The Human Lineage

Foundations of Human Biology

Series Editors:
Matt Cartmill
Kaye Brown
Boston University

Books:

The Growth of Humanity by Barry Bogin

Fundamentals of Forensic Anthropology by Linda L. Klepinger

The Human Lineage by Matt Cartmill and Fred H. Smith

THE HUMAN LINEAGE

MATT CARTMILL
Boston University

FRED H. SMITH
Illionis State University

KAYE BROWN, SERIES EDITOR

A John Wiley & Sons, Inc., Publication

Wiley-Blackwell is an imprint of John Wiley & Sons, formed by the merger of Wiley's global Scientific,
Technical, and Medical business with Blackwell Publishing.

Published by John Wiley & Sons, Inc., Hoboken, New Jersey.
Published simultaneously in Canada.

For general information on our other products and services or for technical support, please contact our Customer Care
Department within the United States at (800) 762-2974, outside the United States at (317) 572-3993 or fax (317) 572-4002.

Wiley also publishes its books in a variety of electronic formats. Some content that appears in print may not be available in
electronic format. For information about Wiley products, visit our web site at www.wiley.com.

Library of Congress Cataloging-in-Publication Data:
Cartmill, Matt.
  The human lineage / Matthew Cartmill, Fred H. Smith.
   p. cm.
  Includes bibliographical references and index.
  ISBN 978-0-471-21491-5 (cloth)
  1. Fossil hominids. 2. Human evolution. I. Smith, Fred H. II. Title.
  GN282.C37 2009
  599.93'8—dc22

                                                    2008055229

Printed in the United States of America.

10 9 8 7 6 5 4 3 2 1

Contents

Preface

Unlike other animals, people wonder how things got to be the way they are; and one of the things we wonder about most is how we got to be so unlike other animals. The science of human origins therefore attracts a lot of interest and attention from a wide range of audiences, from average readers to dedicated researchers. This book about paleoanthropology has been written for readers at the scholarly end of that spectrum, especially for students who have already had a beginning course in the subject. But we have tried to keep it accessible to any educated reader, because we think it is important in today's political and cultural environment to make a substantial account of human evolution available to anyone who has serious questions about it.

As its title implies, this is book about the evolutionary lineage of the human species, *Homo sapiens*. Because it is specifically centered on the *human* lineage, it touches on other aspects of the history of life on earth only to the extent that they bear on human origins. And because this book focuses on reconstructing the human *lineage*, it deals mainly with the facts of paleontology, which furnishes the only direct evidence we have of that lineage. Comparative anatomy, molecular biology, historical geology, archaeology, and other fields of science are brought in mainly to provide necessary background and context for the study of the human fossil record.

Opponents of scientific biology are fond of dismissing that record as a pathetic handful of controversial fragments. If that were so, this book would be a lot shorter. An often-repeated creationist canard insists that all known human fossils would fit on a billiard table. This was probably true in the late 19th century, but it has not been true for a hundred years. Known human fossils number in the thousands and represent the remains of hundreds of individuals. They are more numerous and better-studied than the fossils of any comparable vertebrate group, because the intense interest that people have in the bones of their ancestors has driven them to devote far more effort to collecting and studying fossil humans than (say) fossil horses or herring. Having seen most of the major collections of human fossils in the world's museums, we can assure our readers that those collections can no longer be laid out on a billiard table. It would be hard to cram them all into a boxcar.

The growth of the human fossil record has been especially rapid over the past half-century. In 1959, W. W. Howells could still provide a basic exposition of almost all of the significant human fossils then known in a relatively slim (384-page) volume entitled *Mankind in the Making*. Only three years later, Carleton Coon took 724 pages to present an only slightly more detailed account in his book *The Origin of Races*. Any book that tried to survey today's human fossil collections in the same detail would not fit between a single pair of covers. A recent catalog of most of the major cranial and dental remains of currently known fossil hominins—just photographs and descriptions of skulls and teeth—occupies three quarto volumes totaling more than 1500 pages (Schwartz and Tattersall 2002, 2003, 2005). Most of the decisions involved in writing a paleoanthropology textbook thus concern what to leave out, not what to put in. In making such decisions about the fossil evidence, we have tried to focus not on the details of particular fossils, but on the temporal, regional, or taxonomic patterns that they reveal. Conversely, in covering the theoretical aspects of evolutionary biology, we have stressed those facets of the theory that are deeply involved in current paleoanthropological debates and have skimped on others that are not.

Debates in paleoanthropology are often vigorous and contentious. Some writers would have you believe that such heated debates are both an idiosyncrasy and a shortcoming of our discipline. We think they are wrong on both counts. The scientific enterprise is grounded in the assumptions that all knowledge is provisional and that knowledge increases through the refutation of old ideas and their replacement by new ones. Each generation of scientists makes its mark by overthrowing the received wisdom of the previous generation or transcending its limitations. Active and lively sciences are arenas in which ideas and claims compete for survival. Because reputations and egos are involved in these clashes, arguments are often heated and sometimes intemperate. In these respects, debates over the meaning of fossil skulls are not different from similar debates in other sciences over such issues as the reality of polywater or the planetary status of Pluto.

In grappling with the important debates in our discipline, we have tried to do an even-handed job of laying out the core arguments and key facts that support various currently conflicting interpretations of the fossil record. It will not be difficult to figure out where we stand on most of these issues. But it should also be possible for the reader to understand why others read the facts differently, and even to come to conclusions that differ from ours. In general, we have not tried to articulate the reasoning behind ideas that nobody espouses any longer; but we have included a few such discarded ideas that seem to us to have special historical importance, set off inside boxes and labeled "Blind Alleys."

The first three chapters of this book situate the human fossil record in the larger framework of evolutionary biology and provide the necessary background for what follows. Chapter 1 surveys the development of historical geology, including brief summaries of geological dating techniques and the fossil record of life up through the early radiation of mammals. Chapter 2 lays

out a sketch of the underpinnings of evolutionary theory, with emphasis on its paleontological applications. In Chapter 3, we discuss the mammalian background and evolutionary history of the order Primates. We have undertaken a relatively complete survey of the other living members of our order, because we think that doing so helps put humans in heir proper biological context. Our overview of fossil primates focuses more narrowly on those that have some relevance to the earlier stages of the human lineage, either as potential ancestors or as comparative cases that illustrate relevant concepts and phenomena.

The final five chapters cover the specifically human (hominin) part of the lineage of our species. The initial hominin radiation in Africa, including the origin of the genus *Homo*, is reviewed and discussed in Chapter 4. Chapters 5 and 6 deal respectively with fossils commonly assigned to *Homo erectus* and *Homo heidelbergensis*. This division is not intended as a proxy for a taxonomic distinction or as a presentation of "stages" of human evolution, but only as a heuristic structure to organize the relevant material. A full chapter (Chapter 7) is devoted to the Neandertals, because these are the best-known pre-modern humans and offer unique insights into the pattern of later human evolution. Chapter 8 deals with the emergence and radiation of modern humans—people fundamentally like us. Finally, we provide an appendix detailing the anatomical points and measurements used in the book.

Unless otherwise indicated, all illustrations were drawn by one of us (MC). Some of them are diagrammatic or conceptual, but most were redrawn from photographs, with an eye to both anatomical accuracy and ease of interpretation. Figures for which no published source is credited are based on our own ideas, observations, and photographs.

Most of this book's contents represent other people's work. We have relied heavily on the published work and private thoughts of our colleagues, which have of course greatly enriched our own ideas and interpretations as well. In producing the text and figures for this book, we have tried diligently to give credit to our colleagues for their work and ideas and have attempted to be as thorough as possible in providing citations and references. Despite our best efforts, it is inevitable that we have overlooked some important sources and misinterpreted some others. We accept responsibility for such errors and omissions, and we ask that our colleagues bring them to our attention.

Writing a book like this one brings clearly into focus for us the high quality of research and researchers in paleoanthropology. Of all the pleasures involved in working on this book, none has been greater than our interactions with colleagues. This project could not have been completed without their input, information, and encouragement. Many of them have kindly provided access to fossils and other material critical to the production of this volume. For general assistance we thank: J. Ahern, S. Bailey, O. Bar-Yosef, M. Bolus, D. Boyd, C.L. Brace, G. Bräuer, A. Busby, J. Calcagno, R. Cann, R. Caspari, S. Churchill, R. Clarke, M. Cole, T. Cole, G. Conroy, D. Curnoe, S. Donnelly, A. Durband, A. Falsetti, R. Franciscus, D. Frayer, J. Gaines, J. Gardner, D. Gebo, D. Glassman, A. Grauer, M. Green, L. Greenfield, F. Grine, P. Habgood, T. Holliday, R. Holloway, N. Holton, J.-J. Hublin, K. Hunt, V. Hutchinson, W. Hylander, I. Janković, R. Jantz, D. Johanson, C. Jolly, W. Jungers, R. Jurmain, I. Karavanić, R. Kay, W. Kimbell, J. Kidder, R. Klein, J. Kondrat, L. Konigsberg, A. Kramer, C.S. Larsen, S. Leigh, D. Lieberman, M. Liston, F. Livingstone, D. Lordkipandze, C.O. Lovejoy, A. Mann, J. McKee, N. Minugh-Purvis, J. Monge, S. Myster, L. Nevell, S. Pääbo, M. Parrish, R. Quam, M. Ravosa, J. Relethford, G.P. Rightmire, C. Ruff, D. Rukavina, P. Schmid, D. Serre, J. Simek, T. Simmons, E. Simons, F. Spencer, F. Spoor, C. Steininger, M. Stoneking, A. Thorne, A.-M. Tillier, E. Trinkaus, H. Ullrich, K. Valoch, C. Vinyard, P. Vinyard, A. Walker, C. Wall, C. Ward, S. Ward, K. Weiss, M. Weiss, T. White, M. Wolpoff, B. Wood, and T. Yokley. We would like to credit K. Porter for her initial ideas about dinichism in *Australopithecus*. We are grateful to the following individuals for permission to study fossil hominins, as well as for other forms of assistance: T. Akazawa (Tokyo), B. Asfaw (Addis Ababa), G. Avery (Capetown), H. Bach (Jena), D. Brajković (Zagreb), J. Brink (Bloemfontein), N. Conard (Tübingen), Y. Coppens (Paris), I. Crnolatac (Zagreb), A. Czarnetzki (Tübingen), H. Delporte (Saint-Germain-en-Laye), J. DeVos (Leiden), R. Feustel (Weimar), J.-L. Heim (Pairs/Leipzig), F.C. Howell (Berkeley), W.W. Howells (Cambridge, MA), J. Jelínek (Brno), H.-E. Joachim (Bonn), R. Kraatz (Heidelberg), R. Leakey (Nairobi), A. Leguebe (Brussels), J. Lenardić (Zagreb), R. Macchiarelli (Rome/Poitiers), M. Malez (Zagreb), D. Mania (Halle), G. Manzi (Rome), F. Masao (Dar es Salaam), E. Mbua (Nairobi), J. Melentis (Thessalonika), A. Morris (Capetown), D. Pilbeam (Cambridge, MA), J. Radovčić (Zagreb), Y. Rak (Tel Aviv), M. Sakka (Paris), M. Schmauder (Bonn), R. Schmitz (Tübingen), B. Senut (Paris), C. Stringer (London), J. Szilvássy (Vienna), M. Teschler-Nikola (Vienna), T. Thackeray (Pretoria), M. Thurzo (Bratislava), P. Tobias (Johannesburg), B. Vandermeersch (Paris/Bordeaux), R. von Koenigswald (Frankfurt-am-Main), E. Vlček (Prague), J. Zias (Jerusalem), and R. Ziegler (Stuttgart).

Over the years, our work has been supported by a number of agencies and universities. These are the National Science Foundation (USA), the National Academy of Sciences (USA), the National Institutes of Health (USA), the Wenner-Gren Foundation, the National Geographic Society, the L. S. B. Leakey Foundation, Sigma Xi, the Alexander von Humboldt Foundation, Fulbright Foundation, the John Simon Guggenheim Memorial Foundation, Duke University, the University of Tennessee, the University of Tübingen, Northern Illinois University, Loyola University Chicago, the University of Zagreb, the University of Hamburg, the Croatian Natural History Museum, the Institute of Anthropology (Zagreb), the Institute for Quaternary Geology and Vertebrate Paleontology—Croatian Academy of Sciences and the Arts, the Frank H. McClung Museum, Boston University, and Illinois State University. We are grateful to all of these institutions for their support.

Several of our colleagues—J. Ahern (University of Wyoming), A. Durband (Texas Tech University), R. Franciscus (University of Iowa), P. Habgood (University of Queensland), A. Hartstone-Rose (Pennsylvania State University), I. Janković (Institute of Anthropology, Zagreb), K. Rosenberg (University of Delaware), and D. Schmitt (Duke University)—deserve special thanks for reading and making suggestions on parts of the manuscript. Their help was invaluable. We are particularly grateful to our Series Editor Kaye Brown for her editorial suggestions and other assistance; and we are indebted to Karen Chambers, our editor at Wiley-Blackwell, for her patience and help. Last and most importantly, we want to express our deep gratitude to our families (Kaye Brown, Maria O. Smith, Erica Cartmill, Burton T. Smith, and Maria K. Smith) for all the warm and generous support, patience, advice, understanding, and encouragement that they gave us during the researching and writing of this book.

MATT CARTMILL
FRED H. SMITH

CHAPTER ONE

The Fossil Record

CHANGING IDEAS ABOUT THE CHANGING EARTH

The past is no longer with us. To reconstruct it, we have to look at the traces that it has left in the present. Discerning those traces and figuring out their meanings is not a simple task. It has taken over 300 years for scientists to arrive at the methods that we rely on today in using the present to resurrect the distant past.

Ideas about the remote past began with pre-scientific speculation about the formation of the earth. When our ancestors looked at the landscape around them, it was not obvious to them that it contained a record of slow changes over an immense period of time. Most of the changes in the earth's surface that are noticeable to a human observer are small, swift, and local. After a storm, water running off deforested hills may cut gullies in the soil. Further downstream, that same floodwater may cause a muddy river to overflow its banks and deposit a layer of silt for miles around. At long intervals, an earthquake may shift a piece of land by a few inches, or a volcano may shower a region with ash and spill lava down its slopes.

Changes of this sort have been noticed and commented on since the dawn of history. But few of us ever see all four of these processes—erosion, sedimentation, tectonics, and vulcanism—working to alter the landscape in the course of our own lifetimes. Until about 250 years ago, nobody seems to have thought about how these forces might combine over millions of years to produce the rocks, soils, mountains, and valleys of the earth.

Throughout most of the history of Western thought, speculation about the history of the earth has been constrained by the sacred poetry of the Hebrew scriptures, in which the voice of God from the whirlwind reproves would-be geologists in words of the gravest majesty:

> Where wast thou when I laid the foundations of
> the earth?
> declare, if thou hast understanding.
> Who hath laid the measures thereof, if thou knowest?
> or who hath stretched the line upon it?
> Whereupon are the foundations thereof fastened?
> or who laid the cornerstone thereof;

When the morning stars sang together,
  and all the sons of God shouted for joy?

These lines from the Book of Job represent God as a divine mason building a world with floor plans, foundations, and a cornerstone. Few people took this poetic metaphor literally. But the creation story in Genesis, which tells how God shaped the face of the earth and planted it with grass and trees in the course of a single day, was taken very literally indeed (and still is, in some quarters). Reckoning forward from the first day of that Creation by adding together the ages of patriarchs and the reigns of kings listed in the Old Testament, biblical scholars reasoned that the universe had to be less than 6000 years old. Some Christian authorities fixed on a date of 4004 B.C. for the beginning of the world. Jewish scholars numbered the years of their calendar from a supposed Creation date equivalent to 3760 B.C.

As long as this short time scale was accepted, floods, earthquakes, and volcanoes could not be thought of as having produced the rocks and topography of the earth. The changes wrought by these processes were too small and slow to have brought the present landscape into being in a few thousand years. The minor changes that people could actually see happening in the face of the earth tended to be regarded as blemishes on the original work: signs of decay presaging the approaching end of the world, or marks left by outbursts of God's wrath.

The biggest such outburst was thought to have been the Flood of Noah, in which God had supposedly submerged the whole world under water in order to get rid of sinful humanity (see Blind Alley #1). As evidence of that universal Flood, many people pointed to the seashells and fish bones found in rocks far above the present level of the sea. Those fossil shells and bones seemed to bear out the truth of the Genesis

---

Blind Alley #1: The Flood Story

People all over the world recount tales of disasters long ago that ravaged the earth and threatened humanity. In the ancient Near East, the Sumerians, Babylonians, and Hebrews told stories about a great flood sent by the gods to wipe out the wicked human race. In the Babylonian version, the goddess Ishtar incites the gods to drown the world. Her plan is foiled by the god Ea, who instructs the hero Utnapishtim to build a huge vessel and load his family and all the world's animals aboard it. Safe in the ship, the people and beasts weather a great storm that destroys all other living things. After sending out a dove, a sparrow, and a raven to seek land, Utnapishtim grounds his ark on Mt. Nisir and offers a burnt sacrifice. The gods gather around, repent their acts, and promise never to do it again.

The similar flood story in the Old Testament is another member of this family of Near Eastern legends. But because it was part of sacred scripture, Jews and Christians long accepted it as historical fact. When they found fossil shells in rocks lying hundreds of feet above sea level, they took them as confirmation of a great flood that had drowned the mountains and laid down the fossil-bearing strata. When they dug up ancient human artifacts and remains, they described them as "antediluvian"—that is, dating to a time before the Flood.

The biblical Flood story has implications that can be checked out. If all the fossilized organisms died at the same time a few thousand years ago, then all the fossil-bearing rocks should contain fossils of people and other extant creatures. But they don't. If the Ark saved all the world's animals from destruction, there shouldn't be any extinct species in the fossil record. But there are. (Almost all fossil species are extinct.) If all the animals dispersed from the Ark's landing site on Mt. Ararat, then the world's faunas should grow less and less diverse the further away they are from Turkey. They don't. If enough rain had fallen to cover all the land—raising the seas an additional seven miles—then the total rainfall would have amounted to some 1.4 billion cubic miles of fresh water. How did all that water dry up in 150 days (Genesis 7:24)? Where is it now?

And so on. None of the implications of the biblical Flood story check out. It was not until this old Mesopotamian legend had been set aside that the building of scientific theories about the history of the earth's rocks, animals, and plants could begin in earnest.

story, but they also posed certain problems. For one thing, they were not really shells and bones. Most of them were composed of minerals like those found in the surrounding rocks. Some insisted that the seeming plants and animals imbedded in the rocks were not really the remains of once-living things, but just *lusi naturae*, "games of nature"—freakish, abortive organic forms produced by mysterious creative forces inherent in the stone itself. But the fossils' detailed similarities to bones, shells, and leaves made this hard to credit.

In 1669, Nicolaus Steno, a Danish physician living in Italy, published a book that used fossils to help interpret the history of the earth. Rocks that contain fossils are always organized into **strata** or layers. We can see these layers exposed wherever a quarry or cliff face cuts a vertical section through such rocks. Different layers contain rocks of different composition: Beds of smooth slate may alternate with gritty sandstone or coarse conglomerates containing pebbles of yet another sort of stone. Steno argued that all these layers had been formed when sediments were discharged into lakes or seas by rivers, settled to the bottom, and became cemented together. Fast-moving streams carried coarse sediment that became sandstone or conglomerates; broad, slow rivers carried silt that gave rise to slate and other fine-grained rocks.

The fossils imbedded in these rocks, Steno insisted, truly were the remains of creatures that had lived in or near the seas and lakes. He showed that strange triangular fossils long known as "tongue stones" were identical to the teeth of living sharks and that other fossils found with them were clearly the remains of other sea creatures. The rocks containing such fossils, he argued, must have been laid down on the ocean floor. Other rocks, containing the remains of land plants and animals, had probably been formed from fresh-water sediments.

The layers of sedimentary rock are usually more or less horizontal, as you would expect beds of sediment to be. However, Steno recognized that some of them are tilted or folded. He concluded that the originally flat layers of sediment must have been worked on later by powerful disruptive forces capable of bending or breaking thick beds of solid stone. Steno suggested that these layers had been first lifted above the sea and then undermined by the formation of huge internal caverns. When their roofs fell in, these collapsed caverns became valleys, in which the fallen roof fragments lay as beds of tilted rock. The remaining elevations were left as mountains. Streams running down these mountains then carried silt and gravel down into the valleys, which gradually filled with new sediments.

All these ideas were generally rejected in Steno's own day, but they influenced later geologists (Gillespie

1959, Gohau 1990). Three of Steno's basic postulates—the sedimentary origin of rock strata, the organic nature of fossils, and the rule that overlying strata are younger than those below them—are still fundamental assumptions in reconstructing the histories of the earth, of prehistoric life, and of ancient human activity.

■ NEPTUNE VS. VULCAN

The rise of industrial economies in northern Europe in the 1700s lent new practical importance to theories about the history of the earth. As the prosperity of nations began increasingly to depend on mineral wealth, it was worth money to understand how these resources were distributed in the rocks. It was obvious that there were patterns there to be understood. The coal and metal ores that the new industrial order needed were not scattered around in random pockets. They tended to occur in particular strata, sandwiched in between layers of other sorts of rock. Miners had long ago learned to identify these mineral-bearing strata on the surface and dig down after them into the earth.

Amateur scientists began to trace and study these stratigraphic patterns. In the 1740s, a French physician named Jean Guettard, whose hobby was collecting plants, noticed that certain species of plants were found only where the underlying rocks were of certain sorts, such as limestone or chalk. Mapping the distributions of these plants, he found that their associated rocks were arranged in bands that ran across the map of Europe, sometimes for hundreds of miles. Guettard concluded that these bands represented the exposed edges of superimposed sheets or layers of rock, which had been laid down atop one another as sediments at the bottom of the sea.

If the landscape really had been carved out of stacked-up layers of sedimentary rock, then it might be possible to reconstruct the story of the earth's creation by reading upward through those layers, starting at the bottom with the oldest rocks of all. One of the first people to try this was a German scholar named Johann Lehmann. Lehmann distinguished three major phases in the formation of the earth. He thought that the oldest rocks, which had no fossils in them, had precipitated out of a planetary suspension of liquid mud during a primitive period before the creation of life. This "Primary" period corresponded to Day 1 of the Genesis creation story, when "the earth was without form, and void." Later rocks, containing fossil remains of plants and animals, had been laid down over the Primary rocks during the Flood of Noah. Overlying these "Secondary" strata was a third, superficial layer of recent deposits formed by erosion and volcanic action.

These ideas formed the nucleus of the first systematic theory of earth history, developed and promulgated from the 1780s on by Abraham Werner, a professor at Freiburg in Germany. Werner adopted Lehmann's account of the series of Primary and Secondary rocks. His main contribution was his explanation of volcanoes. Werner taught that the earth's deposits of coal and petroleum had been laid down during the Secondary period. When some of these flammable minerals caught fire, the heat had melted the surrounding rocks and produced volcanic eruptions, spreading sheets of lava in places on top of the Secondary strata. These strata of **igneous rocks** (rocks formed by melting) represented a brief third or Tertiary period of rock formation. Finally, erosion and floods had deposited soil, gravel, and other loose sediments on top of everything else during the most recent period, the Quaternary. Because Werner and his followers thought that almost all geological formations had been formed by the action of water, they came to be known as **Neptunists**, after the Roman god of the sea.

Their opponents argued that certain ancient rocks that the Neptunists called sedimentary were in fact volcanic in origin. For this reason, the anti-Neptunist school were sometimes referred to as **Vulcanists**. Rejecting the notion that different rock-forming processes had operated at different periods of the earth's history, the Vulcanists insisted that the forces seen at work on the landscape today—erosion, sedimentation, volcanoes, and earthquakes—had worked together throughout geological time in the same way they do now and that together they were sufficient to account for all the observable features of the earth.

This doctrine—that the laws and processes that operated in the past are the same ones we see operating in the world around us—is called **uniformitarianism**. The eventual acceptance of the uniformitarian principle was crucially important in the maturation of geology into a genuine science. It also helped to lay the groundwork for Charles Darwin's application of a similar uniformitarianism to the study of biology (Gould 1986). However, scientists today do not regard the uniformitarian principle as entirely valid over the long run. For example, the universe is now thought to have originated in an explosive cosmic event called the Big Bang, which involved processes and forces no longer at work in the world we observe. The formation of the solar planets from condensations of interstellar dust surrounding the newborn sun was another one-time-only process that has no counterpart in today's solar system, though we can get some insight into it by studying other stars younger than our own. The history of the earth and its living organisms has been profoundly affected by asteroid impacts and other catastrophic global events, which we (luckily enough) have never had a chance to witness in the brief span of human history.

But over the span of time that concerns biologists, from the origin of life some four billion years ago down to the present, the uniformitarian principle is true enough for most practical purposes. The small, local forces that we can see reshaping our landscapes today suffice to account for the formation of the rocks of the earth over the vast duration of geological time. They provide the geological background for the history of life.

A BRIEF GUIDE TO SEDIMENTOLOGY

Most of the fossils that paleontologists study are the remains of marine life, buried ages ago in the rain of fine mineral particles and organic detritus that falls ceaselessly through the oceans of the earth to accumulate as mud and sand on the sea bottom. When mud and sand are themselves buried and compressed by overlying sediments, they tend over millions of years to become cemented together by chemical solution and recrystallization, forming hard, dense rock. In much the same way, sugar granules in a bowl will "petrify" into a single stony lump if they get damp and then dry out again. The process takes vastly longer for sand than it does for sugar because silica is far less soluble than sucrose.

The texture of the resulting rock, as well as its suitability as a medium for the preservation of fossils, depends mainly on the size of its constituent particles. The size of those particles depends in turn on how fast the water around them was moving when they were deposited. A swiftly rushing mountain creek will wash away pebbles and gravel together with sand and silt and carry them all downstream. As it flows into flatter country and slows down, it no longer has enough kinetic energy to move the larger bits of rock. One by one, they tumble to the bottom and form deposits of coarse sediment. The smaller particles remain suspended in the water and travel further downstream. Wherever the stream speeds up, it picks up larger particles; wherever it slows down, it drops them. By the time it reaches the ocean as a broad, slow-flowing river, it may have nothing left in it but a thin suspension of clay minerals. Borne out to sea by the currents, this sediment eventually settles to the bottom as a fine-grained mud—the first stage in the formation of shale. Similar processes of sedimentation take place on a smaller scale at the bottom of lakes, or on plains flooded by overflowing rivers.

At the edges of the continents, the action of waves breaking against the rocks has the same sorts of effects that stream runoff has on dry land. High, energetic waves tear loose sizeable chunks of stone and rub them against each other in the surf until all their rough edges are worn away and they become smooth, rounded pebbles. A lot of the rock particles produced by wave action get carried away by ocean currents, in which the particles sort themselves out by size just as they do in a river. Big pieces drop out immediately and form stony bottoms or pebble beaches near the shoreline. Sand travels further, so sandy deposits extend further away from the zone of wave action. Far out to sea, most sediments are fine-grained and may contain a high proportion of organic material. Rocks formed from such sediments are often largely made up of fossil shells and other remains of sea life. For example, the thick beds of chalk that accumulated in many areas near the end of the age of the dinosaurs are composed mainly of the calcareous shells of tiny one-celled animals called foraminifera. In general, the finer the sediment, the more likely it is that the remains and impressions of dead organisms buried in it will be preserved as fossils.

Sediments can also be formed on land by the action of air or ice. Winds blowing across dry, dusty soils can carry small particles away and deposit them as sand dunes, or as deposits of fine silt called **loess**. The particles in such windborne deposits are small, since blowing air is far less energetic than rushing water.

Rivers of flowing ice, on the other hand, can carry huge boulders for long distances. These ice rivers, called **glaciers**, form wherever the snow that falls during the winter does not melt entirely during the summer. In these chilly places, snow builds up year after year, becoming compressed into thick layers of dense ice. The weight of the accumulating ice forces some of it to begin moving away from the areas of buildup, flowing downhill with what is appropriately called "glacial" slowness. As it flows, the moving ice breaks away big and small pieces of the rock beneath it, scouring out characteristic U-shaped valleys like a carpenter's gouge running across the earth.

When it reaches a warmer area where snow melts faster than it accumulates, the glacier melts too. As the rock fragments imbedded in a flowing glacier reach its melting edge, they fall to the ground, depositing a jumble of boulders and gravel known as a **glacial moraine**. If a flowing glacier reaches the sea before it melts, it may form or add to an oceanic ice sheet, which rides on top of the waves (because ice floats in water). The edges of such ice sheets eventually break off as icebergs and drift out to sea. Melting, they may drop big intrusive chunks of continental rock called **dropstones** into the fine sediments of the sea floor many miles from land. Like moraines, dropstones in fine-grained sedimentary rocks are an indicator of the presence of ancient glaciers.

Near the poles, where temperatures remain frigid year-round, sheets of glacial ice may cover the sea permanently and become three or four kilometers thick over land areas. There have been several periods in the earth's history when these polar ice caps spread toward the equator, covering large areas of the temperate zones with ice. These periods are called **ice ages** or **continental glaciations**. The continental ice sheets that form during an ice age swell and shrink periodically, retreating toward the poles during **interglacial** phases and advancing again during **glacial maxima**. The effects that these fluctuations had on the course of human evolution during the last great ice age, the **Pleistocene** epoch, are debated by scientists. We will return to these debates in the later chapters of this book.

At the moment, the only large land areas covered by continental ice sheets are Greenland and Antarctica. But Northern Hemisphere ice sheets extended far south into North America and Eurasia during the last glacial maximum, about 20,000 years ago. Geologists infer the former presence and extent of these ice sheets from signs of glacial erosion and from moraines deposited at the melting edges of long-vanished ancient glaciers. In reviewing the prehistory of human populations in northern Eurasia, it should be borne in mind that some evidence of early human presence at high latitudes may have been destroyed by the erosive forces of continental glaciation.

▌DATING THE ROCKS

The dispute between the Neptunists and the Vulcanists hinged largely on the question of the age of the oldest igneous rocks. This issue could not be entirely settled until there was a generally agreed way of telling which rocks were older. It was easy enough to do this as long as geologists looked at a single locality where rock strata lay on top of each other. In a given geological section, the strata higher up are always younger than those lower down in the section (if you make allowances for occasional folding and twisting). But no single section contained a sample of all time periods and rock types. It was therefore hard to say whether a layer of shale in, say, Scotland was of the same age as a similar-looking shale in Pennsylvania.

The key to this puzzle lay in the fossils imbedded in the sedimentary rocks. In the 1790s, the young English surveyor and engineer William Smith began a systematic study of the rock strata of Somerset, where he had been hired to supervise the construction of a canal. He found that the stacked-up sequence of successive rock types was different in different sections through the earth, so that a bed of red sandstone that was overlain by a layer of gray shale in one canal cut might be covered by a layer of brown sandstone in another. But the sequence of fossils was always the same from cut to cut. Therefore, the distinctive **index fossils** that were restricted to a particular part of one section could be used to match that part up with different rocks in

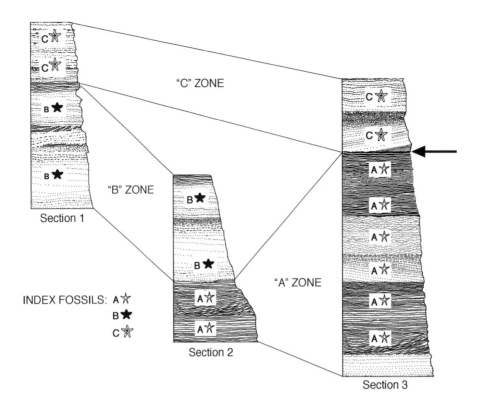

FIGURE 1.1

William Smith's principle of dating rocks by their fossils. If strata containing fossils of the species B are always younger (higher in the section) than those with A, and strata containing C are always younger than both, then these species can thus be used as **index fossils** to determine the relative ages of rocks containing them. The composite **stratigraphic sequence** diagrammed here comprises three successive units, the "A," "B," and "C" zones, though no one section contains all three. Section 3 does not contain a "B" zone, implying that it was a site of erosion rather than deposition during that period. The inferred gap or **unconformity** in this section (*arrow*) is usually also reflected in an interrupted pattern of sedimentation at that time horizon.

other areas and assign them all to the same time period (Fig. 1.1).

Armed with this insight, Smith began to classify and organize the rocks of southern Britain on the basis of the fossils they contained. With the help of a local minister who collected fossils as curiosities, he began in 1799 to publish a series of maps, charts, and books bearing such titles as *A Delineation of the Strata of England and Wales*. Other scientists soon began to build on Smith's work and to extend his system to other lands and strata. Smith lived to see himself hailed on all sides as the pioneering founder of the science of stratigraphy before he died in 1839.

The use of index fossils is the basis of **biostratigraphy**, often referred to as faunal or floral dating. Since no single locality preserves the entire record of life on earth, scientists must correlate strata from different sites in order to determine their relative ages and fit them into a single time sequence. This process begins with the tallying up of the shifting lists of organisms found in successive strata at a single site. This sequence is then compared with similar sequences from other sites in the same area. Of course, two strata from the same time period will not always contain exactly the same suite of organisms, particularly if these strata represent different ecological zones or come from sites that are far apart from each other. Within a region, however, certain key organisms will usually be characteristic of a specific time period. By comparing the biostratigraphic sequences of the sites in a localized area, scientists can work out the time relationships between them and piece them together to form a regional biostratigraphic column. Then by comparing sites that overlap two contiguous regions (or ecological zones), a picture of relative stratigraphic relationships on a broader scale can be formed.

While Smith was walking along his canal cuts in Somersetshire and laying the foundations of stratigraphy in his mind, the Scottish gentleman farmer James Hutton was putting the final touches on his 1795 masterwork, *Theory of the Earth*. In this book, Hutton laid out detailed evidence for believing that a single set of rock-forming processes had operated throughout the history of the earth. Hutton contended that the small-scale geological processes that could be observed acting in the present were the same processes that had fashioned the large-scale features of the earth's topography: Mountains were the result of slow uplift, and valleys and deltas were the result of gradual erosion and deposition by rivers and streams. He argued that the fossil-free "Primary" rocks, which the Neptunists saw as condensations from a worldwide suspension of some nebulous primordial ooze, showed clear signs of having been formed in the same ways as later rocks—either through the consolidation of silt and sand and pebbles washed into the world's waters by waves and streams, or through the solidification of molten stuff belched out of the earth's interior through volcanic vents. The same subterranean heat that had melted those rocks, Hutton suggested, had in some way produced buckling or heaving movements of the crust of the earth, raising former sea beds into the air to form mountains. He surmised correctly that some ancient rocks, which we now call **metamorphic**, represented sedimentary deposits that had been considerably altered by the action of heat and pressure inside the earth. Hutton insisted that all these processes had gone on throughout geological time and were still going on today. His arguments implied that the earth must be much older than was generally thought, because it would have taken an enormous amount of time for these processes to produce the current landforms. God, Hutton concluded, had created the earth as a perpetually working machine like the Newtonian solar system, perfectly designed and exquisitely balanced to keep turning over forever. In a famous phrase, he declared that the study of the earth disclosed "no vestige of a beginning, no prospect of an end."

▮ THE SUCCESSION OF FAUNAS

From the union of Smith's stratigraphic principles with Hutton's uniformitarianism, the science of historical geology emerged during the early 1800s. The canonical expression of this synthesis was the English geologist Charles Lyell's massive compendium, *Principles of Geology*, published in 1830. Lyell's *Principles* had a major influence on Darwin's *Origin of Species* (1859) in at least three respects: it provided an exemplary model (of a big theoretical work supported by masses of empirical detail), it established gradual uniformitarian transformation as the normative mode of large-scale prehistoric change, and it demonstrated that the age of the earth was great enough to allow for the production of life's diversity through the slow, imperceptible processes of evolution that Darwin postulated.

The great task facing the early geologists was the reconstruction of the **geological column**—the overall sequence of the rocks of the earth throughout its history (Fig. 1.2). No one locality preserves the column in its entirety, because there is no place on the face of the earth where sediment has been building up without interruption for four billion years. The longest single exposure is that seen in the cliff faces of the mile-deep Grand Canyon in Arizona, but even this sequence dates back only some 1.75 billion years—and about half of that time span is represented by gaps in the sequence, produced during periods in this region's history when erosion outpaced the deposition of sediments.

The complete geological column therefore has had to be pieced together by collecting fossils and rock samples from all over the world and bringing them back for publication and comparison. Two centuries of dangerous, painstaking, difficult labor and the lives of thousands of scientists have gone into this reconstruction. The job is still far from being completed, but what has been accomplished so far is one of the greatest triumphs of the scientific enterprise.

It became obvious early on in this undertaking that the creation story in Genesis was not a satisfactory account of the history of the earth. First of all, it was clear that in spite of the supposed labors of Noah and his sons, most of the species that had once lived were now extinct. In fact, this was what made Smith's stratigraphic methods work. If all the plants and animals that had ever lived were still alive today, rocks of all ages would contain the same species. Index fossils could be used to order the earth's rocks only because species had finite life spans. The mortality of species was a disturbing discovery.

Different rocks not only contained different species, they contained different *kinds* of species. Only the very youngest rocks preserved the remains of creatures much like those living today. As one traced the history of life downward through the geological column, the fossil animals and plants became more and more alien. Human remains and artifacts, for example, were restricted to the Quaternary deposits at the very top of the column. Below these, the Tertiary rocks contained various sorts of fossil mammals—but the mammals found in lower Tertiary strata were not at all like those of today. Still further back in time, there were no fossil mammals whatever, and the only large land animals were fearsome dragonlike reptiles, sometimes of gigantic size. Before that, no land animals of any sort could be found, and all the earth's organisms had apparently lived in the sea. And at the very bottom of the column, one encountered the fossil-free rocks of the Primary series, whose sediments seemed to bear witness to an unthinkable expanse of time when there had been no living things on the face of the earth.

The scientists of the early 19th century weighed two alternative interpretations of all these uncomfortable facts. One possibility was that ancient species had from

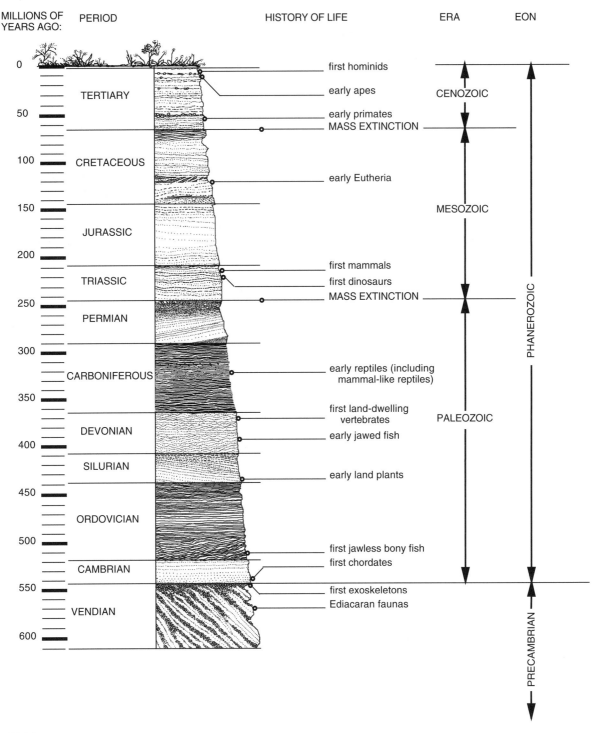

MILLIONS OF YEARS AGO: PERIOD   HISTORY OF LIFE   ERA   EON

0   first hominids
        TERTIARY   early apes
50   early primates   CENOZOIC
                                         MASS EXTINCTION
100   CRETACEOUS
                                         early Eutheria
150   MESOZOIC
        JURASSIC
200   first mammals
        TRIASSIC   first dinosaurs
250   MASS EXTINCTION
        PERMIAN
300
        CARBONIFEROUS   early reptiles (including
                                         mammal-like reptiles)
350
                                         first land-dwelling
        DEVONIAN   vertebrates   PALEOZOIC   PHANEROZOIC
                                         early jawed fish
400
        SILURIAN
                                         early land plants
450
        ORDOVICIAN
500
                                         first jawless bony fish
        CAMBRIAN   first chordates
550
                                         first exoskeletons
        VENDIAN   Ediacaran faunas   PRECAMBRIAN
600

FIGURE 1.2

The geological column, showing the sequence and age of the principal stratigraphic units and some major events in the history of the human lineage.

time to time been obliterated and succeeded by specially created species belonging to new, more advanced types. The French paleontologist Georges Cuvier argued that the abrupt transitions from one stratum to another in the geological column represented brief periods of convulsive change in the earth's surface, in which most of the planet's life forms had been wiped out (Cuvier 1831). After each of these catastrophes, new species had been created to rule the earth during a long, peaceful period of uniformitarian change. Many followers of this doctrine of **catastrophism** did not hesitate to see the hand of God at work in this cycle of extinction and rebirth, and the final catastrophe in Cuvier's scheme was often identified with the Flood of Noah.

There were two major problems with catastrophism. The first was that it violated the uniformitarian principle, by postulating processes of episodic destruction of a kind and magnitude unknown to human experience. The second was that it provided no explanation for the generation of new species. Some catastrophists invoked divine intervention to account for the periodic creation of new forms of life. But this sort of miraculous "explanation" was scientifically unacceptable. (A science of miracles is a contradiction in terms.) Even from a religious standpoint, it was not entirely satisfactory. Why, after all, would an all-powerful God obliterate all his living creatures from time to time and start afresh with new, improved versions, instead of producing the desired product in the first place? Surely he was not creating life over and over because he needed the practice.

The other way out of this dilemma was to adapt Hutton's model of the earth as a perpetually working machine and to suppose that today's life forms had come into being through gradual transformation of the earlier, more primitive species. This so-called **development hypothesis** was articulated in different ways by Charles Darwin's grandfather Erasmus Darwin (1794) and a few thinkers of the early 1800s, including J.-B. Lamarck (1809), E. Geoffroy Saint-Hilaire (1830), and R. Chambers (1844). But the development hypothesis was not highly thought of by most experts, because it too seemed to violate the uniformitarian principle. No one, after all, had ever witnessed one species evolving into another one. Cuvier pointed to the mummies of animals recovered from Egyptian tombs, which were estimated to be 3000 years old but displayed no differences from their modern counterparts (Ferembach 1997). This fact, he argued, shows that species have no tendency to change through time. The development theorists could only reply that 3000 years was evidently not enough time to produce detectable changes, and that it must have taken hundreds or thousands of millions of years to accumulate all the transformations seen in the fossil record. Many people found it hard to accept these enormous expanses of time.

▌ RADIATION-BASED DATING TECHNIQUES

How could this question be settled? The stratigraphic methods used to reconstruct the geological column would not do the job. All they could provide were dates for one stratigraphic unit *relative to others*, based on the principle of superposition (the rule that stacked-up strata had been deposited in chronological sequence, with the ones on the bottom being older than those overlying them). These **relative dates** allowed geologists to determine that, say, rocks containing fossils of trilobites were everywhere older than rocks containing dinosaurs and that strata containing a particular species of trilobites were probably contemporaneous with strata at other localities containing the same species. But biostratigraphy was of no help in determining how many thousands or millions of years had passed since a particular layer of rock had been laid down.

Some tried to attach such **absolute dates** to the geological column by estimating how many millimeters of sediment get deposited each year on the bottoms of today's oceans and then comparing these estimates to the number of meters of rock found in various stratigraphic units. But sedimentation rates, erosion rates, and thicknesses of strata vary too much from place to place to be very useful as clocks. About all that could be said for sure was that it had taken many hundreds of millions of years to build the sedimentary rocks of the earth. To produce reliable absolute dates, geologists needed to find some physical process that proceeds at a constant rate in all times and places, producing changes in the sedimentary rocks that could be measured to determine how much time had elapsed since the rock was first laid down.

Such processes began to be identified in the 1950s, as scientists made use of the phenomena of radioactivity to develop **radiometric dating** techniques. Unstable atoms "decay" by emitting or absorbing subatomic particles and changing into something else. For example, the unstable carbon isotope **carbon-14** (^{14}C) undergoes "beta decay" by emitting an electron (beta particle) and changing into an atom of nitrogen-14 (which is stable). These decay events are unpredictable; but their average rate across a large sample of ^{14}C atoms is constant. It takes 5730 years for 50% of the atoms in a sample of ^{14}C to turn into nitrogen-14. After another 5730 years has gone by, half of the remaining C-14 will have turned into nitrogen, and the sample will be 75% nitrogen—and so on. The period of 5730 years is called the **half-life** of the ^{14}C isotope.

If we found a sealed-up canister labeled "pure carbon-14" and we wanted to know how long it had been sealed up, we could find out in two ways. First, we could take a sample and determine the ratio of nitrogen to ^{14}C in it. (The more nitrogen in the canister, the older the contents must be.) Second, we could measure the rate at which the canister's contents give off beta particles. The lower the rate of emissions, the smaller the percentage of ^{14}C remaining must be—and therefore, the older the contents are.

We do not find sealed canisters of once-pure ^{14}C in ancient archaeological sites, but we find something just as useful: tissues from dead animals and plants. Carbon-14 is produced in the earth's atmosphere at a more or less constant rate through the action of cosmic radiation on carbon dioxide molecules. Plants incorporate the radioactive CO_2 into their tissues, and ^{14}C moves up the food chain from there. When plants and animals die, they stop assimilating ^{14}C from their environments. The ^{14}C remaining in their dead bodies gradually disappears through beta decay, while the stable isotopes of carbon remain unchanged. We can therefore determine how old a site is by determining the ratio of ^{14}C to other carbon isotopes in wood or bones from the site. We can do this either directly (through mass spectrometry) or indirectly (by measuring beta radiation).

Carbon-14 dating (also known as **radiocarbon dating**) has its complications and shortcomings. The rate of production of ^{14}C is not constant, because it varies with the amount of CO_2 in the atmosphere. Moreover, some organisms have physiologies or ways of life that cause them to assimilate less ^{14}C than other organisms do. As a result, not all organisms have the same percentage of ^{14}C in their tissues at death. Carbon-14 dates have to be corrected to take these sources of error into account. And even when all the error factors are compensated for, bone or wood that is older than about 50,000 years has too little ^{14}C left in it to be used for dating. With a time depth of only 50,000 years, ^{14}C dating is useful to archaeologists but has little utility in most paleontological contexts.

Carbon dating has other limitations. Because the death of an organism is required to start the ^{14}C "clock" running, radiocarbon can only be used to estimate an age for bone, wood, or other organic materials. It cannot be used to directly date a mineral sample or a stone tool. Even in dealing with biological materials of suitable age, there is an ever-present risk of postmortem contamination of the sample by carbon compounds of later origin. There are techniques for detecting contamination in samples, but they do not always work. These sources of error introduce uncertainty into any radiocarbon date. A ^{14}C date (or other radiometrically derived date estimate) is therefore always followed by an error estimate—for example, $25,850 \pm 280$ years. The error estimate usually represents the 95% confidence limits on the date, meaning that the estimator calculates that the chances are 95 out of 100 that the true age of the sample lies within the plus-or-minus range.

A relatively new and more precise technique called **accelerator mass spectrometry** (AMS) allows ^{14}C dates to be obtained from much smaller samples, and may in theory allow us to extend the time range for radiocarbon dating back to as much as 100,000 years ago (100 Kya). In practice, AMS has yet to extend the range of ^{14}C dating very much, because contamination is an even more serious problem when tiny amounts of ^{14}C are being used. Differing sample preparation techniques can also alter AMS age estimates by as much as 10% (Higham et al. 2006). Nevertheless, AMS has two major virtues: it can reduce the error estimates on ^{14}C dates when circumstances are favorable, and it allows

the direct dating of irreplaceable fossils without significant damage to them. Used with care, AMS radiocarbon dating has helped to provide a more accurate chronology for the later stages of our biological history.

But what about the earlier stages of that history, where even the most sophisticated radiocarbon techniques are of no use? Fortunately, there are lots of other radioactive isotopes with longer half-lives. One such isotope is potassium-40, which decays into the inert gas argon-40. It has a very long half-life—about 1.25 billion years. In certain minerals, the decay of an imbedded potassium-40 atom leaves the "daughter" argon atom trapped in the rock. If the rock is melted, the argon escapes. The ratio of potassium-40 to argon-40 in a volcanic rock can therefore be used to estimate how long ago the molten rock cooled into a solid form capable of trapping argon. Many rocks of volcanic origin can be dated using this **potassium–argon dating** technique. It is not as precise as radiocarbon dating, but it works over a far greater span of time, from around 2.8 billion years ago down to the present.

The major limitation of potassium–argon dating is that it can only be used on certain volcanic minerals. A fossil itself cannot be directly dated; and its geological context can be dated by potassium–argon only if that context is defined by strata of volcanic origin. In recent years, traditional potassium–argon dating has been largely replaced by the argon-40/argon-39 technique. This technique is more precise and can be applied to mineral samples that mix elements from more than one volcanic event. Argon–argon dating has been particularly useful in East Africa, which has been a hotbed of volcanic activity for the past ten million years. Other parent–daughter pairs of isotopes (and the parents' half-lives) that are used in dating various sorts of rocks include rubidium-87/strontium-87 (48.8 billion years), uranium-235/lead-207 (704 million years), and uranium-238/lead-206 (4.47 billion years). Again, each of these dating techniques has its own limitations and restrictions.

When radioactive atoms give off particles, they often do damage to materials in their vicinity. The ages of some materials can be estimated by measuring the amount of accumulated radiation damage. For instance, fissioning atoms of uranium-238 embedded in the crystals of certain minerals spit out energetic particles that leave tiny **fission tracks** in the crystalline material. Melting the crystal obliterates these tracks. The time elapsed since the material was last melted can therefore be estimated by comparing the number of tracks with the amount of ^{238}U remaining in the material. Sources of error in this method include the production of irrelevant fission tracks by external radiation sources (e.g., cosmic rays) and the erasure of fission tracks by high temperatures that soften the crystalline minerals. When such errors can be ruled out or corrected for, this technique of **fission-track dating** is useful for dating rocks ranging in age over the whole length of the geological column, from the oldest Precambrian strata up to a few thousand years ago.

When crystalline minerals absorb energy from radiation, electrons may become trapped within imperfections in the crystal lattice. Heating the material causes these electrons to escape, producing light. The amount of light produced by heating depends on the amount of radiation that has been absorbed—which in turn depends on the time that has elapsed since the material was formed or last heated. The elapsed time can be estimated by measuring either the light emission when the sample is heated (**thermoluminescent dating**) or the effect of the trapped electrons on the magnetic properties of the material (**electron spin resonance dating**, or **ESR**). These dating techniques can provide estimates for ages of two My (million years) or less. They are often used on fossil tooth enamel. Because buried rocks and fossils receive different amounts of radiation at different sites, these techniques must take account of such variable factors as the amount of uranium dissolved in the local ground water. They accordingly have relatively large margins of error, in the neighborhood of 15%. With ESR, date estimates will vary depending on whether it is assumed that the uptake of uranium occurs soon after burial (**early uptake**, EU) or is taken up at a constant rate after burial (**linear uptake**, LU). EU generally yields lower age estimates than LU. Either model might provide the more accurate date estimate, depending on the details of the particular case. However, LU date estimates usually agree more closely with dates obtained using other radiometric techniques (Schwarcz and Grün 1992, Grün 2006).

For all of these radiation-based techniques, the range of error of the estimated dates can be reduced by applying them to multiple samples of the same material and taking an average. Two or more techniques can also be used to cross-check each other by applying different tests to different materials found in the same stratigraphic level. Within their margins of error (which are usually less than 10%), all these dating techniques give consistent dates across the time span of the geological column. They allow us to infer that trilobites became extinct around 250 Mya (million years ago) and that the dinosaurs died out some 185 My after that.

OTHER DATING TECHNIQUES

Many organic molecules exist in two forms, left-handed (L) and right-handed (D, from Latin *dexter*, "right"), each a mirror image of the other. The amino acids that link together to form proteins in living organisms are all of the L type. As time goes by, some of them will flip over into the D configuration at predictable rates. In a living animal or plant, these right-handed amino acids are continually being removed and replaced by left-handed equivalents. But after death, they begin to accumulate in the degraded proteins of the dead organism at a rate that varies with the temperature of their surroundings. Eventually, the ratio of L to D forms will stabilize at 1:1, producing a so-called "racemic" (equal-ratio) mixture. When amino acids persist in ancient bones, teeth, or wood, the ratio of L to D forms can provide an estimate of the time since death. This technique of dating by **amino acid racemization** (AAR) can be used to estimate the age of organic material as old as 6 My. Because the rate of racemization depends on temperature, these estimates have large margins of error (around 20%). Early AAR dates were generally too old because the racemization rate was at first misestimated (Bada 1985). AAR estimates are also very susceptible to contamination by extraneous amino acids. As a result of all these problems, AAR is generally no longer used on bone. However, AAR dating on ostrich eggshell and some mollusk shell has shown more promise (Miller et al. 1993, Johnson and Miller 1997).

Similarly large sources of error attend other dating methods that depend on fluctuating environmental variables. These include **obsidian hydration dating** (based on the thickness of the hydrated surface layer that forms on buried stone tools), **fluorine dating** (based on the amount of fluorine that organic materials absorb from groundwater) and **uranium-series dating** (which depends on the rate of absorption of uranium isotopes into calcite crystals precipitating out of calcium-rich groundwater). Uranium-series dating incorporates a series of different decay phenomena, each of which has its own half-life and its own sources of error. All these methods can be used with more confidence to determine the *relative* ages of specimens from a single site. For example, one technique that helped to expose the "Piltdown Man" hoax (Chapter 6) was fluorine dating, which showed that the fraudulent fossils had been buried only recently in the gravels where they were found (Weiner et al. 1953).

DATING BASED ON THE CYCLES OF THE EARTH

When sediments are deposited in still water, iron-containing minerals in the sediments tend to line up with the earth's magnetic field. The resulting rocks retain a magnetic signature showing which way the compass needle pointed when they were laid down. When this **remanent magnetism** is measured across a sufficiently long stratigraphic sequence, it becomes apparent that the earth's magnetic field changes polarity at irregular intervals, so that what was previously the north magnetic pole becomes the south pole and vice versa. This produces a succession of "normal" and "reversed"

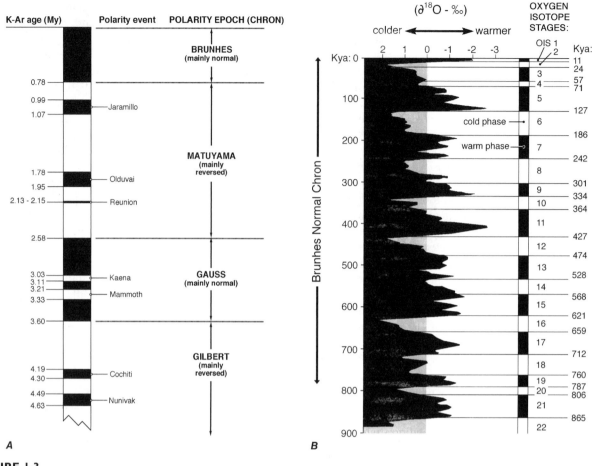

FIGURE 1.3

Geochronological sequences. **A:** Sequence of normal and reversed geomagnetic periods for the late Cenozoic. **B:** Alternation of warm and cold periods across Oxygen Isotope Stages 1 through 21, inferred from oxygen-isotope ratios in fossil foraminifera. The boundaries between stages are drawn on the basis of Milanković cycles. (After Butler 1992, Klein 1999, and Gradstein et al. 2005.)

periods of varying length (Fig. 1.3A), which is the same everywhere in the world. The magnetic "signature" of a rock stratum can therefore help to determine its age.

Inferring a date from a paleomagnetic sequence is no easy matter. Simply identifying a **reversal** (from normal to reversed polarity or vice versa) is not enough, because it must be determined which reversal it is. To do this, a date must be found for some part of the sequence using other techniques. Without such chronological anchors, a local reversal might correspond to any number of different reversals in the global paleomagnetic sequence. For example, the preferred dates for the Gran Dolina site at Atapuerca in Spain are now some 300,000 years older than those originally presented, because it was determined that the local paleomagnetic sequence had to be shifted back one reversal event on the global chart (Chapter 5).

Earth's history has witnessed countless climatic changes, many of which led to major shifts in the types

of organisms inhabiting the oceans and land masses of the planet. Many of the climatic changes that have occurred during the last 65 million years. and probably during earlier periods as well, can be explained by the astronomical theory of M. Milanković (or Milankovitch). Milanković noted that the motion of the earth through space fluctuates in a predictable way with respect to such variables as the shape and position of the earth's orbit, the angle of the axis of the earth's rotation, and the wobble of the earth on that axis as it spins. These **Milanković cycles** (Imbrie and Imbrie 1979, Berger et al. 1984, Berger 1992) all have different periods, and so they continually move in and out of phase with each other. When the cycles line up in a certain way, the earth's atmosphere receives less sunlight, and the planet grows cooler. During the coldest phases of these cycles, mean annual temperatures at high northern latitudes were as much as 16 °C colder than they are today. Even near the equator, mean temperatures dropped some

3-5 °C, and intermediate regions experienced intermediate amounts of cooling (Stanley 1989).

The most recent of these cooling events resulted in a fluctuating advance and retreat of continental ice sheets during the **Pleistocene epoch** (Flint 1971, Van Couvering 2000a,b; Stanley 1989). The Pleistocene (from the Greek meaning "most recent") lasted from around 2 Mya until the latest retreat of the continental ice some 12 Kya. The Pleistocene was characterized by a series of climatic oscillations between cold **glacial** periods and warmer **interglacial** phases. In the colder periods, annual snowfall exceeded annual snowmelt in high latitudes and at high altitudes, resulting in the growth of glaciers. Tremendous amounts of water became tied up in these masses of ice, and sea levels fell correspondingly. At the last glacial maximum, sea level was 100-140 m lower than it is today (Flint 1971). Within the glacial periods, there were brief episodes of warmer temperatures characterized by small-scale ice retreats. These periods are called **interstadials**, and the intervening periods of maximal cold are known as **stadials**. By convention, the term "interglacial" is restricted to the warmest periods of the Pleistocene, when glaciers were completely absent from temperate lowland regions. We are living in one such interglacial period. Geologists exclude this latest interglacial from the Pleistocene and dignify it with its own special name as the **Holocene** or Recent epoch, mainly because it contains us.

In 1909, A. Penck and E. Brückner defined four Pleistocene glacial maxima on the basis of their observations in the Bavarian Alps. These glacial periods were named (from oldest to youngest) the Günz, Mindel, Riss and Würm. This four-phase scheme was subsequently extended to all of Europe. Similar systems were adopted for other northern continents, and the four glacial periods were thought to correspond to "pluvials" (periods of increased rainfall) in Africa. But over the years, it became increasingly clear that this scheme was far too simplistic and that the whole Pleistocene world did not march in time to the same four-beat pattern (Van Couvering 2000b).

The complexity of Pleistocene climate fluctuations is most clearly seen in sediments sampled by drilling into the deep ocean bottoms, where sedimentation is unremitting and the Pleistocene stratigraphic sequence contains few gaps (Shackleton 1967, 1975, 1987, 1995; Bassinot et al. 1994). The distribution of oxygen isotopes in microfossils from these sediments provides an indirect record of planetary temperatures. Ocean water contains two stable isotopes of oxygen, ^{16}O and ^{18}O. Both isotopes combine with hydrogen to form water (H_2O). When seawater is evaporated from the oceans and deposited in glacial ice, those water molecules formed with ^{16}O are evaporated more easily than those with the heavier ^{18}O isotope—and so the proportion of ^{18}O to ^{16}O in the remaining ocean water is increased.

When oxygen gets incorporated into the skeletons of foraminifera and other marine organisms, the $^{18}O/^{16}O$ ratio in these remains reflects the ratio present in the seawater at the time these animals lived. We can therefore measure the amount of water locked up in ice—and therefore estimate the overall temperature of the earth—by determining the $^{18}O/^{16}O$ ratio in the fossil shells trapped in sea-bottom sediments.

The **oxygen-isotope stages** (OIS) for the past 900 Ky are presented in Fig. 1.3B. By convention, the odd-numbered stages represent warm periods, with OIS 1 being the current or Holocene interglacial. Designating the present as an interglacial period may be a bit unnerving, but the expert consensus is that the Pleistocene is not over and that we can expect the glaciers to return. In fact, data from the deep-sea studies indicate that the intensity of each successive glacial advance has tended to increase over the most recent glacial periods (Stanley 1989). This could mean that the next one will be a humdinger—unless the well-documented human influences on today's climate disrupt the Milanković cycles.

The OIS chart reveals 11 glacial periods during the last 900 Ky, but there are also numerous stadials and interstadials in each of the 22 OIS recognized here. It may be significant that none of the warm phases over this period are quite as warm as OIS 1—except for the last major interglacial, OIS 5e. None of the other warm periods reach this level (Fig. 1.3B), and they are of rather short duration (perhaps 10-15 Ky). Maybe the glaciers will be returning sooner than we think! The dates given for the temperature fluctuations shown in Fig. 1.3 of course cannot be inferred from the $\partial^{18}O$ data themselves and must be determined by other methods, including both radiometric and relative dating techniques.

▌THE PROBLEM OF OROGENY

Until the middle of the 20th century, there was a big mysterious piece missing from geological theory. It seemed clear that sedimentary rocks had been formed by the accumulation of mud, sand, and gravel under lakes and seas. But it was not clear how these rocks had been lifted above the waters to become exposed on dry land. The lifting processes, whatever they were, had to be at least as powerful as the processes of erosion, or else the continents would long ago have been worn away by ice, wind, and rain and scattered as sediment beneath the waves.

In the aftermath of an earthquake, it was sometimes found that a block of the earth's crust had been thrust upward a few inches along one side of a crack or **fault** in the rock. Such faults often extended for hundreds of miles. Perhaps sudden sliding or slipping movements along these faults were the cause of earthquakes. Study of the sedimentary rocks on opposite sides of fault lines showed that movement along some faults had continued in the same direction for a long time, so that once-continuous rock formations had become displaced from each other by many vertical feet or horizontal miles. If this process of earth movement went on long enough, it might result in the buckling of the moving crust and the formation of mountains.

It was evident that faults, earthquakes, and **orogeny** or mountain-building all had to be connected somehow. The newest mountains (the ones that are highest, least eroded, and most jagged) invariably are found in areas rich in faults and shaken frequently by earthquakes—for example, along the western edges of North and South America. But what colossal forces could be responsible for shoving vast chunks of the continents around and heaving beds of solid rock up out of the ocean floor to stick up miles above the sea surface?

Some 19th-century geologists conjectured that the earth is still cooling from its original molten state, and that as it cools it shrinks. In theory, the shrinking of the earth might cause the crust of the dwindling planet to wrinkle and buckle like the skin of a drying apple, producing mountains and valleys. But this account did not explain why the mountainous "wrinkles" were found in some regions and not in others. And as more was learned about geophysics, it became clear that the earth is not cooling at all—that it in fact continually gives off more energy than it receives, radiating surplus heat produced in its interior by the decay of radioactive atoms.

▌CONTINENTAL DRIFT

In 1915, a German scientist named Alfred Wegener proposed that all the continents of the earth had at one time been fused together into a single immense land mass. He called this imaginary supercontinent Pangaea, which is Greek for "all earth." Wegener marshaled data from geography, stratigraphy, and paleontology to support his ideas. He showed, for instance, that the Atlantic edges of Africa and South America fit together like pieces of a huge jigsaw puzzle and that when the two were put together, the older stratigraphic rock formations of one continent matched up neatly with those of the other.

Wegener also noted that his theory would help to account for the otherwise puzzling distribution of certain plants and animals. There are some groups of organisms that are found on all the southern continents but do not occur in more northern lands, either alive or as fossils. If the continents had always held their present positions on the globe, these taxa would have had to get into Africa, Australia, and South America by traveling through Eurasia and North America without leaving any traces of their passage. But if those southern continents had been attached directly to each other in the past, their shared fauna and flora would no longer pose a mystery.

Wegener's hypothesis of **continental drift** was listened to with interest and mentioned politely in geology textbooks. But it was not taken too seriously for many years, because Wegener was unable to come up with a plausible mechanism that could have moved the continents around in the way his theory required. The rocks of the earth's crust are of two basic sorts: magnesium-rich igneous rocks such as basalt, which form the bedrock of the ocean basins, and lighter and softer rocks rich in aluminum, which make up the substance of the continents. Wegener's theory seemed to demand that the soft rocks of the moving continents had somehow plowed their way through the harder and denser rocks of the ocean basins, like a ship made of butter sailing through a sea of clay. This was clearly impossible.

The mystery was gradually cleared up in the 1950s and 1960s, as research ships began exploring the topography and stratigraphy of the deep ocean floor. It was found that the basaltic rocks of the ocean basins are divided by fault lines into **crustal plates**. These plates are in motion relative to each other. In some places, adjoining plates are moving apart, leaving gaps or **rifts** in the crust, through which molten rock from the earth's mantle wells up to form new basalts and chains of volcanic mountains. In other places, where plates are colliding, the edge of one plate is riding up over another, heaving the buckling edge of the top plate skyward to form jagged new mountain ranges and forcing the bottom plate down into the mantle to be remelted.

The movements of the continents are now thought to be driven by convection currents produced in the hot, fluid rock of the mantle by the planet's internal heat. As the basaltic rocks move around and around in these great, slow cycles of subduction and re-eruption, they carry the lighter, overlying continents along, tearing them apart and thrusting them together like islands of froth circulating on the surface of a gently simmering pot of soup. The resulting theory of **plate tectonics** has solved the riddle of orogeny, filled the old gap in our understanding of sedimentary processes, and vindicated the insights of Alfred Wegener.

The continual disappearance or **subduction** of crustal material into the planetary interior contributes to the steady loss of ancient rocks. The earth has had a solid crust for over four billion years, but the oldest rocks known are less than 3.8 billion years old. All the older rocks have long since been drawn down into the mantle, or have been eroded away and redeposited as new sediments on the ocean floor. Because rocks do not last forever, older rocks get harder to find as we move down the geological column—and so the fossil record becomes more and more fragmentary the farther back in time we look. We accordingly know far more about the origins of humankind than we do about those of (say) vertebrates; and we know almost nothing at all about the origin of life itself.

LIFE: THE FIRST THREE BILLION YEARS

Although the origin of life is still a mystery, we have a fairly detailed idea today of what it would take to upgrade nonliving carbon-based molecules into self-reproducing systems. Because carbon is one of the most common chemical elements, the rocks and air of the early earth must have contained a lot of simple carbon compounds—carbon monoxide and dioxide (CO, CO_2), hydrogen cyanide (HCN), hydrocarbons, and so on—just as many of the sun's other planets do. Once the ocean had condensed out of the earth's steamy primitive atmosphere, some of these carbon compounds would have dissolved in its waters. Others would have floated in oily wisps on the sea surface. When concentrated in certain favorable sites—by natural distillation in evaporating lagoons, for instance, or by binding to the surface of clay minerals—these compounds could have reacted with each other and combined to produce somewhat more complicated organic molecules, including amino acids, sugars, purines, and pyrimidines. Polymerized amino acids are **proteins**, and polymerized sugars with purines and pyrimidines attached are **nucleic acids**—and proteins and nucleic acids are the key constituents of life.

The major stumbling block in understanding the origin of life is a chicken-and-egg problem. In today's world, the nucleic acid DNA carries the hereditary information in all self-reproducing organisms. DNA needs specific protein catalysts to replicate itself. But those protein catalysts are themselves synthesized by reading out instructions coded in the DNA. It is not clear how instructions for making the proteins needed to make DNA could have gotten encoded in a DNA molecule. The first self-replicating molecules may have been compounds capable of catalyzing their own synthesis—for example, ribonucleic acid (RNA), which carries the hereditary information in some viruses (Orgel 1998). Perhaps proteins and nucleic acids teamed up only belatedly, after a long period of independent evolution. At the moment, we have no way of telling. We may never know until we find and study other earthlike planets. It is hard to understand phenomena, such as the origin of life or of the universe, for which we have a sample size of one.

Once a molecular complex appeared that was capable of making copies of itself, natural selection would have begun to work on those copies, preserving any variations that out-reproduced the others. Naked molecules of nucleic acids that turned up in local puddles of organic chemicals may have been the first self-copying systems. However, they could not spread beyond those puddles until they became enclosed in some sort of envelope that let them take a bit of the puddle along. Resembling a stripped-down version of a modern bacterium, the first living cell would have been a minute droplet of a watery solution of proteins and other chemicals, wrapped in a fatty cell membrane to separate it from the surrounding water. It contained a single strand of DNA, which carried the necessary information to produce several different sorts of proteins. Those proteins catalyzed the cell's metabolic processes (including DNA replication), participated in the cell membrane, and operated to produce more proteins from the genetic instructions coded in the DNA.

The smallest and simplest self-reproducing systems on earth today are almost this simple. These tiny organisms, called **mycoplasmas**, are biochemical cripples that need to get their food and energy by absorbing fairly complex molecules directly from the surrounding fluid—for example, in human lungs, where they cause a form of pneumonia. The earliest cells may have made a meager living in this way by floating around in the thin organic soup of the early ocean, waiting to bump into nourishing molecules of ammonia and sugar.

More advanced microorganisms, including bacteria, are more self-sufficient. A key step in the evolution of self-sufficiency was the development of enzymes that enabled microbes to make their own sugar. Some of them did this, as some bacteria still do, by using light energy from the sun to drive metabolic reactions that combined the smelly gas hydrogen sulfide (H_2S) with CO_2 to yield the sugar **glucose** ($C_6H_{12}O_6$) plus elemental sulfur (S) and water: $12H_2S + 6CO_2 \rightarrow C_6H_{12}O_6 + 12S + 6H_2O$. The main disadvantage of this strategy is that H_2S is not very common. A later modification used different light frequencies to power more complex reactions in which the ubiquitous compound water, H_2O, replaces H_2S in the system, so that oxygen instead of sulfur is given off as an end product: $6CO_2 + 6H_2O \rightarrow C_6H_{12}O_6 + 6O_2$. This new process of **photosynthesis** was a great evolutionary success. Bacteria equipped with it, known as **cyanobacteria**, rapidly became the most abundant life forms in the ancient oceans, and they remain one of the most abundant forms of life today.

Life could not have come into being until around 4 billion years ago, when the earth's surface had cooled enough to let liquid water condense on it. Filaments of carbonaceous matter found in 3.5-billion-year-old Australian chert are interpreted by some as fossils of bacteria (Schopf 1999), but the authenticity of these and other supposed traces of exceedingly ancient life is disputed (Brasier et al. 2002, 2005; Altermann and Kazmierczak 2003, Lepland et al. 2005, De Gregorio and Sharp 2006).

More solid evidence of life appears around 2.8 billion years ago, in the form of indisputable fossils— stacked-up bacterial mats called **stromatolites**—and complex organic compounds in ancient shales. Photosynthesis had probably begun by this time (Olson 2006), but the earth's atmosphere was not yet oxidative. Fine-grained sedimentary rocks older than 2.2 billion years contain granules of pyrites and other easily oxidized minerals. If the air back then had been rich in corrosive oxygen gas, those minerals would have oxidized and decomposed into other compounds while they were being broken up into fine granules of sediment (Knoll 2003). The definitive onset of an oxidative atmosphere is signaled by the deposition around 2 billion years ago of the world's major deposits of iron ore, laid down when reduced (ferric) iron ions that had been dissolved in the primitive ocean combined with oxygen, turned into insoluble ferrous oxide, and precipitated out as vast beds of rust-colored, iron-rich sediments on the floor of the sea (Klein and Buekes 1992).

The release of vast amounts of oxygen gas into the air and water of the earth by cyanobacteria must have had much the same catastrophic effects on microbial life that the chlorination of a swimming pool has. By the time a new equilibrium was finally reached around 2 billion years ago (Strauss et al. 1992), most of the earth's original life forms had been killed off. Those that survived did so only by retreating into deep, dark crevices and crannies where the deadly oxygen could not penetrate. Their modern descendants, the **anaerobic bacteria**, still populate such refuges today, buried deeply in soils and sediments or hiding inside the guts of people and other animals. They occasionally break out and take revenge on oxygen-breathing organisms by infesting their tissues, causing such infections as peritonitis and gangrene.

Atmospheric oxygen was a disastrous setback for the earth's anaerobic organisms, but it presented the cyanobacteria and their descendants with fabulous metabolic opportunities. Earlier life forms had metabolized glucose by breaking each six-carbon glucose molecule apart into two molecules of a three-carbon compound called **pyruvate**. In the new oxygen-rich world, that pyruvate could be further combined with the readily available oxygen to yield carbon dioxide and water. This novel **aerobic** form of respiration brought an almost tenfold increase in the energy generated by metabolizing a sugar molecule. The pyruvate wastes that had been the ashes of the fire of life now became a plentiful new fuel.

One group of microbes managed to cope with the flood of oxygen by enclosing its vulnerable DNA in an inner, protective membrane, forming a cell **nucleus** separated from the surrounding **cytoplasm**. A branch of this group set out on a new evolutionary path by

forming a partnership with some cyanobacteria. They did this by growing larger and bringing the cyanobacteria "indoors," enclosing them within pinched-off inpocketings of the host cell's surface. The imported cyanobacteria brought with them not only their own separate DNA, but also all their valuable enzymatic machinery for aerobic respiration.

These cells appear to have incorporated at least two different strains of "domesticated" bacteria, specialized for doing different jobs. The cyanobacteria that did the job of making air and water into glucose kept their photosynthetic green pigment, chlorophyll, and turned into **chloroplasts** (Chu et al. 2004). Those that handled the task of aerobic respiration, oxidizing the glucose back into CO_2 and water to yield energy, became **mitochondria**. The resulting **eukaryote** (Greek, "true nucleus") cell was a relatively large membranous sac containing the cell nucleus, ribosomal protein factories, and other components suspended in a watery broth of dancing organic molecules, transporting raw materials, waste products, and energy back and forth between the cell contents, the cell surface, and the imbedded chloroplasts and mitochondria (Fig. 1.4).

All later plants and animals are descended from those first eukaryotes. Plants have retained both chloroplasts and mitochondria. We animals and our relatives, the fungi, lost our chloroplasts early on (or are descended from primitive eukaryotes that had not yet acquired them), and so we are not able to make our own food. But almost all the cells in our bodies contain mitochondria descended from the immigrant bacteria of long ago. Even today, after more than a billion years of symbiotic evolution, our mitochondria are still separated from the cell's cytoplasm by a double barrier (their own membranous envelope surrounded by a pinched-off vesicle of cell membrane), reproduce separately from the rest of the cell, and have their own DNA. As we will see later, mitochondrial DNA has played an important part in the study of recent human evolution.

The earliest fossils generally accepted as eukaryotes are simple collapsed spherical sacs from northern China, dated to around 1.8 billion years ago (Zhang 1997). Their fossils are recognizable as eukaryotes because they are several times larger than even the largest bacteria, although they are still one-celled and visible only through a microscope. Fossils that can be seen by the naked eye—coiled-up, millimeter-wide ribbons called *Grypania*—date back even further, to about 2.1 billion years (Han and Runnegar 1992). These ribbons may be fossils of early multicellular algae. But some living cyanobacteria clump together in similar-looking colonies, so it is not clear whether *Grypania* is an early fossil eukaryote or a colonial bacterium.

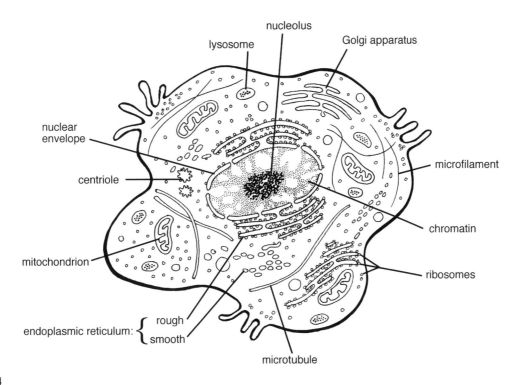

FIGURE 1.4

Diagram of a eukaryotic (animal) cell, showing some of the principal internal structures. Nuclear DNA and proteins form linear complexes called chromosomes, which appear throughout the nucleus as a diffuse material called **chromatin** when the cell is not actively dividing. The **nucleolus** is a center for the assembly of protein-making units called **ribosomes**. These nuclear structures are surrounded by a double-walled **nuclear envelope** confluent with a complexly folded **endoplasmic reticulum**. Part of that reticulum is called "rough" because it is studded with ribosomes synthesizing proteins. The **Golgi apparatus** is another system of folded sacs, within which various chemicals are stored and processed. **Microtubules** and **microfilaments** are structural elements involved in maintaining and changing the shape of the cell. Large organic molecules are hydrolyzed into smaller ones inside digestive sacs called **lysosomes**. **Centrioles** function in cell division. **Mitochondria** are centers for the respiratory processes that provide energy for the other chemical reactions of the cell. (After Campbell 1993.)

▌MULTICELLULAR LIFE

Just as eukaryotic cells originated as colonies of bacteria, so multicellular organisms originated as colonies of eukaryotic cells. Most multicellular organisms, including human beings, develop as **clones**—that is, all their component cells are genetically identical, derived from a single progenitor cell by repeated cell division. In multicellular organisms, that progenitor cell is usually formed by the fusion of two smaller cells called **gametes** (eggs or sperm). Each gamete is derived from one parent. It contains one copy of the total genetic complement, or **genome**, of the species to which the organism belongs. The progenitor cell formed from the fusion of two gametes therefore has two copies of the genome. So do the cells that are cloned from it as the embryo develops. Biologists refer to these two-copy cells as **diploid** and the one-copy gametes as **haploid**. When the new diploid organism eventually produces its own

haploid gametes, the separate genomes derived from its two parents get mixed or **recombined** in the process, so that each gamete carries genes from both parents. When these gametes combine with those from another individual, the resulting new progenitor cell (the fertilized ovum, or **zygote**) will develop into another organism that inherits genes from four grandparents but is genetically different from all of them.

This intricate two-parent reproductive arrangement is called **sex**. It represented a major advance in the mechanics of evolution, because it made it easier for favorable mutations that cropped up in two different individuals to get together in their descendants. Bacteria manage to do something similar in other ways, partly by picking up bits of DNA from dead bacteria and sticking them into their own genomes, but this system will not do for multicellular plants and animals. Although all the cells of a many-celled organism are cloned from a single starter cell, they wind up forming different organs in the adult, with divergent structures and functions. Such organisms therefore need to have a hierarchically arranged genome, in which some of the genes function to switch other genes on or off during development, so that the cells in different organs can develop in different ways in spite of their genetic sameness. Many-celled organisms therefore need to have more delicately adjusted and integrated genomes, and cannot evolve by simply scrounging up genetic information from things they eat. The invention of sex in one-celled eukaryotes provided a more useful and sophisticated method of genetic recombination, which aided and accelerated the evolution of multicellular life.

The earliest multicellular animals must have been tiny and soft-bodied. They have left no traces in the fossil record. Sponges have been recovered from 580-million-year-old rocks in China (Liu et al. 1998). Tiny, globular, many-celled fossils found in slightly younger Chinese rocks appear to represent the early embryonic stages of some multicellular animal (Xiao et al. 1998). Undisputed multicellular animals, in the form of small, nondescript conical shells, occur all over the world in deposits dated to the very end of the **Vendian period** (Fig. 1.2), about 550 Mya. Immediately after that, at the beginning of the **Cambrian period**, these so-called "small shelly faunas" are replaced by a great variety of fossil marine animals that appear seemingly out of nowhere. Because this sudden "Cambrian explosion" of fossils is a worldwide marker of this time horizon in the sedimentary rocks, the boundary between the Vendian and Cambrian periods has been adopted as a line separating the two primary divisions of the geological column: the **Precambrian eon** and the following **Phanerozoic** (Greek, "visible animals") **eon**, roughly corresponding to the "Primary" and "Secondary" rocks of the early geologists.

For a long time, it had been thought that the Precambrian was devoid of multicellular fossils. Their abrupt appearance and great diversity at the beginning of the Cambrian period was perceived as a troubling mystery. Then in 1946, fossils of multicellular organisms were recovered from Vendian deposits around 570 My old at the Ediacara copper mine in South Australia. Since then, geologists have found similar **Ediacaran** faunas in Precambrian rocks from other parts of the world as well.

Unfortunately, these Precambrian discoveries have not cleared up the mystery of the Cambrian explosion. Most of the larger Ediacaran organisms are not likely ancestors of any Cambrian life forms. They appear to have been flattened mats of pasted-together tubes, growing outward from a central core or axis. In cross-section, these things probably looked like an inflatable swim-mat or air-mattress. Nothing much like this body plan is found in any organisms known from later periods. One theory about these creatures holds that the hollow tubes acted like greenhouses, in which photosynthetic

or chemosynthetic bacteria grew in a protected environment and provided their tube-making host with nutrients (McMenamin and McMenamin 1989, Runnegar 1992, Seilacher 1994).

Why was that protected environment necessary? Vendian **trace fossils** hint at an answer. These fossils can be described as worm tracks in mud—traces left by multicellular animals moving in or along the surface layer of the ocean floor. The unknown creatures that left these Precambrian tracks were probably grazing on the thin mat of cyanobacteria and other one-celled organisms that covered the bottoms of the Precambrian seas. The hollow tubes of the Ediacaran "air-mattress" creatures may have protected their internal microorganisms from being eaten by these wormlike or sluglike grazers.

M. and D. McMenamin (1989) have called this late Precambrian phase in the history of life the "Garden of Ediacara," with reference to the Garden of Eden in the Bible. And here at the dawn of multicellular life, there does appear to have been something rather like a Peaceable Kingdom. Ediacaran life consisted mainly of microscopic plants and bacteria living directly or indirectly by photosynthesis. A few larger organisms lived off the bacteria, either by filtering them out of sea water (as those Vendian sponges did), or by slurping them up from the sea floor (the worm tracks), or by culturing them internally and metabolizing their byproducts (the "air-mattress" creatures). But there are no signs that multicellular animals had started eating each other. No jaws, teeth, or defensive structures—no shells or carapaces or spines or spikes—are seen in the Ediacaran fossils. We can assume that the contest between predator and prey had not yet begun, or at any rate had not yet progressed beyond a very low level of intensity. This was a world without weapons or armor, and therefore it was a world without speed and a world without brains.

■ THE CAMBRIAN REVOLUTION

Quite suddenly at the end of the Precambrian, in the space of a few million years of transition, hundreds of genera of armored fossil animals appear, covered with shells, carapaces, spines, spikes and plates of keratin, chitin, and calcium carbonate. Among the most common and familiar of these Cambrian newcomers are the trilobites, which vaguely resemble armored seagoing centipedes. These creatures, once wildly successful but now extinct, represent a major group (phylum) of animals called **arthropods**—segmented animals with external skeletons of chitin and many jointed legs. Other arthropods are still wildly successful today. They include centipedes, spiders, scorpions, and all the innumerable hosts of insects on the land and crustaceans in the sea.

Most of the other major groups of living invertebrates are represented in the Cambrian, including molluscs, brachiopods (lamp shells), echinoderms (starfish and their relatives), and several phyla of worms, as well as a lot of strange-looking creatures of uncertain relationships.

Unlike their Ediacaran predecessors, these Cambrian animals lived by eating other multicellular organisms, killing them and consuming their tissues just as humans and most other animals do today. The advent of the new ecology of predators and prey is signaled by the widespread appearance of some common anatomical features in many Cambrian animals.

The most fundamental of these is **bilateral symmetry**. In contrast to most Precambrian organisms, the primitive members of most of the Cambrian phyla have a head end, a tail end, and left and right sides. This tells us that these creatures were specialized for moving in one particular direction. The head, which is the end of an animal that first encounters new stimuli as it moves forward, is usually distinguished from the other parts of the body by having eyes, feelers, or other special sense receptors. Such heads presumably housed some sort of central nerve ganglion to receive incoming stimuli from those receptors and coordinate the animal's responses to them.

Heads, tails, and sense organs do not necessarily imply a predatory lifestyle. They all must have been present in some form in the wormlike microbe-grazers that left those tracks in the Precambrian mud. But bilateral symmetry and its correlates make it easier to evolve predatory adaptations. The advent of predation in the Cambrian is attested to by the appearance in many Cambrian animals of three key anatomical innovations: jaw-like organs around the head end of the gut for seizing prey, strong propulsive swimming paddles adapted for swift evasion or pursuit, and **defensive armor** against predators. Precambrian fossils are rare not only because Precambrian rocks are rare, but also because few Precambrian organisms had any hard parts. The widespread simultaneous emergence of defensive structures in many Cambrian phyla—mollusc and brachiopod shells, the calcareous plates of echinoderms, trilobite exoskeletons, and so on—bears witness to the onset of new selection pressures caused by the "invention" of predation. With the advent of this new ecology, the Garden of Ediacara was closed, and the harmless air-mattress creatures of the Precambrian swiftly vanished from history. The arms race between the eaters and the eaten has been a major engine of evolutionary change throughout the subsequent history of animal evolution.

Cambrian members of our own phylum, the Chordata or **chordates**, had no bones or other hard parts; but fossils of them have been found recently at two Lower Cambrian sites in China where soft-bodied animals have been preserved in great detail (Chen et al. 1999, Shu et al. 1999). Some of the Chinese fossils appear to represent primitive chordates something like the living lancelet, *Branchiostoma*, a wormlike animal about 5 centimeters long that survives today in shallow seas off the coasts of Asia. *Branchiostoma* and its Cambrian relatives exhibit four main characteristics that we find in all chordates (Fig. 1.5). The first is a rubbery stiffening rod called the **notochord**, a kind of primitive backbone that runs down the back of the animal. The second chordate trait is another cord, made of nerve cells—the **spinal cord**, lying between the notochord and the skin. The third chordate trait is **segmented muscles** attached to either side of the notochord. These muscle segments, or **myotomes**, are the "flakes" that separate from each other when you eat a cooked fish with a fork. In a living fish or lancelet, they wiggle the animal's propulsive tail and send it scooting through the water. Nerve impulses passing through the spinal cord coordinate the contractions of these blocks of muscle with each other and with stimuli that the animal detects in its environment. The fourth chordate trait seen in *Branchiostoma* is a series of holes called **gill slits** in the side walls of the throat. *Branchiostoma* feeds by pumping sea water out through these slits and swallowing any solid leftovers. All vertebrates, including ourselves, exhibit these four chordate characters in one form or another. In humans, the noto-

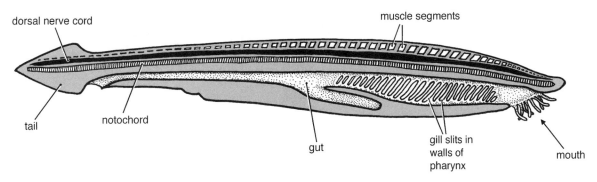

FIGURE 1.5

Major characteristics of the phylum Chordata, diagrammed as they appear in a primitive living chordate (*Branchiostoma*).

chord and gill slits are transient embryonic structures, but the segmented muscles and spinal cord persist throughout life.

The lancelet is a chordate, but it is not a vertebrate in the strict sense. Unlike true vertebrates, it has no jaws, no brain, no eyes or other complex sense organs at the front end—just gill slits and a ring of tentacles around the mouth. Vertebrates are distinguished from more primitive chordates like *Branchiostoma* by two major innovations: the head and the skeleton.

The first of these innovations to evolve was a true **head** with eyes and brain, which was more or less simply added onto the front end of the animal in front of the notochord. Some of the Cambrian chordates from China evidently had heads, though they appear to have lacked eyes (Chen et al. 1999, Shu et al. 1999). More advanced cranial structures are seen in other early chordates called **conodonts**, which were equipped with a pair of big eyes and rows of prey-shredding calcareous spikes inside the mouth (Fig. 1.6A). True vertebrates added a second big innovation—an internal **skeleton** of bone or cartilage, with a brainbox surrounding the brain and a string of vertebral elements forming a primitive backbone around the notochord. The first vertebrate skeletons appear in the **Ordovician** period, which followed the Cambrian (Fig. 1.2). Known only from

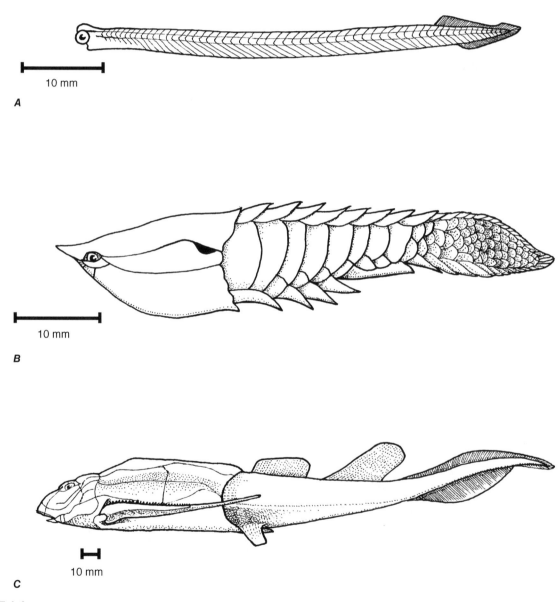

A

10 mm

B

10 mm

C

10 mm

FIGURE 1.6

Paleozoic fishes. **A:** Lower Carboniferous conodont (*Clydagnathus*). **B:** Armored jawless fish (*Anglaspis*). **C:** Early jawed fish with pectoral and pelvic fins (*Bothriolepis*). (**A**, after Sweet and Donoghue 2001; **B–C**, after Moy-Thomas and Miles 1971.)

fragmentary remains, these Ordovician vertebrates appear to have been primitive fishes something like living lampreys, which are found today in rivers and lakes and seas all around the world. Lampreys have no jaws, but they have eyes of the standard vertebrate sort, along with a skull and vertebrae formed from cartilage, and they also have a small but clearly vertebrate brain elaborated out of the front end of the spinal cord.

Most of the jawless fish known as fossils from the Ordovician onward had more rigid and elaborate skeletons, made of bone instead of cartilage. In addition to having a braincase and vertebrae, these early fish had an extensive layer of osseous plates lying beneath the skin, forming a sort of dermal armor that helped ward off attacks from predators. The bony shield over the vulnerable head was especially solid (Fig. 1.6B). We still preserve many of these **dermal bones** in our own skulls.

▌JAWS, FINS, AND FEET

Vertebrate jaws evolved from (or developed as an anterior extension of) the rows of bones that form in between the gill slits. The first vertebrates with jaws appear in the fossil record in the early **Devonian** period, about 400 Mya. These early jawed fish sported another innovation that was crucial for the emergence of humankind: four **paired fins**, comprising a front pair of **pectoral fins** just behind the head and a hind pair of **pelvic fins** back near the anus (Fig. 1.6C). In most of these Devonian fish, the vulnerable eyes and brain were still protected by a head shield of bony armor. But the rest of the body wore a sort of flexible chain-mail coat of small scales made of a bonelike tissue called **dentine**, coated with hard, shiny **enamel**. The scales inside the mouth were pointed, with sharp tips to help hold and tear prey. We still have a set of these enamel-coated scales in our own mouths: our teeth.

The scales on the bodies of most modern fish have been pared down into delicate little translucent chips. Human beings, however, are descended from a group that retained the big, primitive, enamel-covered scales for a long time. These fish are called **sarcopterygians**, from the Greek for "fleshy fins," because their fins were thick and muscular (Fig. 1.7A). It was lucky for us that their fins were robust, because this made it easier for those fins to function as arms and legs when some of these fish began coming out of the water. The larger bones inside the fins of some fossil sarcopterygians can

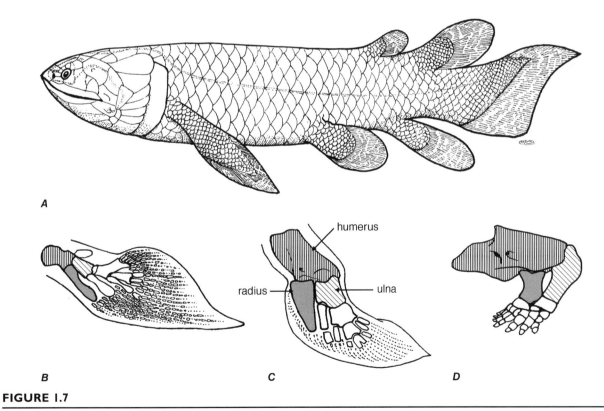

FIGURE 1.7

Fins and limbs. **A:** Devonian sarcopterygian *Holoptychius*. **B–D:** Pectoral fins of *Eusthenopteron* (**B**), a sarcopterygian fish close to the origins of the tetrapod stock; the transitional fish-like Devonian tetrapod *Tiktaalik* (**C**); and the early tetrapod *Acanthostega* (**D**), showing homologies and successive stages in the evolution of the humerus, radius, and ulna. (**A**, after Moy-Thomas and Miles 1971; **B–D**, after Shubin et al. 2006.)

be matched up, bone for bone, with the major bones in our own arms and legs (Fig. 1.7B–D). Surviving sarcopterygian fish include the ocean-dwelling coelacanth *Latimeria* and three species of lungfish that inhabit seasonal bodies of fresh water in the southern continents. Land vertebrates, or **tetrapods** (Greek, "four feet"), evolved from sarcopterygians in the late Devonian. Unlike the rather stiff, flipper-like fins of the earlier sarcopterygians, the modified fins—which we now must call arms and legs—of these Devonian tetrapods stuck out sideways and bent downward to reach the ground. They were both stouter and more limber than the ancestral fins. Their joints, bones, and muscles were big and strong enough to carry the animal's weight when it was out on land, but flexible enough to allow the limbs to swing freely back and forth in the cyclical movements of walking. The little plates of bone that had lain at the bases of the ancestral fins were expanded into **scapulae** (shoulderblades) and **hipbones** (Fig. 1.8), providing bigger areas of attachment for the enlarged limb muscles. The fan of radiating bony fin supports found in the fins of sarcopterygian fish (Fig. 1.7B) became strengthened and simplified into a single stout proximal element (the **humerus** in the arm, the **femur** in the thigh) and a pair of distal elements (the **radius** and **ulna** in the forearm, the **tibia** and **fibula** in the lower leg), terminating in a cluster of wrist or ankle bones (**carpals** or **tarsals**) and a spray of **digits** (fingers or toes). Each digit contained a basal bone called a **metacarpal** or **metatarsal** and a terminal string of small bones called **phalanges**. Most of these early tetrapod elements persist in human limbs.

The viscera of these land-going fish were supported and protected by enlarged ribs curving down from the backbone toward the belly. At the tail end of the trunk, the hipbones grew up to touch the ribs and down to touch each other in the midline of the belly, thus forming a complete bony ring—the **pelvis**—that provided a firm foundation for the hind limb. In most living tetrapods, including humans, the ribs that touch the hipbone fuse together with their vertebrae in the adult to form a single composite bone called the **sacrum** (Fig. 1.8).

Getting around on land is one of the two biggest problems faced by a fish out of water. The other major problem it confronts is water loss. The early land vertebrates evolved eyelids and tear glands to keep their eyes from drying out, but (like the frogs and salamanders, which are their least-changed modern descendants) they retained the moist, glandular skin of their fishy forebears. Therefore, they had to stay in humid surroundings to avoid desiccation. They also had to return at mating time to the water, as frogs do. Out of the water, their ejaculated sperm would not have been able to swim to the eggs and fertilize them, and the jelly-covered eggs would have been unable to develop into water-breathing, tadpolelike larvae.

These early tetrapods were not the first organisms to colonize the land. If they had been, there would have been nothing for them to eat when they got there. Plants had preceded the vertebrates onto land by some 30 My, and various invertebrate groups—including the ancestors of insects—followed the plants onto the land not long afterward. The abundance of edible arthropods on land may have been one of the things that led some sarcopterygian fish to start coming out of the water.

The appearance and spread of land plants altered the earth's atmosphere. Changes in the chemistry of the

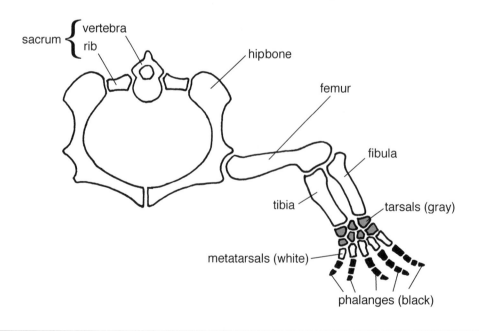

FIGURE 1.8

Diagrammatic anterior view of pelvis and one hind limb of an early terrestrial vertebrate.

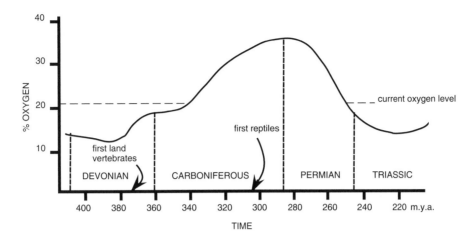

FIGURE I.9

Atmospheric oxygen levels in the late Paleozoic and early Mesozoic. (After Graham et al. 1997.)

sedimentary rocks suggest that levels of oxygen in the air began to rise some 380–400 Mya (Fig. 1.9), at about the same time that plants were colonizing the continents. During the **Carboniferous** period that succeeded the Devonian, great forested swamps spread across the continental lowlands, and land plants underwent an evolutionary radiation. Much of our coal today is mined from deposits laid down in these swamps—which is why this period is called the Carboniferous, from the Latin word for "coal-bearing." The air's oxygen content continued rising until it reached a peak some 280 Mya in the early part of the next period, the **Permian**. It plunged back down again during the Permian to its earlier level, and then it rose more slowly up to an intermediate level like that of the present.

Some biologists think that these fluctuations in atmospheric oxygen were driving major events in animal evolution. They contend that the increasingly oxygen-rich air of the Late Devonian made it easier for vertebrates to move out of the water onto the land at this time (Graham et al. 1995). The high oxygen levels in the Carboniferous—and the high metabolic rates that must have gone with them—also may have made it easier for insects to evolve the machineries of flight. The Carboniferous witnessed the evolution of giant insects, looking like dragonflies with a wingspread a meter across, which probably would not have been able to live in the less oxygenated air of other periods in the earth's history.

▌THE REPTILIAN REVOLUTIONS

During the Permian and the period that followed it, the **Triassic**, the continents slowly drifted together near the equator, forming a single supercontinent that geologists call **Pangaea** in belated tribute to Alfred Wegener. The huge size and low latitude of Pangaea promoted the spread of deserts. And as the world's land masses grew drier and merged together, plant life suffered. The diversity of fossil land plants fell at the end of the Permian—one of only two times that this has ever happened in geological history (Niklas and Tiffney 1994). Oxygen in the earth's atmosphere dropped along with plant diversity throughout the Permian, hitting a low point in concert with the final consolidation of Pangaea during the Triassic (Fig. 1.9).

The coalescence and drying-out of the Permian continents contributed to the success of the next major innovation in the human lineage—the development of an egg that could be laid on land. The new eggs were covered with a tough, leathery shell, stiff enough to support the egg against gravity and impervious enough to keep out bacteria. Because their shells were not entirely watertight, they needed to be laid in damp soil to keep from drying out, as the eggs of many reptiles still do today. This slight permeability of the eggshell had its useful side, however, because it allowed the egg to absorb water from its moist surroundings and grow larger between laying and hatching—again, as the eggs of many modern reptiles do (Carroll 1988, Stewart 1997). The shell also had to be porous enough to let oxygen in and carbon dioxide out, so that the embryo could breathe. All this presents a tricky set of conflicting demands for an eggshell to have to juggle. Some scientists think that the high oxygen levels of the late Carboniferous atmosphere were what made the evolution of this new egg possible (Graham et al. 1997).

The new eggs had a complicated structure (Fig. 1.10). The developing embryo sported a **yolk sac** of the usual vertebrate sort—a bag connected to the gut, containing liquid food to sustain embryonic metabolism and growth. But in the new egg, a second baglike intestinal outgrowth was added on behind this to receive the embryo's urinary wastes. (These now had to be stored

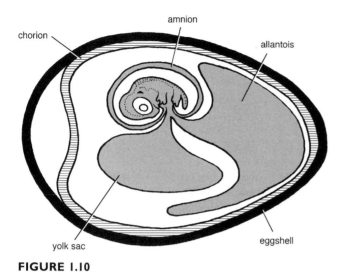

FIGURE 1.10

The amniote egg (diagrammatic).

inside the egg, because there was no longer any surrounding body of water to dump them into.) This new sac, the **allantois**, also served the embryo as a sort of lung, spreading out under the shell to provide a large surface richly supplied with blood vessels for absorbing oxygen from the air and discharging CO_2. Between the allantois and the shell, the entire embryo was wrapped in an outer protective membrane called the **chorion**. Finally, an innermost membrane called the **amnion** enclosed the embryo proper within its own private water bath of amniotic fluid, separating it from all the other apparatus and goings-on inside the eggshell.

Most reptiles and all birds today continue to lay eggs of this sort. So do three species of mammals (two echidnas and one platypus). Other mammals (and a few reptiles) have given up laying shelled eggs, allowing the zygote's early development to take place inside the secure refuge of the mother's egg ducts. But the embryos of these animals still retain the chorion, amnion, allantois, and yolk sac. All mammals, reptiles, and birds are therefore classed together as **amniotes**, after the amnion—whether or not they still lay amniote eggs.

The animals that were laying these eggs were the first **reptiles**, known as fossils from the late Carboniferous. In addition to their famous eggs, the ancestral reptiles displayed two other key adaptations to a fully terrestrial life. The first was **internal fertilization**. In primitive tetrapods, the males had fertilized the eggs as they emerged from the female's reproductive tract, just as male frogs and toads do today. But eggshells make this impossible. If the egg is going to be laid with a shell around it, the male has to introduce sperm directly into the body of the female before the shell forms, so that his gametes can reach the ova to fertilize them. Other animals that lay eggs protected by an impervious

coating—sharks, insects, spiders, and so on—have adopted internal fertilization for similar reasons.

The other key innovation developed in the ancestral reptiles was a **cornified epidermis** covered with a thick layer of dry, dead skin cells. This horny layer curtailed evaporation from the body surface, thus solving the last big problem faced by a fish out of water. The leathery skin and new reproductive adaptations of the early reptiles allowed their descendants to move completely away from ponds and swamps and take up a new style of life in the dryer uplands.

We can trace the separate lineages of birds and mammals all the way back to the Carboniferous period. The first reptiles (Fig. 1.11A) had a solid skull roof, like those of sarcopterygian fish and the primitive land vertebrates. But other Carboniferous reptiles evolved openings in the sides of the dermal armor of the skull. There were two main groups of these reptiles—a group with two holes on each side, and another group with only one hole (Fig. 1.11B, C). The descendants of the early two-holed or **diapsid** reptiles include the dinosaurs and birds, as well as all living reptiles with the possible exception of turtles (Kumazawa and Nishida 1999). The one-holed group, the **synapsids**, included the ancestors of the mammals. The hole in the side of the synapsids' dermal skull roof can still be traced in the human head, where it is represented by the **temporal fossa**—the area full of muscles on the side of the skull, between the dermal bones at the crown of the head and those of the cheekbone (Fig. 1.11).

The first synapsids were still thoroughly reptilian—lumbering, scaly, cold-blooded carnivores with a sprawling alligatorlike posture. They included in their numbers, however, some of the first plant-eating terrestrial vertebrates. Although these synapsid experiments in herbivory turned out to be evolutionary dead ends, they signaled the dawning of a new sort of ecosystem.

On land and sea alike, the earth's major food chains begin with green plants. But in the sea, most green plants are microscopic. Out at sea, beyond the coastal shallows, the ocean bottom lies in permanent darkness, and so photosynthetic plants must float near the surface of the waves. Under these circumstances, one-celled plants have a competitive advantage. Larger, many-celled plants are less efficient at absorbing sunlight (Niklas 1994), and they tend to get shredded by wave action at the sea's surface. Accordingly, most of the plants in the sea are single-celled—and so most of the herbivores are tiny as well. Most of the oceanic animals that we can see without a magnifying glass are animals near the top of the food chain: carnivores that eat other carnivores.

On the land, things are reversed. Small plants are at a disadvantage in capturing sunlight on land, because they have to grow in the shadows of their larger neighbors. Therefore, wherever circumstances permit, the

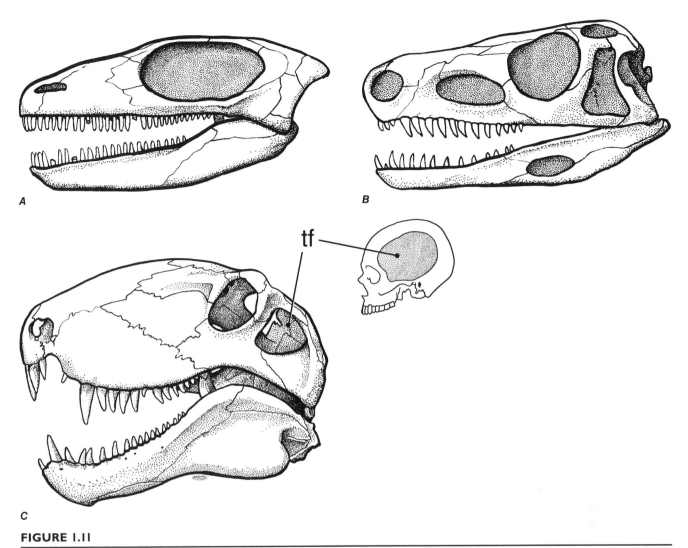

FIGURE 1.11

Skulls of three types of early reptiles: **(A)** the Lower Permian anapsid *Nyctiphruretus*, **(B)** the Lower Triassic diapsid *Euparkeria*, and **(C)** the Lower Permian synapsid *Dimetrodon*. The temporal fossa (tf) in the human skull (inset) is the homolog of the single temporal opening of *Dimetrodon*. Not to same scale. (**A–C**, after Romer 1956.)

surface of the continents is covered with trees and other tall plants, whose spreading crowns are adapted to intercept sunshine and deny it to their competitors. To compete in this battle of shadows, land plants have to invest a lot of their glucose production in making **cellulose** and other stiff, long-chain sugar polymers, which they use to support their towering tissues against the pull of gravity and the buffeting of rain and wind. These polymers contain a lot of energy, as the heat given off by burning wood or paper attests; but it is hard to release by any means short of combustion. Some bacteria can crack these **structural carbohydrates** apart, but animals cannot. Therefore, all animals that eat land plants either specialize in easily digested tissues like fruits or seeds, or else rely on anaerobic microbes in their guts to do the job of breaking down structural carbohydrates into sugars.

Early land plants bore neither fruits nor seeds, and early land vertebrates had no way of digesting cellulose. The terrestrial food chains of the Carboniferous therefore lacked big herbivores. Terrestrial ecosystems of the modern sort, in which big herbivores outnumber big carnivores, began to appear when some groups of Early Permian synapsids evolved the two key adaptations that leaf-eating vertebrates need: (a) **shredding teeth** to reduce plant leaves and stems to a slurry that gut bacteria can work on efficiently and (b) **capacious intestines** to provide space for this silage to sit and ferment. The intestines of the Permian herbivores are not known from the fossils, but we know from their ribs that they had big, bulging abdomens (Fig. 1.12). Most leaf-eating vertebrates, from duckbilled dinosaurs to horses and gorillas, have evolved these two specializations in some form or other (Hotton et al. 1997).

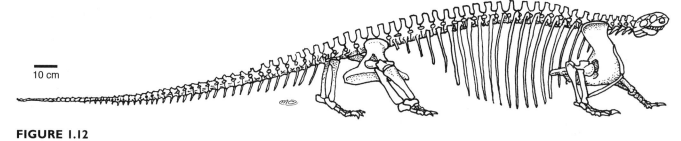

FIGURE 1.12

The skeleton of the plant-eating Early Permian synapsid *Cotylorhynchus*. (After Stovall et al. 1966.)

■ THE TWO GREAT EXTINCTIONS

The synapsid reptiles were the dominant large land animals of the Permian, but their rule was short-lived. The end of the Permian saw perhaps the greatest mass extinction in the history of life on this planet. The causes of this ecological catastrophe remain unclear. The coming-together of Pangaea may have had something to do with it. One theory holds that massive volcanic eruptions that occurred in Siberia at that time released vast amounts of greenhouse gasses into the atmosphere, causing a spike in world temperatures and driving already-stressed animal populations to extinction (Hallam 2004). For whatever reason, most animal species disappeared at this time. It has been estimated that more than 80% of all multicellular animals may have become extinct at the end of the Permian—including most of the synapsid reptiles (Raup 1991).

The surviving synapsids made a modest comeback during the Triassic period that followed the catastrophe. But their ecological space was mostly taken over during this period by the diapsid (two-holed) reptiles, including the ancestors of the dinosaurs. The first dinosaurs made their appearance around 200 Mya, in the late Triassic. The old-fashioned view of dinosaurs was that they were slow, cold-blooded, stupid, lumbering behemoths who died out because they could not adapt to changing times. The current fashion is to think of typical dinosaurs as active, nervous, birdlike creatures with warm blood and feathers. A lot of ink has been spilled in debates over warm blood in dinosaurs, but the point is a moot one, especially for the biggest and most herbivorous dinosaurs. Any very large land animal will have a more or less constant body temperature, because its great bulk takes a long time to warm up or cool off; and any large herbivore will have an elevated body temperature because of all the heat given off by its gently rotting internal compost. And if birds are surviving small dinosaurs, as the consensus has it nowadays, then at least some of the smaller dinosaurs must also have been warm-blooded and birdlike.

By the end of the Triassic, the dinosaurs and other diapsids had replaced the synapsids as the dominant land vertebrates. But one lineage of one-holed reptiles hit on a way to survive in the shadow of the dinosaurs. They became tiny, nocturnal, and warm-blooded, scurrying around in the darkness under the roots and leaves of the forest floor, protected against the chill of the night by a coat of insulating hairs. They were fierce predators—on beetles and worms and such—and they were also fiercely protective mothers, guarding and incubating their eggs and also feeding their hatchlings on fatty secretions from modified sweat glands on their bellies.

These were the early **mammals**, from which we are descended. The suite of innovations that made their way of life possible included a lot of the key inventions in the human lineage. We will look at these mammalian innovations in more detail in Chapter 3. For now, they can all be summed up in two phrases: **small size** and a constantly **high metabolic rate**. Most of the distinctive mammalian features of the teeth, jaws, skin, lungs, and reproductive system are corollaries of these two items.

The mammals stayed small and inconspicuous for the next 135 million years. They did not stop evolving during that period. Their teeth became more complicated and efficient at grinding and slicing food. Some of them evolved keener senses of smell or hearing. Others began eating plants. One group took to bearing live young. But despite these innovations, mammals retained their nocturnal habits and their small size, from a few grams up to no more than five or six kilograms (Hu et al. 2005). During the **Jurassic** and **Cretaceous** periods that followed the Triassic (Fig. 1.2), mammals did not grow big or presume to compete with the dinosaurs. It took an astronomical collision to bring them out into the sunshine and raise them into prominence as the dominant land animals.

The collision involved a medium-sized asteroid, a rocky piece of space debris some 10–15 kilometers in diameter. It struck the earth 65 Mya, near what is now the northern coastline of Yucatan, in Mexico. We can still trace the faint outlines of that impact, which left a crater 200 kilometers across. Sedimentary rocks all around the world from this time horizon carry material thrown up by the impact—dust with enriched levels of the metal iridium, tiny spheres of molten glass, amino acids not found in terrestrial sources, and so on. The incandescent re-entry of these millions of tons of splattered material may have produced a brief but titantic pulse of infrared heat waves, powerful enough to set all the world's forests on fire and cook all the terrestrial animals that were too big to hide under a rock (Robertson et al. 2004). In the strata overlying the thin layer of asteroid debris, no one has ever found an undisputed fossil of a dinosaur.

It is now generally agreed that the Mexican asteroid impact caused the extinction of the dinosaurs at the end of the Cretaceous. Many other groups of animals also vanished at this time, resulting in a second great mass extinction rivalling that at the end of the Permian. The abrupt faunal changes occurring at these two horizons are used to divide the Phanerozoic eon into three so-called **eras**: the **Paleozoic** ("ancient animals") era preceding the Permian extinction, the **Mesozoic** ("middle animals") era following it, and the **Cenozoic** (Greek, "recent animals") era following the asteroid impact (Fig. 1.2).

▌ THE MAMMALS TAKE OVER

The Cenozoic, which is the shortest of the three eras, is sometimes called the Age of Mammals for the same reason that the Mesozoic is called the Age of Reptiles. In both eras, the large land animals belonged respectively to those two groups. The 65-million-year-long Cenozoic is divided into only two subunits: the **Tertiary** period and the later and much shorter **Quaternary** period, which lasted only some two million years. (The words "Tertiary" and "Quaternary," which mean "Third" and "Fourth," are the sole surviving reminder in modern geology of the four-period model of the early Neptunists.)

The increasing brevity of the stages recognized at the top of the geological column reflects the fact that younger rocks are more plentiful than older ones. It also reflects the early geologists' underestimates of the amount of time that elapsed before *Homo sapiens* came on the scene. Those underestimates were rooted in an older, more anthropocentric view of the world than ours. For people who believed deep down that all things were made for man's benefit, it was hard to grasp the fact that the earth was devoid of human beings throughout more than 99.9% of the history of life. We have adopted an anthropocentric focus of our own in writing this book, and most of the rest of it will be concerned with the human lineage during the Cenozoic in general and the Quaternary in particular. But it is worth pausing at this point to note how rapid our success and how vanishingly brief our tenure on this planet thus far has been.

The asteroid impact at the end of the Cretaceous exterminated all the large land animals, but it had less profound effects on plant life. Outside of North America, where the asteroid struck down, latest Cretaceous floras look pretty much like early Cenozoic floras (Hallam 2004). Cenozoic mammals have therefore tended simply to take over the ecological roles for large animals that were left vacant when the dinosaurs went away. The duckbilled, sauropod, and ceratopsian dinosaurs of the Cretaceous have been replaced as large herbivores by such creatures as elephants and hoofed mammals. Raptorial dinosaurs have been supplanted as large terrestrial predators by big cats, wolves, and their kin. Dolphins and seals have slipped into the niches for large marine carnivores once occupied by plesiosaurs and ichthyosaurs. And so on. In addition to taking over all this ecological space from the vanished great reptiles, mammals have also held on to the general niches that they occupied in the Mesozoic, including the roles of small herbivores (multituberculates in the Cretaceous, rodents and rabbits nowadays) and of small nocturnal insect-eaters such as shrews.

Mammals have not, however, made many incursions into the niches for small *diurnal* animals. Most of those niches are still occupied by the surviving diapsids. If you go for a midday walk in the wilds on almost any continent, you will see a number of small diapsids—lizards, snakes, and above all birds—and you may encounter some large mammals; but the only small mammals you are likely to see belong to a few groups (including squirrels, mongooses, and monkeys) that have hit on ways of competing with the birds in the daytime (Charles-Dominique 1975). At small body sizes, the diapsids still rule the day and the mammals rule the night, now as in the Cretaceous.

Cenozoic mammals have of course made some ecological innovations of their own. Grasses and grasslands did not evolve until the Tertiary (Jacobs et al. 1999), and so today's herds of grazing mammals with special teeth for grinding the gritty, siliceous leaves of grass had no equivalents in the Cretaceous. Mesozoic reptiles also never came up with anything like the filter-feeding adaptations of the baleen whales, which short-circuit the sea's food chains by allowing the very largest predators to feed directly on some of the smallest ones. Perhaps the most original mammalian innovation was the nocturnal flight of bats, made possible by the evolution of their leathery wings and the extraordinary sonar-like echolocation that permits them to zoom through the air safely in total darkness. This breakthrough, which probably took place around the beginning of the Cenozoic, has been a great success. So have the specializations of the teeth and jaws that have allowed the rodents to become extremely effective small herbivores, with no real counterpart among non-mammalian vertebrates. Most mammals today are either rodents or bats.

One general feature of Cenozoic mammalian evolution is of special relevance to the human lineage—namely, a trend toward evolving bigger brains. In many groups of mammals, we find that the modern forms have brains that are conspicuously larger than those of their similar-sized relatives in the early Tertiary (Jerison 1973, Martin 1973). This trend shows up in many mammal lineages with widely differing lifestyles, including rodents, carnivores, primates, whales, and hoofed mammals. Brains, as we will see later, are metabolically expensive organs. Perhaps the trends toward brain enlargement that we see in many groups of Cenozoic mammals were made possible by underlying improvements in metabolic efficiency, which we have no way of detecting in the fossil record.

CHAPTER TWO

Analyzing Evolution

PARSIMONY AND PIGEONS

By the middle of the 19th century, the broad outlines of the earth's history were clear to everyone who cared to look. Although reliable chronometric dating techniques still lay a century in the future, the geologists of the 1800s had already established that the earth is millions of years old and that the features of the earth's surface were produced over time by the same geological forces we see at work today. The history of life revealed in the fossil record showed that new life forms have been appearing, and old ones disappearing, throughout earth's history. But how could this shifting parade of new creatures be explained?

There are only two kinds of stories we can tell to explain the appearance of something new in the world: Either the new thing has popped into existence out of nowhere, or it has been produced by reworking something that was already in existence. In the case of species, the first approach is represented by the story of **special creation**, which says that new species are continually being created—presumably by God—as old

ones die out. The second approach yields a story of transformation: some old species change and develop into new species, while others dwindle and disappear forever. This second story is what is now called the theory of **evolution**.

Each of these accounts had its adherents in 1850, but neither had won general acceptance. Both had two serious defects. First, they violated the uniformitarian principle: nobody had ever seen a new species come into existence out of nowhere, nor had they ever seen an old species change into a new one. Second, neither account proposed any mechanism by which new species could come into existence. Special creation was almost by definition a miraculous event—and miracles do not have mechanisms. And although there might in principle be a mechanism that caused one species to evolve into another, nobody in 1850 knew what that mechanism could be.

When we are confronted with two equally good (or equally unattractive) explanations of something, we can often make a choice by invoking the **principle of parsimony**. This rule says that when all else is equal, we

---

The Human Lineage, By Matt Cartmill and Fred H. Smith.
Copyright © 2009 John Wiley & Sons, Inc.

should pick the account that demands the fewest assumptions. The principle of parsimony is only a philosophical axiom, not a law of nature. Neither theory nor experience gives us any reasons for thinking that the simplest explanation is always true. But looking for the simplest explanation is an operating principle that underlies all scientific practice. The uniformitarian principle, for example, is just a special case of the principle of parsimony. Because it is always more parsimonious to posit one law of nature than two, science always assumes that nature operated under the same rules and exhibited the same patterns in the past as it does in the present—unless we have good reasons for thinking otherwise.

Do we observe any patterns in the present that might suggest a mechanism for the origin of species in the past?

Although we do not see new species being created in the world around us, we do see processes that produce changes in existing species. In the 1850s, the best-known examples were found in the practices of animal and plant breeders. Starting with wild stocks adapted for survival in a state of nature, breeders select unusual individuals with rare traits that make them more useful

for human purposes. Over hundreds of years, selection for such traits has produced domesticated strains that differ radically from the original wild stock: corn plants that cannot release seed without human help, cows that give more milk than their calves can drink, pigeons with weird deformities of body and plumage that appeal to the esthetic tastes of pigeon fanciers (Fig. 2.1), and so on.

These artificially selected breeds would not fare very well in the wild, and they would not have come into existence without human intention, maintained generation after generation. But they show that over the long run, the hereditary properties of a plant or animal species can be greatly altered by selecting and breeding individuals that differ from the norm in a certain desired direction. Is there some way in which such a program could be carried out automatically, without the intervention of a guiding intelligence?

▌ DARWIN'S THEORY

During the early 1800s, a number of British authors independently came up with the same theory to explain

FIGURE 2.1

Some of the variations in skull morphology and plumage preserved and fostered among domestic pigeons (*Columba livia*) by pigeon-breeders of the 19th century, illustrated in Darwin's works as examples of artificial selection. **A–D**: Skulls of wild rock-pigeon (**A**) and three artificial breeds: short-faced tumbler (**B**), English carrier pigeon (**C**), and Bagadotten carrier (**D**). **E**, English barb; **F**, English carrier pigeon; **G**, pouter pigeon; **H**, fantail pigeon. (From Darwin 1883.)

FIGURE 2.2

Charles Darwin, as seen by a contemporary caricaturist.

how the blind processes of nature could bring about selective breeding for desirable traits. The most famous of these men was the naturalist Charles R. Darwin (Fig. 2.2). Darwin was not the first to propose the theory, but he arrived at it on his own and worked it out in far greater detail than any of its other discoverers. His 1859 book on the subject, usually referred to as *The Origin of Species*, is a scientific classic that called the world's attention to the theory and its importance in a way that none of his predecessors had. For his work on the theory of natural selection, Darwin is celebrated as one of the three or four most important figures in the history of science.

Darwin's theory can be stated in its essentials very simply. Not all members of a species survive to reproduce, and survival is not wholly a matter of luck. Some individuals do better than others because they have inherited superior features—longer legs, deeper roots, bigger brains, or whatever else makes for survival in that species' way of life. On the whole and in the long run, these superior individuals will leave more offspring than others. Therefore, reasoned Darwin, their hereditary advantages will tend to spread throughout the population. In time, all the members of that species will

have longer legs (or deeper roots, or bigger brains). The species will then have changed from what it was. As long as new and useful variations keep appearing, the species can go on changing forever, adapting to a succession of changing environments. Over hundreds of millions of years, Darwin concluded, this process of **natural selection** could have evolved all the world's organisms from a few ancestral forms, without any need for intelligent planning or divine intervention. The full title of Darwin's 1859 book—*On the Origin of Species by Means of Natural Selection, Or the Preservation of Favoured Races in the Struggle for Life*—encapsulates his theory in just 21 words.

By proposing a mechanical explanation for evolutionary change, Darwin made the idea of evolution intellectually respectable among skeptical scientists. Most biologists soon gave up on the idea of special creation and embraced some form of evolutionary theory. This was due partly to the force of Darwin's arguments. But it also reflected the great explanatory power of the ideas of evolution and natural selection. In addition to explaining the puzzling succession of ancient plants and animals in the fossil record, the new theory cleared up several other fundamental problems:

1. **The Problem of Biogeography.** Different parts of the world have different animals and plants. The Bible story of Noah's Ark implies that all the animals in the world have somehow gotten to where they are today from Mt. Ararat in Turkey. This was not too hard to believe in the Middle Ages, when the world as Europeans saw it centered on the Mediterranean Sea. But starting in the 1500s, Europe's world-spanning voyages of discovery and conquest brought back new knowledge about plants and animals that cast doubt on the biblical account. How could tree sloths, found only in tropical America, have swum the seas or walked through the frozen Arctic to get from Mt. Ararat to Brazil? Had people carried them there? What on earth for? Why would people have wanted to drag along useless beasts like sloths and skunks to the New World, while leaving behind such valuable livestock as horses, sheep, and cattle? Pondering questions like these, European naturalists gradually abandoned the Bible story. However, they still could not explain why there should be sloths in Brazil but not in the Congo, or why each of the little desert islands in the Galápagos archipelago should have its own variety of tortoise slightly different from all the others.

The evolutionary hypothesis solves these problems. Tree sloths evolved in the American tropics. They have not been able to spread beyond that region because they cannot live outside of tropical forests. The various island populations of Galápagos tortoises closely resemble each other because all have descended from a single species that reached the archipelago long ago. They differ from each other because they have evolved in different directions as they spread from island to island.

2. **The Problem of Homology.** To design an airplane, an intelligent designer would not start with a blueprint for a battleship and try to modify the shape of each part to make it serve a different purpose. Yet that seemed to be the program that God had followed in designing living creatures. For instance, most of the parts of a cow's body can be matched up one-to-one with the parts of a human body, right down to the tendons in the fingers and the bones in the skull. But despite their similarities, the corresponding parts do different work in the two species. Men do not fight by banging their foreheads together as bulls do, and cattle do not use their fingers to make tools. Why, then, did the Creator build people using a modified cow blueprint (or vice versa)? Why use a similar blueprint for dolphins, but employ a greatly altered version for sharks and a completely different plan for the bodies of squid—all of which are fast-swimming marine predators?

The evolutionary hypothesis furnishes the answers. Most of the similarities between organisms represent heritages from shared ancestors. Humans, cows, and dolphins are all descended from the primitive placental mammals of the Mesozoic. They have retained such features as warm blood and milk glands from that common ancestry. Sharks do not share these features, because their ancestors never had them. But sharks and mammals do share certain more fundamental features—eyes, spinal cords, brains, backbones—that reflect their common ancestry among Paleozoic vertebrates. And because the last common ancestor of squid and mammals was a simple, wormlike creature back in the Precambrian, no similarities can be discerned today even in their basic body plans. The family resemblances between squid and people are visible only at the molecular level.

Similarities between organisms that are due to shared ancestry, such as the milk glands of people and dolphins or the spinal cords of people and sharks, are called **homologous** traits or **homologies**. Not all similarities are homologies. An example of a nonhomologous similarity is the resemblance of the eyes of squid and octopuses to those of vertebrates. Like our eyes, the eyes of these molluscs have a retina, iris, pupil, cornea, and eyelids, and they function in much the same way as our own. But other molluscs have simpler eyes, and the most primitive living chordates and molluscs do not have eyes at all. These facts suggest that eyes came into being separately in chordates and molluscs. If so, then the striking similarities between the eyes of squid and vertebrates are not homologies.

In cases like this, where similar features appear independently in two different lineages, the similarities are called **convergences**, and the process is known as **con-**

vergent evolution. Convergent evolution is sometimes distinguished from **parallel** evolution. In parallel evolution, homologous structures are modified independently in similar ways in two lineages. For instance, people and dolphins both have big, highly convoluted brains, which have been evolved out of the smaller and simpler brains of the ancestral mammals. The process of brain enlargement has proceeded separately in the human and dolphin lineages. Their brains are homologous *as brains*, but their relatively large size is a nonhomologous resemblance between them, produced by parallel evolution.

It is not always easy to distinguish convergence from parallelism. For example, the development of eyes in most invertebrates seems to be controlled by genetic mechanisms that are homologous with those underlying vertebrate eye development (Pineda et al. 2000). This suggests that the last common ancestor of molluscs and vertebrates may have had eyes of a simple sort. If so, we might want to describe the resemblances between squid eyes and human eyes as parallelisms— homologous features changing in parallel—rather than as convergences. Many biologists choose to ignore this tricky distinction and group all nonhomologous resemblances together as **homoplastic** resemblances or **homoplasies**.

3. **The Problem of the Taxonomic Hierarchy.** Organisms appear to be distributed in nested sets. Among multicellular animals, for instance, 20 or 30 main groups called **phyla** can be distinguished, including chordates, molluscs, and arthropods. Each phylum has a distinctive basic anatomical groundplan, different from those of the others. Within the vertebrate subdivision of the chordate phylum, we can make out seven or more major groups called **classes**: jawless fish, cartilaginous fish, bony fish, amphibians, reptiles, birds, and mammals, each with its own distinctive features. Similarly, there are 20 or so chief subgroups or **orders** of mammals, each of which can be broken down into successively smaller subdivisions called **families, genera** (singular = *genus*), and **species** (singular = *species*). Each of these groupings is called a **taxon** (plural = *taxa*).

There is no obvious reason why the diversity of life should be arranged in this **taxonomic hierarchy** of phylum, class, order, and so on. We can imagine all sorts of composite organisms that would combine useful features from several different phyla or classes—say, a feathered lizard with four legs and two wings that suckles its young. But such composite creatures do not occur. Why do living things appear to be arrayed in discontinuous, nonoverlapping, concentric groups, rather than in a great chain or network of intergrading forms (see Blind Alley #2, p. 33)?

The reason for this becomes clear once we accept that life's diversity has been produced by a tree-like, branching process of evolutionary change. The animal

Blind Alley #2: The Ladder of Nature

If God is perfect Being and utterly self-sufficient, then why did he decide to clutter up the universe by creating lesser, imperfect stuff? Pondering this question, ancient thinkers argued that the universe would not be perfect unless it were *complete* and that it could not be complete unless it included a full range of all possible kinds of beings, descending from God on high down through angels, people, beasts, and plants to mere rocks and dirt at the bottom.

This doctrine came to be known as the Principle of Plenitude (fullness). The envisioned hierarchy of created beings was called the Great Chain of Being or the Ladder of Nature (*scala naturae*). Besides answering the question of why anything should exist, the Principle of Plenitude also explained why so many nasty or useless things had been created (they were needed to fill empty slots in the hierarchy) and why the lower classes should remain subordinate to the aristocracy. By the mid-1700s, many people were beginning to see the Ladder as a kind of moving staircase, sweeping all creatures upward toward human status and beyond. As the German poet Christoph Wieland wrote in 1750,

> God made things thus, that all might rise toward
> The fount of being and adore their Lord.
> All things strive upward toward one lofty end:
> Whatever feels, feels God; and all ascend
> Infinite ladders through eternity,
> Approaching God in rising ecstasy.

This idea laid the groundwork for the first evolutionary speculation. The notion of the Chain of Being was undermined and eventually swept away by Linnaean systematics and evolutionary theory, which arranged living things in the form of a branching tree rather than a linear sequence. Vestiges of a hierarchical vision of nature persist in certain conventional phrases—for example, when primate biologists speak of lemurs as "lower" and monkeys as "higher" primates. The fascinating history of the Ladder of Nature is recounted in A. O. Lovejoy's classic book *The Great Chain of Being* (1936).

phyla represent major branches of that tree. Their subgroups—classes, orders, families, and so on—represent consecutively finer branches (Fig. 2.3). Because the terminal buds on the tree of life never reunite once they have branched off from their neighbors, they are arrayed in nested sets and not a reticulated network. Each of these nested sets is distinguished from other like sets by uniquely shared homologies that reflect the uniquely shared ancestry of its members.

4. **The Problem of Vestigial Structures.** Many organisms have features that serve no purpose in their lives, even though homologous structures are functionally important in other creatures. For example, human ears are immobile. We cannot turn them this way and that to locate sounds, as most other mammals can. Yet we have little muscles attached to our ears, resembling those that move the ears of a cat, a horse, or a lemur. When we contract those muscles, all they do is wiggle our ears a bit. Most of us cannot even do that much. Our ear muscles do nothing useful. Why do we have them? The evolutionary answer is clear. We have them because we are descended from mammals with more typical ears and functional ear muscles. We retain remnants of those muscles, even though wiggling our ears no longer serves any useful purpose.

Human ear muscles and other such evolutionary leftovers are called **vestigial** structures. Such structures are often found in early stages of embryological development, where organs that have lost their functional value

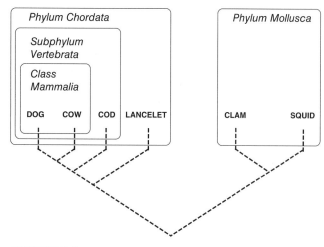

FIGURE 2.3

Any branching tree diagram or **dendrogram** (dashed lines) can be mapped as an **Euler diagram**, a system of nonoverlapping nested sets (boxes). The two figures have the same logic and contain the same information.

may nevertheless be retained in a reduced form because they are implicated in some important developmental process. The human yolk sac and allantois provide examples (Fig. 1.10). Human embryos no longer need to draw food from the yolk sac or deposit wastes in the allantois, because the human placenta has taken over those jobs. But these two sacs, though empty and shrunken in a human embryo, still play a crucial role in development, because their blood vessels grow out to become the veins and arteries of the placenta.

5. **The Origin and Imperfections of Adaptations.** Perhaps the most important problem solved by Darwin was that of the origin of adaptations. Whether we call this a "problem" or not depends on our viewpoint.

Living things are fitted to survive in their environments and ways of life. Animals that eat plants have organs and chemistries suited to grinding up and breaking down plant tissues. Animals that live in seawater have organs and chemistries suited to breathing water and getting rid of excess salt. Plants that grow in deserts can survive with little water, and have organs and tissues that can store and retain whatever water comes along. And so on. No matter what sort of organism we look at, almost every aspect of its anatomy and chemistry testifies to the close fit between the organism, its life habits, and the world. Literally millions of examples of this phenomenon of **adaptation** could be cited.

For a creationist, the fact of adaptation is not a problem but an answer. It provides one reason for believing in a Creator. Organisms and environments are

fitted to each other because a benevolent God created each one with the other in mind. Sharks can live in salty water because God built them to dwell in the sea. Conversely, the sea is salty because God created it to be, among other things, a home for sharks. And all this is comforting. If God has taken so much trouble to provide for the needs of sharks, we can feel sure that he knows our own needs and will provide for them. As the words of the old hymn put it: "His eye is on the sparrow, and I know He watches me."

After drawing up a long list of such marks of divine providence in nature, the Anglican priest William Paley concluded:

> Therefore one mind hath planned, or at least hath prescribed, a general plan for all these productions. One Being has been concerned in all. ... All we expect must come from him. Nor ought we to feel our situation insecure. In every nature, and in every portion of nature which we can descry, we find attention bestowed upon even the minutest arts. The hinges in the wings of an *earwig*, and the joints of its antennæ, are as highly wrought, as if the Creator had had nothing else to finish. We see no signs of diminution of care by multiplicity of objects, or of distraction of thought by variety. We have no reason to fear, therefore, our being forgotten, or overlooked, or neglected. (Paley 1802, pp. 347–348)

But the facts of nature are less comforting than Paley would have us believe. The fit between organism and environment is imperfect. Sometimes it becomes so imperfect that the species can no longer reproduce itself and disappears. The fossil record shows us that extinction is the fate of almost all species sooner or later. And even for those species that survive, adaptation is never perfect. Many adaptations seem like clumsy, imperfect makeshifts unworthy of an omnipotent designer. Even Paley's favorite example of design in nature, the human eye, has some obvious design flaws. For one thing, the focusing mechanism stops working after about 40 years of use. Why can't God make a camera that lasts as long as those designed by human engineers?

Darwin's theory accounts both for adaptations and for their imperfections. In every species, natural selection is continually operating to eliminate the worst-adapted variants and promote the best-adapted ones. This produces long-term evolutionary trends toward better adaptation and apparent design. But every adaptation is a makeshift, produced by modifying something that originally evolved in a different adaptive context. No species is ever perfectly adapted to its present environment and way of life. Environments are continually changing, and the selection pressures of the present can work only on the traits produced by the selection pressures of the past. Therefore, improvements are always possible—though most of them will never come into existence, because the necessary variants will never crop up in the population. Eventually, every species is hit by environmental changes that it cannot adapt to. It then becomes extinct.

The evolutionary theory of Charles Darwin explained all of these puzzling phenomena at a single stroke: the geological succession of fossil organisms, the prevalence of extinction, the geography of life, the distribution of homologies, the taxonomic hierarchy, the existence of vestigial structures, the origin of adaptations, and the imperfection of adaptations. The principle of parsimony demanded the adoption of Darwin's theory. It still does. There is no alternative theory that does all these jobs at once.

PROBLEMS WITH DARWINISM

Although Darwin worked his theory out with inexorable logic and richly detailed examples, it was not complete or satisfactory in its original form. Additional refinements and discoveries were needed to improve the theory and to deal with problems pointed out by Darwin's critics.

1. Nuclear Energy. In 1862, the physicist William Thomson, also known as Lord Kelvin, declared that Darwinian evolution is a physical impossibility. From the rate at which the sun is currently losing heat, Thomson calculated that it could not be as old as the Darwinian theory implied. "The most energetic chemical action we know," he wrote, "… would only generate about 3000 years' heat." Thomson admitted that the sun might be as old as 20 million years, if its heat were derived from the kinetic energy of matter falling into it; but even this was far too short a time to allow for Darwinian evolution of all the earth's life forms from a single-celled ancestor. From the rate at which the earth itself presently radiates heat, Thomson (1864) argued similarly that the earth's surface must still have been a sea of molten lava less than 100 million years ago. Both lines of reasoning, he concluded, prove that life on our planet cannot be as ancient as Darwin's theory requires.

Given Thomson's assumptions, his arguments were flawless. Evolutionists could only stubbornly insist that those assumptions had to be wrong somewhere: the sun and earth must have some other, unknown sources of heat, since the geological record clearly shows that life is hundreds of millions of years old. The evolutionists were proved right in the 20th century when scientists found that the earth's internal heat is maintained by radioactivity and that the sun's energy derives from hydrogen fusion—not from chemical reactions or the kinetic energy of meteor strikes.

Evolutionary biologists cherish this story, because it represents one of the few times in the history of science that we got the drop on the physicists. But although Thomson was wrong, he was not wrong because he was a bad scientist. He was a great scientist, who reached some wrong conclusions because nuclear reactions were not known to science in the mid-19th century. All that he or any other any scientist could do was to try to explain the world on the basis of the knowledge and theories of the time. Because science advances by discarding old interpretations when new information demands it, almost every scientist's ideas will eventually turn out to be wrong or incomplete in one way or another. Thomson was no exception. Neither was Darwin.

2. Particulate Inheritance. In an 1867 review of *The Origin of Species*, Thomson's friend and collaborator Fleeming Jenkin argued that natural selection could not work, because new and better variations would be swamped by interbreeding with inferior competitors.

Suppose, for example, that a dark green individual appeared by chance in a population of yellow frogs. Suppose, furthermore, that its green color helped to camouflage it more effectively from predators. The green frog therefore would live long and leave unusually large numbers of offspring. But because the green frog would find only yellow frogs to mate with, its offspring (by Jenkin's assumptions) would be a color halfway between green and yellow; and *their* offspring (from matings with other yellow frogs) would be still closer to pure yellow—and so on. After a few generations of interbreeding, the descendants of the green frog would be indistinguishable from the yellow-colored frogs, leaving nothing for natural selection to work on. "The advantage, whatever it may be," Jenkin concluded, "is utterly outbalanced by numerical inferiority."

Again, the reasoning was sound, but the assumptions turned out to be mistaken. Jenkin's argument assumes that heredity works by striking a balance between the characteristics of the two parents. This **blending theory** of heredity was the conventional wisdom of the 19th century, and it is still prevalent in popular thinking about inherited traits. But this is not how heredity works.

Two years before Jenkin's review of the *Origin* appeared, an Austrian monk named Gregor Mendel had published an obscure paper that established the foundations of a **particulate theory** of inheritance. Mendel took pea plants that bore green seeds, and he crossbred them with a different strain of peas that bore yellow seeds. He found that the plants of mixed parentage always had yellow seeds—not seeds of some intermediate yellowish-green color, as a theory of blending inheritance would predict. And when he bred the mixed-parentage plants to each other, roughly 25% of their offspring bore green seeds, while the remaining 75% had yellow seeds like those of their parents.

These numbers can be explained by assuming that the hereditary instructions governing seed color come in discrete packets. Each diploid plant inherits two separate color instructions, one from each parent. In a plant that gets "green" instructions from one parent and "yellow" instructions from the other, the yellow color always prevails. Therefore, the original mixed-parentage plants bore only yellow seeds. But because each of those plants carried both "yellow" and "green" instructions, it could pass on either one (though not

both at once) to its offspring via its haploid gametes. If a mixed-parentage plant has a fifty–fifty chance of transmitting either "yellow" or "green" instructions, then the chances are one in four that an offspring of two such plants will receive "green" instructions from both parents (Fig. 2.4). Only offspring with two "green" instructions—25% of the total, on the average—will bear green seeds. The other three-quarters will bear yellow seeds. And those are the ratios that appeared in Mendel's experiment. Mendel carried out other experiments, using different combinations of these and other hereditary traits, and got consistently similar results, showing separate, nonblending inheritance of the different packets of instructions for each trait.

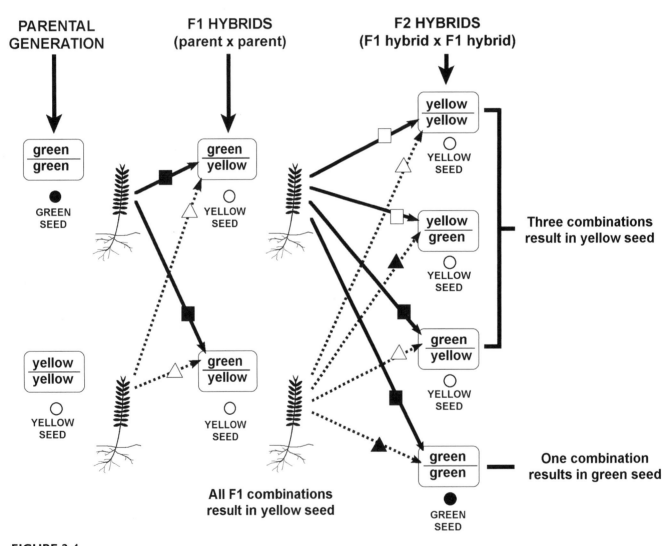

FIGURE 2.4

Fundamental principles of Mendelian genetics, illustrated with reference to the results of Mendel's pea-breeding experiments. Crosses between two strains of peas, one with consistently yellow seeds and one with consistently green seeds (parental generation), always yield first-generation hybrids (F1) with yellow seeds. However, F1 × F1 crosses yield both yellow-seeded and green-seeded plants (F2), in approximately a 3:1 ratio. The Mendelian conclusion: each plant carries two sets of instructions for seed color, only one of which gets read out during development. See text for explanation.

The discrete packets of hereditary instructions revealed by Mendel's work are now referred to as **genes**. Twentieth-century biologists discovered that each gene corresponds to a sequence of DNA involved in the synthesis of a particular polypeptide. When a single gene occurs in multiple variant forms (such as the "yellow" and "green" forms of the gene for seed color in Mendel's peas), the variants are called **alleles** of the gene, and the species is described as **polymorphic** for that gene and the trait(s) it affects. Different alleles, in various combinations, have different effects on the development and functioning of the organism that carries them. Some alleles have more favorable or useful effects than other alleles of the same gene. Individuals that carry less useful alleles tend to reproduce less, and so those alleles tend over time to be eliminated from the **gene pool** of the population. This **differential reproduction** of genetic variants is approximately equivalent to Darwin's "natural selection." In every generation, errors of transcription in the copying of DNA introduce new alleles, which are known as **mutations**. The promotion of the most favorable alleles by differential reproduction, along with the introduction of new mutations through DNA copying errors, drives the process of evolution.

The complement of genes carried by an organism (its **genotype**) is not always reflected in the list of its visible traits (the **phenotype**). The seeds of Mendel's hybrid pea plants had a "yellow" phenotype, but their genotype contained the "green" allele as well. Bred together, such hybrid plants produced some offspring with green seeds. Likewise, a frog carrying a mutation that produces green skin color can transmit that mutation to its descendants, along with all its adaptive advantages, even if all the other members of its species carry a different allele of that gene. Because different alleles remain discrete down through the generations and are not diluted by blending, Jenkin's objections to Darwin are invalid.

3. Sexual Selection. Some animals have heritable traits that would seem to work against their survival. For example, while most animals have rather dull, inconspicuous colors that help them blend into their environments and escape the notice of predators, some species show bright colors or striking patterns that make them stand out from the background. Some of these represent cases of warning coloration, like the black and yellow stripes of many bees and wasps that put predators on notice that these stinging insects are not good to eat. But how can we explain the brilliant red feathers of a male cardinal, or the cumbersome, iridescent tail fan of a peacock, in terms of natural selection? The beautiful, showy plumage of these birds must surely put them at greater risk of being eaten than they would be if they were more drably and sensibly colored, like the females of their species. Is the tail of the peacock the handiwork of a Creator who sometimes puts beauty before utility?

Darwin considered this question in the *Origin* and addressed it at length in a later book (Darwin 1871). Traits of this sort, he argued, render the more showy males more attractive to females of their species. This gives them a reproductive advantage that more than makes up for the increased risk of predation. Over time, the selection pressures imposed by the preferences of female cardinals and peacocks have worked to intensify the conspicuous coloration of the males and make it uniform throughout the species. Darwin labeled this process **sexual selection**.

When we find innate differences between the two sexes of a species (apart from those differences directly related to reproduction), we say that the species exhibits **sexual dimorphism**. Most animal species are sexually dimorphic to some degree. As a rule, the females are larger than the males. This sort of dimorphism is not due to sexual selection. It just reflects the general rule that it takes more materials to make eggs than to make sperm, because eggs are bigger and contain more nutrients.

But other sorts of sexual dimorphism do result from sexual selection. When the males of a species are larger than the females, or bear horns or other special weapons not seen in the females, it is almost always because the males fight or threaten each other for access to females at mating time. This, too, is a form of sexual selection. Males that can deny other males access to females will leave more offspring. Some male traits do two jobs at once: They help them win fights with other males, and they also make them more attractive to females. (Big muscles in human males might be an example of this.) Indeed, we would expect to find a general evolutionary tendency for females to become attracted to males that can win fights. Any preference a female has that makes her pick winners over losers as mates will be favored by natural selection, since it will tend to make her have more successful male offspring—and thus give her own genes an improved chance of long-term survival.

When the sexes differ, why is it usually the males who have the brighter colors, the showier displays, and the more combative dispositions? Why do males generally compete for the females' favor, and not the other way around? Current thinking on this subject begins with A. Bateman's (1948) study of reproductive rates in fruit flies. Bateman found that male fertility was much more variable than female fertility. Most of the females in his experiments had similar numbers of offspring. But a lot of the males had no offspring at all, whereas others had more offspring than even the most fecund females had—up to three times as many for some of the males.

Bateman argued that this difference was ultimately a result of the size difference between eggs and sperm. Because eggs are much larger than sperm and take far more energy and nutrients to produce, a female can produce only a relatively limited number of offspring during her reproductive lifespan. But a male can beget practically infinite numbers of offspring. His reproductive success is limited chiefly by his ability to find eggs to fertilize. Male reproductive rates vary more than those of females because most eggs find a sperm cell but most sperm cells never find eggs. Bateman suggested that this difference between the gametes of the two sexes also explained a key behavioral difference between them—the pattern of what he called "undiscriminating eagerness in males and discriminating passivity in females," seen in many animal species.

In 1972, the evolutionary theorist R. L. Trivers elaborated these ideas into a more general theory of sexual selection. Trivers argued that males and females have different reproductive priorities and different strategies. Because a female invests more time and energy in each offspring than a male does, she can afford to bear only a limited number of offspring. Beyond that, she reaches a point of diminishing returns on her reproductive investment. For example, a hen that tried to lay 30 eggs in a single clutch would be unable to put enough nutrients into the eggs to hatch them all out, or to look after 30 chicks properly if they did hatch. But a rooster can fertilize thirty eggs in a single afternoon with no difficulty. The point of diminishing returns in female reproductive investment differs with different species; but so long as eggs are larger than sperm, it will be lower for females than for males.

Because the optimum number of offspring is lower for a female than it is for a male, female reproductive investment is a relatively limited resource. Therefore, males compete for it. A male's best strategy is to fertilize as many females as possible and to keep other males away from them. This explains why male animals are often more mobile than females—and why they often have special courtship displays to attract females, or bear weapons like horns and antlers to drive off other males. Conversely, a female's best strategy is to choose her few mates carefully, with an eye to general quality and to any specifically male traits that will make her sons successful competitors.

In some animal species, including wolves, human beings, and many songbirds, males too invest a lot of time and energy in the rearing of their offspring. In such species, each sex becomes a resource for the other. Both sexes therefore compete for mates, and sexual dimorphism is small—or even reversed, as in the shorebirds called phalaropes, whose brightly colored females display to court the favors of nest-tending males. Whenever males invest heavily in offspring, we would expect a female to seek to mate with a good provider (to minimize the drain on her own resources) and to try to prevent him from having any other mates (to avoid diluting his exclusive investment in her offspring). Her mate, on the other hand, may benefit from cheating on her—more than she will from cheating on him—because if he gets away with it undetected, he can channel the parental investment of other males into his own offspring.

These predictions seem to echo familiar patterns of human social and sexual behavior. Some social scientists think that Trivers's theory of parental investment explains many aspects of human nature. Others suspect that the theory itself is just a scientific-sounding projection of our culture's economic theory and gender stereotypes. But no matter what we think of the application of Trivers' ideas to human behavior, they have become a generally accepted part of the theory of evolution by natural selection. Biologists invoke these ideas in explaining a wide range of phenomena, from infanticide to sex differences in lifespan and mortality rates.

Sexual selection and natural selection are not really separate phenomena. The two labels merely name different factors that influence reproductive success. Reproductive success is ultimately the only thing that has evolutionary value. In the final analysis, it is irrelevant whether an organism is strong, long-lived, or intelligent. Strength, longevity, and intelligence are useful only insofar as they boost reproductive fitness. The only thing that matters in evolutionary terms is the contribution that a genotype makes to the gene pools of future generations, relative to the average of all genotypes. (This is the technical definition of **fitness**, which is a property of genotypes, not of individuals, in modern evolutionary theory.) Most of the properties of organisms have become what they are because they contribute to lifetime reproductive output and have accordingly been preserved and enhanced by natural selection.

▌THE CONCEPT OF SPECIES

What is meant by the claim that two animals—say, a horse and a donkey—belong to different species? In the creationist framework that prevailed before Darwin, this claim implied that the two had no ancestors in common. It was thought that God created the first pairs of donkeys and horses separately on the sixth day of creation and that the two lines of descent have been entirely separate ever since. Although horses and donkeys can breed together, their offspring are sterile beasts called mules that have no prospect of having progeny of their own. The Bible intimates that God is displeased by the perverse human enterprise of trying to mate horses to donkeys. Orthodox Jews are forbidden by biblical commands from breeding mules, or sowing a field with two kinds of seed at once, or otherwise intermixing and miscegenating different kinds of creatures that God intended to be separate.

This way of thinking about species does not make sense in an evolutionary framework. A horse and a donkey, like any other two organisms, have innumerable ancestors in common, going all the way back to the first living things more than three billion years ago. In current biological theory, horses and donkeys are placed in different species not because they lack shared ancestors, but because they cannot have any shared *descendants* (other than mules, which have no reproductive future). For an evolutionary biologist, the hallmark of species is that they are **reproductively isolated** from each other. Because genes cannot move back and forth between the horse and donkey populations, the evolutionary future of horses (genus *Equus*, species *caballus*) is entirely distinct from that of donkeys (*Equus asinus*). This is what is meant by calling them different species. This defining mark of species status is enshrined in the standard modern definition of species:

Species are groups of actually or potentially interbreeding organisms, which are reproductively isolated from other such groups.

This definition, formulated by Ernst Mayr in 1942, is known as the **biological definition of species** or the **biological species concept** (BSC). The BSC is based on demonstrable patterns of breeding behavior—or more precisely, of reproductive discontinuity. There are many other definitions of species (Kimbel and Martin 1993, Holliday 2003), but almost all of them involve the notion of reproductive isolation in one form or another. We will consider some of these alternatives below.

Although Darwin entitled his book *The Origin of Species*, he had little to say about how species come into being. He was mainly concerned with demonstrating the power of natural selection to effect change *within* an evolving population. This sort of change is now called **anagenesis**. Later evolutionary biologists have looked more carefully into the process of **cladogenesis**, or the splitting of a single species into two. That process is also known as **speciation**.

Speciation requires one part of a single interbreeding population to become reproductively isolated from another. (This follows from the biological definition of species.) If a local population becomes *geographically* isolated—say, by colonizing an island that is hard to get to—it will be unable to exchange genes with the rest of its species. This kind of isolation may be only temporary. When Europeans arrived in Australia, they found native peoples who had apparently been isolated on that island continent for thousands of years. Yet the Australians had not lost the potential for interbreeding with people from the opposite side of the planet, and they soon proceeded to do so. Native Australians belong to the same biological species as the rest of humanity.

But although geographical isolation does not necessarily produce permanent reproductive isolation, it can do so. In fact, it tends to do so if it lasts long enough, because geographically separated populations of a species tend to start diverging from each other genetically. Such divergence can result from divergent pressures of natural selection. A population that is geographically isolated from the rest of its species is likely to be living at the edge of the species' range. If this marginal environment is not typical for the species, it will impose its own unique selection pressures on the isolated population, causing it to become more and more different from the parent stock.

Isolated peripheral populations also tend to diverge genetically because of random errors in allele sampling. This sort of random change is called genetic **drift**. Sampling errors are always largest in small samples, and so drift is most pronounced in small populations. If there are few individuals in a population, or each individual has few offspring, some alleles will be lost due to chance in each generation, because luck will play a relatively large part in determining which individuals reproduce (and also because half of an individual's genome is discarded at random in the making of each gamete). When all but one of the alleles of a gene have been lost, that one is then universal, or **fixed**, in the population. If a population stays small for many generations, drift may eliminate almost all the genetic variation in it. Over time, random fixation of alleles can cause a small, isolated peripheral population to develop genetic peculiarities not seen in the parent population.

Sampling errors can produce large evolutionary effects even in the span of a single generation. A population that is nearly exterminated, so that it is reduced at some point to a mere handful of reproducing individuals, will probably lose most of its genetic variability in the process. Some extant species appear to have passed through this sort of genetic **bottleneck** in the recent past—for example, cheetahs, all of which are nearly genetically identical (O'Brien 2003). A special case of a genetic bottleneck occurs when a species colonizes a new area. The first colonizing individuals are not likely to be a fully representative sample of the genetic variation within the species. If there is no further immigration into the new area, the random sampling error in the makeup of the small colonizing group will result in average differences between the gene pool of the colony and that of the parent population. This so-called **founder effect** is another factor that contributes to the genetic peculiarities of isolated, peripheral populations. Some theorists think that it plays an important role in speciation (Mayr 1982).

In the long run, sampling error and divergent selection pressures can be expected to produce average genetic differences between a peripheral isolate and its parent population. If these processes go on long enough, they can result in significant morphological differences between the two. But regardless of morphological differences, these populations remain parts of the same biological species so long as they remain mutually interfertile. If the two come back into contact with each other and individuals of the two populations mate successfully, the two populations may become genetically confluent again as genes flow in both directions. However, such matings will be fruitless if the two populations have accumulated genetic differences that interfere with the development of hybrid zygotes or make the hybrids infertile (like mules). If the two "populations" cannot interbreed and produce fertile offspring, a reproductive barrier has been established. They then belong to different species.

Such a speciation process, in which the newly forming species are completely separated geographically until gene flow between them is no longer possible, is called **allopatric** (Greek, "different fatherland"). Allopatric speciation is contrasted with **sympatric** speciation, in which two species arise out of one without any geographical separation, and with **parapatric** speciation, in which two populations occupying adjacent territories gradually become reproductively isolated as the flow of genes between them dwindles and finally ceases. Darwin stressed the importance of sympatric and parapatric modes of speciation (Darwin 1859, Chapter 4), and there has been a recent revival of interest in these processes among evolutionary biologists (Mallet 2001). However, the consensus view is that speciation is usually allopatric (Maynard Smith 1966, Bush 1975, Coyne 1992, Futuyma 1998).

∎ EVIDENCE FOR ANAGENESIS AND CLADOGENESIS

All this is theory. Are there empirical reasons for believing that species can change and split?

First, we know for sure that natural selection can produce anagenetic evolutionary changes within a single species. Such changes are often swift enough to be noticed within the lifetime of a human observer, and many have been observed. Bacterial resistance to antibiotics is one well-known example. In the 1940s, when antibiotics first came into use, a shot of penicillin could be relied on to clear up almost any bacterial infection in short order. In the ensuing years, the widespread use of antibiotics placed severe new evolutionary pressures on disease-causing bacteria. Mutations that made bacteria resistant to antibiotics were conserved and promoted by natural selection. As a result, after only a few decades of evolution, many of the early antibiotics have now lost almost all of their power to cure human disease. Similar resistance to human pesticides has evolved in many other sorts of organisms, from malaria parasites to mice. Hundreds of other examples of recent

evolutionary change attributable to Darwinian mechanisms have been observed and studied. Evolution by natural selection is not a mere theory; it is a familiar fact of life.

There is also firm evidence that an interbreeding population can divide into two reproductively isolated species. In certain cases, new species have been observed to come into being overnight. For instance, this can happen when diploid organisms produce aberrant gametes bearing two complete genome copies (2N) instead of the usual single copy (1N). If a diploid (2N) gamete unites with a normal, haploid (1N) gamete, the resulting **triploid** (3N) zygote is usually sterile, because it cannot divide its three genome copies evenly in half to make gametes of its own. However, if the 2N gamete unites with another 2N gamete, the result is a **tetraploid** (4N) zygote. A tetraploid is more likely to be fertile than a triploid is, since its 4N genome can be evenly halved to produce another 2N gamete. The tetraploid is reproductively isolated from its parents, because a backcross to the parent species produces a sterile triploid. But if the tetraploid can fertilize itself (as many plants can), or if one of its gametes can find another 2N gamete from some other individual to combine with, it may give rise to a new, reproductively isolated tetraploid species. This sort of instant speciation is rare among vertebrates, and only one mammalian example is known (Gallardo et al. 1999), but it is a frequent occurrence among plants. Familiar agricultural examples of tetraploid plants include alfalfa and potatoes.

A new species may also come into being overnight when a sperm carrying $(2 + x)$ genome copies from one species encounters an egg bearing $(2 - x)$ copies of the genome of a related species. Such a mating can yield a fertile 4N (tetraploid) hybrid, which is reproductively isolated from both parents (again, because a backcross to either parent species produces a triploid). This sort of speciation is also well-documented among plants, where new tetraploid hybrid species have actually been caught in the act of emergence (Müntzing 1930, Lowe and Abbott 2004). It is the main exception to the previously mentioned rule that the terminal buds (species) on the tree of life do not reunite once they have become reproductively isolated.

∎ THE TEMPO OF SPECIATION

Among animals, instant speciation by genome doubling is rare. But in both animals and plants, reproductive isolation can be produced almost as swiftly by an accidental reshuffling of the **chromosomes** on which the genes are arranged. Many pairs of closely related species are reproductively isolated from each other by such rearrangements, which prevent the chromosomes of hybrid animals from recombining properly in the process of producing haploid gametes. This is what makes mules sterile. Although these horse-donkey hybrids are strong and healthy beasts, their eggs and sperm are defective because of incompatibilities between the chromosomes of horses and donkeys (Chandley et al. 1975). Chromosomal rearrangements that get fixed in a small local population may isolate it in this way from the rest of its species, bringing a new species into existence in a very short span of time.

However, this is not the usual mode of speciation in animals. Chromosomal anomalies do not ordinarily become fixed when they crop up in local populations. Any new chromosomal arrangement that results in reduced fertility will usually be quickly eliminated by natural selection within that local population. Conversely, if a new arrangement does *not* reduce fertility, it cannot reproductively isolate a local population from the rest of the species. Many chromosomal rearrangements have no effect on fertility. A species can maintain several different chromosomal arrangements within a single interbreeding population (Bruere and Ellis 1979).

Among animals, cladogenesis seems as a rule to result not from genome duplication or chromosome rearrangement, but from the more gradual operations of natural selection and drift in small, isolated, peripheral populations. The evidence for this is admittedly indirect. The process cannot be observed and studied directly, precisely because it is gradual and therefore takes many thousands of years to proceed to completion.

Biologists argue about just how many thousands of years are involved. Some think that founder effect and drift work relatively quickly to fix genetic novelties and produce reproductive isolation in small peripheral populations. Others think that speciation is a more gradualistic process, resulting mainly from novel pressures of natural selection that are imposed by the marginal environments where peripheral populations live. This debate is not easy to settle.

In general, we cannot rely on the fossil record to give us information about the tempo of speciation. Speciation itself is rarely captured in the fossil record. Reproductive isolation, the ultimate product of the speciation process, can be detected fairly readily. If two anatomically distinct types of extinct animals persist for some time side by side in the same deposits with no intermediate forms filling the morphological gap between them, we are justified in concluding that they were reproductively isolated from each other (once we rule out the possibility that they represent the males and females of a single dimorphic species). But we rarely have a fossil record that shows a single species gradually splitting into two. What we generally find is that new species pop into existence in the fossil record, seemingly out of nowhere, then evolve along for several million years

undergoing varying amounts of anagenetic change and finally become extinct.

The abrupt appearance of new types in the fossil record might be taken as evidence that the speciation process is very swift. But this inference is not warranted. Abrupt appearances are just what we would expect to see if speciation is usually allopatric. If two incipient species are allopatric, a given fossil locality can preserve only one of them. The larger, more extensive parent population is more likely to be sampled in the fossil record than the small peripheral population that is evolving more rapidly. The two will not be found together in the same deposits until speciation is complete and they can occupy the same territory without coalescing. The eruption of one species into the territory of another may look quite abrupt in the fossil record, but this abrupt appearance does not tell us how much time it took for the two to become reproductively isolated while they were allopatric.

The amount of time that speciation takes can best be estimated by studying living populations. One way to do this is to find cases where populations of a single species have been separated from each other by the recent appearance of barriers that interrupt gene flow—for example, when a large lake dries up to form several small lakes, or when rising waters partly flood a land area so that it is divided into smaller islands, or when a climate change warms up a region and forces a lowland species to retreat up to isolated, persistently cool mountaintops. When the drying, flooding, or warming events can be dated geologically, we can tell how long the populations inhabiting the lakes, islands, or mountaintops have been separated. We can then do breeding experiments to determine whether they have become reproductively isolated.

We can also use a so-called **molecular clock** to estimate how long ago one lineage branched off from another. Whenever we study the genomes of two different species, we can always pick out some sequences of DNA that are homologous—meaning that they have both been copied, over many generations, from a single DNA sequence in the genome of a common ancestor of the two species. Their homologies can be detected by their similar nucleotide sequences. They will also usually have similar functions (e.g., coding for functionally equivalent proteins), and they may occupy similar positions on equivalent chromosomes. In general, the more closely related two species are, the more numerous and similar their homologous DNA sequences will be. If we know the phylogenetic relationships within a group of species, we can look at homologous stretches of their DNA to determine whether they have evolved at a more or less constant rate in all the species throughout their evolutionary history (Fig. 2.5). This often turns out to be true. If

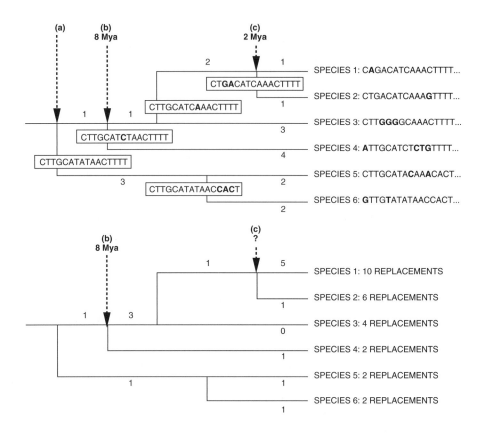

FIGURE 2.5

How a "molecular clock" works. If we know the phylogenetic relationships within a group of species, we can compare homologous stretches of DNA to determine how many nucleotide replacements have accumulated along the lineage leading to each species since they diverged from a common ancestor. If the number is the same for each species, then the rate of change has been constant across all the lineages. The upper diagram represents a hypothetical tree in which this rate is constant; each lineage has accumulated 5 nucleotide replacements (boldface letters) since the time of the last common ancestor at **(a)**. The letters in boxes represent ancestral sequences inferred for each node (branching point) of the tree. If we can date node **(b)** at 8 million years ago (Mya), we can then infer that species 1 and 2 split apart approximately 2 Mya at node **(c)**. The lower diagram represents a tree in which evolutionary rates differ in different lineages; species 1 has accumulated five times as many nucleotide replacements as species 4, 5, and 6 have since the time of their last common ancestor. In such cases, the molecular clock does not keep accurate time, but probabilistic methods can still allow us to derive a divergence-time with attached confidence limits.

the replacement rate has been constant, and if a date can be ascertained for one branching point within the tree, then we can infer the dates of other branch points.

There are some problems with molecular clocks. The biggest problem lies in determining a date for that one branch point, which usually has to be estimated by using paleontological or geological data. The chain of inferences involved in these estimates introduces multiple sources of error into the results (Marko 2002). Errors may also be introduced by the phenomenon of **gene duplication**. Genes often get duplicated within a single genome, and so apparently similar genes in two species may have diverged long before the species themselves split apart. These sources of error can be

minimized by basing our estimates on a wider sample of the genome.

Taken together, biogeographical and molecular data show that the tempo of speciation varies greatly. Under favorable circumstances, it may take only a few thousand years for evolving populations to become reproductively isolated from each other (Fryer and Iles 1972, Johnson et al. 1996). At the other extreme, some populations that have been separated for as long as 20 My can still exchange genes (Stebbins and Day 1967). The average time that it takes for two separated populations to attain reproductive isolation has been estimated to be somewhere between 1 and 3 My (Coyne and Orr 1989, Futuyma 1998, Curnoe et al. 2006, Cartmill and Holliday 2006).

▌SEMISPECIES, HYBRIDS, AND ISOLATING MECHANISMS

No matter whether speciation takes a few thousand years or a few million, it is still far too slow for anybody to witness. But the theory of allopatric speciation gives rise to some predictions that we can test. We would expect to find incipient species in the process of coming into being in the world around us. This implies that we should be able to observe all possible degrees of interfertility, from unimpeded gene flow to perfect reproductive isolation, between closely related adjacent populations.

In fact, this is what we find. Realizing the importance of this fact for his theory, Darwin devoted a whole chapter of the *Origin* to the topic of hybridization between species. He observed that "… when forms, which must be considered as good and distinct species, are united, their fertility graduates from zero to perfect fertility" and that in certain cases interspecies hybrids are more fertile than either parent stock. These facts, he concluded, "… support the view, that there is no fundamental distinction between species and varieties" (Darwin 1859, Chapter 8). Later work has borne out Darwin's conclusions. Interspecies hybridization is a common occurrence. At least 9% of all bird species interbreed with others and produce hybrid offspring (Grant and Grant 1992). Hundreds of pairs of animal species have been observed to produce hybrids in the wild, with varying degrees of reduced fertility (Bush 1975, White 1968, Harrison 1993, Levin 2002).

Most hybridizing species pairs represent once-united populations that have undergone a period of geographical isolation from each other, diverged genetically, and then come back into contact. What happens in such cases depends on the amount and kind of divergence between the two species. If the two are still genetically compatible, they may simply merge with each other and become one species again. However, they can remain discrete indefinitely if some **isolating mechanism** acts to limit the production of hybrids. The most common isolating mechanisms are genomic incompatibilities that prevent hybrid embryos from forming. Such incompatibilities preclude hybridization between most pairs of species. Less fundamental genetic incompatibilities may allow hybrid zygotes to develop but make the hybrids less fit or less fertile than their parents. Even vigorous, healthy hybrids may be rendered sterile by chromosomal incompatibilities, as in the case of mules.

Anatomical and behavioral differences can also act as isolating mechanisms. For instance, two parapatric sister species may differ markedly in size, or have different mating signals, or be active at different times of the day, or feed in different ecological zones. Differences of all these sorts tend to discourage interspecies matings. When such matings do occur in spite of such differences, any hybrids that result may have anatomy or behavior that falls somewhere in between the two optimal types. This often makes the hybrids less well-adapted than either parent (Jiggins et al. 1996).

For all these reasons, the individuals produced by interspecies hybridization are usually relatively few in number and not fully competitive with the members of either parent stock. They tend to be restricted to a relatively narrow **hybrid zone** between the territories of hybridizing parapatric species. But as long as fertile hybrids continue to be produced, a limited amount of gene flow will continue between the two populations. What happens to the alien genes entering each population's gene pool depends on the numbers. If hybrids are at a selective disadvantage, the intrusive genes will be culled by natural selection. If hybrids are rare, if the population is small, or if the forces of selection are weak, then the intrusive genes will tend to disappear due to drift. However, if hybrids are favored by selection, gene flow between the populations will increase and the hybrid zone between them will grow wider. If it becomes wide enough, it will become a mere gradient in allele frequencies within a single indivisible population.

Whenever interspecies hybrids are at a selective disadvantage, females that invest their reproductive resources in hybrid offspring will also be at a disadvantage. (In a state of nature, a mare that accepts a donkey as a mate and rears a mule foal is wasting a lot of time and resources.) The geneticist T. Dobzhansky argued that isolating mechanisms are adaptations, which evolve to prevent females from squandering their resources on unfit offspring (Dobzhansky 1951). This argument sounds intuitively plausible, but there are theoretical problems with it (Butlin 1989, Futuyma 1998). It is impossible for selection to favor a mutation that makes its carriers sterile. Therefore, hybrid sterility cannot be promoted by natural selection. This holds for all isolating mechanisms that limit interspecies gene flow by reducing the fertility or fitness of the hybrids. Such **postzygotic** isolating mechanisms can become established in a population only by chance, or as a side effect of some other, more adaptive change.

Once postzygotic mechanisms have been established, selection may then favor mutations that produce **prezygotic** isolating mechanisms, which act to prevent the formation of hybrid zygotes in the first place—say, by causing mares to spurn the advances of male donkeys. But such mutations will not be adaptive unless the fertility of hybrid zygotes is markedly inferior. And outside of the hybrid zone, prezygotic isolating mechanisms will not carry any adaptive advantage. (A mutation that causes mares to spurn male donkeys is not useful in the absence of donkeys.) Therefore, such mechanisms will not become rapidly or reliably fixed in the whole population.

If there is no selection pressure driving the evolution of isolating mechanisms, then sister populations may remain parapatric for long periods of time without either merging or becoming completely reproductively isolated. The baboons of the genus *Papio* are a case in point. These large, ground-dwelling monkeys are divided into five to nine regional populations of distinctive appearance, which range across much of sub-Saharan Africa (Fig. 2.6). The ranges of these regional forms or types do not overlap. Some of them merge into each other through genotypic and phenotypic gradients. Others are separated by relatively narrow hybrid zones in areas where they contact each other. So far, no isolating mechanisms have been identified that keep these regional populations discrete. As far as is known, all produce fertile hybrids with each other, and genetic studies demonstrate that there is gene flow between at least some of these populations. But the narrowness of some of the hybrid zones, and the slightly differing adaptations and social behaviors of some of the regional forms (Kaplan et al. 1999, Jolly 2001, Jolly and Phillips-Conroy 2003), suggest that there may be factors restricting gene flow. If there are, then these populations may go on exchanging genes indefinitely while still remaining separate. The gelada baboons of Ethiopia appear to have done just that. These distinctive-looking primates, which branched off from *Papio* around 5 Mya and are usually assigned to a genus of their own (*Theropithecus*), still produce fertile hybrids with *Papio* in captivity and probably in the wild as well (Jolly et al. 1997, Jolly 2001).

Populations of this sort, which are partly but not completely reproductively isolated from each other, are sometimes referred to as **semispecies**. They illustrate Darwin's point that there is no fundamental distinction between species and varieties.

▌"RACES"

Darwin's claim is true if we interpret "varieties" to mean semispecies. But it is not true if "variety" is taken to mean what is usually meant by "race." Human "races" are not incipient species. No isolating mechanisms separate any human populations, and no reduction of fertility is found in matings between any two such populations. Therefore, no narrow hybrid zones are seen.

The popular concept of race reflects the fact that human populations in the Old World show average genetic differences correlated with geography. Human pigmentation is darkest in the tropics and lightest in northwestern Eurasia; human scalp hair is tightly curled in southern Africa and straight in northeastern Eurasia; and so on. These and other geographically patterned differences are popularly used to divide the human species up into a varying number of "races." But the populations living in between these extremes display a continuous gradient of skin tones and hair textures, with a good deal of local variation among individuals. Geographical gradients of this sort are called **clines**. Modern study of human genetic variation is grounded in the proposition that "There are no races, only clines" (Livingstone 1962). The division of the human species into racial categories is a cultural construct. It does not reflect the facts of human biology.

Why haven't any isolated human populations—for instance, the natives of Australia—become semispecies? There are two possible reasons: either they have not been isolated long enough, or the geographic barriers separating them from the rest of the species have not been sufficient to prevent gene flow. The second reason has probably been more important than the first in the case of *Homo sapiens*. Large predatory mammals tend to have large individual ranges and to be highly mobile. The geographical ranges of such species are accordingly extensive. Wolves, lions, cougars, leopards, and cheetahs are all single species, united by patterns of gene flow that extended across multiple continents until recently, when human competition broke these populations up into small, isolated remnants. Human beings are also large predatory mammals; and human technology, intelligence, and self-awareness make us even more mobile than wolves and lions. All this makes it uniquely hard to interrupt the flow of human genes, and it makes it correspondingly unlikely that Pleistocene *Homo* ever speciated (Bush 1975, Arcadi 2006).

To draw classificatory distinctions between contiguous populations, we must have some evidence of isolating mechanisms that restrict gene flow between them. This principle needs to be kept in mind in analyzing human "racial" variation. It should also guide us in thinking about earlier stages in human evolution. Three million years ago, our ancestors were apelike, mainly herbivorous animals restricted to the continent of Africa. They were evidently not as mobile or wide-ranging as later humans. Like many other African herbivores, they appear to have been divided into many local species and semispecies. But around two million years ago, ancestral humans belonging to our own genus, *Homo*, began to migrate out of Africa into Europe and Asia, crossing major geographic barriers that had checked the spread of other large mammals. On the face of it, we would not expect these mobile, predatory animals to have fragmented into dozens of local species. It seems prudent to assume that they did not, unless there is solid evidence to the contrary. We would expect, for example, that the archaic, low-browed European humans known as **Neandertals** would not have been reproductively isolated from other human populations and that they would therefore be best described as a subspecies or regional variant of *Homo sapiens* rather than a distinct species (see Chapter 7).

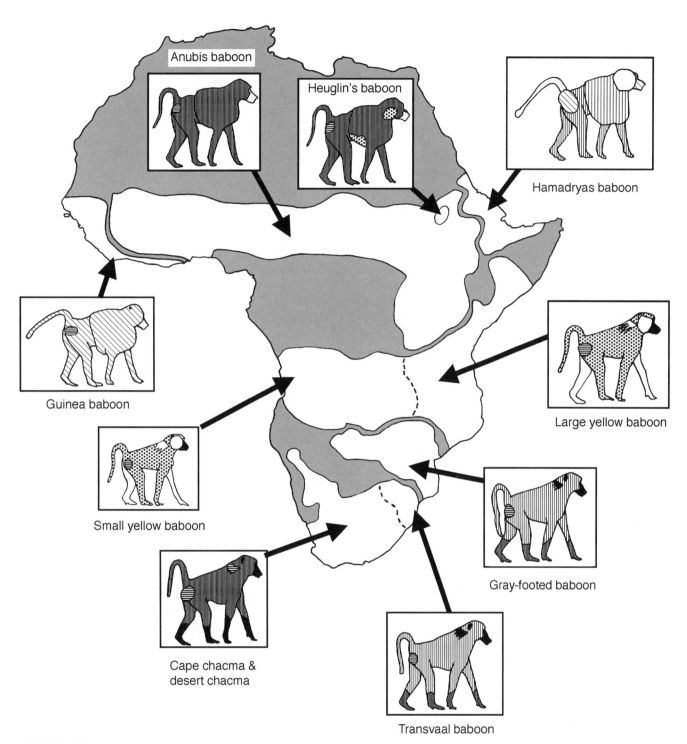

FIGURE 2.6

Semispecies of baboons (*Papio*). Nine or more distinctive regional populations can be distinguished on the basis of such traits as pelage pattern and tail length. Some of these (e.g., the small and large yellow baboons) blur into each other along a clinal gradient, but others (e.g., hamadryas and anubis baboons) are separated by narrow hybrid zones reflecting constraints of some sort on interbreeding between the adjoining populations. Whether we describe such semi-isolated populations as different species or not is a matter of definition. (After Jolly 2001.)

The classification of extinct varieties of early *Homo* as species, semispecies, or "races" has been hotly debated among paleoanthropologists since the 1860s, when T. H. Huxley and W. King clashed over the human status of the Neandertal 1 calvaria (Huxley 1863, King 1864). In the later chapters of this book, we will discuss these debates, but we will suspend judgment on species-level taxonomy. Instead, we will use vernacular tags—what G. G. Simpson (1963) referred to as "N1-level naming terms"—to label distinctive spatial, temporal, and morphological clusters within the genus *Homo*. For example, we will call Neandertals simply "Neandertals," instead of forcing a choice between the taxa *Homo neanderthalensis* (a species) or *Homo sapiens neanderthalensis* (a variety or subspecies).

We adopt this convention for both theoretical and practical reasons. At a theoretical level, we accept Darwin's conclusion that there is no fundamental distinction between species and varieties. When we are dealing with closely related populations, this distinction is often better expressed as a value of a continuous variable, the rate of gene flow between the populations, than by a discontinuous all-or-nothing label. But at a practical level, we recognize that it is often impossible, in the present state of our science, to determine what the rate of gene flow was between two ancient human populations. When one morphological type, or **morph**, supplants another in the fossil record, the transition may represent a genuine extinction event, in which one species outcompeted and replaced another species from which it was reproductively isolated. However, gene flow between semispecies can itself produce effective extinction and give the appearance of replacement. This is most likely when hybridization occurs between a morph that is abundant on the landscape and one that is much rarer. The rarer morph can be driven to "extinction" even if hybrids between the two morphs are not sterile (Levin 2002). The abundant morph simply overwhelms or **swamps** the rare one, either genetically or demographically or both. Even when absolute amounts of gene flow are the same in both directions, the smaller gene pool will be disproportionately affected. In such circumstances, the rare forms will disappear relatively quickly—not because they die off or are outcompeted, but because they become absorbed or **assimilated** by the abundant form. Situations like this are well known in recent human history—for example, in the cases of the pre-colonial natives of New England, Argentina, and Tasmania. We suspect that they were common in our evolutionary history as well.

SPECIES AND FOSSILS

Despite its name, the biological species concept (BSC) does not always fit the biological facts. It cannot be applied to sexless organisms that reproduce by cloning themselves. It is also a poor fit for many plants and other organisms in which the genomes of two fully separate species can be brought together to produce a new hybrid species. Some think that the BSC should not be applied to bacteria or plants (Raven 1976, Futuyma 1998). If so, then most of the biological realm is excluded from the "biological" definition of species. Similar criticisms apply to other species concepts that focus specifically on mating—for example, H. Patterson's (1978) **recognition species concept** (RSC), which defines species boundaries with reference to the presence of a shared "specific mate recognition system."

Species concepts defined in terms of breeding patterns are also hard to apply to fossils (Holliday 2003). Even when we are dealing with multicellular animals, to which the BSC or RSC might apply in theory, breeding tests and mate recognition tests cannot be carried out on extinct organisms. Although extant species distinguished by these criteria are usually morphologically distinctive as well, that distinctiveness is not always reflected in the parts of the anatomy preserved in the fossil record. A deeper theoretical problem becomes apparent when we ask whether two populations from different *time* horizons should be grouped in the same species. Given enough time, evolutionary change within a single evolving lineage can produce greater differences between ancestors and descendants than we ever see encompassed within a single living species. If such ancestors could somehow be brought together with their much-changed descendants, they would probably not be fertile mates for each other. Yet they are in some sense reproductively connected through time. When we have a detailed fossil record of an evolving population, we sometimes find that the beginning and end of the sequence differ enough to be placed into different species, but that there is no gap in the continuum to give us an excuse for drawing a species boundary between the earliest and latest forms (Gould and Eldredge 1977, Rose and Bown 1984, Gingerich 1985, Eldredge 1993). Some people hold that the fossil record of our own genus, *Homo*, fits this description (Brace 1967, 1995; Wolpoff 1999).

Some biologists have tried to deal with this problem by concocting species concepts that involve a time dimension. The **internodal species concept** (ISC) is one example (Fig. 2.7). By this definition, two organisms belong to the same species "by virtue of their common membership in a part of the genealogical network between two permanent splitting events or between a permanent split and an extinction event" (Kornet 1993).

This concept seems clear enough in theory: a species is born when its parent species splits, and it dies when

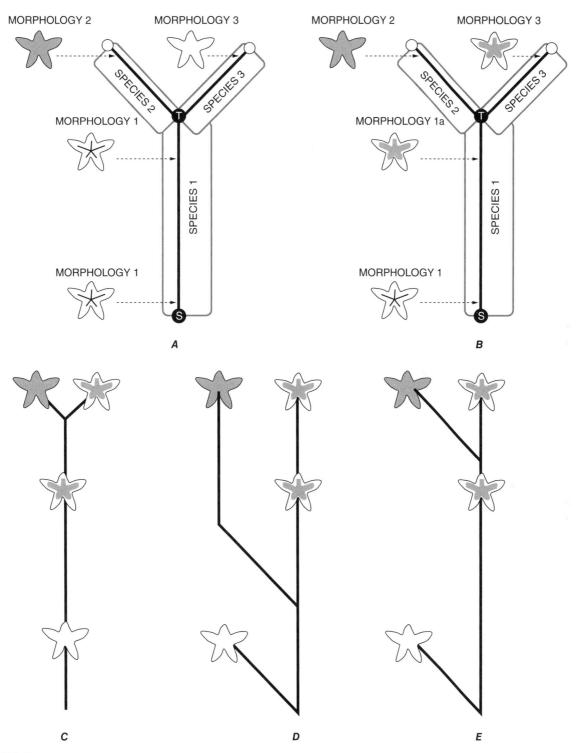

FIGURE 2.7

The internodal species concept. Species are defined as segments of the evolutionary tree delimited by nodes—either splitting events (black circles) or extinction events (white circles). In **A**, species 1 comes into existence with morphology 1 at node S and ceases to exist at node T, when it splits into sister species 2 and 3 with their own distinctive morphologies (2 and 3). In **B**, species 1 undergoes anagenetic change, producing a new morphology (1a); nevertheless, organisms with both morphologies must be classified in species 1, since there has been no splitting event between nodes S and T. After node T, the sister of species 2 continues to preserve morphology 1a; yet the ISC requires that it be recognized as a new species (species 3), even though it has no distinctive traits. Different assumptions about the placement of splitting events in the fossil record (**C–E**) will impose different species boundaries on the same set of fossils.

it goes extinct or splits in two itself. However, there are some problems with it. Because the ISC defines species solely in terms of the geometry of the evolutionary tree, the boundaries it recognizes will not always correspond with the observable differences between fossil specimens. If anagenesis is minimal and if cladogenesis is always accompanied by evolutionary change in every descendant lineage, the ISC works well enough (Fig. 2.7A). But when these assumptions do not hold, the ISC does not work well. It does not allow us to draw a species boundary between an ancestor and a descendant transformed by anagenetic change; all points on a single evolving lineage must be parts of one internodal species, no matter how great the differences between ancestors and descendants. Conversely, if a parent species persists unchanged after a splitting event, it has to be given a new species name (Fig. 2.7B), even though there are no differences between the earlier and later forms.

The ISC also assumes that we have detailed knowledge of the evolutionary tree diagram for the organisms we are trying to classify. But there is almost always more than one plausible family tree that we can draw, and so the ISC rarely provides an unambiguous classification (Fig. 2.7C–E). Finally, what counts as a "splitting event" ultimately has to be defined in terms of reproductive isolation. The ISC thus preserves all the shortcomings of the biological species concept in addition to its own defects. Because it puts a higher value on distribution in time and space than on phenotypic traits in assigning organisms to species, it has limited practical value in classification.

Similar criticisms apply to other attempts to define species in terms of the geometry of the evolutionary tree. A close relative of the ISC is the **evolutionary species concept** (ESC), first defined by G. G. Simpson (1961) and subsequently modified and elaborated by others, particularly E. O. Wiley (1981, 1992; Wiley and Mayden 2000). An "evolutionary species" is defined by Wiley (1981, p. 25) as "a single lineage of ancestor-descendant populations which maintains its identity from other such lineages and which has its own evolutionary tendencies and historical fate." Like internodal species, evolutionary species are time-bounded individuals with a delimited lifespan—discrete entities that originate in one cladogenetic event and end in another one (or in extinction). But the notion that a species has an individual "identity" presupposes a boundary that delimits that species from others. If this boundary is defined in terms of reproductive exclusivity, then the ESC (like the ISC) incorporates the BSC and inherits its deficiencies. And if the boundary is defined in terms of shared morphology, then the ESC (like the ISC) runs afoul of the fact that the distribution of shared traits need not always conform to the geometry of the evolutionary tree.

MORPHOSPECIES

Defining species in terms of evolutionary tendencies, historical fate, and lineage identity is not a simple matter of taking caliper readings from fossil bones. These are nebulous, tricky concepts, and any conclusions we reach concerning them depend on a chain of uncertain inferences from the facts of morphology, geography, and stratigraphy. Not surprisingly, many scientists prefer to approach the definition of "species" by setting aside questions about the geometry of the tree and relying almost entirely on the anatomy of the specimens. This approach underlies the **morphological species concept** (MSC). Under the MSC, a species is defined as *the smallest cluster of organisms that can be distinguished from other such clusters by consistent phenotypic traits.* (The anatomical differences between males and females are always ignored in identifying these clusters, so even the MSC incorporates a minimal breeding-related criterion of species membership.) Such phenotypic clusters, or **morphospecies**, are not equivalent to the clusters recognized as species under the BSC. Some reproductively isolated populations, known as **sibling species,** are not morphologically distinct (Mayr 1976, pp. 509–514; Ayala 1982). Conversely, some morphologically distinct populations are not reproductively isolated. For example, white-tailed deer (*Odocoileus virginianus*) and mule deer (*O. hemionus*) are distinctively different in behavior and anatomy (Fig. 2.8). But in areas where both species occur together, they interbreed and produce hybrids. In those areas, hybrid individuals constitute 8 to 15 percent of all fawns born and represent an ongoing source of gene flow from each species into the other (Stubblefield et al. 1986, Rollins 1990, Derr et al. 1991, Cantu and Richardson 1997, Cathey et al. 1998).

The morphospecies is essentially a pre-Darwinian species concept, which makes no use of the insights of evolutionary theory. It does not take into account either the mechanisms of population genetics or the variation within populations that provides the material for natural selection to work on. Nevertheless, the MSC is the species concept that usually gets employed in the actual classification of newly discovered organisms. When biologists describe and name a new primrose or earwig, they ordinarily have only a few specimens to work with, and so they can say little about the variability or geographical distribution of the new species. They thus have to rely on morphology—consistently distinctive clusters of shared phenotypic traits—in defining that species and contrasting it with other primroses or earwigs (Sokal and Crovello 1992).

Paleontologists too are forced to rely on the MSC, because the fossil record rarely preserves an adequate sample of an extinct species' variation in time, space, and anatomy. We can sometimes compensate for the

FIGURE 2.8

Despite the distinctive forms of their antlers (black) and other differences in morphology and behavior, white-tailed deer (*Odocoileus virginianus*, left) and mule deer (*O. hemionus*, right) interbreed and hybridize in areas where they coexist. Limited amounts of interspecies gene flow result, mainly flowing from *O. virginianus* into the *O. hemionus* gene pool.

MSC's deficiencies by making use of what we know about the living relatives of the fossil. For example, if several specimens of an extinct ape are recovered, the differences between them can be compared with the variation found within each species of living ape, to see whether the fossils can fit comfortably within the range of variation of a single ape species. If they can, then they can all be lumped together in the same morphospecies. If they cannot—and if the fossils form two distinguishable clusters at the same time horizon—they are put into separate species. If they come from different time horizons, they may be taken as representing an ancestral population and a descendant population altered by anagenesis. In such cases, adherents of the morphological species concept may choose to distinguish the earlier and later clusters as different morphospecies. Such ancestor–descendant pairs are called **chronospecies**.

Chronospecies are controversial. Many theorists refuse to recognize such taxa, and most species concepts preclude defining a species in this way. As we noted earlier, chronospecies do not conform to any concepts of species boundaries based on patterns of gene flow, since ancestors and descendants are always in some sense reproductively linked. Adherents of internodal and evolutionary species concepts will refuse to place ancestors and descendants in separate species unless there has been a splitting event somewhere in between them. The word "speciation" is rarely used to describe changes produced by anagenesis within a single lineage, even by those who find it necessary in practice to draw taxonomic distinctions between ancestors and their modified descendants. The theoretical

difficulties with chronospecies come into sharp practical focus whenever new fossil finds fill in the temporal and morphological gap between previously established chronospecies, rendering the species boundary arbitrary. In the human lineage, the distinction between *Australopithecus anamensis* and *Australopithecus afarensis* is a good example of this sort of situation (Chapter 4).

When the differences between two fossil samples are greater than those found within some living species, but not greater than those found within others, paleobiologists differ on what ought to be done. Some believe that it is always best to err on the side of recognizing too many species, so as not to overlook any cladogenetic events. Others think that it is better to recognize as few species as possible, in accord with the logical principle called **Occam's Razor**. This principle is named after the medieval philosopher William of Occam, who insisted that "entities should not be multiplied unnecessarily"— that is, that the number of things we recognize as existing should be kept to a minimum. (This is another version of the Principle of Parsimony.) These two schools of thought are familiarly known as "splitters" and "lumpers." The authors of this book tend to be lumpers, but we recognize that this preference has no empirical basis. Like many areas of contention in science, it is ultimately a philosophical question.

We can sum up these debates over the nature and identification of species by saying that the biological species concept is fundamental to an understanding of both evolutionary process and biological classification. Nevertheless, it does not apply very well to situations involving semispecies, interspecies hybrids, anagenetic evolution, or reproduction by cloning. In constructing scientific theories, it is usually a danger signal when we are unable to define a key concept in a way that covers all the relevant cases. Such situations often alert us to some fundamental wrongness in the concept. The problems surrounding the concept of species suggest that it has only a loose correspondence to the living world and its complexities. This is exactly what we should expect to find if Darwin is right. Trying to sort living things into discrete groups is bound to be difficult, and sometimes futile, if the diversity of life has been produced by slow evolutionary change.

▌ MICROEVOLUTION AND MACROEVOLUTION

Up to this point, we have been dealing with evolutionary phenomena involving individuals, local populations, and species. Evolutionary processes at these basic levels are referred to collectively as **microevolution**. They can be inferred straightforwardly from longitudinal studies of the biology of living populations over the span of a single human lifetime. There is little argument about the reality of microevolution. Even many creationists concede that descent with modification over several generations can transform a species, and that natural causes can bring new species into being (Wieland 1994, Lester 1994). There is less agreement whether such small changes can add up to big ones over long ages of geological time. Some Fundamentalist creationists admit that microevolution occurs, but they insist that large-scale evolutionary change—or **macroevolution**—does not. A horse, they concede, might evolve into a zebra; but it is impossible for a fish to evolve into an amphibian, or for an ape to evolve into a human.

No reputable biologists doubt that macroevolution occurs. However, its causes are disputed. The more orthodox Darwinians (sometimes called "gradualists" by their opponents) think that macroevolution is only accumulated microevolution. In their view, most evolutionary change is anagenetic and gradual, with large changes and long-term trends resulting from long-sustained or intense natural selection of genetic variants within evolving species. The opposing school of thought holds that cladogenesis is required for macroevolution—that evolutionary breakthroughs are concentrated in bursts surrounding speciation events, when founder effect and drift produce abrupt, largely random genomic alterations in isolated peripheral populations. In this view, anagenetic change is typically restricted in scope and brief in duration. Major evolutionary changes and long-term trends are produced mainly by selection *of* species rather than by selection *within* them. Chance processes throw up new species that differ from the parent population in essentially random ways, and inter-

specific competition then determines which species survive. This model of the evolutionary process was forcefully promoted by N. Eldredge and S. Gould, who labeled it "punctuated equilibrium" (Eldredge and Gould 1972, Gould 1982a,b).

The difference between punctuationists and gradualists is not as great as partisans on either side sometimes like to pretend. Both agree on the following:

1. Anagenetic change does take place. It can be either slow or fast, and it can produce improvements in adaptation (Gould and Eldredge 1977).
2. Evolutionary change is sometimes concentrated around speciation events.
3. Cladogenesis can be very rapid—overnight, in cases of speciation by genome doubling. Even when it is driven by the slower processes of selection and drift, speciation may under favorable circumstances take only a few thousand years. This looks instantaneous in the fossil record (Ayala 1985).
4. Natural selection operates at different levels, including that of competing alleles within a gene pool and that of competing species within an ecosystem.

These points are not in dispute. The disputed points concern the grain of the picture. Whether we see evolution as smooth or jerky depends to a large extent on how closely we zoom in on it. Because mutations are essentially random, all evolutionary change begins with sudden fortuitous changes. And because all things that reproduce themselves—alleles, organisms, or species—disappear sooner or later, and some of them disappear sooner than others, the fountain of random novelty is channeled and directed by natural selection at whatever level it operates. There are real and unsettled disagreements about the relative importance of anagenesis and cladogenesis in macroevolution. The fossil record has (once again) been of little help in settling these disagreements, because it does not provide a sufficiently detailed picture to tell us how much of which changes are produced by what processes.

Some theorists have argued that the fundamental evolutionary changes that were involved in the origins of phyla and other higher taxa demanded wholesale reorganization of the genome, made possible either by rare "macromutations" that produced "hopeful monsters" (Goldschmidt 1940) or by the general instability of the genomes of early multicellular animals (Gould 1989, p. 231). These ideas have not won general acceptance. There is no good reason to think that macroevolution requires revolutionary genomic upheavals. Gradual morphological change, representing accumulating microevolution, is observed in at least some fossil populations across geological time; and some rates of anagenesis seen in living populations are high enough to yield rapid evolutionary changes that would look punctuated in the fossil record.

But even if fundamental genetic transformations are not needed to explain macroevolution, that does not imply that they never happen. Single mutations in certain genes can have far-reaching effects on development and morphology, because the genomes of multicellular organisms are hierarchically organized. This hierarchy is necessary for the differentiation of cells into different types. Not all genes are **expressed** (read out to generate RNA and proteins) in every cell of the body. Whether a gene gets expressed in a particular cell or not is determined by the presence in the cell nucleus of proteins called **transcription factors**, which are the products of other genes known as **regulatory genes**. There are many levels of gene regulation. Some regulatory genes act early in development to establish the basic anatomical groundplan of the organism, while others come into play much later and have only minor effects on the phenotype. Mutations in regulatory genes can result in large anatomical changes. Perhaps the best-known examples are the genes in the so-called *box* complex, which control the fundamental differentiation of the embryo around a head-to-tail axis in multi-celled animals. Mutations in these genes may have been involved in the Precambrian origins of some of the major animal phyla (Valentine 2004). As we learn more about the hierarchical organization of animal genomes, we are sure to discover more regulatory gene complexes that have the potential to produce adaptive macromutations.

Following the rediscovery of Mendel's work in the early 20th century, biology went through a phase in which major evolutionary change was attributed to macromutations that brought new types of organisms into existence instantaneously (Bowler 1983). Natural selection was seen as having a mainly stabilizing influence on the course of evolution. This anti-Darwinian episode in the history of evolutionary theory lasted for only a decade or two. Its main legacy was to inspire the myth of the mutant superman, which made its first literary appearance in the 1930s (Stapledon 1935) and has lingered in science fiction and comic books ever since. As we will see later, this image of mutants with superior powers also lingers on in the literature of anthropology. Some linguists and anthropologists contend that language came into being through a macromutation, bringing with it all the distinctively human powers of the mind (Lenneberg 1964, 1967; Chomsky 1964, 1975; Bickerton 1990, Tattersall 1998, Klein 1999; cf. Burling 1992, Hauser et al. 2002).

During the 1920s and 1930s, the two conflicting bodies of evolutionary theory—the Darwinian theory with its focus on natural selection and adaptation, and the early "geneticist" theory with its focus on individual variation and mutation—were reconciled and combined to form the so-called **synthetic theory of evolution (STE)**, also known as **neo-Darwinism** or the **modern evolutionary synthesis**. This synthesis originated largely as a consequence of the emergence of the new science of **population genetics**, in which the genes, alleles, and mutations of the geneticists' theories were analyzed mathematically and studied empirically and experimentally in a statistical framework grounded in the Darwinian concepts of fitness and selection. The title of a classic early treatise on the STE, T. Dobzhansky's *Genetics and the Origin of Species* (1937), summarizes the reconciliation of these once-opposed ways of thinking. The STE asserted that:

- ▸ All evolutionary change is genetic change.

- ▸ Mutation is indispensible for creating new genetic variety (new alleles), but is not itself the primary agent of evolutionary change.

- ▸ Natural selection is the major agent of change, but can only operate on existing variation.

- ▸ Macroevolution is simply an extension of microevolution.

This last point is the one that was most emphatically rejected by the architects of the punctuated-equilibrium model in the 1970s. Eldredge and Gould (1972) and other punctuationists used paleontological data to question the adequacy of the synthetic theory, asserting that the pattern of evolutionary change revealed by the fossil record refuted the assumption of "phyletic gradualism" that they saw as implicit in the neo-Darwinian synthesis. According to the punctuationists, stasis is the rule in lineages traced through geological time, and significant morphological change is uncommon and abrupt. On the basis of these assertions, punctuationists argue that almost all evolutionary change occurs relatively rapidly in small populations and chiefly during cladogenetic events. They conclude that the STE, which mistakenly conflates macro- and microevolution, does not adequately explain the origin of new species and new morphologies.

Paleontology had little role in the emergence of the STE, mainly because the fossil record was seen as too incomplete to contribute much to the development of biological theory. But there were several paleontologists who felt that the STE had to be integrated into paleontology. Foremost among them was the American paleobiologist G. G. Simpson, who wrote extensively on what might be termed the neo-Darwinian theory of paleontology (Simpson 1944, 1953). Simpson's works anticipated several of the major ideas of 1970s punctuationism. However, he saw these ideas as complementing rather than contradicting the tenets of the STE. In his influential 1944 book *Tempo and Mode in Evolution*, Simpson argued that the STE did not presuppose a slow, continuous rate of evolutionary change, of the sort that later punctuationists would label "phyletic gradualism." In many cases, Simpson asserted, an evolving species would reach an **adaptive plateau** or temporary optimum and remain there until environmental or other changes disturbed its adaptive balance. While on the plateau, a species would not change very much, and its morphological pattern would be characterized by relative stasis. Significant change would occur only when the adaptive plateau disintegrated and its occupant had to adapt or go extinct. Simpson argued that such disruptions were likely to occur relatively quickly – that is, during only a small fraction of the lifespan of an evolving lineage—and would often take place in small, peripheral populations. These brief bursts of rapid change would rarely be sampled in the fossil record, and the resulting absence of fossils from intermediate or transitional populations (those that were evolving) would in many cases produce a misleading appearance of discontinuous macroevolutionary transformation. Simpson described this process of **quantum evolution** as one of the dominant patterns or "modes" of evolutionary change. Thus, as one of the major contributors to the neo-Darwinian synthesis, Simpson had already incorporated the central theoretical ideas of punctuationism into that synthesis more than 25 years before they were put forward as revolutionary novelties in the 1970s.

THE POLITICS OF MACROEVOLUTION

Covert political issues provide much of the energy behind the debates over gradualist versus punctuationist models of evolution. Throughout the history of evolutionary thought, these opposing conceptions of normative evolutionary processes have been associated with opposing political subtexts. Darwin believed that the struggle for existence ensures that the strong and fit will win out over the weak and unfit, and he also believed that this is a good thing. He therefore thought that population control and social welfare programs were bad ideas. "Man," Darwin wrote,

> ... has no doubt advanced to his present high position through a struggle for existence consequent on his rapid multiplication; and if he is to advance still higher, it is to be feared that he must remain subject to a severe struggle. Otherwise he would sink into indolence, and the more gifted men would not be more successful in the battle of life than the less gifted. Hence our natural rate of increase, though leading to many and obvious evils, must not be greatly diminished by any means. (Darwin 1871, p. 919)

The beliefs that competition yields progress and that winners deserve to win have been important parts of both Darwinian orthodoxy and capitalist ideology from Darwin's time down to the present. Conversely, liberals and leftists tend to think that cooperation is nobler than competition, that winning and losing are in large measure a matter of luck, and that progress results from revolution at least as often as it does from individual striving to succeed. Biologists on the left are accordingly predisposed to (a) downplay the importance of natural selection and competition and (b) look to macromutations, global catastrophes, and other abrupt, unpredictable transformations as the main engines of macroevolutionary change.

Both sides in this fight are allowing themselves to be seduced by the powerful metaphors of Darwinism. Those metaphors can be suggestive and instructive, but they are only figures of speech. "Natural selection," for example, is an oxymoron, equivalent to "mechanical choice." This sort of "selection" does not involve any actual choosing. Likewise, "competition" in evolution involves no competitors and no contests. It may not be too much of a stretch to use those words in talking about the actions of some nonhuman animals—say, two male antelopes locking horns over the possession of a breeding territory. But to compete, one must have an intention of winning; and species, genotypes, and alleles are not the sorts of things that have intentions. They are merely natural objects that make copies of themselves. Because there is only a limited space available for those copies, the ones that replicate most are more likely to persist. But a species does not win anything by persisting, nor does it lose anything by becoming extinct. Since species have no awareness or intentions, they have no interests to be served, no more than rocks or stars have.

Talking about two species as competing for resources is ultimately a poetic trope, like talking about the sea and the land as contending for mastery over the earth's surface. The sea does not care whether it floods the land or recedes from it, and species, genotypes, and alleles do not care whether they persist or vanish. The metaphor of "competition" in evolutionary biology is useful, but it is also dangerous because it tempts people to apply Darwinian explanations to human society—and conversely, to read their social ideals back into the phenomena of nature.

RECONSTRUCTING THE TREE OF LIFE

To study patterns of evolutionary change, we need to know something about the family tree of the organisms we are looking at. The family tree or genealogy of a group of organisms is called the **phylogeny** of that group. Reconstructing a group's phylogeny is trickier than it might seem. One might think that we could do the job simply by adding up resemblances. For instance, we humans resemble chimpanzees in every way more closely than we resemble goats. This suggests that we are more closely related to chimpanzees than we are to goats. Conversely, goats are in all ways more like sheep than they are like humans; and so we might think that goats are closer relatives of sheep than of humans. And so on.

These particular phylogenetic judgments happen to be correct. But counting resemblances will not do as a general technique for making such judgments. For example, alligators and iguanas resemble each other and differ from chickens in most visible features: scaly featherless skin, cold blood, sprawling quadrupedal gait, long muscular tail, and so on. But alligators are more closely related to chickens than they are to iguanas.

How do we know this? As usual, the fossil record is of some help, but less than one might think. If that record were complete, we could simply trace lines of descent back in it—following the chicken lineage back to the ancestor of all birds, doing similar things for the alligator and iguana lineages, and seeing in what order the three lines of descent join up. But the fossil record is not complete. It has more gaps in it than it has fossils. In the main, what we get from the fossil record is just more organisms: additional pieces to be fitted into the puzzle. Usually the fossils help, but sometimes they raise more questions than they answer.

The general procedure for answering such questions is diagrammed in Fig. 2.9, using imaginary molecular data. Suppose that we want to determine the phylogenetic relationships of species 1, 2, and 3. There are only three possible patterns: either 1 and 2 are most closely related, or 1 and 3 are, or 2 and 3 are. These three possibilities can be represented as nested sets: [(1, 2), 3] or [(1, 3), 2] or [(2, 3), 1]. They can also be represented as branching tree diagrams called **cladograms** (Fig. 2.9A–C). The two representations are equivalent (Fig. 2.3).

Imagine that we have located a stretch of DNA that can be taken as homologous in the genomes of all three species. The nucleotide sequences of the three species differ in only two places. In the first position in the sequence, species 2 and 3 have guanine (G - -) where species 1 has cytosine (C - -). In the third position, species 1 and 2 have thymine (- - T) where species 3 has adenine (- - A). Given these facts, the three possible cladograms imply different patterns of nucleotide substitution during the evolution of the group (Fig. 2.9A–C, black circles).

We might try to use the Principle of Parsimony to choose among these three cladograms, by assuming that the most probable choice is the cladogram that implies the smallest number of nucleotide substitutions. But parsimony considerations are of no help here. Each of the three possible cladograms requires just two changes,

if we make the appropriate assumptions about what the ancestral (primitive) nucleotide sequence was. Figure 2.9D shows why this is the case. If we unfold the three cladograms into an unrooted network, all three are identical. In each case, we encounter two nucleotide substitutions in tracing evolutionary change from point to point along the network. The three cladograms differ only in the starting point, the point at which the network is folded to produce a rooted tree (white circles in Fig. 2.9D).

To choose among the three cladograms in this case, we need external evidence to help us root the tree. This can be obtained by bringing in another species, called the **outgroup**, that we know to be a more distant relative of the species we are looking at. For example, if species 1, 2, and 3 are humans, chimpanzees, and goats—all of which are placental mammals—we might use an opossum or other marsupial as the outgroup. The outgroup is affixed to the network at a point that implies the minimum number of nucleotide substitutions (Fig. 2.9E). The resulting four-species network can now be rooted on the line leading to the outgroup. This provides a root for the tree (Fig. 2.9F) and tells us which of the original three cladograms we should pick.

If we know what the primitive condition is for the variable traits, we can dispense with the outgroup and use parsimony considerations to evaluate the conflicting cladograms directly. For instance, if we know that the nucleotide sequence in the last common ancestor of species 1–3 began with GAA … , then the most parsimonious cladogram is the one that clusters species 1 and 2 together to the exclusion of species 3 (Fig. 2.9A).

We can use other kinds of varying traits in this sort of analysis. For example, we might count the teeth in our three species and then score every species as having either two or three molars in each molar row. In this analysis, "two molars" and "three molars" are taken as alternative **states** of the **character** "number of molar teeth," just as C and G were states of the character "first nucleotide in the sequence." We can then evaluate a cladogram's parsimony by counting the number of state transitions that it requires, just as we counted the number of nucleotide substitutions in our DNA example. Although different characters may yield different answers, we can make our reconstruction as accurate as we wish by increasing the number of characters used in the analysis.

SOURCES OF ERROR IN PHYLOGENETICS

The basic idea behind this technique of phylogeny reconstruction seems simple enough. But as soon as we move beyond molecular data and small numbers of species, large problems emerge.

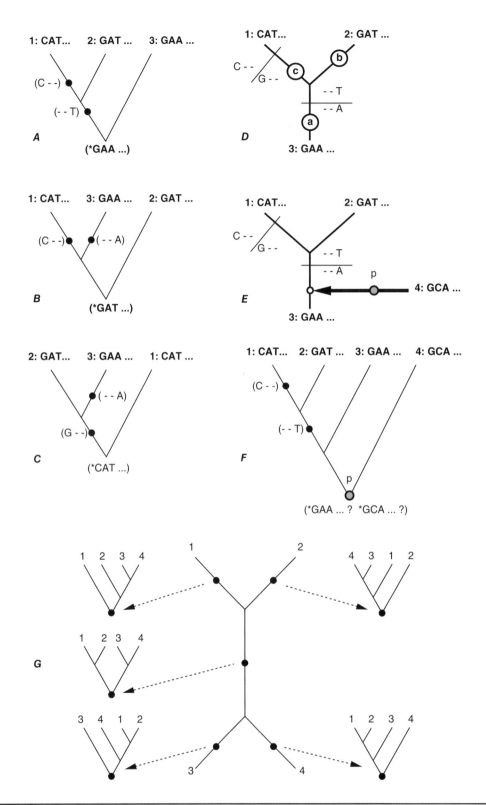

FIGURE 2.9

Phylogeny reconstruction. Three species (1, 2, and 3) can be related to each other in three different ways, diagrammed here as cladograms (**A, B, C**). Three nucleotides at the beginning of a homologous DNA sequence are indicated for each species (CAT … etc.). Each cladogram has different implications for what the ancestral sequence (*asterisks*) must have been and where nucleotide substitutions occurred in the evolution of the group (*black circles*). All three cladograms can be made to work with only two substitutions, if we assume the right ancestral sequence. This is because all three are identical when unfolded into an unrooted network (**D**). Each cladogram corresponds to a different folding point or root on the network (**D**, white circles a–c). To choose among them, we need to compare the sequence in a 4th species known to be an outgroup. When the outgroup is added to the arm of the network where it fits best (**E**, *arrow*), we can root the expanded network at a point on the line leading to the outgroup (**E**, *gray circle p*). In this case, this resolves the three-species problem in favor of cladogram **A** (**F**). If the fourth species is not known to be an outgroup, no resolution is achieved, and the number of possible alternatives increases to 5 (**G**).

The astute reader will have noticed that we rooted the network in our imaginary example by bringing in a known outgroup (Fig. 2.9E,F). In other words, we resolved the phylogenetic relationships of species 1–3 by invoking prior knowledge of their phylogenetic relationship to species 4. How did we get that prior knowledge?

If we do not have such knowledge, adding more species to the analysis only makes matters worse. We can use parsimony to tell us where to splice species 4 into the unrooted network (Fig. 2.9E), but parsimony considerations cannot tell us where to fold that network to produce a rooted tree. If the tree has no root, adding another species to it only yields a more complex unrooted network (Fig. 2.9G). In order to reconstruct phylogeny, we must already know some phylogenetic relationships (Cartmill 1981).

Fortunately, we have other sources of information about phylogeny. The most important source is the fossil record, which tells us a lot about what character states are primitive (or **plesiomorphic**) and which are **derived** or **apomorphic** (nonprimitive). For example, we know that the earliest vertebrates were jawless fish that lived in the sea and breathed water with their gills. Therefore, "having jaws" and "breathing air" are derived character states among vertebrates. Knowing this, we can rule out all of the phylogenies in which these states are taken as primitive. Using the fossil record to help distinguish primitive and derived character states greatly reduces the number of phylogenies we have to look at.

Unfortunately, there are a lot of phylogenies to be examined (Felsenstein 1978, Cartmill 1981). With three species, only three cladograms are possible (Fig. 2.9A–C), but the possibilities increase at a terrific rate as we add more species. If each fork in a cladogram gives rise to two branches, the number of different cladograms that can be drawn for n species is the product of the first $(n-1)$ odd numbers (Fig. 2.10). This number, $(2n-3)!!$ (pronounced "$2n$ minus 3 double factorial"), rises with the lunatic acceleration typical of factorial numbers (Fig. 2.11). More than 13 billion cladograms can be drawn for just one dozen species, and the number of possible phylogenies for (say) the 60-odd species of Asian and African monkeys is larger than the estimated number of atoms in the universe. For analyses involving more than a few species, it is not possible to run an exhaustive search of all the possible cladograms to find the best one.

Are there techniques that can find the best solution without an exhaustive search? Yes and no. The problem of finding the best phylogeny belongs to a class of problems called "NP-complete," for which no reliable shortcuts are known (Garey and Johnson 1979). However, the computer programs used for seeking optimal phylogenies allow the user to introduce phylogenetic knowl-edge derived from other sources—say, by specifying primitive character states or by requiring that certain groups of species be clustered together within the cladogram. Such constraints can keep the search confined within a reasonable time period and still yield reliable results, if the number of species involved is not too great. But without something like a complete search, there is no guarantee that the solution found will be the best one.

Perhaps the thorniest problem that the cladogram-builder faces is choosing the characters to be used in the analysis. When we are dealing with DNA data, this is fairly easy. The characters are the first, second, third, and so on, positions in homologous DNA sequences (though figuring out which sequences are homologous can be tricky), and the possible states of each character are simply the four nucleotides (A, C, G, and T). But when we are dealing with anatomical or behavioral traits, the characters and states that we recognize are not given by nature. They are determined by the words and concepts that anatomists choose to use to describe what they see. Such choices are not phylogenetically neutral. Different ways of describing the same anatomy may have the same information content but result in different assessments of phylogeny (Cartmill 1982).

For example, the members of the mammalian order Primates (which includes lemurs, monkeys, and humans) differ from primitive mammals in having a bridge of bone around the lateral side of the eye. In primitive primates, this **postorbital bar** is a simple arch, composed of outgrowths from two bones (frontal and zygomatic: Fig. 2.12A). In so-called "higher" primates—monkeys, apes, and humans—the two bones that make up the bar, plus a third bone (alisphenoid) from the side wall of the braincase, develop flanges that grow together to form a bony partition (Fig. 2.12C). This partition, the **postorbital septum**, separates the eye from the chewing muscles behind it. One genus of living primates, *Tarsius*, develops flanges that do not meet competely, forming an incomplete septum with a gap in the middle (Fig. 2.12B).

Some experts think that the postorbital septum and other monkeylike features of *Tarsius* show that *Tarsius* is a uniquely close relative of higher primates. Others deny this (Chapter 3). How the septum figures in this argument depends on the words we use in scoring its characters and their states. If we score the septum as "present" versus "absent," it counts as one resemblance between tarsiers and higher primates (Cartmill and Kay 1978, Cartmill 1980). But we can count up to four such resemblances if we score the septal flanges separately for each bone. On the other hand, if we score the septum as "complete" versus "incomplete," it can be made to weigh in on the other side of the scales, as a difference *separating* tarsiers from higher primates (Beard and MacPhee 1994). There are still more possibilities. For

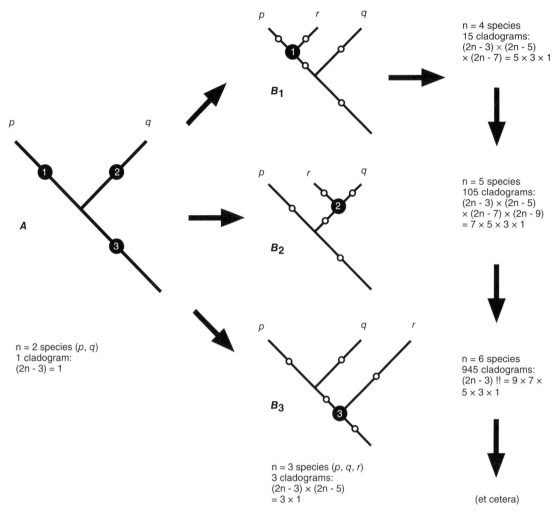

FIGURE 2.10

The number of possible cladograms for *n* species. For two species (**A**), only one cladogram can be drawn. A third species can be added to it in any one of three positions (**A**, *black circles* 1–3). Therefore, three cladograms can be drawn for three species (B₁, B₂, B₃). To each of these three, a fourth species can be added in any one of five positions (**B**, *white circles*); therefore, 15 cladograms (3 × 5) are possible for four species. Each new species grafted on to a cladogram adds one line segment and divides an existing line segment into two, thus creating three topologically different sites for the addition of the next branch—two more sites than were there previously. The number of cladograms that can be drawn for *n* species (*n* > 1) is therefore the product of the first (*n* − 1) successive odd integers: 1, 1 × 3, 1 × 3 × 5, and so on, from 1 to (2*n* − 3). This number is conventionally written as (2*n* − 3)!! The product of this series rises very rapidly (Fig. 2.11). More than 13 billion different cladograms can be drawn for just a dozen species.

instance, in Fig. 2.12, do B and C resemble each other in having a contact between the alisphenoid and zygomatic bones? Or do they differ in the *number* of such contacts, with B having only one and C adding a second contact below the lateral orbital fissure? Or both? Is having an alisphenoid-zygomatic contact the same thing as having a postorbital septum? Or are these two different characters, which deserve to be counted separately in the analysis? Our assessments of the relative merits of conflicting cladograms depend on how we make such

decisions in describing complex morphology (Cartmill 1994, Young and MacLatchy 2004).

Another example, this time from paleoanthropology. Fifty thousand years ago, Europe was inhabited by humans of the type called Neandertals, who differed anatomically from modern humans. Some experts think the Neandertals were a separate species, reproductively isolated from the modern populations that supplanted them. Others suspect that there was gene flow between the two populations, and that certain peculiarities of modern Europeans are traits inherited from Neandertal ancestors. Szilvássy et al. (1986) suggest that one such Neandertal inheritance unique to Europeans may be the proportions of the air-filled cavities called **sinuses** in the frontal and maxillary bones of the European face. In African peoples today, they claim, the frontal sinus is larger than the maxillary; in Asians, the proportions are reversed. Only Europeans resemble Neandertals in having roughly equal-sized maxillary and frontal sinuses (Fig. 2.13A). But if we score each sinus independently as "large" or "not large," we will tally these resemblances differently (Fig. 2.13B,C), since both sinuses are large in some Neandertals and smaller in others.

The fundamental problem in such cases is that there are many different ways to describe any complex object, and all of them omit some of the object's properties.

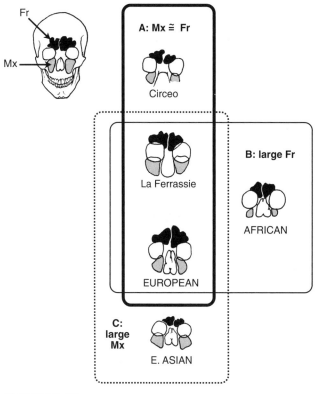

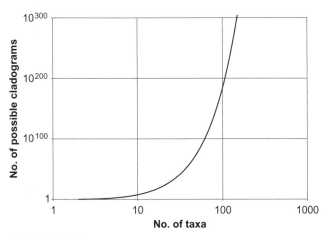

FIGURE 2.11

The number of cladograms possible for *n* species is a double-factorial function of *n*. On this log–log plot, each tick on the vertical axis represents 100 orders of magnitude.

FIGURE 2.13

In Neandertal skulls (Circeo, La Ferrassie), the maxillary air sinuses (Mx) are about the same size as the frontal sinuses (Fr). This can be seen as a resemblance between Neandertals and modern Europeans, distinguishing them from modern African and East Asian peoples. However, other verbal descriptions yield different groupings. See text. (After Szilvássy et al. 1986.)

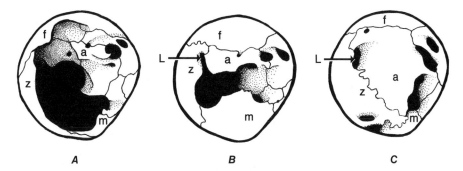

FIGURE 2.12

The inside of the right orbit (eye socket) in three primates: semidiagrammatic front views. In the primitive condition, seen in the bushbaby (*Galago*, **A**), the frontal, zygomatic, and maxillary bones (f, z, m) form a postorbital bar. In the higher-primate arrangement, seen in the squirrel monkey (*Saimiri*, **C**), outgrowths from the alisphenoid (a) and zygomatic combine with flanges of the maxillary and frontal bones to form a complete bony partition, the postorbital septum, that separates the eye from the chewing muscles behind it. The condition in *Tarsius* (**B**) is intermediate. L, lateral orbital fissure. (After Cartmill 1980.)

Unfortunately, every organ of every animal is a complex object. When different people report what they see in the anatomy of an animal, they incorporate their own opinions about what is important and what is not. This can introduce error and bias into the analysis.

Another source of error in phylogenetics is the **weighting** of characters. For example, among mammals, "short tail" and "retractable claws" are both derived character states, different from the ancestral mammalian condition. They deserve to be treated as such in analyzing phylogeny. However, we know that tail length is much easier to alter than basic finger and toe anatomy, and so it should count for less. When we assess (say) the genealogical relationships between rabbits, bobcats, and tigers, it would make sense to count the retractable claws of the two cats more heavily than the short tails of the bobcat and rabbit. But exactly how to go about weighting these traits is not clear. When we are dealing with DNA data, we can make out a case for weighting all nucleotide substitutions equally. But how should we weigh a short tail against a retractable claw? The usual practice is to give every character the same weight and to count every difference between states of a character as a single unitary difference. This seems arbitrary. After all, we can be sure that some of these characters and states involve more extensive genetic changes than others. Yet assigning different weights to them seems just as arbitrary, because their genetic bases are in most cases unknown. Again, our estimates of parsimony often hinge on such choices.

Finally, we should remember that the Principle of Parsimony is not a law of nature. It is merely a rule to be followed in deciding what the safest bet is likely to be. But the safest bet is not always the actual outcome. In some cases, parallel and convergent evolution will produce superfluous changes, piling up homoplastic similarities that trick us into seeing relationships where none exist. We can therefore expect that the most parsimonious cladogram will sometimes be wrong. In the worst cases—for example, in dealing with extinct taxa, where the number of characters we can use is limited—the most parsimonious cladogram will *usually* be wrong (Hawks 2004).

As long as we are looking at small numbers of species with very different degrees of relationship, the probability of coming up with the wrong phylogeny is negligible. No matter how we interpret the evidence or what data set we use, we cannot help but conclude that people are more closely related to donkeys than they are to daisies. But when experts disagree about phylogeny, there is usually a good reason for the disagreement. It may be that one or more of the sources of error noted above has introduced enough static into the analysis to blur the picture.

Our best recourse in such cases is once again to use several independent sources of information, to see whether they give the same result. When anatomy, paleontology, and genetics all point to the same pattern of relationships, we are entitled to conclude that we have arrived at the right answer. And when different sources of information or different modes of analysis give different answers, the appropriate thing to do is to suspend judgment until we know more.

Although parallel and convergent evolution can blur the phylogenetic picture, they often make up for it by clarifying our understanding of adaptation. If certain traits are found together in several groups of organisms, it may be a phylogenetic signal that they have inherited those traits from a common ancestor. But it may also be an adaptive signal, telling us that the two traits work together in some functional way and thus tend to evolve together in many independent lineages. For those who study phylogeny, the adaptive signal is annoying noise that needs to be filtered out. Those who study adaptation feel the same way about the phylogenetic signal. As you might expect, scientists who are mainly interested in phylogeny tend to downplay the importance of adaptation, whereas those who are interested in adaptation tend to be skeptical about phylogenetic reconstructions.

∎ LINNAEAN CLASSIFICATION

Why do we classify organisms? The usual answer nowadays is that classification is another way of expressing hypotheses about phylogenetic relationships. But a deeper answer is that we have to classify things in order to talk about them at all. The words of any human language classify all things into categories. Every language's vocabulary includes a collection of names for types of organisms, which add up to an informal biological classification. These informal systems are not very useful for scientific purposes. They differ from one language to another, and their categories are always vague, ambiguous, and overlapping. (For example, the English word "animal" includes humans in some contexts but not in others.)

To correct these defects, biologists have had to develop a more rigorous system of categories for talking about organisms. The system used today was established in the 1700s by the Swedish naturalist Karl Linne, who wrote under the Latin name of Linnaeus. His system was not intended to express evolutionary relationships. Like other naturalists of his time, Linnaeus believed that God had created all things in essentially their present form. His aim was to discern the logic of God's creation.

We can get some idea of Linnaeus's objectives if we compare his system to the game of Twenty Questions. In that game, one player thinks of something and the other tries to guess what it is by asking 20 yes-or-no questions. It would be a bad strategy to guess particular types or individuals at random ("Is it a frog? The moon? A mushroom? Julius Caesar?"). A better strategy is to start with a large general category and make finer and finer concentric subdivisions, until you arrive at exactly the group or individual to be identified. In effect, Linnaeus's system constitutes a universal strategy for playing Twenty Questions when guessing any natural object. Like the game of Twenty Questions, the Linnaean system began by sorting everything into three categories—animal, vegetable, and mineral—that covered the whole territory of nature. These three categories Linnaeus called **kingdoms**. Each kingdom was divided into classes, each class was divided into orders, and so on down through the taxonomic hierarchy. (That hierarchy of nested sets is also known today as the **Linnaean hierarchy**, in honor of its inventor.) Every taxon was defined by a few key traits or properties common to all its members. For example, the key traits of the class of mammals were "Hairy body. Four feet. Live-bearing, milk-giving females" (Linne 1735).

For Linnaeus, this system was not a game or a mere key for identifying species. His kingdoms, classes, orders and so on were conceived as real entities—**natural kinds**, established by the Creator at the beginning of time. Linnaeus strove to define each taxon in terms of the **essential** properties that made that grouping a truly natural one, and he continually revised his system to try to make his groupings more natural. In the first (1735) edition of his book *Systema Naturae* ("The System of Nature"), Linnaeus classified whales as fish because they lacked the four feet that he saw as part of the mammalian essence. But in later editions, after he had decided that mammary glands and live birth were the really essential things about mammals, Linnaeus changed the name of the class from Quadrupedia ("four feet") to Mammalia ("having mammaries") and moved the whales into it.

One of the more controversial elements of Linnaeus's system was his treatment of human beings. Instead of setting humans up as a kingdom, class, or order of their own as a token of their unique relationship to God, Linnaeus decided on anatomical grounds to classify man as a mere genus (*Homo*) within the same order of mammals as apes, monkeys, and lemurs. The only sign of humanity's special status was the name that Linnaeus gave to this order: the Primates, meaning "chiefs" or "honchos."

▌ EVOLUTIONARY SYSTEMATICS

Linnaeus's system was widely adopted, because it really did seem to express a natural order behind the diversity of life. But it was not clear just what it was that made it seem more natural than many of the possible alternatives. Why, for instance, was it more natural to group

whales with sheep and kangaroos than with salmon and sharks? Did the Creator somehow attach more importance to milk glands and live birth than to feet and hair? Or was that just an arbitrary human perception?

The theory of evolution furnished the answer. The concentric sets of a Linnaean classification were real entities insofar as they corresponded to consecutive branchings of the tree of life (Fig. 2.3). Linnaeus's 1758 classification of whales as mammals was objectively better than his earlier classification of whales as fish, because whales have more recent common ancestors with sheep and kangaroos than they have with sharks or salmon.

This recognition led eventually to the development of a new school of **systematics** (the science of classification) that took evolutionary relationships into account. This **evolutionary systematics** dominated the field for a hundred years, reaching its canonical form in the writings of G. G. Simpson (1961). It was a mixed system, which combined phylogenetic criteria with a lingering Linnaean regard for essential differences between groups. For example, evolutionary systematists recognized that birds were more closely related to crocodiles and dinosaurs than to lizards and turtles. Nevertheless, it was felt that the warm blood, motherly love, and feathers of birds made them a different kind of creature from their crawling, cold-blooded relatives. Rather than demoting the birds to a sub-subset of diapsids within the class Reptilia, evolutionary systematists followed Linnaeus in giving the birds a separate class Aves of their own. The co-equal class status given to birds signaled their major adaptive differerences from reptiles.

There are practical reasons for using a mixed system of this sort. In any cluster of evolving lineages radiating out from a common ancestor, the early, primitive representatives of the diverging lineages are usually more like each other than they are like their more transformed descendants. If classification is going to reflect similarity, these primitive forms have to be grouped together despite their genealogical divergence. Evolutionary systematics did just that, lumping the early members of an evolutionary radiation together with their persistently primitive descendants to form an ancestral **wastebasket taxon**. Thus, among vertebrates, the early amniotes and all their scaly, cold-blooded descendants were dumped into a wastebasket class Reptilia, while their hairy and feathered descendants were placed in separate classes of their own, Mammalia and Aves. Early side branches from the mammal and bird lineages, such as the synapsid reptile *Dimetrodon* (Fig. 1.11C) and the dinosaur *Velociraptor*, were retained in the wastebasket class (Fig. 2.14A).

Taxa of two different sorts were recognized in principle: **grades**, separated by horizontal boundaries between ancestors and descendants (e.g., Reptilia vs. Mammalia), and **clades**, separated by vertical

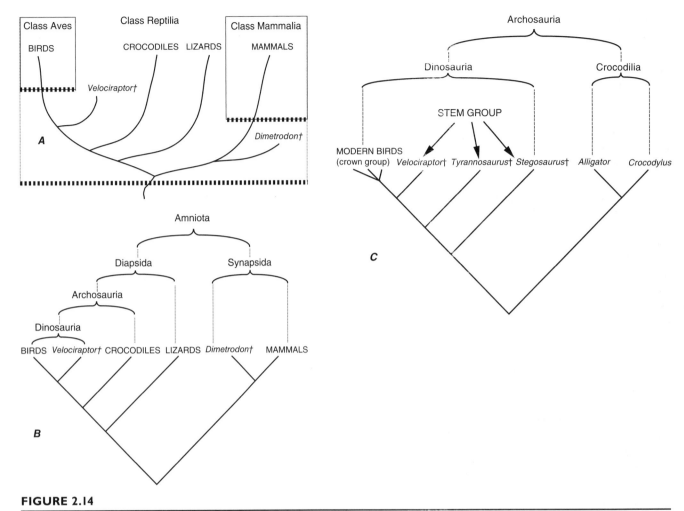

FIGURE 2.14

Evolutionary versus phylogenetic systematics. In evolutionary systematics (**A**), the primitively scaly and cold-blooded members of the bird–mammal–reptile clade (Amniota) are grouped together as a wastebasket taxon Reptilia, separated from birds and mammals by horizontal **grade** boundaries (*dashed lines*) that mark major **adaptive shifts** (in this case, to warm-bloodedness and its corollaries). A similar grade boundary (*lowest dashed line*) separates Reptilia from the amphibian ancestors of the amniotes. In phylogenetic systematics (**B**), the concentric groupings of the classification map precisely onto the branching pattern of evolutionary relationships. Grade boundaries and wastebasket taxa are forbidden, the class Reptilia vanishes, and birds are reduced to a subset of the dinosaurs. Within the Dinosauria, the big, flightless Mesozoic ones popularly called "dinosaurs" can be distinguished as "stem-group" dinosaurs (**C**) from the "crown-group" dinosaurs (the living birds, plus all other descendants of their last common ancestor).

boundaries drawn through branching points on the evolutionary tree. In practice, almost all taxa were separated by grade boundaries from their ancestors, their descendants, or both. However, most evolutionary systematists also required that every taxon be **monophyletic**. This meant that the last common ancestor of all the taxon's members had to be itself a member of that taxon. **Polyphyletic** taxa, which violate this criterion, were forbidden. For example, a taxon containing all and only warm-blooded vertebrates (birds plus mammals) would be rejected as polyphyletic, since the last common ancestor of these animals was a cold-blooded reptile. The taxa recognized by evolutionary systematics thus had to meet criteria of both overall similarity (based on grade boundaries) and of phylogenetic coherence (based on monophyly).

PHENETICS AND CLADISTICS

Trying to combine two sometimes conflicting criteria can be a hard juggling act, and deciding exactly where to draw a grade boundary between ancestors and descendants on an evolutionary tree is a matter of taste. In the early 1960s, competing schools of systematics emerged that got around these problems of evolutionary systematics by relying on a single criterion and getting evolutionary trees out of the picture.

The school of **phenetics** discarded criteria of phylogeny and classified organisms solely on the basis of their overall similarity. The pheneticists insisted that the distribution of shared traits is a matter of observable fact, from which phylogeny can only be secondarily (and dubiously) inferred. Systematics, they argued, should be constructed on the rock of observation and not on the shifting sands of phylogenetic conjecture. This approach dates back to the 1700s. It was given new life by the advent of the digital computer, which made it possible to manipulate large bodies of data incorporating hundreds of facts about organisms. Because the phenetic approach was dependent on computer algorithms and numerical data, it was also known as **numerical taxonomy** (Sokal and Sneath 1963).

The school of **cladistics** or **phylogenetic systematics** (Hennig 1966) has taken the opposite tack, discarding criteria of overall similarity and classifying organisms solely on the basis of the degree of their relationship. Cladists draw a sharp distinction between primitive traits (**plesiomorphies**) and nonprimitive or **derived** traits (**apomorphies**) of organisms. Within a clade, traits that are primitive for that clade carry no information about relationships. (For example, the fact that alligators and lizards are both cold-blooded and scaly does not tell us which is the closer relative of the warm-blooded, feather-covered birds.) Therefore, shared primitive traits or **symplesiomorphies** cannot be used in phylogenetic classification. Wastebasket taxa, which are defined by symplesiomorphies, are labeled as **paraphyletic** and forbidden. Every taxon is required to be **holophyletic**, meaning that it has to contain all and only the descendants of its last common ancestor. (Cladists usually call this "monophyletic.") All taxa are defined exclusively by shared derived traits or **synapomorphies**. Cladists therefore define a species as the smallest cluster of organisms united by one or more synapomorphies. This formulation, or some variation of it, is sometimes called the **phylogenetic species concept.** It is essentially the same thing as the morphological species concept, but with the added proviso that the shared phenotypic traits defining a species have to be synapomorphies. (The term "phylogenetic species concept" is also sometimes applied to what we are calling internodal or evolutionary species concepts: Baum and Donoghue 1995).

The jargon of cladistics sounds forbidding, but it is simple enough conceptually. In our earlier four-species example (Fig. 2.8F), for instance, "T" in the third position is a synapomorphy of a clade comprising species 1 and 2. An "A" in that position is plesiomorphic within this group, so the fact that species 3 and 4 share this symplesiomorphy furnishes no grounds for grouping them together. And a "C" in the first position is a uniquely

apomorphic trait, or **autapomorphy**, peculiar to species 1. Every taxon needs to have at least one such autapomorphy; otherwise, it cannot be recognized as a holophyletic clade.

What this means for classificatory practice is exemplified by Figure 2.14B. The classification has the same geometry as the cladogram. Unlike the evolutionary tree diagrams that underlay the practices of evolutionary systematics (Fig. 2.14A), the cladogram has no time dimension. It is just another way of writing a system of nested sets (Fig. 2.3). Since there is no time dimension, horizontal lines cannot be drawn between earlier and later forms; therefore there are no grade boundaries. All taxa are holophyletic clades. Extinct forms like *Dimetrodon* and *Velociraptor* are grouped with their living relatives. Therefore, the wastebasket taxon Reptilia is replaced by the clades Diapsida and Synapsida, and birds are classified as dinosaurs.

■ PROS AND CONS OF PHYLOGENETIC SYSTEMATICS

By the end of the 20th century, phylogenetic systematics had swept away its competitors and become standard biological practice. The reasons for the triumph of cladistics were mainly esthetic (Cartmill 1999). The elegance and rigor of cladistic methodology were pleasing. Many scientists were attracted by its seeming objectivity, its quantitative treatment of data, and its ingenious formal algorithms for tree-building (originally developed by the numerical taxonomists). And apart from its purely esthetic merits, cladistics had made a real contribution to science by rendering the logic of phylogenetic reconstruction far clearer and more explicit than it had been before.

But phylogenetic systematics also has some significant drawbacks. Because its taxa are defined by genealogical relationships rather than by overall similarity, they sometimes have little descriptive meaning or adaptive significance. For example, molecular data and new fossil evidence have demonstrated that whales are more closely related to hippopotamuses and pigs than to other living mammals (Arnason et al. 2000, Gingerich et al. 2001, Thewissen et al. 2001). The rules of cladistic systematics therefore require that whales be placed in the order Artiodactyla or even-toed hoofed mammals, along with hippos, pigs, giraffes, sheep, and cows. But there is virtually nothing that all these mammals have in common apart from some shared DNA sequences. A killer whale and a sheep are as different in anatomy, ecology, and way of life as it is possible for two placental mammals to be. Classifying them together as artiodactyls conveys no information about their biology.

A more important drawback is that cladistic methodology does not allow any taxon to be characterized by primitive traits (symplesiomorphies). Unfortunately for cladistic theory, some organisms are wholly primitive relative to others. For example, an ancestor is, by definition, plesiomorphic (primitive) in every way relative to its descendants. Therefore, if we find a fossil ancestor of any known organism, we cannot classify it. This seems absurd, but it follows logically from the prohibition of wastebasket taxa.

The inability of cladistics to deal with plesiomorphic fossils may not have seemed like a grave drawback to its inventor, the entomologist Willi Hennig, because he was an expert on the classification of insects. Insects have a skimpy fossil record. But vertebrates and other organisms with readily preserved hard parts have rich fossil records, which include many wholly plesiomorphic species. The need to classify such species represents a major obstacle to the use of strict cladistic systematics in vertebrate paleontology.

In order to deal with extinct organisms and keep biological classifications comprehensive, cladistic systematists have adopted various subterfuges to sneak wastebasket taxa back into the system. Many cladists simply continue to use the old wastebasket names whenever convenient, putting quotation marks around words like "reptile" to show that they are being naughty. Another strategy is to call an extinct group defined by plesiomorphies a "plesion" rather than a class, order, or whatever, and deny it a categorical rank within the Linnaean hierarchy. Some cladists (e.g., Schwartz et al. 1978) extend this convention to include living taxa as well.

Another device for accommodating wastebasket groupings is the distinction between "crown groups" and "stem groups." A **crown group** is a group of living species plus all the descendants, both living and extinct, of their last common ancestor (Runnegar 1992). A **stem group** comprises the set of "all fossils more closely related to a particular crown group than to any other, but more basal than its most basal member" (Budd 2001, p. 333). For example, within the dinosaur-bird-and-crocodile group of diapsids, known as **archosaurs** ("ruling reptiles"), there are only two crown groups: modern birds and the crocodile–alligator group (Crocodilia). Other archosaurs include *Velociraptor, Tyrannosaurus, Stegosaurus*, and all the other big Mesozoic reptiles commonly known as dinosaurs. These extinct creatures are more closely related to birds than to crocodiles. They are therefore stem-group birds—although we can go on calling them dinosaurs, since the larger clade is named Dinosauria (Fig. 2.14C). But no matter whether we call them stem-group birds or stem-group dinosaurs, it is easy to see that a stem grouping of this sort must be a paraphyletic wastebasket relative to the crown group.

Perhaps the most fundamental problem with phylogenetic systematics is that it often demands that we frame premature phylogenetic hypotheses on the basis of insufficient evidence. Using paraphyletic taxa allows us to withhold such speculative judgments, retaining species and groups of uncertain affinities in basal wastebaskets until we have decisive evidence for their affinities. In this book, we will use paraphyletic taxa whenever they seem necessary or useful.

CHAPTER THREE

People As Primates

▌ EARLY MAMMALS

During the Triassic period, the synapsids radiated into a wide variety of forms and ways of life, both predatory and herbivorous (Fig. 3.1). Mammal-like traits developed in several lineages of these Triassic "reptiles." Many of them probably had warm blood and fur, incubated their eggs, and may even have suckled their hatchlings. But despite their initial success and varied adaptations, almost all of them were supplanted by dinosaurs and other diapsid reptiles during the Triassic and Jurassic. Apart from a dubious late survivor found in the earliest Tertiary of Canada (Fox et al. 1992), we know of only one synapsid lineage that made it through the Jurassic period.

That lone surviving synapsid lineage was our own, already represented in the late Triassic by the ancestral mammals. Although the first mammals were more reptile-like in some ways than any mammal alive today, they differed from their synapsid ancestors in several crucial respects.

1. Size. The first mammals were *tiny* animals—as small as living shrews, which can weigh as little as 2.5 grams as adults. One of the biggest early mammals, *Megazostrodon* from the late Triassic of South Africa (Fig. 3.2A), measured about 14 centimeters from head to tail and probably weighed no more than 30 grams (Jenkins and Parrington 1976). At the opposite extreme, *Hadrocodium wui* from the early Jurassic of China (Fig. 3.2B) may have weighed only 2 grams (Luo et al. 2001b; cf. Meng et al. 2003). By contrast, most of the mammals' close relatives among the late synapsids weighed at least 100 times that much.

2. Teeth. Early mammals had differently-shaped teeth in different parts of the jaws, serving different functions: **incisors** in front for nibbling and holding things, a pair of stabbing **canine** fangs behind that, a series of puncturing **premolars** behind the canines, and some more complicated **molar** teeth at the very back of the mouth for slicing and shearing (Fig. 3.3A).

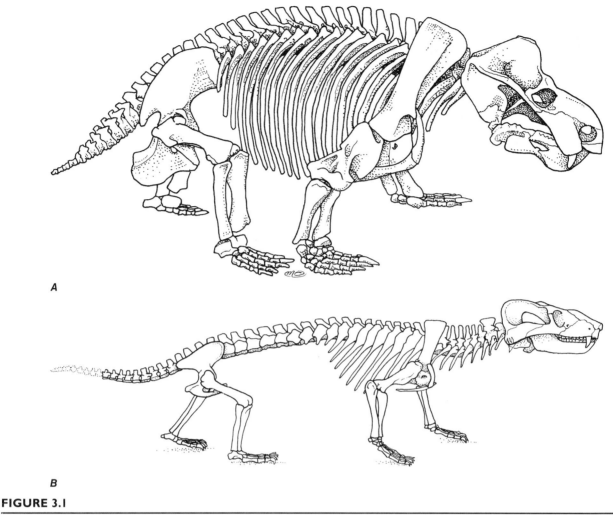

FIGURE 3.1

Triassic synapsids. **A**: The herbivorous dicynodont *Kannemeyeria* (after Carroll 1988). **B**: The mammal-like carnivore *Thrinaxodon* (after Jenkins 1971). Not to same scale.

An enlarged canine tooth is a very old synapsid feature, dating back to the Permian (Fig. 1.11C), and both Triassic mammals and nonmammalian Triassic synapsids sported teeth of several different sorts (Fig. 3.1B). But the **cheek teeth** (molars and premolars) of early mammals differed from those of almost all other synapsids in two ways. First, they were anchored in the jawbones by **multiple roots** like those of a human molar, whereas all the teeth of other synapsids had simple conical roots like those of our incisors and canines. Second, early mammals had **dental occlusion**—that is, the convexities on their lower cheek teeth matched up with the concavities on their upper teeth and vice versa. When the animal chewed, these cusps and notches slid across each other, trapping food particles in the notches and slicing through them like tiny pinking shears (Fig. 3.4).

3. Palate. When a turtle drinks and swallows, water squirts out of its nostrils. This does not happen in a mammal, because a mammal's mouth has a bony and ligamentous roof called the **palate** that separates the mouth from the nasal air passages. The palate allows a mammal to go on breathing through its nose while it processes food in its mouth, and lets it swallow without suffering any nasal leakage.

4. Ear Region. The inner ears of mammals have a more elongated **cochlea** than those of other vertebrates. The cochlea is a tiny membranous tube connected to the semicircular canals that lie buried in the bottom of the skull near each ear. Vibrations in the fluid surrounding the cochlea are detected by auditory nerve endings and transmitted to the brain as sensations of sound. The longer the cochlea is, the more frequencies

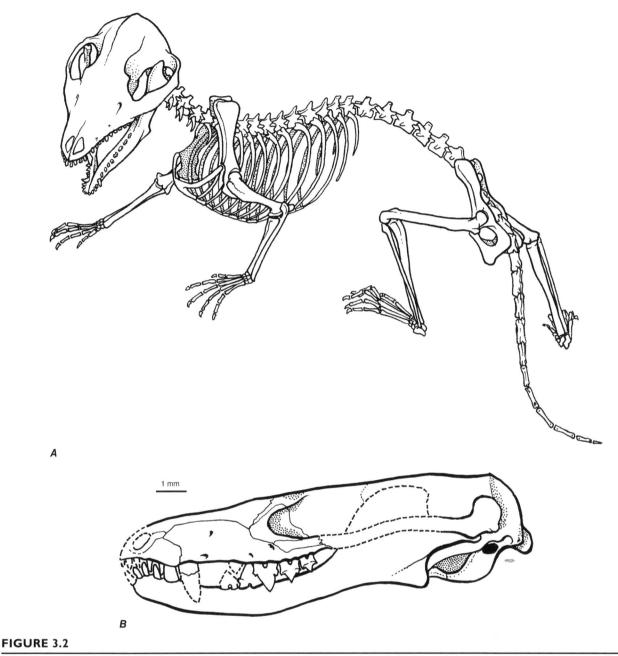

FIGURE 3.2

Early mammals. **A**: Skeleton of a Triassic mammal (composite based on *Megazostrodon* and *Eozostrodon*). **B**: Skull of *Hadrocodium wui* from the early Jurassic of China. (**A**, after Jenkins and Parrington 1976; **B**, after Luo et al. 2001b.)

it can detect and distinguish. The elongated cochleas of early mammals gave them more acute hearing at higher frequencies than their reptilian ancestors had. The cochleas of most modern mammals are even longer—so much so that they have to be curled up into a helix to fit into the available space (Fig. 3.5B).

Primitive synapsid reptiles had short cochleas of the standard reptilian sort, and they probably did not have eardrums or external ear holes (Allin and Hopson 1992). Their sense of hearing was not acute. Like many reptiles today, these mammal ancestors heard mainly loud, low-frequency sounds that reached the cochlea through the

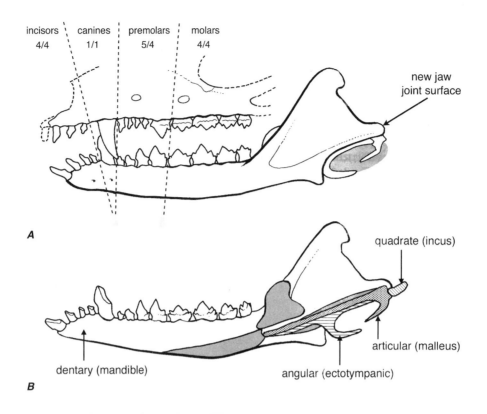

FIGURE 3.3

Jaws and teeth of the Triassic mammal *Morganucodon* (after Crompton 1985). **A**: Lateral view of upper and lower dentition. Gray oval indicates location of eardrum. **B**: Medial view of lower jaw.

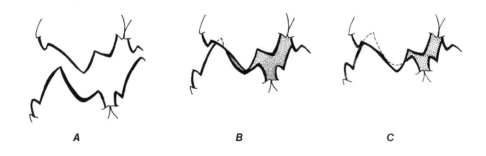

FIGURE 3.4

Molar occlusion in early mammals (diagrammatic). **A**: Upper and lower molars seen in a lateral view (based on *Morganucodon*). When the jaws close, food particles are caught in the narrowing gaps between the cutting edges of the upper and lower molars (*stipple*, **B,C**) and sliced in two. Because the upper and lower edges are set at an angle to one another, the cutting pressure is concentrated at a single traveling point as in a pair of scissors, not spread across the entire blade like that of a cleaver coming straight down on a block.

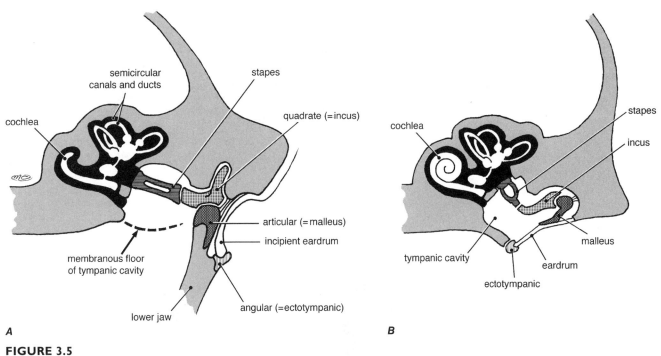

A   **B**

FIGURE 3.5

Diagrammatic cross sections through the ear regions of an Early Triassic synapsid reptile (**A**) and a modern mammal (**B**). (**A**, based on *Thrinaxodon* reconstructions by Allin 1975 and Crompton 1985.)

bones of the skull. The most effective route for such sounds to follow to the cochlea was through the lower jaw. Vibrations that reached the jaw joint were transmitted to the cochlea along a bony rod called the **stapes**, which extended inward from the jaw joint through an air-filled **middle ear cavity** to reach the skull base in the vicinity of the cochlea.

More advanced synapsids improved their hearing ability by making the stapes lighter and thinner, and by thinning the skin over the back edge of the jaw joint to form a rudimentary **eardrum** for picking up airborne sound (Fig. 3.5A). In some of these animals, the jaw muscles had already begun to evolve toward a mammal-like configuration, with a forward-pulling **masseter** muscle and a backward-pulling **temporal** muscle behind it. These changes redirected the muscle forces so that they could be brought to bear more directly on objects held between the teeth. This reduced the chewing stresses carried by the **quadrate** and **articular** bones that formed the jaw joint (Fig. 3.6). As the loads on that joint diminished during later synapsid evolution, the tooth-bearing **dentary** bone grew larger and the other bones in the lower jaw dwindled in size (Fig. 3.6D, E).

The early mammals carried these trends still further. To begin with, they developed a new jaw joint on each side between the enlarged dentary bone and the **squa-mosal** bone in front of the ear. The first mammals thus had two jaw joints on each side: the old one between the quadrate and articular, and the new one between the dentary and squamosal (Figs. 3.3B and 3.6F). The new jaw joint freed the quadrate and articular from having to carry any of the stresses of chewing. The quadrate and articular therefore became smaller and lighter, more specifically adapted to the demands of hearing, and more loosely attached to the jaws (Crompton and Parker 1978, Crompton and Hylander 1986, Allin and Hopson 1992).

In adult mammals today, the quadrate and articular are completely detached from the jaws (though a transient reptile-like arrangement appears briefly in the embryo). The dentary is the only bone left in the lower jaw, and so the dentary-squamosal joint is the only joint between the skull and lower jaw on each side. The mammalian dentary is called the **mandible**, and the new jaw joint is called the **temporomandibular** joint. The articular and quadrate bones of mammals are renamed the **malleus** and **incus**. They are tiny bones lying completely within the middle-ear cavity, where they are usually attached to the surrounding bones of the skull only by ligaments (Fig. 3.5B). Their sole function is to transmit sound waves from the eardrum to the stapes. Another bone in the reptilian lower jaw, the angular, has been refashioned to form a bony ring called the

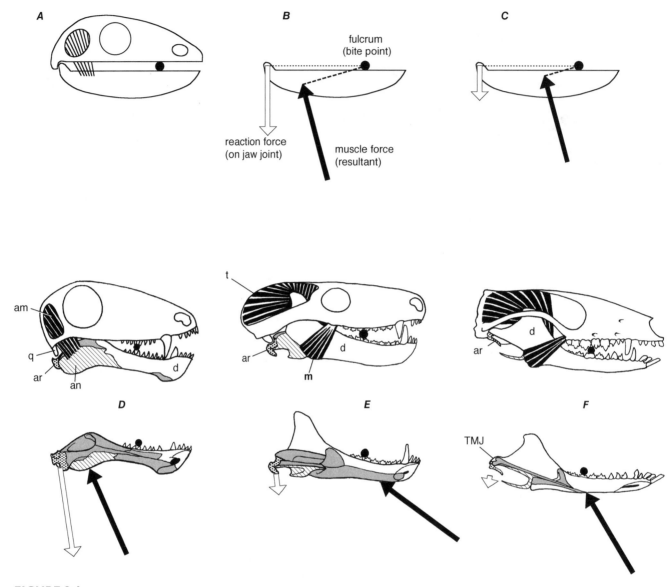

FIGURE 3.6

Evolution of mammalian chewing muscles and ear ossicles (diagrammatic). **A**: Attachments and orientation of the jaw muscles in a primitive reptile. The lower jaw can be analyzed as a lever with its fulcrum at the point where food is being bitten (*black dot*). When the forces on the lower jaw are in static balance (**B**), the magnitude of the summed jaw-muscle force vector (*black arrow*) multiplied by its lever arm around the bite point (*heavy dashed line*) must equal the jaw-joint reaction force times its own lever arm (*white arrow, light dashed line*). If the summed muscle forces pass closer to the bite point (*black arrow*, **C**), the force borne by the jaw joint will be proportionately reduced (*white arrow*, **C**). In the evolutionary progression from early synapsids (**D**) to later synapsid reptiles (**E**) and early mammals (**F**), the jaw adductor muscle (am) differentiated into a backward-pulling **temporalis** muscle (t) and a smaller, forward-pulling **masseter** (m). These changes brought the combined jaw-muscle forces to bear more directly on the bite point (*black arrows*, **E,F**), thus enhancing bite force while reducing the reaction force and lowering the stresses on the non-tooth-bearing bones in the lower jaw. As more and more of the jaw-muscle force was transferred to the teeth, the tooth-bearing **dentary** bone (d) became progressively larger, while the other lower-jaw bones (*gray tone*) grew smaller and eventually disappeared—except for the **articular** (ar, *stipple*) and **angular** (an, *hachure*), which were drawn into the evolving ear apparatus and became the malleus and ectotympanic respectively. The development of a new temporomandibular jaw joint (TMJ) between the dentary and the braincase freed the **quadrate** (q) from load-bearing and allowed it to become another ear bone, the incus (cf. Figs. 3.3B and 3.5).

ectotympanic, which encircles the mammalian eardrum.

Some diapsid groups, including certain lizards and the ancestors of birds, also evolved eardums and keen hearing; but they achieved this differently, by attaching the stapes directly to the eardrum. Their quadrate and articular bones did not get involved in the ear. These two bones still form the jaw joint in all living reptiles and birds.

5. Vertebrae and Breathing. Between the skull in front and the sacrum at the rear, the vertebral column of mammals is divided into three functionally different sections: the **cervical** (neck), **thoracic** (chest), and **lumbar** (abdominal) vertebrae. These three kinds of vertebrae differ most obviously in their ribs. Early reptiles had prominent curving ribs on most of the vertebrae of the neck and trunk. But in mammals, such ribs are seen only in the thoracic region. In the cervical and lumbar regions, the ribs are reduced to short bony projections from the sides of the vertebrae.

This reduction of the cervical and lumbar ribs is seen in the earliest fossil mammal skeletons (Jenkins and Parrington 1976). The reduced lumbar ribs of the Triassic mammals suggest that they had already developed a modern mammalian pattern of breathing. In mammals, the drawing of air into the lungs is driven by the **diaphragm**, a dome-shaped muscular partition stretched across the abdominal end of the rib cage. The bulging surface of this curved sheet of muscle protrudes into the rib cage. When the diaphragm contracts, it flattens out and shoves the abdominal viscera tailwards. This increases the volume of the space enclosed by the ribs, and air rushes into the lungs to fill that extra space. Lacking lumbar ribs, the mammalian abdomen can stretch more freely on inhalation to accommodate the guts displaced by the contracting diaphragm.

The lumbar ribs were also reduced in some of the mammal-like synapsid "reptiles" of the Triassic (Fig. 3.1B). This is one more piece of evidence suggesting that some of these animals had high metabolic rates and warm blood.

6. Vertebrae and Locomotion. In moving around on land, early terrestrial vertebrates continued to employ a sinuous, fish-like side-to-side wiggling of the spine. This sort of undulation is still important in the locomotion of most living salamanders and lizards, which retain a primitive sprawling posture with the limbs sticking out sideways. When these animals walk, their side-to-side undulations help swing their limbs forward and back to provide propulsive thrust (Fig. 3.7A). Some amphibians and reptiles—for example, snakes—have abandoned their limbs altogether and rely on a version of the fish wiggle to get around on the land.

But other reptilian lineages have evolved in the opposite direction, reshaping their bones and muscles to produce a more erect posture in which the feet are planted underneath the body and the limbs propel the animal forward by swinging forward and back in arcs that roughly parallel the midline plane of the body. This sort of posture and locomotion has evolved at least three times: in the chameleon family of lizards, in the dinosaur group of diapsid reptiles (including the ancestors of birds), and in the synapsid ancestors of the mammals.

A mammal-like locomotor posture was already present in pre-mammalian synapsid "reptiles," and the limb bones of early mammals differ little from those of late synapsids. But the backbones of the first mammals show certain innovations, reflecting new patterns of locomotion not present in their synapsid ancestors. The joints between the lumbar vertebrae of early mammals are reoriented so that their opposing faces roughly parallel the body's midline plane. This makes the lumbar part of the vertebral column stiffer overall, but leaves it free to flex and extend in the midline plane—that is, to bend down toward the belly or up toward the back. The up-and-down bending movement of the spine in the midline is an important aspect of mammalian locomotion. The fast running gait of most living mammals is a **bound** or **gallop**, in which the animal extends its back when it pushes off with its hind legs. It then flies through the air with all four feet off the ground, lands on its forefeet alone, and flexes the back to bring the hind feet far forward before they touch down (Fig. 3.7B). This cycle then repeats (Gambaryan 1974). The cyclical flexing and extending of the back adds to the distance covered in each bound, and thus increases the animal's speed. The characteristically mammalian lumbar intervertebral joints seen in the fossil skeletons of early mammals suggest that they had evolved a mammalian style of fast running (Jenkins and Parrington 1976).

This conclusion is bolstered by the orientation of their vertebral **spinous processes**, the bony projections that jut out from the backs of the vertebrae. The spinous processes can be felt on the human back as a row of bumps in the midline, running from the neck down to the pelvis. In typical reptiles, these processes all point in roughly the same direction (Fig. 3.1). But in living mammals that have a galloping or bounding gait, the spinous processes tend to line up with the pull of the big back muscles that extend the vertebral column at the beginning of the gait cycle. This means that the processes at the head end of the trunk slant back toward the pelvis, and those at the pelvic end slant forward toward the head. Early mammals show these regional differences in the angles of the spinous processes. This is further evidence that they had evolved a mammalian running gait, and that in time of need they bounded along with an up-and-down wiggle like a squirrel in a

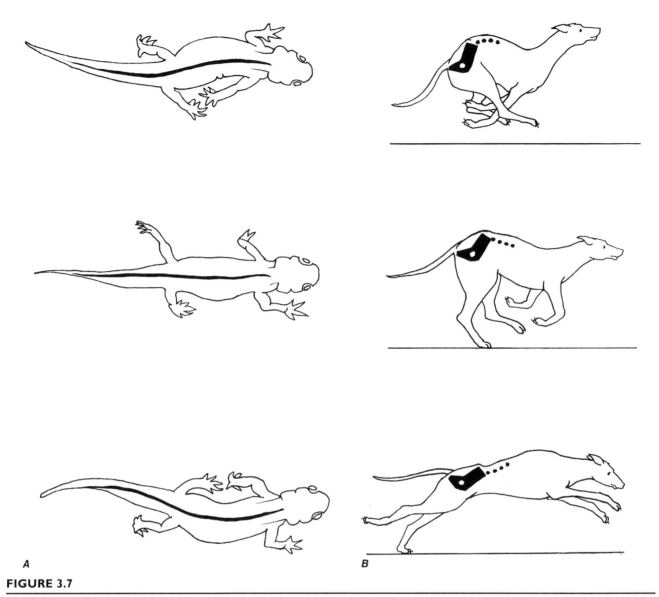

FIGURE 3.7

Locomotion patterns in (**A**) a walking salamander, seen from above, and (**B**) a galloping dog. Black areas in **B** indicate the flexion and extension of the lumbar vertebrae and pelvis during the gallop. (After Gray 1959.)

hurry, rather than executing a side-to-side wiggle like a lizard scooting for cover.

7. **Brain and Braincase.** The brains of even the most mammal-like of the other Triassic synapsids were no larger than those of typical reptiles. But the early mammals appear to have had brains three or four times as big for their size (Crompton and Jenkins 1978). This expansion was most marked in the hind end of the brain, which is concerned with the coordination of movements and the inner-ear senses of hearing and balance, and in the front part of the brain, which was at first mainly concerned with the sense of smell.

Brain enlargement in early mammals was accompanied by a solidification of the braincase (Fig. 3.8).

In typical reptiles, the roof, the floor, and the hind and front ends of the braincase are bony, but there is a big gap in the middle where the sides of the brain are covered over only by membranes and jaw muscles. In early mammals, flanges of bone grew inward from the edges of this gap, largely filling it in and helping to protect the vulnerable midsection of the enlarged brain. In later mammals, the hole was filled in completely, enclosing the brain inside a safe, solid bony box. The new side walls of the braincase also provided added surfaces for the attachment of jaw muscles. Parallel trends toward brain enlargement have independently produced a solid, mammal-like braincase in birds, presumably for similar reasons.

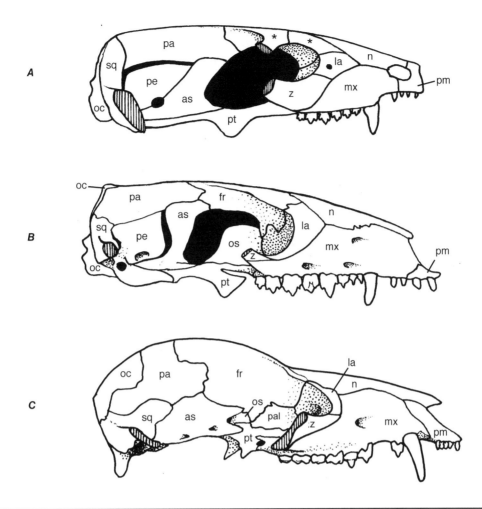

FIGURE 3.8

The gap in the side wall of the reptilian braincase (*black area*) was progressively filled in by bone during the evolution of Mesozoic mammals. **A**: Side wall of the braincase in a Triassic synapsid reptile (diagrammatic composite of *Probainognathus* and *Thrinaxodon*: after Romer 1970a and Parrington 1946). **B**: The Jurassic mammal *Morganucodon* (after Kermack et al. 1981). **C**: The extant opossum *Didelphis* (after Carroll 1988). Vertical hachure indicates cut surfaces of the zygomatic arch, removed to display the lateral side of the braincase. Abbreviations: as, alisphenoid; fr, frontal; la, lacrimal; mx, maxilla; n, nasal; oc, occipital; os, orbitosphenoid; pa, parietal; pal, palatine; pe, petrosal; pm, premaxilla; pt, pterygoid; sq, squamosal; z, zygomatic. Asterisks indicate reptilian bones lost in the mammalian skull.

▌ALLOMETRY

Early mammals had relatively bigger brains than their synapsid ancestors. However, it is hard to say exactly how much bigger, because there is very little overlap in body size between the tiny early mammals and the non-mammalian synapsids. This causes problems in interpretation. Smaller animals tend to have more brain volume per unit of body weight, and the relationship between brain size and body size is not a simple one.

That relationship is an example of the larger phenomenon called **allometry**—the patterned change of body proportions and ratios when body size changes. Most of the allometric differences between different-sized animals reflect the mathematical fact that the ratio of surface area to volume—the amount of outside surface per cubic unit of inside contents—changes with size. It is easy to see why this must be so if we think about a simple uniform shape like a cube. A 1-inch cube has 6 square inches of surface for its one cubic inch of volume. A 2-inch cube has four times as much surface ($6 \times 2^2 = 24$ square inches), but eight times as much volume ($2^3 = 8$ cubic inches). Thus, the 2-inch cube has only half as much surface per unit volume ($24 \div 8 = 3$) as the 1-inch cube ($6 \div 1 = 6$). More generally, multiplying all the linear dimensions of something by a factor of n gives it n^2 times the original surface area and n^3 times the original volume, so it has only $(n^2/n^3) = 1/n$ times the original surface-to-volume ratio. This is called the **square-cube law**.

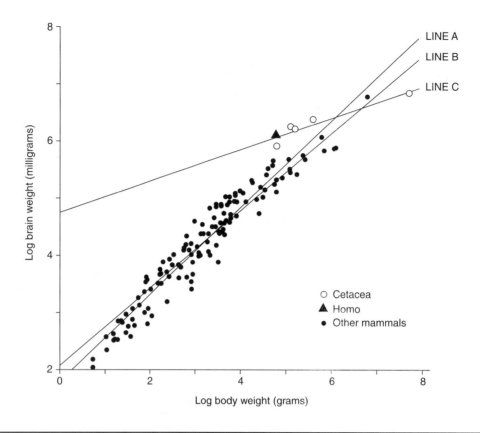

FIGURE 3.9

Log–log plot of brain weights and body weights in 142 species of living mammals (data from Cartmill 1990). Lines A and B are drawn through the bivariate average with slopes of ¾ and ⅔, respectively. Line C is the least-squares line fitted to the data for dolphins and whales (Cetacea).

To maintain **functional equivalence**, an animal that grows larger or smaller therefore needs to alter some of its body proportions to make up for the inevitable change in surface-to-volume ratios. For example, a larger animal may need to have disproportionately larger gills or lungs so that each cubic inch of its flesh continues to be connected to about the same amount of surface area for the exchange of respiratory gasses. Large land animals may need to move more carefully or have relatively thicker bones than their smaller relatives, because a bone's breaking strength depends on its cross-sectional area whereas the weight it has to bear depends on the animal's volume. And so on.

What does all this have to do with brain size? As we might expect, body weight varies roughly as the cube of body length in closely related animals of different size. But brain weight does not. It varies roughly as the *square* of length. To put it another way, brain weight is roughly proportional to (body weight)$^{2/3}$ (i.e., the square of the cube root of body weight). Why this should be is unclear. Some think that brain size needs for some unknown reason to vary with surface area to maintain functional equivalence (Jerison 1973). Others (Martin 1981, 1996, 1998; Armstrong 1983) argue that brain size is tied to metabolic rate, which (for some equally unclear

reason) happens to be roughly proportional to (body weight)$^{3/4}$.

If we graph mammalian brain and body weight data on logarithmic coordinates (which convert exponential curves into straight lines), we can see that an exponent of either ⅔ or ¾ fits the scatter of data reasonably well (lines A and B, Fig. 3.9). But both lines tend to diverge from the cloud of data at its ends, because the cloud is slightly curved and not strictly linear.

As a result of this curvature, different ranges of body size yield different empirical estimates of the allometric constant—that is, the slope of a line drawn through the scatter of data points. If we look at just the smallest mammals, we get a high estimate of the slope; if we look at the largest mammals, we get a much lower one.

For example, the datum for human beings (*Homo*) falls further above both lines A and B in Fig. 3.9 than any of the other data. This suggests the flattering conclusion that human beings are uniquely brainy mammals when we take allometry into account. But if we were to base our estimates of brain-body allometry solely on the data for mammals that are larger than we are—say, on the whales and dolphins (Cetacea), which lie at the upper end of that curved scatter—we would come up with a very different picture. Judged by the whale–dolphin standard (line C, Fig. 3.9), human brains are just average in size, and almost all other mammals are shockingly pea-brained.

This example illustrates how risky it is to extrapolate empirical estimates of allometry into ranges of body size above or below the range that the estimates were based on. We face this problem in trying to say how much bigger the brains of early mammals were than those of other synapsids, because we do not know of any other synapsids that were so small. If we had some independent theory that would predict how brain size ought to vary with body size to maintain functional equivalence, we could estimate relative brain size by comparing deviations from the values predicted by the theory. But as yet we have no such theory. All we can say for sure is that early mammals had brains larger than those of *living* reptiles that are equally tiny.

This sort of problem arises whenever we try to compare the relative sizes of organs in animals of different body weight. We will see later how such problems complicate our attempts to make inferences from fossil bones to the activity patterns of the ancestral crown-group primates. Similar problems confront us when we try to draw conclusions about the biology of the bipedal apes from which we are descended, some of which were much smaller than the modern human average.

ALLOMETRY AND EARLY MAMMALS

Cladistic analysis of the anatomical facts about living animals reveals some things about the early mammals that we cannot infer directly from their fossil remains. We know that the first mammals still laid shelled eggs, because egg-laying is primitive for amniotes and some living mammals still lay eggs. The ancestral mammals probably had warm blood and fur, and skin with glands in it that secreted an oily fluid to help keep their hairs supple and shiny, because the common ancestors of living mammals must have had these things. The first mammals probably had smaller eyes and dimmer vision than their ancestors, because the apparatus of vision in living mammals appears to have lost some features that are found in other amniotes. For example, most reptiles and birds can tell red from green, and they can change the shape of the lenses of their eyes to focus at different distances, which is called **accommodation**. Typical mammals, however, are more or less color-blind and have fixed-focus eyes. (The monkey-like ancestors of human beings re-evolved color vision and accommodation, but in a different form.) Evidently, early mammals did not need or have keen eyesight. The senses of touch, hearing and smell were more important to them.

Everything that we know or can infer about early mammals suggests that they were shrew-like creatures that had managed to survive in a reptile-dominated world by becoming diminutive, warm-blooded, largely nocturnal predators that lived and fed in dense cover. They were uniquely suited to this special way of life. Their high internal temperatures gave them an advantage over cold-blooded reptiles during the cool night-time hours. Their reliance on senses other than vision in hunting also gave them an edge over the vision-guided diapsids in darkness and shadows. They made the most of this advantage by developing especially acute nonvisual sense receptors: an expanded cochlea to detect the high-frequency rustle of insect movements, mobile **external ears** to help locate the direction of those sounds, special tactile whiskers or **vibrissae** bristling from the tip of the questing snout, and enlarged olfactory organs for sniffing out concealed prey in the dark.

The early mammals faced little competition for their new niche because it was a difficult way of life to maintain, bordering on the physically impossible. A tiny animal finds it hard to sustain a constantly elevated body temperature. The square-cube law works against it. As noted earlier, a huge land animal like a big dinosaur can hardly help being warm-blooded, because it cools off and heats up very slowly and has a great many cubic feet of heat-generating internal tissues for every square foot of heat-radiating skin. Things are reversed at the other end of the allometric scale. A small shrew has far too much outside for its pathetic 3 or 4 grams of inside. Every calorie of energy that it gets from its food is carried away almost at once by radiation from its skin and evaporation from its lungs. To stay warmer than its surroundings, a shrew has to eat almost continuously around the clock every day of its short, frenzied life. Denied food for a few hours, it will starve to death—a victim of the implacable laws of geometry and physics.

Faced with these allometric problems, early mammals were under intense pressure to come up with ways of processing both food and oxygen as rapidly and efficiently as possible. Most of the distinctive peculiarities of mammals originally evolved to serve those ends, either directly or indirectly. It is easy to see the connection when we look at such mammalian features as the diaphragm (to help pump air in and out faster), fur (to minimize heat loss), a **four-chambered heart** (to prevent the oxygen-rich blood from the lungs from mixing with the oxygen-depleted venous blood coming back to the heart from the rest of the body), and dental occlusion (to speed digestion by slicing and grinding food into smaller bits). It is less obvious, but equally likely, that the fundamental facts about the mammalian life cycle—the mammalian patterns of birth and death—also started out as allometrically imposed adaptations to a shrew-like way of life.

DEATH AND MOLAR OCCLUSION

All organisms die sooner or later, but not all organisms have fixed lifespans. One of this book's authors has a

450-gram pet turtle that was probably about 50 years old when it was acquired, judging from its size and the growth rings on its horny carapace. Some 40 years have passed since then, during which the turtle's mammalian owner has grown slower, grayer, and feebler. Every 450-gram mammal that was alive forty years ago has died long since. But the 90-year-old turtle is as strong and perky as ever. Barring fatal accidents or disease, it can go on more or less indefinitely. So can many other vertebrates. People and other mammals cannot.

The lifespan of most mammals is limited by their teeth. Most vertebrates with teeth have an endless supply of them. As soon as the tooth of a fish or a lizard wears out, it drops out of the jaw and is replaced by a new one growing in from underneath. Typical fish and reptiles can keep replacing their teeth indefinitely because they do not chew their food and do not have precise dental occlusion. Their teeth are simple conical fangs, which do their jobs of grabbing and stabbing well enough even when they have erupted incompletely, or if they erupt in a position a bit different from the one occupied by the old tooth. But the upper and lower molar teeth of mammals have reciprocal surfaces that need to fit against each other with precision. If a new molar grew in to replace an old one, it would be of little use until it had erupted into full occlusion. Even when it got there, it would not fit properly at first, because it would lack the ground-down **wear surfaces** that had developed reciprocally between its predecessor and the teeth that opposed it during chewing.

Mammals therefore do not replace their complicated molars, and they replace their other, simpler teeth—incisors, canines, and premolars—only once, if at all (Fig. 3.10). In those living mammals that have the most complete dental replacement, a first wave of teeth erupts on each side of the upper and lower jaws, beginning with the central incisor in front and proceeding more or less straight back to the last molar (with variations in different taxa). A second wave then starts in front again, proceeds back to the last premolar, and stops. The two waves overlap in time, and the last tooth in the first wave may not come in until sometime after the second wave of eruptions is completed. But once that last molar has erupted, the animal has no more teeth in reserve. It will die after its so-called "permanent" (unreplaced) teeth wear out. Most mammals die well before that happens; but in a few large, long-lived mammals such as elephants, death may come by starvation in toothless old age. This would be a regular occurrence in human beings as well, were it not for dentures and soup.

Even with dentures, the human lifespan is still finite. A mammal's doom is ultimately written in its cell-replicating machinery. After about 40 successive cycles of mitotic cell division, mammalian cells in tissue cultures stop dividing, and the same thing evidently happens in the living body. It is not clear why mammals

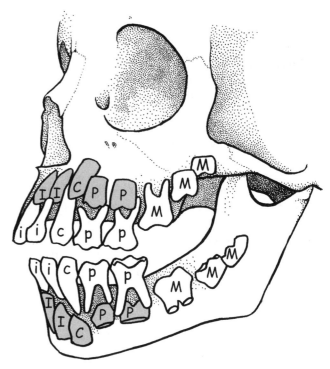

FIGURE 3.10

Tooth replacement in *Homo sapiens* (diagrammatic), showing first (white) and second (gray tone) series of teeth. Lowercase letters indicate replaced ("milk") incisors (i), canines (c), and premolars (p). M, molar.

(and birds) have this cellular limitation on their span of life, while many other vertebrates do not (Girondot and Garcia 1998, Nakagawa et al., 2004). Perhaps it is an inevitable side effect of being warm-blooded. Perhaps it confers some evolutionary advantage in a larger context. Perhaps it has no adaptive significance one way or the other for mammals, given that most aging mammals are facing death by dental attrition anyway.

Future advances in cellular and molecular medicine may someday repeal the limits on human mitosis and offer us the prospect of an indefinitely extended life. In the meantime, we can comfort ourselves by noting that it is probably better on the whole to be a mortal human being than an immortal turtle.

▌ALLOMETRY, MOTHERHOOD, AND MILK

The early mammals were as small as a warm-blooded animal can possibly be. But their newly hatched offspring were smaller still. Therefore, their hatchlings were not warm-blooded. Like newborn shrews today (Nagel 1989), the babies must have allowed their body temperature to rise and fall with the temperature of their surroundings whenever their mother was not around.

Such minuscule creatures would have had a hard time surviving if they had begun life by simply breaking out of the egg and wandering off in search of food, as the hatchlings of most reptiles do. Newly hatched reptiles also have a hard time surviving, but their mothers can compensate for high infant mortality rates by laying larger egg clutches. This was not an option for early mammals. They were living so close to metabolic bankruptcy that they could not afford to lay a lot of large-yolked eggs all at once. Their only recourse was to make a smaller number of eggs and then invest additional energy in each offspring more gradually, by helping the hatchling to survive until it grew big enough to survive on its own (Crompton and Jenkins 1979).

Mammalian motherhood came into existence to address these problems. Early mammals must have laid and incubated their eggs in a protected nest, probably in a burrow in the ground. (Burrowing could have evolved out of the common reptilian habit of burying the eggs in a shallow hole.) Like the newborn babies of mice and many other mammals today, the newly hatched young of early mammals were **altricial**, meaning that they emerged into the world in a semi-fetal condition— pink, hairless, blind, and unable to walk. The mother fed them at first with **milk** secreted from two rows of **mammary glands** on her belly—enlarged, modified versions of the skin glands used in scent communication. Having no need to chew, the newly hatched young had no teeth. Later on, when their eyes, ears, teeth, and thermoregulatory machineries were adequately developed, the mother began leading them out of the nest to learn to find food on their own.

The newly hatched young of the earliest mammals may have simply lapped up milk oozing from the mother's mammary glands. But mammals soon evolved a quicker, more efficient way of refueling the growing babies. The ducts of the mammary glands came to empty into raised **teats** or nipples, permitting the young to nurse faster and more efficiently by getting their mouths entirely around the faucet and sucking on it.

Suckling was not a trivial innovation. It demanded a reorganization of the reptilian mouth and throat. This reorganization depended on the development of a **palate** separating the mouth from the nasal passages. We can see why the palate is important in suckling if we look at a human baby born with a so-called cleft palate, in which a developmental accident has left a reptile-like gap between the right and left sides of the roof of the mouth. When such a baby tries to suck milk, air rushes into the mouth through the nose and breaks the suction. When the baby finally manages to get some milk into its mouth and tries to swallow it, part of the milk goes back out the same way, dribbling from its nose like water from the nostrils of a drinking turtle.

The mammalian hard palate helps prevent this by separating the nose from the mouth. The ideal arrange-

ment would have been to extend this partition between the food and air passages all the way down to the lungs and stomach. Unfortunately, evolution works not by jumping to the optimal setup, but by making continual minor improvements in the inherited materials; and the inherited materials did not permit this ideal arrangement to evolve. The early land vertebrates had nostrils that lay *above* the mouth; but their lungs were connected to the *ventral* (belly) side of the gut. The food and air passages therefore had to cross at some point. Later land vertebrates have inherited this intersection. To manage the traffic at this crossing and make suckling possible, early mammals developed a moveable flap, the **soft palate**, that could be raised or lowered at the back of the mouth. When a baby mammal suckles, this flap is pulled down and forward across the rear opening of the mouth cavity. This allows the nursing baby to breathe without letting air into its mouth to break the suction of the tongue. When the nursling swallows, the flap is pulled up to close off the back of the nasal passages and keep the milk from escaping through the nose. (Breathing therefore has to stop briefly during swallowing, as you can prove for yourself by trying to do both at once.)

A nursling's mouth can also leak milk or air between its lips. This would be a serious problem if the baby had stiff, horny lips like those of living reptiles. The evolution of the teat in mammals was accompanied by changes in the lips, which became soft, moist, and muscular so that they could form an adequate seal when pursed around a nipple.

The muscles that invaded the lips came from the muscles of the second branchial arch. In a reptile, this band of muscle forms a sphincter around the gut tube in the neck (Fig. 3.11A). When it contracts, it squeezes the **pharynx** (the part of the gut between the tongue and esophagus) and helps to move swallowed food down toward the stomach. These muscles still develop in the neck in the mammalian fetus, but then migrate forward under the skin to the face, where they eventually form a complex web of **facial muscles** (Fig. 3.11B). These muscles do several distinctively mammalian jobs. In suckling, they act to pucker the lips and tense the cheeks. They also move the external ears and wiggle the nose and whiskers. In people and many other mammals, some of them are specialized for moving the skin of the face to produce **facial expressions**.

The job of squeezing the pharynx is handed over in mammals to a series of new **pharyngeal constrictor** muscles derived from the last three branchial arches (Fig. 3.11B). Other similarly derived muscles also work during swallowing to pull the tongue backwards and draw the soft palate down across the back of the mouth. None of this apparatus—the hard palate, the soft palate and its muscles, the facial muscles, or the pharyngeal constrictors—exists in any modern reptiles. Most of it

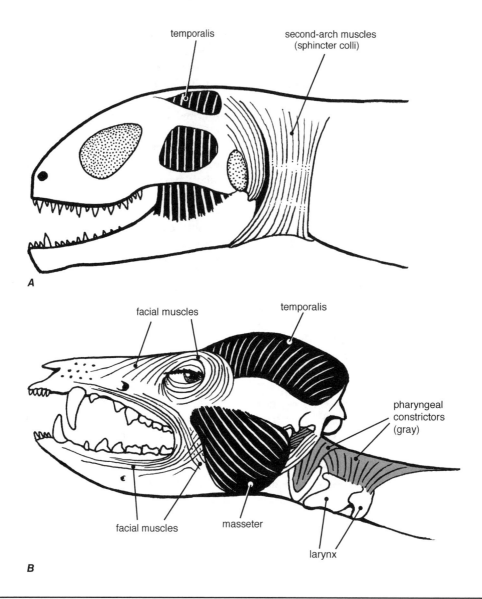

temporalis

second-arch muscles
(sphincter colli)

A

facial muscles

temporalis

pharyngeal
constrictors
(gray)

facial muscles

masseter

larynx

B

FIGURE 3.11

Pharyngeal and facial muscle evolution. **A**: The second-arch muscles in a reptile (*Sphenodon*) form a muscular sphincter around the pharynx. **B**: The equivalent muscles (facial muscles) in a mammal, plus the new mammalian constrictors of the pharynx. (**A**, after Romer 1970b.)

presumably evolved as a concomitant of suckling (K. Smith 1992).

▌RESPIRATION AND THE PALATE

Mammals may have inherited some of these features from their synapsid ancestors. Some of those mammal-like "reptiles" already had a bony palate separating the mouth from the nasal passages (Strauss et al. 1992). This does not necessarily mean that they had begun feeding their young on milk. The palate serves more than one

function. In addition to facilitating suckling, it also helps to solve certain respiratory problems that result from having an elevated body temperature. It may have evolved originally for that purpose.

Warm-blooded animals face difficulties with respiratory water loss. The air expelled from any animal's lungs is always saturated with water vapor. Because warm air can hold more water vapor than cool air, a warm-blooded animal loses more water vapor from its lungs in every exhalation. Warm-blooded animals also have high metabolic rates, and thus have to breathe faster than cold-blooded ones. As a result, mammals tend to become dehydrated just by breathing. This is not a problem for cold-blooded vertebrates. A lizard with a body temperature the same as that of the surrounding air actually gains more water per hour from metabolizing its food (e.g., by oxidizing glucose into CO_2 and H_2O) than it loses by breathing. But in mammals, the reverse would be the case (Hillenius 1992) were it not for a special structure in the mammalian nose. This structure, called the **maxilloturbinal**, acts as an air conditioner, humidifier, and water recycling system.

The maxilloturbinals in most mammals are a pair of elaborate bony scrolls attached to the side walls of the nasal passages in front, behind the nostrils and above the palate (Fig. 3.12A,B). When a mammal inhales, incoming air flows through these scrolls. In doing so, it picks up water from the expanse of moist mucous mem-

brane covering the folds of the maxilloturbinals. The evaporation of this water cools the membrane and saturates the inflowing air with water vapor. Since the inhaled air is now laden with water, little more can be lost to it by additional evaporation when it reaches the lungs.

When a mammal exhales, the warm, saturated air flows back out across the cooled maxilloturbinals, and its temperature drops. This causes some of the exhaled water vapor to condense on the maxilloturbinal mucosa, ready to be picked up again in the next intake of breath. This cycle cuts down on respiratory water loss. [Although the human maxilloturbinals are greatly simplified and reduced (Fig. 3.12D) you can still feel a similar process happening on the inside of your own nasal passages when you breathe deeply.] When evaporative cooling is

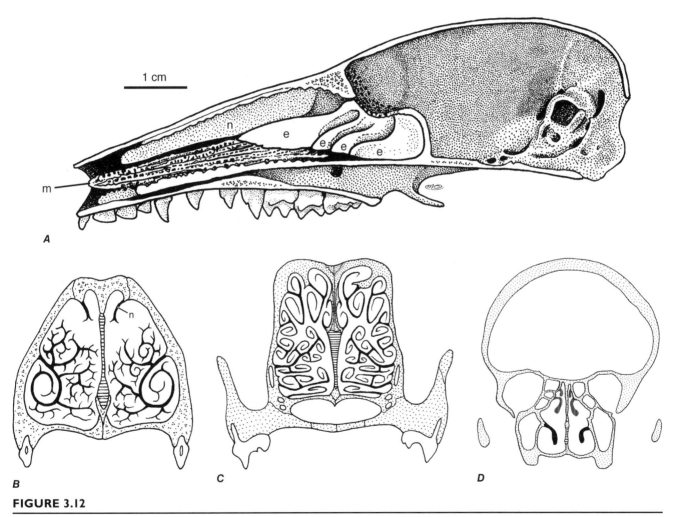

FIGURE 3.12

The mammalian nasal fossa. **A**: Hemisected skull of an unspecialized marsupial (*Caenolestes*), viewed from the medial aspect. **B**: Cross section through the maxilloturbinals and nasoturbinals of a dog. **C**: Cross section through the ethmoturbinals of a dog. **D**: Cross section through the maxilloturbinals (inferior conchae: black) and ethmoturbinals (superior and middle conchae: hachure) in *Homo sapiens*. Abbreviations: e, ethmoturbinal; m, maxilloturbinal; n, nasoturbinal. (**A**, after Osgood and Herrick 1921; **B,C**, after Miller et al. 1964; **D**, after Grant 1972.)

needed, most mammals increase evaporative water loss by **panting**, breathing in through the nose and out through the mouth (Schmidt-Nielsen et al. 1970). Air exhaled through the mouth does not pass across the maxilloturbinals, so it carries away more water vapor—and heat along with it—at every pant.

To make this system work to best effect, there needs to be a bony palate separating the nose from the mouth. The palate confines the air flow to the nose, forcing it to pass through the narrow passages between the maxilloturbinal surfaces (instead of taking a path of less resistance around the maxilloturbinals) and thereby producing more efficient heat exchange. When heat loss is needed, the hard and soft palates allow exhaled air to be shunted away from the maxilloturbinals into the mouth. The palate also walls off the nasal apparatus from all the powerful crunching, grinding, and dismemberment that goes on in the mouth. Protected by the palate, the maxilloturbinal scrolls can become as delicate and elaborate as necessary to provide the right amount of air-conditioning surface (Fig. 3.12B).

A second set of delicate bony scrolls is found in the mammalian nose above and behind the maxilloturbinals. These scrolls, the **nasoturbinals** and **ethmoturbinals** (Fig. 3.12A), are covered with mucous membrane containing the receptor cells for the sense of smell. These complex bony surfaces may carry a lot of olfactory surface in large mammals—nearly 8 square meters in a big, keen-nosed dog (Fig. 3.12C; Moulton 1977, Doty 2001). Equipped with all this scent-detecting machinery, mammals have a uniquely acute sense of smell. The remnants of similar olfactory turbinals can be made out in the fossil skulls of some early mammals (and also in some of the other late Triassic synapsids), testifying to the importance of olfaction in their way of life.

Birds, the other group of small warm-blooded vertebrates, have independently come up with their own ways of dealing with some of these problems, evolving four-chambered hearts, high-turnover respiratory apparatus, nest-building and incubation of eggs, fluffy feathers (instead of fur), gizzards (instead of dental occlusion), and maternal feeding of blind, naked babies with insects or regurgitated stomach contents (instead of milk). The independent appearance of such mammal-like traits in birds supports the notion that these features are functionally linked to warm blood and small body size—though both birds and mammals appear to have inherited some of them from larger-bodied synapsid or diapsid ancestors.

Although mammals stayed small throughout the Mesozoic, they continued to evolve. Toward the end of the Mesozoic, they began to radiate into lineages leading to the mammals of today. We are descended from one such lineage that made important innovations in two main areas: the molar teeth and the reproductive apparatus.

THE TRIBOSPHENIC MOLAR

The molars of primitive mammals (Fig. 3.3) were serrated blades looking something like the teeth of sharks, with three main cusps set in a fore-and-aft row. The total length of the shearing edges in such a linear array is limited by the depth of the notches between the cusps and by the length of the jaws. Throughout mammalian history, mammals that need to chop up their food finely have evolved ever more elaborate nonlinear molar cusp arrangements, allowing them to escape those limitations and pack in more shearing edges per unit of jaw length. This process has demanded the evolution of more precise dental occlusion.

The first step in getting more shearing edges was to fold the three linearly arranged cusps into a "V" shape. In the Mesozoic mammals that took this step, the upper and lower molars were folded in opposite directions (Fig. 3.13A), with the apices of the upper V's pointing in toward the **lingual** (tongue) side of the tooth row and those of the lower V's pointing out toward the **buccal** (cheek) side. The straight line of shearing edges was converted into a zigzag, shortening it in a fore-and-aft direction and allowing more molars to be fitted into the available space without lengthening the jaw. (*Morganucodon* had only four molars, but the first mammals with this zigzag arrangement managed to fit in seven or eight.) When the jaws closed, the lower molars moved upward and lingually (medially, toward the midline of the body), coming to rest in the triangular spaces between the upper teeth, so that the cutting edges of each lower molar sheared across the back edge of the upper tooth in front of it and the front edge of the upper tooth behind it (Fig. 3.13B). This sort of reversed-triangles arrangement is seen in several groups of Jurassic mammals.

The medially directed sweep of the lower molars, and the empty spaces between the triangles in each jaw, afforded opportunities for further improvements. In the next stage of this process, the transverse width of the upper molars was augmented by the addition of the **protocone**, a new cusp on the lingual side, and the **stylar shelf**, a crest-bearing extension on the buccal side (Fig. 3.13C). Both additions provided new shearing edges for the **trigonid** (the lower triangle) to move across during its lingual phase of movement. The spaces between the lower molars were filled in by another crest-bearing extension, the **talonid**, projecting from the rear edge of the trigonid. Appearing first as a simple shearing edge, the talonid expanded to become a small basin ringed with cusps. The protocone came to rest in this basin at the end of its medial sweep, producing a crushing action like that of a pestle in a mortar (Fig. 3.13C).

The continuation of these evolutionary trends produced the **tribosphenic** molar pattern (Bown and

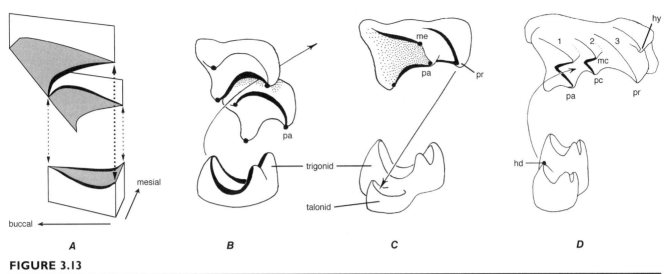

FIGURE 3.13

Evolution of molar occlusion (diagrammatic). **A**: Basic pattern of reversed "V"s or triangles. **B**: Dental occlusion in an early mammal (symmetrodont) with interlocking triangular molars. When the teeth are brought together (*arrow*), the shearing edges (*heavy black curves*) on the lower triangle (the **trigonid**) sweep medially across reciprocally curved edges on the upper triangle. **C**: The addition of a **stylar shelf** (*stippled area*) and **protocone** (pr) to the original three cusps (*black dots*) augments the total length of the shearing edges (*heavy black curves*) on the upper molar. At the end of the medial sweep, a crushing action is produced when the protocone comes to rest in the basin of the heel or **talonid** of the lower molar (*arrow*). **D**: The tribosphenic pattern seen in the ancestors of all living mammals. The edges of the trigonid now shear across three shearing edges (1–3) on each side of the upper triangle (the **trigone**). The two additional edges are borne by two new cusps (paraconule, pc; metaconule, mc). The outer cusp in the talonid (hypoconid, hd) now sweeps (*curved arrow*) through its own series of notches (*heavy black curves*) in the center of the upper molar, further augmenting the amount of shearing. Other abbreviations: me, metacone; hy, hypocone.

Kraus 1979), the ancestral arrangement from which the patterns of molar occlusion in all living mammals have been derived. In the tribosphenic lower molar (Fig. 3.13D), the talonid was expanded to furnish more crushing surface. The **hypoconid** cusp on the outer edge of the talonid became the apex of a new lower triangle, which sheared through a series of V-shaped notches in the center of the upper molar. As the jaws were opened again, the lower jaw moved a bit further medially, dragging the hypoconid down the lateral side of the protocone for a grinding action.

This whole series of molar modifications was a natural evolutionary development from the original reversed-triangles zigzag. In fact, it was so logical and natural that it appears to have evolved independently three times: once in our own Mesozoic ancestors, separately in the ancestors of today's egg-laying monotremes, and a third time in a Jurassic beast called *Shuotherium* that developed a talonid on the *front* of the lower molars rather than the back (Chow and Rich 1982, Luo et al. 2001a, Kielan-Jaworowska et al. 2002).

The combination of shearing, crushing, and grinding functions in a single occlusal package made the tribos-

phenic molar not only an important evolutionary advance in itself, but also a versatile point of departure for evolving new dietary adaptations. Carnivorous mammals have tended to accentuate the shearing features of their tribosphenic molars. Conversely, mammalian lineages that specialize in eating plant tissues have evolved molars that emphasized crushing and grinding functions at the expense of puncturing and shearing. A common first step in adapting to plant-eating has been the addition of a **hypocone** (Fig. 3.13D), a new cusp that projects from the rear and inner part of the upper molar **cingulum** (the bulging base of the tooth's crown, near the gum line). The hypocone grinds into the central depression in the lower molar's trigonid, in much the same way that the adjoining protocone grinds into the talonid basin. Further specialization for plant-eating almost always involves lowering the trigonid to the level of the talonid and making the hypocone taller and fatter, so that both upper and lower molars eventually present a more or less square array of several cusps at about the same height. You can study one of the results of this squaring-off process by looking at your own squared-off molars in a mirror.

LIVE BIRTH AND PLACENTATION

In addition to their tribosphenic molar teeth, our Mesozoic mammalian ancestors came up with another key innovation. Instead of laying shelled eggs like a typical reptile or a platypus, they began retaining their eggs inside the mother's oviducts until the embryos had developed far enough to come out, breathe air, and drink milk. This habit of **viviparity** or "live birth" has some obvious advantages. It makes eggshells unnecessary, and thus saves the energy and materials that would go into making them. It allows the mother to keep her eggs warm inside her body and protect them from predators 24 hours a day, even while she runs around foraging. This must have been a big help to a tiny Mesozoic mammal that had to eat more or less continually around the clock.

But retaining the eggs internally poses certain problems. The body of the offspring is foreign tissue, genetically different from that of the mother, and it is likely to be attacked by her immune system if she keeps it inside her too long. An embryo developing inside the mother also faces respiratory problems, since there is not much oxygen available in her oviducts. Divergent solutions to these problems have evolved in the two clades of living mammals that practice live birth. The pouched mammals or **marsupials**, including opossums and kangaroos, manage by keeping intra-uterine life very short. Their tiny babies emerge from the womb in a semi-embryonic condition (lacking hind limbs and fully mammalian ear apparatus) and crawl up the mother's belly to reach the pouch, where they fasten onto a nipple to drink milk and finish their development. The largest extant marsupial, the red kangaroo, which may grow to weigh as much as a human adult, weighs only about three-quarters of a gram at birth (Sharman and Pilton 1964).

The other viviparous group of living mammals is the **eutherians**, the group to which humans and other primates belong. Fetal eutherians complete more of their development while inside the womb than marsupials do. The eutherian fetus is able to do this because it has evolved ways of tapping into the mother's bloodstream to get food and oxygen and to unload CO_2 and other metabolic wastes. The mother's blood does not enter directly into the fetus's own arteries and veins; that would provoke a massive immune response from the mother and result in the death of her offspring. Instead, a special organ called the **placenta**, formed partly from the tissues of the fetus and partly from those of the mother, develops in the lining of the uterus and mediates the contact between maternal and fetal blood. Eutherian mammals are sometimes referred to as "placental mammals," after the placenta. (Various sorts of transient placental structures are also developed briefly in marsupials.) At birth, the eutherian placenta is expelled along with the fetus and the leash of blood vessels, the **umbilical cord**, that connects them to each other. The mother then bites through the cord and thriftily eats the placenta.

The placental not-quite-contact between the maternal and fetal bloodstreams differs in intimacy in different groups of eutherian mammals. In the least intimate arrangement, the outermost layer of fetal tissues (the chorion and its derivatives) makes contact only with the membrane (epithelium) lining the inside of the uterus. This **epitheliochorial** (epithelium + chorion) type of placenta is found in a wide range of living eutherians, including pangolins, dugongs, hoofed mammals (including whales), and relatively primitive primates such as the lemurs of Madagascar. In the so-called "higher" or **anthropoid** primates—monkeys, apes, and people—the contact between maternal and fetal blood is more intimate. The tissues of the uterus break down completely where they touch the chorion, so that the outermost embryonic membranes come to be directly bathed in small pools of the mother's blood. This **hemochorial** type of placenta (Greek *haima*, "blood" + chorion) is also found in a wide range of eutherian groups, including manatees, hyraxes, armadillos, rodents, and such relatively primitive mammals as shrews and hedgehogs.

At least one of these two placental types must have evolved several times in parallel. Until recently, it has generally been thought that the epitheliochorial placenta was primitive, because all placentation begins with a superficial contact between maternal and fetal tissues, and because a similar minimal contact must have characterized the earliest stages in the evolution of the placenta (Luckett 1975, 1976, 1977). It has also been argued that the hemochorial type of placenta must be the more advanced or derived type because it facilitates a more efficient exchange of nutrients and wastes across the placenta. Some have suggested that the efficiency of the hemochorial placenta was an important factor in the evolution of the higher primates, because it allowed for faster prenatal brain growth and thus permitted the evolution of exceptionally large brains in large anthropoids with long gestation periods (Elliot and Crespi 2008). The main problem with this argument is that dolphins, the other group of big mammals with relatively huge brains (Fig. 3.9), have an epitheliochorial placenta.

The hemochorial placenta has certain disadvantages. It causes the mother to bleed—sometimes fatally—when the placenta detaches from the uterus at birth, and it exposes the developing fetus to more direct challenges from the mother's immunological defenses. What, then, could explain its parallel evolution from the epitheliochorial type in various groups of mammals?

Inclusive-fitness theory suggests that there is a delicately balanced competition between the mother and the developing fetus (Haig 1993, Crespi and Semeniuk 2004). In some sense, each of them needs the other. The fetus will die if the mother dies; and from an evolutionary standpoint, the mother will have lived in vain if she has no offspring. But their interests are not identical. The mother's own life is worth more to her in evolutionary terms than the life of any one of her children. A female mammal can sometimes increase her lifetime reproductive success by compromising or sacrificing her present offspring to improve her own chances of having more offspring in the future. For example, although mammalian mothers will generally put up a fight to defend a baby from predators, few mammalian mothers will fight to the death to do so—and any genes that promote such potentially suicidal mother-love will not be conserved by natural selection. A pregnant mammal may even spontaneously abort or resorb her own fetus when times are hard and food is in short supply. By doing so, she avoids wasting precious resources on an offspring that would have had a reduced likelihood of surviving to transmit her genes to succeeding generations.

Conversely, the fetus's life is worth more to the fetus than is the life of any one of its mother's future offspring. On the average, it will share only 25% of its variable genes with one of its future half-siblings (if begotten by a different male). Selection will therefore tend to favor any fetal adaptations that promote the survival of a fetus at the cost of the survival of up to three of its mother's future offspring. The fetus can thus sometimes increase its selective fitness by pre-empting maternal time, nutrients, and energy that the mother might otherwise have invested in the production of three later babies. By doing so, it increases its own reproductive chances at the expense of its mother's. The hemochorial placenta may have originated as an essentially parasitic adaptation, which allows a developing fetus to make more efficient use of the resources of the mother's body by evading some of her immunological defenses and gaining a degree of control over her tissues through the secretion of fetal hormones. If so, it might have evolved many times in parallel because its selective advantage to the fetus sufficiently outweighed the disadvantages to the mother.

However, this argument cuts both ways. The epitheliochorial placenta can be interpreted as the derived state, evolved in several different groups as a *maternal* defense against invasion and hormonal manipulation by the fetus. It has recently been argued that when the distribution of placental types is mapped onto cladograms generated by molecular data, the most parsimonious conclusion is that the hemochorial placenta is a eutherian symplesiomorphy, and that an epitheliochorial placenta has evolved three or four times in parallel (Vogel 2004, Wildman et al. 2006). As we will see, this issue has implications for our understanding of primate phylogeny.

CRETACEOUS MAMMALS

Because early tribosphenic molars tend to look much alike and Mesozoic mammal fossils are generally fragmentary, experts disagree about which fossils represent eutherians. The earliest mammal generally recognized as a eutherian is *Eomaia scansoria*, from Lower Cretaceous deposits of China dated to around 125 Mya (Ji et al., 2002). *Eomaia* is known from a wonderfully complete skeleton (Fig. 3.14), including even an impression of its fur. In life, it probably would have looked something like a large mouse—until it opened its mouth to reveal a set of teeth resembling those of an opossum, not those of a rodent.

The upper dental battery of *Eomaia* included five peg-like incisors, one slender canine fang, five blade-like premolars, and three molars on each side. The count in the lower jaw was similar, but with only four incisors (Fig. 3.15). Anatomists call this tally the **dental formula**,

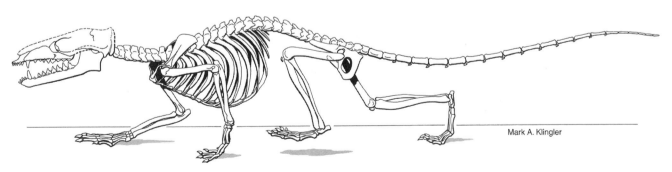

FIGURE 3.14

Skeleton of *Eomaia scansoria*. (Reconstruction and drawing by Mark A. Klingler. Reprinted, with permission, from Ji et al. 2002.)

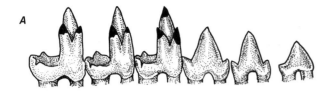

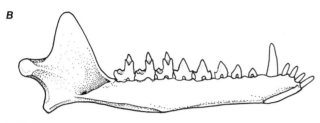

FIGURE 3.15

Dentition of *Eomaia scansoria*; medial views of lower P3-M3 (**A**) and mandible with complete dentition (**B**). (After Ji et al. 2002.)

and they write it like a fraction: 5.1.5.3/4.1.5.3. *Eomaia's* dental formula appears to be the primitive one for eutherian mammals. With rare exceptions, evolutionary changes in the dental formula involve the loss of teeth, so almost all living eutherians have smaller numbers in their dental formulas than their ancestors had. For example, the dental formula of the earliest primates was 2.1.4.3/2.1.4.3, seven teeth shy of *Eomaia's* complement on each side. That of their descendant *Homo sapiens* is only 2.1.2.3/2.1.2.3 (Fig. 3.10). Thus, in the course of our evolution, a total of 22 teeth were lost from the ancestral eutherian dental battery.

The molars of *Eomaia* were sharp, insect-slicing teeth of a primitive tribosphenic sort, with trigonids that towered high above the talonids (Fig. 3.15A). *Eomaia* had a long snout, and its orbits were small and laterally oriented. All this suggests that it retained the ancestral mammals' way of making a living, sniffing around on the ground at night and poking its nose into crevices and leaf litter in search of insect prey. Its relatively long, slender digits hint that it had some climbing ability as well. However, there is no reason to think that it was any more arboreal than (say) a chipmunk. All small forest-dwelling animals need some climbing ability, if only to navigate the fallen logs and sticks that cover the forest floor (Jenkins 1974). The fossil evidence indicates that *Eomaia* still preserved a way of life much like that of the ancestral mammals.

Later Cretaceous eutherians started to diversify and radiate into new ecological territory. Experts disagree about the time at which the major groups of eutherian mammals began to diverge from each other. Most of the extant eutherian orders, including rodents, bats, carnivores, insectivores, and our own order Primates, do not appear in the fossil record until after the departure of the dinosaurs at the end of the Cretaceous. But fragmentary fossils from the Late Cretaceous of Uzbekistan appear to represent primitive hoofed mammals (Archibald 1996, Cifelli 2001). If so, then the adaptive radiations of other eutherian groups must have been well under way at that time. Estimates using "molecular clocks" derived from genetic data suggest that the major groups of eutherians separated from each other even further back in the Cretaceous (Bromham et al. 1999, Huchon et al. 2002), well before any of them shows up in fossil form. This seems plausible, since the first species of any higher taxon is unlikely to be preserved as a fossil. We would therefore expect some time to elapse between the origin of a taxon and the appearance of its earliest fossil representative. The amount of time that can be expected to elapse, however, is a matter of debate (Martin 1993, Gingerich and Uhen 1994, Tavaré et al. 2002).

From a broad ecological standpoint, the most important Cretaceous radiation was not that of the eutherian mammals, but that of the modern flowering plants, the **angiosperms**. In the late Jurassic, when the first angiosperms appear in the fossil record (Sun et al. 1998), the world's forests were made up chiefly of conifers, cycads, ferns, and other non-flowering plants with relatively inedible tissues. By the end of the Cretaceous, angiosperms had come to dominate most terrestrial floras.

One reason for the angiosperms' success was that they evolved ways of co-opting animals into helping them reproduce, by offering little dabs of food as a bribe. The flowers of many angiosperms are baited with bright colors, sweet perfumes, and sugary nectars. The bait entices insects and other animals into entering male flowers, picking up a dusting of sperm-bearing pollen, and carrying it around to the female flowers of neighboring plants. Many angiosperms also produce tasty fruits containing small, hard seeds built to survive passage through vertebrate guts. Animals that eat the fruits carry the seeds away and deposit them later in their droppings, thus helping to disperse the plant's offspring.

Angiosperms evolved flowers and fruits for their own benefit. But these reproductive adaptations also afforded a lot of ecological opportunities for animals. The major beneficiaries of the angiosperm radiation were the insects. Bees evolved out of wasps during the Cretaceous, taking advantage of the new niches opening up for pollen- and nectar-eaters (Engel 2000). A host of other insect lineages, including moths and butterflies, swiftly developed adaptations for feeding on these and other angiosperm tissues (Grimaldi 1999).

Cretaceous mammals, too, began exploiting the new vegetarian opportunities that were opened up by the radiation of the angiosperms. The first major group of herbivorous mammals to take advantage of these opportunities was the **multituberculates**. The large front teeth and multicusped molars of these extinct creatures give them a superficial resemblance to rodents, and many of them probably resembled rats and mice in their general appearance and diets. But they were far less closely related to rodents than *Eomaia* was, or than even a platypus is (Cifelli 2001). They represent a wholly separate line of descent from the mammals of the Jurassic. Multituberculates radiated into a variety of adaptive types, from woodchuck-sized herbivores to squirrel-like tree-climbers with prehensile tails (Clemens and Kielan-Jaworowska 1979, Krause and Jenkins 1983). They peaked in diversity in the **Paleocene epoch** at the beginning of the Tertiary, declined during the following **Eocene** epoch, and disappeared in the early part of the next epoch, the **Oligocene**. Their decline coincided with a marked fall in global temperatures (Fig. 3.16), which wiped out subtropical forests in high latitudes

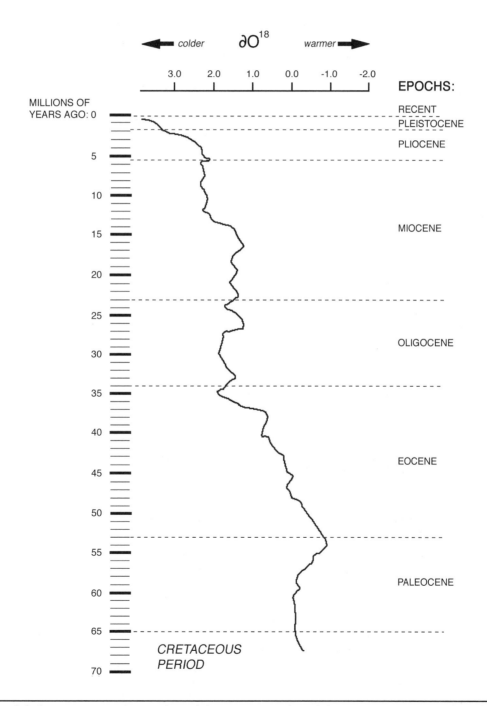

FIGURE 3.16

Fluctuations in global temperature during the subdivisions (epochs) of the Cenozoic era, as inferred from the ratios between oxygen-16 and oxygen-18 isotopes in the fossil shells of foraminifera. (After Hallam 2004.)

during the Oligocene and may have contributed to their extinction.

Marsupials also entered the trees during the Cretaceous. Stem-group marsupials are known from the early Cretaceous of China, about 125 Mya (Luo et al. 2003), and from mid-Cretaceous deposits of North America some 27 My later (Cifelli and de Muizon 1997). By the end of the Cretaceous, marsupials were abundant in North America and probably present in most of the other continents (Cifelli 2000, Krause 2001). Comparative anatomy indicates that the last common ancestor of the living marsupials had thumb-like big toes: divergent, grasping, and clawless. Its hindfoot probably looked much like that of a small, arboreal opossum. Similar feet have been convergently evolved in certain other lineages of climbing mammals, including our own primate ancestors. This sort of grasping foot is adaptive for moving around on twigs and other relatively slender supports that are too small for a climbing animal to anchor itself by digging in its claws (Cartmill 1972, 1992; Rasmussen 1990). The ancestral crown-group marsupials, like many small marsupials today, were probably arboreal creatures that moved and foraged in the terminal branches of trees (Cartmill et al. 2002, 2007).

The tree-climbing multituberculates of the Cretaceous appear to have entered the trees to feed on the new plant tissues—fruit, pollen, nectar, and so on—that the angiosperm radiation made available. The ancestral marsupials were also arboreal, but not primarily herbivorous. The teeth of most Cretaceous marsupials are of a persistently primitive, bug-crunching tribosphenic type, with tall trigonids and small talonids on the lower molars and a wide, crest-bearing stylar shelf on the uppers. These features resemble their equivalents in early eutherians (Fig. 3.15) and in insect-eating marsupials of today. They contrast with the low trigonids, expanded talonids, and reduced stylar shelves seen in extant fruit-eating marsupials and eutherians. It looks as though the first marsupials took to the trees chiefly to exploit the insects that abounded in the blossoming branches of the new angiosperm forests of the Cretaceous. As we will see, some think this was also true of the ancestral primates.

THE ORDER PRIMATES

There are over 200 species of living primates. Most primates are largely or exclusively arboreal animals, which spend their time feeding, moving, and sleeping in trees in tropical and subtropical forests of Asia, Africa, and the Americas. Despite being largely restricted to low-latitude forests, primates have been a moderately successful group of mammals. Unfortunately, the long-term survival of most primate species is now in doubt because one atypical primate that has adapted to life on the ground, *Homo sapiens*, is busily cutting down the world's tropical forests to make way for the more open country that it finds congenial and useful.

Primates are united by a number of derived features that were not present in the early eutherians. The most obvious of these distinctive primate synapomorphies are traits of the hands, feet, and visual apparatus (Fig. 3.17).

1. **Primate Hands and Feet.** The palms and soles of mammals serve the same sort of function in locomotion that pneumatic tires serve in the movement of a car: they provide a tough, elastic friction surface that enhances traction and helps to absorb the shocks of the road. In primitive eutherian mammals, the animals' weight is borne by hairless skin, stretched over bulging **pads** of specialized fatty connective tissue that protrude from the palms, soles, and tips of the digits (Fig. 3.17C). The hairless skin provides traction, while the fatty pads act as shock absorbers. In the ancestral eutherians, all 20 fingers and toes were tipped with pointed cylindrical **claws**, used in digging, grooming, and climbing. The claws of the hand also helped to secure and hold insects and other food items.

In the ancestral primates, the hind feet were remodeled to serve as prehensile organs, adapted to grasping the branches of trees. In almost all primates, the first digit of the foot—the "big toe" or **hallux**—is furnished with strong muscles and sticks out at a marked angle to the other toes, so that it can be opposed to them in a powerful grip. Some primate lineages, including our own, have developed a similarly divergent and opposable thumb as well (Fig. 3.17D). But in most nonhuman primates, the foot is even more specialized for grasping than the hand is. In all primates, the claw of the hallux has been flattened out into a **nail**. Most primates have flattened nails on their other digits as well. The fatty pads of the palms and soles are also flattened out and more or less fused together in primates, forming a soft, pliable, high-friction surface covered with curving "fingerprint" ridges.

2. **Primate Eyes and Orbits.** Compared with other mammals, primates have unusually keen vision. As noted earlier, the ancestral mammals evidently had small eyes and a dim sense of sight. Primates, however, have moved back toward the sharper-eyed reptilian condition, re-evolving large eyeballs, color vision, and mechanisms for focusing the retinal image. The visual centers of the brain are big and complicated in even the most primitive living primates (Fig. 3.17B); and in the so-called "higher" primates (monkeys, apes, and humans), they attain levels of size and complexity not achieved by any other mammals (Allman 2000, Kirk 2003).

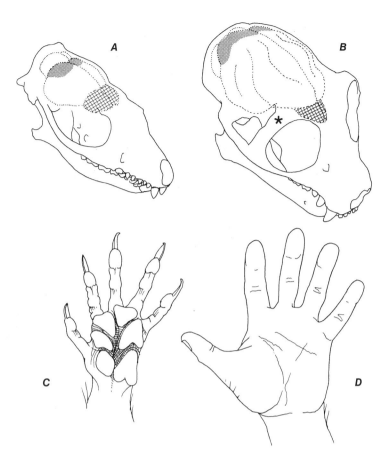

FIGURE 3.17

Major adaptive differences between primates and more primitive mammals. **A**: Skull of a primitive mammal, the hedgehog *Erinaceus*. The brain (*dotted outline*) is small, the visual part of the cortex (*stipple*) is restricted, and the olfactory bulbs (*hachure*) are large. The eye sockets face as much sideways as forwards, and are not encircled in bone. **B**: Skull of the strepsirrhine primate *Otolemur*. The brain is enlarged, the visual cortex is expanded, and the olfactory bulbs are reduced. The eye sockets are larger and face more toward the front of the head than those of *Erinaceus*, and they are enclosed by a complete bony ring (*asterisk*). **C**: Hand of the tree shrew *Tupaia*. All five digits are about equally divergent, and all are tipped with long, pointed claws. The palm bears separate, protruding pads like those on the feet of a dog. **D**: Hand of a primate (*Homo*). The palmar pads are fused together into a broad, soft surface. The first digit (thumb) diverges widely from the other four, and all digits are tipped with flattened, shieldlike nails. (From Cartmill 1992.)

In most nonprimate mammals, the eyes point in different directions, and each eye socket faces more or less toward its own side of the head. In all primates, both eyes point in the same direction, so that the sector of the world seen by the left eye overlaps widely with that seen by the right. This parallel orientation of the two eyes is reflected in the eye sockets or **orbits** of the primate skull (Heesy 2004), which are likewise rotated around to face more anteriorly. The eyes and orbits of primates also tend to be drawn together toward the middle of the face, compressing the organs of smell in between them. (In some small primates, the two eye sockets actually touch each other in the midline.) The lateral edge of the eye socket, which is a band of tough fiber in most mammals, becomes a bony **postorbital**

bar in primates, so that a primate's eye is completely surrounded by a ring of bone (Fig. 3.17A,B).

3. **The Primate Ear.** The **petrosal** bone forms a dense protective capsule around the mammalian cochlea and semicircular canals. In primitive mammals, the eardrum lay in a nearly horizontal position underneath the petrosal, and the **tympanic** or middle-ear **cavity** was a narrow, air-filled space intervening between the bone and the eardrum (Fig. 3.18C). Most mammals have evolved away from this primitive arrangement by enlarging the tympanic cavity. The inflated tympanic cavity is almost always enclosed in a rigid shell of bone, to prevent it from collapsing under the pressure of surrounding tissues. This bony shell is called the **auditory**

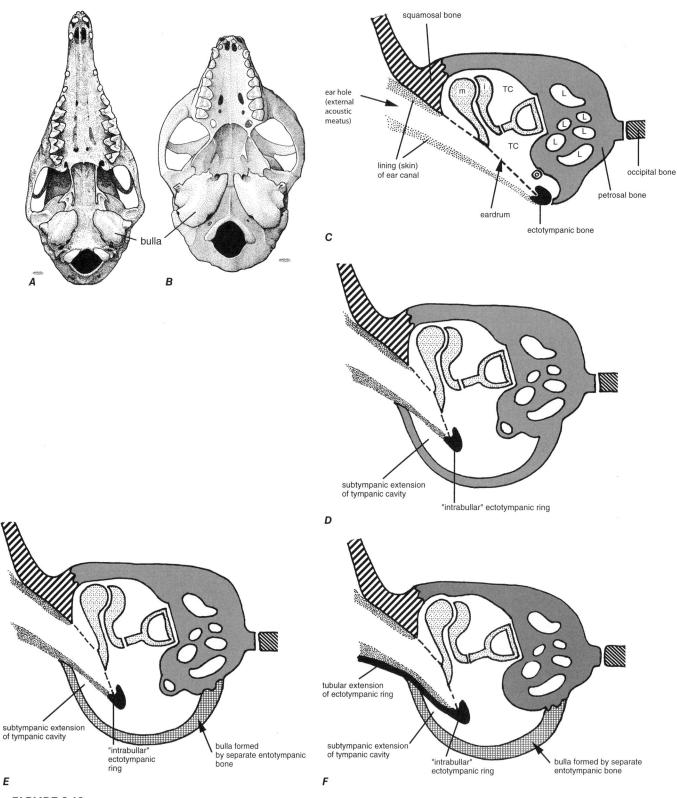

FIGURE 3.18

The auditory bulla. **A,B**: Underside of the skull of the tree shrew *Tupaia tana* (**A**) and the mouse lemur *Microcebus murinus* (**B**), showing the entotympanic bulla of **A** and the similar-looking petrosal bulla of **B**. (Not to same scale.) **C–F**: Schematic cross (coronal) sections through the ear regions of a primitive mammal (**C**), a lemur (**D**), a tree shrew (**E**), and a plesiadapiform (**F**). Abbreviations: i, incus; L, bony labyrinth; m, malleus; TC, tympanic cavity.

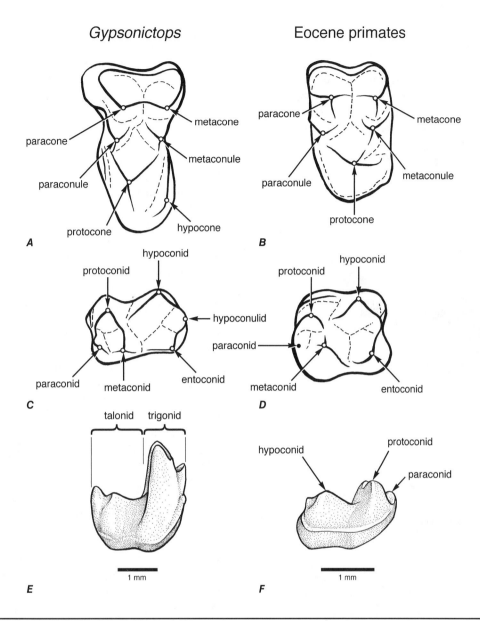

Gypsonictops   Eocene primates

FIGURE 3.19

Primate molar synapomorphies. **Top row**: Diagrammatic occlusal views of the upper left second molars of the Cretaceous eutherian *Gypsonictops hypoconus*, illustrating primitive eutherian morphology, and the Eocene omomyoid primate *Tetonoides*, illustrating primitive euprimate morphology. **Middle row**: Lower right M2s of the same species. **Bottom row**: Lateral views of the lower right M2s of *Gypsonictops* and the primitive Eocene omomyoid *Teilhardina belgica* (**F**), illustrating the reduction of trigonid height in early primates. (**A,B**, after Kay and Hiiemae 1974; **C,D**, after Kay 1977; **E**, after Lillegraven 1969; **F**, after Szalay 1976.)

bulla. In most small mammal skulls, the bullae protrude conspicuously from the underside of the skull like a pair of big bony bubbles (Fig. 3.18A,B). Primates are unique among mammals in having a bulla that forms entirely as an outgrowth of the petrosal bone (Fig. 3.18D). In other mammals, a bulla is formed by outgrowths from other bones of the cranial base, or by one or more special bulla bones called **entotympanics** (Fig. 3.18E).

4. **Primate Molars and Diets.** Most living primates feed mainly on plant materials. This habit is reflected in their molar teeth. In primates, the W-shaped zigzag of crests borne on the stylar shelf of the upper molars is substantially reduced by comparison with primitive eutherian mammals (Fig. 3.19A,B). In the lower molars, the trigonid is lowered relative to the surface of the talonid, and its front cusp, the **paraconid**, is reduced (Fig. 3.19C–E: cf. Fig. 3.15A,B). These changes have the effect of enhancing the crushing functions of the tribosphenic molar at the expense of its piercing and slicing functions. Similar changes have taken place in parallel in many lineages of plant-eating mammals. They are

seen in most living primates, and they appear to be a shared heritage from the primate ancestry (Szalay 1968, 1972).

The degree to which these changes are expressed in a primate species depends on its diet. Those primates that feed to a significant extent on insects and other animal prey have molars with sharper cusps, higher trigonids, and more shearing surfaces than those of related species with diets centered on fruit or gums (Kay 1984; Fig. 3.20). The molars of primates that specialize in

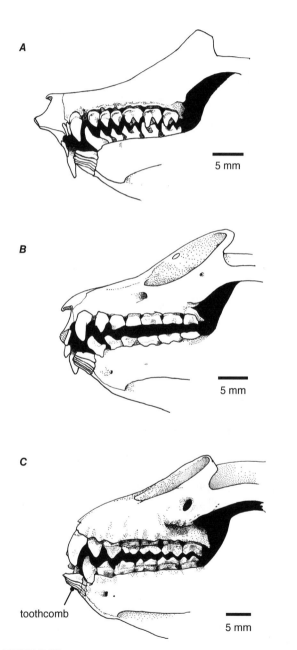

FIGURE 3.20

Dentitions of three strepsirrhine primates. **A**: An insect-eater, *Loris tardigradus*. **B**: A fruit-eater, *Cheirogaleus medius*. **C**: A leaf-eater, *Hapalemur griseus*. (After James 1960.)

eating leaves are somewhat intermediate, with moderately tall cusps and moderately enhanced shearing functions. Leaf-eaters, like insect-eaters, need effective molar shearing, because leaves and insects have some structural properties in common. Both leaves and insects have soft, digestible stuff on the inside, encased in tough, flexible envelopes of less digestible polymers (cellulose or chitin). These polymer casings need to be first punctured to release their contents, and then chopped into tiny bits to increase their surface area and speed their enzymatic and bacterial digestion in the intestines.

Despite the functional and structural similarities between the molars of leaf-eaters and insect-eaters, we can still tell the teeth of leaf-eating, insect-eating, and fruit-eating primates apart. The flat-topped crushing molars of fruit-eaters are the easiest to spot. Leaf- and insect-eaters overlap in morphology, but they differ in body size. Insect-eaters have to be small animals, because insects are small and seldom occur in large quantities in a single place. Apart from a few specialized termite-eating animals—aardvarks, anteaters, pangolins, and the like—big mammals cannot catch enough insects to rely on them as a dietary staple. Leaf-eating mammals, on the other hand, have to be fairly big. They need to have long gut passage times, to give their gut bacteria time to digest their fibrous food during its journey through the intestines; and they need to have relatively low metabolic rates, because leaves are a low-energy food. Small mammals, which have quick gut passage times and high metabolic rates, therefore cannot live on leaves. (Some tiny rodents with exceptionally efficient shearing teeth are exceptions to this rule.) All primates that rely on leaves as their main source of energy or protein have average body weights of 800 grams or more, whereas most primates that feed chiefly on insects have weights of 200 grams or less. The 500-gram midpoint of this gap dividing the two groups is sometimes known as **Kay's Threshold**, after its discoverer (Kay 1975, 1984).

5. Primate Reproduction. Primitive eutherian mammals bore litters of altricial young—blind, pink, and helpless, like newborn dogs or rats. Primate babies are not altricial, but **precocial**: that is, they are born with their eyes and ears open, and with skin that is already pigmented and hairy. Primates also have much smaller litters than primitive mammals had. In humans and most other primate species, a single baby is the norm. This is why almost all female primates have only two nipples, instead of the 8–12 that we see on a bitch or the 6–8 found on a female shrew.

These differences reflect a difference in reproductive strategy. Dogs, shrews, and other mammals that bear altricial babies typically have short gestation periods and low birth weights for their adult body size. They can thus turn out offspring at a high rate, because they make a relatively small investment in each individual baby (Martin 1990). This strategy allows their populations to expand rapidly when circumstances permit. However, their blind, pink, helpless, and extremely edible young need to be protected from predators—say, by being concealed in a nest or burrow, or by being carried around in a pouch. If a mammal cannot protect her babies in these ways, she cannot manage to have big litters of blind, pink babies. She has to bear a smaller number of precocial young, which are less helpless and can follow her around from the moment of birth. Mammals that bear such young include most eutherian herbivores that cannot build protected nests—for instance, hoofed mammals and elephants. Baby whales and dolphins also have to be precocial, because they have to be able to swim from the moment of birth to keep from drowning.

Primates follow an in-between sort of reproductive strategy. Although primate newborns are precocial, they are at first too feeble and ill-coordinated to keep up with the mother in fleeing from predators. Most primate mothers solve this problem by carrying their babies around with them. The newborn uses its grasping hands and feet to cling to the fur of its mother, leaving the mother's own hands and feet free for running, feeding, or fighting. The baby-carrying task is unusually burdensome for human mothers, because the human newborn is unable to cling to its mother by itself. To get food, a woman who is burdened with an infant must therefore leave her baby in the care of other adults while she forages, or carry it around in her own arms (or in some technological substitute like a backpack)—or else rely on someone else to bring food to her and the baby. As we will see, this has implications for human social organization.

THE LIVING STREPSIRRHINES

Almost all of the living primates fall into one of two major clades. One is the higher primates or **Anthropoidea**, which comprises monkeys, apes, and human beings. The other group, the toothcomb primates or **Strepsirrhini**, comprises the lemurs of Madagascar and the lorises and galagos of tropical Africa and southern Asia. If all primates were either anthropoids or strepsirrhines, primate classification would be simple. Unfortunately, one genus of living primates, *Tarsius*, does not fit comfortably into either major clade. Some people group it with the anthropoids in a suborder **Haplorhini** (Table 3.1, System A). Others group *Tarsius* and the strepsirrhines together in a suborder **Prosimii** or prosimians (Table 3.1, System B).

The strepsirrhines are distinguished from the anthropoids mainly by having retained a lot of primitive euthe-

rian features that have been lost or modified in anthropoids. The primitive character of strepsirrhines is apparent in their faces (Fig. 3.21). Like that of a dog, the snout of a lemur is tipped with a moist tactile area, the **rhinarium**, surrounded with rows of sensory vibrissae. Strepsirrhine snouts are long and filled with typical mammalian nose equipment (Fig. 3.12), including convoluted maxilloturbinals for air-conditioning and rows of ethmoturbinals bearing an expansive array of olfactory receptors. The olfactory bulbs in a typical lemur's brain are nearly as big as those of other arboreal mammals, and its brain is no larger than those of many other modern mammals of similar size (Jerison 1973).

These shared primitive retentions (symplesiomorphies) provide no grounds for grouping all strepsirrhines together as a clade. However, the living strepsirrhines share a few nonprimitive traits as well. One such trait may be their epitheliochorial placenta, if this is interpreted as non-primitive among eutherians (Wildman et al. 2006). Another apparent synapomorphy of this group is the reshaping of their lower incisors and canines into a narrow, comb-like array of long, slim teeth sticking directly forward out of the front of the lower jaw (Fig. 3.20). This so-called **toothcomb** is used like a comb in grooming the fur. Some strepsirrhines also use jabbing movements of the toothcomb in feeding—for example, in prising edible gummy secretions off of tree trunks. A third possible synapomorphy of the living strepsirrhines is the presence of a claw (rather than a nail) on the second toe of the hind foot, which is used for scratching. However, this **toilet claw** is also found in tarsiers.

Most living strepsirrhine species are restricted to the island of Madagascar off the southeastern coast of Africa. Like other isolated land masses, Madagascar has a peculiar native fauna, in which a small number of taxa have radiated to fill a large number of ecological roles. The ecological niches for large herbivores were filled on Madagascar by the evolution of gigantic flightless birds and big lemurs. Some of these extinct lemurs weighed up to 200 kilograms—as much as an adult male gorilla (Godfrey and Jungers 2003).

Much of Madagascar's unique wildlife disappeared abruptly a few thousand years ago, following the first human settlement of the island (Godfrey and Jungers 2002). However, the surviving fauna still includes a wide variety of strepsirrhine primates, which have evolved to fill niches occupied elsewhere in the Old World tropics by monkeys and apes. The living primates of Madagascar are known collectively as **lemurs** and are grouped together as the infraorder **Lemuriformes** (Table 3.1). They fall into five main clades, often classified as families. Typical **Lemuridae**, in the genera *Lemur, Eulemur,* and *Varecia*, are medium-sized herbivores (1.0–3.6 kg) with long tails and long, foxy faces (Fig. 3.21). The lemurid family includes diurnal animals,

TABLE 3.1 ■ A Partial Classification of the Crown-Group Primates (Euprimates)

SUBORDERS (SYSTEM A)	SUBORDERS (SYSTEM B)	INFRAORDER	SUPERFAMILY	FAMILY	GENERA
STREPSIRHINI	PROSIMII	ADAPIFORMES	ADAPOIDEA	Adapidae	†Adapis, †Adapoides, †Cryptadapis, †Leptadapis, †Palaeolemur (etc.)
				Sivaladapidae	†Sivaladapis, †Guangxilemur, †Indraloris, †Sinoadapis (etc.)
				Notharctidae	Notharctinae: †Notharctus, †Cantius, †Pelycodus, †Smilodectes (etc.)
					Cercamoniinae: †Cercamonius, †Aframonius, †Donrussellia, †Periconodon (etc.)
		LEMURIFORMES (toothcomb prosimians)	PLESIOPITHECOIDEA?	Plesiopithecidae	†Plesiopithecus
			LEMUROIDEA	Megaladapidae	†Megaladapis
				Palaeopropithecidae	†Palaeopropithecus, †Mesopropithecus, †Babakotia, †Archaeoindris
				Archaeolemuridae	†Archaeolemur, †Hadropithecus
				Lemuridae	Lemur, Hapalemur, Eulemur, Varecia, †Pachylemur
				Cheirogaleidae	Cheirogaleus, Microcebus, Mirza, Allocebus
				Lepilemuridae	Lepilemur
				Indriidae	Indri, Propithecus, Avahi
				Daubentoniidae	Daubentonia
			LORISOIDEA	Lorisidae	Loris, Nycticebus, Perodicticus, Arctocebus, †Karanisia (etc.)
				Galagidae	Galago, Galagoides, Euoticus, Otolemur, †Saharagalago (etc.)
		TARSIIFORMES	OMOMYOIDEA	Omomyidae	Omomyinae: †Omomys, †Shoshonius, †Ekgmowechashala? (etc.)
					Anaptomorphinae: †Teilhardina, †Tetonius, †Anaptomorphus (etc.)
					Microchoerinae: †Necrolemur, †Microchoerus, †Nannopithex, †Pseudoloris
			TARSIOIDEA	Tarsiidae	Tarsius, †Xanthorhysis? †Afrotarsius?
HAPLORHINI	ANTHROPOIDEA	(BASAL AND PUTATIVE ANTHROPOIDS OF UNCERTAIN AFFINITIES)		Proteopithecidae	†Proteopithecus, †Serapia
				Parapithecidae	†Parapithecus, †Apidium, †Qatrania, †Biretia (etc.)
				Amphipithecidae?	†Amphipithecus, †Pondaungia, †Siamopithecus, †Myanmarpithecus?
				Eosimiidae?	†Eosimias, †Phenacopithecus?
		PLATYRRHINI (New World monkeys)	CEBOIDEA	Callitrichidae	Callithrix, Saguinus, Callimico (etc.)
				Cebidae	Cebus, Saimiri
				Atelidae	Ateles, Brachyteles, Lagothrix, Alouatta, †Protopithecus (etc.)
				Aotidae	Aotus, †Tremacebus (etc.)
				Pitheciidae	Pithecia, Cacajao, Callicebus, †Homunculus (etc.)
		CATARRHINI (Old World monkeys, apes, humans)	PROPLIOPITHECOIDEA	Oligopithecidae?	†Catopithecus, †Oligopithecus
				Propliopithecidae	†Propliopithecus, †Aegyptopithecus, †Moeripithecus
			PLIOPITHECOIDEA	Pliopithecidae	†Pliopithecus, †Epipliopithecus, †Laccopithecus, †Crouzelia, †Platodontopithecus, †Dionysopithecus, †Egarapithecus, †Plesiopliopithecus, †Anapithecus
			PROCONSULOIDEA	Proconsulidae	†Proconsul, †Afropithecus, †Kenyapithecus, †Nyanzapithecus, †Turkanapithecus, †Mabokopithecus, †Rangwapithecus, †Kamoyapithecus, †Griphopithecus? †Nacholapithecus?
				Sugrivapithecidae*	†Sivapithecus, †Gigantopithecus, †Ankarapithecus, †Lufengpithecus?
				Dendropithecidae	†Dendropithecus, †Micropithecus, †Simiolus
			HOMINOIDEA	Morotopithecidae?	†Morotopithecus
				Oreopithecidae?	†Oreopithecus
				Dryopithecidae?	†Dryopithecus, †Pierolapithecus
				Hylobatidae?	Hylobates, Symphalangus
				Pongidae?	Pongo, †Khoratpithecus?
				Hominidae	Homininae: Pan, Homo, †Australopithecus, †Orrorin, †Ardipithecus, †Sahelanthropus, †Otavipithecus? †Samburupithecus?
					Gorillinae: Gorilla
			CERCOPITHECOIDEA	Victoriapithecidae	†Victoriapithecus, †Prohylobates
				Cercopithecidae	Cercopithecus, Miopithecus, Allenopithecus, Chlorocebus, Cercocebus, Erythrocebus, Papio, Theropithecus, Macaca (etc.)
				Colobidae	Colobus, Presbytis, Semnopithecus, Rhinopithecus, Nasalis (etc.)

Gray tone denotes extinct groups; daggers (†) indicate extinct genera. Question marks signify uncertainty concerning placement of taxa in the next higher grouping.

FIGURE 3.21

The face of a lemurid (*Varecia* juvenile).

nocturnal animals, and some species that are active at all hours. Most of them live in small family groups up in the trees. However, *Lemur catta*, a stripy-tailed diurnal lemurid often seen in zoos, spends more time on the ground (15–20% of daily activity) and has larger social groups (up to 30 animals) than other living strepsirrhines (Jolly 1966, 2003). The dwarf and mouse lemurs in the family **Cheirogaleidae** are small (30–500 g) and largely solitary animals that prowl around at night through the fine branches of the forest canopy and marginal undergrowth, where they feed on fruit, insects, and plant secretions. The largest living lemurs (up to 9.5 kg) are found in the family **Indriidae**—long-legged jumpers with human-looking body proportions and with teeth and guts adapted for eating leaves. *Lepilemur*, a somewhat similar leaping folivore, is only distantly related to the indriids and is usually put into its own family **Lepilemuridae**.

The fifth lemur family, **Daubentoniidae**, is represented by a single genus and species—*Daubentonia madagascariensis*, the aye-aye, a rare and strange beast with rodent-like gnawing incisors instead of a toothcomb. It feeds on wood-boring grubs, which it detects with its huge, bat-like ears, uncovers by gnawing with its enlarged incisor teeth, and hooks out with an elongated, spidery middle finger. It thus occupies the ecological niche of the woodpeckers, which do not occur in Madagscar (Cartmill 1974a). Perhaps because of its unique feeding adaptations, the aye-aye has relatively the largest brain of any strepsirrhine, well within the monkey range (Stephan 1972).

The other living strepsirrhines inhabit continental Africa and Asia. They comprise two adaptive types: **lorises**, found in the forests of equatorial Africa and southeastern Asia, and **galagos** or bushbabies, found in forests and woodlands of sub-Saharan Africa. Lorises are stubby-tailed, short-eared animals with vise-like, powerfully gripping hands and feet. They move stealthily around in bushes and the fine branches of the forest canopy, feeding on fruits, insects, small vertebrates, and plant secretions. Galagos have a similar range of dietary habits, but they are more acrobatic, leaping animals. Their elongated ankle bones make their hind limbs look something like rabbit legs ending in monkey hands. Unlike lorises, galagos have long tails (useful as balancing organs in leaping) and big, mobile ears that can fold up like an accordion.

Lorises and galagos are grouped together as the infraorder **Lorisiformes**. They resemble each other, and differ from the Lemuriformes, in the construction of the ear region. In lemuriforms, the tympanic cavity expands sideways after birth, swelling out laterally below and past the lower edge of the eardrum. A lemur's eardrum and the ectotympanic bone that encircles it thus appear to be enclosed inside the bulla (Fig. 3.18D). In a lorisiform, the tympanic cavity is smaller and does not expand laterally, and so the eardrum and ectotympanic wind up being attached to the bulla's lateral edge (Fig. 3.22A). The small tympanic cavity in lorisiforms is augmented by two accessory air spaces linked to the main cavity: a mastoid cavity that expands into the back end of the petrosal, behind the eardrum, and a medial accessory cavity that expands into the medial edge of the petrosal. Another distinctive peculiarity of the lorisiform ear region is the loss of the internal carotid artery. It is replaced by an enlarged ascending pharyngeal artery, which passes around the medial edge of the petrosal to enter the braincase. A similar specialization is seen in the cheirogaleids of Madagascar (Cartmill 1975a, Yoder 1992).

ANTHROPOID APOMORPHIES: EARS, EYES, AND NOSES

Higher primates or anthropoids—monkeys, apes, and humans—have their own distinctive arrangement of the air-filled spaces of the middle ear. The anthropoid arrangement is rather like that of lorisiforms, with a mastoid cavity and an ectotympanic attached directly to the lateral edge of the bulla. However, anthropoids lack the medial accessory cavity. Instead, they have an **anterior accessory cavity**, which grows from the back end of the Eustachian tube out into the front edge of the petrosal bulla (Fig. 3.22F). The bony septum between the anterior accessory cavity and the main middle-ear cavity contains the internal carotid artery (MacPhee and Cartmill 1986). In most anthropoids (including humans), both the anterior and mastoid cavities branch as they grow, giving rise to masses of little bone-enclosed air cells that swell out the front and back ends of the petrosal.

Anthropoid eyes are also distinctive. There are two types of light-sensitive receptor cells in the vertebrate retina: rods and cones. Rods, which can detect light at lower levels than cones can, predominate in all mammalian retinas. Although cones are less sensitive, they come in several varieties that respond differently to different wavelengths (colors) of light. Cones therefore give their possessors color vision. The retinas of all anthropoids are mainly composed of rods out at the edges—which is why we cannot distinguish colors in peripheral vision (or in dim light, when our cones stop working). However, the retinas of almost all anthropoids have a small central area, the **fovea**, in which all the receptors are cones. Each cone in the fovea has its own separate neuronal "wire" running back to the brain. Images that fall on the fovea are therefore seen in color and in great detail.

No other mammals have a fovea of this type. It gives large anthropoids sharper vision than almost any other animals on the planet. The only organisms known to excel humans and apes in keenness of sight are eagles (Kirk 2003).

As if to compensate for their acute vision, anthropoids have a feebler sense of smell than most mammals. The olfactory parts of their brains are far smaller than those of primitive mammals and most strepsirrhines, relative both to body size and to brain size (Stephan and Andy 1969, Stephan 1972). Anthropoids have markedly reduced numbers of olfactory nerve fibers, ethmoturbinals, and square inches of olfactory receptor surface (Fig. 3.11D). Their nostrils are surrounded with ordinary skin instead of the doglike rhinarium seen in strepsirrhines. With their dull noses, sharp eyes, color vision, diurnal activity patterns, arboreal habits, noisy social groups, big brains, keen intelligence, and appetite for fruit, anthropoids have been characterized as mammals that have converged with parrots (Brace 1976).

The visual specializations of anthropoids are reflected in their skulls. As noted earlier, primates differ from primitive mammals in having a postorbital bar (Fig. 3.17). This stiff bony arch serves to keep the temporalis muscle's contractions from bending the protruding outer rim of the eye socket backwards (Cartmill 1972, Ravosa et al. 2000). A similar stabilizing bar has been evolved for similar reasons in other mammals whose

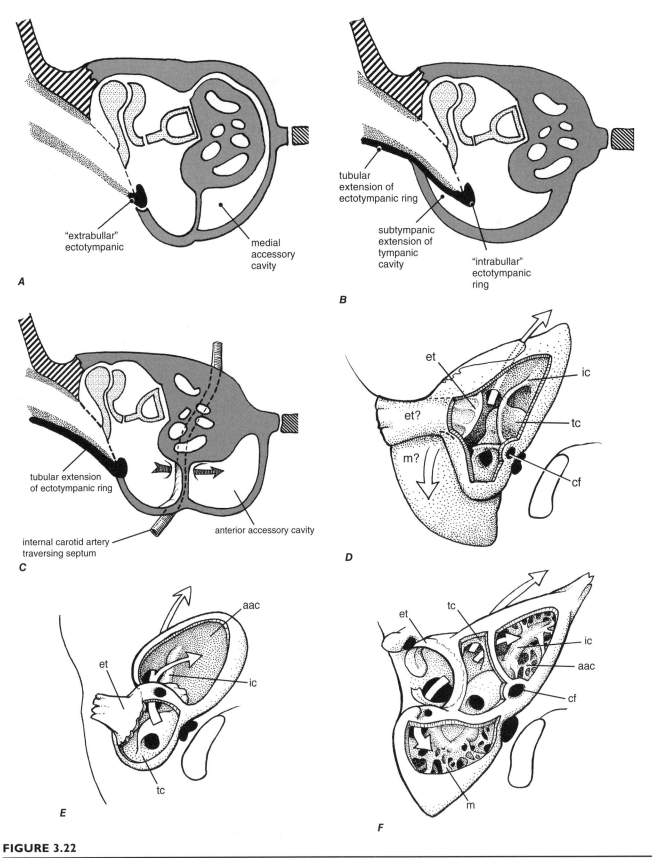

FIGURE 3.22

Primate ear regions (diagrammatic). **A–D**: Schematic cross sections through the middle ear cavities of (**A**) a lorisiform, (**B**) an omomyoid, and (**C**) a tarsier. **D–F**: Schematic ventral views of the right bullae of (**D**) an omomyoid, (**E**) a tarsier, and (**F**) an anthropoid, opened to show connections (*white arrows*) between air-filled spaces. The question marks in **D** indicate uncertainty about which bone forms the bony ear tube (external acoustic meatus) in omomyoids and how the mastoid air cells are connected to the middle ear. Abbreviations: aac, anterior accessory cavity; cf, carotid foramen; ic, internal carotid artery (in bony tube); m, mastoid air cells; tc, tympanic cavity. (**D–F**, after MacPhee and Cartmill 1986.)

eyes or orbits project from the surface of the head (Heesy 2005). Anthropoids have gone a step beyond this by developing a complete bony wall, the **postorbital septum**, between the eye socket and the temporal muscle (Fig. 2.12C). The septum prevents the contracting temporalis muscle from jiggling the eye socket's contents. Were it not for the septum, temporalis contractions would disrupt the acute vision that goes along with with the anthropoid fovea (Heesy et al. 2007). A similar but less complete septum is found in tarsiers, which also have a retinal fovea (Fig. 2.12B).

Among the lemurs of Madagascar, sexual dimorphism is minimal and females generally dominate the males, pushing them aside or driving them off whenever competition arises. But in most anthropoid species, the males are on average larger and more muscular than the females, and they tend to be dominant in social interactions. This sort of sexual dimorphism probably represents another anthropoid synapomorphy (Simons et al. 1999).

Living anthropoids share several other derived traits, including a need for vitamin C in the diet, a marked general increase in brain size, the fusing together of the two halves of the lower jaw in adults, and the reduction of some of the bony bumps on the femur and humerus (third trochanter, humeral epicondyles) for the attachment of limb muscles. But none of these traits can be counted as an anthropoid synapomorphy. The first one is shared with tarsiers as well, and the others were not yet present in the earliest fossil anthropoids (Simons 1990, 2004; Bush et al. 2004, Seiffert et al. 2004).

▌TARSIERS

The genus *Tarsius* comprises six species of small (80–160 g) nocturnal primates found on the islands surrounding the Celebes Sea (Sulawesi, Borneo, and the southeastern Philippines). Tarsiers have huge, bugged-out eyes, enlarged fingertip pads, and elongated hind limbs and ankle bones adapted for leaping, all of which gives them a vaguely frog-like appearance despite their long tails and big ears. Like frogs, tarsiers feed exclusively on insects and other animal prey. A tarsier catches its victims by jumping on them, seizing them with its long fingers, and biting them with its conical, canine-like incisor teeth (Fig. 3.23). The high, sharp cusps and shearing edges of tarsiers' molars reflect their insectivorous habits.

In most of these respects, tarsiers seem to be uniquely specialized. But other traits of tarsiers suggest a special relationship to anthropoid primates. This was first noted by the German anatomist A. Hubrecht (1897), who discovered that tarsiers resemble anthropoids in certain distinctive embryological features, including the reduction of the allantois to a vestige, the development of a

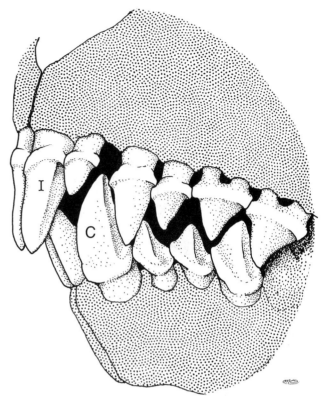

FIGURE 3.23

Antemolar dentition of *Tarsius*. C, lower canine; I, upper central incisor.

body stalk, and the formation of a hemochorial placenta with no participation by the yolk sac (Luckett 1975, 1976; Fig. 3.24). As noted above, the hemochorial placenta itself may be a primitive eutherian feature. However, some of the unique features of the placentation process shared by anthropoids and tarsiers appear to be synapomorphies (Cartmill 1994).

Tarsiers also share several of the distinctive anthropoid features of the sense organs. Apart from anthropoids, they are the only mammals with a postorbital septum (Fig. 2.12) and a retinal fovea (Ross 2004). Like anthropoids, tarsiers lack the reflecting layer in the retina, the **tapetum lucidum**, that makes the eyes of strepsirrhines shine brightly in the beam of a flashlight at night. The tapetum serves to enhance the light-gathering capacity of the eye, ensuring that any photons that are not absorbed by a retinal receptor cell on the way in bounce back and have a second chance to be detected on the way out. Tarsiers also resemble small anthropoids in having a large **interorbital septum**. This means that the left and right eye sockets of a tarsier contact each other in the midline, obliterating most of the space available for olfactory apparatus. As a result, the olfactory organs of tarsiers are much reduced, like those of anthropoids. Like anthropoids, tarsiers have

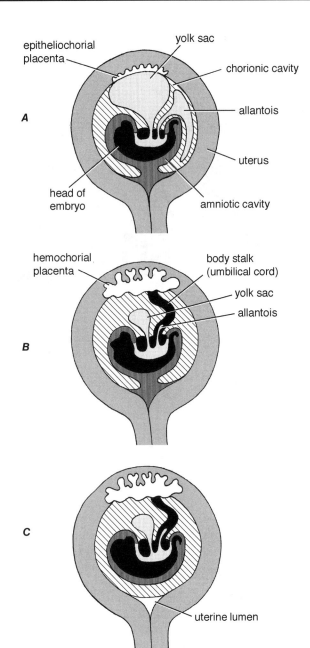

FIGURE 3.24

Placentation in primates (diagrammatic). **A**: lemur; **B**: tarsier; **C**: anthropoid.

lost the rhinarium and have hairy skin around their nostrils. They also have a rather anthropoid-like middle ear arrangement, with an anterior accessory cavity separated from the main cavity by a bony partition containing the internal carotid artery (Fig. 3.22E and F). Like anthropoids, but unlike nearly all other mammals and all other primates, tarsiers cannot make their own vitamin C and must ingest it in their diets to stay healthy (Pollock and Mullin 1987).

These and other derived traits that *Tarsius* shares with anthropoid primates have led most systematists to classify tarsiers and anthropoids together as Haplorhini. However, the degree of relationship and the relevance of modern tarsiers to the origin of anthropoids are still debated (Kay et al. 2004b). Genetic studies yield conflicting results. Some of them link tarsiers to anthropoids, while others link tarsiers to the Strepsirrhini (Eizirik et al. 2004, Schmitz and Zischler 2004). As we will see, the fossil record has not settled the issue.

PLATYRRHINES: THE NEW WORLD ANTHROPOIDS

The living Anthropoidea are divided into two groups. One comprises the monkeys and apes of Africa and Asia—including humans, who evolved in Africa but have subsequently colonized the whole globe. The other anthropoid group is the monkeys of the New World. The American monkeys are distinguished from their Old World cousins by a few synapomorphies. Most of them share a distinctive arrangement of the bones in the side wall of the braincase (Fig. 3.25), and their lower premolars are single-rooted, instead of having two roots like those of primitive mammals and most primates. But most of the distinguishing features of New World monkeys are merely retained primitive traits. Their dentition (2.1.3.3 (or 2)/2.1.3.3 (or 2)) retains an anterior premolar that is lost in the Old World anthropoids (2.1.2.3/2.1.2.3). The American monkeys have a simple ring-shaped ectotympanic bone surrounding the eardrum (Fig. 3.22F). This too is a primitive feature, contrasting with the more derived tube-shaped ectotympanic seen in today's Old World anthropoids. The nose of American monkeys preserves some primitive traits as well, including a well-developed **vomeronasal organ**— a separate scent-sensing pocket inside the nasal passages, used by most mammals for detecting social and sexual odorants (Maier 1980). In crown-group Old World anthropoids, the vomeronasal organ is reduced or absent, and sexual receptivity is advertised by behavior and visual signals (Dixson 1983, Rossie 2005). The nostrils of New World monkeys are also primitive in facing more laterally than those of Old World anthropoids. These nasal differences are memorialized in the names of the two anthropoid infraorders: **Platyrrhini** (Greek, "flat-nosed") for the American monkeys, and **Catarrhini** ("downward-nosed") for the Old World monkeys, apes, and humans (Table 3.1).

Isolated in South America throughout most of the past 25 million years, the platyrrhines have undergone a modest evolutionary radiation. Most of them are small- to medium-sized (1–3 kg) fruit-eaters. All are mainly or wholly arboreal, and almost all are diurnal. The exception to this last rule is the owl monkey (*Aotus*), the only nocturnal higher primate. It differs from other anthropoids in having an all-rod retina with no fovea. However, like other anthropoids (and tarsiers), it has no tapetum

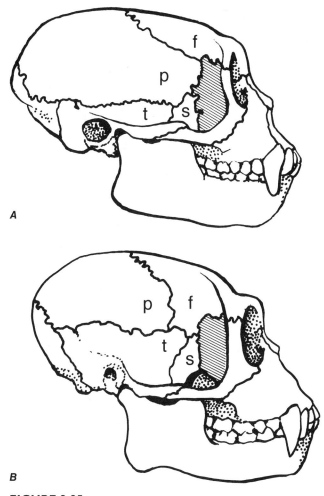

A

B

FIGURE 3.25

Side wall of the braincase in a New World monkey (*Cebus albifrons*, **A**) and an Old World anthropoid (*Hylobates lar*, **B**). In New World anthropoids, the zygomatic bone (hachure) usually contacts the frontal (f), parietal (p), and sphenoid (s). In Old World anthropoids, it is typically separated from the parietal by a frontal contact with the sphenoid or temporal (t). Compare Fig. 3.34. Not to same scale. (Modified from Cartmill 1980.)

lucidum. Its eyes are unusually large, affording increased light-gathering capacity to make up for the absence of the tapetum. The owl monkey is probably a secondarily nocturnal descendant of a diurnal anthropoid ancestor (Cartmill 1980, Martin 1990, Heesy and Ross 2004).

Several clades can be discerned within the living Platyrrhini (Harada et al. 1995, von Dornum and Ruvolo 1999, Steiper and Ruvolo 2003, Cartmill in press), although their relationships to one another are not wholly clear. Perhaps the most distinctive of these platyrrhine clades is the marmoset and tamarin group, often distinguished as a family, **Callitrichidae**. These diminu-

tive monkeys are the smallest of the anthropoids; an adult pygmy marmoset (*Cebuella*) can weigh as little as 100 grams. They have lost their third molars, giving them a unique dental formula of 2.1.3.2/2.1.3.2. Callitrichids are exceptional among primates in two other ways as well: they have sharp, squirrel-like claws rather than nails on all their digits (except the hallux), and they usually give birth to two babies at a time. To reduce the burden of bearing twins, callitrichid mothers farm out some of the usual duties of a mother primate to other family members. The mother of course gestates and suckles the babies, but they are carried around after birth by their father and older siblings, who may also help nourish the infants by sharing food with them after weaning (Brown and Mack 1978, Ingram 1978, Goldizen 1986). Among primates, this sort of systematic delegation of parts of the mother's job to other family members is peculiar to callitrichids and human beings.

Another platyrrhine subgroup that has a particular relevance to human evolution is the family **Atelidae** (Fig. 3.28C). These animals are bigger than other platyrrhines—5–10 kg in the living forms, and up to 25 kg in some extinct genera (Cartelle and Hartwig 1996). In typical atelids, the forelimbs are elongated, the fingers are curved and hooklike, and the shoulder bones—the shoulderblade or **scapula** and collarbone or **clavicle**—are altered in ways that allow them to hold their arms over their heads more easily (Erikson 1963). In all these respects, the atelids have converged evolutionarily with the living apes (Hominoidea) of the Old World—the superfamily to which human beings belong. Unlike any Old World primates, the atelids have evolved long, powerful prehensile tails, which serve as a sort of third hand in suspensory locomotion.

CERCOPITHECOIDS: THE OLD WORLD MONKEYS

The higher primates of the Old World today comprise two subgroups, usually distinguished at the superfamily level: the Old World monkeys or **Cercopithecoidea**, and the ape-human group or **Hominoidea**. Like the spider monkeys of the New World, the common ancestors of the crown-group hominoids seem to have been adapted for suspensory locomotion. All the living hominoids (including people) share anatomical traits that appear to have originated as specializations for arm-swinging. The cercopithecoids, on the other hand, are dedicated quadrupeds.

Cercopithecoid monkeys are widely distributed throughout tropical Africa and Asia, ranging north into temperate regions of Japan and China. Their teeth are distinctive. The front lower premolar (P_3) is an elongated blade that acts as a whetstone, against which the dagger-like upper canine is honed to a keen cutting

FIGURE 3.26

A male baboon (*Papio*) "yawning" to display his canines.

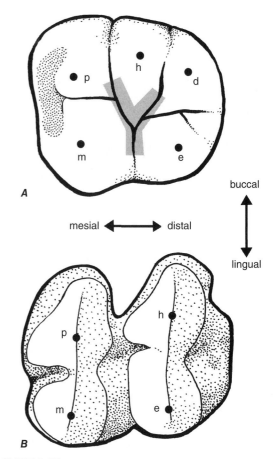

FIGURE 3.27

Right second lower molars of (**A**), a hominoid (*Pongo*) and (**B**), a cercopithecoid (*Macaca*). Gray tone indicates the "Y" frequently formed by the sulci of hominoid lower molars. Cusp homologies: d, hypoconulid; e, entoconid; h, hypoconid; m, metaconid; p, protoconid. Not to same scale. (After James 1960.)

edge. These canines are fearsomely well-developed in the males, especially in baboons and other large terrestrial cercopithecoids (Fig. 3.26). The molars of cercopithecoids bear four principal cusps, with transverse ridges or **lophs** joining the two front and two back cusps (Fig. 3.27B). This **bilophodont** arrangement yields more shearing surface per tooth than the cusp setup in hominoids does (Kay 1977).

Living cercopithecoids are divisible into two families: the leaf monkeys, or **Colobidae**, and the guenon–baboon–macaque group, the **Cercopithecidae** (Groves 1989). Colobids are sometimes called "leaf monkeys" because they are specialized for eating leaves. Like many other leaf-eating mammals, they have teeth with exaggerated shearing features for shredding leaves, and big stomachs with multiple compartments for digesting the resulting slurry. Colobids generally avoid eating a lot of ripe fruit, perhaps because the simple sugars in ripe fruit have a disruptive effect on the leaf-fermenting bacteria in their stomachs (Kool 1993, Jablonski 2002). The colobids' ruminant-like guts make them efficient browsers and allow them to outnumber cercopithecids in forests where they occur together (Andrews 1981).

Cercopithecids lack the colobids' specializations for processing leaves, and fruit is important in their diets. They have their own distinctive peculiarity of the gut—namely, outpocketings of the mouth lining called **cheek pouches**, into which a foraging monkey can cram food items to prevent other monkeys from appropriating them before they can be eaten. The smallest and least specialized of the living cercopithecids are the **guenons** (*Cercopithecus, Miopithecus, Allenopithecus, Chlorocebus*). These monkeys and their larger cousins, the **mangabeys** (*Cercocebus, Lophocebus, Rungwecebus*), are mainly arboreal fruit-eaters that inhabit the canopy layers of tropical African forests.

Some other cercopithecids spend most of their time on the ground. One of these terrestrial forms is the patas monkey (*Erythrocebus*). Built for speed, the patas monkey is a specialized guenon with long, slender legs that give it the ability to run like a greyhound in fleeing from predators. The **baboons** (*Papio, Theropithecus*) are more familiar and successful terrestrial primates. These big, long-faced, short-tailed monkeys (Figs. 2.6 and 3.26) are common in open country in Africa, where they move around in large groups (up to 200 individuals) feeding on everything from grass rhizomes to baby gazelles. The baboons' close relatives, the **macaques** (*Macaca*), extend from North Africa into tropical Asia and Malaysia as far east as Sulawesi, where they coexist with the most northerly of the Australian marsupials. As the only cercopithecids in Asia, the macaques have had the evolutionary opportunity to radiate into a wider variety of ecological and morphological types—large versus small, long- versus short-tailed, arboreal versus terrestrial—than we usually find within a single primate genus.

∎ HOMINOIDS: THE LIVING APES

The other surviving catarrhine superfamily is our own group, the Hominoidea. Modern hominoids (including humans) share a cluster of skeletal features that seem to have originated as adaptations to hanging and swinging from overhead branches (Keith 1923). These features are traits of the crown group (they are found in all living apes and humans) but not the stem group (they are absent in primitive fossil apes). The major ones are the following:

1. *Remodeled Shoulder Girdle.* Hominoids have a long, stout clavicle or collarbone that thrusts the scapula around to the back, onto the dorsal side of the flattened thorax (Fig. 3.28). The hominoid scapula is narrowed from side to side and elongated in a head-to-tail direction, and skewed to direct the point of the shoulder up toward the head (Fig. 3.29). In a quadrupedal monkey, the **glenoid cavity** (the socket of the shoulder joint) faces down toward the ground (Fig. 3.28); in hominoids, it faces laterally and

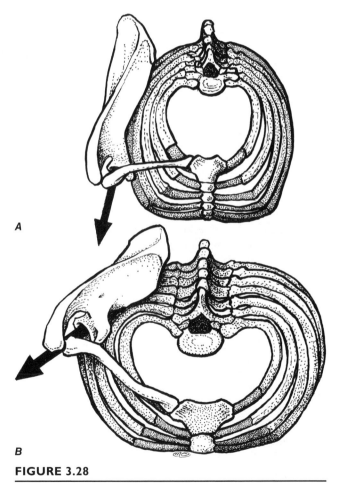

FIGURE 3.28

Shoulder girdle and thorax in a cercopithecoid monkey (baboon, **A**) and hominoid (human, **B**), seen looking from the head end toward the pelvis. Not to same scale. (After Schultz 1969.)

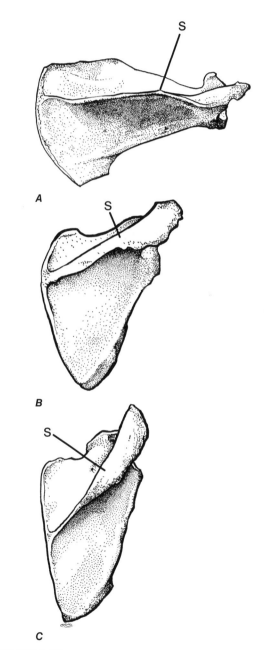

FIGURE 3.29

Scapulae of a baboon (**A**), a human (**B**), and a chimpanzee (**C**). Dorsal views; not to same scale. S, scapular spine. (**A** and **C**, after Swindler and Wood 1973.)

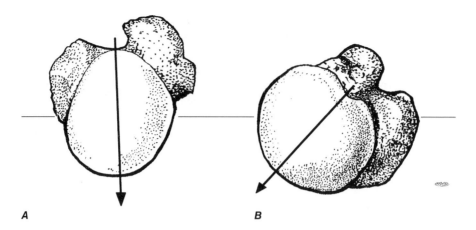

FIGURE 3.30

Humeral torsion. Proximal right humeri of a cercopithecoid (*Papio*, **A**) and hominoid (*Homo*, **B**), viewed from above looking downward along the long axis of the shaft, showing the difference in the orientation of the humeral head (*arrows*) relative to the axis of elbow flexion and extension (*horizontal line*). (**A**, after Swindler and Wood 1973; **B**, after Grant 1972.)

headwards—which is how the arm needs to be oriented in swinging along an overhead support. As a result of these changes, an ape's shoulder is fixed in a sort of permanent shrug, and so apes tend to look as though they have no necks. The remodeling of the scapula is less pronounced in humans than in other living hominoids (Fig. 3.29). Our glenoid cavity therefore faces more directly sideways and less headwards, and the human shoulder is held correspondingly lower, which gives us a more distinct neck than an ape's.

2. ***Humeral Torsion.*** In a quadrupedal mammal, the ball-shaped head of the humerus faces backwards and dorsally to meet the downward-facing glenoid cavity on the scapula. But in apes and people, the glenoid faces more laterally, and the humeral head is twisted around medially to meet it (Fig. 3.30).

3. ***Remodeled Thorax.*** In cercopithecoids and most other mammals, the thorax is deeper (from back side to belly side) than it is wide (from left to right). These proportions are reversed in hominoids (Fig. 3.28). This has the effect of redistributing the body's mass more evenly around all sides of the vertebral column, from which much of the weight is suspended in an animal hanging by its arms.

4. ***Reduction of the Epaxial Muscles.*** When a typical mammal gallops, the epaxial or "deep back" muscles that run from head to tail along the dorsal side of the vertebral column contract as the animal shoves off with its hind legs. Their contraction arches the back and extends the lumbar part of the vertebral column (Fig. 3.7). This arching swings the head end of the body forward as the galloping animal flies into the air, adding propulsive thrust and increasing the

length of the stride. But in an animal swinging through the trees by its arms, the epaxial muscles contribute little to propulsion. Those muscles are accordingly much smaller in hominoids than they are in typical quadrupedal mammals (Fig. 3.31A,B). This reduction of the epaxial muscles is reflected in the position of the transverse processes of the lumbar vertebrae, which lie on the ventral side of the epaxial muscles (Fig. 3.31C,D). The change in the epaxial muscles' function is also reflected in the angulation of the spinous processes of the lumbar vertebrae, which are not angled toward the head as they are in mammals that gallop (Fig. 4.12A).

5. ***Shortened Lumbar Region.*** The lumbar vertebrae of hominoids are reduced in number—from 6–8 in a typical galloping mammal to only 4 or 5 in humans and apes. The lumbar region of apes is shortened for the same reason that the epaxial muscles are reduced: lumbar flexion and extension no longer contribute to propulsion in a suspensory style of locomotion. Having a shorter lumbar region may also make it easier for an animal swinging along suspended by its arms to control the movements of its dangling pelvis and hind limbs. The vertebrae subtracted from the lumbar region appear to have become incorporated into the sacrum, so that the sacrum of most hominoids incorporates 5 or 6 fused vertebrae, as opposed to only 3 or 4 in quadrupedal monkeys (Keith 1923, Schultz 1961).

6. ***Straightening-out of the Forelimb.*** Typical quadrupeds have a curved humerus and ulna, built to stand the stresses of standing and running on all fours with the elbow flexed. Hominoids have a straighter humerus and ulna, and the elbow joint

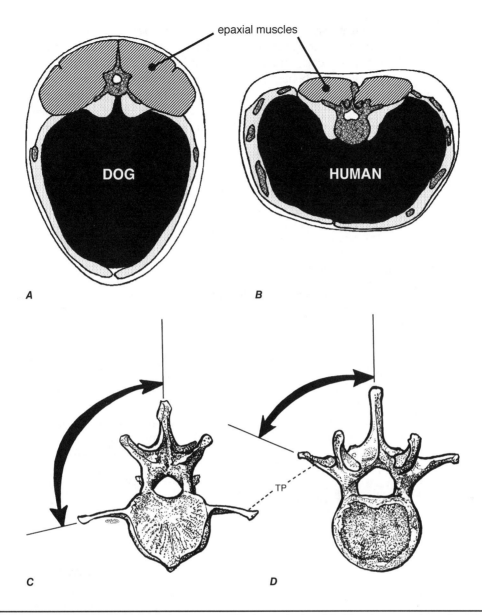

FIGURE 3.31

Diagrammatic cross sections of a dog (**A**) and a human (**B**) through the third lumbar vertebra, showing the comparative size of the epaxial muscles that extend the.lumbar spine. Lumbar vertebrae of a cercopithecoid (*Papio*, **C**) and a modern hominoid (*Pan*, **D**), demonstrating the reduction in the epaxial sector of the vertebral musculature (*curved arrows*). TP, transverse process. (**A**, after Popesko 1977; **B**, after Grant 1972; **C,D**, after Swindler and Wood 1973.)

between them is remodeled in a way that allows apes (and people) to straighten the elbow out completely in extension and keeps the joint stable throughout its enhanced arc of movement. All these changes reduce the bending stresses on the bones and ligaments around the elbow joint in an ape hanging below a branch.

7. ***Increased Supination.*** Mammals that have the ulna firmly connected to the carpal (wrist) bones at the base of the hand cannot turn their palms to face directly upward. A human or ape, however, can turn its palm to face straight up (**supinated**), or down (**pronated**), or anywhere in between, by swiveling

the radius (and the hand with it) back and forth around the stationary ulna. In great apes and people, the ulna is connected to the carpals only indirectly, by soft tissues. This loose connection allows the hand to rotate more freely around the ulna. The hominoid carpal bones and elbow joint are also remodelled to allow the radius to rotate more freely. An ape hanging from a branch needs this sort of rotatory mobility, in order to turn this way and that and to reach out to grab new branches with diverse orientations. Since an arm-swinging animal must have the strength to spin its whole body around when it supinates and pronates the forearm that it is hanging from, the muscles that produce those movements are relatively larger in apes than in quadrupedal monkeys. The radius and ulna of apes are markedly curved away from each other, which increases the space between the bones to accommodate bigger finger-flexing muscles (for supporting the body weight) and improves the leverage and torque of the muscles of supination and pronation (Oxnard 1963, Aiello et al. 1999)

8. *Loss of Tail.* In hominoids, the tail is no longer needed for balance in quadrupedal running atop branches. It has been reduced to a vestigial internal nubbin, the tailbone or **coccyx**. The old tail muscles

have been reworked to form a thin conical sheet of muscle called the **pelvic diaphragm**. Anchored to the vestigial tail and the inside of the pelvis, the pelvic diaphragm supports the rectum and uterus from underneath and helps to keep them from prolapsing out through their openings. An animal that hangs or stands upright needs this sort of added support for the pelvic viscera (Keith 1923).

All the living hominoids share these eight specializations of their **postcranial** (non-head) skeleton. Hominoids are also distinguished from the cercopithecoids

TABLE 3.2 ■ Intermembral Indices of Primates

Locomotion	Extant Primates (n)	Mean Inter-Membral Index (MIM)	Range	Extinct Primates	MIM (est.)
Arm-Swingers	Symphalangus (31)	148	142–155		
	Pongo (103)	144	135–150		
	Hylobates (55)	129	121–138		
	Gorilla (178)	117	110–125		
	Pan sp. (128)	107	102–114		
	Pan paniscus	102			
	Brachyteles (7)	105	102–110 ←	†Australopithecus cf. africanus (Stw 573)?	"subequal"
	Ateles (34)	105	99–109		
	Lagothrix (22)	98	93–100		
	Alouatta (34)	98	92–105		
Quadrupeds	Papio (22)	95	92–100		
	Nasalis (12)	93	89–97		
	Loris (8)	92	89–98		
	Erythrocebus (7)	92	86–99		
	Arctocebus (2)	90	90–90		
	Nycticebus (7)	90	87–94 ←	†Proconsul heseloni	88
	Macaca (116)	89	83–95	†Australopithecus afarensis (AL 288-1)	88
	Perodicticus (6)	88	86–90		
	Cercocebus (2)	86	82–90		
	Cercopithecus (37)	84	79–81		
	Cebus (15)	81	77–85		
	Colobus (8)	79	77–83		
	Presbytis (87)	78	73–84		
	Callithrix (38)	76	73–78		
	Saguinus (11)	75	73–78		
	Aotus (12)	74	72–75		
	Daubentonia (6)	71	70–72		
	Lemur (17)	70	68–72		
	Homo sapiens	69	68–70		
Leapers	Propithecus (10)	64	63–66		
	Indri (13)	64	61–66		
	Euoticus (5)	63	62–64		
	Galago (10)	62	51–71		
	Avahi (4)	56	55–56		
	Tarsius (8)	55	53–59		

Index = length of humerus + radius, as a percentage of femur + tibia. Daggers (†) indicate extinct taxa.
Sources: Coolidge (1933), Napier and Napier (1967), Aiello and Dean (1990), Harrison (2002), Clarke (2002b).

by their molar teeth. In contrast to the bilophodont molars of cercopithecoids, the molars of hominoids typically have four or five low, rounded cusps separated by a variable pattern of **sulci** or grooves. The hominoid arrangement is sometimes called a **Y-5** molar pattern, because the branching sulci on the lower molars commonly form a transversely oriented letter "Y" on the occlusal surface (Fig. 3.27).

Within the hominoids, humans and gibbons stand out as natural groups set off from the other apes by their unique locomotor adaptations. The gibbons or **hylobatids** are highly specialized **brachiators**, or arm-swingers. All the hominoids share adaptations to this sort of locomotion, but the gibbons have perfected it. Gibbons are treetop acrobats that swing from branch to branch through the air like circus aerialists. The other living apes are more cautious in their locomotion, only occasionally taking all four hands and feet off the support at once in jumping or swinging. This difference is just what we would expect on the basis of the square-cube law. Adult gibbons weigh from 5 to 12 kg. Other living apes are much larger (33–175 kg), which is why they are called "great apes." Because the great apes are so large, they have more cubic inches of ponderous mass per square inch of supporting structures than the smaller gibbons. They are therefore inherently more fragile, and tend to be correspondingly careful in their arboreal movements.

As part of their adaptation to brachiation, gibbons have evolved exceptionally long forelimbs. All the living apes have relatively long arms, but a gibbon's **intermembral index**—the ratio of forelimb length to hindlimb length—is uniquely high (Table 3.2). Gibbons also have unusually long, curved, hook-like fingers.

Adult gibbons are intolerant of other adults of the same sex, and so they tend to wind up living in small family groups consisting of a single mated pair and their subadult offspring. This sort of one-male, one-female social arrangement is called "monogamy," by analogy with monogamous human marriages. (It isn't a very precise analogy. Married human couples seldom live in isolation and drive away all visitors, as gibbon pairs do.) Like most other so-called monogamous mammals, gibbons show little sexual dimorphism in body size or canine weaponry. Such differences between males and females are most pronounced in primates with **polygynous** mating patterns, in which one male mates with many females and competes aggressively with other males for the opportunity to do so. The intense male-male competition in such situations favors the bigger and more formidable males and thus selects for greater dimorphism (Plavcan and van Schaik 1997b).

PONGIDS AND HOMINIDS

Most systematists divide the rest of the living hominoids into two families, Pongidae and Hominidae. The type genus of the family Pongidae is the orangutan genus, *Pongo*. Orangutans are large (40–80 kg) arboreal herbivores covered with long orange hair. They inhabit the Indonesian islands of Borneo and Sumatra, where they and their rain forest habitat are being rapidly wiped out by deforestation. Orangutans are clever with their hands (they are notorious in zoos as lock-pickers and escape artists), and they make and use a few simple tools in the wild (van Schaik et al. 2003). Oddly enough for such intelligent animals, orangutans are largely solitary as adults. Typically, a single full-grown male will patrol and try to defend a large territory (up to 700 hectares) encompassing the smaller, overlapping territories of several adult females. The male will mate with these females whenever they come into estrus, and try to prevent other males from doing the same (van Schaik and van Hooff 1996). Orangutans show the pronounced sexual dimorphism that might be expected from this mating pattern: fully grown males may weigh more than twice as much as adult females (Table 4.4). Male orangutans are additionally distinguished by flat, fleshy pads or "flanges" protruding from their cheekbones.

Flanged male orangs usually do not tolerate the presence of other flanged males, but they do allow subadult males (which look like females) to hang around at a respectful distance. Some of these supposed subadults are actually adult males with arrested development, who take the opportunity to force copulations on the females whenever the resident flanged male is absent or looking the other way. Retaining an immature appearance into adult life may represent an alternative mating strategy, which gives these males a better chance of mating with the local females whenever the flanged male leaves them unguarded. In zoos, unflanged adult males grow flanges and get bigger when there are no other flanged males around (Maggioncalda 1995), and a similar succession probably takes place in the wild whenever a flanged male grows feeble or dies.

The type genus of the remaining hominoid family, Hominidae, is the human genus, *Homo*. Before the advent of cladistics, the family Hominidae was reserved for humans and their close fossil relatives. The family Pongidae was generally used as a wastebasket taxon, in which all the great apes were lumped together in a catchall category of non-gibbon, nonhuman hominoids (Marks 2005).

In recent years, this practice has fallen out of favor. Molecular evidence has convinced almost all of the experts that the apes of Africa—chimpanzees and gorillas—are more closely related to us than orangutans are (Ruvolo 1997, Deinard et al. 1998, Wimmer et al. 2002, Wildman et al. 2003, Marks 1994, 2005; cf. Schwartz 1984, Grehan 2006). The rules of cladistic systematics therefore require that the African apes be grouped with humans (rather than with orangutans) as members of the human family, Hominidae. This usage is becoming

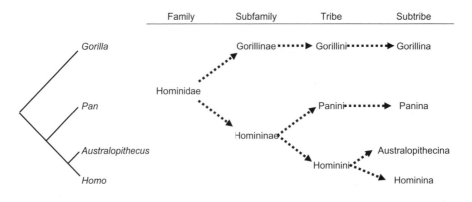

FIGURE 3.32

The current consensus concerning the relationships of the living African hominoids (cladogram, **left**) and the cladistic classification (**right**) that results from it.

increasingly prevalent. Human beings are now often distinguished from the African apes at a lower taxonomic rank, as the tribe Hominini or the subtribe Hominina (Mann and Weiss 1996, Wood and Richmond 2000). We will follow this usage by referring to people and their close fossil relatives as "hominins" rather than "hominids." Hominins include modern humans and fossil *Homo*, and also several extinct species of *Australopithecus* and related genera from the Pliocene and Pleistocene of Africa (Fig. 3.32).

The living African apes are not bipeds, but they exhibit another unique form of terrestrial locomotion called **knuckle-walking** (Fig. 3.33). When chimpanzees and gorillas travel on the ground, they do not walk on the palms of their hands, but on the backs of their flexed fingers. Because these apes have long, curved fingers adapted for suspensory arm-swinging, they would risk sprains or fractures if they stuck their fingers out and walked around on their palms as most quadrupedal monkeys do. They avoid that risk by walking on the backs of their curled-up fingers. The skin there is thickened and specialized to carry weight and transmit propulsive thrust, like the skin on the palms and soles.

Gorillas are the largest living primates. Adult females weigh around 75–90 kg (Table 4.4); a fully adult male can weigh as much as 200 kg. As we might expect of such big animals, adult gorillas are predominantly terrestrial (they are too ponderous to do much brachiating in the trees) and herbivorous, with a diet that includes a large daily intake of stems, leaves, and other low-energy plant tissues (Kuroda et al. 1996, Watts 1996, Yamagiwa et al. 1996).

Gorillas live in small family groups usually consisting of one dominant "silverback" adult male and two to four adult females, plus their subadult offspring. The female offspring tend to leave and join other family groups, presumably to avoid inbreeding. The male offspring also move out, going off to hang around the fringes of other

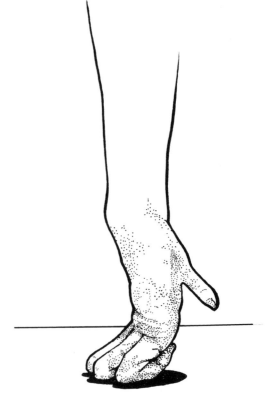

FIGURE 3.33

Hand of knuckle-walking chimpanzee, *Pan troglodytes*. (After Tuttle 1967.)

family groups and trying to lure females off to start new families of their own. Although larger, multi-male groupings headed by a few closely related adult males are sometimes seen (Stewart and Harcourt 1986), the usual social pattern of gorillas is polygyny, in which one or two males defend females and territory against the incursions of nearby males who might replace them. Sexual dimorphism is therefore extreme in gorillas, and the silverback males are territory-defending machines—big, scary beasts, equipped with powerful muscles, fearsome canine teeth, and impressive chest-beating threat displays. Many taxonomists divide gorillas into a Western species (*G. gorilla*) and an Eastern species (*G. beringei*), which differ in some aspects of their morphology, locomotor repertoire, and social behavior (Bradley et al. 2004, 2005).

BONOBOS AND CHIMPANZEES

There are also two living species in the chimpanzee genus, *Pan*. The bonobo or "pygmy" chimpanzee, *Pan paniscus*, inhabits Central African rain forests south of the Congo River. The more familiar *Pan troglodytes* or "common" chimpanzee lives in forests and woodlands north of that river, and ranges east into extreme western Tanzania. The nicknames of these two species are not really appropriate, since body size is as large in "pygmy" chimpanzees as it is in some populations of "common" chimpanzees, and "common" chimpanzees are endangered animals that are by no means common. We will refer to *P. paniscus* as the bonobo and reserve the term "chimpanzee" for *P. troglodytes*.

The social patterns of bonobos and chimpanzees are more human-like than those of the other apes. They may be uniquely relevant for understanding our own ancestry, since *Pan paniscus* and *P. troglodytes* appear to be our closest living relatives (Fig. 3.32). Both species of *Pan* associate in large "communities" of from 20 to over 100 individuals, including several adults of both sexes. Each community has a more or less exclusive territory. Males tend to remain in their native territory when they grow up; adolescent females usually emigrate to join other communities nearby. The individuals comprising a community almost never assemble in one place at the same time. Rather, they band together in shifting, transient groupings called "parties," which may consist of a few individuals or dozens of one or both sexes (Boesch 1996).

This description applies thus far to both chimpanzees and bonobos. But the two differ markedly in some other aspects of their behavior. Chimpanzees (*P. troglodytes*) might be described fancifully as "patriarchal" or "militaristic" in their behavior and social organization. Male chimpanzees usually displace or dominate females in competitive encounters. Adult males band together to patrol the boundaries of the community's territory, attacking any interlopers they encounter. Such patrols may even invade the territories of smaller neighboring communities to hunt down and kill their residents (Goodall et al. 1979). Perhaps as a corollary of this sort of "warfare," male bonding is strong and conspicuous in chimpanzees. *Pan troglodytes* is the only nonhuman primate in which adult males within multi-male groups tend to groom each other, associate together to the exclusion of females, and cooperate in such activities as hunting animal prey—which plays an important role in chimpanzee social hierarchies and relationships (Stanford 1999, 2001a,b). Unrelated female chimpanzees, on the other hand, show little evidence of mutual attraction and rarely cooperate with each other (Nishida and Hiraiwa-Hasegawa 1986). Female chimpanzees sometimes kill and eat each other's babies. Infanticide and cannibalism are even more common among males (Goodall 1986).

By contrast with chimpanzees, bonobos are thought by many observers to be sweeter, more matriarchal creatures whose motto is "make love, not war." Female bonobos tend to dominate males. Bonobos have much lower levels of interpersonal and intergroup aggression than chimpanzees have. Female bonobos stay sexually receptive for a longer percentage of their menstrual cycle than female chimpanzees do, and bonobos' levels of sexual activity of all sorts—especially of homosexual acts between adult females—are much higher. Female bonding is correspondingly close and intense among bonobos; male bonding is not. Male bonobos do not band together to police and defend their territories. Neighboring bonobo communities often mingle, with grooming and copulations occurring peacefully across community lines. Although bonobos can be quite aggressive, they rarely hunt, and infanticide and cannibalism have never been observed (Nishida and Hiraiwa-Hasegawa 1986, Kano 1996, Takahata et al. 1996, White 1996, de Waal 2001).

The two species of *Pan* also seem to differ in their technological skills. Chimpanzees make and use a wide variety of simple artifacts, including leaf "hats" for shedding rain, wadges of chewed leaves for sopping up drinking water from tree crevices, and stone anvils for cracking nuts. The most common chimpanzee implements are trimmed sticks and twigs, which are used for everything from grooming to pulling edible termites out of their nests. Because chimpanzee tool use is learned rather than instinctive, and because the inventory of these tools differs from one chimpanzee group to another, many scientists do not hesitate to refer to chimpanzee technologies as a nonhuman form of material culture (Whiten et al. 1999, McGrew 2001a).

Tool use is especially common and most highly elaborated in those chimpanzee groups that have the most "social tolerance," as measured by the time animals spend together in various friendly interactions. The primate biologist C. van Schaik and his co-workers (van Schaik et al. 1999, 2003) argue that such interactions provide a necessary condition for learning how to use tools. They suggest that the relatively solitary habits and low social tolerance of orangutans may explain why these apes, which are skilful makers and users of tools in captivity, use so few tools in the wild. Van Schaik conjectures that early hominins became better tool users than orangutans or any other ape because "there was a social change that made them tolerate each other" (Vogel 1999).

This theory leads us to expect that bonobos, with their lower levels of aggression and intense female bonding, would be even more technologically inclined than chimpanzees. Puzzlingly enough, the reverse seems to be the case. Although bonobos are able to make and use tools in captivity (Toth et al. 1993), they have so far not been observed to do so much in the wild, and their repertoire of "cultural" behaviors seems relatively impoverished (Hohmann and Fruth 2003).

Not surprisingly, these differences between the two species of *Pan* have developed a mythological resonance in the scientific literature. Scholars who see hunting, aggression, and male bonding as important factors in human evolution like to point to *Pan troglodytes* as a model for ancestral hominin society and culture (Wrangham and Peterson 1996, Stanford 1996, 2001a,b). Anthropologists of a more pacifist or feminist inclination prefer to take *Pan paniscus* as the point of departure for the evolution of human behavior (Zihlman 1979, Zihlman et al. 1978, Tanner 1981, de Waal 2001). But both models may be inappropriate. As we will see in the next chapter, early hominins appear to have been more sexually dimorphic than bonobos or chimpanzees. This fact suggests that our Pliocene ancestors may not have been much like either species of *Pan* in their social organization and mating patterns.

In this connection, it seems worth pointing out that we should not invoke the behavior of our ancestors as justifications for our own actions. Whether those ancestors were more like bonobos or chimpanzees, our descent from them implies nothing about what is natural or proper for human beings—no more than our descent from fish implies that we should live in the water and breathe through our ears (Cartmill 2002).

▌HUMANS VS. APES: SKULLS AND TEETH

There are many significant anatomical differences between humans and the extant African apes. Many of them are differences from the neck down, reflecting the bipedal habits of *Homo* versus the quadrupedal knuckle-walking of *Pan* and *Gorilla*. We will look at these postcranial differences in the next chapter.

But several crucial differences between people and apes are found above the neck. Of all the distinctive features of our species, perhaps the most adaptively important is our big brain. Although great apes are among the brainiest of all animals, a human brain is more than three times as big as an ape's of the same body size—over 1200 cc, versus less than 400 cc for a chimpanzee or bonobo (Tables 4.1 and 6.1).

The effect that the huge human brain has on the architecture of the human skull is evident at a glance. Compared to that of a chimpanzee, a human skull is all braincase and no jaws (Figs. 3.34, 3.40, 3.36, 8.1). The ape's big, protrusive facial skeleton thrusts forward well beyond the front end of the braincase; the smaller human face is pulled back under the bulging front end of a much larger braincase. In the ape, the front end of the braincase sports projecting brow ridges or **supra-orbital tori** that roof over and protect the protruding eye sockets. The front end of the human braincase overlaps the orbits almost completely, and the bony brow ridges are small or negligible.

Other differences between *Pan* and *Homo* in the braincase and face are evident when we look at a midline section through the skull (Fig. 3.36A,B). In the ape's skull, the **foramen magnum** (Latin, "big hole"), through which the Spinal cord emerges from the skull, faces backward and downward. So do the **occipital condyles**, the paired joint surfaces alongside the foramen magnum where the skull articulates with the uppermost cervical vertebra (Fig. 3.35C). But in the human skull, the foramen magnum faces *forward* as well as downward, and the occipital condyles are shifted forward along with it.

It has been argued that our big brain is responsible for these peculiarities of the human skull. The enlarged human braincase sticks out in back just as it does in front, protruding backward well past the foramen magnum. Some experts think that the fore and aft expansion of the human brain causes the basicranium (floor of the braincase) to flex in the middle, thrusting the foramen magnum downward and forward (Fig. 3.36C–E; Biegert 1963, Ross and Ravosa 1993). Others reject this conclusion, arguing that brain size is not correlated with basicranial flexion throughout human growth and development (Jeffery and Spoor 2002). No doubt, the brains of humans are big, globular, and flexed by comparison with the brains of primitive mammals (cf. Fig. 3.12A). But so are the brains of apes; and the amount of difference in flexion that we find between apes and humans, or between human infants and adults, depends on the lines and angles we choose to draw in making our measurements (Ross et al. 2004).

Whatever its connection with brain enlargement may be, the forward rotation of the human foramen magnum and occipital condyles is unequivocally associated with our upright posture. The forward shift of the condyles in *Homo* puts the cervical vertebrae roughly underneath the head's center of mass, thereby balancing the human head more precisely on the vertical human neck. The head of an ape, by contrast, hangs forward in front of its shoulders, and so its weight has to be continually supported by the epaxial neck muscles attached to the back of the skull. These muscles are accordingly larger in apes than in humans, and their area of attachment

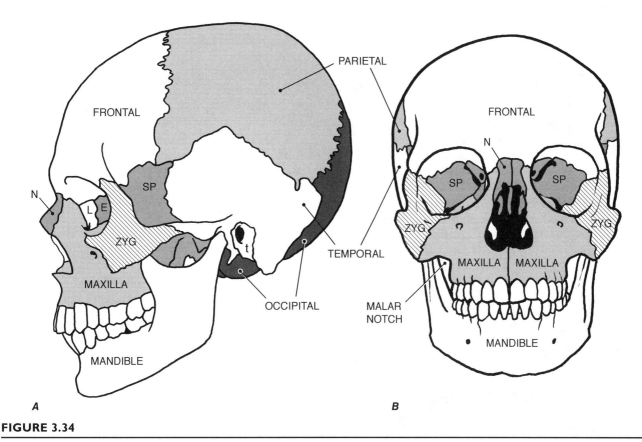

FIGURE 3.34

Modern human (*Homo sapiens*) skulls, showing the boundaries of the constituent bones from the lateral (**A**) and anterior (**B**) aspects. Abbreviations: E, ethmoid; L, lacrimal; N, nasal; SP, sphenoid; t, tympanic (= ectotympanic) part of temporal; ZYG, zygomatic (malar). (**A**, after Rohen et al. 1998; **B**, after Clemente 1997.)

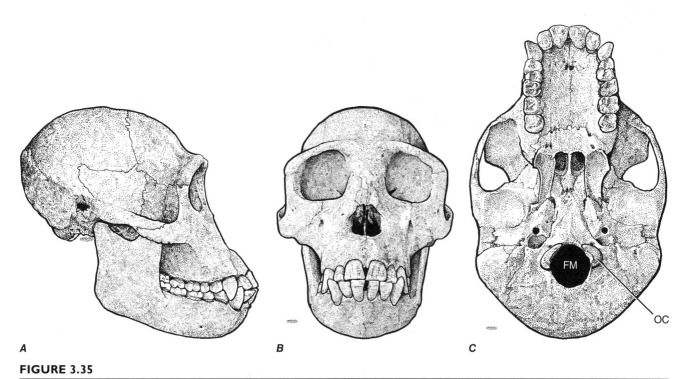

FIGURE 3.35

Skull of female chimpanzee (*Pan troglodytes*, BMNH 1864.12.17): Lateral (**A**), frontal (**B**), and basal (**C**) views. FM, foramen magnum; OC, occipital condyle.

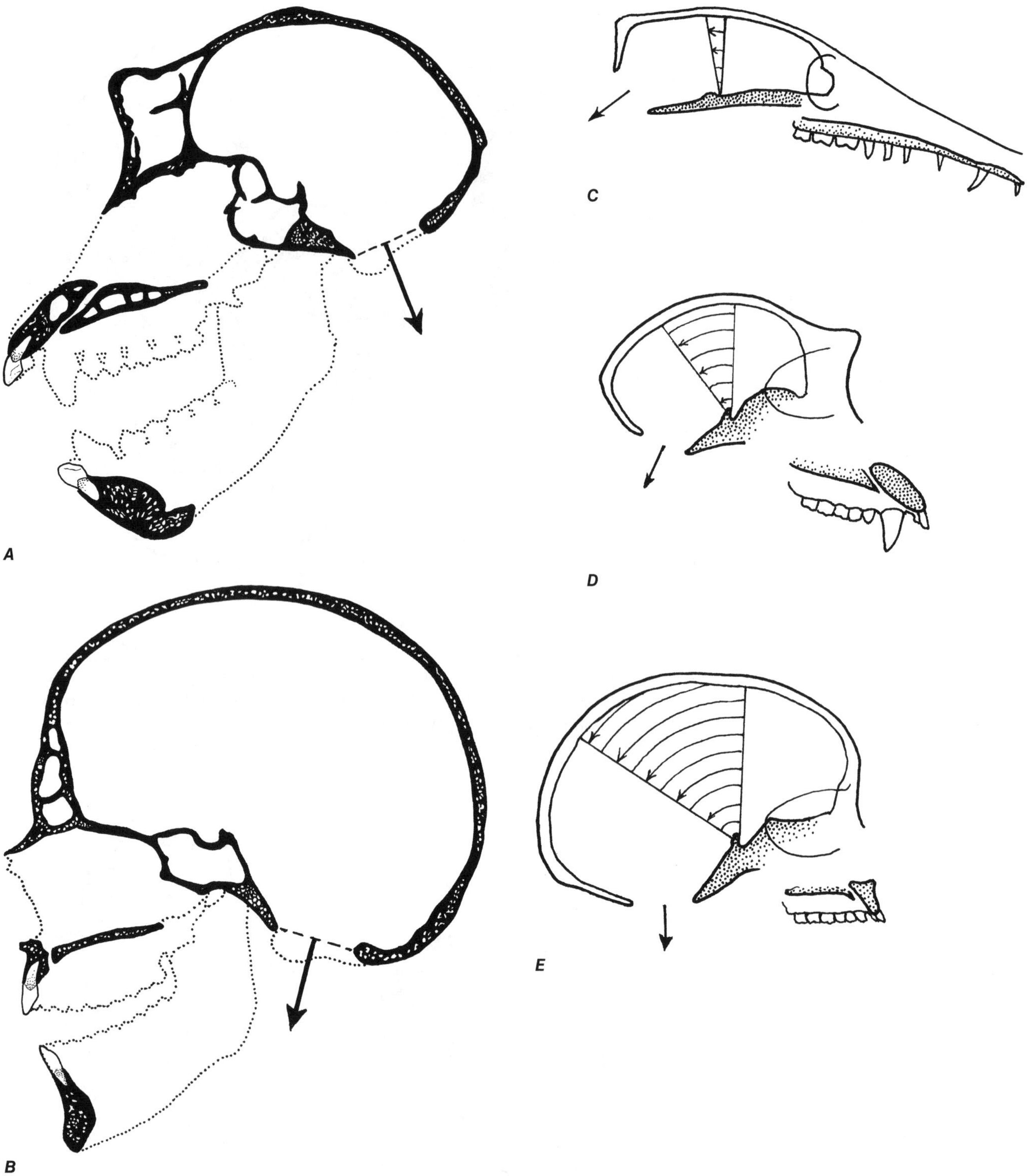

FIGURE 3.36

Basicranial flexion. **A,B**: Hemisected skulls of a male chimpanzee (**A**) and a male human (**B**). **C–E**: Hypothetical effect of progressive brain enlargement on basicranial flexion and foramen magnum orientation (*arrows*) in a tree shrew (*Tupaia tana*, **C**), a chimpanzee (**D**), and a human (**E**). (**A**, after Schultz 1969; **B**, after Pernkopf 1963; **C–E**, after Biegert 1963.)

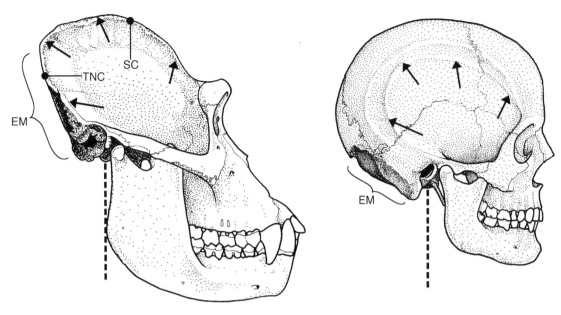

FIGURE 3.37

Skulls of male *Gorilla gorilla* (*left*) and *Homo sapiens* (*right*). In the gorilla, most of the weight of the head lies in front of a vertical line through the occipital condyles (*dashed line*). The area for attachment of the epaxial muscles of the neck (EM) therefore needs to be more extensive to stabilize the head. The gorilla's big temporalis muscle extends beyond its small braincase (*black arrows*). This results in the formation of a bony sagittal crest (SC) between the left and right temporal muscles on top of the skull, along with a temporonuchal crest (TNC) between the areas of temporal and neck muscle attachment in back. (After Elliot 1913 and Grant 1972.)

extends further up the back of the skull. The larger the ape, the heavier the head and the bigger the neck muscles. The massive neck muscles that are needed to hold up the ponderous head of a big male gorilla produce a flaring **nuchal crest** protruding from the gorilla's skull at the upper end of the area of neck-muscle attachment. The human equivalent, the **superior nuchal line** (Fig. 8.7), is far more delicate and located much lower. Humans can get by with such puny neck muscles because the human head is lighter and better balanced— and also because the rearward expansion of the human braincase provides the epaxial neck muscles with a long lever arm in producing rotation around the forward-shifted occipital condyles (Fig. 3.37).

Because an ape's jaws are bigger relative to the size of its braincase than a human's are, the attachments of its jaw muscles also extend relatively further up the sides of its braincase. In big apes, the temporal muscles may extend all the way up to the midline of the skull roof. This induces the growth of a bony keel, the **sagittal crest** (Fig. 3.37), from the midline of the skull, to furnish additional attachment surface in the midline for the two temporal muscles.

Sagittal crests are almost always present on adult male gorillas and frequently seen in male orangutans and chimpanzees. Their presence always reflects some combination of relatively small brains and big jaws. Sagittal crests are accordingly more common in bigger

mammals, which have relatively small brains and big jaws for allometric reasons. And because male anthropoids usually have bigger jaws and more robust muscles than females of their species have, sagittal crests are more often seen in males than in females. In big gorilla males, all these factors work together, and so cranial cresting is especially pronounced. The sagittal crest in apes is tallest at the rear of the skull, reflecting the large size of the posterior part of the temporal muscle. The temporalis may extend all the way back to the neck muscles, causing a **temporonuchal crest** to grow up between them. When a sagittal crest is present as well, it merges the midline with the temporonuchal crests (Fig. 3.37).

Some early hominins also had cranial crests (Chapter 4). This does not mean that they were especially close relatives of chimpanzees and gorillas, but only that they had relatively small brains and big jaws. Crests develop in any mammal whenever the jaw muscles need more attachment space than is present on the braincase. The crests may disappear in that mammal's descendants if the evolving brain grows larger, the jaw muscles get smaller, or body size dwindles and the ratio of brain to jaw size changes allometrically. The size relationships between brain and jaws are largely determined by genetic factors, but the cranial crests are not under independent genetic control. In a classic experiment, S. Washburn (1947) showed that rats that had their temporalis muscles cut did not develop crests, while a control group from the same strains (fed the same diet) developed a normal cresting pattern. Impressive as they are, cranial crests themselves tell us nothing about patterns of evolutionary relationship.

Human dentitions differ strikingly from those of apes. Because our front teeth—canines and incisors—are smaller than those of human-sized apes, our dental arcade is shorter overall and narrower in front. This gives the human arcade a parabolic shape (in occlusal view) that contrasts with the U-shaped arcade of apes (Figs. 3.35 and 8.7). Although the two front molars of humans are about the same size as a chimpanzee's (Schuman and Brace 1954) and usually preserve the primitive hominoid cusp pattern (Fig. 3.27), our third molars may be smaller and simpler, or even missing altogether. The canines of apes are primitively protruding and fang-like, with long roots, whereas human canines are relatively small and stubby. The large crowns of ape canines have a distinct point at the apex and a distinctive blade on the distal edge, extending from the apex to the crown base. This blade, sometimes referred to as the **distal cutting blade**, is not present on human canines (Fig. 8.7). Although our canine teeth are no longer of much use in a fight, their still slightly pointed crowns and persistently long roots make them the teeth of choice for hard cutting use—say, in biting through a thread or opening a packet of nuts. The upper incisors of apes are strongly **procumbent**—that is, they slope forward as well as downward, so as to position their working surfaces out in front of the big lower canines. Human incisors are more vertically implanted. The upper central incisor is larger than the lateral one in both humans and apes, but the size differential is much less in humans.

Human dental enamel is thicker than that of any African ape. Our thick enamel correlates with high rates of dental wear. Among preindustrial human populations, hard chewing of tough, abrasive foods caused all the teeth to wear down flat, producing a smooth and uniform occlusal surface. The canines did not protrude at all in adults, and the upper and lower canines and incisors met each other squarely in an edge-to-edge bite. In addition to the heavy wear on the occlusal surfaces of the teeth, the grinding of each tooth against its neighbors whenever it moved slightly in its socket produced **interstitial wear** between adjacent teeth (Wolpoff 1971c). This made each tooth get steadily shorter in a mesiodistal direction and drift **mesially**—that is, along the arc of the dental arcade toward the midline (Fig. 3.38). The dental arcade thus grew shorter throughout life.

In most human populations today, there is less abrasive grit in the diet and less dental wear. The molars thus tend to retain their cusps, like those of African apes,

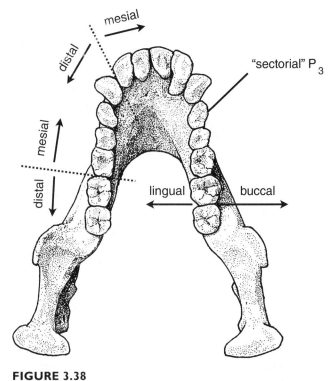

FIGURE 3.38

Mandible of female chimpanzee (*Pan troglodytes*, BMNH 1864.12.17), occlusal view.

instead of wearing off flat. Because interstitial wear is slight in modern populations, the M2 sometimes fails to move mesially far enough to get out of the way of the late-erupting M3. Blocked from erupting, the M3 remains partly buried in the jaw and becomes a so-called "impacted wisdom tooth." The resulting abscesses and infections can be life-threatening. Another side effect of reduced dental wear in modern human populations is an **overbite**, in which the upper canines and incisors occlude in front of the lower ones (Fig. 8.1) instead of meeting them edge-to-edge. The overbite is not found in fossil humans. It is a diagnostic sign of a diet rich in such things as white bread and cupcakes.

In apes and other nonhuman catarrhines, the large upper canines occlude *behind* the lower canine and in front of the elongated, blade-like lower front premolar (P₃). Ape P₃ crowns usually bear one very large cusp, and the tooth is rotated mesially (toward the incisors). This **sectorial premolar** (Fig. 3.38) is thereby positioned to wear against the back edge of the upper canine, producing a wear facet that sharpens the canine's distal cutting blade. The even bigger sectorial P₃ of Old World monkeys hones their upper canines into razor-like slashing blades (Fig. 3.26). The human P₃, by contrast, has evolved into a molar-like grinding tooth (Fig. 8.5). In preindustrial peoples, its occlusal surface is continuous with the flat wear surface of the rest of the teeth.

PRIMATE ORIGINS: THE CROWN GROUP

Some of the synapomorphies of primates have no special adaptive value. For example, primates are the only mammals that have an auditory bulla formed from the petrosal bone; but other mammals have bullae composed of other bones, which do the same job just as well. The fact that primates have an osseous bulla is adaptively significant, but (as far as we know) the unique composition of the primate bulla is not.

But other primate synapomorphies appear to be telling us something about the ecology and adaptations of the ancestral primates. For example, the prehensile hind feet of nonhuman primates are clearly adapted for grasping branches. It seems reasonable to conclude from this that the ancestral primates, like almost all living primates, lived in trees.

In the early 20th century, most scientists simply left it at that. It was generally thought that arboreal life accounts for most of the peculiarities of primates, because the pressures of arboreal life automatically select for reduced claws, grasping extremities, special-ized visual systems, and other primate-like traits in tree-dwelling mammals (see Blind Alley #3). The problem with this **arboreal theory** of primate origins is that most arboreal mammals do not in fact have or need these sorts of features (Cartmill 1972, 1974b,c, 1992). Therefore, living in trees is not a sufficient explanation for the evolution of primate features.

We can seek a better explanation by looking for primate-like features in nonprimate mammals. Since none of the primate synapomorphies is primitive for mammals, their occurrences in other groups of mammals must represent evolutionary convergences with primates. Such convergences often originate as independent adaptations to the same selection pressures. Studying their distribution and correlates can therefore help us to determine what those shared selection pressures are.

Among nonprimate mammals, grasping adaptations of the hind foot (long digits, a divergent big toe, reduction of claws, etc.) are found in most arboreal marsupials and some climbing rodents. The most primate-like hind feet of all are found in small arboreal marsupials that move and forage among twigs and vines in tropical forests, out at the tips of branches where fruit grows and insects gather to feed on flowers, fruit, and nectar. There is growing agreement among the experts that

---

Blind Alley #3: The Arboreal Theory

The arboreal theory of primate origins, worked out by scientists in the 1920s and 1930s, held that the distinctive peculiarities of Primates were the natural result of living in trees. When early mammals started climbing trees, it was claimed, they needed stereoscopic vision to judge distances in jumping from branch to branch. So they evolved big eyes and moved them to the front of the head to allow for 3-D vision. They scaled back their sense of smell, which was of little use in the treetops. Their brains grew bigger to handle the new visual input and to furnish the sharp wits they needed to navigate the maze of branches. They swiveled their thumbs and big toes around to oppose the other digits in grasping branches, flattened out their sharp claws into broad nails to give their fingertips a more secure grip, and, in short, turned into primitive primates, something like a modern lemur. Carrying all these trends a bit further eventually converted the lemur into a monkey; and from there it was only a short evolutionary step to a tenured professor. Or so the story ran.

The trouble with this story is that most tree-dwelling mammals don't look like primates. Squirrels, for example, have big noses of the standard mammalian sort. (There are, in fact, lots of things to smell up in trees.) Their eyes point in different directions. Squirrels grip tree trunks and branches not with their thumbs and big toes (which are no more opposable than a dog's), but by digging in their claws. Those claws give them a grip on trees that primates, with their flat nails, can't match. And since squirrel anatomy is very well suited to arboreal life, why would arboreal life have driven the early primates to abandon it for something different?

Arboreal life by itself can't explain the peculiarities of primates. Those peculiarities must have originated as adaptations to a particular, specialized way of life in the trees. Today's small bush-dwelling marsupials, which have primate-like grasping feet, give us an idea of what that way of life may have been like.

grasping, hand-like hind feet, in primates and marsupials alike, evolved to provide more stable and secure locomotion in this precarious "fine-branch milieu" (Charles-Dominique and Martin 1970, Cartmill 1970, Lemelin and Schmitt 2007). Most other arboreal mammals have more primitive, clawed hands and feet, because these are more useful for grasping thicker supports (Cartmill 1974c).

Cartmill (1970, 1972) offered a related explanation for the distinctive visual traits of primates—the convergence of the eyes and orbits, the enclosure of the eyes in stabilizing bony rings and cups, and the enlargement of the retinas and the visual parts of the brain. Among nonprimates, such traits are well-developed in cats and other predators that rely on their eyes for detecting and ranging prey (Fig. 3.39; Heesy 2008). Cartmill proposed that the primate visual traits had originated as predatory adaptations, like the comparable specializations seen in cats and owls. He concluded that the last common ancestor of the living primates had been a small, insect-eating prosimian, resembling a modern mouse lemur or loris, that foraged by night in the terminal branches and shrub layers of tropical forests. The apomorphies of primate hands, feet, and eyes (Fig. 3.17) had originated as adaptations to this way of life.

This **visual-predation** theory was an advance over its predecessors, but there are some problems with it as well. For one thing, it does not account for the synapomorphies of the primate dentition (Fig. 3.19). The reduction of the stylar shelf and paraconid, the lowering of the trigonid, and other features that distinguish primate teeth from those of more primitive eutherians suggest that plant tissues were important elements in the diets of ancestral primates (Szalay 1968, 1972; Silcox et al. 2005).

The paleontologist F. S. Szalay and his co-workers dismiss the visual-predation theory on dental grounds. Szalay argues that the grasping hallux and some other primate apomorphies originated as adaptations to an acrobatic, jumping form of locomotion that he calls **grasp-leaping** (Szalay et al. 1987, Szalay 2007), in which the grasping hind feet provide increased security during takeoffs and landings and the visual specializations of primates allow for more precise estimates of the nature and distance of potential landing sites. One problem with this account is that other arboreal leapers—for example, flying squirrels and other gliding mammals—get along perfectly well without primate-like apomorphies. Another is that moving the two eyes together toward the middle of the face actually reduces the range over which stereoscopic vision can be used to estimate distances, because it diminishes the differences between the left and right visual fields.

Another picture of what early primates were like has been put forward by the primatologist R. Sussman (1991; Rasmussen and Sussman 2007). Like Cartmill, Sussman

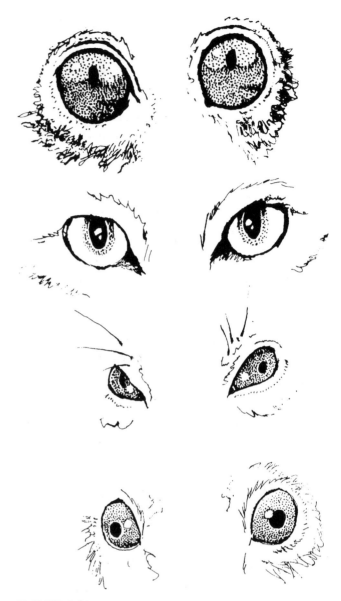

FIGURE 3.39

Orientation of the eyes in various mammals. **From top to bottom**: A lorisiform primate (*Nycticebus coucang*), a small Asian cat (*Felis margarita*), a megachiropteran fruit bat (*Nyctimene rabori*), and a phalangeroid marsupial (*Pseudocheirus peregrinus*). In the loris and the cat, both eyes point in the same direction; the bat and the phalanger are somewhat wall-eyed. (From Cartmill 1992.)

thinks that ancestral primates were small terminal-branch foragers. But like Szalay, he thinks that they were primarily fruit-eaters. Sussman argues that early primates developed grasping feet to facilitate foraging for fruit, nectar, and other angiosperm products out in the terminal branches, and that primate visual specializations evolved to help early primates locate and manipulate food items "... of very small size (e.g., fruits, flowers, *and* insects), at very close range, and under low light conditions." There is no doubt that grasping feet can evolve in climbing herbivores (Kirk et al. 2003), but Sussman's explanation of primate visual specializations seems less plausible. Among prosimians and most non-primate mammals, specializations of the eye and brain for keen vision are associated with visual predation, whereas fruit- and nectar-eaters have lower-resolution eyesight (Arrese et al. 1999, 2002; Tetreault et al. 2004; but cf. Pettigrew 1986). Fruit-eating does not demand acute vision, since fruits do not run or hide from their "predators." After all, angiosperm fruit and nectar were evolved precisely in order to encourage animals to eat them.

Each of these theories about primate origins—the visual–predation theory, the grasp–leaping theory, and Sussman's angiosperm-products theory—has its own strengths and weaknesses. It may be possible to combine them in a way that conserves their strengths and eliminates their weaknesses. Perhaps each applies to a different phase of early primate evolution. To determine this, we need to know more about the distribution of those synapomorphies among stem-group primates.

FOSSIL PRIMATES: THE STEM GROUP

Crown-group primates, also known as "primates of modern aspect" or **euprimates** (Greek *eu-*, "proper, true"), first show up in the Eocene. All the skeletal synapomorphies of euprimates—the petrosal bulla, grasping hind feet, claw loss, specialized visual apparatus, and so on—are found together as a cluster in these Eocene animals. The extinct animals commonly regarded as stem-group primates show up earlier, in the latest Cretaceous, and continue through the Eocene. Most of them were not much like euprimates in their appearance or adaptations. These earliest "primates" are primates mainly by courtesy of the conventions of Hennigian systematics; that is, they seem to be more closely related to euprimates than anything else is. If they are, then the rules of cladistics require that they be classified with euprimates.

These extinct animals are the **Plesiadapiformes**. Depending on which cladograms you accept, they are either a suborder of Primates or a more distantly related group that belongs in a special order of its own (Wible and Covert 1987, Hooker et al. 1999, Bloch and Boyer 2002, Silcox et al. 2005). Some have argued that the so-called "flying lemur" *Cynocephalus*, a peculiar gliding mammal of southeast Asia, is the closest living relative of plesiadapiforms (Beard 1991, 1993; Kay et al. 1992). The plesiadapiforms take their name from the widely distributed Paleocene and Eocene genus *Plesiadapis* (Fig. 3.40A). When this extinct animal was first described in 1877, it was noted that its molars resembled those of the lemur-like Eocene euprimate *Adapis*. It was accordingly named *Plesi-adapis* ("close to *Adapis*") and classified as a primate. *Plesiadapis* and related fossil forms were at first known almost entirely from teeth and jaws, and they were grouped in the order Primates because of their primate-like molars.

But as fossil skulls and feet of plesiadapiforms began to turn up, it became apparent that these animals lacked most of the other hallmarks of the crown-group primates. At least some (and perhaps all) plesiadapiforms had an entotympanic bulla (Fig. 3.18F; Kay et al. 1992, Bloch and Silcox 2001). They all had small, primitive brains and big snouts. Their eyes were generally small and faced in different directions, and their orbits were not enclosed in complete bony rings (Fig. 3.40). Most of them appear to have had digits tipped with sharp claws rather than flattened nails (Bloch and Boyer 2007, Bloch et al. 2007). Although many plesiadapiforms appear to have been tree-dwellers (Kirk et al. 2008), their last common ancestor probably did not have euprimate-like grasping feet. Plesiadapiforms also have peculiar dental apomorphies of their own, including a reduced dental formula, small canines, and big, specialized front teeth. These apomorphies make it difficult to fit any known plesiadapiforms (with the possible exception of the poorly known early genus *Purgatorius*) into the direct ancestry of euprimates. Although the Eocene plesiadapiform *Carpolestes* had grasping hind feet with a divergent, nail-tipped hallux (Bloch and Boyer 2002, 2007), it has no other distinctively primate-like features, and its cranial and dental apomorphies exclude it from euprimate ancestry (Kirk et al. 2003, Silcox et al. 2005, Bloch and Silcox 2006). Its primate-like hind feet probably represent an evolutionary convergence.

Although *Carpolestes* is not a stem euprimate, it nevertheless demonstrates that grasping hind feet could have evolved in the primate ancestry before the distinctive euprimate specializations of the eyes and brain came into being. The first primate peculiarities to evolve were probably the molar apomorphies—reduced stylar shelf, lowered trigonid, and so on—that are shared by euprimates and plesiadapiforms and were presumably present in their last common ancestor (Cartmill 1975b). That ancestor may have been a somewhat squirrel-like arboreal mixed feeder, like some of its plesiadapiform descendants. Primitive euprimates derived from such an ancestry may have been either *Carpolestes*-like creatures feeding on fruit and nectar out in the terminal branches, as in Sussman's model, or visually predatory mixed feeders, as Cartmill proposes. Either sort of feeding habit could have selected for the euprimate grasping specializations of the hind feet. In Cartmill's model, the predator-like apomorphies of the euprimate visual system (Nekaris 2005, Heesy 2008) also originated at this stage. However, they might have come into being secondarily, as an adaptation for eating the insects

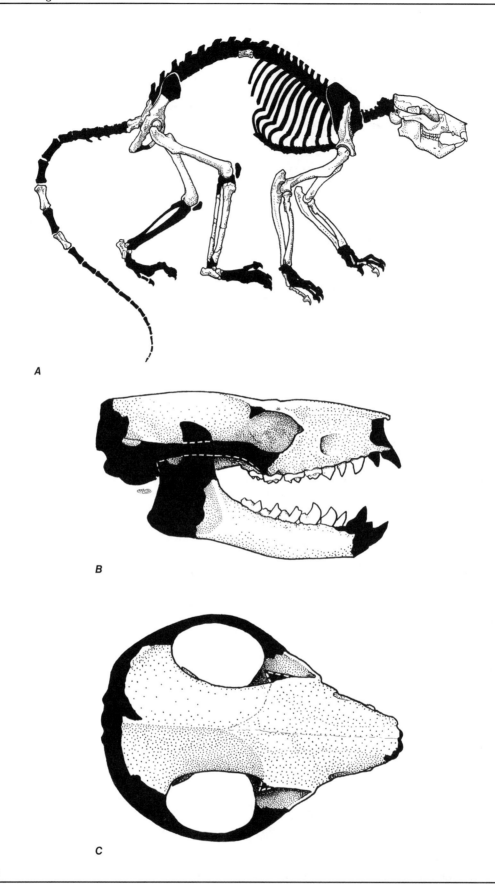

FIGURE 3.40

Plesiadapiforms. **A**: Skull and skeleton of *Plesiadapis*; reconstruction based mainly on *P. tricuspidens* (late Paleocene, Europe). **B,C**: Lateral and dorsal views of the skull of *Palaechthon nacimienti* (middle Paleocene, North America). (**A**, after Simons 1964a; **B,C**, after Kay and Cartmill 1977.)

that were also out in the terminal branches feeding on angiosperm products (Rasmussen 1990, Ravosa and Savakova 2004). The "grasp-leaping" features of the foot skeleton that Szalay emphasizes could also have evolved during a late, visually predatory phase in the process of euprimate origins (Gebo 2004, Bloch and Boyer 2007, Bloch et al. 2007). To sort out these possibilities, we need more fossil finds of early euprimate relatives to help us determine the sequence in which the various primate apomorphies were acquired.

When we look at a crown group of animals, we usually find that the group is distinguished from its living relatives by a suite of shared adaptive traits. These traits are often taken at first to be an **adaptive complex** of features, which evolved together to yield the characteristic new way of life seen in the crown-group animals. In Simpson's evolutionary systematics, such inferred **adaptive shifts** provide most of the rationale for drawing horizontal taxonomic boundaries between ancestors and descendants. But as our knowledge of the fossil record increases, we often discover that the various apomorphies of the crown group evolved at different times for different reasons, taking on changed or enhanced functions when new apomorphies were added later.

The multistage model of euprimate origins sketched above illustrates this pattern of **mosaic evolution**. We find such mosaic patterns elsewhere in the story of human evolution—for example, in the origins of mammals and of "higher" or anthropoid primates and in the evolution of the distinctive apomorphies that separate humans from the living apes.

▌THE FIRST FOSSIL EUPRIMATES

Altiatlasius koulchii from the late Paleocene of North Africa is probably the earliest fossil euprimate found so far (Gingerich 1990, K. Rose 1995). However, it is known only from isolated teeth. It may prove to be a plesiadapiform (Hooker et al. 1999).

The oldest fossils of undoubted euprimates appear in the earliest Eocene. From the very beginning, these early euprimates include representatives of two separate clades, sometimes distinguished as the superfamilies **Adapoidea** and **Omomyoidea** (Covert 2002). Most investigators think that the adapoids represent the earliest radiation of the strepsirrhines (Gebo 2002), and that the omomyoids represent a parallel radiation of stem haplorhines (Gunnell and Rose 2002). Both groups show up almost simultaneously in North America, Europe, and Asia about 55 Mya. One genus, *Teilhardina*, appears in both Europe and Asia at this time. Although details of its dental morphology ally *Teilhardina* with the omomyoids, some contemporary adapoids are very similar to it. In fact, one of the most primitive adapiforms known, *Donrussellia* from the early Eocene of France (Godinot 1992, Rose et al. 1994, Gebo 2002), was originally described as a species of *Teilhardina* (Russell et al. 1967, Szalay 1976). The dental similarities between the earliest adapoids and omomyoids (Fig. 3.41A,B) suggest that *Teilhardina* is a good model for the common ancestor of the two groups (Rose and Bown 1991, Ni et al. 2004).

The most primitive species of *Teilhardina* is *T. asiatica* from the early Eocene of China (Ni et al. 2004). *T. asiatica* was a diminutive animal, with a body weight estimated at 28 grams—about the same as that of the smallest living primate, the pygmy mouse lemur *Microcebus berthae* (Atsalis et al. 1996, Yoder et al. 2001). Unlike plesiadapiforms (other than *Purgatorius*), *T. asiatica* retains all four premolars on each side in both the upper and lower dentition. Its dental formula of 2.1.4.3/2.1.4.3 is primitive for euprimates. The cheek teeth of *T. asiatica* are sharp-cusped, with well-developed shearing edges, resembling those of living primates that eat insects. The small size and dental anatomy of *T. asiatica* imply that insects were an important part of its diet. Its eye sockets are fully euprimate—larger and more frontally directed than those of known plesiadapiforms, as well as separated from the temporal muscles behind by a bony postorbital bar.

On the basis of these facts, the describers of *Teilhardina asiatica* endorse the idea that the ancestral euprimate was a small visual predator. However, they report that the fossil's eye sockets are smaller than would be expected for a nocturnal primate of its size. They conclude that *T. asiatica* was probably active during the daytime rather than at night, and that the ancestral euprimate may also have been diurnal. A recent study of visual photopigment genes in primates (Tan et al. 2005) reaches a similar conclusion.

In rebuttal, the primatologist R. D. Martin (2004) defends the conventional view of primitive euprimates as nocturnal by invoking the assumption that adaptation is a gradual process. If it is gradual, then it must often lag behind what would be optimal for a species in its current situation. When a species adopts a new habit, its morphology will at first be more suited to the way of life of its ancestors than to its own novel lifestyle. Martin argues that *Teilhardina asiatica* may have been a case of this sort—a newly nocturnal animal, whose eyes were still too small to be optimal for nighttime activity. It was not until later in the Eocene, Martin suggests, that nocturnal omomyoids evolved big eyes well-suited to a nocturnal way of life.

The supposed lag between the initial adoption of a new way of life and the subsequent evolution of adaptations to it is often referred to as **phylogenetic inertia**. How much of the anatomy of an organism reflects its own way of life, and how much reflects the adaptations of its ancestors? This issue comes up over and over in

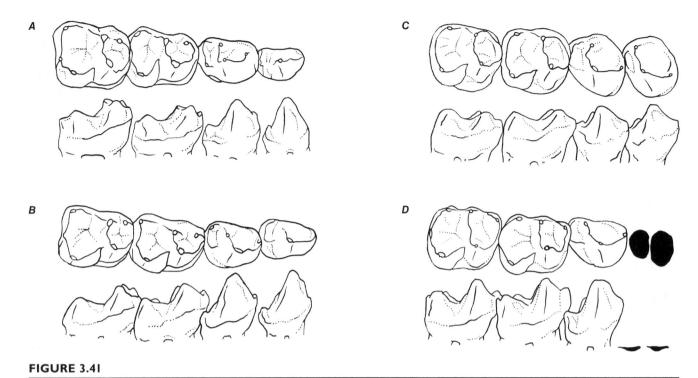

FIGURE 3.41

Occlusal views of the lower right cheek teeth (P₃-M₂) of fossil euprimates. **A**: A primitive adapoid, *Cantius frugivorus* (Early Eocene, North America). **B**: The primitive omomyoid *Steinius vespertinus* (Early Eocene, North America). **C**: The South American anthropoid *Neosaimiri fieldsi* (Miocene, Colombia). **D**: *Eosimias sinensis* (Middle Eocene, China). (After Beard et al. 1994.)

the scientific literature on primate and human evolution. For example, early human relatives in the genus *Australopithecus* walked on their hind legs, but retained many apelike features of their limb bones (Chapter 4). From those apelike features, some experts infer that these creatures must still have been spending a lot of time in the trees (Oxnard 1975a,b, Stern and Susman 1983). But others say that *Australopithecus* was fully terrestrial, and that its apelike features are functionally meaningless leftovers from a simian ancestry, preserved through phylogenetic inertia (Latimer 1991).

Nobody doubts that phylogenetic inertia is a real phenomenon. Adaptation is never perfect, and the morphological novelties that an evolving population comes up with are always limited by the materials that the process of evolution has to work with. But it is not clear how much the *speed* of evolution is restricted by phylogenetic inertia. Observations of evolving populations in the modern world suggest that adaptive change under natural selection is often surprisingly rapid (Grant 1986, Futuyma 1998). We might therefore expect that the lag time between the adoption of a novel behavior and the appearance of morphological adaptations for it should be relatively short on the geological time scale. It seems

hard to believe, for example, that our ancestors in the genus *Australopithecus* walked around exclusively on their hind legs for millions of years while retaining nonadaptive apelike features of their limb bones.

Martin offers another, less controversial objection to the interpretation of *Teilhardina* as diurnal. He points out that *T. asiatica* is smaller than any of the modern primates with which it has been compared. Therefore, the claims made about its relative orbital diameter involve taking a curve calculated from measurements of bigger animals and extrapolating it to tiny animals. It may be misleading to do this, for the same reason that it is misleading to estimate the relative brain size of humans on the basis of cetacean data (Fig. 3.9).

Perhaps the most important single fact about *Teilhardina asiatica* is its extremely small size. Several other early euprimates were no larger, and some appear to have been smaller still, weighing no more than 10–15 grams (Gebo et al. 2001). Primates this small would have been subject to some of the same adaptive constraints as shrews. Scientists have just begun to explore the implications of this possibility for the origin and early evolution of the euprimates (Gebo et al. 2000b, Gebo 2004).

EOCENE "LEMURS" AND "TARSIERS"

Most of the known Eocene euprimates are either adapoids or omomyoids. The two groups differed in their typical ways of life. Although the first adapoids were small animals like *Donrussellia*, later adapoids were mostly medium-sized arboreal herbivores reminiscent of some of the lemurids and indriids of Madagascar (Covert 1986, Gebo 2002). Most omomyoids, by contrast, were little fruit-and-insect-eating prosimians with locomotor habits something like those of today's galagos and cheirogaleids (Strait 2001, Covert 2002, 2005). Both adapoids and omomyoids are relevant to the human evolutionary story, because each is suspected of having given rise to the first monkeys.

The Eocene adapoids are arrayed in three families: Adapidae, Sivaladapidae, and Notharctidae. The **Adapidae** comprise a half-dozen genera of mainly medium-sized primates from the middle and late Eocene of Europe and China. Those adapids known from the most complete remains were heavily built, slow-climbing leaf-eaters with robust skulls and chewing machinery (Fig. 3.42A). The **Sivaladapidae** were a long-lasting but exclusively Asian group. Known almost solely from teeth and jaw fragments, they eventually culminated in some largish (~7 kg) genera with deep, monkey-like jaws found in Chinese deposits from the late **Miocene** epoch, around 7 Mya (Fig. 3.16). The third and largest adapoid group, the **Notharctidae**, comprises some 20 genera classified into two subfamilies: the mainly North American Notharctinae and the less well-known Cercamoniinae of Europe and Africa. Some North American notharctines are represented by complete skeletons, which show them to have been agile, lemur-like arboreal animals (Fig. 3.42B; Dagosto 1993).

The other main group of Eocene primates, the omomyoids, are mainly small, big-eyed animals that are often classified as "tarsioids." Most authorities lump them into a single family, **Omomyidae**. Widespread and diverse, the omomyids radiated in North America, Europe, Africa, and Asia throughout the Eocene and lingered on into the early Oligocene in the northern continents. A possible omomyid, *Ekgmowechashala*, survived into the Miocene of North America. About half of the 30-odd omomyid genera are distinguished from the rest by their big, procumbent central incisors, small canines, and reduced anterior premolars (Fig. 3.42C). These buck-toothed omomyids are commonly set apart as the subfamily Anaptomorphinae.

The relationships of all these Eocene primates are disputed. The long snouts, small brains, and large olfactory apparatus of the adapoids give them a generally lemur-like appearance, which has led many experts to classify them as strepsirrhines related to the living tooth-comb primates. But most of the traits that contribute to the adapoids' lemurish look are just primitive features retained from the ancestral euprimates. Such traits are of no help in figuring out phylogeny. A few features of the foot, including a broad and flexible **talocrural** joint between the uppermost tarsal bone (the **talus**) and the bones of the leg, are shared uniquely by adapoids and tooth-comb primates, and may be strepsirrhine synapomorphies (Dagosto 1988, Gebo 1986, 2002).

Some adapoids show certain monkey-like peculiarities of the jaws and anterior teeth. These include vertically implanted, spatulate incisors, sexually dimorphic canines, and a fusion of the primitive joint (the **mandibular symphysis**) between the left and right halves of the mandible. These features were once hailed as signs that anthropoids evolved from adapoids (Gingerich 1980, 1984; Rasmussen and Simons 1988). But we now know that primitive anthropoids had unfused mandibular symphyses (Simons 2004), and so the fused symphysis cannot be an adapoid-anthropoid synapomorphy. It must be a convergence. The other resemblances between some adapoids and anthropoids may represent convergences as well.

The relationships of the omomyids are equally unclear. Most experts regard them as haplorhines, with affinities to living tarsiers and anthropoids. However, the grounds for this conclusion are skimpy. Some experts interpret the big central incisors of anaptomorphines as a synapomorphy linking tarsiers to omomyids, but the morphologies of these teeth are rather different in the two groups (Figs. 3.23, 3.42C). The limb bones of some omomyoids exhibit somewhat tarsier-like specializations for leaping (Dagosto et al. 1999), but similar specializations have been convergently arrived at in galagos. One feature that omomyids share with tarsiers and early anthropoids was a relatively stiff, steep-sided talocrural joint. If the flexible talocrural joint of adapoids and tooth-comb strepsirrhines is a primitive retention, then a stiffer and tighter joint may be a synapomorphy that connects omomyids to anthropoids and tarsiers. But the stiff "haplorhine" talocrural joint may itself represent the primitive euprimate condition (Dagosto and Gebo 1994, Gebo et al. 2000a).

No omomyoids show the distinctive peculiarities of the orbit and bulla shared by tarsiers and anthropoids (Figs. 2.12 and 3.22). The omomyoid ear region is basically like that of adapoids, except that the ectotympanic ring was apparently drawn out into a bony ear tube in many omomyoids (Fig. 3.22B,D). The omomyoid ear has been described as lemur-like (Hürzeler 1948) because the inner end of the ectotympanic tube (where the eardrum sits) is "intrabullar" (Fig. 3.22B). However, it has also been described as tarsier-like (Simons 1961a) because the ectotympanic is tubular, like a tarsier's, and most of the tube is "extrabullar" (Fig. 3.22B,C). This example illustrates once again how the same anatomy can be described in different words

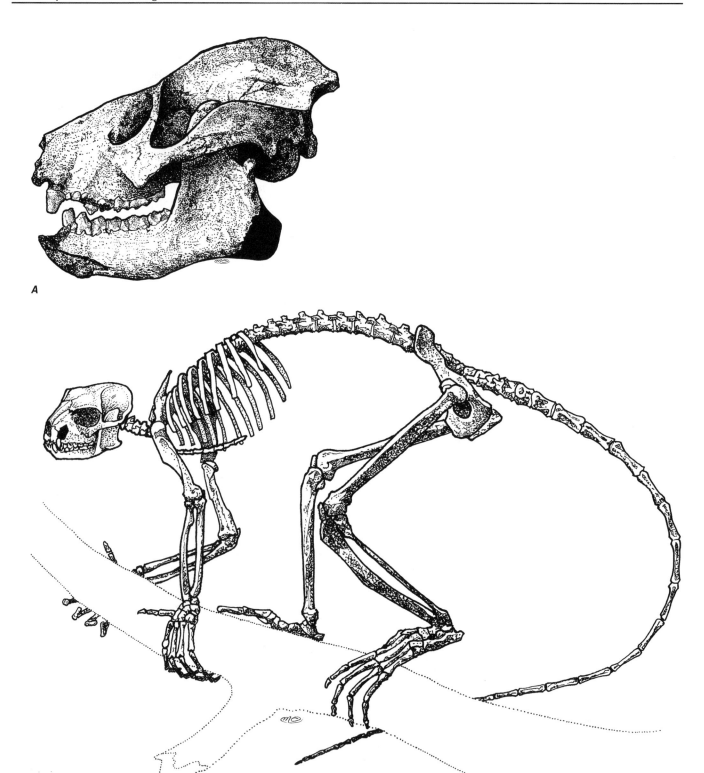

A

B

FIGURE 3.42

Eocene primates. **A**: Skull of *Leptadapis magnus*, an adapine adapid from the middle Eocene of France (after Piveteau 1957). **B**: Skeleton of *Smilodectes gracilis*, a notharctine adapid from the North American middle Eocene (after Simons 1964a); **C**: Reconstructed skull of *Tetonius homunculus*, an anaptomorphine omomyid from the early Eocene of North America (after Szalay 1976). **D**: Reconstructed skull of *Shoshonius cooperi*, an omomyine omomyid from the early Eocene of North America (after Beard and MacPhee 1994).

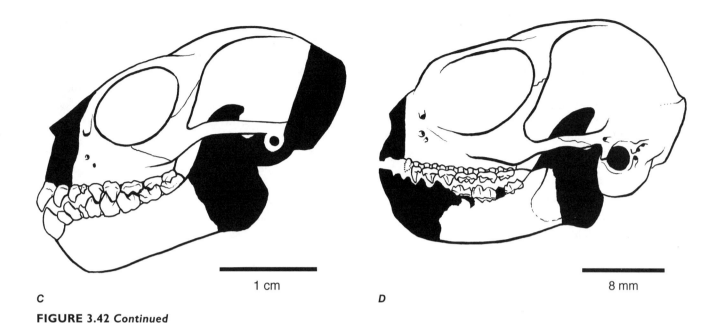

C

1 cm

D

8 mm

FIGURE 3.42 *Continued*

that express and reinforce different ideas about evolutionary relationships.

Over the past 50 years, various investigators have singled out one or another omomyoid as being especially tarsier-like and transferred it to the modern family Tarsiidae. However, different features of the teeth and skeleton are cited in each case, and no Eocene primate exhibits the whole "tarsioid" kit. The list of candidate Eocene tarsiids includes *Necrolemur* from Europe (Simons 1972), *Shoshonius* from North America (Fig. 3.42D; Beard et al., 1991), *Xanthorhysis* from China (Beard 1998), and *Afrotarsius* from Egypt (Simons and Bown 1985, Rasmussen et al. 1998). Some tiny but otherwise strikingly tarsier-like molars and a fragmentary maxilla from the middle Eocene of China have been assigned to the genus *Tarsius* itself (Beard et al. 1994, Rossie et al. 2006). This Eocene "*Tarsius*" was smaller than any living species of *Tarsius* (Gebo et al. 2000b), and we might well assign it to a different genus or family if we had more of its skeleton. But if a true tarsiid was already present in China during the Eocene, then most of the isolated tarsier-like features scattered here and there among omomyoids elsewhere in the world must be convergences or parallelisms.

These issues of classification are hard to resolve, because most of the key apomorphies that unite modern anthropoids and tarsiers as "haplorhines" are features of soft anatomy and biochemistry that leave no traces on fossil skeletons. Some experts suggest that the adapoid "lemurs" of the Eocene may in fact have belonged to the haplorhine clade (Rasmussen 1986). Others argue that the omomyid "tarsioids" were anatomically strepsirrhine, with a wet nose like a lemur's (Beard 1988). The affini-

ties of these Eocene primates may not be settled to everybody's satisfaction until we learn more about the fossil antecedents and evolutionary history of *Tarsius*.

THE FIRST ANTHROPOIDS

Monkey-like primates are also found in the Eocene. The oldest unmistakable anthropoids come from late Eocene strata of the Jebel Qatrani Formation in the **Fayum Depression** of northern Egypt. Searched over with indefatigable persistence and skill by the paleontologist E. L. Simons and his co-workers since 1961, these deposits have yielded a treasure trove of uniquely important fossil primates. The Fayum primates include not only representatives of the usual Eocene groups, but also the tarsier-like *Afrotarsius* (Simons and Bown 1985), two lorisoids (*Saharagalago* and *Karanisia*) that are the earliest tooth-comb primates known (Seiffert et al. 2003), the strange and unique prosimian *Plesiopithecus* (Simons 1992, Simons and Rasmussen 1994), and 18 species of early anthropoids belonging to at least four different extinct families.

All four families of Fayum anthropoids are represented by dental, cranial, and postcranial remains. The known skulls of these animals show two diagnostic anthropoid traits: (1) a complete postorbital septum and (2) a distinctively anthropoid ear region, with an inflated mastoid and an anterior accessory cavity bounded by an intrabullar septum enclosing the internal carotid artery (Fig. 3.22F; Cartmill et al. 1981, MacPhee and Cartmill 1986, Simons 1987, 1990, 1997; Kay et al. 1997). Nevertheless, the Fayum anthropoids are more primitive than any living anthropoids. Their braincases were smaller than those of similar-sized modern monkeys, and the jaws of the earlier and more primitive ones retained an unfused mandibular symphysis into adult life (Beard 2002, Simons 2004, Simons et al. 2007).

In some respects, the early anthropoids from Egypt are more like New World monkeys (Platyrrhini) than like the Old World anthropoids of today. Their ectotympanics were ring-shaped like a platyrrhine's, not tubular like those of modern catarrhines. Their postcranial bones generally resemble those of unspecialized quadrupedal platyrrhines or prosimians, with few or none of the apomorphies that characterize living catarrhines (Gebo et al. 1994, Seiffert et al. 2004). Two of the four Fayum anthropoid families, the **Proteopithecidae** and **Parapithecidae**, retained three premolars above and below, giving them a platyrrhine dental formula of 2.1.3.3/2.1.3.3. However, all these "platyrrhine" traits are merely primitive features. These two families are probably stem-group anthropoid offshoots that lie outside the crown group altogether (Miller and Simons 1997, Ross et al. 1998, Seiffert et al. 2004, 2005). The earliest Fayum primate, *Biretia*, is generally regarded as a parapithecid. Originally described on the basis of isolated teeth from Algeria (Bonis et al. 1988), *Biretia* is now known from a late Eocene site in the Fayum. Cranial fragments of the Fayum species suggest that *Biretia* was a nocturnal animal with big, tarsier-like eyes (Seiffert et al. 2005). A later and less primitive parapithecid, the Oligocene parapithecid *Parapithecus grangeri*, had a small, low braincase, a big snoutlike face, and divergent orbits, giving it a strikingly prosimian appearance (Fig. 3.43A) reminiscent of some of the lemurs of Madagscar.

A third Fayum family, the **Oligopithecidae**, is another basal anthropoid group of uncertain affinities. The earliest and best-known oligopithecid is *Catopithecus browni* from the late Eocene L-41 site, dated on paleomagnetic grounds to around 35.7 Mya (Kappelman et al. 1992). Known from teeth, jaws, crushed skulls, and a few postcranial remains, this little monkey is often taken as a model of the ancestral anthropoid. Its anterior teeth are definitively anthropoid, with interlocking, sexually dimorphic canines and vertically implanted spatulate incisors (Fig. 3.43B). Its molar teeth are primitive for anthropoids: the upper molars are simple, three-cusped teeth with an incipient hypocone, and the lowers retain a high trigonid with a slight but distinct paraconid on M_1 and a less distinct paraconid bump on M_2 (Fig. 3.43C). The small size of *Catopithecus* (around 600 g) and the well-developed shearing features of its molars suggest that insects were an important part of its diet (Kirk and Simons 2001). The oligopithecids are sometimes classified as stem catarrhines (Propliopithecoidea: Table 3.1), mainly because they have a catarrhine dental formula (2.1.2.3/2.1.2.3) and a sectorial P_3 (Rasmussen 2002, Seiffert et al. 2004, Ross and Kay 2004). A few postcranial features of *Catopithecus* have also been put forward as catarrhine synapomorphies (Seiffert et al. 2004).

The fourth and final family of Fayum anthropoids, the **Propliopithecidae**, also have a catarrhine dental formula and are generally accepted as primitive catarrhines. Like those of most later catarrhines, their lower molars have a low trigonid and no paraconid (Fig. 3.43D). Five species of three genera are known, all from early Oligocene sites. The best-known propliopithecid, *Aegyptopithecus zeuxis*, was the largest of the Fayum anthropoids; large males weighed around 7.5 kg (Gingerich 1977). *A. zeuxis* appears to have been an unspecialized arboreal quadruped rather like the living howler monkeys (*Alouatta*) of the New World, to which it is often likened in size and postcranial anatomy (apart from the howlers' prehensile tail). Two fairly complete skulls of *A. zeuxis* have been described. The larger skull, presumably that of a big male, has a long face and a small braincase, which make it look almost as prosimian-like as *Parapithecus* (Fig. 3.43E). A recently discovered female skull, however, has a more gracile, anthropoid-like face (Simons et al. 2007). The mandible is fully anthropoid in both sexes, with a fused, vertically deep symphysis. Just as *Catopithecus* comes close to our ideas of what an ancestral anthropoid would have looked like, *Aegyptopithecus* provides a plausible model ancestor for later catarrhines.

Some earlier and more fragmentary fossils may represent even more primitive anthropoids. *Algeripithecus* and *Tabelia* from the early Eocene of Algeria are known only from isolated teeth (Godinot 1994), and their anthropoid status is uncertain. *Pondaungia, Amphipithecus*, and some related primates from Eocene sites in Myanmar and Thailand, grouped together as the family **Amphipithecidae**, are regarded as anthropoids or anthropoid relatives by some authors (Jaeger et al. 1998, Ducrocq 1999, Kay et al. 2004a, Kay 2005). Known mainly from teeth and limb bones, amphipithecids appear to have been slow-climbing arboreal animals. Their jaws were vertically deep, like those of anthropoids, but the two halves of the lower jaw remained unfused. *Pondaungia*, at least, had vertically oriented incisors and large canines like an anthropoid's. Although a talus attributed to *Amphipithecus* shows a tight, uniaxial talocrural joint of the haplorhine type (Marivaux et al. 2003), other details of amphipithecid cheek teeth and limb bones suggest that they may have been specialized adapoids (Szalay and Delson 1979, Gebo et al. 1999, Ciochon et al. 2001, Beard 2002, Ciochon and Gunnell 2004). Frontal bones attributed to *Amphipithecus* lack a postorbital septum (Shigehara et al. 2002, Takai et al. 2003), but some experts suspect that these frontals belonged to a nonprimate mammal or a reptile (Beard et al. 2005).

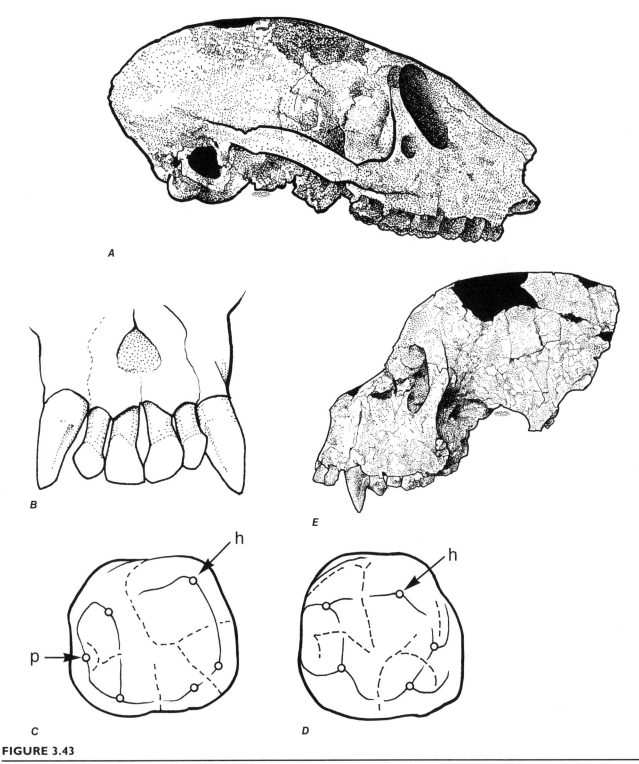

FIGURE 3.43

Fayum anthropoids. **A**: Skull of the parapithecid *Parapithecus grangeri* (after Simons 2004). **B**: Reconstructed upper front teeth of *Catopithecus browni* (after Simons and Rasmussen 1994). **C**: Lower right M2 of *C. browni* (after Simons 1989). **D**: Lower right M2 of *Aegyptopithecus zeuxis* (after Kirk and Simons 2001: reversed). Compare Figs. 3.19, 3.41. **E**: Skull of *A. zeuxis* (after Simons 1972). Abbreviations: h, hypoconid; p, paraconid.

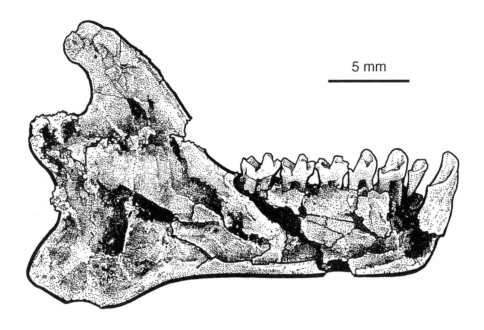

FIGURE 3.44

Eosimias centennicus mandible, lateral view. (After Beard and Wang 2004.)

Another candidate for the title of "earliest anthropoid" is *Eosimias* from the middle Eocene of China, a small primate with sharp-cusped cheek teeth like those of primitive euprimates and an anthropoid-like anterior dentition (Figs. 3.41D and 3.44). Ten other species of *Eosimias*-like primates, mostly unnamed and undescribed, have been found in the same deposits in China (Gebo and Dagosto 2004) and in the middle Eocene of Burma (Takai et al. 2005). These species are grouped together with *Eosimias* as a family, **Eosimiidae**. Some of these species are known only from unassociated tarsal elements, which resemble those of omomyoids but show some anthropoid-like apomorphies (Gebo et al. 2000a, Gebo and Dagosto 2004). The affinities of *Eosimias* and its relatives are debated. Because their cheek teeth look somewhat like those of tarsiers, people who think that anthropoids are related to omomyids and tarsiers tend to accept *Eosimias* as a pre-Fayum anthropoid (Kay et al. 1997, 2004b; Beard 2002, 2004). Conversely, people who suspect that anthropoids evolved from adapids tend to dismiss *Eosimias* as an "Eocene tarsioid" (Rasmussen 2002; cf. Godinot 1994, Tattersall and Delson 1999, Gunnell and Miller 2001).

Because *Eosimias* has dental resemblances to both anthropoids and tarsiers, it tends to pull tarsiers into the anthropoid clade in cladistic analyses, so that the modern *Tarsius* gets grouped with anthropoids to the exclusion of both omomyoids and adapoids (Kay et al. 1997, 2004b; Seiffert et al. 2005, Takai et al. 2005). This grouping was originally proposed (Cartmill and Kay 1978) to account for the apomorphies of the orbit and ear region that *Tarsius* shares uniquely with crown-group anthropoids (Figs. 2.12 and 3.22). However, a petrosal bone from China that has been attributed to *Eosimias* looks like a typical omomyoid petrosal, with none of the anthropoid-like peculiarities seen in *Tarsius* (Beard and MacPhee 1994, Ross and Covert 2000). No matter whether *Eosimias* turns out to be an early anthropoid or an omomyoid, it shows at any rate that adapoids were not the only middle Eocene primates with anthropoid-like front teeth. Vertically implanted incisors may in fact be a primitive trait inherited from the ancestral euprimates (Gingerich et al. 1991, Covert and Williams 1991, 1994).

All these supposed basal anthropoids illustrate how difficult it can be to identify "missing links" in the context of mosaic evolution. The synapomorphies of the anthropoid crown group, which seem so clear and distinctive when we look at living primates, melt away one by one as we work backward through stem-group taxa toward the ancestry of the anthropoid clade. The Fayum anthropoids lack the big brains of modern anthropoids. Some of them still had unfused or incompletely fused mandibular symphyses. Amphipithecids may have lacked a postorbital septum. *Eosimias* may have had an omomyoid type of bulla, not the anthropoid type. We can argue that the absence of these anthropoid features excludes these pre-Fayum groups from the anthropoid clade—or that it shows in what sequence the mosaic of anthropoid synapomorphies was acquired. Here again, it will take new fossil discoveries to resolve these questions and shed more light on the origins and ancestral adaptations of higher primates.

ANTHROPOID RADIATIONS

Planetary temperatures fell steadily throughout the Eocene (Fig. 3.16), leading to a drop in plant biodiversity rivalling that seen at the end of the Permian (Niklas and Tiffney 1994). During the succeeding Oligocene and Miocene epochs, the world remained relatively cool, though it was warmer overall than it is today. Polar ice sheets covered most of Antarctica by the Early Miocene, and they appeared and spread in other high-latitude areas—Greenland, Siberia, Alaska, Scandinavia—during the second half of the Miocene, from 10 to 5 Mya (Denton 1999). Subtropical forests disappeared from North America and northern Eurasia, to be replaced by cooler, drier forest communities and the world's first grasslands (Jacobs et al. 1999). These ecological changes wiped out primates in the higher latitudes (along with the last multituberculates and plesiadapiforms). But closer to the equator, primates continued to flourish and evolve.

Outside of Madagascar, almost all of the surviving post-Oligocene primates were anthropoids. Anthropoids from Africa somehow managed to reach South America during the Oligocene. Exactly how they did this is a mystery, since South America at this time was surrounded by seawater for hundreds of miles in every direction. Some think that the crossing might have been accomplished by a pregnant female "rafting" across the South Atlantic on a floating mass of vegetation torn loose by storms along the African coast (Ciochon and Chiarelli 1980).

However they got to South America, the New World monkeys appear to have split off from catarrhines about 34 Mya (Ross and Kay 2004). They first appear in the fossil record some 8 My later, in the late Oligocene of Bolivia (Fleagle and Tejedor 2002). Platyrrhines radiated during the Miocene, producing several diverse subgroups in South America and the Greater Antilles. Along with other South American fauna, they spread northward into tropical North America around 3 Mya, when the two American continents first became joined together (Marshall 1988, MacPhee and Horovitz 2002, Hartwig and Meldrum 2002).

In the Old World, there is a 10-million-year gap in the anthropoid fossil record after the Fayum. Fossil catarrhines reappear in the latest Oligocene, around 24 Mya (Leakey et al. 1995b). Like platyrrhines, catarrhines diversified into several groups during the Miocene. One of the earliest offshoots from the catarrhine stem was the **Pliopithecoidea**, an extinct group of monkey-like primates known from several Miocene genera of Europe and China (Table 3.1). The pliopithecoids share few synapomorphies with other catarrhines. Although pliopithecoids have the catarrhine dental formula, their ectotympanics do not form a long bony tube of the modern catarrhine type (Zapfe 1958, 1960), and they retain some primitive postcranial details that have been lost in the crown group (Andrews and Delson 2000). They appear to be stem-group catarrhines derived from an ancestor resembling *Aegyptopithecus* (Begun 2002a).

The other anthropoids known from the Old World Miocene are definitely catarrhines of the modern sort, with tubular ectotympanics like those of living apes and Old World monkeys. Most of these Miocene catarrhines are "dental apes"—monkey-like quadrupeds that resemble modern hominoids in the cusp pattern of their molar teeth (Fig. 3.27) but lack some or all of the hominoid specializations for arm-swinging. Most of them are known only from jaws and dentitions, and they are therefore distinguishable mainly on the basis of size and variations on the general theme of "ape teeth": Y-5 molars, sectorial P_3's, and big canines. When different experts describe these variations in their own words and run them through the machinery of cladistic systematics, they come up with different cladograms and classifications (Young and MacLatchy 2004). We will sidestep these disputes by lumping all the quadrupedal "dental apes" together into a paraphyletic superfamily **Proconsuloidea** (Harrison 1987, 2002). In what follows, the superfamily Hominoidea and the word "hominoid" are reserved for animals that are known or reasonably suspected to have had the suspensory locomotor specializations characteristic of the crown group (Table 3.1). We will use the informal term "Y-5 primates" as a label for the whole collection of proconsuloids and hominoids—dental "apes" plus real apes—taken together.

Although we have swept most of the Miocene "apes" into a single taxonomic wastebasket, they are a diverse collection of primates. We will sketch their diversity by describing six key taxa that are known from skulls and postcranial bones as well as teeth, and we will mention other taxa that may have phyletic affinities to each of these six animals.

1. *Proconsul*: A Primitive Proconsuloid. *Proconsul heseloni* is a medium-sized arboreal primate found in early Miocene deposits of Kenya, dated to about 18 Mya. Several partial skeletons have been recovered, along with a fairly complete skull of an adult female (Fig. 3.45A). *P. heseloni* females probably weighed around 10 kg (Harrison 2002). Because of its African origin and its apelike dentition, *Proconsul* was at first thought to have a special relationship to the modern African apes.[1] But subsequent studies (Le Gros Clark and Leakey 1951, McHenry and Corruccini 1983, Harrison 2002) have shown that *P. heseloni* differed from modern hominoids

1. The Roman-sounding title "Proconsul" was given to the genus as a pun on the name of Consul, a chimpanzee in the London Zoo (Hopwood 1933).

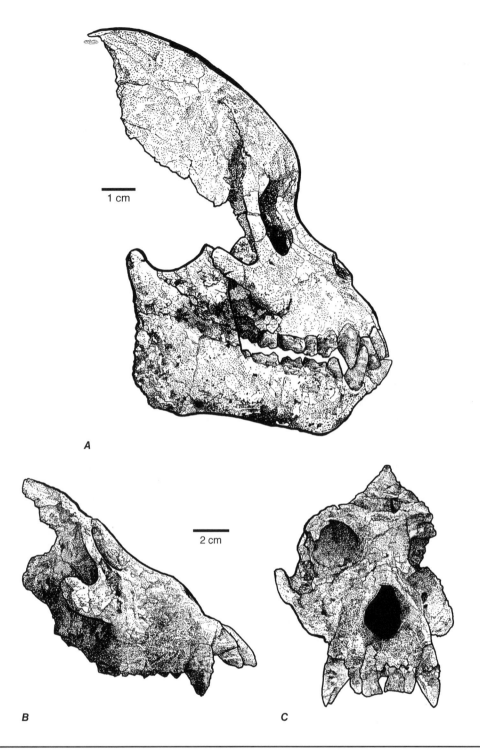

1 cm

2 cm

A

B   C

FIGURE 3.45

Proconsulids. **A**: Skull of *Proconsul heseloni*. **B,C**: Skull of the chimpanzee-sized *Afropithecus turkanensis*. (**A**, after Simons 1972; **B,C**, after Leakey et al. 1988.)

in most of the key postcranial traits that distinguish hominoids from quadrupedal monkeys—the shape of the thorax and scapula, the orientation of the humeral head, the mobility of the forelimb joints, and so on (Figs. 3.29, 3.30, 3.31). The morphology of the carpal bones shows that the supination of the hand was restricted in *P. heseloni*, as in typical quadrupedal monkeys (Morbeck

1975, Ward 1998; *contra* Lewis 1972). Like most quadrupedal primates today, but unlike apes, *P. heseloni* had legs that were longer than its arms (Table 3.2). Remains of a larger species of *Proconsul, P. nyanzae*, reveal that its lumbar vertebrae were elongated and that their transverse processes lay in a monkey-like ventral position (Ward 1993), indicating that *Proconsul* retained the long, flexible lumbar spine and big epaxial musculature of a bounding quadruped (Fig. 3.46). All these facts suggest that *Proconsul* moved around in the trees somewhat like a modern colobid monkey, walking and running on all fours with occasional bouts of leaping and hanging (Walker 1997).

Proconsul is now generally regarded as a primitive crown-group catarrhine—a plausible model for the common ancestor of all the Y-5 primates (proconsuloids and hominoids), and perhaps for the cercopithecoids as well. A dozen or so other Miocene genera from Africa are here provisionally grouped together with *Proconsul* in a paraphyletic family **Proconsulidae** (Table 3.1). Several of these animals were as big as living chimpanzees or orangutans; big males of the species *Proconsul major* may have weighed as much as a female gorilla. The known postcranial bones of these larger proconsulids mostly resemble those of *P. heseloni* (Leakey and Walker 1997, McCrossin and Benefit 1997, Ward 1998, Nakatsukasa et al. 1998, Harrison 2002), implying that they too had quadrupedal locomotor habits. Quadrupe-

dal animals of this size are not likely to have been exclusively arboreal, and some of the largest proconsulids may have been ground-dwelling forms.

The skulls of these big proconsulids (Fig. 3.45B,C) have relatively smaller braincases and larger faces than those of similar-sized great apes today. Unlike the living African apes, they had thick dental enamel. Their **nasal** bones—the bones that form the roof of the nasal opening (Figs. 3.8, 3.34, 3.35)—jut forward high up, above the level of the orbit's lower rim. This primitive trait gives the big proconsulids a snouty appearance reminiscent of *Aegyptopithecus* (Fig. 3.43E). The nasal openings of proconsulids were narrow at the bottom, like those of cercopithecoids or the Fayum anthropoids but unlike those of modern great apes (Fig. 3.35).

2. *Morotopithecus*: The First Hominoid? *Morotopithecus bishopi* is another large-bodied early Miocene catarrhine from Africa with Y-5 molars. It had the snouty face and diamond-shaped nasal opening of a large proconsulid, but it may have been more like a modern ape

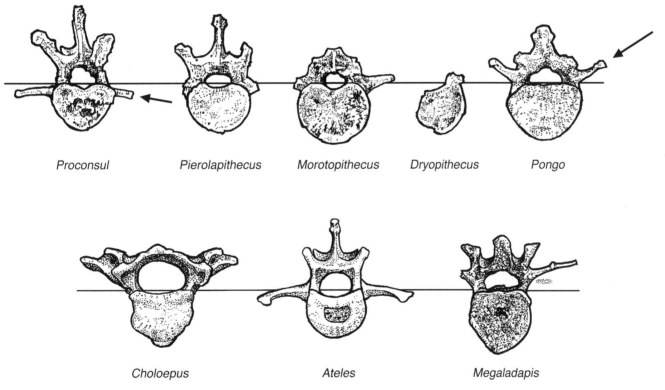

Proconsul   *Pierolapithecus*   *Morotopithecus*   *Dryopithecus*   *Pongo*

Choloepus   *Ateles*   *Megaladapis*

FIGURE 3.46

Lumbar vertebrae of some clambering and swinging arboreal mammals. **Top row**: Mid-lumbar vertebrae of the proconsulid *Proconsul nyanzae* (early Miocene of Kenya), the dryopithecids *Pierolapithecus catalaunicus* and *Dryopithecus laietanus* (late Miocene of Europe), *Morotopithecus bishopi* (early Miocene of Kenya), and a modern orangutan, *Pongo pygmaeus* (after Moyá-Solá et al. 2004). **Bottom row**: Lumbar vertebrae of a two-toed sloth (*Choloepus*), the spider monkey *Ateles*, and the extinct lemur *Megaladapis* (after Shapiro et al. 2005). The vertebrae are co-aligned on horizontal black lines passing through the dorsal edges of the vertebral bodies. Black arrows indicate the position and orientation of the lumbar transverse processes in *Proconsul* and *Pongo*. Not to same scale.

in its locomotion and body form. The main evidence for this claim is the ape-like shape of a mid-lumbar vertebra attributed to *Morotopithecus*. The transverse processes of this vertebra are intermediate in position between those of apes and quadrupedal monkeys (Fig. 3.46). This indicates that the epaxial muscles were reduced in size and no longer working to extend the back powerfully in galloping and bounding. A fragmentary scapula attributed to this species has also been described as distinctively apelike in the shape of its glenoid cavity (MacLatchy et al. 2000; cf. Johnson et al. 2000). These and other features of the postcranium suggest that suspensory clambering and swinging predominated over quadrupedal running in the locomotor behavior of *M. bishopi* (Sanders and Bodenbender 1994, Gebo et al. 1997).

3. *Sivapithecus*: An Ancestral Orangutan? Y-5 catarrhines are also found in later Miocene faunas all across southern Asia, from Turkey to China. Most of these Asian "dental apes" have robust jaws and heavily worn, thick-enameled molar teeth. They exhibit marked sexual dimorphism, which has caused problems for systematists. For many years, the two sexes of the species *Sivapithecus sivalensis*, found in the late Miocene of India and Pakistan, were placed by most experts in two different *families*: the males were identified as apes (Pongidae), while a few particularly gracile females were assigned to a different genus, "*Ramapithecus*," that was thought to be a hominin (Lewis 1934, Simons 1961b, Simons and Pilbeam 1965; see Blind Alley #4, p. 136).

A partial skull of a male *S. sivalensis* found in Pakistan (Fig. 3.47) bears a striking resemblance to that of an orangutan both in overall form and in some anatomical details, including the delicate rib-shaped brow ridges, the narrow space between the orbits, and the level floor of the anterior end of the nasal cavity (Ward and Brown 1986). When this skull was first described (Pilbeam 1982), it was taken as proof that *S. sivalensis* was an ancestor or close relative of the living orangutan (Ward and Kimbel 1983, Ward and Pilbeam 1983). But when postcranial remains of *Sivapithecus* were found, they proved to exhibit certain primitive, monkey-like features associated with quadrupedal locomotion. Although the distal humerus is ape-like, the finger bones are relatively short and straight, and the humeral shaft has a backwards-canted (retroflexed) proximal end—implying that the (unknown) humeral head also faced backwards, as in quadrupedal monkeys (Fig. 3.30). These features suggest that *Sivapithecus* walked quadrupedally atop branches like *Proconsul*, rather than hanging and swinging beneath branches on extended arms, as orangutans and other modern hominoids do (Pilbeam et al. 1990, Pilbeam 1997, 2002; Ward 1997, Rose 1997, Madar et al. 2001).

We can interpret this finding in two ways. If the postcranial specializations shared by the living great

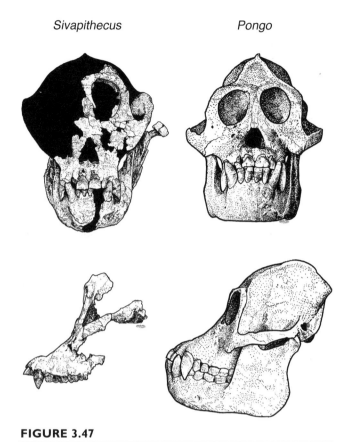

Sivapithecus   *Pongo*

FIGURE 3.47

Facial skeleton of *Sivapithecus sivalensis* (**left**) and *Pongo pygmaeus* (**right**). (**Left**, after Kelley 2002. **Right**, after Elliott 1913.)

apes are synapomorphies, then *Sivapithecus* is not in the orangutan lineage, and so the traits of the facial skeleton shared by *Sivapithecus* and *Pongo* must be either primitive retentions or convergences. Some experts favor this conclusion (Pilbeam et al. 1990, Pilbeam 1996, 2002; Gebo et al. 1997). But others prefer the opposite conclusion—that the postcranial peculiarities shared by the living apes evolved independently in the orangutan lineage and that *Sivapithecus* is an ancestral orangutan despite its primitive humerus and short, uncurved phalanges (Andrews 1992, Begun et al. 1997, Ward 1997, Larson 1998).

Other Miocene proconsuloids from eastern Eurasia have been put forward as candidate proto-orangutans. *Ankarapithecus* from Turkey shares some of the orangutan-like cranial traits seen in *Sivapithecus* (Andrews and Cronin 1982, McCrossin 2005). The Chinese form *Lufengpithecus* has more orang-like molar teeth than *Sivapithecus* has, but its skull is not as much like that of an orangutan (Schwartz 1990, 1997; Kelley 2002). *Khoratpithecus* from Thailand also has orang-like teeth, and it appears to have resembled orangutans in lacking a jaw-opening muscle (the anterior digastric) found in all other catarrhines (Chaimanee et al. 2004). Nothing is known about its facial or postcranial skeleton.

All these supposed orangutan relatives are of approximately the same age (late Miocene, around 10 Mya). Because they all resemble *Pongo*, albeit in different ways and to different degrees, they have all been placed in the orangutan family Pongidae by some authorities. But if *Sivapithecus* and its relatives are pongids, then the arm-swinging adaptations shared by the living apes must have evolved at least three separate times: once for orangutans, once for gibbons, and once for hominids. It seems equally likely that orangutan-like facial morphology evolved twice—once in the orang lineage, and convergently in the ancestry of *Sivapithecus*. We provisionally interpret *Sivapithecus* and most of the other Miocene pseudo-orangutans as comprising an extinct family of specialized proconsuloids, the **Sugrivapithecidae**[2] (Table 3.1). However, *Khoratpithecus* may prove to be a true member of the orangutan family.

The sugrivapithecids are represented in the Pleistocene by the last and largest proconsuloid of all, *Gigantopithecus blacki*. Known only from huge teeth and jaws found in China and Vietnam, *G. blacki* has been estimated to have weighed nearly 550 kg—almost three times as much as a male gorilla (Ciochon 1991). This horse-sized catarrhine was the largest primate that ever lived. A smaller, more *Sivapithecus*-like species, *G. giganteus*, occurs in the late Miocene of India and Pakistan (Kelley 2002).

4. *Dryopithecus*: A Suspensory Primate from Western Europe. The Black, Caspian, and Aral Seas of today are remnants of what was once a continuous, shallow ocean, the **Tethys Sea**, stretching from the Alpine region eastward into central Asia (Fig. 3.48). The Tethys was connected at its western end with the Mediterranean, leaving western Europe isolated at the tip of a long northern peninsula of the Eurasian continent. During the second half of the Miocene, Y-5 catarrhines

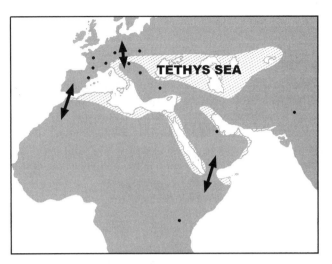

FIGURE 3.48

Paleogeography of Europe in the mid-Miocene, around 17 million years ago. Black dots indicate middle Miocene sites that have yielded proconsuloids and hominoids; black arrows indicate possible faunal interchange routes. (After Bernor 1983.)

colonized this isolated region and underwent a modest evolutionary radiation in the forests of Europe. That European radiation is part of the human story, because it produced the extinct apes that are the most probable fossil ancestors of the modern African hominids.

Dryopithecus laietanus is a medium-sized ape (~25 kg) found in late Miocene deposits of Spain, dated to around 10 Mya. It is known from teeth, jaws, skull fragments, and a partial skeleton that preserves many of the limb bones. The forearm and hand of *D. laietanus* were strikingly elongated relative to its femur, making it the earliest primate known to have had ape-like limb proportions. The primitive contact between the ulna and the wrist bones was reduced, indicating an increased ability to supinate the forearm. The anatomy of the lumbar vertebrae shows that the epaxial muscles of *D. laietanus* were at least as reduced as those of *Morotopithecus* (Fig. 3.46). In short, the postcranial skeleton of *D. laietanus* suggests that it was a suspensory, arm-swinging animal with locomotor anatomy and behavior much like those of modern apes (Moyà-Solà and Köhler 1996, Begun 2002b).

Other late Miocene apes of Europe, here grouped together as a family **Dryopithecidae**, fit the same general picture. The elbow joint of *Dryopithecus brancoi* from Hungary had articular surfaces that facilitated supination of the radius and allowed the elbow to be straightened out completely when the animal hung by its arms (Morbeck 1983). A humeral shaft known for the French species *D. fontani* is missing the upper end, but what is preserved indicates that the humeral head was twisted medially, like that of a modern ape

2. Most recent authors call this family "Sivapithecidae," but this violates an obscure law of taxonomy. The names of family-group taxa (families, tribes, and their sub- and supergroups) are constructed by taking the name of a typical or "type" genus in the group and pasting on various endings: *-oidea* for a superfamily, *-idae* for a family, and so on (cf. Fig. 3.37). The International Code of Zoological Nomenclature (Art. 40) requires that a family-group taxon keep the name first given it, even if its type genus has been sunk into another genus (Mayr 1969). The first person to set up a separate family-group taxon for any of the *Sivapithecus* group of fossil apes named it for a genus, "*Sugrivapithecus*," that was later sunk into *Sivapithecus*—so "Sugrivapithecidae" has priority (Szalay and Delson 1979).

(Begun 2002b). In the related genus *Pierolapithecus* from Spain, the epaxial muscles were reduced (Fig. 3.46), and the ulna had completely lost its contact with the carpals (Moyà-Solà et al. 2004). In all these respects, these European apes resemble living hominoids and differ from the quadrupedal "dental apes" of the African and Asian Miocene. The dryopithecids also have thin dental enamel like today's African apes, whereas sugrivapithecids, orangutans, and most of the few late Miocene "dental apes" known from Africa have thick dental enamel.

Summarizing all this evidence, the paleontologist D. Begun postulates that Europe was colonized about 17 Mya by a thick-enameled "dental ape"—perhaps the poorly known *Griphopithecus*, which occurs at this time horizon in both Asia Minor and western Europe. From some such ancestry, the thin-enameled, arm-swinging dryopithecids evolved, migrating into Africa 8 My later to give rise to the ancestors of humans, chimpanzees, and gorillas (Begun 2002b). This is a plausible story that synthesizes a lot of diverse facts. The main problem with it is that it assumes that orangutans evolved independently from a *Sivapithecus*-like stock. It therefore entails parallel evolution of the hominoid suite of postcranial specializations.

5. *Dendropithecus* and the Hylobatid Problem. The gibbons (Hylobatidae) do not fit very well into what we know about the evolution of the hominoids. These small apes are thoroughly specialized for arm-swinging, and they have often been put forward as a model of what the last common ancestor of the crown hominoids might have looked like. However, no hylobatids are known in the fossil record before the first appearance of *Hylobates* itself, which pops up out of nowhere in the middle Pleistocene of southeastern Asia (Szalay and Delson 1979, Kelley 2002). Molecular-clock analyses indicate that the hylobatids branched off from the other living apes between 15 and 18 Mya (Pilbeam 2002, Rauum et al. 2005). But there is no plausible gibbon ancestor among the known Miocene catarrhines of that date, or among the later proconsuloids of Asia. A poorly preserved late Miocene fossil jaw from China is thought by some to represent a very small species of the crown-group hominoid *Dryopithecus* (Kelley 2002), otherwise known only from Europe. If it really is a dryopithecid, it might be a gibbon ancestor. Unfortunately, it is more likely to be a pliopithecoid (Harrison 2005).

Several other fossil catarrhines have at one time or another been identified as Miocene hylobatids, mainly because they were gibbon-sized animals that had long, slender limb bones, gracile mandibles, and long canines, just as gibbons have. Nevertheless, they can all be unequivocally excluded from the gibbons' ancestry. At least one of them, *Epipliopithecus* from the late Miocene of Spain, is a pliopithecoid, not a crown-group catar-

rhine. Its forelimb bones were long, but its hindlimb bones were longer, and its shoulder, wrist, and elbow joints were like those of quadrupedal monkeys, Despite its long arms and legs, its vertebral column was that of a galloping quadruped, with a primitively long, flexible lumbar region, a short sacrum, and an external tail (Zapfe 1959, 1960; Ankel 1965). The other candidate proto-gibbons comprise four genera of small early Miocene catarrhines from East Africa that have been variously classified as pliopithecoids or specialized proconsuloids (Szalay and Delson 1979, Delson 2000b, Begun 2002a). The best-known of these African forms, *Dendropithecus macinnesi*, is known from several partial skeletons. Like *Epipliopithecus*, *Dendropithecus* appears to have been a long-limbed quadruped with no special resemblance or relationship to gibbons.

6. The Enigma of *Oreopithecus*. *Oreopithecus bambolii* is known from the latest Miocene of Italy, around 6 Mya. It is one of the best-known Miocene hominoids, and also one of the most puzzling. If proconsuloids are dental apes, then *Oreopithecus* is just the reverse—a "postcranial ape," which would be unhesitatingly acclaimed as a crown-group hominoid if its teeth were missing. From the neck down, *Oreopithecus* resembled the living great apes in many details. Its arms and digits were elongated, its thorax was broad, its humeral head was medially directed, its wrist and elbow joints were capable of full supination and elbow extension, its lumbar vertebrae were short and few, and it lacked a tail (Sarmiento 1987, Begun 2002b).

Despite all the ape-like features of its postcranium, the skull and teeth of *Oreopithecus* are a curious mixture of autapomorphies and primitive features unexpected in a European ape of this late time period. Its lower molars sport a peculiar central cusp, the centroconid, which is joined to the buccal cusps by sharp crests that sheared against reciprocal crests on the upper molars (Butler and Mills 1959). The only other fossil primate with lower molars much like this is the Fayum anthropoid *Apidium*, which is not even a catarrhine. The P_4 and first two lower molars of *Oreopithecus* still bore a paraconid cusp—a primitive mammalian feature retained in basal anthropoids and some pliopithecoids, but lost in *Apidium* and all crown-group catarrhines (Figs. 3.19, 3.28: Szalay and Delson 1979, Begun 2002a,b; Beard 2002, Rasmussen 2002). The only known skull of *Oreopithecus* has been crushed flat, and its endocranial volume cannot be measured; but the braincase appears to have been exceptionally small, with extensive sagittal and temporonuchal crests for the attachment of massive temporal muscles (Szalay and Berzi 1973, Harrison 1989, Harrison and Rook 1997). *Oreopithecus* has a tubular ectotympanic, and there is little doubt that it is a catarrhine of some sort. But there is no agreement about just what sort it is. Its locomotion has been variously reconstructed as sloth-like, ape-like, and human-like, and it has been placed in almost every possible catarrhine group, from the cercopithecoids to the hominins (Hürzeler 1968, Szalay and Delson 1979, Sarmiento 1987, Köhler and Moyà-Solà 1997, Moyà-Solà and Köhler 1997b, Begun 2002b, Rook et al. 2004, Wunderlich et al. 1999).

The cheek teeth of *Oreopithecus* are similar in some ways to those of the Early Miocene proconsulid *Nyanzapithecus* from Kenya, and it was once argued that the two are phyletic sisters (Harrison 1986). However, this idea was subsequently given up as requiring too much convergent evolution in other parts of the skeleton (Harrison and Rook 1997). The European dryopithecids were near neighbors of *Oreopithecus* in both space and time. They appear to have shared with it an apelike suite of adaptations to arm-swinging. It seems safest to regard *Oreopithecus* as an aberrant offshoot of the dryopithecids (Harrison and Rook 1997). Its moderately large size (around 30 kg), sharp-crested molars, and powerful jaw muscles suggest a specialized diet of leaves and other fibrous plant tissues low in food value (Kay and Ungar 1997). This might account for its small braincase as well. Some speculate that *Oreopithecus* may have evolved from a dryopithecid trapped on a Mediterranean island where both predators and food resources were scarce and it was no longer advantageous to maintain a big, calorically expensive brain (Begun 2002b).

How we choose to reconstruct the story of ape evolution depends on what we think the chances are that the hominoid suite of postcranial specializations could have evolved in parallel in different groups (Larson 1998). One way to assess this probability is to look at other arboreal mammals that have evolved similar specializations for hanging underneath branches. There are at least three such groups: the tree sloths, the big subfossil "sloth lemurs" (**Paleopropithecidae**) of Madagscar, and the spider-monkey group of New World monkeys (Atelidae). All these animals have converged with apes in some ways. They all have elongated forelimbs and exceptionally limber joints adapted for suspensory modes of locomotion. Their lumbar vertebrae are reduced in number. Some species in each of the three groups have the dorsally shifted lumbar transverse processes that signal a reduction of the epaxial muscles (Johnson and Shapiro 1998, Narita and Kuratani 2005, Shapiro et al. 2005). So did the big extinct lemuriform *Megaladapis*, which appears to have been a non-leaping quadruped that moved around in the trees with the caution befitting such a large animal (up to 75 kg; Godfrey and Jungers 2002). Certain apelike anatomical traits have also evolved among the lorises as adaptations to a similarly cautious arboreal quadrupedalism (Cartmill and Milton 1977).

All these convergences support the notion that the postcranial peculiarities of hominoids are functionally correlated with apelike clambering or suspensory locomotion in the trees. But convergence is not identity. Even when we can legitimately use the same words to describe the hominoid traits and their convergently evolved equivalents in other tree-dwelling mammals, some of the anatomical details are always different. The postcranial skeletons of modern hominoids cannot be confused with those of spider monkeys, sloths, sloth lemurs, lorises, or *Megaladapis*, despite the convergences between these groups (Fig. 3.46).

If the derived traits shared by the living hominoids are true synapomorphies, then something like the following picture of ape evolution emerges. A proconsulid radiation in the early Miocene of Africa gave rise to two radiations of later Miocene apes outside of Africa: the thick-enameled, quadrupedal sugrivapithecids in Asia and southeastern Europe, and the thin-enameled, arm-swinging dryopithecids of western Europe. The living hominoids of Africa, and probably those of Asia as well, have inherited their shared postcranial specializations for arm-swinging from a dryopithecid ancestry.

The biggest problem with this picture is the gibbons. If the molecular clocks have been calibrated correctly, gibbons split off from the other living apes too long ago to be descendants of any known dryopithecid (Yoder and Yang 2000, Young and MacLatchy 2004, Rauum et al. 2005). Gibbons also retain slightly more primitive states of some of the diagnostic hominoid characters than other living apes do. For example, gibbons preserve a small contact between the ulna and the carpal bones, and they usually have one more lumbar vertebra than the great apes (Zapfe 1960, Schultz 1961, Lewis 1972). The medial torsion of the humeral head is less pronounced in gibbons than in the other living hominoids (Larson 1998). All these facts suggest that gibbons might have branched off from the ape ancestry early on, from an early Miocene proto-hominoid (*Morotopithecus*?) that had not yet developed all the hominoid peculiarities of the postcranium.

The few and fragmentary ape fossils recovered from the later Miocene of Africa neither confirm nor refute this picture. Most of them seem to have been thick-enameled quadrupeds like the earlier large proconsulids of Africa or the later sugrivapithecids of Asia (Ward and Duren 2002, Ishida et al. 2004). However, at least one of these creatures, *Nacholapithecus*, resembled modern apes in lacking an external tail (Nakatsukasa et al. 2003). *Otavipithecus*, from the middle Miocene (~13 Mya) of Namibia, has relatively thin enamel and has been proposed as a possible ancestor of the living African hominoids (Pickford et al. 1997). So has *Samburupithecus*, known from a partial lower jaw from the later Miocene (~9.5 Mya) of Kenya (Ishida and Pickford 1997). Some isolated, thick-enameled cheek teeth from early Ethiopian rift deposits (>10 Mya) have been assigned to a new genus *Chororapithecus*, which their discoverers think may be an offshoot from the gorilla lineage (Suwa et al. 2007). Finally, two African hominoids dated to the very end of the Miocene, around 6 Mya, have been put forward by their discoverers as ancestral hominins. We will describe these two genera (*Orrorin* and *Sahelanthropus*) in the next chapter.

In the present state of our knowledge, any reconstruction of ape evolution can only be tentative. A single fossil discovery of the right sort—say, a sugrivapithecid with a lot of humeral torsion, or an undoubted *Dryopithecus* from China—would shake all the parsimony-based cladograms into different configurations (Young and MacLatchy 2004). This sort of thing has already happened more than once following successive finds of new *Sivapithecus* specimens. We can expect it to happen many more times as new fossils and new taxa are recovered from the Miocene.

The latest major anthropoid radiation, and in some ways the most successful, was that of the Old World monkeys, the Cercopithecoidea. Cercopithecoids appear to have originated from early African catarrhines during the late Oligocene gap in the Old World fossil record. Molecular evidence points to a date about 25 Mya for the cercopithecoid–hominoid split (Pilbeam 2002). The earliest fossil cercopithecoids, known from early and middle Miocene sites in North and East Africa, are assigned to their own paraphyletic family, **Victoriapithecidae**. The victoriapithecids already had the bilophodont molars that are the hallmark of the cercopithecoids (Fig. 3.27), but they retained small braincases and long faces somewhat like those of *Aegyptopithecus*. Like later cercopithecoids, victoriapithecids appear to have been adapted for quadrupedal running, with hip, ankle, and elbow joints that were more stable and less limber than those of other catarrhines. The limb bones of these early cercopithecoids, the microscopic wear on their teeth, and the other fauna found at their fossil sites combine to yield a picture of the victoriapithecids as semi-terrestrial animals inhabiting woodlands and forest margins, where they foraged for seeds and tough fruits (McCrossin et al. 1998, Benefit and McCrossin 2002).

This was an unusual way of life for early Miocene primates, but it turned out to be the key to the cercopithecoids' eventual success (Jablonski 2002). World temperatures, which had remained more or less stable through the first half of the Miocene, began to fall once more around 14 Mya (Fig. 3.16). During the later Miocene, dryer and more seasonal forests became prevalent throughout much of the Old World (Janis 1993), and late Miocene primates adapted to living in them. In Asia, the new environments were exploited by proconsuloids, producing the radiation of sugrivapithecids—wholly or partly terrestrial quadrupeds with thick-enameled, heavily worn teeth and massive jaws suited to feeding on hard objects (Kay 1981, Kay and Ungar 1997). But in Africa, the cercopithecoids were already adjusted to this sort of semi-terrestrial existence, and they moved into the new environments as fast as they became available.

Cercopithecids and colobids, the two modern families of cercopithecoids, first appear in the late Miocene of Europe and Africa. By the end of the Miocene, they considerably outnumbered the Y-5 primates. Cercopithecoids soon spread eastward into southern Asia, where they occur in Pliocene sites extending from Iran all the way over to China and Japan. The Asian sugrivapithecids disappeared at this time horizon, apparently driven into extinction by continuing climatic deterioration (Nelson 2003) and by competition from the cercopithecoid invaders. The ancestors of the living apes (and of *Gigantopithecus*) must have survived in the Pliocene forests of Asia and Africa, but no trace of them has yet been found in the fossil record. Perhaps they were restricted to environments where fossilization did not occur—for example, upland areas with high rainfall.

During the Pliocene and Pleistocene, the cercopithecoids radiated to produce dozens of arboreal and terrestrial species, including huge baboons weighing more than 60 kg (Delson 2000a, Jablonski 2002), while the arm-swinging hominoids dwindled to a few survivors in wet tropical forests. After the Miocene, the only well-documented fossil apes are those belonging to a specialized African group that challenged the cercopithecoids on their own turf, moving down to the ground and evolving a striding, bipedal form of terrestrial locomotion paralleling that of the early dinosaurs. These striding hominoids were our own ancestors and close cousins, the "man–apes" of the genus *Australopithecus* from the Pliocene and Pleistocene of sub-Saharan Africa.

CHAPTER FOUR

The Bipedal Ape

■ BEING HUMAN VS. BECOMING HUMAN

We lay a lot of stress on the distinction between human beings and beasts. In our laws and customs, the animal–human boundary separates persons from things that can be made into soap and sandwiches. Because this distinction is so important for us, the stories that we tell about human origins center on what we think are the crucial differences between people and other species.

Our upright, bipedal posture is one important anatomical marker of human status (which is why we almost always imagine space aliens as bipeds), but most of the things that we cherish as essential to our humanity are mental and behavioral attributes. We see big brains and tool use as the main things that make us human—and so we tend to think that these traits, at least in some incipient form, have always distinguished our lineage from those of the apes.

When scientists first began to apply the theory of evolution to our own species, they too tended to think of big brains and technology as the fundamental apomorphies of the human lineage. Nobody expected bipedality to have evolved long before big brains and tools. Although Darwin suggested that the shift to bipedality was the key event in the ascent of man from the apes, he thought that the main value of standing upright lay in freeing up the hands for using tools and weapons (Darwin 1871, pp. 435 ff.). For Darwin and other early Darwinians, big brains, small canines, tool use, and bipedality were inseparable parts of a single adaptive complex. It was assumed that these traits had all evolved along together in a humanlike direction from the beginning of the human story.

When early hominins began to come to light, these assumptions warped scientists' views of the fossils. These extinct creatures had braincases no larger than those of some living apes. Their bones were not found in association with stone tools. Most of the experts therefore initially dismissed them as a group of aberrant apes—a sterile evolutionary sideline, ultimately pushed into extinction by the true human ancestors that were thought to be lying out there somewhere in the stratigraphic column, just waiting to be discovered. Because the apelike early hominins did not show the key human

The Human Lineage, By Matt Cartmill and Fred H. Smith.
Copyright © 2009 John Wiley & Sons, Inc.

traits of big brains and stone tools, they were for many years read out of the story of human evolution.

THE TAUNG CHILD

In 1922, the South African anatomy professor E. P. Stibbe was fired from his job at the University of the Witwatersrand after he was seen entering a Johannesburg movie theater with a woman who was not his wife. (The threshhold for campus sex scandals was lower in those days.) A "Help Wanted" message went out to the British medical establishment in London, which pressed forward a young Australian doctor named Raymond Dart as Stibbe's replacement. Dart was at first reluctant to leave his cozy situation in London for a chancy post in a colonial backwater, but he finally yielded to his advisors' urgings and set off for Johannesburg. It turned out to be a fabulous career move.

The Johannesburg area is dotted with the filled-in remains of ancient caves. Like many caves, these had formed long ago as solution cavities etched out of the limestone bedrock by the dissolving action of acid rainwater. They were subsequently filled in again by calcium carbonate precipitating out of the water trickling through the underground caverns. The rock fragments, soil, and trash on the cave floors—including the bones of ancient animals—were cemented together by the precipitating minerals, forming conglomerates called **breccias**. From the beginning of the industrial era in South Africa, mining companies have been blasting out, grinding up, and cooking these minerals to convert the calcium carbonate into lime (calcium oxide) for use in industry and agriculture.

In 1923, when Dart arrived in South Africa, the Northern Lime Company was quarrying and processing ancient deposits of carbonate minerals at a place called Taung, 250 miles from Johannesburg. Miners occasionally found fossils in the rock and kept them as curiosities. When Dart was given one such souvenir, the skull of an extinct baboon, he asked the geologist R. B. Young to fetch him some more fossils from the site. Young went to Taung in November, 1924 and returned with another fossil skull. Removal of its encasing breccia revealed the face and mandible and a partial **endocast**—a natural cast of the inside surface of the braincase—of a very interesting primate (Figs. 4.1 and 4.2A).

The new fossil was no baboon. It looked like the skull of a baby ape. That fact by itself would have made it an important specimen, since no living or fossil apes had ever been found in South Africa. But Dart perceived certain humanlike features in the fossil infant not found in any known ape. Deciding that he had been handed a historic opportunity to describe the proverbial

FIGURE 4.1

The infant skull from Taung: The type specimen of *Australopithecus africanus*. (After Johanson and Edgar 1996.)

"missing link" between man and the apes, Dart began working night and day on the Taung child. It took him less than six weeks to prepare the skull, write it up, and mail his manuscript off to the British science journal *Nature*.

Dart's paper was published on February 7, 1925. In it, he named the Taung anthropoid *Australopithecus africanus*, meaning "the southern African ape." He argued that the small size and apical wear of the Taung baby's canine teeth placed it squarely on the human branch of the hominoid family tree. Features of the preserved endocast hinted that the foramen magnum had been centrally located underneath the skull, implying an upright posture and bipedal habits. Dart estimated that the cranial capacity would have been 525 cm^3 when the child reached adulthood, which is somewhat larger than those of most apes. The faint traces of the brain's surface convolutions on the endocast suggested to Dart that the brain of the Taung infant had some distinctively human-like features—for example, a posterior displacement of the lunate sulcus, implying a relative

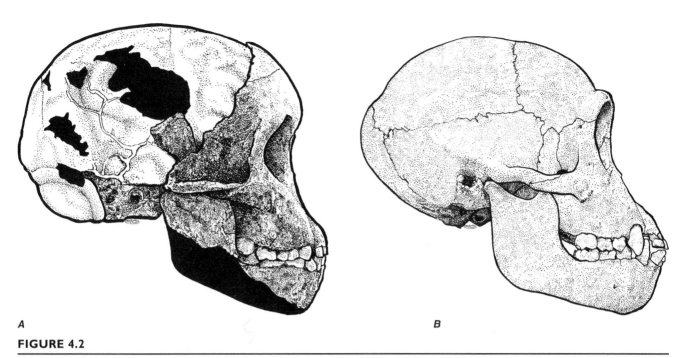

A   B

FIGURE 4.2

The Taung skull (**A**) compared with that of an infant chimpanzee (*Pan troglodytes*, **B**) at a similar stage of dental development (first permanent molar erupted). (**A**, after Johanson and Edgar 1996)

expansion of anterior cortical areas associated with "higher" brain functions.

Like Darwin, Dart thought that a move out of the forest onto the savanna had been a major factor in the development of upright bipedality in our ancestors. *Australopithecus*, Dart argued, represented the first stage in this process: an ape driven to develop certain human-like adaptations because it lived in an environment unsuited to apes. In his view, the Taung area had been grassy, open country in the time of *Australopithecus*, just as it is today. To survive in such a setting, *Australopithecus* must have been more humanlike than any living ape in its anatomy, ecology, and behavior. Summing up, Dart hailed *Australopithecus* as "a creature well advanced beyond modern anthropoids in just those characters, facial and cerebral, which are to be anticipated in an extinct link between man and his simian ancestor." The missing link was no longer missing—or so it seemed to Dart.

Unfortunately, the Taung child was not the best possible specimen to base such claims on. Its face, jaws, and dentition were largely complete, but most of the rest of the skull was missing. Most of the rear two-thirds of the braincase was represented only by the endocast. And because it was only a baby—roughly equivalent in dental development to a 4-year-old chimpanzee—it was hard to be sure how human-like it would have looked if it had grown to adulthood.

Dart's baby met with an icy reception. In the very next issue of *Nature*, four leading experts on fossil man attacked Dart's claims, dismissing *Australopithecus* as merely an extinct great ape, "in the same group or sub-family as the chimpanzee and gorilla" (Keith 1925). Dart's own former advisor G. Elliot Smith called it "an unmistakable anthropoid ape that seems to be much on the same grade of development as the gorilla and the chimpanzee without being identical with either" (Smith 1925). The experts admitted that the Taung child was a peculiar ape, with some suggestively human features. Its braincase was high-vaulted, its brow ridges nonexistent, and its jaws oddly small for a baby ape (Fig. 4.2B). But the final verdict was that "in size of brain this new form is not human but anthropoid" (Keith 1925). Therefore, it could not represent an early stage in the big-brained human lineage.

This remained the orthodox opinion for the next twenty years, as more and more fossils of *Australopithecus* came to light from the Transvaal caves. "The fundamental human characteristic, that is, the great development of the brain, the basis of all our psychological evolution, was never fulfilled in them," wrote the French paleoanthropologist H. Vallois (Boule and Vallois 1952). "Because they lacked humanlike brains," concluded the leading American physical anthropologist E. Hooton (1946), "they remained apes, in spite of their humanoid teeth."

▍ *AUSTRALOPITHECUS* GROWS UP

No more man–ape fossils turned up at Taung. But other cave sites proved more fruitful. In August 1936, the paleontologist R. Broom began exploring a cave complex at Sterkfontein, 70 miles northwest of Johannesburg (Fig. 4.3). Within a few days, Broom and his co-workers had discovered a partial skull of an adult *Australopithecus* (Broom 1936a). Further excavation at Sterkfontein, which continues to this day, has yielded hundreds of bones and teeth of other "missing link" man-apes resembling the infant from Taung.

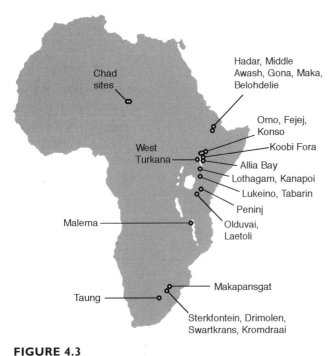

FIGURE 4.3

Early hominin sites of Africa.

Broom (1938) put the man-apes from Sterkfontein into a new genus *Plesianthropus* (Greek "close to human"), mainly because the mandibular symphysis had a different shape from that of the Taung infant. The lower deciduous canines of one Sterkfontein juvenile (Sts 24) also show a small cusp on the front edge that is lacking in Taung. Unimpressed by these differences, most later workers have lumped the Sterkfontein fossils into *Australopithecus africanus*. A more northern cave site, Makapansgat (Fig. 4.3), has also yielded remains of this species.

The Sterkfontein and Makapansgat fossils answered the question of what the Taung baby would have looked like when it grew up. The most complete skull of *A. africanus* is Sts 5, the famous "Mrs. Ples" (Plesianthropus) from Sterkfontein (Fig. 4.4). Despite its nickname, Sts 5 may be the skull of an adult male (Rak 1983, Thackeray et al. 2002). Other, less complete skulls of this species look much like Sts 5, with differences due to individual variation and sex (Fig. 4.5).

At first glance, the adult *A. africanus* skulls look ape-like, with a small brain and a lower face that sticks out far in front of the braincase. Looking at these skulls, many experts at the time thought that Dart's critics had been right to classify the Taung child as an ape. Hooton (1946) dismissed the new Sterkfontein fossils with a limerick:

Cried an angry she-ape from Transvaal,
"Though old Doctor Broom had the gall
To christen me *Plesi-
anthropus*, it's easy
To see I'm not human at all."

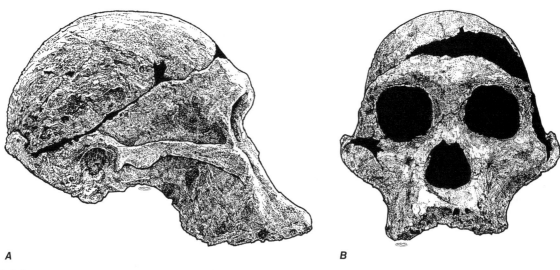

A   *B*

FIGURE 4.4

Lateral (**A**) and frontal (**B**) views of the Sts 5 skull ("Mrs. Ples") of *Australopithecus africanus* from the cave site at Sterkfontein. (After Johanson and Edgar 1996.)

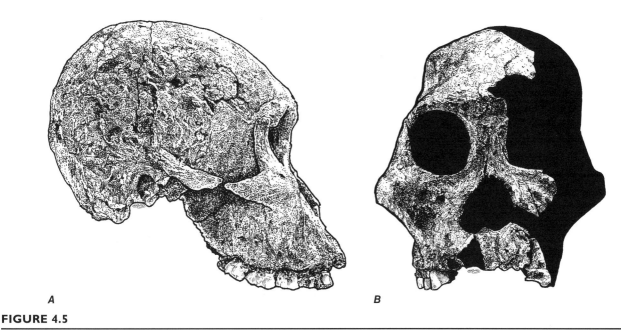

A   B

FIGURE 4.5

Lateral (**A**) and frontal (**B**) views of the Sts 71 skull of *Australopithecus africanus* from Sterkfontein. (After Johanson and Edgar 1996.)

But first impressions can be misleading. Take that small brain, for example. The braincase of Sts 5 has a volume around 480 cc. Estimated cranial capacities for 6 adult *A. africanus* skulls from Sterkfontein and Makapansgat range from 428 to over 500 cc, with an average around 431 (Tobias 1971, Holloway 1975, Conroy et al. 1998, Holloway et al. 2004). In a literal sense, this is an ape-sized brain—no bigger than that of a male orangutan, and smaller than the male gorilla average of some 530 cc. Scientists who prefer to see *Australopithecus* as essentially ape-like have always stressed the fact that it had a brain smaller than a male gorilla's. But male orangutans and gorillas are much larger animals than *A. africanus* (Table 4.1), which was probably no bigger than the living bonobo (McHenry 1992). A plot of body weight against brain volume (Fig. 4.6) shows that *A. africanus* had a larger brain than would be expected for a living ape of its estimated size. To put it another way: *Australopithecus* had a bigger brain than any animal of its size known from all the previous history of life on earth.

Likewise, although the face of *A. africanus* is apelike in its general proportions, the teeth set in that face are very different from those of any living ape (Fig. 4.7). The most conspicuous differences involve the permanent canines, which are small and incisor-shaped—much like ours and unlike the big stabbing fangs found in most apes. Like the milk canines of the Taung child, the permanent maxillary canines of adult *A. africanus* specimens are worn down from the tip as in humans, rather than from the rear as in apes. Because the lower canines are smaller than those of apes, there is no gap

TABLE 4.1 ■ Average Brain and Body Weights of Living Apes and Early Hominins

Species	Body Weight (kg)	Brain Volume (cc)	Log Body Weight	Log Brain Volume
Pan troglodytes male	51.5	398.5	1.71	2.6
P. troglodytes female	40	371.1	1.6	2.57
Pan paniscus male	45	356	1.65	2.55
Pan paniscus female	33	329	1.52	2.52
Pongo male	81	434.4	1.91	2.64
Pongo female	37	374.5	1.57	2.57
Gorilla male	168	534.6	2.23	2.73
Gorilla female	83	455.6	1.92	2.66
A. africanus male	41	485	1.61	2.63
A. africanus female	30	428	1.48	2.63
A. afarensis male	45	430	1.65	2.63
A. afarensis female	29	380	1.46	2.58
A. boisei male	48.6	530	1.69	2.72
A. robustus male	49.8	530	1.70	2.72

Sources: Tobias (1971), Holloway (1972, 1975, 1988, 2000); McHenry (1988, 1992, 1994a); Fleagle (1999), Asfaw et al. (1999), White (2000).

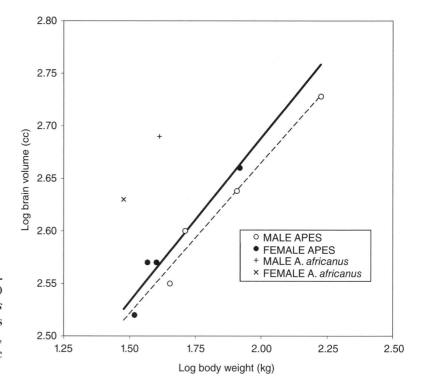

FIGURE 4.6

Adult body weight versus brain volume (cc) in living great apes and *Australopithecus africanus*: log–log plot and least-squares regression line. (Data from Tobias 1971, Holloway 1975, McHenry 1992, and Fleagle 1999.)

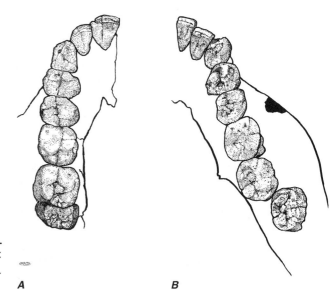

FIGURE 4.7

Permanent dentition of *Australopithecus africanus*. **A**: Occlusal view of upper teeth (Sts 52a from Sterkfontein). **B**: Occlusal view of lower teeth (Sts 52b).

or **diastema** between the upper permanent canine and the lateral incisor, where an ape has a space to accommodate the big lower canine when the jaws are closed. (The Taung child does show a small diastema here in front of the upper milk canine.) The upper permanent canines are also small, so the canine facet on the anterior lower premolar is reduced. That premolar (P_3) therefore

has a rounder outline in occlusal view than the P_3 of any ape (Fig. 3.38) and tends to have at least two cusps of roughly equal size. In apes, the P_3 generally has a single cusp, and is mesially rotated to allow the upper canine to fit in front of it on the buccal side. This morphology allows the P_3 of an ape to have an important shearing or honing function in its occlusion with the big upper

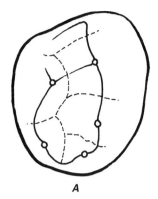
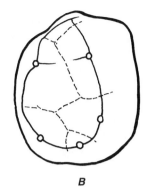

FIGURE 4.8

Lower right anterior deciduous premolars (dp3) of a human child (**A**), the Taung infant (**B**), and a juvenile chimpanzee (**C**). (After Robinson 1956.)

canine, keeping that canine's distal blade sharp by wearing it down from the rear. But the P$_3$ of *A. africanus* functions as a crushing and grinding tooth. This difference is even more conspicuous in the deciduous P$_3$, which is strongly sectorial in apes but molariform in *A. africanus* and humans (Fig. 4.8). The dental enamel of *A. africanus* is thick like that of *Homo*, not thin like the enamel of *Pan* and *Gorilla*. In adult life, the teeth of *A. africanus* wore down to a smooth, flat occlusal surface, with a pronounced wear gradient from front to back, so that the cusps of the first molars were almost completely ground off by the time the third molars erupted (Mann 1975). In all these features of the dentition, *A. africanus* resembles preindustrial humans and differs from the other living African hominoids.

Subsequent studies of the fossil teeth from South Africa supported Dart's claims about the basic "humanness" of the *A. africanus* dentition (Le Gros Clark 1950, Robinson 1956). By itself, the dental evidence might not have been enough to prove a close phylogenetic relationship between *Australopithecus* and *Homo*. Many other mammals, including some Miocene sugrivapithecids, also have relatively heavy dental wear, thick enamel, and small canines (Wolpoff et al. 2006; see Blind Alley #4, p. 136). The human affinities of *A. africanus* were not conclusively established until the discovery of fossil postcranial bones in the late 1940s. Those bones showed that *Australopithecus* had walked upright on its hind feet, unlike any other nonhuman mammal.

Although many mammals occasionally stand or walk on their hind legs, and some hop bipedally whenever they go fast, humans are the only mammals that are habitual striding bipeds. To find anything comparable to human locomotor behavior, we have to look to the diapsids. The primitive archosaurs were also striding bipeds, and their descendants the birds have perfected a bipedal habit derived from the dinosaur model. Nevertheless, human bipedality is quite different from that of a dinosaur or a bird.

All bipeds confront the problem of having a much smaller support base than quadrupeds have. It is harder

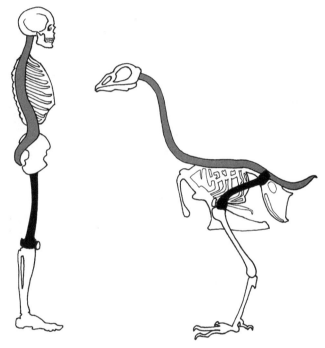

FIGURE 4.9

Bipedality of human and bird compared. The bird balances on its hind feet by moving them forward to a position under the center of mass. The human body is balanced by swinging the trunk upright to bring the center of mass directly above the hind feet. (**B**, after Koch 1973.)

to balance on two feet than on four. To stay balanced, a biped needs to rearrange its body and its limbs so that its center of gravity lies over the contact points of its hind feet. Birds have done this by bringing their knees up alongside the rib cage and shifting their hind feet headward, so that a line connecting the two feet passes directly under the body's center of gravity (Kummer 1991). Humans have adopted just the opposite solution to the balance problem: we fling our hind legs straight out backward and stand up on them (Fig. 4.9; Alexander 2004). The entire length of the body—and the center of

Blind Alley #4: *Ramapithecus*

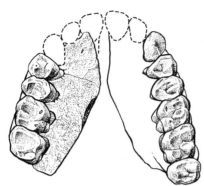

"RAMAPITHECUS PUNJABICUS"

In 1910, Guy Pilgrim described a fragmentary lower jaw from the late Miocene of India. Its Y-5 molars identified it as a "dental ape" of the same general sort as the European *Dryopithecus*, so Pilgrim named it *Dryopithecus punjabicus*. During the 1930s, G. E. Lewis studied some other Miocene primates from British India. One piece of upper jaw with four teeth in it had a short face and a small, empty canine tooth socket. Lewis named this fossil *Ramapithecus brevirostris* (Latin, "short-faced ape of Rama") and suggested that it might have a special relationship to the human ancestry.

Twenty-five years later, Lewis's work sparked a revolution in the study of human origins when Elwyn Simons read it and reexamined the Miocene apes from India. Putting Lewis's *Ramapithecus* maxilla together with its own mirror image, Simons (1961b) argued that it must have had a dental arcade of a distinctively human shape. In 1964, Simons lumped it together with Pilgrim's "*D. punjabicus*" and a thick-enameled contemporary from Africa, *Kenyapithecus*, as a single species—*Ramapithecus punjabicus* (Simons 1964b). He identified this composite creature as a 15-million-year-old human ancestor, virtually identical to *Australopithecus* in its known parts. Studying various bits and pieces attributed to *Ramapithecus*, other scientists found evidence of many human-like traits, including diet, chewing behavior, ground-dwelling habits, and tool use.

Not everybody was convinced. The loudest dissent came from the molecular biologists V. M. Sarich and A. Wilson (1967). Their attempts to calibrate a molecular clock of primate evolution convinced them that the common ancestor of humans and chimpanzees must have lived far more recently than *Ramapithecus*. Therefore, the Miocene fossils from India and Africa could not have a special relationship to the human lineage. "One no longer has the option," wrote Sarich (1971), "of considering a fossil older than about eight million years as a [hominin] no matter what it looks like."

Sarich's blunt words evoked cries of outrage from the morphologists. But the consensus today is that Sarich and Wilson were right and the teeth-and-bones experts were wrong. *Ramapithecus* appears to have been an artificial construct, manufactured by sticking a small canine from an African ape onto maxillae from two gracile females of the Asian sugrivapithecid *Sivapithecus* (Fig. 3.52A: Greenfield 1979, Kay 1982, Kay and Simons 1983). In retrospect, "*Ramapithecus*" looks like one more expression of a recurrent wish to think that humans diverged from the apes a lot further back in the past than we actually did.

---

gravity with it—is thus balanced vertically over the hind feet.

Our ancestors' transition to this upright posture was made easier by the adaptations to uprightness that they inherited from the arm-swinging apes. These included the evolution of the pelvic diaphragm and the ventral displacement of the vertebral column toward the body's line of gravity (Chapter 3). But standing and walking on the hind feet alone presented some special challenges that no other apes have ever had to deal with. These challenges were met by the evolution of a suite of distinctive anatomical features found in the human vertebral column, skull, pelvis, legs, and feet. Many, though not all, of these features are found in *Australopithecus*.

▌BIPEDAL POSTURE AND THE VERTEBRAL COLUMN

Supported by the fore and hind limbs and their limb girdles, a typical mammal's vertebral column functions something like the main cable slung between two pillars of a suspension bridge. In a quadrupedal posture, this vertebral "cable" is not compressed by the weight of the body. But when such an animal rears up on its hind feet, all the weight that was formerly carried by the forelimbs is transferred to the vertebral column. The human vertebral column does not carry 100% of the body's weight. Our legs hold up their own weight, which amount to about one-third of total body mass (Zihlman and Brunker 1979), and the weight of our internal organs is partly supported from underneath by the pelvis and pelvic diaphragm. But the weight of the thorax, upper limbs, head, and neck—plus the weight of anything that a human being may be carrying—has to be passed downward through the thoracic, lumbar, and sacral vertebrae to the weight-bearing hind limbs. This places almost all of the human vertebral column under compression.

To bear the stresses of this compression and keep them as small as possible, our vertebral column has undergone some significant alterations. The most obvious is its sinuous shape. Viewed from the side, the

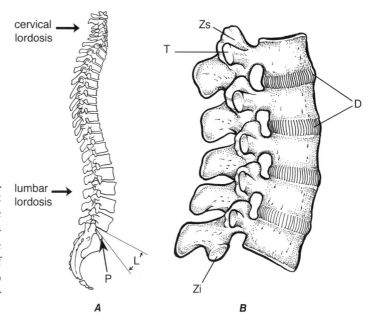

FIGURE 4.10

Curvatures in the human vertebral column (right lateral views). **A**: The column as a whole. **B**: The lumbar vertebrae and intervertebral disks. Abbreviations: D, disk; L, lumbosacral angle; T, transverse process of first lumbar vertebra; P, promontory of sacrum; Zi, inferior zygapophysis (articular process) of L. 5; Zs, superior zygapophysis of L. 1. (After Bogduk and Twomey 1987.)

vertebral column of a gorilla or chimpanzee traces a nearly straight line that slopes at a 45-degree angle from the skull down to the coccyx. By contrast, the human vertebral column describes a series of serpentine curves (Fig. 4.10A). The thoracic and sacral parts of the human column curve forward, while the cervical and lumbar regions are bent backward. The **lordosis** (dorsally concave curvature) in our cervical vertebrae throws the head backward so that it is nearly balanced atop the neck, and the lumbar lordosis throws the thorax back so that it is nearly balanced over the sacrum. The balance of the human head is further improved by the forward and downward relocation of the occipital condyles and foramen magnum (Figs. 3.36 and 3.37). The lumbar lordosis is made possible in part by an increase in the average number of lumbar vertebrae, from four in *Pan* to five in modern *Homo* (Schultz 1961). This increases both the length and flexibility of the lumbar part of the column (Lovejoy 2005a).

Despite these modifications, the balance is not perfect. The centers of gravity of the human head and thorax still lie slightly anterior to the cervical and lumbar parts of the vertebral column. Therefore, the epaxial muscles in the neck and the small of the back have to contract continuously to prevent the head and thorax from toppling forward when we stand or sit upright. (This is why we tend to fall forward when we doze off sitting up.) But the job of those muscles is made easier by the cervical and lumbar curvatures. By reducing the amount of force that the epaxial muscles have to exert to keep things balanced, those lordoses reduce the stress on the vertebral column as a whole—which has to bear all the compression generated by epaxial muscle con-

tractions, as well as that imposed by the body's weight.

The cervical and lumbar lordoses in the human vertebral column are produced mainly by the intervertebral disks in these regions, which are wedge-shaped—thicker in front than they are in back. In the lower lumbar region, a similar "wedging" is evident in the bony bodies of the vertebrae as well (Fig. 4.10B). This wedging accentuates our lumbar lordosis. It is not usually seen in the lower lumbar vertebrae of apes (Robinson 1972, Aiello and Dean 1990, Latimer and Ward 1993; Cartmill, unpublished data; but cf. Rose 1975).

Although our lumbar lordosis helps solve the problem of bipedal balance, it creates new problems. Because the forelimb bears no weight in a biped, all the weight of the upper part of the body has to be transmitted downward to the lower limb. Each vertebra from the skull down to the sacrum therefore has to carry more weight than the one above it. Human lumbar and upper sacral vertebral bodies are enlarged to handle the load (Figs. 4.10 and 4.11). Nevertheless, the stresses at the bottom of the column sometimes become too much for the lumbar vertebrae to bear. Human beings are accordingly plagued by lower back problems: compression fractures of the lumbar vertebrae, tears and ruptures of their intervertebral disks, strained and pulled back muscles, and other pains, traumas, and ailments of the lumbar region. And although our lumbar lordosis helps with balance, it also intensifies the local stresses on the heavily burdened lumbar vertebrae—as you might expect a bend in a pillar to do. The angle between the lowest lumbar vertebra and the top of the sacrum is particularly sharp (L, Fig. 4.10), so "slipped disks" and

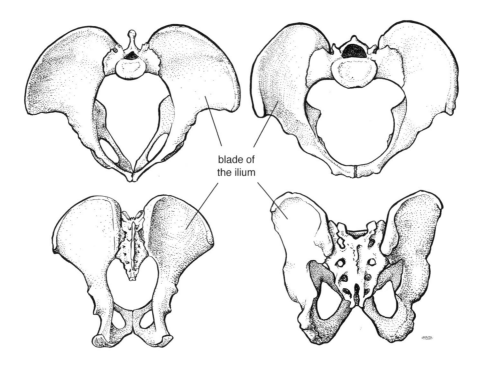

FIGURE 4.11

Pelves of *Gorilla* (**left**) and *Homo* (**right**) seen from the cranial (**top row**) and dorsal (**bottom row**) aspects, showing differences in size of the sacrum and in the orientation and curvature of the iliac blades. (Gorilla pelvis after Schultz 1961.)

other degenerative changes tend to concentrate in this area.

In short, the human lower back is not much of an advertisement for intelligent design! Rather, it exemplifies the Darwinian principle that no adaptations are perfect, because natural selection can act only on the imperfect materials that history has given it to work with.

▋ BIPEDAL POSTURE AND THE PELVIS

With the possible exception of the foot, the pelvis is probably the most distinctive part of the human skeleton. Some small monkeys have vaguely human-looking skulls, with big braincases and short faces, but no other living mammal has a pelvis like ours.

The human difference is most obvious in the ilium, the part of the hipbone that extends up above the hip socket. In quadrupedal mammals, including most primates, the ilium is a stout bony rod. It slants forward and upward from the hip socket to reach the sacrum, which it contacts through the **sacroiliac joint** (Fig. 4.12A). In a standing human, all the weight of the upper part of the body is transmitted downward through the lumbar vertebrae, sacrum, and sacroiliac joint to the hipbone, and thence from the hip sockets down into the bones of the lower

limb. To maintain a bipedal stance with a minimum of muscular exertion, the sacrum and the weight it is carrying need to be more or less balanced over the hip sockets. This could be achieved simply by tipping the pelvis backward (Fig. 4.12B). However, the long ilium would still give the weight acting through the sacrum a correspondingly long moment arm around the hip joint. This long moment arm would make it harder to control the hip joint when the trunk was tipped in any direction.

The human ilium has adapted to the bipedal stance by becoming shorter and wider. Making the ilium vertically *shorter* pulls the sacroiliac joint down toward the level of the hip socket, reducing the moment arm of the trunk's weight around the hip joint. Making the ilium *wider* on its dorsal edge thrusts the sacroiliac joint backward, thereby shifting the hip joint forward relative to the line of gravity (Fig. 4.12C)—so far forward that the gravitational vector of the weight of the trunk actually passes slightly behind the hip socket in a standing human, so that the body tends to tip backward at the hip joint (Fig. 4.13). This backward tipping (extension of the hip joint) is checked in humans not by muscle contractions, but by a thick **iliofemoral ligament** running across the front of the hip joint. As the pelvis tips backward, this ligament becomes taut and checks the backward pitch of the trunk (Fig. 4.13). The attachments of this strong ligament to the bone (Fig. 4.14A) commonly produce a

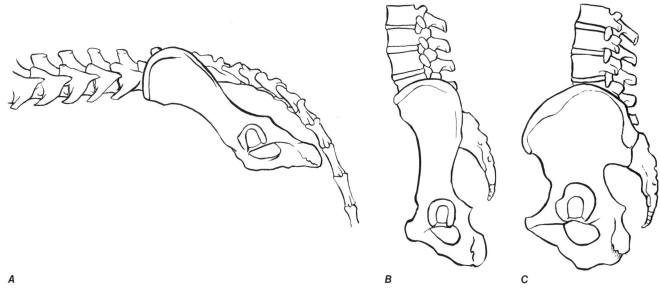

A   B   C

FIGURE 4.12

Remodeling of the human pelvis. **A**: The hipbone and associated vertebrae in a dog. **B**: A similar hipbone superimposed on an upright (human) vertebral column with the sacroiliac joint positioned over the hip socket. The elongation of the ilium results in unnecessary instability. **C**: The human configuration, with a short, broad ilium. (**A**, after Miller et al. 1964.)

faint roughening on the front of the femur, the **intertrochanteric line** (Fig. 4.14B)—a human peculiarity not seen on the thighbones of apes. The pelvic attachment of the iliofemoral ligament is similarly marked by a bony prominence, the **anterior inferior iliac spine** (Fig. 4.15), which is much smaller or absent on apes' pelves (Le Gros Clark 1954, 1967). These adaptations for the stabilization of the hip joint are peculiarly human traits, associated with upright posture. The exaggerated backward sweep of the human ilium produces a deep concavity on the back edge of the hipbone, the **greater sciatic notch**, through which the major nerves of the lower limb emerge from the pelvis, the corresponding concavity in a quadruped's pelvis is much shallower (Fig. 4.15).

The hip joint also has to be stabilized against flexion. This is managed not by ligaments but by contraction of muscles—especially of the big **gluteus maximus** muscle that runs down from the hipbone to the femur in back. The dorsal broadening of the human ilium shifts the origin of gluteus maximus backward relative to the hip socket, giving it a more effective lever arm in extending the hip joint. The lower half of gluteus maximus has been lost in humans, so that its mass is wholly concentrated behind the hip joint; but what is left is greatly enlarged, so that the human muscle is actually some 60% larger than its more extensive counterpart in human-sized apes (Sigmon 1975, Lieberman et al. 2006). All these changes make our gluteus maximus a uniquely powerful and dedicated extensor of the hip. Although it shows no consistent pattern of activity during upright

standing and normal walking, it plays a crucial role in other bipedal activities, acting powerfully to swing the femur into an extended position (e.g., in running, climbing stairs, or in rising from a squatting posture) or to check flexion at the hip (e.g., in leaning forward). In some studies, gluteus maximus has been found to contract briefly whenever the heel touches down in walking, preventing the decelerating body from jackknifing forward at the hip (Wootten et al. 1990, Lovejoy 2005a,b). The redeployment of this muscle was an important factor in the evolution of our upright posture and locomotion (Tuttle et al. 1975, Lovejoy 1988).

The cranio-caudal shortening of the human ilium moves the sacroiliac joint down closer to the hip joint, which reduces the work that gluteus maximus and other hip muscles have to do in habitual bipedality. But this shortening also has an undesirable consequence. Because it draws the sacrum closer to the pubic symphysis (Fig. 4.16), it reduces the size of the pelvic opening through which the head of the fetus must pass during childbirth. Given the large braincase of the human newborn, this makes for an exceedingly tight squeeze. This squeeze is relieved to some extent by the dorsal broadening of the ilium (which shifts the sacrum backward, away from the pubis) and by three peculiarities of the human sacrum itself:

1. The human sacrum is **increased in breadth** (Fig. 4.16). This spreads the hipbones further apart in back and makes the pelvic aperture wider from side to side.

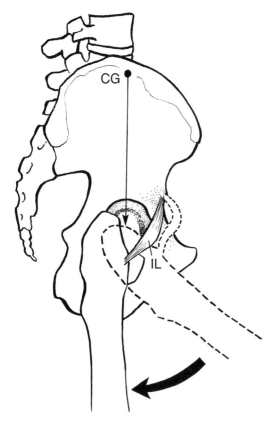

FIGURE 4.13

Stabilization of the extended human hip joint. A line (**vertical arrow**) drawn straight downward through the center of gravity of the upper body (*black dot*, CG) passes just behind the center of the hip joint. The trunk therefore tends to pitch backward around the hips. This movement (*hip extension*) is checked by the iliofemoral ligament (IL), which is slack in the flexed position (*dashed outline*) but becomes taut as the hip moves into extension (*curved arrow*).

2. The human sacrum is also **short**, containing only five fused vertebrae as opposed to a norm of six in *Pan* and *Gorilla* (Schultz 1961). The shortening of the sacrum pulls the coccyx up out of the pelvic aperture, making that aperture deeper from front to back than it would be if we had a long, narrow sacrum like an ape's.

3. Finally, the human sacrum is **less vertical** than the rest of our vertebral column. Its oblique orientation (about 45° out of vertical) has the effect of swinging its tail end up and back, thus further enlarging the anteroposterior (front-to-back) diameter of the pelvis.

Despite these compensating modifications, vaginal delivery in *Homo sapiens* is still a far more difficult, painful, and dangerous process than it is in any ape. During labor, a human fetus's head and shoulders have to twist and rotate in various directions as it squirms its way through the birth canal. The fetus's big braincase cannot pass through the canal without deformation of the skull and brain. To allow the braincase to deform more easily, it retains cranial **fontanels** or soft spots—unossified membranous gaps between the dermal bones of the skull roof—instead of having a more fully ossified skullcap like a newborn chimpanzee (Abitbol 1996). Sometimes the human fetus gets hopelessly stuck during labor, resulting in the death of both baby and mother unless a surgeon intervenes. In combining a relatively huge brain with upright bipedality, the human species is pushing the limits of the mammalian body plan and the eutherian reproductive strategy.

The oblique orientation of the human sacrum helps to ease vaginal delivery, but there is an orthopedic price to be paid for the obstetrical benefit. Reorienting the

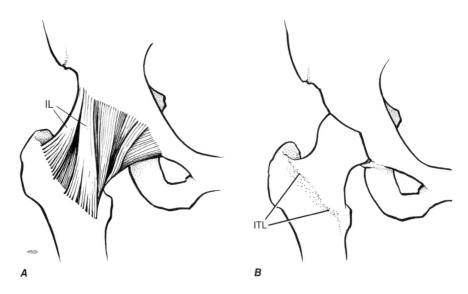

FIGURE 4.14

A: Anterior view of the fibrous capsule of the human right hip joint, showing the iliofemoral ligament (IL). **B:** The intertrochanteric line (ITL).

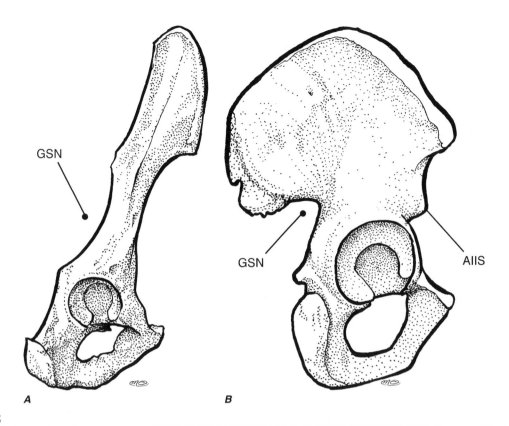

FIGURE 4.15

Lateral views of the hipbone of a chimpanzee, *Pan troglodytes* (**A**), and a human (**B**). AIIS, anterior inferior iliac spine; GSN, greater sciatic notch. (After Swindler and Wood 1973.)

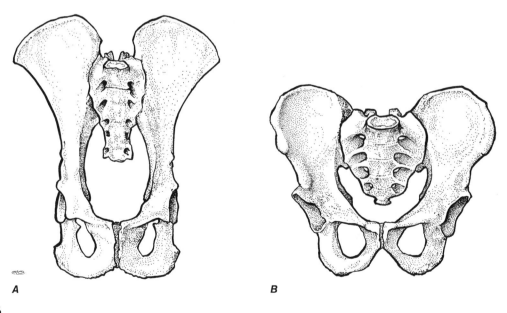

FIGURE 4.16

Pelvis (hipbones + sacrum) of a chimpanzee (**A**) and human (**B**). Ventral views. (After Swindler and Wood 1973.)

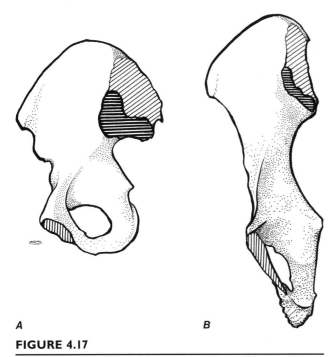

A   **B**

FIGURE 4.17

Semi-diagrammatic medial views of the hipbone of a chimpanzee, *Pan troglodytes* (**A**), and a human (**B**), showing the larger sacroiliac joint (*dark horizontal hachure*) and attachment area for the dorsal sacroiliac ligaments (*diagonal hachure*) in the latter. (After Swindler and Wood 1973.)

sacrum introduces a sharp bend, the **lumbosacral inflection**, between the last lumbar and first sacral vertebrae. As already noted, this inflection makes this a particularly weak spot in the vertebral column, susceptible to tears and ruptures.

The oblique slant of the sacrum also places the top of the first sacral vertebra anterior to the sacroiliac joint. In erect postures, the body's weight therefore presses downward on the front of the sacrum and tends to make the sacrum rotate around the sacroiliac joint, thrusting the sacrum's head end downward and its tail end upward. This is prevented by the helical shape of the sacroiliac joint surface, and also by powerful ligaments that hold up the head end of the sacrum and tie it down to the ischium at its tail end. The **dorsal sacroiliac ligaments** that hold up the sacrum at its head end attach to an expanded area of the medial surface of the ilium above and behind the sacroiliac joint. This area is much larger in humans than in an ape—as is the area of the sacroiliac joint itself (Fig. 4.17).

▌BIPEDAL LOCOMOTION: KNEES

When a terrestrial mammal walks or runs, each of its limbs goes through a repeating cycle of movements.

The limb is lifted off the ground and swung forward (the **swing phase**), then placed down and drawn backward (the **stance phase**), then picked up again, and so on. Whenever the hind leg swings forward, the knee is **flexed** (bent) to keep the foot from dragging as it swings forward. During the stance phase, when the knee is bearing part of the animal's weight, the knee must be **extended** (straightened out) and *prevented* from flexing, so that it can help to support the body.

In our ordinary locomotion, we humans walk with our knee fixed in an extended position during stance phase. At the end of stance phase, the ankle flexes to thrust the body forward and upward, so that we bob up and down as we walk, falling to a low point at the end of each swing phase and rising to a high point in mid-stance (Fig. 4.18A). Like a pendulum swinging back and forth, we are continually converting kinetic energy (of movement) into potential energy (of height) and back again. This sort of **inverted-pendulum** walking helps to conserve energy (Cavagna et al. 1977, Griffin et al. 2004).

This description applies equally well to bipedal birds (Gatesy and Biewener 1991) and to most quadrupedal mammals, which also have an inverted-pendulum walking gait (McMahon 1985). But it does not apply to most of the nonhuman primates. Typical primates walking on all fours flex their knees and elbows as they approach the middle of the stance phase, continually changing the length of the weight-bearing limbs so as to keep their bodies at a constant height above the support (Schmitt 1999, d'Août et al. 2002, 2004). This levels out the fluctuations in their vertical reaction forces. Human beings sometimes walk this way, too (Fig. 4.18B). When we tiptoe, we unconsciously take on a monkey-like pliancy, gliding along with hips and knees flexed to minimize the peak levels of vertical force—for instance, when we are trying to sneak into the house at 3 A.M. without making the floorboards creak.

This sort of yielding or **compliant** walking gait has some benefits for an arboreal quadruped. Flexing the knees and elbows at midstance keeps an arboreal animal's center of mass lowered, thus improving its balance and decreasing its chances of falling off its branch. A compliant gait also helps to save energy in arboreal locomotion. For an animal walking along a horizontal branch, fluctuations in the vertical reaction force will tend to make the branch bounce up and down, generally out of phase with the animal's footfalls. A stiff-legged gait therefore is counterproductive in the trees. It does not yield an efficient, pendulum-like storage and recovery of energy, as it does for an animal walking on solid ground. Most nonhuman primates do not appear to use inverted-pendulum gaits in arboreal locomotion. In laboratory studies, they show more crouching postures,

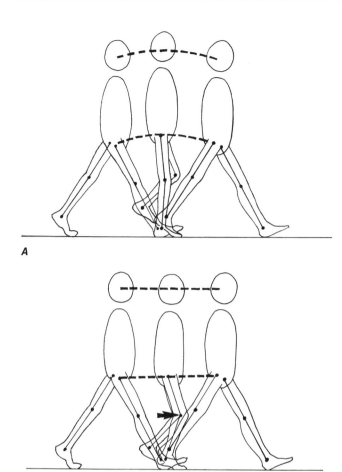

A

B

FIGURE 4.18

Inverted-pendulum and compliant gaits. When the knee is held in the extended position throughout stance phase (**A**), the body arcs up and down like an inverted pendulum (*dashed curves*), changing potential energy (at the high point) into kinetic energy (at the low point) and back again. Flexing the knee as it nears mid-stance (*arrow*) produces a yielding, compliant gait (**B**). Knee compliance results in a flatter trajectory (*dashed lines*) and a smoother "ride," but it requires more muscular work to stabilize the flexed, weight-bearing knee.

more compliant limb deployment, and less fluctuation in reaction forces when they walk on elevated poles than they do when they walk on the ground (Schmitt 1993, 1998, 1999).

Freed from the constraints of arboreal life, human beings have readopted a stiff-kneed gait (except when we are trying to be sneaky). The human knee joint is kept stiff during stance phase chiefly by ligamentous mechanisms, which save on muscular effort. At the end of swing phase, the human knee joint is brought into a fully extended position by the contraction of the big **quadriceps femoris** muscle on the front of

the thigh (Wootten et al., 1990). When the knee is fully extended, several strong ligaments around and inside the knee joint all become taut at once. This brings the joint into a so-called "close-packed" position, preventing any sliding or rotation of the femur on the tibia as long as the quadriceps continues to contract. The **collateral ligaments** on either side of the knee joint are especially important in stabilizing the extended knee. In humans (as in other mammals with a stiff-kneed walk), these ligaments are attached to the lower leg bones and to the big bony "knuckles" or **condyles** at the lower end of the femur in such a way that the ligaments lie well behind the axis of rotation of the knee joint. Our femoral condyles are elliptical—longer from front to back than they are high—and so the collateral ligaments become abruptly taut when the knee reaches complete extension (Fig. 4.19C,D).

The elliptical shape of the human condyles has two other useful effects. When the knee is *flexed* (Fig. 4.19C), the elongated condyles hold the kneecap further away from the axis of rotation of the flexed joint (Lovejoy 2007), increasing the lever arm of the quadriceps muscle (which is attached to the tibia via the kneecap). When the knee is *extended* (Fig. 4.19D), the elongated inferior surface of the condyles comes into contact with the top of the tibia, providing a larger surface area for weight-bearing (Lovejoy 1975). The anteroposterior elongation of the condyles makes the distal end of the human femur about as long from front to back as it is wide, giving it a roughly square outline when seen from below (Fig. 4.20H).

In apes, by contrast, the condyles are more nearly circular in side view, and so the inferior aspect of the distal femur has a more rectangular outline (Fig. 4.20A,B). The collateral ligaments of a chimpanzee's knee are attached close to the joint's axis of rotation, and do not become abruptly taut at the end of extension (Sonntag 1924, pp. 157–158). As a result, there is no preferred, close-packed weight-bearing position for the chimpanzee knee (Lovejoy et al. 1999). This reflects the fact that chimpanzees have a compliant walk, in which the angle of knee flexion changes continually during stance phase, while humans have a stiff-legged walk. This difference is also discernible on the inside of the knee joint. The cartilages or **menisci** inside the joint are more specifically configured in humans to fit the close-packed position of the knee joint, and they are tied down more tightly to the top of the tibia, where they leave more extensive ligamentous markings on the bone than those of apes (Tardieu 1986).

The human knee also looks distinctive from in front. When a chimpanzee stands bipedally, its flexed knees are widely separated (Tardieu 1991). In a standing human, the knees almost touch, and the tibias run

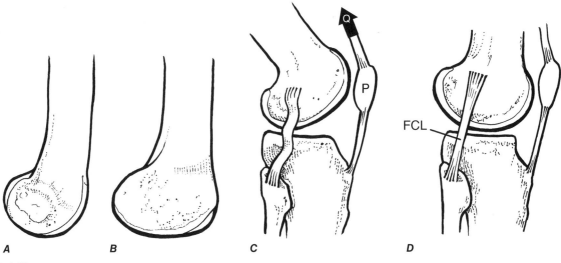

FIGURE 4.19

Distinctive features of the human knee (right side, diagrammatic lateral views). **A**, **B**: Distal femora of a human (**A**) and a chimpanzee (**B**), showing difference in shape of the femoral condyles. **C**, **D**: Human knee joint. When the knee is flexed (**C**), the elongation of the condyles increases the moment arm of the knee-extending muscle, quadriceps femoris (*black arrow* "Q"). When the knee is extended in the close-packed, weight-bearing position (**D**), the collateral ligaments of the joint become taut and the surface area in contact with the tibia is maximized. FCL, fibular collateral ligament; P, patella (kneecap). (**A**, **B**, after Aiello and Dean 1990; **C**, **D**, after Cartmill et al. 1987.)

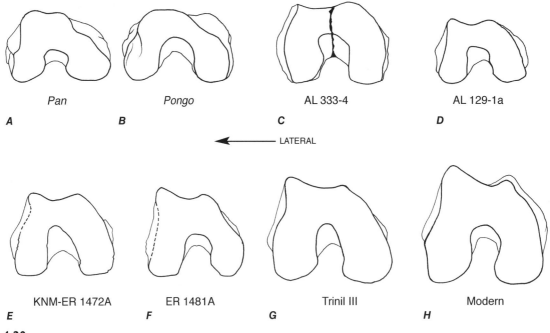

FIGURE 4.20

Diagrammatic distal views of the right femur in great apes and hominins. **Top row**: Apes and *Australopithecus afarensis* (large morph, AL 333-4; small morph, AL 129-1a). **Bottom row**, *Homo*: Two Koobi Fora femora, Trinil femur from Java (reversed), and modern human. (**D**, after Stern and Susman 1983; **A–C**, **E–H**, Tardieu 1983.)

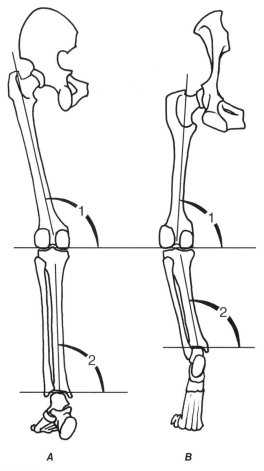

FIGURE 4.21

Diagrammatic posterior views of left hindlimb skeleton of a human (**A**) and a typical quadrupedal mammal (lion, **B**). Angle 1, between the long axis of the femur and a horizontal line drawn across the bottom of the femoral condyles, is greater than 90° in the human (valgus knee) but not in the quadruped. Angle 2, between the long axis of the tibia and the plane of the joint surface between the tibia and the top of the talus, is significantly more than 90° in the quadruped but approximates 90° in the human. (**B**, after Ellenberger et al. 1956.)

straight down from knees to feet. This affords better balance and more effective weight transmission in bipedal walking when the body's support point shifts from one foot to the other. Since our knees are close together while our hip joints are separated by the breadth of the pelvis, the shaft of the femur slants medially downward to the knee (Fig. 4.21). In medical jargon, the human knee is **valgus** (knock-kneed), unlike the knee joints of the African apes. Some orangutans also have a slightly valgus knee (Kern and Straus 1949), but their distal femora are otherwise like those of a chimpanzee (Fig. 4.20A,B). The valgus-ness of the knee can be measured by the angle between the long axis of the

femur's shaft and a line drawn tangent to the two big knuckles or condyles at the bottom of the femur. In *Gorilla* and *Pan*, these two lines are almost perpendicular to each other, deviating from a right angle by an average of only 1° to 2°. In humans, however, this deviation (the **bicondylar angle**) rises to an average of about 8°.

The difference in limb posture between people and the African apes is also reflected in the tibia (Fig. 4.21). In a standing human, the long axis of the tibia is vertical and therefore perpendicular to the horizontal joint surface at the ankle between the tibia and the uppermost tarsal bone, the **talus**. In a chimpanzee, this **talocrural** joint surface lies at an oblique angle to the axis of the slanting tibia (Stern and Susman 1983, 1991; Latimer et al. 1987).

▌BIPEDAL LOCOMOTION: THE HIP JOINT

When a quadruped lifts one hind foot off the ground to swing it forward, the weight of the swinging limb tends to depress the unsupported side of the pelvis. If you walk along behind a walking dog or horse, you can watch its pelvis swiveling left and right as it tips toward the swing-phase side (Fig. 4.22A,B). This pelvic tipping is kept under control by **deep gluteal** muscles (gluteus medius and minimus) that come off the lateral face of the ilium and run laterally down to the **greater trochanter** that sticks out from the upper end of the femur. In a walking dog, the gluteal muscles on one side begin to contract as that hind foot comes down, and continue to contract when the hind foot on the other side is being lifted (Tokuriki 1973). This contraction helps check the pelvic drop on the unsupported side by producing a counter-torque (medial rotation of the femur) around the stance-side hip joint (Fig. 4.22C, arrows 2 and 4). Because the deep gluteal muscles also pull forward on the greater trochanter, they may contribute to propulsive retraction of the femur as well (Fig. 4.22C, arrows 1 and 3). A similar but less pronounced cycle of gluteal contraction is seen in chimpanzees knuckle-walking on all fours (Stern and Susman 1981).

Similar things happen in a walking human: the pelvis tips toward the unsupported (swing-phase) side (Fig. 4.23A,B), and the deep gluteal muscles on the other side are brought into play to counteract this pelvic drop. But because we walk with our hips extended, the deep gluteal muscles have to be arranged differently to do that job. In a human, these muscles run sideways across the top of the extended hip joint. This alters the axis around which they rotate the femur when they contract. When the hip is extended and the foot is off the ground, our deep gluteal muscles act to swing the thigh out sideways (abduction). But when the foot is planted on

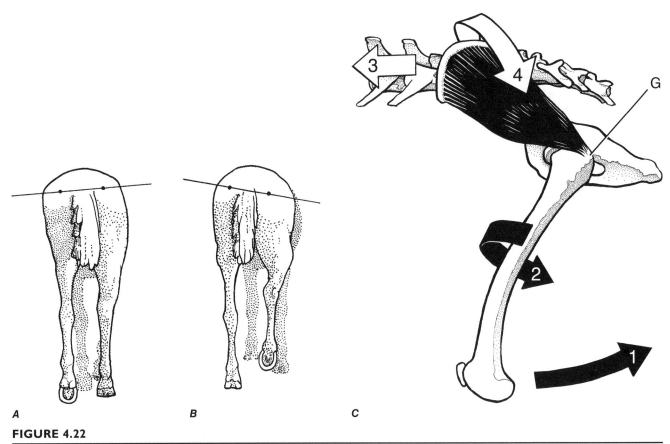

FIGURE 4.22

Pelvic tilt in quadrupedal walking. **A**, **B**: Pelvic tilt in a walking horse, seen from behind (after Muybridge 1887). The drop of the pelvis on the unsupported (swing-phase) side is indicated by a line drawn through equivalent markings on the horse's spotted coat in two successive photographs. **C**: Arrangement and action of the deep gluteal muscles diagrammed in a quadruped (dog). Arising from the dorsal and lateral surface of the iliac blade, these muscles pull anteriorly and medially on the greater trochanter (**G**), producing retraction (arrow 1) and medial (inward) rotation of the femur (arrow 2). When the foot is on the ground (stance phase), these forces move the pelvis rather than the leg (*white arrows*), producing propulsive thrust (arrow 3) and torquing the pelvis toward the supporting foot (arrow 4). This pelvic torque checks the tipping of the pelvis toward the unsupported side during swing phase. (After Miller et al. 1964.)

the ground and supporting the body, they act on the pelvis, preventing it from tipping toward the opposite side when that foot is lifted (Fig. 4.23C). When we walk, the deep glutei on the right side contract when the left foot is off the ground and vice versa, just as they do in a quadruped; but in us, they contract throughout stance phase, act almost solely as abductors, and do not contribute to propulsive thrust.

To make the line of action of the deep gluteal muscles pass directly across the top of the hip joint, the iliac blade of *Homo* has been reshaped so that it curves further around the abdomen toward the belly. The human ilium therefore extends more ventrally and faces more laterally than that of a quadruped (Fig. 4.11). The reoriented ilium is another skeletal correlate of human-style bipedal locomotion.

▌ BIPEDAL LOCOMOTION: FEET

The peculiar human foot is at least as distinctive as our bony pelvis. All other primates have a prehensile hind foot that looks and works something like a human hand, with a thumb-like, divergent hallux that opposes the other toes in grasping tree branches. The human foot has been transformed into a much more rigid structure, which is wholly given over to weight-bearing and has lost its ability to grasp. This transformation has involved a complicated set of changes and adjustments in almost all of the more than two dozen small bones that make up the skeleton of the foot (Fig. 4.24). The most obvious and important of these changes can be summed up under four main headings:

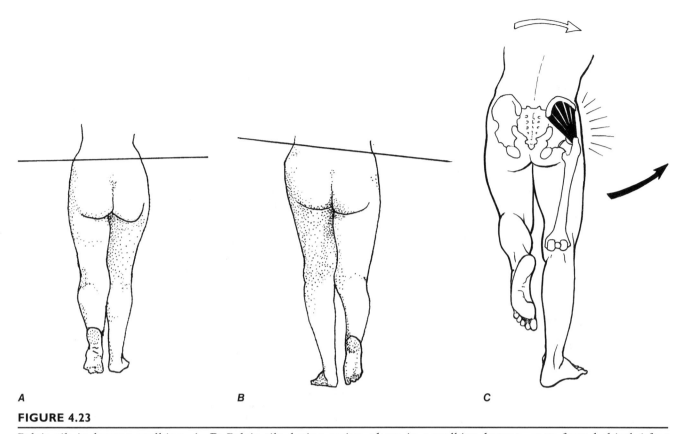

A   *B*   *C*

FIGURE 4.23

Pelvic tilt in human walking. **A**, **B**: Pelvic tilt during swing phase in a walking human, seen from behind (after Muybridge 1913). The line indicating the tilt of the pelvis toward the unsupported (swing-phase) side is traced from a rod fastened to Muybridge's model. **C**: Arrangement and action of the deep gluteal muscles in *Homo*. The deep gluteal muscles run directly over the hip joint on its lateral side. When they contract, they pull medially on the greater trochanter and abduct the hip (*black arrow*). When the foot is on the ground and the opposite foot is lifted, they act instead to resist pelvic tipping toward the unsupported side (*white arrow*). (After Cartmill et al. 1987.)

1. **Formation of a Longitudinal Arch.** The skeleton of the human foot is kinked upward in the middle, forming an arch running the length of the foot. When we stand, about half of the weight borne by the foot is transmitted to the ground through the heel and the rest through the heads of the **metatarsals** (the proximal bones in the five digits), which underlie the ball of the foot. Acting through the talus, the body's weight tends to depress the talus and force the two ends of the arch apart. The primary mechanism preventing this is the **plantar aponeurosis**. This flat tendon deep to the skin of the sole is attached to the bony heel in back and to the metatarsal heads and proximal phalanges in front. By preventing these two attachments from moving apart, the plantar aponeurosis acts to maintain the curvature of the arch in the same way that a bowstring maintains the curvature in a strung bow. Downward displacement of the talus is prevented by ligaments at the top of the arch, especially the so-called "**spring ligament**" underlying the rounded head of the talus (Fig. 4.24A). Whenever heavy loads place dangerous

stresses on these ligaments, muscles in the sole of the foot contract to help maintain the integrity of the longitudinal arch. When we stand on tiptoe, the increased stress on the arch is additionally countered by an automatic ligamentous mechanism. As the **metatarsophalangeal joints** (between the metatarsals and the phalanges of the toes) extend, the proximal phalanges swing up above the plane of the sole into a **dorsiflexed** (bent-upward) position, winding the attached fibers of the plantar aponeurosis around the metatarsal heads like a guitar string around a tuning peg and increasing the tension in the aponeurosis (Fig. 4.24A). This is known in the anatomical literature as the "windlass effect." The tension thus produced in the aponeurosis stores energy, which is recovered when weight is released on the foot and the toes flex during the transition from stance to swing phase. The recovery of energy stored in the plantar aponeurosis and "spring" ligament helps to propel the body forward and makes human walking more energetically efficient. If these ligamentous structures are stretched and lose their springiness,

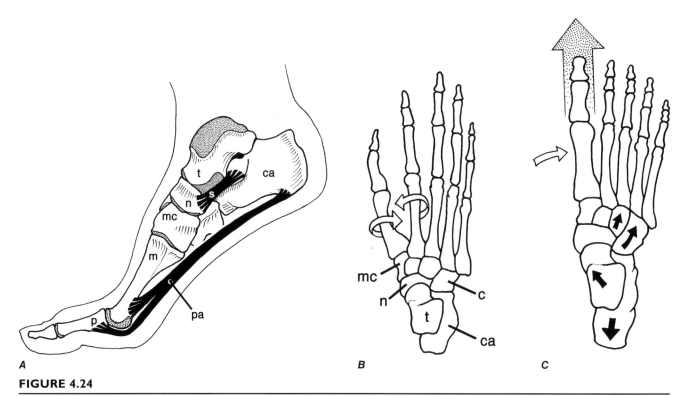

FIGURE 4.24

Skeleton of the foot (diagrammatic). **A**: Medial aspect, showing the longitudinal arch of the human foot and some of the mechanisms that maintain it. **B, C**: Dorsal views of the bones of the right foot in a chimpanzee (**B**) and a human (**C**). Apomorphies of the human foot include the rotation of the metatarsal heads (*arrows* in **B**), the adduction of the hallux (*white arrow* in **C**), the hypertrophy of the hallux (*stippled arrow* in **C**), the reduction of the phalanges of the other four toes, and the general elongation and enlargement of the tarsal bones (*black arrows* in **C**). Abbreviations: c, cuboid; ca, calcaneus; mc, medial cuneiform; m, first metatarsal; n, navicular; p, proximal phalanx of hallux; pa, plantar aponeurosis; s, "spring" (plantar calcaneonavicular) ligament; t, talus. (A, after Basmajian 1982; **B, C**, after Aiello and Dean 1990.)

muscular effort is needed to support the arch in standing and walking. The resulting condition, known as **flatfoot**, is debilitating. It used to disqualify people for military duty, because a soldier with no "spring in his step" cannot march far without suffering pain and muscular fatigue.

2. Enlargement and Adduction of the "Big" Toe. The grasping hallux of nonhuman primates is relatively short, though it is strong and powerfully muscled. But most humans, unlike other primates or any other mammals, have a hallux that is longer and more robust than the other toes (Fig. 4.24C). Its size reflects its importance in human walking: our hallux provides the final propulsive push-off at the end of stance phase, and its metatarsal head carries as much weight as those of the other four toes put together.

To stabilize our hefty hallux in weight-bearing and propulsion, its metatarsal is permanently **adducted** (drawn up alongside the other digits) and bound to the adjoining metatarsal by a strong ligament (Morton 1922). The joint between our first metatarsal and the tarsal

bone at its base, the **medial cuneiform**, has a flat articular surface that is far less mobile than the helically curved joint at the base of an ape's thumblike hallux. These changes render the human hallux incapable of rotating away from the other toes to oppose them in grasping. Since opposition is no longer an option, the heads of all our metatarsals have been reoriented so that they face uniformly downward toward the ground (Fig. 4.24B,C). All five toes of the human foot therefore flex in the same direction and act in essentially the same way in locomotion.

3. Enlargement and Stabilization of the Tarsal Bones. Half of the weight carried by each foot in standing is borne by the bony heel. When we walk, the heel is the first part of the foot to touch down at the beginning of stance phase—and it comes down with a jarring shock, because the knee is locked into full extension at that moment. The human heel and the tarsal bone it springs from, the **calcaneus**, are enlarged and reshaped to bear these increased loads (Latimer and Lovejoy 1989).

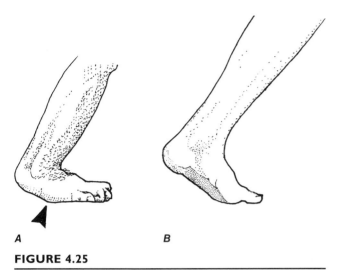

FIGURE 4.25

The midtarsal break (*arrow*) near the end of stance phase in the bonobo foot (**A**) is not seen in the more rigid human foot (**B**). (**A**, after d'Août et al. 2002; **B**, after Muybridge 1913.)

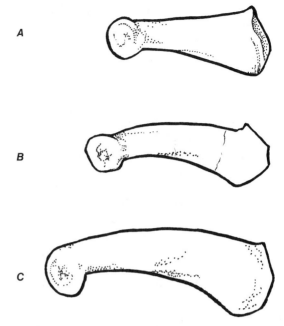

FIGURE 4.26

Proximal phalanx of the middle toe in (**A**) human, (**B**) *Australopithecus afarensis* (AL 333-115h), and (**C**) chimpanzee. (After Susman et al. 1985.)

Although our other tarsal bones normally transmit no weight directly to the ground, they too are enlarged in humans (Fig. 4.24). Taken together, they form the vault of the foot's longitudinal arch. Because that arch needs to be rigid, the tarsal bones and their joints have been reshaped in the human foot to reduce mobility. In the foot of a chimpanzee, these joints allow a lot of bending along a line—the so-called **midtarsal break**—between the calcaneus and talus in back and the other tarsals in front (Elftman and Manter 1935, Meldrum and Wunderlich 1998). When a chimpanzee lifts its heel in walking (Fig. 4.25), the foot bends along this line so that the weight borne by the heel is transferred to the ground through the tarsal bones, and only later on shifted all the way forward to the ball of the foot (the metatarsal heads). In the arched human foot, there is no midtarsal break and the shift to the anterior support point is instantaneous. This ability to shift our support point instantly from one end of the foot to the other is probably an advantage in balancing on one foot, which is something that we have to do far more often than other mammals do.

The upper (talocrural) joint surface of the human talus is not only larger than that of a human-sized ape; it is also shaped differently. When an ape walks, the skewed shape of its tibiotalar surface makes the upper end of the tibia swivel out to the side during stance phase, as the ankle moves from the flexed to the extended position. In humans, the tibiotalar joint surface has a more cylindrical shape. This makes the knee stay close to the midline throughout the stance phase (Latimer et al. 1987).

4. *Reduction of the Phalanges.* Apes have long curved toes for the same reason that they have long curved fingers: their hands and feet alike are used in grasping and hanging from branches in the trees. We can neither grasp with our toes nor hang by them, and so our toe phalanges are short and straight (Fig. 4.26). The powerful finger- and toe-flexing tendons of apes are tied down to the shafts of the phalanges by tendinous bands called **flexor sheaths**, which prevent the tendons from pulling away from the bones like bowstrings when the digits are flexed. The attachments of the flexor sheaths are marked by ridges on either side of the proximal phalanges. These ridges are much less pronounced in humans than in apes. This difference reflects the relative feebleness of our digital flexor muscles, which are no longer called upon to support and propel the body's weight in knuckle-walking or arboreal hanging and swinging.

Most mammals have hands that look quite a lot like their feet. The embryological development of the hands and feet is governed by a set of regulatory genes that affect the distal parts of the forelimb and hindlimb in similar ways (Shubin et al. 1997). But when the hands and the feet need to take on divergent forms and functions, as they have in the course of human evolution, the developmental linkage between hand and foot anatomy can be overridden by natural selection acting on alleles further down in the hierarchy of the genome. The peculiarities of the human foot bear witness to the fact that such low-level genetic changes, accumulated over time by natural selection, can add up to a large evolutionary effect.

AUSTRALOPITHECUS STANDS UP

Halted during the Second World War, excavations at the South African cave sites began again in 1946. On August 1, 1947, a nearly complete pelvis of *A. africanus* was found at Sterkfontein (Broom and Robinson 1950), together with the upper end of a femur and most of a vertebral column. This partial skeleton, Sts 14 (Fig. 4.27), is still one of our major sources of information about the locomotor anatomy of *Australopithecus*. Other postcranial bones have continued to come to light at Sterkfontein ever since.

The postcranial bones of *A. africanus* differed strikingly from those of other primates in distinctively human ways. The iliac blade was short and broad. It curved posteriorly to form a distinct greater sciatic notch and bore a prominent anterior inferior iliac spine in front. Only the upper two sacral vertebrae were preserved, but they were broad, like a human's; and what was left of the sacroiliac joint surface indicated that the sacrum was tipped forward into a human-like orientation. The rotation of the sacrum and the "wedging" of the last two lumbar vertebral bodies suggested that there must have been an overall lumbar lordosis, with a fairly sharp lumbosacral inflection. Two distal femora, found elsewhere in the Sterkfontein deposits, showed that *A. africanus* had a valgus knee joint and human-like bicondylar angle (Robinson 1972). "From considerations such as these," wrote the anatomist W. Le Gros Clark, "it is a reasonable, and indeed an inevitable, inference that the australopithecines had become adapted to an erect bipedalism" (Le Gros Clark 1967, p. 97).

The new fossils helped to bring about a revolution in scientific thinking about *Australopithecus* (Howells 1985, Cartmill et al. 1986, Cartmill 1993). Up to this time, scientists had regarded an enlarged brain as the hallmark of humanity. But the new material from South Africa proved that "… differences in the brain between apes and man … were attained *after* full human status

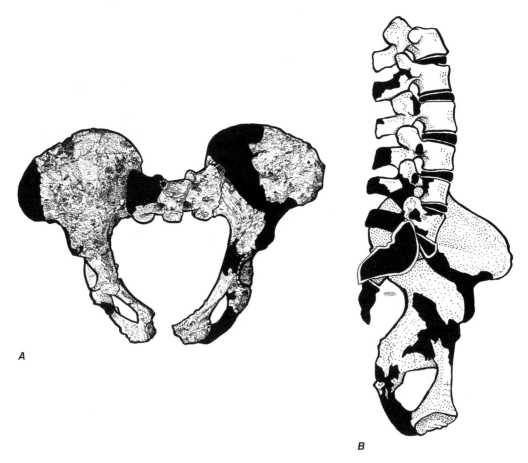

A

B

FIGURE 4.27

The pelvis of *Australopithecus africanus* (Sts 14) from Sterkfontein. **A**: Anterior view of the hipbones and sacrum. **B**: Semi-diagrammatic right lateral view of the pelvis and lumbar vertebrae. (After Pilbeam 1972.)

had been achieved in the limbs and trunk" (Washburn 1951).

With this fact in mind, paleoanthropologists soon came to see bipedality and small, incisor-like canine teeth, not big brains and intelligence, as the essential human properties. The new human hallmarks were interpreted as correlates of tool use: bipedality had been favored by selection because it freed the hands for holding weapons, and hand-held weapons had replaced big canines in fighting. Since *Homo* and *Australopithecus* alike were bipeds with reduced canines, it followed that both must have been tool-using animals that shared an "essential humanity" (Brace 1962). "All species now identified as [hominins]," wrote the paleoanthropologist A. Mann (1972), "shared with modern man a dependence on human cultural behavior, and for that reason they should all be referred to as 'human'." One influential school of thought argued from this premise that there could never have been more than one species of hominins at a time (Blind Alley #5, p. 188). And because some Miocene hominoids had small canines, most of the experts on fossil hominins in the 1960s and 1970s were convinced that a separate human lineage, with at least some incipient form of our adaptive peculiarities, could be traced back into the Miocene (see Blind Alley #4, p. 136).

Many scientists dissented from one aspect or another of this postwar consensus. But there was near-universal agreement on two points: *Australopithecus* had been a biped of some sort, and its teeth and postcranial skeleton showed that it was more closely related to *Homo*—either as an ancestor or a close evolutionary cousin—than any other primate known at the time. These two conclusions are still accepted by almost everybody in the field.

THE SKULL OF *AUSTRALOPITHECUS AFRICANUS*

The skull is the most obviously apelike part of the *A. africanus* skeleton. The cranial vault is low, and the forehead recedes sharply. In a rear view, the maximum breadth of the cranium falls low down on the cranial base, reflecting the small size of the brain relative to the jaws (Fig. 4.84A). *A. africanus* also shows pronounced **alveolar prognathism**, meaning that the lower part of the face (the part containing the **alveoli** or sockets of the upper teeth) sticks far out in front of the braincase (Figs. 4.4 and 4.5). Because the forebrain is relatively small and the face is big, there is a marked, chimpanzee-like **postorbital constriction**—that is, an emphatic narrowing of the braincase behind the eye sockets (Le Gros Clark 1967). In all of these respects, *A. africanus* remains primitive and apelike (Fig. 3.35).

However, a closer look reveals key differences from the apes. Some of these differences relate to posture and the carriage of the head. In *Australopithecus*, the foramen magnum and the occipital condyles are shifted downward and forward into a more horizontal position underneath the braincase. As you might expect from this fact, *A. africanus* has a more flexed cranial base than apes have, and its head was more nearly balanced over its neck than that of an ape. If we draw a vertical line through the occipital condyles, the part of the Sts 5 skull lying in back of that line is about 48% as long as the part in front (Le Gros Clark 1967). This value is intermediate between the corresponding averages of 25% for *Pan* and 80% for modern *Homo* (cf. Figs. 3.36, 3.37, and 4.83). The **nuchal plane**—the area where the neck muscles attach to the back of the skull—is more horizontal in *A. africanus* than in apes, so that not much of it is visible in rear view. The entire nuchal plane of apes is visible from the rear. Like most other large catarrhines, the African apes (*Pan* and *Gorilla*) have a conspicuous bony ridge, the **supraorbital torus**, protruding from the frontal bone above the eye sockets (Figs. 3.35, 3.36, and 3.37). *Australopithecus africanus* lacks a well-defined torus, although there is some thickening along the supraorbital margin.

Apes and *A. africanus* both exhibit extensive alveolar prognathism, because the lower face extends considerably farther forward than the upper face. But in *A. africanus,* the cheekbones are also shifted forward relative to the jaw joint (Rak 1983). This shift, which is carried even further in later species of *Australopithecus,* allows the jaw muscles to exert stronger chewing forces through the cheek teeth (p. 154). Increased loading of the posterior teeth of *A. africanus* is also implied by their large size, their heavy wear, and the anatomy of the jaw joint. In *A. africanus* as in later hominins, the **mandibular fossa** or jaw socket on the temporal bone, where the condyle of the mandible sits, bears a dense bony bump at its front edge called the **articular eminence** or **articular tubercle**. In humans, the condyle slides forward onto this bump when the jaws are parted, and most of the reaction force passing through the condyle during biting is borne by the thickened bone of the tubercle (Hylander 1975a). Hinton (1981) argues that the presence of this bump is functionally correlated with increased loading of the posterior teeth, and that its absence in apes and earlier hominins implies relatively smaller loads on their molars and premolars.

Both *A. africanus* and the African apes have thickened ridges of bone on each side of the face, running up from the alveolar part of the maxilla along the sides of the nasal opening. In the apes, these are called **canine jugae**, and they contain the long, thick roots of the big upper canines. The thinner but similar-looking structures in *A. africanus* are made of solid bone and do not enclose the relatively short canine roots. They were christened **anterior pillars** by Rak (1983), who believes they constitute part of a buttressing system for carrying the enhanced loads on the posterior teeth. Rak originally saw these pillars as a defining apomorphy of *A. africanus* and a precursor to the more extensive facial buttressing seen in later species of *Australopithecus*. However, subsequent work by McKee (1989) showed that the presence of an anterior pillar is variable among *A. africanus* specimens and is not a defining feature of the taxon.

AUSTRALOPITHECUS ROBUSTUS

By 1950, Dart and Broom had between them managed to name a different species of *Australopithecus* at each of the five South African cave sites. Three of these—*Australopithecus africanus* at Taung (Dart 1925), *Plesianthropus transvaalensis* at Sterkfontein (Broom 1938), and *Australopithecus prometheus* at Makapansgat (Dart 1948)—are regarded by most current authorities as different names for the same species, *A. africanus*. But the fossils from two other sites, Swartkrans and Kromdraai, were something else. Their distinctive appearance had led Broom (1938) to assign them to their own genus and species, *Paranthropus robustus*. He later put the Swartkrans fossils into a different species of the same genus, *Paranthropus crassidens* (Broom 1949). A few other scientists have accepted this distinction, on the grounds that the premolars and molars from Kromdraai are generally more primitive than those from Swartkrans in certain details of their crown morphology (Howell 1978, Grine 1985, Jungers and Grine 1986, Grine and Martin 1988). However, most paleoanthropologists think that both these sites sample one species of *Paranthropus* or *Australopithecus*, for which the species name *robustus* has priority. Recent finds at the new site of Drimolen appear to bridge the minor dental differences between the Swartkrans and Kromdraai fossils (Keyser et al. 2000). Those differences may simply be anagenetic differences between successive samples of a single evolving lineage.

In what follows, we will lump the species "*crassidens*" into *robustus*, and the genus "*Paranthropus*" into *Australopithecus*. But *A. robustus* is clearly not the same animal as *A. africanus*. The two differ chiefly in the skull and teeth. In an influential 1956 monograph, Broom's student J. Robinson tallied up the cranial and dental differences between *A. africanus* and *A. robustus* and analyzed their significance. Both species have larger molars than modern humans, and are often described as **megadont** (<Gk., "big-toothed"). But the molars of *A. robustus* are especially big, with an average

TABLE 4.2 ■ Dental Breadths in Early Hominins[a]

	"Robust"*Australopithecus*				"Gracile"*Australopithecus*			Early *Homo*	
	A. robustus	A. boisei	A. garhi	Stw 252	A. anamensis	A. afarensis	A. africanus	Habilines	Ergasters
Mandible									
I_1	6.1 (0.9, 11)	6.8 (0.8, 7)	—	—	7.8 (—, 2)	7.6 (0.2, 4)	6.6 (0.7, 3)	6.8 (0.7, 3)	6.3 (—, 2)
I_2	7.0 (0.4, 8)	6.9 (0.9, 4)	—	—	8.4 (0.5, 3)	7.6 (0.5, 5)	7.7 (0.7, 3)	7.4 (0.7, 4)	6.8 (0.2, 4)
C	7.9 (0.6, 12)	8.8 (0.8, 9)	—	—	10.2 (1.0, 5)	10.4 (1.2, 12)	10.2 (1.0, 7)	9.0 (—, 2)	9.5 (—,1)
P_3	11.7 (0.9, 17)	12.9 (0.9, 7)	—	—	10.9 (1.3, 3)	10.6 (0.8, 21)	11.3 (0.7, 6)	9.5 (0.9, 6)	10.6 (0.5, 5)
P_4	12.9 (1.0, 16)	14.7 (1.1, 14)	—	—	10.7 (0.8, 4)	11.0 (0.8, 18)	11.9 (0.5, 6)	10.5 (0.5, 6)	11.0 (—,2)
M_1	13.8 (0.8, 23)	14.7 (1.0, 11)	—	—	12.0 (—, 2)	12.6 (0.8, 18)	13.1 (1.0, 11)	12.0 (0.8, 6)	11.2 (0.7, 7)
M_2	15.0 (1.0, 22)	17.0 (1.3, 14)	—	—	13.1 (—, 2)	13.5 (1.0, 20)	14.2 (0.8, 10)	13.4 (1.1, 5)	12.2 (0.6, 5)
M_3	14.7 (1.0, 20)	16.7 (1.4, 18)	—	—	12.5 (—, 2)	13.4 (0.9, 14)	13.9 (0.7, 11)	13.1 (0.9, 7)	12.2 (0.4, 6)
Maxilla									
I^1	7.4 (0.4, 13)	—	—	8.0 (—, 1)	8.6 (0.4, 3)	8.3 (0.6, 6)	8.2 (0.2, 3)	8.0 (—, 2)	—
I^2	6.7 (0.7, 8)	6.5 (0.9, 5)	6.9 (—, 1)	7.7 (—, 1)	—	7.6 (0.6,8)	6.6 (0.7, 5)	6.6 (0.9, 4)	—
C	9.4 (0.7, 19)	8.9 (0.9, 6)	12.9 (—, 1)	11.4 (—, 1)	9.7 (—, 2)	10.9 (1.0, 12)	9.6 (0.5, 6)	9.3 (0.8, 6)	—
P^3	14.1 (0.7, 16)	15.4 (1.2, 8)	16.0 (—, 1)	14.1 (—, 1)	12.6 (0.7, 3)	12.4 (0.6, 9)	12.5 (0.8, 14)	11.5 (0.5, 8)	12.0 (—, 1)
P^4	15.3 (0.8, 18)	16.3 (1.0, 7)	16.0 (—, 1)	15.3 (—, 1)	11.9 (2.0, 3)	12.0 (0.5, 7)	13.3 (0.7, 9)	11.7 (0.7, 8)	—
M^1	14.9 (0.7, 20)	16.3 (1.2, 8)	—	15.2 (—, 1)	11.9 (—, 1)	13.2 (1.0, 9)	13.7 (0.7, 12)	12.9 (0.6, 11)	—
M^2	16.0 (0.9, 17)	18.2 (10.2, 8)	—	17.0 (—, 1)	14.0 (—, 2)	14.7 (0.6, 7)	15.3 (1.0, 15)	14.4 (1.1, 9)	13.7 (0.5, 3)
M^3	17.0 (0.7, 19)	18.9 (2.7, 4)	16.9 (—, 1)	18.0 (—, 1)	13.5 (0.5, 3)	14.2 (0.1, 7)	15.5 (1.3, 14)	14.8 (1.4, 11)	—

[a]Mean (standard deviation, sample size). All measurements are in centimeters. Only labio-lingual or bucco-lingual dimensions are used, because these are not affected by interproximal wear. For *A. garhi*, *A. anamensis*, and Stw 252, the indicators of central tendency treat measurements from right and left sides of the same specimens as independent data, but the sample sizes represent number of individuals (Clarke 1998; Leakey et al. 1995a, 1998; Asfaw et al. 1999). Other data are taken from Wood (1991), where sample configuration is detailed.

summed molar area (breadth × length) about 17% larger than the *A. africanus* in Robinson's sample. Likewise, the premolars are big and molarized in both species, but more so in *A. robustus* (Table 4.2). On the other hand, the canines and incisors of *A. robustus* are considerably smaller than those of *A. africanus*. In short, *A. robustus* typically has very big cheek teeth but relatively tiny, human-sized front teeth (Fig. 4.28), whereas *A. africanus* (Fig. 4.7) has teeth that are bigger overall than a human's but have more nearly humanlike relative proportions.

These differences are correlated with differences in the skull, which are most obvious in males (Fig. 4.29). Compared to the more delicately built, "gracile" skull of *A. africanus* (Figs. 4.4 and 4.5), the skull of *A. robustus* is—well, *robust*: more heavily and powerfully built, with marked ridges and crests for the attachment of chewing muscles, and with stout bony buttresses for transmitting and dissipating the bite forces that those big muscles produced. As in *A. africanus*, those buttresses often include thickened "anterior pillars" framing the nasal opening. However, the skulls of the two species differ in at least six major ways:

1. *A. robustus* shows less **alveolar prognathism** than *A. africanus*, meaning that the lower, tooth-bearing (alveolar) part of its face does not stick out so far in front of the eye sockets. Its dental arcade is pulled backward under the braincase, and the rear upper molar (M^3) lies further back relative to the jaw joint (Rak 1983).

2. However, the upper jaw of *A. robustus* is vertically taller than that of *A. africanus*. As a result, its teeth lie further below the level of the jaw joint, and the ascending ramus of its mandible is higher (Fig. 4.30).

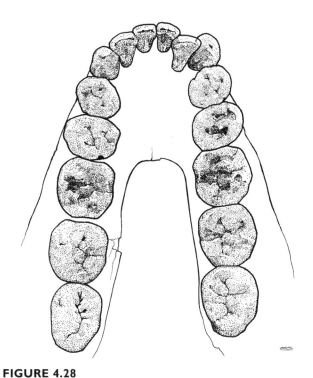

FIGURE 4.28

Mandibular dentition of *Australopithecus robustus* from Swartkrans (SK 23). (After Robinson 1956.)

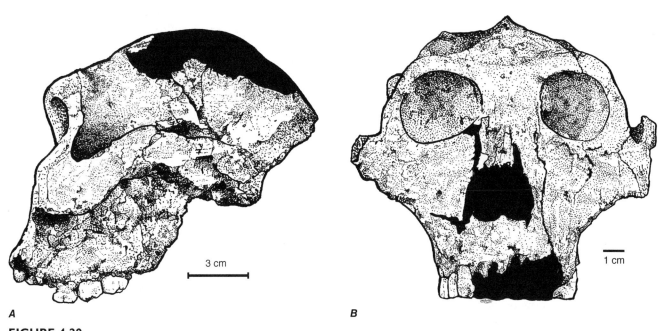

A

B

FIGURE 4.29

Male skull of *Australopithecus robustus* from Swartkrans (SK 48). **A**: Lateral view. **B**: Frontal view. (**A**, after Wolpoff 1980; **B**, after Schwartz and Tattersall 2005.)

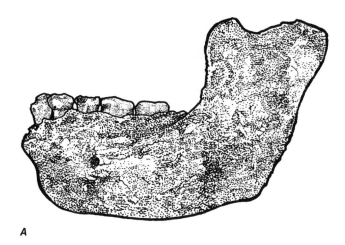

A

1 cm

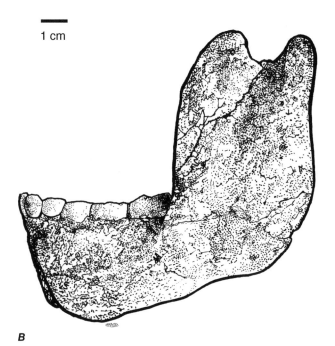

B

FIGURE 4.30

Mandibles of (**A**) *Australopithecus africanus* (MLD 40) and (**B**) *A. robustus* (SK 23). (After Day 1965.)

3. The temporal muscles of *A. robustus* were more massive than those of *A. africanus*—especially in males, where they extended up to the top of the skull and produced a sagittal crest. (Female *A. robustus* appear to have lacked a sagittal crest: Fig. 4.31.) Unlike those of great apes and many other mammals (Fig. 3.37), the sagittal crest of *A. robustus* was restricted to the top of the skull, and did not extend back to the epaxial muscles at the back of the neck. This is partly due to the fact that *A. robustus*, like *A. africanus*, had a downward-facing foramen magnum (presumably because both species were upright bipeds), so the epaxial muscles of the neck attached further down on the skull than they do in an ape (Fig.

3.37). The restriction of the sagittal crest to the top of the skull in *A. robustus* also reflects a particular enlargement of the anterior part of the temporalis muscle. Some *A. africanus* specimens, presumably males, have temporal lines that are almost touching (Wolpoff 1974, Thackeray et al. 2002), so a sagittal crest may have occurred as an occasional variant in *africanus* as well.

4. To accommodate the enlarged anterior temporalis, the robust zygomatic arches of *A. robustus* enclosed a correspondingly expanded space on either side of the braincase. This space (the **temporal fossa**) was expanded in two directions: *medially*, by a narrowing of the braincase in back of the eye sockets (a marked postorbital constriction), and *anteriorly*, by a forward displacement of the orbital openings and the anterior root of the zygomatic arch relative to the braincase. This forward displacement of the upper face gives *A. robustus* a strikingly vertical facial profile, and makes it seem to have even less prognathism than it actually has (see #1, above). In *A. robustus*, the entire face is protrusive, not just the alveolar region.

5. The anterior end of the zygomatic arch of *A. robustus* is shifted so far forward that it protrudes anteriorly beyond the rest of the upper face, so that the nasal bones sit in the center of a concavity when viewed from above (Fig. 4.31B). This is called a "**dished face.**"

6. Finally, *A. robustus* had a larger braincase than *A. africanus*, and probably a larger brain (Table 4.1). (Unfortunately, the only *A. robustus* specimen on for which we have a good cranial-capacity measurement SK 1585, at 530 cc: Holloway 2000). Yet despite its larger braincase, *A. robustus* has less of a forehead. Seen in profile, its frontal bone slants almost horizontally backward from the supraorbital torus. This change in the slope of the forehead is due at least in part to the forward displacement of the eye sockets in *A. robustus*.

The biomechanical significance of these differences between the two species of *Australopithecus* is diagrammed in Fig. 4.32. In *A. robustus*, the increased size of the anterior temporalis and the forward shift of the anterior end of the zygomatic arch (where the masseter muscle is attached) shift the resultant of the jaw-muscle force to a more anterior point on the tooth row (Fig. 4.32B). This shift is augmented by pulling the dental arcade backward. More bite force is thus exerted on the premolars, which are correspondingly bigger and more molarized in *A. robustus* than in *A. africanus* (Table 4.2). The taller ascending ramus of *A. robustus* raises the jaw joint further above the tooth row. Elevation of the jaw joint is a common feature of mammals

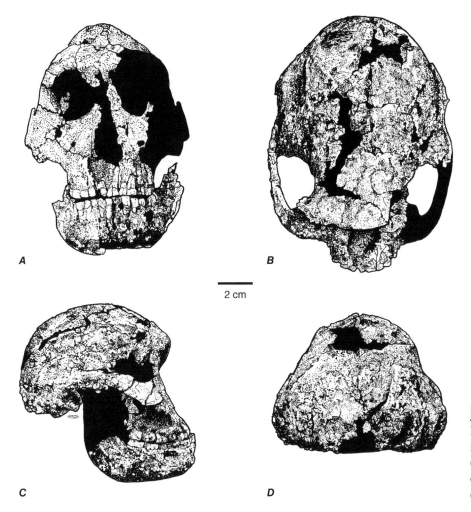

A

B

2 cm

C

D

FIGURE 4.31

Female skull of *Australopithecus robustus* from Drimolen (DNH 7). (**A**) Frontal view, (**B**) vertical view, (**C**) lateral view, (**D**) occipital view. (After Keyser et al. 2000.)

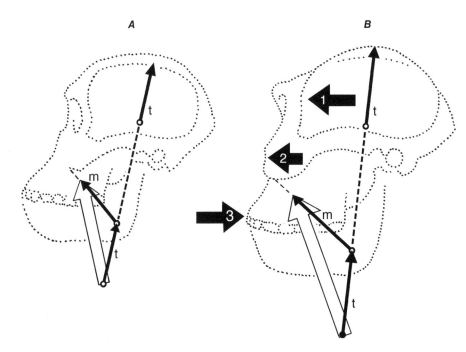

A

B

FIGURE 4.32

Biomechanics of the *Australopithecus* face (diagrammatic). By comparison with *A. africanus* (**A**), the skull of *A. robustus* (**B**) has an enlarged anterior part of the temporalis muscle and temporal fossa (*black arrow* 1), a forward-shifted anterior root of the zygomatic arch (*black arrow* 2), and a dental arcade pulled backward under the braincase (*black arrow* 3). These changes alter the vectors of the temporalis muscle force (*thin black arrows* "t") and the masseter muscle force (*thin black arrows* "m") relative to the dentition. The resultant muscle force (*white arrows*) is thereby increased and redirected to a more anterior point on the tooth row.

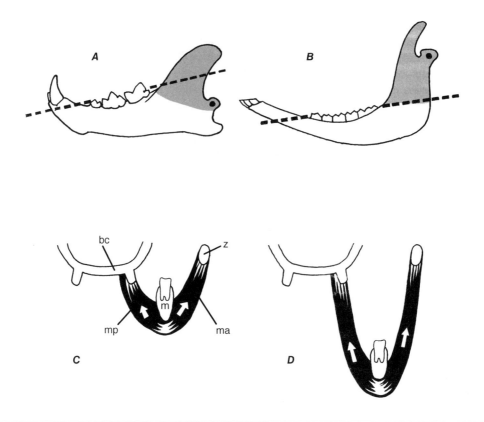

FIGURE 4.33

Effects of condylar elevation. Carnivores (**A**, lion) and herbivores (**B**, cow) characteristically differ in the height of the mandible's ramus (*gray tone*) and the relationship of the mandibular condyle (*black dots*) to the occlusal plane of the cheek teeth (*dashed lines*). Increasing the height of the ramus brings all the teeth into occlusion more nearly simultaneously when the jaws come together, and it distributes crushing action more evenly along the length of the tooth row. **C**, **D**: diagrammatic cross sections through the medial pterygoid and masseter muscles. Increasing condylar elevation (**D**) makes the two muscles pull more nearly in the same direction, with correspondingly less wasted effort. Abbreviations: bc, floor of braincase; m, mandible; ma, masseter; mp, medial pterygoid muscle; z, zygomatic arch. (**A**, **B**, after Ellenberger et al. 1956.)

that eat tough or fibrous vegetation (Fig. 4.33B). It gives the masseter and medial pterygoid muscles a longer lever arm around the joint's center of rotation, and it makes more effective use of the forces they exert on the mandible by bringing them more nearly into parallel with each other and with the bite force on the teeth (Maynard Smith and Savage 1959; Fig. 4.33C,D). Elevating the jaw joint also has the effect of limiting the gape between the front teeth, so it is not seen in animals that kill prey by biting (Fig. 4.33A). A similar complex of changes—elevated jaw joint, more vertical face, enlarged anterior temporalis, increased postorbital constriction, anterior displacement of the masseter origins, incisor reduction and molar enlargement—were evolved in the extinct Malagasy lemur *Hadropithecus,* probably for similar adaptive reasons (Jolly 1970b, Tattersall 1973).

From the differences in the skull and teeth between *A. africanus* and *A. robustus,* Robinson inferred that the two species must have had different ecologies. *A.*

robustus, he concluded, was a gorilla-like herbivore adapted to feeding on leaves, stems, and other low-quality plant tissues; *A. africanus* had a mixed diet more like those of modern humans (Robinson 1954b, 1956, 1968). This contention became known as the **dietary hypothesis.** A competing theory holds that the cranial and dental differences between the two species are merely allometric, reflecting the larger body size of *A. robustus* rather than differing ecologies (Tobias 1967a). However, careful estimates of body size based on postcranial measurements (Table 4.1) have found no significant difference between *A. africanus* and *A. robustus.* Both seem to have been chimpanzee-sized animals averaging around 45-50 kg—roughly a hundred pounds—with the largest males weighing about twice as much as the smallest females (Jungers 1988a, McHenry 1988). Most later authors have therefore accepted some version of Robinson's dietary hypothesis (Grine 1986, Teaford and Ungar 2000).

MAN-APES, JUST PLAIN APES, OR WEIRD APES?

What did all this imply for human evolution? For Robinson, the implications were clear: *A. africanus* was a close relative of *Homo* (and he later reassigned it to that genus), whereas *A. robustus* lay on a specialized, ultimately sterile side branch of the human evolutionary tree (Robinson 1967, 1968). But some other scientists rejected that interpretation.

Both types of *Australopithecus* resembled living apes in retaining various primitive features lost in *Homo*—for example, those small, ape-sized brains. And even though a big brain was no longer seen as a uniquely important marker of human affinities, other primitive traits of *Australopithecus* were put forward as reasons for thinking that these extinct African creatures had no special relationship to humanity.

The attack on the hominin pretensions of *Australopithecus* was led by the formidable British anatomist S. Zuckerman. Zuckerman (1954) thought that humans had diverged from apes in the Oligocene and that *Australopithecus* was more closely related to gorillas and chimpanzees than it was to *Homo*. Zuckerman and his students at the University of Birmingham went after *Australopithecus* with every weapon at hand, from classical comparative anatomy to complex statistical analyses made possible by a new research tool, the digital computer. Their efforts met with little success. Zuckerman first argued that the metrics of the *Australopithecus* dentition were ape-like in crucial respects (Ashton and Zuckerman 1950), but his argument collapsed when a simple mathematical error was found in it (Yates and Healey 1951). He then insisted that because "*Paranthropus*" *robustus* from South Africa had a sagittal crest, it must have had an apelike nuchal crest as well. "The implication," he wrote, "is thus clear that *Paranthropus* carried its head on its vertebral column far more in the manner of a gorilla than of a man" and therefore could not have been an upright biped (Zuckerman 1954). This argument, too, collapsed when R. Holloway (1962) demonstrated that some gibbons have a sagittal crest but no nuchal crest. Later fossil finds showed that this was true of "*Paranthropus*" as well.

The most fundamental objection to Zuckerman's arguments came from Le Gros Clark. Anticipating Hennig's distinction between symplesiomorphic and synapomorphic resemblances, Le Gros Clark (1958, 1959) insisted that only shared derived traits—what he called "characters of independent acquisition"—are useful for inferring phylogeny. We would expect, he said, that an early hominin would look like an ape in most of its characters, because the last common ancestor of humans and apes also looked like an ape. But even if Zuckerman and his colleagues could show that *Australopithecus* had remained primitive and ape-like in most of its characters, such "characters of common inheritance" (symplesiomorphies) could never prove anything about relationships. To place a fossil on the hominin branch of the evolutionary tree, all that was needed was to show that it shared *some* hominin synapomorphies. And that, Le Gros Clark declared, was clearly the case for *Australopithecus*.

Le Gros Clark's reasoning made it clear that the persistently apelike traits of *Australopithecus* could not be invoked to exclude it from the human family. But it could be ruled out of direct human *ancestry* by autapomorphies—nonprimitive, derived traits unique to *Australopithecus*. And there appeared to be some apomorphies that *A. robustus* and *A. africanus* shared uniquely with each other. In a few respects, they were more humanlike than even early *Homo*: for example, in the extreme molarization of the front lower deciduous premolar, or in the reduction of the canine in *A. robustus* (Robinson 1956). In certain other respects, they were neither apelike nor humanlike, but exhibited peculiar apomorphies of their own.

One apparent autapomorphy of *Australopithecus* was its postcanine megadonty. Both *A. africanus* and *A. robustus* had larger cheek teeth than early fossil *Homo* (Table 4.2) or modern *Pan*. This fact made it easy to see *A. africanus* as the midpoint on a sterile, megadont evolutionary sideline, which had arisen from an ancestor with small molars (like those of *Pan* and *Homo*) and terminated in the specialized dead end of *A. robustus*. If that were true, then Zuckerman was right after all, and the true early human ancestors still remained to be found.

POSTCRANIAL PECULIARITIES

The most perplexing autapomorphies of *Australopithecus* were features of the postcranial skeleton. Fragmentary pelves of *A. robustus* from Swartkrans and Kromdraai showed humanlike modifications for bipedality, much like those found in *A. africanus*. The ilium of *A. robustus* was short and broad, with a deep greater sciatic notch and a prominent anterior inferior spine. A faint intertrochanteric line was discernible on the front of the proximal femur (Robinson 1972). A lower lumbar vertebra of *A. robustus* from Swartkrans had a wedged vertebral body, suggesting a lumbar lordosis. It looked as though both the "robust" and "gracile" types of *Australopithecus* had been upright and bipedal. But they both showed some significant differences from the human pattern.

One such difference lay in the size of the vertebral bodies. Compared to humans of similar stature, both species of *Australopithecus* had small vertebral bodies—especially in the lower back (Fig. 4.34)—and a small sacroiliac joint (Sanders 1990, 1998, McHenry 1991a). The small size of these weight-bearing joint surfaces

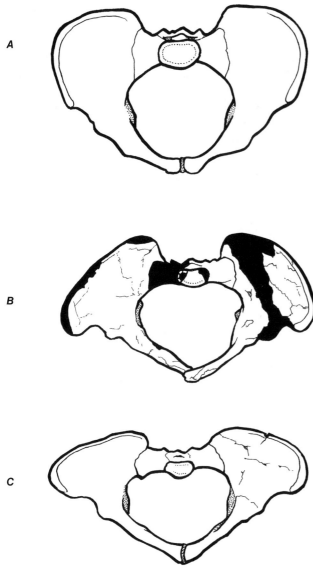

FIGURE 4.34

Pelves of (**A**) *Homo sapiens*, (**B**) *Australopithecus africanus* (Sts 14), and (**C**) *Australopithecus afarensis* (AL 288-1, "Lucy"), viewed from the cranial aspect. Not to same scale. (**A**, **C**, after Stern and Susman 1991; **B**, after Robinson 1972.)

suggested that the lower back of *Australopithecus* had not been as well adapted to the stresses of bipedality as ours.

Another difference from humans was the angulation of the iliac blade. Human ilia curve around the sides of the abdomen toward the belly, so that the outer surfaces of the iliac blades face as much laterally as posteriorly. This reorientation allows the deep glutei to act more effectively as abductors to check pelvic tipping in bipedal locomotion. But the iliac blades of *Australopithecus* faced more posteriorly, somewhat like a gorilla's (Figs. 4.11 and 4.34). There were also differences in

the construction of the hip joint. Like apes, *Australopithecus* had a relatively small **acetabulum** (hip socket) and a correspondingly small femoral head. But unlike those of either apes or humans, its femoral head was mounted on a curiously elongated femoral neck (Fig. 4.35).

These differences between *Australopithecus* and *Homo* were troubling for those who wanted to see *Australopithecus* as a humanlike biped. It was all very well to say that its apelike features were just "characters of common inheritance" with no phylogenetic implications. But it seemed as though they must have had *functional* implications. The small weight-bearing surfaces of the vertebral column and hind limbs of *Australopithecus* appeared poorly adapted to bearing the peculiar stresses and shocks of upright bipedality. Were its forelimbs still playing a role in its locomotion? And why did its femur have such a long neck?

In 1973, the anthropologist C. O. Lovejoy proposed an elegant answer to some of these awkward questions. Lovejoy and his coworkers (Lovejoy et al. 1973, Lovejoy 1974, 1975, 1988) argued that the flaring ilia and long femoral neck of *Australopithecus* had enhanced the leverage of the deep gluteal muscles (Fig. 4.36A). Having a longer lever arm, the deep glutei of *Australopithecus* would not have had to exert so much force to prevent pelvic tipping during bipedal walking. This meant that the fulcrum of the gluteal lever system, the hip joint, would have had to bear less reaction force. Therefore, the femoral head and the acetabulum—the weight-bearing surfaces of that fulcrum—could be correspondingly smaller than ours, and so they were. *Australopithecus* (said Lovejoy) differed from humans not because it was a defective biped, but because it was actually a better and more efficient biped than we are.

Why had humans abandoned this efficient ancestral arrangement? Lovejoy attributed this change to brain growth. To permit vaginal delivery of a big-brained fetus, early *Homo* had to enlarge the transverse diameter of the birth canal. That enlargement had displaced the acetabula laterally (Fig. 4.36B), reducing the length of the femoral neck and shifting the fulcrum of the lever system toward the line of action of the deep glutei. This shift reduced the mechanical advantage of the glutei. To make up for that reduction, the glutei had to pull harder. The acetabulum, femoral head, and other weight-bearing structures of the hipbone and hip joint are enlarged in *Homo* to handle that stronger muscle pull.

LOUIS LEAKEY AND EAST AFRICA

Despite Lovejoy's analysis, some scientists still refused to accept *Australopithecus* as a human ancestor. Perhaps the most stubborn of these holdouts was the African paleoanthropologist Louis Leakey. Born in 1903 to

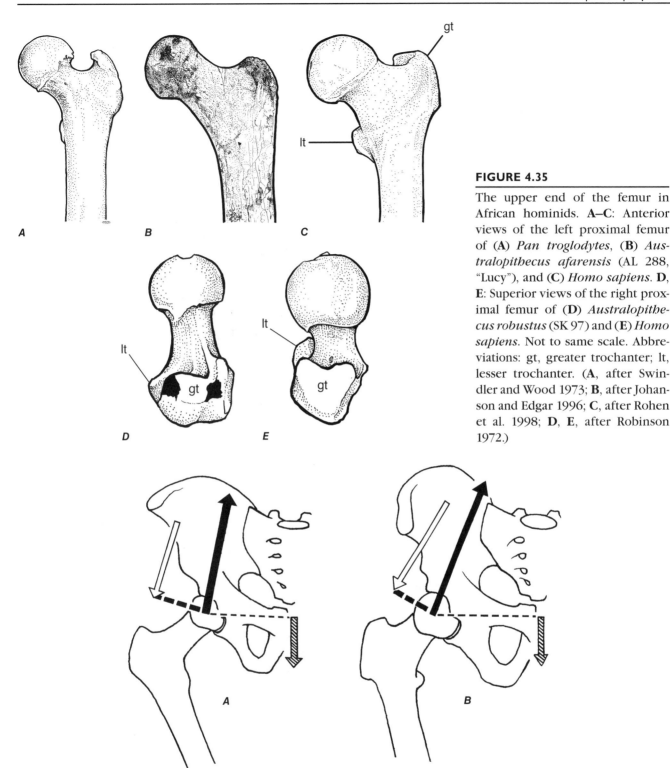

FIGURE 4.35

The upper end of the femur in African hominids. **A–C**: Anterior views of the left proximal femur of (**A**) *Pan troglodytes*, (**B**) *Australopithecus afarensis* (AL 288, "Lucy"), and (**C**) *Homo sapiens*. **D**, **E**: Superior views of the right proximal femur of (**D**) *Australopithecus robustus* (SK 97) and (**E**) *Homo sapiens*. Not to same scale. Abbreviations: gt, greater trochanter; lt, lesser trochanter. (**A**, after Swindler and Wood 1973; **B**, after Johanson and Edgar 1996; **C**, after Rohen et al. 1998; **D**, **E**, after Robinson 1972.)

FIGURE 4.36

C. O. Lovejoy's analysis of pelvic biomechanics in *Australopithecus* (**A**) and *Homo* (**B**), diagrammed in anterior views with the same pelvic width (bi-iliac breadth). Lengths of arrows represent magnitudes of forces. When the left foot is lifted, the weight of the body tends to tip the pelvis to the left, rotating it around the right hip joint. The torque thus produced is equal to the body-weight vector (*hatched arrow*) times its distance from the right hip joint (*light dashed line*). The deep glutei prevent tipping by exerting an equal and opposite torque, which equals the summed gluteal muscle force on the pelvis (*white arrow*) times *its* distance from the hip joint (*heavy dashed line*). In *Homo*, lateral displacement of the hip sockets increases the moment arm of the body-weight vector and decreases that of the gluteal-muscle force vector. Having a shorter moment arm, the human glutei must work harder at an identical body weight to prevent pelvic tipping. Therefore, the reaction force at the hip joint (*black arrow*), which is equal and opposite to the vector sum of the other two forces, is greater in modern humans (**B**). The femoral head of *Homo* is enlarged to carry the increased load.

British missionary parents in Kenya, Leakey remained an outsider to the scientific establishment throughout his life. Most of his long, productive career was spent advocating fringe theories of human evolution and seeking evidence for them in the prehistory of East Africa. His theories were usually mistaken, and some of his evidence was insubstantial. But sometimes—not often, but sometimes—Leakey was right and the scientific consensus was wrong.

One of Leakey's pet theories was the idea that human ancestors were practically unknown in the fossil record. When Leakey was a young man, this was not a fringe idea. During the first half of the 20th century, many orthodox scientists thought that all the low-browed, simian-looking fossil hominins known from Europe, Africa, and Asia were only our retarded evolutionary cousins, not our ancestors, and that there was a true human lineage of great antiquity still waiting to be discovered. Leakey too believed in the reality of an ancient and separate human lineage, which he thought had split off from other hominins back in the Miocene (Leakey 1953). What set Leakey apart from most of the other proponents of this notion was his conviction that the earliest evidence of the true human lineage would be found in Africa (see Blind Alley #6, p. 206).

On this last point, Leakey proved to be right. East Africa has turned out to be our most important source of knowledge about early hominin evolution. There are two reasons for this. First, hominins originated in Africa and have been restricted to Africa for most of their evolutionary history. Second, the processes of plate tectonics have made East Africa a unique catchment basin for early hominin fossils, as well as for volcanic deposits that allow us to determine how old those fossils are.

For the past 20 million years, continental drift has been slowly tearing Africa apart. The main rift between the separating plates runs from Israel in the north down to Kenya in the south. It then splits into eastern and western rift systems, which continue southward around opposite sides of Lake Victoria. As these plates draw away from each other, the planetary crust in between them attenuates and cracks. The rift lines tend to subside, resulting in depressed areas called **rift valleys** that fill with sea water along the coast and with fresh water inland. The resulting bodies of water in the African rift system include the Sea of Galilee, the Red Sea, the Gulf of Aqaba, and the chain of great lakes running down through East Africa from Lake Turkana in the north to Lake Malawi in the south.

As the rifts widen and the rocks crack and shear, adjoining blocks of crust shift vertically along the fault lines between them, leaving some elevated areas protruding above the general subsidence. Silt and gravel eroding from these elevations are carried by streams and rivers into the low-lying areas, where they are deposited as freshwater sediments. Volcanoes open up along the fault lines, contributing basaltic lavas and compacted layers of volcanic ash called **tuffs** to the strata accumulating in the rift valleys and their lakes. When these volcanic deposits cool to atmospheric temperatures, radiometric clocks are set going in some of the minerals they contain. We can therefore measure the age of the lavas and tuffs from isotope ratios, fission-track counts, electron spin resonance, and other absolute-dating metrics. These dates furnish upper and lower boundaries for the age of fossils preserved in the water-deposited sediments lying between the volcanic strata.

Over the course of the past 40 years, absolute dates of increasing sophistication and accuracy have been recovered from hominin-bearing sites in the northern and eastern rift valleys of East Africa. The resulting stratigraphic time scale is sketched in Fig. 4.37. The hominin sites in South Africa and Chad contain few minerals that yield absolute dates, so the ages estimated for these sites are based mainly on correlations with dated fossil faunas from East Africa.

▌OLDUVAI GORGE

In 1931, when Louis Leakey mounted his first expedition to Olduvai Gorge in Tanzania (Fig. 4.3), radiometric geochronology was undreamed of. Leakey was drawn to Olduvai not because of its datable sediments, but because of work done there before World War I, when Tanzania was a German colony. During that time, German scientists had mapped Olduvai, sketched its stratigraphy, collected its terrestrial fossils, and even found a human skeleton of modern type buried in its Middle Pleistocene sediments (Hopwood 1932). When Leakey arrived at Olduvai, he was pleased to find that some of those sediments contained early Acheulean stone tools—crudely worked hand-axes and flakes, resembling the so-called "Chellean" tools found in the oldest levels of human occupation in Europe. All these factors made Olduvai look like the perfect place to search for the true and ancient human lineage that Leakey and others expected to find.

Leakey's search met with one discouragement after another. The "Middle Pleistocene" skeleton from Olduvai turned out to be an intrusive modern burial. A fragmentary modern-looking braincase found near Kanjera in western Kenya, touted by Leakey as the maker of the "Chellean" tools, was largely ignored by other scientists because it was a surface find, of uncertain geological context. Middle Pleistocene strata at Kanam, another Kenyan site, yielded a chunk of mandible with what looked like a projecting bony chin—a modern human trait, not seen in any other jawbone of such antiquity. Leakey (1935) hailed the Kanam mandible as "the most ancient fragment of true *Homo* yet to be discovered anywhere in the world." But this discovery, too, fell flat

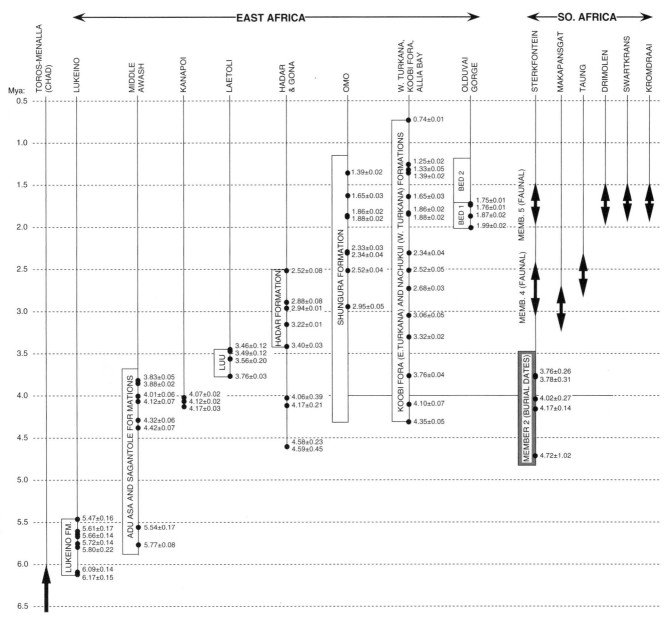

FIGURE 4.37

Absolute dates (*dots*) and relative dates (based on index fossils and paleomagnetism: *double-beaded arrows*) for some Miocene, Pliocene, and Pleistocene African sediments containing fossil hominins. LUU, Laetolil Upper Unit. (Data from McFadden 1980, Delson 1988, Drake and Curtis 1987, Walter et al. 1991, Brown and McDougall 1993, F. Brown 1994, Leakey et al. 1998, Semaw 2000, WoldeGabriel et al. 2001, Haile-Selassie 2001, Sawada et al. 2002, Partridge et al. 2003, Alemseged et al. 2005, Semaw et al. 2005, Walker et al. 2006, White et al. 2006, Egeland et al. 2007, Grine et al. 2006.)

when that vaguely human-like chin region proved to be the product of a pathological bony growth (Tobias 1960, Weiner et al. 2006).

For the next quarter-century, Leakey and his co-workers kept returning at intervals to Olduvai Gorge, collecting animal fossils and ancient artifacts. The hominin fossils that they turned up during this period—some parietal bits, some loose teeth, and a piece of jawbone—were few and disappointing. But their surveys helped establish that the lower levels of the Olduvai deposits contained the earliest and most primitive stone tools yet found. These **Oldowan** tools (named from the German spelling of "Olduvai") consist largely of simple flakes knocked off of quartz cores, plus a smaller percentage of cores and hammerstones used in flake production (Toth and Schick 1986, de la Torre 2004).

On July 17, 1959, Leakey's wife Mary found a broken skull associated with Oldowan tools at the FLK site in Bed I at Olduvai. When reassembled, the skull proved to comprise the nearly complete occiput, face, and upper dentition of a new type of fossil hominin (Fig. 4.38). This skull, Olduvai Hominid 5 (OH 5), was rushed into print by Leakey (1959) and described as the type specimen of a new genus and species, *Zinjanthropus boisei*. Leakey's decades of persistence at Olduvai had finally paid off.

The OH 5 skull was remarkable not only for its beautiful preservation but also for its extraordinary morphology (Tobias 1967a). In most respects, it appeared to be a "hyper-robust" version of *Australopithecus robustus* from South Africa. All the distinctive features of *A. robustus* were present in it and generally exaggerated. It had the forward-shifted orbital openings, the dished face (Fig. 4.39), the protruding cheekbones, the strong postorbital constriction, and the powerful cranial crests and ridges of the robust South African species; and it

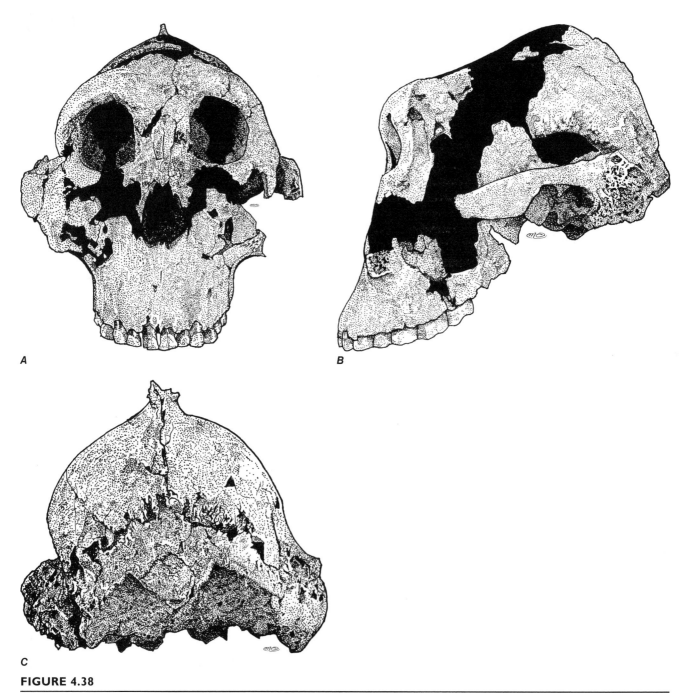

A

B

C

FIGURE 4.38

The type specimen of *Australopithecus boisei* (OH 5), viewed from the front (**A**), side (**B**), and back (**C**). (After Tobias 1967a.)

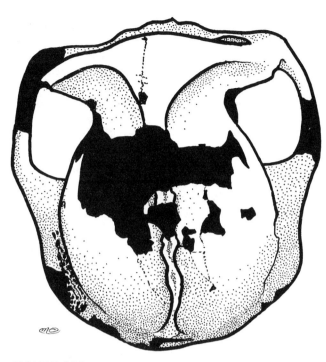

FIGURE 4.39

Semidiagrammatic vertical view of the OH 5 skull of *Australopithecus boisei*, showing the "dished" face and the extensive sagittal crest. (After Tobias 1967a.)

had an even taller upper jaw and bigger temporal fossa. Its cheek teeth (Fig. 4.40A) were extraordinarily large—more than twice as big as would be predicted for an ape of comparable body size (McHenry 1988).

Most of the experts at the time recognized *"Zinjanthropus" boisei* as a variation on *A. robustus*, and they declined to accept it as a genus separate from South African *Australopithecus* (or *Paranthropus*). But Leakey had a reason for stressing the distinctiveness of the new form. Since it had been found associated with some of the oldest stone tools then known, he naturally assumed that it was the maker of those tools. And if "Zinjanthropus" was the first stone-tool maker, Leakey dared to surmise that it might have a special phylogenetic relationship to human beings—that it was, in fact, the true and ancient human ancestor he had been seeking for so many years.

Leakey's hopes did not stay pinned on OH 5 for long. Early the next year, Leakey's son Jonathan discovered fragments of a different, more human-like hominin (OH 6) at another Bed I Olduvai site. Leakey's family and field workers began searching for more complete remains of this creature. Their search was rewarded in 1960 by the discovery in Bed II of what appeared to be the remains of a single hominin individual, comprising the parietals, the bones of a hand, an upper molar, and half of a gracile

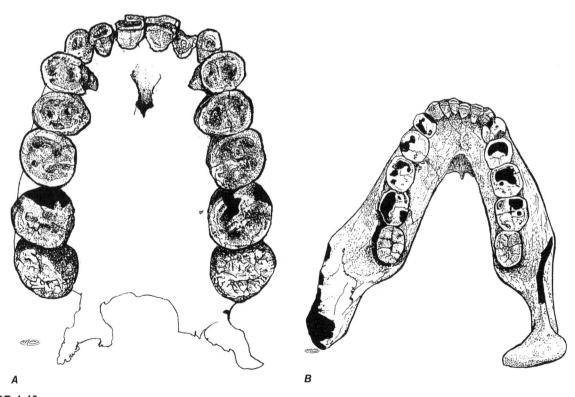

A   *B*

FIGURE 4.40

Dentition of *Australopithecus boisei*. (**A**) Upper dentition (OH 5); (**B**) lower dentition (Peninj). Not to same scale. (After Tobias 1967a, 1991.)

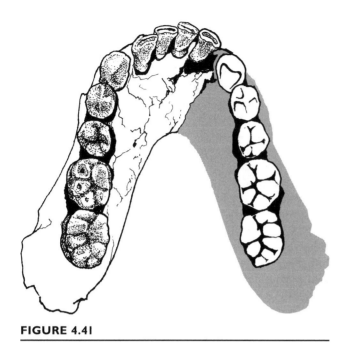

FIGURE 4.41

The mandible of OH 7, the type specimen of *"Homo habilis."* The right side of the mandible (*gray tone*) and the associated canine and cheek teeth are reconstructed as a mirror image of the left. (After Tobias 1991.)

(though still subadult) mandible. This specimen, OH 7 (Fig. 4.41), showed no signs of the super-robust adaptations to powerful chewing seen in OH 5. Although the anterior teeth of OH 7 were about as big as those of OH 5, its molars and premolars were much smaller and transversely narrower in shape. The P_3 was no bigger than the canine and lacked the molar-like occlusal conformation seen in *Australopithecus*. The P_4 too was less molarized. In all these respects, OH 7 resembled early *Homo* and was more primitive than OH 5 or either of the South African *Australopithecus* species. Its parietals hinted at a larger cranial capacity than the 530 cc estimated for the "Zinjanthropus" skull (Tobias 1967a, Blumenscheine et al. 2003).

In short, OH 7 appeared to be Leakey's long-sought early *Homo*: a gracile, relatively large-brained toolmaker that lacked the megadonty and other autapomorphies of contemporary *Australopithecus* (Leakey 1961). Leakey et al. (1964) put OH 7 together with several other non-robust hominin fossils from Olduvai, classified the collection as a new species of *Homo*, and gave it the species name *habilis*—a Latin word meaning "handy," with reference to the tool-making dexterity supposedly discernible in the OH 7 hand bones. Abandoning his earlier claims for the Kanam jaw and the "Zinjanthropus" skull (OH 5), Leakey proclaimed that *H. habilis*—not "Zinj" itself—had made the tools and butchered the animals found in association with OH 5 at the FLK I site.

Since 1962, continuing discoveries of early hominins from Olduvai Gorge and other African sites have made it clear that at least two sorts of bipedal apes—gracile, somewhat *Homo*-like forms and more specialized robust forms with heavy-duty chewing machinery—lived side by side in East and South Africa throughout the Late Pliocene and Early Pleistocene, from around 2.7 million to 700,000 years ago. Beyond this, practically everything else is still being debated. There is no current scientific consensus on the number, distinctiveness, relationships, classification, ecology, or behavior of the Plio-Pleistocene hominins of Africa. The African fossil record presents us with a confusing array of *Australopithecus*-like creatures, some of which are almost indistinguishable from chimpanzees, while others grade insensibly into early *Homo*. Our survey of these bipedal apes will take them up in temporal order, beginning with the Miocene precursors of *Australopithecus* and ending with the late Pliocene and early Pleistocene forms that have been attributed to *Homo*.

◼ *SAHELANTHROPUS*: THE OLDEST HOMININ?

Several East African localities older than 5 million years have yielded fossils that are currently in the running for the title of "first hominin." The best-known, most ancient, and most puzzling of the three is *Sahelanthropus tchadensis*. Remains of this species have been recovered from several sites in the Djurab Desert, in the northern part of the Republic of Chad, dated on faunal evidence to around 7 Mya (Brunet et al. 2002, 2005). Today, these sites lie in the midst of a forbidding desert. Seven million years ago, they stood at the edge of a shallow lake containing fish, turtles, and crocodiles; but even back then, the surrounding country was already a dry desert, attested to by deposits of wind-frosted sand in the ancient lake sediments (Vignaud et al. 2002). Evidently, *Sahelanthropus* was not a rainforest animal.

Sahelanthropus is known from a nearly complete skull, two partial mandibles containing the crowns of P_3–M_1, and some other bits of teeth and jaws. The skull is about the size of a small chimpanzee's, but it looks nothing much like that of any living primate. Exactly what it looked like is not entirely clear, because it was crushed and cracked into hundreds of pieces and the upper jaw had suffered a substantial amount of plastic deformation. But most of the individual pieces of the other skull bones were not much deformed, and they remained in roughly their original relative positions. A virtual reconstruction was undertaken by doing a CT scan of the fossil and then fiddling with the individual "bone fragments" in the computer until they fitted together as precisely as possible (Zollikofer et al. 2005).

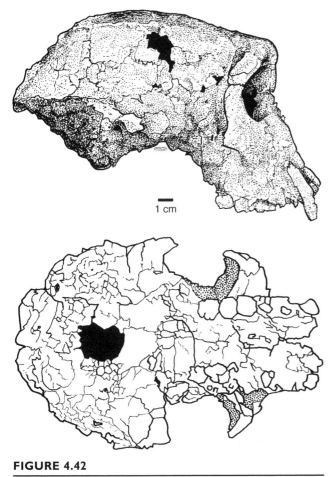

FIGURE 4.42

Virtual reconstruction of the skull of *Sahelanthropus tchadensis*: lateral and diagrammatic basal views. (After Zollikofer et al. 2005.)

This reconstruction (Fig. 4.42) shows *S. tchadensis* as an apelike creature with a small, low braincase but a surprisingly vertical face. The upper canine has a long root, but its crown is short, projects little, and has a worn-down tip (an **apical wear facet**). Although the P$_3$ is unknown, its wear facet on the upper canine is restricted to the canine's distal edge. This implies that the canine-sharpening "sectorial " action of the catarrhine P$_3$ had been essentially lost in *Sahelanthropus*, as in all later hominins (Brunet et al. 2005). In its cross-sectional area, the canine is intermediate between the large canines of chimpanzees and the reduced canines of *Australopithecus*. The dental enamel of *Sahelanthropus* is similarly intermediate in thickness between that of early *Australopithecus* and the thinner enamel of chimpanzees and gorillas (White et al. 1994). Some of these seemingly hominin-like dental traits, however, can be matched among Miocene apes (Wolpoff et al. 2006).

Although nothing is known about the postcranial skeleton of *Sahelanthropus*, its cranial base seems to

have been remodeled for upright posture to some degree. As reconstructed, the foramen magnum is positioned further forward than that of any ape, although the edge of the area of attachment for the neck muscles is marked by an ape-like nuchal crest high up on the back of the skull. Because the crest has a downturned lip, the discoverers of *S. tchadensis* infer that the neck muscles must have been running vertically up the back of an upright neck, and conclude that it must have been a biped. However, the nuchal plane of the reconstructed skull faces as much backward as it does downward—like that of a chimpanzee, and unlike the more horizontal neck-muscle attachment surfaces seen in *Australopithecus* and *Homo* (cf. Figs. 3.40 and 4.4).

The most puzzling thing about the reconstructed skull of *Sahelanthropus* is that it is in some ways more human-like than the Pliocene hominins that came after it (Wood 2002). It has huge brow ridges, bigger than those of an adult male gorilla. This is an ape-like feature—the brow ridges of early *Australopithecus* are relatively more delicate—but it is also a resemblance to some early *Homo* (Chapter 5). The short, vertical face of the *S. tchadensis* reconstruction seems more derived toward the human condition than the jutting, prognathic faces of early *Australopithecus*. Perhaps most surprisingly, the reconstructed foramen magnum of *S. tchadensis* lies further forward, and is tilted more anteriorly, than those of early *Australopithecus* (Zollikofer et al. 2005; cf. Fig. 3.36). This would seem to imply that *Sahelanthropus* had a more advanced upright posture than some of the East African hominins that succeeded it over the course of the subsequent three million years. That proposition is hard to swallow. In view of the shattered and somewhat deformed condition of the skull, many paleoanthropologists are suspending judgment on these supposed attributes of *S. tchadensis* until more fossils are found.

▌MIO-PLIOCENE ENIGMAS: *ORRORIN* AND *ARDIPITHECUS*

Outside of South Africa and the Djurab region of northern Chad, almost all the known fossils of early hominins have been recovered from sites in the northern and eastern rift valleys of East Africa. Two hominid species from Rift Valley sites, *Orrorin tugenensis* and *Ardipithecus ramidus*, are almost as ancient as *Sahelanthropus*. More fragmentary than the fossil from Chad, they nonetheless offer additional hints of very early adaptations to bipedality in the human lineage.

Orrorin tugenensis has been found at four localities in the Lukeino Formation in the Tugen Hills area of western Kenya, all dated to around 5.9–5.7 Mya (Fig. 4.3; Sénut et al. 2001, Sawada et al. 2002, Galik et al. 2004). The known fossils comprise a piece of lower jaw with two molars in it, the nearly complete upper two-thirds

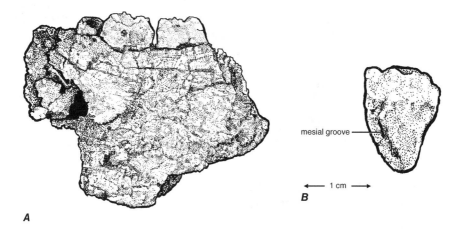

FIGURE 4.43

Teeth of *Orrorin tugenensis*. (**A**) Holotype mandible (BAR 1000'00); (**B**) lingual view of upper right canine crown (BAR 1425'00). Not to same scale. (After Sénut et al. 2001.)

or so of a femur, parts of two other femora, a distal radius, the proximal phalanx of a finger, several jaw fragments, and some isolated teeth. Like the skull of *Sahelanthropus*, the teeth attributed to *Orrorin* (Fig. 4.43) display a mix of ape and hominin traits. Their enamel is nearly as thick as that of *Australopithecus*. The upper canine is smaller than those of most apes—as small in cross section as the canines of early *Australopithecus*—but its crown shape is decidedly apelike, with a trenchant point and a groove running down the mesial and lingual edge. This so-called **mesial groove**, which is common among male monkeys and apes, is also retained in some other very early hominins (Haile-Selassie et al. 2004b, White et al. 2006). In addition, a small remnant of a distal cutting blade is present. Although the *Orrorin* canine surely did not function like that of an ape, its overall morphological pattern is persistently ape-like.

Perhaps the most significant piece in the scrappy collection of *Orrorin* remains is one of the femora (BAR 1002'00). Its lower end is missing, so there is no way of telling whether it had a valgus knee. But it shows some other human-like features (Fig. 4.44). Its lesser trochanter is large and medially directed (Pickford et al. 2002), like that of most modern humans and unlike those of *Australopithecus* and chimpanzees (Robinson 1972; Fig. 4.35). What this means functionally is not clear. The pit between the trochanters in back, the **trochanteric fossa**, is shallower than those of apes. A shallow fossa is characteristic of humans and ground-dwelling monkeys, and it has been described as a sign of reduced hip mobility and terrestrial habits (Bacon 1991). Like *Australopithecus*, *Orrorin* has a long femoral neck; but the head of the femur is relatively larger than that of *Australopithecus*. The fuzzy CT images that have been

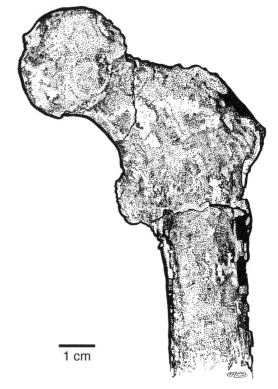

FIGURE 4.44

Left proximal femur of *Orrorin tugenensis* (BAR 1002'00). Anterior view. (After Sénut et al. 2001.)

published (Pickford et al. 2002, Galik et al. 2004) seem to show that the cortical bone surrounding the marrow cavity in the neck of the *Orrorin* femur is thickened along its lower edge and thin along its upper edge—a human-like trait, which some see as an adaptation for habitual loading of the hip joint in an extended position in bipedal walking (Ohman et al. 1997, Galik et al. 2004; cf. Rafferty 1998).

The forelimb skeleton of *Orrorin* appears to have been rather apelike. The humerus is large and robust, with a long crest extending up the lateral side of the shaft above the elbow. A large elbow-flexing muscle (*brachioradialis*) attaches to this crest in apes. Its large size in *Orrorin* implies a continued need for powerful elbow flexion, perhaps in connection with locomotion in the trees. The lone finger bone is curved like that of an ape, which also suggests some persistent use of the hands in hanging and swinging.

Ardipithecus, the third entry in the "first hominin" sweepstakes, was first described from surface finds found at Aramis, an early Pliocene site (~4.4 Mya) west of the Awash River in the Middle Awash region of Ethiopia. These fossils were originally announced as a new species of *Australopithecus, A. ramidus,* by T. White et al. (1994). A year later, the same authors published a terse note placing *A. ramidus* in a genus all its own (White et al. 1995). Slightly older fossils of *Ardipithecus ramidus* (4.5-4.3 Mya) have been found in the nearby Gona area (Semaw et al. 2005). Other sites in the valley of the Awash River have yielded even older *Ardipithecus* remains, dating back to 5.2–5.8 Mya (Haile-Selassie 2001). These oldest finds have been assigned to a different species, *Ardipithecus kadabba* (Haile-Selassie et al. 2004b). Both species of *Ardipithecus* seem to have ranged south into Kenya; a jaw fragment found there at Lothagam and dated to around 5.5 Mya may belong to *Ardipithecus kadabba*, while some later (<5.0 My) scraps of bone—a distal humerus

from the Tugen Hills and a piece of mandible containing two worn molars from nearby Tabarin—may represent a southern population of *Ar. ramidus* (Pickford et al. 1983, Hill 1985, Ward and Hill 1987, Wood and Richmond 2000, Deino et al. 2002, Schwartz and Tattersall 2005).

Most of the features claimed to distinguish *Ardipithecus* from *Australopithecus* are primitive, chimpanzee-like dental traits, including larger and more projecting canines, less megadont molars, and a P_3 with a single cusp and a single root (White et al. 1994, White 2002). *Ardipithecus* has thinner enamel than any other hominin—only some 0.1 mm thicker than the chimpanzee maximum. A mesial groove and distal-blade remnant are found on some isolated upper canines assigned to *Ar. kadabba* (Hailie-Selassie et al. 2004b). However, *Ardipithecus* differs from chimpanzees in having larger molars, smaller upper central incisors, and a less well-developed canine honing mechanism. In these respects, *Ardipithecus* is intermediate between chimpanzees and early *Australopithecus* (Fig. 4.45). A fragment of juvenile mandible contains a deciduous M_1 that is virtually identical to a chimpanzee's.

A fragmentary skull base of *Ardipithecus* from Aramis has been described as showing telltale signs of upright posture, or at least of an anteriorly displaced foramen magnum. The main reason for thinking this is that a line drawn between the foramina for the internal carotid arteries intersects the front edge of the foramen magnum (White et al. 1994). Unfortunately, this turns out not to be a trait that reliably distinguishes humans from chimpanzees (Schaefer 1999, Ahern 2005).

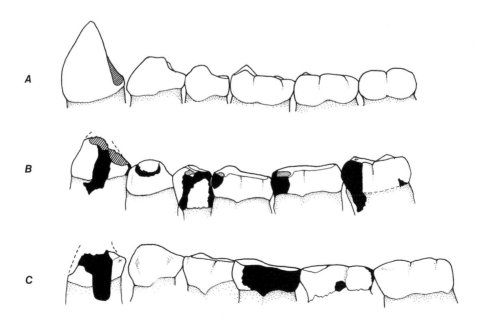

FIGURE 4.45

Left lower canine and cheek teeth of *Ardipithecus*, compared. Lateral views. (**A**) Female chimpanzee, *Pan troglodytes* (reversed); (**B**) *Ardipithecus ramidus* ARA-VP-1/128; (**C**) *Australopithecus afarensis*, Laetoli Hominid 4 (reversed). (After White et al. 1994.)

A few postcranial remains have been attributed to *Ardipithecus*. They are generally ape-like, but a fourth-toe proximal phalanx referred to *Ar. kadabba* has a dorsally tilted joint surface for the fourth metatarsal. This trait has been claimed as a correlate of bipedal habits by some (Latimer and Lovejoy 1990b, Haile-Selassie 2001) and rejected by others (Duncan et al. 1994). A badly crushed partial skeleton of *Ardipithecus*, reportedly discovered in 1994, remains undescribed and shrouded in secrecy (White et al. 1995, Gibbons 2002, 2004).

Currently, each of these three contenders for the title of "first hominin"—*Sahelanthropus, Orrorin,* and *Ardipithecus*—has its own group of proponents who emphasize its importance over that of the other candidates. Some of the *Orrorin* group argue that *Sahelanthropus* is not a hominin but an ape, as demonstrated by its big supraorbital torus, nuchal crest, and large canines (Sénut and Pickford 2004, Wolpoff et al. 2002, 2006). The discoverer of *Sahelanthropus,* M. Brunet, replies that these are persistently primitive features, to be expected in a stem or ancestral form (Brunet 2002, Brunet et al. 2005). The *Orrorin* team dismisses *Ardipithecus* as an ancestral chimpanzee or gorilla (Sénut et al. 2001). In response, the discoverer of *Ar. kadabba* says that the upper canine of *Orrorin* lacks apomorphies seen in all true hominins, and argues that *Orrorin* may be a chimpanzee ancestor (Haile-Selassie 2001). Members of the *Ardipithecus* team disparage the published radiographs of the femoral neck of *Orrorin* and demand that its discoverers present better evidence (Ohman et al. 2005). The *Orrorin* group retorts that it would be nice to see *any* published evidence concerning the 1994 *Ardipithecus* skeleton (Eckhardt et al. 2005).

We assume that T. White, the custodian of that skeleton, knows what he is talking about when he says that "There is still insufficient fossil evidence to determine whether there were one, two, or more hominid species lineages between 5 and 7 million years ago in Africa" (Gibbons 2005). None of these disputes can be resolved until that issue is settled. If all these Late Miocene hominids turn out to belong to the same species, as has been suggested (Hailie-Selassie et al. 2004b; cf. Lovejoy 2005a), its Linnaean binomial would be *Ardipithecus tugenensis*, since the name "*O. tugenensis*" predates "*Ar. kadabba.*"

▍ *AUSTRALOPITHECUS ANAMENSIS*

The oldest fossils that all the experts acknowledge as hominin come from Kanapoi and Allia Bay, near Lake Turkana in northern Kenya (Fig. 4.3). The first of these fossils to be discovered was a distal humerus found in 1965 (Patterson 1966, Patterson and Howells 1967). Isolated hominin molars and premolars were recovered

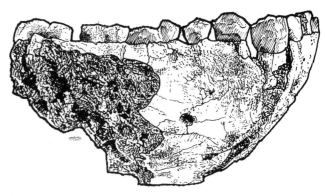

FIGURE 4.46

Lower jaw of the type specimen of *Australopithecus anamensis* (KNM-MP 29281A) from Kanapoi, lateral view. The canines and incisors of this fossil have been rotated downward and forward by postmortem cracking and distortion.

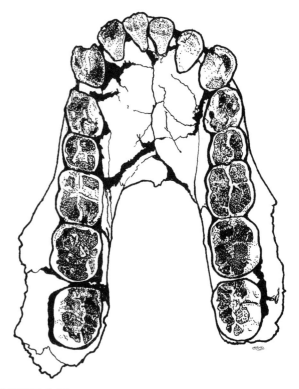

FIGURE 4.47

Australopithecus anamensis mandible (KNM-MP 29281A), occlusal view.

from two sites at Allia Bay in 1982 and 1988, and additional searching at Kanapoi in 1994 turned up more complete material including a proximal tibia, skull fragments, and a mandible with a complete dentition (Figs. 4.46 and 4.47). All these hominin fossils have been assigned to a single species, which has been christened *Australopithecus anamensis* and dated to the lower Pliocene, from 3.9 to 4.2 million years ago (Fig. 4.37;

Leakey et al. 1995a, 1998; Ward et al. 1999, 2001; Wood and Richmond 2000, White 2002). Other remains attributed to *A. anamensis* have recently been recovered from this time horizon in the Middle Awash area in Ethiopia, extending its known range a thousand kilometers to the north and east (White et al. 2006).

Australopithecus anamensis is the earliest species of *Australopithecus*. In several respects, it is more human-like than *Sahelanthropus, Orrorin,* or *Ardipithecus*. It has thicker dental enamel, which is generally thought to be a hominin synapomorphy (Beynon and Wood 1986, Grine and Martin 1988, G. Schwartz 2000). Its molars are enlarged, though not as big as those of later species of *Australopithecus* (Table 4.2). Its canine teeth are reduced by comparison with those of African apes. Like the canines of later hominins, they have worn-off tips and project little beyond the adjoining teeth, with which they form a single functional wear surface (Fig. 4.46). Not much is known about the skull, but a fragmentary temporal bone found near the Kanapoi mandible has an **external acoustic meatus** (opening of the bony ectotympanic ear tube) with an oval cross section—another hominin feature not found in *Ardipithecus* or living apes.

But in other ways, *A. anamensis* was persistently apelike. Although its incisors and canines are smaller than those of a chimpanzee, they are still fairly large. In most later hominins, further reduction of the front teeth resulted in a narrowing of the jaws in front, yielding a characteristic parabola-shaped dental arcade (Figs. 3.38, 4.40, and 4.41). In *A. anamensis*, the large size of the front teeth results in a U-shaped arcade with parallel rows of cheek teeth, as in living apes (Figs. 3.38 and 4.47) and *Sahelanthropus* (Fig. 4.42). A partial maxilla from Kanapoi (KP 29283) resembles those of apes in having canine jugae and a protrusive anterior alveolar region (nasoal vedar clivus). The P₃ of *A. anamensis* was also less molarized and more apelike than those of later hominins, with a single central cusp and elongated shape much like the P₃ of *Ardipithecus*. A big upper canine from the Middle Awash, attributed to *A. anamensis*, retains a mesial groove (White et al. 2006).

Like apes and *Ardipithecus*, *A. anamensis* lacks a distinct articular eminence at the front edge of the mandibular fossa (Ward et al. 1999). Its lower jaw appears to have been less stout and bulbous than those of later species of *Australopithecus*. At the level of the first molar, the jawbone has a breadth/height ratio greater than those of living apes, but smaller than those of later *Australopithecus* species (Ward et al. 2001). All these features of the jaws suggest that *A. anamensis* may not have been biting as hard on whatever it ate as later *Australopithecus* species did. This is also hinted at by the shape of its mandibular symphysis. Slimmer and more steeply angled (and therefore more ape-like) than those of later *Australopithecus*, it appears to have been

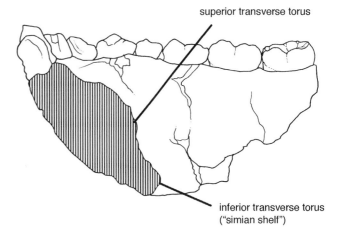

superior transverse torus

inferior transverse torus ("simian shelf")

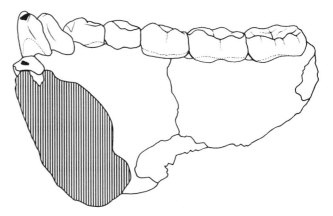

FIGURE 4.48

Medial view of the right lower jaw and teeth of *Australopithecus anamensis* (**top**: KNM-KP 29281A) and the later *A. afarensis* from Hadar (**bottom**: AL 400-1), showing differences in the shape of the mandibular symphysis (*vertical hachure*). The angulation of the *A. anamensis* symphysis has been exaggerated by postmortem cracking and distortion. (After Ward et al. 1999.)

less well adapted for transmitting powerful chewing forces from one side of the jaw to the other (Fig. 4.48). Nevertheless, the teeth are heavily worn, especially near the front of the tooth row.

Postcranial remains attributed to *A. anamensis* show a similar mix of ape-like and hominin features. The tibiotalar joint surface lies at right angles to the long axis of the tibial shaft (Fig. 4.49). This is a correlate of a valgus knee (Fig. 4.21), and suggests that this earliest species of *Australopithecus* was already adapted to a bipedal gait. In the sum of its metric characters, the distal humerus from Kanapoi falls in a broad zone of overlap between African apes and *Homo* (Lague and Jungers 1996). A prominent ridge on the dorsal edge of the distal end of this humerus resembles a similar ridge seen in great apes. In *Pan* and *Gorilla*, this ridge functions to prevent over-extension of the loaded wrist joint when these apes walk on their

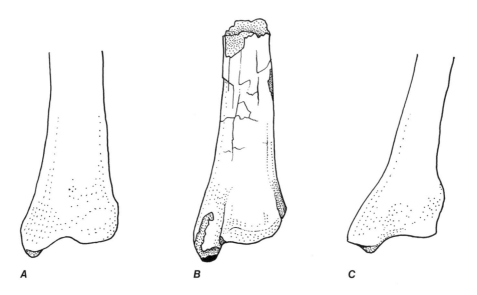

FIGURE 4.49

The lower end of the left tibia (anterior view) in (**A**) *Homo sapiens*, (**B**) the Pliocene hominin *Australopithecus anamensis*, and (**C**) *Pan troglodytes*. (After Ward et al. 1999.)

knuckles (Richmond and Strait 2000). A phalanx from the third digit of the hand shows robust muscle markings and pronounced longitudinal curvature—ape-like features, associated with the function of the hand in arm-swinging locomotion. A distal wrist bone (capitate) from Kanapoi shows that the index finger of *A. anamensis* lacked the ability to rotate slightly found in the human hand (Leakey et al. 1998). This ability, which is shared by all later hominins but is not present in apes, is associated with precise thumb-to-fingertip gripping and manipulation.

From comparisons of these postcranial remains with those of later *Australopithecus* for which we have more complete skeletons, we can estimate that large males of *A. anamensis* may have weighed up to 50 kg—as much as a modern chimpanzee or human (Leakey et al. 1995a, Ward et al. 2001). A nearly complete radius from the locality of Sibilot, 20 km from Allia Bay (Heinrich et al. 1993), is thought to represent *A. anamensis*. This radius is as big as that of a 6-foot-tall man; but *A. anamensis* was probably not that tall, since it presumably resembled later *Australopithecus* in having relatively long arms (compared to leg and trunk length).

Hominin fossils are known from three other Ethiopian sites (Fejej, Maka, Belohdelie) at around this time horizon or slightly later. These may represent either *A. anamensis* or its better-known descendant *Australopithecus afarensis* (Asfaw 1987, Fleagle et al. 1991, White et al. 1993, 2000; Kappelman et al. 1996, Van Couvering 2000, Kimbel et al. 2004). This may prove to be a distinction without a difference. A recent analysis of fossils of both species, ranging from the earliest *A. anamensis* (Kanapoi: 4.17 Mya) to the most recent *A. afarensis* (Hadar: 2.9 Mya), suggests that *anamensis* gradually changed into *afarensis* without any morpho-logical leaps or discontinuities (Kimbel et al. 2006). If so, then any species boundary drawn between them is arbitrary.

AUSTRALOPITHECUS AFARENSIS

With the possible exception of *A. africanus* from South Africa, *A. afarensis* is better known than any other species of *Australopithecus*. Its remains have been found at sites all up and down the Eastern Rift Valley, from Hadar in the north to Laetoli in the south (Fig. 4.3), ranging in age from 3.6 to 2.9 Mya (Kimbel 1988, White 2002). We have over a hundred specimens of *A. afarensis*, representing dozens of individuals and most of the parts of the skeleton (McHenry 1994c)—including the skull and partial skeleton of a 3-year-old infant from Dikika in Ethiopia (Alemseged et al. 2006) and the famous AL 288-1 find from Hadar, which comprises nearly half the skeleton of an adult female nicknamed "Lucy" (Fig. 4.50).

A. afarensis can be characterized as a more primitive version of *A. africanus*. Like that South African species, *A. afarensis* has a prognathic face, a relatively flat skull base, and a domed cranial vault (Fig. 4.51). However, the braincase of *A. afarensis* seems to have been slightly smaller. Estimates of its endocranial volume range from 380 to 550 cc, compared to a range of 428 to 600 for *A. africanus* (White 2002, Holloway et al. 2004). Because *A. afarensis* had larger jaw muscles than *A. africanus* relative to the size of its brain, it had more pronounced crests for jaw-muscle attachment on the outside of the skull, including a temporonuchal crest (Kimbel et al. 1984) and a small, posteriorly placed sagittal crest in some individuals (Boaz 1988, Kimbel 1988, Kimbel et al.

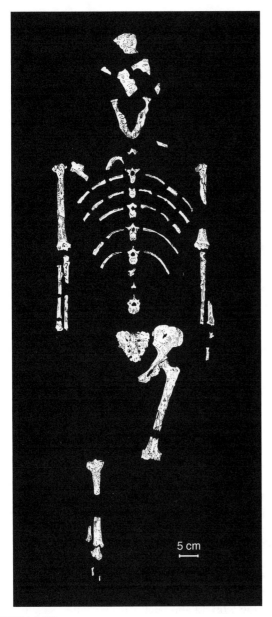

FIGURE 4.50

Partial skeleton of *Australopithecus afarensis*, AL 288-1 ("Lucy"), from Hadar, Ethiopia. (After White 2002.)

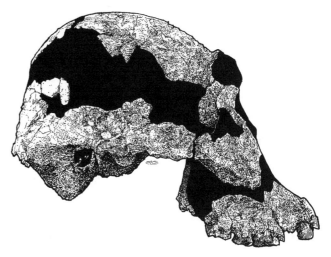

FIGURE 4.51

Skull of *Australopithecus afarensis*, AL 444-2, from Hadar, Ethiopia. (After Johanson and Edgar 1996.)

1994). As in apes, the rear and upper parts of the temporal bone are hollow, inflated by air-filled extensions of the middle ear cavity (Kimbel et al. 1985, 2004).

In these respects, the skulls of all species of *Australopithecus* conform to two general (though not invariable) rules of mammalian skull construction:

1. ***The more the lower face projects in front of the zygomatic arches, the larger the posterior temporalis becomes.*** Mammalian jaw muscles are often arranged so that the vector of their resultant forces passes through or near the part of the dental arcade where the greatest occlusal forces occur.

Therefore, a forward-thrusting dental arcade is commonly accompanied by an enlarged posterior temporalis muscle. Conversely, shifting the teeth backward tends to go along with an enlargement of the *anterior* part of the temporalis. (Moving the anterior end of the zygomatic arch forward has similar correlates; cf. Fig. 4.32). This rule has a lot of exceptions, especially among specialized herbivores in which other jaw muscles have largely supplanted the temporalis. But it applies reasonably well to primates in general (Tattersall 1973) and to *Australopithecus* in particular.

2. ***The larger the jaws are relative to the brain, the more extensive the air sinuses of the skull become.*** The inner surface of the braincase closely follows the surface of the brain, while its outer surface furnishes attachments for the jaw muscles. When the jaw muscles are large and the brain is small, the outside of the skull has to be big relative to the inside. In such cases, the excess space between the inner and outer surfaces is usually occupied by air-filled **sinus cavities** that develop as outpocketings from the nasal passages and the middle ear. In apes and in *A. afarensis*, a system of air cells connected with the tympanic cavity extends throughout most of the temporal bone. In later, larger-brained hominins, there is less disparity between the inner and outer surfaces of the braincase, and the air cells are increasingly restricted to the **mastoid** part of the temporal bone, behind the ear opening (Kimbel et al. 1985).

The dentition of *A. afarensis* is intermediate in most respects between that of its predecessor *A. anamensis* and that of *A. africanus*. The canines and incisors of *A. afarensis* are much like those of *A. anamensis* (Fig.

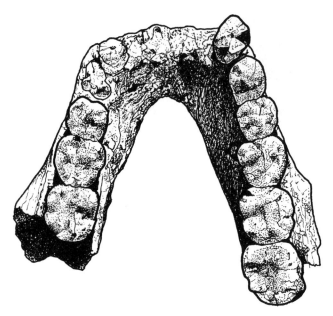

FIGURE 4.52

Lower dentition of *Australopithecus afarensis* (Laetoli Hominid 4). This is the type specimen of this species. (After Johanson and Edgar 1996.)

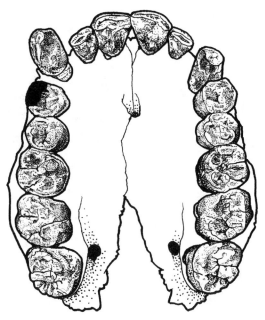

FIGURE 4.53 *Australopithecus afarensis* maxilla from Hadar (AL 200-1a). The protrusion of the right upper canine in this specimen is due to postmortem dislocation, but the apelike gaps (diastemata) in front of the canines are authentic. (After Johanson and Edgar 1996.)

4.48), but its molars are larger, continuing the trend in *Australopithecus* toward postcanine megadonty (Table 4.2). Larger than those of later *Australopithecus*, its upper canines have long roots surrounded by canine jugae (Johanson and White 1979) and tend to retain remnants of the distal cutting blade. Like that of *A. anamensis*, the P_3 of *A. afarensis* remained semi-sectorial, with a strongly dominant buccal cusp rotated toward the incisors, reflecting a derivation from an ancestor with a fully sectorial P_3. However, the premolars of *A. afarensis* are more molarized than those of *A. anamensis*—especially the P_3, and most especially in the later sample from Hadar (Hunt and Vitzthum 1986, Kimbel et al. 2004, 2006; cf. Figs. 4.45 and 4.48). By one estimate, the area of Lucy's lower cheek teeth is 2.8 times what would be expected for a modern hominoid of her body size (McHenry 1984). This enlargement of the cheek teeth relative to the incisors and canines is correlated with the appearance of a parabola-shaped dental arcade in some individuals of *A. afarensis* (Fig. 4.52). Yet the arcade retains an ape-like "U" shape in other individuals, and a small diastema often persists in front of the upper canine (Fig. 4.53; Kimbel et al. 1982).

The "Lucy" skeleton (Fig. 4.50) is a key source of information about body build and locomotor adaptations in *Australopithecus*. This is the only *Australopithecus* specimen yet described that preserves the long bones of both the upper and lower limbs for the same individual. Lucy's limb proportions are intermediate between those of a chimpanzee and a modern human's;

compared to her legs, her arms were longer than ours but shorter than a chimpanzee's (Table 3.2).

Like the later *Australopithecus* species from South Africa, "Lucy" had a femur with a small head and long neck (Fig. 4.35B), though these apomorphies are less pronounced in *A. afarensis* than they became in later *Australopithecus* from South Africa (Stern and Susman 1983). Lucy's pelvis resembles those of other hominins in having a broad and backwardly rotated sacrum and a short, wide ilium, with a deep greater sciatic notch in back and a marked anterior inferior spine on the front of its upper edge. The area of attachment for dorsal sacroiliac ligaments (Fig. 4.17) was larger than in *Pan*, but smaller than in *Homo*, and the iliac blade faced more posteriorly than it does in *Homo* (Stern and Susman 1983). Perhaps the most striking thing about Lucy's pelvis is its exaggerated transverse width. This is particularly apparent in the flaring iliac blades (Fig. 4.34), but it affects the whole pelvis. The pelvic aperture—the birth canal—is exceedingly wide relative to its front-to-back depth. Obstetricians find that this sort of pelvic conformation makes for particularly difficult and dangerous labor and delivery in modern humans. It has been speculated that this was also true of *A. afarensis* (Tague and Lovejoy 1986, Abitbol 1996).

Although the shape of Lucy's pelvis raises questions about her locomotion (see below), it seems clear that

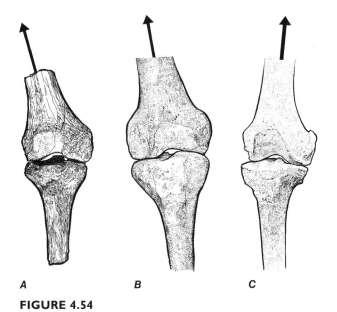

FIGURE 4.54

Anterior views of the right distal femur and proximal tibia of (**A**) *Australopithecus afarensis* (AL 129a-b); (**B**) *Homo sapiens*; and (**C**) *Pan troglodytes*, showing the valgus knee joint in the two hominins. The black arrows indicate the long axes of the femoral shafts. (**A**, after Johanson and Edgar 1996.)

she was a striding biped of a more or less human sort. The postcranium of *A. afarensis* shows dozens of human-like apomorphies thought to be associated with upright bipedality (McHenry 1994c). These include such telltale traits as relatively short toe phalanges, an adducted and enlarged first metatarsal, an expanded heel process of the calcaneus, a tibiotalar joint surface perpendicular to the long axis of the tibia, and a valgus knee (Fig. 4.54).

By an extraordinary stroke of luck, we have direct evidence for the presence of a striding biped at Laetoli, where a volcanic ashfall dated radiometrically to 3.6–3.75 Mya preserves tracks made in the fresh ash by the feet of many species of ancient birds and mammals, including Pliocene three-toed horses (*Hipparion*) and sabertoothed "tigers" (Leakey and Hay 1979, Hay and Leakey 1982). Mammals of two different sorts crossed this surface walking on two legs only (Tuttle et al. 1991, Day 1991). One appears to have been a large bearlike carnivore that took a few short steps on its hind legs (Tuttle 1987, 1990); but the other was a human-like biped, represented by the tracks of three individuals. Some of the experts who have studied these prints regard them as essentially like our own, with a thoroughly human pattern of force transmission involving a fully adducted hallux, a strong longitudinal arch, and short, straight toes (Day 1985, Day and Wickens 1980, Tuttle 1985, Tuttle et al. 1990, 1991). Others see the Laetoli tracks as the imprints of still

partly ape-like feet, which retained a semi-abducted hallux and long, curved toes (Stern and Susman 1983, Clarke and Tobias 1995). This sort of disagreement permeates everything that has been written about the locomotion of *Australopithecus*.

LUCY'S LOCOMOTION: THE VIEW FROM STONY BROOK

Ever since the battles between Le Gros Clark and Zuckerman in the 1950s, there have been two conflicting schools of thought about early hominin locomotion. One school follows Le Gros Clark and Lovejoy in seeing *Australopithecus* as an essentially human upright biped, almost as well adapted to walking around on its hind legs as we are. The other school believes that early *Australopithecus* still spent a lot of time up in the trees, and that its adaptations to bipedality were imperfect compromises aimed at accommodating both arboreal and terrestrial modes of locomotion. This second school has its roots in the work of R. Ciochon and R. Corruccini (1976), R. Tuttle (1981), Zuckerman's student C. E. Oxnard (1973, 1975a,b), and the French morphologists C. Tardieu (1979) and B. Sénut (1980, 1981a,b). However, its most forceful proponents have been the anatomists J. Stern, R. Susman, and W. Jungers from Stony Brook University in New York.

We can call these two schools of thought the "Lovejoy school" and the "Stony Brook school" for short. The opening manifesto of the Stony Brook group was a 1983 paper in which Stern and Susman argued for three propositions:

1. *A. afarensis* retained significant adaptations for arboreal locomotion.
2. When it walked bipedally, *A. afarensis* did not fully extend its knees and hips.
3. *A. afarensis* males were more terrestrial and bipedal than the females.

Stern and Susman were somewhat tentative about items #2 and #3, but they had no doubts about #1. One bone they found particularly telling was the **pisiform**, which underlies the so-called "heel" of the human hand on the ulnar edge of the wrist. Like most mammals, African apes have a large rod-shaped pisiform, which really does function as a sort of heel. Sticking out at right angles to the forearm, it provides a stout bony lever arm for the powerful flexor carpi ulnaris muscle, which flexes the wrist to help support and propel the body in knuckle-walking and in arboreal hanging and climbing. In humans, the forelimb is normally not involved in locomotion, and the pisiform is reduced to a pea-sized nubbin (L. *pisiformis* = "pea-shaped"). The pisiform of *A. afarensis* is an elongated, rod-shaped bone like those of chimpanzees or gorillas (Jouffroy 1991). In these early hominins, Stern and Susman concluded, wrist flexion was still involved in supporting and propelling the body.

Stern and Susman pointed to other apelike features of the *A. afarensis* hand that bolstered this conclusion. The metacarpals of the four fingers (digits 2–5) have large heads with articular surfaces that extend far onto the dorsal side, as in chimpanzees. This suggests that the joints between the metacarpals and phalanges were highly mobile and adapted to carrying a lot of force. The proximal phalanges of the *A. afarensis* hand are longer than ours and curved like those of *Pan* and *Gorilla*, with pronounced, ape-like flexor-sheath ridges. The tips of the distal phalanges are narrow and pointed like those of apes, whereas in humans they are flattened and widened into spade-shaped **apical tufts** to carry the broad, expanded pads of our manipulative fingertips. "When all the elements of the *afarensis* hand are considered together," wrote Stern and Susman, "one is struck by the morphologic similarity to apes."

The other element of the *A. afarensis* forelimb that seemed to Stern and Susman to betray a lingering commitment to arboreal locomotion was the scapula. This broad, thin bone is easily crunched up after death by bone-chewing scavengers, and only damaged fragments of the glenoid end of the scapula were known at that time for any species of *Australopithecus*. But the surviving bits of the *A. afarensis* scapula suggested to Stern and Susman that the glenoid socket had faced in a more cranial direction than ours, like that of an ape (Fig. 3.29)—presumably as an adaptation for suspensory arm-swinging.

Stern and Susman admitted that the pelvis and lower limb of *A. afarensis* were radically transformed in a humanlike way from the ancestral condition, and that it must have been a more or less upright biped when it came down from the trees and walked on the ground. The foot bones from Hadar showed that *A. afarensis* had a "plantigrade, bipedal foot with strong plantar ligaments and reduced potential for intertarsal mobility," implying a human-like arched foot with no midtarsal break. The heel process of the calcaneus was long and stout, modified for bearing the increased forces of bipedality (Fig. 4.24C). The metatarsal heads showed that the toes were able to bend far upward (dorsi-flexion)—a human characteristic, associated with the so-called "windlass effect" that tenses the plantar aponeurosis when we shove off at the end of stance phase (Fig. 4.24A). And both the fossil bones and the footprints from Laetoli demonstrated that the hallux of *A. afarensis* was adducted.

But the phalanges of the toes of *A. afarensis* were relatively longer than ours, fully as curved as an ape's (Fig. 4.26), and ape-like in various details of their anatomy. And the Laetoli footprints showed impressions of varying lengths for the lateral toes—sometimes sticking out beyond the big toe, sometimes not. This sug-

gested to Stern and Susman that these long toes might have been variably curled under in bipedal walking, as they often are in chimpanzees walking on their hind legs (Fig. 4.25). Even the adduction of the big toe, they conjectured, might have been variable. A damaged medial cuneiform from Hadar appeared to have had an ape-like convexity of its distal articular surface, where the first metatarsal articulates (Fig. 4.24), hinting that *A. afarensis* had retained some limited capacity for abducting its big toe to enhance the grasping ability of the foot.

From the apelike lateral flare and dorsal orientation of the iliac blades of the "Lucy" pelvis (Fig. 4.34C), Stern and Susman inferred that only the most anterior part of the deep gluteal muscles could have acted as an effective abductor of the hip joint. They argued that those muscles in *A. afarensis* acted to check pelvic tipping during walking by exerting a medially rotating torque on the flexed femur, as in quadrupeds (Fig. 4.22C, arrows 2 and 4) or chimpanzees walking bipedally (Stern and Susman 1981)—not by exerting an abducting torque on the extended femur, as they do in *Homo* (Fig. 4.23C). Therefore, *A. afarensis* must have walked with a flexed hip, as quadrupeds do when they walk on their hind legs. Stern and Susman thought they could see signs of this in the acetabular surface of the hip joint. The anterior horn of this crescent-shaped articular surface, through which propulsive thrust is exerted at the end of stance phase in human walking, is less extensive in "Lucy" than it is in modern *Homo* (Fig. 4.55). To Stern and Susman, this suggested that Lucy's hip was not fully extended at the end of stance phase.

Stern and Susman's third claim, that the two sexes of *A. afarensis* had different locomotor patterns, drew upon earlier work by Tardieu. Tardieu (1979) had argued that the three distal femora known from Hadar fell into two distinct types or morphs. The two small ones (AL 288-2, "Lucy," and AL 129) had essentially apelike articular surfaces, with nearly circular condyles and a rectan-

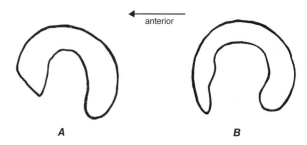

FIGURE 4.55

Outlines of the crescent-shaped articular surfaces of the left hip socket in (**A**) *Australopithecus afarensis* (AL 288-1ao, "Lucy") and (**B**) *Homo sapiens*. (After Stern and Susman 1983.)

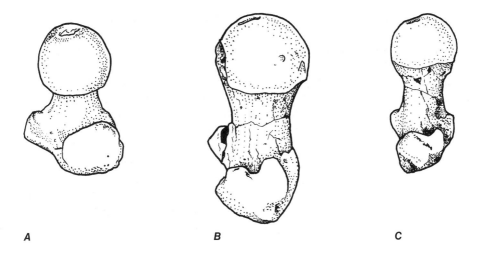

FIGURE 4.56

Superior views of the right proximal femur of (**A**) *Pan troglodytes*, (**B**) *Australopithecus afarensis* AL 333-3 (large morph), and (**C**) *A. afarensis* AL 288-1ap (small morph, "Lucy"). Compare Fig. 4.35. (After Stern and Susman 1983; reversed.)

gular distal outline (Fig. 4.20D). The largest one (AL 333-4) had humanlike elliptical condyles and a square distal outline (Fig. 4.20C). Tardieu concluded that these groups represented two different species. The smaller morph, she suggested, was a persistently primitive hominin that had retained more arboreal locomotor behavior, while the larger morph was a more advanced terrestrial biped with closer affinities to humans.

Stern and Susman noted other anatomical differences between the larger and smaller postcranial bones of *A. afarensis* that seemed to reflect the same sort of differences in locomotor behavior. The femur of the larger morph, they claimed, resembled ours in having a relatively large head and a short neck (Fig. 4.56); the small morph's femur looked more like those of South African *Australopithecus*. The small morph had a more valgus knee than a human's (Fig. 4.54), due to its wide pelvis and short legs, while the large morph had a more human-like knee angle, implying more human-like limb proportions. In the small morph, the distal articular surface of the tibia faced downward and backward, as it does in apes; in the large morph, it faced downward and *forward*, like a human's (Fig. 4.57).

The fossil teeth and jaws of *A. afarensis* from Hadar are also dimorphic, but the differences are no greater than can be seen within single species of living apes (White et al. 1981, 1983; Cole and Smith 1987). Unlike Tardieu, Stern and Susman therefore preferred to interpret the two Hadar morphs as females and males of a single species, *A. afarensis*. But why would the females have been less terrestrial than the males? Stern and Susman thought this might have been mainly a matter of allometry:

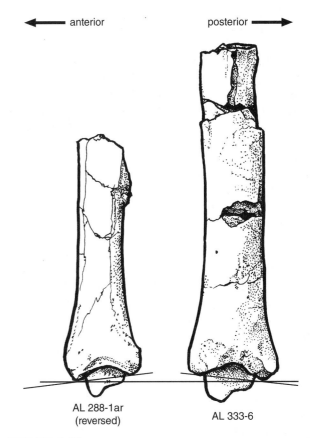

anterior ← → posterior

AL 288-1ar
(reversed)

AL 333-6

FIGURE 4.57

Lateral views of left distal tibiae of *Australopithecus afarensis* AL 288-1ar (small morph, "Lucy") and AL 333-6 (large morph). (After Stern and Susman 1983.)

Male orangutans and male gorillas spend more time on the ground and, in the case of gorillas, near to the ground than do females. Larger males, by virtue of sheer mass, may be compelled (or allowed) to spend more time on the ground than smaller, more vulnerable females (with young) and subadult animals. It is possible that the more pongid-like morphology of the smaller, female Hadar hominid reflects an increased arboreal component in its locomotor profile over the larger males. (Stern and Susman 1983)

Since sex differences in locomotor anatomy are negligible in both orangutans and gorillas, Stern and Susman concluded that "the degree of sexual difference in locomotor behavior in *A. afarensis* was greater than in any living ape."

LUCY'S LOCOMOTION: THE REBUTTAL

The Lovejoy school responded with a string of detailed studies of the bones and joints of the *Australopithecus afarensis* hindlimb. These studies aimed to show two things: that *A. afarensis* and *Homo sapiens* shared unique specializations for upright, straight-kneed, striding bipedalism and that these specializations made both species incompetent in the trees. The talocrural joint surface and axis of rotation of *A. afarensis*, said Lovejoy and his collaborators, was reoriented at right angles to the tibial shaft (Fig. 4.21), and the articular surface of the talus was nearly cylindrical. These changes converted the talocrural joint into a simple hinge—well-suited for bearing vertical weight and for flexion and extension movements, but lacking the range of motion needed for tree-climbing (Latimer et al. 1987). Like the human calcaneus, the calcaneus of *A. afarensis* bore a long, stout, laterally expanded and vertically oriented heel process, adapted to withstand the large stresses generated when the heel strikes down in a stiff-legged bipedal walk (Latimer and Lovejoy 1989). Relative to other skeletal measurements, the phalanges of Lucy's toes were only slightly longer than our own (White and Suwa 1987, Latimer and Lovejoy 1990b) and not at all comparable to the long, finger-like toes of a chimpanzee. The joint surfaces of the toe phalanges and distal metatarsals of *A. afarensis* were dorsally extensive and thus adapted to transmitting their greatest loads during extension, just as ours are (in the shove-off at the end of stance phase: Fig. 4.24A). By contrast, those of apes are shaped for carrying maximal loads during flexion (as in hanging from a branch).

All these descriptions might be debated, and they have all subsequently been called into question (Stern 2000). But for Lovejoy and his co-workers, the anatomy

of Lucy's big toe settled the issue beyond any possibility of argument. Even if the tarsometatarsal joint surface at the base of the hallux was slightly more curved than that of a human, it was nevertheless flatter than any ape's; and other features of the metatarsal and medial cuneiform showed that the hallux was permanently adducted. That made the foot useless as a grasping organ. A primate that was still spending a lot of time up in the trees could not have evolved in this way. "An opposable great toe," concluded Latimer and Lovejoy (1990a), "is the *sine qua non* of primate arboreality. Its absence is direct and virtually absolute confirmation that climbing behavior was negligible in *A. afarensis*."

Like the debate between Le Gros Clark and Zuckerman thirty years earlier, the debate between the Lovejoy and Stony Brook camps turns on the implications of the primitive, ape-like traits of *Australopithecus* (Ward 2002). The issue in question in this case is the importance of phylogenetic inertia. How much of the anatomy of an organism reflects its own way of life, and how much reflects the adaptations of its ancestors? To the Stony Brook group, it seems clear that the persistent apelike traits of *A. afarensis* must have continued to serve a function, presumably in arboreal locomotion. Otherwise, they would have been transformed by the selection pressures imposed by a strictly terrestrial way of life (Duncan et al. 1994, Stern 2000). But to Lovejoy and his colleagues, it is obvious that the human-like changes in the postcranial skeleton of *A. afarensis* must have severely hampered its ability to get around in the trees. They could have evolved only under a set of selection pressures that rendered arboreal adaptations valueless. Any apelike postcranial characteristics that persisted in *A. afarensis* must be vestigial traits, of so little functional importance that they had not yet been eliminated by natural selection. As Lovejoy's frequent collaborator Bruce Latimer put it:

Natural selection, when viewed within [an] animal's historical context, can be said to have a direction vector. The direction of the evolutionary change is of utmost importance in the interpretation of "intermediate" characters. ... When the directionality of the anatomical alterations is considered, it is highly unlikely that significant climbing behaviors were included in the locomotor behavior of the Hadar hominids. (Latimer 1991)

LUCY'S LOCOMOTION: PERSISTENT QUESTIONS

This dispute continues unresolved in the scientific literature. Three years after Latimer wrote the words quoted above, the paleoanthropologist H. McHenry (1994c) listed 54 persistently ape-like (plesiomorphic) traits and 46 distinctively human-like (apomorphic) traits of the *A. afarensis* postcranium, each of which was interpreted differently by partisans of different viewpoints. Since then, the list has grown even longer as more scientists have weighed into the fight, and virtually every observation has been called into question by one side or another (Stern 2000).

One persistent set of questions centers on the vertebral column. Compared to modern humans of similar size, *Australopithecus* had small, ape-like lumbar and first sacral vertebral bodies (Robinson 1972, McHenry 1991a, Schmid 1991, Sarmiento 1998, Sanders 1998, Häusler 2002, Dobson 2005; Figs. 4.16, 4.34, and 4.58). It would seem that the bodies of these vertebrae were not as well adapted as those of *Homo* to carrying the weight of the upper part of the body. This perplexing fact can be given at least four different spins:

1. *Australopithecus* was a mainly quadrupedal animal, like the living African apes. Even when it came down to the ground, it still spent a lot of time standing and walking on all fours (Sarmiento 1996, 1998; Sarmiento and Marcus, 2000).

2. *Australopithecus* was a biped, but it behaved in a way that placed relatively less stress on its vertebral column than humans experience—say, because it walked less energetically or more compliantly, or because it carried loads in its arms less often (Schmitt et al. 1996, Sanders 1998, Schmitt 2003).

3. The lumbar vertebrae of *Australopithecus* were carrying as much weight as those of a similar-sized human, but not as much of that weight was borne by the vertebral bodies and intervertebral disks as in *Homo*. Rather, the zygapophyses—the processes that bear the joints between the dorsal parts of adjacent vertebrae, which in humans mainly check and guide intervertebral movement (Fig. 4.10)—were loaded under compression, and carried a large part of the weight (Sanders 1990, 1995, 1998).

4. The lumbar part of the vertebral column of *Australopithecus* carried as much weight as that of a human of comparable size, but it did not have to carry as much compressive stress. Because it was less sharply curved, or better balanced, less force from the surrounding muscles—and correspondingly less compression of the column—was required to stabilize it.

This last interpretation has a few facts in its favor. Of the Plio-Pleistocene hominin specimens that have been described, only three are complete enough to allow us to count the lumbar vertebrae: two of *Australopithecus africanus* from Sterkfontein (Sts 14 and Stw 431) and one of early *Homo* from Nariokotome in Kenya (Fig. 4.27B; Fig. 5.18). All three of them appear to have had six lumbar vertebrae (Robinson 1972, McHenry 1992, Walker and Leakey 1993c, Latimer and Ward 1993, Sanders 1995, 1998; Sarmiento 1998). There is some debate about this, because the uppermost lumbar vertebra in each of these three individuals has certain thoracic-type features (Häusler et al. 2002, Toussaint et al. 2003). Nevertheless, there is general agreement that all these vertebrae were functionally part of the lumbar series, with lumbar-type joint surfaces and movements (Lovejoy 2005a). Modern humans normally have five functionally lumbar vertebrae, and the living great apes have averages of only three or four (Schultz 1961). What seems to have happened in hominins is that one or two of the uppermost sacral vertebrae, which would otherwise have been left sticking up above the sacroiliac joint when the ilium was shortened (Fig. 4.16), have been freed up from the sacrum and incorporated into the lumbar series. If six lumbars was the norm in *Australopithecus*, then early hominins probably lengthened the lumbar part of the column by two vertebrae when they became upright bipeds.

Having more lumbar vertebrae means that a given amount of lordosis can be accommodated with less abrupt angles between adjacent vertebrae. (Modern humans with six lumbars have less "wedged" lower lumbar vertebral bodies than those with five: Latimer and Ward 1993). And having smaller intervertebral angles presumably reduces stresses on the individual intervertebral disks and ligaments. This may in turn reduce the epaxial muscular effort needed to maintain the lordosis. If so, then having a gentler lumbar curvature than modern humans would have allowed *Australopithecus* to stand erect with less compressive stress on the lumbar vertebrae—which could therefore retain smaller vertebral bodies, like those of an ape. All this is speculation; but it may be significant in this connection that the Nariokotome *Homo* skeleton, which has six lumbar vertebrae, also has unusually small lumbar vertebral bodies for its size (Latimer and Ward 1993). If future fossil finds confirm the hypothesis that six lumbars was the norm in *Australopithecus* and early *Homo* (Wood and Richmond 2000), that discovery will raise another question: why have modern humans retreated to an average of five?

In support of his claim that *Australopithecus* was a chimpanzeelike quadruped, Sarmiento (1998) notes that the **sacral promontory**—the upper front edge of the first sacral vertebral body (Fig. 4.10)—is more sharply angled in humans (53–79°) than in African apes (73–94°). The more acute angle in *Homo* is a corollary of our sharp lumbosacral inflection. In the sacra of Sts 14 and "Lucy," the angle is blunt and ape-like (84–85°), and so Sarmiento infers that these early hominins lacked a human-like lumbar lordosis. But the 5° difference between *Australopithecus* and the human maximum might be just another side effect of *Australopithecus*'s

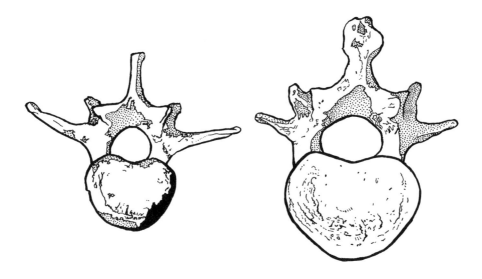

FIGURE 4.58

Second lumbar vertebrae of *Australopithecus africanus* (**left**: Sts 14) and a small female *Homo sapiens* (**right**), viewed from below. (After Robinson 1972.)

having six lumbar vertebrae, with correspondingly reduced intervertebral angles.

The lumbar vertebrae of *Australopithecus* have strikingly elongated transverse processes (Fig. 4.58). This fact can also be interpreted in different ways. It might mean that *Australopithecus* had relatively large epaxial muscles (Schmid 1991). But it is more commonly interpreted as an adaptation to improve the leverage of the trunk-controlling muscles that attach to these bony projections. What this means functionally is not clear, because there are a lot of those muscles and they perform a wide range of different functions including extending and flexing the vertebral column, bending it to either side, and flexing the hip.

Some pieces of ribs are preserved in the "Lucy" skeleton (Fig. 4.50). From these fragments and the associated pelvis, Schmid (1983) tried to reconstruct the shape of Lucy's trunk. Schmid concluded that it was narrow at the top and wide at the bottom, like an ape's. The narrowness of the upper thorax of great apes has been interpreted as a suspensory adaptation (Hunt 1991). An ape's narrow upper thorax and diagonally skewed scapula (Fig. 3.29) allows the shoulder joint to swing up close to the midline when the animal hangs by its arms or climbs upward (Fig. 4.59). This transfers weight more directly from the vertebral column to the forelimb, and reduces lateral sway when the animal changes handholds. Schmid's reconstruction of Lucy's ribcage as apelike thus fits the theory that Lucy still had an ape-like scapula and spent a lot of time in the trees.

The baby *A. afarensis* skeleton from Dikika (Alemseged et al. 2006) preserves a nearly complete scapula, which has an outline almost exactly like a juvenile gorilla's (Fig. 4.60A–D). This discovery might seem to settle

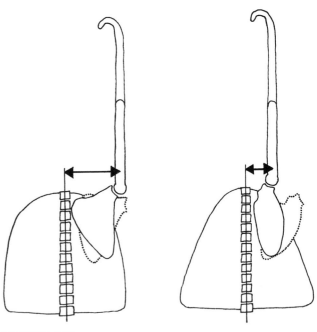

FIGURE 4.59

Schematic torso and right forelimb of human and chimpanzee, viewed from behind in a suspensory posture. The cone-shaped rib cage, narrow scapula, and cranially facing glenoid of the ape allow the scapula to rotate farther from the resting position (*dotted outlines*) when the arm is raised, thus bringing the shoulder joint closer to the vertebral column and the line of gravity (*double-headed arrows*). (After Hunt 1991.)

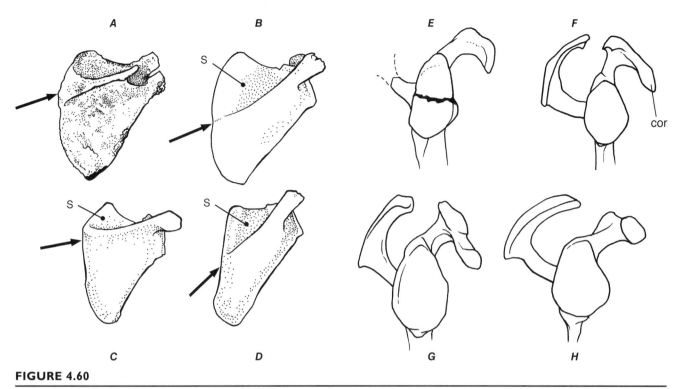

FIGURE 4.60

The scapula of *Australopithecus*. **A**: Dorsal view of right scapula of juvenile *A. afarensis* (DIK-1-1). **B–D**: Similar views of scapulae of gorilla (**B**), modern human (**C**), and chimpanzee (**D**), from juveniles at similar stages of dental development, showing position and orientation of the scapular spine (*arrows*) and size of the supraspinous fossa (**s**). **E–H**: Lateral views of the glenoid region of right scapulae of **E**, *Australopithecus africanus* (Sts 7); **F**, chimpanzee; **G**, orangutan; and **H**, modern human, showing the downward prolongation of the coracoid process (cor) in the first three. Not to same scale. (**A–D**, after Alemseged et al. 2006; **E–H**, after Sénut 1981a.)

the question of Lucy's locomotion in favor of the Stony Brook camp. But like many other parts of the *A. afarensis* skeleton, the Dikika scapula permits more than one interpretation. In humans, the **scapular spine** that projects from the dorsal surface of the scapula is less tilted up toward the head than it is in apes. This difference reflects the less cranial orientation of the human glenoid (Fig. 3.29). The human **supraspinous fossa** (the concavity above the scapular spine) is relatively smaller than those of apes. This difference reflects the reduction of the human supraspinatus muscle, which occupies the fossa and acts to help raise the arm above the head and to stabilize the shoulder joint under load. The Dikika scapula has a more horizontal spine and a smaller supraspinous fossa than the scapulae of baby apes. A proponent of the Lovejoy school might see this as evidence that the *A. afarensis* forelimb was no longer used habitually in apelike locomotion in the trees. Most of the Dikika skeleton remains to be prepared and described, so there are almost certainly more surprises in store.

A fragmentary *A. africanus* scapula from Sterkfontein (Sts 7) sends a similarly ambiguous signal concerning the **coracoid process**. This bony projection (Fig.

4.60E-H) is an attachment site for three arm muscles: biceps brachii and coracobrachialis, which lie in the upper arm and flex the shoulder and elbow, and pectoralis minor, which pulls the scapula toward the ribcage and can act to help support the body in hanging and swinging from the arms. In apes, but not in humans, the coracoid process is prolonged downward, lining it up with the pull of pectoralis minor. The coracoid of the fragmentary Sterkfontein scapula has an intermediate morphology (Sénut 1981a). Like other ape-like features of the *Australopithecus* skeleton, this fact can (again) be read in two ways, as reflecting either phylogenetic inertia or a persistent role for the forelimb in arboreal locomotion.

The exaggerated width of Lucy's pelvis has implications not only for the shape of her trunk, but also for the mechanics of her bipedality. Because the pelvic aperture is widened along with the rest of Lucy's pelvis (Tague and Lovejoy 1986), the **interacetabular distance** between her left and right hip joints is increased. The wide span between the two hip joints would have given the body's weight a longer lever arm around the hip joint during the swing phase of the locomotor cycle. This longer lever arm would increase the tendency of the pelvis to drop toward the unsupported side, canceling out any mechanical advantage that Lucy's deep gluteal muscles might have derived from the lateral flare of her hipbones. Jungers (1991) calculates that Lucy's deep glutei were in fact less mechanically effective than a human's, due to the increased width of her pelvis. If so, then Lovejoy's mechanical analysis (Fig. 4.36) cannot explain the small head and long neck of the *A. afarensis* femur (Ruff 1998).

Jungers has compared Lucy's skeletal dimensions with those of comparable-sized modern humans (Ituri Pygmies) and African apes (bonobos). His data (Jungers 1982, 1988b, 1991; Jungers and Stern 1983) indicate that Lucy's differences from modern humans in her body proportions are not simply allometric side effects of her small size and short stature (about 1 meter), as some have argued (Wolpoff 1983). Lucy's skeleton is human-like in relative humerus length, but she falls squarely with the apes in having a short femur (when we control for body size). These facts are reflected in her intermediate, quadruped-like intermembral index (Table 3.2).

From his Ituri Pygmy data, Jungers (1991) also concludes that Lucy's femoral head was significantly smaller than would be expected for a human of her body size, as most previous researchers (including Lovejoy) have said. But other studies by C. Ruff (1988, 1998) reach the opposite conclusion—that human femoral head diameter is positively allometric (growing relatively larger as body size increases) and that a modern human as small as Lucy would have a femoral head just as small as hers. These contrary conclusions rely on different methods and differently structured samples, and neither one seems to provide a definitive answer to this important question.

Why would early *Australopithecus* have reduced its upper limb length, as in humans, but retained short legs like those of an ape? There are theoretical reasons for thinking that large arboreal primates need relatively long arms in climbing tree trunks (Cartmill 1974c, Jungers 1977). Preuschoft and Witte (1991) infer that Lucy's relatively short legs would have been advantageous in vertical climbing—another sign of a persistent commitment to locomotion in trees. Tardieu (1991), however, thinks that short legs were advantageous in the early stages of bipedal adaptation because they kept the body's center of mass low and thus made it easier for *A. afarensis* to balance on its hind legs. Jungers and Stern (1983) suggest that shortening the upper limb would have had a similar beneficial effect on body-mass distribution and balance in bipedal walking.

We still lack a plausible explanation for the exaggerated width of the *A. afarensis* pelvis. Schmid (1991) sees Lucy's flaring iliac blades as adaptive for suspensory locomotion, because they would have provided increased attachment surfaces for the back muscles (latissimus dorsi and quadratus lumborum) that help to suspend the lower part of the body from the elevated arm. But if this sort of pelvic shape is an arm-swinging adaptation, why would it have evolved in a short-armed, largely bipedal ape and not in specialist brachiators? An alternative account is proposed by the paleoanthropologist Y. Rak (1991), who argues that an enhanced distance between the hip joints could have helped make up for Lucy's short legs, by increasing her stride length when one side of the pelvis rotated forward around the other during swing phase. But we doubt that the resulting slight gain in stride length would have been worth it, since increasing the interacetabular distance would increase the leverage of the body's mass around the stance-phase hip joint, thus requiring more work to check pelvic tipping toward the unsupported (swing-phase) side.

Lovejoy et al. (1999) argue that it is a mistake to try to explain morphology in this particularistic, "atomized" fashion. They postulate that the differences between the pelves of *Pan* and *A. afarensis* might have been produced by a change in some growth-controlling genetic mechanism that governs pelvic shape ratios, with the result that overall pelvic length decreased and breadth increased—a process they liken to applying a simple linear distortion in a digital image-manipulation program. In this way, reducing the length of the pelvis (to make it more stable in bipedal locomotion) might have automatically produced an increase in the distance between the hip sockets. The exaggerated width of Lucy's pelvis may be a useless or even slightly disadvantageous side effect of the adaptively valuable decrease in the length of the ilium.

There is nothing wrong with this sort of explanation in principle. It has long been recognized that changes favored by natural selection can bring with them other changes that are in themselves useless or undesirable. The phenomenon was documented by Darwin, who wrote in Chapter 5 of the *Origin* that "... the whole organisation is so tied together during its growth and development, that when slight variations in any one part occur, and are accumulated through natural selection, other parts become modified." But explaining a particular morphology with reference to what Darwin called "the laws of correlation" requires that some reason be produced for thinking that the supposed correlation is real. Such correlations can be demonstrated in the genetics lab (by showing that mutations producing one sort of change entail the other) or by comparative anatomy (by showing that one sort of change is often attended by the other for no apparent functional reason). If we knew that, say, certain mutations that shorten the hipbone in mammals also make the whole pelvis wider, then the account offered by Lovejoy and his colleagues would have some substance. But we do not, and so it remains insubstantial. In fact, since our own pelves are short without having the exaggerated breadth seen in *A. afarensis*, we know that the length and breadth of the pelvis can vary independently under the control of natural selection.

The Laetoli footprints are no longer accessible to direct observation—they were reburied in 1996 to protect them from erosion—but they continue to provoke controversy. Some observers see at least one of the prints (G1/34) as strikingly chimpanzee-like, with a divergent hallux and long, curled toes (Deloison 1991). The South African paleoanthropologist R. Clarke has renounced his earlier conviction that the Laetoli footprints were fully human-like (Clarke 1979), and now sees them as retaining certain apelike features, including a somewhat divergent hallux (Clarke and Tobias 1995). R. Tuttle argues just the reverse—that the Laetoli prints are like those of modern humans and could not have been made by the long-toed feet of *A. afarensis*. He concludes from this that there must have been a second, more human-like species of hominin at Laetoli (Tuttle 1981, 1985, 1990). White and Suwa (1987) tried to refute Tuttle by showing that a reconstruction of Lucy's foot fits perfectly into the fossil tracks. The problem here is that Lucy's foot consists of a talus and two phalanges (Fig. 4.50: Johanson et al. 1982a). The White and Suwa reconstruction of her foot is therefore based largely on a fossil foot (OH 8) from Olduvai that is usually assigned to *Homo habilis*. This has not convinced anybody in the Stony Brook camp.

Did the foot of *A. afarensis* have a longitudinal arch? This question, too, is debated. The shape of the talonavicular joint in Lucy's foot (Langdon et al. 1991) and the strong markings for the attachment of the "spring" ligament (Fig. 4.24A) on the navicular bones from Hadar (Latimer and Lovejoy 1989) suggest that the midtarsal region had already acquired a humanlike rigidity. However, the navicular of *A. afarensis* also has a large, ape-like tubercle projecting from its lower surface. In African apes, this bony process transfers weight downward from the talus to the sole of the foot during the midtarsal break. Its large size in the Hadar fossils leads some to think that there was a similar weight transfer—and therefore a midtarsal break, and therefore no longitudinal arch—in the foot of *A. afarensis* (Sarmiento and Marcus 2000, Harcourt-Smith and Aiello 2004).

Did Lucy walk with bent knees and flexed hips, as Stern and Susman suggested? This is another open question. Tardieu (1986) concludes that the attachments of the knee menisci to the tibia were loose and ape-like in *A. afarensis*, implying an imperfectly stabilized knee joint—but this is another one of those ape-like traits that might be blamed on phylogenetic inertia, and the evidence for it in *A. afarensis* is debated (Lovejoy 2007). Crompton et al. (1998) contend that walking with bent knees and hips is energetically inefficient and would have been quickly eliminated by natural selection in the earliest bipedal hominins. No doubt, humans find it awkward to walk this way, but this may not have been

true of *A. afarensis*. Almost all mammals stand and walk with their knees and hips flexed. So do most bipeds, because most bipeds are birds (Fig. 4.9). Rejecting Crompton's conclusions, Schmitt (1999, 2003; Schmitt et al. 1996) argues that ape-like compliant walking, with a bent knee and hip at mid-stance, would have been retained in the early stages of hominin bipedality. Since compliant walking increases stride length and decreases impact forces in humans, Schmitt thinks that keeping the knees and hips bent might have been advantageous for a short-legged early hominin that had small, poorly stabilized hindlimb joints like those of an ape.

The hominins that left their footprints at Laetoli took considerably shorter stride lengths, relative to foot length, than humans do (Tuttle et al. 1991). This seems to contradict Schmitt's picture of Lucy as taking long, bent-kneed strides. However, it may simply mean that the Laetoli bipeds had short legs and long feet. If Schmitt is right, a compliant walking gait might even have helped to compensate for the short legs.

The anatomy of the distal femur poses some problems for the Stony Brook school. The long bones of mammalian limbs form in the fetus as rods of cartilage and then ossify in three separate bony pieces: a shaft or **diaphysis** in the middle, and an **epiphysis** bearing the joint surfaces at each end. The diaphysis remains separated from its epiphyses by plates of cartilage. These **epiphyseal plates** persist until the bones finish growing, whereupon the three pieces of bone fuse together. In the distal femora of juvenile chimpanzees and many other baby mammals, the epiphyseal plate and the adjoining surfaces of the bony diaphysis and epiphysis are folded into interdigitating ridges and valleys. These interlocking surfaces presumably help to keep the epiphysis from being pulled off the diaphysis by the shearing forces generated in bent-kneed postures. But in human juveniles, where the knee is held in an extended position when it carries the body's weight, the shearing forces on the joint are correspondingly reduced—and the epiphyseal plate of the distal femur is almost flat. The immature *Australopithecus* femora from Hadar have flat, human-like epiphyseal plates (Tardieu and Preuschoft 1996). This suggests that *A. afarensis*—or at any rate the large morph, which appears to have had a more stabilized and humanlike knee joint than the small morph (Fig. 4.20; Tardieu 1983, 1986)—was walking with extended knees.

The paleontologist L. MacLatchy (1996) studied and measured the hip joint in humans, chimpanzees, *A. afarensis*, and the two South African species of *Australopithecus*. From the shape and size of the articular surfaces, she concluded that the human hip joint is specialized for carrying forces pressing forward and upward on the walls of the socket when the thigh is adducted and extended (our usual upright posture). The more mobile and flexible hip joint of chimpanzees is built to carry loads in a wider variety of postures and to rotate more freely in various directions, especially in abducting the flexed hip (as in clambering up a tree trunk). Of these three species of *Australopithecus, A. robustus* from Swartkrans appears to have been most like *Homo*. Fossils from this site include a humanlike hip socket (SK 3155) and two femoral heads, of which one is fully human-like (SK 82) and the other is intermediate between humans and chimpanzees (SK 97). However, one or more of these fossils may represent early *Homo*, which also occurs in the Swartkrans breccias. The hip joint of *A. africanus* (Sts 14) has an acetabular articular surface that is almost as human-like as that of SK 3155, but is peculiarly narrow in back near the greater sciatic notch. The small and large morphs of *A. afarensis* differ. The small morph (Lucy) has an essentially chimpanzee-like hip joint, but with more restricted abduction. The hip socket of the large morph is not known, but its femoral head (AL 333-3) is quite human-like, with expanded anterior and superior surfaces. MacLatchy's data generally support the idea that *Australopithecus* walked bipedally with an extended hip, as the Lovejoy school contends; but they also support the Stony Brook contention that the small morph of *A. afarensis* was more arboreal than either the large morph or modern humans.

Like humans, but unlike African apes, *A. afarensis* has relatively thick cortical bone along the lower edge of its femoral neck and very thin bone along the upper edge. This morphology is interpreted by Lovejoy and his colleagues as an adaptation for resisting the stresses of bipedal locomotion with an extended hip (Ohman et al. 1997; cf. Lovejoy 2005b). Similar conclusions have been drawn from the *Orrorin* femur (Galik et al. 2004). However, somewhat human-like bone-density patterns are seen in some quadrupedal monkeys and prosimians (Rafferty 1998). The thickening seen in *A. afarensis* suggests that Lucy's locomotion was not like that of any African ape, but it does not necessarily imply that she walked like a human being.

One of the most convincing reason for thinking that *A. afarensis* was still active in the trees is its long, curved phalanges (Fig. 4.26). Such phalanges are mechanically better suited for hanging from a branch than straight phalanges are, because the curvature reduces bending stresses on the bones (Oxnard 1969, 1973; Hunt 1991). The Lovejoy school sees the curved phalanges of the hand and foot of *A. afarensis* as simian leftovers, preserved by phylogenetic inertia. But this position is undermined by Stony Brook studies, still largely unpublished (Paciulli 1995, Richmond 1997), that purport to show that baby apes do not develop markedly curved phalanges until they start swinging and clambering around in the trees. This implies that phalangeal curvature reflects use, not genetic predisposi-

sition. If so, then *A. afarensis* must have still been using its foot for grasping things—presumably tree branches.

In summarizing this vast, messy, and inconclusive debate over the locomotion of early *Australopithecus*, the words that McHenry wrote in 1991 still seem appropriate:

> The host of "ape-like" traits seen in these early hominids probably implies that their bipedalism was kinematically and energetically different than modern humans and may imply that they were more efficient tree-climbers than modern humans. This arborealism was different from ape-like tree-climbing, however, because the hindlimb was specialized for bipedality and had lost essential climbing adaptations such as hallucial divergence.

The meaning of the postcranial differences between the large and small morphs of *A. afarensis* depends on whether we view these as two different hominin lineages or as the respective males and females of a single species. We will return to this question after we finish surveying the diversity and relationships of the species of *Australopithecus*.

AUSTRALOPITHECUS BAHRELGHAZALI?

Seven years before *Sahelanthropus* came to light, M. Brunet's explorations in Chad uncovered an anterior mandible and an isolated P³ of *Australopithecus* (Fig. 4.61) near Koro Toro, 150 km from the *Sahelanthropus* site. Dated to around 3.5 Mya on the basis of associated fauna, these remains were assigned to a species of their own, *Australopithecus bahrelghazali* (Brunet et al. 1995, 1996). The new species was said to differ from its contemporary *A. afarensis* in having a more vertical

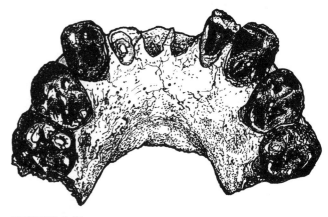

FIGURE 4.61

Australopithecus bahrelghazali, type specimen (KT12/H1). Occlusal view. (After Johanson and Edgar 1996.)

mandibular symphysis and lower premolars with three roots rather than two. However, more recent finds of *A. afarensis* suggest that this is not a real difference between the two (White et al. 2000). Until more material of the Chad *Australopithecus* is found, it seems prudent to regard it as a local variant of *A. afarensis*. It is still an important find, because it shows that *Australopithecus* ranged to the west of the Rift Valley.

AUSTRALOPITHECUS PLATYOPS?

In 1999, a fragmented and badly distorted *Australopithecus* skull was recovered from 3.5-million-year-old deposits west of Lake Turkana in Kenya. Together with two partial upper jaws and a temporal bone from slightly younger levels in the same beds, it has been described as a new genus and species of hominin, *Kenyanthropus platyops* (Leakey et al. 2001). The zygomatic arches of this skull arise far forward on the face, and its maxilla seems to have been short and deep, so that it appears to have an unusually tall and flat lower face for an early *Australopithecus*. In its facial proportions and profile, this skull, KNM-WT 40000, has been likened to the later, more humanlike KNM-ER 1470 skull, which is usually classified as *Homo rudolfensis* (Lieberman 2001).

These descriptions have been called into question by T. White (2003). White points out that the "Kenyanthropus" skull has been pulled apart into thousands of little bony bits by the expansion of the matrix in which it is imbedded. As a result, the size and shape of the original skull can only be estimated. The distortion of the braincase is particularly hopeless. White suggests that the "Kenyanthropus" material may represent a mere local or individual variant of *A. afarensis*. Until better fossils from this place and time turn up, it seems safest to suspend judgment on the taxonomy of these specimens.

White criticizes the naming of the new genus on other grounds. He protests that some fossil hunters are erecting new taxa on the basis of relatively trivial size and shape distinctions, of the sort that can easily be encompassed within the ranges of variation found within single species of living apes. Normal variation within living species, White observes, is considerably greater than is generally recognized. If more paleoanthropologists appreciated this, they would split less and lump more.

In a rebuttal to White, the primatologist J. Schwartz (who is renowned as an enthusiastic splitter) argues that splitting a single fossil species artificially into two species is a harmless error, since the two will wind up side by side on the cladogram as sister species and thus will not distort the overall geometry of the cladogram. In Schwartz's view, lumping two populations that are really separate species is a worse error, which risks losing phylogenetic and morphological information. Therefore,

when in doubt, we should play it safe and choose splitting over lumping (Schwartz 2003).

We concur with White, not Schwartz, on this issue. Our preference for lumping is partly a matter of philosophy: we accept the principle of Occam's Razor (Chapter 2), which says that the number of entities (such as species) that one recognizes should be kept to a minimum. The contrary principle, that entities should be multiplied whenever possible, strikes us as a recipe for chaos. And because we are skeptical about the objectivity and replicability of cladistic methodology, we are not so confident as Schwartz is that a species artificially split into two entities will wind up as a single branch on the cladogram.

Experience does not support Schwartz's optimism about the benign effects of making unnecessary taxonomic distinctions. "Kenyanthropus" is a case in point. Having recognized *K. platyops* as a valid taxon, the paleoanthropologist D. Lieberman (2001) draws a cladogram in which *Kenyanthropus* winds up as the sister group, not of *A. afarensis*, but of all other hominins younger than 4 Mya. This option would not arise if Lieberman adopted White's suggestion and sank *K. platyops* into *A. afarensis*.

We suspect that these early hominins are in fact oversplit, and that more extensive sampling in time and space would disclose continuities between populations that are currently recognized as distinct species. Until more is known about the range of spatial, temporal, and intra-population variation in these creatures, we favor striking a compromise between lumping and splitting, by provisionally recognizing all the *Australopithecus* species that have been named but grouping them all into a single paraphyletic genus, *Australopithecus* (Wood and Richmond 2000).

AUSTRALOPITHECUS GARHI

In 1997, a partial skull of an *Australopithecus* was recovered at Bouri, west of the Awash River in the Middle Awash area of Ethiopia. The skull was found in the Hata Member of the Bouri Formation, overlying a tuff dated to about 2.5 Mya by argon–argon dating (de Heinzelin et al. 1999). The paleoanthropologist B. Asfaw and his co-workers (1999) have designated this skull, BOU-VP-12/130, as the type specimen of a new species, *Australopithecus garhi*. Fragments of two other skulls, a nearly complete lower jaw, some other mandibular bits and pieces, and various postcranial remains from the same general place and time (the Lake Turkana Basin, from 2.7 to 2.3 Mya) may represent the same taxon.

The type skull (Fig. 4.62) is probably that of a male. In general shape, it looks much like those of *A. afarensis*: it has a small sagittal crest, moderately pronounced postorbital constriction, and a projecting, prognathic face. Its main distinctions from *A. afarensis* are features of the dentition (Fig. 4.63), especially its huge upper canines. Although these teeth have the reduced crown height and apical wear typical of *Australopithecus* canines, their buccolingual breadth is two standard deviations larger than the average for *A. afarensis* and a full six standard deviations above the *A. africanus* mean (Table 4.2).

In other respects, the dental metrics of *A. garhi* deviate from those of *A. afarensis* in a "robust" direction. Compared to the earlier species, it has smaller incisors, larger, more molar-like premolars, and much

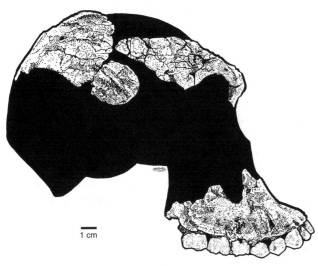

FIGURE 4.62

The BOU-VP-12/130 skull, the type specimen of *Australopithecus garhi*. (After Asfaw et al. 1999.)

larger molars—as big as those of *A. robustus* or *A. boisei*. But in spite of the expansion of the chewing surfaces of its cheek teeth (here extending even to the canine), it lacks the tall, "dished" face and other skull specializations that go along with megadonty in *A. africanus* and *A. robustus*. Its cranial capacity has been estimated at around 450 cc, which is comparable to estimates for male *A. afarensis* and *A. africanus* (Table 4.1).

A partial skeleton, BOU-VP-12/1, was found about 300 meters away from the *A. garhi* skull. These bones probably came from a single individual. They may or may not belong to the same species as the skull. They include parts of a humerus, radius, ulna, and femur, but no teeth or skull bones. All the limb bones are missing one or both ends. As far as can be told from the preserved parts, this individual had a humerus as long as that of the "Lucy" skeleton of *A. afarensis*, and an even longer forearm, so that its upper limb was longer overall than Lucy's. Some think that the femur was also significantly longer than Lucy's and that the BOU-VP-12/1 skeleton may have had a more human-like intermembral index than *A. afarensis* (Asfaw et al. 1999). Others argue that the Bouri skeleton is too fragmentary to tell us anything about its overall limb proportions (Reno et al. 2005a).

The discoverers of *A. garhi* conjecture that it may represent a late survivor of an earlier stock that gave rise to both *A. africanus* from South Africa and the *Australopithecus*-like early "*Homo*" species from East Africa (Asfaw et al. 1999). (A similar notion had been put forward earlier on the basis of fragmentary dental remains from the Omo Shungura Formation: Hunt and Vitzthum 1986, Suwa et al. 1996). However, *A. garhi* is

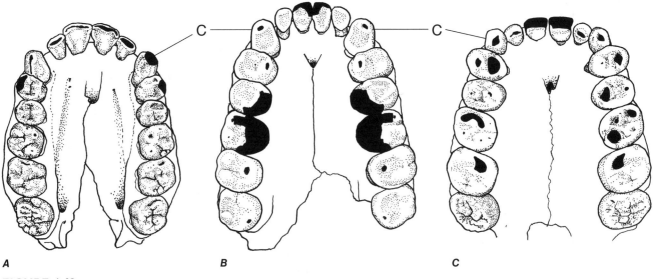

A   B   C

FIGURE 4.63

Maxillary dentition of (**A**) *Australopithecus afarensis* (KNM-MP 29281A) (**B**) *A. garhi* (BOU-VP-12/130), and (**C**) *A. boisei* (OH 5). Where possible, missing or distorted elements have been restored as mirror images of their intact counterparts from the opposite side. C, upper canines. (After Asfaw et al. 1999.)

too late in time to be an actual ancestor of these species; and its uniquely specialized dentition rules it out as a structural ancestor, at least of *A. africanus*. It may represent a species that had its own singular adaptive niche and eventually disappeared without leaving any descendants.

Some bipedal ape in the vicinity at this time appears to have been using stone tools. A few washed-out and redeposited Oldowan cores and flakes have been found downslope from Hata Member sediments of similar age, and cutmarks left by stone tools are found on fossil bones of *Hipparion* and antelopes from localities near the *A. garhi* sites (de Heinzelin et al. 1999). The tool-user may have been *A. garhi* and/or the BOU-VP-12/1 hominin, or some other hominin species yet to be uncovered in these deposits.

In dealing with the supposedly distinctive features of these and many other early hominin fossils, paleoanthropologists often find themselves simply having to take the claims and descriptions of the discoverers on faith, because the specimens are kept locked up for many years off limits to critical investigators. In the short run, there may be good reasons for this sort of secretiveness (Gibbons 2002). But it cannot be tolerated for long, because it obstructs the independent inquiry and replication of observations that are fundamental to the scientific method. Once we are expected to take someone else's word for some finding, we are no longer doing science. As J. Schwartz and I. Tattersall (2003: p. x) complain:

> … the type specimens of a number of species named and described several years ago remain even now off-limits to the general paleoanthropological community due to their describers' steadfast resistance to any independent verification.

Such species include the potentially important *Ardipithecus ramidus*, *Australopithecus garhi*, and *Australopithecus bahrelghazali*. … Until access to such specimens is permitted to researchers outside the closed describing cliques, these species (and specimens) must be regarded by the rest of us as hypothetical constructs.

█ AUSTRALOPITHECUS AETHIOPICUS

In 1967, the paleoanthropologists C. Arambourg and Y. Coppens described an *Australopithecus* mandible from Member C of the Shungura Formation in the Omo River valley of southern Ethiopia, dated to around 2.7 Mya. The mandible, Omo 18-1967-18, was a dubious choice for the type specimen of a new species, because it had no crowns left on any of its teeth. Nevertheless, Arambourg and Coppens (1967, 1968) assigned it to a new genus and species, *Paraustralopithecus aethiopicus*, because they thought that it showed a distinctive combination of characters. The remaining tooth roots and sockets showed that it must have had diminutive front teeth and big molars imbedded in a massive jawbone, like *A. robustus* and *A. boisei*; but the shape of the jaw hinted at a more prognathic, less vertical face than those later robust forms.

These inferences were vindicated 20 years later, when a robust *Australopithecus* skull with a big, protruding face (Figs. 4.64 and 4.65) was recovered from 2.5-million-year-old deposits west of Lake Turkana. Nicknamed "The Black Skull" for its color, this specimen, KNM-WT 17000, was originally attributed to *A. boisei* (Walker et al. 1986). The consensus nowadays is that it is sufficiently different to be assigned to a species of its own, along with the 18-1967-18 mandible and a few

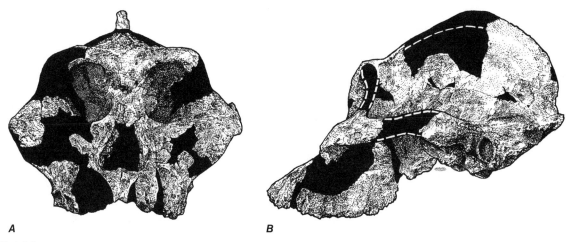

A   *B*

FIGURE 4.64

Skull of *Australopithecus aethiopicus* (KNM-WT 17000), anterior (**A**) and lateral (**B**) views. (After Tattersall and Schwartz 2001.)

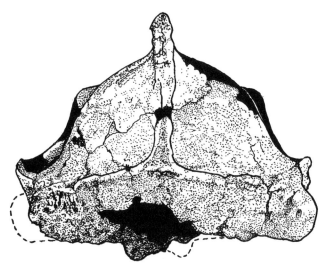

FIGURE 4.65

Skull of *Australopithecus aethiopicus* (KNM-WT 17000), occipital view. (After Walker and Leakey 1988.)

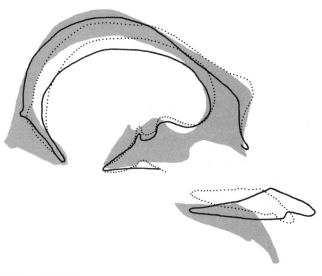

FIGURE 4.66

Diagrammatic midline sections (midsagittal craniograms) of the reconstructed skulls of *A. aethiopicus* KNM-WT 17000 (*solid line*), *A. boisei* KNM-ER 406 (*dotted line*), and *A. boisei* OH 5 "Zinj" (*gray tone*), oriented on the foramen magnum to show differences in basicranial flexion. (After Walker and Leakey 1988.)

other "hyper-robust" jaw and skull fragments from Kenyan and Ethiopian localities dating from 2.7 to 2.3 Mya. However, nobody has shown any enthusiasm for retaining "Paraustralopithecus" as a separate genus. The official name of this species is therefore *Australopithecus* (or *Paranthropus*) *aethiopicus*.

Australopithecus aethiopicus shares many robust-type apomorphies with *A. robustus* and *A. boisei*, including a "dished" midface, a forward shift of the anterior root of the zygomatic arch, a sharply receding forehead, big chewing-muscle attachments, enlargement of the postcanine teeth, molarization of the premolars, and relatively small canines and incisors (Kimbel et al. 1988). The things that distinguish WT 17000 from skulls of the two later robust species are all primitive or plesiomorphic traits. The most conspicuous is its alveolar prognathism—the ape-like forward thrust of its lower face. There is a lot of variation in this trait within living ape species. Nevertheless, if "Zinj" and the "Black Skull" were classified together in one species, the range of variation in facial slopes within that species would be almost twice as great as that seen in any living species of great apes (Kimbel and White 1988).

Another primitive trait of WT 17000 is its small braincase volume, estimated to have been around 400 cc. This is almost 25% less than OH 5 (Table 4.1), even though these two individuals appear to have been about the same size. Because WT 17000 had a small braincase and big jaw muscles, it has massive sagittal and temporal crests. And because its face is prognathic, it had a large posterior temporalis muscle, with robust and expansive attachments to the rear end of the braincase. The result is a very primitive-looking

occipital region (Fig. 4.65): wide and startlingly low, with a gorilla-like fusion of the sagittal, temporal, and nuchal crests. WT 17000 also lacks a well-developed articular tubercle at the front of its mandibular fossa. Its "flat" fossa is another primitive trait shared with apes and earlier *Australopithecus* species. The later robust forms (*A. boisei* and *robustus*) have distinct tubercles. (Kimbel et al. 1998)

Perhaps as a side effect of its prognathic face or its small brain size (or both), the cranial base of WT 17000 is less flexed than that of OH 5. Midsagittal craniograms (midline tracings) of the two skulls oriented on the plane of the foramen magnum show that the base of the braincase of "Zinj" is kinked up in the middle, and the palate is bent downward, by comparison with WT 17000 (Fig. 4.66). This is often cited as a difference that distinguishes *A. aethiopicus* from *A. boisei*. However, the anterior end of the "Zinj" cranial base is a reconstruction (Tobias 1967a); and the only other complete male *A. boisei* skull known, KNM-ER 406 (Fig. 4.67), is no more flexed than WT 17000. The only consistent differences between the craniograms of the two species are the smaller brain and longer, more projecting face of WT 17000. If we had more skulls of both, we might find that they overlapped in these metrics as well.

Dental differences between *A. aethiopicus* and *A. boisei* are difficult to assess, since there are no crowns left on the teeth of either the type mandible of *A. aethiopicus* or the WT 17000 skull. But other, more complete

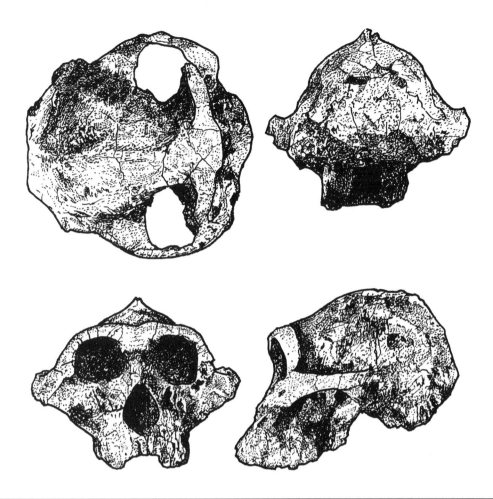

FIGURE 4.67

Skull of male *Australopithecus boisei* from Koobi Fora (KNM-ER 406), viewed from the top, back, front, and side. (After Wood 1991.)

teeth and partial jaws of robust-type *Australopithecus* from Omo (Shungura Formation) probably represent *A. aethiopicus*. They indicate that this species had jaws fully as big as those of *A. boisei*, but a less specialized dentition, with relatively bigger front teeth and smaller cheek teeth. Fragmentary jaws and teeth of robust *Australopithecus* known from around 2.3 Mya in both Ethiopia and Kenya are intermediate between *aethiopicus* and *boisei* in both age and morphology. They appear to document an anagenetic but rather swift and jerky evolution of *A. aethiopicus* into *A. boisei* at this time horizon (Walker and Leakey 1988, Suwa 1988, Suwa et al. 1996).

AUSTRALOPITHECUS BOISEI

In its skull and dentition, *A. boisei* is better known than the other two "robust" species of *Australopithecus*. Fossils attributed to this species have been recov-

ered all over East Africa, from southern Ethiopia in the north (Omo, Konso) to Malema on the shore of Lake Malawi in the south (Fig. 4.3). They occur across a time span of approximately a million years, from approximately 2.5 Mya at Malema and Omo to around 1.4 Mya from Bed 2 at Olduvai Gorge (OH 38) (White 1988, Schrenk et al. 1993, Wood et al. 1994, Suwa et al. 1996, 1997; Sandrock et al. 1999, Alemseged et al. 2002, Schwartz and Tattersall 2005). Throughout its existence in space and time, *A. boisei* lived side by side with, and competed successfully with, more gracile hominins usually assigned to our own genus, *Homo*. The last surviving *A. boisei* actually overlapped with early *Homo erectus* (also known as *H. ergaster*) in Africa (Leakey and Walker 1976). The discovery of the contemporaneity of these two hominin species was a fatal blow to the so-called single-species hypothesis (see Blind Alley #5, p. 188).

The collection of fossils assigned to *A. boisei* is diverse, and distinguishing it from the other two "robust"

Blind Alley #5: The Single-Species Hypothesis

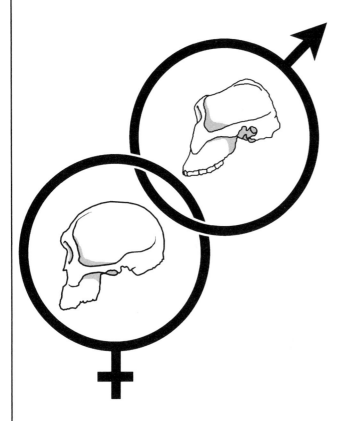

During the 1960s and early 1970s, an influential school of paleoanthropologists contended that there had never been more than one species of hominins at a time. The reasoning behind this "single-species hypothesis" ran as follows:

1. All hominins possess culture.

2. All organisms with culture occupy the same ecological niche.

3. Two sympatric species cannot long occupy the same ecological niche, because one will eventually outcompete the other and either drive it into extinction or force it into a different niche.

4. Therefore, there cannot be two sympatric hominin species.

In principle, this conclusion allows for allopatric speciation events. But one of the sister species thus produced would have to be short-lived, since one or the other would swiftly become extinct whenever the two came back into contact. Proponents of the single-species hypothesis preferred to think that all hominins at a single time horizon were always members of the same species. Therefore, only chronospecies could be distinguished in the single indivisible human lineage. Morphological differences between contemporaneous hominins had to be interpreted as sexual dimorphism, regional differences between local "races," or both.

For example, single-species theorists at first regarded the two species of *Australopithecus* from South Africa as successive phases in the linear course of human evolution, with the earlier, more gracile *A. africanus* evolving first into *A. robustus* and then into the genus *Homo* (Brace 1967). When distinctly different gracile and robust forms were found together in the same time horizon at Olduvai Gorge, the single-species story was amended: now the gracile types from East Africa were regarded as females, and the robust types were interpreted as males of the same species (Wolpoff 1971d).

This interpretation collapsed, and the single-species hypothesis finally died with it, when the KNM-ER 3733 skull was found (Leakey 1973a). Not even the most committed single-species theorist was willing to aver that this large, big-brained creature (Fig. 5.15) and its smaller-brained contemporary ER 406 (Fig. 4.38) were respectively a female and male of one species.

The four-point argument sketched above is logically flawless. However, the truth of its conclusion (4) depends on that of its premises (1–3), all of which are suspect. The main problem with premise 3, which is known as "Gause's Law," is that it is not falsifiable, because it does not predict how long it will take for the extinction (or adaptive divergence) to happen. Since all species eventually either become extinct or evolve into something different, the predicted changes will always take place if we just wait long enough. And we now know that the limited sort of "culture" that early hominins may have possessed, involving socially learned patterns of object modification, is found in several other species of brainy animals, including chimpanzees, orangutans, and even some birds (Hunt 2000, Weir et al. 2002). Chimpanzees and crows would have no trouble coexisting, even though both use tools of a sort.

The only form of "culture" that might be seen as defining an ecological niche is a fully human culture grounded in the use of language. And having this sort of culture frees us from the compulsion of Gause's Law, since language allows us to make deliberate, consultative decisions about what we will choose to do—including allowing a competing species to survive.

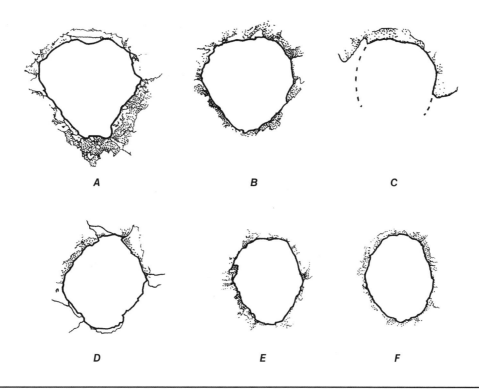

FIGURE 4.68

Outlines of the foramen magnum in (**A**) *Australopithecus boisei* KNM-ER 406, (**B**) *A. boisei* OH 5, (**C**) *A. aethiopicus* KNM-WT 17000, (**D**) *Australopithecus cf. boisei* KNM-ER 407, (**E**) *A. africanus* Sts 5, (**F**) *A. africanus* Sts 19. All are displayed with the occipital (rear) pole toward the bottom of the page. **D** is an endocranial view; the rest are basal views. Not to same scale. (**A**, after Wood 1991; **C**, after Walker and Leakey 1988; **B**, **D–F**, Schwartz and Tattersall 2005.)

species—*A. aethiopicus* and *A. robustus*—is not as easy as it might seem. One seemingly constant autapomorphy of *A. boisei* is the distinctively expanded talonid on its P$_4$ (Suwa 1988, Wood et al. 1994, Suwa et al. 1996). The foramen magnum is strikingly heart-shaped in ER 406 and slightly so in the OH 5 "Zinjanthropus" skull (Fig. 4.68A,B). Some see signs of a similar morphology in the ER 17000 skull (Fig. 4.68C), and call this a synapomorphy of an *aethiopicus-boisei* clade (Leakey and Walker 1988, Kimbel et al. 1988, McHenry 1996a). Others feel that what is left of the foramen magnum in ER 17000 is not heart-shaped enough to count as "heart-shaped" (Wood 1991, p. 263). The fragmentary skull KNM-ER 407 (Fig. 4.68D), which is usually identified as a female *A. boisei* (Wood 1991), has a roughly diamond-shaped foramen magnum. The foramen magnum of the Sts 5 skull of *A. africanus* might be described as heart-shaped or at least vaguely pentagonal, but this description seems inappropriate for other *A. africanus* specimens (Fig. 4.68E,F). The early *Homo* skull KNM-WT 15000 (Fig. 5.12A) has a foramen magnum that has been called heart-shaped (Strait et al. 1997). All this demonstrates once again how tricky it can be to frame, compare, and tally verbal descriptions of complex objects.

A. boisei typically differs from *A. robustus* in having a more exaggerated expression of some of the "robust" apomorphies; but the "typical" morphology is not seen in all *boisei* individuals. *A. boisei* characteristically has a wider, "visor-like" face, with a zygomatic arch that curves out laterally in a smooth circular arc around the temporal fossa as seen from above (Rak 1983), whereas *A. robustus* has a more distinct flexure between the anterior and lateral parts of the arch, and exhibits a shallow triangular depression between the zygomatic bone and the lateral edge of the nasal opening (Figs. 4.29B, 4.31, and 4.67). However, a partial skull of a male *A. boisei* from Konso in Ethiopia resembles *A. robustus* in these features (Suwa et al. 1997). The anterior attachment of the masseter muscle to the zygomatic arch is typically further forward in male *A. boisei* than in *A. robustus*, but the females of the two species do not differ (Rak 1988). Sexual dimorphism appears to have been marked in the skulls of both species, with female crania being less robustly constructed and having less pronounced crests for jaw-muscle attachment (Figs. 4.31 and 4.69).

A. boisei and *A. robustus* share a number of apomorphies not seen in the earlier and more primitive *A. aethiopicus*. These include a larger cranial capacity,

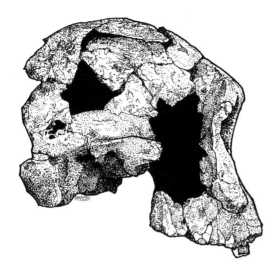

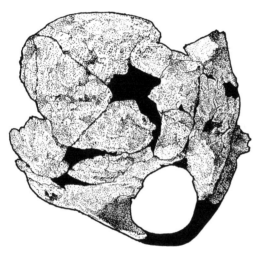

FIGURE 4.69

Skull of female *Australopithecus boisei* from Koobi Fora (KNM-ER 732). Lateral and superior views. (After Wood 1991.)

increased height of the mandibular symphysis, a deeper mandibular fossa with a more pronounced articular eminence, a taller and less prognathic face, a thicker bony palate (thickest in *A. boisei*), a more anteriorly concentrated sagittal crest, smaller incisors and canines, more molarized premolars, and ultra-thick dental enamel (Wood et al. 1994). These shared features are often regarded as synapomorphies demonstrating a special phyletic relationship between *A. boisei* and *A. robustus*. But except for the larger size of the braincase, they are all parts of a single functional complex, concerned with the generation and dissipation of uniform high occlusal forces across all the postcanine teeth. This functional correlation makes it more likely that the complex could have evolved independently in East and South Africa. Most of the features shared by *A. boisei* and *A. robustus* can be seen as extreme expressions of evolutionary trends that operated throughout the history of the genus *Australopithecus*. For example, *A.*

africanus differs from its predecessors *A. afarensis* and *Ardipithecus ramidus* in much the same ways in which *A. robustus* differs from *africanus*: a deeper mandibular fossa, more vertical symphysis, smaller incisors, bigger cheek teeth, thicker enamel, and so on (Fig. 4.45, Table 4.2).

The robust faces and sagittal crests of *A. boisei* and *A. robustus* make them look somewhat apelike; but their dished, vertical faces, hypertrophied molars, and reduced front teeth make it clear that these resemblances are superficial and misleading. The canines of *A. robustus* are smaller on average than those of *A. africanus*, and those of *A. boisei* are smaller still. Unlike apes, both species have small incisors with vertically implanted roots.

The postcranial skeleton of *A. boisei* is essentially unknown. Although a number of unassociated hominin limb bones are known from *A. boisei* localities, they cannot be confidently assigned to that species, because gracile hominins more closely related to *Homo* occur in the same deposits. A fragmentary skeleton (KNM-ER 1500) from the Koobi Fora area east of Lake Turkana includes some limb bones as well as a toothless chunk of lower jaw that some think represents a female *A. boisei*. However, Wood (1991) argues that the mandibular fragment is not diagnostic, and could belong to either a small female *A. boisei* or to one of the gracile *Homo*-like forms. The associated limb bones are badly weathered, but they appear to resemble the "Lucy" skeleton in the relative sizes of the upper and lower limb elements (Grausz et al. 1988).

FITTING IN SOUTH AFRICA: THE PROBLEM[S] OF STERKFONTEIN

There is a lot of uncertainty surrounding the dating of the fossil hominins from South Africa. The stratigraphy of the South African caves is complex, involving a long sequence of successive erosions and redepositions of carbonate minerals into and out of the groundwater trickling through the caverns. Attempts to reconstruct the history of the caves are further complicated by the fact that large parts of the deposits at most of the major sites were destroyed by mining operations. Because the cave deposits are not volcanic, most radiometric dating techniques cannot be applied to them. The ages of these deposits have been estimated mainly on the basis of comparisons between the fauna they contain and faunas from datable sediments in East Africa (Fig. 4.37). Such faunal comparisons suggest that the *A. africanus* sites are all Pliocene, ranging in age from about 3 My (Makapansgat) to 2.8–2.4 My (Sterkfontein and Taung). The three principal *A. robustus* sites are later (Pleistocene) in age, with faunal dates of some 2.0–1.5 Mya (Drimolen), 1.9–1.6 Mya (Kromdraai Member B), and 1.8–1.5 Mya (Swartkrans). Two lesser cave sites have also yielded a few fossils of *A. robustus*: coopers Farm, at 1.9–1.6 Mya, and Gondolin, at around 1.3 Mya (Vrba 1985, Delson 1988, McKee et al. 1995, Grine 2000, Keyser 2000, Berger et al. 2003, Schwartz and Tattersall 2005).

The age of the Sterkfontein breccias is the subject of a lively current debate. The debate is worth sketching briefly, because the site is uniquely important and the debate illustrates the range of techniques and problems involved in dating the South African sites.

The cave deposits at Sterkfontein are divided into six main stratigraphic units, labelled Members 1 through 6 in chronological order. Member 1, the oldest unit, appears to have formed mainly before the cave opened up to the surface. It contains both **flowstones** (undisturbed accumulations of solid carbonate minerals, like the slab that forms at the base of a stalagmite) and a basal layer of rocky breccias with no fossils in them, plus a few bone fragments at the top showing that the cave had developed a small surface opening (Partridge 1978). Member 3 is still largely unexcavated.

Early hominin remains have been found in Members 2, 4, and 5. Member 5, the youngest of these three units, contains crude Acheulean tools in its upper strata along with a few fossils that some attribute to early *Homo* (Clarke 1985b). Its older breccias contain more primitive material, including Oldowan-type tools and three isolated teeth (StW 566, 569, 584) attributed to *A. robustus* by some researchers and to *A. africanus* by others (Kuman and Clarke 2000, Guatelli-Steinberg and Irish 2005). This mix is comparable to what we see in Bed I at Olduvai. This suggests that the oldest part of Sterkfontein Member 5 dates to 2 Mya or less (Fig. 4.37; Vrba 1985), which is compatible with the faunal evidence (Kuman and Clarke 2000).

Almost all of the *A. africanus* fossils from Sterkfontein were found in Member 4, which contains no tools and few useful index fossils (Berger et al. 2002, Partridge 2002). Faunal age estimates for the lower boundary of Member 4 range from 2.1 to 2.8 Mya (Vrba 1985, Kuman and Clarke 2000, Clarke 2002a). A key issue in dating them is the presence of the modern horse/zebra genus *Equus*, which is not known to occur anywhere else in Africa prior to 2.36 Mya. Several *Equus* fossils have been found in Member 4 breccias. But except for one tooth, all of them were found out of context, in heaps of rubble tossed aside by miners. Those who favor an older date (around 2.8 Mya) for the bottom of Member 4 dismiss these *Equus* fossils as contaminants dislodged from Member 5 (Kuman and Clarke 2000, Clarke 2002a). Their opponents (Berger et al. 2002) point to the *Equus* fossils as proof that the Member 4 breccias are younger than that, ranging in age from no more than 2.5 My at the bottom to perhaps as little as 1.5 My at the top.

Member 2 has yielded only a few *Australopithecus* fossils, but one of them is uniquely important because it appears to be the virtually complete skeleton of a single individual. Most of this skeleton (Stw 573) is still lying deep underground, imbedded in hard, rock-studded breccia from which it is being slowly and carefully removed (Clarke 1998, 2002b). The fossil fauna from this part of Member 2, the so-called Silberberg Grotto, is unlike typical faunas from South African cave sites in four ways: (1) It consists mainly of baboons and cats, (2) there are no fossilized carnivore droppings, (3) very few (less than 4%) of the fossil bones show any signs of having been chewed on, and (4) many of the bones are undisturbed, lying in or near their original relative positions. The scientists working at Sterkfontein think that this fauna accumulated when agile climbing animals—baboons and cats, plus one unlucky *Australopithecus*—climbed or fell into the Silberberg Grotto, found themselves trapped, died of hunger or thirst, and turned into dried-out mummies with all their bones intact (Pickering et al. 2004). The limited damage observed on the skeletons may have been inflicted by starving cats vainly trying to eat the skin and bones of previous victims. This mechanism of bone accumulation has preserved at least one early hominin fossil of great value, but it has produced a restricted and peculiar faunal sample that has not been of much use in dating Member 2 (McKee 1996, Tobias and Clarke 1996).

Other dating techniques have been tried. Breccias do not retain a paleomagnetic signature, but solid, undisturbed layers of carbonate flowstone do. There are five such layers in the Silberberg Grotto, which preserve a record of five reversals of geomagnetic polarity. Partridge and co-workers (1999) interpret these as a sequence beginning in the upper Gilbert chron between 4.19 and 3.6 Mya, and ending in Member 3 in the upper Gauss chron (Fig. 1.3). But Berger et al. (2002) shift everything upward by four reversal events, reading the sequence as beginning in the Kaena subchron (<3.11 Mya) and ending in the Olduvai subchron (<1.95 Mya). By the first reading, the Stw 573 skeleton is 3.58–3.22 million years old—as old as the earliest *A. afarensis* from East Africa. The second reading makes it almost a million years younger.

Three absolute dating techniques have been used on the Sterkfontein deposits. An electron spin resonance study on the enamel of fossil antelope teeth from Member 4 yielded an unhelpfully wide range of possible dates, from 2.87 to ~1 Mya (Schwarcz et al. 1994). Recent uranium-series dates for the Member 2 flowstones suggest an age of about 2.2 My for Stw 573 (Walker et al. 2006). But much older dates have been obtained for Member 2 using **burial dating**. This technique relies on the fact that cosmic radiation striking rocks at the earth's surface produces two radioactive isotopes (aluminum-26 and beryllium-10) that have half-lives of around 1.0 and 1.8 million years, respectively. Because cosmic rays cannot penetrate far into the earth, production of these isotopes ceases when a surface rock becomes buried at depths greater than 20 meters. Since ^{26}Al and ^{10}Be decay at different rates, the length

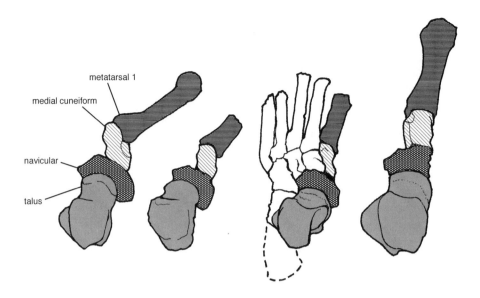

FIGURE 4.70

Left foot bones of (L. to R.) *Pan troglodytes*, Stw 573 (as reconstructed by Clarke and Tobias 1995), OH 8, and *Homo sapiens*. (After Clarke and Tobias 1995.)

of time that a deeply buried rock has been buried can be estimated by comparing the amounts of the two isotopes that remain in it. If a surface rock drops into the opening of a limestone cavern—for example, by falling into the deathtrap of the Silberberg Grotto—and then becomes imbedded in the accumulating breccia, it can furnish an estimate of the breccia's maximum age. Burial dating on an intrusive rock found adjacent to the Stw 573 skeleton yields an age estimate of 4.17 ± 0.35 My. Dates on four other rocks from different areas and levels of Member 2 are compatible with this estimate (Partridge et al. 2003). This implies an age for Stw 573 of about 4 My, which puts this skeleton at about the same time horizon as the *anamensis-afarensis* transition in East Africa.

Given these early dates, we might expect *Australopithecus* fossils from Member 2 to be particularly primitive. Member-2 breccias outside the Silberberg Grotto have yielded a partial cranium of *Australopithecus*, isolated postcranial bones, and some tooth and jaw fragments. All of these are basically like later *A. africanus* from Member 4 (Partridge et al. 2003). However, Stw 573 may be more primitive in some ways. Its zygomatic arch appears to be stouter, taller, and more robust than those of Member-4 *A. africanus* skulls, and its temporal lines appear to extend further up on the sides of the braincase (Clarke 1998). These facts suggest that it had more massive jaws relative to the size of its brain. The upper and lower limbs of Stw 573 are said to be "subequal" in length (Clarke 2002b). This implies that Stw 573 had relatively longer arms than *A. afarensis* (Table 3.2).

The foot is the only part of the Stw 573 skeleton that has been freed from the breccia and studied. Clarke and Tobias (1995) contend that it had a "semi-divergent" hallux that was still useful for grasping branches in arboreal locomotion (Fig. 4.70). Other experts who have studied the Stw 573 foot bones reject this conclusion (Harcourt-Smith and Aiello 2004, McHenry and Jones 2006).

If Stw 573 is distinctly more ancient and more primitive than *A. africanus* from Member 4, it may represent a different species—either a new species yet to be named, or one of the *Australopithecus* species of comparable age (*A. anamensis* or *A. afarensis*) known from eastern Africa. That would bring the tally of early hominin species found at Sterkfontein to as many as four: two (possible early *Homo* and *A. robustus*) from Member 5, another (*A. africanus*) from Member 4, and a fourth, unnamed species for Stw 573 in Member 2. And that may not be the end of the list. Some experts believe that Member 4 contains multiple species. Stw 252, a very fragmentary subadult cranium from Member 4, combines big incisors and canines with unusually large, *robustus*-sized premolars and molars. This combination is reminiscent of *A. garhi* (Table 4.2), which lived at about the same time in East Africa. Clarke (1988) has argued that this specimen, together with Sts 71 (Fig. 4.5) and some other Member 4 fossils with big cheek teeth, belongs to a new, unnamed species of *Australopithecus* that was ancestral to the later *A. robustus*. Two other fragmentary specimens from Member 4 (the Stw 183 maxilla and the Stw 255 temporal bones) also show certain *robustus*-like traits (Lockwood and Tobias 2002) and might be claimed as more evidence for the presence in Member 4 of a *robustus* ancestor. However, isolated *robustus*-like character states crop up here and there in other Member 4 fossils as well (Ahern 1998, Lockwood and Tobias 2002). For now, we prefer to assume that all the Member 4 hominins from Sterkfontein constitute a single species, *A. africanus*, which exhibited a lot of intraspecific variation.

FITTING IN SOUTH AFRICA: SOME *ROBUSTUS* QUESTIONS

Like Member 5 at Sterkfontein, the *A. robustus* sites of South Africa (Swartkrans, Kromdraai, Drimolen, Coopers, Gondolin) are faunally dated to the Early Pleistocene, less than 2 Mya. Also like Member 5, these sites contain Oldowan tools and fossils that have been assigned to two hominin species: *A. robustus*, and a second species with smaller molars that is generally regarded as some form of early *Homo*. We confront the same question here that we ran into in dealing with *A. boisei* in East Africa: when there are two hominin species in the same deposits, how do we tell which cranial and dental remains go with the tools and the postcranial bones?

None of the postcranial fossils from these sites are unequivocally associated with cranial parts or teeth of either *Homo* or *A. robustus* (Susman 1988a, White 2002). Some of the hindlimb bones show morphologies—small sacral vertebral bodies, a wide, flaring ilium, a small acetabulum, a small femoral head and long femoral neck, a posteriorly positioned lesser trochanter (Fig. 4.35)—thought to be distinctive of *Australopithecus*. These fossils are generally thought to represent *A. robustus* (Robinson 1972, McHenry 1975, 1988; McHenry and Corruccini 1978, Aiello and Dean 1990, Gommery et al. 2002). But other postcranials from the *robustus* sites are more humanlike, and it is hard to say which species they belong to.

The SK 50 right hipbone is a case in point. This specimen has always been considered as belonging to *A. robustus*. Most of the iliac blade is missing and postmortem crushing has distorted the ilium and acetabulum. The acetabulum is small, as in *Australopithecus* (MacLatchy 1996); and Day (1986b) believes the iliac blade would have had an *Australopithecus*-type dorsal orientation, and thus would not have positioned the deep glutei laterally as in *Homo*. Nevertheless, SK 50 resembles *Homo* in being much larger than Sts 14 and other earlier *Australopithecus* specimens, and its distortion makes it unclear exactly what its ilium looked like.

The hand bones from Swartkrans have provoked the most debate. There are some two dozen of these, mainly metacarpals and phalanges, representing several indi-viduals. The stratigraphic levels in which they were found (Members 1–3) have yielded all the skulls, jaws, and teeth of *A. robustus* found at Swartkrans, along with several cranial and dental specimens attributed to early *Homo*. R. Susman (1988a,b, 1993, 1994, 1998) has described the Swartkrans phalanges as human-like, without the marked curvature and robust flexor-sheath attachments seen in *A. afarensis*. Like those of humans, the Swartkrans thumb metacarpal SKX 5020 is relatively long, with a large, broad head at one end and a large, mobile carpometacarpal joint at the other. A distal thumb phalanx, SKX 5016, bears a flat, expanded apical tuft and muscle insertion scars of a distinctively human-like sort—traits thought to be associated with precise manipulation, found in *Homo habilis* (OH 7) and modern humans but not in chimpanzees. Susman concludes that the Swartkrans hominins that left these remains must have had a big, nimble thumb and dextrous hands adapted to making tools.

None of this would be controversial if these hand bones were attributed to early *Homo*. However, Susman reasons that most of them must have come from *Australopithecus robustus*, which comprises over 95% of the hominin teeth and jaws from these breccias (Grine 1989). One first metacarpal, SK 84, exhibits a peculiar "beaked" head, of a sort otherwise known only from the early *Homo* skeleton WT 15000 from East Africa (Chapter 5). The SK 84 metacarpal, Susman argues, must come from early *Homo* too. Therefore, the different-looking—but functionally human-like—SKX 5020 metacarpal must represent *A. robustus*. He concludes from all this that at least some of the tools found in Members 1–3 were made and used by *A. robustus*.

Susman's arguments have not persuaded most of the experts. Although the relative breadth of the first metacarpal head is large and humanlike in SKX 5020, it is just as large in mountain gorillas (Hamrick and Inouye 1995), and so it is not a reliable indicator of tool-making. Most of Susman's critics have focused on his attributions of hand bones to particular species. Since the "beaked" head seen in the WT 15000 and SK 84 metacarpals is not found in any other hominins, Walker and Leakey (1993c) suggest that these bones may come from some other sort of animal. Conversely, Trinkaus and Long (1990) show that in their overall dimensions, all the thumb metacarpals from Swartkrans can fit comfortably within the range of variation seen in modern humans. There is also no *a priori* reason to think that the two hominin species at Swartkrans must be present in the same proportion in the cranial and dental remains as in the postcranial remains. Different processes of accumulation may have been involved. For example, baboon finger bones pass undigested through carnivore guts and accumulate where carnivores defecate, whereas skull bones do neither (Pickering 2001). If the hand bones and cranial bones were accumulated by different processes, the relative frequencies of *Homo* and *Australopithecus* may well be quite different in the two samples. Finally, even if we were to grant that a 95% predominance of *A. robustus* skulls and teeth in the Swartkrans breccias "signals a like predominance of postcranial remains from that taxon" (Susman 1991), that would not imply that a particular, non-randomly chosen thumb bone from Member 1 has a 95% chance of belonging to that species. (If 99% of the animals in your house are arthropods, the *a priori* probability that one chosen at random will be an arthropod is 0.99; but the probability that *you* are an arthropod is zero, no matter what the percentages are.)

But whatever we think of Susman's specific arguments concerning the SKX 5016 and 5020 thumb bones, his general conclusions seem inescapable: most of the hominin hand bones from Swartkrans probably represent *A. robustus*, and they are more human-like than those of *A. afarensis*. Some of the same humanlike features are seen in phalanges from Sterkfontein attributed to *A. africanus* (Ricklan 1987, 1990; McHenry 1994c, Susman 1998). This suggests that all three species of Plio-Pleistocene hominins in South Africa—*A. robustus, A. africanus*, and early *Homo*—may have been tool-users.

The archaeological evidence from Swartkrans is also suggestive. Oldowan tools occur at Swartkrans in Members 1 and 2—and also in Member 3, where no *Homo* has yet been found. Members 1–3 also contain peculiar bone "artifacts" or natural tools: pieces of ungulate long bones, whose fractured ends have been scratched and polished in a way that can be experimentally duplicated by using a bone fragment as a digging stick (Brain 1981, Brain and Shipman 1993). Similar bone "tools" are found in Member 5 at Sterkfontein (Brain 1985). This implies the presence of a tool-user at both sites who was delving in the ground for something. That tool-user might have been early *Homo*. But it might equally have been *Australopithecus robustus*, digging up nourishing but tough roots and rhizomes to be processed by those big, flat teeth and powerful jaws.

Apart from the hand, the postcranial skeleton of *A. robustus* does not appear to have been particularly humanlike. The body of a last lumbar vertebra from Swartkrans (SK 3981b) resembles that of *A. africanus* in having a "wedged" shape but a relatively small lumbosacral articular surface (Robinson 1972). The fragmentary and distorted hipbones from Swartkrans, usually attributed to *A. robustus*, are human-like in the shape of the acetabular joint surface (MacLatchy 1996) but seem otherwise to have been much like those of *A. africanus*, with wide, laterally flaring ilia and small hip sockets (McHenry 1975, 1994c). From a first metatarsal found at Swartkrans, Susman and Brain (1988) infer that *A. robustus* had a humanlike foot with a stout plantar aponeurosis and a dorsiflexed position of the metacarpophalangeal joint at toe-off (Fig. 4.24A). But this fossil is not much different in these respects from the first metatarsal of *A. afarensis* (McHenry 1994c), and it is not certain which Swartkrans hominin it belongs to.

When it comes to the skull, *A. robustus* does share some apomorphies with early *Homo* that are not seen in *A. afarensis* or *A. africanus* (Tobias 1988, Rak et al. 1996, Strait et al. 1997, Kimbel et al. 2004). There are at least six of these:

1. A shorter, more vertical face.
2. Smaller canine teeth.
3. A larger brain.
4. A more flexed cranial base, with a more anterior position and more horizontal orientation of the foramen magnum.
5. A more transverse orientation of the long axis of the **petrous** (petrosal) part of the temporal bone (Fig. 4.71).
6. A **superior orbital fissure** (through which several nerves pass from the brain into the eye socket) that takes the form of an elongated slot rather than a small, round foramen.

These six traits are also seen in *A. boisei* from eastern Africa. Are they just convergences, or are they synapomorphies that signal a derivation of the human lineage from these "robust" forms?

Some think that the growth of the two bones that enclose the superior orbital fissure is developmentally correlated with elongation of the anterior cranial base and face (Lieberman et al. 2000). If so, then traits 1 and 6 may be different aspects of a single evolutionary change and should count as one resemblance, not two. It has also been suggested that traits 3–5 should not be tallied separately, because they are all part of a single functional complex correlated with a change in the shape of the **cerebellum**. This part of the brain, which is a major center for the control and coordination of voluntary movements, fills the rear hollow of the braincase surrounding the foramen magnum and presses against the vertical rear face of each petrous temporal. In *Homo, A. robustus*, and *A. boisei*, the cerebellum is large, vertically tall, and displaced forward ("anteriorly tucked") under the posterior (occipital) lobes of the cerebral hemispheres (Holloway 1988). Dean (1986, 1988) speculates that these changes in the brain were driven by selection for improved coordination of hand movements in toolmaking, and that the petrous temporal had to rotate into a more transverse position in these species to make room for the enlarged cerebellum behind it. The other changes in the cranial base might have happened as part of the same process.

However, other fossils show that these changes are not reliably correlated with one another. The WT 17000 skull of *A. aethiopicus* has a transversely oriented

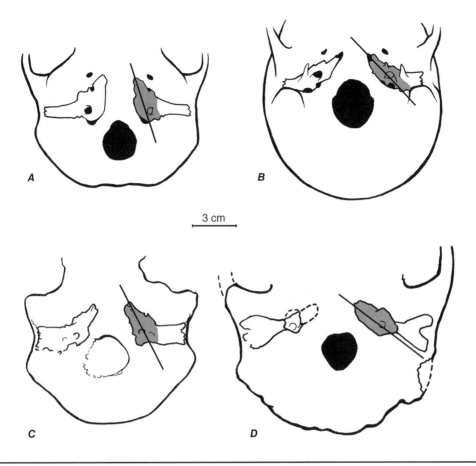

FIGURE 4.71

Diagrammatic basal views of hominid skulls. In a chimpanzee (**A**), the long axis (*black line*) of the petrous part of the temporal bone (*gray tone*) has a roughly fore-and-aft orientation. It is more nearly transverse in *Homo sapiens* (**B**). (**C**) *Australopithecus africanus* (MLD 37/38); (**D**) *A. boisei* (OH 5). (**A**, after Elliot 1913; **B**, after Clemente 1997; **C**, **D**, after Tobias 1967a.)

petrous, but its cranial base is not flexed and its cerebellum appears to have been primitive (Strait et al. 1997). Some *A. boisei* also retain an unflexed cranial base (Fig. 4.66). Conversely, *A. afarensis* appears to have had an "anteriorly tucked" cerebellum without any transverse reorientation of the petrous temporal (White et al. 1983, Kimbel et al. 2004). These facts make it hard to interpret all these cranial-base changes as functional correlates of each other, or of some adaptive mode—tool use, for example, or more perfected bipedalism—that might have evolved in parallel as a single linked complex in the late "robust" species and early *Homo*. We cannot rule out the possibility that some of them are true synapomorphies, carrying a phylogenetic signal about the origins of our own genus.

THE PHYLOGENY OF *AUSTRALOPITHECUS*

In the mid-20th century, when the list of fossil hominins comprised only the genus *Homo* and the two South African types of *Australopithecus*, there were few competing theories about their relationships. After all, only

three cladograms can be drawn for three taxa. Nevertheless, all three of the obvious possibilities had their proponents in the 1950s and 1960s:

1. Zuckerman and his school thought the *Australopithecus* species were a dead-end side branch of human (or ape) evolution and that the true human lineage was an outgroup that had branched off some 30 million years earlier.

2. Robinson, Le Gros Clark, and most of the other experts held that *A. africanus* and *Homo* were sister taxa or an ancestor–descendant pair. According to this theory, it was *A. robustus* that was the outgroup—"a specialized type, somewhat divergent from the main line of (hominin) evolution" (Le Gros Clark 1964).

3. A third view (Brace 1967) placed all three taxa on a single evolutionary lineage running from *A. africanus* through *A. robustus* to *Homo*. This hypothesis would be represented as a cladogram by making *A. africanus* the outgroup of a *Homo*-plus-*robustus* clade.

Even when additional hominin taxa began to emerge from Olduvai Gorge, the same three views could be, and were, stretched to accommodate the new discoveries. This was managed by regarding *A. boisei* as a form of *A. robustus*, and by treating *Homo habilis* as either early *Homo* or a late, "advanced" variant of *A. africanus*, depending on which cladogram you preferred. The fact that *A. boisei* and *H. habilis* had lived at Olduvai at the same time was a blow to the single-lineage theory (#3), but its backers kept it alive for a while by claiming that the two forms were respectively males and females of a single species (see Blind Alley #5, p. 188).

The recognition of a fifth and even more ancient hominin species, *A. afarensis* (Johanson et al. 1978, Johanson and White 1979), raised the number of possible cladograms to 105 and encouraged theorists to stake out new positions. The discoverers of *A. afarensis* hailed it as the last common ancestor of all later hominins (Johanson and White 1979, White et al. 1983). They saw the other *Australopithecus* species as an extinct sideline, characterized by specialized megadonty and powerful jaws. *A. africanus*, they claimed, was a primi-

tive representative of this robust clade, while *Homo* had arisen directly from *A. afarensis* through a separate line of descent. Many later authors have accepted this basic premise, though their cladograms have grown more complex and diverse as the number of *Australopithecus* species has increased (Fig. 4.72). But others have preferred to interpret the similarities between *Homo* and the "robust" species of *Australopithecus* (larger brain, more transverse petrous temporal, more vertical face, and so on) as synapomorphies linking *Homo* to the *boisei-robustus* clade (Fig. 4.73).

Other interpretations have been put forward. In his study of the OH 5 skull of *A. boisei*, Tobias (1967a) noted the unusual arrangement of its cranial **venous sinuses**, the big veins that drain the blood from the brain. In living apes and modern humans, most of the veins inside the braincase empty into the **transverse** sinuses, which run between the cerebrum and cerebellum at the back of the skull (Fig. 4.74A). This arrangement is also visible in *A. africanus*. But in OH 5 and other *A. boisei*, the transverse sinuses are largely replaced by a different set of channels, the **occipitomarginal** (O/M) sinus system

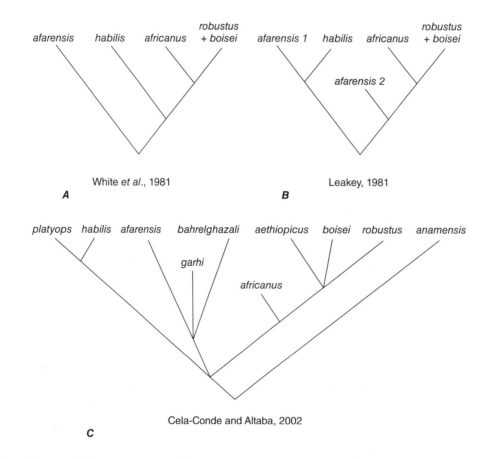

FIGURE 4.72

Some cladistic hypotheses of the relationships between *Australopithecus* species and early *Homo* (*habilis*), in which *A. africanus* is removed from the ancestry of *Homo* to that of a *robustus-boisei* clade. To facilitate comparison, species names have been standardized and all tree diagrams have been rendered as cladograms with no time dimension or ancestor–descendant pairs.

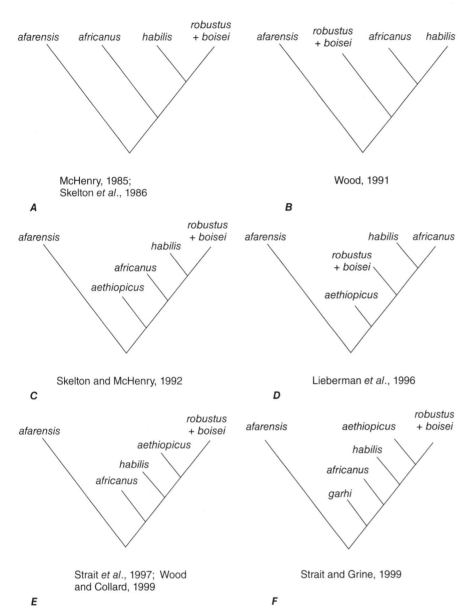

FIGURE 4.73

Some cladistic hypotheses in which *Homo* (*habilis*) is seen as more closely related to one South African *Australopithecus* species than to the other. Conventions as in Fig. 4.72.

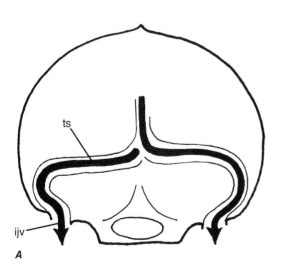

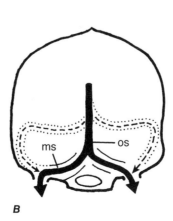

FIGURE 4.74

Cranial venous sinus patterns: diagrammatic occipital views. **A**: Typical *Homo* pattern, in which venous blood drains laterally (*black arrows*) through left and right transverse sinuses (ts) into the internal jugular veins (ijv). **B**: Pattern typical of "robust" *Australopithecus*, in which the transverse sinuses (*dashed arrows*) are largely replaced by a midline occipital sinus (os) and two marginal sinuses (ms). (After Falk and Conroy 1983.)

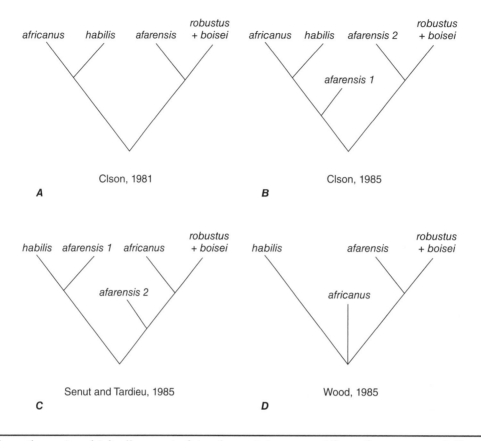

FIGURE 4.75

Some cladistic hypotheses in which all or part of *A. afarensis* is removed from the ancestry of *Homo* to that of the *robustus-boisei* clade. Conventions as in Fig. 4.72.

(Fig. 4.74B). The O/M arrangement appears to be typical of *A. boisei* and *A. robustus* and also of *A. afarensis* (Falk and Conroy 1983, Kimbel 1984, Falk 1986, 1988; Conroy et al. 1990, Strait et al. 1997, Kimbel et al. 2004). This has been proposed as a reason for transferring *A. afarensis* to the *boisei-robustus* clade (Fig. 4.75). More than a dozen other cranial and basicranial apomorphies supporting such a grouping have been put forward (Olson 1981, 1985; Kimbel et al. 2004), though some of them have not held up under critical scrutiny (Kimbel et al. 1985).

Olson (1985), who first suggested that some of the *A. afarensis* specimens from Hadar belonged to the *robustus* clade, thought that other Hadar fossils (including the "Lucy" skeleton) represented a different, more gracile form that was ancestral to *Homo* (Fig. 4.75B). He accordingly divided *A. afarensis* into two species, corresponding roughly to the large and small morphs recognized by Tardieu, and placed one in the genus *Paranthropus* and the other in *Homo*. Since then, others have also split *afarensis* into two species, with varying taxonomic results (Figs. 4.72B and 4.75C). Sénut (1996) prefers to put the Laetoli fossils into a genus of their own, *Praeanthropus*. Even among those who accept *A. afarensis* as a single species, the genus *Praeanthropus* has been gaining currency lately as a convenient taxonomic

wastebasket for the earliest and most primitive hominins, including *Orrorin* and the basal species of *Australopithecus* (Strait et al. 1997, Wood and Collard 1999, Cela-Conde and Altaba, 2002, Cela-Conde and Ayala 2003, Grine et al. 2006).

The place of *A. africanus* in all this has become increasingly unclear as the number of *Australopithecus* species has grown (Kimbel et al. 2004). *A. africanus* shares some apomorphies with the later "robust" species. Many of them are related to megadonty, but some of them—a human-like distal thumb phalanx, a deep mandibular fossa, a less ape-like distal radius, an enlarged brain, small canines, and a generally less ape-like skull—can also be read as resemblances to early *Homo* (Wood and Richmond 2000). Some experts contend that *A. africanus* is ancestral to both *A. robustus* and *A. boisei* (Rak 1983, 1985; Grine 1985, Senut and Tardieu 1985, Skelton and McHenry 1992, McHenry 1996a, Wolpoff 1999, Cela-Conde and Altaba 2002). Others have suggested that it is ancestral only to *A. robustus* and that *A. boisei* evolved independently from *A. aethiopicus* (Walker et al. 1986, Walker and Leakey 1988).

Several investigators think that *A. africanus* spent more time in the trees than *A. afarensis* did. The main support for this idea comes from the distribution of bone dimensions in the two species. Both species are known from more than 50 limb bones. For these samples, the ratios of forelimb to hindlimb joint dimensions are on average significantly greater in *A. africanus* than in *A. afarensis* (McHenry and Berger 1998a,b; Green et al. 2007). The relatively larger forelimb joints of *A. africanus* suggest that its forelimb carried correspondingly greater loads, hinting at a more arboreal pattern of activity. A fragmentary *A. africanus* skeleton from Sterkfontein, Stw 431, which preserves the elbow joint and a partial hipbone from the same individual, also appears to have had bigger forelimbs (relative to hindlimbs) than the "Lucy" skeleton of *A. afarensis* (McHenry and Berger 1998a). It has been claimed that *A. africanus* had a more ape-like tibia than *A. afarensis* (Berger and Tobias 1996) and smaller, less human-like lumbar vertebrae (Dobson 2005). And as noted earlier, some think that the Stw 573 skeleton of *A. africanus* from Sterkfontein Member 2 had a divergent big toe capable of grasping tree branches. On the other hand, the Stw 431 hipbone has been interpreted as revealing a more human-like pattern of bipedalism in *A. africanus* than in *A. afarensis* (Häusler 2002).

We will know far more about the postcranial differences between *A. afarensis* and *A. africanus* when the rest of the Stw 573 skeleton has been prepared and studied. If the postcranium of the South African species proves to be more apelike than that of *A. afarensis*, we will be obliged to conclude that the human-like cranial apomorphies of *A. africanus* are merely homoplasies, evolved in parallel with those of *Homo* in a creature that retained a more arboreal mode of life—or else that the more humanlike postcranial traits of *A. afarensis* are parallelisms (McHenry and Berger 1998a). Both readings might be right. Either one implies that there must have been a great deal of parallel evolution in a human-like direction going on among the Pliocene species of *Australopithecus*.

A lot of the disagreement over the anatomy and phylogeny of these Plio-Pleistocene hominins is driven by an underlying competition to determine whose *Australopithecus* fossils are going to be recognized as ancestral to *Homo*. Will it be *afarensis*, because all later *Australopithecus* species are too megadont? Or one of the South African species, because they share apomorphies with *Homo* not seen in *afarensis*? Or *platyops*, because of its early date and vertical face? Or *garhi*, because of the limb proportions of the BOU-VP-12/1 skeleton and those equid bones with cutmarks on them? Or some sequential combination of two or more of these species? There is no consensus on the answers to these questions.

The task of reconstructing the phylogeny of these early hominins can be made considerably simpler if we leave early *Homo* out of the picture for the time being and confine our attention to *Australopithecus*. Nine species of *Australopithecus—aethiopicus, afarensis, africanus, anamensis, bahrelghazali, boisei, garhi, robustus, and platyops*—figure in the current literature of paleoanthropology (Fig. 4.72C). More than two million different cladograms (15!! = 2,027,025; see Chapter 2) can be drawn for these nine species (Fig. 2.10). How can this be reduced to a usefully small number of candidate phylogenies?

The first step in approaching this sort of problem is to reexamine the **alpha taxonomy** of the sample to be classified—that is, its fundamental division into **observational taxonomic units** (OTUs). In this case, the OTUs are the nine species. We might find it necessary to increase the number of OTUs—for example, by dividing *A. robustus* into two species, *robustus* and *crassidens*, for the Kromdraai and Swartkrans fossils respectively (Grine 1985, Grine and Martin 1988). (As noted earlier, the Drimolen finds seem to render this particular division unnecessary.) But if possible, we will try to simplify our task (and conform to the principle of Occam's Razor) by reducing the number of OTUs. In the case of *Australopithecus*, we can eliminate two of the nine OTUs from consideration right away, by recognizing that the available fossil evidence does not justify granting separate species status just yet to either *"Kenyanthropus" platyops* or *A. bahrelghazali*.

We can simplify our list of OTUs further if we depart from strict cladistic methodology and allow ourselves to use stratigraphic information. *A. anamensis* is the earliest species of *Australopithecus* known. In all respects in which it is known to differ from later *Australopithecus* species, it is relatively primitive (plesiomorphic). It seems reasonable to assume that it is the ancestor of all the later species. *A. anamensis* therefore provides the root for our evolutionary tree, and we are left with only six other OTUs to worry about. Furthermore, there is convincing evidence that *A. anamensis* evolved directly into *A. afarensis* (Kimbel et al. 2004, 2006). For purposes of reconstructing phylogenetic relationships, we can therefore treat these two species as a single ancestral unit. This brings the number of problematic OTUs down to five: *africanus, garhi*, and the three "robust" species *aethiopicus, boisei*, and *robustus*.

This sort of reasoning, in which both temporal ordering and degree of apomorphy are taken into account in inferring ancestor–descendant relationships, is sometimes called **stratocladistic**. It has been shown experimentally to be more reliable than strict cladistic reconstructions of phylogeny (Fox et al. 1999). Stratocladistic reasoning differs from **stratophenetic** inference, in which evolutionary relationships are read directly out of the facts of stratigraphic proximity and phenetic similarity revealed by the fossil record. Pure stratophenetic inference is unreliable, because the fossil record is incomplete, and the oldest fossils of a species are never as old as the species itself. For this reason, stratigraphic precedence is no proof of ancestry. The stratophenetic evidence needs to be supplemented with cladistic evidence.

But when one morphospecies is wholly stratigraphically prior and cladistically plesiomorphic (primitive) relative to another morphospecies that is geographically contiguous and phenetically similar, the inference that the older one evolved into the younger one is warranted. That inference is clinched in those instances where the two are linked by transitional populations that are intermediate between them in all these respects. All these conditions appear to be met in the case of *A. anamensis* and *A. afarensis*. Therefore, the two can be treated as a single unit for the purposes of phylogenetic analysis.

Similar stratocladistic reasoning can be used to establish *A. aethiopicus* as the direct ancestor of *A. boisei*. The two are geographically and stratigraphically contiguous in East Africa. The features that distinguish *A. aethiopicus*—the more prognathic face, smaller cheek teeth, smaller P_4 talonid, and so on—are plesiomorphic relative to the comparable character states in *A. boisei*. There are also stratophenetically and stratocladistically intermediate samples of teeth and jaws from Member G of the Shungura Formation in southern Ethiopia (~2.3 Mya), which appear to document a transition from *A. aethiopicus* (2.5 Mya) to the derived morphology seen in the later *A. boisei* (<2.0 Mya) (Suwa et al. 1996). *A. aethiopicus* and *A. boisei* can therefore also be provisionally put together as a single ancestor–descendant unit.

If the *A. anamensis/afarensis* unit is the ancestor of all later *Australopithecus*, then it is ancestral to both *A. africanus* and the *A. aethiopicus/boisei* unit. Since *africanus* and *aethiopicus* were morphologically very different and overlapped in time, they represent two separate lines of descent and probably had separate origins from *A. anamensis/afarensis*. Putting all these conclusions together reveals a nearly complete evolutionary tree diagram for the genus *Australopithecus*, diagrammed in Fig. 4.76. Only two OTUs remain problematic: *A. garhi* and *A. robustus*.

Australopithecus garhi is a contemporary of *A. africanus* and *A. aethiopicus*, but it is clearly different from both. Its unique canine enlargement makes it an unlikely ancestor for later species of *Australopithecus* (unless Stw 252 represents a connection between *garhi* and *A. robustus*). There are therefore three plausible ways to add *A. garhi* to the evolutionary tree: (1) as the sister of *A. africanus*, (2) as a separate line of descent from *A. anamensis/afarensis*, or (3) as the sister of *A. aethiopicus*. At the moment, there are no compelling grounds for choosing among these alternatives (Fig. 4.76, lines 1–3). Likewise, *A. robustus* shares some apomorphies with *A. africanus* and others with the *A. aethiopicus/boisei* lineage, and can be plausibly linked to either of

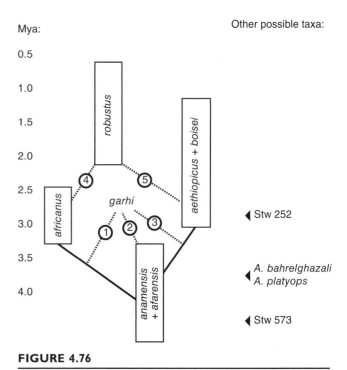

FIGURE 4.76

Stratocladistic diagram of inferred relationships between seven species of *Australopithecus*. Boxes indicate known temporal ranges; dashed lines represent alternative (numbered) possible phyletic connections.

these earlier species, depending on how we choose to describe, score, and weight character states (Fig. 4.76, lines 4 and 5). This leaves us with six viable evolutionary trees (alternatives 1 + 4, 1 + 5, 2 + 4, 2 + 5, 3 + 4, and 3 + 5). This is not an entirely satisfactory resolution. However, 6 is fewer than 2,027,025.

It is instructive to compare the foregoing with a cladistic approach to the same problem. If we are going to play by strict cladistic rules, we are not entitled to treat the ancestor-descendant pairs (*anamensis + afarensis*) or (*aethiopicus + boisei*) as single units in our analysis. The later member of each of these pairs exhibits apomorphies not seen in the earlier member, and some of these might prove to be synapomorphies linking the later member to some other derived OTU. Therefore, we must start with at least 7 OTUs, not 5; and in principle, all the presumptive species of the genus *Homo* should be added to the analysis, since they belong to the same clade. This puts us back where we started.

If we limit our strict cladistic analysis to seven species of *Australopithecus* and grant that *A. anamensis* is wholly plesiomorphic relative to all the other OTUs, then *anamensis* becomes the first or basal branch on our cladogram, and we can devote our attention to the other 6 OTUs. This move eliminates more than 90% of the options. However, it still leaves us with 945 alternative cladograms to evaluate. We could proceed to do this in the usual way, by (1) making choices between alternative ways of describing and scoring various sorts of morphologies, (2) drawing up a list of the resulting character-state distributions, (3) making the assumption that evolutionary change from one state of any character to another is equally probable for all states and all characters, and (4) setting a computer to work evaluating the 945 possible cladograms to find the one—or two, or more—that require the fewest character-state transformations.

For reasons sketched in Chapter 2, we believe that procedure (1) is not wholly objective or replicable among investigators. Assumption (3) is known to be false. We are therefore not persuaded that the usual cladistic methodology represents an improvement over a more straightforward approach that makes direct reference to the facts of geography and stratigraphy as well as to the distribution of shared apomorphies.

In the final analysis, the phylogenetic relationships of the various species of *Australopithecus* can be cleared up only by finding new fossils that clarify the evolutionary connections of *A. robustus, A. africanus,* and *A, garhi.* As far as the placement of the two South African species is concerned, we agree with Suwa et al. (1996) that "… this can ultimately be resolved in a stratophenetic manner only by the accumulation of time successive hominid samples in South Africa." Asfaw et al. (1999) sum up similar conclusions involving *A. garhi* in these words:

The … currently unresolved polychotomy stems from the fact that those characters most widely used in early hominid phylogenetic systematics are predominantly related to masticatory adaptation and are known to be both interdependent and susceptible to parallel evolution. … The atomization of such morphological complexes has led to lengthy trait lists, but the valence of the individual "characters" is clearly compromised. Such exercises have been useful in establishing the extensive homoplasy present among early hominids, but such confirmation only accentuates the precarious nature of phylogenetic reconstructions based on an incomplete and highly fragmentary fossil record.

WHAT DID *AUSTRALOPITHECUS* EAT?

Despite all these ongoing disagreements over the boundaries, functional anatomy, and relationships of the various species of *Australopithecus*, the overall picture of early hominin evolution is clear enough. Bipedal apes evolved in Africa around the end of the Miocene. By 3.5 Mya, they were found throughout most of the continent, from Chad to South Africa. They underwent a modest evolutionary radiation into several species. The earliest forms had chimpanzee-like skulls, but with more downward-facing foramina magna, bigger cheek teeth, and smaller canines with apical wear and reduced C/P_3 honing. Later species of *Australopithecus* tended to develop more humanlike hands and feet, bigger brains, flatter faces, and still smaller canines. However, these trends in the direction of humanity were correlated with parallel trends toward megadonty, an apomorphic specialization that is generally thought to exclude the later *Australopithecus* species from the human lineage. As far as is known, all the species of *Australopithecus* had limb proportions somewhere in between those of humans and apes, and the details of the upper limb skeleton suggest that "… the structure and function of the upper body … was different from that of modern humans" (Alemseged et al. 2006). When they walked on the ground, they walked bipedally; but their bipedality was unlike ours in some respects.

This is quite a lot to know, but we would like to know more about these bipedal apes. How much like us were they in their diets, social organization, and behavior? Did they use tools? Did they fight with weapons? Did they hunt? Did they live in large, multimale groups as chimpanzees and humans do, or in smaller family units like those of gorillas? Was there anything distinctively human in the way they acted, reproduced, and went about making a living?

Given that the experts on these matters have been arguing inconclusively for over 20 years about whether these creatures walked with their knees bent, it is not surprising that there is no scientific consensus on these more speculative issues. Most of the variables involved are not reliably correlated with bony anatomy. And once we start asking questions about things that are "distinctively human," we have moved wholly beyond the scope of correlation, because we are dealing with a data set that has only one member (Cartmill 1990, 2002). Nevertheless, a lot of thought, research, and ingenuity have been expended in repeated efforts to answer these tantalizing questions.

It ought to be relatively easy to reconstruct the diet of *Australopithecus*, since what mammals eat does have implications for the morphology of their skulls and teeth. But it is hard to find extant parallels for *Australopithecus*. The only living mammal that has teeth much like those of *Australopithecus* is *Homo sapiens*; and the apomorphies of human teeth are not a useful guide to feeding behavior, because it is not clear what sort of feeding behavior those apomorphies are associated with. Our teeth have become adapted over the past several hundred thousand years to a feeding regime in which the food is extensively pre-chewed and pre-digested—by chopping, grinding, mashing, cooking, fermentation, and so on—before it gets put in the mouth. All this **extra-somatic** (outside the body) or technological (cultural) preparation of food makes it possible for humans to live on an extraordinarily wide range of diets, from an all-fruit regimen like that of a conscientious Jain to a virtually all-meat diet like that of a traditional Eskimo, without evolving any special dental adaptations. Because there is no such thing as *the* human diet, it is hard to make inferences from the peculiarities of human teeth to the adaptive context that produced them in the first place.

The fact that the teeth of early hominins are not much like those of any living nonhuman primate does suggest some negative conclusions about their diets. Fruit-eating anthropoids generally have big incisors, used for biting into and pulping fruits (Hylander 1975b), and leaf-eating primates have high-cusped cheek teeth with enhanced shearing surfaces. Over the three-million-year history of the genus *Australopithecus*, evolution worked steadily to reduce both the size of the incisors and the shearing features of the molars. We can infer that these creatures did not specialize in eating either leaves or fruit—though they undoubtedly ate some of both, just as all other large anthropoids do.

Big, flat-topped cheek teeth like those of *Australopithecus* are poorly adapted to shearing through leaves, stems, and other tough, fibrous plant tissues. They are better suited to crushing resistant but brittle plant tissues that can be digested without being finely chopped. Most modern humans subsist chiefly on plant tissues of this sort, particularly seeds (rice, wheat, corn, legumes, etc.) and starchy root vegetables (potatoes, manioc, cassava, and the like).

Some authors have suggested that early hominins had similar diets (Jolly 1970a, Hatley and Kappelman 1980, Kay 1981, Laden and Wrangham 2005). There are two main ways of testing these suggestions. The first is to look at the microscopic wear patterns on the occlusal surfaces of the teeth. When crushed between the teeth, leaves and other fibrous but flexible plant tissues tend to leave fine parallel scratches in the enamel. If the leaves are blades of grass, which contain tiny, gritty silica inclusions called **phytoliths**, the fine scratches will be supplemented by microscopic pits where trapped phytoliths have been dragged across and driven into the surface. Dusty or sandy food items—dug-up roots, for example—will leave similar pits, whose diameter depends on the particle size of the adhering grit. Larger, harder inclusions or casings on plant tissues will leave larger pits when crushed between the teeth. Muscle and other soft animal tissues tend to produce a generally polished enamel surface with few microwear features.

The other main way of testing morphological inferences about the diets of extinct animals is to look for the characteristic traces that an animal's food leaves in the chemical makeup of its skeleton. One such trace is left by carbon isotopes. In the tropics, most grasses use a type of photosynthesis known as C_4 photosynthesis, which produces carbon compounds with a higher $^{13}C/^{12}C$ ratio than the so-called C_3 photosynthesis used by most other terrestrial plants. This isotopic difference between grasses and other plants carries over into the tissues of the herbivores that feed on them. Therefore, we can distinguish tropical mammals that eat grass (grazers) from those that eat the tissues of trees and shrubs (browsers) by measuring the $^{13}C/^{12}C$ ratios in the enamel of their teeth.

A second dietary indicator in dental enamel is the ratio of strontium to calcium. These two chemically similar elements get absorbed from soil by plants and are passed along to the animals that eat those plants. Leaves typically contain less strontium than other plant parts, so leaf-eaters have lower Sr/Ca ratios in their enamel than herbivores that eat fruits, roots, or seeds. Because vertebrates tend to retain the calcium they eat and excrete most of the strontium, herbivorous vertebrates have lower Sr/Ca ratios than their food plants have; and predators that eat those herbivores have even lower ratios. As you might expect, the carnivores that prey on leaf-eating vertebrates have the lowest ratios of all. The Sr/Ca ratios in plants vary with the proportions of the two elements found in the local rocks and soil, so what counts as a "high" or "low" ratio has to be judged relative to the entire fauna at a particular site.

When all three sorts of evidence—from $^{13}C/^{12}C$ ratios, Sr/Ca ratios, and microwear—are put together, almost every sort of specialized mammalian diet writes its own distinctive signature in the dental enamel (Table 4.3). We do not yet have evidence of any of these sorts for most of the species of *Australopithecus*, but what we do have is instructive. Microwear on the molars and incisors of *A. afarensis* looks surprisingly like that seen on the teeth of gorillas, with lots of fine parallel scratches and some small pits (Ryan and Johanson 1989, Grine et al. 2006). This rules out a specialized dependence on fruit or hard seeds, which would leave broader scratches and more pits. But it also rules out a specialized concentration on leaves, which would leave finer scratches and fewer pits. *A. africanus* has similar molar microwear (Kay and Grine 1988), exhibits $^{13}C/^{12}C$ ratios in between those of grazers and browsers (Sponheimer and Lee-Thorp 1999, van der Merwe et al. 2003), and shows high Sr/Ca ratios. This combination of signals suggests a diet focussed on grass seeds, grass rhizomes, and bulbs (Lee-Thorp and Sponheimer 2006).

The other *Australopithecus* species for which we have all three sorts of evidence is *A. robustus*. Its molar enamel exhibits more microwear features per square millimeter than that of *A. africanus*, with a higher ratio of pits to scratches, suggesting higher rates of wear and more feeding on hard or brittle objects (Grine 1986, 1987; Kay and Grine 1988). Its incisors have the same *proportion* of scratches to pits as those of *A. africanus*, but both sorts of features occur in lower *numbers* in *A. robustus*, indicating that its reduced incisors were used less in feeding (Ungar and Grine 1991). The $^{13}C/^{12}C$ ratios in *A. robustus* enamel are higher than those seen

TABLE 4.3 ■ Signatures Left by Various Diets on Mammalian Teeth, Compared with Similar Parameters for *Australopithecus* **Species**

Dietary Specialization	Enamel Microwear	$^{13}C/^{12}C$	Sr/Ca
Plant Tissues			
Soft, fleshy fruit	Few features	Low	High
Tree leaves	Scratches	Low	Low
Fruit seeds, nuts	Pits	Low	High
Tubers	Pits	Low	High
Grass leaves	Scratches	High	Low
Grass roots and stems	Pits	High	High
Grass seed	Scratches	High	High
Animal Prey			
Grazers (leaves)	Few features[a]	High	Very low
Grazers (roots/stems)	Few features[a]	High	Low
Browsers (tree leaves)	Few features[a]	Low	Very low
Fruit-eaters	Few features[a]	Low	Low
Mixed feeders	Few features[a]	Moderate	Low
Australopithecus			
A. afarensis	Scratches: few pits	—	—
A. africanus	Scratches: few pits	Moderate	High
A. robustus	Pits: fewer scratches	Moderate high	Moderate low

[a]Unless diet includes bone-crunching behavior.

Sources: Kay and Grine 1988, Ryan and Johanson 1989, Sillen 1992, Sponheimer and Lee-Thorp 1999, van der Merwe et al. 2003, Wood and Strait 2004, Grine et al. 2006, Lee-Thorp and Sponheimer 2006, Ungar et al. 2006.

in *A. africanus*, implying a greater consumption of foods derived from C$_4$ vegetation (Sponheimer and Lee-Thorp 1999). Its Sr/Ca ratios fall toward the low, "folivore" end of the scale—higher than those seen in strict leaf-eaters (or their predators), but lower than those seen in omnivorous mixed feeders like baboons (Sillen 1992, 1993, Sillen et al. 1995).

Taken by themselves, the low Sr/Ca ratios and the powerful herbivore-type jaws of *A. robustus* might support Robinson's claim that this creature was a specialized browser on fibrous vegetation. But that conclusion is contradicted by the increased pitting of its molar enamel, which implies an increased consumption of seeds, nuts, or tubers. The relatively high $^{13}C/^{12}C$ ratio found in *A. robustus* enamel indicates that either grasses or the flesh of grass-eating animals (including invertebrates) must have been important in its diet. However, its flat, low-crowned teeth are ill adapted for chewing the abrasive leaves of grass, and its molars are less pitted than those of baboons that eat a lot of gritty grass roots (Kay 1985, Daegling and Grine 1999). Its low Sr/Ca ratios are also incompatible with a specialized preference for grass rhizomes (Sillen 1992). Grass seeds and grass-eater flesh are the most likely sources of the elevated ^{13}C levels (Lee-Thorp et al. 2000, 2003).

In short, the evidence indicates that *A. robustus* was an opportunistic mixed feeder, which ate many different types of food from many different sources without being specialized for any one of them. How, then, can we explain its peculiarly specialized chewing apparatus—its huge molars, powerful jaws, ultra-thick enamel, dished face, and so on?

One current theory holds that these are specializations for dealing with **fallback foods**. Some primates that appear to have teeth and jaws suited to certain sorts of foods have been found to focus on those foods only during lean seasons, when nothing better is available. For example, the Malagasy lemur *Propithecus* has teeth and guts that are adapted to chewing and digesting leaves and other fibrous vegetation; but it feeds mainly on fruits when it has the opportunity (Yamashita 1998, Norscia et al. 2006). Likewise, all the African apes eat mainly ripe fruits when they are available, and fall back on less tasty and digestible browse—leaves, stems, and piths—when there is little fruit to be had (Wrangham et al. 1996, 1998; Kuroda et al. 1996, Yamagiwa et al. 1996, Tutin et al. 1997, White 1998, Hashimoto et al. 1998, Doran and McNeilage 2001). When we say that *Gorilla* is a folivore and *Pan* is a frugivore, we mean that *Gorilla* can live entirely on leaves for a longer period of time than *Pan* can (Wrangham et al. 1998), not that *Gorilla* prefers leaves to fruit.

Laden and Wrangham (2005) suggest that hominin divergence from the other African apes was driven by a switch in fallback foods, from the fibrous tissues of forest herbs and trees to bulbs, tubers, and other **underground storage organs** (USOs) of savanna plants. They propose that this shift drove the early-hominin evolutionary trends toward large jaws, premolar molarization, megadonty, and thick enamel. These trends opened up grassland areas (which are particularly rich in edible USOs) to hominin exploitation, and they culminated in the late "robust" species of *Australopithecus*.

This theory accounts for the high Sr/Ca ratios in the dental enamel of *A. africanus*. But it does not comport well with the low levels of enamel pitting and the moderately high $^{13}C/^{12}C$ ratios in that species, which suggest that grass seeds were important in its diet. The evidence from *A. robustus* enamel does not point to a USO-based diet, but it is probably compatible with that possibility—if we assume that USOs were mainly fallback foods for *A. robustus*, with most of its diet coming from other sources (to explain the low Sr/Ca ratios) and that *A. robustus* ate a lot of grass-derived tissues (resulting in the high $^{13}C/^{12}C$ ratios).

The USO theory has several points in its favor. It suggests an adaptive advantage for terrestriality in early hominins. It explains the association between *A. robustus* and those long-bone "digging sticks" at Sterkfontein and Swartkrans. And it accounts for the apparently parallel evolution of exaggerated megadonty in *A. robustus* and *A. garhi*—and maybe in *A. boisei* or *A. aethiopicus* as well—as the natural outcome of an increased reliance on USOs.

However, the USO theory is hard to reconcile with the belief that the robust species of *Australopithecus* were particularly close relatives of early *Homo*. If megadonty conferred an advantageous ability to fall back on emergency rations during hard times, it seems strange that natural selection would have worked to abolish this advantage at the base of the *Homo* lineage. Some early *Homo* species had molars and premolars approaching those of "robust" *Australopithecus* in size (see below), but most of them had considerably smaller cheek teeth (Table 4.2). If the USO theory is true, humans might have evolved from some persistently primitive *Australopithecus* species, which lacked marked adaptations for exploiting USOs and was therefore at a disadvantage in competing with *A. robustus* on the savanna—until it began using stone tools, enabling its descendants to break out of the *Australopithecus* niche altogether.

The pelvic anatomy of *Australopithecus* also furnishes some hints about diet. In 1972, Robinson inferred from the flaring ilia of the Sts 14 pelvis that *A. africanus* must have had "wide hips and a bulging abdomen" (Robinson 1972, p. 234). Drawing similar inferences from the "Lucy" skeleton of *A. afarensis*, P. Schmid (1991) argued that the broad pelvis and long lower ribs of *Australopithecus* imply "a well-developed and protuberant abdomen as common in pongids." If so, then these animals must have had capacious guts, suited to a low-quality diet containing a lot of structural carbohydrates. Early *Homo* appears to have been different, with a narrower pelvis more like those of modern humans (Ruff and Walker 1993). This suggests that early *Homo* had

smaller intestines—and, by inference, a higher-quality diet—than *Australopithecus*.

Some argue that the difference in relative brain size between *Australopithecus* and *Homo* also reflects differences in diet. Different sorts of tissues have different metabolic rates: more energy is needed to keep a pound of mammalian brain, intestines, heart, or kidneys alive and functioning than is needed to maintain a pound of bones or fat. Because any animal that is not growing bigger or fatter burns up all the calories contained in its food, evolution cannot simply increase the size of a metabolically expensive organ without finding the calories to pay for it in some other part of the animal's energy budget. In mammals, intestines and brains are both particularly expensive tissues to maintain. Aiello and Wheeler (1995) contend that the size of the two is negatively correlated in anthropoid primates: for a given body size, leaf-eaters (which need to have big guts) generally have larger guts and smaller brains than fruit-eaters. Aiello and Wheeler suggest that modern humans can afford to have exceptionally large brains only because we have much smaller intestines than would be expected for a human-sized anthropoid, and they also suggest that we can afford to have peculiarly small intestines because we feed mainly on high-calorie foods—plant seeds, starchy USOs, animal flesh and fat—and because we practice cooking and other extrasomatic food preparation techniques that do a lot of the work of digestion that would otherwise have to be done by our guts.

This so-called **expensive-tissue hypothesis** makes sense in terms of the hominin fossil record. It explains why relative brain size and megadonty are positively correlated in *Australopithecus*: increasing specialization of the teeth and jaws might have allowed more effective use of new food resources (USOs?) and provided fuel for a larger brain. And it suggests an underlying reason for the co-occurrence of markedly bigger brains and stone tool use in early *Homo*—namely, that the new tools afforded increased access to higher-quality dietary items (animal flesh?), allowing the gut to become smaller and freeing up part of its energy budget to be invested in a larger brain.

If the expensive-tissue hypothesis is true, we should find correlations between brain weight, gut size, and dietary quality in many other animals. The few studies that have sought to test these correlations in nonprimates have yielded mixed results (Isler and van Schaik 2006). Other variables—ontogeny, phylogeny, life history, energetic costs of different sorts of locomotion, and so on—enter in to complicate the picture, making it hard to come up with a decisive test.

The validity of the expensive-tissue hypothesis is still a matter for debate. But its basic premise is undisputed: brains are expensive organs, and larger brains have to be paid for out of an animal's energy budget, either through savings on other organs or through extra caloric intake. Understanding the history of brain enlargement in the human lineage will depend not on figuring out how a bigger brain would have proved advantageous to our ancestors—there are a great many possible benefits to becoming brainier—but on discovering how they managed to find the extra calories to pay for it.

AUSTRALOPITHECUS AND THE ECOSYSTEM

Most living apes inhabit forests, or at least woodland communities with a lot of trees in them. Although some populations of chimpanzees (*Pan troglodytes*) live on the fringes of the savanna (Moore 1996), this is marginal habitat for these animals. Chimpanzees in these fringe habitats make little use of the grasses and other C_4 plants that dominate savanna ecosystems (Sponheimer et al. 2006), and they have to repair to trees and wooded areas for sleeping, for refuge from predators, and for access to the ripe tree fruits that still constitute most of their diet (McGrew et al. 1988).

Humans, by contrast, thrive in open country. For modern humans, it is the closed-canopy tropical forests that are the marginal habitat, where survival is made possible largely by felling trees and making clearings to farm root crops. It is an open question whether human hunters and gatherers have ever been able to live in rain forests without a steady influx of edible USOs acquired in trade from their farming neighbors (Bailey et al. 1989, Bailey and Headland 2004, Bahuchet et al. 2004).

At some point, the human lineage must have shifted its preferred habitat from tropical forests to more open country. Because all the South African cave sites lie today in dry savannas and open woodlands, it was assumed for a long time that *Australopithecus* too was a savanna-dwelling animal, and that the shift to a grasslands ecology had taken place at the very beginning of the hominin evolutionary career. In his initial paper on the Taung infant, R. Dart argued that *Australopithecus* had gotten started down the road to humanity precisely because it was compelled to make a living in a harsh environment. The other apes, Dart wrote, remained primitive because they had things easy in their equatorial forests, where "… Nature was supplying with profligate and lavish hand an easy and sluggish solution … of the problem of existence" (Dart 1925). But for an ape living in the Transvaal, survival "… constantly and increasingly demanded the operation of choice and cunning … to find and subsist upon new types of food and to avoid the dangers and enemies of the open plain." This new way of life "… evoked the thinking and planning powers of the anthropoid, and, with these powers, caused the transformation from anthropoid to man" (Dart 1926).

There is a political subtext in these words, which express the conviction of many Europeans at the time that the dark-skinned natives of the tropics are shiftless and lazy because life in tropical forests is undemanding (see Blind Alley #6, p. 206). This idea is colonialist hogwash. In fact, closed-canopy tropical forests are exceptionally demanding habitats for humans and other terrestrial mammals, because almost all the primary production of edible plant tissues is going on 50 feet overhead, where it is directly accessible only to flyers and tree-dwellers. There is generally less mammalian biomass per hectare in African rain forests than there is on the savanna, and what there is consists largely of elephants (which can feed on understory trees) and arboreal primates adapted to living and foraging in the canopy (Prins and Reitsma 1989, White 1994).

Yet though the savanna is a rich and productive habitat, few primates have managed to adapt to it. The main exceptions are swift-running terrestrial cercopithecines—baboons, vervets, patas monkeys—that rely on varying combinations of wariness, agility, threats, social organization, and speed to discourage or elude predators on the ground. And whatever the locomotor behavior of early hominids was like, they were surely not swift runners, because modern humans are not.

In this respect, humans are different from other bipeds. All other animals that run on their hind legs appear to have taken up bipedality in order to go faster. Large flightless birds, which have inherited their bipedality from dinosaur ancestors, can hit top speeds of 18 meters per second or 40 mph (Blanco and Jones 2005). Many lizards and even some insects get up on their hind legs when they need to run very fast (Alexander 2004). But humans are relatively slow bipeds, with maximum running speeds of around 10 m/sec. It seems like a safe bet that early hominins, with their relatively short legs, were even slower runners than we are (Bramble and Lieberman 2004).

It has long been assumed that these slow-footed bipedal apes, having no fangs, talons, or other built-in weapons, would have needed some special advantages—some combination of technology and teamwork—to survive on the Pliocene savannas alongside such formidable African predators as lions, leopards, cheetahs, sabertoothed cats, hyenas, and African hunting dogs. Two divergent conclusions have been drawn from this assumption. One conclusion, propounded by Dart and many others, is that the new environment forced these apes to evolve human-like mental and behavioral abilities. The other conclusion, originally put forth by C. Darwin (1871, Chapter 2), is that hominins must have already acquired some human-like advantages before they could leave the trees and move out into open country. These ideas are not mutually exclusive, and both undoubtedly contain part of the truth. But they emphasize different causal factors. Stories of the first, Dartian type see early hominins as apes trapped in an increasingly hostile environment, to which they were

Blind Alley #6: The Retarded Tropics

Throughout recorded history, northern Europe has been a cold place, cool in the summer and frozen in the winter. When Europeans began colonizing low-latitude countries, they were impressed by the balmy climates of the tropics. In the European imagination, the tropics came to be seen as an Edenic realm of perpetual summertime, where food could be obtained year-round simply by lying under the trees and waiting for the coconuts and mangos to drop.

This image (along with the vexing reluctance of tropical peoples to work hard for the profit of foreign conquerors) contributed to a widespread European view of the tropics as a backward region, where the natives "… are by nature shiftless and lazy, having no climatic incentive to work or accumulate goods or exercise forethought" (Hanson 1933).

These notions infected scientific thinking about human evolution. During the heyday of European colonialism, many scientists attributed the major advances in human prehistory to our ancestors' having left the tropics for more

stimulating climes. Today's apes, wrote Raymond Dart's mentor G. E. Smith, had remained apes because they had lingered on in equatorial jungles, "living in a land of plenty, which encouraged indolence in habit and stagnation of effort and growth" (Smith 1927, p. 40). Dart himself thought that *Australopithecus* had been thrust down the road toward humanity by its harsh Transvaal environment, while the other apes had been arrested by their undemanding life in tropical forests (Dart 1925). "In tropical and semi-tropical regions where natural food fruits abound, human effort—individual and racial—immediately ceases," proclaimed H. F. Osborn, the leading American paleontologist of the 1920s (Osborn 1926). Osborn held that *Homo sapiens* must have evolved in Central Asia, "… because to the south conditions of life were less rigorous, food was more easily obtained, and the milder sub-tropical climate was less stimulating to discovery and invention" (Osborn 1927). W. W. Howells was still repeating this line 20 years later:

> Where did man evolve? … Asia is the automatic choice of most people. … The chances favor some region north of India and the Himalayas, since in the more heavily forested south apes would probably have remained apes. (Howells 1946, pp. 106–107)

As recently as 1962, C. Coon contended that Europe and northern Asia had been the ancestral homelands of modern humans, because these regions "had challenging climates and ample breeding grounds." Coon insisted that "If Africa was the cradle of mankind, it was only an indifferent kindergarten. Europe and Asia were our principal schools" (Coon 1962, pp. 656, 663).

Racial stereotypes and colonialist prejudices, as well as a series of unrealistically late dates attributed to various African fossils, long impelled many scientists to dismiss Africa as a laggard backwater in human prehistory. Today's consensus is just the reverse—that Africa was the locus of every major breakthrough in hominin evolution. But it is worth remembering that the current consensus may embody current biases, which we ourselves can scarcely perceive. The surprisingly primitive morphology and early dates of the Dmanisi hominins (p. 279) should remind us that we have a lot left to learn about the role of tropical and subtropical Asia in the hominin story.

compelled to adapt or die. Stories of the second, Darwinian type see these creatures as apes that evolved a new way of life in the forest, which allowed them to colonize open-country environments later on (McKee 1999).

The Dartian type of story has probably been the more popular among scientists. Many have attributed the origins of the hominin clade to the cooling and drying trends seen in the global climate of the Late Cenozoic

(Fig. 3.16). These climatic trends, it has been said, brought the grasslands of eastern and southern Africa into existence and left our ape ancestors trapped in shrinking forests—from which the only escape was into the savanna, on two feet. Some see these changes ultimately as side effects of continental drift. By the early Miocene, the drift of Antarctica across the South Pole had resulted in the formation of a continental ice sheet. The sequestration of this ice must have lowered sea levels and altered oceanic circulation, and probably precipitated or accelerated the mid-Cenozoic trends toward cooler and dryer world climates (Denton 1999). It has also been argued that rifting of the African plates during the Miocene initiated an ecological split between East and West Africa by opening up the Rift Valley. According to the so-called "East Side Story" proposed by French paleoanthropologist Y. Coppens, the Miocene uplifting of the western shoulder of the Rift Valley cast eastern Africa into a downwind rain shadow (like that which produces the Great Basin deserts in North America). To the west of this line, Africa remained forested, and the local hominids evolved into gorillas and chimpanzees. To the east, forests dwindled into small remnants and grassy woodlands, and the local hominids evolved into *Australopithecus* to cope with their changed circumstances (Coppens 1994, 1999; Boaz 1994; cf. Kortlandt 1972).

The discovery of *Sahelanthropus* and *Australopithecus bahrelghazali* in Chad, far to the west of the supposed dividing line, has moved Coppens to abandon the East Side Story (Brunet et al. 2002). But that story might be salvaged by interpreting these creatures as open-country animals that evolved in East Africa and then spread westward along the southern edge of the Sahara. A more serious problem for any story that traces hominin origins to the drying-out of eastern Africa is that some early hominins appear to have lived in East African areas with high rainfall.

Trying to figure out the ecology of terrestrial paleontological sites is a tricky and dubious business. Fossilization usually takes place in sediments that accumulate at the bottoms of lakes and rivers, and the wet margins of these watercourses are often populated by moisture-loving animals and plants not found in the surrounding countryside. The fossils preserved in the sediments can therefore yield an unrealistically wet and forested picture of the local ecology. Nevertheless, it does look as though some of the earliest hominin sites in East Africa lay in wooded areas. The *Orrorin* site has been interpreted as supporting an open woodland with patches of closed-canopy forest (Pickford and Sénut 2001). Two Middle Awash sites in Ethiopia—the Aramis site that has yielded *Ardipthecus ramidus*, and the Asa Issie site that preserves early *Australopithecus anamensis*—abound in primates, rodents, bats, and antelopes belonging to genera that live in forests today (WoldeGabriel et al. 1994, 2001; White et al. 2006). By one interpretation, Member 4 at Sterkfontein represents a similar moist habitat (Clarke and Tobias 1995, Kuman and Clarke 2000). On the basis of these and other paleoecological reconstructions, some have concluded that hominins got their start in wooded country and moved out into grasslands only later, as eastern and southern Africa continued to dry out and forested areas became fewer and farther apart (Lovejoy 1981, Berger and Tobias

1996, WoldeGabriel et al. 2001). This account of things harmonizes with the idea that early *Australopithecus* was still largely arboreal.

However, the obituaries for the "savanna hypothesis" that were widely voiced in the late 1990s (Shreeve 1996) seem to have been premature. Some of the earliest hominins may have lived in forests or semi-forested areas, but others apparently did not. *Ardipithecus kadabba* appears to have inhabited "... a sub-humid and seasonally dry climate in a landscape covered by C_3-dominated woodland and grassy woodland" (Semaw et al. 2005). A recent study of the Member 2 fauna from Sterkfontein suggests that the area was a dry, open woodland at the time of deposition, around 4 Mya (Pickering et al. 2004). The *A. anamensis* sites at Kanapoi and Allia Bay in Kenya appear to have had a similar ecology at around the same time (Coffing et al. 1994, Wynn 2000, Ward et al. 2001). And if *Sahelanthropus* is a hominin, then the earliest hominin known lived in an oasis in an arid, windswept grassland bordering on a sandy desert (Vignaud et al. 2002). On the basis of what we know today, the safest conclusion is that early hominins "were apparently not restricted to a narrow range of habitats" (Leakey et al. 1995a).

Perhaps the best evidence for the importance of open country in the evolution of these bipedal apes is the fact that there is relatively little to eat on the ground in a closed-canopy tropical forest. It seems unlikely that our ancestors would have lost their tree-climbing abilities in such a forest. If *Australopithecus* was as fully committed to terrestrial life as the Lovejoy school of thought contends, it must have been largely an open-country animal. On the other hand, if the Stony Brook picture of its locomotion is correct, the early species of *Australopithecus* would have been able to both climb trees and walk bipedally on the ground, allowing them to exploit both arboreal and terrestrial food resources. Such an animal could have lived as an inhabitant of forest and woodland fringes, moving into open areas to feed and retreating to the trees for other sorts of food and for shelter (Ciochon and Corruccini 1976). Many living mammals, from langur monkeys to whitetailed deer, thrive in this sort of forest-edge habitat.

∎ TWO SPECIES OR TWO SEXES?

Robinson's original (1954b, 1956) "dietary hypothesis" posited an ecological difference between "robust" and "gracile" types of *Australopithecus*. The "gracile" type, *A. africanus*, was supposed to be a mixed feeder with a taste for meat. The "robust" type, *Australopithecus* (or *Paranthropus*) *robustus*, was depicted as a more specialized folivore (Wood and Ellis 1986). This difference would be an example of what is called **niche partitioning**, in which two sympatric species with the same general way of life evolve in divergent directions so as to reduce competition for shared resources.

We now think that *A. africanus* and *A. robustus* were never direct competitors, because they lived at different times (Fig. 4.76). However, Robinson's hypothesis is still accepted in a modified version. Throughout eastern and southern Africa, over a time range from around 2.5 to 1.4 million years ago, the late-surviving "robust" types of *Australopithecus*—*A. robustus* and *A. boisei*—lived sympatrically with more humanlike "gracile" types usually assigned to the genus *Homo*. The ecological niches of these two types of hominin were evidently different enough to allow them to coexist side by side almost indefinitely.

The proposition that two hominin species can coexist was vigorously contested at one time (see Blind Alley #5, p. 188). Although this proposition is not in question nowadays, arguments continue over the number of hominin species that were present at various times and places. In the first published description of the *A. afarensis* fossils from Ethiopia, Johanson and Taieb (1976) suggested that *three* hominin species were found in the Hadar sample:

1. A very small, gracile *Australopithecus*, represented by the "Lucy" skeleton, the AL 129 knee (Fig. 4.54A), and some other bits and pieces.

2. A larger, "robust" *Australopithecus* species, represented by a heavily pneumatized temporal bone (AL 166-9) and a femur (AL 211-1) of the *Australopithecus* type but much bigger than Lucy's.

3. An early member of the genus *Homo*, represented by the AL 200-1 maxilla (Fig. 4.53) and some other fossil jaws, all having canines and incisors too big to fit comfortably into any of the three previously described *Australopithecus* species (*africanus, robustus,* and *boisei*).

During the next two years, as more fossils emerged from the Hadar sites, the distinctions between these three groupings grew blurred and a different picture began to emerge. Although all the Hadar femora were of the *Australopithecus* type, with long necks and posteriorly located lesser trochanters, they differed considerably in size, with the largest specimens being 50% bigger than the smallest in some dimensions (Fig. 4.56). There was a similar range of sizes in other skeletal elements—for example, between the dimensions of the biggest mandible (AL 333w-32 + 60) and those of the "Lucy" mandible (White and Johanson 1982). Yet the teeth in the Hadar jawbones were not so disparate in size. M_2 breadths, for example, ranged between 12.5 and 14.6 mm, a difference of about 17% (Johanson et al. 1982b). The biggest of the lower canines, which we would expect to be particularly dimorphic teeth, dif-

fered in breadth from the smallest by only 33%. This is comparable to the intraspecific ranges of canine breadths seen in *A. africanus* and *A. robustus* (Wolpoff, 1975b). These data implied that the relatively large canines seen in the AL 200-1 maxilla were not a distinctive marker of its human affinities. Rather, they were a primitive, persistently apelike feature characteristic of the Hadar sample as a whole. Johanson et al. (1978) accordingly lumped all the fossil hominins from Hadar, together with the geologically older specimens that had previously been found at Laetoli in Tanzania (White 1977), into a single taxon and christened it *Australopithecus afarensis,* with Laetoli Hominid 4 (Fig. 4.52) as the type specimen.

There was still a lot of variation in this collection of fossils, and many experts continued to suspect that it included more than one species (Zihlman 1985, Tobias 1988). One school of thought held that the cranial fossils from Hadar came from an early member of the robust "Paranthropus" clade, whereas "Lucy" and some other small Hadar fossils represented a smaller "gracile" species ancestral to *Homo* (Olson 1981, 1985; Falk and Conroy 1983, Falk 1988; Fig. 4.75). Others contended that the *larger* fossils from Hadar had special affinities to *Homo*, as witnessed by their more human-like postcranial remains (Tardieu 1983, 1986; Sénut and Tardieu 1985; Figs. 4.20 and 4.57). Häusler and Schmid (1995) argued that the pelvis of the AL 228 "Lucy" skeleton is too small to have passed a newborn *A. afarensis*—and therefore, "Lucy" must be a male. And since AL 288 is one of the smallest individuals known from Hadar, recognizing it as a male implies that the *A. afarensis* sample must be a mixture of bones from two species of different body size. This argument, however, relies on a dubious estimate of the size of the baby's skull at birth (Wood and Quinney 1996).

The defenders of the original concept of *A. afarensis*—as a single species ancestral to all later hominins—fought back vigorously against these multiple-species ideas. Nothing about the teeth of the Hadar hominins, they insisted, suggests that they represent more than one species (Kimbel et al. 1985, White 1985, Grine 1985, Kimbel and White 1988, White and Johanson 1989). The cranial features linking *afarensis* to *robustus* were dismissed either as primitive (plesiomorphic) or as too variable to be used as phylogenetic markers (Kimbel 1984, Kimbel et al. 1985). And the differences between the large and small morphs of *A. afarensis* from Hadar were interpreted as differences between large males and small females of the same species (Johanson and White 1979). A slightly different position, proposed by Cole and Smith (1987), held that (a) the variation in the *A. afarensis* sample approaches the level where two separate species might be recognized and (b) the *afarensis* material may preserve a record of a single population in the process of splitting.

The consensus that emerged from this debate was that *Australopithecus afarensis* was a single species, but with a high degree of sexual dimorphism (Frayer and Wolpoff 1985, Johanson and White 1979, Johanson 1980, Stern and Susman 1983, Jungers 1988a). Johanson and Edgar 1996, White 2002). Although the size range in the relative breadth of the canines was smaller in *A. afarensis* than it is in some species of living apes, it was nevertheless larger than it is in modern humans or other *Homo* (Wolpoff 1975b). There were equally striking size differences in the postcranial bones, implying that sexual dimorphism in body size had also been pronounced. Most estimates for size dimorphism in *A. afarensis* based on postcranial metrics yield a **dimorphism index** of around 125, meaning that linear measurements for presumed males average 125% of similar measurements for females. This is about what we see in gorillas and orangutans, and considerably above the index values of 103–116 found in humans and chimpanzees (Table 4.4). *A. africanus* and *A. boisei* also seem to have been more dimorphic in body size than humans are (Table 4.4), though sample sizes are smaller for these species and the estimates are correpondingly shakier.

Among other anthropoid primates, body-size dimorphism is strongly correlated with canine dimorphism (Plavcan and van Schaik 1992). Both kinds of dimorphism are also strongly associated with two behavioral variables. The first is the amount of terrestrial activity in the species. In anthropoid species that live on the ground, males tend to be larger than females and to have relatively big canines, for threatening and defending against predators. The second behavioral variable correlated with dimorphism is the intensity of sexual competition between males. In species where the males habitually threaten and attack each other to gain access to females, the males tend to have larger and scarier canine fangs than the females do. But in species where there is little intermale competition for females—either because they live in mixed-sex groups where female choice governs mate selection (e.g., *Pan*) or because males and females form exclusive "monogamous" pairs (e.g., *Hylobates*)—there is little dimorphism in either canine or body size.

Australopithecus afarensis is a puzzling exception to these rules (Plavcan 2001). It has a low, chimpanzee-like level of canine dimorphism, and its canines are seemingly too puny and short to have been of much use in stabbing and slashing. However, it appears to have been far more dimorphic in body size—almost as much as gorillas, by some people's reckoning (Lockwood et al. 1996; Table 4.4). These seemingly contradictory facts have spawned various conflicting conclusions. From its marked body-size dimorphism, some infer that *A. afarensis* must have had a **polygynous** mating system—that is, one in which several females may mate exclusively with one male, and intermale competition for females is correspondingly intense (Leutenegger and Shell 1987, McHenry 1996b, Harmon 2006). But this notion is hard to reconcile with the reduction of the size and dimorphism of the canine teeth in *A. afarensis*. Others conclude that *A. afarensis* must have had a unique, unknown pattern of social and sexual behavior, with no direct analog among living primates (Plavcan and van Schaik 1997a). Lovejoy and his co-workers (Reno et al. 2003) claim that everybody else has been estimating dimorphism the wrong way, and they conclude that dimorphism in *A. afarensis* was actually much like that seen in modern humans. The methods and assumptions underlying this conclusion are debated (Plavcan et al. 2005, Reno et al. 2005b).

The ongoing controversy over sexual dimorphism in *Australopithecus* has a political subtext. To understand this controversy, we need to trace its historical roots in the cultural and political turmoil of the 1960s and 1970s.

HUNTING, GATHERING, AND DIMORPHISM

In 1949, R. Dart began publishing a string of articles and books proposing that *Australopithecus africanus* had made its living as a predator. As Dart saw it, an ancestral shift from a diet of vegetation to a diet of meat was responsible for all the distinctive traits of humanity, including most of the nastier aspects of human nature (Dart 1949, 1953, 1955, 1957; Dart and Craig 1959). Dart's ideas were picked up by others and elaborated into a global theory of human origins and evolution that reigned in the textbooks of physical anthropology for the next 20 years (Cartmill 1993). This theory is sometimes referred to as the **feedback model** or the **hunting hypothesis**. In brief, it ran like this:

Apes are primarily fruit- and leaf-eaters; but humans are descended from an aberrant ape that took to hunting animal prey as a major part of its diet. Like other apes, the baby proto-humans clung to their mothers and took a long time to grow up. Perennially hampered by these dependent offspring, the females could not be effective predators. The job of hunting therefore fell to the males.

A predatory lifestyle required some fundamental changes in ape behavior. Several man-apes had to act in concert to bring down a large quarry, so the males had to learn to cooperate and coordinate their actions. Because they lacked natural equipment for killing and butchering, the males had to make and use crude tools—pointed sticks, broken rocks, and the like—to handle these jobs. And to use these tools effectively, they had to free up their hands by becoming fully bipedal.

TABLE 4.4 ■ Estimated Body Sizes (kg) and Measurements of Sexual Dimorphism for *Australopithecus* Species and Living Hominoids

	M (kg)	F (kg)	Body Weight Dimorphism (M, % of F)	Postcranial Dimorphism (Linear Metrics: M, % of F)	Lower Canine Dimorphism (B-L Width: M, % of F)	Upper Canine Dimorphism (Crown Height: M, % of F)	Source
Homo sapiens	—	—	—	109	—	—	Zihlman (1985)
	—	—	(165)	—	—	—	Jungers (1988a)
	68.2	50	136	—	—	—	Jungers (1988a)
	64.9	53.2	122	—	—	—	McHenry (1991c)
	47.9	40.1	120	—	107	108	Plavcan and van Schaik (1997)
	—	—	—	116	—	—	Reno et al. (2003)
	—	—	113–122	110–115	—	—	Plavcan et al. (2005)
	—	—	—	110–115	—	—	Lee (2005)
	—	—	—	—	109–112	—	Lee (2005)
	—	—	—	112	—	—	Harmon (2006)
Pan paniscus	46.3	33.2	139	—	—	—	Jungers (1988a)
	47.8	33.1	144	—	—	—	McHenry (1991c)
	45	33	136	—	118	138	Plavcan and van Schaik (1997)
	—	—	—	103	—	—	Harmon (2006)
Pan troglodytes	—	—	(223)	—	—	—	Jungers (1988a)
	56.6	40.1	141	—	—	—	Jungers (1988a)
	54.2	39.7	137	—	—	—	McHenry (1991c)
	43	33	130	—	132	143	Plavcan and van Schaik (1997)
	—	—	—	105–111	—	—	Reno et al. (2003)
	—	—	127–130	105–108	—	—	Plavcan et al. (2005)
	—	—	—	105–107	—	—	Lee (2005)
	—	—	—	—	124–128	—	Lee (2005)
	—	—	—	108	—	—	Harmon (2006)
Gorilla gorilla	—	—	—	127	—	—	Zihlman (1985)
	—	—	(302)	—	—	—	Jungers (1988a)
	164.3	75.5	218	—	—	—	Jungers (1988a)
	164.3	75.5	218	—	—	—	Jungers (1988a)
	157.9	75.4	209	—	—	—	McHenry (1991c)
	159.2	97.7	163	—	138	173	Plavcan and van Schaik (1997)
	—	—	203	125–126	—	—	Reno et al. (2003)
	—	—	238–247	124–125	—	—	Plavcan et al. (2005)
	—	—	—	124–130	—	—	Lee (2005)
	—	—	—	—	153–159	—	Lee (2005)
Pongo pygmaeus	—	—	—	146	—	—	Zihlman (1985)
	—	—	(278)	—	—	—	Jungers (1988a)
	81.2	37.2	218	—	—	—	Jungers (1988a)
	78.8	38.8	203	—	—	—	McHenry (1991c)
	86.3	38.7	223	—	147	169	Plavcan and van Schaik (1997)
	—	—	219	124	—	—	Plavcan et al. (2005)
	—	—	—	122	—	—	Harmon (2006)
Hylobates syndactylus	11.4	11.3	101	—	—	—	Jungers (1988a)
	11.3	11.3	100	—	—	—	McHenry (1991c)
	10.9	10.6	102	—	113	118	Plavcan and van Schaik (1997)
Australopithecus afarensis (estimated)	—	—	—	(136)	—	—	Zihlman (1985)
	—	—	(223)[a]	—	—	—	Jungers (1988a)
	—	—	(266)[c]	—	—	—	Jungers (1988a)
	44.6[b]	29.4[b]	152[b]	—	—	—	McHenry (1991c)
	70.5[c]	39.5[c]	178[c]	—	—	—	McHenry (1991c)
	60.1[a]	35.6[a]	169[a]	—	—	—	McHenry (1992)
	—	—	—	—	125	118	Plavcan and van Schaik (1997)
	—	—	—	117–122	—	—	Reno et al. (2003)
	—	—	148–243	125–129	—	—	Plavcan et al. (2005)
	—	—	—	129–142	—	—	Lee (2005)
	—	—	—	—	125	—	Lee (2005)
	—	—	—	126	—	—	Harmon (2006)

TABLE 4.4 ■ *Continued*

	M (kg)	F (kg)	Body Weight Dimorphism (M, % of F)	Postcranial Dimorphism (Linear Metrics: M, % of F)	Lower Canine Dimorphism (B-L Width: M, % of F)	Upper Canine Dimorphism (Crown Height: M, % of F)	Source
A. africanus	—	—	(223)[a]	—	—	—	Jungers (1988a)
(estimated)	—	—	(266)[c]	—	—	—	Jungers (1988a)
	52.8[a]	36.8[a]	143[a]	—	—	—	McHenry (1992)
	40.8[b]	30.2[b]	135[b]	—	—	—	McHenry (1992)
	—	—	—	—	113	113	Plavcan and van Schaik (1997)
A. robustus	—	—	(155)[a]	—	—	—	Jungers (1988a)
(estimated)	—	—	(210)[c]	—	—	—	Jungers (1988a)
	49.8[a]	40.3[a]	124[a]	—	—	—	McHenry (1992)
	40.2[b]	31.9[b]	126[b]	—	—	—	McHenry (1992)
	—	—	—	—	118	131	Plavcan and van Schaik (1997)
A. boisei	—	—	(210)[a]	—	—	—	Jungers (1988a)
(estimated)	—	—	(240)[c]	—	—	—	Jungers (1988a)
	76	42	181[a]	—	—	—	McHenry (1992)
	48.6	31.5	154[a]	—	—	—	McHenry (1992)

[a]Calculated from hominoid (human + ape) regression formulae.
[b]Calculated from human regression formulae.
[c]Calculated from (nonhuman) ape regression formulae.
B-L, buccolingual; M, male; F, female. Numbers in parentheses represent the sample maximum expressed as a percentage of the sample minimum, which is usually an overestimate of population dimorphism (male mean as % of female mean).

Cooperative hunting and tool use were learned behaviors, not instinctive. To learn them, the young proto-humans had to have bigger brains and a longer learning period, and therefore a more prolonged period of infant brain growth and juvenile dependency. This extended "childhood" was made possible by the appearance of another distinctively human behavior: the males began helping the females to feed and care for their dependent offspring.

The incentive for the males to do this was the sexual favors of the females, who kept the males continually interested by hiding their estrous cycles. Natural selection ensures that male mammals will be most attracted to females just following ovulation, when conception is most likely—*if* the males can tell when that happens. Like other female mammals, human females are most interested in sexual activity at this point in the cycle (Adams et al. 1978); but they are able and willing to copulate at any point in the cycle, and they have lost the conspicuous changes in the external genitals that signal the time of maximum fertility in chimpanzees or baboons. As a result, male humans have a hard time telling when females are ovulating, and so they remain attracted to them more or less constantly. Their sexual attentions are diffused throughout all the phases of their mates' menstrual cycles, including times of pregnancy.

With her mate helping to feed her and their young, the female could devote more time to tending babies, and her growing offspring could devote more time to

brain development and learning. (Bipedality was an advantage here, because it freed up the hands for bringing food home to the females and for carrying the increasingly helpless young.) All these trends brought about the appearance of the human nuclear family, consisting of a hunting male, a sexy but sedentary female, and a string of immature offspring busily learning to make tools and take care of their younger siblings. Supporting that lengthening string of offspring made more demands on the males' hunting abilities and called for more effective tools, hunting techniques, and teamwork. This in turn required more prolonged learning, which meant even bigger brains, longer infancies, and more dependent infants—and so on round and round, onward and upward to the genus *Homo*, with everything human flowing from that crucial shift to a more predatory lifestyle.

By this account, a sexual division of labor, "in which the male specialized as hunter, the female as domestic" (Etkin 1954), marked the human lineage from the start. The politically sensitive reader will observe that this theory projects the idealized domestic arrangements of 1950s America back into prehistory and identifies them as the engines of human progress. It might be disrespectfully referred to as the Flintstone model of human origins, in which Fred goes off to work every day to put food on the table while Wilma stays home with the kids and does the housework.

The feminist movement that emerged during the 1960s was not disposed to accept the Flintstone arrangement as the archetypal human condition. The hunting hypothesis came under increasing suspicion from academics of a liberal persuasion (who predominate in anthropological circles). There were empirical grounds for this skepticism. The flat-crowned, megadont cheek teeth of *Australopithecus* differ from those of the African apes in ways that suggest adaptation to munching hard, resistant plant materials, not animal flesh. Field studies of the !Kung "Bushmen," a modern South African non-agricultural group, revealed that most of their subsistence was obtained by female gathering rather than by male hunting (Lee 1968, 1969). Detailed studies of the bones from the *Australopithecus* sites in South Africa showed that the cave faunas had been accumulated by non-hominin carnivores carrying the remains of kills back to their lairs. The man-apes, it seemed, had been mainly prey, not predators (Brain 1981). The archaeologist L. Binford (1981, 1985, 1987) argued that there was little evidence for hunting of large prey by hominins much before the appearance of modern humans. And observations of chimpanzees in the field revealed that they kill and eat a small but significant number of medium-sized mammals, all without the benefit of tools, language, bipedality, or other distinctively human traits (Goodall 1963, 1986; Teleki 1973, 1981). These findings cast doubt on the old stories that assigned a primary role to hunting in human origins.

During the 1970s, anthropologists N. Tanner and A. Zihlman elaborated a new, female-centered model of early hominin society, which stood the hunting hypothesis on its head (Tanner and Zihlman 1976, Tanner 1981, Zihlman 1976, 1978, 1981, 1987). The "Man the Hunter" story of human origins, protested Tanner and Zihlman,

> … stresses the role of men almost to the exclusion of women. In that tradition men are portrayed as protectors of children and women with whom they are assumed to be attached sexually in a pair-bond relation. Men bring back meat, the presumed major food source, to the waiting dependents at camp. This view links males with technology and the provision of basic sustenance and assumes stable, long-term sexual bonds. It promotes the idea of male aggression as necessary for hunting and for protecting the weak and passive females and children and assumes male dominance over females inherent to the hunting way of life.

The alternative story put forward by Tanner and Zihlman started from the premise that chimpanzees are the closest human relatives and should be taken as models of the ancestral hominin way of life. The primitive hominins were assumed to have been chimpanzee-like inhabitants of tropical forests and woodlands, with a diet consisting almost entirely of fruit and other plant parts, supplemented with a small amount of animal prey. Like chimpanzees, the ancestral hominins made and used simple tools for various purposes in procuring food and shelter. And like chimpanzees, they lived in large mixed-sex groups divided into temporary parties of shifting composition, with mother–offspring groupings forming the most stable long-term subunits. "We propose," wrote Tanner and Zihlman (1976), "that the human adaptation commenced as ape populations moved away from the dense African forests and began utilizing resources in the relatively more open country of eastern Africa." The innovation that made this possible was the invention of **gathering**—that is, the collection and subsequent sharing of such portable, high-quality food items as eggs, honey, edible insects (McGrew 2001b), seeds, and USOs. These foods were collected out in the savanna during the daytime (when predator pressure was lowest) and shared at the end of the day between mothers and offspring bivouacked at safe locations near clusters of trees, into which the group could flee if a predator showed up.

Gathering required the invention of several sorts of tools, particularly digging sticks and containers for carrying gathered food. "Because of nutritional requirements of pregnancy and nursing, and overt demands from hungry children," wrote Tanner, "women had more motivation for technological inventiveness, for creativity in dealing with the environment, for learning about plants, and for developing tools to increase productivity and save time" (Tanner 1981, p. 222). Gathering also favored the development of bipedality, since walking on the hind legs freed up the hands for collecting while carrying a food parcel. Thus, all the new human-like behaviors were pioneered by the adult females. The less sociable males continued to forage for themselves, occasionally contributing larger vertebrate prey to their mothers' feasts. Unrelated males were allowed to associate with the female collectives if they ingratiated themselves by sharing food, grooming, protecting the group, and playing with the babies. Females preferred such accommodating males as companions and sexual partners; therefore, the most sociable males were favored by sexual selection. And because the most sociable males are the ones that look most like females (perhaps because they have low testosterone levels), "selection for more sociable males reduced sexual dimorphism in dentition" (Tanner and Zihlman 1976). The feminization of the male canine teeth resulted in canine reduction in the species as a whole. Added together, all these changes converted a *Pan*-like creature into an early *Australopithecus*. In short, mother-centered, food-sharing groups of female gatherers—not nuclear families dependent on a hunting male—were the driving force behind the origin of the key human traits of behavior and anatomy.

By 1980, cooperative female gathering and females' mate choices were starting to figure importantly in anthropological accounts of the causes of human origins (Dahlberg 1981). With exceptionally bad historical timing, Lovejoy (1981) chose this moment to propose a theory that attributed hominin origins to a sexual division of labor, in which the male served as breadwinner and the female as domestic in a monogamous nuclear family.

Lovejoy dismissed the importance of hunting, noting that "there is no evidence whatever that early [hominins] hunted." His version of the feedback theory focussed instead on the reproductive implications of male provisioning of females. Hominoids, argued Lovejoy, had dwindled and almost disappeared from tropical faunas during the Miocene because they had been replaced by the faster-breeding cercopithecoids. Old World monkeys mature faster than modern apes or humans, and they space their births more closely than apes do. Resources permitting, they can therefore multiply at a faster rate. Lovejoy suggested that this was the main reason why cercopithecoid monkeys had outcompeted and replaced the hominoids during the Miocene. The one ape that effectively resisted the cercopithecine onslaught was the ancestral hominin. It had managed this, Lovejoy thought, by recruiting the foraging activities of the males in the service of female reproduction. This was accomplished through the creation of the nuclear family, centered on a mated pair bound together through frequent sex and the male's helping to feed the female and their offspring. With the male working to feed his mate and their young, the female could afford to space her pregnancies closer together and thus reproduce at a faster rate. Males that provisioned their families in this way, as well as females who attracted such males, would therefore have been favored by natural selection.

Other distinctive peculiarities of hominins, argued Lovejoy, could be explained with reference to this story. Bipedality had initially been favored because it allowed the foraging males to use their hands to carry food back to their mates and offspring. Increased tool use began with the males' recruiting "simple and readily available natural articles to enhance carrying ability," which eventually led to the manufacture and use of stone tools by early *Homo*. Females had evolved concealed estrus in order to attract the males at the bottom of the dominance hierarchy. The smaller size and less robust canines of these low-ranking males rendered them less competitive in mating with desirable, visibly estrous females. They were therefore more willing to play the provisioning game to be allowed to have sex with females that were not visibly estrous. Female preference for these wimpy but provident males brought about an overall reduction of canine size and sexual dimorphism (Lovejoy 1993).

But it is not to a male's advantage to help feed a female's offspring unless he can be reasonably sure that the babies he is bringing food to are his own. Males who divert a lot of their own resources to another male's offspring will be selected against. Therefore, early hominins must have been monogamous, with both members of the pair generally refraining from having sex with anyone else. "Pair bonding," argued Lovejoy (1981), "was fundamental and crucial to early [hominin] reproductive strategy."

Lovejoy's model and the Tanner–Zihlman model had a lot of elements in common: the unimportance of hunting, the importance of concealed estrus, the origin of bipedality as a carrying adaptation, the invention of food-carrying artifacts for bringing food back to a home base, and canine reduction through female choice of small-toothed males. But the spin was different. In the "Woman the Gatherer" story, bipedality, food-carrying technology, and family provisioning are all female innovations, and the males are dragged along in a human direction by "bipedal, tool-using, food-sharing, and sociable mothers choosing to copulate with males also possessing these traits" (Tanner and Zihlman 1976). In Lovejoy's story, the human innovations originate in male behaviors, with the females and young sticking close to a home base and limiting their daily ranges to avoid competition with the wide-ranging, adventuresome male foragers.

Some critics detected the flavor of gender bias in Lovejoy's male-centered account (Ranieri and Washburn 1981, Zihlman 1987, Wolfe et al. 1982). Others assailed his facts and figures. It was argued that birth spacing in pre-agricultural humans was longer than Lovejoy assumed; that the math underlying his model was flawed; that his explanation of concealed estrus was unworkable; and that hunting was the only way to make male provisioning of mothers and young truly effective (Harley 1982, Hill 1982, Isaac 1982, Wolfe et al. 1982, Wood 1982). Lovejoy (1982b) fended off these attacks with statistics and calculations, but few anthropologists were won over to his account of hominin origins. Two main objections continue to be urged against it (Wrangham 2001). The first is that the monogamous pair-bonding assumed by the model would be unreliable if the male parent were away from the home base all day, leaving the female and offspring unguarded against seducers, rapists, or infanticides. The second is that monogamous, pair-bonded primates do not show the high degree of sexual dimorphism attributed to in *Australopithecus*. And if dimorphism really was more extreme in *Australopithecus* than in either humans or chimpanzees, that fact constitutes a fatal objection to the Tanner-Zihlman model as well.

For all these reasons, the amount of sexual dimorphism in early hominins remains controversial.

Everyone agrees that *A. afarensis* is highly variable in size—maybe as variable as orangutans or gorillas, but at any rate more variable than humans or chimpanzees. The ongoing disagreements are over the causes of this variability. Although most investigators attribute it to size differences between males and females, others attribute most of it to temporal and geographic variation within a population whose average body size was increasing over time (Lockwood et al. 2000, Reno et al. 2003, 2005b). This dispute will probably be resolved only by future discoveries that will increase the size of the fossil sample. In the meantime, it seems reasonable to think that at least the later "robust" species of *Australopithecus* were more sexually dimorphic than *Pan* or *Homo*, because the morphological differences between male and female skulls appear to have been greater in these creatures than they are in humans or chimpanzees (Figs. 4.29, 4.31, 4.38, and 4.69).

▌DINICHISM: A POSSIBLE SYNTHESIS

As noted earlier, some experts still suspect that *A. afarensis* is highly variable because it is a hodgepodge of mixed remains from two sympatric hominin species. The chief reason for thinking this is that many different researchers have found that the postcranial remains of the larger *afarensis* specimens are more human-like in some ways than those of the smaller individuals. But these findings might also be taken as an indication that Lovejoy's story of hominin origins is basically correct—that in *A. afarensis*, males and females had different social roles and ecological niches, with the males being wide-ranging foragers who brought food back to safer core areas where females were sequestered with the infants. Such a difference might select for sexual dimorphism in both body size and locomotor adaptations. This point has been made by the Stony Brook group:

> … within such a framework, it is inconceivable to us that, if the larger male *A. afarensis* were off foraging during the day, leaving the diminutive females with their offspring to fend for themselves, that the latter would have survived without recourse to the trees… It seems that the only possibilities of survival, especially if the Hadar hominids were monogamous, would have rested in their being at least part-time arborealists. (Susman et al. 1985)

A synthesis of parts of Lovejoy's theory, the Tanner-Zihlman theory, and the Stony Brook position on *A. afarensis* locomotion might be attempted along these lines. Let us assume that (as Lovejoy suggests) early hominins were able to compete with cercopithecoids

more effectively than other apes because they hit upon a reproductive strategy that allowed for closer birth spacing. This strategy involved male foraging at some distance from a relatively safe core area containing the females and young, to whom the males brought food at the end of their daily round. As Susman et al. (1985) argue, the females would have been safer if they were "part-time arborealists" who carried on much of their daily activity in the trees and seldom strayed far from arboreal refuges. The males ranged further afield into more dangerous environs. They were correspondingly larger and more robust (hence the sexual dimorphism). They were also more specifically adapted to terrestrial locomotion—and to bipedal locomotion in particular—to facilitate food-carrying on the homeward trip. Conversely, the females retained a greater degree of arboreal competence, though both sexes were bipedal on the ground (as indicated by the Laetoli footprints).

Lovejoy's (1981) theory postulated that perfected terrestriality evolved in forests or woodland-savanna mosaics, not in open country. But as noted earlier, there is not much to eat on the ground in a closed-canopy tropical forest. Wide-ranging male foragers would have fared better by venturing into open areas in search of more diverse food sources, including "… fruits, berries, nuts, seeds, underground tubers, and roots … a wider range of young animals than in the tropical forests … and termite hills—the last a visible source of attraction from far away" (Tanner and Zihlman 1976). Digging up USOs and carrying food back to the core area would have stimulated the invention of several sorts of artifacts. Sticks and stones might have been used as weapons in killing prey or driving competing scavengers away from carnivore kills, although this sort of foraging was probably infrequent and restricted to relatively small game until the advent of the genus *Homo* (O'Connell et al. 2002).

We can refer to this division of labor as **sexual dinichism**. We do not know what selective forces might have brought such a system into being, but it seems clear that male provisioning of offspring could have had the sorts of selective advantages claimed by Lovejoy. It is true that foraging males would not have been able to guard their mates against the sexual advances of other males, and so pair-bonding could not have been a 100% reliable guarantee of paternity (Wrangham 2001). But this is not a fatal objection. To make the system work, a male's foraging effort would not need to benefit his own offspring exclusively. Benefitting them *disproportionately* could provide a selective advantage, if the disproportion were high enough.

Several factors could have helped to ensure this disproportion, even though the male was unable to monitor his mate's sexual activity while he was off looking for food. First, the core-area females could have formed a foraging and baby-tending collective, of the sort envisioned in the Tanner–Zihlman model. Such a collective would have provided a group defense against predators and extragroup males, and could also have served as a deterrent to male seducers or rapists from within the group. Second, though "adulterous" copulation on the sly would no doubt have been part of the social landscape, as it is in human beings and most so-called monogamous vertebrates (Schülke et al. 2004), this need not have placed cuckolded male provisioners at a selective disadvantage—if extrapair conceptions were infrequent enough, and if all the males within the group were closely related. Inclusive-fitness theory suggests that a male who provisions an offspring begotten on his mate by his father or brother would not simply be wasting his effort, since such an offspring would be carrying a large percentage of the cuckolded male's genes. Among chimpanzees and bonobos, males continue to reside in the group where they were born, while females tend to emigrate from their home group and find new homes in neighboring groups. Widespread in human societies as well, this **patrilocal** male residence pattern is apparently an inheritance from the last common ancestor of *Pan*, *Homo*, and *Gorilla* (Bradley et al. 2004). As such, it would also have been present in the earliest hominins. It would have eased the negative selective consequences of provisioning an occasional bastard offspring.

As critics of Lovejoy's model have observed, no arrangement like this is known among other primates, but certain elements of it are seen elsewhere. In a few carnivorans (canids, hyaenas) and in many pair-living birds, foraging males help to provision nest-guarding females. A particularly close parallel noted by Lovejoy (1981) is provided by the toucan-like African and Asian birds called hornbills (Bucerotidae), in which the nesting female is physically walled up inside a tree cavity and fed through a small hole by the male until the chicks are fledged. It is suggestive that this breeding behavior correlates with an unusually large brain in adult hornbills (Diamond and Bond 2003). A possible precursor of dinichism is seen in chimpanzees, where the males do most of the hunting of mammalian prey, making over 90% of the kills in some areas, and eat most of the meat. The small amount of meat that the hunters share with others is doled out for "political" purposes—to kin, to high-ranking or estrous females, or to animals that have given the hunters meat in the past (Stanford 2001a,b). It is not hard to imagine moving from a pattern of donating meat to desirable females and close kindred to a pattern of increasing food-sharing with a mate and her offspring. It should be stressed, however, that sexual dimorphism is minimal in all these mate-provisioning animals. No precise modern analog is known for the behavioral ecology sketched above for early hominins.

EXPLAINING HOMININ ORIGINS

This story, like all its alternatives, is speculative and resistant to empirical testing. Its chief merit is that it explains the apparent coupling in early hominins of slight canine dimorphism with high degrees of dimorphism in the rest of the skeleton. Its main shortcoming is the absence of a parallel case—a highly dimorphic species in which males provision females and young—among other animals. The absence of parallel cases is a significant defect, but it does not furnish automatic grounds for rejecting this or any other explanatory account of human origins. Human beings are unique in many ways, and we need to be open to the possibility that no parallels can be found in other animals for the combination of factors underlying the divergence of our ancestors from the apes.

Nevertheless, an explanation that holds for only a single case is necessarily *ad hoc*. We may have to live with the possibility that some aspects of human evolution will always resist causal explanation. We have already faced this possibility in dealing with the origins of hominin bipedalism. Other bipeds got up on their hind legs to run faster. Our ancestors did not, and so *ad hoc* explanations of their shift to bipedality are the only sort we can have.

Many such explanations have been offered. Rose (1991) reviews 19 different accounts of the causes of human bipedality that have been put forward in the scientific literature, grouped under four headings:

1. **Pre-emption of the forelimb** for nonlocomotor jobs (throwing; carrying food, infants, and/or tools).
2. **Social behavior** (threat displays, sexual displays, aggression, vigilance, evasion).
3. **Feeding behavior** (bipedal gathering, scavenging, or predation, either on the ground or in the trees—or even in the water, in the so-called "aquatic ape" theory (Morgan 1982)).
4. **Other**—including thermoregulation, biomechanical necessity (e.g., for a long-armed gibbon-like ancestor), locomotion on slippery substrates, iodine deficiency, and various combinations of other listed factors.

This list could be expanded. Some of these supposed causes are demonstrably wrong, inadequate, or silly. But many of them are not. Long analytical chapters could be (and have been) written, evaluating the pros and cons of each variant explanation in order to sieve out some relatively plausible factor as the probable first cause of human bipedality. We have not written such a chapter, because choosing among *ad hoc* explanations strikes us as a matter of taste. In modern human behavior and biology, bipedality serves several of these purposes; it allows us to carry things, throw things, swing weapons, run long distances (Carrier 1984), swim better than other hominoids, gesture with both hands, see farther than we could on all fours, and so on. The bipedality of our ancestors must have conferred several simultaneous advantages as well. But it is virtually impossible, six million years after the fact, to determine which of these advantages got the process of bipedalization started. Claims about the causal priority of, say, carrying food versus gesturing are untestable, because there are no parallel cases in other animal lineages that can be studied to check out the correlations between the supposed causes and their effects.

In the case of hominin bipedality, it is not even clear what the effects were that we are seeking to explain, because we do not yet know how our ancestors moved around before they took to their hind legs. As Day (1986b) put it, "We really have no clear idea of what form of locomotion, in what creature, preceded, or was immediately pre-adaptive for, upright posture and bipedal gait." Some believe that the pre-bipedal hominins were stout-wristed, semi-terrestrial knuckle-walkers resembling chimpanzees (Corruccini 1978, Zihlman and Brunker 1979, Zihlman 1990, Begun 1993, Gebo 1992, 1996; Richmond and Strait 2000, Collard and Aiello 2000, Corruccini and McHenry 2001). Others envision them as orangutan- or gibbon-like arboreal clamberers and swingers, with long arms and limber joints (Stern 1975, Tuttle 1969, 1974, 1975, 1981, 1994; Fleagle et al. 1981, Reynolds 1985, Schmitt and Larson 1995). Questions of phylogeny are at issue here (Richmond et al. 2001). If *Pan* is the phyletic sister group of the hominins, as most experts currently believe (Fig. 3.32), then either

(a) *Pan* and *Gorilla* evolved knuckle-walking independently of each other, or else—

(b) people are descended from knuckle-walkers.

Proponents of a clamberer/swinger model accept proposition (a). Those who think the ancestral hominins were chimpanzee-like prefer (b). Others reject the consensus phylogeny and regard *Pan* as the sister group of *Gorilla*. This move allows knuckle-walking to be read as a synapomorphy of a *Pan*–*Gorilla* clade. The fossil record is not of much help here, because all the known hominid postcranial fossils are those of bipedal hominins. It has been argued that a few chimpanzee-like anatomical features of the wrist and elbow in *Australopithecus* and *Homo* originated as adaptations to knuckle-walking; but other interpretations are possible for most of these traits (Richmond et al. 2001).

The adoption of bipedality in the human lineage cannot be explained without determining what sort of locomotion came before bipedality, because the sequence of events here is what we are trying to explain.

But getting the story straight is only part of an explanation. Causal explanation links a cause with an effect by telling a story and invoking a law—showing that the cause came immediately before the effect (this is the story-telling part) and that events like the supposed cause can be expected to result in effects of that sort (which is a natural law). The expectation or law can be checked out only by showing that it holds true in other, similar cases (Cartmill 2002). This is why scientists do experiments to test their explanations. Parallel or convergent evolution often provides natural experiments to test evolutionary explanations. But because human bipedality is unique, there are no parallel cases. Explanations of its origin are thus necessarily conjectural. We can rule some of them out by showing that they are inconsistent with the facts, but we cannot show reasons for thinking that the proposed cause actually does have the proposed effect.

What is true of human bipedality is probably true of some other human peculiarities that have no parallels in other animal lineages—for example, the enlargement and adduction of our big toe. But for at least some other hominin apomorphies, we can escape from the trap of human uniqueness by reconceptualizing the trait in a way that makes it no longer unique (Cartmill 1990). The expensive-tissue hypothesis is an example of this. If we ask, "What could cause a human to have a brain three times as big as that of a similar-sized chimpanzee?" then we are limited to awestruck conjecture, because there are no other human-sized animals that have a brain anywhere near the size of ours. But if we ask more broadly framed questions about the correlates of brain enlargement in other animal groups, we may discover regularities in nature that apply to the human case and explain its peculiarities. The expensive-tissue hypothesis offers an explanation of human brain enlargement in terms of the high quality of the human diet, and implies that these variables should be correlated in other animals as well. This implication is testable. Whether the expensive-tissue hypothesis proves true or not, it provides a model of the way in which reorganizing our concepts can liberate us from the dead end of *ad hoc* explanation.

Another recent example of this sort of reconceptualization involves the problem of canine reduction in hominins. Most students of human evolution, from Charles Darwin onward, have tried to explain this phenomenon in one of three ways: (1) as an adaptation to allow freer movements of the jaw in chewing, (2) as a corollary of disuse resulting from the use of weapons in fighting, or (3) as a social adaptation of some sort—for instance, in the female-choice explanations favored by Tanner, Zihlman, and Lovejoy. Explanation (1) has been shown to be empirically false, because big canines do not preclude human-like jaw mobility (Greenfield 1990). Explanation (2), which has been the most popular, makes little sense when looked at closely (Jolly 1970a). Humans, despite their short canines, still bite each other in fighting, and early stone tools would not have rendered canine fangs useless in a no-holds-barred combat. (Which would you rather fight: a big man carrying a handaxe, or a big man with a handaxe *and* three-inch canine teeth?) Explanations of type (3) may be correct, but they are untestable. Any evolutionary novelty at all can always be explained by saying that females found it attractive. In living species, such claims can be checked out by behavioral experiments. But inferences cannot be drawn from such experiments to the remote past. Even if experiments proved that modern women generally find feminine-looking men more attractive, that would tell us nothing about the mate preferences of female *Australopithecus*.

Hylander and Vinyard (2006) have recently attacked the problem of canine reduction in a more productive way, by asking what limits the length of primate canines. One limiting factor is jaw gape. To be useful in biting, the canine teeth have to be fairly widely separated from each other when the mouth is wide open (Fig. 3.26). Jaw gape is limited chiefly by the jaw muscles. When these muscles have been stretched to their utmost, the mouth cannot be opened wider. Moving the attachments of these muscles forward can have advantageous effects on the bite force exerted through the post-canine teeth (Fig. 4.32), but it also restricts the opening of the mouth, because the anteriorly shifted jaw muscles become taut at smaller angles of jaw rotation. Hylander and Vinyard suggest that the forward shifting of the jaw muscles in early hominins made it necessary to reduce the length of the canines, because they could not be used in biting unless they became shorter. Again, the truth of this account is irrelevant to its main virtue, which is that it explains an apparent human peculiarity in terms of a regularity that should apply in many other cases. If this explanation is correct, we should find a reliable correlation between jaw-muscle anatomy and canine length in other groups of mammals—another testable expectation.

None of this implies that all human peculiarities can be explained with reference to general regularities seen throughout mammalian evolution. Some human apomorphies almost certainly had unique causes. We are not entitled to expect that the causes of events will always be of the sort we prefer to have, and in some cases an *ad hoc* explanation will be the correct one. The problem is that we cannot identify which cases those are. We may have to settle for such explanations of many evolutionary events; but we should never be content with them, because the number of *ad hoc* explanations that can be dreamt up for any event is endless.

▌PRIMITIVE *HOMO*—OR "ADVANCED" *AUSTRALOPITHECUS*?

The first fossils that everyone accepts as *Homo* show up in the African fossil record around 1.8 Mya. The genus appears soon afterward in other low-latitude areas of the Old World. These wide-ranging early humans are usually grouped into the species *Homo erectus*. Many scholars prefer to consign the oldest fossil *Homo* from Africa and western Asia to a separate species, *Homo ergaster*. We will deal with these fossils in the next chapter under the taxonomically noncommittal headings of "Erectines" and "Ergasters."

But as the theory of evolution might lead us to expect, there are older African fossil hominins that seem to be intermediate between *Australopithecus* and *Homo* in their morphology. Many of these fossils have been referred to a third species of *Homo*: *H. habilis*. Researchers have been debating the reality of this species ever since it was first described In 1964. Some experts think that the "*H. habilis*" fossils represent primitive forms of *Homo*. Others regard them as "advanced," humanlike versions of *Australopithecus*. For the time being, we will evade this taxonomic issue as well, by referring to them all as "Habilines."

The first fossils to be identified as intermediate forms connecting *Australopithecus* and *Homo* came from the Transvaal caves. In 1949, R. Broom and J. Robinson reported the discovery of a gracile mandible from Swartkrans, SK 15, that looked more human to them than anything previously found at the Transvaal cave sites. Pointing to the small molars and slender mandibular body of SK 15 as key differences from any known *Australopithecus*, Broom and Robinson assigned it to a new genus and species, *Telanthropus capensis*. Robinson (1953b, 1954a) hailed "Telanthropus" as an ancestor of the genus *Homo*, and eventually sank it into *H. erectus*.

The "Telanthropus" specimens were fragmentary and unconvincing. In 1960, theories about Pliocene *Homo* in Africa took on more substance with the discovery of OH 7 in Bed I at Olduvai Gorge (Leakey 1961). OH 7, comprising a mandible, some hand bones, and parts of both parietals, was almost as fragmentary. It was also immature, roughly equivalent in dental development to a 12- to 13-year-old child. Nevertheless, it seemed clear that this gracile skeleton was not the same sort of creature as the other Bed I hominin, the hyper-robust *A. boisei*. Louis Leakey and his co-workers (Leakey et al.1964) assigned the "pre-Zinj child" and a partial foot skeleton (OH 8) found nearby to a new species, *Homo habilis*.

The definition of this new species, and its assignment to the genus *Homo*, revolved around three supposed differences between OH 7 and known species of *Australopithecus*. The first was a bigger brain. Cranial capacity estimates for OH 7 ranged between 642 and 723 cc, with a central tendency of 681 cc (Tobias 1964)—significantly greater than any estimates for *Australopithecus*. Second, the OH 7 mandible (Fig. 4.41) was described as having smaller cheek teeth and less molarized P3's than even the most gracile *Australopithecus*. Finally, the OH 7 hand bones were interpreted as having features associated with a so-called "precision grip," in which the thumb tip is fully opposed to the tip of the index finger (Napier 1961, 1962). The toolmaking skill supposedly reflected in the enlarged brain and nimble hand of OH 7—and in the stone tools found throughout the Olduvai deposits—was touted as a key part of the taxonomic and adaptive distinctiveness of the new species.

During the next 10 years, more fossil hominins that were clearly not *A. boisei* emerged from the lower levels of Olduvai Gorge. All were assigned to *H. habilis* by Leakey and his coworkers. The most important of these new *H. habilis* finds were the OH 24 partial cranium from Bed I (Fig. 4.77: Leakey et al. 1971); the OH 13 fragmentary cranial vault with an associated maxilla and mandible from lower Bed II (Fig. 4.78: Leakey et al. 1964); and the OH 16 fragmentary cranial vault and associated dentition, also from lower Bed II (Fig. 4.79: Leakey et al. 1964 Tobias 1991).

There was general agreement that the gracile hominins from Olduvai represented a different species from *A. boisei*. But not everyone agreed that it was a *new* species. Many thought that the new material could be accommodated within either *Australopithecus africanus* or the established primitive species of *Homo, H. erectus*. Robinson (1960, 1965, 1966) argued that OH 7 and the other Bed I Habilines represented *A. africanus*, while the Bed II specimens were *H. erectus*—which he regarded as the immediate descendant of *A. africanus* (Robinson, 1972). Brace et al. (1973) also concluded that OH 7, 16 and 24 could not be differentiated from *A. africanus*, and that OH 13 fitted comfortably into *H. erectus*. M. Wolpoff (1969, 1970a) raised doubts about the large brain size claimed for *H. habilis*, insisting that cranial capacities could not be reliably estimated from the fragmentary skulls attributed to the new species.

In the early 1970s, the Leakeys' son Richard and a team of collaborators started looking for fossil hominins in the Plio-Pleistocene deposits of the Koobi Fora region, east of Lake Rudolf in northern Kenya. Like Olduvai Gorge, the Koobi Fora sites soon began to yield jaws and teeth of both *A. boisei* and more gracile hominins (Walker and Leakey 1978, Wood 1991). A nearly complete cranium, known by its museum number KNM-ER 1470 (Fig. 4.80), came to light In 1972. This find proved that some of the "gracile" forms from Koobi Fora had had much larger brains than any known *Australopithecus* (Leakey 1973a). The first estimate of the cranial capacity of the 1470 skull was an impressive 810 cc, and its age was pegged at around 3 Mya, based on an initial

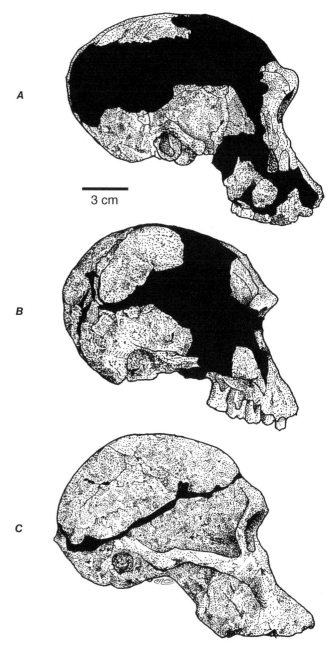

3 cm

FIGURE 4.77

Two crania sometimes assigned to early *Homo*, OH 24 (**A**) and Stw 53 (**B**), compared to the Sts 5 cranium of *Australopithecus africanus* (**C**). OH 24 is from Olduvai Gorge, Tanzania; Stw 53 and Sts 5 are from Sterkfontein, South Africa. Although these three skulls are similar, OH 24 has a larger braincase (Tables 4.1 and 4.6), a shorter, more flexed cranial base (Fig. 4.83), and a more *Homo*-like frontal/supraorbital region. (After Tobias 1991.)

age estimate of 2.6 Mya for the overlying KBS tuff. Even when the age of the KBS tuff was revised downward to 1.88 My (see below), and the skull's estimated cranial capacity was lowered to 752 cc (Table 4.5), the 1470 cranium still seemed to validate L. S. B. Leakey's long-

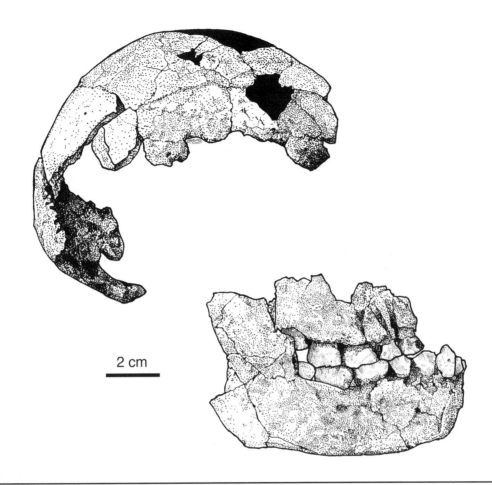

2 cm

FIGURE 4.78

OH 13 cranial vault, maxilla and mandible from Olduvai Gorge, also known as "Cinderella." This is one of the latest and most Erectine-like specimens assigned to the Habilines. (After Tobias 1991.)

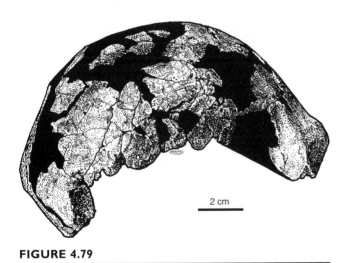

2 cm

FIGURE 4.79

Reconstruction of the OH 16 cranial vault, known as "Olduvai George." Louis Leakey reported that this was a complete skull when he first found it. But it was late in the day, and when he returned the next day to excavate it, Masai cattle had trampled it into bits. The reconstruction of this specimen is open to question. (After Tobias 1991.)

standing belief in an ancient *Homo* lineage separate from *Australopithecus*. To be sure, there were still a few inconvenient facts that did not fit into the picture. As R. Leakey's co-worker A. Walker (1976) noted, 1470 has a face and palate that rival those of *A. boisei* in size. This seemed wrong for a skull attributed to *H. habilis*, a species supposedly distinguished by its small teeth.

The 1470 find was followed by further discoveries of relatively large-brained crania (Table 4.5) and human-like postcranial bones from the Koobi Fora Formation (Leakey 1971, 1972, 1973a, 1973b, 1974; Leakey and Wood 1973, 1974; Wood 1991). The skull KNM-ER 1813 (Fig. 4.81), which came to light In 1973, was a perfect match for the *H. habilis* material from Olduvai in most respects, particularly in the small size of its cheek teeth. But its cranial capacity was a strikingly low 510 cc (Table 4.5), within the range of robust *Australopithecus* (Holloway 1988). The differences between ER 1470 and 1813 suggested that there were not one but two ancient "*Homo*" lineages—one with an *Australopithecus*-sized brain and a *Homo*-like face, and another with the opposite combination. Most experts now recognize these as different taxa. ER 1813 is generally retained in *H. habilis*;

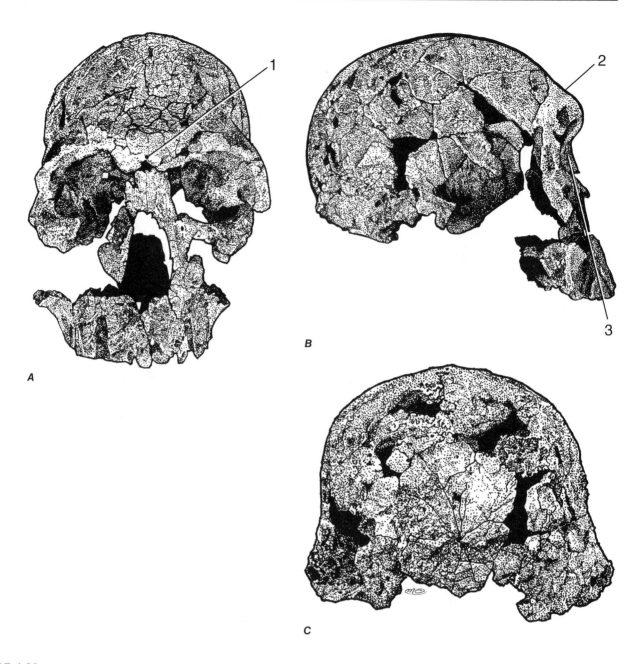

A

B

C

FIGURE 4.80

The KNM-ER 1470 skull from Koobi Fora, Kenya. (**A**) Frontal view, (**B**) lateral view, (**C**) occipital view. Although there is a swelling at glabella (1), the absence of a distinct supraorbital torus is evident (2). Also evident in lateral view is the anterior inclination of the malar (infraorbital) region from its superior to inferior aspects (3). (After Leakey et al. 1978.)

but 1470 is accepted as the type of a separate species, *Home rudolfensis* (Alexeev 1986, Wood 1999a).

Postcranial remains were also found in the Koobi Fora deposits. Some of these—the 1472 femur, the 1481 femur and partial tibia, and the 3228 hipbone (Leakey 1973a, 1976)—were more *Homo*-like than the corresponding parts known for any *Australopithecus*. It seemed reasonable to attribute these to "*Homo*" *habilis*. But this attribution was cast into doubt by the 1986 discovery of a partial skeleton, OH 62, from Bed I in Olduvai

Gorge. OH 62 comprises a lower face with much of the upper dentition, plus various postcranial remains. The teeth and facial anatomy of OH 62 resembled those of *H. habilis* from Olduvai, and so D. Johanson and his colleagues assigned it to that species (Johanson et al. 1987). But they inferred from the associated long bones that *H. habilis* had body proportions resembling those of the "Lucy" skeleton of *Australopithecus afarensis*. Although the sources of error involved in the reconstruction make it impossible to be sure (Korey 1990),

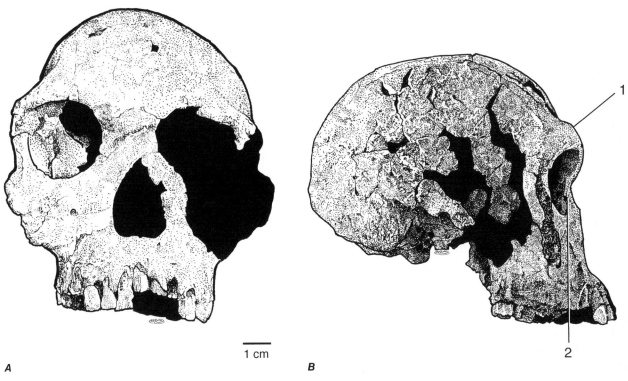

1 cm

A   **B**

FIGURE 4.81

The KNM-ER 1813 skull from Koobi Fora, Kenya. (**A**) Frontal view, (**B**) lateral view. This specimen exhibits an incipient supraorbital torus (1) and a more vertical orientation of the malar region (2) than ER 1470 (Fig. 4.80). Not to same scale. (After Leakey et al. 1978.)

TABLE 4.5 ■ Cranial Capacity Estimates for Selected Specimens Assigned to "Homo" habilis or "Homo" rudolfensis

Specimen	Site	Cranial Capacity (cc)
OH 7	Olduvai Gorge	687
OH 13	Olduvai Gorge	650
OH 16	Olduvai Gorge	638[a]
OH 24	Olduvai Gorge	590
KNM-ER 1470	Koobi Fora	752
KNM-ER 1590	Koobi Fora	>800
KNM-ER 1805	Koobi Fora	582
KNM-ER 1813	Koobi Fora	510
KNM-ER 3732	Koobi Fora	600–650

[a]From Tobias (1991). Other data from Holloway (2000).

"Homo" habilis seems to have retained a body build like that of *A. afarensis*, with relatively long arms and a small body mass of around 30 kg (Wood 1992). The disparity between the *Homo*-like hindlimb bones from Koobi Fora and the *Australopithecus*-like limb and body proportions of OH 62 muddied the concept of *Homo habilis* even further.

Several finds from other sites of about the same age have been attributed to early *Homo* (Table 4.6). Most of these are isolated teeth, but they also include two frag-

TABLE 4.6 ■ Major Specimens Assigned to "Homo" habilis or "Homo" rudolfensis in East Africa

Site	Crania	Mandibles	Postcrania
Baringo-Chemeron[a] (KNM-BC)	1	—	—
Hadar (AL)[b]	666-1	—	—
Koobi Fora (KNM-ER)	807, <u>1470</u>, 1478, <u>1590</u>, 1805, 1813, <u>3732</u>, 3735, <u>3891</u>	819, 1482, 1483 1501, 1502, 1506, <u>1801</u>, <u>1802</u>, 1805, 3734	<u>813</u>, <u>1472</u>, <u>1481</u>, 3228, 3735
Olduvai (OH)	6, 7, 13, 14, 16, 24, 52, 62, 65[c]	7, 13, 37, 62	7, 8, 10, 35, 43, 48, 49, 50, 62
Omo (Omo)	L894-1, 75-14	75-14, 222-2274	—

[a]Hill et al. (1992).
[b]Kimbel et al. (1997) assign AL 666-1 to "Homo aff. habilis."
[c]Blumenscheine et al. (2003). See text.
Underlining indicates specimens assigned to *H. rudolfensis*. Isolated teeth and Erectine specimens are not included. Specimens appear in more than one column if multiple elements are preserved. The Omo, Hadar, and Baringo-Chemeron specimens attributed to *Homo* are not assigned to a specific species. List and attributions after Wood (1992) unless otherwise indicated.

mentary crania (L894-1 and 75-14) from the Omo Shungura Formation in Ethiopia, a partial cranium (Stw 53) from Member 5 at Sterkfontein, and the "Telanthropus" material from Swartkrans (Howell and Coppens 1976, Boaz and Howell 1977, Hughes and Tobias 1977, Tobias 1978, Wood 1992, Clarke 1994a). There is some sketchy evidence of the genus *Homo* at two other South African *A. robustus* sites: a child's jaw and some postcranial bits from Drimolen, and a lower molar from Gondolin (Menter et al. 1999, Keyser et al. 2000).

All these supposed early *Homo* remains are dated to less than 2 Mya, but there are some signs that the genus is more ancient than that. Isolated teeth from Omo-Shungura, found with Oldowan tools and dated to between 2.41 and 2.34 Mya (Howell and Coppens, 1976, Howell et al. 1987, Brown 1994), have been identified as early *Homo* by some experts (Howell et al. 1987). Others regard these Omo teeth as *Australopithecus* (Wolpoff 1999, Hunt and Vitzthum 1986). A temporal fragment (KNM-BC 1) from the Chemeron Formation in Kenya has been radiometrically dated to 2.4 Mya and assigned to *Homo* on the basis of a few apomorphies, including a deep, medially placed mandibular fossa and sharp crests on the petrous and tympanic bones (Hill et al. 1992, p. 720). But some features of this fossil, including its large mandibular fossa and big, circular tympanic opening, are more like their counterparts in *A. boisei* (Strait et al. 1997). A maxilla, AL 666-1 (Fig. 4.82), found at Hadar at a horizon containing Oldowan

tools and directly underlying a tuff dated to 2.33 Mya, has been ascribed to an undetermined species of *Homo* on the basis of the broad, square shape of its palate, its weak prognathism, and various features of the maxillary sinus, nasal floor, and dentition (Kimbel et al. 1996, 1997). Like other Habilines, the AL 666-1 maxilla has big front teeth and moderately large cheek teeth (Table 4.7). An adult mandible with a complete dentition from Uraha in the Malawi Rift Valley has been assigned to the large Habiline species, *H. rudolfensis*, because of its resemblance to a Koobi Fora jaw (KNM-ER 1802) thought to represent that species (Schrenk et al. 1993, Bromage and Schrenk 1995). The Uraha mandible is correlated to deposits at Omo dated to 2.35 Mya. This makes it older than any of the Koobi Fora specimens, which are all less than 2 My old (see below). It has accordingly been argued that *H. rudolfensis* is the earliest form of the genus (Bromage and Schrenk 1995; cf. Suwa et al. 1996).

In trying to sort out all this material, scientists have focused on three questions:

1. Can all these fossils be encompassed within the single species *H. habilis*?

2. If not, how should they be parceled out into different species (Tables 4.6 and 4.8)?

3. Do any of these specimens really belong in the genus *Homo*?

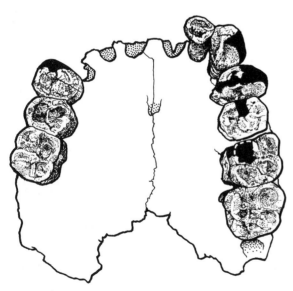

FIGURE 4.82

Dentition of the AL 666-1 maxilla from Hadar, Ethiopia, attributed to early *Homo* and dated to slightly over 2.33 million years ago. Occlusal view. (After Johanson and Edgar 1996.)

TABLE 4.7 ■ Dental Breadths for Selected Habiline Specimens and AL 666-1

	I1	I2	C	P3	P4	M1	M2	M3
Mandible								
OH 7	6.6	7.5	10.11	9.8	10.7	12.5	13.7	—
OH 13	—	—	7.9	8.7	9.9	11.6	12.0	12.4
OH 16	7.0	7.6	10.0	—	10.9	12.8	15.1	14.4
Maxilla								
OH 13	—	—	—	12.1	12.5	12.8	14.0	13.1
OH 16	8.8	7.9	—	12.1	12.1	13.8	16.0	16.8
OH 62[b]	—	—	9.9	—	—	—	(16.0)	(15.5)
OH 65[c]	8.0	7.6	9.9	12.6	13.0	13.5	14.3	14.0
ER 1590	7.7	—	12.7	13.4	13.6	15.0	17.6	—
ER 1805	—	—	10.5	11.5	—	13.4	15.1	14.1
ER 1813	—	6.1	8.6	11.3	11.5	13.0	13.9	13.7
AL 666-1[d]	—	7.7	10.2	12.5	12.6	12.4	14.4	—

[a]Tobias 1991.
[b]Johanson et al. 1987.
[c]Blumenscheine et al. 2003.
[d]Kimbel et al. 1997.
All measurements are labio-lingual or bucco-lingual dimensions in mm. If both sides of a specimen are present, right-side elements are used preferentially. Data from Wood (1991) unless otherwise noted.

TABLE 4.8 ■ Features Supposedly Distinguishing *"Homo"* habilis from *"Homo"* rudolfensis

Feature	*"Homo"* habilis	*"Homo"* rudolfensis
Mean endocranial vol. (cm³)	610	751
Overall cranial vault form	Enlarged occipital contribution to the sagittal arc	Primitive condition
Frontal bone	Incipient supraorbital torus	Torus absent
Parietal bone	Coronal > sagittal chord	Primitive condition
Face—overall	Upper face > midface	Midface > upper face
Nose	Margins sharp and everted; evident nasal sill	Margins less everted; no nasal sill
Malar surface (anterior face of zygomatic)	Vertical, or near vertical	Sloping anteriorly downward
Palate	Foreshortened	Large
Upper teeth	Probably two-rooted premolars	Premolars three-rooted; absolutely and relatively large anterior teeth
Mandibular fossa	Relatively deep	Shallow
Mandibular corpus	Moderate relief on external surface; rounded base	Marked relief on external surface; everted base
Lower teeth	Buccolingually narrowed post-canine crowns; reduced talonid on P4; M3 reduction; mostly single-rooted lower premolars bifid or twin plate-like P3 roots	Broad post-canine crowns; relatively large P4 talonid; no M3 reduction; twin plate-like P4 roots,
Limb proportions	*Australopithecus*-like[a]	uncertain[b]
Forelimb robusticity	*Australopithecus*-like[a]	uncertain[b]
Hand	Mosaic of ape-like and modern human-like features	uncertain[b]
Hindfoot	Retains climbing adaptations	Like later *Homo*[c]
Femur	*Australopithecus*-like	Like later *Homo*[c]

[a]Based on OH 62.
[b]No definite association of appropriate skeletal elements.
[c]Assumes the association of KNM-ER 1472, 1481, and 813 with *"Homo"* rudolfensis.
Source: Adapted from Wood (1992: 786). Specimen assignment as in Table 4.6.

Because the answers to these questions are still unclear (Stringer 1986, Wood 1991, 1992; Tobias 1991, Lieberman et al. 1996, Wolpoff 1999, Wood and Collard 1999), we will refer to these fossils collectively as "Habilines."

DATING AND GEOLOGICAL CONTEXT OF THE HABILINES FROM OLDUVAI, OMO, AND KOOBI FORA

The Bed I deposits at Olduvai contain a series of volcanic tuffs that have been radiometrically dated to a span of time ranging from 1.99 Mya at the bottom (Tuff IA) to 1.75 Mya at the top (Tuff IF). Fossil hominids are found throughout most of this sequence (Egeland et al. 2007). This relatively short span of time encompassed a period of extensive climatic and ecological change following Tuff IC times (1.84 Mya), when the habitat became more open and the climate grew hotter and dryer (Walter et al. 1991). Oldowan tools, and evidence of their use in the processing of carcasses and bones, are common throughout the upper part of Bed I (Leakey 1971b, Toth and Schick 1986). The oldest of the major Habiline specimens, OH 24, is bracketed in age between Tuff IA (1.99 My) and Tuff IB (1.8/1.87 My). OH 7, 8, and 62 come

from higher in the sequence, between Tuff IB and Tuff IC/1D (Blumenscheine et al. 2003). OH 65 is intrusive into Tuff IC and dates to 1.84–1.79 Mya. Later Habilines have been found in the overlying Bed II. OH 16 derives from the Lemuta Member (1.66 Mya), while OH 13 comes from above the Lemuta Member but below Tuff IID (1.48 Mya). OH 13 and 16 are associated with Oldowan tools (White 2000).

The Shungura Formation west of the Omo River in southern Ethiopia has also been carefully and extensively dated (Feibel et al. 1989, Brown 1994). Omo 75-14, a mandible with some cranial fragments, has been attributed to early *Homo*. It derives from below the G4 submember, which is dated to 2.12 Mya (Howell and Coppens 1976, Suwa et al. 1996). G. Suwa and his colleagues argue that Omo 75-14 shares a number of dental features with the larger representatives of early *Homo*, which are conventionally assigned to *H. rudolfensis*. A later specimen of early *Homo* from the Shungura Formation is a gracile partial cranium (L894-1) that Boaz and Howell (1977) liken to known skulls of the smaller Habilines. This specimen dates to slightly over 1.88 Mya (Brown 1994).

By far the largest sample of early *Homo* remains comes from the site of Koobi Fora in northern Kenya (Table 4.6). Early *Homo* fossils are found in three successive divisions of the Koobi Fora Formation: the Upper Burgi Member (2.0–1.88 Mya), the KBS Member (1.88–1.6 Mya), and the Okote Member (1.6–1.39 Mya) (Feibel et al. 1989, Wood 1991). All the late specimens from the Okote Member are generally accepted as *H. erectus*, except for a nondescript piece of lower jaw (KNM-ER 819) that some assign to *Australopithecus boisei* (Wood 1991; Table 4.6). Specimens from the KBS Member include two fragmentary crania (KNM-ER 1478 and 3891), a partial skull with a braincase and upper and lower jaws (KNM-ER 1805), and the partial cranial vault of a juvenile (KNM-ER 1590) associated with some loose teeth. The 1805 and 1590 specimens come from just above the KBS Tuff at the bottom of the KBS Member (1.88 Mya), and are estimated to be 1.85 My old.

The most important cluster of early *Homo* fossils from Koobi Fora derives from just below the KBS Tuff at the top of the oldest of these stratigraphic units, the Upper Burgi Member. They fall into an age range from 1.88 to 1.9 Mya. These finds include:

1. An occipital fragment, KNM-ER 2598, that has been assigned to *Homo erectus* (Wood 1991, Wolpoff 1999).

2. Four important specimens assigned to other species of early *Homo*—the crania KNM-ER 1470, 1813, and 3732, and the partial skeleton ER 3735.

3. Several mandibles—KNM-ER 1482, 1483, 1501, 1502, 1801, 1802, and 3734.

4. A few postcranial remains—KNM-ER 1472, 1481, and 3735.

An even earlier Koobi Fora specimen that figures importantly in arguments about early *Homo* is a virtually complete hipbone, KNM-ER 3228, which Wolpoff (1999) attributes to *H. erectus*. This specimen derives from just above a datable tuff (the Lokalalei Tuff) in the lower part of the Upper Burgi Member, and is estimated to be 1.95 My old.

▌ HABILINE SKULLS

Scientists' initial ideas about the cranial anatomy of the Habilines came from the early discoveries at Olduvai. The parietals of the type specimen of *H. habilis*, OH 7, revealed only two basic facts about the new species: it lacked a sagittal crest, and it appeared to have had an expanded braincase. OH 13 and 16 (Figs. 4.78 and 4.79) are almost as fragmentary and not much more informative. OH 24 is also heavily reconstructed (Fig. 4.77), but it preserves the cranial base and face. In the original description of *H. habilis*, Leakey and his colleagues (Leakey et al. 1964) asserted that *Homo habilis* would have lacked the facial "dishing" seen in robust *Australopithecus*, in which the zygomatic bones jut forward on

either side beyond the nasal region. However, OH 24 turned out to have a dished face (Anonymous 1971). This awkward fact bolstered the contention of some that *Homo habilis* was not clearly separable from *Australopithecus*. Nevertheless, OH 24 differs from any earlier *Australopithecus* in several respects. It has a more distinctly defined supraorbital torus and a deeper mandibular fossa. In basal view, it shows signs of a more flexed basicranium, with a more central foramen magnum and a shorter anterior cranial base than earlier "gracile" types of *Australopithecus* (Fig. 4.83). Of course, the same thing can be said of some robust *Australopithecus* crania.

Two nearly complete crania from Koobi Fora, ER 1470 and 1813, provide a clearer picture of Habiline cranial anatomy. The ER 1470 skull demonstrates unequivocally that some Habilines had bigger brains than any *Australopithecus* (Table 4.5). Otherwise, however, 1470 does not differ much from some *Australopithecus*. The cranial vault looks very much like an enlarged version of *A. africanus*. Viewed from the rear (Fig. 4.84), the 1470 skull shows the "bell curve" silhouette characteristic of *Australopithecus*, reflecting a braincase that is narrow relative to the width of the cranial base. Like *Australopithecus*, 1470 has thin vault bones and lacks the distinct supraorbital torus seen in undisputed early *Homo*.

Walker (1976) has shown that the size and form of the 1470 face are also *Australopithecus*-like. Walker compared 1470's face/cranium index (upper facial height as a percentage of cranial length) with similar figures for various representatives of *Australopithecus* and *Homo*. He found index values ranging from 51 to 64.5 in *Australopithecus*, and from 30 to 45 in *Homo* (including other early members of the genus). The index value for ER 1470 is 59—completely out of the *Homo* range, but squarely in the middle of the *Australopithecus* range and approaching the values seen in robust *Australopithecus*.

The lower part of the face (below the orbits) of the 1470 skull is prognathic as well as tall. The facial breadth of 1470 cannot be measured, because the lateral parts of the zygomatic bones are missing along with the rest of the zygomatic arch (Fig. 4.80A). But what is left of the cheek region shows that the facial surface below the orbit inclines forward as it descends from the orbit's lower rim. The lower edge of the cheekbone intersects the alveolar (tooth-socket) part of the maxilla at a very low position, reflecting the great depth of the lower face (Fig. 4.80B).

The palate of 1470 is also very large, and the preserved roots of the upper cheek teeth indicate that the missing crowns must have been expansive. As B. Wood (1991) notes, the palate is actually larger than those of such robust *Australopithecus* as KNM-ER 405, KNM-ER 406 (Fig. 4.67), and OH 5 (Fig. 4.40A). Wood also points

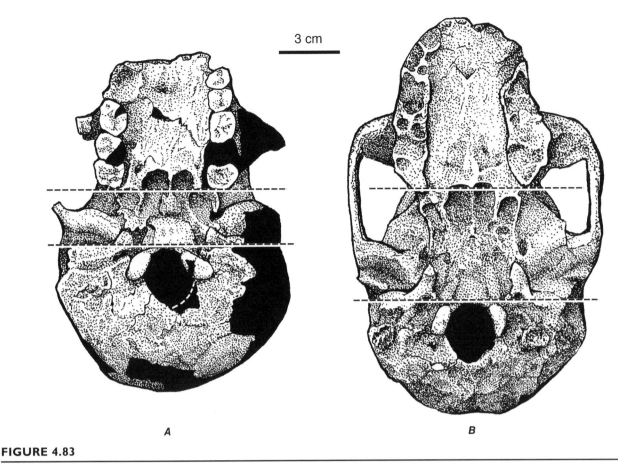

3 cm

FIGURE 4.83

Basal views of OH 24 (**A**) and Sts 5 (**B**). OH 24 has a shorter and more flexed cranial base, resulting in a more anterior position of the foramen magnum relative to the dentition (horizontal dashed lines). However, similar morphology is seen in some other Sterkfontein specimens and in later robust *Australopithecus* (Wolpoff 1999). (After Tobias 1991.)

out that despite its size, the 1470 face differs in some key respects from robust *Australopithecus*: it does not have the "dished" face seen in all three robust species, and it lacks the thickened, stress-bearing "anterior pillars" of bone that usually frame the nasal opening in *A. robustus* and *A. africanus* (Rak 1983).

The KNM-ER 1813 cranium (Fig. 4.81) presents some striking contrasts to 1470. (The comparisons drawn in Table 4.8 are largely based on the differences between these two skulls.) In its cranial form, 1813 is the epitome of what a Habiline should be. Compared to 1470, it has a more distinct supraorbital torus, a broader upper face, a smaller palate with relatively small posterior teeth, a more vertical orientation of the infraorbital (zygomatic) region, and a shorter anterior cranial base (Wood 1991). Although similar in some ways to *A. africanus* (Rak 1983), ER 1813 is far more *Homo*-like in its facial and dental features. It would be an ideal representative of *H. habilis* if it did not have a pea-sized braincase (Table 4.5).

Three other Upper Burgi crania from Koobi Fora deserve special notice. KNM-ER 1590, a partial cranial vault and maxillary dentition from the KBS Formation (Leakey 1973b, 1974; Wood 1991), appears to be the same sort of creature as 1470. Its cranial capacity cannot be estimated accurately, but it appears to have been even larger than that of 1470 (Table 4.5). Its cranial vault bones are at least as large as 1470's, with no indication of cranial cresting. Unfortunately, the face (including the brow region) is almost totally missing, but the cheek teeth are huge (see below)—another similarity to 1470.

A second Upper Burgi specimen, KNM-ER 1805, seems to represent the opposite, small-toothed and small-brained pole of the Habiline spectrum. It comprises a cranial vault (lacking the anterior frontal bone and part of the base), a lower face with most of the teeth left, and a partial mandible (Leakey 1974, Wood 1991). Its cranial capacity is only slightly larger than that of 1813 (Table 4.5). The small brain of 1805 may help to explain its surprising possession of cranial crests. Apart from robust *Australopithecus*, 1805 is the only hominin skull younger than 2 My that exhibits cranial cresting. It has a strong compound temporonuchal crest and a slight

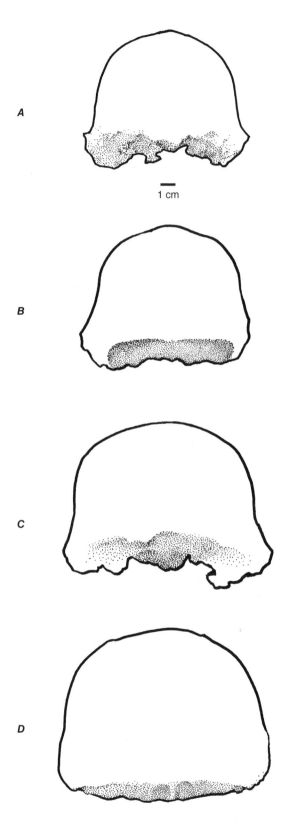

1 cm

FIGURE 4.84

Diagrammatic occipital views of the braincases of (**A**) *Australopithecus africanus* from Makapansgat (MLD 37/38) and three Koobi Fora skulls attributed to early *Homo*: KNM-ER 1813 (**B**) KNM-ER 1470 (**C**), and KNM-ER 3733 (**D**). *Australopithecus* and the two Habilines (**B**, **C**) exhibit a bell-shaped rear silhouette that contrasts with the more convex outline of the Erectine/Ergaster skull (**D**). Compare Figs. 4.31, 4.38, 4.65, and 4.68. (**A**, after Schwartz and Tattersall 2005; **B**, **C**, after Leakey et al. 1978; **D**, after Wood 1991.)

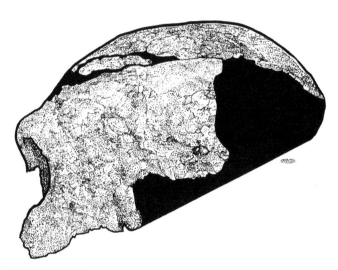

FIGURE 4.85

KNM-ER 3732 (from Koobi Fora, Kenya), lateral view. Note the incipient supraorbital torus and the downward and anterior slope of the anterior face of the zygomatic (malar) region. Compare Figs. 4.80 and 4.81. (After Wood 1991.)

sagittal crest over the posterior end of the braincase. The anterior extent of the sagittal crest cannot be determined because of postmortem damage. Its teeth are comparable in size to those of other Habilines (Table 4.7), which it resembles in other aspects of vault shape and morphology. Various experts have tentatively attributed this enigmatic specimen to just about every possible hominin taxon, including *Homo erectus* (Howell 1978, Wolpoff 1978, 1984), its African variant *H. ergaster* (Groves and Mazak 1975), *Australopithecus boisei* (Tobias 1988), and *Australopithecus* cf. *africanus* (Falk 1983, 1986). Rejecting these possibilities, B. Wood (1991, 1992; Wood and Collard 1999) continues to include 1805 among the Habilines, which seems like the best place for it at present.

A third puzzling find from the Upper Burgi Member, KNM-ER 3732 (Fig. 4.85), is a skull preserving a partial cranial vault, including most of the supraorbital region and the right lateral face (Leakey 1974, Wood 1991). Only a rough estimate of cranial capacity can be made for this fragmentary specimen, but it is thought to be at least 100 cm^3 smaller than 1470 or 1590 (Table 4.5). Like 1813, it has a distinct supraorbital torus—but its infraorbital region slants forward, as in the 1470 skull. We will return to this specimen below.

HABILINE TEETH

In the original description of *Homo habilis*, smaller teeth were one of the features cited as distinguishing the new species from *Australopithecus*. But this claim was dubious even for the type specimen, OH 7. Admittedly, the posterior teeth of OH 7 and many other Habilines are much smaller than those of robust *Australopithecus*, and Habiline cheek teeth tend on average to be somewhat smaller than their counterparts even in *A. africanus*. OH 13 (Fig. 4.86) is very similar to African *Homo erectus* (Ergasters) in molar size. But except for the lower premolars, dental breadths in OH 7 are close to the *A. africanus* means; and the M2 and M3 breadths of another Olduvai Habiline, OH 16, actually exceed those means (Tables 4.2 and 4.7).

Subsequent discoveries of *habilis*-like hominins from Koobi Fora have dispelled the original image of *H. habilis* as having small cheek teeth. Habilines of the 1470 type have some of the largest molars and premolars of all fossil hominins. ER 1470 itself preserves no molar crowns, but the similar-looking ER 1590 skull has molars and premolars that are generally about three standard deviations above the mean for *A. africanus*—and even exceed the mean for *A. boisei* (Table 4.2). Yet despite their size, the posterior teeth of 1590 do not exhibit the crown complexities seen in *A. boisei* (Fig. 4.40). The very large maxillary canine of 1590 is another conspicuous difference from any "robust" *Australopithecus*. ER 1805 exhibits a comparable tooth-size pattern, though its cheek teeth are less massive than those of 1590. Even ER 1813 is not tiny-toothed (Table 4.7).

In short, the dental morphology of Habilines is just as complex and puzzling as their patterns of cranial morphology and capacity. The early claims made about the small teeth of *H. habilis* are hard to defend, even if only the Olduvai specimens are considered. Still, these hominins are clearly not robust *Australopithecus*, and their dental morphology is not specialized enough to rule them out of an ancestral position in the human genus. Habiline dental morphology appears to extend all the way back to AL 666-1 at 2.33 Mya, although the pattern seen in that specimen is not greatly different from that of *A. africanus* (Figs. 4.7 and 4.82; Tables 4.2 and 4.7).

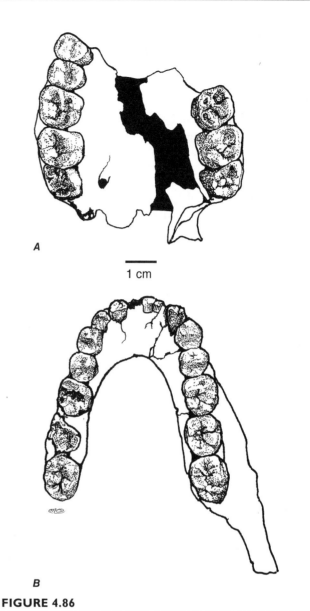

FIGURE 4.86

The maxillary (**A**) and mandibular (**B**) dentition of OH 13. Occlusal views. (After Tobias 1991.)

HABILINE POSTCRANIAL REMAINS

L. S. B. Leakey and his colleagues (Leakey et al. 1964) assigned OH 7 to the genus *Homo* on the basis of the postcranium as well as the skull and teeth. Their original diagnosis of *H. habilis* included features of the OH 7 hand and the OH 8 foot that were thought to be more humanlike than their counterparts in *Australopithecus*. They noted that the OH 7 thumb is well-developed and fully opposable, and concluded that the hand would have been capable of both a power grip (as in grasping a baseball bat) and a precision grip (as in using a screwdriver)—implying the manual dexterity that gives *H. "habilis"* its name. However, the OH 7 hand also retains certain primitive traits not seen in humans, including markedly curved middle phalanges. And although the OH 8 foot is strongly arched and has a robust, adducted hallux, it too retains some primitive, ape-like features, particularly in the shape of the talus (Lewis 1981, Wood 1992).

In 1982, Susman and Stern reassessed the morphology of the OH 7 hand, the OH 8 foot, and the OH 35 partial tibia and fibula. They reaffirmed the humanlike form of the carpometacarpal joint at the base of the thumb in OH 7 and concluded that the OH 8 foot is fundamentally like those of fully bipedal hominins. They based this latter conclusion partly on the humanlike form of the OH 8 calcaneocuboid joint. This joint exhibits a "locking mechanism" that limits supination of the foot (rotation of the sole toward the midline of the body) and stabilizes the foot as a weight-bearing, locomotor organ (Gebo and Schwartz 2006). Monkeys and apes have a more flexible calcaneocuboid joint that allows for greater supination—for example, in grasping the trunk of a tree between the two feet while climbing. Susman and Stern also argued that the lower ends of the OH 35 leg bones have several uniquely human features, and that the tibiotalar joint surface is clearly that of a habitual biped. They concluded that all three of these specimens represent the same taxon, *H. habilis*, and may even have come from the same individual. Although the curved phalanges and other features of the OH 7 hand bones suggested to them that this creature may have retained some climbing ability, its hind limbs were those of a functionally obligate biped.

The same conclusion is implied by several hindlimb bones from Koobi Fora—particularly the ER 813 talus, the ER 1472 and 1481 femora, the ER 1481 partial tibia, and the ER 3228 hipbone. Collectively, this sample constitutes the "Koobi Fora legs." The earliest of these, the ER 3228 hipbone, will be discussed later as a possible early Ergaster (Chapter 5). If it is a Habiline, then Habilines (or one group of them) are essentially identical to Ergasters and Erectines in hip morphology. The ER 813 talus has recently been analyzed by Gebo and Schwartz (2006), who argue that it and the suspected Habiline tali from Olduvai and Omo (OH 8 and Omo 323-76-898) are fundamentally like the modern human talus and different from those of apes or earlier *Australopithecus*.

The Koobi Fora femora (ER 1472 and 1481) are long, humanlike bones with slender shafts, large heads, and short necks (Leakey 1973a, Day et al. 1975, Kennedy 1983b). Their knee-joint surfaces are also human-like in shape and relative size (Fig. 4.20). Like more recent human femora, they have a less sharp bend between the neck and shaft (neck-shaft angle) than the femora of *Australopithecus*. The sharp neck-shaft angle seen in *Australopithecus* is probably a corollary of having relatively short femora that have to slope sharply medially as they run down from widely separated hip sockets to touch each other at the knee. The wider angle seen in

the Koobi Fora femora therefore suggests more human-like proportions of the hindlimb and pelvis. Despite their slender shafts, the ER 1472 and 1481 femora have a relatively thick outer (cortical) layer of bone, like the femora of *Homo erectus* (Kennedy 1983b).

In all of these features, these Koobi Fora limb bones differ from known *Australopithecus* and resemble *Homo*, including the nearly complete adolescent male Ergaster skeleton KNM-WT 15000 from Nariokotome in Kenya (Chapter 5, Fig. 5.18). Taken as a whole, the Koobi Fora legs testify to a pattern of hindlimb anatomy, body proportions, and locomotor behavior that resembles later *Homo* and differs from what we see in *Australopithecus*. A strong argument can be made that whichever early *Homo* group the Koobi Fora legs derive from is the ancestor of later people.

But in dealing with these fossils, we run into the same problem we encountered with the postcrania from Swartkrans: when there are multiple hominin species in the same deposits, we cannot tell which legs go with which skulls and teeth unless we find them associated as parts of the same individual. Unfortunately, there are no hipbones, tali, or complete femora unequivocally associated with diagnostically Habiline cranial remains, and so we cannot say for sure which Koobi Fora postcranials belong to Habilines rather than to an early Ergaster—or even to *Australopithecus boisei*, as we noted earlier in connection with the fragmentary ER 1500 skeleton.

The shakiness of such assumptions was brought home by the discovery of OH 62. This specimen from Olduvai Gorge comprised 302 relatively small fragments, many of which were pieced together to reassemble much of the lower face (maxilla with teeth), the right humerus shaft, the right ulna, and the proximal ends of the left femur and right tibia (Johanson et al. 1987). The maxilla and teeth mark this specimen as a Habiline. The estimated molar breadths (Table 4.7) are well above the mean for Habilines, falling close to OH 16 and ER 1590, but the canine is not significantly bigger than the Habiline average (Table 4.2). The nasal opening has an everted and sharp margin, which is one of the features that distinguishes *Homo* from *Australopithecus*, and the maxilla lacks anterior pillars. D. Johanson and his colleagues noted its strong similarity to Habiline specimens, including both ER 1470 and ER 1813.

The postcranial bones of OH 62 are a different story. They indicate a small hominin, perhaps slightly smaller than "Lucy" (AL 288-1), with long arms and short legs like hers. Johanson and his co-workers estimated that the humerus was about 95% as long as the femur (Johanson et al. 1987)—relatively even longer than Lucy's, and approaching the relative lengths seen in some African apes. The neck-shaft angle is low, as in *Australopithecus*. A more fragmentary Habiline skeleton from Koobi Fora, KNM-ER 3735, appears to have had similar limb proportions (Leakey and Walker 1985).

This similarity of OH 62 to *afarensis* came as a startling surprise, given its late date (1.8 Mya) and the humanlike features of all the other limb bones that have been attributed to Habilines. Naturally, other authors have questioned the conclusions of Johanson and his team. Asfaw et al. (1999), who have inferred more humanlike limb proportions for the BOU-VP-12/1 limb bones associated with *Australopithecus garhi*, are not disposed to accept the evolutionary reversal implied by the estimates for OH 62. They argue that the OH 62 remains are too fragmentary to yield reliable limb-length indices. But Reno and his team (Reno et al. 2005a) say the same thing about the BOU-VP-12/1 skeleton.

Häusler and McHenry (2004) argue that small apes and people have relatively long forelimbs for allometric reasons and that when allometry is taken into account, OH 62 and ER 3735—and even the AL 288-1 "Lucy" skeleton—actually have limb proportions that fall within the range of recent humans (although OH 62 also falls within a zone of overlap between humans and chimpanzees at very small body sizes). They conclude that limb proportions in early *Homo* and "Lucy" are not ape-like when viewed in this light. However, Häusler and McHenry grant that OH 62 apparently had a relatively longer (and therefore more ape-like) ulna, relative to humeral length, than later *Homo*.

For all their imperfections, OH 62 and the even more fragmentary ER 3735 are important because they are the only specimens in which hindlimb bones are directly associated with Habiline cranial remains. The Koobi Fora legs, which lack such associations, appear to represent a creature with a more humanlike kind of bipedal locomotion. If Häusler and McHenry are right, this difference may be wholly due to allometry, and OH 62 and ER 3735 may be mere size variants of the pattern seen in the Koobi Fora legs. But just as in the case of *A. afarensis*, we cannot feel too confident about this conclusion, since it contradicts those that Jungers (1991) drew from a different approach to the allometry of limb length.

◼ ADVANCED *AUSTRALOPITHECUS*: THE FRUSTRATIONS OF VARIATION

As specimens assigned to *Homo habilis* accumulated, it became increasingly clear that there was a lot of variation in individual features within the Habiline sample. In 1986, Stringer (1986a) concluded that there was evidence for as many as three African species of early *Homo*: *Homo habilis* (including ER 1470, ER 1590, OH 7, and OH 24), *Homo* cf. *erectus* (including OH 13, OH 16, and ER 1805), and another species (possibly *H. ergaster*) represented by ER 992 and ER 1813. Although Stringer was not certain about the reality of that third species, he suggested that we should expect a radiation of early *Homo* lineages, evolving in parallel in a human-

like direction at different rates and with emphasis on different features.

In his analysis, Stringer compared the coefficient of variation (CV) of endocranial volume in various groupings of Habiline specimens with CVs for other extant and extinct hominoid samples. He showed that the entire Habiline sample has a CV of 12.4, which is larger than those found in any of the comparative samples he used. In order to get CVs in the range of those from the other samples in Table 4.9, Stringer had to exclude either 1470 (and 1590) or 1813 from the Habiline sample (Fig. 4.87). As Stringer notes, even higher CVs of endocranial

TABLE 4.9 ◼ Coefficients of Variation for Endocranial Volume in Living and Fossil Hominoid Samples

Species	Volume (cm³)	s.d.	N	CV
Gorilla gorilla	522	57	39	10.9
Pan troglodytes	395	35	33	8.9
Pan paniscus	343	33	40	9.5
Homo sapiens (recent)	1475	149	13	10.1
Early *Homo*:				
Ngandong	1151	99	5	8.6
Neandertals	1510	150	6	9.9

Source: Data from Stringer (1986a).

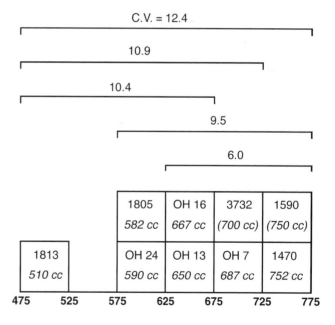

CRANIAL CAPACITY

FIGURE 4.87

Coefficients of variation for various combinations of endocranial volumes of Habilines (revised after Stringer 1986a). Some of these estimates differ from those presented in Table 4.5.

volume—as high as 13.1—have been recorded in other samples of *Gorilla* (Tobias 1971, Miller 2000). Still, it is clear that endocranial volume and other features are highly variable in Habilines. If all these fossils represent a single species, it must have been a markedly dimorphic one—or one in the process of splitting.

In his systematic analysis of the Habilines from Koobi Fora, B. Wood (1991) also concluded that the entire sample was too variable to fit into the single species *Homo habilis*. He proposed a division of the sample into two taxa: *H. habilis* and an unnamed new *Homo* species, for which he later recognized that the name *Homo rudolfensis* (Alexeev 1986) had taxonomic priority (Wood 1992). Wood documented the anatomical patterns differentiating these two early *Homo* species (Table 4.8) and assigned diagnostic specimens to each (Table 4.7). In Wood's schema, all of the Olduvai specimens and many of those from Koobi Fora fall into *H. habilis*, while some 10 Koobi Fora specimens are assigned to *H. rudolfensis*. Kramer et al. (1995) reached similar conclusions from an analysis of craniofacial variation in the Habilines.

In most of the characters tallied by Wood, *H. rudolfensis* seems more *Australopithecus*-like than *H. habilis*. These characters include nasal form, orientation of the malar area, posterior tooth size and form, aspects of vault contours, and absence of a distinct supraorbital torus. A cladistic analysis by Lieberman et al. (1996) also found several derived characteristics shared between the *H. rudolfensis* sample and *Australopithecus*, while the *H. habilis* group emerged as the sister group of the Erectines. The only features that Wood (1991) found more "advanced" (human-like) in *H. rudolfensis* than in *H. habilis* were its larger cranial capacity and certain characteristics of the foot and femur. These postcranial characteristics, however, are questionable, since they depend on an assignment of the Koobi Fora legs (minus the ER 3228 hip) to *H. rudolfensis*. There are no clear associations of *H. rudolfensis* crania with postcranial remains in the fossil record. The only reason for thinking that the more *Homo*-like postcrania from Koobi Fora represent *H. rudolfensis* is the fact that the OH 62 skeleton of *H. habilis* seems to have had a more primitive lower limb. But if Häusler and McHenry are correct (see above), this may be simply an allometric difference; and it is not clear in any case that the Koobi Fora legs represent a Habiline rather than an Ergaster.

Not everyone agrees that variation in the Habilines requires recognition of two species. Tobias (1991) and Wolpoff (1999) believe that the entire Habiline sample can be accommodated in a single species. J. Miller (1991) asserts that the Habiline CV of 12.7 for cranial capacity is no greater than those of *Gorilla gorilla* or modern humans, for which he reports maximal CVs of 13.1 and 13.5, respectively. Miller also argues that the 95% confidence interval for the Habiline CV is so large that no biologically meaningful inferences can be drawn from it. In a later study, Miller (2000) showed that variation in other cranial metrics of Habilines does not conclusively establish the presence of more than one taxon. However, if Habiline variation is not beyond that of other samples, it is always at the extreme of intraspecific variability—as even Miller's data show.

Even if we accept that Habilines need to be divided into two species, drawing the line between them is not a simple job. So long as we confine our attention to the extreme forms ER 1470 and 1813, it is relatively easy to come up with a list of distinctly diagnostic features (Table 4.8). It is unlikely that ER 1470 and 1813 sample the same species, unless that species is more dimorphic than gorillas (Lieberman et al. 1988). But most other Habilines do not fall so neatly into one morph or the other. An excellent example of this is KNM-ER 3732 (Fig. 4.85). This specimen is generally included in the 1470 group. The anterior face of its zygomatic bone slopes forward and downward, as is expected for *H. rudolfensis*. The preserved portion of its vault suggests a cranial capacity on a par with that of ER 1470 (although R. Holloway estimates it at only 600–650 cm³; see Table 4.5). However, 3732 has a distinct supraorbital torus, not an amorphous supraorbital swelling like that of 1470. In this feature, it is more like the 1813 group. Other specimens show similarly ambiguous morphologies, especially OH 24 and the enigmatic ER 1805. The bottom line is that despite the clarity of Table 4.8 and the distinctiveness of ER 1470 and 1813, it is not a simple matter to divide the remaining Habiline specimens into unequivocally defined subsets.

A final complication has been introduced by the discovery of OH 65, a lower face and palate with a complete dentition found in the little-explored western part of Olduvai Gorge and dated to around 1.8 Mya (Blumenscheine et al. 2003). Most of its dental and cranial features—its broad, vertical face, its massive tooth roots, and the contours of its maxilla—resemble their counterparts in ER 1470. It suggests the presence of a second Habiline species, presumably *H. rudolfensis*, at Olduvai. However, its discoverers argue on the basis of the size of the OH 7 parietals that OH 7 belongs to the same species as OH 65 and ER 1470. Since OH 7 is the type specimen of *Homo habilis*, this conclusion implies that the species everybody has been calling *H. rudolfensis* should properly be called *H. habilis*. If so, then a new species name needs to be coined for the small-toothed, small-brained form represented by ER 1813 and most of the other Habilines from Olduvai.

ADVANCED *AUSTRALOPITHECUS*: BACK TO SOUTH AFRICA

Four South African sites—Swartkrans, Gandolin, Drimolen, and Sterkfontein—have yielded fossils that have been attributed to early *Homo*. The non-*Australopithecus* hominins from Member 1 at Swartkrans are probably *Homo erectus* (Chapter 5). While the supposed *Homo* from Drimolen and Gondolin are too few and fragmentary to be assigned to a species, their geological age means they could be Erectines as well. A contrary opinion is offered by Curnoe (2006), whose analysis of molar size and morphology leads him to conclude that these specimens—particularly the SK 15 mandible—generally resemble the Habilines and *A. robustus*, not early Erectines.

The Sterkfontein "*Homo*" specimens, a small series of fragmentary fossils from Members 4 and 5, seem even less likely to represent Erectines. Most Sterkfontein hominins are conventionally assigned to *Australopithecus africanus*, but some investigators think that a few of them show affinities to *Homo*. Stw 151 is a handful of loose teeth and skull bits found together in a Member 4 solution cavity and laboriously pasted together to form the jaws, teeth, and basicranium of a baby hominin that died at about 5 years of age. Its supposed resemblances to *Homo* (Moggi-Cecchi et al. 1998) have not been generally accepted. Sts 19, a cranial base with associated facial fragments from Member 4, has been said to share with *Homo* a suite of 12 basicranial synapomorphies not found in *Australopithecus* (Kimbel and Rak 1993), including a more transverse orientation of the petrous temporal (Fig. 4.71). However, Ahern (1998) finds that 9 of these characters are polymorphic in gorillas, chimpanzees, Habilines, and/or other *A. africanus*. He concludes (as have others) that *A. africanus* is simply a more variable species than many paleoanthropologists appreciate—and that this is true of other extant and extinct hominid species as well.

The most important of these Sterkfontein "*Homo*" specimens is Stw 53 (Fig. 4.77), a heavily reconstructed cranium recovered in 1976 (Hughes and Tobias 1977) from what is apparently a pocket of infill between Members 4 and 5 (Kuman and Clarke 2000). Several researchers, from A. Hughes and P. Tobias to D. Curnoe (2001, 2002; Curnoe and Tobias 2006) have presented detailed arguments for assigning Stw 53 to early *Homo*. If these arguments are sound, this specimen establishes the presence of *H. habilis* in South Africa. Curnoe and Tobias (2005) find that Stw 53 shares 11 derived conditions with Habilines (but only two with Erectines). M. Wolpoff (1999), however, asserts that Stw 53 most closely resembles *A. africanus* specimens from Sterkfontein Member 4 in such features as its shallow mandibular fossa, the form of its mastoid region, its vault shape as seen from the rear, and the presence of anterior pillars in its face. Kuman and Clarke (2000) list five additional reasons for affiliating Stw 53 with *Australopithecus*:

1. The cranial capacity fits in the *Australopithecus* range.

2. The front of the braincase is narrow, like that of *Australopithecus*.

3. The nasal skeleton is flattened, lacking the eversion of *Homo*.

4. The posterior teeth are very large, as in *Australopithecus*.

5. The M^3 is larger than M^2, whereas in *Homo* the reverse is true.

In addition, the supraorbital region of Stw 53 seems less well-defined than those of other specimens of early *Homo* (excepting 1470) and looks much like those of the other *Australopithecus* fossils from Member 4.

ADVANCED *AUSTRALOPITHECUS* OR EARLY *HOMO*? PHYLOGENETIC ISSUES

In 1999, two important assessments of the Habilines agreed, for somewhat different reasons, that Habilines were more correctly placed in *Australopithecus* than in *Homo*. One of these assessments was offered by Wolpoff (1999), who argued that the earliest Erectines occur at the same time (just under 2 Mya) as the first unequivocal Habilines. Therefore, the human lineage does not include the classic Habiline specimens. In Wolpoff's view, the Habilines exhibit features that evolved in parallel with those that identify the *Homo* lineage, but they are not directly involved in its ancestry. Demonstrating that the majority of Habiline features lie within the *Australopithecus* range of variation, Wolpoff refers to the Habilines as "*Homo*-like *Australopithecus*."

The other reassessment of the Habilines, by Wood and Collard (1999), reviewed the definition of the genus *Homo* and concluded that fossil specimens should be assigned to this taxon only if they resemble *Homo* and differ from *Australopithecus* in the following six characteristics:

1. The estimated body mass should approach that of *H. sapiens* more closely than it does that of any *Australopithecus*.

2. The same thing should be true of the body proportions.

3. The locomotor pattern should be a humanlike obligate bipedalism, with limited climbing adaptations.

4. The relative size of the jaws and teeth should be more like those of modern humans than like those of *Australopithecus*.

5. Ontogenetic development should approach the modern human pattern, with extended juvenile periods of growth and development.

TABLE 4.10 ■ **Comparisons of *Homo* (or Purported *Homo*) Species with *Homo sapiens* and *Australopithecus***

Species	Body Size	Body Shape	Locomotion	Jaws/Teeth	Development	Brain Size
H. rudolfensis	—	—	—	A	A	A
H. habilis	A	A	A	A	A	A
H. ergaster	H	H	H	H	H	A
H. erectus	H	—	H	H	—	I
H. heidelbergensis	H	—	H	H	—	A
H. neanderthalensis	H	H	H	H	H	H

H, character state as in *H. sapiens*; A, state more like that in *Australopithecus*; I, intermediate state. A dash indicates that data are not available.
Source: Revised from Wood and Collard (1999, p. 70).

6. Endocranial capacity should be both absolutely and relatively more like those of other members of *Homo* than like those of *Australopithecus*.

The assessments of Wood and Collard are summarized in Table 4.10. Their study differs from Wood's (1991, 1992) previous analyses in that the Koobi Fora legs are not included with *H. rudolfensis*. If they were so included, *H. rudolfensis* would be scored as *Homo*-like ("H") in the characters associated with locomotion and body size. One might quibble with some of the other assessments as well. But the overall pattern is clear: Habilines are more like *Australopithecus* in every character, while even the earliest Erectines (Ergasters) are almost uniformly grouped with *Homo*. Wood and Collard admit that including Habilines in *Australopithecus* would make the craniodental morphology of that genus extremely variable, and would almost certainly make *Australopithecus* paraphyletic. Nevertheless, their analysis leads them to refer the species *habilis* and *rudolfensis* to *Australopithecus* as the most realistic and taxonomically conservative option available.

An additional reason for referring the Habilines to *Australopithecus* is offered by Foley (2001), who notes that the presence of multiple Habiline species in the same time and place "... might be taken as evidence for an essentially australopithecine grade of adaptation." In later phases of human evolution, there is no conclusive evidence for the coexistence of sympatric species of *Homo*. The apparent co-occurrence of four hominin species (*boisei, habilis, rudolfensis,* and *ergaster/*

erectus) at Koobi Fora, and perhaps at Olduvai, intimates that the first three of the four were making a living in a way that did not place them in direct competition with *Homo erectus*. If the Habilines had occupied the high position on the ecological food chain characteristic of later *Homo*, Foley argues, they would have been rarer and less speciose.

If Wood (1991) and Wolpoff (1999) are correct in attributing specimens like the ER 3228 hipbone and the ER 2598 occipital to *Homo erectus* (see Chapter 5), then the first Erectines are just as early as the diagnostic Habilines, and specimens like ER 1470, ER 1813, OH 7, and OH 62 are almost certainly our cousins, not our ancestors. But the oldest fossils that have been attributed to *Homo* (Uraha, AL 666-1, Baringo-Chemeron 1, and the Omo Shungura E-G remains) may well represent the common ancestry both of the Habilines and of the genus *Homo* in the strict sense. The craniodental morphology of these specimens closely approaches that of the larger, more *Australopithecus*-like Habiline, *A. rudolfensis* (Howell and Coppens 1976, Howell et al. 1987, Wood 1991, 1992; Schrenk et al. 1993, Bromage and Schrenk 1995, Suwa et al. 1996). They may mark the transition from something like *Australopithecus garhi* to early *Homo*, as Asfaw et al. (1999) suggest. This conjecture needs to be supported by more and better fossil evidence. For now, the phyletic relationships between *Homo* proper and the various species of *Australopithecus*—including *A. habilis* and *A. rudolfensis*—remain speculative and indeterminate.

CHAPTER FIVE

The Migrating Ape: *Homo erectus* and Human Evolution

▌THE "MUDDLE IN THE MIDDLE"

Australopithecus and its Plio-Pleistocene relatives (*Orrorin, Ardipithecus,* "*Kenyanthropus,*" and the Habilines) are restricted in their distribution. Not only are they geographically limited to Africa, they also seem to have been restricted ecologically to sheltered and protected areas near permanent water sources (Rogers et al. 1994, Cachel and Harris 1995, 1998). Around 1.8 Mya (and probably at least 150,000 years earlier), a radically new human form belonging to the genus *Homo* proper shows up in Africa. This new hominin appears to have spread throughout Africa, except for the dense rain forests and the most inhospitable deserts, by 1.5 Mya. And early on—perhaps as early as 1.8–1.7 Mya—it extended its range beyond Africa into Eurasia, becoming the first of our ancestors to leave the natal continent.

Most of the earliest non-African fossils of this new, differently designed primate have been recovered from tropical and subtropical regions around the eastern end of Asia. These low-latitude areas may have been prime real estate for the new genus because the local ecologies resembled its original home in the African tropics. But the new hominins were not confined to tropical regions. Some of the oldest remains of *Homo* found outside Africa come from Dmanisi in western Asia, at about 42° north latitude. And by around 400,000 years ago (400 Kya), the genus was found in temperate, mid-latitude regions all across Eurasia, up to 35°N (Fig. 5.1). The ability of these migrating apes to adapt to such a wide range of environments was due to some fundamental biological and behavioral changes. These included a marked increase in brain and body size, changes in body proportions, modifications in physiology, a broader and higher-energy dietary regime than that of *Australopithecus* (with more meat and tubers on the menu), and generally more sophisticated technological adaptations.

What triggered the spread of *Homo* out of its African homeland is not yet clear from the archaeological or paleontological evidence. This spread appears to have been more a matter of diffusion than of directed migration, involving many waves of movement out of (and

---

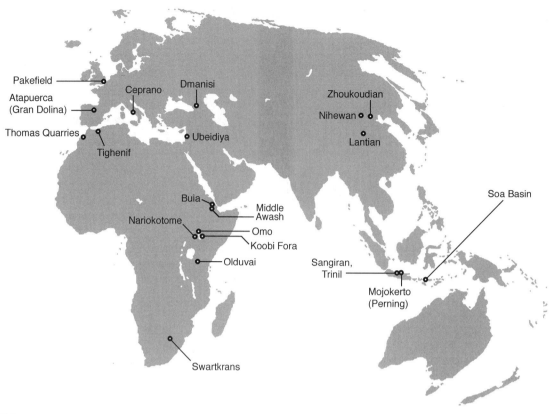

FIGURE 5.1

Map of the major Erectine and Ergaster sites mentioned in the text.

almost certainly back into) Africa. But whatever its causes and details may have been, the "Wanderlust" of these early members of the genus *Homo* marks the onset of that wide-ranging mobility that has typified our genus throughout its history, making humans the most widely distributed of all terrestrial mammals. Beginning with early *Homo*, we humans can be characterized as the migrating ape.

For most of the second half of the 20th century, all of the early humans discussed in this chapter were lumped into a single species, *Homo erectus*. There were few disagreements about the definition of this species. It comprised essentially all "Middle Pleistocene" hominins from Asia, Europe and Africa. The study of these fossils was a relatively uncontroversial aspect of paleoanthropology, providing a welcome oasis of cool consensus between all the heated arguments about *Australopithecus* and the equally heated debate over modern human origins. Most scientists believed that *H. erectus* had evolved from some advanced Habiline and given rise to archaic forms of *H. sapiens* in a unilineal, nonbranching pattern of descent.

Today, however, there is as much debate surrounding the fossils once assigned to *H. erectus* as there is for any other part of human evolutionary history. The debate began to heat up in 1975, when K. Butzer and G. Ll. Isaac

commented on the difficulty of sorting out relationships and affinities in this heterogeneous collection of fossils, and coined the phrase "the muddle in the middle" to describe the Middle Pleistocene span of human evolution. In the same year, C. Groves and V. Mazák established a new species, *Homo ergaster*, for the KNM-ER 992 mandible of early *Homo* from Koobi Fora. By the mid-1980s, other leading experts (Stringer 1984, Andrews 1984, Wood 1984, Groves 1989) had accepted the notion that the earliest (African) representatives of the Erectine group lacked many of the defining characteristics of *H. erectus* and should be placed in a different species. The name *H. ergaster* is now widely applied to these early Erectines from Africa, and sometimes to the Dmanisi specimens from Georgia as well (Gabunia et al. 2000b; but cf. Gabounia [Gabunia] et al. 2002). As we will see later (Chapter 6), many authors now assign the earliest European fossils of *Homo* to yet another species, *Homo heidelbergensis* (Tattersall 1986). Thus, a set of specimens that were all previously lumped into *H. erectus* by near-universal consent is now commonly partitioned into three different species of *Homo*, and there is constant lobbying to recognize even more (Bermúdez de Castro et al. 1997, Tattersall and Schwartz 2001, Gabounia et al. 2002). On the other end of the spectrum of opinions, Wolpoff and his colleagues argue that even the species *H. erectus* is one species too many, and contend that all these early humans should be formally sunk into our own species, *H. sapiens* (Wolpoff et al. 1994a, Wolpoff 1999).

It may seem strange that learned and sensible people can look at the same collection of fossils and see such radically different things. But these opposed positions reflect different views of the evolutionary process and the nature of species in evolutionary history. Those who want to split *H. erectus* into multiple species tend to see evolution as punctuated, with evolutionary change being concentrated in "genetic revolutions" surrounding speciation events (see Chapter 2). They accordingly want to recognize lots of human species, to provide for all the speciation events needed to produce the big evolutionary changes that occurred in the human lineage during the Pleistocene. Adherents of the single-taxon idea, on the other hand, see no evidence of any sharp discontinuities in the fossil record of the genus *Homo*. They think that humans evolved gradualistically as a single biological species united by continuing gene flow, from the emergence of the genus *Homo* down to the present. And if human populations have always constituted a unitary web of interconnecting lineages, any division of this web into successive chronospecies must be arbitrary. It follows that all living and fossil humans should be lumped into a single species, *Homo sapiens*.

Most of the rest of this book will be devoted in one way or another to trying to resolve this issue, with reference to the facts of paleontology and the propositions of evolutionary theory. Although we will put forward our own pet theories about where the truth lies on this spectrum of opinion, we encourage you to study the fossil evidence and other information discussed in this and the following chapters, and to try to judge for yourself which pattern best explains the data. This decision is not as easy as some partisans would have you think.

To avoid imposing our own preconceived notions on the reader, we will avoid formal taxonomic designations and use the term "Erectines" to refer to all fossil hominins that could be included in *H. erectus* in the broad sense. We adopt this convention simply as a heuristic means of referring to the appropriate specimens. In the same vein, we will use "Ergasters" to refer to early Erectines (~1.9–1.5 Mya), with no implications for alpha taxonomy. In what follows, the terms "early Erectine" and "Ergaster" should be considered synonymous. We will follow a similar procedure in subsequent chapters as well.

A BRIEF HISTORY OF *HOMO ERECTUS*: 1889–1950

The Indonesian island of Java figures importantly in the history of paleoanthropology, because it was there that the first specimens of *H. erectus* were discovered. As the first non-modern human remains to be found outside of Europe, they had a fundamental impact on the history of ideas about human evolution.

Several early Darwinians had speculated that Asia might have been the place where humans evolved from their ape ancestors. The eminent German biologist E. Haeckel, who was never one to shy away from speculation, even coined a taxonomic name for the imaginary "missing link" that he surmised would be found to connect humans with apes—particularly with the gibbon, which Haeckel thought shared important features with humans. Haeckel (1866) called his hypothetical ape-human form "*Pithecanthropus*" (ape–man), and later (1868) added the species nomen "*alalus*" (without speech).

Haeckel's speculations about *Pithecanthropus alalus* and human origins in Asia fired the imagination of a young anatomist and physician at the University of Amsterdam, Eugen (or Eugène) Dubois. Dubois was so eager to find Asian fossils pertaining to human evolution that he joined the Dutch army for an eight-year term as a medical surgeon in order to be posted to the Dutch East Indies (Indonesia), where he hoped to find Haeckel's missing link. Amazingly, he did just that (Theunissen 1988, 1997; Howell 1994a, Shipman 2001).

Dubois arrived in Indonesia in 1887. In 1889, his searches on Java turned up an early human skull similar to one (Wadjak I) that had been found nearby a year earlier. But as Dubois recognized, both these skulls were robust but not particularly primitive (Dubois 1922). They were not the Haeckelian missing link he had come to Indonesia to find.

Later that year, Dubois began to work in central Java, concentrating on exposures of Pleistocene sediments along the Solo River. Indonesian workers in his employ soon found a fragmentary mandible at the site of Kendung Brubus. This was probably the first find of *H. erectus*, though it was not immediately recognized as such (Tobias 1966). In 1891, his workers recovered a simian-looking calotte with a shelf-like supraorbital torus, receding frontal, and sharply angled occipital bone, together with three teeth, from deposits near the village of Trinil (Fig. 5.2). Dubois initially thought these might be the remains of a fossil chimpanzee (Dunsworth and Walker 2002). But in 1892, he turned up a left femur in what he thought were the same deposits. This femur was essentially identical to a modern human's. Dubois put the primitive, apelike cranium (Trinil 2) and the modern femur (Trinil 3) together and identified them as "Java Man," a transitional form between apes and humans (Dubois 1894). With a nod to Haeckel, Dubois named this composite creature *Pithecanthropus erectus*, meaning "erect ape-man."

Over the course of the following 8 years, four more fragmentary femora were recovered from Trinil

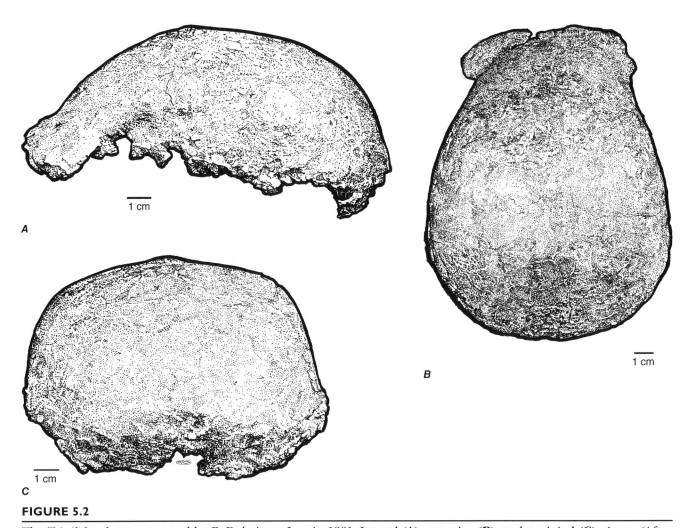

FIGURE 5.2

The Trinil 2 calotte recovered by E. Dubois on Java in 1891. Lateral (**A**), superior (**B**), and occipital (**C**) views. (After Schwartz and Tattersall 2003.)

deposits and assigned to *Pithecanthropus*. Dubois returned in 1895 to Europe, where he became embroiled in lively debates about the validity of *Pithecanthropus* (Bowler 1986, Gould 1990). Some textbook accounts claim that Dubois was so disheartened by the negative response to his ideas that he hid the fossils away from other scientists and retreated into scientific isolation until after World War I. But this story is a myth (Theunissen 1988, 1997). Dubois refused access to his fossils more as an act of defiance than of defeat. During his withdrawal from the *Pithecanthropus* debate, he embarked on a detailed study of brain evolution in mammals. His studies convinced him that the modern human brain evolved in two saltatorial "jumps", from ape to *Pithecanthropus* and then from *Pithecanthropus* to modern human, doubling its neurons at each jump (Dubois 1899a,b).

Dubois' ideas about brain evolution ultimately led him to describe *Pithecanthropus* as a sort of giant gibbon (Dubois 1935). Some have interpreted this as an admission that *Pithecanthropus* was not a human ancestor after all. But Dubois drew the gibbon analogy primarily to emphasize how *Pithecanthropus* fitted perfectly into Haeckel's model as an intermediate step in human brain evolution from a gibbon ancestor, confirming it as the "missing link" between apes and modern people. As Gould (1990: 22) put it: "... Dubois used the proportions of a gibbon to give *Pithecanthropus* a brain at exactly half our level, thereby rendering his man of Java, the pride of his career, a direct ancestor to all modern humans. He argued about gibbons to exalt *Pithecanthropus*, not to demote the greatest discovery of his life."

For his discoveries in Java and his broad, innovative approach to their analysis, Dubois deserves to be recognized as one of the founders of scientific paleoanthropology (Shipman and Storm 2002). But ironically, his correct insights about the Erectine morphological pattern were probably based in part on a mistaken association between fossils of very different ages. There is increasing suspicion that the Trinil femora belong to a more recent population than the Trinil calotte, because the femora do not show the characteristic differences from modern human femora that are seen in undoubted Erectines (Kennedy 1983a, Day 1995). Bartstra (1983) suggests that Dubois failed to recognize that late Pleistocene sediments were also present at the Trinil locality and thus inadvertently commingled femora from that period with the early Pleistocene cranial remains of "Pithecanthropus."

The next significant discovery to play a role in the *Homo erectus* story was a complete mandible found at the village of Mauer near Heidelberg, Germany in 1907 (Fig. 6.11). Named *Homo heidelbergensis* by Schoetensack (1908), this massive, chinless jaw was usually included in *H. erectus* throughout most of the last century, and was often cited as a typical example of a *H. erectus* mandible. But as we will see later (Chapter 6), many experts now prefer to assign the Mauer mandible to the separate species originally created for it.

Like Dubois, the Canadian anatomist Davidson Black became convinced by his readings that Asia had been the cradle and nursery of humankind, and he set off for China to search for proof and to serve as professor of anatomy at the Peking Union Medical School (Black 1933). In 1927, excavations under his direction recovered a human molar from a site known as Longgushan (Dragon Bone Hill) near the village of Zhoukoudian. Convinced that this and other teeth traced to the site were evidence for early humans in China, Black erected the taxon "*Sinanthropus pekinensis*" for these dental remains. Further excavation under the field direction of W. Pei turned up two partial mandibles, a calotte, and other cranial fragments in 1928 and 1929. By the end of excavations in 1937, some 45 cranial, postcranial, and mandibular remains, as well as 147 isolated teeth, had been recovered from what was designated Locality 1 at Zhoukoudian (Figs. 5.3 and 5.4).

Black died in 1933 and was replaced in 1935 by Franz Weidenreich, who was charged with analyzing the human remains from Zhoukoudian. Under the threat of a looming Japanese invasion, Weidenreich left China and moved to the American Museum of Natural History in New York, where he produced a series of monographs on the Zhoukoudian (ZKD) material (Weidenreich 1936a,b, 1937, 1941, 1943). Lamentably, the Zhoukoudian fossils were no longer available for direct study, because they were lost at the outset of World War II. Given to the American Embassy in Beijing for safekeeping, they had been evacuated to the port serving Beijing on December 7, 1941—the day of the Japanese bombing of Pearl Harbor. The Japanese Army promptly captured the fossils and their guardian contingent of Marines from the U.S. embassy, and the bones disappeared. There are a lot of conjectures about what happened to them (Shapiro 1974, Jia and Huang 1990), but

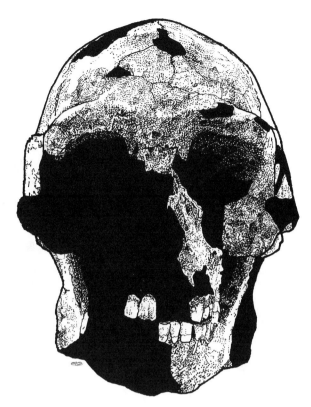

FIGURE 5.3

Reconstruction of the L-3 Zhoukoudian (China) Erectine skull by I. Tattersall and G. Sawyer (1996), anterior view. Note the sagittal keeling and the relatively horizontal orientation of the lateral part of the zygomaticoalveolar margin (ZAM). (After Johanson and Edgar 1996.)

the most likely account is that of the colonel in charge of the Beijing Marine detachment, who thought that they were simply thrown away by Japanese soldiers (Moore 1953). Although Weidenreich's monographs were based not on the ZKD fossils themselves, but on the notes, photographs, drawings, and casts he had made in China and brought with him to the United States, his masterful studies have made them the most thoroughly documented Erectine sample known.

While Zhoukoudian was yielding its treasures, more fossils were emerging from Java. Between 1931 and 1933, several *erectus*-like fossil skulls were recovered from Ngandong, another site on the Solo River. Originally viewed as representing a more "advanced," humanlike form than "*Pithecanthropus*," the Ngandong fossils are now regarded by some as a late-surviving population of *H. erectus* (see Chapter 6). In 1936, von Koenigswald reported the discovery of a juvenile Erectine skull (Perning 1) from Modjokerto in central Java. This find was followed between 1937 and 1941 by the discovery of three Erectine braincases (Pithecanthropus II–IV) and several jaws and teeth from the Sangiran Dome, also in central Java (von Koenigswald 1940, 1975).

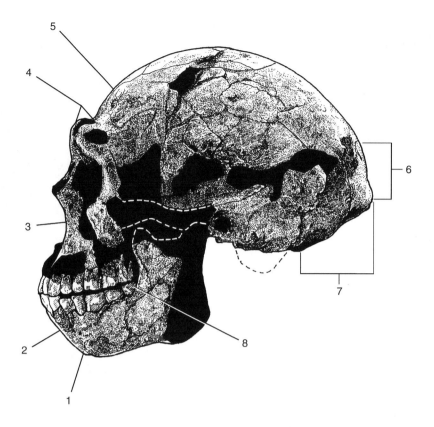

FIGURE 5.4

Reconstruction of the L-3 Zhoukoudian (China) Erectine skull by I. Tattersall and G. Sawyer (1996), lateral view. This reconstruction exhibits many characteristic features of Erectine skulls, including: (1) absence of a chin or mental eminence; (2) receding mandibular symphysis; (3) facial prognathism and an everted nasal margin, the latter feature suggesting a pronounced external nasal structure (Franciscus and Trinkaus 1988); (4) well-developed supraorbital torus and supratoral sulcus; (5) low, receding frontal squama (forehead) with a frontal boss; (6) relatively short occipital plane and prominent nuchal torus; (7) relatively long nuchal plane; and (8) absence of a retromolar space between the distal M3 and the anterior margin of the ramus. (After Johanson and Edgar 1996.)

Before being interned by the Japanese in Indonesia, von Koenigswald brought his Javan fossils to China to compare them with the Zhoukoudian material. Both he and Weidenreich agreed that the early human remains from China and Java were fundamentally similar and probably should be thought of as "races" of the same species (von Koenigswald and Weidenreich 1939). Nevertheless, the generic names "*Pithecanthropus*" and "*Sinanthropus*" continued to be used as distinctive labels for the Indonesian and Chinese material, respectively—for example, in Weidenreich's ZKD monographs.

This practice fell into disrepute during the 1950s following the Cold Springs Harbor Symposium on Human Evolution in 1950, a historic meeting that had a transforming impact on the study of human evolution. At that symposium, Ernst Mayr (1950) endorsed the views of Weidenreich and von Koenigswald, arguing that the genera "*Pithecanthropus*" and "*Sinanthropus*" should be sunk into the genus *Homo*. Krogman (1950) proposed that all Middle and Late Pleistocene humans should be placed in either *H. erectus* or *H. sapiens*. In

the aftermath of the Cold Springs Harbor meeting, lumping came to replace splitting as the normative practice in anthropological systematics. By the 1960s, Krogman's broad definition of *H. erectus* was almost universally accepted (Howell 1960, Howells 1966, Le Gros Clark 1964, Simpson 1963), and it continued to be so into the 1980s (Howells 1980, Wolpoff 1980b).

LATER DISCOVERIES IN AFRICA AND EURASIA

The first Erectine-like form discovered in Africa was the Kanam jaw, found in Kenya in 1932 (Leakey 1935). Because the specimen had small anterior teeth and seemed to have a protuberant bony chin, L. S. B. Leakey offered it in support of his recurring claims about an ancient lineage of modern human ancestors in Africa (Chapters 4 and 6). However, subsequent analysis has shown that the specimen is much younger than the Lower Pleistocene age attributed to it by Leakey and that its supposed "chin" is a bony tumor (Tobias 1960, 1962; Oakley 1975).

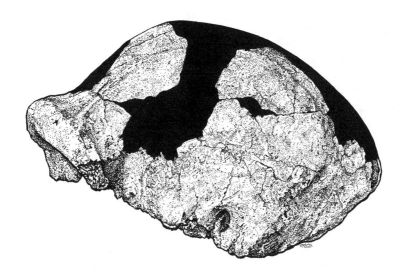

FIGURE 5.5

The OH 9 ("Chellean Man") cranium from Upper Bed II, Olduvai Gorge (Tanzania). (After Delson 1985.)

The next Erectine-like specimen to be found in Africa was the "Telanthropus" mandible, SK 15, recovered at Swartkrans in 1949. This mandible and later "Telanthropus" finds (Clarke 1994a) are often included with the Erectines (Walker 1981, Rightmire 1990, Clarke 1994a,b; Klein 1999, McCullom et al. 1993, Antón 2003), although not everyone agrees with this assessment (Wolpoff 1968a, 1970b; Grine et al. 1993, Curnoe 2001, 2006).

In 1954, evidence of Erectines surfaced for the first time in North Africa, when Arambourg (1954, 1955) excavated three mandibles and a juvenile parietal at Ternifine (now Tighenif) in Algeria. Although he recognized the strong similarity of the remains to those of the Asian Erectines, Arambourg nonetheless created a new taxon, "*Atlanthropus mauritanicus*," for these ancient North Africans. A year later, he placed additional finds from Sidi Abderrahmen (Morocco) in his new taxon (Arambourg and Biberson 1956).

Unmistakable Erectines began to turn up in East Africa in the 1960s. In 1961, L. S. B. Leakey described an *erectus*-like cranium (OH 9) from site LLK II at Olduvai Gorge (Fig. 5.5). Characteristically, Leakey argued that this "Chellean Man" was more advanced and humanlike than other Erectines. But subsequent studies by Rightmire (1979a, 1984a 1990) and Day (1971, 1995) have shown that OH 9 and a series of other remains from Beds II and IV at Olduvai, including the OH 28 hipbone and femur (M. Leakey 1971a), fit comfortably into *H. erectus*.

In the 1970s, the Koobi Fora Research Project began to find fossils that could be included in the Erectine group. The first discoveries were mandibles, including KNM-ER 992 (Leakey and Wood 1973); but the field season of 1974–1975 yielded the cranium KNM-ER 3733 and the hipbone KNM-ER 3228 (Leakey 1976, Leakey and Walker 1976). These two specimens are extremely important. ER 3733 is the key cranial specimen on

which the concept of *H. ergaster* is based, while ER 3228 is the oldest known specimen (1.95 Mya) that can plausibly be called Erectine. Later Erectines from Koobi Fora include the ER 3883 cranial vault, the ER 1808 partial skeleton, and several less complete specimens (Dunsworth and Walker 2002). Finally, at the site of Nariokotome III on the western shore of Lake Turkana, the most complete Erectine skeleton known was discovered in 1984 (Brown et al. 1985). Approximately 85% complete, the Nariokotome skeleton (KNM-WT 15000) has provided a wealth of information about Erectine paleobiology (Walker and Leakey 1993a). Three more recently discovered cranial fossils from Africa also appear to have Erectine affinities: the Buia cranium from Eritrea (Abbate et al. 1998), the Daka skull from the Middle Awash of Ethiopia (Asfaw et al. 2002), the Olorgesailie frontal and temporal from Kenya (Potts et al. 2004), and the small KNM-ER 42700 calvaria (Spoor et al. 2007b).

Between 1950 and 2000, more Erectine specimens were recovered from China and Java. The new Chinese finds include a few fragments from Zhoukoudian (Woo and Chao 1959, Wu 1985), the Gongwangling skull from Lantian (Woo 1966), and a series of archaic human skulls from Hexian and Yunxian that could be described as late Erectines (Chapter 6). Excavations on Java have recovered a number of Erectine remains from Sambungmacan, in the central part of the island, as well as the Sangiran 17 or "Pithecanthropus VIII" cranium, which is the most complete skull we have of an Asian Erectine (Sartono 1971, 1975).

New Erectine-like specimens were also found in Europe and western Asia during the second half of the 20th century. The earliest European finds were cranial fragments from the sites of Vértesszöllös in Hungary (Thoma 1969) and Bilzingsleben in Germany (Vlček 1978, Mania and Vlček 1987). The Erectine affinities of these enigmatic bits and pieces were long debated, but a nearly complete skull from Ceprano in Italy has since provided more convincing evidence for the presence of *H. erectus* in Europe (Ascenzi et al. 1996, 2000; Clarke 2000). Some fragmentary Erectine-like fossils from Spain, the TD6 and TE9 remains from Atapuerca, have been assigned to yet another species, "*Homo antecessor*" (Bermúdez de Castro et al. 1997, 1999a; Carbonell et al. 2008). This supposed species, proposed as the common ancestor of Neandertals and modern humans, has not been widely accepted. Burial dating, paleomagnetism, and faunal correlations all suggest an age of 1.2-1.1 My for the oldest of these Spanish fossils.

Considerably older Erectine fossils come from "the gates of Europe" (Dean and Delson, 1995), at the site of Dmanisi in Asian Georgia on the southern side of the Caucasus. In 1991, a human mandible was found there along with Oldowan-like tools and extinct fauna, and assigned a tentative date of 1.7 to 1.8 Mya (Gabunia and Vekua 1995). It was initially argued that this specimen's closest morphological affinities were with later Erectines, rather than earlier hominins (Bräuer and Schultz, 1996). But subsequent work at the site has uncovered more mandibles and four skulls and postcrania, which resemble the early Erectine material from Koobi Fora and Nariokotome in many ways but are more primitive in others (Gabunia et al. 2000b, Vekua et al. 2002, Rightmire et al. 2006, Lordkipanidze et al. 2007). The surprisingly early dates proposed for these fossils imply that Erectines spread out of Africa very soon after their emergence, and intimate that we may have a lot to learn about the initial human colonization of the European subcontinent.

ERECTINE CHRONOLOGY AND GEOGRAPHIC DISTRIBUTION

Although the oldest Asian Erectines may be nearly as ancient as the earliest ones from Africa, few experts doubt that the genus *Homo* originated in Africa. All the earlier hominins were restricted to Africa, and our evolutionary connection to the African great apes is unmistakable. As we might expect, the oldest Erectine fossils known come from East Africa, though the picture is muddled by disagreements about taxonomy and dating.

The dating of the East African sites is basically secure, because it combines evidence from biostratigraphy, paleomagnetic correlation, and radioisotopic dating of volcanic tuffs that can be correlated from site to site by chemical fingerprinting (Haileab and Brown 1992, Feibel et al. 1989, Brown et al. 1985, Brown and McDougall 1993, Brown 1994). These dating techniques have established that the oldest East African Erectines are those from Koobi Fora. However, there is some question about exactly which specimens are Erectines and which

are Habilines (Spoor et al. 2007b). Undoubted early Erectines include the ER 3733 skull, at 1.78 Mya, and a partial skeleton and fragmentary skull (ER 1808 and 730) at about 1.7 Mya. An occipital bone (KNM-ER 2598) dated to 1.9 Mya may represent an earlier Erectine. Even earlier is the ER 3228 hipbone, which is about 1.95 My old. Although this specimen's morphology is similar to those of such later *H. erectus* specimens as OH 28, it might perhaps belong to a Habiline—maybe to *Australopithecus rudolfensis* (Ch. 4). If ER 3228 really is an Erectine, it pushes the age of this group back to almost 2 My. The Nariokotome III skeleton (WT 15000) is about 1.6 My old, and the ER 3883 cranial vault dates to 1.58 Mya. The ER 992 mandible and some other fragmentary Erectine remains from Koobi Fora are between 1.5 and 1.6 My old, and the ER 42700 cranial vault has an estimated geological age of 1.55 My (Spoor et al. 2007b). South African dates are less secure, but faunal correlations point to an age of 1.8-1.6 My for the possible Erectine material from Member 1 at Swartkrans (Vrba 1985, 1995). The fossils from Swartkrans Members 2 and 3 are only slightly younger.

The Erectine remains from Dmanisi (Georgia) and the Perning 1 (Modjokerto) juvenile skull from Java have both been assigned dates of ~1.8 Mya (Gabunia et al. 1999, 2000b; Swisher et al. 1994), which would make them as old as the oldest undoubted African Erectines. However, there may be problems with these dates (see below). Two other specimens from Java, the crushed skulls Sangiran 27 and 31, are dated by Swisher and his colleagues to 1.66 Mya.

The next oldest Erectines are again from Africa. At Olduvai Gorge, OH 9 (Fig. 5.5) is now correlated to the top of Bed II, at about 1.4 Mya (Leakey and Hay 1982). From Ethiopia, a fragmentary cranium from Omo K (P996-117) and a relatively complete mandible from Konso (KGA 10-1) have been dated to about the same age as OH 9 (Howell 1978, Asfaw et al. 1992). Slightly later are a partial skull from Buia in the Danakil (Afar) Depression of Eritrea (Abbate et al. 1998), a complete calvaria from the Dakanihylo (Daka) Formation in the Middle Awash of Ethiopia (Asfaw et al. 2002), and frontal and temporal bones from Olorgesailie, Kenya (Potts et al. 2004). These fossils have been assigned dates of about 1 Mya on the basis of paleomagnetic correlations to the Jaramillo Subchron (Buia) and the later Matuyama Reversed Chron (Daka), and association with dated volcanic tuffs (Olorgesailie).

The dating of the Erectine specimens from Indonesia is particularly problematic. None of the major specimens were actually excavated *in situ*, but rather were collected and brought in by locals. Therefore, their precise stratigraphic positions are unknown—except for a single tooth, excavated by Kramer et al. (2005). And although Java has a lot of datable volcanic tuffs, particularly at Sangiran, these have been extensively

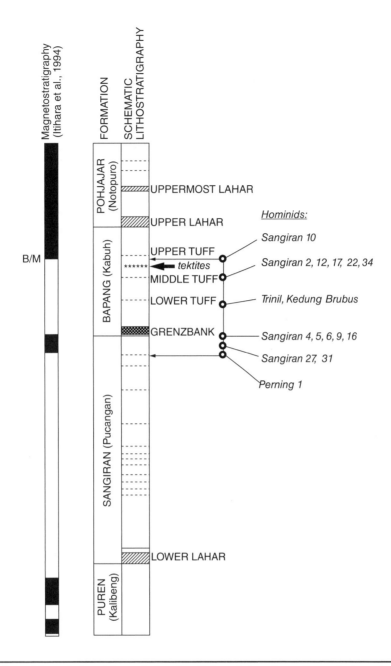

FIGURE 5.6

Schematic of the stratigraphic relationships of the fossil hominins from Java, Indonesia. (After Langbroek and Roebroeks 2000.)

reworked by erosion and deposition, with the result that segments of older tuffs are often redeposited in younger layers. De Vos and Sondaar (1994, De Vos et al. 1994) point out this latter problem with respect to the early dates reported by Swisher and colleagues at Perning and Sangiran, which are contradicted by both biostratigraphic and paleomagnetic data. Tektites found at several horizons between the Middle and Upper Tuffs of the Kabuh Formation have been reliably dated to between 800 and 700 Ky (Langbroek and Roebroeks 2000). These dates, along with the paleomagnetic evidence (Itihara et al. 1994), support the bracketing of all

the Sangiran fossils between the onset of the Jaramillo (normal) Subchron at 1.1 Mya and the Brunhes/Matuyama boundary at 780 Kya (Fig. 5.6). This bracket would encompass the human remains from both the Bapang (Kabuh) Formation and the older Sangiran (Pucangan) Formation. Hyodo et al. (2002) support this dating bracket on the basis of magnetostratigraphic correlations between Indonesian and Chinese sites. If this chronology is correct, the earliest humans in Java would be about 1 My old rather than 1.8 My.

Huffman and colleagues have used new archival materials and fieldwork to locate the site where the

Perning 1 juvenile calvaria was recovered in 1936 (Huffman 2001, Huffman et al. 2006). On the basis of their assessment of stratigraphy and taphonomy, Huffman et al. (2006) demonstrate that the skull was found ~20 m above the level from which Swisher et al. (1994) obtained the 1.8-My date. This would support a younger date for the Perning specimen. The equally controversial dating of the Ngandong sample, which some include in *H. erectus*, is discussed in Chapter 6.

Claims have also been made for the early appearance of humans in China. Longgupo cave in south-central China has yielded a fragmentary mandible (containing a first molar and fourth premolar) and an isolated upper lateral incisor, which Huang et al. (1995) claim are human (cf. Larick and Ciochon 1996). On the basis of paleomagnetic correlations, they assign an extremely early date—around 1.9 Mya—to these specimens. However, others contend that the Longgupo mandible is that of an ape, very similar to *Lufengpithecus* in corpus and dental form (Etler 1996, Wolpoff 1999), and that the context of the incisor is uncertain. As things stand, Longgupo does not provide compelling evidence for so early a human presence in China (Schwartz et al. 1996). Suggestions that human molars from the imprecisely dated sites of Jianshi and Badong represent Chinese australopithecines are discounted by Zhang (1984, 1985b), who notes that they are morphologically most similar to Indonesian Erectine teeth. The Jianshi molars were found in association with *Gigan-*

topithecus teeth (Gao 1975), suggesting a very late survival of these huge primates in southern China.

The earliest firm indications of human presence in China come from stone tools found in the Nihewan Basin of north-central China, which are dated to about 1.36 Mya (Zhu et al. 2001, Schick et al. 1991). A cranium recovered from the Gongwangling site near the town of Lantian (Fig. 5.1) is associated with similar stone tools and may be of comparable age. The paleomagnetic evidence at Gongwangling is equivocal, allowing for dates of either 1.15 My (An and Ho 1989, Hyodo et al. 2002) or 780 Ky. The Chenjiawo mandible, from another site near Lantian, is younger—perhaps 600 Ky old. The continued presence of Erectines in Africa during this time span is demonstrated by remains from Tighenif (Ternifine), at about 700 Kya (Geraads et al. 1986), and from Beds III and IV at Olduvai Gorge (including OH 12 and 28), conventionally dated to between 780 and 620 Kya (Leakey and Hay 1982). However, a recent revised paleomagnetic sequence for Olduvai Gorge by Tamrat et al. (1995) would put these specimens at about 1.1 Mya.

Zhoukoudian (ZKD) is still the best-known and most productive Erectine site in China. Located about 50 km southwest of Beijing, the site consists of a number of ancient caves. The deposits in the "Lower Cavern" at Locality 1, which contain several concentrations of human and animal bones and archaeological remains, are over 40 m thick and divided into 17 layers. Three layers with high concentrations of stone tools were designated as "cultural layers" (Fig. 5.7). The oldest human

Layer	Climatic cycles	Human remains	TIMS date/ polarity	Oxygen isotope stage	ESR dates
1	1 cold			10	
2	warm		400± 8	11	
3	2 cold	Locus H 3		12	282±45
					312-386
4	warm	Upper CL		13	
5			>600		
6	3 cold			14	
7	warm			15	
8-9		Loci L 1-3			418±48
	4 cold	Lower CL		16	
10	warm	Bottom CL		17	
11			Locus E 1		578±66
12					669±84
13	5 cold			18	
14	warm		Brunhes/ Matuyama boundary (780 Ky)		

FIGURE 5.7

Stratigraphy and dating of the "Lower Cavern," Locality 1 at Zhoukoudian, China. The TIMS date for layer 2 is from Shen et al. (2001), and the ESR dates are from Huang et al. (1991) and Grün et al. (1997). CL, cultural layer. (Adapted from Zhou et al. 2000.)

fossils come from Layer 11 and the most recent ones from Layer 3, but the vast majority comes from Layers 8–10.

The dating of the ZKD sequence continues to be somewhat uncertain. There are two competing chronologies. The first, based on electron spin resonance (ESR) dating, brackets the sequence from Layers 2 through 12 between 280 Ky and 669 Ky, with the most important fossil layers (8–10) falling between 418 Kya and 578 Kya (Huang et al. 1991, Grün et al. 1997). The second chronology, proposed by Shen et al. (2001), is based on a thermal-ionization mass-spectrometry (TIMS) date of 400 ± 8 Ky for Layer 2—which is more than 100 Ky older than the ESR age estimate for this layer. Shen and colleagues also report a mean age estimate for Layer 5 of more than 600 Ky, although this figure is questionable because of high error factors in the individual dates. If the TIMS chronology is accurate, then most of the ZKD Erectine remains are bracketed between 600 Ky and the Brunhes/Matuyama boundary at 780 Ky.

Zhou et al. (2000) have recently estimated the paleoclimatic cycles implied by several aspects of the ZKD cave sediments (e.g., fossil mammals, spores and pollen, trace elements) and correlated them with the $\delta^{18}O$ curve of deep-sea cores and the Chinese loess sequence. Their results support the older (TIMS) age for Layers 2–10 (Fig. 5.7). The implied time span of 250 Ky for the sequence of hominin-bearing strata at the site conforms to G. Pope's (1988) inference of an approximately 200-Ky span on the basis of the fossil fauna. However, these chronologies conflict with the suggestion that the hominin deposits accumulated during a single interstadial (warm period) (Aigner 1986). At this point, it is not certain which of these schemes is closer to the truth.

In Africa, China, and Indonesia, there are additional fossil remains that could well be classified as *H. erectus*. These include the Hexian, Nanjing, and Yunxian finds (all from China); Sambungmacan, Hanoman 1 (Widianto et al. 1994) and Ngandong in Indonesia; and a series of fragmentary remains from North and East Africa. All of these remains have been claimed to show more "advanced" morphology than that of typical Erectines, and so they will be discussed in Chapter 6. We will address the issue of *H. erectus* in Europe later in this chapter.

CRANIAL VAULT MORPHOLOGY OF *HOMO ERECTUS*

In order to assess competing definitions of *H. erectus*, we need to establish baseline morphology for undoubted Erectines, using samples and specimens that virtually all the experts would place in *H. erectus*—which means the fossils from Indonesia and China. We will start by applying this program to cranial morphology and then

follow it in subsequent sections dealing with other aspects of Erectine anatomy. The basic references for our discussion of the cranial morphology of *H. erectus* are Weidenreich (1943), Le Gros Clark (1964), Howells (1980), Hublin (1986), Rightmire (1990), Dunsworth and Walker (2002), Wolpoff (1999), Stringer (1984), Wood (1984, 1994), Schwartz and Tattersall (2003), and Antón (2003).

Not all the cranial features discussed below are equally expressed in all Erectine specimens. Cranial features vary within any population, and the larger the sample of a population is, the more variants it will capture. Differences in time and space also contribute to variation within a fossil sample. For example, some of the differences between the Erectines from Zhoukoudian and those from Java may represent regional differences, while others may reflect the fact that the ZKD sample is probably some 400 Ky younger than any of the Indonesian specimens. Sexual dimorphism can also contribute significantly to variation within and between samples. However, we will have little to say about it in this chapter, because there is little or no basis for assigning a sex to most Erectine fossils.

In spite of all these complicating factors, we have enough specimens from Indonesia and China to draw a solid picture of the basic gestalt of Erectine cranial form. Seen in lateral view (Figs. 5.4, 5.8, and 5.9), Erectine skulls exhibit a low, moderately long cranial vault (Table 5.1), characterized by a receding forehead (the **frontal squama**). One regional difference between Java and China is immediately apparent: the Indonesian specimens tend to have flatter, more steeply receding frontals (Figs. 5.8 and 5.9) than those from Zhoukoudian. Conversely, the ZKD specimens have more development of a frontal bulge or **boss**, giving the forehead a somewhat steeper appearance (Fig. 5.4). The skulls of Erectines are often described as long and low; however, Wood (1984) demonstrates that various cranial-shape indices do not separate Asian *H. erectus* from African Erectines, or even from specimens of *Australopithecus habilis* and *A. rudolfensis*.

Although indices of cranial *shape* may not distinguish Erectines from Habilines, the Erectines tend to have larger linear *measurements* of the cranial vault, reflecting their larger brains (see below). The largest-brained pre-Erectine skull is KNM-ER 1470, and its maximum cranial length, breadth, minimum frontal breadth, and bi-parietal breadth are 161 mm, 129 mm, 79 mm, and 108 mm, respectively. Comparison with the equivalent Erectine values in Table 5.1 bears witness to the absolute increase in braincase size seen in early *Homo*. The Erectine supraorbital torus projects forward strongly, and is separated from the frontal squama by a depression or **supratoral sulcus**. Less obvious, but still prominent in Erectines, are the posterior projection of the **nuchal torus** (the thickening of the vault where the

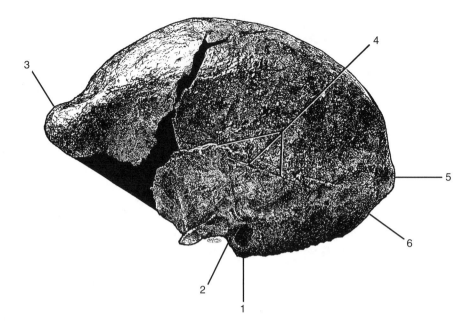

FIGURE 5.8

The "Pithecanthropus II" calvaria (Sangiran 2) from Java. Important features illustrated here are: (1) the thick tympanic ring; (2) the pronounced postglenoid process; (3) a thick, projecting supraorbital torus; (4) a relatively straight (non-arched) squamosal suture; (5) the position of inion and opisthocranion; and (6) the position of endinion (internally). See text for details. (After Johanson and Edgar 1996.)

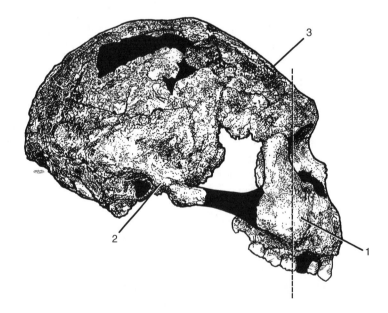

FIGURE 5.9

Sangiran 17 ("Pithecanthropus VIII") from Java, Indonesia; lateral view. Characteristic Erectine features of this fossil include: (1) total facial prognathism, with both the midfacial and lateral facial regions positioned anterior to the anterior end of the braincase (*vertical dashed line*); (2) a thickened postglenoid process; and (3) a long, low frontal squama lacking a frontal boss. This last feature is specifically characteristic of Indonesian Erectines. (After Johanson and Edgar 1996.)

TABLE 5.1 ■ Selected Cranial Vault Measurements for Erectine Specimens (Measurements in mm)

Specimen/Sample	Maximum Cranial Length	Max Cranial Breadth	Bi-parietal Breadth	Breadth Index (Parietal/Base, %)	Minimum Frontal Breadth	Auricular Height
ZKD mean[a]	193.6	141	136.2	96.60	87.2	98.4
ZKD range	188–199	137–143	133–139	—	81.5–91	95–105
Hexian[b]	190	160	145	90.63	101	95
Sangiran 2[c]	—	137	123	89.78	83	90
Sangiran 4[c]	—	146	124	84.93	—	–83
Sangiran 10[d]	184	139	126	90.65	82	—
Sangiran 17[d]	208	160	120	75.00	103	95
Sambung. 1[b]	199	151	146	96.69	105	105
Sambung. 3[e]	178.5	127.5	126.6	99.29	101	103
Ngand. mean[a]	209	146	146	100.00	105	107.4
Ngand. range	193–219.5	138–156	140–152	—	101–108	105–111
ER 3733[f]	183	144	122	84.72	92	90
ER 3883[f]	181.5	136	121	88.97	88	91
WT 15000[f]	175	141	119	84.40	90.5	—
OH 9[g]	206	150	139	92.67	88	–88
Ceprano[h]	198	161	—	—	106	105
D-2280[i]	177	136	118.5	87.13	74.5	—
D-2282[i]	167	125	116	92.80	65	—
D-2700[i]	153	125	115	92.00	66	—

[a]Weidenreich (1943).
[b]Courtesy of M. Wolpoff.
[c]Measured by FHS.
[d]Jacob (1973).
[e]Márquez et al. (2001).
[f]Walker and Leakey (1993b).
[g]Rightmire (1979a).
[h]Ascenzi et al. (2000).
[i]Vekua et al. (2002).

neck muscles attach) and the development of the **supramastoid crest** (ridge) over the external acoustic meatus (ear canal) and mastoid process. Regional differences between the northern Chinese and Indonesian samples of Erectines are evident in many measurements and anatomical details, and in the overall form of the cranium as reflected by multivariate analysis (Antón 2002, Kidder and Durband 2004).

The Erectine supramastoid crest is essentially continuous with the **angular torus**, a thickening of the posterior and inferior corner (angle) of the parietal bone. The parietal bone itself tends to be relatively short (as measured along the **sagittal suture** between the two parietals) and relatively flat. Parietal shape and the presence of an angular torus are often cited as unique defining features (autapomorphies) of *H. erectus* (see Stringer 1984, Rightmire 1990). In side view, the occipital bone appears strongly angled. The **occipital angle**, the angle between the nuchal and occipital planes (Howells 1973b; see Appendix), is estimated to be ≤107° in *H. erectus* (Stringer 1984). This angle is more obtuse in later *Homo*. The Erectine mastoid process is relatively

small, and inclines medially in rear view; but unlike those of some later humans, it projects below the other structures on the cranial base. The **squamosal suture** (at the upper edge of the temporal bone) runs a relatively straight course from the frontal bone back to the occipital. In later humans, this suture arches upward in lateral view.

The strong anterior projection of the Erectine supraorbital torus is also evident from the front (Figs. 5.3 and 5.10). The torus is relatively thick from top to bottom, and its thickness tends to be more or less constant all along the torus. Again, there are regional differences. The Indonesian forms tend to exhibit a shelf-like torus with relatively little depression above **glabella** (the most anterior midline point on the frontal bone), whereas the Chinese (especially the ZKD) specimens generally have a better-developed supraglabellar depression. A distinctive thickened ridge of bone is visible on the midline of the frontal squama, usually beginning just above the supratoral sulcus. This midline ridge, referred to as a **sagittal keel**, often extends onto the parietals. This is another feature often cited as a defin-

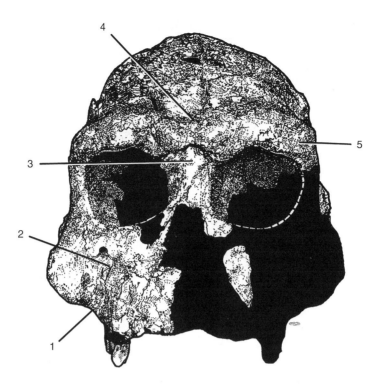

FIGURE 5.10

Frontal view of Sangiran 17. This specimen illustrates the following features: (1) an oblique zygomaticoalveolar margin (ZAM) lacking a malar notch; (2) a relatively flattened infraorbital region lacking a canine fossa; (3) broad interorbital area; (4) a shelf-like supraorbital torus with a weak supraglabellar depression; and (5) a prominent supraorbital torus of near-constant vertical thickness all the way across the orbit, from the medial to the lateral side. (After Johanson and Edgar 1996.)

ing characteristic of *H. erectus* (Stringer 1984, Rightmire 1990). This keeling is also observable on the Zhoukoudian endocasts (Begun and Walker 1993).

Sagittal keeling and the nuchal torus are clearly seen in a rear view of the Erectine skull (Fig. 5.11). The nuchal torus is usually most strongly developed medially. This medial development of the torus precludes the presence of a separate **external occipital protuberance**. (In modern humans, this midline bump on the occipital bone is the attachment point of the **nuchal ligament**, which connects the spinous processes of the cervical vertebrae to the back of the head and helps support the weight of the head when the neck is flexed.) At its lateral ends, the nuchal torus is less pronounced and generally continuous with the angular torus/supramastoid ridge. Often a slight sulcus separates the medial part of the nuchal torus from the occipital plane above it.

The maximum breadth of the Erectine cranium is located down on the cranial base, at the level of the supramastoid ridges. The maximum breadth of the vault above the base is located not far above this point, on the upper temporal squama. This gives the skull a distinctive pentagonal shape from the rear (Fig. 5.11) that is sometimes described as "gabled." The breadth of the skull across the parietals just at or below the temporal

1 cm

FIGURE 5.11

Sangiran 4 ("Pithecanthropus IV") from Java (Indonesia), posterior view. The maximum breadth of the cranium is located on the cranial base, at the level of the supramastoid ridges. Biparietal breadth is markedly less (Table 5.1). The nuchal torus is most well developed medially. Compare Fig. 5.25 (After Schwartz and Tattersall 2003.)

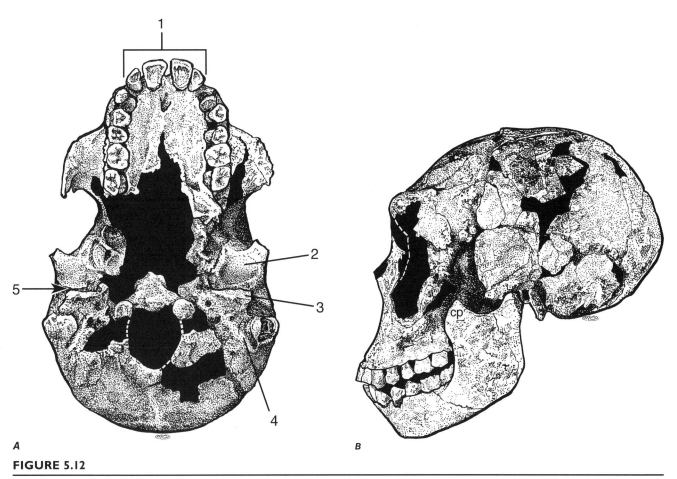

A

B

FIGURE 5.12

The Nariokotome cranium (KNM-WT 15000) from northern Kenya in basal view (**A**) and lateral view including the mandible (**B**). In basal view, note that the maxillary incisors are shovel-shaped (1). Other features indicated are: (2) a relatively deep mandibular fossa, with a well-developed articular tubercle (anterior rim) and a medial wall formed entirely from the temporal bone; (3) a sharp petrosal crest; (4) a wide digastric sulcus; and (5) a distinct fissure separating the postglenoid process from the tympanic plate. cp, coronoid process. (**A**, after Walker and Leakey 1993b; **B**, after Dunsworth and Walker 2002.)

lines is noticeably less than at the cranial base (Table 5.1), which constitutes a difference from most later humans.

Viewed from above (Fig. 5.2B), Erectine crania show degrees of postorbital constriction similar to those seen in the Habilines. As measured by the postorbital constriction index (minimum frontal breadth as a percentage of maximum skull breadth across the zygomatic arch), the narrowing of the braincase behind the orbits in Asian Erectines does not differ significantly from that seen in *Australopithecus rudolfensis* and *A. habilis* (Wood 1984). But as noted earlier, linear measures of the cranial vault tend to be larger in Erectines than in *Australopithecus* because Erectines had bigger brains.

Perhaps the most distinctive area of the Erectine skull is the occiput (Figs. 5.2C, 5.11). Because of the marked development of the nuchal torus, the most posterior point on the skull, **opisthocranion**, is always located on that torus in Erectines. The **superior nuchal line**, a ridge marking the upper edge of the attachments of the neck muscles in back, also extends to the posterior/inferior aspect of the nuchal torus. Thus, the anatomical point called **inion** (where the superior nuchal line crosses the midline) essentially coincides with opisthocranion (Figs. 5.8, 6.7). In modern humans, inion always lies below opisthocranion. This relatively high positioning of the superior nuchal line on the skull vault is an aspect of the characteristic Erectine expansion of the **nuchal plane**, the flattened neck-muscle attachment surface at the back of the skull. The nuchal plane runs from the rear end of the foramen magnum (**opisthion**) up to inion, and extends laterally to the grooves for the digastric muscles on the medial side of the mastoid process (Figs. 5.4 and 5.12). In Erectines, but not in most later humans, the nuchal plane is larger than the **occipital plane**, which extends from inion up to **lambda** (the rear end of the **sagittal suture** between the two parietals; see Appendix). The distinctively large relative size

TABLE 5.2 ■ Lengths of the Occipital and Nuchal Planes in Erectines and Later Humans (Measurements in mm)

Sample	Occipital Plane	Nuchal Plane	Index
Sangiran mean	45.7	61.3	134.3
Range	45–46	60–64	130.4–142.2
ZKD mean	49.6	59.6	120.5
Range	48–52	58–63	109.4–131.3
Ngandong mean	57.2	52.4	92.1
Range	52–64	47–53	80.7–101.9
Neandertal mean	59.4	46.5	75.3
Range	56–67.8	43.2–50.1	63.4–81.1

The occipital plane is measured from inion to lambda; the nuchal plane, from inion to opisthion; and the index is the nuchal-plane length expressed as a percentage of occipital-plane length.

of the Erectine nuchal plane (Fig. 5.4) is demonstrated by the data in Table 5.2. The relatively high position of inion and the expansion of the nuchal plane in Erectines are also reflected in the displacement of inion above **endinion**. Endinion is the central point on the **cruciate eminence** (the ridge below the grooves for the transverse venous sinuses) on the internal aspect of the occipital. As Figure 5.8 illustrates, endinion is located much lower than inion in Erectines, whereas these points tend to lie on the same plane in more recent humans. All this, along with the generally marked areas of muscle attachment, indicates that the nuchal muscles—the epaxial neck muscles that hold up the head—were quite robust in Erectines.

These well-developed cranial superstructures of Erectine skulls—the supraorbital, nuchal, and angular tori, the supramastoid ridge, and the sagittal keel—give the impression of a reinforcement or buttress system for the cranium. *H. erectus* braincases are also often described as having generally thicker bones than those of other humans. The thickening of the cranial base is especially marked along the nuchal plane (Fig. 6.7). It may be related to the expansion of the nuchal plane itself, representing an adaptation to stresses generated by the apparently powerful neck musculature. However, it is hard to imagine what Erectines might have been doing with their neck muscles that would have distinguished them from other humans and led to this peculiar thickening of the nuchal plane. Weidenreich (1939a, 1943) suggested that the general thickening of the cranial vault in Erectines might have evolved as a protection against blows to the head. To support this notion, he noted a high frequency of cranial injuries, both healed and unhealed, in the Zhoukoudian sample. P. Brown (1994) has likewise argued that this thickness has to do specifically with interpersonal violence during the Erectine period.

D. Lieberman (1996) offers a different explanation. His experimental studies on pigs and armadillos lead him to think that thickening of the cranial vault is an incidental side effect of high overall levels of cortical bone deposition stimulated by exercise before the skeleton has finished growing. In his view, vault thickening is not a highly heritable feature, and is thus not useful for taxonomic or phylogenetic purposes. Lieberman compared bone thickness at **bregma** (the anterior end of the sagittal suture) and at the parietal eminence or boss (the most laterally bulging part of the parietal bone) among three groups of pre-modern humans: Erectines (including Ngandong and Salé), Neandertals, and "other" archaic *Homo* (mainly Heidelbergs). He found that in Erectines, bone thickness at bregma is significantly thicker than in Neandertals, but not than in the other archaics. At the parietal eminence, there was no statistically significant difference among any of the samples. These findings imply that the thickness of Erectine vaults is not as distinctive as is often claimed.

Other, smaller details of the cranial base are often considered characteristic of *H. erectus* (Figs. 5.9 and 5.12). One such detail is an unusually marked **digastric groove**, a broad fissure on the medial edge of the mastoid process that serves as the origin of the digastric muscle. Medial to that groove, there is a ridge called the **occipitomastoid crest** (or juxtamastoid eminence), which is more prominent in *H. erectus* than in modern humans, though not as well developed as it is in Neandertals (Chapter 7). In many Erectine specimens, the mastoid process is separated from the petrosal crest of the temporal bone by a distinct fissure, termed the **mastoid fissure** (Andrews 1984, Bräuer and Mbua 1992). The socket of the jaw joint (the mandibular or glenoid fossa) is generally rather deep in Erectines, with a pronounced articular eminence in front of it and a marked posterior margin. According to Stringer (1984), the medial wall of the fossa is formed entirely from the temporal bone in Erectines, with no sphenoid contribution. The fossa narrows medially; and as Figure 5.12 demonstrates, there is a marked fissure at its medial end between the **tympanic plate** (the flattened front surface of the ectotympanic tube) and the large **postglenoid** process (Andrews 1984, Bräuer and Mbua 1992). The tympanic plate itself is long and robust and exhibits a sharp petrosal crest inferiorly (Fig. 5.12). Spoor and Zonneveld (1994) note that the bony labyrinth, which houses the inner-ear mechanisms for hearing and balance inside the petrous portion of the temporal, exhibits the same form as that of modern humans (cf. Fig. 7.14).

❙ CRANIAL CAPACITY AND THE BRAIN

Asian Erectines have significantly more capacious braincases, and presumably larger brains, than any *Australo-*

TABLE 5.3 ■ Asian Erectine Cranial Capacities (in cm³)

Site/Specimen	Cranial Capacity	Reference
Trinil 2	940	Holloway (1981a)
Sangiran 2	813	Holloway (1981a)
Sangiran 4[a]	908	Holloway (1981a)
Sangiran 10	855	Holloway (1981a)
Sangiran 12	1059	Holloway (1981a)
Sangiran 17	1004	Holloway (1981a)
Gongwangling	780	Woo (1966)
Zhoukoudian II	1030	Weidenreich (1943)
Zhoukoudian III	915	Weidenreich (1943)
Zhoukoudian V[b]	1140	Weidenreich (1943)
Zhoukoudian VI	850	Weidenreich (1943)
Zhoukoudian X	1225	Weidenreich (1943)
Zhoukoudian XI	1015	Weidenreich (1943)
Zhoukoudian XII	1030	Weidenreich (1943)
Hexian[b]	1025	Wu and Dong (1985)
Total sample mean	972.6	
Indonesia mean	929.8	
Kabuh mean	934.2	
China mean	1001.1	
Zhoukoudian mean	1029.3	

[a]Sangiran 4 is the only Indonesian specimen not from the Kabuh Formation.
[b]Zhoukoudian V and Hexian are sometimes considered "advanced" compared to the remainder of the Chinese sample.

pithecus. The mean endocranial volume estimated for the largest-brained *Australopithecus* species, *A. rudolfensis*, is around 752 cm³ (Wood, 1992), whereas Asian Erectines have an average cranial capacity of 972.6 cm³ (Table 5.3). Whether this represents a *relative* increase in brain size depends on estimates of body size in the two groups, which remain problematic and debated (see below; Begun and Walker 1993).

The later Erectines appear to have bigger brains on the average than the earlier ones; the sample means are 934.2 cm³ for the Indonesian specimens and 1029.3 cm³ for the Zhoukoudian sample. To those who see human evolution as gradualistic, this difference signals a significant trend toward brain expansion throughout the Erectine phase of the human lineage (Wolpoff 1984, 1999). To others, it does not (Rightmire 1981, 1986, 1990). Other factors besides time may explain all or part of the difference between the earlier and later samples. One such factor is geography. Cranial capacities in Australasia remain somewhat low in later periods of human evolutionary history, even in recent samples (Freedman et al. 1991, Beals et al. 1984, Holloway 1980, 1981a, Wolpoff 1999, Smith 2002). As Freedman and his colleagues point out, this is probably due to the "thermodynamic effect" (Beals et al. 1984), whereby cranial capacities tend to be larger at higher latitudes in colder climates. Here again, we run into the problem of estimating body size, which has to be taken into account

in assessing the meaning of differences in brain size. Since no postcranial fossils can be unequivocally attributed to Javan Erectines (see below), we cannot tell how much of the difference in average brain size between the Indonesian and Zhoukoudian samples is due to allometry. As will be seen later, we encounter similar problems in trying to compare Erectine and Habiline brain volumes.

Endocasts of the inside surfaces of Erectine braincases tell us something about the external morphology of their brains. Unfortunately, trying to say anything about brain structure from endocasts is a very tricky and imprecise enterprise. The detailed assessments of fossil sulci and gyri that were published in the first half of the 20th century are no longer thought to contain much information about patterns of neural organization in early humans. Thus Weidenreich's (1936a) study of the Zhoukoudian endocasts is rarely cited nowadays, whereas his descriptions of other aspects of Zhoukoudian anatomy are still classic references for analytical and descriptive studies.

Holloway (1981a, 1982, 2000), who has studied the endocasts of many Erectines, offers some cautious general observations on what those endocasts can tell us about their brains and behavior. Holloway notes that Erectine endocasts, unlike those of apes, generally show asymmetric protrusions or **petalia** of the right frontal and left occipital lobes. This is not surprising, since this is a distinctive pattern seen in all hominins, from *Australopithecus* down to modern *H. sapiens* (Holloway and de La Costa-Lareymondie 1982). The cortical area known as **Broca's cap** or Broca's area tends to be well-developed in Erectines—another human-like feature, which has traditionally been thought to have something to do with language abilities. However, recent studies indicate that the area around Broca's cap may be more generally related to higher-level control of motor functions, not specifically to linguistic or speech functions (Begun and Walker 1993).

What little can be told from the Erectine endocasts thus suggests that, except for their small size and narrow frontal lobes, there is little to differentiate them from those of modern humans. However, the endocasts do not provide any real insight into the internal "wiring" of the brain itself. Coqueugniot et al. (2004) reached some different conclusions from their study of the infant Erectine braincase from Perning (Modjokerto). Comparing it with a large series of extant humans and chimpanzees, they concluded that the Perning 1 individual was about 1 year old at death—at which time its endocranial capacity was already about 72–84% of that of an average adult Erectine. Having such a mature brain so early in life implies that the period of postnatal brain growth must have been short. This suggests an ape-like pattern of brain development, different from the protracted brain growth (and correlated prolonged helplessness) seen in modern human infants. From these conclusions, Coqueugniot's team inferred that *H. erectus* probably did not have spoken language or "cognitive skills comparable to those of modern humans." Leigh (2006) argues that the *absolute* rate of brain growth in Perning 1 appears to have been about the same as in modern humans. But even if it was, it is clear from the small braincase of adult Erectines that brain growth must have stopped earlier in infancy than it does in human babies today.

FACES AND MANDIBLES OF ASIAN *HOMO ERECTUS*

The supraorbital torus is the only part of the Asian Erectine face that is well known. Sangiran 17 has a fairly complete but partly reconstructed facial skeleton. Three other Indonesian specimens—the Sangiran 4 palate (Fig. 5.13), Sangiran IX, and the Sangiran 27 crushed face—provide some additional information. From China, Gongwangling preserves the palate and lower face up to the nasal floor, and the Zhoukoudian sample included some facial pieces. Weidenreich (1943) described only six facial fragments, apart from upper facial remains attached to calvariae (Table 5.4). He noted that many other smaller facial pieces were associated with badly crushed skulls like X and XI, but these fragments were too small for precise identification or reconstruction.

The known facial specimens of Asian Erectines resemble each other in several features, including an emphatically columnar lateral orbital margin and a broad upper face. Overall, Asian Erectines have generally broad (but not particularly tall) faces and broad anterior palates. The upper face is especially broad, reflecting the broad eye sockets and wide interorbital region. The large nasal opening is framed by broad, projecting nasal bones and everted lateral margins. From these features of the nasal region, Franciscus and Trinkaus (1988) infer that Erectines had a protruding, human-like external nose with downward-facing nostrils, unlike the ape-like flat nose and forward-facing nostrils inferred for *Australopithecus*. They argue that the turbulence generated by exhaling into this external nose (which is generally cooler than the nasal cavity inside the skull) would enhance condensation of moisture in expired air. The condensed moisture could then be captured by inspired air, providing an effective mechanism for moisture conservation in an arid environment. By this account, the protruding human nose thus reclaims some of the functions originally served by the primitive mammalian maxilloturbinals, which are greatly reduced in anthropoids (Chapter 3). The increased breadth of the Erectine nose would have allowed greater amounts of air to be inspired per unit time, allowing these early humans to maintain high activity levels for longer periods (Trinkaus 1987a). This

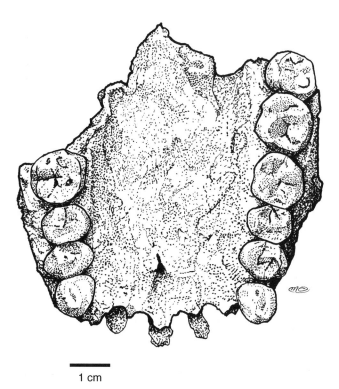

1 cm

FIGURE 5.13

Maxilla of Sangiran 4 in occlusal view. Note the broad anterior palate, the large right canine, and the relatively small M[3]. (After Schwartz and Tattersall 2003.)

TABLE 5.4 ■ Facial Pieces from Zhoukoudian
(after Weidenreich, 1943)

Description	Locus	Association	Important Features
Frontal process of left maxilla	L	Skull X	Columnar lateral orbital pillar.
Fragment of left zygomatic bone	L	Skull X	Angled zygomaticoalveolar margin (ZAM). Malar notch.
Fragment of left maxilla with P[3]–M[3]	L	Skull XI	Angled ZAM. Malar notch.
Right half of palate	L	Skull XI?	
Fragment of left maxilla with I[1], P[3]–M[3]	O	Skull XIII	Canine fossa. Angled ZAM. Malar notch.
Fragment of left maxilla with P[3], M[1-3]	UC[a]	Skull XIV?	

[a]From the Upper Cave, but thought to be associated with the Erectine sample.

may have been advantageous in view of the Ergasters' increased brain and body sizes (see below).

There is a lot of variation in other aspects of the facial anatomy of Asian Erectines. On the Sangiran 17 face, as reconstructed by Wolpoff (Thorne and Wolpoff 1981), the zygomaticoalveolar margin or ZAM (the inferior margin of the bony cheek, seen in frontal view) runs obliquely downward from the zygomatic arch to the dental alveoli, without any concavity or angulation (Fig. 5.10). This Javanese face thus lacks the **malar notch** or *incisura maxillaris* (Fig. 3.34), a notch seen in the ZAM of later humans between the masseter attachment laterally and the alveolar process medially (Pope 1992). Sangiran 17 also lacks a human-like **canine fossa**, which is a broad, groove-like depression in the bony facial surface (**infraorbital plate**) just lateral to the nasal opening. But in China, the Gongwangling cranium has both an angled ZAM and a canine fossa, and the Zhoukoudian remains have angled ZAMs, malar notches, and canine fossae (Fig. 5.3). Wolpoff (1999) attributes these differences to the time difference between the samples and the fact that Sangiran 17 is, in his view, a male and the Zhoukoudian material is probably female. However, he identifies some other features of the Sangiran 17 specimen as regional features that persist in later humans from Australasia (Thorne and Wolpoff 1981, Wolpoff et al. 1984, Frayer et al. 1993). Since Sangiran 17 is the only Indonesian specimen on which these features are preserved, it is impossible to tell which aspects of its facial skeleton represent regional features and which are due to individual variation.

Erectines had prognathic faces. The Sangiran 17 cranium, Weidenreich's (1943) reconstruction of the L2 specimen from Zhoukoudian, and Tattersall and Sawyer's (1996) reconstruction of the L3 ZKD cranium provide some insights into the nature of this prognathism. All three specimens suggest a condition known as **total facial prognathism**, which combines three distinct sorts of facial projection:

▶ *Upper midfacial prognathism* (forward projection of the area around **nasion**, the upper end of the suture between the nasal bones).

▶ *Alveolar prognathism* (forward projection of the alveolar process under the nose).

▶ *Lateral facial prognathism* (forward projection of the cheek and anterior zygomatic arch; see Fig. 5.9).

In Erectines, all three of these components of the facial skull project well forward in front of the braincase, producing a characteristically flat but protrusive face. Again, there is variation in these features, even in the available sample of just three faces. The Sangiran 17 specimen has swept-back zygomatics and marked alveolar prognathism (Wolpoff 1999, Thorne and Wolpoff 1981), and the L3 reconstruction suggests a good deal

of alveolar prognathism (Fig. 5.4), but these features are built on a face that is fundamentally prognathic overall. The Erectine face is less protrusive than those of apes or *Australopithecus*; but compared to more recent humans, Erectines preserve a relatively primitive pattern of facial prognathism.

Erectine mandibles are known from both Indonesia and China. Several lower jaws from Java come from the Pucangan Formation right around the "Grenzbank" (Fig. 5.6). Weidenreich (1946) referred four of these, Sangiran 5, 6, 8, and 9, to his taxon *Meganthropus paleojavanicus*, to which we will return later. There are also mandibles from the Kabuh Formation, including Sangiran 22 and Dubois' Kendung Brubus specimen. Zhoukoudian yielded a series of 11 mandibles in various states of completeness, six of which are regarded as subadults (Weidenreich 1936b). Some additional ZKD specimens were recovered after World War II (Wu 1985). Also from China is the Chenjiayao mandible, from a site near Lantian (Woo 1964). This specimen, estimated to be about 600 Ky old, consists of a complete corpus (body) and several teeth. It is particularly interesting because it is the earliest specimen in human evolutionary history to exhibit congenital absence of the wisdom teeth.

In general, Erectine mandibles are characterized by taller but less thick mandibular bodies (corpora) than those of *Australopithecus*, although there is considerable variation. Because mandibular body height tends to correlate with overall body size, the taller mandibular bodies of Erectines reflect their general increase in body size compared to earlier hominins (see below). The outer surface of the Erectine symphysis slopes backward from the alveolar process to the base, and there is no evidence of a distinct chin structure. On the inner surface of the jaw in the midline, the alveolar plane also slopes backward (as would be expected with a receding symphysis), and a variably developed **superior transverse torus** (Fig. 4.48) is present. Mental foramina are often multiple, ranging in location from under the M_1 to under the septum between the sockets of P_3 and P_4. The rami tend to be tall, with pronounced coronoid processes that often rise higher than the condyle (Fig. 5.12B). Viewed laterally, the anterior margin of the ramus normally bisects the M_3, so there is no **retromolar space** (a space between the distal aspect of the M_3 and the ramus, found in Neandertals). For their size, the mandibles tend to have broad rami. Their breadth probably reflects the skull's marked lateral facial prognathism, which increases the anteroposterior length of the temporal fossa and the ramus along with it (Fig. 7.16).

THE ERECTINE DENTITION

The teeth from Zhoukoudian number about 150, which makes them the largest sample of Erectine teeth yet recovered. On average, the ZKD premolars and molars are larger than those of more recent *Homo* (Table 6.4) but markedly smaller than those of late *Australopithecus*, including the *Homo*-like forms *A. habilis* and *A. rudolfensis* (Tables 4.2 and 6.5). The buccolingual breadths of most of the ZKD cheek teeth fall within the range of recent humans (Wolpoff 1971b). Unlike those of earlier hominins, the ZKD third molars tend to be smaller than the M1 and M2 (which are roughly equal in size). The ZKD canines exceed the modern range and are larger on average than those of *Australopithecus*, although they are smaller than the canines of some later archaic *Homo*. The ZKD incisors, however, tend to be smaller than those of earlier African Erectines (Brown and Walker 1993) or most later archaic human samples (Table 6.5). These data show that the evolution of dental size in human evolutionary history cannot be summarized simply by contrasting the shrinking cheek teeth with the rest of the dentition, because the evolution of the canine follows a different pattern from that of the incisors. Weidenreich (1937) thought that the variation in the ZKD teeth reflects substantial sexual dimorphism in the sample. However, ZKD tooth-size variation is no greater than in other Middle and Late Pleistocene *Homo* (Table 6.5).

The most morphologically distinctive of the Zhoukoudian teeth are the upper incisors, which have often been described as "shovel-shaped." This is an imprecise term that has been applied indiscriminately to several different morphologies. The ZKD upper incisors combine marked marginal (lateral) ridges with a straight incisal margin connecting the marginal ridges (Fig. 5.14). This pattern differs from the "shovel-shaped" incisors of South Asians (which have less pronounced marginal ridges) and those of Neandertals (which have a more marked basal tubercle and a more convex incisal margin). However, it resembles the pattern seen in modern North Asians. The ZKD-North Asian "shoveling" pattern is therefore often cited as evidence for regional continuity between Middle Pleistocene and modern human populations in China (Frayer et al. 1993, Crummett 1994, 1995; Wolpoff et al. 1994a).

Except for the incisors, there are no significant size differences between the teeth of early African Erectines and Zhoukoudian (Brown and Walker 1993, Brown 1994). However, there are differences between these samples and Indonesian Erectine teeth. The cheek teeth of the Javan specimens from the Pucangan Formation are larger on average than those of the earlier (African) or later (Zhoukoudian) Erectines, although the ranges overlap. The Javan molars also tend to be mesiodistally longer. Some of the bigger Pucangan teeth have been assigned to "Meganthropus" (see below), but both Lovejoy (1970) and Kramer and Konigsberg (1994) have shown that these teeth do not differ systematically in size from other Erectine teeth. The deciduous molars from Java do not differ in morphology from those of

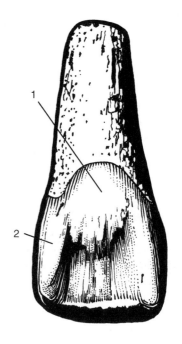

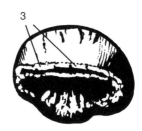

FIGURE 5.14

Maxillary central incisor of a Zhoukoudian Erectine. Note the well-developed basal tubercle (1) and marginal ridges (2), as well as the relatively straight course of the incisal margin (3). (After Weidenreich 1937.)

other Erectines (Grine 1984). Finally, the upper incisors of Indonesian Erectines have straight incisal margins like those from ZKD, but differ in having less developed marginal ridges and basal tubercles. This pattern is also seen in later samples from South Asia and has been cited as another piece of evidence for regional genetic continuity (Crummett 1994, 1995).

ERECTINE POSTCRANIAL REMAINS

The only postcranial elements unequivocally associated with Asian *H. erectus* are those from Zhoukoudian. Eleven postcranial pieces—a segment of clavicular shaft, a hand bone, two humeral pieces, and seven femoral fragments—were recovered during the pre-World War II excavations at the site and published by Weidenreich (1941). More humeral and tibial pieces were recovered following World War II (Wu 1985). All the femora differ from those of modern humans in three features: an anterior-posterior flattening of the shaft, a thickening of the cortical bone, and an absence of the ridge (the **pilaster**) marking the major muscle attachments on the back of the thigh. These three features are also seen in the OH 28 femur and other Erectine specimens from Africa. They appear to distinguish Erectines from modern *Homo* (Day 1971, 1984, 1995).

The most complete femur from Zhoukoudian, ZKD IV (M4), is a right shaft that lacks both ends but preserves the lower neck. Weidenreich (1941) estimated that the intact bone would have been 41 cm long, from which a body mass of 56–58 kg has been inferred (Ruff et al. 1993). If these estimates are correct, this individual would have had relatively short limbs. This might be seen as a cold-climate adaptation, obeying the rule (Bergmann's Rule) that mammals living in cold climates evolve less spindly bodies than their warm-climate relatives (so as to minimize surface-to-volume ratios and conserve heat). If so, the M4 femur would be the earliest evidence of human body-shape adaptation to nontropical conditions. But this is a shaky inference from a reconstructed femur, which may have been a few centimeters longer than Weidenreich's estimate.

The femora from Trinil present problems. As noted, it is not clear where they were found within the Trinil deposits. Although Trinil 3 (femur I) is complete, it has a pathological bony outgrowth below the neck, probably produced by the ossification of muscle and tendon (*myositis ossificans*) in response to some injury (Ortner and Putschar 1981). Trinil 6 (femur II) is slightly less complete, and the other femoral pieces (Trinil 7-9) are only shaft segments. Trinil femora 6-9 were discovered by Dubois in collections after he was back in Holland, and no documentation has survived giving their exact contexts. All of these specimens, including Trinil 3, were studied by M. Day and T. Molleson (1973; Day 1984), who demonstrated that their morphology and histology cannot be distinguished from those of recent humans. A more detailed comparative analysis by G. Kennedy (1983a) demonstrates that the Trinil femora fit with later *Homo* and not with other Erectine femora in their overall cortical thickness, degree of anterior-posterior shaft flattening, and pattern of trabeculae in the neck. Although chemical analysis does not clearly demonstrate that these femora come from a later time period than the Trinil 2 skullcap does (Day 1984), the case that they represent *H. erectus* is not very convincing.

EARLY AFRICAN ERECTINE SKULLS AND THE ERGASTER QUESTION

The early East African Erectine remains comprise a relatively few specimens from two Kenyan sites: Koobi Fora

and Nariokotome III, from the east and west sides of Lake Turkana respectively. As noted earlier, these date to between 1.95 and 1.5 Mya. Nariokotome III has yielded only one major specimen, the 85% complete adolescent skeleton of KNM-WT 15000, which preserves a nearly complete cranium and associated mandible (Brown et al. 1985, Walker and Leakey 1993a). Koobi Fora provides a partial adult skeleton, KNM-ER 1808 (Walker et al. 1982), plus some additional limb bones (see below). Koobi Fora has also produced two informative crania, KNM-ER 3733 (Fig. 5.15) and 3883, and other cranial fragments including the fragmentary ER 730 skull and the ER 2598 occipital (Leakey and Walker 1976 1985; Walker and Leakey 1978). There are a number of partial mandibles from Koobi Fora, among them the type specimen of *H. ergaster*, ER 992. Klein (1999) has recently suggested that all later African remains traditionally assigned to *H. erectus* should be classified in *H. ergaster*, which would extend the time range of this species to about 600 Ky. However, most workers restrict *H. ergaster* to the earliest African material, and this convention will be followed here. Table 5.5 lists other specimens that we include in the "Ergaster" category, in addition to those mentioned above.

Even the staunchest supporters of a separate species status for *H. ergaster* agree that the cranial anatomy of early African *Homo* shares a great deal with that of the Asian Erectines (Andrews 1984, Stringer 1984, Wood 1984, Klein 1999, Tattersall and Schwartz 2001). The overall shape of the skull, the anatomy of the occiput and cranial base, reduced masticatory apparatus, facial prognathism, and the patterns of tooth sizes are among the most obvious similarities. All of the major Ergaster cranial specimens were originally classified as *H. erectus* (Brown et al. 1985, Leakey and Walker 1976, Walker and Leakey 1978) before anyone suggested that they ought to be put in a separate species. Claims that Ergasters have higher cranial vaults than Asian Erectines (Tattersall and Schwartz 2001) are not borne out by the pertinent data (Table 5.1). The Ergasters are characterized, on average, by the smallest cranial capacities among Erectines (Tables 5.3 and 5.6). However, we might expect that the earliest Erectines would have the smallest brains; and Ergaster braincases are still considerably bigger than those of *Australopithecus* (a 14% increase over *A. rudolfensis*). The faces of WT 15000 and ER 3733 exhibit more angled zygomaticoalveolar (ZAM) margins than the Zhoukoudian sample, and also lack malar notches and canine fossae (Fig. 5.16). However, Indonesian Erectines—specifically Sangiran 17 (Fig. 5.10)—are more similar to the Ergasters in these respects than the ZKD skulls are.

Much has been made of the supposedly thinner supraorbital torus of Ergasters. But this is known for certain only in ER 3733. ER 3883 has a thicker torus, and WT 15000 probably would have had one (judging from the preserved lateral portions) if he had lived to grow up. Even in ER 3733, the form of the torus (with

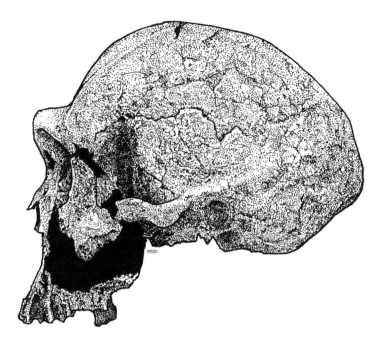

FIGURE 5.15

Lateral view of KNM-ER 3733 (Koobi Fora, Kenya), the earliest dated major cranial fossil of an Erectine (Ergaster). In lateral view, the total facial prognathism characteristic of Erectines is clearly evident. (After Leakey and Leakey 1978.)

TABLE 5.5 ■ Additional Specimens of Early _H. erectus_ from Lake Turkana, Kenya

Specimen	Preserved Elements	Stratigraphic Member
ER 164	Cranial fragment, vertebrae, phalanges	KBS
ER 731	Mandible fragment	Okote
ER 741	Tibia	Okote
ER 803	Partial skeleton	Okote
ER 806	Isolated teeth	Okote
ER 807	Maxillary fragment	KBS
ER 808	Isolated teeth	Okote
ER 820	Juvenile mandible	Okote
ER 1466	Frontal fragment	Okote
ER 1821	Parietal fragment	KBS
ER 2592	Parietal fragment	Okote
ER 3892	Frontal fragment	Okote
ER 5428	Talus	KBS
WT 15000	Skeleton	Natoo
WT 16001	Cranial fragments	Natoo

Age ranges (Mya): KBS 1.88–1.65; Okote and Natoo 1.65–1.39 (Brown 1994; Brown and McDougall 1993).
Source: After Walker (1993).

TABLE 5.6 ■ African Erectine Cranial Capacities (in cm^3)

Site/Specimen	Cranial Capacity	Reference
KNM-ER 3733	848	Holloway (1983, 2000)
KNM-ER 3883	804	Holloway (1983, 2000)
KNM-WT 15000	909	Begun and Walker (1993)
KNM-ER 42700	691	Spoor et al. (2007b)
OH 9	1067	Holloway (1983, 2000)
OH 12	727	Holloway (1983, 2000)
Buia	775[a]	Abbate et al. (1998)
Daka	995	Asfaw et al. (2002)
Tighenif 4	~1300	Arambourg (1955)
Total sample mean	901.8	
Circum-Turkana (Ergaster) mean	813.0	
Non-Ergaster Africa mean	972.8	

[a]Abbate et al. estimate cranial capacity between 750 and 800 cm^3.

its relatively even thickness medially and laterally; see Fig. 5.16) and the presence of a distinct supratoral groove represent similarities to Asian Erectines. The mandibular fossa and digastric sulcus of ER 3733 are also like those of their Asian counterparts.

Ergasters and Asian Erectines also share distinctive features of occipital morphology. All three relatively complete Ergaster skulls have (or appear to have had) larger nuchal than occipital planes, an inion that essentially coincides with opisthocranion, an elevation of inion above endinion, and distinct nuchal tori. The Koobi Fora crania appear to have less angled occipitals than the Asian Erectines have; but this is uncertain,

TABLE 5.7 ■ **Comparative Data on Facial Flatness and Prognathism**

Sample	Skull Length	ZMR	PRR	ZMR Index	PRR Index	ZMR/PRR Index	SSA[a]
ER 3733	175.0	84	116	48	66.3	72.4	142
Petralona	209.2	77	120	36.8	57.4	64.2	119
Kabwe I	206.0	75	121	36.4	58.7	62.0	116.4
Neandertal mean	212.9	75.8	120	35.6	56.4	63.2	110.6
Recent mean	177.8	70.4	97.5	39.6	54.8	72.2	128.8

[a]SSA (subspinale angle) data from Stringer (1983) in degrees. Other data from Smith and Paquette (1989, p. 184). Other abbreviations: PRR, prosthion radius (from the anterior end of the maxilla in the midline to the biauricular plane); ZMR. zygomaxillary radius (the horizontal distance from the front of the cheek to the biauricular plane drawn vertically through the two auditory meatuses); ZMR Index, zygomaxillary radius as a percentage of maximum cranial length; ZMR/PRR Index, zygomaxillary radius as a percentage of prosthion radius. All measurements defined in Howells (1973b): see Appendix.

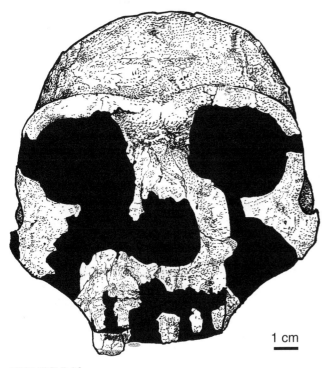

1 cm

FIGURE 5.16

KNM-ER 3733 (Koobi Fora, Kenya), frontal view. Note the thin but prominent supraorbital torus, the broad interorbital area, and the rather oblique orientation of the zygomaticoalveolar margin (ZAM). (After Schwartz and Tattersall 2003.)

because the crania are damaged in ways that preclude exact measurement of the angle (Stringer 1984). The ER 2598 occipital, which preserves the occipital plane, nuchal torus, and superior nuchal plane, has a sharp occipital angle and an inion that lies at a significantly higher position than endinion. ER 2598 also has thicker cranial bones and a more pronounced nuchal torus than the three more complete Ergaster crania. Wood (1991) judges the specimen to be specifically Erectine-like.

These features support Wolpoff's (1999) claim that ER 2598 at ~1.9 My, represents the earliest dated specimen to exhibit an Erectine (or in Wolpoff's terms, early *H. sapiens*) cranial morphology.

Like other Erectines, ER 3733 and WT 15000 have relatively broad upper faces, with a broad nose and everted nasal margins (Figs. 5.15 and 5.16). The face of Ergasters is also relatively flat (as viewed from above). This is reflected metrically in the **subspinale angle** (SSA, Table 5.7). The larger this angle is, the less the lower edge of the nasal fossa sticks out in front of the cheekbones (see Appendix). The estimated SSA of ER 3733 is 142°, compared to means of 128.8° for modern humans and only 116° for the beak-nosed Neandertals (Stringer 1983). Because the 3733 face is partly reconstructed (Fig. 5.16), there is some uncertainty about the precision of this estimate, but it is highly likely that 3733 would have been closer to the modern values than to the Neandertals. A different analysis by Smith and Paquette (1989) also demonstrates that the lateral face in ER 3733 was placed relatively far forward. This is determined by measuring the **zygomaxillary radius** (ZMR) and the **prosthion radius** (see Appendix). A larger ZMR relative to overall cranial length (ZMR Index) indicates a more forwardly placed lateral face. The relative protrusion of the midsagittal and lateral parts of the face is measured by the ratio of ZMR to prosthion radius (ZMR/PRR Index, Table 5.7), with a lower value reflecting a more projecting midsagittal face. These metrics show that ER 3733, like other Erectines, has a face that is relatively flat (high ZMR/PRR index) but is markedly prognathic overall (high ZMR and PRR indices), unlike the true orthognathic face of modern humans.

Most of the mandibles attributed to the Ergasters are not associated with diagnostic cranial remains, and some of them may belong to Habilines. Even the type specimen of *H. ergaster*, ER 992 (Fig. 5.17), is very similar to specimens generally classified with the Habilines. However, the ER 730, ER 1808, and WT 15000 mandibles are certainly Ergasters, associated with defin-

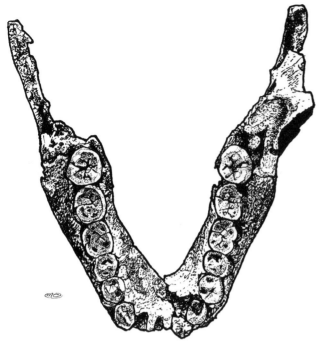

FIGURE 5.17

KNM-ER 992 (Koobi Fora, Kenya), the type specimen of *Homo ergaster*. The canine is large, and the molars are small compared to the mandibular corpus. (After Johanson and Edgar 1996.)

itive cranial elements. Their morphology does not differ systematically from that described above for Asian Erectines: their symphyses are receding and devoid of chin structures, the mandibular corpuses tend to be relatively tall and not thick, and the rami are relatively broad and bisect the M_3s (so that retromolar spaces are lacking). Like those of late *Australopithecus*, the M_3s of Ergasters are often the largest molars, and all the molars in these jaws tend to have very complex occlusal morphology (Fig. 5.17). However, overall the posterior teeth are reduced compared to late *Australopithecus*. The anterior teeth are relatively large, though not larger than those of some *Australopithecus*. The upper incisors of Ergasters exhibit slight to moderate degrees of shoveling. As can be seen in the Nariokotome youth, the maxillary incisors have only weakly developed basal tubercles and marginal ridges, and their incisal margins are straight to slightly curved (Fig. 5.12). Both metrics and morphology tend to place WT 15000 with Erectines, but many of its features are also shared with *A. habilis* (Brown and Walker 1993, Brown 1994, Tobias 1991).

The case for recognizing *Homo ergaster* and *H. erectus* as separate species hinges on claims that each group has apomorphies not found in the other. In 1984, three influential articles put forward such claims.

Stringer (1984) provided a list of 27 cranial features that might be used to define *H. erectus*, but concluded that 12 of these features were found only in the Asian representatives of the taxon (1984). Wood proposed 32 possible autapomorphies of the Asian group, including the morphology of the occipital (nuchal) torus, presence of an angular torus, presence of a post-toral sulcus on the frontal, proportions and shape of the occipital, and a relatively large occipital arc. Both Wood and Stringer found that the Ergaster specimens (basically ER 3733 and 3883) did not share these supposed apomorphies of Asian Erectines.

Andrews (1984) evaluated these lists of features and concluded that most were primitive retentions or were not specific enough to define *H. erectus* in a precise sense. However, Andrews did identify seven features that he felt were autapomorphies of Asian Erectines, allowing a cladistic definition of *H. erectus* that would restrict the taxon to Asia. These features are:

1. Frontal bone keeling.
2. Keeling of the parietal bones.
3. Marked thickness of the cranial vault bones.
4. Angular tori.
5. Separation of inion and endinion.
6. Mastoid fissure.
7. Postglenoid/tympanic-plate fissure.

But more recent studies cast doubt on these supposed autapomorphies of Asian Erectines. Bräuer and Mbua (1992; Bräuer 1994) studied the distribution of these seven features in a large series of Asian and African Erectines, later archaic forms of *Homo* from Africa and Asia, and several species of *Australopithecus* (including the Habilines). They concluded that these traits exhibit a pattern of continuous variation in the samples they investigated, and could not simply be scored as "present" or "absent." Not all seven features are present in every specimen of Asian Erectine; and all of them frequently occur in African Erectines—including the Ergasters— and in earlier and later hominins. Rightmire (1990, 1994) reached similar conclusions using a somewhat different list of features that included the form of the supraorbital torus and the capacity of the braincase. In a principal-components analysis of cranial measurements, Wu et al. (2005) found close affinities between Ergasters and Asian Erectines. Finally, Kramer (1993) compared *H. erectus,* including both African and Asian forms, to a mixed-species sample of Koobi Fora hominins dating from 2.0 to 1.7 Mya (including robust *Australopithecus*, Habilines, and Ergasters) and to a sample of recent *H. sapiens* drawn from 30 populations. His analyses show that the degree and pattern of variation in the *H. erectus* sample resembles that of the single-species sample of *H. sapiens,* rather than that of the mixed-species sample. In each of these studies, the authors conclude that separation of the early African Erectines into a separate species (*H. ergaster*) is not warranted.

EARLY AFRICAN ERECTINE POSTCRANIAL MORPHOLOGY

The two partial skeletons known for the African Ergasters provide considerable information on the body form and adaptation of these early humans, and both have some interesting anomalies as well. KNM-ER 1808 preserves about 50% of the skeleton of a probable female (Ruff and Walker 1993). Although the cranium and mandible are fragmentary, they resemble other Ergaster specimens in their supraorbital torus form, cranial thickness, and mandibular and dental anatomy. However, the ER 1808 long bones exhibit extensive pathological thickening, produced by the subperiosteal deposition of coarse woven bone. Several possible causes for this pathology have been suggested, including vitamin-A poisoning caused by the eating of carnivore livers (Walker et al. 1982), the consumption of honeybee eggs, pupae, and larvae (Skinner 1991), and the infectious disease known as yaws (Rothschild et al. 1995). If this last diagnosis is correct, the ER 1808 skeleton would represent by far the earliest known incidence of a treponemal disease, a class of infections that also includes venereal syphilis. All of these pathogenic factors can cause sickness in children without any effect on the skeleton (Ortner and Putschar 1981), so it is not clear that any of them are the actual cause of the observed pathology.

The other, more complete Ergaster skeleton is KNM-WT 15000 from Nariokotome, sometimes referred to as the "strapping youth" (Fig. 5.18). It comprises some 85% of the skeleton of a male, whose age at death is estimated at 11 years from his teeth and at 13–13.5 years on the basis of his skeleton (B. Smith 1993). When he died, WT 15000 already had achieved a height of 160 cm, and would probably have topped out at around 185 cm (6 feet) if he had lived to grow up (Ruff and Walker, 1993).

The WT 15000 skeleton differs in several ways from those of a comparable modern adolescent. Some of these differences look like primitive retentions from an *Australopithecus* ancestor: relatively long femoral necks, small vertebral bodies, long and straight vertebral spines, and the presence of six lumbar vertebrae (Latimer and Ward 1993, Pearson 2000a, Walker and Leakey 1993c). The meaning of some other differences is less clear. The **vertebral canal**, the passage where the spinal cord runs through the stacked-up vertebrae, is narrower in WT 15000 than it would be in a modern human, particularly in the neck and thorax. This has been interpreted as a sign of a similar **stenosis** or nar-

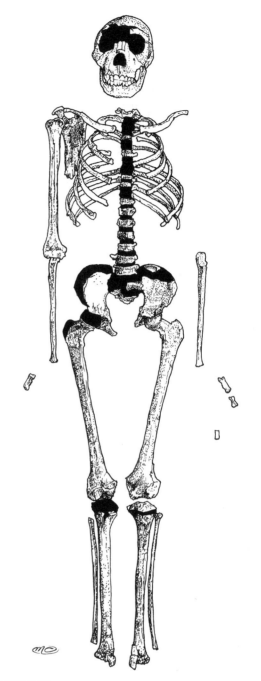

FIGURE 5.18

KNM-WT 15000 (Nariokotome III), an 85% complete skeleton of an adolescent male Ergaster from the west side of Lake Turkana, Kenya. Limb proportions and a linear trunk indicate adaptation to a tropical environment (see text). (After Johanson and Edgar 1996.)

rowing of the cervical and thoracic segments of the spinal cord, indicating a relative lack of fine motor control over the muscles that those segments innervate (MacLarnon 1993). The muscles that would be affected include the intercostal muscles of the thorax and the muscles of the abdominal wall, which act to help drive air in and out of the lungs. From these facts, some argue that early Erectines may not have had humanlike vocal control, and therefore probably lacked articulate language (Walker 1993, MacLarnon 1993). However, other abnormalities in the WT 15000 vertebral column, and the reported absence of similar stenosis in five vertebrae associated with the D2700 Erectine skeleton from Dmanisi, suggest that the narrow vertebral canal in WT 15000 may reflect some sort of pathology (Latimer and Ohman 2001, Meyer 2005, Meyer et al. 2006). A detailed assessment of what this might be has yet to be published. For now, the cause and implications of the vertebral stenosis in WT 15000 remain unknown.

Ruff and Walker (1993) have estimated both stature and body weight for a number of East African specimens dating to between 1.89 and 0.7 Mya (Table 5.8). In making these estimates, they assumed that all individuals had tropical limb proportions (see below), and used equations for predicting stature in South African Bantu (Feldesman and Lundy 1988). Body weight was predicted using equations based on multiple-regression equations of body weight on stature and latitude in living populations (Ruff 1991). If ER 736, 737, and 1808 all represent Ergasters along with WT 15000, their average stature and body weight would be 174.5 cm and 60.3 kg. This represents a 19% increase in stature and 31% increase in body weight compared to the earlier Koobi Fora specimens in Table 5.8. From hip-bone morphology, we know that OH 28 and probably ER 1808 are females, while WT 15000 is a male. Although this is a perilously small sample, it intimates that there was moderate sexual dimorphism in size in the African Erectines.

The Nariokotome skeleton provides a significant amount of additional information on body form. Judging

TABLE 5.8 ■ Estimated Body Sizes of Selected Human Specimens from East Africa

Specimen	Date (Mya)	Femur Length	Stature	Body Weight
ER 1472	1.89	40.0	149	47
ER 1481	1.89	39.5	147	46
ER 3728	1.89	39.0	145	45
ER 736	1.70	50.0	180	62
ER 1808	1.69	48.0	173	59
ER 737	1.60	44.0	160	52
WT 15000	1.53	51.7	185	68
OH 34	1.0	43.0	158	51
OH 28	0.7	45.0	163	54

Revised from Ruff and Walker (1993, p. 259). Femur (bicondylar) length and stature are given in cm, body weight in kilograms. Data for WT 15000 are estimated adult values. ER 1472, 1481, 3228, and 3728 may represent "Homo" (Australopithecus) rudolfensis.

from the form of the clavicle and ribs, the thorax appears to have been barrel-shaped like a human's, not funnel-shaped like an ape's, although it may have widened out at the bottom like a bell due to the lateral flare of the pelvis (see below). The axillary or lateral border of the scapula resembles modern humans, and differs from some other early *Homo*, in having an **axillary groove** on its ventral aspect and a distinct buttress or crest on its dorsal aspect. Since both of the femora and tibiae, one humerus, one radius, and most of the vertebrae are preserved, WT 15000 tells us a lot about limb and body proportions in early Erectines. If the Nariokotome boy had lived to grow up, his humerus would have been only about 63% as long as his femur. Comparable percentages in other hominids are more than 95% in African apes, around 90% in OH 62, 85% in AL 288, and 70–73% in recent humans. In short, the Nariokotome boy had strikingly long lower limbs. His long legs would have enhanced both the speed and efficiency of his bipedal locomotion. From this and other features of the early Erectine skeleton, Bramble and Lieberman (2004) argue that these humans were adapted for endurance running, and that this was a key element in the adaptive success of the genus *Homo*.

The lengths and mass distributions of the limbs determine their natural periods of oscillation—that is, the frequency with which they tend to swing back and forth, considered as pendulums hanging from the hip and shoulder joints. We tend to walk at about the natural pendular period of our lower limbs, and swing our upper limbs at the same frequency with the sides reversed, so that each arm swings backward while the leg on that side is swinging forward and vice versa. This synchrony conserves energy by transferring angular momentum back and forth between the swinging arms and the counter-rotating pelvis. Because our upper limbs are shorter than our lower limbs, they have a shorter inherent period of oscillation. It takes a slight muscular effort to keep them swinging in synchrony with our natural walking rhythm.

Why, then, are our upper limbs not relatively longer (as they were in *Australopithecus*), so that all our limbs have the same natural frequency? Perhaps because we often carry heavy things in our hands. When we do this, the added weight at the end of the upper limb lengthens the natural period of its swing. (Compare this with sliding the weight downward on a pendulum clock or upward on the arm of a mechanical metronome.) A recent analysis concludes that human arms, including those of the Nariokotome skeleton, are of the right length to maximize the weight that can be efficiently carried in the hands while keeping the upper and lower limbs synchronized in walking (Wang et al. 2003). By this analysis, bipedal walking while empty-handed may have been slightly less efficient for the Nariokotome youth than for *Australopithecus*; but carrying things was less energetically costly. This suggests that the efficient long-distance transportation of heavy loads— food, babies, tools, and weapons—may have been more important for early *Homo* than it had been for previous hominins.

The limb proportions of Nariokotome indicate specifically tropical adaptations. Modern human limb proportions vary clinally with ambient temperature (Trinkaus 1981, 1983a; Holliday 1997a,b). Tropically adapted populations have relatively longer and thinner limbs than populations adapted to colder climates, and this difference appears early in childhood (Eveleth and Tanner 1976). Tropically adapted groups also have relatively longer distal limb elements (tibia and radius, as compared to femur and humerus) than groups in colder climates. These patterns reflect the wider generalization known as Allen's Rule (Allen 1877)—that warm-blooded animals in hot climates have skinnier, more attenuated extremities and bodies than their cold-climate relatives, giving the tropical forms a higher surface-to-volume ratio for more effective dissipation of body heat. WT 15000's brachial index (100 × radius/humerus) and crural index (100 × tibia/femur) are "hyper-tropical," falling just above the sample mean range for living African peoples (Table 5.9). This strongly suggests that Ergasters and modern Africans shared physiological as well as anatomical adaptations to a tropical environment.

Christopher Ruff (1991, 1993, Ruff and Walker 1993) has analyzed relative body breadth as an aspect of climatic adaptation. Ruff argues that the human body can be modeled as a cylinder, with stature representing the height and some measure of body breadth representing the diameter of the cylinder. From this model, it follows that the body's surface-to-volume ratio is determined chiefly by its relative breadth. Using bi-iliac breadth to represent body breadth (because it can sometimes be measured directly on skeletal remains), Ruff shows (Table 5.9) that WT 15000's bi-iliac breadth/stature ratio falls slightly below the range of means for recent African populations and well below those for Europeans. Thus WT 15000 again falls out as "hyper-tropical" in adaptation.

The general gestalt of the Nariokotome skeleton is fundamentally modern (Walker 1993), and there is no reason to think that the locomotion of this early human would have differed in any significant way from that of people today. Nevertheless, the skeleton has some archaic features. As noted earlier, some of these may represent retentions from *Australopithecus*. The Nariokotome femora also show a complex of non-modern features—absence of the pilaster, thick cortical bone, **medullary stenosis** (narrowed medullary canals), and **platymeria** (anterior–posterior flattening) in the upper shaft—that are seen in other Ergaster femora (ER 736 and 737) and in the Erectine femora from Zhoukoudian (Walker and Leakey 1993c).

Nariokotome is the only Erectine skeleton that preserves both hipbones, but two other important early *Homo*-like hipbones are known from East Africa. The left hipbone of OH 28 (Fig. 5.19) dates to at least 700 Kya.

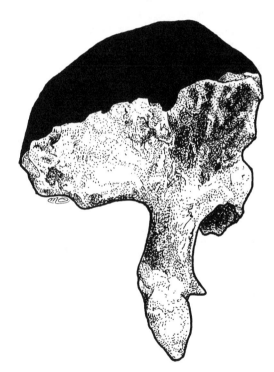

FIGURE 5.19

The OH 28 hipbone from Bed IV, Olduvai Gorge (Tanzania). (After Dunsworth and Walker 2002.)

TABLE 5.9 ■ Limb Lengths and Body Breadth Indices for WT 15000, Compared to Means of Recent and Fossil Human Samples

Specimen or Sample	Brachial Index	Crural Index	Bi-Iliac Breadth	Bi-Iliac Index (breadth/stature)
WT 15000	79.9	88.0	26.6	.144
Recent Africans	76.4–78.7	82.8–85.8	23.1–26.3	.148–.174
Recent Europeans	72.9–74.0	78.4–83.1	27.4–29.8	.160–.188
Neandertal mean	73.2	78.7	31.8[a]	.187[a]

[a]Kebara 2 only. Data from Ruff and Walker (1993), except for Neandertal indices (see Table 7.16). The values for WT 15000 are the estimated adult dimensions.

Although it has been dented and splintered by carnivore chewing, it preserves much of the ilium (except for the most superior part) and the ischium. Day (1971), who first described this bone, noted that its overall morphology was that of a fully bipedal hominin, but he also observed some distinctive features. These include a laterally flared ilium, a very prominent acetabulo-cristal buttress or vertical iliac pillar, a relatively thin iliac blade between the thick buttresses, and a relatively small auricular surface (for the sacroiliac joint). The wide greater sciatic notch indicates that the OH 28 individual was female. The hipbones of WT 15000 preserve similar portions of the bone, but are slightly less complete than the OH 28 hipbone. On the basis of its narrow greater sciatic notch, Walker and Ruff (1993) consider WT 15000 a male. The Nariokotome boy's hipbones also differ from those of OH 28 in being less robust (with less prominent pillars and tuberosities of the ilium and ischium), probably because he died before reaching maturity.

A third hipbone from the Lake Turkana sites, KNM-ER 3228 (Fig. 5.20), is an adult specimen lacking the pubis and ischial ramus. Older than the other two, ER 3228 comes from the lowest part of the Upper Burgi Member at Koobi Fora and is dated to ca. 1.95 Mya. It is not associated with cranial material, so its taxonomic status is

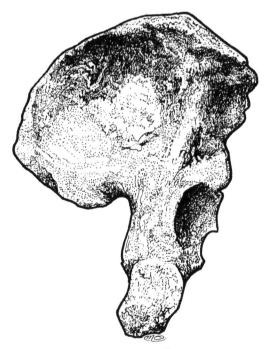

FIGURE 5.20

KNM-ER 3228, a 1.95-My-old hipbone that may represent the earliest Erectine fossil. In its overall size, pronounced iliac flare, and strong iliac buttress, it is markedly similar to other Erectine hipbones (OH 28 and WT 15000: cf. Fig. 5.19). (After Dunsworth and Walker 2002.)

unclear. There are no hipbones known for either *Australopithecus habilis or A. rudolfensis*. (In fact, no postcranial remains of any sort can be definitely attributed to *A. rudolfensis*.) However, the ER 3228 hipbone is notably larger than two femora (ER 1472 and 1481) that come from the same time span (Table 5.8) and might belong to one of these taxa. And in its morphology, ER 3228 is very similar to OH 28 and Nariokotome. The similarities to OH 28 were noted by R. Leakey (1976), who attributed ER 3228 to the genus *Homo*. Walker and Ruff (1993) state that ER 3228 differs from WT 15000 only in its more robust vertical iliac pillar and muscle/ligament attachments—differences that (again) probably reflect the subadult age at death of WT 15000.

Like WT 15000 and OH 28, ER 3228 does exhibit some similarity to *Australopithecus*, particularly in its widely flaring ilia. But its acetabulum is large and human-like, and it has thickened weight-bearing iliac pillars like those of OH 28. All these features probably reflect increased body size, which McHenry (1994a) estimates at 61.6 kg for the ER 3228 individual. This estimate is considerably larger than most estimates for *Australopithecus* (Table 4.4), including the Habiline species (McHenry 1994a, Fleagle 1999, Wood and Collard 1999), and conforms to those for other African Erectines (Table 5.8). The size of ER 3228 and its morphological similarity to the hipbones of later African Erectines suggest that this specimen represents the earliest known Ergaster.

The Nariokotome pelvis has prompted speculation about the birthing process in early *Homo*. Walker and Ruff (1993) reconstruct the pelvis as relatively broad (from side to side) and shallow (from front to back). There is some guesswork involved in the front-to-back measurement, since the pubic bones and the anterior rami of the ischial bones are not preserved. However, Ruff and Walker think that the pelvic inlet probably had a flattened oval shape like that of *Australopithecus* (Fig. 4.34), and that this shape would have been maintained throughout the pelvic inlet, even at the most interior part. Ruff and Walker used data on growth and sexual dimorphism in pelvic inlet dimensions in recent humans to estimate these parameters in WT 15000 as if it were a female that followed a *Homo sapiens* growth trajectory after age 11. If their reconstruction is accurate, it suggests that the birthing process in early *Homo* was similar to Lucy's (Rosenberg 1992, Rosenberg and Trevathan 1996, 2001; Ruff 1995). This means that the head of the neonate would have entered the birth canal facing laterally, so that the long axis of the head would have corresponded to the widest part of the oval-shaped pelvic inlet (Tague and Lovejoy 1986). And because the birth canal seems to have had a transversely oval shape throughout, the baby's head probably would not have rotated in the inlet to face posteriorly just before emerging, as it does in later humans (Ruff 1995). Unlike other primate neonates (which emerge facing the mother), Ergaster babies would therefore emerge facing to the side. In this position, the mother would have great difficulty clearing the infant's breathing passages or guiding the infant out of the birth canal, because there is a danger of injury to the baby's spine if the mother grabs the baby and tries to guide it from this angle. Rosenberg and Trevathan (2001) conclude from all this that Ergaster females (and *Australopithecus* females as well) probably cooperated in childbearing, helping each other to give birth.

Walker and Ruff (1993) also tried to estimate the maximum brain size of a newborn Ergaster from the WT 15000 pelvis. Given their reconstructed pelvic inlet breadth of 120 mm for WT 15000 (reconstructed as an adult female) and reducing this by an extra 10 mm for soft tissue in the canal, they calculated that the maximum neonatal head length that could be accommodated would be 110 mm. From data on modern newborns, they estimated that the brain weight of a full-term Ergaster fetus would have been 200–240 grams. If the newborn Ergaster had followed a great-ape growth pattern from this starting point, adult brain weight would have topped out between 440 and 600 grams. This is far smaller than the actual brain size of an adult Ergaster (Table 5.6). The discrepancy suggests that Ergasters had already developed the uniquely human trait of prolonging the fetal brain-growth rate into the first postnatal year. Such a pattern would allow the brain to grow to 2.5 times its birth weight of 200 grams during the first year, and to add another 200 grams before adulthood. These assumptions yield a more realistic adult brain weight of ~900 grams. (The estimated cranial capacity for WT 15000 is 909 ml.) Ruff and Walker conclude that the human pattern of brain growth—what Martin (1983) called the "secondary altricial" pattern—is an inheritance from the earliest days of the genus *Homo*.

All these ideas about parturition, midwifery, and brain growth in Ergasters are intriguing. They are also highly speculative, because they involve estimating the dimensions of an adult female's birth canal from one reconstructed pelvis of a juvenile male.

■ EARLY ERECTINE ADAPTATIONS: ANATOMY AND PHYSIOLOGY

Erectines had bigger brains than any earlier hominins, but they also had bigger bodies. It is not clear whether the brain-size increase was greater than would be expected from the increase in body size. D. Begun and A. Walker (1993) present four different calculations of Erectine **encephalization quotients**, which express an animal's observed brain size as a ratio of that expected for its body weight on the basis of brain-body allometry.

TABLE 5.10 ■ Comparison of Four Calculations of Encephalization Quotients for Habilines, *Homo erectus*, and Modern *H. sapiens*

Group	Jerison	Bauchot	Martin 1	Martin 2
Habilines	4.30	18.40	3.47	2.32
Erectines	4.40	19.09	4.29	2.98
Moderns	7.15	29.80	5.08	3.60

Each EQ is calculated using the allometric equation used by the author indicated. Other *Australopithecus* species fall markedly below the Habilines, except in the Martin 2 calculation. See Walker and Begun (1993) for detailed references.
Source: Adapted from Begun and Walker (1993, p. 349).

In two of these four calculations (Table 5.10), Erectines are only slightly more encephalized than Habilines; in the other two, Erectines are roughly midway between Habilines and modern humans in relative brain size. Studies by H. McHenry (1994a,b) reach similar conclusions. One problem with all these comparisons is that we have no definitely identified postcranial remains of the larger-brained Habiline species, *A. rudolfensis*. Although Ergaster brains are some 14% bigger than those of *A. rudolfensis*, we cannot tell how much of this difference is due to body size alone, because the body size of *A. rudolfensis* is unknown. If the ER 3228 hipbone (with an associated body-size estimate of 61.6 kg) belongs to *A. rudolfensis*, then Ergasters were about the same size as *rudolfensis*, and their bigger brains represent a real increase in encephalization. But if the ER 1472 and 1481 femora belong to *A. rudolfensis*, then it was a smaller animal (Table 5.8) and may have been equally encephalized despite its absolutely smaller brain.

The increase in body size seen in the early Erectines is striking (Table 5.8), particularly if ER 3228 is a member of this group. An increase as marked as this is not likely to have been produced by random factors; it must have been driven by rather strong selection pressures. Body-size increase may have had two main adaptive advantages for early *Homo*. First, if Erectines ate more meat than their predecessors, which seems likely from the archaeological record and other indicators (see below), then size would have mattered. Although increased carnivory does not necessarily demand larger size, larger size can confer advantages in both hunting and scavenging. Other things being equal, bigger predators can take and defend larger prey, even if that "prey" is a carcass. For a tool-using predator like early *Homo*, increased size would have made it possible to put more force behind the use of spears, clubs, and other implements for hunting and defensive purposes.

A second, less obvious but perhaps more significant advantage to increased body size is its adaptive value in hot, arid environments. Increasing body size in such an environment would seem to run counter to Bergmann's Rule (Bergmann 1847), that tropical climates tend to favor smaller body size in mammals (because smaller size increases the surface-to-volume ratio and facilitates the dissipation of body heat: Beall and Steegmann 2000). But if we accept Ruff's contention that the surface-to-volume ratio in the human body (modeled as a cylinder) and its allometric relationships (between that ratio and body size) depend mainly on the body's cross-sectional area, then augmenting body size by increasing trunk length will not have much of a negative impact on thermoregulation in tropical humans so long as body breadth is kept consistently narrow. And augmenting body size would have had two big additional advantages for early Erectines. First, increasing body size while maintaining a linear body build helps to conserve moisture in hot, arid environments (Wheeler 1991b, 1992b, 1993). Other things being equal, the rate of water loss relative to overall weight drops off as body size goes up (Table 5.11). It has been calculated that under high-temperature conditions (35–40 °C), with metabolic rate, food intake, and excretory losses held constant, a 70-kg human with a body form like that of early *Homo* (WT 15000) would need to drink only some 82–85% as much water as an *Australopithecus*-shaped Habiline (Wheeler 1993, p. 25). Thus, increasing body size would have allowed early *Homo* to travel farther per unit of water consumed—an important factor in allowing this migrating ape to range more widely and exploit a broader ecological niche. This adaptive advantage would have been further enhanced by the water-conserving changes in nose morphology seen in early *Homo*.

Second, increasing body stature would have yielded benefits in locomotion. Longer limbs result in a longer stride, and increase in stride length allows an animal to cover more ground in a given amount of time. (For this reason, thoroughbred horses, greyhounds, and human track stars tend to have relatively long legs and lanky builds by comparison with horses, dogs, and humans less selected for racing.) Increased speed in walking and running would have obvious advantages for a mobile and wide-ranging creature like early *Homo*. It has been suggested that this stride-length increase would also have been energetically more efficient, and that it evolved to allow human hunters to follow migrating herds of herbivores (Sinclair et al. 1986). Although these conclusions have been criticized as overly simplistic (Lovejoy 2005a,b), a longer stride would surely have had value in many different hunting strategies (for example, in the relentless tracking of a wounded quarry). This value would have helped to promote size increase in Erectines.

Brain size, body size, and subsistence strategies are intertwined with one another in a complicated way. W. Leonard and M. Robertson (1994) calculate that simply increasing body size from that of *A. habilis* to that suggested for early Erectines would have resulted in a 40–45% increase in total energy consumption. And big brains are metabolically expensive. In humans today, the brain utilizes over 25% of the body's energy budget, although it makes up only about 2.5% of body mass. Humans expend two or three times as much energy in brain metabolism as would be expected for an average primate of the same body weight, and five times as much as the average human-sized mammal (Martin 1989, Leonard 2000). Yet despite their big, metabolically costly brains, modern humans do not have higher metabolic rates than would be expected for anthropoids of our body size (Leonard and Robertson 1997b). Evidently, brain enlargement has been "funded" by making cuts in other metabolically expensive body tissues. As noted in Chapter 4, most of the savings appears to have come from reduction in the size of the human gut, made possible by improvements in the caloric content and digestibility of the human diet (Aiello and Wheeler, 1995). Judging from the shape of the Nariokotome skeleton, Erectines had less protruding abdomens (and therefore smaller guts) than apes or *Australopithecus*, and so they must have had diets of correspondingly improved quality (Clutton-Brock 1980, Aiello and Wheeler 1995, Leonard and Robertson 1994, 1997a, Leonard 2002).

Much of that improvement probably came from an increased consumption of animal flesh (Leonard 2000). As any calorie-counting dieter can tell you, meat contains more energy than most other foodstuffs: 100–600 kilocalories per 100 grams (depending on the organs being eaten), as opposed to 20–240 kcal for edible leaves, fruits, and roots (Pennington 1989). This does not mean that humans had to become anything like strict carnivores to support their ravenous central nervous systems.

TABLE 5.11 ■ Thermoregulatory Water Loss, Drinking Requirements, and Potential Ranging Distances Calculated for Active Humans of Differing Body Weights at 35–40 °C

Body Weight (kg)	Sweat Loss (% of Weight)	Total Water Loss (% of Weight)	Water Required (% of Weight)	Range (in km)[a]
35	2.36	3.71	2.4	11
52.5	1.97	3.19	2.0	19
70	1.84	2.98	1.9	24
87.5	1.68	2.76	1.7	30

[a]Assumes a 4% dehydration rate over a 12-hour time period. Differing dehydration rates would change these parameters.
Source: Data from Wheeler (1992b).

For example, seeds and nuts are fully as nutritious as meat (300–550 kcal per 100 g), and provide the main dietary staples for most humans today. Humans are in fact distinguished from other primates by their uniquely wide variety of diets and subsistence strategies (Leonard 2002). But because our large brains are coupled to small guts, a workable human diet has to center on foods that are both high in energy and relatively easy to digest. And since Erectines appear to have had relatively larger brains and smaller guts than *Australopithecus*, they must have been eating more of such foods than their predecessors—and probably using new technologies to relieve their guts of some of the work of processing and digesting food.

The skull and teeth of Erectines bear independent witness to this change. Erectine chewing muscles, facial buttresses, and cheek teeth are markedly reduced compared to those of *Australopithecus*. Brace (1967) has long attributed this reduction of the chewing teeth and supportive architecture to a relaxation of natural selection, brought about by dietary shifts and technological improvements related to food preparation. Although there is still considerable overlap in these dimensions between *Australopithecus* and early *Homo* (Wolpoff 1971b), the reduction in the average size of Erectine cheek teeth is unmistakable (Table 5.12). This reduction tends to be most marked in the most posterior and latest-erupting molars (M3s) and premolars (P4s). Erectines also had thinner mandibular corpuses (relative to their heights) and much smaller temporal fossae than *Australopithecus* had. All these changes imply that occlusal forces on the cheek teeth were reduced in Erectines.

Similar inferences have been drawn from the characteristic Erectine increases in the size of the mandibular fossa and the breadth of the face. M. Wolpoff (1999) argues that these increases reflect a change in the positions of attachment for the masseter and temporalis muscles, which yielded a larger side-to-side or transverse component in chewing movements—implying more emphasis on grinding and less on crushing in mastication. These changes, too, can be logically viewed

as correlates of a diet requiring less powerful occlusal forces.

The softer Erectine diet must have featured increased amounts of meat. But a meat-rich diet creates problems of its own. Hunting and scavenging require wider ranges than collecting and gathering, and covering these wider ranges requires greater caloric expenditures. Moreover, lean meat is hard for humans to digest. This would have presented a problem in dry seasons, when prey animals grow lean and do not furnish human predators with the body fat needed to promote digestion. This problem can be addressed by smashing the bones of prey and extracting the fatty marrow (Klein 1999), or by the seasonal use of plant tissues as fallback foods. O'Connell et al. (1999) have emphasized the importance of tubers and other underground storage organs as dietary resources in recent hunters and gatherers and possibly also for Erectines. USOs might have been consumed more intensely in the dry seasons by early Erectines, which would have ameliorated the problems associated with meat-eating during such periods.

No matter what assumptions we make, it is a virtual certainty that early humans could not have afforded a specialized focus on hunting and scavenging. They must have retained a solid base of gathering and collecting as well (O'Connell et al. 1999, 2002; Hawkes et al. 1997, 1998; Lancaster 1975, 1978). This suggests that Erectines had developed—or perhaps improved upon from an earlier stage of hominin evolution, as Lovejoy's (1981) theories imply—a human-like sexual division of labor, summed up in the phrase "man the hunter, woman the gatherer." This sort of two-track subsistence strategy must have cut in at some point in human evolution, because it characterized the modern human hunters and gatherers known to recent history. The anatomical evidence we have reviewed indicates that this strategy may already have been present in early Erectines, but this conclusion needs to be tested against the archaeological record.

Before we go on to deal with knapped stones and animal bones, we need to consider another related aspect of human anatomy and physiology. The anatomical peculiarities of the human skin—the loss of body hair, the development of extensive eccrine sweat glands, and a general increase in pigmentation—probably first appeared in early African *Homo*. This conclusion is speculative, since skin almost never leaves any traces in the paleoanthropological record. But these apomorphies seem likely to have originated as part of the adaptive shift to a modern human foraging pattern.

In hot environments, mammals use a variety of strategies to get rid of excess heat. Many dig burrows or depressions in the soil, to which they retreat during the hottest part of the day to dump heat into the earth by direct **conduction** from some naked or lightly furred part of the body surface (e.g., a dog's belly). Large mammals, which have low surface-to-volume ratios that

TABLE 5.12 ■ Comparison of Mandibular Dental Breadths in the Posterior Dentitions of Late *Australopithecus* and Erectines (Measurements in mm)

Tooth	A. habilis	A. rudolfensis	A. robustus	Erectines
P$_3$	11.7	12.4	12.5	10.3
P$_4$	11.7	11.4	13.7	10.1
M$_1$	12.8	13.1	14.5	11.8
M$_2$	14.4	14.5	15.5	12.3
M$_3$	13.5	14.3	14.9	11.7

make them especially susceptible to metabolic heat buildup, often immerse themselves in waterholes or mud wallows where they can lose heat by **convection** to the surrounding liquid (and indirectly into the surrounding earth). Some mammals have enlarged ears or other skin surfaces that can help to cool the body by **radiation**. All these options require, however, that the animal retreat into a favorable microenvironment to cool itself when its body temperature rises. They are all available to, and are adopted by, reptiles and other ectotherms as well as mammals.

The distinctively mammalian (and avian) mechanism for losing excess heat is cooling by **evaporation**, which can carry heat away from the body even when the air temperature is higher than the animal's temperature. Most mammals rely largely on the maxilloturbinal surface for evaporative cooling. We use this mechanism to some extent, but the reduced turbinals that we have inherited from our anthropoid ancestors (Fig. 3.12D) do not provide enough evaporative surface to cool off a heat-stressed human. Our ancestors got around this limitation by evolving novel adaptations that allow water, in the form of sweat, to be evaporated from the entire surface of the body.

There are two types of sweat glands in mammals. **Apocrine** glands, associated with hair follicles, secrete sweat containing the remnants of ruptured cells. The decomposition of this detritus produces the characteristic musky smells of mammalian skin, yielding distinctive odors for different species. (Compare the smells of wet human hair and a wet dog.) In humans, apocrine sweat glands are concentrated in the pubic region, in the armpits, and around the nipples. **Eccrine** sweat glands lie closer to the skin surface and open to it directly through pores (Folk and Semken 1991). These glands, which secrete a watery, salty fluid, are unusually numerous in humans, where they are spread densely and relatively evenly over the whole body surface. The evaporation of eccrine-gland exudate is our main recourse for cooling off under conditions of heat stress.

In order for sweat to carry heat directly away from the skin, it must evaporate directly from the skin. This implies that the skin must not be covered by an extensive and continuous coat of body hair, as it is in most mammals (Newman 1970, Wheeler 1984, 1985, 1991a, 1992a). While body hair can act as insulation from external heat, it also holds metabolic heat in. Evaporation from sweat-soaked fur can carry away a certain amount of heat (as it does, for example, in a sweating horse). But skin-surface sweating is more effective, and a naked skin allows cooling by radiation as well as by evaporation. Human body hair is reduced to allow for these effects. We actually have as many body hairs per square inch as chimpanzees (Harrison and Montagna 1969), but most of the hairs are very thin and do not

protrude much beyond the skin surface, so the skin is left largely exposed. Under hot conditions, our reliance on evaporative cooling results in a rapid loss of water, which has to be replenished by frequent or copious drinking. All this explains why *Homo sapiens* is, in R. Newman's (1970) phrase, "such a sweaty and thirsty naked animal."

These apomorphies of human skin carry information about our ancestors' way of life. The loss of body hair and the multiplication of eccrine sweat glands in *Homo* presumably represent adaptations to increased daytime activity levels in a hot climate. They would have allowed relatively hairless early humans to tolerate higher environmental temperatures and levels of metabolic heat than their hairier, apelike ancestors (Wheeler 1992a). Evaporative cooling by sweat, however, does not work very well when the relative humidity is high. The human reliance on sweating is therefore more likely to have evolved in the relatively dry environments of eastern and southern Africa than in the humid tropical forests of West Africa. Earlier hominins (*Australopithecus*) may have been considerably hairier than the Erectines and later humans, since they apparently lived in less open habitats and lacked the human body size and form seen in early *Homo* (Cachel and Harris 1995, 1996, 1998).

Hair helps to shield the skin of diurnal mammals from sun exposure. Fur loss could not have proceeded very far in early humans without an alternative way of protecting the skin from the damaging effects of the intense tropical sun. Humans, like elephants and some other semi-hairless tropical mammals, are defended against such exposure by their densely pigmented skin. The pigment occurs in the form of **melanin** particles, located at the base of the outer skin layer (the epidermis). The amount of melanin produced by skin cells is controlled by both genetic and environmental factors. In humans today, skin color is a genetic polymorphism under the control of several different genes, which also determine melanin concentrations in the hair and the iris of the eye. Untanned human skin varies in color from pale pink to near-black; but most of us can grow darker by tanning, depositing more melanin in skin areas that are regularly exposed to the damaging effects of sunlight. The most injurious wavelengths in sunlight lie in the 280- to 400-nanometer range, just beyond the violet end of the visible spectrum. The melanin particles in the epidermis absorb this ultraviolet radiation (UVR) and prevent its penetration further into the body. Unblocked UVR can lead to several negative consequences. The most obvious of these is skin cancer, but the prevention of sunburn is also important because even mild burning has a significant negative effect on the rate of sweating and thus on thermoregulation. Too much UVR can also lead to extensive degrading of folate, a derivative of folic acid that is critical to development of the embryonic neural tube and to spermatogenesis

(Jablonski and Chaplin 2000). The combination of these factors would almost surely result in strong selective pressures favoring the evolution (or retention) of darkly pigmented skin, to block the penetration of tropical UVR into Erectine bodies.

But why go to all this trouble? Early Erectines would not have needed to worry about moisture conservation, heat gain, and sunburn if they just stayed out of the sun and restricted activity to cooler periods of the day when the sun and heat are less of a problem. Most mammals follow some such strategy. Nocturnal mammals do not have to deal with solar radiation at all; but a nocturnal lifestyle was not an option for early humans, who (like all catarrhines) had retinas that were ill-suited to nighttime activity. An alternative option would have been to adopt a **crepuscular** activity cycle that peaked during dawn and dusk. However, this is the time when most large African carnivores are active. Crepuscular foraging would have placed early humans in competition with lions, leopards, and hyenas. Worse yet, it would also have exposed them to direct predation by these formidable killers. But the role of a diurnal hunter/scavenger was pretty much there for the taking by early Erectines. The available evidence from comparative anatomy and physiology strongly suggests that they succeeded by meeting the physiological challenges of this diurnal niche. A diurnal pattern of subsistence activity—incorporating intensive gathering, collecting, hunting, and scavenging—defined a Paleolithic human feeding strategy that remained characteristic of humans throughout the Pleistocene, and persisted among hunters and gatherers into the ethnographic present.

▌ EARLY ERECTINE ADAPTATIONS: THE ARCHAEOLOGICAL EVIDENCE

The oldest stone tools, dated to about 2.6 Mya, are Oldowan choppers, scrapers, and flakes that come from several Ethiopian sites in the Awash region (Semaw 2000). These tools are some 500,000 years older than the first fossils that may represent Erectines, and nearly a million years older than the earliest undoubted Erectines (~1.7 Mya). The current evidence thus suggests that the most ancient stone tools may have been made by Pliocene hominins that we assign to the genus *Australopithecus*.

However, all later occurrences of the Oldowan Industry fall within the time span of the African Erectines. The possible early Erectines ER 2598 and 3228, dated to around 1.95 Mya, are fully as old as the most ancient Oldowan tools from Olduvai Bed I and Koobi Fora. Later, more sophisticated types of Oldowan tools—the Developed Oldowan of Upper Bed II at Olduvai, and the Karari Industry at Koobi Fora—begin around 1.7 Mya, when unequivocal Erectines (e.g., ER 3733, 1808, 730)

are known. Admittedly, none of these tools are associated exclusively with Erectines. Oldowan and Developed Oldowan tools turn up in association with several other hominin species: robust australopithecines in East and South Africa, possible late-surviving *A. africanus* at Sterkfontein, *A. garhi* in East Africa (by inference from the cutmarks on animal bones from Bouri), and the two Habiline species (*Australopithecus habilis* and *rudolfensis*). It is tempting to think that stone tools are the exclusive hallmark of the genus *Homo*, but that proposition cannot be demonstrated by the archaeological record.

But no matter who else may have been making stone tools, the early Erectines certainly were doing so. In addition to their association with Oldowan cultural material, Erectines are the only ancient hominins (>1 My) regularly found with Acheulean tool assemblages (Barut Kusimba and Smith 2001). The only other hominins associated with the Acheulean are robust *Australopithecus*, at the Peninj site in northern Tanzania (Figs. 4.3, 4.40) and at Swartkrans (where possible Erectines are also present).

After 1.6/1.5 Mya, the Oldowan and Developed Oldowan disappear as discrete entities, although the tool types associated with them persist as a part of later assemblages. The first definite Acheulean assemblages occur around 1.4 Mya at Konso (Asfaw et al. 1992), but some think that the Acheulean goes back even further, to the end of the Oldowan (Cachel and Harris 1998, Barut Kusimba and Smith 2001). The Developed Oldowan differs from the Oldowan mainly in its inclusion of "protobifaces" and crude bifaces (handaxes) and in differing percentages of certain other "tool types." As S. Cachel and J. Harris (1995) note, it is often difficult to decide whether certain assemblages should be considered Developed Oldowan or early Acheulean. It is therefore reasonable to suspect that the latter tradition developed from the former, and that all these tool types represent a single line of cultural evolution. Although M. Leakey (1971b, 1975) did not believe that the Acheulean arose from the Developed Oldowan, other archaeologists, from Bordes (1968) to Klein (1999), have seen a natural progression and continuity from Developed Oldowan to Acheulean in many artifact forms and stone-working techniques. In short, it is possible (although far from certain) that the entire Early Stone Age (ESA) in Africa, from the Oldowan to at least the earlier part of the Acheulean, was produced by a single hominin, *Homo erectus/ergaster*.

No matter how many hominin species were involved, it seems clear that the main objective of stone-tool making throughout this period was the production of sharp cutting edges (Toth and Schick 1986, Semaw 2000). These were useful for many tasks besides the processing of carcasses. Use-wear analysis of flakes from Koobi Fora reveals that they were used for working several materials, including both animal and plant tissues (Keeley and Toth 1981). These facts provide further evidence for an increasingly broad pattern of resource exploitation by early *Homo*.

Does archaeology provide evidence for increased amounts of meat in the Erectine diet? In Africa throughout the Early Stone Age (ESA: Oldowan + Developed Oldowan + Acheulean), local concentrations of animal bones are associated with stone tools and other indications of human activity. The problem is that it is not easy to determine what causes produced these associations. In some cases, water transport—particularly high-energy transport by rapidly moving rivers and streams—can produce fortuitous aggregations of bones and stones. Such cases are relatively easy to identify, because bones and stones transported in this way generally exhibit telltale signs of having been knocked about. But even when these bone-and-stone associations are not fortuitous, their ecological meaning is hard to ascertain. Some have contended that ESA bone/artifact assemblages are best explained as chance associations, or as the result of human scavenging at other predators' kill sites (Binford 1981, 1985, Binford et al. 1988). Others argue that at many sites, humans appear to have been responsible for bringing the bones and tools together (Bunn 1981, 1991, 1994; Bunn and Kroll 1986, Kroll and Isaac 1984, Kroll 1994, Shipman 1988, 1989; Potts 1988, Potts and Shipman 1981).

Whether or not humans were the primary predators or the agents of transportation, it seems clear that early humans equipped with ESA technology had increasing access to meat, marrow, brains, and other animal tissues that would have improved the quality of their diet. From the Oldowan into the Acheulean, there is a general upward trend in the variety of stone tools made and the quality of their manufacture, and a similar upward trend in the occurrence of animal bones (including those of large mammals) associated with evidence of human activity (Leakey 1971b, Monahan 1996, Cachel and Harris 1996, 1998). Thus, the archaeological record strongly suggests that meat was becoming an increasingly important part of the human diet throughout the course of the ESA. The relative importance of hunting and scavenging are still being debated. But the distinction between these activities is less important than it might seem, for all predators will steal kills from others if they get the chance.

The inferences we can make about early humans' ecology are limited, because our sample of their technology is biased and incomplete. Early tool users often must have utilized bones, sticks, and other objects without modifying them, and such "natural tools" are usually not identifiable as tools. Artifacts made of materials more perishable than stone must also have been a critical part of early human technology, but these are rarely preserved in the archaeological record. Wooden

implements are known from a few late ESA sites. At Kalambo Falls in Tanzania, the Acheulean levels have yielded what appears to be a wooden club, wooden pieces that may have been used to haft stone tools, and several sharpened pieces of wood (Clark 1969). A presumed spear point made of wood was recovered at Clacton-on-Sea in England (Oakley et al. 1977), and wooden pieces polished by use were excavated at Gesher Benot Ya'akov in Israel (Belitsky et al. 1991). These rare survivals hint that wood was extensively used by archaic humans. The sharpened and polished pieces from these sites may have been digging sticks rather than weapons. However, the late Lower Paleolithic (equivalent to ESA) site of Schöningen in Germany has yielded three virtually complete throwing spears, all over 2 m long, that were almost certainly hunting implements (Thieme 1996, 1997, 1999, Dennell 1997). They are associated with a large series of horse remains, many of which were clearly butchered by stone tools. Dated to around 400 Kya, Schöningen is probably associated with post-Erectine humans, but it proves that early members of the genus *Homo* were hunting large mammals.

Utilized bone fragments have also been found at various sites dating to the Erectine period. These include the site of Bilzingsleben in Germany and the African sites of Olduvai Gorge, Swartkrans, and Kalambo Falls. We have already (in Chapter 4) mentioned the more than 60 utilized bones from Swartkrans. Experimental studies suggest that the polish on many of these bones results from their use as digging implements (Brain and Shipman 1993). This serves a reminder that early humans must have continued to get most of their food from plants. Unfortunately, little is known about the vegetable items on their menus. Although plant remains are preserved at such sites as Kalambo Falls, Zhoukoudian, Gesher Benot Ya'akov, and Bilzingsleben, we cannot say for sure whether any of them represent food plants. But evidence for digging implements, and indications of microwear on ESA tools commensurate with the working of plant material (Keeley and Toth 1981), provide indirect indications that plant tissues were important to ESA peoples.

As noted earlier, it is not certain whether the bone "digging sticks" from Swartkrans were utilized by *Australopithecus robustus* or early *Homo*. Underground storage organs (USOs) of plants may have figured in the diets of both species. In a comparative analysis of strontium/calcium levels in the fossil hominin bones from Swartkrans, Sillen et al. (1995) found that the SK 80 *Homo* had higher levels of strontium relative to calcium than any of the seven *A. robustus* specimens examined. This is surprising. Diets with more meat content generally have proportionally less strontium (because the strontium gets filtered out by more and more vertebrate kidneys as it nears the top of the food chain). If early

Homo had been more of a predator than *A. robustus*, we might expect its bones to have a lower Sr/Ca ratio. The high Sr/Ca ratio in the SK 80 Erectine may reflect a biased loading of strontium in the plant part of its diet. Sillen and co-authors note that the USOs of plants that grow near Swartkrans today have high Sr/Ca ratios. Given the evidence for digging use on the Swartkrans bone implements, these authors suggest that Swartkrans *Homo* may have been using those tools and exploiting USO resources more than *A. robustus* was. However, another possible *Homo* specimen from Swartkrans, the young SK 27 skull, has a Sr/Ca ratio that falls into the robust australopithecine range at the site. This may reflect dietary variation in early *Homo*, or the specimen's young age, or both.

Taken together, all these sources of information point to a very inclusive subsistence strategy for ESA *Homo*, in which new technologies were employed to secure a wide range of food items by hunting, scavenging, gathering, and collecting. The archaeological evidence overall suggests a marked increase in the breadth and quality of human diets at this time—which conforms to the inferences drawn above from the fossil bones, skulls, and teeth of early *Homo*.

The use of fire in food preparation would have profoundly affected this dietary shift. Fire would allow for the processing of both plant and animal foodstuffs, some of which might not be edible or easily digested when raw. The control of fire was also a prerequisite for the spread of *Homo* into temperate environments, probably even within Africa. Unfortunately, archaeological proof of early human control of fire has been elusive. The earliest possible evidence for human-controlled fire comes from site FxJj 20 at Koobi Fora, dated to ~1.5 Mya, where reddened patches of sediment suggest a controlled fire that burned for several days (Bellomo 1994). Fragments of burned clay of about the same age have been found in association with Oldowan tools at the Kenyan site of Chesowanja (Gowlett et al. 1981). However, similar effects can be produced by natural brushfires, particularly when roots burn underground (Isaac 1984). There is no indication at either site that the fire area was defined by lining the fire pit or encircling it with stones. The evidence for human use of fire at this early date thus remains equivocal.

Member 3 at Swartkrans has produced more convincing evidence of the use of fire around 1 Mya. Brain and Sillen (1988) analyzed apparently burned bone from this level and found structural alterations of the bone tissue that would have required higher temperatures than those normally generated in bush fires. They argued that these bones must have been burned as the result of human-controlled fire (see also Brain 1993). Again, however, there are no indications of distinct hearth areas. Control of fire is also reported for several Eurasian sites, and among these Zhoukoudian is usually cited as the most convincing. However, the evidence for human-controlled fire at ZKD has recently been strongly questioned (see below). Science has yet to come up with a conclusive test that will enable us to identify the earliest stages in the human control of fire.

■ PATTERNS OF DEVELOPMENT AND EVOLUTIONARY CHANGE IN ERECTINES

Growth and development in early Erectines appear to have differed from the pattern seen in modern humans (Leigh 2001). Our best evidence for this comes from the juvenile Erectine skeleton from Nariokotome, WT 15000. B. H. Smith (1993) has shown that if we assume a human-like pattern of ontogeny for WT 15000, different sources of information yield conflicting estimates of his age at death. Analysis of the stages of formation of individual teeth (which is generally regarded as the most accurate technique) estimates his age at 10.5 to 11 years, whereas estimates derived from the stage of epiphyseal formation/fusion and from stature put his age at 13–13.5 years and 15 years, respectively. Smith believes that WT 15000's imperfect conformity to human growth standards, reflected in the spread of these age estimates, may imply that Erectines did not grow up precisely as modern humans do.

What might the difference in growth pattern be? B. H. Smith (1993) suggests that it may relate to a unique human characteristic, the adolescent growth spurt. At the beginning of sexual maturity, chimpanzees have attained a higher percentage of their adult size than humans have (Gavan 1953). Humans catch up with a marked increase in growth rate during adolescence (hence the term "adolescent growth spurt"). Bogin (1990) believes that this delay in body-size maturation is the result of selection to extend the period during which children can learn from parents before they signal their adulthood by achieving adult stature. From the Nariokotome skeleton, Smith infers that early Erectines lacked the adolescent growth spurt, and would thus have attained their adult size earlier in development than modern humans do.

Among mammals in general, and among primates in particular, adult brain size correlates with virtually every aspect of a species' life history (Sacher 1975, B. H. Smith 1989a,b; Barrickman et al. 2008). By estimating brain size from fossil remains, we can use regression equations derived from living primates to estimate life-history parameters for extinct species. In a study of 21 extant primate species, B. H. Smith (1989a, 1991) established that a species' brain size correlates strongly with the age at emergence of the first permanent molar. From this correlation, Smith estimates that early Erectines (with an estimated adult brain volume of 810 cm^3) would be expected to erupt their first molars at 4.5 years and their second molars at about 9 years. Their lifespan would potentially extend some 15 years beyond that of a chimpanzee. These estimates describe a species with a life history that is intermediate between those of *Pan* and modern *Homo*.

Smith's calculations are supported by a recent comparative study of hominin tooth development by Dean et al. (2001). This study examined daily incremental growth markings in dental enamel to calculate rates of enamel formation in 13 fossil hominins, including robust and gracile *Australopithecus* and five early Erectine specimens (ER 808, ER 820, WT 15000, SK 27, and Sangiran 4). All of these early hominins exhibit a faster trajectory of enamel growth than modern humans. With this faster trajectory, early Erectine first molars would have erupted at around 4.5 years—the same age that Smith arrived at from her brain-size data. This figure is closer to the M1-eruption age for the great apes (~3.5 years) than to that for modern humans (~6 years). Dean and his co-workers estimate that growth was completed in Erectines at 14–16 years of age. Their data do not preclude the existence of an adolescent growth spurt, but they do show that the Erectine growth curve was different from that of recent humans.

Both punctuationists and gradualists have studied evolutionary rates within the *H. erectus* lineage and tried to show that they fit their preferred models of evolutionary modes and mechanisms. For example, it might seem at first glance (Tables 5.3 and 5.6) that earlier Erectines had smaller cranial capacities than later ones. This suggests that there was a gradual, anagenetic increase in brain size over the time span of *H. erectus*. However, Rightmire (1981) used regression analysis of several measurements (cranial capacity, biauricular breadth, M_1 breadth and mandibular-robusticity index) to show that no significant pattern of change with time could be documented. He divided the Erectine sample into six successive groups, plotted the group average for each metric character against the average date for the specimens included in the group, and found that none of the resulting regressions differed significantly from zero. Rightmire concluded that Erectines represented a period of stasis in human evolution.

This conclusion fits punctuationist expectations. But it contradicts an earlier analysis by Bilsborough (1976), who divided the Erectine sample into only two subsamples—early and late—and used both univariate and multivariate statistics to demonstrate significant differences between the two. Wolpoff (1984) found that dividing the Erectine sample into *three* time spans also reveals significant change over time in a number of cranial, dental and mandibular features, which mirrored rates of change in other vertebrate lineages over similar time spans. However, Wolpoff included some specimens in his Erectine sample that Rightmire did not (e.g., Hexian and Vértesszöllös) and excluded some that Right-

mire used (most importantly, Salé). Wolpoff also argued that least-squares regression was not appropriate, and instead analyzed his data in terms of evolutionary rates. A subsequent debate over what specimens and techniques ought to be used (Rightmire 1986, Wolpoff 1986a) had no clear winner, and the issue of evolutionary stasis in *H. erectus* remains undecided.

▌ EARLY ERECTINE RADIATIONS IN AFRICA

The oldest archaeological sites in the Lake Turkana Basin are restricted to relatively sheltered and protected closed-habitat settings near lake margins (Rogers et al. 1994, Harris and Capaldo 1993). But by 1.8/1.7 Mya, archaeological traces are found in a wider variety of habitats—along stream channels, in floodplains, and even in the dryer upland parts of the basin (Cachel and Harris 1995, 1998). Africa seems to have experienced an arid climatic pulse at this time (de Menocal 1995), reflected in an expansion of savanna vegetation (Cerling 1992) and savanna-adapted bovids (Vrba 1995). Drier climates during this period are also evinced by changes in lakes and watercourses. For example, the proto-Omo River became a series of braided smaller channels, thereby breaking up the surrounding gallery forest (Feibel 1995). This climate change may have been one of the factors that encouraged ancestral Erectines to move into more varied habitats and to venture further away from permanent water sources, thus helping to drive the changes in adaptation that characterize the genus *Homo*.

Outside of the Turkana Basin, the earliest Erectine fossils come from South Africa, where specimens from Sterkfontein (Member 5), Swartkrans (Members 1 and 2), Gandolin, and Drimolen have been attributed to *Homo*. As discussed in Chapter 4, the Sterkfontein specimens are probably *A. africanus*, and the affinities of the supposed *Homo* material from Gandolin and Drimolen are unclear. But Members 1 and 2 at Swartkrans have produced a small series of remains that do not appear to represent *Australopithecus robustus*, by far the most common hominin at the site. SK 15, a gracile mandible preserving M_{1-3}, was the first of this group to be recognized (Broom and Robinson 1949). Broom and Robinson initially called the specimen *Telanthropus capensis*, but Robinson transferred it to *Homo erectus* in 1961. Several other cranial fragments from Swartkrans have since been proposed as representatives of the genus *Homo*. These include SK 27, a crushed juvenile skull preserving some teeth; SK 45, a small section of mandibular corpus with M_1 and M_2; SK 80, a complete face with parts of the left lower vault and cranial base; and SK 2635, a fragmentary palate with five teeth (Clarke 1994b). A few isolated teeth from Swartkrans have also been assigned to *Homo*. As noted earlier (Chapter 4), some of the

Swartkrans hand bones may represent this genus as well. Most of these specimens purportedly come from Member 1, which is dated to ~1.8 to 1.6 Mya; but SK 15 is thought to be slightly younger, having been found in a chunk of Member 2 breccia projecting into Member 1 (Clarke 1994a,b). Acheulean/Developed Oldowan tools are also found in Members 2 and 3 at Swartkrans (Brain et al. 1988), implying for many scholars the presence there of the genus *Homo*. However, there are questions about the context of both the SK 15 and 45 mandibles because of uncertainties about R. Broom's record-keeping practices (Wolpoff 1999).

It is clear that SK 15 and 45 are not robust australopithecines. The anatomy of SK 15 is comparable to that of known Erectines (especially ER 992), although it is a bit smaller in size. Its molars appear to be also Erectine-like in size and form, although Curnoe (2006) argues that their crown shape and form are more like those of Habilines (see Ch. 4). The SK 45 mandible bears similar-sized teeth, but its corpus decreases notably in height from the level of the M_1 to that of the M_2—a pattern typical of more modern mandibles. This distinctive difference from other Erectine mandibles makes it likely that SK 45 is intrusive into Member 1 and comes from a later, more modern human. The context of the cranial pieces is not at issue. Although SK 27 (Clarke 1977) has a Sr/Ca ratio like that of Swartkrans *A. robustus* (see above), it is considered a possible *Homo* because it appears to have a large cranial capacity (perhaps >700 cm^3), a relatively large canine and lateral incisor, molars at the small end of the Swartkrans size range, *Homo*-like cranial base features (Clarke 1994b), and no sagittal crest. The absence of the crest might reflect the specimen's young age at death (~9 or 10 years). But its age would not explain the other *Homo*-like features of SK 27—particularly the co-occurrence of large anterior and small posterior teeth, which is the opposite of the *A. robustus* pattern.

Anatomically, the most informative specimen is SK 80 (Fig. 5.21). R. Clarke reconstructed the skull from two different pieces, a palate (SK 80) originally assigned to "Telanthropus" and a fragmentary cranium (SK 847) once thought to be a robust *Australopithecus* (Clarke et al. 1970). Attributed to "early *Homo*" by Clarke and Howell (1972), the resulting composite (Fig. 5.21) has also been thought to represent a small *A. robustus* (Wolpoff 1968a, 1970b). However, SK 80 bears a strong similarity in several features to the undoubted Erectine

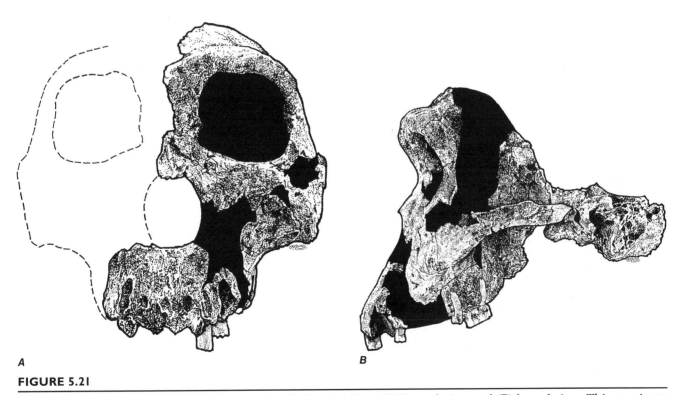

A   *B*

FIGURE 5.21

The SK 80 specimen from Swartkrans (Member 1), South Africa. (**A**) Frontal view and (**B**) lateral view. This specimen is a composite of the SK 80 maxilla and the SK 847 craniofacial specimen. The former specimen was classified as *Homo* and the latter as a robust *Australopithecus* until it was discovered that they fit together. The reconstructed specimen has been variously classified as *Homo habilis* (Curnoe and Tobias 2005), a small female robust *Australopithecus* (Wolpoff 1968a, 1970b), and *Homo erectus* (Clarke 1994a,b). (After Johanson and Edgar 1996.)

ER 3733 (Walker 1981, Clarke 1994a,b, Dunsworth and Walker 2002). The form of its nasal rim and spine indicate a *Homo*-like nose (Franciscus and Trinkaus 1988), and analysis of subnasal morphology (McCollum et al. 1993) indicates that SK 80 does not conform to the pattern seen in *A. robustus*. The overall shape of the face, particularly in the zygomaticomaxillary and infranasal regions, is very similar to that of ER 3733. Finally, the supraorbital torus of SK 80 is Erectine-like in form, and is separated from the frontal squama by a supratoral sulcus. In all of these respects, SK 80 differs conspicuously from robust *Australopithecus*—and also from its contemporary, the supposed "Homo habilis" skull Stw 53 (regarded here as an *A. africanus*) from Sterkfontein Member 5 (Walker 1981). It is not certain, however, that SK 80 is not a Habiline. A multivariate analysis by Grine et al. (1993) groups it with Habilines rather than with Erectines, and a recent comparative study (primarily of the molars) comes to the same conclusion (Curnoe 2006).

In his studies of the hand bones from Swartkrans, Susman (1988b, 1993, 1994, 1998) concluded that their possessor had a well-developed precision grip between the thumb and index finger—and thus had a dextrous, human-like hand capable of making stone tools. However, Susman argued that almost all of these hand bones represent *A. robustus*, mainly because their remains comprise over 95% of the diagnosable cranial and dental remains at the site. As noted earlier (Chapter 4), Susman's arguments have not convinced other experts. At present, it is not certain which species left the humanlike hand bones from Swartkrans Members 1–3, or made the stone tools found in the same deposits.

After ~1.6 Mya, stone tools begin to be found throughout much of Africa, and these artifacts are often accompanied by fossil human remains. These remains were once uniformly assigned to *H. erectus*, but their classification is now a matter of debate. Some of them show certain traits more typical of Asian Erectines than of earlier Ergasters. As mentioned previously, Klein (1999) argues that all of these should be lumped into *H. ergaster*, mainly because he rejects the idea that *H. erectus* could have migrated from its Asian homeland back to Africa and replaced the local Ergasters. But as in later human evolution, there are alternatives to a migration-and-replacement model. Stringer (1984) has suggested that the specifically *H. erectus*-like characteristics of such African specimens as OH 9 and Bodo may be due either to increased gene flow between Asia and Africa during the later Middle Pleistocene, or to parallel evolution of thick skulls and other Asian *erectus* features in African *Homo* populations. On the basis of the recently described Daka calvaria from Ethiopia (Fig. 5.22a), Asfaw et al. (2002) support the view that African and Eurasian Erectines represent **demes**—differentiated

1 cm

A

1 cm

B

FIGURE 5.22

The Erectine calvaria from Daka (Middle Awash Valley, Ethiopia). The lateral view (**A**) shows the projecting supraorbital torus and distinct supratoral groove. In posterior view (**B**), the skull exhibits the enhanced biparietal breadth (relative to the cranial base) characteristic of later Erectines, resulting in more nearly vertical side walls of the braincase. Compare Figs. 5.2, 5.11, 5.26, and 5.27. (After Asfaw et al. 2002.)

local populations (cf. Howell 1999)—of a single wide-spread species, *H. erectus*. Some other workers have posited a similar continuum of Erectine evolution in Asia and Africa (Bräuer and Mbua 1992, Bräuer 1994, Rightmire 1988, 1990, 1994). On the other hand, Manzi et al. (2003) offer a different analysis of Daka and question the connections between African and Asian Erectines. Migration from Asia back into Africa is certainly a theoretical possibility. But at present, we see no compelling reasons for choosing among migration, gene flow, parallelism, or some combination of these processes to explain the similarities between Asian Erectines and some late Erectines from Africa.

The earliest of these post-Ergaster African Erectines are the OH 9 partial calvaria from Upper Bed II at Olduvai Gorge, Tanzania (Fig. 5.5) and two Ethiopian fossils—a relatively complete mandible from Konso (a.k.a. Konso-Gardula), and a very thick parietal fragment from Omo Member K (L 996-17). All are ~1.4 My old. The Konso specimen, KGA 10-1, is associated with Acheulean tools and exhibits such specifically *erectus*-like features as a reduced M_3, a robust mandibular corpus, a broad vertical ramus, and the absence of a retromolar space. Although unassociated mandibles are hard to classify, there is no reason to doubt that Konso is an Erectine. Asfaw et al. (1992) note that it is one of the largest Erectine mandibles known and would make a good fit for the large Olduvai Hominid 9 cranium. Also associated with Acheulean artifacts, the OH 9 specimen has a prominent supraorbital torus, a supratoral sulcus, a marked angular torus, and an angled occipital (Fig. 5.5). It has been most completely described by Rightmire (1979a, 1990), who considers it *H. erectus*. Although neither the occipital nor nuchal planes are complete enough to measure, it is clear that the latter was expanded as in other Erectines. Rightmire thinks that the occipital and nuchal planes were roughly equal in size. The nuchal torus is not as well developed as in many other Erectines, but opisthocranion coincides with inion. Of the seven proposed *H. erectus* apomorphies investigated by Bräuer and Mbua (1992), OH 9 lacks only the frontal keel. (Parietal keeling cannot be assessed on this specimen.) Rightmire also suggests that the cranial base was relatively flat and that the facial skeleton must have been markedly prognatic.

Later Olduvai Erectines, including OH 28, 22, 51 and 12, were once dated to ~700 Kya. However, new paleomagnetic correlations for Olduvai Bed III/IV assign an older date to these specimens, around 1.1 Mya (Tamrat et al. 1995). We have already discussed the OH 28 hipbone, and the associated femur fits the Erectine pattern (Day 1971). OH 12, from site VEK IV at Olduvai, is a small, fragmentary cranium with one of the smallest cranial capacities attributed to an Erectine (Table 5.6). Otherwise, the specimen exhibits characteristic Erectine features, particularly on the occipital and temporal

(Bräuer and Mbua 1992, Rightmire 1979a, 1990, Antón 2004). The fragmentary facial pieces of OH 12 approach KNM-ER 3733 in shape (Antón 2004), further blurring the distinction between Ergasters and later, more "typical" (Asian-like) African Erectines. OH 22 is the right half of a mandible, and OH 51 is a small piece of mandibular body. They are very similar to OH 23, a segment of mandibular body from the Masek Beds preserving P_4–M_2 (Rightmire 1990), which is of about the same age (if the new Olduvai correlations are correct). As described by Rightmire (1990), the morphology of all these mandibles and their teeth is fully commensurate with their inclusion in the Erectine group.

Another African Erectine from this same time horizon (~1 My), the complete calvaria from Daka, has been described as exhibiting typically Erectine features shared by both Asian Erectines and Ergasters (Asfaw et al. 2002). However, the Daka specimen is more "advanced" than most other Erectines in two ways: It has a distinctly double-arched supraorbital torus, and its cranial vault has its maximum breadth higher up, so that the side walls of the braincase look almost parallel when viewed from the rear (Fig. 5.22). The small frontal bone from Olorgesailie (~1 My) also has a double-arched torus. The temporal bone found with it has a shallower mandibular fossa than is typical for Erectines, but most other features of the Olorgesailie specimens are clearly Erectine (Potts et al. 2004). A cranium of similar date from Buia in the Afar (Danakil) depression of Eritrea is described by Abbate et al. (1998) as exhibiting features generally attributed to *H. ergaster* or *H. erectus*. But its cranial capacity is low, even for an early Erectine—around 750–800 cm³. Like Daka, Buia exhibits maximum cranial breadth high up on the parietals, but this "advanced" feature may be due to postmortem distortion. A thick parietal fragment from Gomboré II (Melka Kunturé) in Ethiopia falls in the Erectine range for vault thickness and dates to between 1 Mya and 780 Kya (Chavaillon et al. 1974). Certain later East African specimens, including the Bodo cranium and the Baringo mandibles, have many features in common with these later African Erectines, and a case can be made for classifying them as Erectines (Wood 1999b). However, their dates and morphology suggest affinities with later, post-Erectine archaic humans (Heidelbergs), and we will take them up in the next chapter.

The earliest human remains known from North Africa are the fossils from Tighenif (Ternifine) in Algeria, excavated by C. Arambourg in 1954 and 1955 (Arambourg 1954, 1955) and dated to around 700 Kya on the basis of faunal correlations and paleomagnetism (Geraads et al. 1980, 1986). Three adult mandibles, three upper molars (two deciduous, one permanent), and a juvenile parietal were recovered in association with Acheulean artifacts (Arambourg 1963, Balout et al. 1967). The Tighenif mandibles have receding, chinless symphyses, broad rami, no retromolar spaces, relatively tall bodies, big anterior teeth, and M_3s that are reduced in breadth compared to the other molars. This morphology is commensurate with that of other Erectine mandibles. J. Schwartz (2004) sees similarities between the Tighenif teeth and those from the Gran Dolina. The Tighenif 3 mandible is exceptionally large, with particularly broad and tall rami and prominent coronoid processes. The parietal (Tighenif 4) is thin, lacks an angular torus, and has a very weakly developed midline keel. The weak development of these features is probably due to this individual's subadult age at death. It appears to have had one of the largest cranial capacities of any Erectine, estimated at around 1300 cm^3 (Table 5.3).

Tillier's (1980) study of the Tighenif upper molars suggests that their closest relationships are with other early North African humans from three later sites in Morocco—Sidi Abderrahmen, the Thomas Quarries, and Salé (Hublin 1985). The finds from Sidi Abderrahmen and the Thomas Quarries have some morphological features that could place them with the Erectines. But because all these fossils are thought to show more "advanced" features as well, and because their dates lie at or after the end of the Erectine time span, we will deal with them in Chapter 6.

OUT OF AFRICA I: THE ERECTINE RADIATION

For a long time, it was thought that the earliest hominin site outside of Africa was 'Ubeidiya in the Jordan Valley, dated to ~1.4 Mya by faunal correlation to European sequences and by the similarity of the stone tools found there to Developed Oldowan/Acheulean lithics from Olduvai Upper Bed II (Bar-Yosef and Goren-Inbar 1993, Belmaker et al. 2002). However, recent dates and discoveries suggest that Erectines had spread all across the Asian land mass, from the southern Caucasus to Indonesia, by 1.8 Mya (Swisher et al. 1994, Gabunia et al. 1999) and had penetrated into southwestern Europe by 1.65 Mya (Gilbert et al. 1998). Some of these early dates remain controversial, but they are supported by some molecular-clock readings. Templeton (2005) investigated 25 DNA haplotype regions among extant human populations, using a variety of statistical techniques. On the basis of at least three of these haplotypes, Templeton concludes that the first human populations left Africa about 1.9 Mya (Fig. 8.36).

But by what routes did Erectines migrate out of Africa? Why did they go? And if these early dates are correct, how were they able to spread so far so fast? The most obvious route for the spread of early Homo out of Africa lies along what O. Bar-Yosef has called the "Levantine Corridor," around the northwestern rim of the Red Sea and up through what is now the Sinai. The Red Sea is shallow here, and the floor of its northern arms may

have been dry land at times during the Pleistocene when sea level was at its lowest, some 80 meters below its present height. During these same periods, the shallow southeasternmost part of the Red Sea, where today it meets the Gulf of Aden, would also have been dry land. This land would have provided a southern corridor along which early Erectines could have spread eastward into tropical Asia and Indonesia, while the Levantine Corridor offered a route for dispersal north and west into Europe.

Erectines might also have dispersed from North Africa into Europe further west, via Mediterranean islands like Corsica and Sardinia. Any western route would have required crossing open water, since sea level never fell far enough during the Erectine period to expose a complete land bridge joining Africa to Europe across the Mediterranean Sea. The narrowest place for a crossing would have been at the Strait of Gibraltar at the western end of the Mediterranean. The deepest part of the strait, which today lies at a depth of 183 m, persisted as a sea barrier throughout the Pleistocene. But during peak glacial advances, when sea level was at its lowest, the gap would have shrunk to a distance of 8–10 km, making the shores of Europe visible to Erectines standing on the beach in North Africa. There is evidence that Indonesian Erectines crossed open-water barriers (see below), so it is possible that this happened in the Mediterranean as well. Still, the most likely routes for Erectine movement into Europe are those around the eastern Mediterranean.

Why did the Erectines leave Africa at this time? The answer may simply be, "Because they could!" The biological and cultural innovations documented above for early Homo gave the new genus the ability to live in more varied environments and to range more widely than Australopithecus could. This ability is reflected in the Erectines' expansion into all but the most extreme areas of Africa. That expansion was surely not a conscious effort on the part of these humans to set out in search of new horizons, but rather the natural result of a more intense search for resources. As Gabunia et al. (2001) have noted, that search would lead to increasingly larger home ranges, especially at higher latitudes. Leonard and colleagues (Antón et al. 2002a, Leonard et al. 2002, Leonard 2002) estimate that the foraging territories would have had to increase in size as much as tenfold to support the Pleistocene human subsistence strategy.

From a biological perspective, Homo sapiens is a uniquely generalized species. Generalized species tend to exploit wider geographic and ecological ranges, show more intraspecific variation, and exhibit less of a tendency to branch off new species than do more specialized species, which exploit narrower ecological niches (Vrba 1980, 1984; Eldredge 1989). Modern humans have carried these tendencies to an extreme, and the archaeological and paleontological data indicate that early

Homo was more like us in all these respects than any of the known species of *Australopithecus*. The selection pressures that produced early Erectine morphology must have been both causes and effects of this shift to a more generalized adaptation. As technology and other aspects of human cultural adaptation became more sophisticated, humans would have been able to live in more varied ecological settings and exploit an ever-widening resource base. The result would have been a classic feedback loop, with the Erectine biocultural adaptations allowing for (or driving) a quest for expanded resources, while that quest in turn imposed selection pressures that favored greater niche breadth and more generalized adaptations.

Assuming that all the early Erectine dates are correct and represent the earliest stages of human occupation in their several regions, the genus *Homo* took only about 150 Ky to reach Indonesia, 250 Ky to reach the southern Caucasus, and 300 Ky to reach Europe proper after it first appeared on the African scene. The swiftness of its spread to Indonesia seems hard to believe. Perhaps the dates from Java are wrong. Perhaps we have not yet found the earliest Erectines in Africa. But as we have noted, tropical Asia may not have presented a difficult ecological adjustment for these early humans, and the Erectine adaptive strategy must have both conferred and demanded considerable mobility. Wobst's (1976) model of population structure for Pleistocene hunter/gatherers suggests that early human groups would have had to be very mobile and cover large ranges for a number of reasons, including the need to find suitable mates. Considering all this, it is not so remarkable that Erectines spread very quickly into suitable climatic zones—particularly since the world beyond Africa was virgin territory, where they faced no resistance from competing hominin species.

In a series of papers on *Homo erectus*, Cachel and Harris (1995, 1996, 1998) suggest that Erectines were such successful colonizers and able to spread so quickly and broadly because they acted as a "weed species." Plant and animal "weeds" that quickly colonize new regions thrive on environmental disruption caused by a variety of agents, and benefit by moving into vacant niches in regions that lack species with a similar adaptation. This phenomenon is often observed when invading species colonize islands (MacArthur and Wilson 1967) and in comparable situations on the mainland. Erectines and their spread fit this concept of a "weed species." The climatic cycles of the late Pliocene and Pleistocene caused considerable environmental disruption, which would have offered various opportunities to technology-bearing humans. But the fact that no organism with a comparable adaptation existed in the newly colonized areas would have afforded a still greater advantage. The new Erectine biocultural adaptation opened a wholly novel ecological niche, in which no

other organism could successfully compete; and as that adaptation became even more sophisticated, the niche of the genus *Homo* would have become broader, opening more and more ecological zones and geographical regions to human occupation.

Cachel and Harris evoke the concept of "character release" to explain the regional variations in morphology and technology that characterized Erectines once they become widely spread in the Old World. According to Van Valen (1965), lack of competition in new environments may allow species to alter aspects of their behavior and morphology that would be more constrained if the organisms were under stronger selective pressure. The release of stabilizing selection in this way would lead to increased variability in both anatomical and behavioral features. Cachel and Harris suggest that this phenomenon may be responsible for the regional differences seen in early Erectines from Africa, Java, and China. Since character release can be regarded as a type of genetic drift, features that are not stabilized by selection might be expected to vary randomly, with different variants becoming fixed by chance in different regional populations. This appears to be what we see in Erectines. Their regional variation tends to affect anatomical details (which one might expect to reflect the effects of drift) and not the general gestalt of the Erectine pattern, which must have been controlled by more stringent and consistent selection pressures.

In addition to the early fossils from Georgia and Indonesia, a lot of archaeological evidence also bears witness to the Erectine migrations northward and eastward out of Africa. Acheulean/Developed Oldowan tools have been found at the 1.4-My-old site of 'Ubeidiya in the Jordan Valley, along one of the probable northward migration routes. Four human cranial fragments and two teeth from this site (Tobias 1966) have been shown to be younger than the artifact-bearing deposits (Molleson and Oakley 1966). However, Belmaker et al. (2002) have recently described a human lower incisor from 'Ubeidiya that unquestionably does come from the Lower Paleolithic deposits. Morphometrically, this tooth fits within the Erectine range, and Belmaker et al. provisionally assign it to the Ergasters.

To the east, Acheulean/Developed Oldowan artifacts are commonly found in South and mainland Southeast Asia. The vast majority of these are surface finds with no reliable geoarchaeological context, but there are a few exceptions. The oldest of these is the site of Riwat in northwestern Pakistan, which has yielded three flaked quartzite pebbles and two flakes, dated to ~2 Mya on the basis of paleomagnetic and other geological evidence (Dennell et al. 1988). It is not certain that these broken rocks are artifacts. The deposits containing the possible tools are consolidated gravels, in which such flaking might have occurred naturally (Klein 1999).

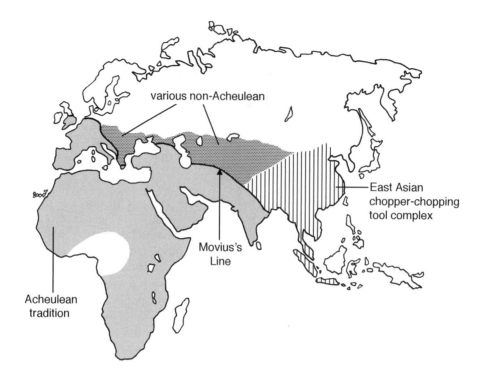

FIGURE 5.23

Geographical distribution of Acheulean technology versus that of the East Asian chopper-chopping-tool complex and other non-Acheulean traditions of the late Early Paleolithic. Movius's Line largely separates the Acheulean from the others, although Acheulean assemblages have also been found to the east of this line (see text). (After Schick 1994.)

However, Dennell et al. (1994) have argued that other stones from the deposits do not show flaking, and that the supposed tools and their flaking pattern stand out noticeably. Unmistakable tools are found at younger sites (<1.0 My) in South Asia. These include Ban Mae Tha in northern Thailand, where three flaked pebbles are dated by radiopotassium and paleomagnetism to 800 to 600 Kya (Pope et al. 1986); Dina in northeast Pakistan, where Acheulean handaxes are dated to sometime between 780 and 400 Kya (Rendell and Dennell 1985); and Bori in west-central India, where Acheulean tools are dated by radiopotassium to 670 Kya (Mishra et al. 1995). These sites are too young to be involved in the initial radiation of Erectines into this part of the world, because even the most conservative estimates indicate habitation of Java by ~1 Mya. But they show that Erectines were widespread in parts of Asia other than Indonesia and China during the late Lower and Middle Pleistocene.

In the 1940s, Harvard archaeologist H. Movius (1944, 1948) proposed a boundary between regional Paleolithic cultures known today as the "Movius Line" (Fig. 5.23). This concept was based on Movius's work on the Lower Paleolithic of Asia, particularly at Zhoukoudian. He pointed out that hand axes were absent at ZKD, where the stone-tool assemblages were characterized by

choppers and chopping tools. Movius noted that such a chopper/chopping tool tradition (CCT) characterized Asia from northern India (and Pakistan) to East and Southeast Asia. Although there were regional variants within the CCT, including the Patjitanian (Pacitinian) in Java, all were distinguished from the hand-axe traditions of southern India, Europe and Africa (Fig. 5.23) by the absence of hand axes and of the Levallois technique, a prepared-core technique developed in the Acheulean (Movius, 1969).

Movius's concept has been criticized on a number of grounds (Yi and Clark 1983, Zhang 1985a). Bifaces (hand-axes) are now recognized at several East Asian sites, including Lantian, Dingcun, and Kehe (all in China) and Chon-Gok-Ni (in South Korea). Hou et al. (2000) have reported Acheulean-like bifaces dated to 800 Ky in the Bose Basin of southern China. Furthermore, some of the regional traditions that Movius assigned to the CCT do not appear to date from the Lower Paleolithic time span. Bartstra (1984) has shown that the Pacitinian tools from the type locality actually derive from Late Quaternary alluvial fill, and there is evidence that others are also more recent (Hutterer 1985). In addition, it is now recognized that many sites in *western* Eurasia also lack bifacial tools and Levallois technology (Yi and Clark 1983). These include pivotal sites in central Europe

(Vértesszöllös, Bilzingleben), northern Europe (Clacton) and Pakistan (Riwat).

Schick (1994) argues that the Movius Line still retains some integrity as a concept, particularly when it is also extended to the west (Fig. 5.23). For example, she points out that the East Asian bifaces are not typical Acheulean handaxes, but generally cruder and thicker. This seems to be the case for the Bose Basin tools as well. Schick also argues that the idea of a CCT should be replaced by referring to these cultures as what Clark (1977) labels as "Mode 1" industries. These crude industries, which begin with the earliest Oldowan tools in East Africa, are characterized by roughly made cores lacking any standarized shape and big flakes used without any reworking. Regional traditions cannot be defined in Mode 1 industries, because they are too primitive technologically. Schick believes that the more advanced "Mode 2" Acheulean industries, including the Levallois technique, spread relatively slowly out of Africa into western Eurasia, and perhaps never fully made it past the East Asian "barrier" defined by the Movius "Line." As she notes, the arrested spread of the Acheulean is not easily accounted for by the conventional explanations—lack of suitable raw materials for bifaces, preeminent use of one type of raw materials for tools, or different functional requirements in the eastern Asian environments. Schick suggests that the eastward spread of Acheulean industries may have been impeded by geographical and ecological bottlenecks that restricted movements of people and technologies in certain places at certain times.

Cachel and Harris (1996, 1998) offer a different explanation. Extending the idea of "character release" to technology, they suggest that whether a particular group of early humans made bifaces, or used raw materials other than stone, may largely reflect whatever tool-making traditions were initially established in that group. This sort of cultural drift or "founder effect" could produce random differences between regional industries, in much the same way that random errors in gene-pool sampling contribute to genetic differences between regional populations.

Even if regional differentiations in genetics and in culture are produced by analogous processes, the results of the two processes should not be equated. It is significant that neither Cachel and Harris nor Schick attribute the regional differences in Erectine cultural traditions to genetic differences in regional hominin type. As Schick specifically notes, the presence and late continuation of Mode 1 industries in East Asia demonstrates the inadvisability of equating any Paleolithic cultural tradition with a hominin taxon. It seems clear that the division between Acheulean and CCT industries does not correspond with a division between two species of *Homo*. But however we interpret it, the Movius Line retains some validity as a real if indistinct boundary

between the Acheulean and CCT. It challenges our ability to understand the meaning of the distribution of Lower Paleolithic technology.

INDONESIAN ERECTINES AND THE SPECTER OF "MEGANTHROPUS"

The initial migrations of Erectines onto Java must have occurred during glacial maxima, when the sequestering of seawater in continental and alpine glaciers resulted in a lowering of sea levels. Many Southeast Asian islands lie on a single common expanse of continental shelf, the Sunda Shelf, and are thus separated by relatively shallow expanses of water (Fig. 8.26). During maximal glaciation, when sea level was lowered ≥100 m, the Sunda Shelf would been an expanse of dry land covering almost 4 million square kilometers (Bellwood 1992, 1997), and Erectines from Asia could have walked to Java without getting their feet wet. Micropaleontological and oxygen-isotope data from deep sea cores indicate that maximal exposure of the Sunda Shelf occurred about 3 Mya, 1.25 Mya, and 65–45 Kya. However, a sea level drop of only 20 m would have sufficed to join both Sumatra and Java to the Asian mainland. This level of reduction in sea level occurred frequently during the past 320 Ky and was probably common throughout the Pleistocene (Chappell and Shackleton 1986), but the exact periodicity prior to 500 Ky is not clear.

As discussed previously, it is not certain whether the earliest Erectines on Java date to slightly more than 1 Mya or all the way back to some 1.8 Mya. The specimens providing the earlier dates are Perning 1 (the "Modjokerto" skull) and Sangiran 27 and 31. All three derive from the Pucangan Formation (characterized by the so-called Djetis fauna; see Fig. 5.6), and all have been involved in taxonomic debates.

Perning 1 is represented by a complete calvaria found in 1936. It was initially described by von Koenigswald (1936), who subsequently classified it as "Pithecanthropus modjokertensis" (see von Koenigswald 1956). The skull comes from a young individual, aged about 4–6 years by most estimates. Coqueugniot et al. (2004) argue that it was only around 1 year of age. Antón (1997) notes that, despite its young age, Perning 1 already exhibits Erectine morphology (supraorbital and nuchal tori, angled occipital, overall vault shape) of the sort seen in adults, though less pronounced. The specimen has a cranial capacity of 650 cm³. It would probably have attained an endocranial volume of 750 cm³ at adulthood, but that would still make it the smallest-brained Erectine individual known from Java (Table 5.3). It has even been suggested that Perning 1 might be a Javanese *Australopithecus*; but its morphology, as detailed by Antón, clearly shows that this is not the case.

Sangiran 27 and 31 are fragmentary, crushed skulls. In fact, Sangiran 27 is still imbedded in matrix. T. Jacob (1980) and S. Sartono (1982) described these crania as belonging to "Meganthropus" (see also Tyler 1994, 1995, 2001), along with two other fragments that do not have "Sangiran" numbers assigned to them. According to Tyler (2001), these specimens have unusually thick braincases and exhibit relatively high vaults, exceedingly pronounced nuchal tori, and small endocranial capacities compared to later Pucangan Erectines. Tyler suggests that although the "Meganthropus" material does not warrant generic distiction, it is different enough to be placed in a species of *Homo* separate from *erectus*—presumably *H. "palaeojavanicus."* However, other investigators (Kramer 1994, Wolpoff 1999) find no significant differences between these specimens and other Javan Erectines. Sangiran 31 appears to be quite similar to Sangiran 4, an undoubted Erectine from the Pucangan, although the nuchal torus and sagittal keeling are more pronounced and the cranial vault seemingly higher in Sangiran 31. A recent reanalysis of all the cranial remains attributed to *"Meganthropus,"* including Sangiran 31, shows that no feature on any specimen falls out of the Indonesian Erectine range (Durband 2003). One possible exception is the purported presence of a sagittal crest on Sangiran 31 (Tyler 1994, 2001); but even if cresting exists on this specimen, crests are also present in at least one other specimen that may represent early *Homo*, ER 1805 from Koobi Fora.

The other major evidence for the distinctiveness of "Meganthropus" comes from jaws and teeth, all from the Pucangan Formation. The most distinctive of these is the type specimen found in 1941, Sangiran 6, a piece of a very large and exceptionally thick mandibular corpus retaining the P_3, P_4, and M_1 (Fig. 5.24). A possibly associated left posterior corpus with M_2 and M_3 was recovered in 1950. Other mandibles (Sangiran 5, 8, and 9) have also been assigned to "Meganthropus" (Pope 1997a). Weidenreich (1946) regarded "Meganthropus" as a key component of his theory that humans had evolved from a "giant" progenitor (in his view, *Gigantopithecus*). Robinson (1953a) considered these fossils as evidence for a robust *Australopithecus* in Java. Tobias and von Koenigswald (1964) noted similarities between Sangiran 6 and OH 7, but they also found clear similarities between other Habilines and Javan Erectines. Lovejoy (1970) concluded that the "Meganthropus" specimens overlap too much with other *H. erectus* in such features as premolar morphology, form of the mandibular symphysis, and overall size to justify putting them in separate species. Kramer (1994) found that all of the Sangiran "Meganthropus" mandibles showed clear affinities to *Homo* rather than to *Australopithecus*. However, Tyler (1994) argues that when the Kendung Brubus mandible is added to the total Indonesian mandibular sample, variation is too great to be encompassed within a single

species. One univariate study found marked similarities between Sangiran 6 and *Australopithecus* (Orban-Segebarth and Procureur 1983). But multivariate statistical analyses of the same specimen (Kramer and Konigsberg 1994) show that while there is a superficial size similarity with *Australopithecus*, the overall pattern of size and shape clearly places Sangiran 6 with *Homo*. In short, there is no compelling reason to remove any of the "Meganthropus" specimens from *H. erectus*.

The Sangiran 4 partial skull from the Pucangan Formation is often presented as the typical example of early crania from Java. Found in 1938, it comprises the rear of a calvaria and a palate/lower face (Fig. 5.25). Weidenreich (1945a) assigned this specimen to "Pithecanthropus robustus," another stage in his evolutionary progression of humans from "giant" ancestry (Weidenreich 1946). Von Koenigswald (1950) placed the specimen in "P. modjokertensis" along with other Pucangan remains, including Perning 1. Although the calvaria has always been accepted in the Erectine fold, the palate was once identified as that of an aberrant orangutan (Krantz 1975). The palate is large, with a broad lower nose. The broad anterior palate indicates that large incisors were present. The canines are also large, but human-like in morphology. Although it is impossible to be certain, it is likely that no canine fossa was present. The anatomy and size of the Sangiran 4 palate and lower face suggests a morphological pattern similar to that of Sangiran 17, with no feature that is distinctively ape-like.

The younger Kabuh (Bapang) Formation, characterized by the so-called Trinil faunal assemblage, has yielded a larger series of cranial remains, among them the Trinil 2 type specimen of *H. erectus*. Most of these crania come from the Sangiran Dome and exhibit variations on the Erectine cranial theme discussed above. As noted previously, no major Indonesian Erectine specimen was actually excavated under controlled scientific conditions. However, a few recent finds, such as the human incisor from Bukuran (in the Kabuh Formation), have been excavated under such conditions, confirming the association of Erectines with the Kabuh deposits (Baba et al. 2000). Most Trinil specimens lack faces (Trinil 2, Sangiran 2, 3, 10 and 12), but two have partial faces (Sangiran 17 and IX). There is considerable variation in these specimens, mainly in size (Table 5.1). This variation has generally been attributed to sexual dimorphism. Overall, the Kabuh sample exhibits a 10% increase in cranial capacity compared to Ergasters, reflected in generally more expanded cranial vaults. Compared to the Ergasters, the Kabuh crania also have thicker vaults and more pronounced tori, but a slightly reduced nuchal area. Sangiran 17, as reconstructed by M. H. Wolpoff, is particularly significant for the issue of regional continuity in Australasia (Thorne and Wolpoff 1981, 1992; Frayer et al. 1993, Wolpoff 1999), because it

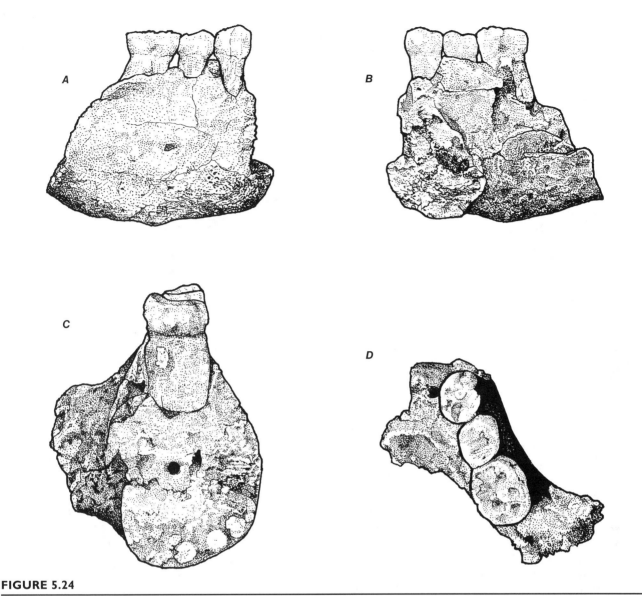

FIGURE 5.24

Four views of the Sangiran 6 "Meganthropus" mandible from Java (Indonesia). The jaw retains P_3–M_1. (**A**) Buccal view, (**B**) lingual view, (**C**) posterior view, (**D**) an occlusal view. Not to same scale. (After Schwartz and Tattersall 2003.)

is the only archaic face complete enough to compare with modern Australasian faces.

One curious aspect of the Erectine period in Indonesia is the relative paucity of stone tools. Whereas Erectines are generally associated with either Acheulean or Oldowan-type artifacts in the remainder of the inhabited Old World, no fully convincing stone tools are known from this period in Indonesia. Crude, rather amorphous cores and flakes have been recovered at Sambungmacan (Jacob et al. 1978) and at Ngebung (Bartstra 1985, Sémah et al. 1992), but there are no associations for these tools, and their age and makers are uncertain. Bartstra (1985) believes the Sambungmacan tools are from a much later time period. Pope (1988, 1989; Pope and Cronin 1984) has posited that bamboo

may have been used in lieu of stone tools by Javan Erectines. Bamboo is easily workable and becomes very strong when dried. It also forms sharp, quite functional cutting edges. However, it probably takes stone tools to work bamboo. Certainly, no bamboo tools have yet been found in the Javan record. For now, the existence of a Lower Paleolithic "bamboo industry" in Java is only a plausible speculation.

The most tantalizing recent archaeological discoveries in Indonesia come not from Java but further east, on the small island of Flores. At the sites of Mata Menge and Boa Leza in the Soa Basin on Flores, purported stone artifacts have been found in association with remains of the extinct elephant *Stegodon*, thought to have migrated to Indonesia at the same time as Erectines (Sondaar

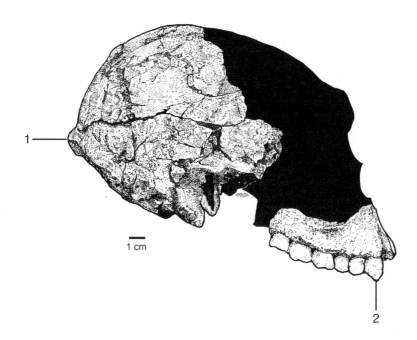

1

2

1 cm

FIGURE 5.25

The Sangiran ("Pithecanthropus") 4 skull from Java (Indonesia). Note the prominent nuchal torus (1) and large maxillary canine (2). (After Schwartz and Tattersall 2003.)

et al. 1994, Morwood et al. 1998, 1999). Fission-track and paleomagnetic dating estimate the age of these bones and tools as just older than the Brunhes/Matuyama boundary—about 880 Kya (Morwood et al. 1998). But Flores lies beyond the edge of the Sunda Shelf, on the Australian side of Wallace's Line (Wallace 1890), the great boundary between Asian-derived faunas to the northwest and Australia-derived faunas to the southeast (Fig. 8.26). If the Flores lithics are truly artifacts and correctly dated, then Erectines must have been able to cross an approximately 50-kilometer stretch of open water between the islands of Bali and Lombok, which were not connected during Pleistocene periods of lowered sea level, to reach Flores. This implies that Erectines were capable of limited water travel, here and presumably at other places in the world as well. Even more exciting is the discovery of diminutive Erectine-like fossils from Liang Bua on Flores (Morwood et al. 2004). These possible late-surviving Erectines are discussed in Chapter 6.

CHINESE ERECTINES

Although there are a number of Erectine fossils from China in addition to the well-known specimens from Zhoukoudian (Etler and Li 1994, Wu and Poirier 1995), the ZKD fossils always dominate discussions of Erectines in China. This may not be appropriate. S. Antón (2002) suggests that the ZKD sample is not a good proxy for Asian Erectines in general, because it manifests some

idiosyncratic patterns of cranial form. J. Kidder and A. Durband (2004) also report that the ZKD crania exhibit a unique metric pattern, not shared with African or Indonesian Erectines. Specifically, the ZKD calvariae have a cranial vault that is broad in the middle but narrow at the frontal and occipital ends, while the other samples (including other Chinese specimens) have relatively broader frontals and occipitals. Kidder and Durband take this not as a marker of a taxonomic boundary, but as just another example of regional variation within the Erectines.

Apart from its human remains, which have been previously discussed, the ZKD site also has been cited as providing archeological evidence for various human behaviors during the Middle Pleistocene, including systematic hunting focused on specific prey, collection of plant remains, controlled use of fire, the practice of cannibalism, and the production of relatively sophisticated stone artifacts (Wu and Li 1983, Zhang 1985a). In recent years, all but the last of these have been challenged. The stone tools are often described as fitting into Movius's chopper-chopping tool tradition, but most of them in fact are flakes. Bifaces do not occur at ZKD.

The question of how the remains accumulated in the cave is critical to all of the behavioral questions. Two species of deer make up a large percentage of the extensive fauna from the site, and their bones have been interpreted as the product of human deer hunting. However, several other species are also strongly represented, including hyenas (≥2000 bones) and horse. Stiner (1992) contends that the pattern of bone accumulation at ZKD is characteristic of hyena dens. Binford and Ho (1985) observed that the ZKD stone tools are not always found in association with human remains or in any functional association with animal bones. They suggested that both hyenas and wolves acted as accumulating agents for both human and faunal remains at Locality 1. Binford and Ho also argued that the four ash layers were not evidence for human-controlled fire, because there are no distinct hearths and (apart from some stone tools) no evidence of any human activity associated with the supposed remains of fires. Another detailed analysis of the ZKD sedimentary sequence posits that the "ash layers" are largely finely laminated, water-deposited silts that washed in from outside the cave (Goldberg et al. 2001). At the moment, the evidence for controlled use of fire at ZKD is very much in doubt.

Other Erectine sites in China have yielded small fragments of human bone or isolated teeth (Wu and Poirier 1995). Right and left upper central incisors, recovered as surface finds from the site of Yuanmou, are morphologically quite similar to those from ZKD, but with less developed marginal ridges. Reassessment of the magnetostratigraphy at Yuanmou suggests that these teeth are about 700 Ky old, not >1 My old as once claimed (Hyodo et al. 2002). Their morphology is commensurate with this age estimate. Other sites yielding loose teeth are Yunxian (Meipu), Yunxi, Yiyuan (also cranial fragments), Nanzhao, Luonan, Wushan (also a mandible fragment), and Xichuan. The morphology of all these specimens is consistent with classification in *H. erectus,* as is that of the previously discussed Lantian fossils (Wu and Poirier 1995).

The site of Tangshan (Nanjing) yields a more complete cranium, Nanjing 1, dating to >500 Kya (Zhao et al. 2001). This specimen exhibits an Erectine-like occipital morphology and a vault form that groups with other Erectines in multivariate statistical analyses (Antón 2002, Wu et al. 2005). Like the Zhoukoudian specimens, Nanjing 1 has a malar notch and an angled zygomaticoalveolar margin. Finally, cranial and other remains from the sites of Yunxian and Hexian exhibit fundamentally Erectine morphology. However, since they supposedly show some more advanced features as well, these specimens will be considered in the next chapter.

DMANISI—HUMANS AT THE PERIPHERY OF EUROPE

In 1991, the year it became an independent country, Georgia provided paleoanthropology with one of the most important discoveries of early *Homo.* The village of Dmanisi, located in the uplands on the southern side of the Caucasus Mountains, lies about a kilometer from the confluence of two rivers, overlooking the route of the former Silk Road through the Caucasus. Excavation

of a medieval storage pit in the ruins of fortifications at Dmanisi exposed underlying Plio-Pleistocene materials in the 1980s, and excavation of those deposits in 1991 yielded a mandible (D 211), Oldowan-type tools, and archaic faunal remains (Gabunia 1992, Gabunia and Vekua 1995). Continuing excavations at the site have yielded considerably more material. In 1997, a right third metatarsal (D 2021) was recovered (Gabunia et al. 2000b), and in 1999 two crania (D 2280 and 2282) joined the Dmanisi family (Gabunia et al. 1999, 2000b). A second, much larger mandible (D 2600) was unearthed in 2000, followed by a third mandible (D 2735), a third cranium (D 2700), and other bones in 2001 (Gabounia et al. 2002, Gore 2002, Vekua et al. 2002). A fourth, edentulous cranium (D 3444) and an associated mandible (D 3900) were found between 2002 and 2004 (Lordkipanidze et al. 2005); and additional postcranial remains, some associated with some of these crania and mandibles, were described in 2007 (Lordkipanidze et al. 2007).

The initial find, the D 211 mandible, was described as similar to early *Homo erectus* by Gabunia and Vekua (1995), although Bräuer and Schultz (1996) considered it more similar to later archaic *Homo* specimens. After the discovery of the first two crania, the Dmanisi hominins were placed in *"Homo ergaster"* by Gabunia et al. (2000b). It is now claimed that some of the Dmanisi remains might belong in their own species, *"Homo georgicus"* (Gabounia et al. 2002). This claim is strongly influenced by the D 2600 mandible, which is much larger than the other specimens and has a rather enigmatic morphology (Skinner et al. 2006). However, the other specimens, and perhaps D 2600 as well, can be accommodated within the Erectine pattern (Vekua et al. 2002, Rightmire et al. 2006, 2008).

The deposits containing the human fossils, tools, and faunal remains derive from a distinctive, loamy sand layer that lies on the top of a spur of the Masavera Basalt. This basalt has been dated by ^{40}Ar/^{39}Ar to 1.85 ± 0.01 Mya (Gabunia et al. 2000a,b), establishing a maximum age for the human remains and fauna. The Masavera Basalt exhibits normal polarity, which together with the argon date places it within the Olduvai Subchron (1.77–1.93 Mya) of the Matuyama Reversed Chron. The level containing the human remains (Level A2) was originally reported to exhibit reversed polarity (Gabunia et al. 1999). This suggested the remains could date anywhere from about 1.77 Mya (the end of the Olduvai) to about 1.1 Mya, the beginning of the Jaramillo Normal Subchron in the Matuyama Reversed Chron. However, a new analysis of the A2 deposits concludes that they too exhibit normal polarity (Gabunia et al. 2000b). Level A2 lies conformably on the lower normal deposits (Level A1/Masavera Basalt). This fact, taken together with the new polarity assessment, supports placement of this level and the Dmanisi hominins within the Olduvai

Subchron. The vertebrate fauna, particularly the presence of certain rodents, also points to a latest Pliocene-earliest Pleistocene date for the A2 level (Gabunia et al. 2000a,b; Vekua et al. 2002), and the Oldowan-like, Mode 1 lithics are also commensurate with this early date. No bifaces or Developed Oldowan tools have yet been found in the Dmanisi deposits.

Although the Dmanisi hominids conform to the general pattern seen in other Erectines, their overall morphology is exceptionally primitive. For example, the cranial capacities of the D 2280, D 2282 and D 3444 crania are respectively, only 775, 650, and 625 cc (Gabunia et al. 2000b, Lordkipanidze et al. 2005). D 2700 is even smaller, with a cranial capacity of only ~600 cm³ (Vekua et al. 2002). These values are well below the African Ergaster mean of 854 cm³ (Table 5.6), and overlap broadly with the range seen in Habilines (Table 4.5). The primitiveness of the Dmanisi fossils comports with the surprisingly early dates estimated for the A2 level and its contents. The age and primitive morphology of the Dmanisi specimens imply that they are the product of a very early migration out of Africa. While attributing the Dmanisi hominins to *H. erectus*, Vekua et al. (2002) note that the primitiveness of these fossils suggests that the first humans to leave Africa may have represented an even earlier and more primitive stage in the evolution of the genus *Homo*. The short stature of the Dmanisi people (see below) has also been noted as a possibly primitive trait lost in the earliest African Ergasters (Lordkipanidze et al. 2007; cf. Lieberman 2007). It has even been speculated that the migration went the other way—that *Homo* originated in Asia, and that the Dmanisi fossils may represent a survival of the ancestral stock that gave rise to the early Ergasters in Africa (Dennell and Roebroeks 2005).

Gabunia (or Gabounia) et al. (2000a) have used faunal and paleobotanical evidence to show that hominin occupation of the Dmanisi site took place in a mosaic environment of open steppe and gallery forest located near a lake or pond. This sort of environment, which would have provided a wide array of rich resources for these early humans, is exactly what early Erectines might be expected to prefer. Gabunia and coworkers indicate that the climate would have been quite warm during the time the Dmanisi people were living there. Even today, much of Georgia is surprisingly warm (which is what allows it to produce such crops as tea and coffee). Despite the presence of these Erectines in a warm spot at the periphery of Europe at this early date, there is no good evidence for human penetration into Europe proper until at least half a million years later (see Gabunia et al. 2000a). We will return to this issue below.

Perhaps the least primitive of the Dmanisi specimens is the D 211 mandible. Its corpus is low and not as thick as most other approximately contemporary mandibles. Its mandibular symphysis is weakly receding, with a slight anterior bulging on the lower alveolar process. There is no evidence of a mental trigone (Chapter 8) or any other structure associated with a true bony chin. The mental foramina are located at the P_3/P_4 septum. The incisors and canine are relatively small. Their breadths are smaller than those of either Mauer or Tighenif, and lie well below the Zhoukoudian means. The posterior teeth are moderate in size, and the molar size pattern is $M_1 > M_2 > M_3$. All the teeth of D 211 are moderately to lightly worn. The second mandible, D 2600, is much larger in overall dimensions and seemingly more primitive (Gabounia et al. 2002). It does share some features with D 211: its symphysis is weakly receding, there is no evidence of chin structure, and the mental foramina lie at the premolar septum. However, the corpus is higher and thicker. The anterior edge of the ramus bisects the M_3 (so there is no retromolar space), and the coronoid process appears to have been higher than the condyle. The teeth are much more heavily worn than those of D 211, and they increase in size going backward; Gabounia and co-workers describe the molar size sequence as $M_1 < M_2 < M_3$. The canines and their roots are larger than those of D 211. The P_3 of D 2600 has two roots, rather than one as in D 211. D 2600 also manifests an unusual alveolar bulging, extending from the front edge of the canine to the front edge of the first molar. The third Dmanisi mandible, D 2735, is intermediate in size but also has a receding symphysis, high coronoid process, absence of retromolar spaces, and mental foramina located under the premolars (Vekua et al. 2002). It is associated with the D 2700 cranium. The fourth mandible (D 3900) is associated with the D 3444 cranium (Lordkipanidze et al. 2005). Almost all of its teeth were lost during life, and the alveolar bone surrounding their sockets was largely resorbed. The resorption of the incisor alveoli gives the misleading impression that the base of the mandibular symphysis was projecting and vaguely chin-like. This jaw and the associated cranium demonstrate that some combination of diet, food sharing, and food preparation techniques allowed some individuals of early *Homo* to survive into toothless old age (Lordkipanidze et al. 2005). Although the Dmanisi mandibles are highly variable, Gabounia et al. (2002) attribute most of the variation to sexual dimorphism, and regard all these specimens as representing a single species.

Published descriptions of the Dmanisi skulls (Gabunia et al. 1999, 2000b; Vekua et al. 2002, Lordkipanidze et al. 2005, Rightmire et al. 2006) also demonstrate considerable variation with an overall similarity to East African Ergasters, particularly in the cranial vault (Figs. 5.26 to 5.28). As might be expected from their small cranial capacities, all three crania have their maximum breadth on the cranial base; the upper biparietal breadth

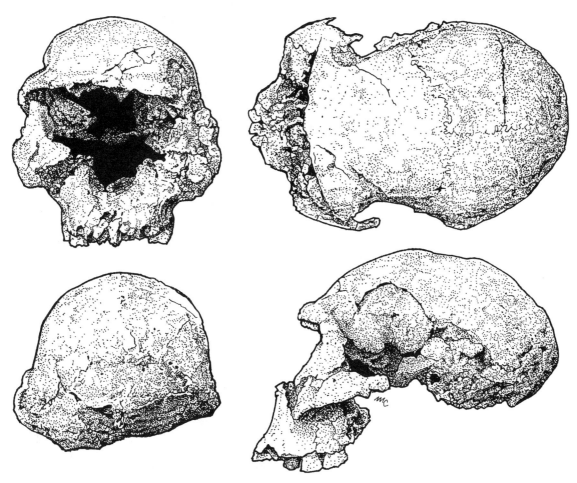

FIGURE 5.26

Four views of the D 2282 cranium from Dmanisi, Georgia. The Dmanisi crania exhibit Erectine features, but many are quite primitive. The supraorbital torus extends straight across the glabellar area, as in ER 3733; and the superior and lateral views show the total facial prognathism characteristic of Erectines and Ergasters, as well as the exaggerated alveolar prognathism found in all the Dmanisi crania. (After Gabunia et al. 2000b.)

is markedly less (Table 5.1). Vault bones are moderately thick. The course of the squamosal suture is not arched, and occipital morphology matches that described for Erectines earlier in this chapter (except that the nuchal torus is not as prominent). D 2282 and D 2700 (Figs. 5.26 and 5.27) have prominently projecting supraorbital tori, but these are relatively thin like the torus of ER 3733. D 2280 (Fig. 5.28) has a projecting but thicker torus, more like that of ER 3883. The faces of the D 2700, D 2282, and D 3444 crania, however, show some interesting differences from those of ER 3733 and WT 15000 (Figs. 5.15, 5.16, and 5.18). They exhibit obvious lateral facial prognathism like other Erectines, but they appear to have more alveolar prognathism. The anterior root of the zygomatic arch is located far to the rear, alongside M^1—a position usually correlated with lateral facial retreat (Chapter 7). But in this case it reflects an unusually anterior position of the palate, even by comparison with other Ergasters. Most of the other facial

structures (e.g., the orbits) are comparable in size to those of other Erectines, but noses appear to have been narrower in the Dmanisi specimens, at least in D 2282 (Gabunia et al. 2000b). Nevertheless, the upper part of the nasal aperture of D 2700 exhibits the Erectine nasal features noted by Franciscus and Trinkaus (1988). Finally, the inferior zygomaticoalveolar margin (ZAM) in the Dmanisi specimens is notably less oblique in orientation than in East African forms (Fig. 5.26). In this regard, the Dmanisi specimens are more like later Erectines from Asia. On the basis of the cranial form, Rightmire et al. (2006) suggest that the Dmanisi hominins are probably best assigned to *Homo erectus* (which for them includes the Ergasters), although the D 2600 mandible is still problematic. S.-H. Lee (2005) reaches a similar conclusion based on variation in Dmanisi cranial capacities.

Postcranial remains from at least four individuals have also been found at Dmanisi (Lordkipanidze et al.

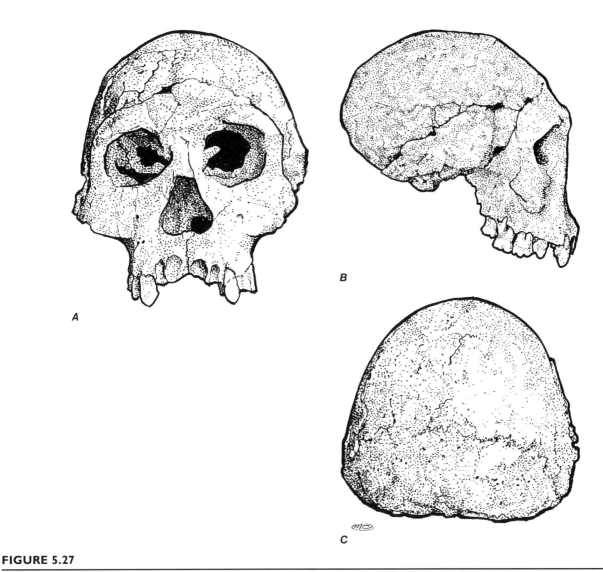

FIGURE 5.27

(**A**) Anterior, (**B**) lateral, and (**C**) posterior views of the D 2700 cranium from Dmanisi. (After Vekua et al. 2002.)

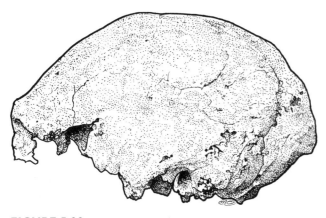

FIGURE 5.28

Lateral view of the D 2280 cranium from Dmanisi. (After Dunsworth and Walker 2002.)

2007). Subadult postcranials associated with the D 2700 cranium and the D 2735 mandible include a clavicle, both humeri, several ribs and vertebrae, a femur, and four hand and foot bones. A similar collection of adult bones (clavicles, scapula, humerus, femur and tibia, and several foot bones) may all belong to the same individual as the odd D 2600 mandible. Two other adults are represented by a few foot bones. From the long-bone measurements, Lordkipanidze and his team conclude that the Dmanisi hominins were only 145–166 cm tall and weighed some 40–50 kg. Their short stature and small bodies make them look more like earlier hominins than the tall, lanky early Ergasters from Africa. They also resemble *Australopithecus*, and differ from later *Homo*, in retaining a rather cranially oriented scapular glenoid. The torsion of the Dmanisi humeri (Fig. 3.30) is less than in African apes and humans (145–150°), but greater than in quadrupedal monkeys (85–100°)—about 110–115°, as in gibbons, some orangutans, the Nariokotome Ergaster, or "*Homo floresiensis*" (Morwood et al. 2005, Larson et al. 2007). This too may be a primitive retention from an *Australopithecus* ancestry (Larson 1998).

In most other respects, including limb proportions and the overall form of the clavicle, femur, and tibia, the Dmanisi postcrania are more like those of later *Homo*. The foot appears to have been at least as modern as the OH 8 Habiline foot (Lieberman 2007), with an adducted hallux and a longitudinal arch. Lordkipanidze and colleagues suggest that the small body size of the Dmanisi humans could be either a plesiomorphic feature or an indication of different ecological adaptations. Despite their short stature, the Dmanisi hominins had relatively long hindlimbs, like Ergasters; and it is argued that their biomechanical efficiency during long-range walking and energy storage/return during running were equivalent to those of modern humans (Lordkipanidze et al. 2007).

THE INITIAL OCCUPATION OF EUROPE

Beginning in the 1960s, a number of sites in Europe were touted as demonstrating an early human presence on that continent. The site of Le Vallonet Cave in southern France was particularly championed as proof that humans equipped with Oldowan tools had occupied Europe by ~1 Mya (de Lumley 1975, 1976; de Lumley et al. 1988). Other similar sites, all dated to between ~1 My and 780 Ky (the Brunhes/Matuyama Boundary), include the Kärlich A and B_a localities in Germany (Würges 1986, 1991, Vollbrecht 1992), Soleihac in the Massif Central of central France (Thouveny and Bonifay 1984, Bonifay 1991), and the Stránská Skála cave sites in the Czech Republic (Valoch 1984, 1987; Musil and Valoch 1968). However, Roebroeks and van Kolfschoten (1994) note that in all of these sites the only indications of human activity are presumed lithic artifacts, all of which are found in contexts suggesting that the "artifacts" were probably produced by natural flaking. From detailed assessments of the material from each site, they demonstrate that none of these sites provides convincing evidence for a human presence in Europe at this early date.

In a few sites, the problem is the dating itself. For example, the central Italian site of Isernia La Pineta has produced over 11,000 unquestioned lithic artifacts and other indications of human activity (Peretto 1991, Mussi 1995, 2001). Radiopotassium dating suggests an age of ~730 Ky for the archaeological remains, but the context of the dated material relative to the artifacts is not certain (Mussi 1995, 2001; Villa 1996). The deposits have reversed magnetic polarity, suggesting an age of >780 Ky; but Roebroeks and van Kolfschoten (1994)

emphasize that the fauna suggests a younger age. They note particularly that the deposits contain the modern water vole (genus *Arvicola*) rather than its more primitive ancestor (genus *Mimomys*). Since the transition from *Mimomys* to *Arvicola* in Europe is known to have occurred between 500 and 600 Kya (von Koenigswald and van Kolfschoten 1996), the La Pineta deposits must be younger than 600 Ky. Many of the other sites mentioned above, however, yield remains of *Mimomys*, supporting earlier dates for the faunal complexes at those sites.

At Prezletice (Czech Republic) and Šandalja (Croatia), two possibly human teeth found with supposed stone artifacts have been claimed to demonstrate an early human presence in central Europe (Fridrich 1989, Malez 1976). However, the artifacts from these sites are questionable (Roebroeks and van Kolfschoten 1994, Valoch 1995), and the "human" teeth have turned out to be a bear and deer incisor, respectively. More recently, fragmentary fossils of early humans have been reported from several sites in the Orce Basin in Spain. The most productive of these, Venta Micena, has yielded a cranial-vault fragment and shafts of what are identified as human adult and juvenile humeri (Gilbert et al. 1989, 1991, 1994, 1998). Paleomagnetic and faunal studies indicate an age of ~1.65 My for the deposits containing these remains, which would make them the earliest fossil evidence of humans in Europe (Martínez-Navarro and Palmquist 1995, Palmquist 1997). But apart from the fossils, there is no other convincing evidence of human presence at the site (Dennell and Roebroeks 1996). Unfortunately, the purported human remains are not convincing either. The cranial vault probably comes from a young horse (Moyá-Solá and Köhler 1997b), and paleontologists who have seen the humeri do not consider them human (Dennell and Roebroeks 1996, Klein 1999). Thus Venta Micena, like Šandalja and Prezletice, does not provide persuasive evidence for such an early presence of humans in Europe.

W. Roebroeks and colleagues have argued that Europe was essentially uninhabited by humans until after 600/500 Kya (Roebroeks and van Kolfschoten 1994, 1995; Dennell and Roebroeks 1996). At least north of the Pyrenees and Alps, older sites claimed to show evidence of human presence contain only pseudo-artifacts that were probably produced by natural processes (Roebroeks 2002, Tuffreau and Antoine 1995). After this time horizon, undoubted tools are relatively common throughout Europe, along with other archaeological evidence of human activity and associated human fossils. Later sites that fit this description include Boxgrove (England), Mauer (Germany), Fontano Ranuccio and Visgliano (Italy), Torralba/Ambrona (Spain), Kärlich G and Miesenheim I (Germany), and the earliest sites in the Somme Valley of Northern France (Roebroeks and van Kolfschoten 1994, 1995).

However, recent evidence from sites in Spain indicates that earlier habitation was possible in at least the warmer, southern parts of Europe. In the Guadix Baza area in southern Spain, such sites as Barranco Lèon and Fuente Nueva-3 have produced relatively large (≥100 pieces each) samples of convincing artifacts, associated with faunal remains and paleomagnetic evidence suggesting a pre-Jaramillo Subchron age of >1 My (Turq et al. 1996, Orms et al. 2000, Roebroeks 2002). No human fossil remains have emerged from these sites; but in the 1990s, human fossils were found in the ATD6 level from the Gran Dolina site at Atapuerca and dated to ~800 Kya (Carbonell et al. 1995, Bermúdez de Castro et al. 1999a). In light of these finds, Roebroeks (2002, 2005) has backed off from his earlier claim that no human occupation can be demonstrated in Europe until after 600/500 Ky. Although Roebroeks stresses that the Spanish sites are "unique, loner" examples of early human presence in Europe, he acknowledges them as convincing evidence of at least a sporadic human presence in the southern reaches of Europe before 600/500 Kya. Subsequently, evidence of human occupation in the form of crude tools was recovered from the Cromer Forest-bed Formation at Pakefield in England (Parfitt et al. 2005). Various indirect dating techniques place the age of these deposits in the early part of the Brunhes Chron, around 700 Kya. Roebroeks (2005) concedes that even if these dates are slightly too old, the evidence indicates that hominins were inhabiting areas far to the north relatively soon after they entered the European continent.

The Spanish locale of the earliest occupation sites in Europe intimates that these first Europeans might have entered the Iberian Peninsula via the Strait of Gibraltar. As noted earlier, the distance across the Strait was reduced to less than 10 kilometers during glacial maxima. The evidence from Flores indicates that Erectines could have negotiated this distance across open water, even in the teeth of treacherous currents. The possibility of human movement across the Strait will be revisited in Chapter 8.

The most ancient human remains yet found in Europe were recently recovered from the TE9 level at a northern Spanish cave site, the Sima del Elefante in the Sierra de Atapuerca (Carbonell et al. 2008). Found in association with a few Oldowan-type Mode 1 tools, the Sima del Elefante fossils comprise an isolated P_4 and parts of a mandible (ATE9-1) preserving the symphyseal region, the right lateral incisor and two damaged canines, and the alveoli and some roots of the teeth from I_1 to M_1. Paleomagnetic and faunal evidence date this level of the cave deposits to the latter part of the Matuyama Chron. This is corroborated by burial dates on intrusive rocks from the TE9 level itself (1.22 ± 0.16 My) and the underlying TE7 unit (1.13 ± 0.18 My).

A larger sample of early European *Homo* comes from another Atapuerca site, Gran Dolina, where the ATD6 Aurora stratum has so far yielded almost 80 human fossil specimens along with 238 lithic artifacts and a large faunal sample (Carbonell et al. 1995, Bermúdez de Castro et al. 1997, 1999a; Schwartz and Tattersall 2002). This material was derived from careful, systematic excavations of some 6 meters of deposit, so there is no question regarding the context of these remains. Like those at Sima del Elefante, most of the artifacts from ATD6 and older strata at Gran Dolina are small flakes and crudely flaked pebbles, made from a variety of locally available raw materials (Carbonell et al. 1999). The faunal sample contains typical early Pleistocene animals, particularly the primitive water vole (genus *Mimomys*), demonstrating that the ATD6 level is older than 600/500 Ky (Cuenca-Bescós et al. 1999). Originally, paleomagnetic correlation suggested an age of ≤500 Ky (Carbonell and Rodríguez 1994), but the faunal evidence and subsequent ESR and U-series dating of the ATD6 stratum indicates an older age. Ungulate teeth from the ATD6 level have yielded age estimates of 840 ± 42, 844 ± 30, and 846 ± 56 Kya (Falguères et al. 1999). Unfortunately, dates for some other levels of Gran Dolina are less internally consistent. But taken together with the negative (Matuyama) polarity of the ATD6 level (Parés and Péres-Gonzáles 1999), the available indicators support an age of around 840 Ky for the Gran Dolina human fossils and artifacts.

Although the human fossils from these early Atapuerca sites are rather fragmentary, their discoverers believe that they present a unique combination of modern and archaic traits and have referred them to a new human species, *Homo antecessor*—the purported last common ancestor for the Neandertals of Europe and a modern-human lineage originating in Africa. Their supposedly derived or modern-looking traits include features of the mandibular symphysis, infraorbital region, and frontal bone. The ATE9-1 mandible has a shallow, gracile corpus and a rather vertical-looking symphysis, with a slight concavity below the alveoli in front and a slight bump at the lower edge in the midline. Carbonell et al. (2008) interpret this bump as a "mental trigone"—an incipient bony chin, a trait generally regarded as distinctively modern (Chapter 8).

The fossils from Gran Dolina include the type material of "Homo antecessor." They comprise cranial remains, isolated teeth, vertebrae, ribs, and limb-bone fragments from at least six individuals. The postcranial bones have been described as being generally more similar to those of modern humans than Neandertals (Carretero et al. 1999, Lorenzo et al. 1999). However, the comparisons made thus far involve traits that do not consistently distinguish Neandertals from modern humans. For example, the tuberosity of the ATD6-43 radius is fully anterior in orientation, rather than anteriomedial as in many Neandertals (Trinkaus and Churchill 1988). But approximately a third of all Neandertal radii also have an anteriorly oriented tuberosity. Carretero et al. (1999) estimate the length of the same radius as 257 mm, which falls 1.7 standard deviations above a Neandertal mean of 233.2 (SD = 14.1). This is interpreted to suggest that the forearms of the Gran Dolina folk did not show the relative shortening characteristic of Neandertals. Again, however, this value lies well within the Neandertal range of variation. Stature estimates based on the ATD6-43 radius, the ATD6-50 clavicle, and a metatarsal are 172.5 cm, 174.5 cm, and 170.9 cm, respectively. These figures fall solidly in the overlap area of stature estimates for the later (Sima de los Huesos) Atapuerca humans, Neandertals, and early modern Europeans (Vandermeersch and Trinkaus 1995).

Gran Dolina has yielded 30 human teeth, belonging to at least six individuals, which have been studied in detail by Bermúdez de Castro and colleagues (1999b). Their analysis shows that most of the morphological features of the ATD6 teeth are primitive for *Homo*, and therefore of little use in determining their phylogenetic relationships within that genus. The posterior teeth are relatively large and resemble the teeth of African early *Homo* more than they do European Middle Pleistocene humans. The ATD6 sample exhibits a $P_3 > P_4$ size sequence, expansion of the incisors compared to earlier humans, and some reduction of the mandibular canines. In these features, the ATD6 sample is generally like both Ergasters and Middle Pleistocene Europeans. Published pictures of the lateral upper incisor of the ATD6-69 facial specimen reveal a shoveling morphology recalling later European specimens rather than Asian and African Erectines.

The cranial and mandibular specimens from Gran Dolina are listed in Table 5.13. The larger pieces come from subadults (ATD6-5, -14, -15, and -69), and many of the smaller fragments may be subadults as well because they show no sutural stenosis (fusion). Only a few pieces are clearly from adults (Table 5.13). The juvenile remains have figured prominently in the assessment of the phylogenetic position of the Gran Dolina hominins. Bermúdez de Castro and colleagues (1997, 1999a; Rosas and Bermúdez de Castro 1999) have relied heavily on specimens ATD6-5, 14, 15, and 69 in assigning the Gran Dolina specimens to the new species "*Homo antecessor*." The ATD6-15 frontal bone exhibits a projecting but relatively thin (from top to bottom) supraorbital torus (Fig. 5.29). This is typical of a subadult supraorbital torus. The thin vault bone also points to a subadult status for this specimen. Arsuaga et al. (1999b) estimate an age of 10 to 11.5 years at death for ATD6-15, although the frontal sinuses seem a bit large for this age. Arsuaga and his team describe the torus as being double-arched like those of Neandertals, rather than horizontal and shelf-like across the orbits as in many Erectines. They also stress the specimen's minimum frontal breadth, which is very broad (95–100 mm) for a subadult—about the same as that of the later adolescent cranium 6 from

TABLE 5.13 ■ Cranial and Mandibular Specimens from Atapuerca (Gran Dolina)

ATD6 Number	Brief Description
5	Right mandibular corpus of a juvenile with erupted M_1 and M_2. M_3 crown fully formed in crypt.
14	Left maxilla of a child with dc and dm^1. I^1, I^2, and C in crypts.
15	Left and central frontal bone of a juvenile with small adhering pieces of right nasal, right parietal, and right maxilla (frontal process).
16	Right temporal (mastoid angle). Exhibits cutmarks.
17	Associated pieces of temporal (17a) and sphenoid (17b). Temporal part exhibits of degenerative joint disease in the mandibular fossa area.
18	Left adult temporal (petrous portion).
19	Adult facial fragment with the zygomatic body and adhering maxilla.
20	Fragment of left parietal and temporal squama.
38	Left juvenile zygomatic bone.
57	Right temporal (mastoid process).
58	Left zygomatic with adhering maxilla.
69	Left zygomatic and maxilla, plus right maxilla, with adhering fragments of vomer and sphenoid. Right I^2-M^1, left P^3 and M^1 are erupted (right C and P^4 only partly). Left M^2 and M^3 are in crypts.
77	Right occipital condyle with hypoglossal canal.
84	Left adult zygomatic arch fragment (temporal and zygomatic bones).
89	Basilar part of adult occipital.

Source: After Arsuaga et al. (1999b) and Rosas and Bermúdez de Castro (1999). See also Schwartz and Tattersall (2002).

the Sima de los Huesos (100 mm). They imply that the cranial capacity of ATD6-15 might be comparable to the 1220 cm^3 estimate for SH 6.

The ATD6-57 temporal bone exhibits a small mastoid process that does not project below the occipitomastoid crest (juxtamastoid eminence), a trait characteristic of Neandertals and some Heidelbergs. The petrous portion of the temporal (ATD6-18) shows fusion of the styloid process to the basicranium, a primitive feature in *Homo* that some think was lacking in Erectines. The squamosal suture appears to be more arched than is the case in most Erectines, making it more like those of Heidelbergs and later archaic and modern folk. Other cranial vault pieces (Table 5.13) tend to be rather small and tell us little about the affinities of the Gran Dolina remains.

The middle and lower facial anatomy of the Gran Dolina people is best preserved in another subadult specimen, ATD6-69 (Figs. 5.29 and 5.30), which has been dentally aged to between 10 and 11.5 years (Bermúdez de Castro et al. 1997). Arsuaga and colleagues (1999b) describe this specimen as exhibiting a fully modern pattern of mid-facial topography—including a

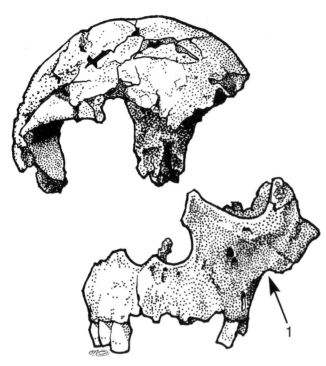

FIGURE 5.29

The ATD 6-69 juvenile partial face from the Gran Dolina at Atapuerca, Spain. This is the most complete specimen of *"Homo antecessor"* and one of the earliest hominin specimens from Europe. The angled zygomaticoalveolar margin and malar notch are visible in this frontal view. (After Arsuaga et al. 1999b.)

FIGURE 5.30

Lateral view of the ATD 6-69 lateral facial fragment (Gran Dolina, Sima de Atapuerca, Spain) showing the canine fossa. (After Arsuaga et al. 1999b.)

coronal orientation of the infraorbital plate, the presence of a canine fossa, an infraorbital plane that slopes posteriorly from the orbit, and an angled zygomaticoalveolar (ZAM) margin. The zygomatic root is at the front edge of M^1, indicating a flat face (Fig. 5.29), and the Spanish team think that the specimen probably exhib-

ited total facial prognathism. The root of the zygomatic is found rather high up on the alveolar process. Other, more fragmentary facial pieces are similar to ATD6-69 in morphology. The fully adult ATD6-58 zygomatic/maxillary specimen has a canine fossa, and the ATD6-38 specimen possibly has one as well. The ATD6-19 adult facial fragment has an angled ZAM. Its internal nasal anatomy has been claimed to lack the Neandertal apomorphies described by Schwartz and Tattersall (1996) and Arsuaga et al. (1997c). As we will see later (Chapter 7), these apomorphies have been disputed. The ATD6-69 specimen also exhibits an anteriorly placed incisive canal that runs almost vertical to the clivus. The mandible fragment, ATD6-5, shows a generalized morphology, differing little from other Lower and Middle Pleistocene *Homo* in Europe and Africa. However, Rosas and Bermúdez de Castro (1999) believe that the unusual position of the mylohyoid groove uniquely groups this specimen with African Erectine mandibles.

As early as 1995, Carbonell and colleagues had intimated that the Gran Dolina humans did not fit comfortably into any known species of *Homo*. The 1997 creation of the taxon *"H. antecessor"* for these remains (Bermúdez de Castro et al. 1997) was met with rather widespread skepticism (see Gibbons 1997). More detailed analysis of the remains (Bermúdez de Castro et al. 1999a) has not dispelled this skepticism. Bermúdez de Castro et al. (1997) list 29 defining characters for the new species, the most salient of which are listed below. Information in parentheses has been added by us.

1. "Modern" midfacial morphology exhibited by ATD6-69 and confirmed by other smaller fragments (but probably with total facial prognathism).
2. Modern human interior nasal anatomy, including the pattern of crests on the anterior nasal floor.
3. Absence of any derived Neandertal traits.
4. A relatively broad frontal bone, suggesting a cranial capacity of ~1220 cm^3, and a double-arched supraorbital torus.
5. Arched squamosal or temporal suture.
6. Primitive dental traits in posterior dentition.
7. Some size expansion of the anterior teeth.
8. Fusion of the styloid process to the cranial base.
9. Position of the mylohyoid groove on the mandible.
10. (Non-Neandertal limb proportions).
11. (European pattern of shoveling in maxillary incisors).
12. Anterior position and vertical orientation of the incisive canal.

For Bermúdez de Castro and colleagues, the Gran Dolina humans present a complex mosaic of primitive

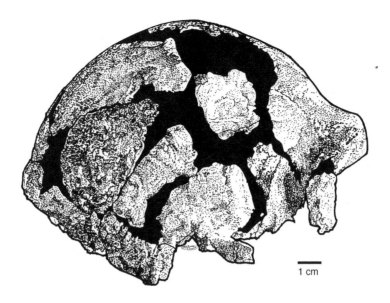

1 cm

FIGURE 5.31

The Ceprano (Italy) cranium in lateral view. Note the Erectine-like conformation of the nuchal torus and the supraorbital region, including the supratoral sulcus.

and advanced traits. The case for recognizing *Homo antecessor* as a species rests on a twofold claim: that the Gran Dolina humans are primitive enough to be the last common ancestors of both Neandertals and modern humans, but not primitive enough to be retained in *H. erectus*. According to them, the traits break down as described in the following sentences. The primitive traits in the foregoing list are 1, 2, 6, 9, and 10. Thus a "modern" human-like midfacial morphology, with a canine fossa, angled ZAM, and so on, is seen as primitive for *Homo*. Apomorphies separating the Gran Dolina humans from Erectines are found in features 4, 5, 8, and 12. Features 4, 5, 7, and 12 are derived traits that link Antecessors with both Neandertals and modern humans. Feature 10 is a plesiomorphy that implies that Antecessors have not become a part of the "specialized" Neandertal lineage. But #11 may indicate otherwise.

However, whether this combination of features really defines *H. antecessor* as a separate species is still open to question. The ATD6-15 supraorbital torus is no more double-arched than those of some Erectines—for example, the Zhoukoudian tori. Other suggested derived traits, like "some expansion of the incisors," are not very compelling. Some believe that too much of the evidence for "*H. antecessor*" is based on juvenile material, particularly ATD6-15 and ATD6-69 (Gibbons 1997). Finally, the material is quite fragmentary, and there is simply too much about the anatomy of the Gran Dolina humans that remains a mystery.

Another potentially ancient (800 Ky) European fossil comes from the central Italian site of Ceprano, where roadway construction in 1994 turned up a cranial vault

missing much of its base and part of its top but preserving a virtually complete frontal, right parietal, temporals, and much of the occipital (Fig. 5.31). The specimen apparently derives from Level 9 of the stratigraphic sequence at Ceprano, but it is not associated directly with artifacts or other remains (Ascenzi et al. 1996). A primitive Acheulean industry is found in the overlying Level 7, and these strata underlie a volcanic level dated by radiopotassium to ~700 Ky (Ascenzi et al. 1996). More recent discussion of the chronology of the Ceprano basin suggests that only radiopotassium dates on the order of 0.9 to 1.0 Mya underlie the specimen (Manzi et al. 2001). All these facts are compatible with a date of 800 Kya. But because of the circumstances of the find and questions about the specimen's geological context, a reliable date for it has so far proven elusive.

The Ceprano cranium has been described in detail by Ascenzi and his co-workers (1996, 2000; Manzi et al. 2001; see also Clarke 2000, Schwartz and Tattersall 2002). Initial publications by the Ascenzi team emphasized the Erectine-like features of the specimen, even designating it, in 2000, as the earliest *H. erectus* from Europe. This view is supported by Clarke's (2000) analysis of the specimen as well. The main similarities to Erectines are found in the pattern of thickness and projection of the (massive) supraorbital torus, and in the well-developed supratoral sulcus. Because there is a distinct supraglabellar depression, the supraorbital torus area is very similar to the Zhoukoudian tori. Other Erectine-like features include presence of an angular torus, thick cranial vault bones (especially on the cranial base), a strongly angled occipital bone, and a pronounced nuchal torus (with inion and opisthocranion at the same point). Also there is probably a high position of inion relative to endinion and a very pronounced mastoid/nuchal plane region, with the cranium's maximum breadth located at the supramastoid level (on the cranial base). Metric data for the Ceprano cranium are given in Table 5.1. A cranial capacity estimate of 1057 cm^3 (Ascenzi et al. 2000) places the specimen near the mean for Erectines (Table 5.3).

However, other features may suggest a more advanced hominin form. The vault is relatively short compared to its breadth. The braincase is nearly as broad as the cranial base, so the sides of the vault are relatively vertical. Sagittal keeling is absent on both the frontal and parietals, and the temporal squama is described as arched. On the basis of such advanced (or at least non-Erectine) features and a multivariate analysis of cranial form and discrete traits, Manzi et al. (2001) have argued that Ceprano should be placed in a new taxon (possibly *H. antecessor*) to demonstrate its transitional nature between Erectines and Heidelbergs.

Gilbert et al. (2003) challenge this conclusion, on the grounds that the phenetic analysis used by the Manzi group does not distinguish primitive from derived characters and thus does not accurately reflect phylogenetic relationships. Gilbert and his co-workers also argue that the Manzi group have scored some character states incorrectly. For example, Ceprano is scored as having a "short" cranial vault, while such Erectines and Ergasters as ER 3883, 3733, Sangiran 2, Daka, Dmanisi 2280 and 2282, and three Zhoukoudian crania are scored as having a "long" cranial vault. The problem is that all of these Ergaster/Erectine specimens have a vault length that is shorter than Ceprano's!

As noted earlier, the number of unquestioned human archaeological sites and human fossils in Europe increases markedly after 600/500 Kya. Most of these sites are discussed in the next chapter, because the associated human remains generally do not fit classic definitions of *H. erectus*. However, some of these later remains—for example, the fragmentary specimens from Bilzingsleben (Germany)—do not differ significantly from Erectines in anatomical pattern, particularly in the pronounced supra-orbital torus and occipital morphology. Thus, the question of whether there are Erectines in Europe is far from resolved, and many of the early fossils from the continent suggest that we cannot yet exclude Erectine influence totally from the initial homesteading of Europe.

If the dating is correct, Europe was the last of the Old World continents to be occupied by humans—even though Dmanisi, at the edge of Europe, was inhabited quite early. What made Europe itself so difficult to conquer? Europe lies further north than most of the regions of Asia inhabited by Erectines. Was it simply the cold of the Pleistocene temperate zone that kept early humans out of Europe, or did other factors play a role?

One factor that probably inhibited the northward spread of early *Homo* was skin color. As noted earlier in this chapter, Erectines adapting to life in the tropics would have had darkly pigmented skin in order to limit the absorption of ultraviolet radiation (UVR) from the sun. Too much absorption of UVR leads to skin cancer, damage to human thermoregulatory ability, and reproductive problems. But when humans moved out of the tropics, they ran into an opposite set of problems related to skin pigmentation. Absorption of adequate amounts of UVR is necessary for the synthesis of vitamin D, which in turn is necessary for calcium absorption and normal patterns of growth of the skeletal system. Jablonski and Chaplin (2000) note that deficiencies of this vitamin can result in rickets, which can cause relative immobilization, pelvic deformities that can preclude normal childbirth, and even death. They also point out that requirements for vitamin D$_3$ are elevated during pregnancy and lactation because of the mother's need to absorb and mobilize more calcium to provide for her offspring's developing skeleton. Vitamin D$_3$ also apparently triggers antimicrobial actions that target the tuberculosis bacillus (Liu et al. 2006).

According to W. Loomis (1967), depigmentation of human skin would be a necessity for humans to successfully adapt above 40° N latitude. Above this "pigment line," which runs through present-day northern Turkey, Greece, Italy and Spain, dark pigmentation would preclude absorption of sufficient UVR. Without adequate dietary sources of vitamin D, dark-skinned humans north of this line would risk all the ills associated with vitamin D deficiency. Most of East Asia inhabited by early Erectines is south of 40° N latitude, and so pigmentation was not a critical issue for them; but early humans probably could not thrive in Europe until their skin-pigment levels had been reduced by natural selection. This may explain why it took so long for humans to successfully "invade" Europe.

▌ MAJOR ISSUES: A SUMMING UP

Taxonomy

There is a general trend today to recognize more species within the genus *Homo*. Fossil remains that would have been grouped together as *H. erectus* ten years ago are now sometimes partitioned among as many as five species, including *H. ergaster*, *H. antecessor*, *H. georgicus*, and *H. heidelbergensis* in addition to *H. erectus* proper. Although we are not yet approaching the abundance of fossil human genera and species proclaimed in the taxonomies of the early 20th century (Campbell 1965, Smith 2002), it is certainly a fact that the classification of the genus *Homo* is becoming increasingly complex once again. It is also a fact that many of the new taxa are not as precisely defined or as strongly supported by available data as they need to be to win general acceptance by students of human evolution.

Brace (1988, 1989) has argued that this multiplication of taxa is rooted in a Neoplatonic, essentialist concept of species that is willfully blind to the messy realities of evolutionary biology. Others insist that Brace's view is the one that is out of touch with biological reality. As we noted earlier, a preference for taxonomic splitting over lumping is logically connected to the idea that most evolutionary change occurs during speciation events. Some scholars who accept this idea (Tattersall 2000, J. Schwartz 2000) have argued that human paleontology has lagged behind biology in general by failing to recognize the linkage between evolutionary change and speciation. In their view, a multiplication of species within the genus *Homo* is needed to bring paleoanthropology into line with other biological sciences.

Is there any way of using empirical facts to resolve this philosophical dispute? One way is to determine how many species there are nowadays of other mammals resembling *Homo* in ecology and size. Conroy (2002) compared species diversity in early *Homo* with that in other mammals that fall within the body size range estimated for early *Homo*. Species diversity in mammals is generally tied to overall body mass, with larger forms exhibiting less taxonomic diversity. Conroy found that mammals as large as early *Homo* generally do not exhibit large numbers of species. He argues that recognition of more than two synchronic species in early *Homo* would put the human lineage at odds with the pattern for similar-sized mammals. Foley (1991) suggests that once *Homo* achieved its status as a broad exploiter of many different sorts of resources, including meat, it was unlikely that the human lineage would speciate extensively. Broad resource exploitation would make later hominins a classic example of a generalized species, and such species show very low rates of speciation. Foley argues that speciation events would have been far less common in *Homo* than in earlier hominins with narrower ecological niches. This would be especially true in nontropical regions, where speciation rates tend to be reduced for all mammalian lineages.

We suggest that increasing human dependence on culture would also have inhibited speciation—not because culture is an ecological niche, as some theorists have suggested (see Blind Alley #5, p. 188), but precisely because culture is a way of escaping from any ecological niche. By allowing humans to opportunistically exploit any resources that become available, cultural adaptations lead to more encompassing exploitation of all resources and therefore to greater niche breadth—which would be expected to inhibit speciation. As noted earlier, early humans appear to have been very mobile and wide-ranging animals, and the extensive gene flow associated with such mobility would also inhibit speciation. Arcadi

(2006) has further argued that Pleistocene *Homo* would have had a high tolerance for an unusually wide range of habitats, which again would not facilitate speciation. Thus, biological theory, observed patterns of species diversity in living mammals, and the facts of fossil human morphology all raise serious doubts about the recent proliferation in the number of species of *Homo* recognized by many paleoanthropologists.

Of course, saying that we would expect little speciation within *Homo* on theoretical grounds does not refute the existence of multiple species. But this admission cuts in both directions. When a new taxon is created, it is incumbent upon its creators to demonstrate that it is distinct from those previously described. It will not do to simply assert that theory demands the presence of more taxa; the facts of biology must support their presence. The distinctiveness of *H. ergaster*, *H. antecessor*, *H. georgicus*, and even *H. heidelbergensis* (see Chapter 6) from *H. erectus* is far from unequivocally demonstrated by the pertinent fossil record. Increased knowledge may prove that some or all of these taxonomic designations are valid, but none of them are conclusively supported by currently available facts.

We have used the terms "Ergasters" and "Erectines" throughout this chapter in an effort to identify the specific samples under discussion without becoming mired in these taxonomic issues. Basically, we identify these groupings with what the late F. C. Howell (1996, 1999) called **paleodemes**. This term denotes a grouping of fossils that are of "proximate geographic distribution and closely similar age, a level between the individual specimen(s) and a higher taxonomic category" (Howell 1999, p. 294). Howell defined some 20 such paleodemes within the fossil record of the genus *Homo*. One or more paleodemes may represent a single species, but one can recognize these demes without necessarily agreeing on their formal taxonomic assignment. We will use such designations as Ergasters, Erectines, Heidelbergs, and Neandertals in this same manner. Of course, we will address the formal taxonomic issues as well.

Dates and Additional Evidence

As far as we can tell, *Homo* must have evolved from some late Pliocene Habiline in Africa. The very early dates offered for the oldest fossil *Homo* in Georgia and Java are therefore perplexing. The KNM-ER 3733 cranium establishes the presence of Erectines in Africa by 1.78 Mya, but the dates from Dmanisi and Perning are just as old. There are some indications that Erectines were already prowling the Lake Turkana Basin 175 Ky earlier. Although both the ER 2598 occipital and ER 3228 hipbone (which date to 1.9–1.95 Mya) may be Erectines, we need more diagnostic remains to be sure. Emergence of Erectines in Africa at close to 2 Mya would provide sufficient time for members of this group to have migrated to Georgia and Java by 1.8 Mya. If Erectines were present in Africa by 2 Mya, *A. habilis* and *A. rudolfensis* are less likely to have played a role in the origin of the genus *Homo*, because Erectines would be just as early as the oldest diagnostic remains from these other taxa. Earlier remains (>2 My) that have been suggested to represent *Homo*, such as the Baringo-Chemeron temporal and the AL 666 maxilla from Hadar, may be the ancestors of Erectines rather than members of the *A. rudolfensis/A. habilis* clade(s).

It should also be noted that the 1.8-Mya dates for Erectines at Dmanisi and Perning are still questionable, as is the 1.66-Mya date for Erectines at Sangiran. There is no *a priori* reason to doubt that Erectines could be this old in either locale, particularly if they were already in Africa by ~2 Mya. The recent reassessments of magnetostratigraphy at Dmanisi lend additional support to the early date there. However, the situation in Java is more complex, and there continue to be sound reasons to suspect that the early dates from Java are not accurately representing the true ages of the human remains.

We might understand more about Erectine radiations into Eurasia if we had a firmer grasp of the timing of subsequent movements of humans into new areas in the Old World. It would be helpful to know more about Erectine migration routes and their ability to cross water barriers. Although there seems to be a growing consensus that humans entered southern Europe by 700/800 Kya, more evidence documenting this early presence would be useful. Even if those early dates hold up, it still seems odd that humans would have inhabited Georgia for almost a million years before they managed to extend their range westward into Europe. The recently described Kocabaş skullcap from Turkey (Kappelman et al. 2008), provisionally dated to around 500 Kya (Chapter 6), suggests that more discoveries of early *Homo* remain to be made in southeastern Europe.

Evolutionary Patterns

The fossil record clearly documents geographical variation in Erectines. Although all Erectines share a basic anatomical pattern, aspects of that pattern differ systematically as a function of whether the remains come from Australasia, North Asia, or Africa. However, it is not clear that the differences between Asian and African or early European forms are great enough to warrant their placement in different species. Antón (2003) has suggested that the regional morphs of Erectines might best be considered as allotaxa, like the regional morphs of African baboons (Fig. 2.6). Accepting this view amounts to an admission that the limitations of the fossil record make it practically impossible to resolve the perennial arguments about species boundaries in early humans. According to both Antón and Jolly (2001), downplaying these unresolvable issues should allow a clearer focus on patterns of adaptation and other biological questions. We tend to agree!

CHAPTER SIX

The Big-Brained Ape: Regional Variation and Evolutionary Trends in the Middle Pleistocene

The most important differences between people and other animals lie in our mental capacities, and the physical matrix of the human mind is the human brain—which is much larger than those of other hominoids. These facts led early Darwinians to see brain enlargement as the key factor in the emergence and evolution of humans. As the anatomist Arthur Keith (1947) put it, "The essential mark of man lies neither in his teeth or his postural adaptations, but in his brain, the organ of his mentality." We have already recounted how the ape-sized brain of *Australopithecus* was for many years regarded as grounds for excluding this transitional form from the human lineage. Conversely, some scientists could not accept the original Neandertal specimen as representing an archaic phase of human evolution, because its braincase was fully as big as a modern human's.

These judgments are grounded in two suppositions: that there is a boundary between apes and humans, and that it can be defined in terms of brain size. Early in the 20th century, the pioneering neuroanatomist G. E. Smith pegged that boundary at 1000 cm³, which he thought was the minimum amount of brain matter needed for normal human neurological functions (Smith 1927).

Keith (1947) put the bar somewhat lower, claiming that it took a cranial capacity of only 750 cm³ to sustain a human-grade mentality. Keith referred to this boundary between "apehood and manhood" as the **cerebral Rubicon**—a reference to the River Rubicon in northern Italy, proverbial as a metaphor for historic turning points ever since Julius Caesar started a civil war by leading his armies across it.

Was there a decisive and transforming event in the evolution of the human brain, as Keith's metaphor suggests? As we have seen, cranial capacities in *Australopithecus* generally do not exceed those of living apes, which can reach as much as 700 cm³ in a big male gorilla. But even early *Australopithecus* specimens appear to have had larger brains than those of apes of similar body size today (Chapter 4). And around 1.8 million years ago, we begin to find fossil hominins with braincases that are even more capacious than those of big gorillas (Tables 4.5 and 5.6). Some of these creatures, including advanced *Australopithecus*-like forms (Habilines) as well as true early *Homo* (Ergasters), have endocranial volumes that lie on the human side of Keith's "Rubicon," though not yet within the range of normal modern human adults. In short, increased encephalization appears to be an ancient trend in the human lineage, extending back before the appearance of the genus *Homo*.

But having reached the intermediate levels seen in early *Homo*, hominin cranial capacities remained more or less static for over a million years (Rightmire 1981, Ruff et al. 1997, Wolpoff 1999). The phrase "more or less" here skirts an inconclusive debate of the usual sort between punctuationists, who think stasis is the evolutionary norm, and gradualists, who do not. The data do not decisively demonstrate an absolute stasis in brain size during this period (Leigh 1992, Lee and Wolpoff 2003), but they suggest an essentially flat curve of absolute cranial capacity against time from 1.8 Mya up to ~400 Kya (Fig. 6.1). After that date, average brain size increases in a steady linear fashion until about 100 Kya, when a second period of stasis sets in.

The notion of a cerebral Rubicon defined in terms of absolute brain size is no longer regarded as valid, because it does not consider allometric factors. Absolute size matters in brains, but it is not the whole story. A brain in the human size range does not guarantee human mental capacities. If it did, elephants and whales would be mental as well as physical giants. When we compare pairs of species that have comparable brain structure and mental abilities, but differ in body size—say, house cats versus tigers, or sheep versus cows—we find that the larger animals have brains that are absolutely larger, but *relatively* smaller (as a percentage of body weight) than those of their smaller relatives. Evolving cranial capacities therefore need to be interpreted in the light of concomitant changes in body size, and so body mass has to be taken into account in tracing the history of human brain evolution.

As we saw in Chapter 4, inferring body mass from the fragmentary remains of extinct animals like *Australopithecus* can be a tricky business. Luckily, the job becomes easier when we restrict our attention to the genus *Homo*. Since ancient humans appear to have

had modern human body proportions, we can use regression equations derived from modern humans to infer ancient human body mass from skeletal measurements. Using estimates obtained in this way, Ruff et al. (1997) find only two significant changes in human body mass over the past 1.8 million years. The first is a marked increase between 90 and 75 Kya in late "archaic" humans—meaning Neandertals and their primitive-looking contemporaries (see below). This size increase distinguishes these people from their predecessors—and also from early "modern" humans, a category that includes some Near Eastern near-contemporaries of Neandertals (Skhūl/Qafzeh) as well as later people associated with the early Upper Paleolithic in Europe. The second big change is a significant reduction in body mass beginning about 35 Kya and continuing down to the present day. Neither of these changes correlates with the absolute increase in human brain size between 400 and 100 Kya, and so that increase in braininess cannot be read as a mere side effect of increasing body size. In fact, encephalization quotients (EQs) calculated from these body-mass data and brain-volume estimates (Table 6.1) show that when we correct for allometry, *relative* brain size increases significantly only after 400 Kya. By this time, humans have clearly become *the* big-brained ape. If there is a decisive event in the history of human brain evolution, it is not the crossing of some imaginary Rubicon signaling the acquisition of a critical brain mass, but the change in the *rate* of brain evolution that sets in about 400 Kya.

OF "ARCHAIC *HOMO SAPIENS*" AND *HOMO HEIDELBERGENSIS*

Because brain-size averages start to fall consistently into the modern human range after 400 Kya, many paleoanthropologists in the 1960s and 1970s regarded the fossil human remains from around 400 Kya as the earliest

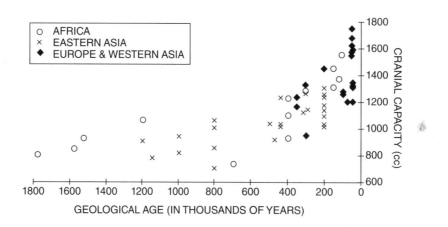

FIGURE 6.1

A plot of hominin cranial capacity over time beginning 1.8 Mya. (Modified from Ruff et al. 1997.)

TABLE 6.1 ■ Estimated Body Mass, Cranial Capacity, and Brain Mass for Humans over the Past 1.8 Million Years

Sample	Temporal Range (Kya BP)	Body Mass (kg)[a] (Mean ± SE (n))	Cranial Capacity (cc) (Mean ± SE (n))	Brain Mass (g) (Mean ± SE (n))	EQ Sample Means (Mass)
Living worldwide	—	58.2 ± 1.0 (51)	1349	1302	5.288
Pecos Pueblo	—	55.5 ± 1.2 (29)	1308 ± 23 (29)	1263 ± 22 (29)	—
Late Upper Paleolithic	10–21	62.9 ± 0.9 (71)	1466 ± 35 (23)	1412 ± 33 (23)	5.406
Early Upper Paleolithic	21–35	66.6 ± 1.3 (33)	1517 ± 30 (15)	1460 ± 28 (15)	5.352
Late archaic *H. sapiens*	36–75	76.0 ± 1.4 (17)	1498 ± 45 (14)	1442 ± 42 (14)	4.781
Skhul-Qafzeh	90	66.6 ± 2.2 (10)	1501 ± 45 (6)	1444 ± 42 (6)	5.293
Early Late Pleistocene	100–150	67.7 ± 2.4 (10)	1354 ± 41 (8)	1307 ± 39 (8)	4.732
Late Middle Pleistocene	200–300	65.6 ± 5.1 (6)	1186 ± 32 (17)	1148 ± 30 (17)	4.257
Middle Pleistocene	400–550	67.9 ± 6.4 (5)	1090 ± 38 (12)	1057 ± 36 (12)	3.818
Late Early to early Middle Pleistocene	600–1150	58.0 ± 4.3 (3)	856 ± 52 (7)	835 ± 50 (7)	3.400
Early Pleistocene	1200–1800	61.8 ± 4.0 (5)	914 ± 45 (5)	890 ± 43 (5)	3.458

[a]See source for details of sample composition and methods for deriving brain mass from cranial capacity and the calculation of encephalization quotients (EQ). For most of this sample, especially those prior to 35 Kya, brain and body data cannot be derived from the same individuals.
Source: Ruff et al. (1997).

representatives of our own species, *Homo sapiens*. Most of the pertinent fossils known at the time came from European sites, including Swanscombe (England), Steinheim (Germany), Arago (France), and Petralona (Greece). The skulls from these sites had human-sized braincases, but their cranial morphology was certainly not modern. Like Erectines, they tended to have persistently low cranial vaults, receding foreheads, large and projecting faces, and pronounced supraorbital tori. These people were therefore routinely labeled as "archaic *Homo sapiens*"—admitted to membership in our species because of their almost modern-sized brains, but set off as "archaic" because of their primitive-looking cranial morphology.

During the 1970s, the growth of cladistic systematics and the spread of punctuationist models of macroevolution made this terminology and way of thinking increasingly unpopular. S. J. Gould, N. Eldredge, S. Stanley, and others revived and vigorously championed an older view of human phylogeny as a "bush, not a ladder," in which multiple dead-end lineages had repeatedly branched off the human stock and been supplanted by later offshoots. The branches on these phylogenetic bushes were conceived of as internodal species, beginning in a speciation event and ending in another speciation or an extinction. Such concepts suited the theoretical needs of both punctuationism and cladistics. From the 1980s on, these trends led to calls for the recognition of more hominin species (e.g., Eldredge and Tattersall 1982, Tattersall, 1992) and encouraged paleoanthropologists to make ever more and finer taxonomic distinctions within the genus *Homo*.

Tattersall (1986) particularly criticized the concept of "archaic *Homo sapiens*," a category that he saw as taxo-nomically meaningless and without parallels in other domains of biology. Tattersall suggested resurrecting the taxon *Homo heidelbergensis* for the earliest European *Homo*, as distinguished from the later Neandertals (*H. neanderthalensis*) and modern humans (*H. sapiens*). The recognition of *H. heidelbergensis* as a distinct species also received a boost from the growing perception that none of the European fossils represented *Homo erectus* in the strict sense (see Chapter 5). The two species *H. heidelbergensis* and *H. neanderthalensis* partitioned the same taxonomic space that had been covered in Europe by the "archaic *H. sapiens*" category. With the additional of appropriate African specimens (e.g., Kabwe, Bodo, Ndutu and Elandsfontein), *H. heidelbergensis* was resurrected and reconceptualized as the last common ancestor of two later human lineages: the Neandertals in Europe and the first modern humans in Africa (Rightmire 1990, 1996, 1998).

Distinguishing these three groupings as separate species would seem to make for greater taxonomic precision. But as we have noted in previous chapters, imaginary precision is not a satisfactory substitute for messy biological reality. It is important to remember that there is no unanimity among scholars concerning the diagnosis, membership, or even existence of *H. heidelbergensis*. A precise morphologically based definition of this taxon has yet to be offered. Some experts (Rightmire 1990, 1996, 1998; Tattersall and Schwartz 2001) include all African "archaics" and the oldest European ones in *H. heidelbergensis*, but regard later European specimens (e.g., Swanscombe) as early representatives of the Neandertal clade. Klein (1999) prefers to assign all of the European fossils to Neandertals and all African ones to *H. sapiens*. Bermúdez de Castro et al. (1997) restrict

H. heidelbergensis to the "pre-Neandertals" of Europe (see also Rosas and Bermúdez de Castro 1998). McCarthy et al. (2007) hold that there is too much cranial variation, even within Africa, to allow a single species to encompass all specimens currently placed in *H. heidelbergensis*, and that at least two species need to be distinguished within this grouping.

The reality and boundaries (if any) of the taxon *H. heidelbergensis* can be assessed only after reviewing the pertinent fossil record. In doing so, we will focus on specimens and sites between ca. 400 Ky and 200 Ky in age. These temporal boundaries encompass the entire "archaic *sapiens*" group (excluding the so-called "classic Neandertals") from Europe and Africa, as well as some arguably comparable fossils from Asia and Australasia. To avoid prejudging the taxonomic issue, we will refer to these fossils collectively under the informal label of **Heidelbergs**.

The Heidelbergs are the first hominins to exhibit a human level of encephalization in both relative and absolute terms. But there is no way of knowing whether this breakthrough in brain size was accompanied by any changes in the internal organization or "wiring" of the brain. The human brain is not only larger than those of apes, but also more complex (Martin 1990, Deacon 1997, Holloway 2000, Holloway et al. 2004). Unfortunately, we cannot judge levels of neurological complexity in ancient brains, because all our knowledge of them is derived from casts of the inside of the braincase. Such endocasts can reveal information only about the external topography of the brain. In hominins, they do not even do that very well, because the fit between the brain's surface and the inside of the braincase is rather loose in humans (and other large mammals), and the impression left by the brain on the skull is correspondingly fuzzy. In short, we have no direct morphological evidence bearing on the organization of the brain in fossil humans.

In principle, some inferences about brain reorganization might be drawn from changes in early human behaviors revealed by the archaeological record. But the acceleration in human brain-size evolution that began some 400 Ky ago was not accompanied by any detectable changes in such behaviors. In the western parts of the Old World, the period between 400 and 200 Kya continues to be characterized by Acheulean cultural assemblages containing bifacially worked tools. These assemblages are not different from those associated with late African Erectines (Barut and Smith 2001, Bordes 1968, Clark 1977, Gamble 1986, Klein 1999). Later Acheulean tools are slightly more sophisticated; but the change is not particularly striking, and it does not seem to reflect any fundamental behavioral shifts (Gamble 1986, Klein 1999). Further east, biface assemblages are rare (though not entirely absent) throughout this period, and non-biface assemblages of a vaguely

Oldowan sort predominate (Fig. 5.23). Similar non-biface tool "cultures," designated by such labels as Clactonian, Tayacian, Taubachian, and Buda, are found in Central and Eastern Europe and at certain localities further west (de Lumley 1975, Svoboda 1987). Some think that the geographical separation of these two types of assemblages may reflect a dual source for the colonization of Europe: an African source bringing Acheulean traditions, and an Asian source with nonbiface traditions carried by more Erectine-like people (Rolland 1992, Klein 1999). Others argue that the differences between the two traditions may instead reflect local differences in ecology and in the raw materials available for making stone tools (Villa 1983). The non-biface assemblages from northern and eastern Eurasia are sometimes considered "transitional" to the succeeding (post-Acheulean) Middle Paleolithic, but the fact is that they generally parallel the Acheulean temporally rather than occurring after it.

It is not clear exactly what the distribution of biface versus nonbiface assemblages means in any adaptive or populational sense. All we can say for sure is that some European Heidelbergs are found with Acheulean tools, while others are associated with nonbiface tool kits. Whatever behavioral changes underlay the new morphology seen in these big-brained apes were rooted in the Acheulean and its contemporary traditions, and must have involved aspects of human behavior that leave no trace in the archaeological record.

◼ EARLY MODELS OF LATER HUMAN EVOLUTION

At this point in our story, we need to sketch the history of 20th-century thought about the origin of modern humans. Although the fossils considered in this chapter are humans of a distinctly archaic sort, their interpretation has been profoundly influenced by models of modern human origins, and many of them have played important roles in the debates surrounding those models. Like it or not, scientific interest in all the phases of human evolution in the Pleistocene has been driven and channeled by a natural human preoccupation with the question of how we ourselves came into existence. Conflicts between competing models of modern human origins therefore underlie a lot of seemingly unconnected arguments about human evolution—not only about the first truly modern-looking human remains in the Late Pleistocene, but also about the meaning of earlier hominin fossils and the way the processes of evolution have worked throughout the history of the genus *Homo*. To understand those arguments, we need to understand those models.

During the first half of the 20th century, paleoanthropologists debated several conflicting models of modern human origins. Vallois (1954, 1958) grouped them under three general headings: Pre-Sapiens, Pre-Neandertal, and Neandertal-Phase models. These widely used labels obscure the significant variation found within each group (Spencer and Smith 1981). All the **Neandertal-Phase** models, for example, depicted the European Neandertals as ancestors of modern humans (Brace 1964, Bowler 1986, Spencer 1984, Smith 1997b). However, some proponents of this idea, especially Weidenreich (1946, 1947a,b), saw the Neandertals as specifically ancestral only to today's European and West Asian populations, whereas A. Hrdlička (1927, 1930) believed that Neandertals gave rise to early modern humans in Europe and that these early moderns then spread from Europe into the rest of the world (Spencer and Smith 1981). Both stories are "Neandertal-Phase" models in Vallois's sense. But Weidenreich's version is a precursor of today's **multiregional-evolution** model, which posits regional genetic continuity between archaic and modern humans in various parts of the Old World (Wolpoff and Caspari 1997). Hrdlička's version resembles today's **single-origin** models, which view modern humans as originating in one small area and then spreading outward to replace more archaic populations.

The **Pre-Neandertal** models held that modern humans evolved from some archaic human ancestor(s) somewhere in the Old World. This ancestry might have included the early, "progressive" Neandertals (early European and Near Eastern specimens), but it excluded the later and more specialized "classic" Neandertals of Europe. Such models were propounded by the American scientists W.W. Howells (1959) and F. C. Howell (1951, 1957) and the Italian anthropologist S. Sergi (1953). Broadly speaking, their ideas can be seen as precursors of today's leading single-origin theory, the **Recent African Origin** model, although the focus on Africa as the center of modern human origins came later.

Pre-Sapiens models have no current proponents or derivatives, but they had a powerful historical influence on the interpretation of many of the fossils considered in this chapter. Pre-Sapiens theorists (Keith 1915, Boule 1921) believed that modern *Homo sapiens* had a long lineage separate from those of Neandertals, Erectines, and other archaic humans. They argued that this lineage was documented by certain ancient fossils that showed distinctively modern anatomical features, particularly those involving the skull and brain. Most of these fossils were eventually shown to be incorrectly dated modern humans of no great antiquity. But some of the most ancient of these pre-Sapiens fossils, the remains from Piltdown in southern England, turned out to have been concocted as a deliberate hoax.

The historical details of the Piltdown affair are well-known and notorious (Spencer 1984, 1986, 1990, 1997b). We summarize them here because of the impact Piltdown had on the study of human evolution for much of the 20th century. In 1911, several cranial fragments were unearthed in a gravel pit at Piltdown by C. Dawson, who showed them to A. Smith Woodward at the British Museum. Working together, they recovered more pieces of the cranial vault and a right partial mandible (with two molars) in 1912, found in association with fossil animal bones and possible tools. Between 1913 and 1915, a lower canine and pieces of a second skull came to light, but no other parts of the face or cranial base of Piltdown 1 were found.

The pieces of Piltdown 1 were broken in such a way that no direct articulation between the cranium and mandible was possible. Among other things, the condyle of the mandible was missing. But it was clear that the skull, as reconstructed by Woodward, had a relatively modern shape, although the vault was unusually thick and the initial cranial capacity estimate of 1070 cm^3 was on the small end of the modern range. A second reconstruction of the vault (Fig. 6.2) by A. Keith made the skull look even more modern, and raised the cranial capacity to 1400 cm^3.

The Piltdown mandible, on the other hand, was very ape-like, with a backward-sloping symphysis bearing a so-called "simian shelf" (a protruding bony buttress on the lower rear edge; see Fig. 4.48). Yet the lower molars had flat, seemingly humanlike wear surfaces unlike those of living apes. When one of us (FHS) examined the Piltdown specimens in 1973, the thing that stuck out most was the peculiar pattern of wear on the molar teeth. They were worn almost perfectly flat, and the edges between the sides and occlusal surfaces were quite sharp. No primate teeth wear like this. It may seem surprising that this anomaly escaped expert attention for so long. But hindsight is always 20/20, and comparative anatomists today know much more about patterns of primate dental wear than their predecessors knew in the early 20th century.

To their credit, some leading scientists (including W. K. Gregory, D. Gorjanović-Kramberger, and A. Hrdlička) never accepted Piltdown and provided insightful criticisms of its anatomy (Spencer 1984, 1990, 1997b; Radovčić 1988). However, it was readily accepted by many other experts for a variety of reasons. Probably the main reason for Piltdown's broad acceptance was that (like other candidate Pre-Sapiens specimens) it catered to one of the most cherished preconceptions about human evolution held at that time—namely, the primacy of the role of the brain in human evolution. Because the brain was viewed as the most distinctive and important human feature, it was generally thought that evidence of the brain's distinctiveness should be discernible even in the earliest stages of the human lineage. All of the supposed Pre-Sapiens fossils—even the purportedly earliest one, Piltdown—had relatively high, rounded cranial vaults. These skulls, and presumably the underlying brains, were more like those of modern humans than were the low-vaulted brainpans of

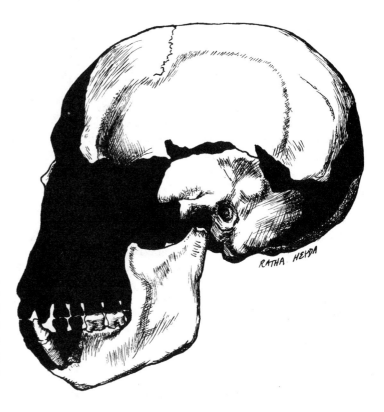

FIGURE 6.2

The Piltdown skull from Sussex, England, as reconstructed by Arthur Keith. This famous hoax was concocted by combining a relatively thick modern skull with the mandible of an orangutan. Pieces were deliberately broken so as to obscure the lack of articulation in critical areas (e.g., at the temporomandibular joint). (Drawing by R. Heyda.)

Neandertals and "Pithecanthropus." It was concluded from all this that modern human brains and skull form were just as ancient as those of "lowbrows" like Neandertals, and that living humans had an older and more intellectual ancestral lineage than the Neandertal-Phase and Pre-Neandertal models implied. At bottom, the chief charm of Piltdown was that it relieved scientists of the distasteful need to admit Neandertals and Erectines into their own family trees.

As noted in Chapter 4, Piltdown also was a significant factor in the initial dismissal of *Australopithecus* from human ancestry. The bipedal apes from the South African cave sites had ape-sized braincases. They did not fit the preconceived picture of the primacy of the brain, which Piltdown so conveniently corroborated. In the context of Piltdown, even the human-like jaws of the *Australopithecus* fossils were seen by some as ruling them out of human ancestry. Because Piltdown had a big brain, it had to be a human ancestor; and because Piltdown had an ape-like mandible and canine, it proved that these primitive features had persisted for a long time in the line leading to *Homo sapiens*. Therefore, the seemingly more humanlike jaws and teeth of *Australopithecus* had to be mere convergences and autapomorphies. (The term "autapomorphy" had not yet been invented during the vogue of the Pre-Sapiens model, but the concept was familiar under other labels.) And as autapomorphies, these human-convergent traits excluded *Australopithecus* from the

human lineage—in the eyes of the Pre-Sapiens theorists, at any rate.

But as knowledge of human prehistory accumulated, the Pre-Sapiens model grew increasingly tattered. By the 1950s, such purported Pre-Sapiens fossils as the Grimaldi skeletons from Italy and the Ipswich and Galley Hill remains from England had been shown to be intrusive recent burials in older deposits—not modern-looking contemporaries of the Neandertals, as the Pre-Sapiens theorists had claimed (Brace 1964, Spencer 1984, 1986). The Grimaldi fossils turned out to be associated with the Upper Paleolithic, and they are still an important (though relatively recent) part of the human fossil record. But the others are not. New geochemical techniques helped to establish these facts. The Galley Hill remains, for example, were revealed as intrusive and recent when it was found that they had absorbed less fluorine from the groundwater than the genuinely ancient animal bones associated with them on the 100-foot terrace of the Thames (Oakley and Montagu 1949).

The revised dates for these late Pre-Sapiens remains focused suspicion on their supposed ancient ancestor, Piltdown. Initial attempts to apply the fluorine test to the Piltdown materials were inconclusive, but J. Weiner was able to demonstrate that the Piltdown jaw and teeth had been artificially modified. The results of an improved fluorine test showed conclusively that the Piltdown cranial pieces were not contemporaneous with the jaw. These findings led to the distressing suspicion that at least parts of the Piltdown assemblage were bogus (Weiner et al. 1953). More intensive analysis of the cranial pieces and other specimens demonstrated that the Piltdown finds were an elaborate hoax (Weiner 1955, Spencer 1990, 1997b; Gee 1996). The fraudulent "fossils" had been concocted by taking a modern ape jaw, filing the teeth down with a metal rasp to make them look more human, mixing them together with pieces of a pathologically thick modern human skull, and staining the lot to resemble long-buried fossils.

There has been considerable speculation about the identity of the culprit or culprits. Dawson was certainly involved, but it is not clear whether others were in on the plot or what anyone's motives may have been (Spencer 1990, 1997b; Gee 1996). What is clear is the lessons Piltdown conveys about science in general and paleontology in particular. When Piltdown was created, the human fossil record was not extensive, and the pre-Neandertal part of it was particularly sketchy. It was relatively easy to slip a fraudulent fossil into the picture, because no one really knew what genuinely ancient hominins looked like. But as more fossils of *Homo* and *Australopithecus* were recovered during the 1930s and 1940s, Piltdown began to stand out like the proverbial sore thumb. All the other fossils were telling scientists that humanlike dentitions came before humanlike brains, with Piltdown as the only exception. The increasingly bad fit between Piltdown and the growing knowledge base of paleoanthropology ultimately prompted the studies by Weiner and K. P. Oakley that exposed the hoax.

The story of Piltdown teaches us two important lessons about the nature of science. The first is that empirical science is ultimately self-correcting. In the long run, enhanced knowledge can be expected to lead to more accurate understanding of anything, be it history, physics, cosmology, or paleoanthropology. But the second, equally important lesson is that we can never be confident that we know enough *now*, in the short run, to feel entirely sure of our knowledge. Sometimes increased knowledge leads us through even greater periods of confusion before we reach a more accurate understanding. Humility in confronting the natural world and a decent respect for the opinions of others are virtues appropriate to the conduct of science. Unfortunately, scientists are no more naturally virtuous than anybody else.

The demise of Piltdown removed the last of the original Pre-Sapiens specimens from the scientific scorecard. However, the Pre-Sapiens theorists were not ready to concede the game. Following in the footsteps of his mentor Boule, Vallois became the primary champion in the 1950s of a revised Pre-Sapiens model. In place of the original, discredited Pre-Sapiens specimens, Vallois put forward new candidates, arguing that the Swanscombe skull from England (see below) and the remains from the French site of Fontéchevade (Chapter 7) established the presence of a Pre-Sapiens lineage in Europe (Vallois 1954, 1958). Vallois's interpretations of these specimens kept the Pre-Sapiens model going well into the 1970s.

Although the Pre-Sapiens model as such has no living proponents, some aspects of it have remained alive in later theories. The original classification of the Habilines as early *Homo* was rooted in L. S. B. Leakey's Africa-centered version of the Pre-Sapiens theory. Some experts (e.g., Howell 1960) identify specimens from the time period covered in this chapter as representatives of a non-Neandertal lineage living alongside the Neandertals in Europe. This claim can be thought of as a sort of minimalist limiting case of the Pre-Sapiens model.

In recent years, discussions of the origin of modern *Homo sapiens* have been dominated by two models: the **Recent African Origin** model (**RAO**) and the **Multiregional Evolution** model (**MRE**). The ongoing debate over these contending models has been summarized in many books and articles (Smith and Spencer 1984, Trinkaus 1989, Mellars and Stringer 1989, Bräuer and Smith 1992, Nitecki and Nitecki 1994, Clark and Willermet 1997). Both models have connections of one sort or another to the earlier models classified by Vallois, though these connections are not always simple or obvious. Unlike almost all earlier models, both RAO and MRE are based on information drawn from genetic studies as much as on the fossil record itself, and both emphasize strong connections to biological theory. Over their 20-year history, these models themselves have evolved, so that both models as they are currently articulated differ from the versions first proposed in the mid-1980s (Smith 1997a, 2002; Stringer 2002, Trinkaus and Zilhão 2002).

■ THE RECENT AFRICAN ORIGIN MODEL

The RAO model is most closely associated with the English paleoanthropologist Christopher Stringer, who outlined its basic tenets in a classic 1988 paper written with his colleague Peter Andrews. The ideas of Stringer and Andrews were based both on the fossil record and on new information about mitochondrial DNA (mtDNA) variation in living human populations (Cann et al. 1987). In its original form, the RAO model posited that modern humans arose in Africa as a new species (*Homo sapiens*) around 100 Kya. The new species soon spread into Eurasia—first into the Near East and then to the rest of Asia and Europe—where it replaced the local archaic peoples (see Fig. 6.3) without interbreeding with them. Although Stringer (1994) conceded that some slight degree of interbreeding was a theoretical possibility, he thought that genetic admixture between modern and archaic humans had been insignificant

FIGURE 6.3

Map of the major sites discussed in this chapter.

(Stringer 2002, Stringer and Bräuer 1994, Bräuer and Stringer 1997).

In the ensuing 20 years, RAO has met with widespread acceptance. It has become a popular explanation for the apparent morphological and genetic homogeneity of recent humans. As presented in 1988, it incorporated three central claims:

1. RAO represented the origin of modern *Homo sapiens* as single-centered or **monocentric**, with the whole set of distinctively modern apomorphies coming into being as a suite in a restricted geographical region (in this case, Africa).

2. RAO was a **total-replacement** model, depicting modern *Homo sapiens* as taking over the ranges of earlier humans and replacing them without interbreeding.

3. RAO was a **speciation** model; that is, it proposed that modern human morphology originated through a speciation event.

Like most new ideas, the RAO model of Stringer and Andrews had its roots in earlier thinking. Most of the Pre-Neandertal and Pre-Sapiens models had depicted the origins of modern humans as monocentric, as did Hrdlička in his version of the Neandertal-Phase model.

W. W. Howells and F. C. Howell (1994b, 1996, 1999), the leading American proponents of the Pre-Neandertal model, also promoted more or less monocentric theories about modern human origins, although both were often vague on the location of the "center." In a series of papers and books spanning more than three decades, Howells (1959, 1973a, 1974, 1976, 1993) argued for the expulsion of Neandertals from modern human ancestry and intimated that modern humans had originated somewhere to the East of Europe. Howells's most important statement of his position was a paper published in 1976, in which he combined his conclusions about modern human cranial homogeneity (Howells 1973b) with the pertinent fossil record and the available information about the population genetics of living humans. This landmark article laid the groundwork for the RAO and all other subsequent models that claim a recent monocentric origin for modern humans.

The patterns of genetic polymorphism seen in living human populations suggested to Howells that the center for modern human origins might have been in Africa. L. S. B. Leakey (1934, 1936), and R. Protsch (1975) had put forward similar ideas earlier, but their arguments were founded on questionable dates attributed to certain African fossils. G. P. Rightmire (1976) focused attention on later human fossil remains from Africa and suggested cautiously that the first modern humans might have appeared there (see also Rightmire 1984b). But it was the German paleoanthropologist Günter Bräuer who first unequivocally argued that Africa was the place of origin of modern humans, a conclusion based on his own extensive analysis of the pertinent African fossil record (Bräuer 1984a,b, 1989). Like Howell, Howells, and Stringer, Bräuer accepted the theoretical possibility that "hybridization" could have occurred between modern humans and more archaic forms. But his later writings (Bräuer 2003, 2006; Stringer and Bräuer 1994, Bräuer and Stringer 1997, Bräuer and Broeg 1998, Bräuer et al. 2004, Bräuer et al. 2006) indicate that he does not regard interbreeding as a significant factor in the peopling of Eurasia by early modern humans.

Early studies of nuclear DNA polymorphisms were interpreted as pointing to an African origin for modern humans (Nei and Roychodbury 1982). However, the strongest genetic evidence for this claim was provided by mtDNA. From their survey of the distribution of mtDNA variants in modern humans, Cann et al. (1987) drew three conclusions:

1. All the mtDNA variants in living humans can be traced back to a single mtDNA haplotype that existed in Africa as recently as 90 Kya.

2. The ancestors of all living non-African peoples emigrated from Africa at about this time.

3. No signs of interbreeding between these emigrants and indigenous archaic peoples in Eurasia can be discerned in the genetic data.

By this interpretation, the mtDNA evidence provided support for all the major elements of Stringer's RAO theory: monocentrism, speciation, total replacement, and recent African origin. Because of the Biblical parallels to the single "mother" of us all implied by the mtDNA data (more precisely, a single group of mothers with the same mtDNA haplotype), this perspective became widely known as the "African Eve theory."

Later researchers have offered significant refinements and changes to the original RAO model. M. Lahr (1996, 1997; Lahr and Foley 1994, 1998) postulates a serial replacement of Eurasian archaics, involving multiple waves of progressively more and more modern populations coming out of Africa one after another. Others suggest that a weaker "Out of Africa" model, involving variable amounts of gene exchange with indigenous archaic populations, might be an appropriate fit to the data (Harpending and Relethford 1997, Relethford 2001). From his analysis of 25 DNA haplotypes in extant human populations, Templeton (2002, 2005) argues for a "mostly out of Africa" genetic model, involving not one but three migration events from Africa. In addition to the spread of early modern people from Africa, which he sees as occurring ~130 Kya (Chapter 8), Templeton

detects traces of two earlier series of migrations. The first was the initial movement of hominins out of Africa, some 1.9 Mya (Chapter 5). The second, around 650 Kya, may represent the radiation of early archaic *Homo sapiens* (a.k.a. *Homo heidelbergensis*) from Africa into Eurasia. If Templeton is correct in detecting signs of the last two of these migrations in the genetics of human populations today, then the wave of early moderns that left Africa around 130 Kya must have interbred with the archaics that they encountered. If they had not, then all traces of the earlier migration would have been wiped away by the last one.

Although RAO is a theory about the origin of modern humans, it bears on broader issues in evolutionary theory. The speciation event assumed by RAO fits the punctuated-equilibrium model of macroevolution, in which the role of anagenesis is minimized and evolutionary change is concentrated in the "genetic revolutions" attending the formation of new species. To be sure, the evidence for the RAO model of modern-human emergence provides no logical grounds for assuming that similar things happened during earlier stages of human evolution. But many feel that the RAO model validates the expectations of punctuationist theory, and that if the origin of modern humans involved a speciation event, then other such events must surely have occurred many times during the preceding two million years of human evolution. S. J. Gould, one of the architects of the punctuated-equilibrium model, embraced RAO as a paradigmatic example of how macroevolution works, and argued that it provided evidence for the occurrence of multiple speciation events in the evolution of the genus *Homo* (Gould 1987). The species thus produced would have included *H. heidelbergensis*, *H. erectus* in the strict sense (restricted to Asian Erectines), *H. ergaster*, *H. neanderthalensis*, and perhaps others. Thus, the RAO model has been both a product of, and at the same time a major contributor to, the growth of the multiple-species approach to the hominin fossil record.

THE MULTIREGIONAL EVOLUTION MODEL

The Multiregional Evolution model (MRE) put forward by M. Wolpoff, A. Thorne, and X. Wu in 1984 is a more detailed and sophisticated version of Weidenreich's "trellis" model of human evolution. Like Weidenreich (1946, 1947a,b), the proponents of MRE contend that humans in the genus *Homo* have always been members of a single species. Although regional populations have developed distinctive regional peculiarities, they have always been connected by enough gene flow to prevent any of them from becoming reproductively isolated (Fig. 6.4). The genus *Homo* has thus evolved as an interconnected web of regional lineages extending back to

MODERN POPULATIONS

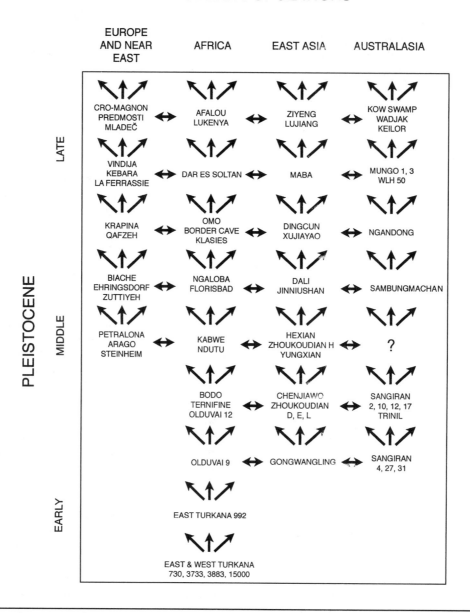

FIGURE 6.4

Multiregional Evolution in schematic form. The arrows indicate patterns of gene flow. (Modified from Frayer et al. 1993.)

the Erectines. As expounded by Wolpoff (1999), MRE is not monocentric; it holds that modern human anatomy did not evolve as a complex in a single region, but emerged through the consolidation of traits originating in different regions. The length of time that it took for a fully modern anatomical pattern to emerge in a particular region would have depended on several factors, including the rate and pattern of gene flow, the local selective environment, and genetic drift. Although local populations might occasionally have been replaced by immigrants, some archaic populations survived in every geographical region to become contributors to the modern human gene pool. The extent of archaic contribution to early modern populations would have varied from region to region.

Despite assertions to the contrary (Howells 1993), MRE does not imply the parallel evolution of modern humans independently in different regions of the Old World (Smith 1997a, Wolpoff et al. 2000). Like Weidenreich's "trellis," MRE posits genetic exchanges not just within but also between regional lineages (Fig. 6.4). Naturally, the amount of gene flow between regions would have been variable over time and space, and certain regions may have been effectively isolated at times. Early formulations of MRE (Wolpoff et al. 1984, Wolpoff 1989, 1992) seemed to imply that the movement of genes between regions probably did not involve systematic movements of populations. However, more recent statements of the theory (Wolpoff et al. 2001, 2004) concede that migrations—for instance, the movement of modern people into Europe around 35 Kya—may at times have been important in fostering the spread of new genes.

Thus, like RAO, MRE has evolved as a model over the past 20 years. The two models have in fact tended to converge, with some recent versions of RAO admitting the possibility of limited admixture between archaics and modern humans, and current statements of the MRE allowing for some migration and local replacement (but not speciation). Although many students of modern human genetics defend some version of RAO, others (Templeton 1993, 1997, 2002; Relethford 2001) feel that the mtDNA data actually support MRE, or at least something different from classic RAO.

A. Thorne's "center and edge" hypothesis has played a large part in the development of MRE. Thorne (1976, 1981; Thorne and Wolpoff 1981, 1992) applied two well-known biological principles to the study of fossil humans. The first is that **polytypism**—the presence of multiple modal peaks of variation within a species—will evolve in any species that has a wide range encompassing diverse ecological communities. The second is that populations on the peripheries of a polytypic species' range are usually less variable than those in the core region where the species evolved. Because the core region is the environment to which the species has been adapting the longest, it normally represents the optimal environment for the species. In such optimal environments, stabilizing selection is reduced and more variants can flourish.

From these principles—and from the plausible assumption that Africa was at the geographical center of the human story, whereas East Asia and Australasia were "edges"—Thorne reasoned that human variation would be expected to be highest in Africa. In the more peripheral environments of Eurasia, suboptimal conditions would have imposed new, harsher selection pressures, reducing variation and leading to the fixation of locally adaptive traits. And because those marginal environments would initially have been occupied by small groups, founder effect and drift would result in the random fixation of other variants. The combined effects of divergent selection and sampling error would be expected to produce distinctive regional types—geographical races—at the periphery. Thorne argued that this pattern had always characterized human populations, starting with the initial spread of Erectines out of Africa. Thorne's reasoning was not only supported by evolutionary theory; it also explained some aspects of the fossil record (e.g., variable Erectine morphology in

Africa vs. distinctive regional morphs in Java and China). The center-and-edge model accordingly became a fundamental tenet of the MRE proponents.

RAO and MRE have different implications for the systematics of the genus *Homo*. RAO encourages its supporters to see the genus *Homo* as containing more species than traditionally thought, whereas MRE leads logically to the conclusion that all members of *Homo* should be lumped into one species (Wolpoff et al. 1994a). As we might expect, proponents of RAO models are continually on the lookout for autapomorphies in regional groups that can be interpreted as signals of reproductive isolation. Conversely, MRE supporters are eager to find long-lasting regional traits that demonstrate morphological continuity within specific regions over hundreds of thousands of years. Both groups want to identify regional peculiarities of morphology in fossil *Homo*, but they differ in the durations and taxonomic meanings that they assign to the persistence of those regional traits. Not surprisingly, these divergent views lead to different readings of fossil human anatomy.

■ EUROPEAN HEIDELBERGS

Human evolution in Europe appears to have been a rather late and peripheral part of the story of the human lineage. But because Europe has more fossil-hunters per square mile than any other part of the world where fossil humans are found, we know more about fossil humans from Europe than from anywhere else. Before the 1960s, virtually all the human fossils known from the late Middle Pleistocene came from Europe. The most important exception, the partial skeleton of "Rhodesian Man" discovered in 1921 near Kabwe (Broken Hill) in Zambia, was long thought to be less than 100 Ky old (Clark 1970), and was usually dismissed as irrelevant to human evolution in the Middle Pleistocene. For these reasons—and also because of the long-standing habit Europeans have of thinking of themselves as the vanguard of human progress—European fossils have until recently exerted a disproportionate influence on scientific thinking about later human evolution.

The European fossil humans spanning this time range include all the specimens from Europe that have been identified as *Homo heidelbergensis* (Rightmire 1990, 1998). Some of these fossils are fragmentary, and many of their dates are vague. Despite these drawbacks, the European Heidelberg sample provides a reasonable view of the anatomical character of humans from this time span. However, this sample exhibits some regional peculiarities, which suggest a special phyletic connection with later Europeans—especially with the Neandertals. Regardless of what they think of the validity of the taxon *H. heidelbergensis*, or of the merits of MRE versus RAO models, most paleoanthropologists agree that the

European Heidelbergs are Neandertal ancestors (Stringer et al. 1979, Wolpoff 1980b, Hublin 1982, Smith 1985, Rightmire 1990, 1998; Arsuaga et al. 1997b, Dean et al. 1998, Tattersall and Schwartz 2001).

Petralona

Petralona is a massive cranium found accidentally in either 1959 or 1960 (accounts differ) by shepherds in Clemontsi Cave near the Greek village of Petralona, not far from Thessalonika. It is probably not the most ancient Heidelberg specimen from Europe, and it was not the first to be discovered. But it is a good starting point for a survey of human remains from this period, because it is an excellent example both of their morphology and of the problems involved in their interpretation.

The Petralona cranium (Figs. 6.5 and 6.6) is complete except for the incisors, right canine, and right P³. Claims that an associated skeleton was also found (Poulianos 1971) have not been substantiated. Because Petralona was not excavated systematically, the skull's exact context within the cave will never be known. No artifacts are associated with it, but undiagnostic artifacts were reported from elsewhere in the cave. Dating of the specimen by faunal correlation is inconclusive (Cook et al. 1982). Latham and Schwarcz (1992) used uranium-series dating to determine a minimum age of 200 Ky for

matrix in the cave that matched matrix retained on the cranium. ESR dating places the age of the Petralona skull between 250 and 150 Ky (Grün 1996).

The skull exhibits a mosaic of Erectine and Neandertal traits (Murrill 1981, Stringer et al. 1979, Stringer 1974a, 1983). The most primitive, Erectine-like part of the skull is the rear of the braincase. The occipital bone is characterized by a well-developed nuchal torus, separated from the occipital plane by a moderate sulcus. The nuchal region is broad, thick, and pneumatized by extensions of the mastoid air cells. As in Erectines (Fig. 6.7), inion lies high up on the back of the skull and coincides with opisthocranion. Although the nuchal plane is relatively large, the occipital plane is a bit larger still, reflecting the large size of the braincase (Table 6.3). However, the maximum breadth of the cranium lies at the level of the broad cranial base (Fig. 6.5B), as in Erectines.

Several other features contribute to the primitive gestalt of this specimen. In lateral view (Fig. 6.5A), Petralona displays a low cranial vault with a receding forehead. Cranial capacity measures 1230 cm³, which is over two standard deviations below Holloway's (1985) mean for European Neandertals (1507 ± 116 cm³, $N = 10$) and slightly below a mean of 1261 ± 119 cm³ ($N = 7$) for the European Heidelberg sample as a whole (Table 6.2). This relatively small cranial capacity is surprising given

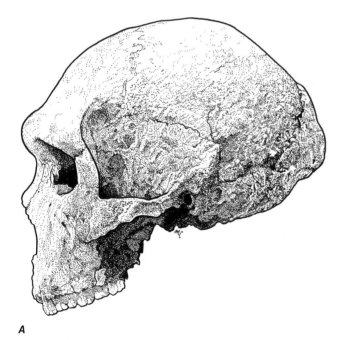
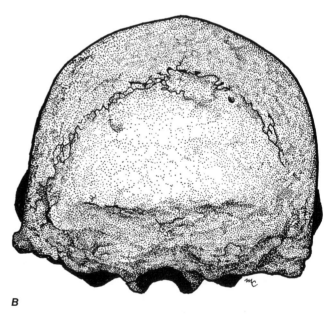

A   *B*

FIGURE 6.5

The cranium from Petralona (Greece). In lateral view (**A**), Petralona exhibits facial prognathism, a prominent nuchal torus and. The Neandertal-like inflation of the infraorbital region is evident, resulting in the absence of a canine fossa. The posterior view (**B**) reveals that maximum cranial breadth is still on the cranial base, but on the less distorted left side it is evident that the biparietal breadth is not much less. The elongation of the occipital plane and the medially developed nuchal torus are also evident. (**A**, after Johanson and Edgar 1996.)

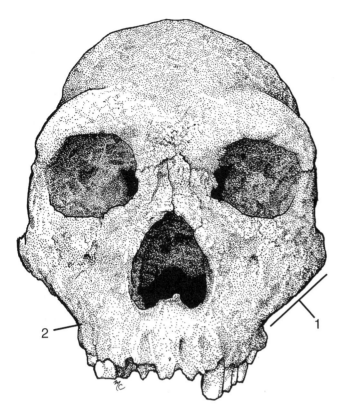

FIGURE 6.6

Frontal view of the Petralona cranium (Greece). Significant morphological features include (1) an oblique zygomaticoalveolar margin, with its medial root positioned low on the alveolar process (2); a broad face and nasal aperture; a broad anterior palate; a prominent supraorbital torus with incipient lateral thinning; and a broad interorbital area.

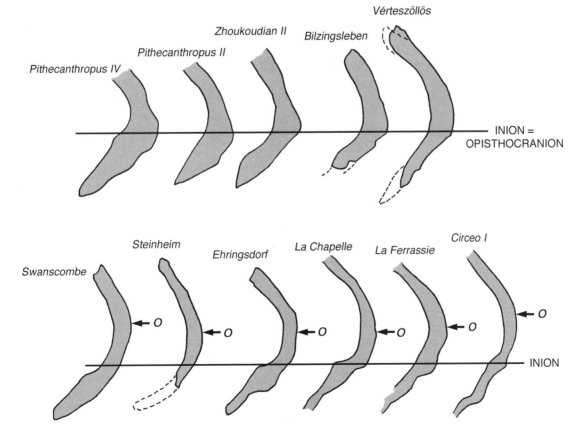

FIGURE 6.7

Mid-sagittal (midline) profiles of occipital bones from Middle Pleistocene *Homo* (**top**) and Neandertals (**bottom**). The solid lines represent the position of inion, which coincides with opisthocranion in all the top-row specimens. In the bottom row, an O with an arrow marks the position of opisthocranion. Note also the average difference between the two groups in the thickness of the cranial vault. (After Vlček 1978.)

TABLE 6.2 ■ Cranial Capacity Estimates for European and African Heidelbergs and Contemporary Asian Specimens

Specimen	Cranial Capacity (cm^3)
Arago 21	1160
Bodo	1250
Kabwe 1	1325
Petralona	1230
Reilingen	1450
Saldanha	1225
Sima de los Huesos 4	1390
Sima de los Huesos 5	1125
Sima de los Huesos 6	1140^a
Steinheim	1160^b
Swanscombe	1325
Hexian PA 830	1025
Nagwi	870
Narmada	1260
Ngandong 1 (Solo I)	1172
Ngandong 6 (Solo V)	1251
Ngandong 7 (Solo VI)	1013
Ngandong (Solo IX)	1135
Ngandong 13 (Solo X)	1231
Ngandong 14 (Solo XI)	1090
Sambungmacan 1	1035
Sambungmacan 3	918^c
Sambungmacan 4	1006^c
Yunxian EV 9001	1200
Yunxian EV 9002	1200

^{a}Adolescent.
^{b}From Howell (1960).
^{c}From Baba et al. (2003). All other estimates are from Holloway et al. (2004).

the skull's large overall size, which exceeds those of other European Heidelbergs and compares well to the largest Neandertal skulls. On the other hand, Petralona's endocranial volume is well above the Erectine average (Table 5.3). The vault is somewhat distorted, as reflected in the asymmetries that can be seen in a rear view (Fig. 6.5B). Nevertheless, the braincase is clearly expanded in breadth (biparietal) relative to the cranial base (Tables 5.1 and 6.3), resulting in a more vertical orientation of the side walls than in Erectines.

In several respects, Petralona looks forward to Neandertal morphology. Like Neandertals, it exhibits some flattening of the vault at lambda and some shelving of the posterior nuchal plane, producing an incipient occipital protuberance or "bun" that bulges outward from the back end of the skull (Fig. 6.5A). However, this "occipital bunning" is not as pronounced as it is in many Neandertals (Chapter 7). In facial view (Fig. 6.6), the Petralona skull is even more Neandertal-like. The face is broad and long, and the palate is very broad, especially at the front end. The supraorbital torus (SOT)

TABLE 6.3 ■ Selected Cranial Measurements (in mm) for African and European Heidelbergs

Measurement	Petralona	Swanscombe	Kabwe 1	Ndutu
Maximum cranial length	209	—	203	178
Maximum cranial breadth	156	—	(153)	137
Bi-parietal breadth	149	130	138	133
Auricular height	115	—	114	—
Opisthion–inion length	62	52	(55)	53
Inion–lambda length	64	58	(56)	57
Occipital index^a	97	90	(98)	93

^{a}The occipital index is the length of the nuchal plane, measured from the back edge of the foramen magnum (opisthion) to inion, expressed as a percentage of occipital-plane length (from inion to lambda). All measurements were taken by FHS.

arches slightly over each orbit, as in Neandertals, rather than forming a continuous bar of the sort seen in many Erectines. Although Petralona's midorbital SOT thickness value of 23.7 mm is above the Neandertal average (Table 7.6), its torus clearly follows the post-*erectus* pattern in exhibiting thinning of the lateral portions of the structure. The lateral thinning and double-arching of the SOT are less marked than they are in most Neandertals, but a trend in the Neandertal direction is evident. Also like Neandertals, Petralona has a column-like lateral orbital margin continuous with the lateral end of the SOT, a very large nasal aperture, and inferior cheekbone margins (zygomaticoalveolar margins, ZAM) that slant sharply downward from the zygomatic arch to end low down on the upper jaw, close to the teeth. And as in typical Neandertals, the maxillary air sinus is expanded, resulting in a vertically oriented, inflated, puffy-looking infraorbital region that lacks a canine fossa or any other sort of concavity. In short, the face of Petralona exhibits a morphological buttressing system that is markedly similar to that seen in Neandertal skulls (Smith 1983, Rak 1986, Trinkaus 1987b).

Though the face is generally Neandertal-like, the pattern of facial prognathism is somewhat different. Viewed from above, Neandertal skulls have a "prow-like" midfacial region, with the nasal region thrusting forward in the midline and the cheekbones sweeping sharply backward on either side. Stringer (1983) quantified this peculiarity by measuring the **subspinale angle** (SSA). This is defined as the angle between lines drawn from subspinale (an anatomical point just below the nasal spine at the bottom of the nose opening) back to the zygomatics on each side (see Appendix). Petralona has an SSA of 119°—flatter than in Neandertals ($\bar{x} = 110.6 \pm 4.3°$), but deviating in a Neandertal-like direction from the even flatter faces of Erectines and modern humans ($\bar{x} = 128.8 \pm 5°$). The prominence of Petralona's midface results more from posterior migration of the cheekbones than from a Neandertal-like forward protrusion of the nasal region (Trinkaus 1987b, Smith and Paquette 1989).

All in all, Petralona appears to represent an intermediate stage in the evolutionary transition from the total facial prognathism of Erectines to the prow-like, midsagitally prognathic face of Neandertals. However, Petralona lacks any trace of the **suprainiac fossa**, an oval depression just above the nuchal torus in the midline, which is a distinctively Neandertal trait (Santa Luca 1978). And although the mastoid processes are missing, it is clear that they would have been large and would have projected well below the base of the skull—which is not the case in Neandertals (Chapter 7). Thus, while Petralona is morphologically on its way to Neandertalhood, it is not there yet.

Bilzingsleben

Bilzingsleben is a complex and important Middle Pleistocene site in eastern Germany. Its age is uncertain, but it seems clear that humans were occupying the site at least 400,000 years ago (Schwarcz 1992). The numerous stone tools from the site include no bifaces, and the assemblage is considered non-Acheulean (Mania and Vlček 1987). The site has yielded cranial and dental fragments belonging to at least two human individuals. Bilzingsleben 1 is an adult represented by three pieces of very thick frontal bone (including the central portion of an impressive supraorbital torus), an occipital lacking only the lower portion of the nuchal plane, two other vault pieces, and an upper molar. A second individual, Bilzingsleben 2, is represented by the lateral part of a left supraorbital torus, a left parietal, and a left temporal bone.

The most distinctive aspects of the Bilzingsleben human remains are the supraorbital torus (SOT) and occipital. The occipital is strongly Erectine-like—so much so that Vlček (1978) initially described Bilzingsleben 1 as *H. erectus* and reconstructed it using OH 9 as a model. Like that of the Petralona skull, it has a well-developed nuchal torus and an inion that coincides with opisthocranion (Fig. 6.7). But Bilzingsleben 1 is even more primitive in having a sharp occipital angle and a very short occipital plane. The short occipital plane is reflected in the high position of inion, which lies on the same horizontal plane as internal inion (endinion)—a distinctive characteristic of Erectines (Vlček 1978). The Bilzingsleben 1 supraorbital torus fragment is similar to the corresponding part of the Petralona skull, but with slightly less of a dip in the midline, so that the torus

looks more like a bar and less like a double arch. There is no clear supratoral sulcus. Bilzingsleben 2's most distinctive anatomical feature is the thick, projecting lateral supraorbital torus fragment (Mania et al. 1993, 1995). This specimen is generally similar to the lateral aspect of the SOT in Petralona, Arago, and the Sima de los Huesos skulls (see below) in that the supratoral sulcus is less developed than is usually the case in Erectines. In both Bilzingsleben individuals, the bones of the cranial vault are quite thick.

In general, the Bilzingsleben crania seem more archaic than Petralona, particularly in occipital morphology and possibly in the central supraorbital region. Apart from Ceprano and Dmanisi, Bilzingsleben shows the most Erectine-like cranial morphology known from the European subcontinent. However, there is no reason to assume that the missing parts of the Bilzingsleben skulls would have looked particularly Erectine. In Petralona and other European Heidelbergs like Arago and Sima de los Huesos, the greatest similarities to later humans are seen in the face below the orbits—an anatomical region completely unknown at Bilzingsleben.

Swanscombe

In the 1930s, A. Marston discovered a virtually complete occipital and left parietal in a gravel seam on a terrace of the Thames River, near the English town of Swanscombe (Marston 1936, 1937). When J. Wymer relocated this deposit in 1955, he was lucky enough to find the other, less well-preserved parietal of the very same skull.

The Swanscombe skull was initially dated to the Hoxnian interglacial (now OIS 13, ~500 Kya), and its morphology was considered to be fundamentally modern. This morphological assessment was based primarily on G. Morant's 1938 study of the specimen and on W. Le Gros Clark's (1938) cranial-capacity estimate of 1325 cm³, which placed the brain size of the specimen within the modern human range. Because of its antiquity and supposedly modern morphology, Swanscombe became one of the key fossils promoted by Vallois (1954) as evidence of the existence of a Pre-Sapiens lineage in Europe (Brace 1964). However, subsequent reassessments of Swanscombe, beginning with that of Ovey (1964), have rejected both the ancient date and the modern morphology. The gravel seam yielding the Swanscombe skull is correlated to the base of the Upper Middle Gravels in the Thames Terrace sequence. Fauna (including rodents and mollusks) and geological evidence from this level indicate a relatively cool phase and a possible correlation with OIS 7 or early OIS 6 (Cook et al. 1982). More recent geomorphological studies (Bridgeland 1994, Stringer and Hublin 1999) suggest that the terrace containing this site correlates to OIS 11, around 400 Kya (Fig. 1.3B). Thermoluminescence dating and a uranium-series date on bone from the Swanscombe site yield respective age estimates of 272 Ky (Szabo and Collins 1975) and 200-230 Ky (Bridgeland et al. 1985). Given the depositional environment at Swanscombe, it is also possible that the skull is not in primary association with the Acheulean artifacts and fauna used to date it (Conway et al. 1996). About all we can say is that the specimen is probably between 200 and 400 Ky old.

Swanscombe (Fig. 6.8) is much less robust than Petralona, despite its apparently larger cranial capacity. In lateral view, Swanscombe appears rather different from the comparable parts of the Petralona and Bilzingsleben skulls. The cranial contours of Swanscombe are more rounded. The low position of its inion with respect to both opisthocranion and endinion (Fig. 6.7) reflects a relatively reduced nuchal area and an expanded braincase (relative to the cranial base) by comparison to Petralona and earlier members of *Homo* (Table 6.3). Unlike those of Petralona and Bilzingsleben, Swanscombe's nuchal torus is not prominent and does not project strongly posteriorly.

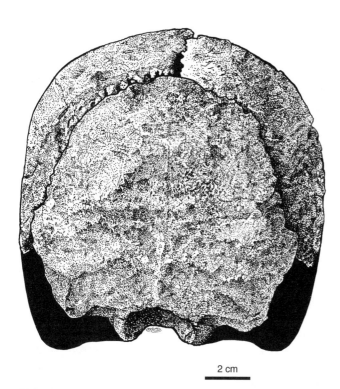

FIGURE 6.8

The Swanscombe (England) cranium in rear view. Absence of the temporal bones makes it difficult to visualize the shape of the vault from the rear, but maximum breadth would probably have been at the level of the inferior part of the parietal bones. Thus, the occipital silhouette is approaching the typical Neandertal pattern. A faint suprainiac fossa is also present. (After Day 1965.)

Nevertheless, the broad occiput of Swanscombe and the dimensions and shape of its parietals are distinctly archaic (Sergi 1953, Wolpoff 1980b). The vault is actually quite low, with a basion-bregma height of 126 mm, falling right at the Neandertal mean. The "modern" features of the occipital noted above are not unique to recent humans; they are also characteristic of Neandertals. Other archaic traits seen in Swanscombe include a pneumatized nuchal plane, a small but distinct nuchal torus, and—as Figure 6.8 demonstrates—an incipient suprainiac fossa (Stewart 1964, Hublin 1982, 1988a,b; Santa Luca 1978). Like Neandertals, Swanscombe appears to have had strongly developed occipitomastoid crests. The temporal bones are missing, but the contours of the lateral occipitals at the occipitomastoid suture, as well as distinct V-shaped sulci seen just lateral to the foramen magnum, demonstrate that the occipitomastoid crests would have been very marked. This is another Neandertal apomorphy, which results from the lowering of the occipitomastoid crest region relative to the foramen magnum (Smith et al. 2006). Finally, multivariate metric analyses (Corruccini 1974, Stringer 1974b, 1978) consistently group Swanscombe with Neandertals rather than with more modern humans.

In short, Swanscombe is a Neandertal ancestor and already exhibits some distinctive Neandertal peculiarities. For Klein (1999), these apomorphies qualify it for inclusion in the species *H. neanderthalensis*. However, it lacks some other Neandertal apomorphies. It does not have a Neandertal-like occipital bun. The broadest point on the skull would have been on the cranial base, as in Erectines, rather than halfway up the sides of the braincase, as in Neandertals. The contour of Swanscombe from the rear exhibits the same basic shape seen in Petralona, without the latter's distortion. We therefore prefer to group Swanscombe with the Heidelbergs. But we decline to draw any taxonomic conclusions from this preference, because the real lesson to be learned from Swanscombe is that the boundary between European Heidelbergs and Neandertals is vague and arbitrary.

Steinheim

In 1933, a largely complete cranium, lacking much of the left side, the central base, and the anterior teeth and their sockets, was found in a commercial gravel pit near the village of Steinheim in Germany (Fig. 6.9). The specimen was excavated and initially described by F. Berckhemmer (1933), who considered it to date to the Riss glacial maximum (OIS 10), about 350 Kya. Others (Adam 1954) have assigned it to the preceding Holsteinian interglacial period (OIS 11, ~400 Kya). The faunal remains from the site are said to be similar to those from Swanscombe, but as at Swanscombe the association between the fauna and specimen is not entirely secure

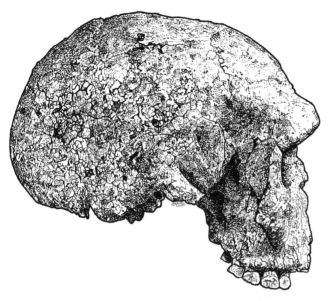

FIGURE 6.9

Lateral view of the Steinheim cranium (Germany). Although the specimen appears to have a rather modern vault shape and weak prognathism, these features result from postmortem distortion. (After Johanson and Edgar 1996.)

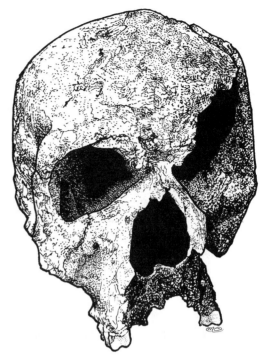

FIGURE 6.10

The Steinheim cranium (Germany) in frontal view. Distortion of the specimen is evident on its left side. The right side of the face is relatively non-distorted. The supraorbital torus and moderately angled zygomatico-alveolar margin are not affected by distortion. (After Johanson and Edgar 1996.)

(Cook et al. 1982). Although questions of dating and context persist, an attribution of Steinheim to the time span between 200 Kya and 400 Kya seems reasonable.

The Steinheim cranium figured prominently in the paleoanthropological literature of the mid-20th century, when Howell (1960) and others saw Steinheim and Swanscombe as representing a generalized archaic-human stock, potentially ancestral to both Neandertals and modern humans (Brace 1964). But no one has studied Steinheim systematically since Wienert's initial analysis in 1936. The main reason for this neglect is the extreme post-depositional plastic distortion of the specimen. The distortion is particularly evident in an inferior view, where it can be seen that the left side of the dentition is almost at the skull's midline. Its occipital and the lower parts of its parietals are pushed inferiorly and anteriorly, producing unnaturally rounded lateral occipital contours. Other aspects of its deformation help make the vault deceptively modern-looking when viewed from the rear or side (Fig. 6.9). Because of all this distortion, we believe that no contour on the Steinheim cranial vault preserves its original form. The widely cited cranial-capacity estimate of 1150 cm^3 (Howell 1960) should therefore be considered questionable.

The only portion of the Steinheim skull that is not distorted is the right side of the face and the supraorbital region (Fig. 6.10). The thick, projecting supraorbital torus thins laterally, and forms a distinct double arch over the orbits. In these features, Steinheim approaches the condition seen in Neandertals rather than the more primitive Erectine form. Steinheim may also exhibit an incipient suprainiac fossa, but this is weakly developed at best. M. Wolpoff (1980b) argues that Steinheim also resembles Neandertals in having marked midfacial prognathism. This is certainly true, although the plastic deformation makes it impossible to quantify. The face, like the vault, is relatively small and gracile, but resembles Neandertals in having a relatively broad nasal aperture and interorbital region. However, the zygomaticoalveolar margin (ZAM) is more angled than a Neandertal's (or Petralona's), and it intersects the alveolar part of the maxilla higher up—and unlike Neandertals, Steinheim has a canine fossa. This is uncommon in specimens from this time period, but some of the Atapuerca skulls—including AT404 from Sima de los Huesos and AT D6-69 from Gran Dolina (see below)—also have this feature.

Mauer

In 1907, workers at the Graferain Quarry at the German village of Mauer, near Heidelberg, found a primitive-looking human mandible in a layer of Pleistocene river sediment known as the "Lower Sands." Various age indicators for these deposits, including TL dates and faunal correlations, suggest an age for the mandible between 475 Ky and 700 Ky (von Koenigswald 1992, Wagner 1996). Some relatively crude bifacial and flake tools have been reported from these deposits (Fiedler 1996), but these may have come from a more recent stratum in the quarry (Müller-Beck 1996).

The Mauer mandible (Fig. 6.11) was complete when discovered, but the left premolars are now missing. It was initially described by O. Schoetensack (1908), who established it as the type specimen of *Homo heidelbergensis*. Many of its features are similar to those of Erectine mandibles described in Chapter 5. The symphysis is retreating, with no evidence of any basal projection, and the corpus is thick—even by comparison with its considerable height. Internally, the alveolar plane slopes markedly and the symphysis is reinforced by two horizontal shelves of bone (internal transverse tori) with a sulcus between them, as in earlier *Homo* and *Australopithecus* (Fig. 4.48). The form of the symphysis and the great overall length of the mandible suggest that the missing skull would have had a prognathic face, like other crania from this period. Like other contemporary and older mandibles, Mauer also has a markedly broad ramus (relative to total length), resulting in the absence of a retromolar space. The extreme breadth of the ramus suggests a pattern of total facial prognathism, in which the entire face was situated far in front of the braincase and the coronoid process was shifted forward with it (Trinkaus 1987b). Czarnetzki et al. (2003) suggest that the broad ramus in Mauer reflects an anomalous insertion pattern of the temporalis muscle. However, other Heidelberg mandibles—Arago 2, for example—also have broad rami, though not as broad as Mauer's.

The outer front surface of the mandible bears a hole, the **mental foramen** (Fig. 6.11), through which the sensory nerve to the chin area emerges from its canal in the bone. In humans, the position of this foramen generally reflects facial prognathism: the further the face sticks out in front of the braincase, the further back the mental foramen lies in relation to the dental arcade. The mental foramen of the Mauer jaw lies under the rear edge of P_4. This is further back than the usual position in modern humans (under the P_3/P_4 contact), reflecting the pattern of overall prognathism seen in other features of the Mauer mandible. But the mental foramen of Neandertals usually lies still further back, under M_1 (Trinkaus 1993a).

Rosas and Bermúdez de Castro (1998) believe that Mauer possesses specific Neandertal apomorphies in the breadth of the lower incisors, morphology of the premolars, position of the mental foramen, and retromolar space. However, the position of the mental foramen does not constitute a Neandertal apomorphy (Smith

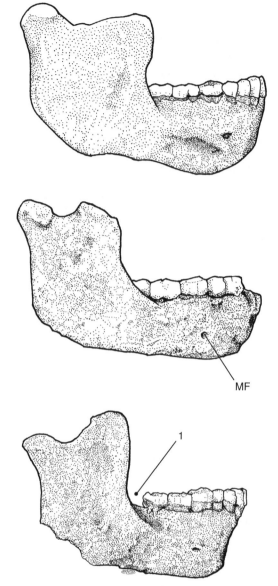

FIGURE 6.11

Lateral views of mandibles from Mauer (**top**), Arago 13 (**middle**), and Régourdou (**bottom**). The latter two specimens are from France; Mauer is from Germany. Régourdou is a Neandertal, while the other specimens are Heidelbergs. All three mandibles have receding mandibular symphyses and no evidence of mental eminences. The Neandertal mandible exhibits a retromolar space (1)—a distinct gap between the M_3 and the anterior edge of the mandibular ramus. The Heidelbergs have broader mandibular rami and lack retromolar spaces. MF, mental foramen.

1976, Trinkaus 1993a), and the foramen position in Mauer falls in the range of overlap between Neandertals and recent humans. Furthermore, the definition of the retromolar space used by Rosas and Bermúdez de Castro does not conform to that used by other workers. By the usual definition (a gap between the distal M_3 and the anterior margin of the ramus), Mauer clearly lacks a retromolar space (Fig. 6.11).

Like other Erectine and Heidelberg mandibles, Mauer has posterior teeth well within the modern human range, but anterior teeth that continue to be quite large. Mauer also exhibits taurodontism, an expansion of the molar pulp cavities characteristic of the later Neandertals (Skinner and Sperber, 1982). However, it lacks the conspicuous Neandertal apomorphies of the mandibular foramen, coronoid process, and mandibular incisure crest (see below and Chapter 7). Thus, like other European remains from this time period, the Mauer jaw reveals a mosaic of anatomical features, some of which look backward to the Erectines while others foreshadow the Neandertals.

Boxgrove

The Boxgrove site (West Sussex, England) has yielded two human lower incisors and a 294-mm-long left tibial shaft lacking both articular ends. The tibia (Boxgrove 1) was carefully excavated by M. Roberts and colleagues. It is reliably associated with Acheulean artifacts and a rich mammalian fauna, the latter indicating an OIS stage 13 correlation at between 478 Kya and 512 Kya (Roberts et al. 1994, Stringer 1996, Stringer et al. 1998).

The Boxgrove tibia gives us some information about human body size and build in northern Europe at this time. Using a regression analysis based on measurements of Middle/Late Pleistocene and recent human tibias, Stringer et al. (1998) estimate that the complete Boxgrove tibia would have had a biomechanical length of 373 mm. From this estimate, a body mass of 80 kg can be predicted. In cross section, the tibia's central marrow cavity is surprisingly large for the amount of cortical bone. Nevertheless, Trinkaus et al. (1999b) calculate that the shaft would have been stronger than modern tibias, as were those of other Pliocene and early Pleistocene *Homo*. They also infer from the metrics of the shaft that the Boxgrove individual would have had body proportions intermediate between those of arctic and tropical human populations today. This suggests that this high-latitude fossil may document an early stage in the adaptation of human body form to cold climates (Holliday 1997b, Ruff et al. 2002). Similar intermediate proportions seem to characterize the Zhoukoudian postcranials and the large series of Spanish specimens from Atapuerca (see below). However, indications of similar body form are also present in the Ngandong specimens from Java, where adaptation to cold is unlikely. All these inferences are essentially speculative, since none of these specimens is complete enough to be a genuinely reliable indicator of its body's limb proportions.

Atapuerca—Sima de los Huesos

The Sima de los Huesos (SH) or "Pit of Bones" is a long, sloping chamber at the base of a 10-m vertical shaft in the complex cave system of the Sierra de Atapuerca in northern Spain. To date, over 5000 human fossils representing every part of the skeleton have been recovered from this remarkable site, making SH one of the largest fossil hominin samples ever recovered from a single locality. The recovery of these remains is a tribute to the dedication of J.-L. Arsuaga, J.-M. Bermúdez de Castro, E. Carbonell, and their co-workers, who have also guided the interpretation of the SH finds.

At least 33 human individuals are represented in the SH sample (Arsuaga et al. 1997b). All the other animals from the site are carnivores, mainly bears (166 individuals of *Ursus deningeri*). Because the cave contains no herbivore bones and has no entrance other than the vertical shaft, it cannot have been a carnivore den. Arsuaga and his co-workers believe that the human bodies were probably thrown down the shaft by other humans and that the bears and other carnivores were drawn by the smell of carrion to the cave, where they were trapped and died. They also argue that all of the human bones probably come from the same temporal horizon, because bones from very different parts of the excavated deposit can be articulated (refitted) with each other. ESR and U-series dating on human and bear bones from the cave indicate a minimum age of 350 Ky (Bischoff et al. 1997, 2003). This fits with the faunal dates for the carnivores and micromammals (Garcia et al. 1997; Cuenca-Bescós et al. 1997). However, high-resolution uranium-series dating suggests a much older age for the SH hominins, around 600 Kya (Bischoff 2007). No used tools or other evidence of cultural behavior are associated with the SH humans, but a single unused quartzite handaxe was found in the pit. Dubbed "Excalibur" by the Spaniards, it may represent evidence of ritual behavior at SH.

More than 75% of all known postcranial remains of Middle Pleistocene humans come from SH. At the end of the 1995 field season, the SH tally stood at 54 vertebrae, 30 scapulae and clavicles, 47 arm bones, 246 wrist/hand bones, 25 pelves, 29 femora, 14 patellae, 35 lower leg bones, and 204 foot/ankle bones. The large number of hand and foot bones strengthen the claim that the SH bodies were complete when thrown down the shaft.

Taken as a whole, the SH postcrania exhibit some primitive traits linking them to the Erectines, but there are also a number of features that appear Neandertal-like (Pearson 2000a). The lower limbs at SH have thick shafts with high percentages of cortical bone and small medullary cavities. The femoral specimens lack pilasters, have a poorly developed *linea aspera*, and exhibit platymeria of the proximal and middle shaft. One (Femur X) is estimated to be 470–480 mm long (Arsuaga et al. 1999a), which is at the upper reaches of the Neandertal range (430-484 mm). Femoral necks from SH tend to be long and anteroposteriorly flattened. Like those of Neandertals, the SH tibias are generally short; an estimated length of 345 mm for Tibia II puts it well below the Boxgrove length estimate and within the Neandertal range. These data suggest that the lower-limb proportions of the SH humans conformed to a Neandertal pattern (Chapter 7, Table 7.16).

The pelvic sample from SH is particularly intriguing. There are 18 specimens, including a virtually complete pelvis (Pelvis 1) of a large male (Arsuaga et al. 1999a). Pelvis 1 is markedly broad, and the body mass estimated from its maximum breadth (across the ilia) is 93.1-95.4 kg—the largest body-mass estimate for any fossil human (Ruff et al. 1997). The vertical diameter of Pelvis 1's acetabulum is 60 mm, slightly above a Neandertal mean of 57.9 ($N = 9$). This corroborates the large body-size estimate and suggests an increased body mass in the SH sample compared to earlier *Homo*. Like Arago 44 and such earlier Pleistocene hipbones as OH 28 and KNM ER 3228, the SH pelves exhibit strong lateral iliac flare and pronounced iliac pillars (Arsuaga et al. 1999a). The Spanish researchers view the elongated femoral necks and marked iliac flare of the SH skeletons as bio-mechanical compensations for the very broad pelvic girdle—a primitive feature retained from *Australopithecus*. Several SH pubic bones have an elongated, thinned superior ramus. This feature has been called a Neandertal autapomorphy (Arsuaga et al. 1999a); but it is also seen in both earlier and later hominins, and is more likely to be another primitive character (see Chapter 7). An elongated pubis may have persisted in European Heidelbergs and their Neandertal successors because it produces a deeper, more barrel-shaped trunk, which would be adaptively valuable in cold climates (Smith 1991, 1992b).

The SH pelves resemble Neandertals and modern humans, and differ from those of the WT 15000 Erectine and earlier hominins, in having an inlet that changes its shape below the midplane of the birth canal, becoming rounder down near the outlet. In the SH pelves, the pelvic inlet is typically almost as deep from front to back (pubic symphysis to coccyx) as it is broad from side to side. Rosenberg and Trevathan (2001) conclude from this that birthing in the SH females would have involved the same pattern of neonatal head rotation in passing through the birth canal that we see in modern humans. This constitutes the earliest evidence for the establishment of "rotational" parturition. However, Rosenberg and Trevathan feel that ancient females would have continued to benefit from assistance during the birthing process, just as women do today.

Carretero et al. (1997) find that the SH upper-limb bones share many morphological and metrical similarities with those of Neandertals. Some of these shared

TABLE 6.4 ■ Humeral Indices

	Midshaft Index[a]	Robusticity Index[b]
SH	75.1 ± 4.5 ($N = 8$)	19.6 (Humerus II)
Neandertal	75.8 ± 6.3 ($N = 22$)	20.4 ± 1.4 ($N = 6$)
EUP[c]	81.8 ± 6.3 ($N = 13$)	18.4 ± 1.3 ($N = 7$)
Skhūl/Qafzeh	86.3 ± 4.1 ($N = 8$)	17.7 ± 1.3 ($N = 5$)
Recent human range	70.0–83.5	17.6–22.8
	(56 samples)	(85 samples)

[a] $100 \times$ minimun/maximum midshaft diameter.
[b] $100 \times$ minimun circumference/maximum length (of shaft).
Source: Adapted from Carretero et al. (1997).

traits are symplesiomorphies—for example, a long, narrow glenoid fossa, craniocaudally compressed clavicular shafts, low humeral torsion angles, and a narrow, "closed" deltoid insertion running parallel to the long axis of the humerus. These shared primitive traits tell us nothing about the affinities of the two groups. The Spanish scientists note that humeral midshaft flattening and humeral robusticity in the SH sample seem similar to the Neandertal pattern (Table 6.4), but are not conclusive indicators of Neandertal affinities. However, Carretero and his co-authors interpret some features seen in the SH sample as probable synapomorphies of a specifically European *Homo* clade that ends with Neandertals. These supposed synapomorphies include a laterally oriented scapular glenoid fossa, a transversely oval humeral head, and a dorsal axillary border of the scapula (see Chapter 7). As Carretero and his team are quick to point out, the functional meaning of most of these traits is unknown.

A large sample of teeth has been recovered at SH. Bermúdez de Castro and Nicolás (1997) identify five children (ages 4-12), 14 adolescents (ages 14-20), 10 young adults (ages 22-30), and two mature adults (ages 30-40) in the sample of jaws and isolated teeth. Metric patterns for these teeth (Table 6.5) are complex. The incisors tend to be smaller than those of Neandertals, but large compared to the Zhoukoudian Erectines. Canines in both SH and Zhoukoudian are relatively larger than those of Neandertals. The SH posterior teeth vary in size, but in general fall into the modern human range (Table 6.5). Bermúdez de Castro et al. (1993) observes that both Neandertals and the SH humans had front teeth (I1-P3) that were big compared to their back teeth (P4-M3). He also notes that the relatively small size of the SH lower molars is associated with a simplification of the cusp pattern (high frequency of hypoconulid absence). These facts suggest that selective pressures favoring big posterior teeth were relaxed at SH for some reason, or that directional selection was beginning to favor smaller posterior teeth in these people.

Although the SH incisors are smaller than those of Neandertals, they share some Neandertal peculiarities.

TABLE 6.5 ■ Dental Metrics for Selected Middle and Late Pleistocene Samples

Tooth	Sima de los Huesos Mean, SD (N), Range	Krapina Mean, SD (N), Range	Neandertals Mean, SD (N), Range	Zhoukoudian Mean, SD (N), Range
I_1	6.5 ± 0.4 (5) 5.4–5.6	7.7 ± 0.5 (7) 7.0–8.2	7.2 ± 0.3 (11) 7.0–8.0	6.3 ± 0.5 (5) 5.8–6.8
I_2	7.3 ± 0.4 (9) 6.7–8.1	8.3 ± 0.5 (10) 7.0–9.2	7.7 ± 0.6 (13) 6.7–9.0	6.9 ± 0.3 (8) 6.4–7.3
C^-	8.7 ± 0.8 (9) 7.3–9.8	9.4 ± 0.7 (11) 8.0–10.2	8.8 ± 0.8 (20) 7.5–10.2	9.1 ± 0.3 (8) 8.2–10.4
P_3	8.9 ± 0.8 (8) 8.3–10.0	9.1 ± 0.7 (12) 7.9–10.3	8.8 ± 0.7 (29) 7.3–10.2	9.8 ± 0.6 (13) 8.9–10.8
P_4	8.7 ± 0.9 (10) 7.2–10.1	9.7 ± 0.6 (10) 9.2–11.3	8.9 ± 0.8 (32) 7.1–10.3	10.5 ± 0.8 (8) 9.6–11.8
M_1	10.5 ± 0.5 (10) 9.7–11.4	11.4 ± 0.8 (11) 10.0–12.6	10.8 ± 0.6 (34) 9.6–12.0	12.0 ± 0.8 (14) 10.6–13.0
M_2	10.0 ± 0.5 (11) 9.2–10.8	11.7 ± 0.6 (11) 11.0–12.7	10.9 ± 0.7 (26) 9.6–12.3	12.1 ± 0.7 (11) 11.1–13.0
M_3	10.0 ± 0.7 (10) 8.7–11.3	10.8 ± 0.7 (14) 9.7–12.0	10.9 ± 1.1 (26) 8.7–13.0	11.1 ± 0.7 (10) 10.0–12.4
I^1	7.7 ± 0.2 (11) 7.5–8.1	8.9 ± 0.5 (19) 8.1–9.7	8.2 ± 0.4 (18) 7.4–9.0	7.8 ± 0.3 (6) 7.5–8.2
I^2	7.9 ± 0.3 (4) 7.5–8.1	8.6 ± 0.6 (15) 7.7–9.5	8.3 ± 0.5 (15) 7.7–9.0	8.1 (3) 8.0–8.2
C^-	9.9 ± 0.8 (6) 8.8–10.7	9.8 ± 0.7 (16) 8.1–11.2	9.5 ± 0.6 (21) 8.6–10.6	10.1 ± 0.4 (5) 9.8–10.6
P^3	11.2 ± 0.6 (4) 10.5–11.8	11.4 ± 0.6 (11) 8.5–11.9	10.4 ± 0.6 (19) 9.0–11.4	11.6 ± 0.8 (7) 10.5–12.6
P^4	10.8 ± 0.6 (5) 9.9–11.4	10.8 ± 0.4 (14) 10.3–11.4	9.9 ± 0.6 (20) 9.0–11.2	11.4 ± 0.6 (11) 10.3–12.5
M^1	11.9 ± 0.6 (6) 10.9–12.6	12.3 ± 0.4 (11) 11.0–13.2	11.9 ± 0.4 (22) 11.1–12.5	12.5 ± 0.7 (7) 11.7–13.7
M^2	12.8 ± 1.0 (5) 11.3–13.8	12.4 ± 1.0 (17) 10.5–14.0	12.3 ± 1.2 (19) 10.0–14.5	12.6 ± 0.5 (8) 11.9–13.4
M^3	11.7 ± 1.0 (5) 10.1–13.0	11.5 ± 1.4 (6) 8.7–12.5	12.0 ± 1.0 (16) 10.0–14.0	10.5 ± 0.8 (10) 10.4–12.5

Only labio-lingual or bucco-lingual dimensions are used, because these are not affected by interproximal wear. Measurements in mm. Krapina data are from Smith (1976); other data are from Bermúdez de Castro et al.(1993). The Neandertal sample excludes Krapina and includes both European and Near Eastern specimens.

The incisal edges of the crowns are curved, the edges have well-developed marginal ridges, and the basal tubercles are very prominent. All this is typical for European Neandertals (Crummett 1994) and unlike what we see in Pleistocene humans from East Asia (Chapter 5). At least some SH individuals had taurodont molars (Rosas 1987)—another Neandertal characteristic (Fig. 7.20)

The mandibles from SH present a mosaic of features (Rosas 1987, 1995). They vary a lot in size, but all have relatively thick, robust mandibular bodies and primitively chinless and sloping symphyses. Rosas (1995) notes that several specimens exhibit an incipient basal projection—that is, a faint bump in the chin area, of the sort that Weidenreich called a *mentum osseum*. Rosas describes two of the larger mandibles (AT 605

and AT 300) as possessing a mental **trigone**, a distinctive complex of ridges and depressions generally associated with recent human chins (Fig. 8.2). However, examination of these specimens shows that they do not have mental trigones in the modern human sense. The typical morphology of SH mandibles is well illustrated by AT 888 (Figs. 6.12 and 6.13). All the SH jaws have receding symphyseal profiles, although some of the larger specimens are less receding than the others.

One SH specimen has a mental foramen placed very anteriorly, under the P_3/P_4 contact, as in modern humans; but 50% of the SH mandibles have foramina located further back, under M_1—a frequency similar to that found in Neandertals (Trinkaus 1993a). Unlike other Heidelberg mandibles, eight SH specimens exhibit a

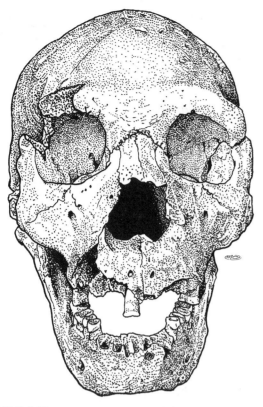

FIGURE 6.12

The Sima de los Huesos 5 cranium (AT 700) from the Sierra de Atapuerca (Spain), articulated with a mandible from the same site (AT 888 + AT 721) that may belong to the same individual. The facial morphology of this specimen closely approaches that of Neandertals.

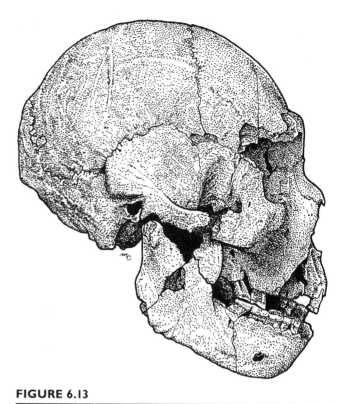

FIGURE 6.13

The Sima de los Huesos 5 cranium (AT 700) and the AT 888/AT 721 mandible in lateral view.

retromolar space, resulting in part from their relatively narrow rami. In these features, the SH mandibles resemble Neandertals more than most Heidelbergs do, particularly the early ones.

The paleoanthropologist Y. Rak (Rak et al. 1994, Rak 1998) has identified a complex of mandibular traits that supposedly distinguish Neandertals from other *Homo*. These include a coronoid process that sticks up above the level of the condyle, pterygoid tubercles, and an **incisure crest** (running forward from the condyle to the notch or incisure behind the coronoid process: Fig. 8.5) that intersects the condylar neck centrally in front. All these features are variable in the SH specimens. For example, the large mandible AT 605 conforms to the purported Neandertal condition in the first two of these features, but has a lateral incisure crest. Mandible AT 607, on the other hand, is Neandertal-like in the last two features but not the first. Only one specimen (AT 83) has a horizontally oval mandibular foramen, another possible apomorphy of Neandertals (Chapter 7).

As of 1997, the SH cranial sample comprised nine skulls, seven partial facial skeletons, and 19 other cranial

pieces, representing at least 13 individuals (Arsuaga et al. 1997c). The best-known finds are the three skulls (SH crania 4, 5, and 6) that were described in 1993 by Arsuaga and colleagues. The most complete of the three is SH 5 (Figs. 6.12 and 6.13), but the other two are almost as informative.

Cranium 5 (AT 700) exemplifies several features that are typical (though not invariate) in the SH cranial sample. Its supraorbital torus is strongly projecting. The height of the torus tends to thin evenly across the orbit from the midline to its lateral end, as in the Petralona and Arago skulls. In some other SH specimens, the torus is thinnest over the center of each orbit, as it is in Neandertals. Like other SH specimens, cranium 5 has a receding forehead and features related to incipient occipital bunning (Chapter 7)—for example, flattening around the lambdoidal region (Fig. 6.12). Cranium 4 (AT 600) is larger, with a higher vault (Table 6.6), a less retreating forehead, and more rounded occipital contours, as in the Swanscombe skull. The adolescent (ca. 14-year-old) cranium 6 (AT 764) also has a high, rounded vault (Table 6.6), but this may be due to its subadult status. Like cranial shape, cranial capacity is variable in the SH sample (Arsuaga et al. 1993), ranging from 1125 cm³ (Cranium 5) up to 1390 cm³ (Cranium 4). This is a large range, which encompasses all the variation found in other European Heidelbergs (Table 6.2).

TABLE 6.6 ■ Cranial Measurements for Sima de los Huesos

Specimen	Maximum Cranial Length	Maximum Cranial Breadth	Auricular Height	Minimum Frontal Breadth	Maximum Frontal Breadth
Cranium 3	—	—	—	102	115
Cranium 4	(201)	164	121	117	126
Cranium 5	185	146	108	106	119
Cranium 6[a]	186	136	121	100	—

[a]Denotes subadult.

TABLE 6.7 ■ Occipital and Nuchal Plane Measurements from the Sima de los Huesos Sample[a]

Specimen	Nuchal Plane Length	Occipital Plane Length	Index
Cranium 1	—	64	—
Cranium 3	—	74	—
Cranium 4	46	67	69
Cranium 5	50	61	82
Cranium 6[b]	35	66	53
Occipital IV	39	(65)[c]	60

[a]All lengths in mm. The index equals 100 × the ratio of the nuchal to the occipital plane.
[b]Subadult.
[c]Estimated.
Source: Data from Arsuaga et al. (1997c).

All the SH occipitals have nuchal tori that are moderately well-developed but poorly defined laterally. Variable oval depressions above the torus represent various degrees of suprainiac fossa development. Unlike Neandertals, the SH crania generally have rather large mastoid processes that project distinctly below the cranial base, rather than angling toward the midline. No SH mastoids have an anterior mastoid tubercle (Chapter 7), another feature that has been claimed as a Neandertal synapomorphy (Santa Luca 1978). As in other Heidelbergs and modern crania, the prominence of the mastoid area in the SH skulls results from the absence of the inferiorly projecting occipitomastoid crest region seen in Neandertals. The position of inion relative to opisthocranion is low on the SH skulls. As a result, their occipital planes are larger than the nuchal planes, and so their occipital indices are low. The exceptionally low occipital index for cranium 6 (Table 6.7) may be another reflection of the specimen's subadult age at death. These low index values reflect an expansion of the braincase relative to the cranial base, which is characteristic for Pleistocene *Homo* after around 400 Kya. This expansion is also reflected in the maximum cranial breadths, which lie rather high up, at the mid-parietal level (*contra* Arsuaga et al. 1993, 1997c).

The facial parts of the SH skulls also present a variable mosaic of Erectine and Neandertal features. Cranium 5, which retains a complete face, has primitively columnar lateral orbital margins, and foreshadows Neandertals in its broad nasal aperture and vertically elongated facial dimensions (Fig. 6.13). These characteristics are also evident in the adolescent cranium 6 and in other, more fragmentary facial pieces. The SH infraorbital regions are somewhat inflated, generally resulting in the absence of a canine fossa. (However, at least one adult specimen, AT 404, has a distinct canine fossa.) The zygomaticoalveolar margin (ZAM) of the inflated maxilla tends to slope obliquely down toward the alveolar process, intersecting it at a low position. Again, there is variation, with some specimens (e.g., AT 1100) exhibiting a more angled or curved ZAM. Arsuaga et al. (1997c) describe the internal nasal morphology of SH 5 as a precursor of the purportedly apomorphic Neandertal condition (see Chapter 7) defined by Schwartz and Tattersall (1996).

Cranium 5 is mid-sagittally prognathic, resulting in the most "beaked" midface in the sample (Arsuaga et al. 1997c). This is evident in the very anterior placement of the upper midface (nasion area) and the low subspinale angle of 111° (see Table 5.7). The infraorbital plate slopes markedly backward from the nose, approaching the Neandertal condition (Rak 1986). But most other SH specimens appear to exhibit more total facial prognathism, like Petralona and earlier members of the genus *Homo*.

The excellent cranial preservation at SH is reflected in the presence of ear ossicles in five individuals. In a study of the bones of the middle and outer ears in five of the SH skulls, Martínez et al. (2004) conclude that, like modern humans (but unlike apes), these hominins had acute hearing in the 2 to 4 kHz range. This is a human apomorphy, generally thought to be an adaptation to facilitate the reception and decoding of acoustic information in human speech. Martínez and his coworkers think that the anatomy of the ear region in the SH fossils may signal the presence of a human capacity for language in the European lineage leading to Neandertals.

TABLE 6.8 ■ Traits Shared Between the Sima de los Huesos Fossils and Other Human Groups

Trait	Erectines	Neandertals	Moderns
Vault broadest at base	+		
High total prognathism	+		
Laterally thick supraorbital torus	+		
Mandibular morphology	+	+	
Lower limb robusticity	+	+	
Cranial capacity range	+	+	+
Supraorbital torus shape		+	
Incipient suprainiac fossa		+	
Midface projection		+	
Lateral occipital profile		+	+
Tympanic morphology		+	+
High cranial vault		+	+
Temporal squama shape		+	+
Rear parietal profile	+		+
Adult occipitomastoid morphology			+

Source: After Stringer (1993).

Arsuaga et al. (1997c) view the SH sample as representing an intermediate stage on an evolutionary trajectory from the earlier Erectine condition to what they see as the highly derived morphology of Neandertals in Europe. This intermediate position is reflected both in the mosaic pattern of individual traits and in the Spanish team's multivariate analyses of craniofacial form. Stringer (1993), who agrees with this judgment, tallies up the features that the SH sample shares with earlier and later *Homo* (Table 6.8), and concludes that the SH humans are close kin to Neandertals. As noted earlier, some experts actually assign them to *Homo neanderthalensis*. However, Arsuaga and his team accept the taxon *Homo heidelbergensis* (for European specimens only) and assign the SH sample to this species.

Other European Heidelbergs

Several other sites yield human remains that fall into this time period. Human crania have been recovered at the sites of Apidima (Greece), Arago (France), and Reilingen (Germany). The Apidima skulls are less well preserved than Petralona, but appear to be very similar, particularly in facial anatomy (Coutselinis et al. 1991). The Arago cranium (Arago 21) comes from a carefully excavated cave site, but its age is not clear (Stringer et al. 1984). This cranium preserves the face, frontal, and parietals, all of which are smaller than the corresponding parts of Petralona. The face is clearly distorted (Fig. 6.14), but the details of its morphology are similar to those described for Petralona. From the side, Arago 21 appears to show the same sort and degree of progna-

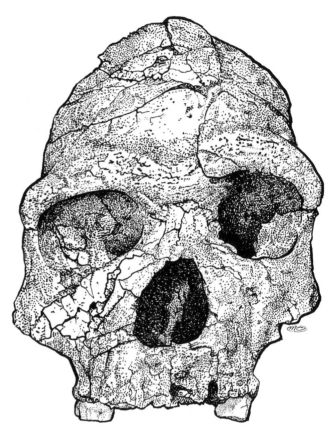

FIGURE 6.14

Arago 21 (France) is markedly distorted on the right side, but exhibits morphology on the left side that is similar to that of Petralona. (After de Lumley 1981.)

thism ascribed to other European Heidelbergs (Fig. 6.15), but its distortion and incompleteness make precise measurements impossible. An educated guess as to its cranial capacity (Holloway 1982) puts it at 1166 cm³.

The Reilingen specimen was dredged up from a Pleistocene gravel pit that has also yielded Middle Pleistocene fauna. Although the association of the faunal remains with the cranium is not certain, Ziegler and Dean (1998) think that a Middle Pleistocene age for the specimen is highly likely. A. Czarnetzki (1989), who first studied the specimen, considered it an advanced Erectine. More recent assessments indicate Neandertal affinities (Condemi 1996, Dean et al. 1998). The specimen preserves the rear of a cranial vault (occipital, parietals and a temporal). Like Neandertals, it exhibits a suprainiac fossa and distinct occipital bun (including the inferior shelf). Its estimated cranial capacity of 1430 cm³ (Table 6.2) exceeds those of other Heidelbergs. On the other hand, the mastoid process projects further below the skull than in typical Neandertals, mainly because the occipitomastoid crest is less well developed. The maximum breadth of the Reilingen braincase is low down, on the temporals; but the biparietal breadth is almost as great. In short, the specimen has a silhouette

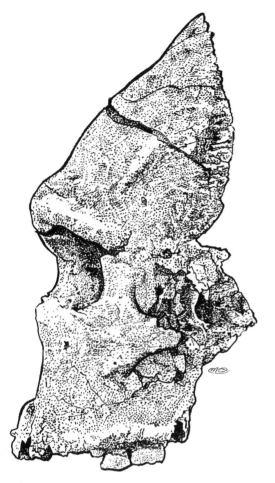

FIGURE 6.15

The relatively undistorted left side of Arago 21 (France) shows the specimen's facial prognathism, prominent supraorbital torus (lacking a supratoral sulcus), and absence of a canine fossa. (After de Lumley 1981.)

in occipital view that resembles those of other Heidelbergs.

In addition to the lower jaws from Mauer and Sima de los Huesos, Heidelberg mandibles have been found at Arago and the Middle Pleistocene site of Bau de l'Aubesier (Lebel et al. 2001). The Arago mandibles are similar morphologically to the Mauer jaw (Fig. 6.11), but exhibit considerable size variation. The Arago 13 mandible is slightly smaller than Mauer and is probably a male; the even smaller Arago 2 mandible is probably that of a female. Arago 13 also has unusually large teeth, but the Arago dental sample as a whole is about average for Middle Pleistocene humans. The Bau de l'Aubesier site, also in France, has yielded a single edentulous mandible (Aubusier 11) that exhibits extensive periodontal pathology. Lebel and Trinkaus (2002) believe that this specimen's masticatory function was so strongly impaired that the individual could not have survived without assistance from others. However, equal degrees of pathology can be found in non-human primates, which

do not exhibit such patterns of social care (DeGusta 2002, 2003). Lebel and Trinkaus also describe an upper molar (Aubesier 10) that exhibits carious lesions, a relatively rare occurrence in fossil human remains.

The Arago site also preserves some postcranial remains, including a crushed hipbone (Arago 44), a proximal femur and two other femoral fragments, a piece of a fibula, a metatarsal, and a distal manual phalanx. The hipbone is close morphologically to Pelvis 1 from the Sima de los Huesos sample. Marchal (1999) studied fossil human pelves from the Ergasters up through the probable early Neandertal specimen Le Prince 1, and concluded that pelvic morphology (including that of Arago 44) remained quite stable in human evolution from around ~2 Mya up to 100 Kya. The Arago femora have the thick cortices and other strengthening features that appear to be characteristic for Mid-Pleistocene *Homo* (Lovejoy 1982a, Ruff et al. 1993).

The cave site of Vértesszőllős in Hungary has yielded three fragmentary teeth and an adult occipital. Originally dated to about 400 Kya, these specimens have been placed at about 225 or 185 Kya by uranium-series dating of travertines from the site (Schwarcz and Latham 1984). But as usual, the association between the dated minerals and the human remains is not certain. The two deciduous teeth—a canine and a molar—are described by Thoma (1967) as being similar to those from Zhoukoudian. The Vértesszőllős occipital bone, also initially described by Thoma (1966), is complete except for the basal part, in front of the foramen magnum. It exhibits some Erectine features (nuchal torus, opisthocranion coinciding with inion; see Fig. 6.7). On the other hand, its occipital plane is larger than the nuchal plane, and there is incipient development of a suprainiac fossa. Cranial capacity appears to have been large, around 1325 cm^3 (Thoma 1969, Wolpoff 1971a). Apart from this large cranial-capacity estimate, the morphology of the Vértesszőllős remains is relatively primitive and generally similar to that seen in other Heidelbergs (Table 6.8).

Finally, Kappelman et al. (2008) describe an anterior calotte recovered from travertines at the site of Kocabaş in western Turkey. This specimen preserves the frontal bone (missing the glabellar region) and parietals of an adult individual. The travertines from this site have been assigned thermoluminescence dates ranging from 510 ± 0.05 Kya to 490 ± 0.05 Kya (Kappelman et al. 2008). Kappelman and his coworkers attribute the specimen to *Homo erectus* on the basis of its supraorbital torus form and its position on a bivariate plot of maximum biparietal and minimum frontal breadths. However, Erectines overlap with Heidelbergs in these respects; and given the specimen's estimated age, including it among the latter seems reasonable. This attribution helps to connect the African and European Heidelberg samples, and strengthens the argument that comparable West Asian specimens (see below) can also be considered Heidelbergs.

AFRICAN HEIDELBERGS

In 1921, a massive, unusual skull was recovered from the Broken Hill Mine near Kabwe in Northern Rhodesia (now Zambia). This skull, christened "Rhodesian Man" in the press, was the first post-Erectine archaic human found outside of Europe. Originally placed in its own species, "Homo rhodesiensis," by A. Smith Woodward in 1921, the Kabwe skull soon came to be widely regarded as an African variant of the Neandertals (Hrdlička 1930, Howells 1959). Coon (1962) pointed to more archaic aspects of Kabwe's morphology, and interpreted it as representing an intermediate stage between the Erectine and Neandertal phases of human evolution. In 1975, Murrill (1975, 1981) noted strong similarities between the palates of Petralona and Broken Hill. More recently, Rightmire (1990, 1996, 1998) and other researchers have emphasized similarities between the European and African fossil human skulls that date from 700 to 200 Kya. At present, it is common practice to lump all these specimens, including Kabwe, into the species *Homo heidelbergensis* (Rightmire 1998, Tattersall and Schwartz 2001). As noted above, some workers still prefer to retain the African specimens in their own species.

All the problems that hamper understanding of European Heidelbergs are even worse for the African Heidelberg sample. Africa is much bigger than Europe; yet there are only a handful of human fossils in Africa that might represent the Heidelberg group, and these are geographically widely scattered around the continent. And while the chronological framework for the European forms leaves much to be desired, almost all of the African forms have contextual problems that render estimation of their age even more problematic.

Kabwe

Known until the 1980s as Broken Hill, this site has yielded the largest and most informative collection of Heidelberg remains from Africa. In addition to the well-known skull, Kabwe 1 (E 686), the site has produced a partial maxilla of a second, smaller skull (Kabwe 2 or E 687), a frontal bone, and a large series of postcranial specimens (Hrdlička 1930), along with Paleolithic artifacts and Middle Pleistocene faunal remains. Unfortunately, Kabwe epitomizes all the problems that beset the African Heidelberg sample. The remains were not recovered through systematic excavation, but were just turned up by lead-mining operations (Smith Woodward 1921). Although all material reportedly was recovered from the same cave chamber, it is not certain that all the remains are contemporaneous. Attempts to date the

Kabwe human fossils directly have not been successful, so their geological context can only be estimated from their uncertain archaeological and faunal associations.

In 1970, J. D. Clark argued that the Paleolithic artifacts associated with the Kabwe fossils suggested an age of between ca. 40 Ky and 60 Ky—a surprisingly recent age for a skull with so archaic a morphology. This recent date for Kabwe helped create the impression during the 1960s and 1970s that Africa was something of a backwater area during the later phases of human evolution (Coon 1962). However, in 1973, R. Klein showed that the faunal material from Kabwe was more primitive than faunas from the Middle Stone Age of South Africa—which at that time was thought to have begun about 125 Kya. More recent assessments of African Pleistocene biostratigraphy suggest an age in excess of 300 Ky for the Kabwe fauna (Klein 1999). Chemical analysis of the human remains indicates that all contain a high proportion of lead, combined with smaller and varying amounts of zinc (Clark et al. 1950). This suggests, but does not prove, that all of the Kabwe human specimens are of similar age. In short, there is continuing uncertainty concerning the context, pattern of association, and age of the Kabwe remains.

The Kabwe 1 skull is very massive, with a strikingly prominent supraorbital torus, long face, and relatively low cranial vault (Figs. 6.16 and 6.17). Cranial capacity measures $1325\,cm^3$ (Holloway et al. 2004), which is

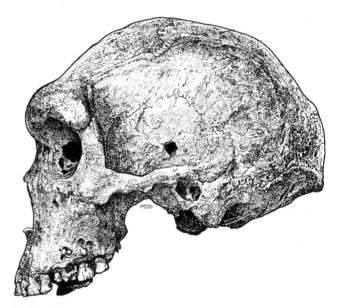

FIGURE 6.16

Lateral view of the Kabwe 1 (Broken Hill) cranium from Zambia. In addition to its facial prognathism, prominent supraorbital torus, and archaic vault contour, the pathologies on this specimen are evident in this view. These include the destruction of the teeth by carious lesions, apical abscesses, and an enigmatic lesion of the temporal squama. (After Pycraft et al. 1928.)

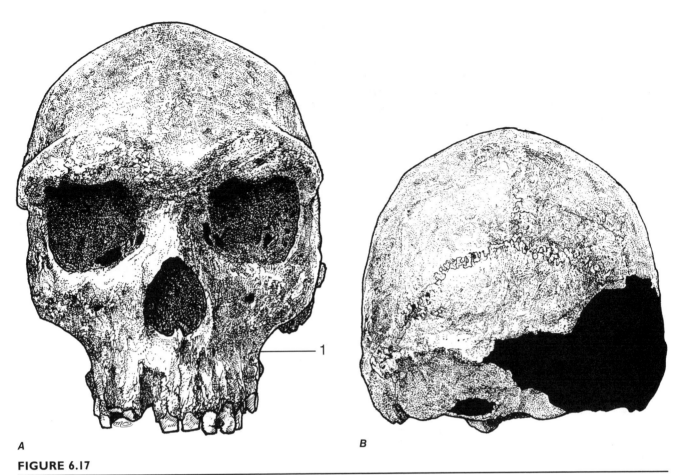

FIGURE 6.17

Kabwe (Broken Hill) 1 from Zambia in frontal view (**A**) and rear view (**B**). The archaic features of the face, including the massive supraorbital torus and broad interorbital region, are visible. The nasal aperture is smaller than in European Heidelbergs. The zygomaticoalveolar margins (ZAM) are more angled and have their root higher up on the alveolar process (1) than in most European Heidelbergs. The rear view shows that the maximum cranial breadth is still at the level of the supramastoid ridges (cranial base), but the biparietal breadth is almost as great. (After Pycraft et al. 1928.)

similar to that of European Heidelbergs. Despite its primitive appearance, the skull's vault morphology reflects this brain expansion in ways much like those seen in its European counterparts. In a rear view, the biparietal breadth of Kabwe 1 is virtually equal to that of its cranial base (Fig. 6.17), and the occipital plane of its occipital bone is probably slightly larger than the nuchal plane (Table 6.3). Postorbital constriction—the transverse narrowness of the frontal "waist" between the braincase and the face—is less than in Erectines. All of these features contrast with the Erectine condition and signal an expansion of the vault relative to the cranial base.

Kabwe 1's frontal bone is long and low, with a distinct sagittal keel that is reminiscent of that seen in Erectines. The supraorbital torus is strongly projecting and continuous over the nasal region, which is another Erectine feature. The torus is also quite thick (in its superior-inferior dimension), and although it is thickest medially,

its dimensions are fairly even across the entire structure (Smith 1992a). The second Kabwe frontal (E897) is more gracile but similar in the morphology of the squama (behind the torus).

In contrast, the occipital bone of Kabwe 1 differs markedly from those of Erectines (except for a prominent, apparently continuous nuchal torus). As in European Heidelbergs, inion is located lower than opisthocranion, reflecting a relative reduction of the nuchal plane. The expanded occipital plane is more vertically positioned than in Erectines (Fig. 6.16). The occipitomastoid crests are well developed, so that the moderately large mastoid processes appear not to project inferiorly. Although a cursory look at the lateral view of the skull suggests the presence of occipital bunning (Fig. 6.16), that impression actually stems from the presence of a slight depression just above the lambdoidal suture and not from lambdoidal flattening (Chapter 7). Because the lower part of the occipital plane is missing

(Fig. 6.17), we do not know whether this skull had a suprainiac fossa.

The Kabwe 1 face has a substantial level of midsagittal facial prognathism with moderately retreating and backswept zygomatics, as in Petralona and other European Heidelbergs. Like those European fossils, the Kabwe 1 skull is intermediate in this respect between Erectines and Neandertals (Stringer 1983, Smith and Paquette 1989). In such indicators of mid-facial prognathism as subspinale angle and zygomaxillary index, it falls much closer to Neandertals than to Erectines—closer, in fact, than Petralona does (Table 5.7). The Kabwe 1 face is very long but only moderately broad (Fig. 6.17). Its nose is large, but not as prominent as in many European specimens. The maxillary alveolar process is elongated and the anterior palate is broad, suggesting that the missing anterior teeth had broad crowns and long roots (Smith and Paquette 1989). Buttressing of the face is reflected in the broad intraorbital region, the rather oblique zygomaticoalveolar margin (ZAM), and the columnar-like lateral orbital pillars. Although the infraorbital region is less inflated than in many European Heidelbergs, there is no canine fossa.

While the facial form of Kabwe 1 is broadly similar to those of European Heidelbergs, there are some differences. As noted above, the nose is less prominent and the supraorbital torus is more primitive. Although the Kabwe 1 ZAM is oblique in orientation, it intersects the alveolus rather high above the alveolar margin (Fig. 6.17). Most of the European Heidelbergs have a lower, more Neandertal-like point of intersection. The Kabwe 2 maxillary fragment is similar to Kabwe 1 in its ZAM and features of its palate, but it is considerably smaller—and unlike Kabwe 1, it has a distinct canine fossa.

The proportions of the teeth and palate of Kabwe 1 are much like those of Petralona (Murrill 1975). The teeth are heavily eroded by both occlusal wear and carious lesions (cavities). Some of these lesions have destroyed almost the whole crown of the tooth (Fig. 6.16)—an unusual finding in pre-agricultural humans. The pulp cavities of several teeth have been penetrated by dental decay, and the bone around their root tips has been eaten away by spreading bacterial colonies, producing **apical abscesses**. The septic teeth and jaws of the Kabwe 1 individual surely had a serious impact on his health, and may have contributed to his death.

Most of the postcranial bones from Kabwe do not differ markedly from the modern human condition. Stringer (1986b) noted that the larger of the two Kabwe hipbones (E 719) bears a thick buttress of bone above the acetabulum, a feature normally found only in Erectines and some other Heidelbergs (Arago 44). Marchal's (1999) survey of Lower and Middle Pleistocene hipbones results shows that the Kabwe hipbones (E 719 and E 720) resemble European specimens from the same time period and retain a strong similarity to Erectine

hipbones from Africa (OH 9 and ER 3228). Although the Kabwe femur fragments are morphologically modern in external morphology, they have unusually thick cortical bone (Kennedy 1984). They lack the primitive elongation of the femoral neck seen in African Erectines. The complete left tibia from Kabwe (E 691) is one of two postcranial elements supposedly discovered in close proximity to Kabwe 1. Unlike the femora, it retains some persistently Erectine features (Pearson 2000a). Its length (416 mm) suggests an attenuated body build adapted to the tropics, and it contrasts with those of European Neandertals and some recent cold-adapted modern human populations, which have short distal limb segments. The differences between the long Kabwe tibia and the apparently short tibias from Boxgrove and Zhoukoudian hint that latitudinal differences in body form, evident in later Pleistocene human skeletons, were beginning to appear between 300 Kya and 500 Kya.

Bodo and Ndutu

The Bodo skull is the face and anterior cranial vault of a very robust individual from the Middle Awash Valley of Ethiopia (Kalb et al. 1982). The skull, a second cranial fragment, and a distal humerus were associated with Acheulean tools and a mid-Pleistocene fauna dated to around 600 Ky by biostratigraphy and $^{40}Ar/^{39}Ar$ dating (Clark et al. 1994). Ndutu is a partial skull found in 1973 near the edge of Olduvai Gorge in Tanzania, in association with Acheulean tools and fragmentary faunal remains (Clarke 1976). The Ndutu site lies just below a volcanic tuff that may correspond to an Olduvai tuff dated to 400 Kya (Leakey and Hay 1982). Ndutu has been carefully described by Rightmire (1983, 1984b). Due to complex political issues, Bodo has never been completely described; but Conroy et al. (1978) provide a preliminary assessment, and Rightmire (1996) has published the most thorough analysis to date of this impressive specimen.

Bodo and Ndutu represent near-extremes of variation in the African Heidelberg sample. Ndutu's estimated cranial capacity is only some 1090 cm³, while Bodo's is 1250 cm³ (Holloway et al. 2004). Bodo (Fig. 6.18) has a massive face, with a huge nose and an inflated infraorbital region that precludes the presence of a canine fossa. Ndutu has a much smaller face and nose, and what remains of the maxilla seems quite similar to that of Kabwe 2—especially in having a canine fossa. Bodo has a thick, projecting supraorbital torus; Ndutu's is also projecting, but much thinner. The tori of both skulls are relatively even in thickness all the way across, which seems to be typical of African Pleistocene *Homo* from this time period and later (Smith 1992a).

Ndutu clearly shows the expansion of the occipital plane and the relative broadening of the braincase (Table

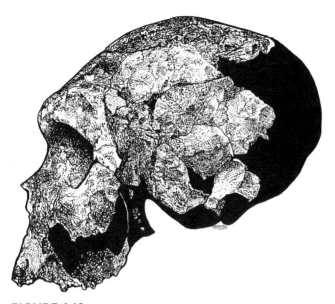

FIGURE 6.18

The Bodo cranium from Ethiopia in lateral view. Note the massive face and supraorbital torus, as well as the total facial prognathism. (After Johanson and Edgar 1996.)

6.3) that distinguish Heidelbergs from Erectines. Its occiput exhibits all the features seen in Kabwe that reflect expansion of the vault relative to the base (increased biparietal breadth, lowered inion, etc.). Ndutu has a nuchal torus and no indication of occipital bunning or a suprainiac fossa. The Bodo skull and a second cranial piece from Bodo (a parietal) seem more primitive in several features. Both exhibit an angular torus, a thickening of bone at the lower back corner of the parietal commonly seen in Erectines (Asfaw 1983). The Bodo face shows less swept-back zygomatics ("zygomatic retreat": Chapter 7) and a more straight, continuous supraorbital torus than Kabwe or other African Heidelbergs. In these features, Bodo seems persistently Erectine-like. This may be because it is more ancient, but some aspects of the face and supraorbital torus may be related to the effects of overall size or sexual dimorphism (Wolpoff 1999). Rightmire (1996) interprets the Bodo cranium as indicating a speciation event that marks the origin of *H. heidelbergensis*. We will come back to this idea later.

African Heidelberg Mandibles

Mandibles from this period are known from both East and South Africa. Near Lake Baringo in Kenya, two nearly complete mandibles (BK 67 and BK 8518) were found in association with Acheulean tools in deposits dated to <660 Kya (Wood 1999b). In South Africa, the Cave of Hearths at Makapansgat has yielded the right half of a small mandible preserving the P_3 and the first

two molars. The associated artifacts suggest a mid-Pleistocene age (Volman 1984). Finally, a small piece of mandibular ramus discovered as a surface find near Elandsfontein, on the southwest coast of South Africa (Fig. 6.3), is thought to be associated with Acheulean artifacts and mid-Pleistocene fauna. Klein (1988) suggests an age of somewhere between 500 Ky and 200 Ky for the Elandsfontein material.

The rami of the Baringo and Elandsfontein jaws appear to be relatively broad, which implies that the temporal fossae of the missing skulls must have been long from front to back. This in turn implies a relatively long distance between the jaw joint and the back of the orbit, suggesting a pattern of total facial prognathism, as in the Bodo skull, rather than the sort of zygomatic retreat seen in Kabwe 1. Both Baringo mandibles have receding mandibular symphyses and lack any indication of a chin. The Cave of Hearths specimen is damaged at the symphysis, but also seems to follow this pattern. BK 8518 from Baringo has two internal buttresses at the symphysis; BK 67 and Cave of Hearths lack any internal buttressing. The teeth in all these mandibles are on the small side for Heidelbergs.

Other African Heidelbergs

The Elandsfontein site in South Africa (a.k.a. Saldanha or Hopefield) has also yielded a calotte that is much like that of Kabwe 1 but smaller (Rightmire 1984b, Bräuer 1984a). A rather gracile partial ulna from Baringo has a coronoid process that is short relative to its olecranon process, a primitive feature found in Erectines and other archaic forms of *Homo* (Churchill et al. 1996). A large proximal femur from Berg Aukas in Namibia is more primitive than the Kabwe femora in some respects (Grine et al. 1995) and may belong to the Heidelberg group; however, it lacks a secure context and is undated. In contrast, a large femur from Kenya, KNM-ER 999, seems fundamentally modern—for example, it exhibits the modern pattern of anterior-posterior reinforcement of the shaft, rather than the medial-lateral pattern seen in Erectines (Trinkaus 1993b, Pearson 2000a). The ER 999 femur has been dated to 301 ± 96 Kya by gamma-ray spectrometry (Bräuer et al. 1997), but the large standard error makes this estimate highly questionable.

Finally, from Eyasi in Tanzania comes a series of fragmentary remains of at least three individuals. These specimens, recovered by the Kohl-Larsen expeditions in 1935 and 1938, include surface finds that have suffered heavy abrasion. Acheulean tools found with these remains, together with the geology of the area, suggest an age of about 200 Ky for the site (Masao 1992, Mehlman 1984, 1987). The most complete skull (Eyasi 1) was analyzed by H. Weinert (Weinert et al. 1940), who emphasized its primitive aspects and offered a fanciful reconstruction based more on Erectine material from

China and Java than on the Eyasi fossils themselves. The supraorbital torus of Eyasi 1 is moderately thick and projecting, and the two occipitals (1 and 2) have nuchal tori. In these features, the Eyasi remains are like smaller versions of Kabwe 1. Interestingly, Eyasi 1 has a depression above the nuchal torus that is very similar to the suprainiac fossae of Neandertals (Trinkaus 2004; see Chapter 7).

North Africans

A few finds from North Africa fall in the Heidelberg time span. Most of them were found near the coast of North Africa in Morocco (Fig. 6.3), but one of these specimens (the Wadi Dagadlé maxilla) derives from the Red Sea coast, far to the south and east of the others. The affinities of all these fossils are in doubt, and most experts do not group them with the Heidelbergs. Some of them show clear similarities to Erectines; others may represent post-Heidelberg stages of human evolution in Africa.

The best known of these North African specimens is the skull from Salé in Morocco. It consists of a basicranium and the posterior part of the cranial vault, lacking the face and brow ridges (Hublin 1985). ESR dating on fauna from the site suggests an age of either 389 or 455 Ky. The skull is very small, with an estimated cranial capacity of only 880 cm^3 (Holloway 1981b). Some aspects of the skull are appropriately archaic: there is a well-defined postorbital constriction, a low frontal profile, and a well-developed sagittal keel on both the parietals and frontal. The specimen has a very modern-looking posterior vault, but the significance of this is complicated by an abnormal position of the superior nuchal line well below the moderate nuchal torus. The effect of this pathology on overall vault form is disputed, but Salé certainly cannot be considered an anatomically normal representative of its population. The modern-looking aspects of its anatomy may be products of its obvious pathology.

The Thomas Quarries and Sidi Abderrahman localities are faunally dated to a Mid-Pleistocene span that is younger than the 700-Ky-old Tighenif site (Geraads et al. 1980). One estimate for the Thomas Quarry site puts it at 400 Kya (Hublin 1985). These estimates would place both sites in the appropriate time range for the Heidelberg group. Thomas III, an upper facial fragment with a bit of frontal, has a prominent supraorbital torus, but the torus thinness and surface morphology suggest it is a subadult. Some associated teeth include Erectine-sized molars and a large lateral incisor. Thomas I and Sidi Abderrahman are robust mandibles with large posterior teeth.

Two later but persistently primitive-looking specimens may also belong to this broadly defined grouping. The Kébibat specimen comprises several cranial pieces,

including an occipital and parts of a maxilla and mandible, and a nearly complete dentition (Hublin 1985). The occipital, which is the only interpretable cranial piece, has moderately thick bone and a weak nuchal torus. Saban (1977) reports impressions of endocranial blood vessels that conform to a primitive pattern. Both the anterior and posterior teeth are large, falling in the Erectine size range. The second specimen is a megadont, robust partial palate from Wadi Dagadlé, Djibouti. Despite their primitive morphology, the Kébibat and Wadi Dagadlé fossils are at most 250 Ky old and 200 Ky old respectively (Boins et al. 1984, Hublin 1985).

ASIAN HEIDELBERGS?

As in North Africa, there are a few Asian finds that fit in the Heidelberg time frame and exhibit comparable morphological patterns. Those from western and southern Asia are mostly too few and fragmentary to tell us much about regional norms and variation in this area and period. Most of their dates are vague. Fossils from East and Southeast Asia are more numerous, but their dating is still problematic. These eastern fossils generally show few of the morphological changes that distinguish Heidelbergs from their Erectine precursors, and they are usually considered part of the Erectine group or clade.

Mugharet El-Zuttiyeh

The Zuttiyeh specimen from Israel (Fig. 6.19) is the most likely candidate Heidelberg from western Asia. It consists of a complete frontal bone and the right lateral portion of the face, including the zygomatic and parts of the sphenoid and maxilla. When Zuttiyeh was discovered in 1925, it was hailed as the second "Neandertal-like" specimen (Kabwe being the first) to be found outside of Europe. Excavated near the base of a cave near the Sea of Galilee (Fig. 6.3), the Zuttiyeh skull is sometimes referred to as the Galilee skull. It was originally attributed to a Mousterian level (Turville-Petre 1927), but a later analysis (Gisis and Bar-Yosef 1974) concluded that it derived from the underlying Acheuleo-Yabrudian level. The Acheuleo-Yabrudian is a biface-rich tradition that seems to be a technological precursor to the Mousterian (Bar-Yosef 1994, 1995; Klein 1999). Although there are no reliable dates from the Zuttiyeh site—uranium-series dates from the site are generally considered contaminated (Schwarcz 1994)—correlation with other Acheuleo-Yabrudian sites suggests that Zuttiyeh is between 250 Ky and 350 Ky old (Bar-Yosef 1992, 1995).

Initially, the Zuttiyeh skull was described as a Neandertal (Keith 1927, Hrdlička 1930), but later studies have shown that the face is relatively flat, with no evidence of Neandertal-like midsagittal facial prognathism (Van-

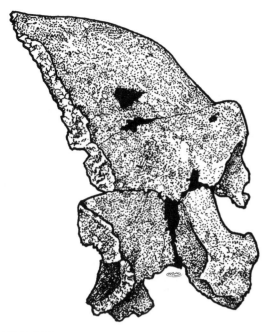

FIGURE 6.19

The Zuttiyeh anterior craniofacial skeleton from Israel. Also known as "Galilee Man," Zuttiyeh resembles European and African Heidelbergs in its total facial prognathism (indicated by the forward placement of the lateral face) and its prominent, arched supraorbital torus. (After Delson 1985.)

dermeersch 1985, Smith 1985, Simmons et al. 1991, Sohn and Wolpoff 1993). Vandermeersch (1982, 1985, 1989) interpreted this facial flatness as evidence of Zuttiyeh's phyletic affinities to the later, more modern-looking Skhūl and Qafzeh fossil skulls, which also come from Israel (Chapter 8). But a subsequent analysis (Simmons et al. 1991) has shown that the Zuttiyeh face is both flat *and* prognathic. Like many Heidelbergs, it exhibits total facial prognathism—a primitive condition compared to both Skhūl/Qafzeh and Neandertals.

Zuttiyeh has a very projecting and thick supraorbital torus, separated from the frontal squama by a well-excavated supratoral sulcus. The torus thins laterally, but it shows no signs of the thinning over the middle of each orbit that characterizes Neandertals and some other late Pleistocene humans (Chapter 7). (In fact, Zuttiyeh does show a sort of mid-orbital thinning on the left side only, but this is the product of a healed injury; Simmons et al. 1991.) Post-orbital constriction is marked, and the frontal is relatively narrow, with a prominent frontal boss. The lateral orbital margin is columnar—another generally primitive feature. The preserved part of the zygomatic, however, seems relatively gracile. Overall, this morphology compares reasonably well to those seen in African and European Heidelbergs.

Sohn and Wolpoff (1993, Wolpoff 1999) argue that Zuttiyeh shares certain distictive character states with East Asian forms, particularly with the Erectines from Zhoukoudian. The resemblances mainly involve the overall shape of the anterior vault and some midline features above the nose, including a depression at glabella (as seen from above) and a flat superior nasal region. These shared traits suggest to them that some significant amount of gene flow connected East and West Asia during the Mid-Pleistocene.

Other West Asian Candidates

Two other Levantine sites have yielded fragmentary fossils from this period. Segments of two femoral shafts are known from Gesher Benot Ya'acov (Geraads and Tchernov 1983), and another femoral shaft and a molar were excavated from Level E at Tabūn (McCown and Keith 1939). The Tabūn fossils are associated with Acheuleo-Yabrudian tool assemblages and thus are of comparable age to Zuttiyeh (Bar-Yosef 1995). The Gesher Benot Ya'acov fossils are probably somewhat earlier, around 500 Ky or slightly more, judging from their archaeological associations (Geraads and Tchernov 1983). All these femora have thick cortical bone and are platymeric—that is, broader from side to side than from front to back (Ruff et al. 1993). This pattern is seen in all Heidelberg femora, but it is also characteristic of Erectine and Neandertal femora (Fig. 7.26). These West Asian femora might fit into any of these groups, but temporally they fall in the Heidelberg span.

South Asia

Virtually no fossil human remains are known from India, but one partial skull from the Narmada Valley in north central India (Fig. 6.3) may belong to the Heidelberg group. Found in 1982, this specimen preserves much of the cranial vault, lacking part of the left side of the frontal and all the face except the right lateral orbital margin (Fig. 6.20A). The associated fauna and Acheulean artifacts suggest a late Middle and early Late Pleistocene age, but there is no direct dating of the specimen (Kennedy et al. 1991, Kennedy 2000). Early studies of the Narmada skull (Sonakia 1985, de Lumley and Sonakia 1985) aligned it with *Homo erectus*, largely because the maximum cranial breadth lies down low on the cranial base. On the other hand, the cranial capacity is $1260\,cm^3$ (Table 6.2), the occipital plane is larger than the nuchal plane (opisthocranion is higher than inion), the occipital contour is rather rounded, and the vault is relatively high. From these facts, Kennedy et al. (1991) argue that Narmada is more advanced than the Erectines.

East Asia

The situation in China during this period is perplexing. As noted in Chapter 5, Zhoukoudian (ZKD) provides the

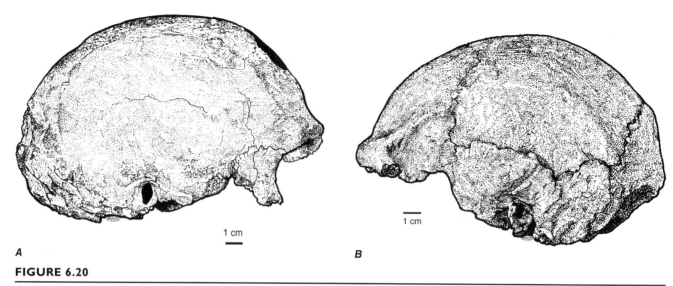

FIGURE 6.20

The Narmada cranium from India (**A**), and the Poloyo (Sambungmachan 3) cranium from Indonesia (**B**). (**A**, after Schwartz and Tattersall 2003; **B**, after Delson et al. 2001.)

best sample of Chinese Erectines by far. However, the ZKD sample appears to be atypical in some ways (Antón 2002, Kidder and Durband 2004), and its chronology is in dispute. If the ESR dating is correct, ZKD spans a time period from 312 Kya to 578 Kya, with the Locality 1 Erectine sample falling at 418 Kya—a range of dates similar to those attributed to the early Heidelbergs from the West (Steinheim, Mauer, Sima de los Huesos, Bodo, Ndutu). The Locality 1 specimens, however, do not exhibit the expansion of the braincase relative to the cranial base that characterizes Heidelbergs. This could mean that Erectine morphology persisted later in eastern Asia than in the West, as some think it did in Australasia (see below). Alternatively, it may be that the TIMS dating scheme is correct and Locality 1 is older than 600 Ky (Fig. 5.7).

A few specimens from ZKD are stratigraphically younger than the Locality 1 sample. Level 3 is dated by ESR to 312-386 Kya (Grün et al. 1997) but could be more than 410 Ky old (Zhou et al. 2000). This level yielded a partial cranium and mandible from Locality H. Parts of the H3 skull were recovered in the 1930s, but the frontal bone was one of the major pieces found after World War II (in 1966). Like the other ZKD specimens, H3 lacks a face; but it preserves the supraorbital torus, most of the occipital and temporals, and parts of the parietals. Qiu et al. (1973) suggest that the H3 cranium is more advanced in some features than the more ancient remains from Locality 1. These features include a broader braincase (particularly in the frontal and occipital) and reduced cranial tori (a thinner supraorbital torus and a less pronounced nuchal torus). As Wolpoff (1999) points out, the individual features of H3 all fall within the Locality 1 range, and its "fundamental similarities" are

to this sample. Nevertheless, Wolpoff and others see the differences between H3 and the older ZKD fossils as reflecting advancement toward modernity.

On the other hand, the various indicators of relative braincase expansion in H3 do not differ significantly from the Locality 1 sample. Qiu and colleagues note that the distance between inion and endinion is reduced, but the occipital still conforms to the Erectine pattern. Given these facts, and given that no single feature of the H3 skull falls out of the Locality 1 range, it becomes difficult to see it as anything but a particularly late example of persistent Erectine morphology at Zhoukoudian.

Other Chinese specimens from about the same time period, from the sites of Hexian and Yunxian (Fig. 6.3), seem to fit this pattern as well. The fossil human level from Hexian (Level 2) has yielded a fairly complete cranium (Fig. 6.21) and a number of other cranial pieces. Age estimates for this level range from 620 to 153 Kya, but a combination of ESR and U-series dating provides an estimate of 412 ± 25 Kya (Grün et al. 1998). The Hexian (Longtandong) specimens have been described as being quite similar to the H3 skull from Zhoukoudian. They also resemble the ZKD Locality 1 crania, and have a generally Erectine gestalt (Etler 1996, Wu and Poirier 1995, Pope 1988, 1997b). Estimated cranial capacity for Hexian PA 830 is 1025 cm³, which is within the ZKD range (Table 5.3). Although these specimens do exhibit some of the same "progressive" or "advanced" features seen in ZKD H3, including broader frontals and some changes in the occipital, the overall form of the occipital is clearly Erectine-like. Wu and Poirier (1995) report a distance of 22 mm between inion and endinion, imply-ing a big nuchal plane of the typical Erectine sort

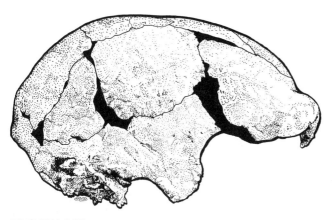

FIGURE 6.21

Lateral view of the Hexian skull from China. (After a photograph of a cast in Schwartz and Tattersall 2003.)

(Chapter 5). The Hexian teeth are larger than those from ZKD, but their morphology is Erectine, and the mandible would be lost in the ZKD sample.

Two crushed and distorted crania from a Middle Pleistocene terrace at Yunxian (Li and Etler 1992, Etler 1996, Wu and Poirier 1995) may be roughly contemporary with Hexian. Chronometric dating on fauna from the terrace points to an age exceeding that of Zhoukoudian Locality 1 (Chen et al. 1997), but the skulls' association with the dated fauna is not certain. Both of the Yunxian specimens, EV 9001 and the larger EV 9002, preserve relatively complete faces—a rarity in early fossil humans from this period in Asia. Both faces are broad and generally large. Both have well-developed, Erectine-like supraorbital tori that appear to be separated from the frontal squamae by supratoral sulci. Both appear to exhibit total facial prognathism. Both have malar notches, angled zygomaticoalveolar margins, canine fossae, a high position for the root of the zygomatic arch, and large noses. The flat upper faces seen in the Yunxian skulls resemble those of the Zhoukoudian sample, and their occipital morphology and dental morphometrics do not differ from those seen in other Erectines. Unfortunately, the vaults of the Yunxian crania are so distorted that not much can be said about them. The cranial capacity of 1200 cm^3 estimated for both skulls (Table 6.2) should accordingly be taken with several grains of salt.

Australasia

Australasia comprises the islands of Southeast Asia as well as greater Australia (Chapter 8, Fig. 8.24). As noted in Chapter 5, dating of the Javan Erectines is generally problematic due to the complex local geology, the reworking of volcanic tuffs, and the uncertain stratigraphic provenience of most of the specimens. These problems are no less evident in the late Middle and Late

Pleistocene. They are exemplified by the material from the sites of Sambungmacan and Ngandong, located along the Solo River in central Java.

Sambungmacan Locality 1 yielded a calvaria and a right tibia shaft (Jacob 1975, Baba and Aziz 1992). Both specimens have some archaic morphological features, but their stratigraphic context and geological age is uncertain. The tibia (Sambungmacan 2) is a central shaft fragment that has very thick cortical bone and a small medullary cavity. The specimen is similar in overall size to, but relatively narrower than, the Zhoukoudian Erectine femora. The calvaria (Sambungmacan 1) resembles earlier Erectine individuals in Indonesia, particularly in its very flat forehead (lacking a frontal boss), the development of the cranial buttressing system, the relatively straight contour of the supraorbital torus, and the form of the supratoral groove. Its cranial capacity is 1035 cm^3, which is within the range of the Indonesian Erectines from the Kabuh Formation (Table 5.3).

In 1977, another cranium was discovered by miners near the village of Poloyo in the Sambungmacan District. This specimen, now known as Sambungmacan 3 (Sm 3), comprises a cranial vault lacking the face and much of the base (Fig. 6.20B). It was bought by an antiquities dealer, and examined and briefly described in Jakarta. It then disappeared. The specimen surfaced again in New York in 1999, where it was noticed by H. Galiano, a conscientious owner of a natural history shop. Known details of the story are recorded by Márquez et al. (2001), but it is not certain how the specimen found its way to New York. The skull has now been returned to Indonesia. A third Sambungmacan calvaria (Sm 4) was recovered in 2001 during the collection of sand for construction purposes, at a site just 100 m upstream on the Solo River from the locality where Sm 3 was found. According to Baba et al. (2003), the Sm 4 specimen exhibits no evidence of extensive water transport, though it does evince several healed head injuries.

Like those of most human fossils from Java, the context and geological age of both Sm 3 and 4 are uncertain. There is no evidence that they are contemporary with the other Sambungmacan specimens or each other. However, a recent comparative study of Sm 3 found that its closest similarities were with Sm 1 and the (probably later) Ngandong sample, rather than with earlier Indonesian Erectines (Delson et al. 2001). A comprehensive morphometric study of Sm 4 (Baba et al. 2003) comes to similar conclusions, but emphasizes the intermediate status of the Sambungmacan specimens as transitional forms linking earlier Indonesian Erectines to the later Ngandong crania (see below).

Several distinctive features shared by the Sambungmacan and Ngandong fossils suggest a genetic continuity between them. They are particularly similar in the supraorbital torus, which forms a straight horizontal bar

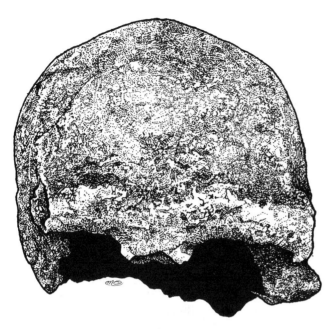

FIGURE 6.22

The Ngandong 5 (Java, Indonesia) cranium in frontal view. Note the massive, shelf-like supraorbital torus, which is actually thickest laterally. (After Santa Luca 1980.)

FIGURE 6.23

The Ngandong 5 (Java, Indonesia) cranium in rear view. The vault shape and prominent nuchal torus are evident. The maximum cranial breadth is still at the level of the cranial base, but the biparietal breadth closely approaches it. (After Santa Luca 1980.)

across the frontal (Fig. 6.22). In both groups, the tori are thickest laterally and exhibit lateral **frontal trigones**. This feature, described by Wolpoff (1999) as a knob-like backward-facing triangle on the posterior aspect of the lateral supraorbital torus, is better developed in the Ngandong sample and lacking in earlier Indonesian Erectines. The Sambungmacan and Ngandong samples may also uniquely share a distinctive course of the squamotympanic fissure (the suture between the ectotympanic and squamosal parts of the temporal bone), which in these fossils traverses the roof of the mandibular fossa instead of running across the fossa's rear edge (Durband 2002; but see Delson et al. 2001). Compared to typical Indonesian Erectines, the Sambungmacan and Ngandong crania are more Heidelberg-like in having more vertical foreheads, reduced postorbital constriction, less angled occipitals, and more globular vaults (Delson et al. 2001, Márquez et al. 2001, Antón et al. 2002b, Baba et al. 2003). Like the Heidelbergs to the West, the Sambungmacan and Ngandong crania have a bigger cranial vault relative to the cranial base than the Kabuh Erectines had (Table 5.1). The biparietal breadth is about the same as cranial-base breadth, the sides of the skulls are relatively straight when viewed from the rear (Fig. 6.23), and the occipital plane of the occipital bone is expanded and more vertically oriented than in the Kabuh fossils. All these vault features reflect an increase in relative brain size in these later specimens from Java.

Sm 1 and 3 have less developed nuchal tori than Ngandong and thus are more like earlier Erectines in this feature. Baba and colleagues, however, note that Sm 4's torus development is more like Ngandong's. The Sambungmacan crania also exhibit midline keels on the frontal bone, a feature found in both Ngandong and earlier Erectines. Estimated cranial capacity is 918 cm^3 for Sm 3 and 1006 cm^3 for Sm 4 (Baba et al. 2003). The mosaic of features in the three Sambungmacan crania seems to place them in a morphologically intermediate position between typical Erectines and the Ngandong people, but closer to the latter. This conclusion is supported to some extent by a recent multivariate analysis of Indonesian Erectines by Antón et al. (2002b). Their factor plots show that the Ngandong specimens overlap somewhat with the Kabuh Erectines but form a reasonably distinct cluster, and that the two Sambungmacan crania included fall closer to the Ngandong cluster than to any other group.

Ngandong

Between 1931 and 1933, the remains of 12 ancient human crania and two tibias were recovered by Indonesians working for the Dutch Geological Survey. These specimens were found in deposits associated with the High or 20-meter Terrace of the Solo River and were associated with an extensive fauna. A calvaria discovered in 1987 near Nagwi is now generally included with the Ngandong sample (Widianto and Zeitoun 2003), as is a small piece of a hipbone. All these crania lack faces, and only a few preserve much of the cranial base (Figs. 6.22–6.25). Because they were found along the Solo River, these remains are often referred to as the Solo remains. Their initial describer, W. Oppenoorth, erected a separate species, "Homo (Javanthropus) soloensis," for the Ngandong specimens in 1932 (Santa Luca 1980). The Ngandong fossils present perplexing problems in terms of both taxonomy and chronology. Though relatively ignored for many years, they have become a focus of the current debate over human evolution in the late Pleistocene.

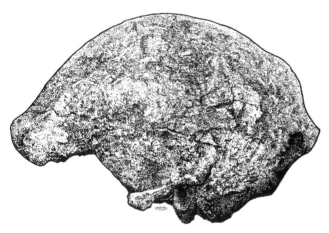

FIGURE 6.24

The Ngandong 5 (Java, Indonesia) cranium in lateral view. Note the very prominent nuchal torus, the long and low frontal bone, and the absence of a supratoral sulcus. (After Santa Luca 1980.)

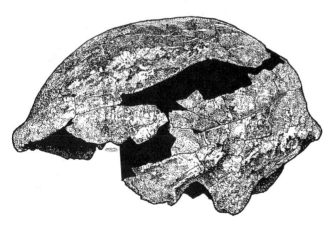

FIGURE 6.25

The Ngandong 9 (Java, Indonesia) cranium in lateral view. The features of this specimen closely resemble those noted for Ngandong 5. (After Santa Luca 1980.)

The Ngandong remains have been generally attributed to the Late Pleistocene on the basis of the geology of the locality and the associated fauna. As Santa Luca (1980) notes, however, the taphonomic pattern and surface characteristics of the animal bones from Ngandong differ markedly from those of the humans. Many of the nonhuman bones were found in articulation, and they show no evidence of extensive water transport or sorting. The humans, on the other hand, exhibit several signs of both, including the pattern of skull-part preservation (only the most durable structures remain), the type of surface damage, and the disassociation of elements. Thus it does not appear that the fauna and humans derive originally from the same location, which makes it difficult to argue that they are contemporaneous. There is a uranium-series date on one Ngandong skull of >300 Kya and a fission-track date from the Notoporo Beds (often considered equivalent to Ngandong in age) of <250 Kya (Jacob 1978, Sémah 1984).

Some younger dates have been obtained using other chronometric dating techniques on faunal remains. Fossils from the base of the High Terrace where the crania were found have yielded a uranium-thorium date of 101 ± 10 Kya (Barstra et al. 1988). More recent and controversial ESR and uranium-series dates on nonhuman teeth from the site suggest that the Ngandong remains fall between 53 and 27 Kya (Swisher et al. 1996). Since the actual site is gone, the contemporaneity of these teeth with the human remains is open to question, as noted by Santa Luca. Grün and Thorne (1997) also point out that the human fossils are heavily mineralized and have a ceramic-like appearance. They also have a consistent, distinctive, coffee-brown coloration. The animal bones associated with the dated teeth are much less dense, and have a gray color with manganese stains. These differences, and the taphonomic evidence cited by Santa Luca, strongly suggest that the human and faunal elements may derive from different depositional environments. Although it is still not clear whether the sample is very recent or 200–300 Ky old, there is little convincing evidence favoring the very recent dates.

The Nagwi specimen was an accidental discovery and lacks a definitive geological context. Morphometric analysis of the specimen points to stronger affinities with the Ngandong sample than with earlier Erectines (Widianto and Zeitoun 2003). Nagwi's lack of a datable context is particularly unfortunate, as it might have helped to answer the questions surrounding the geological age of the Ngandong sample.

The morphology of the Ngandong and Nagwi (hereafter just "Ngandong") crania is distinctive. Their frontal bones have long, relatively flat squamae. The impressive supraorbital tori form a continuous, relatively straight bar across the orbits; and, as already noted, the lateral part of the torus is often the most pronounced (Fig. 6.22). The cranial vaults are low and angular. The nuchal-muscle attachments and nuchal tori in back are strikingly robust. These features make the Ngandong sample stand out from all other Middle or Late Pleistocene hominins. There are strong similarities between them and earlier Erectine remains in the overall form of the cranium (Santa Luca 1978)—especially in side view—and in the morphology of their endocranial casts (Holloway 1982). All these features support the view of the Ngandong humans as late-surviving Erectines. But in spite of their striking robustness and primitive appearance, the Ngandong skulls exhibit advances related to brain and braincase expansion. Although Nagwi's cranial capacity is only 847 cm^3, the average for the rest of the Ngandong sample is 1149 cm^3 (range 1013-1251: Holloway 1980; Table 6.2). This is a 20% increase over the mid-sex average of 954 cm^3 for the Erectines from the Kabuh Formation (Holloway 1982). The Ngandong cranial capacities might be considered small for Late Pleistocene hominins, even if the earlier dates are correct. But it must be remembered that even in recent humans, cranial capacities in tropical populations tend to congregate toward the small end of the modern range, due to thermodynamic effects of climate and latitude (Chapter 8). Although the Ngandong crania exhibit more pronounced nuchal tori than typical Erectines (Fig. 6.25), the relative length of their nuchal planes is smaller (Table 5.2). In Erectines, the nuchal plane is always larger than the occipital plane; Ngandong tends toward the reverse proportion, found in Heidelbergs and other late Pleistocene humans. The pneumatization of the temporal bone in the Ngandong crania is also similar to the modern pattern, rather than to that seen in the Erectines (Balzeau and Grimaud-Hervé 2006).

One of the two tibias (Ngandong 13) is relatively complete. It is short like the Boxgrove tibia, but more slender. The other (Ngandong 12) is less complete, preserving much of the shaft of a far more robust individual. Ngandong 13 does not appear to have had a very "tropical" body build, but the fragmentary evidence makes it impossible to be sure. The Ngandong 17 hipbone fragment is similar to African and European hipbones dating between ~2 Mya and 100 Kya (Marchal 1999).

Several studies have been devoted to the Ngandong fossils. Weidenreich (1946, 1951) saw the Ngandong humans as a morphological link between Indonesian Erectines and some modern Australians. For him, they occupied a grade corresponding to European Neandertals. Von Koenigswald (1958) went so far as to dub them "tropical Neandertals." Santa Luca (1980) undertook a cladistic analysis of the skulls and concluded that they were strongly linked to earlier Indonesian Erectines. He argued that the Ngandong sample should be included taxonomically within *Homo erectus*, a conclusion reached by many other authors both before and since (Howells 1959, Coon 1962, Rightmire 1990, 1992, 1994;

Klein 1999, Tattersall and Schwartz 2001). Traits that Santa Luca sees as uniting Ngandong with *H. erectus* include a horizontal supraorbital torus, the presence of an angular torus (a thickened ridge on the posterior/inferior aspect of the parietal), the conjoined mastoid and supramastoid crests, and the form of the nuchal torus. An alternative view, grounded in the MRE perspective, sees most of the features connecting Ngandong to earlier Indonesian Erectines as regional Australasian characteristics, not as defining features of *H. erectus* (Smith 1985, Wolpoff 1999). Supposed regional features shared by these Javan populations include the supraorbital torus form, the flat frontal squamae, frontal-bone keeling, and the presence of a **prebregmatic eminence** (a cross-like elevation at the intersection of the coronal and sagittal sutures). The expansion of the brain in the Ngandong humans, and the concomitant changes seen in their vault morphology, also argue against grouping them with the Erectines. But however one interprets them, there is general agreement that the morphological similarities of the Ngandong sample to earlier Javan Erectines reflect regional continuity in the hominin populations spanning this time range in Australasia.

Liang Bua

If the idea of possible Erectines surviving at Ngandong until 27 Kya seems surprising, the story of the pint-sized hominins at Liang Bua Cave, on the Indonesian island of Flores, is mind-boggling! Brown et al. (2004) described a partial skeleton of a remarkable specimen, Liang Bua 1 (LB 1), including a cranium and mandible, a fragmentary pelvis and right femur, a right tibia, and other bones of an adult individual. Postcranial elements of other individuals were recovered, but only the LB 1 specimen has a skull. The next year, a second mandible (LB 6), additional postcranial remains of LB 1 (right humerus and ulna, left fibula and ulna), and other postcranial bones were described (Morwood et al. 2005). These discoveries have elicited strong opinions and fierce controversy, but it is not yet clear who these people were or what explains their unique morphology.

The cranium of LB 1 (Fig. 6.26) exhibits a moderately developed supraorbital torus, facial prognathism (particularly alveolar prognathism), and a low cranial vault. The maximum breadth of the vault lies at the level of the cranial base. In the frontal view, prominent ridges can be seen running up the side of the nasal aperture. Unfortunately, the mid-face is missing, but the overall facial form of LB 1 closely approaches that of the small early Erectine D2700 from Dmanisi (Fig. 5.27). Multivariate metrical analysis (Nevell et al. 2007) and morphological comparisons of the vault show strong similarities to Erectines. The mandible has a retreating

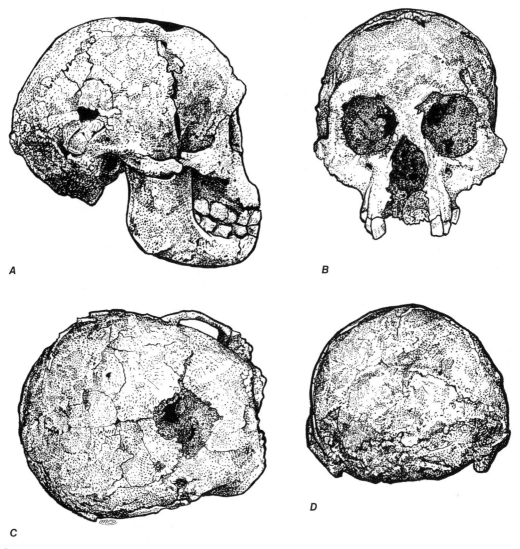

FIGURE 6.26

The Liang Bua 1 skull from Flores, Indonesia: lateral (**A**), frontal (**B**), occipital (**C**), and superior (**D**) views. Note that some critical areas of the cranium are missing. Similarities to Erectines are noted in such features as the receding symphysis and broad ramus of the mandible, the pattern of (total facial) prognathism, the form of the lateral supra-orbital torus, and the shape of the vault as seen from the rear. (After Brown et al. 2004.)

symphysis, with no evidence of a chin. It lacks a retro-molar space and has a relatively broad ramus.

Brown and his colleagues initially considered LB 1 to be very similar in morphology to Indonesian Erectines, though much smaller. But they also noted similarities to *Australopithecus* in the pelvis and hip: a marked iliac flare, a large bicondylar angle of the femur, and a long femoral neck with a low neck-shaft angle. The subsequent discovery of the right arm bones for LB 1 reinforced this picture. The humerofemoral index of LB 1 is 85.4, which falls outside the recent human range but

is identical to the index for the "Lucy" skeleton of *Australopithecus afarensis* (Morwood et al. 2005) and that for *A. garhi* (Argue et al. 2006). LB 1's tibia and a second tibia from another individual (LB 8) share an unusual midshaft cross-sectional shape; like chimpanzee tibias, they lack a distinct anterior margin (the sharp edge at the front of the human shin). However, this may be an allometric feature, reflecting the uniquely small size of these people. Liang Bua 1 stood only about 3′7″ (1.09 m) in height, smaller than any other known adult hominin. By inference from the dimensions of the tibia, LB 8 appears to have been even shorter, around 3′4″ (1.01 m) in stature. On the basis of the new postcranial evidence and further analysis of the LB 1 skull, Brown et al. (2004) repudiated their previous stand and argued that the Liang Bua people were probably not derived from Erectines. Given the primitive features of the postcranium, they suggested that these people might be descended from a radiation of even more primitive hominins into Australasia.

In addition to its astonishingly small stature and unique postcranial morphology, LB 1 is also remarkable in having a cranial capacity of just 417 cm³—smaller than those of almost any *Australopithecus* (Falk et al. 2005). Even more amazing is the fact that LB 1 apparently lived only 18,000 years ago! This date, derived from AMS radiocarbon dating of two charcoal samples associated with the LB 1 skeleton, is supported by thermoluminescence dates (Morwood et al. 2004). More recent dates based on a child's left radius (LB 4) imply that some of the Liang Bua hominin remains are no more than 12 Ky old (Morwood et al. 2005), and Morwood et al. (2004) have suggested that still other Liang Bua specimens may be as old as 95 Ky. If all these dates are correct, Liang Bua preserves a surprisingly long record of these diminutive people from a single site.

Fauna associated with the Liang Bua hominins include a variety of small vertebrates (some with charred bones), the giant lizard called the Komodo "dragon" (*Varanus komodoensis*), another large lizard, and a dwarfed species of elephant (*Stegodon*). The dwarf elephants are represented by 17 individuals, most of which are juveniles. It has been widely assumed that all these bones represent prey hunted by the Liang Bua people. The excavations at Liang Bua have yielded a series of relatively sophisticated stone tools, including various blade tools (Morwood et al. 2004). It is surprising that such tools could be produced by a hominin that has one of the smallest cranial capacities known for of any member of the human tribe. If these tools were made by people with skulls like that of LB 1, it becomes hard to argue that there is any linkage between human tool-making ability and brain size. However, only a few of these tools are associated with the hominin skeleton, while a large number were found with *Stegodon* remains in another part of the cave. Thus, the assumption that LB 1-type hominins were making the tools (and hunting the lizards and elephants) is open to question. Similar tool assemblages are found in association with modern humans at other Indonesian locations (Lahr and Foley 2004); and modern humans, not the Liang Bua people, may have been the ones who made and used them at Liang Bua as well.

Many experts continue to support the initial contention (Morwood et al. 1998, 1999) that the Liang Bua homininns represent a dwarf species derived from Erectines who reached Flores over 800 Kya. Islands often produce peculiar animals. Chance processes, analogous to founder effect and drift, ensure that few species colonize islands and that those that do are easily exterminated by random accidents. Small land masses are unable to support as many species as larger ones. As a result, island faunas are always taxonomically impoverished compared to those of continents. The small animals that reach islands often evolve into giant species that fill the ecological niches of the missing large herbivores and carnivores. We noted this phenomenon earlier in the case of the giant herbivorous birds and lemurs of Pleistocene Madagascar. The Komodo dragon furnishes a predatory example. Conversely, large-bodied animals that reach islands often become smaller, because the island does not provide enough resources for them to maintain adequate populations at their original body size. Such **insular** or **endemic dwarfs** include the tiny white-tailed deer of the Florida Keys and the dwarf elephants and mammoths found on Malta, Sicily, and the California Channel Islands during the Pleistocene. The restricted dietary resources of the small island of Flores make it a perfect place for endemic dwarfism to evolve, and the dwarfed *Stegodon* species found at Liang Bua demonstrates that this in fact occurred in some species of large mammals. There is no reason to think that Pleistocene *Homo* would have been exempt from these evolutionary pressures. The discovery of the LB 6 mandible and small postcranial elements from other hominin individuals provide persuasive reasons for thinking that a population of tiny endemic-dwarf humans inhabited Flores as recently as 10,000 B.C.

All this sounds convincing, but urgent and perplexing questions persist. How did the ancestors of these humans get to Flores in the first place? The island lies on the Australian side of Wallace's Line, and any terrestrial mammal probably would have had to cross at least 20 km of open water to reach Flores from Asia (Chapter 5). How could Erectines have managed this? Argue et al. (2006) suggest that local tectonic events may have opened dry-land connections across Wallace's Line from time to time, but there is no clear evidence for such land bridges. Did the cultural evolution that took place between the Soa Basin sites at ~800 Kya on Flores (Chapter 5) and the more advanced lithic assemblage at Liang Bua occur in isolation, or was there external influence? If the latter, did this have any biological impact on Flores homininns? Were the tools even made by LB-1-type humans? How were these homininns able to maintain an effective breeding population for as much as 800 Ky on Flores unless there was external input? And why is the braincase of LB 1 so tiny—far smaller than would be expected for a dwarf *Homo* of its body size (Schoenemann and Allen 2006, Martin et al. 2006a,b)?

These unanswered questions lead many experts to suspect that there is something fundamentally wrong with the Liang Bua evidence. Some believe that the LB 1 specimen is pathological. Certain features of its morphology, including the tiny braincase, the form of the supraorbital region, and the absence of detectable cranial sutures even in CT scans (Brown et al. 2004), suggest the pathological condition known as **microcephaly** (Gk., "small head"). Microcephaly is correlated with cranial stenosis (premature closure and obliteration of the cranial sutures), which results in an abnormally small cranial vault and brain and usually produces mental retardation. In microcephalic modern humans, there is generally a projecting, "beaked" appearance to the nose and midface (Ortner and Putschar 1981, p. 304). This part of the LB 1 cranium is missing (Fig. 6.26), but the surrounding anatomy does not seem to indicate a beaked midline. Moreover, microcephalics usually have a distinct chin, even when there is alveolar prognathism; yet the symphyseal region in LB 1 is clearly receding. It looks primitive, not pathological. There are dozens of clinical syndromes that involve microcephaly as a symptom, and some of them also involve small stature and abnormalities of the limbs, jaws, and teeth. Not all of these syndromes have been explored or ruled out as potential explanations for the unique morphology seen in the Liang Bua hominins. But proponents of the "microcephalic" interpretation of that morphology have not been able to point to a pathological modern skull that looks exactly like LB 1.

Much of the debate over microcephaly at Liang Bua has centered on brain morphology. An analysis of a virtual endocast of LB 1 by Falk et al. (2005; cf. Balter 2005, Falk et al. 2006, 2007) revealed expanded areas of the frontal lobes matched only in nonpathological modern endocasts, and an overall brain shape that differs markedly from that of the one microcephalic examined. A more detailed subsequent analysis shows that the LB 1 endocast lacks the projecting cerebellum and relatively narrow frontal lobes shared by nine microcephalics of various types and geographic origins (Falk et al. 2006). However, another study reports matching LB 1's endocast shape with that of a microcephalic individual (Weber et al. 2005). In a comparative study of the LB 1 endocast, Holloway et al. (2006) found features on the frontal and temporal lobe that are not present on normal modern or Erectine endocasts, and concluded that microcephaly cannot be ruled out. And as noted above, LB 1's brain/body ratio does not conform to allometric relationships established for either modern or ancient *Homo*. This finding implies either that LB 1 is the product of selection pressures unlike those affecting any other human population, or that the specimen is pathological.

The discovery of the second adult mandible (LB 6), which resembles LB 1 in size and many features of its morphology, is damaging to the microcephalic interpretation. The postcranial remains certainly show that all

these people were small; but the second jaw suggests that they all had small crania as well. These findings undermine the contention that LB 1 is a unique, pathological specimen (Lieberman 2005). Argue et al. (2006) used metric and morphological analyses to compare Liang Bua with australopithecines, Erectines (including Ergasters), and a varied sample of modern humans. They found no evidence that LB 1 is a microcephalic, at least of the one type that they examined. LB 1 also did not show any close similarities to either Erectines or normal recent humans, including such diminutive people as African pygmies and Andaman Islanders. A multivariate study by Nevell et al. (2007) yielded similar results: LB 1 differs from recent humans in the same ways that other archaic crania do, but it is not specifically like Asian Erectines. What all this means, however, is unclear. If LB 1 is anomalous because of abnormalities of growth and development, morphometric studies on it would not be expected to have much taxonomic significance (Eckhardt et al. 2007).

In support of the "pathological" interpretation of Liang Bua, Jacob et al. (2006) report that LB 1, despite its archaic-looking morphology, falls within the recent human range of variation in over 100 cranial features. They assert that the LB 1 specimen is characterized by "marked craniofacial and postcranial asymmetries and other indications of abnormal growth and development" (Jacob et al. 2006, p. 13421). Jacob and his co-workers argue that some of the unusual features of the specimen (negative chin and rotated premolars) are features often found in pygmoid people living in the Liang Bua region today. In their view, the Liang Bua people represent a pygmy population of *Homo sapiens*, and LB 1 is simply a pathological specimen of this population. This analysis has been strongly criticized by Larson et al. (2007), who claim that the limb proportions and stature of LB 1 lie completely outside the range of modern humans, including pygmoid populations from the Flores region and the smallest pygmoid populations worldwide. They also contend that Jacob and his team greatly exaggerated the asymmetries and supposed pathologies of the skeleton, and that LB 1's level of asymmetry falls within the normal human range.

If LB 1 is not a modern microcephalic dwarf, what is it? The Liang Bua hominins bear little specific resemblance to known Erectines or to modern humans of like antiquity. From the primitive skull features and small brain of LB 1, Lahr and Foley (2004) conclude that the LB 1 specimen is not a pygmy modern human; but they also conclude from the shape and proportion of the skull and teeth that it is certainly *Homo*. Nevertheless, they acknowledge some similarities to *Australopithecus* in the pelvis and femur. The unique limb proportions and hipbone morphology of the Liang Bua hominins, and the fact that their wrist morphology has been described as rather apelike (Tocheri et al. 2007), provide some support for the idea that they might have evolved from pre-Erectine ancestors (Brown et al. 2004, Falk et al. 2005, Argue et al. 2006).

Whether a new species or higher-level taxon is appropriate for these fossils remains to be seen. The body/brain size ratio of LB 1 is unlike anything else seen in the genus *Homo*, but unusually small brains are found in some other insular mammals. (We have already seen a similar explanation invoked to account for the small brain of *Oreopithecus*.) Long-term adaptation to tropical forest conditions, on both islands and continents, has resulted in the emergence of pygmy-like populations of modern humans in several widely separated parts of the world. Similar conditions might have produced similar results in a late-surviving population of Erectines, or of more advanced archaic types like the Ngandong people. For now, the dwarfed-Erectine hypothesis may seem the most plausible, but other ideas cannot yet be rejected, including the possibility of pathology. More than one of the competing explanations may contain elements of the truth. It is possible, for example, that the Liang Bua humans were a population of dwarfed Erectines *and* that the LB 1 specimen is a pathological variant of that population with an abnormally small brain.

If *Homo floresiensis* turns out to be a valid taxon, then at least one other species of hominin existed alongside modern humans (*Homo sapiens sensu stricto*). Much of the heat in the arguments about Liang Bua flows from the ongoing debate between the RAO and MRE theories of modern human origins. Some RAO partisans hope, and some MRE proponents fear, that recognizing a separate species of *Homo* on Flores will support giving separate species status to other late-surviving archaics, including the Ngandong people and the Neandertals. But both the fear and hope are unjustified. The occurrence of one speciation event does not affect the likelihood of another. Even if *H. floresiensis* is real, the MRE model could still be correct in all particulars. At most, the late persistence of archaic humans on Flores might make the survival of the Ngandong people to ~27 Kya on Java seem more plausible. But we do not need Flores to reach that conclusion, since we already know that "classic" Neandertals survived to almost that date in Europe alongside more modern humans in Africa (Chapter 8). The significance of the little people from Flores, and their relevance to the story of modern human origins, will become clear only with the recovery of more specimens. One skull is never enough!

▌ SUPRAORBITAL TORI, CHINS, AND PROJECTING FACES

In any anatomical comparison in which different features are being tallied and contrasted, there is a tendency to think of all the items in the tally as separate entities produced independently by evolutionary forces.

But in most cases, anatomical features change in concert with other features, either because they are correlated functionally or because they are **pleiotropies**—joint side effects of a single genetic change. For example, in this and the previous chapters we have repeatedly emphasized the expansion of the cranial vault in the course of human evolution. Several trends in the dimensions of the skull reflect this expansion: an increase in the parietal breadth of the braincase (relative to cranial-base breadth), a broadening of the frontal bone (both posteriorly at the coronal suture and just behind the eye sockets), and an increase in the height of the braincase. These trends begin in the Erectines and continue into the Heidelbergs. Other related trends, such as the expansion of the occipital plane of the occipital bone (relative to the nuchal plane), first show up in the Heidelbergs. But all of these changes are byproducts of the expansion of the cerebral cortex relative to the brainstem, which causes the upper part of the braincase to get bigger relative to the rest of the skull. Although they are often discussed independently, they represent different aspects of a single evolutionary trend.

Other correlated evolutionary trends in the human genus involve the skeleton of the face. Erectines and Heidelbergs are distinguished from modern humans by their massive, protruding supraorbital tori, their prognathic faces, and their receding, chinless mandibular symphyses. These features, too, are not independent of each other. In one way or another, the loss of all these primitive traits in modern *Homo sapiens* reflects the changing morphology of the teeth.

Fossil and living apes have projecting faces because their teeth and jaws are large and long relative to the size of their braincases. The jaws are large not only because the teeth themselves are large, but also because the roots that hold large teeth in the jaws must be very large as well. The canines of apes have particularly long roots, and their jaws need to be large to hold these roots. Because of the way mammalian faces grow (Enlow 1975), a larger upper face requires more growth in both forward and downward directions. Thus the big upper face needed to accommodate the large roots of big canine teeth must be both longer and more forwardly projecting (prognathic) than it would be if the canines were smaller. Throughout the evolution of the genus *Australopithecus*, there is a trend toward reduced prognathism, resulting partly from reduction of the anterior teeth. Erectines are even less prognathic than the more primitive species of *Australopithecus*, particularly in the alveolar region, but they retain the complex of total facial prognathism in a reduced form. Heidelbergs often show some reduction in *lateral* facial prognathism—in other words, their cheekbones are not positioned as far forward as they are in Erectines—but they can still be characterized as totally facially prognathic. Basically, this represents a retention of a primitive complex in both Erectines and Heidelbergs.

The persistent prognathism of all these early humans is the main reason why they lack chins. If the maxillary dentition is positioned generally more forward, then the mandibular dentition must also be positioned anteriorly to maintain proper occlusion. This is accomplished structurally by moving the mandibular alveolar process (the portion of the mandible in which the teeth are imbedded) more anteriorly relative to the ramus and the base of the mandibular corpus. One result of this is a symphyseal profile that inclines posteriorly from the superior part of the bone to the base—in other words, a receding mandibular symphysis. With such a profile, a basal anterior concentration of bone would not manifest itself as a chinlike projection, even if it were present structurally.

However, there is just as much functional need for reinforcement of the mandibular symphysis in chinless forms like the Erectines or Heidelbergs (or for that matter in *Australopithecus* or apes) as there is in modern humans with protuberant chins. Experimental studies show that stresses produced on the anthropoid mandible during chewing and biting tend to concentrate at the symphysis, making it a structural weak point (Hylander 1984). In large anthropoids, the symphysis is buttressed to withstand these stresses. In the primitive hominid arrangement, seen in modern apes and retained in early *Australopithecus* (Fig. 4.48), the symphysis slopes downward and backward from the projecting lower teeth toward a basal thickening, which is continuous with the body of the mandible. One or two transverse bars of bone called **transverse mandibular tori** may run across the symphysis in back to increase its structural strength. (An inferior transverse torus that juts backward beyond the rest of the symphysis is called **a simian shelf**.) Throughout the evolution of the genus *Homo*, the teeth and their roots become progressively smaller. As a result, the face grows less prognathic and the upper dentition gradually shifts backward toward the braincase. The lower teeth and their alveoli of course follow the upper teeth, to maintain occlusion. But the mandibular symphysis remains stout, and its basal, non-alveolar part does not move backward as far or as fast as the lower teeth do. (The less marked retreat of the lower part of the symphysis may reflect a need to maintain enough space inside the mandibular arch to accommodate the tongue muscles that originate from its inner surface.) Because the lower part of the symphysis retreats more slowly than the upper part, the symphysis becomes more vertically oriented as midsagittal prognathism diminishes. In the final stages of this process, the thick basal part of the symphysis is left sticking out in front, forming a **mental eminence** or bony chin. Interestingly, Dobson and Trinkaus (2002) have shown that

Middle and Late Pleistocene mandibles lacking chins are just as well equipped to resist the stresses at the mandibular symphysis as are later mandibles with chins. The process is not quite so simple as this description implies, and the modern human chin has some anatomical features that have nothing to do with the structural integrity of the symphysis (Chapter 8). But the essential story is the one told above: the teeth retreat backward until the bony buttress of the symphysis is left jutting out beneath them.

The significance of the supraorbital torus, which Schaaffhausen (1858) described as a "most remarkable peculiarity," is more obscure. This projecting bar or double arch of bone across the tops of the eye sockets is unique to anthropoids, but it is not seen in the skulls of all anthropoids (Fig. 3.43E). It is a conspicuous difference between the skulls of early and modern *Homo*, although tori of a modest size occur in some modern human populations (Russell 1985, Vinyard and Smith 2001). The tori of some hominoids, including gibbons, orangutans, some *Australopithecus*, and even some Ergasters (e.g., ER 3733), are little more than thin shelves or ridges (Figs. 3.47, 4.4, 5.16). However, the supraorbital tori of chimpanzees, gorillas, and most Erectines and Heidelbergs are very thick and visibly prominent (Figs. 3.40A, 3.37, 5.5, 5.8, 6.5 and 6.16).

Many explanations of the distribution and anatomical variation of supraorbital tori have been offered (Smith and Ranyard 1980, Russell 1985), but none is entirely satisfactory. Some of the proposed functions of the torus, such as keeping hair out of the eyes or providing sunshades, are just silly. Several authors, from Schaaffhausen (1858) to Tappen (1973, 1980), have explained the supraorbital torus as a sort of visor that evolved to protect the eyes and brain from blows to the head, supposedly as an adaptation to the rigors and perils of the early human way of life. Weidenreich (1943, 1946), for example, viewed the supraorbital torus as a part of an overall buttressing system that reinforced the entire skull against traumatic injuries. This view owes a lot to the image of early humans as violent, club-swinging cavemen. That image is not wholly unwarranted. More than one early human skull does in fact show signs of healed injuries to the bony ridges surrounding the eye. But the protective-buttress theory does not explain most of the facts about the torus and its distribution. It does not tell us, for example, why gorillas have huge, projecting tori while orangutans get by with delicate supraorbital ridges. It also does not explain why the torus is greatly reduced or absent in our own skulls. After all, people have not stopped hitting each other in the head as their brains have increased in size. On the contrary, those enlarged brains have allowed us to invent ever more ingenious and damaging methods of hitting one another in the head; and the increased volume of the modern human brain makes it more susceptible to traumatic injury, simply as a corollary of the square-cube law.

Some have tried to explain the massive supraorbital tori of Middle and Late Pleistocene hominins as buttresses of a different sort, evolved to protect the integrity of the skull against its own internal forces generated by the jaw muscles during chewing and biting. The most cited of these explanations is the bent-beam model proposed by Russell (1985). Russell suggests that strains generated from chewing and biting, especially on the front teeth, are concentrated in the supraorbital region. In her model, the torus acts as a load-bearing beam across the top of the eye sockets to resist the downward pull of the jaw muscles below and behind the orbits. Unfortunately for Russell's theory, experimental studies of mastication in nonhuman primates do not show the predicted pattern of bone strain in the supraorbital region (Hylander et al. 1991a,b; Vinyard and Smith 2001).

Moss and Young (1960) hypothesize that in primate skulls that combine prognathic faces with small, low cranial vaults and receding foreheads, the supraorbital torus is needed to provide a skeletal bridge between the orbit and the braincase. In a primitive mammal, whose orbits face more laterally (Fig. 3.17C), a protruding torus is not needed, because the upper edge of the orbit more or less follows the contours of the side wall of the snout and braincase. In a modern human skull (Figs. 3.34, 3.36B, and 8.1), a torus is also unnecessary, because the expanded frontal end of the braincase juts forward over the orbits and covers them. But in a chimpanzee, where the forward-facing orbit projects far beyond the front end of the brain (Fig. 3.35A), a skeletal superstructure is needed to protect the eyeball in this exposed position. This need is filled by inflating the frontal sinus to produce an air-filled bar of bone projecting above the eye sockets (Fig. 3.36A)—the supraorbital torus.

This so-called **spatial model** of the torus's function sounds plausible enough, and it is supported by metric and experimental studies (Ravosa 1988, 1991a,b; Vinyard and Smith 2001, Fiscella and Smith 2006). But questions persist. Why does the orbit of a chimpanzee have to stick out far beyond the front end of the braincase? The eye would work just as well if it were pulled back underneath the frontal end of the brain—as it is in orangutans, which lack a projecting torus (Fig. 3.47C,D). No doubt, an eye that protrudes far beyond the braincase needs a roof of some sort; but this answer begs the question. Neither visual function nor the biomechanics of mastication requires that the eye of a chimpanzee must lie far anterior to the braincase, or that the opening of the orbital cone has to face directly forward. We can appreciate this latter point if we look at the skull of a cat, whose eyes face directly forward but whose orbits face more upward and sideways, tilted at an angle of about $45°$ to the midline plane of the head.

One missing piece in this three-dimensional jigsaw puzzle is the anterior part of the temporalis muscle. A more vertical face generally correlates with a relatively larger anterior temporalis; and one way of making space for a larger anterior temporalis is to shift the orbit forward relative to the braincase. Some such difference in the architecture of the facial skeleton and the jaw muscles may explain the differences between gracile and robust *Australopithecus* in the protrusion of the supraorbital region (Fig. 4.32). It may help to explain at least part of the difference between the tori of chimpanzees and orangutans as well. But it does not appear to be helpful in explaining the spectacularly big tori of such beetle-browed early *Homo* skulls as Ceprano, Kabwe, and Petralona, whose brow ridges stick out much further in front of the temporal fossa than they do in front of the braincase (Figs. 5.29, 6.5, and 6.16).

The functional and developmental causes that contributed to the variable development of our ancestors' protruding brow ridges are not yet clear. In 1980, Smith and Ranyard concluded that although the main function of the torus is to provide a structural bridge between the orbit and the braincase, other factors—particularly biomechanical ones—probably impact its development and overall form. This is still about all we can say for sure. But we can say with some confidence that those ridges have been reduced to vestigial bumps in our own skulls because their protective and load-bearing functions were taken over by the expanding forehead, which moved forward and grew more vertical as the brain grew forward above the orbits and the facial skeleton shrank and withdrew beneath the braincase. The later stages in the evolution of the human supraorbital torus are secondary effects of both brain enlargement and reduced prognathism.

MAJOR ISSUES: SPECIATION, MIGRATION, AND REGIONAL CONTINUITY

The fossils discussed in this chapter represent the first humans with an average level of encephalization that falls into the modern human range. All the experts agree that these Heidelbergs are descended from the earlier, smaller-brained Erectines surveyed in Chapter 5. It is also generally agreed that in at least parts of the Old World, the local Heidelbergs represent intermediate stages in evolutionary lineages leading from the local Erectines to later regional populations. In Europe, for example, the earliest Heidelbergs (Petralona, Mauer, Bilzingsleben, Vértesszöllös) are the most Erectine-like, whereas Neandertal apomorphies appear more frequently in later specimens (~400–200 Kya). Scientists with such widely disparate views as Wolpoff (1980a,b, 1999) and Hublin (1982, 1988a,b) concur in seeing these European Heidelbergs as transitional forms between

Erectines and the Neandertals. Several authors (Stringer et al. 1979, Dean et al. 1998) break the European Heidelbergs and Neandertals into a sequence of temporal and evolutionary grades leading toward full "Neandertalhood" (Table 6.8).

The Middle Pleistocene fossils from China and Australasia are generally held to show signs of regional continuity with earlier Erectine populations in their respective regions. But the supposed markers of regional continuity are used in very different ways by proponents of different models. In Java, for example, the resemblances between the Kabuh and Ngandong samples are invoked by proponents of replacement models as grounds for classifying Ngandong as late-surviving *Homo erectus*. Others see these features simply as signs of regional continuity, and regard Ngandong as an East Asian archaic *Homo sapiens* (or perhaps a local form of Heidelberg). But neither school of thought questions the descent of the Ngandong hominins from earlier Javan Erectines. The Chinese material provokes similar agreements and conflicts. Although possibly more advanced than typical Erectines, the later Zhoukoudian remains—along with those from Hexian and Yunxian—may, in fact, be best considered as Erectines rather than Heidelbergs. Here again, their lineal and genetic continuity with earlier Chinese Erectines is generally conceded. The points at issue concern the connections between these regional lineages and their modern successors. We will take these issues up in the last two chapters of this book.

The markers of regional continuity are different in Europe, China, and Australasia. For example, a flattened frontal squama and laterally thickened supraorbital tori are characteristic of all the Javan fossils, whereas Neandertal features (e.g., suprainiac fossae, occipital bunning, and various aspects of Neandertal facial form) are more distinctively European. There are no specifically regional features that demonstrate the descent of the African Heidelbergs from African Erectines. Nevertheless, the Africans are generally distinguished by the absence of the distinctive features found in other regional populations. All these facts go to show that morphologically distinct regional populations—variously described as geographical races, paleodemes, semispecies, or species—were in place throughout most of the Old World by this time in human evolutionary history.

While there are regional differences in morphological details, some general trends are manifest everywhere: brains get bigger, cranial vaults expand, and faces get smaller. Explanations of this fact reflect differing views of human evolutionary history during the Middle and Late Pleistocene. On the one hand, Tattersall (1996) argues that these trends could easily have gone on in parallel in separate regional species of *Homo*. Such closely related species would have had very similar genomes and morphology, and thus would be expected

to have responded to similar selection pressures in similar ways. In this case, body size would probably also have had to change in parallel in the separate regional lineages (Table 6.1). All this is certainly possible. Parallel evolution is more common in later human evolutionary history than is generally thought (Trinkaus 1992), and it may have been at work in this instance. Although brain size and skull form changed in similar directions in different regions, they apparently did so at different speeds, with eastern Asia and Australasia lagging somewhat behind the rest of the Old World. This disparity can be interpreted as a signal of parallel evolution, proceeding at different rates in reproductively isolated regional populations. But an obvious alternative interpretation is that similar evolutionary trends in different regions reflect the interconnection of the regional populations through gene flow. By this interpretation, different rates of change in different regions might reflect local differences in selection pressures, or the additional time needed for new mutations to spread from one end of the Old World to the other.

There is a third possibility. It may be that the Heidelbergs, like the Erectines, had a particular region of origin, and that their eventual distribution across the entire Old World resulted from population movements out of this source area. Templeton's (2005) interpretation of modern human DNA haplotypes (Chapter 8) points to a wave of migrations from Africa around 650 Kya (Fig. 8.36). Such a dispersal event might reflect the spread of Heidelbergs from a homeland in Africa. If this is what happened, then Zhoukoudian, Ngandong, and other late Erectine-like forms with certain "progressive" traits may represent the results of interbreeding between indigenous Erectines and incoming African migrants. However, there is little in the way of fossil evidence that supports a migration out of Africa at ~650 Kya, with the possible exception of the Kocabaş anterior calotte from Turkey (Kappelman et al. 2008).

There is general agreement that early *Homo* (Ergasters) originated in Africa and then spread into Eurasia around 1.8 Mya. Most paleoanthropologists (but not all) think that this "Out of Africa 1" event was followed by a similar "Out of Africa 2" dispersal of modern humans from Africa around 100–130 Kya. Templeton's ideas imply that this last dispersal was in fact "Out of Africa 3," and that there was an additional, intermediate wave of out-migration from Africa that spread Heidelberg-grade morphology across the Old World. As noted earlier, any lingering trace of this second dispersal event in the DNA data would imply that the third dispersal must have involved admixture between the African invaders and more primitive indigenous populations. Unfortunately, the fossils we have from this period are too few and too imprecisely dated to refute any of these alternative theories about their population biology. More information is needed—on both ancient bones and modern genes—before we can say whether the new morphologies seen in these fossils originated through parallel evolution, gene flow, or migration with admixture.

There is also general agreement that the large-brained, big-faced humans (Heidelbergs) known from the Middle Pleistocene of both Europe and Africa were the ancestors of their immediate successors in those two regions. In 1979, C. B. Stringer and colleagues (Stringer et al. 1979) suggested a three-stage grade structure for the evolution of recent humans. In their scheme, most Heidelbergs fall into Grade 1. A few of the specimens described in this chapter are regarded as transitional to Grade 2, which includes early Neandertals in Europe (e.g., Saccopastore and Ehringsdorf) and some primitive-looking "modern" specimens from Africa (e.g., Florisbad and Omo 2). Zuttiyeh, which could be considered a Heidelberg, is placed on the line between Grades 2 and 3. Grade 3 is divided into two subgroups: "classic" Neandertals from Europe and the Near East (Grade 3a), and early modern humans (Grade 3b). Although no specific taxonomy is suggested, the implication is that a single interbreeding population evolved from Grade 1 to Grade 2 and then split into two new species: the Neandertals in Europe, and early modern humans in Africa.

More recently, Hublin (1998) and Dean et al. (1998) have offered a similar but more detailed picture of human evolution in Europe. In this version, which they call the "Accretion Model" (Table 6.9), four stages are recognized. Stage 1 (early pre-Neandertals) and Stage 2

(pre-Neandertals) encompass the European Heidelbergs. Stage 3 is basically the same as Stringer's Grade 2, and Stage 4 comprises the "classic" Neandertals. These stages progress through gradual accumulation or accretion of Neandertal features, beginning with facial features and then extending to the rear and base of the vault, until full Neandertal morphology is attained in Stage 4. The distinctive character of this European lineage is explained as the result of genetic drift acting on a small founder population that was geographically isolated from the rest of the Old World. That isolation would have been created by the onset of glacial conditions, probably at the peak of isotope stage 12, when Europe was walled off by glaciers and tundra to the north and east and by other sorts of barriers around the Mediterranean to the south.

Hawks and Wolpoff (2001a) recognize the existence of a distinctive European lineage leading to the Neandertals in Europe, but deny that it was isolated from other human populations. Hawks and Wolpoff examined six metric traits supposed in the Accretion model to be distinctive of the European lineage: mastoid height, nasal breadth, orbit shape, foramen-magnum shape, lambdoid-depression index, and relative maxillary-incisor breadth. If Europeans were isolated from the remainder of the world, then European fossils could be expected to show significant differences from comparable non-European population samples in these metric features. But Hawks and Wolpoff found such differences in only 22% of their comparisons, mostly in mastoid height. They also found that variation in the European

TABLE 6.9 ■ The Four-Stage Neandertal Accretion Model

Stage Number, "Old" Names	Derived Features Present	Specimens
1. Early Pre-Neandertals, early archaic *H. sapiens*	Convex, receding infraorbital profile, wide nuchal torus.	Arago, Mauer, Petralona
2. Pre-Neandertals, late archaic *H. sapiens*	Bilaterally protruding nuchal torus, incipient to well-defined suprainiac fossa, strong juxtamastoid eminence, incipient *en bombe* shape, occipital plane convexity, double-arched supraorbital torus, incipient midfacial prognathism.	Bilzingsleben,[a] Vértesszőllős,[a] Atapuerca SH, Swanscombe, Reilingen
3. Early Neandertals	Full suprainiac fossa, *en bombe* shape from rear, small mastoids, elongate temporal bone, anterior mastoid tubercle, increased dolichocephaly, more occipital convexity.	Ehringsdorf, Biache I, La Chaise Suard, Lazaret, La Chaise Abri Bourgois-Delaunay, Saccopastore, Krapina,[b] Shanidar[c]
4. "Classic" Neandertal	Exaggerated occipital plane convexity, suprainiac fossa, high midfacial prognathism, large piriform aperture, rounded circumorbital morphology, post-toral sulcus deepens.	Neandertal, Spy, Monte Circeo, Gibraltar Forbes Quarry,[d] La Chapelle-aux- Saints, La Quina, La Ferrassie, Shanidar (late group), Le Moustier, Amud

[a]Morphologically most similar to stage 1; placed in stage 2 because of dating.
[b]Most of the Krapina sample belongs here; also hard to judge because of fragmentary state or remains.
[c]Early Shanidar group; also often hard to judge due to fragmentation.
[d]Some morphology hard to judge.
Source: Adapted from Dean et al. (1998).

sample did not systematically decrease over time (a trend that some supporters of the Accretion model associate with isolation), and that many features did not follow the accretion pattern of becoming more Neandertal-like over time.

Hawks and Wolpoff assert that their data do not support the level of European isolation required by the Accretion model, but instead show that the European "lineage" was connected to other regional populations via gene flow during this period. They reject the idea that a unique process of "accretion" has to be invoked to explain the European pattern of morphological evolution. An additional problem with the Accretion model is that the remains listed do not always fall clearly in the specific stage indicated. For example, it has been argued for many years that the early Neandertals from Krapina (Stage 3) are no less "classic" in morphology than the later specimens placed in Stage 4 of the Accretion model (Smith 1976, 1982).

Is *Homo heidelbergensis* a valid taxon? This is not an easy question, and opinions vary widely. G. P. Rightmire (1990, 1996, 1998) believes that this species originated in Africa and that it includes both the European and African Heidelbergs (but not the Asian forms) discussed above. His concept of *H. heidelbergensis* as the last common ancestor of modern humans and Neandertals has gained wide acceptance. Rightmire also attempts the most comprehensive diagnosis yet offered for this taxon, based primarily on the morphology of the Bodo skull (Rightmire 1996). He lists the following as diagnostic features of the species:

1. Thick cranial bones.
2. Large, projecting supraorbital torus.
3. Midline keeling and bregmatic eminence.
4. Angular torus.
5. Position of the petrous portion of the temporal and a crevice-like *foramen lacerum* (a persistent bit of unossified basicranial cartilage at the front end of the petrosal).
6. Low braincase and flattened frontal profile.
7. Expanded cranial capacity.
8. Broader mid-vault, with incipient parietal bossing.
9. Division of the supraorbital torus at midorbit, with reduction of the lateral part of the torus.
10. High contour of the temporal squama.
11. Incisive canal running vertically (through the roof of the mouth).
12. Vertical sides of piriform aperture (the external nose hole in the skull).

The first six of these features are common in Erectines, while the others are more "advanced" character states. But none of these 12 traits is unique to Heidelbergs. As Rightmire (1998) clearly states, it is the combination of features that defines the group. A quick review of all the specimens described in this chapter will show that there are no unique defining characters for Heidelbergs—just various mosaics of features that are not the same for all regions, or even for all specimens within a region. Taxa of this sort, defined by possessing some variable and unspecified preponderance of the properties on a list, are sometimes referred to as **polythetic** (Beckner 1959). The polythetic character of "*H. heidelbergensis*" is not necessarily grounds for rejecting it as a valid species. Most extant species are polythetic groupings. (If you doubt this, try making a list of properties that are both necessary and sufficient for membership in the species *Homo sapiens*.) But such groupings do not conform to the requirements of cladistic systematics, which demand that every taxon be defined by unique synapomorphies.

If we forget about the rules of cladistics and accept the use of grade-based taxa, then there is nothing wrong in principle with defining *H. heidelbergensis* as an intermediate evolutionary grade, separated from an ancestral species defined by symplesiomorphies and one or more descendant species defined by synapomorphies. But it is not clear that such a taxon is any more biologically meaningful than the vague concept of "archaic *H. sapiens*." It is also not clear that the species *H. heidelbergensis* has any practical utility in classification. For example, any line drawn between Heidelbergs and Neandertals in Europe will always be somewhat arbitrary. Do Swanscombe and Sima de los Huesos belong with the Heidelbergs or with the Neandertals? They certainly have features in common with other Heidelbergs and with Neandertals, but they are not typical representatives of either. What, apart from a false semblance of precision, do we gain by drawing species boundaries across a polythetic spectrum of transitional forms? Perhaps a vague label is the most appropriate one for a vague category.

Tattersall and Schwartz (2001) argue that the taxon *H. heidelbergensis* deserves to be accepted even if it is not precisely delimited in terms of either defining characteristics (synapomorphies) or membership. To do otherwise, they claim, would obscure the "bushiness" of human evolution and foster old-fashioned unilineal thinking about hominin phylogeny. They insist that getting the alpha taxonomy right is necessary before moving on to more complex phylogenetic inferences. But it is well to remember that imprecisely defined "species" are not necessarily going to lead to improved accuracy in the later stages of analysis. Tattersall (1986) further argues that formal Linnaean names like *H. heidelbergensis* should be used in place of descriptive labels like "archaic *H. sapiens*" in order to bring paleoanthropological terminology in line with general biological practice. This is a legitimate consideration, but we feel that the conventions of biological nomenclature ought to have a lesser claim on biologists than the facts of biology. We see no reason to think that recognizing more species is always a better reflection of biological reality in human evolution (cf. Smith 1994).

If *H. heidelbergensis* is accepted, it is perhaps best thought of as a chronospecies, defined more by a span of time than by a particular set of properties. Bermúdez de Castro et al. (1997) adopt such a definition for this taxon, which they restrict to the European Pre-Neandertals. But again, it is not clear that defining a taxon on this basis has any specifically biological (as opposed to geological) meaning. For this reason, and because there is no agreement on the taxon's definition or membership, we have grouped the human fossils from this time span under the informal heading of "Heidelbergs." Used in this way, "Heidelberg" represents a non-taxonomic heuristic category that identifies certain archaic *H. sapiens* in much the same way as the term "Neandertal" identifies specific European and West Asian samples. It provides adequate precision of reference without committing us to any hypotheses about species boundaries. Specifically, it comprises the set of all fossil hominins that exhibit modern human levels of brain size (and correlated characters reflecting cranial-vault expansion) while retaining primitive traits otherwise characteristic of smaller-brained early *Homo* (the Erectines). As such, the term "Heidelberg" can be used with little hesitation for the European, Near Eastern, and African specimens discussed in this chapter, probably including those from North Africa. Whether the term should be extended to samples and specimens further east in Asia is a matter of choice; but in making that choice, we need to recognize that those Asian remains do exhibit some of the same trends (brain and vault expansion) that characterize the European and African specimens.

CHAPTER SEVEN

Talking Apes: The Neandertals

More than once in this book, we have laid out the pros and cons of some disputed point and then gravely declared that more fossils need to be found to settle the matter. The imperfections of the paleontological and archaeological evidence limit our ability to answer many of the questions we have about the human lineage. One might think that if only we had a really good record for some period and region—the remains of hundreds of human individuals of various ages and both sexes, including some complete skeletons, maybe even preserving some actual DNA, found in association with thousands of animal bones and artifacts documenting different sorts of activity at different sites—then, we might be able to find answers to all of the fundamental questions about ancient human biology and behavior that have baffled us under less ideal conditions. Or so one might think.

The case of the Neandertals should give us cause for reflection. We have all these things for the Neandertals, and yet the fundamental questions persist. More and better-preserved fossils, better contextual and chronological information, and much richer behavioral data are available for these ancient aborigines of Europe than for

any earlier population of fossil hominins. We even have some limited knowledge of Neandertal genetic structure. But the same large questions that were fought over in Darwin's time—Were the Neandertals members of our own species? Are any living humans descended from Neandertals? Did they have language and other uniquely human capacities of thought and behavior?—continue to be debated in scientific meetings today as hotly as they were 140 years ago.

Throughout the first hundred or more of those 140 years, the Neandertals played a central role in scientific discussions of the later phases of human evolution. They took on a special importance from the very start of those discussions in the 1860s, simply because they dwelt in Europe—which is where the scientific study of prehistory began and where prehistorians first started digging into the earth in search of ancient human relics. Neandertal fossils from Europe (Table 7.1) were recognized and described long before any ancient hominins were found elsewhere (on Java in 1891). Early findings and suppositions about Neandertals made a correspondingly deep initial impression on both popular and scien-

The Human Lineage, By Matt Cartmill and Fred H. Smith.
Copyright © 2009 John Wiley & Sons, Inc.

TABLE 7.1 ■ Neandertal Fossil Discoveries, 1829–1909

Site/Specimen (Country)	Year[a]	Context[b]	Description
Engis 2 (Belgium)[c,d]	1829/30	F, P, (?)	Child's cranium with several teeth
Forbes Quarry (Gibraltar)[d]	1848	N	Adult cranium
Kleine Feldhofer Grotte (Germany)	1856	N[e]	Calotte, 15 postcranial elements[f]
La Naulette (Belgium)	1866	F, P	Adult, edentulous mandible
Šipka (Czech Republic)[g]	1880	F, P	Partial mandible of a child
Spy (Belgium)	1886	F, P	Skulls and postcrania of 2 adults, 1 child
Bañolas (Spain)	1887	N	Mandible with teeth
Taubach (Germany)	1887	F, P	1 deciduous molar, 1 permanent molar
Krapina (Croatia)[g]	1899	F, P	Fragmentary remains of several individuals
Ochoz (Czech Republic)[g]	1905	F	Adult mandible with teeth
La Chapelle-aux-Saints (France)	1908	F, P	Skull and partial postcranium
Weimar-Ehringsdorf (Germany)	1908	F, P	Cranial, mandibular, and postcranial remains
Le Moustier (France)	1908	F, P, (?)	Skull and partial postcranium[h]
La Quina (France)	1908	F, P	Skulls and postcrania of several individuals
La Ferrassie (France)	1909	F, P	Several partial skeletons

[a]Year of initial discovery of human remains at the site; some sites yielded additional remains in subsequent years (e.g., Krapina, La Quina, La Ferrassie, Weimar-Ehringsdorf).

[b]Context of the remains: F, Pleistocene fauna; P, Paleolithic artifacts; (?), some question concerning the associations; N, no associations.

[c]Sometimes designated Engis 1.

[d]Not recognized as Neandertals until after 1856.

[e]New excavations have established faunal and artifact associations (see text).

[f]New excavations have recovered additional pieces of this skeleton and remains of at least two other individuals.

[g]Then part of the Austro-Hungarian Empire.

[h]Lost during World War II. Cranium rediscovered in 1965 (see Hesse and Ullrich 1966). The Le Moustier 2 infant skeleton, found in 1914 and also presumed lost, was rediscovered in France in 1996 (Maureille 2002).

tific thinking about "primitive" humans. By the 1920s, "Neanderthal" (with an H) had entered common parlance as an adjective meaning "crude, boorish, or old-fashioned"; and the stoop-shouldered, club-dragging image of the "cave man," derived from early reconstructions of Neandertals, had entered popular consciousness as the stereotypical picture of Stone Age Man.

The Neandertals have also had a disproportionate impact on the study of prehistory because they left more fossils than any of their predecessors. There are several reasons for this. For one thing, Neandertals were the first humans who buried their dead (at least sometimes). Unburied bodies usually disintegrate swiftly under the assaults of sun, rain, and bone-cracking scavengers, but the deliberate burial of a body increases the chances that its bones will be preserved. Moreover, many Neandertals inhabited lands containing limestone caves, which they lived and died in and used as garbage dumps. Such caves are ideally suited to preserve bones and other debris in the accumulating sediments and minerals laid down by percolating groundwater. And because caves are known to be promising places to look for ancient remains, they are more likely than other sorts of occupation sites to be sought, found, and excavated by prehistorians. (Ancient open-air sites are not particularly rare, but they are usually discovered by

accident.) Finally, Neandertals lived relatively recently, over a span of time extending from just over 200 Kya to about 30 Kya. Their remains therefore outnumber those of their predecessors, simply because they have been exposed for less time to the destructive forces of nature. This is true even at the molecular level. Because Neandertals lived so recently, scientists have succeeded in extracting mitochondrial DNA from 13 Neandertal specimens and nuclear DNA from two. It is no coincidence that almost all of these DNA samples come from late-surviving Neandertals, dated to 40 Kya or less.

As ancient Europeans, Neandertals have also loomed large in the literature of paleoanthropology because Westerners tend to assume that Europe has always held center stage in the human drama. We now know enough about the prehistory of other regions to realize that this was a false assumption. Although other parts of the Old World have still not been as thoroughly picked over in search of Paleolithic sites as Europe has, some areas (such as Israel, South Africa, and parts of East Africa) are getting close. Our wider knowledge of the ancient world has led most scientists today to view Africa, not Europe, as the source of the most important innovations in the evolution of the human genus. The Neandertals of Europe had more "advanced" or modern-looking contemporaries in Africa; and it is to these African people, not to the Neandertals, that we must look for the main source of our own genetic heritage (Chapter 8). But even if Europe and the Neandertals were peripheral to the central thread of the human story, they are nonetheless critical to our understanding of the historical context of modern human origins. The unusually detailed knowledge we have of the Neandertals is also important for what it can tell us about the processes at work in other geographic regions and less well-documented phases of human evolution.

Perhaps the most important marker of the boundary between people and animals is the ability to talk. The language abilities of Neandertals, and thus their human status, have been debated ever since the first studies of Neandertal remains in the mid-19th century. Some researchers have argued that Neandertals were unable to produce the full range of sounds needed for efficient encoding of speech (Lieberman and Crelin 1971). Others contend that fully modern language emerged only about 50 Kya, possibly as a single genetic mutation (Klein and Edgar 2002). Although few have claimed that Neandertals lacked language altogether, it is often asserted that whatever speech capacities they had must have been subhuman.

However, the preponderance of current evidence suggests otherwise. For example, a recent analysis of Neandertal DNA recovered from El Sidrón Cave in northern Spain shows that Neandertals shared the uniquely derived human form of the FOXP2 gene (Krause et al. 2007a). This gene encodes a transcription-factor protein that is extremely conservative in mammals, differing in only one amino acid between chimpanzees and mice. Modern humans, however, show two additional amino-acid substitutions not shared with other mammals. Because people carrying mutations in this gene suffer impairments in various aspects of speech production and grammatical ability (Lai et al. 2001), it has been suggested that the distinctive human form of FOXP2 was selected for as an adaptation to language (Enard et al. 2002). Its presence in both Neandertals and modern humans indicates that that adaptation took place before the two groups diverged. While this does not prove that Neandertals had fully developed modern human language capacities, it is one strong piece of evidence that they did. We will examine other evidence bearing on this question later. For now, it is sufficient to note that Neandertals were almost certainly "talking apes," and probably not the first ones.

NEANDERTALS—EARLY DISCOVERIES AND IDEAS (1829–1909)

The first Neandertal to be recognized as a primitive type of human was found in 1856 by laborers at a limestone quarry in Germany's *Neanderthal* or Neander Valley (German *Thal* "valley" = English *dale*), a scenic spot named in honor of a local choirmaster and author who wrote under the pen name of "Neander." (The last syllable of *Neanderthal* is pronounced like the first syllable in English *tolerate*. Later German spelling reforms eliminated the silent H, but the old spelling persists in everyday English. Both spellings are pronounced the same way.) Two other Neandertal fossils, the skull of an 8- or 9-year-old child from Engis in Belgium and an adult skull from Forbes Quarry on Gibraltar (Table 7.1), had in fact been found before 1856 and kept as curiosities (Busk 1865, but they were not recognized as Neandertals until much later (Spencer 1997a, Smith 1997c).

The 1856 skeleton was recovered while the walls of Neander's pretty little valley were literally being leveled to obtain construction-grade limestone. The story of the specimen's recovery and recognition has been pieced together by several prehistorians (Hrdlička 1930, Schmitz 1997, 2006, Schmitz and Thissen 2000, Schmitz et al. 2002). The original specimen—a cranial vault and 15 well-preserved pieces of the postcranial skeleton (Fig. 7.1)—was found in a small cave, the Kleine Feldhofer Grotte (G., "Little Feldhofer Cave"). Some pieces were found *in situ* as workmen were shoveling out the cave fill in preparation for quarrying of the limestone matrix. W. Beckershoff, co-owner of the quarrying company, spotted the bones (which he took for fossils of the extinct cave bear) and told the workers to save them. A search soon turned up other pieces, including the cranial vault (Fig. 7.2), in discarded fill that had been dumped onto the valley floor 60 feet below; but the workers seem to have retrieved only the largest and most easily identified bones. These were recognized as human some weeks later by J. Fuhlrott, a local teacher and natural historian, who turned them over to H. Schaaffhausen, Professor of Anatomy at the University of Bonn, for analysis. Schaaffhausen and Fuhlrott made the first scientific report on the specimen in 1857, and the debate on the Neandertal problem began soon afterward.

Three competing interpretations of the original Neandertal specimen (hereafter Neandertal 1) and similar fossils emerged during the late 19th century (Brace 1964, Spencer 1984, 1986; Bowler 1986, Smith 1997b, Schmitz 2005). The first was the interpretation championed by Schaaffhausen (1858, 1888) and T. H. Huxley (1863). They recognized that the Neandertal 1 skull had some primitive features, including a low vault and a heavy brow ridge, and considered it to be of considerable antiquity. However, because of its large brain (~1525 cm³), they did not regard it as a missing link between humans and apes, nor as belonging to a different species from *Homo sapiens*. For Huxley and Schaaffhausen, it was unthinkable that a primitive type of human could have a modern-sized brain. Rather, they

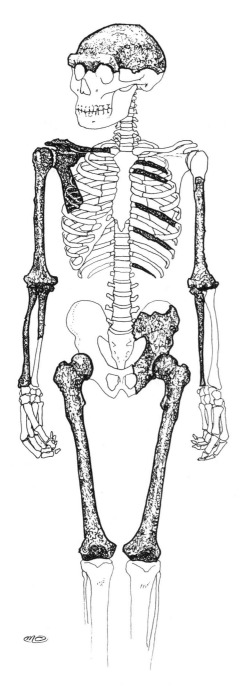

FIGURE 7.1

The original Neandertal 1 skeleton, recovered in 1856 from the "Kleine Feldhofer Grotte" near Düsseldorf, Germany. The shaded elements are the 16 pieces found in 1856. (After Schmitz and Thissen 2000.)

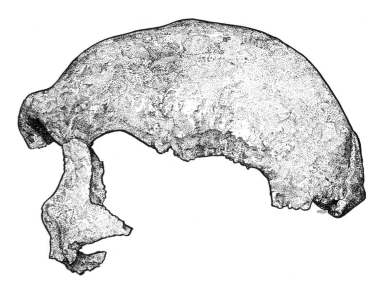

FIGURE 7.2

The original Neandertal calotte with the NN 34 left zygomatic bone. NN 34 was excavated in 2000 from the cave fill thrown out of the "Kleine Feldhofer Grotte" in 1856. The specimen exhibits distinctly Neandertal zygomatic features and fits perfectly onto the calotte.

saw the skull from Neander's valley as representing a low, savage, and barbarous "race" of aborigines that inhabited Europe prior to the coming of the ancient Celts and Germans. Both Schaaffhausen and Huxley held evolutionary views, and they certainly did not avoid the controversies associated with the idea of evolution. But that big brain kept them and most other scholars at the time from seeing Neandertal 1 as anything but a backward "race" of *Homo sapiens.*

This explanation of Neandertal 1 conformed to 19th-century habits of thinking about prehistory. European scholars of the late 1800s tended to see large-scale human history as a succession of ethnic replacements in which one "race" was supplanted by others moving in from elsewhere. This was a natural way of thinking for Europeans of the time, who often viewed their own colonial adventures in this light and pictured themselves as the predestined successors of more primitive folk in the colonized lands of Africa, Australia, and America. In Europe itself, some scholars defined an indigenous population called the "Cannstatt race," (Quatrefages de Breau and Hamy 1882) which had been subjugated first by invading Celts, then by the Romans, and next by Germanic tribes, with each succeeding wave of peoples overwriting the cultural and racial signature of the one before it. Both ancient remains and modern diversity were interpreted in these terms. A Welshman with a dark complexion, for example, might be pointed to by students of this sort of racial "science" as a surviving specimen of the Celtic stock, while the big nose and curly hair of his cousin was attributed to Roman "blood" and their blonde, blue-eyed grandmother was hailed as a Nordic descendant of the Anglo-Saxons.

Nowadays, we would interpret these facts in the light of genetics and simply say that the population of Wales, like all other European populations, is polymorphic with respect to pigmentation and hair texture. But in 1882, when Quatrefages de Breau and Hamy concocted the concept of the Cannstatt race, ideas of racial hierarchy and succession were prevalent among antiquarians and prehistorians. Neandertal 1 (along with Forbes Quarry and an assortment of other, mostly robust modern specimens) was identified by many scholars as part of that imaginary "Cannstatt race." The Cannstatt construct artificially and erroneously mixed Neandertals with modern humans, supporting the perception that Neandertals were only extreme variants of a "savage" modern-human ethnic group rather than a potentially ancestral form of human (Schwalbe 1904, 1906; Smith 1997b). This perception not only conformed to European ideas about racial hierarchy and succession; it also allowed prehistorians who were uncomfortable with Darwinian ideas to avoid thinking about the processes by which one population might over long periods of time change into something quite different.

A second, even more clearly antievolutionary school of thought interpreted the strange morphology of Neandertal 1 as pathological. The champion of this school was the German pathologist and anthropologist R. Virchow (Brace 1964, Smith 1997d). Virchow (1872) claimed that Neandertal 1's strange appearance was brought on by a unique combination of trauma and disease (including fractures, rickets, and arthritis). Virchow noted that the Neandertal 1 individual had survived to an advanced age (judged by the extensive fusion of the cranial sutures). Believing that this was possible only in a relatively civilized society, Virchow contended that the Neandertal specimen could not represent a "primitive" race. To help prove his contention, Virchow tried to show that people with equally "low," primitive-looking skulls could be found among living northern Germans. There was no reason, he concluded, to consider the Neandertaler to be anything but a pathologically abnormal member of a "civilized" race.

Interpreting Neandertal 1 as a pathological case was defensible as long as only the one specimen was known. Many new disease syndromes and pathologies were being identified in the late 19th century, and it was reasonable to suspect that the seemingly primitive features of the Neandertal 1 skull might be the result of some unknown combination of maladies. Virchow's impressive credentials as a pathologist lent this interpretation an air of authority. His view of Neandertal 1 was popular with those who opposed the idea of human evolution, and can still be found in creationist books and Web sites (often with citations of Virchow's writings).

But as proof accumulated that specimens like Neandertal 1 were very ancient and not unique, the tide began to turn against both the racial-succession and pathology interpretations. Partial Neandertal mandibles from La Naulette (Belgium) and Šipka (Czech Republic) were found in 1866 and 1880, respectively, alongside Pleistocene fauna and primitive stone tools (Dupont 1866, Schaaffhausen 1880). These were only lower jaws, and their affinities to Neandertal 1 were not obvious to everybody—particularly after Virchow (1882) offered pathological explanations of both of these specimens as well. But more convincing specimens were not long in coming. In 1886, two relatively complete Neandertal skeletons were excavated from the cave of Spy (Belgium), in direct association with Pleistocene fauna and Mousterian tools (the cultural complex generally associated with Neandertals). The distinct similarity of the Spy skeletons to Neandertal 1 was made clear in their initial descriptions (Fraipont and Lohest 1886, 1887). With the Spy, La Naulette, and Šipka specimens added to the Neander Valley material, Virchow's position became increasingly less tenable. The Gibraltar specimen was also recognized as a Neandertal in the early 1900s (Schwalbe 1906, Sollas 1907). The evidence that all these remains were of considerable antiquity undermined Virchow's interpretation, as did Schwalbe's (1901) careful demonstration that the Neandertal skull's morphology was not due to pathology or to advancing age.

This conclusion could have been reached much earlier if the original Neandertal site had been properly excavated. During the 1990s, the German archaeologists R. W. Schmitz and J. Thissen studied the available documents dealing with the Feldhofer site and pinpointed the area where the discarded refuse from the cave had been dumped, on what had been the valley floor back in 1856 (Fig. 7.3). They began digging there in 1997. Some four meters below the present ground level, Schmitz and Thissen (2000) found Micoquian tools (a form of Middle Paleolithic closely related to Mousterian), Pleistocene animal remains, and pieces of human bone. One of the small human fragments fitted exactly onto the left femur of Neandertal 1, proving that they had indeed found the Kleine Feldhofer Grotte cave fill. Further excavations in 2000 yielded two cranial pieces that fitted onto the Neandertal 1 skull (Fig. 7.2), plus 65 other human bones from a minimum of three individuals (Schmitz et al. 2002, Schmitz 2003, Smith et al. 2006). If all this evidence of remote antiquity and multiple individuals had been recovered from the Neander Valley in the first place, ideas like Virchow's might never have materialized.

The third interpretation of Neandertal 1 that emerged in the latter half of the 19th century was that it was an extinct, primitive relative of *Homo sapiens*. The Irish scientist W. King, one of the earliest proponents of this idea, redescribed the cranium in 1864 and concluded that it belonged to a non-modern human species, which he named *Homo neanderthalensis*. King suggested that the Neander Valley creature might better be separated from *Homo sapiens* at the genus or even family level,

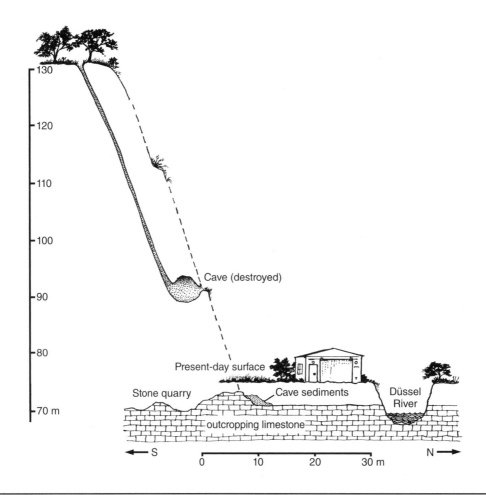

FIGURE 7.3

Schematic diagram of the original location of the "Kleine Feldhofer Grotte." In 1856, the cave was 20 m up the valley wall from the valley floor, represented here by the top of the "outcropping limestone." The "cave sediments" thrown from the cave in 1856 are represented here by the deposits shown on the valley side of a limestone berm. This berm was left as a natural levee to keep the Düssel River from overflowing its banks. It was from these deposits that R. Schmitz and J. Thiessen recovered the additional Neandertal skeletal remains, Middle Paleolithic tools, and Late Pleistocene fauna in 1997 and 2000. The "present-day surface" is ~4 m above the level of the floor in 1856. (After Schmitz and Thissen 2000.)

because the shape of the cranial vault showed that it must have had subhuman mental capacities resembling those of living apes.

King's views won few adherents in the 1800s. But in the first decade of the 20th century, expert opinion began to favor the idea that Neandertals were a primitive species, distinctly different from modern humans. From his metric analyses of the available fossils, the German anatomist G. Schwalbe (1901, 1904, 1906) concluded that Neandertals were distinctive enough to be placed in a species of their own, which he identified with E. Haeckel's (1868) imaginary "*Homo primigenius.*" (Like "*Pithecanthropus,*" this name had been coined by Haeckel as a label for one of his hypothetical ancestral forms—in this case, for the supposed last common ancestor of living humans.) Similar ideas were urged almost simultaneously by the geologist and pale-

ontologist D. Gorjanović-Kramberger, who had directed and reported on the exciting discovery of hundreds of fragmentary Neandertal remains from the cave site of Krapina in his native Croatia (Gorjanović-Kramberger 1906). Both Schwalbe (1906) and Gorjanović-Kramberger (1906) boldly contended that Neandertals were the logical ancestors of modern people. Although Schwalbe conceded that the Neandertals might be late-surviving primitive forms rather than direct human ancestors, he insisted that in a "… pure zoological sense *Homo primigenius* is an intermediate between *Homo sapiens* and *Pithecanthropus erectus*" (Schwalbe 1906, p. 14). Gorjanović-Kramberger, however, never wavered in his contention that Neandertals were directly ancestral to modern humans (Smith 1976, Radovčić 1988). These two early 20th–century scientists deserve credit for developing the first systematic arguments for a Neandertal phase in human evolutionary history.

In 1908 and 1909, a flood of important Neandertal remains from French sites—La Chapelle-aux-Saints, Le Moustier, La Quina, La Ferrassie—were added to the Neandertal family register (Table 7.1). The impact of these accumulating discoveries, along with the publications of Schwalbe and Gorjanović-Kramberger, ushered in a brief period during which most scientists accepted the idea of a Neandertal ancestry for modern people. This period was brought to an abrupt close a few years later by M. Boule's analyses of the moderately complete Neandertal skeleton from La Chapelle-aux-Saints. In his publications on La Chapelle between 1911 and 1913, Boule accepted the view that Neandertals were a separate species; but he denied their position as modern human ancestors. Ever since Boule, the Neandertal problem has centered around these two issues: Were Neandertals members of our species, and are any living people descended from them?

▎ IDEAS ABOUT NEANDERTALS—FROM BOULE TO THE 21ST CENTURY

Boule argued that several postcranial features of the La Chapelle skeleton indicated that Neandertal posture and locomotion had been somewhat ape-like. These features included:

▶ Massive musculature and bone structure of the hip region, which Boule believed had prevented the hip from being fully extended.

▶ A retroverted (backwardly angled) proximal articular surface of the tibia, suggesting a bent-kneed stance.

▶ A generally ape-like foot structure, with a divergent hallux.

▶ A single-arched vertebral column (lacking the "S" shape of modern humans), with long, straight, posteriorly projecting vertebral spines.

Boule concluded from all this that the Neandertals had not stood fully erect, and had walked with bent knees and hips (rather like the 1980s Stony Brook model of *Australopithecus afarensis*). He also pointed to the low cranial vault, receding forehead, projecting face, prominent supraorbital torus, and absence of a chin as distinctly primitive characteristics. Despite the large cranial capacity of the La Chapelle skull (1625 cm³), Boule held that its low vault and various features of its endocast (Boule and Anthony 1911) indicated inferior mental abilities.

Boule's studies were thorough and of generally high quality (Trinkaus 1985, Trinkaus and Shipman 1993). However, subsequent work has shown that all of his postcranial interpretations were mistaken (Straus and Cave 1957, Stewart 1962a, Trinkaus 1985). Whereas Virchow had misread primitive features of Neandertal 1 as pathologies, Boule erred in the opposite direction, misreading the results of degenerative joint disease and other pathologies in La Chapelle as primitive traits. Boule recognized the pathologies of the La Chapelle skeleton, but he mistakenly thought that they were not particularly significant to interpreting its anatomy. Perhaps his major shortcomings were his disregard of the extensive skeletal variability in living people, and his peremptory dismissal of studies that refuted many of his interpretations.

Boule's misinterpretations were also conditioned by his preconceptions about the evolutionary process (Hammond 1982). His studies of other mammalian groups and the influence of his mentor, A. Gaudry, had led him to view evolution as involving extensive branching. (In this regard, his views foreshadowed current ideas about the necessary "bushiness" of the human family tree.) Given the primitive morphology of the Neandertals and the short time separating them from the earliest modern humans, Boule concluded that the Neandertals must have been a side branch, not an ancestor, of the modern human line.

Luckily for Boule, his studies of La Chapelle roughly coincided with the emergence of such purported Pre-Sapiens fossils as the Grimaldi skeletons from Italy and the Ipswich, Galley Hill, and Piltdown remains from England (Chapter 6). These converging influences led a lot of anthropologists and anatomists to abandon unilineal, Neandertal-phase accounts of human phylogeny and adopt some version of the Pre-Sapiens scheme (Bowler 1986, Spencer 1984, 1986; Trinkaus and Shipman 1993, Smith 1997a,b).

Although Boule's theories and other Pre-Sapiens models remained influential, other ideas emerged as Neandertal-like fossils started coming to light in other parts of the world—in Zambia (Kabwe), Palestine (Zuttiyeh, Skhūl, and Tabūn), South Africa (Florisbad), and Java (Ngandong). These accumulating finds contributed to the growth of two new perceptions. The first was that the Neandertals of Europe were not singularly different from other fossil humans of comparable age. By the 1950s, the Neandertals and other big-brained but low-browed early *Homo* were commonly grouped together as representing an Old-World-wide "archaic *H. sapiens*" stage in human evolution (Brace 1967; see Chapter 6). The second new perception was that later human evolution had not been centered in Europe, and that models of human evolution needed to have a wider geographic scope.

This second perception resulted in large measure from the fossil discoveries made between 1929 and 1934 at the Mt. Carmel sites of Skhūl and Tabūn. The

Mt. Carmel fossils comprised a large series of "Neandertal-like" specimens, all of which were thought to date to ~100 Kya. Keith and McCown (1937) initially distinguished two separate groups in the Mt. Carmel remains: Neandertals from Tabūn (Level C and below), and more modern types from Skhūl. However, after further analysis, McCown and Keith (1939) concluded that the Tabūn and Skhūl specimens graded imperceptibly into each other, forming a single variable population. Although McCown and Keith were not entirely clear about the biological meaning of all this variation, they described the Mt. Carmel sample as being in the "throes" of evolutionary change. Since the Mt. Carmel fossils were then thought to be earlier than the European Neandertals, the logical interpretation was that they represented an Asian population beginning to evolve toward modern humans from a non-European Neandertal-like ancestor. Thus the Mt. Carmel remains could be viewed as demonstrating an origin of modern humans that bypassed the European Neandertals.

The Mt. Carmel sample provided much of the basis for the Pre-Neandertal model (Chapter 6) elaborated by F. C. Howell (1951, 1952, 1957). Howell argued that modern humans and European Neandertals had descended separately from early "progressive" Neandertals. In Howell's model, the "Classic" European Neandertals had evolved from an earlier, less extreme Neandertal type—represented by such last-interglacial (OIS 5e) fossils as Krapina, Saccopastore (Italy) and Ehringsdorf (Germany)—by developing more specialized adaptations to the colder climate of Europe during the Würm glacial advance (OIS 5d through 3). Outside of Europe, perhaps in the Near East, "progressive" Neandertals of the older, less specialized sort (including the Skhūl people) had gone on to evolve into modern humans.

With the unmasking of the original pre-Sapiens fossils as intrusive or fraudulent (Chapter 6), the popularity of Pre-Sapiens models waned in the 1950s and 1960s, and variations on Howell's Pre-Neandertal model became increasingly popular. Similar views were propounded at the same time by S. Sergi (1953) and adopted by many other paleoanthropologists, including G. Bräuer, C. B. Stringer, J.-J. Hublin, B. Vandermeersch, and W.W. Howells. As noted in Chapter 6, the Pre-Neandertal perspective is a precursor of today's Recent African Origin model for modern human origins (Smith 1997a,b, 2002).

Alongside the Pre-Neandertal model, several versions of the Neandertal-Phase model continued to find defenders, from the 1920s down to the present. These included A. Hrdlička's (1927, 1930) "Neanderthal Phase of Man" concept and Weidenreich's (1946, 1947a,b) trellis model, discussed in Chapter 6. In the 1960s, C. Coon and C.L. Brace articulated views rooted in the ideas of Weiden-

reich. Coon (1962) believed that regional human lineages (for him, "races") had evolved from *H. erectus* into *H. sapiens* largely in parallel in the major geographic regions of the Old World, with each lineage crossing the threshold to modernity independently of the others. Coon's ideas were controversial because he argued that the different lineages had crossed that threshold at different times. This claim implies that a single interbreeding population must have been divided for a time into two different species—an implication that many critics saw as biologically absurd. Criticism of Coon focused particularly on his contention that the northern lineages had crossed the threshold first, and by his hints that the northern "races" in Europe and Asia today are somehow more highly evolved than the modern human populations of Africa and Australasia. These suggestions led Coon's critics to accuse him of racism (see Shipman 1994, Wolpoff and Caspari 1997). Yet although Coon emphasized parallel evolution in these lineages, he also clearly accepted Weidenreich's contention that these lineages were always interconnected by gene flow, so that regional evolution in one area really did not occur in *total* isolation from other lineages.

As noted in Chapter 6, Weidenreich's "trellis" model provides the theoretical underpinning for the modern Multiregional Evolution model. A similar position has long been argued by Brace (1967, 1995), who contends that all modern humans passed through a Neandertal-like stage in their region of origin. For him, the major driving force in the modernization process was the relaxation of natural selection brought about by increasing technological sophistication. Once technology had reached a level at which the robust morphology of archaic humans was no longer necessary for survival, relaxed selection led to the emergence of modern dental, cranial, and postcranial anatomy. Brace argued that this process would be expected to occur independently in different regions. Thus, modern morphology had eventually become universal without any need for "modernizing" genes to spread from any particular center. However, Brace followed Weidenreich much more closely than Coon had in stressing the continual strong interconnection of regional lineages throughout the world. Because of this historical interconnection, and because of the clinal gradients of genetic and morphological variation that this interconnection has produced in modern humans, Brace concludes that human races are artificial constructs and have no biological basis.

New fossils of Neandertals have continued to accumulate. The remains of around 400 Neandertal individuals are currently known, ranging from relatively complete skeletons to fragments of a single bone. These remains span the entire age range from babies to relatively old individuals, although it appears that few Neandertals lived much beyond their reproductive years (Trinkaus 1995a). This extensive sample tells us a lot about patterns of Neandertal growth, development, and adult variation. Since the 1920s, our understanding of the Neandertals has been enhanced by a long series of discoveries from European cave sites— Gánovce in Slovakia (in 1926), Subalyuk in Hungary (1932), Guattari (Monte Circeo) in Italy (1939), Arcy-sur-Cure (Grotte du Renne) in France (1949), Régourdou and Hortus in France (1957 and 1960), Kůlna in the Czech Republic (1965), Vindija in Croatia (1974), and Saint-Césaire in France (1979). The Vindija, Hortus, Arcy, and Saint-Césaire specimens have provided especially important evidence about the anatomy and behavior of the latest Neandertals in Europe (Vandermeersch 1997, Wolpoff 1999, Smith 2002, Bailey and Hublin 2006a). More recently, Rosas et al. (2006b) have reported over 1300 fragmentary Neandertal specimens from a minimum of eight individuals at the Spanish site of El Sidrón. Possible Neandertals of considerable significance have been recovered from Altamura in Italy (1993) and Mezmaiskaya in Russia (1993). The Altamura specimen is encased in cave flowstone and cannot yet be excavated or removed from the cave (Delfino and Vacca 1993), but the skeleton appears to be relatively complete, and its known morphology is commensurate with that of a Neandertal or Neandertal ancestor. Mezmaiskaya is significant because the second published Neandertal mitochondrial DNA sequence was derived from the bones of a very young child from this site (Ovchinnikov et al. 2000).

The context and affinities of the Mezmaiska child have been questioned (Hawks and Wolpoff 2001b), but other sites in West and Central Asia have yielded undoubted Neandertal fossils (Trinkaus 1984a, 1995b). In addition to Tabūn, the most significant of these come from Teshik Tash in Uzbekistan (1938), Shanidar Cave in Iraq (1953), Amud Cave (1961) and Kebara Cave (1965) in Israel, and Dederiyeh Cave in Syria (1993). Taken together, these specimens define the geographic range of Neandertals outside of Europe. They inevitably lead us to ask how Neandertals came to inhabit these regions. Did they evolve *in situ* from more archaic people, as a single population spanning the entire. Neandertal range in Eurasía, or did they enter Asia secondarily as migrants from a European center of origin?

Bar-Yosef (1992, 1994, Bar-Yosef and Vandermeersch 1993) contends that the Asian Neandertals were late immigrants from Europe. According to Bar-Yosef's "Late Migration Hypothesis," the Neandertals were forced to the South because they lacked adequate cultural buffering to deal with the intolerable cold of European glacial maxima (Bar-Yosef 1988; cf. Aiello and Wheeler 2003). Bar-Yosef believes that this migration could have occurred

any time after oxygen isotope stage (OIS) 5d (~105 Ky), but was most likely during OIS 4, beginning ~74 Kya.

It seems reasonable to think that human movement from the North into the Levant could have occurred more than once during the Middle and Late Pleistocene. This possibility suggests the existence of a long-term fluctuating boundary between Neandertals and other (presumably more modern) people to the South, who either evolved in the Levant or migrated there from the South and/or East. Fogarty and Smith (1987) have shown that multiple migration routes would have been open between Europe and Asia Minor during parts of the Middle and Late Pleistocene glacial cycles, and that Neandertals could have migrated south into the Near East. However, these same connections would also have allowed gene flow into Europe. The fact that migration was possible in many directions poses a problem for the view that total geographic isolation of Europe underlay the evolutionary differentiation of the Neandertals (and their precursors, the Neandertal-like European Heidelbergs). Neandertal populations were no doubt *relatively* isolated during parts of the later Pleistocene, but they were probably never completely cut off from human populations outside of Europe.

The degree to which Europe was genetically isolated at this time is the crux of the ongoing debates over the later phases of human evolution. There is a spectrum of expert opinion on this question. The Recent African Origin and Multiregional Evolution models represent the opposing poles of this spectrum, but there are intermediate models of the relationship between Neandertals and early modern humans. One such intermediate, already discussed in Chapter 6, is G. Bräuer's Afro-European Sapiens or Replacement with Hybridization model. Another is the Assimilation model, which was first articulated in 1989 by F. Smith and colleagues (Smith et al. 1989a). The Assimilation model arises out of earlier work by E. Trinkaus and Smith (1985; Smith 1985) and shares many features with the MRE model, particularly a focus on the importance of gene flow in later human evolution and the rejection of a speciation event to explain modern human emergence. However, the Assimilation and MRE models differ on two major points (Smith 2002). First, the Assimilation model depicts the basic morphological apomorphies of modern humans as having emerged as a complex in a single region of origin, probably in East Africa (Chapter 8). Second, the Assimilation model finds evidence of regional European continuity in several details of morphology, but not in the fundamental gestalt or bauplan of the skeleton. For example, some early modern human skulls and postcrania from Europe exhibit certain morphological details that are hard to explain if Neandertals are totally excluded from their ancestry (Smith et al. 2005). However, their overall morphological pattern is easily identifiable as non-Neandertal, implying that the Neandertal contribution to the European gene pool was probably far less than some supporters of the MRE model have intimated. The fit between all these models and the available data will be assessed in the next chapter.

In addition to debating the taxonomy and phylogenetic relationships of Neandertals, scientists during the past quarter-century have devoted a great deal of attention to functional and adaptive models of Neandertal anatomy, ontogenetic studies focusing on Neandertal growth and development, reconstruction of various aspects of Neandertal behavior and technology, analysis of Neandertal diet, and the patterns of Neandertal distribution in time and space. Perhaps the most surprising field of Neandertal studies to emerge in the last decade has been the analysis of mitochondrial DNA (mtDNA), which has been isolated from several individuals. In 1997, M. Krings and colleagues, working in the laboratory of S. Pääbo, published the first such sequence, extracted from the right humerus of the original Neandertal 1 skeleton (Krings et al. 1997). This was followed by a second sequence from another part of the Neandertal 1 mitochondrial genome (Krings et al. 1999). Later researchers have recovered mtDNA sequences from twelve other Neandertal individuals, including: Mezmaiskaya (Ovchinnikov et al. 2000), three specimens from Vindija (Krings et al. 2000, Serre et al. 2004), La Chapelle-aux-Saints (Serre et al. 2004), Engis 2 (Serre et al. 2004), a second individual (Neandertal 2) from the Neander Valley (Schmitz et al. 2002), El Sidrón (Lalueza-Fox et al. 2005), Les Rochers-de-Villenueve 1 (Beauval et al. 2005), and Scladina Cave (Orlando et al. 2006). All of these sequences cluster at the very periphery of the modern human range. They have accordingly been interpreted as proving that Neandertals are a different species from *H. sapiens* and as providing support for the Recent African Origin model. Nuclear DNA recovered from a Vindija specimen in 2006 is also cited as evidence for the specific distinctiveness of the Neandertals (Green et al. 2006, Noonan et al. 2006). Proponents of the Recent African Origin model point particularly to the genetic data from Europe as providing compelling evidence for the replacement of indigenous archaic people—in this case the Neandertals—by intrusive modern humans who migrated into the continent just over 30 Ky ago (Tattersall and Schwartz 2000, Stringer 1990a, 1994, Stringer and Andrews 1988, Stringer and Bräuer 1994, Stringer et al. 1984, Bräuer and Stringer 1997, Lahr 1996, Pearson 2000a,b, Bräuer 2006).

▌NEANDERTAL CHRONOLOGY AND DISTRIBUTION

Picking a date for the first appearance of Neandertals is difficult—not because there are not enough dated

specimens, but because there is no consensus on what specimen first shows enough typical Neandertal morphology to be considered a "true" Neandertal. As noted in Chapter 6, Neandertal features began to appear in European specimens almost as soon as Europe was inhabited. Some workers consider even such early specimens as Petralona to be Neandertals in a broad sense. Wolpoff (1980b, 1999), Stringer et al. (1979), and Dean et al. (1998) all agree that Neandertal features gradually accumulate in European hominins during the Middle and early Late Pleistocene, although they invoke different evolutionary models to explain this accumulation (cf. Rosas et al. 2006a). All also agree that the European fossils from the last interglacial (OIS 5e) are Neandertals (Janković 2004). These include the Krapina sample (Smith 1976), dated at 130 Ky (Rink et al. 1995), the two Saccopastore crania (Stringer et al. 1984, Condemi 1992, Bruner and Manzi 2006), and the slightly older specimens from the Bourgeois-Delaunay Cave at La Chaise (France). Condemi (2001) has conclusively established the Neandertal affinities of the remains from the French site, which are dated to ≥151 ± 15 Ky by uranium/thorium.

How much further back into OIS stages 6 and 7 can "typical" Neandertals be traced? One of us (Smith 1984) has argued that the human remains from Fischer's and Kämpfe's quarries at Ehringsdorf, near Weimar (Germany), exhibit a typically Neandertal morphological pattern. If this is accepted (and it certainly is not accepted by everyone; see Vlček 1993), then the Ehringsdorf specimens would be the earliest unequivocal Neandertals. ESR and uranium-series dating put the deposits containing these fossils at ~205 Ky, during OIS 7 (Blackwell and Schwarcz 1986, Grün and Stringer 1991). Distinctively Neandertal anatomy is also seen in two cranial specimens from Biache-Saint-Vaast in France (Stringer et al. 1984, Guipert et al. 2007), which are dated to between 190 Ky and 159 Kya on the basis of thermoluminescence of associated burned flints (Aitken et al. 1986), and may be as old as 250 Ky (Guipert et al. 2007).

There are other European Neandertals that appear to date back to OIS stages 6 and/or 7, but their dating is less secure. For example, the fragmentary remains from the Abri Suard at La Chaise appear to be similar to Biache (Hublin 1980) and are chronologically older than the Bourgeois-Delaunay remains from the same cave complex (Condemi 2001). Other human remains that probably derive from OIS 6 (~200 Ky to ~130 Ky), but are not as securely dated, include the Montmaurin mandible, the Lazaret parietal, and the Fontéchevade crania (all from France). Also from this time horizon are a crushed hipbone from Prince Cave (Italy), which derives from a level dated to 160–110 Kya by uranium-series dating, and a partial calotte from the German site of Ochtendung, associated with a thermoluminescence date of 170 Ky (von Berg et al. 2000). All of these specimens have features that indicate Neandertal affinities, and there is no reason to doubt their attribution to Neandertals.

Neandertals undoubtedly populated Europe during the last part of OIS 5 (Mellars 1996), but most Neandertal remains are younger, with dates that range between the onset of OIS stage 4 (~74 Kya) and the end of stage 3 (~35 Kya). Neandertal fossils directly dated between ~40 Kya and ~35 Kya are listed in Table 7.2. A few of these late dates are suspect because they are based on radiocarbon dating, which is of marginal utility for sites this old, even when AMS (accelerator mass spectrometry) technology is employed. The application of uranium-series, electron-spin resonance, and thermoluminescence dating to materials from deposits older than ~40 Ky has therefore been critical in establishing a firmer chronological background for the Neandertals (Aitken et al. 1993). For example, the Bañolas (Banoyles) mandible has been recently dated by this combination to 66 ± 7 Kya, placing it within OIS 5a (Grün et al. 2006), and the El Sidrón site has been similarly dated to ~43 Kya (Rosas et al. 2006b).

The latest-dated Neandertals in Europe are not restricted to southwestern Europe, as some have suggested (Hublin et al. 1995, 1996; Tattersall 1995a, Zilhão and d'Errico 1999), but are actually rather widely spread in Europe (Table 7.2). The most recent dates, just under 30 Kya, in fact come from the Mezmaiskaya infant in the

southern Caucasus of Russia and from two Neandertals from the G1 level at Vindija in Croatia. But these dates may be underestimates. Use of a more refined sample-preparation technique, known as ultrafiltering, has produced slightly older AMS radiocarbon dates for the Vindija specimens (Higham et al. 2006), and the other radiocarbon dates in Table 7.2—including the AMS dates—are probably also a few thousand years too young. This is certainly the case for the most recent supposed Neandertal from Iberia, an upper premolar from Figueria Brava (Schwartz and Tattersall 2002) dated to ~30 Kya by standard ^{14}C technology. The affinities of this isolated tooth are difficult to establish conclusively, and it is attributed to a Neandertal largely on the strength of its association with Mousterian artifacts. Other, even more recent "Neandertal" dates from the Iberian Peninsula (Hublin et al. 1995, Zilhão and d'Errico 1999) are derived from archaeological levels containing Mousterian artifacts but no human fossils. In Europe, Mousterian tool assemblages are reliably associated with Neandertals, and it may be reasonable to infer the late survival of Iberian Neandertals from the presence of Mousterian cultural assemblages (Gamble 1986, Klein 1999, Mellars 1996, Stringer and Gamble 1993). But even if this inference is sound, the Iberian radiocarbon dates are still probably underestimates; and other dates show that Neandertals survived just about as long in several other parts of Europe as well.

TABLE 7.2 ■ Selected European Neandertals Dating to ~42 Ky or Less

Site (Level)	Date	Technique	Dated Material[a]	Reference
Vindija (G1)	28,000 ± 400	AMS	Parietal (Vi 208)	Smith et al. (1999)
Vindija (G1)	29,100 ± 400	AMS	Mandible (Vi 207)	Smith et al. (1999)
Mezmaiskaya	29,195 ± 965	AMS	Rib	Ovchinnikov et al. (2000)
Figueira Brava	30,900 ± 700	^{14}C	Mollusk shell	Antunes and Cunha (1992)
Vindija (G1)	32,400 ± 800	AMS*	Parietal (Vi 208)	Higham et al. (2006)
Vindija (G1)	32,400 ± 1800	AMS*	Mandible (Vi 207)	Higham et al. (2006)
Vindija (G1)	33,000 ± 400	AMS	Cave bear	Karavanić (1995)
Arcy (Xb)[b]	33,820 ± 720	^{14}C	Stratum	Hedges et al. (1994)
Zafarraya	34,400 ± 2000	Th/U	Ibex tooth	Hublin et al. (1995)
La Quina (1)	35,250 ± 530	^{14}C	Animal bone	Henri-Martin (1964)
Saint-Césaire (8)	36,300 ± 2700	TL	Burned flints	Mercier et al. (1991)
Vindija (G3 ?)	38,310 ± 2130	AMS	Tibia fragment	Serre et al. (2004)
Neandertal 2	40,052 ± 409	AMS	Humerus (NN 1)	Schmitz et al. (2002)
Le Moustier	40,300 ± 2600	TL	Burned flints	Valladas et al. (1986)
Neandertal 1	40,394 ± 512	AMS	Humerus (N1)	Schmitz et al. (2002)
Les Rochers-de-Villenueve	44,152 ± 817	AMS	Femur	Beauval et al. (2005)
Vindija (G3)	>42 Ky	AMS	Foot bone (Vi 205)	Krings et al. (2000)

Only dates directly on or directly associated with fossil human remains are included. Dates on archaeological assemblages without associated fossil human remains are not included. Abbreviations: AMS, accelerator mass spectrometry (^{14}C); AMS*, accelerator mass spectrometry on ultrafiltered samples; ^{14}C, standard radiocarbon; Th/U, thorium–uranium; TL, thermoluminescence.
[a]Bones unless otherwise identified derive from human remains.
[b]Grotte du Renne.

Although Mousterian artifacts in Europe may always imply the presence of Neandertals, such tools occur with other, more modern-looking hominins in the Near East (see below). Conversely, some late Neandertals in Europe appear to have been making more advanced sorts of stone tools. For example, the Neandertal fossils from Saint-Césaire and Arcy-sur-Cure are not associated with the (Middle Paleolithic) Mousterian, but with the early Upper Paleolithic industry known as the Châtelperronian (Lévêque and Vandermeersch 1981, Hublin et al. 1996, Hublin and Bailey 2006). At Vindija, the same level (G1) that yielded the late Neandertal fossils contains a cultural assemblage with both Mousterian and Upper Paleolithic characteristics (Karavanić and Smith 1998, 2000). These so-called "transitional" assemblages—that is, assemblages that combine Mousterian with some aspects of early Upper Paleolithic technology and tool typology—are often referred to as the Initial Upper Paleolithic (IUP) (Churchill and Smith 2000a).

These post-Mousterian industries raise important questions concerning the nature of interaction between Neandertals and early modern peoples in Europe (Conard et al. 2004). The dates currently assigned to the latest European Neandertals imply that they may have coexisted with the first modern people to migrate into Europe. Churchill and Smith (2000a) suggest that the overlap may have lasted as long as 8–10 Ky, if currently available dates are correct. But some of the dates invoked by Churchill and Smith have been shown subsequently to be erroneous (Table 8.12). The late dates for Neandertals should be regarded as minimum estimates, since ^{14}C dates on samples older than about 20 Ky tend to give age estimates that are too young (Evin 1990). Tiny amounts (≤1%) of modern carbon contamination in a 40-Ky sample can make it appear to be 6–7 Ky younger than it actually is. A more systematic problem involves what Conard and Bolus (2003) identify as the "Middle Paleolithic Dating Anomaly." Radiocarbon dating techniques generally assume that the rate of production of atmospheric ^{14}C production has been constant, but there is evidence that this assumption is not always valid. Data from fossil North Atlantic planktonic foraminifera indicate anomalous peaks in ^{14}C production rates during the period between 40 and 30 Kya. This fluctuation could produce temporal offsets of as much as 6–10 Ky in radiocarbon dates, so that an apparent ^{14}C date of 30 Ky would correspond to an actual age of 35–38 Ky. Many ^{14}C dates from this time period may thus be a few thousand years too young. Conard and Bolus argue that this anomaly may exaggerate the duration of the period of overlap between Neandertals and early modern Europeans. If Conard and Bolus are right, the overlap may have been no more than two or three millennia, rather than the longer span suggested by Churchill and Smith. But if the dates for the latest Mousterian sites must be set back in time, then by the same token so

must the earliest dates for the Upper Paleolithic assemblages associated with modern humans (Zilhão and d'Errico 2003a). Applying the corrections suggested by Conard and Bolus to both sets of dates may thus simply shift the period of coexistence between Neandertals and modern humans in Europe to an earlier date, without significantly shortening its duration.

In western and central Asia, establishing a chronological framework for Neandertals depends on what specimens are included with the Neandertals, and on which of two competing chronological schemes are adopted (Fig. 7.4). As noted in Chapter 6, the Zuttiyeh partial cranium and the femora from Tabūn E and Gesher Benot Ya'acov in Israel are plausible antecedents of the Near Eastern Neandertals, both anatomically and chronologically (500–350 Ky). These fossils are too fragmentary to classify accurately, but they are associated with Lower Paleolithic tools of an Acheulean type and can be regarded as Near Eastern Heidelbergs. The earliest Mousterian tools in the Levant come from Level D at Tabūn, which has yielded conflicting dates—around 265 Kya by thermoluminescence, but only around 120–170 Ky by electron spin resonance, depending on which uptake model is used (Table 7.3). The next oldest human remains from this region, and the oldest ones associated with a Mousterian cultural complex, come from Tabūn Level C, which is dated to ~170 Kya by TL and to ~120 Ky by ESR (assuming a linear uptake). Here and elsewhere in the Levant, TL dates are systematically older than ESR dates for a given level (Table 7.3). The TL dates for Tabūn D in particular are much earlier than any dates for other Mousterian sites in the region (Bar-Yosef 1998). These facts justify a suspicion that TL dates from the Levant may be systematically in error for some reason (Porat et al. 2002).

The human fossils from Tabūn Level C include two important finds: (a) a partial skeleton with a complete skull (Tabūn C1) and (b) an impressive mandible (Tabūn C2). The C1 skeleton is widely accepted as a Neandertal, but its stratigraphic context is unclear and various estimates of its age span more than 200 Ky. Although some experts assign it to Level C (Quam and Smith 1998, Grün and Stringer 2000, Grün 2006), others suspect that it comes from the older Level D (implying an age of up to 260 Ky) (Jelinek 1982, 1992) or from the younger Level B, which dates to 86–103 Ky by ESR (Bar-Yosef and Callander 1999). Direct uranium-series dating of the C1 skeleton (using gamma-ray spectrometry) has yielded an extremely low estimated age of 34 ± 5 Ky (Schwarcz et al. 1998)—less than half the ESR estimate for Level B (Grün et al. 1991). However, this result has been called into question because U-series dating of bone often produces large over- or underestimates (Millard and Pike 1999; see also Alperson et al. 2000). U-series dating on Tabūn C1 teeth, a more reliable procedure, indicates that the specimen is at least 112 Ky old (Grün and

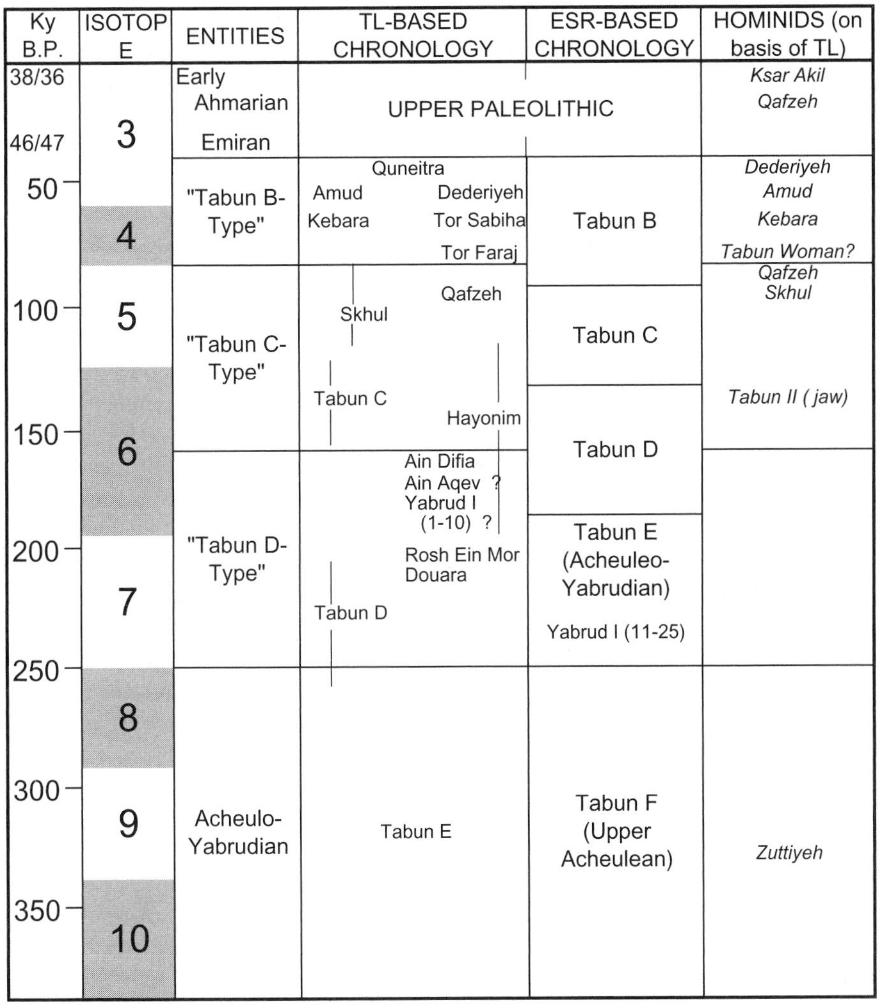

Ky B.P.	ISOTOPE	ENTITIES	TL-BASED CHRONOLOGY	ESR-BASED CHRONOLOGY	HOMINIDS (on basis of TL)
38/36	3	Early Ahmarian	UPPER PALEOLITHIC		Ksar Akil / Qafzeh
46/47		Emiran			
50	4	"Tabun B-Type"	Quneitra / Amud Dederiyeh / Kebara Tor Sabiha / Tor Faraj	Tabun B	Dederiyeh / Amud / Kebara / Tabun Woman?
100	5	"Tabun C-Type"	Qafzeh / Skhul / Tabun C / Hayonim	Tabun C	Qafzeh / Skhul
150	6			Tabun D	Tabun II (jaw)
200	7	"Tabun D-Type"	Ain Difia / Ain Aqev ? / Yabrud I (1-10) ? / Rosh Ein Mor Douara / Tabun D	Tabun E (Acheuleo-Yabrudian) / Yabrud I (11-25)	
250	8				
300	9	Acheulo-Yabrudian	Tabun E	Tabun F (Upper Acheulean)	Zuttiyeh
350	10				

FIGURE 7.4

The chronostratigraphy of the Levantine Neandertals and early modern human fossils and archaeological complexes. Note the differences between the ESR- and TL-based chronologies. (After Bar-Yosef 1994, 2000.)

TABLE 7.3 ■ Electron Spin Resonance (ESR) and Thermoluminescence (TL) Dates (in Ky) Associated with Hominin Fossils or Middle Paleolithic Artifacts from the Near East

Site/level	TL	ESR/EU	ESR/LU	Source
Tabūn Unit I (C)	171 ± 17			Mercier et al. (1995)
Tabūn Unit II (D)	212 ± 22			Mercier et al. (1995)
Tabūn Unit V (D)	244 ± 28			Mercier et al. (1995)
Tabūn Unit IX (D)	265 ± 27			Mercier et al. (1995)
Tabūn Unit X (E/D)	270 ± 22			Mercier et al. (1995)
Tabūn B		86 ± 11	103 ± 18	Grün et al. (1991)
Tabūn C		102 ± 17	119 ± 11	Grün et al. (1991)
Tabūn CI		112 ± 29	143 ± 37	Grün and Stringer (2000)
Tabūn D		122 ± 20	166 ± 20	Grün et al. (1991)
Qafzeh (XVII-XXII)	92 ± 5 (average)			Valladas et al. (1988)
Qafzeh (XV-XXI)		96 ± 13 (avg.)	115 ± 15 (avg.)	Schwarcz et al. (1988)
Kebara (Unit X)	61.6 ± 3.6			Valladas et al. (1987)
Kebara (Unit X)		60 ± 6	64 ± 4	Schwarcz et al. (1989)
Skhūl	119 ± 18			Mercier et al. (1993)
Skhūl		81 ± 15	101 ± 12.6	Stringer et al. (1989)
Amud B1/6		43 ± 5	48 ± 6	Schwarcz and Rink (1998)
Amud B2/8		59 ± 8	70 ± 8	Schwarcz and Rink (1998)

All fossils are associated with Mousterian artifacts. EU, early uptake; LU, linear uptake.
Source: Rvised from Bar-Yosef (1998).

Stringer 2000), and other elements that probably belong to this same skeleton have been unequivocally recovered from Level C (Trinkaus 1993d, Quam and Smith 1998). A Level-C age for C1 seems likely.

The C2 mandible's derivation from Level C has never been questioned, but its affinities are debated. Rak (1998) claims that the specimen (Fig. 7.34) represents a modern human, with no significant archaic features. Conversely, V. Stefan and E. Trinkaus (1998a) consider it an archaic (but not specifically Neandertal) human, with no significant *modern* features. Quam and Smith (1998) believe the C2 jaw exhibits a mosaic of archaic (*including* specifically Neandertal) and more modern features. If nothing else, this debate demonstrates the difficulties of making sense out of unassociated mandibles.

The uncertainties surrounding these specimens make it difficult to put a starting date on "Neandertalhood" in the Near East. Mousterian tools, which in Europe reflect the presence of Neandertals, date back in Asia to late OIS stage 7, around 215 Kya (Porat et al. 2002). If the TL dates from Tabūn D are included, this is pushed back to ~265 Ky (Bar-Yosef 1998, 2000). But in the Near East, the Mousterian is not always associated with Neandertals. The Skhūl/Qafzeh remains, which are fundamentally modern and dated to ~80–100 Ky, are also found with Mousterian cultural assemblages (McCown and Keith 1939, Vandermeersch 1981, Bar-Yosef 1994, 1998, 2000).

Later Near Eastern Neandertals appear to fall into a time range from around 70 to 45 Kya (Table 7.3). In the Levant, this series includes the specimens from Kebara in Israel (at 60–64 Ky) and Dederiyeh in Syria (estimated at ~50 Ky). The latest dated Neandertals in the Near East come from the Wadi Amud at 43–48 Ky (Bar-Yosef 1998, 2000; Bar-Yosef et al. 1992). Further to the East, in Iraq, the Shanidar Neandertals were found below a stratum bearing a standard ^{14}C date of 46 Ky, and some of them could be several tens of thousands of years older (Trinkaus 1983b, 1984a). Unfortunately, there are no chronometric dates for the Crimean or Teshik Tash Neandertals, so it is not possible to determine when Neandertals arrived in extreme Eastern Europe or central Asia. However, there is evidence of Neandertals—or at any rate, distinctively Neandertal mtDNA variants—in Siberia by ~37 Kya (Krause et al. 2007b).

The geographic distribution of Neandertals, based on the presence of fossils attributed to this group, is illustrated in Fig. 7.5. Neandertal fossils are widely spread in Europe, from Gibraltar (Forbes Quarry, Devil's Tower) in the southeast to the present-day island of Jersey (La Cotte de St. Brelade, 49°N) and north central Germany (Saltzgitter-Lebenstedt, 52°N) in the north. Neandertals probably lived even further north in Europe (Roebroeks et al. 1992), at least during warmer periods, but subsequent continental glacial activity has destroyed any evidence of this. Neandertal remains are also found extending through the southern parts of central Europe, across southern Russia to the Teshik Tash site in Uzbekistan at 67° East longitude. In the Near East, the Neandertal remains from Kebara are the furthest south, at about 34° North latitude. A Neandertal range estimated from archaeology would be considerably larger. Mousterian sites extend into more of Europe (for example, into the Alps) as well as Siberia and North Africa. Middle Paleolithic-like tools and evidence of cut marks on mammoth remains are also present in the Russian Arctic (67°N) at ~37 Kya (Pavlov et al. 2001). However, some Mousterian tools were the handiwork of more modern-looking people—not only at Skhūl and Qafzeh, but also in North Africa, where Mousterian tools are associated with non-Neandertal humans at Jebel Irhoud and Haua Fteah (Hublin 1993, Smith 1985). Since the Neandertals' geographical range must have fluctuated in response to the changing climates of the later Pleistocene, their overall distribution (Fig. 7.5) probably covers a larger territory than they inhabited at any one time.

NEANDERTAL MORPHOLOGY—THE CRANIAL VAULT

The braincases of Neandertals reflect the continuing enlargement of the brain during the late Middle and early Late Pleistocene (Table 6.1). Neandertal brains were as large as those of living humans, and all the trends associated with cranial-vault expansion seen earlier in the European Heidelbergs (Chapter 6) are continued and extended in the Neandertal braincase. But despite their fully human-sized brains, Neandertals had cranial vaults that differ from those of recent humans in a number of obvious ways. The descriptions that follow generally apply to both European and Asian Neandertals, but differences between the two groups—for example, in occipital bunning—will be specifically noted.

Neandertal cranial vaults are longer, wider, and lower than those of modern humans (Table 7.4). The low height of the Neandertal vault is obvious to visual inspection (Figs. 7.6 and 7.7; cf. Figs. 8.22 and 8.32). This difference can be quantified using the **vertex radius** (see Appendix) and the **vault-height index** (vertex radius as a percentage of maximum cranial length). Average values of both the radius and the index are lower for Neandertals than for recent Europeans (Table 7.4). The lower index values in Neandertals partly reflect the greater length of their skulls. But early modern humans from Europe and West Asia (European Early Upper Paleolithic and Skhūl/Qafzeh, Table 7.4)

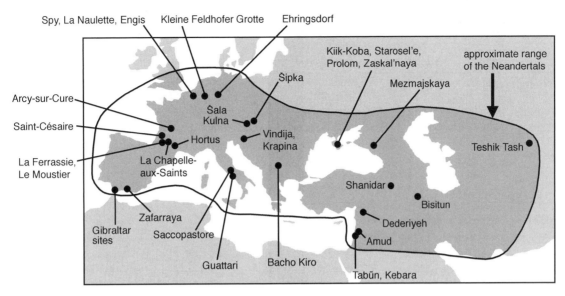

FIGURE 7.5

Geographic distribution of Neandertals, showing the location of major sites mentioned in the text.

have equally long skulls, and still have significantly larger vault-height indices. The data in Table 7.4 also show that Asian Neandertals tend to have vault-height indices intermediate between those of European Neandertals and early moderns, and that some early moderns (e.g., Skhūl 4, Mladeč 1, Mladeč 5) fall in the Neandertal range.

If Neandertal vaults are lower than those of living humans, how can they contain brains as big as ours (Table 7.9)? The lesser height of the Neandertal vault is offset by the greater width of the braincase, which is reflected in both maximum cranial breadth and minimum frontal breadth (Table 7.4). In effect, the modern human braincase protrudes forward over the eye sockets, while the Neandertal braincase bulges out sideways above the ears. In modern humans, cranial breadth is greatest high up on the parietals (Fig. 8.6); in Neandertals, maximum cranial breadth lies lower on the parietals, just above the squamosal suture (Fig. 7.8). The increased cranial breadth in Neandertals and its low position give the braincase a distinctive oval silhouette in rear view. This vault shape, conventionally described as *"en bombe"* (French, "bomb-like"), is regarded by some as a uniquely Neandertal apomorphy (Hublin 1980).

Frontal Bones

Neandertal frontals, like those of Erectines and Heidelbergs, are persistently primitive in having big supraorbital tori and a relatively flat frontal squama (forehead) that slopes sharply backward. In a modern human skull, the braincase is more uniformly curved in lateral view, giving it a near-circular outline with a higher, more bulging and more vertically oriented forehead (Fig. 8.1). The outward bulge of the modern forehead is reflected in a smaller **frontal angle** (see definition in Appendix) than those seen in Neandertals and earlier hominins (Table 7.5). Using a differently defined frontal angle, Suzuki (1970a) found that the Neandertal mean differs from those of living human groups by >4 s.d. Again, the sample means for early moderns (European Early Upper Paleolithic and Skhūl/Qafzeh) fall in between those of living humans and Neandertals.

In Neandertal skulls, the frontal sinuses are generally restricted to the supraorbital region near the midline, extending as far laterally as the center of each orbit (Fig. 2.13; Vlček 1967, Heim 1978, Smith and Ranyard 1980). Unlike those of many modern humans (Fig. 3.36B), Neandertal frontal sinuses do not extend into the frontal squama. Nevertheless, Neandertals had far more voluminous frontal sinuses than modern humans have. The Le Moustier 1 Neandertal skull has been said to have a small frontal sinus of the modern type (Vlček 1967), but its small size is probably due to the subadult status of this individual, who died at around 14 years of age (Ahern and Smith 2004). In fully adult Neandertals, the sinus is much larger and multichambered, and often becomes so expanded that the walls of the supraorbital torus near the midline are paper-thin and translucent (Smith and Ranyard 1980).

TABLE 7.4 ■ Selected Cranial Vault Measurements for Late Pleistocene European and West Asian Hominins

Specimen	Maximum Cranial Length	Maximum Cranial Breadth	Vertex Radius[a]	Minimum Frontal Breadth	Vault-Height Index[b]
European Neandertals					
Spy 1	202	145	117	114	58
Spy 2	200	153	120	110	59
La Quina 5	201	139	116	101	57
Neandertal 1	201	147	—	110	—
La Ferrassie 1	208	159	118	112	57
La Chapelle	211	157	117	111	55
Guattari 1	204	155	113	106	55
Saccopastore 1	181	142	110	110	60
European Neandertal mean	201	150	116	109	57
(N, σ – 1)	(8, 8.9)	(8, 7.3)	(7, 3.3)	(8, 4.0)	
Asian Neandertals					
Tabūn C1	183	141	108	98	59
Amud 1	215	154	126	115	59
Shanidar 1	208	154	127	110	60
Asian mean (N = 3)	202	150	120	108	59
European Early Upper Paleolithic (EUP)					
Mladeč 1	203	145	119	100	59
Mladeč 2	185	139	119	103	64
Mladeč 5	208	153	119	108	57
Brno 2	205	134	132	101	64
Dolní Věstonice 3	187	132	118	94	63
Cro Magnon 1	203	151	129	104	64
Cro Magnon 2	192	140	—	98	—
Cro Magnon 3	204	154	—	98	—
EUP mean	197	143	123	100	63
(N, σ – 1)	(9, 9.5)	(9, 8.1)	(7, 5.8)	(9, 4.1)	
Asian Early Moderns (Skhūl/Qafzeh)					
Skhūl 4	206	148	119	106	58
Skhūl 5	193	146	127	102	66
Skhūl 9	213	145	—	96	—
Qafzeh 6	195	147	125	110	64
SQ mean	201	146	123	104	63
(N, σ – 1)	(4, 9.4)	(4, 1.3)	(3, —)	(4, 6.0)	
Post-Pleistocene Europeans					
Ofnet mean[c]	182	138	121	94	66
(N, σ – 1)	(11, 4.2)	(12, 5.6)	(10, 2.7)	(9, 3.8)	
Norse Male mean[d]	188	142	122	—	64
(N, σ – 1)	(55, 5.3)	(55, 4.7)	(55, 4.2)		
Zalavar Male mean[d]	185	141	122	—	66
(N, σ – 1)	(54, 5.8)	(54, 4.0)	(54, 3.7)		
Berg Male mean[d]	180	148	121	—	67
(N, σ – 1)	(56, 7.4)	(56, 5.5)	(56, 3.6)		

All measurements are by FHS except for Shanidar 1 (Trinkaus 1983b) and the recent European Norse, Zalavar, and Berg samples (Howells 1973b).

[a]Vertex radius as defined by Howells (1973b).

[b]Vault-height index = $100 \times$ vertex radius/maximum cranial length.

[c]Ofnet is a series of Epipaleolithic or Mesolithic crania from southern Germany (Saller 1962).

[d]Since females exhibit even more "modern" patterns in these values, only male values are given here.

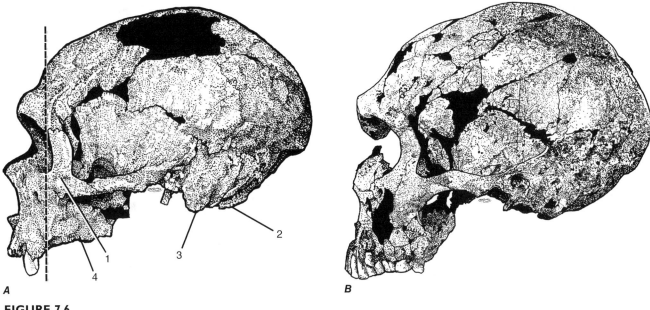

A   **B**

FIGURE 7.6

Lateral views of the La Chapelle-aux-Saints (**A**) and La Ferrassie 1 (**B**) Neandertal crania from France. Neandertals continue to exhibit marked midfacial and alveolar prognathism, but the lateral face (1) is located posterior to the anterior end of the braincase (*vertical dashed line*). Other features illustrated are (2) a prominent occipitomastoid region located in a relatively inferior position, which results in the appearance of (3) a nonprojecting mastoid process. La Chapelle's posterior teeth were lost before death, and the alveolar process is totally resorbed in this area on both the maxillae and mandible (4).

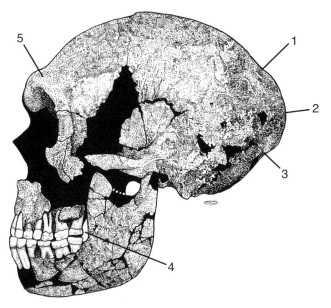

FIGURE 7.7

The La Quina H5 skull and mandible (France) in lateral view. Illustrated are the components of occipital bunning: (1) flattening in the lambdoidal region; (2) a relatively vertical occipital plane; and (3) the presence of an infratoral shelf, exaggerated by the inferior position of the occipitomastoid area (cf. Fig. 7.6). Other features indicated are the retromolar space (4) and the prominent supraorbital torus without a supratoral sulcus (5). (After Martin 1923.)

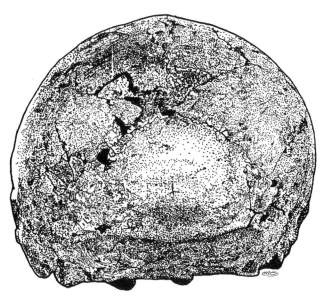

FIGURE 7.8

The La Ferrassie 1 cranium from France, rear view. The maximum breadth of the Neandertal cranium is located on the braincase (lower parietals) rather than the cranial base. The suprainiac fossa is weakly developed in this specimen.

TABLE 7.5 ■ Midsagittal Cranial Angles

Sample Mean	Frontal Angle	Parietal Angle	Occipital Angle
European Neandertals	139	145	109
(N, σ − 1)	(11, 2.3)	(9, 3.8)	(3, —)
Asian Neandertals (N = 3)	140	138	113
EUP	129	137	116
(N, σ − 1)	(17, 4.5)	(19, 3.5)	(11, 5.7)
SQ	132	137	
(N, σ − 1)	(4, 1.6)	(4, 3.6)	
Norse	132	133	115
(N, σ − 1)	(55, 3.7)	(55, 3.6)	(55, 5.0)
Zalavar	129	133	116
(N, σ − 1)	(54, 3.6)	(54, 3.3)	(54, 4.6)
Berg	128	133	117
(N, σ − 1)	(56, 3.8)	(56, 4.2)	(56, 5.5)

Measurements (in degrees) from Trinkaus (1983b) for the fossil samples and from Howells (1973b) for the Berg, Zalavar, and Norse samples. For the latter three samples, only male data are given because the modern female values are even more removed from those for the fossils. Angle definitions are from Howells (1973b; see also Appendix).

The thinness of the torus's walls near the nose tends to discredit the theory that it functioned as a protective visor. Because the walls of the torus are so thin, torus size is more closely correlated with frontal-sinus volume in Neandertals than in recent humans (Vinyard and Smith 1997). Torus size and sinus volume were probably causally linked during Neandertal ontogeny. Heim (1978) suggests that projecting supraorbital tori probably result from large sinuses. But if the size of the torus reflects prognathism, as the so-called spatial model suggests (Chapter 6), then the large sinuses may be the effect rather than the cause.

The protuberant supraorbital torus is the most conspicuous feature of the front part of the Neandertal braincase (Figs. 7.6, 7.7, 7.9–7.11). Unlike the straighter, more bar-like tori of some Erectines and Heidelbergs, the Neandertal torus has a dip in the middle (a superior depression above glabella) that divides the bar into separate arches over the two eye sockets. The torus of Neandertals is separated from the frontal squama by a supratoral sulcus, as in most other archaic humans; but the sulcus is generally not as pronounced as it is in Erectines, perhaps because the frontal part of the braincase was bigger in Neandertals.

Some of the peculiarities of the Neandertal supraorbital torus are reflected in its metrics. Its **thickness**, measured from top to bottom, is greatest over the nasal edge of the eye socket and dwindles laterally. This is also true of many Heidelbergs; but the lateral thinning of the torus is more pronounced in Neander-

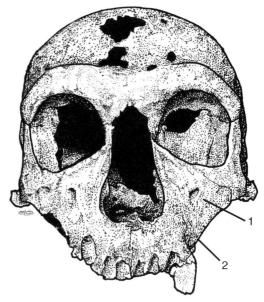

FIGURE 7.9

The face of the La Chapelle-aux-Saints Neandertal (France). Neandertals are characterized by large faces and nasal apertures, broad interorbital regions, and broad anterior palates. Supraorbital tori are both projecting and thick, but they tend to thin slightly laterally. The zygomaticoalveolar margin is markedly oblique and has its root low on the alveolar process (1). There is no evidence of a canine fossa (2). (After Martin 1923.)

tals, and the tori of Neandertals do not thin evenly as those of most Heidelbergs do. Rather, Neandertal tori are thinnest over midorbit, and thicken again where the torus connects to the lateral orbital margin (Table 7.6). The supraorbital tori of the Early Upper Paleolithic modern humans are thinner overall than those of Neandertals, especially at midorbit. The same pattern is found in recent human skulls (Cole and Cole 1994).

The Neandertal supraorbital torus is not only vertically thick, but also horizontally protuberant—that is, it sticks out in front of the braincase farther than the brow ridges of modern humans do. Its horizontal **projection**, defined as the anteroposterior distance between the inner surface of the braincase and the front of the torus (Smith and Ranyard 1980), is greatest laterally and smallest medially. This is what we would expect to see in an anthropoid skull: since the front edge of the braincase curves backward to either side and the orbits face roughly forward, the distance between the braincase and the bar over the orbital openings naturally increases laterally. Torus projection in Neandertals is everywhere greater than in modern humans, but the difference between the two groups is slight in the lateral plane and

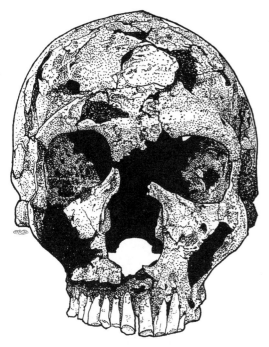

FIGURE 7.10

The La Ferrassie 1 Neandertal face (France), frontal view. This specimen shows the same basic morphological pattern seen in La Chapelle (Fig. 7.9). Note the extensive occlusal wear on the anterior teeth, which have lost practically all their enamel. All that is visible of these teeth in this illustration is their roots, exposed because the bone of the alveolar process has been broken away.

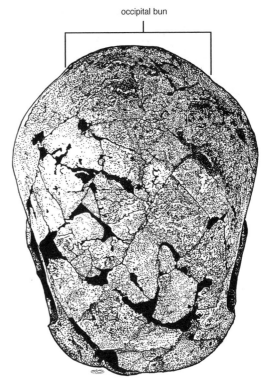

occipital bun

FIGURE 7.11

La Ferrassie 1 (France), superior view. Expansion of the anterior braincase in Neandertals results in a reduced degree of postorbital constriction, similar to that seen in recent humans.

TABLE 7.6 ■ Dimensions of the Supraorbital Torus of Neandertals and Comparable Structures in Early Modern Humans[a]

	Projection			Thickness		
Sample	Lateral Plane[b]	Midorbital Plane[c]	Medial Plane[d]	Lateral Plane	Midorbital Plane	Medial Plane
Central European Neandertals (N, SD)	22.6 (14, 1.8)	22.4 (17, 2.9)	20.2 (5, 2.0)	11.8 (17, 1.6)	9.9 (19, 1.9)	17.0 (5, 2.8)
Krapina (N, SD)	24.3 (8, 1.4)	23.9 (11, 1.2)	20.3 (4, 2.3)	12.5 (11, 1.6)	10.7 (13, 1.8)	17.6 (4, 3.0)
Vindija (N, SD)	22.1 (5, 1.8)	19.0 (5, 3.1)		10.6 (5, 0.5)	8.6 (5, 0.6)	
Western European Neandertals (N, SD)	24.4 (8, 2.4)	23.0 (8, 2.6)	22.5 (7, 2.1)	12.5 (9, 1.5)	11.1 (9, 0.7)	20.1 (8, 2.0)
Central European EUP (N, SD)	20.3 (9, 2.6)	16.1 (9, 3.4)	13.0 (9, 3.0)	8.1 (11, 1.4)	5.4 16.6 (11, 1.7)	(11, 3.3)

[a]Means (in mm) of measurements were taken by FHS.
[b]Measured just inside the lateral orbital wall.
[c]Measured at the thinnest part of the torus, just lateral to the middle of the orbit.
[d]Measured just lateral to the medial orbital margin.

TABLE 7.7 ■ Presence of Selected Discrete and Metric Traits in the Skulls of Late Pleistocene Humans from Europe and Western Asia

Sample	Suprainiac Fossa	Anterior Mastoid Tubercle	Horizontal-Oval Mandibular Foramen	Nasion Projection[a]
European Neandertals	100 (23)[b]	34.8 (23)	52.6 (19)	29.3 (11)
EUP	38.5 (26)	20.0 (25)	18.2 (22)	21.9 (16)
Late UP	23.7 (37)	0 (19)	6.7 (30)	19.3 (23)
Skhūl/Qafzeh	14.2 (7)[c]	40.0 (5)	0 (2)	12.4 (3)
European Mesolithic	19.3 (161)	0 (179)	1.9 (161)	19.4 (114)

Data from Frayer (1992b) except where noted. Incidences of discrete traits are expressed as percentages; nasion-projection data are in mm. Numbers in parentheses indicate sample sizes. Definitions for the features are given in the text and in Frayer (1992a,b).
[a]Projection of nasion anterior to the bi-fmt plane. This is the plane that extends between the two frontomalaria temporalia (see Appendix).
[b]Frayer (1992b) considers one Neandertal from Krapina (Kr 11.5) to lack a fossa. However, FS believes that this specimen does not preserve the area necessary to make a judgment on presence or absence of the fossa.
[c]Frayer (1992b) identifies suprainiac fossae in both Qafzeh 6 and Skhūl 9; FS identifies a fossa only in the latter specimen.

very marked in the medial plane (Table 7.6). To put it another way, the Neandertal supraorbital torus is very protuberant toward the midline (where it contains the big frontal sinus), but is swept backward toward the sides, so that it roughly parallels the curved front edge of the braincase. This reflects the posterior displacement of the lateral parts of the face, or "zygomatic retreat," typical of Neandertal skulls (see below: Fig. 7.16). The marked projection of the Neandertal torus toward the midline is thus one aspect of the distinctive prow-like midsagittal protuberance of the Neandertal face. This is seen also in the strong projection of nasion (the upper end of the suture between the two nasal bones) in Neandertals compared to later Europeans (Table 7.7).

The supraorbital torus and the enclosed frontal sinus took their final, impressive form late in Neandertal ontogeny, just prior to the attainment of full adulthood (Smith and Ranyard 1980, Wolpoff et al. 1981, Ahern and Smith 2004), in connection with the eruption of the last permanent teeth (Oyen et al. 1979). Neandertal children do not exhibit such prominent supraorbital tori, and even the late adolescent Neandertal Le Moustier 1 lacks the prominent torus of an adult (Vlček 1970, Smith and Ranyard 1980, Minugh-Purvis 1988). Nevertheless, the beginning of a torus can be palpated and visually identified even in the youngest known Neandertals, which distinguishes them from modern human children of similar age.

Occipital Bones

The occipital bone at the back of the human braincase roughly models the size and shape of the posterior end of the brain. Like modern humans—and unlike the Erectines (Table 5.2)—Neandertals have an occipi-

tal plane (the part of the bone above inion) that is longer in a midsagittal section than the nuchal plane (the part below inion), reflecting the enlargement of the brain's occipital cortex. But because Neandertals have a broader braincase than modern humans, their occipital bones are usually broader as well, and so the *area* of the nuchal plane (and thus the attachment area for the neck muscles) is generally greater in Neandertals than in more recent humans. Compared to Erectines, Neandertals have less pronounced nuchal tori, normally restricted to the medial half of the bone and not marked superiorly by a supratoral sulcus. Rak et al. (1994) contend that Neandertals are also distinguished by an elongated foramen magnum. This is consistent with other indications that the cranial base is elongated in Neandertals (Smith 1991). However, like many other supposed peculiarities of Neandertals, an elongated foramen magnum is found in some other human populations and cannot be counted as a Neandertal autapomorphy.

Some other features of the Neandertal occipital bone are more distinctive. The most significant of these are the presence of an inferiorly projecting **occipitomastoid crest** or eminence, a **suprainiac fossa**, and an **occipital bun**. The occipital bun is a posterior protuberance of the back of the vault (Figs. 7.7 and 7.11), produced by a combination of all of the following features:

► A marked flattening of the parietals and the occipital plane around the anatomical point of lambda (the rear end of the suture between the two parietals). This **lambdoidal flattening** extends to about the midpoint on the course of the lambdoidal suture (between the parietals and the occipital) on each side and continues onto the posterior part of each parietal.

▶ A relatively vertical lower part of the occipital plane, from the nuchal torus below to the onset of the lambdoidal flattening above.

▶ The presence of a **nuchal shelf**, a narrow flat area that runs essentially the entire breadth of the nuchal plane just below the nuchal torus.

▶ A nuchal plane that is horizontally oriented anterior to the nuchal shelf.

▶ A more inferior position of the part of the nuchal plane anterior to the shelf, compared to more recent humans (see below).

Because all these features must be present for an occipital protuberance to count as a true occipital bun, the presence or absence of a bun cannot be scored on a skull unless the posterior cranial vault is virtually complete. Only 11 of the known adult European Neandertal skulls are complete enough to make an unequivocal judgment about bunning. Of these 11, nine (81.8%) have fully developed occipital buns. The nine are La Quina 5, La Chapelle, Ehringsdorf H, Gánovce, Biache 1, Neandertal 1, Guattari 1, Spy 1, and Spy 2 (with a weak bun). The other two—Saccopastore 1 and La Ferrassie 1—possess some, but not all, of the full complement of bunning features.

Neither cranial size nor geological age seems to correlate with the presence of an occipital bun. The early and small Saccopastore 1 and the large and late La Ferrassie 1 (Fig. 7.6) both lack this feature, whereas Biache 1, one of the earliest and smallest Neandertals, has a well-developed occipital bun (Fig. 7.32). Moreover, occipital buns are characteristic of only European Neandertals. The three Asian Neandertals that are complete enough to assess the presence of bunning—Tabun C1, Shanidar 1, and Amud 1—all lack most of the components that define the feature. This difference between European and Asian Neandertals is reflected in their occipital angles (Table 7.5). The relatively low angles found in European Neandertals are due to the presence of the bun, which compresses the rear vault in a superior–inferior direction and thus decreases the occipital angle. Values of this angle for Asian Neandertals are higher (but not as high as the mean values seen in moderns). This difference is chiefly due to the absence of bunning in the Asian skulls, although the estimate for the Shanidar 1 skull from Iraq is also unnaturally inflated by postmortem distortion (Trinkaus 1983b).

It has been hypothesized that the effect of bunning on the occipital angle is the reason for the evolution of the occipital bun—that the bun developed in order to direct the nuchal plane downward, thus providing a more horizontal surface for the attachment of the nuchal muscles at the back of the neck (Brose and Wolpoff 1971, Wolpoff 1980a, Smith 1983). By this analysis, the occipital bun is a sort of functional equivalent of the

nuchal crest of more primitive hominids. This may be a side effect of occipital bunning, but it seems unlikely to be the prime reason for the structure's existence.

Trinkaus and LeMay (1982) argue that the cranial vault generally responds to changes in brain growth, and that occipital buns are no exception. They suggest that bunning is a side effect of posteriorly directed growth of the occipital pole of the brain, which produces a localized posterior bulge in the upper part of the occipital bone. Trinkaus and LeMay note that anteriorly directed growth of the front end of the brain ceases in modern humans at the end of the third postnatal year, when the brain has attained about 80% of its adult size. The remaining 20% of brain growth is concentrated more toward the rear. Therefore, a spurt in brain growth late in ontogeny might be expected to affect the occipital pole of the brain disproportionately, and the resulting response of the surrounding bone at the rear of the cranial vault might yield an occipital bun. Neandertal ontogenetic patterns are compatible with this idea. The youngest Neandertal individuals, including the roughly 4-year-old Subalyuk 2 Neandertal, lack strong indications of bunning (Pap et al. 1996). Slightly older Neandertals—for example, the ~6-year-old specimens Engis 2 (Fraipont 1936, Tillier 1983) and Krapina 2 (Gorjanović-Kramberger 1906, Smith 1976)—exhibit more distinct, but still incipient, occipital buns. However, the ~14-year old Le Moustier 1 juvenile has only a weak development of bunning features.

Trinkaus and LeMay do not suggest a functional reason for the supposed late growth spurt of the brain in Neandertals. It is unlikely to reflect any structural uniqueness of the Neandertal brain, because occipital buns are also found in most of the early modern human skulls from Europe (and in a few from North Africa). This distribution is sometimes cited as evidence for gene flow from Neandertals into early modern populations in this part of the world (Chapter 8). Gunz and Harvati (2006) tentatively accept the Trinkaus–Le May explanation for bunning, but contend that the degree of bunning, even in early modern specimens, can be predicted based on overall skull form. According to them, this invalidates its use as a distinct morphological feature.

The origin of the term "occipital bun" merits a brief mention. It is a translation of the French phrase *chignon occipital*, first applied by Boule to the La Chapelle specimen. *Chignon* means a "bun" in the sense of a woman's hairstyle, in which the hair is worn in a tight roll on the back of the head, and in which Boule saw a fanciful resemblance to the protuberance of the Neandertal occiput. The literal English translation confuses many students, who struggle to recognize some similarity between this Neandertal feature and the bread around their hot dogs.

The suprainiac fossa is a shallow depression located just superior to and bordering the nuchal torus. The

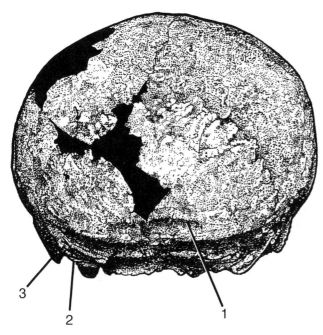

FIGURE 7.12

La Chapelle-aux-Saints (France), posterior view. The maximum breadth of the braincase falls low on the parietals, yielding a typically oval (*"en bombe"*) shape of the vault viewed from the rear. Also note the well-developed, horizontally elongated suprainiac fossa (1), the inferior position of the occipitomastoid region (2), and the relatively nonprojecting and medially inclining mastoid process (3).

superior nuchal line (the attachment of the fascia covering the nuchal muscles) defines the fossa's inferior boundary. (The superior boundary of the fossa is not precisely delimited.) In Neandertals, this fossa generally takes the form of a broad oval, with the long axis of the oval running parallel to the torus below it (Fig. 7.12). However, it is sometimes more nearly circular, and may be divided by a low vertical ridge near the midline. The surface bone in the fossa itself is characterized by distinct pinprick-like pitting, not found on the surrounding bone, which gives the fossa an impression of rugosity. This fossa was first noticed by Gorjanović-Kramberger (1902), and was briefly discussed by Weidenreich (1940) as the *fossa supratoralis*. Apart from these mentions, it was largely ignored until the late 1970s, when the French paleoanthropologist J.-J. Hublin redirected scientific attention to this distinctive structure.

The reason for the fossa's existence is debated (Hublin 1978, 1980; Caspari 1991). We suspect that it is connected with the absence of an external occipital protuberance in Neandertals. In modern humans, this protuberance represents the attachment point for the nuchal ligament. Pinprick pitting resembling that seen

in the suprainiac fossa is sometimes seen above the occipital protuberance in modern crania, especially during childhood (Heim 1982, Frayer 1992a). This suggests that the nuchal ligament in Neandertals attached below the suprainiac fossa. Evidently, the ligament was more diffuse and fan-like than ours, with attachments extending laterally along the underside of the nuchal torus rather than being concentrated in the midline. Caspari (1991) posits that the suprainiac fossa is an area of resorptive bone that develops in response to bending forces acting along the nuchal torus. A broad attachment of a fan-like nuchal ligament to the underside of the torus would explain the absence of the external occipital protuberance, and might produce the pattern of tension and suprainiac resorption proposed by Caspari.

Both Hublin (1978, 1980) and A. Santa Luca (1978) consider the suprainiac fossa to be a defining characteristic of Neandertals, although Hublin acknowledges that the fossa is also found in low frequencies in modern humans. All European Neandertal occipital bones that preserve the pertinent anatomical region show suprainiac fossae (Table 7.7).

The occipitomastoid crest or eminence is located lateral to the foramen magnum and just medial to the occipitomastoid sulcus. The surface of this eminence bears a distinct ridge in Neandertals, possibly associated with the attachment of the epaxial muscle obliquus capitis superior. But the distinctive feature of the Neandertal skull is not this ridge, but a general lowering (more inferior position) of this part of the basicranium, extending even over onto the temporal bone. This lowering reduces the degree to which the temporal bone's mastoid process protrudes below the cranial base around it. The Neandertal mastoid process is usually described as smaller and less projecting than that of modern humans (Vallois 1969, Heim 1976, Hublin 1988c, Stringer et al. 1984, Condemi 1988, 1992), but this description is misleading. The distance from the tip of the mastoid process to the top of the external acoustic meatus (from mastoideale to porion) is actually about the same in Neandertals as in Upper Paleolithic and more recent human skulls (Table 7.8). The mastoid process only appears to be shorter and smaller in Neandertals (cf. Figs. 7.6 and 8.34) because the whole occipitomastoid region medial to it projects inferiorly almost as far as the mastoid process does. The projection of the mastoid below the surrounding bone (measured on the medial edge of the mastoid, from the floor of the digastric groove; see Fig. 7.12) therefore averages several millimeters less than in more recent humans (Table 7.8). It is not clear why the occipitomastoid surface has been displaced downward in Neandertals. It may be related to the same pattern of brain growth that produces an occipital bun. However, even when a bun is not present in

TABLE 7.8 ■ Mastoid Process Projection in Neandertals and Modern Humans[a]

Sample/Specimen	Mastoid Projection[b]	Porion-Mastoidale[c]
Neandertal mean	6.6	32.6
(N, σ − 1)	(22, 2.5)	(18, 4.2)
EUP mean	9.1	32.0
(N, σ − 1)	(7, 3.9)	(7, 32.9)
Mladeč 1	11.0	32.0
Mladeč 5	8.7	34.3
Brno 2	11.0	34.0
Pavlov	16.8	39.6
Dolní Věstonice 3	5.5	28.0
Recent human mean[d]	9.1	30.2
(N, σ − 1)	(99, 2.6)	(99, 3.1)
Recent human male mean	10.6	32.0
(N, σ − 1)	(48, 2.1)	(48, 2.6)

[a]Measurements (in mm) are from Smith (1982).
[b]Measured medially from the floor of the digastric sulcus.
[c]Measured from just above the external auditory meatus (porion) to the tip of the mastoid process (mastoidale).
[d]Sexes pooled. The modern humans in this table are Proto-Historic Arikara Native Americans.

Neandertals, the inferior projection of the occipitomastoid crest always is.

Temporal Bones

Neandertal temporal bones exhibit several arguably distinctive features, both externally and internally (Gorjanović-Kramberger 1906, Vallois 1969, Smith 1976, Santa Luca 1978, Hublin 1978, 1988c; Stringer et al. 1984, Condemi 1988, 1992; Hublin et al. 1996, Ponce de Léon and Zollikofer 1999, Spoor et al. 2003, Harvati and Weaver 2006 a,b). The most significant of these are:

▶ A gracile, weakly projecting mastoid process.

▶ The presence of anterior mastoid tubercles.

▶ Pronounced development of the supramastoid crests.

▶ Broad digastric sulci (reflecting the increased breadth of the cranium).

▶ A relatively high position of the external acoustic meatus relative to the mandibular fossa.

▶ The morphology of the mandibular fossa.

▶ The form of the bony labyrinth.

K. Harvati (2003; Harvati and Weaver 2006b) has shown that multivariate metric analyses distinguish Neandertal temporal bones from those of both early modern and recent humans. Many of her metric variables reflect the features listed above.

As noted above, the weak projection of the Neandertal mastoid process should be viewed as a peculiarity of

the occipital bone, not of the mastoid process itself. The other supposed peculiarity of the Neandertal mastoid, the **anterior mastoid tubercle**, was described by Santa Luca (1978) as a rounded protuberance found just posterior to the external acoustic meatus. As far as is known, this tubercle is not associated with the attachment of any muscle or tendon. Santa Luca found this structure on all of the specimens he used as his "core" Neandertal group, including the Krapina 3 cranium, and both he and Hublin (1978) consider the feature as apomorphic for Neandertals. But others (Frayer 1992b; FS, personal observations) note that all adult temporals from Krapina lack this tubercle, and that its incidence among Neandertals overall is not appreciably greater than in other Late Pleistocene human samples (Table 7.7).

The socket of the jaw joint (the mandibular or glenoid fossa) is typically large in Neandertals, with well-developed medial and posterior walls (Gorjanović-Kramberger 1906, Smith 1976). Unlike those of many modern humans, the medial wall of the fossa is generally formed entirely from the temporal bone in Neandertals, with no significant sphenoid contribution. The posterior wall is formed by a **postglenoid process** that is thicker and more projecting than those of most modern humans. Behind this process, the bony tympanic ring surrounding the external acoustic meatus also tends to be thick in Neandertals.

In contrast, the anterior margin of the mandibular fossa is generally less well marked in Neandertals than in recent populations, because Neandertals tend to have a less pronounced articular eminence (Fig. 7.13). This bump at the anterior edge of the jaw socket becomes the articular surface for the mandibular condyle whenever the jaws are parted, when the condyle moves forward and downward along the tubercle's posterior surface. (You can feel this motion of the condyle in your own head by placing a finger in front of your ear hole and opening your mouth.) As the cheek teeth come together during the power stroke of mastication, the condyle glides and rotates back up into the center of the mandibular fossa. Pressure exerted on the temporal bone by the condyle—which is to say, all the chewing force that is not transmitted to the upper jaw through the teeth—is therefore concentrated toward the center of the mandibular fossa during molar chewing in centric occlusion, but is shifted onto the articular eminence when the mouth is opened to bite with the anterior teeth (for example, in biting into an apple).

In a systematic study of articular-eminence variation in humans, Hinton (1981) concluded that a weakly developed tubercle generally reflected heavy use of the anterior dentition. Its reduction in Neandertals would thus seem to imply heavy anterior dental loading. Two other lines of evidence point to the same conclusion. The first is the extraordinary amount of anterior dental wear seen in many Neandertals (see below). The second

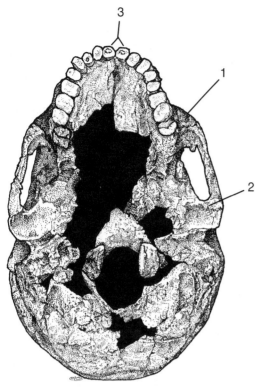

FIGURE 7.13

La Ferrassie 1 (France), basal view. Note the relatively posterior position of the lateral face (1), the poorly defined articular tubercle of the glenoid fossa (2), and the extensive wear of the anterior teeth, often resulting in exposure of the pulp cavity (3).

is the high incidence of degenerative joint disease on the articular surfaces of Neandertal mandibular fossae and condyles (Gorjanović-Kramberger 1906, Alexandersen 1967, Smith 1976), reflecting heavy stress on the jaw joint. This stress could not have resulted entirely from anterior dental loading (Antón 1994), but it seems clear that such loading must have contributed significantly to it.

The **bony labyrinth** of the Neandertal inner ear is also distinctive (Spoor and Zonneveld 1998, Spoor et al. 1994). The bony labyrinth comprises a system of little cavities that enclose the semicircular canals and cochlea inside the petrosal (petrous) part of the temporal bone and provide a sort of loose-fitting external cast of those inner-ear structures. Labyrinth morphology is established prenatally, and can tell us things about an animal's locomotion, posture, or taxonomic relationships (Spoor and Zonneveld 1998, Spoor et al. 2007a). High-resolution CT scans (Fig. 7.14) reveal that in Neandertals the anterior and posterior semicircular canals are relatively smaller than in modern humans, the posterior canal has a lower position relative to the plane of the lateral canal, and a line through the ampullae of the anterior and posterior canals forms a higher angle with the plane of the lateral canal (Spoor et al. 2003). These morphological differences appear early in ontogeny; the Neandertal pattern is found in specimens as young as two years of age (Spoor et al. 2002). Hublin and colleagues (1996) regard the Neandertal pattern as a derived (autapomorphic) feature, because the labyrinth of the

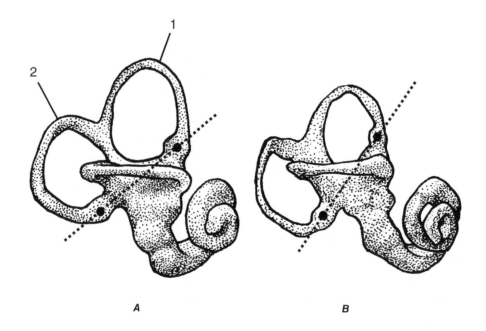

A   *B*

FIGURE 7.14

Virtual (CT) endocasts of the bony labyrinth of the inner ear in modern humans (**A**) and Neandertals (**B**): lateral aspect. The anterior (1) and posterior (2) canals are larger in modern humans. Moderns also have a higher position of the posterior canal and have a smaller angle of the ampullar line (*dashed line*) relative to the lateral canal. (After Spoor et al. 2003.)

Erectines is more like that of moderns (Spoor and Zonneveld 1994). Like many other distinctive Neandertal features, the Neandertal type of bony labyrinth is also found in European Heidelbergs (Spoor et al. 2003).

It is not clear what these differences in pattern mean. They presumably do not reflect postural differences, because Neandertals were certainly habitual, fully upright bipeds. Some have speculated that the peculiarities of the Neandertal labyrinth may reflect unspecified differences in brain development, in head movements, or in locomotion (Hublin et al. 1996, Spoor et al. 2003). Although this distinctive labyrinth form has been found in 15 Neandertal skulls (Spoor et al. 2003), the Le Moustier 1 specimen from France has a pattern more like that seen in modern humans (Ponce de Léon and Zollikofer 1999). Thus, as is the case with many other features, this supposedly distinctive Neandertal apomorphy is not present in all individuals.

Brains

Neandertal brain size, as measured by cranial capacity, is often said to be greater than that of modern humans. But consideration of the full range of Neandertal endocranial volumes gives a slightly different impression (Table 7.9). The mean for Würm European Neandertals (after OIS 5d to OIS 3) is actually smaller, though not significantly so, than the Early Upper Paleolithic (EUP) mean in Europe, and the same relationship exists in West Asia. All of these groups, Neandertals and moderns alike, tend to have larger cranial capacities than *recent* humans (Tobias 1971, Ruff et al. 1997, Holloway 2000). This fact reflects post-Pleistocene decreases in average body mass (Table 6.1). When body size is taken into account, there is no evidence that Neandertal brains differ significantly in size from those of modern humans. However, pre-Würm Neandertals do appear to have smaller brains than the later (Würm) Neandertals have. This difference may be the result of evolutionary change. It probably does not represent a latitudinal difference, because the most southerly Neandertals (the ones from West Asia) have the largest mean cranial capacities (Table 7.9).

Similarity in size does not guarantee similarity in structure. The Neandertal brain was somewhat different in shape and proportions from the brains of modern humans, and some have speculated that there may have been corresponding differences between the two groups in neural organization and mental abilities. Unfortunately, endocasts of fossil hominin braincases provide only vague and general hints about brain structure. R. Holloway, who has studied more fossil hominin endocasts than anyone else and has done everything possible to draw inferences from the anatomy of the skull to the anatomy of the brain, finds nothing in Neandertal endocasts to suggest significant differences from modern

humans in neural structure or function (Holloway 1981b, 1985; Holloway et al. 2004; cf. Kochetkova 1978). Like the Erectines, Neandertals exhibit a modern human pattern of hemispheric asymmetry. They show a human-like development of Broca's and Wernicke's areas. Although these assessments of surface topography tell us little about the internal organization of the brain, the available indicators provide no real support for the claim that Neandertal brains were functionally different from our own.

One possibly significant difference noted by Holloway is that Neandertal brains appear to have been somewhat larger than those of most recent humans in the anterior part of the occipitomastoid area of the cerebral cortex. This is in the area where Trinkaus and LeMay's postulated neural growth spurt (related to occipital buns) is supposed to have occurred. If this increase is real, it may signal an expansion of the primary visual striate cortex in Neandertals—which might indicate that they possessed enhanced visual and spatial abilities. A possible parallel among recent humans is seen in Native Australians, who have expanded visual striate cortices and make higher average scores than Europeans do on tests of visual–spatial abilities (Kleklamp et al. 1987).

NEANDERTAL FACES

The most obvious fact about Neandertal faces (Figs. 7.9, 7.10 and 7.15) is that they are *big*, in all three dimensions: height, breadth, and length. The faces of some Heidelbergs are just as massive, but later humans pale in comparison. For example, upper facial height in Neandertals is significantly greater than in Early Upper Paleolithic (EUP) humans or in a sample of three recent European populations (Table 7.10). The faces of the early moderns from Skhūl/Qafzeh (mean = 73.2 mm, $N = 5$, SD = 3.7) are a bit taller than the modern European average (around 68 mm in males), but still fall well below the Neandertal mean (85.6). Overall upper facial breadth (represented by bifrontal breadth in Table 7.10) follows a similar pattern. The expanded vertical height and transverse breadth of Neandertal faces are also reflected in alveolar height (Table 7.10), nasal height, nasal breadth (Table 7.11), and most other facial height and breadth measurements.

Like Erectines and Heidelbergs, Neandertals had broad interorbital areas and thick, columnar lateral orbital margins. In modern humans, the frontal (postorbital) process of the zygomatic bone comprises two plates: an **orbital plate** that contributes to the postorbital septum, and a **facial plate** that faces forward and forms the lower part of the postorbital bar. The two meet at the lateral margin of the orbit but diverge as they extend away from that margin, producing a V-

TABLE 7.9 ■ Estimated Adult Cranial Capacities for European and Asian Neandertals and Early Modern Humans.

Specimen	Group	Cranial Capacity	Specimen	Group	Cranial Capacity
Ehringsdorf H	European Neandertal (pre-Würm)	1450	La Chancelade	"	1530
			Combe Capelle	"	1570
			Cap Blanc	"	1434
Biache-St. Vaast	"	1200	Bruniquel 2	"	1555
Lazaret	"	1250	Predmostí 3	"	1580
Krapina 2	"	1450*	Predmostí 4	"	1250
Krapina 3	"	1255	Predmostí 9	"	1555
Krapina 6	"	1205	Predmostí 10	"	1452
Saccopastore 1	"	1245	Brno 2	"	1600
Saccopastore 2	"	1300	Mladeč 1	"	1540
Gánovce	"	1320	Mladeč 2	"	1390
Guattari 1	European Neandertal (Würm)	1360	Mladeč 5	"	1650
			Dolní Věstonice 3	"	1285[a]
Engis 2	"	1362*	Dolní Věstonice 13	"	1481[a]
Spy 1	"	1305	Dolní Věstonice 14	"	1538[a]
Spy 2	"	1553	Dolní Věstonice 15	"	1378[a]
La Chapelle-aux-Saints	"	1625	Dolní Věstonice 16	"	1547[a]
La Ferrassie 1	"	1640	Pavlov	"	1472[a]
Neandertal 1	"	1525	Sungir 1	"	1464
La Quina 5	"	1172	Skhūl 1	West Asian early modern	1450*
La Quina 18	"	1200*			
Gibraltar 1	"	1200	Skhūl 4	"	1554
Gibraltar (Devil's Tower)	"	1400*	Skhūl 5	"	1520
Le Moustier 1	"	1565*	Skhūl 9	"	1590
Tabūn C1	West Asian Neandertal	1271	Qafzeh 6	"	1600
			Qafzeh 9	"	1508[b]
Amud 1	"	1740	Qafzeh 11	"	1280*
Shanidar 1	"	1600			
Shanidar 5	"	1550	European Neandertal mean (N, SD)		1350 (16, 158)
Teshik Tash	"	1525*	Pre-Würm European Neandertal mean (N, SD)		1278 (8, 81)
Cro-Magnon 1	European early modern	1730	Würm European Neandertal mean (N, SD)		1423 (8, 187)
			Asian Neandertal mean (N, SD)		1540 (4, 196)
			Early Upper Paleolithic mean (N, SD)		1504 (21, 114)
Cro-Magnon 3	"	1590	Skhūl/Qafzeh mean (N, SD)		1554 (5, 41)

All values (in cm^3) are taken from Holloway (2000) or Holloway et al. (2004) unless otherwise indicated. These data differ from those presented in Table 6.1, in large part because the data are grouped differently and somewhat different sample constructions are employed. European Neandertals are divided into pre-Würm (end of OIS 7 through OIS 5e) and Würm (after OIS 5e) subsets. Asterisks indicate subadult data, not included in determinations of means.
[a]From Vlček (1993).
[b]From Vandermeersch (1981).

shaped trough between them in back that forms part of the front end of the temporal fossa. But in Neandertals and other archaic humans, these two plates are not as distinct, and the sulcus between them is largely filled in with bone, forming a thick postorbital bar or **lateral orbital pillar** with an increased cross section (Fig. 8.20).

Despite these differences between Neandertals and modern humans in the bony structures around the orbit, the orbital cavities themselves do not differ as much as is often claimed. Although Neandertals had broader orbits than recent Europeans have (Table 7.10), early modern humans were more like Neandertals than their modern descendants are. Average orbital breadths in the

EUP and Skhūl/Qafzeh (mean = 43.4, N = 4, SD = 2.5) samples lie within a single standard deviation of the Neandertal mean. B. Maureille and F. Houët (1997) find significant differences in size between Neandertal and recent human orbits (no early moderns were compared), but no differences in *shape*. This finding contradicts previous claims (e.g., Heim 1976) that Neandertals have rounder orbits.

The front teeth of Neandertals are very big (Table 6.5). The alveolar part of the Neandertal maxilla is therefore both tall and wide, to accommodate the long roots and broad crowns of the incisors and canines (Smith and Paquette 1989). The large size of the Neandertals'

TABLE 7.10 ■ Neandertal, Early Modern, and Recent Facial Metrics[a]

Sample[b]	Upper Facial Height	Alveolar Height	Interorbital Breadth	Orbital Breadth	Bifrontal Breadth[c]
Neandertal mean	85.6	25.5	28.1	45.9	112.7
(N, SD)	(10, 5.1)	(10, 3.1)	(12, 2.3)	(10, 3.2)	(12, 3.0)
EUP mean	65.8	18.7	25.7	42.0	102.6
(N, SD)	(8, 2.7)	(8, 3.6)	(11, 3.4)	(8, 3.6)	(9, 6.3)
Norse mean	68.9	—	22.3	40.4	99.0
(N, SD)	(55, 3.4)		(55, 2.5)	(55, 1.4)	(55, 2.9)
Zalavar mean	68.5	—	21.4	40.0	98.1
(N, SD)	(54, 4.2)		(54, 2.4)	(54, 1.2)	(54, 2.1)
Berg mean	67.9	—	22.9	40.1	99.6
(N, SD)	(56, 4.2)		(56, 2.5)	(56, 1.4)	(56, 3.3)

[a]All measurements are in mm.

[b]Berg, Zalavar, and Norse data are from Howells (1973b) and are for males only. Shanidar 1 and 5 data from Trinkaus (1983b). All other measurements taken by FS. EUP, Early Upper Paleolithic.

[c]Bifrontal breadth is the maximum facial breadth at the zygomaticofrontal suture (see Howells 1973b).

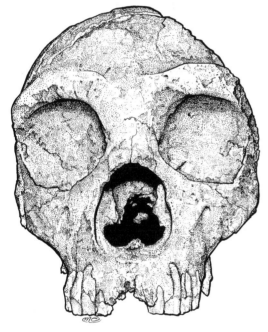

FIGURE 7.15

The Gibraltar Forbes Quarry Neandertal, frontal view. Although small in size, Gibraltar 1 displays the same facial morphology seen in larger Neandertals. The nose exhibits the internal nasal rim and lateral wall swellings emphasized as Neandertal apomorphies by J. Schwartz and I. Tattersall (1996).

TABLE 7.11 ■ Neandertal, Early Upper Paleolithic, and Recent European Nasal Metrics[a]

Sample/Specimen	Nasal Height	Nasal Breadth	Nasal Index
Neandertal mean	61.6	33.6	55.4
(N, SD)	(10, 4.2)	(12, 2.5)	(9, 3.7)
La Chapelle	62	34	54.8
Gibraltar 1	55	34	61.8
Shanidar 1	62	30	48.4
Vindija 225	—	28	—
Vindija 259	—	26	—
EUP mean	52.2	24.5	47.2
(N, SD)	(10, 4.1)	(10, 1.4)	(10, 47.2)
Cro Magnon 1	52	23	44.2
Mladeč 1	50	25	50.0
Mladeč 8	—	31	—
Norse mean	51.9	25.4	48.9
(N, SD)	(55, 2.7)	(55, 1.5)	
Zalavar mean	51.4	25.4	49.4
(N, SD)	(54, 2.9)	(54, 1.5)	
Berg mean	51.7	25.5	49.3
(N, SD)	(56, 2.9)	(56, 2.0)	

[a]All measurements are in mm. Nasal index = 100 × nasal breadth/nasal height. Measurements are by FS except for Shanidar (Trinkaus 1983b) and the recent human samples (Berg, Zalavar, and Norse: Howells 1973b).

front teeth is reflected metrically in their alveolar height (Table 7.10) and in various measurements of the breadth of the lower face and anterior palate (Smith 1983, 1984). For example, the mean distance from the anterior midpoint on the palate (prosthion) to the back edge of the upper canine is 24.9 mm in Neandertals (SD = 1.1), but only 22.4 mm in Early Upper Paleolithic Europeans. The broad anterior palate of Neandertals may be one of the reasons why they have such wide nasal openings (Table 7.11).

Although large, the Neandertal face is not uniformly robust. There is strong internal buttressing around the nose and below and lateral to the orbits, but the cheek area itself is rather gracile. The infraorbital region of

TABLE 7.12 ■ Relative Upper Facial Projection[a]

Sample	NAR	NAR Index
Neandertal mean	110.4	53.9
(N, SD)	(8, 4.6)	(7, 2.10)
EUP mean	100.0	51.9
(N, SD)	(8, 8.4)	(8, 2.7)
Norse mean	95.3	50.5
(N, SD)	(55, 3.4)	
Zalavar mean	94.7	51.1
(N, SD)	(54, 3.6)	
Berg mean	94.9	52.6
(N, SD)	(56, 3.8)	
African mean[b]	92.8	50.1

[a]Nasion radius (NAR) is defined in Howells (1973b), which is the source for the recent human data. The Pleistocene samples were measured by FS, except for values for Shanidar (Trinkaus 1983b). The NAR index (100 × NAR/maximum cranial length) indicates the relationship of NAR to overall size.
[b]Means of five sample means in Howells (1973b).

Neandertals is classically described as looking "puffy," and it is in fact inflated with air by greatly expanded maxillary sinuses (Heim 1978). That expansion results in some distinctive morphological peculiarities. Like the walls of the supraorbital torus enclosing the frontal sinus in Neandertals, the bony front wall of the bloated maxillary sinus is exceedingly thin and often translucent. This wall, known as the **infraorbital plate**, is flat or even slightly convex in the Neandertal skull. In modern humans, the infraorbital plate is markedly concave, producing a sunken vertical furrow or depression below the orbit. This depression, the **canine fossa**, extends from the infraorbital foramen down to the lower edge of the bony cheek (the zygomaticoalveolar crest or margin). It is characteristic of modern human skulls and is also found in some Erectines and a few Heidelbergs. But in Neandertals (and most Heidelbergs), the canine fossa has been obliterated by the inflation of the "puffy" infraorbital region. The loss of this fossa is regarded by some as an important and uniquely derived (autapomorphic) peculiarity of Neandertals (Maureille and Houët 1997). However, it needs to be understood in the context of the entire Neandertal facial complex and not as an isolated character state.

Neandertals, like some Erectines and Heidelbergs, have a zygomaticoalveolar crest or margin (ZAM) that slopes down in a more or less straight line from the front of the zygomatic arch to the alveolar plane of the maxilla. In most modern human skulls, the ZAM is concave or inflected, with a roughly horizontal lateral part that forms a distinct angle with the more vertical medial portion (Fig. 8.3). The oblique ZAM of Neandertals (Figs.

7.9, 7.10 and 7.15) has been compared to the architectural structures known as "flying buttresses," which prop up and support the thin vertical walls of Gothic cathedrals. However, the absence of an inward bend of the ZAM, like the absence of a canine fossa, may be just a side effect of the "puffy" inflation of the Neandertal maxilla.

The infraorbital foramen, which opens in the middle of the infraorbital plate of the maxilla, transmits nerves and blood vessels that help to supply the skin of the face. This foramen is much larger in Neandertals than it is in modern humans. Coon (1962) thought that this was an adaptation to cold—that the foramen had been enlarged in Neandertals in order to accommodate larger blood vessels, increasing blood flow to the face and helping to prevent frostbite. However, modern human populations adapted to cold do not have particularly large infraorbital foramina (Steegmann 1970). The foramen may be large in Neandertals simply because a big face needs a big blood supply.

External Nose

The **nasal** or **piriform aperture** (nose opening) in the skull is exceptionally large in Neandertals, in keeping with the general size of the face. On average, Neandertal nasal apertures are both broader and taller than those of any modern humans from Europe, including the Upper Paleolithic sample (Table 7.11). The nasal apertures of Neandertals are also framed by projecting nasal bones, very prominent anterior nasal spines, and markedly everted upper lateral walls. These features and the sheer size of the opening suggest that a living Neandertal would have sported a truly impressive schnoz of Durante-like proportions.

Within the Hominidae, a projecting external nose is a peculiarity of the genus *Homo* (Chapter 5). It seems to have originated in early *Homo* as an adaptation for regulating and conserving temperature and moisture in respired air, in connection with an increase in daily ranges and activity budgets (Trinkaus 1987a, Franciscus and Trinkaus 1988). The big noses of Neandertals would have made them champions at conserving heat and water in this way—an ability that would have been adaptively valuable during cold, dry glacial maxima in northwestern Eurasia. Coon (1962) suggested that such respiratory functions explained both the big noses and the marked upper facial prognathism of Neandertals. In periglacial environments, Coon argued, the big Neandertal nose would have been invaluable in warming inspired air, so that it would not provide a cold shock to the brain stem above the nasal part of the pharynx. Increased prognathism in the upper part of the face would also have helped in warming inhaled air, by increasing the distance that the air had to travel before it encountered the braincase. Adaptations for warming and moistening inhaled air may have had other advantages for the Neandertals, such as reducing the incidence of upper respiratory infections.

This analysis sounds plausible, and studies have shown that in late Pleistocene *Homo*, metric variation in the face is more tied to climate than neurocranial variation is (Harvati and Weaver 2006a,b). However, comparisons with modern humans living in cold, dry climates cast serious doubts on Coon's interpretation. Such modern humans do not have Neandertal-like features of the nose, particularly its relative breadth. Some of the peculiarities of Neandertal nasal anatomy may have little or nothing to do with nasal function. For example, there is a strong correlation between nasal breadth and anterior palate breadth in humans (Glanville 1969), and it may be that the Neandertal nasal aperture is broad simply because of the space demands of the anterior teeth.

Prognathism

Like more primitive *Homo*, Neandertals had prognathic faces. But Neandertal faces are prognathic in a distinctive way. Erectines (and, to a lesser extent, Heidelbergs) exhibit total facial prognathism, in which all three components of the face—the dental part (alveolar prognathism), the nasal part (nasal or upper facial prognathism), and the cheek and lateral orbital region (lateral facial prognathism)—have a more forward position relative to the braincase than they have in modern humans. Alveolar prognathism in Neandertals is about the same as, or a little less than, that seen in Heidelbergs (Trinkaus 2003); but upper facial prognathism appears increased

in Neandertals compared to Erectines and most Heidelbergs, and lateral facial prognathism is reduced (i.e., the lateral parts of the face have a more posterior position). The resulting combination of protrusion in the midline and swept-back cheekbones to either side produces the distinctive "prow-like" or "beak-like" appearance of the Neandertal face, reflected metrically in uniquely low values of the subspinale angle (Table 5.7).

As intimated earlier (Chapter 6), one explanation for the oblique, backward-sloping orientation of the Neandertal cheek below the orbits is that the lateral parts of the face are moved backward or retreating relative to the midsagittal part of the face (Smith 1983, Trinkaus 1987b, 2003; Smith and Paquette 1989). An alternative view attributes this Neandertal peculiarity not to the retreat of the sides of the face, but to increased protrusion of the middle of the face (Rak 1986, Demes 1987, Spencer and Demes 1993, Maureille and Houët 1997, Antón 1994). The difference between these two perspectives is significant. If the Neandertal pattern of prognathism is produced by lateral facial retreat, then it can be seen as morphologically intermediate between the total facial prognathism of Erectines and the retreating, non-prognathic faces of recent humans (Fig. 7.16). By this reading, the Neandertal pattern would be distinctive but not necessarily autapomorphic (uniquely derived). But if the Neandertal pattern is caused by exaggerated forward protrusion of the upper face in the midsagittal plane, then it becomes reasonable to argue that the Neandertal facial form is autapomorphic. If so, the Neandertals are correspondingly more likely to have been a specialized side branch of the human lineage, with no living descendants (Rak 1986).

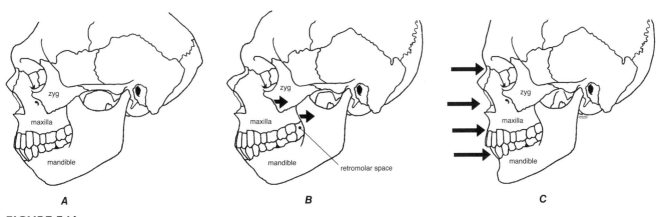

FIGURE 7.16

The zygomatic-retreat model of Neandertal facial morphology. In primitive *Homo* (**A**), the entire face is anteriorly positioned relative to the braincase. In Neandertals (**B**), the lateral parts of the face are displaced posteriorly (*arrows*), resulting in a protrusive, prow-like nasal and midfacial skeleton, a narrowed ramus of the mandible, and a retromolar space. In modern humans (**C**), the central part of the face and dental arcade are also pulled back (*arrows*). As a result, the nasal region, brow ridges, and maxilla become less prominent, and the retromolar space disappears.

Both of these interpretations find some support in the data. It seems obvious to visual inspection (Figs. 7.6 and 7.7) that the lateral parts of the face are more posteriorly placed in Neandertals than in either Erectines or later humans. This impression is confirmed by cranial measurements, which show that lateral facial projection, scaled against either maximum cranial length or alveolar projection (ZMR and ZMR/PRR indices, Table 5.7), is demonstrably less in Neandertals than it is in ER 3733 or recent *Homo sapiens*. These data support the "lateral facial retreat" model. But it is also true that the nasal region (at nasion) projects further in front of the braincase in Neandertals than it does either in modern humans (Tables 7.7 and 7.12) or in Erectines and Heidelbergs (Smith and Paquette 1989, Arsuaga et al. 1997c, Trinkaus 2003). This aspect of Neandertal facial form does appear to be apomorphic.

Appearances, however, can be deceiving. The position of the bridge of the nose relative to the cranial base depends on the size of both the upper facial skeleton and the anterior end of the braincase. Trinkaus (2003) suggests that the Neandertal nasal region is uniquely protrusive because Neandertals are unique in having big brains, big noses, *and* big supraorbital tori. Their big braincase thrusts nasion further forward than in Erectines and Heidelbergs, while their big nasal and supraorbital skeleton holds nasion further forward than in modern humans. Adding these two factors together puts nasion in a uniquely anterior position. Since brain enlargement appears to have preceded reduction of the upper facial skeleton in human evolution, we might expect to find exaggerated nasal protrusion in human populations that have undergone the first of these transformations but not the second.

Even if we interpret the protrusive nasal region of Neandertals as an autapomorphy, it does not follow that Neandertals were a separate species. All distinctive regional populations have autapomorphies. If they did not, they would not be regionally distinctive. For example, high frequencies of light-colored hair and blue eyes are a distinctive apomorphy of recent European populations. But this fact does not exclude blonds from *Homo sapiens*. The distribution of alleles for pigment reduction forms a cline with no sharp boundaries, both in space (with pigment-reducing alleles declining in frequency as we move outward from Europe) and through time (with those alleles declining in Europe, and increasing in temperate-zone areas elsewhere, as a result of population movements over the past 300 years).

There is some evidence for a similar clinal distribution of upper facial projection. The skulls of the earliest modern humans from Europe (EUP, Tables 7.7 and 7.12) are more Neandertal-like than other modern humans, including Skhūl/Qafzeh, in measurements reflecting upper facial projection (nasion projection and nasion radius). Later European samples maintain this differ-

ence from recent Africans, even when the nasion radius is scaled to cranial length (NAR index, Table 5.12). Thus, if upper midsagittal facial prognathism is considered derived for Neandertals, its continued presence in later European samples is evidence for some degree of regional genetic continuity in Europe across the Neandertal/EUP boundary.

Internal Nose

Schwartz and Tattersall (1996) identify three features of the internal nose which they claim make Neandertals unique among primates, and possibly among terrestrial mammals. These features are: (1) an interior rim or flange just inside the nasal aperture, which terminates bilaterally about halfway up the side of the nose in a wide, blunt projection from the medial nasal wall (Fig. 7.15); (2) a protuberant swelling of the lateral nasal wall into the posterior nasal cavity; and (3) the absence of an ossified roof over the groove in the lacrimal bone through which tears drain from the eye into the nose. All three of these features are clearly evident on Gibraltar 1 (Forbes Quarry), but Schwartz and Tattersall also recognize at least one of them on Spy 1, Kůlna, La Chapelle, La Ferrassie 1, Saint-Césaire, Engis 2, Roc du Marsal, Gibraltar-Devil's Tower, and Pech du l'Azé. All the non-Neandertals that they examined, including most importantly Skhūl 5, Cro-Magnon 1, and Kabwe 1, are reported to lack all three of the Neandertal-specific nasal features. Although Schwartz and Tattersall correctly note that these supposed autapomorphies do not prove that Neandertals were a separate species, they think that they support that conclusion. From the findings of Schwartz and Tattersall, Laitman et al. (1996) go on to infer that Neandertals had a highly specialized upper respiratory tract, reflecting significant behavioral differences from modern humans.

The Schwartz and Tattersall study was critically evaluated by Franciscus (1999), who examined a much larger sample of 200 Pleistocene specimens (including 25 Neandertals) and over 500 recent humans from varied localities in the western Old World. Franciscus's observations and his summary of the previous literature contradicted the study of Schwartz and Tattersall in every particular. Feature 3 had previously been shown to be common in a diverse sample of modern skulls. It is also absent in the Krapina Neandertals (Yokley 1999). For feature 2, Franciscus showed that the internal breadth of the nasal cavity is not relatively reduced in Neandertals, which it would have to be if the reported swelling were an apomorphy. Finally, Franciscus identified Schwartz and Tattersall's feature 1, the internal nasal rim, with the so-called turbinal crest (*crista turbinalis*) described by Gower (1923). The development of this crest is variable in recent human populations, but the extreme condition seen in Neandertals (Gower's

Stage 5) is seen in frequencies of >10% in recent Bantu, western European, Near Eastern, and North African samples.

Even in light of Franciscus's analysis, the internal nasal anatomy of Neandertals is still distinctive. No recent human sample has an incidence of the turbinal crest remotely approaching the 68% reported by Franciscus for Neandertals. The most fully developed form of the crest (Stage 5) is characteristic of Neandertals, but it is absent in Middle and Late Pleistocene Africans, Skhūl/Qafzeh, and European Heidelbergs, and occurs in only 6.7% (1 of 15 specimens) of early modern humans from Europe. And although Franciscus interprets the medial projection from the turbinal crest as a consequence of the exaggerated breadth and eversion of the external rim of the Neandertal nasal aperture, this analysis gets rid of one apomorphy only by explaining it as a side effect of another.

In short, there is an overall pattern of internal nasal form that is far more common in Neandertals than in other groups. But no element in that pattern appears to be either unique to or uniform in Neandertals, as Schwartz and Tattersall claimed. And there are other problems with the Schwartz/Tattersall study. For instance, the best example of feature 1 is unquestionably Gibraltar 1 (Fig. 7.15). This specimen had a nasal cavity filled with hard breccia that was cleaned out with a pneumatic drill. Franciscus (1999) suggests that the medial projection from its turbinal crest—which is much more pronounced in Gibraltar 1 than in other Neandertals—may be the result of some "pneumatic sculpting" during preparation, and may not represent real anatomy. Finally, Schwartz and Tattersall score the presence or absence of some of their three features in specimens that entirely lack the anatomical region where the feature is found—for example, Spy 1, Kulna, Gibraltar-Devil's Tower, and Skhūl 5. Given all this, claims that these features document a uniquely Neandertal morphology or imply uniquely specialized behavior and physiology remain unconvincing.

In a subsequent study, Franciscus (2003) documents the relationship between the floor of the nasal fossa (the upper surface of the hard palate) and the lower edge of the nasal aperture in *Homo*. He distinguishes three patterns: (1) **level**, in which there is a smooth transition between the nasal margin and nasal floor, and both are on the same horizontal plane; (2) **sloped**, in which the transition is smooth but the nasal floor is lower than the nasal margin; and (3) **bilevel**, in which there is a distinct step down from the nasal margin to the nasal floor. As might be expected from their generally exaggerated and everted nasal margins, Neandertals exhibit a high frequency (80%) of the bilevel pattern. However, it also occurs in substantial frequencies in the Skhūl/Qafzeh fossils (40%), the African Middle Pleistocene (50%), European Heidelbergs (29%), and recent Bantu (19%),

and is present in lower percentages in most other fossil and recent samples. The significance of the bilevel arrangement and its distribution is unclear. It does not appear to be related to cold adaptation, at least in recent humans. Franciscus thinks that it may be a developmental correlate of pronounced prognathism, which would certainly fit the Neandertal case.

NEANDERTAL MANDIBLES

Mandibles are the most common surviving Neandertal skull element other than teeth, and mandibular anatomy has played a large role in debates on the taxonomic and phylogenetic status of Neandertals. In assessing the meaning of the distinctive features of the Neandertal mandible, it is important to understand the relationships between cranial and mandibular anatomy. The anatomy of the lower jaw is largely the product of underlying structural, functional, and developmental factors that are manifested in other ways in the rest of the skull. Most of the anatomical traits of the Neandertal mandible (Fig. 7.17) are predictable consequences of three other Neandertal characteristics that we have already mentioned: large front teeth, marked midsagittal prognathism, and zygomatic retreat.

Because Neandertals have a highly prognathic face (at least near the midline), their upper front teeth lie far in front of the braincase. It follows that the lower front teeth and their alveoli (root sockets) must also lie far forward. Because the alveolar process at the upper edge of the mandibular symphysis lies far forward, the thick lower part of the symphysis does not protrude beyond

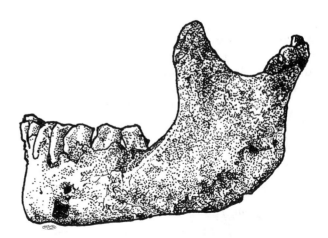

FIGURE 7.17

The Neandertal mandible from La Ferrassie 1 (France). This specimen has a relatively vertical mandibular symphysis, but no evidence of structures associated with a chin. The size of the retromolar space is exaggerated by the extreme wear on the M_3.

TABLE 7.13 ■ Comparative Metrics of the Mandibular Symphysis[a,b]

Sample	Symphyseal Angle[b]	Symphysis Height	Symphysis Thickness (Base)	Symphysis Thickness (Alveolar Process)
Neandertals	98.7	35.1	14.9	11.3
(N, SD)	(16, 5.7)	(18, 4.0)	(16, 1.3)	(16, 1.5)
Vindija	87.0	30.8	15.2	10.3
(N, SD)	(3, —)	(3, —)	(3, —)	(3, —)
EUP	78.2	32.5	15.1	9.8
(N, SD)	(10, 6.7)	(11, 4.2)	(12, 1.5)	(11, 1.4)

[a]Symphyseal angles are in degrees; other measurements are in mm.
[b]Measurements were taken by FS. Symphyseal angles were measured following Olivier (1969). Sample sizes and means are given for all samples; standard deviations are given where appropriate.

TABLE 7.14 ■ Metrical Comparisons of Neandertal Mandibles[a]

Sample	Mandible Superior Length	Corpus Height[b]	Corpus Breadth[b]	Minimum Ramus Breadth
Neandertals	109.3	32.9	15.7	41.5
(N, SD)	(15, 6.7)	(23, 3.3)	(23, 1.8)	(15, 2.7)
EUP	99.9	32.2	12.1	38.2
(N, SD)	(13, 7.3)	(12, 4.1)	(11, 1.4)	(13, 3.4)
Skhūl/Qafzeh	117.6	37.1	14.9	43.1
(N, SD)	(3, —)	(3, —)	(3, —)	(3, —)

[a]Sample means and sizes are given in all cases. Standard deviations are given where appropriate. All measurements are in mm.
[b]Taken at the level of the mental foramen.
Source: Data from Trinkaus et al. (2003a).

it. As a result, there is no bony chin (Fig. 7.17), and the symphysis has a receding rear surface that slopes downward and backward. In all these respects, Neandertal mandibles are persistently primitive (Chapter 6).

Measurements of the mandibular symphysis (Table 7.13) demonstrate the Neandertals' receding symphyses (reflected in high symphyseal angles) and the big roots of their large front teeth (reflected in the thickness of the alveolar process). The combination of a receding symphysis and long anterior dental roots in Neandertals probably explains their great symphyseal height (Smith and Paquette 1989). The thickness of the base of the symphysis in Neandertals does not differ significantly from that in the EUP samples, showing that both groups have equally well developed buttresses at the base of the mandible. Although Neandertals lack a projecting chin, some of them have more vertical symphyses than others and a slight but noticeable protuberance of the symphyseal base in front. This is particularly true of some of the later Neandertals, including the Vindija sample from Croatia (Wolpoff et al. 1981, Smith 1984) and the La Quina 9 and Hortus 4 mandibles from France (Stefan and Trinkaus 1998b, Wolpoff 1999). These incipient basal projections represent what Weidenreich (1936b) called a ***mentum osseum***. Neandertal basal projections are defined by the appearance of a slight ***incurvatio mandibulae*** or mandibular incurvature (Fig. 8.1), a concavity of the front surface of the symphysis below the alveolar process. This concavity gives the impression of some forward projection of the base (Fig. 7.17). However, most Neandertal mandibles lack the mental trigone and related features that characterize the modern human chin (Fig. 8.2). Some believe that these differences preclude any genetic continuity between

late Neandertals and early modern humans in western Asia (Rak 1998, Rak et al. 2002, Schwartz and Tattersall 2000). Others argue that some of the distinctive anatomical features of the modern chin are detectable, though not prominent, in certain Neandertal mandibles (Trinkaus 2002, Trinkaus et al. 2003a, Mallegni and Trinkaus 1997, Stefan and Trinkaus 1998b, Quam and Smith 1998, Wolpoff 1999).

Unlike the rest of the facial skeleton, Neandertal mandibles are not particularly large by comparison with those of early modern humans. In most dimensions, they are intermediate between the EUP and Skhūl/Qafzeh specimens (Table 7.14). One conspicuous exception is the transverse breadth of the mandibular corpus, which is greater in Neandertals, especially when compared to corpus height. The index of robusticity (100 × corpus breadth/height) at the position of the mental foramen is 48.1 for Neandertals, but only 37.9 for EUP and 40.4 for Skhūl/Qafzeh.

In Erectines—and in Heidelbergs that preserve the Erectine pattern of total facial prognathism—both the dentition and the anterior part of the temporalis muscle occupy an anterior position relative to the braincase. Therefore, the front edge of the mandible's coronoid process (where the anterior temporalis attaches) lies further in front of the jaw joint in these archaic humans than it does in a modern human skull. As a result, the rami of their mandibles are broader from front to back than modern ones are. The Neandertal face, however, is prognathic near the midline but not at the sides (the "zygomatic retreat" configuration). Compared to Erectines, Neandertals have a temporal fossa that is shifted backward relative to the dentition; and the anterior temporalis and the front edge of the mandibular ramus are shifted backward with it (Trinkaus 1987b). This shift has two effects (Fig. 7.16). First, it makes the ramus narrower, reducing its relative breadth to modern proportions (Table 7.14). The breadth of the ramus, expressed as a percentage of the overall length of the mandible, is 37.9 for Neandertals and 38.1 for the EUP sample. Second, the backward shift of the front edge of the ramus relative to the teeth opens a small gap between the distal edge of M_3 and the ramus (cf. Figs. 6.11 and 7.16). This gap, the **retromolar space**, is often considered a distinctive Neandertal characteristic (Stringer et al. 1984); but it is not present in all Neandertals, and it is also seen in some early modern mandibles from Europe and western Asia (Smith 1976, 1984, Franciscus and Trinkaus 1995, Frayer 1992b, Quam and Smith 1998, Trinkaus 1987b). Trinkaus et al. (2003a,b) found a retromolar space in 75% of Neandertals ($N = 28$)—but also in 40% of the Skhūl/Qafzeh people ($N = 5$) and in 22.9% of a EUP sample ($N = 24$).

Another aspect of the Neandertal mandible that reflects their pattern of midsagittal facial prognathism and lateral facial retreat is the location of the mental foramen (or foramina, since Neandertals often have multiple ones). This foramen tends to maintain a constant position relative to the braincase, regardless of how far the face sticks out in front. In Neandertals, it therefore tends to be positioned further back along the jaw (relative to the lower tooth row) than in modern humans (Smith 1976, Quam and Smith 1998). Among Neandertals, 92.6% have mental foramina positioned posterior to the P_4/M_1 septum ($N = 27$). However, 33.3% of Skhūl/Qafzeh ($N = 6$) and 19.2% of EUP mandibles ($N = 26$) also have such a posterior placement (Trinkaus 1993a, Trinkaus et al. 2003a,b). Like the retromolar space, the posteriorly placed mental foramen is not a marker of a specifically Neandertal mandibular morphology. Although a distinctive Neandertal pattern is evident, it is a secondary effect of the Neandertal pattern of prognathism—which can be seen either as a Neandertal autapomorphy or as a point on an evolutionary trajectory leading to modern morphology (Fig. 7.16).

Some discrete (nonmetric) traits of the Neandertal mandible have been interpreted as autapomorphies. One such trait is the **horizontal-oval (H-O)** pattern of the mandibular foramen on the inner surface of the ramus (Stringer et al. 1984). This feature, first noted in the Krapina Neandertals (Gorjanović-Kramberger 1906, Smith 1976), is defined by the presence of a broad band of bone that incorporates the mandibular lingula and extends horizontally across the entrance of the mandibular nerve into the mandible. Fig. 7.18 illustrates the H-O morphology and contrasts it with the "normal" condition usually found in other *Homo*. The H-O trait, which has also been referred to as "lingual bridging" (Lebel and Trinkaus 2002, Trinkaus et al. 2003a), is reported in 53% of Neandertals ($N = 19$), 18% of EUP

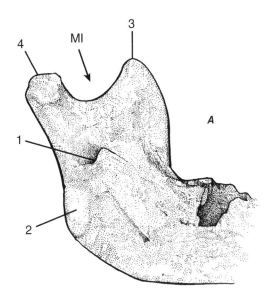

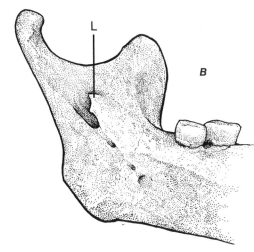

FIGURE 7.18

The Krapina 63 Neandertal mandible from Croatia (**A**) compared to that of a recent human (**B**). Krapina 63 exhibits a horizontal-oval mandibular foramen (1), a pterygoid tubercle (2), and an extension of the coronoid process (3) above the level of the condyle (4). The latter two features are considered characteristic of Neandertals by Y. Rak (1998, Rak et al. 2002). L, lingula; MI, mandibular incisure. (After a drawing by M. O. Smith in Smith 1978.)

($N = 22$), 6% of later Upper Paleolithic Europeans ($N = 30$), and <2% of large Mesolithic and Medieval Hungarian samples (Frayer 1992b). The trait is absent in the Skhūl/Qafzeh sample and in known mandibles of Heidelbergs. Smith (1978) interprets this feature as providing an expanded, reinforced attachment area for the attachment of the sphenomandibular ligament, which runs from the skull base to the lingula and helps to stabilize the mandible during use of the teeth. It is not known whether the H-O trait is genetically determined or develops as a response to unusual patterns or levels of dental stress.

Rak et al. (1994, 2002; Rak 1993, 1998) have noted three other discrete traits of the Neandertal mandible that they propose as unique, defining autapomorphies of *H. neanderthalensis*. The first of these is the **medial pterygoid tubercle**, a clearly demarcated plateau at the corner of the jaw that serves as the insertion point for fibers of the medial pterygoid muscle (Fig. 7.18A). Although Rak and co-workers claim that this feature is unique to Neandertals, other studies (Richards and Plourde 1995, Quam and Smith 1998) have found that it occurs at lower frequencies in the EUP (10%) and in six recent human samples (up to 24%). Secondly, Rak and his colleagues claim that the **incisure crest**, the crest defining the outline of the mandibular incisure or notch (Fig. 7.18), intersects the anterior condylar neck medially in Neandertals but more laterally in all recent humans (Fig. 8.5). However, Quam and Smith (1998) found that the supposedly distinctive Neandertal pattern was absent in 30% of their Neandertal sample and present in 8% ($N = 38$) of a recent modern sample from West Asia. Several researchers have suggested that in both Neandertals and modern humans, the placement of the incisure crest depends on the size of the lateral condylar tubercle of the fibrous capsule of the jaw joint (Suzuki 1970a, Smith 1976, Trinkaus 1995b). In a detailed and comprehensive recent study, Jabbour et al. (2002) found that hypertrophy of the lateral condylar tubercle is a strong predictor of medial placement of the crest. Although they observe that Neandertals do tend to have a higher frequency of more medial placement of the crest, with 78% of Neandertals exhibiting their two most medial stages ($N = 9$), they also found medial crest placement in Upper Paleolithic and recent humans (22%, $N = 9$ and 62%, $N = 102$, respectively).

Finally, Rak and colleagues note that the contour of the mandibular incisure when viewed from the side is a more or less symmetrical curve in modern humans, with the deepest point close to the center of the arc (symmetrical pattern). In Neandertals, however, the deepest point is close to the neck and the crest rises sharply to the coronoid process (Fig. 7.18). But once again, this supposed autapomorphy of the Neandertal mandible is neither uniform among Neandertals nor unique to them. The asymmetrical "Neandertal" pattern is also found in some European Heidelberg mandibles (Rak et al. 2002). And although the "modern" symmetrical pattern is indeed characteristic of EUP (88%, $N = 17$) and Skhūl/Qafzeh (100%, $N = 2$) mandibles, it is also found in 29% ($N = 14$) of Neandertals (Trinkaus et al. 2003a). The difference between the two patterns reflects the height of the coronoid process, which rises high above the level of the mandibular condyle in Neandertals but is essentially on the same horizontal plane in

modern humans. Since the coronoid is the point of insertion of the temporalis muscle, it seems likely that this difference too is related to the mechanics of lower-jaw use.

As the cited studies demonstrate, there is a distinctive pattern to Neandertal mandibular rami. But none of the elements of that pattern qualifies as a uniquely derived characteristic of Neandertals (Wolpoff and Frayer 2005). All the distinctively Neandertal traits—the H-O mandibular foramen, the medial pterygoid tubercle, the medial position of the incisure crest, and the asymmetrical mandibular notch—are probably related to the function of the mandible in mastication and nonmasticatory behaviors. All these bony features mark the attachments of muscles and ligaments that produce or limit jaw movement (Smith 1978, Richards et al. 2003). They represent parts of a single functional complex, not independent synapomorphies to be tallied up as separate traits in a cladistic analysis.

The traits in this complex suggest that Neandertals were making hard demands on their teeth and jaws. This inference is supported by the anatomy of the mandibular fossa (see above), by the high incidence of degenerative joint disease on Neandertal condyles (Alexandersen 1967, Smith 1976, Trinkaus 1983b), and by the anatomy and wear patterns of Neandertal teeth.

▮ NEANDERTAL DENTITION

The front teeth are the most distinctive part of the Neandertal dentition. The unworn incisors and canines of Neandertals are large, even by comparison with those of more primitive humans. This is especially true of the earliest Neandertals. The Neandertals of OIS 5e, particularly those from Krapina, had larger incisors and canines in both jaws than their Heidelberg predecessors from Sima de los Huesos (Table 6.5). However, their successors, the so-called "classic" Neandertals of the later Würm, had smaller front teeth (Smith 1976, Wolpoff 1979); and even among the Würm Neandertals, the very latest ones (dating to ≤36 Ky) have slightly smaller front teeth than the others (Wolpoff 1979). In this regard, Neandertals became less and less distinctive as time went on. Of course, there was further size reduction in modern humans; but whatever factors caused that reduction seem to have been already set in motion during the Neandertal period.

The large size of the front teeth is not always evident in a Neandertal skull, because the big crowns of these teeth may be completely worn away. Neandertal incisors and canines are usually very heavily worn, even compared to the cheek teeth in the same individual (Fig. 7.12; Smith 1983, Trinkaus 1983b). Wear on the incisors and canines often exceeded the capacity of the teeth to withstand it. For example, in both La Ferrassie 1 from

France (Heim 1976, Wallace 1975) and Shanidar 1 from Iraq (Stewart 1959, 1977, Trinkaus 1983b), the crowns of the anterior teeth were functionally worn away altogether at the time of death, and wear was continuing on the roots! In some cases, the wear was so severe that it laid open the soft tissues in the pulp cavity, exposing the individual to life-threatening infection. We do not know what caused these high attrition rates (see below); but whatever it was, it would have placed a high selective premium on large incisor and canine crowns that could resist wear and protect the vulnerable parts of the front teeth for as long as possible.

Similar considerations may explain the distinctive "shovel" shape of Neandertal incisors, especially in the upper jaw (Gorjanović-Kramberger 1906, Hrdlička 1920). As noted in Chapter 5, the shoveling morphology seen in European Neandertals is not the same as that seen in high frequencies in Asian populations today (Cadien 1972). Crummett (1994, 1995) argues that the European pattern involves larger basal tubercles and a greater curving of the incisal margins of the crowns, which she attributes to a need to pack as much occlusal surface as possible into the available space (Figs. 5.14 and 7.19). The resulting increase in the area of the occlusal surface may represent an adaptation to counter the rapid attrition of the anterior teeth in Neandertals (Hrdlička 1920).

It has also been suggested that the shovel-shaping of the incisors and the large size of the anterior teeth of Neandertals would have enabled them to bear and exert more powerful bite forces, and that the distinctive Neandertal morphology evolved for this purpose (Smith 1983). This view is supported by a comparative analysis of Neandertal and Eskimo craniofacial morphology by Spencer and Demes (1993). Eskimos leading a traditional lifestyle used their anterior teeth extensively and loaded them heavily for para- and nonmasticatory purposes. Although Eskimo cranial anatomy bears little obvious similarity to that of Neandertals (Hylander

1977), Spencer and Demes concluded that the distinctive morphologies of both groups have the effect of enhancing the leverage of the jaw muscles for producing bite force on the front teeth (Spencer and Demes 1993). It has also been noted that the midfacial prognathism of Neandertals yields a more vertical orientation of the anterior dental roots than does the total facial prognathism of earlier *Homo*—which would have further improved the Neandertals' ability to generate and resist intense bite forces on their front teeth (Smith 1983, Smith and Paquette 1989). The enhanced length of the roots of these teeth also makes sense as an adaptation for resisting force.

Of course, the million-dollar question here is; Why are Neandertal anterior teeth so big and so heavily worn? Since the early 1960s, Brace (1962, 1964, 1967, 1979, 1995) has argued that the extensive wear seen on these teeth reflects their use for purposes that would have been served by tools in later human populations. In this view, the severe attrition resulted from the habitual use of the teeth as a vise for holding, pulling, and twisting materials that were being worked by other means. The front teeth may have been used directly for working such substances as hides and wood and may even have been used during certain phases of stone-tool manufacture. Ungar et al. (1997) note that the pattern of anterior dental wear in Neandertals most closely matches Inuit (Eskimo) populations among modern samples they examined, and that the degree of wear among Inuit is directly related to the oral preparation of animal hides.

It also has been suggested that the wear on Neandertal front teeth could have been produced by their use in food preparation. In most Neandertals, anterior dental wear was heaviest on the outer (labial) edges of the teeth, which are thereby rounded off to produce a "beveling" effect (Ungar et al. 1997). Wallace (1975) argued that the labial beveling on the incisors of La Ferrassie 1 was probably due to the gripping and pulling of grit-

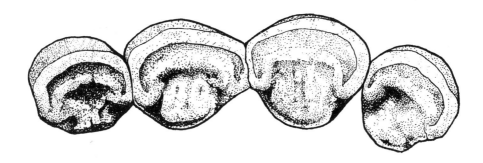

FIGURE 7.19

Four maxillary incisors from the Krapina Neandertal site (Croatia) in occlusal view. The marked basal tubercles and marginal ridges, evident in both the central and lateral teeth, give Neandertal maxillary incisors a shovel-shaped appearance. However, the incisal margin is distinctly arched or curved, in contrast to the straighter margin seen in East Asian shovel-shaped incisors (cf. Fig. 5.14).

laden food over the incisors, rather than to the use of the teeth as tools. This may be true as far as it goes, but food preparation of this sort is itself a form of tool use, outside the realm of normal chewing (Hylander 1977). The microwear features on Neandertal incisors—the pattern of striations on the occlusal surfaces, the high incidence of enamel chipping and other microfractures, and the microscopic striations on the front surfaces of the crowns (resulting from using the teeth as a vise while cutting objects with a stone tool)—all support the claim that Neandertal anterior teeth were involved in far more strenuous activities than routine biting and chewing (Smith 1983, Trinkaus 1983b, Wolpoff 1999, Smith et al. 2006).

But even if Neandertals used their front teeth as tools, it does not follow that such uses were the driving force behind the evolution of Neandertal dental morphology and facial architecture. The big front teeth of Neandertals no doubt would have facilitated these behaviors, but they may have evolved for other reasons. Several authors have surmised that the distinctive features of the Neandertals' anterior dentition may not be adaptations produced by selection, but side effects of other Neandertal peculiarities. Antón's (1994) study of facial biomechanics in Neandertals and modern humans leads her to conclude that Neandertals actually could not bite as hard on their front teeth as modern humans can. Antón believes that the derived features of the Neandertal face are not adaptations but pleiotropic side effects of an overall pattern of craniofacial development peculiar to Neandertals, which probably developed in Europe as the result of founder effect and genetic drift. This analysis fits nicely with the "accretion model" discussed in Chapter 6. In Antón's view, as well as in the nose-centered view of Coon (1962), Neandertal front teeth are enlarged simply in order to fill the space in the anterior palate caused by the expanded face and big schnoz.

Others have conjectured that the relatively large anterior teeth of Neandertals may be side effects of growth patterns involving the entire body. Trinkaus (1978, 1983b) notes that front teeth show a stronger size correlation with body mass than cheek teeth do, and suggests that Neandertals may have had big front teeth simply as a consequence of an elevated lean body mass. Smith (1991) posited that Neandertal craniofacial form may have been the result of selection for accelerated prenatal growth in order to produce bigger babies, which would have had a better chance of surviving the highly vulnerable early months of life in a cold environment (Roberts 1978). Several of the distinctive features of the Neandertal skeleton can be interpreted as consequences of this hypothetical growth pattern. As we will see later, supposed differences between Neandertals and modern humans in ontogeny and life history are a matter of controversy.

At present, there is no consensus about the causes of the peculiarities of the Neandertal face. But whatever the explanation proves to be—selection, drift, pleiotropy, or some combination—the "teeth as tools" model of Neandertal dental *function* is still likely to be correct. The heavy wear and other attrition-related features of Neandertal front teeth are facts, and anterior dental loading and use must have been heavy and frequent. No matter whether the behavior resulted in selection for the morphology, or the morphology was produced for other reasons and then made the behavior possible, Neandertal craniofacial architecture would still have been adapted to the generation and dissipation of anterior dental loading, as has been repeatedly suggested (e.g., Brace 1979, Smith 1983, Rak 1986, Demes 1987). We should remember in this connection that new morphology always comes into existence for reasons unrelated to behavior, appearing either as a random mutation or as a side effect of something selected for in a different context. The novel trait becomes adaptive only secondarily, if and when its possessors adopt behaviors that take advantage of it.

Neandertals' posterior teeth (molars and premolars) are less distinctive than their front teeth. They generally fall into the size range of modern human cheek teeth (Brace 1967, Wolpoff 1971b, Smith 1976, Frayer 1978), which means that they are reduced compared to earlier *Homo* (Table 6.5). However, the early (OIS 5e) Neandertals from Krapina again stand out as rather megadont, even by comparison with the Zhoukoudian Erectines or the Neandertal-like Heidelbergs from Sima de los Huesos (Table 6.5).

A few morphological features distinguish Neandertal posterior teeth from those of more recent humans. According to Bailey (2002), Neandertals are distinctive in having asymmetrical P_4 crowns and a mid-trigonid crest (MTC) connecting the protoconid and metaconid on the lower molars. On the upper molars, Bailey (2004) finds that Neandertals have smaller metacones and larger hypocones than modern humans, and a somewhat different crown shape with more centrally placed cusps. But most of these features also occur among early modern humans. While the MTC is found on more than 90% of Neandertal lower molars, it is also present in smaller frequencies (1 of 3, or 33%) on the M_1s of the early moderns from Skhul/Qafzeh. The Neandertal pattern of metacone/hypocone size (though not the overall shape difference) is also seen in a small sample of early modern human upper molars ($N = 6$). The significance of these features is unclear, but they are offered by Bailey as further evidence for the derived character of Neandertal morphology. Stringer et al. (1997) similarly conclude from a cladistic analysis of nine discrete molar features that the Krapina Neandertals always form a distinctive outgroup compared to a series of modern human populations. Furthermore, they

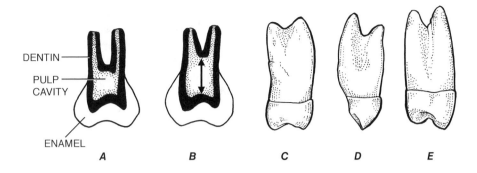

DENTIN —
PULP
CAVITY —
ENAMEL —

A   B   C   D   E

FIGURE 7.20

Taurodont maxillary molars. **A**: Diagrammatic vertical section of a human molar showing the pertinent aspects of dental anatomy. **B**: Section through a stylized Neandertal taurodont molar, showing the expanded pulp cavity (*arrows*). **C–E**: External views of taurodont teeth from Krapina (Croatia). (Revised after Klein 1999.)

argue that the Krapina teeth are more similar to African than to European teeth, and interpret this finding as evidence against genetic continuity between Neandertals and later Europeans.

Another purportedly distinctive feature of the Neandertal postcanine dentition is the incidence and extent of development of **taurodontism** in the lower molars. Taurodontism is a marked expansion of the pulp cavities in the deciduous and permanent mandibular molars (Fig. 7.20). Its *raison d'être* is not understood. First reported for Neandertals by Gorjanović-Kramberger (1907), taurodontism is characteristic of many Neandertals, but by no means all (Skinner and Sperber 1982). And as Gorjanović-Kramberger himself pointed out in 1908, taurodontism is not restricted to Neandertals. It is found in Mauer, Pontnewydd, and other potential Neandertal ancestors (Wolpoff 1999), and it occurs among living humans as well. For example, modern Khoisan people from southern Africa have sample incidences of taurodontism comparable to those of some Neandertal samples (Constant and Grine 2001). Like many other supposed autapomorphies of Neandertals, taurodontism is common in Neandertals, but it is not unique to them.

▌BODY SIZE AND PROPORTIONS

Body-mass estimates for Neandertals ("Late archaic *H. sapiens*," Table 6.1) are greater than those for other *Homo*—including late Middle Pleistocene hominins, early modern Europeans, early modern Western Asians, and worldwide averages for living humans (Ruff et al. 1997). Evidently, Neandertals had relatively high lean body masses (Table 6.1). Conversely, estimates of Neandertal stature indicate that they were shorter than any of these other people (Table 7.15). This combination of short stature and heavy weight means that the average

TABLE 7.15 ■ Stature Estimates for Selected Middle and Late Pleistocene Hominin Samples[a]

Sample	Femur/Stature Ratio	Trotter–Gleser Formula
	Stature Estimates (cm) Based on:	
Erectines (N = 4)	168.07	169.07
European Neandertals (N = 6)	161.2	165.04
West Asian Neandertals (N = 6)	162.48	165.43
Skhūl-Qafzeh (N = 5)	178.54	175.7
Early Modern Europeans (N = 26)	171.33	171.12

[a]The femur/stature ratio estimates are calculated using the formula [stature = 100 × femur length (cm) /26.74]. The Trotter–Gleser estimates use the formula presented by Trotter and Gleser (1958).
Source: All data are from Feldesman et al. (1990).

Neandertal was exceptionally stocky and robustly built.

Although Neandertals have low, thoroughly human-like intermembral indices (Trinkaus 1983b), their limbs have relatively short distal segments—that is, the tibia and fibula are unusually short compared to the femur, and the radius and ulna are short compared to the humerus (Fraipont and Lohest 1887, Hrdlička 1930, Trinkaus 1981, 1983a; Holliday 1997a,b; Ruff et al. 2002, Pearson 2000b). Coon (1962) suggested that Neandertals evolved these limb proportions as an adaptation to cold climates, an idea that has been widely accepted by other scholars (Trinkaus 1981, 1983a; Holliday 1997a,b; Pearson 2000b). Coon's idea fits the

TABLE 7.16 ■ Brachial and Crural Indices for Neandertals and Modern Human Populations[a]

Sample	Crural Index	Brachial Index
European Neandertals		
Mean	78.7	73.2
SD (N)	1.6 (4)	2.5 (5)
European Early Modern		
Mean	84.9	77.7
SD (N)	1.7 (18)	2.0 (20)
European Mesolithic		
Mean	85.5	77.5
SD (N)	2.6 (10)	1.9 (10)
West Asian Neandertals		
Mean	78.1	77.2
SD (N)	1.2 (5)	2.1 (4)
Skhūl-Qafzeh		
Mean	86.0	75.2
SD (N)	2.1 (4)	6.8 (3)
Koniag Eskimo		
Mean	80.5	75.3
SD (N)	2.3 (20)	2.6 (20)
Recent Europeans		
Mean	82.9	75.1
SD (N)	2.4 (243)	2.5 (240)
Recent North Africans		
Mean	85.0	78.6
SD (N)	2.3 (133)	2.4 (136)
Recent Southern African		
Mean	86.1	79.6
SD (N)	2.2 (66)	2.5 (67)

[a]Brachial index = 100 × radius length/humerus length; crural index = 100 × tibia length/femur length (Trinkaus 1981). Data from Holliday (1997b) and Ruff et al. (2002).

general principle known as **Allen's Rule**, which says that warm-blooded animals living in cold places tend to have relatively short limbs in order to decrease the body's surface-to-volume ratio and thus reduce heat loss (Allen 1877). In humans, it is the distal bones of the limbs that are the most shortened elements. Modern human populations generally conform to Allen's Rule, with a strong correlation between mean year-round ambient temperature and various measures of body proportions (Roberts 1978, Trinkaus 1981, Holliday 1997a). The data in Table 7.16 show that brachial and crural indices, for example, are significantly smaller on the average in Europeans and Eskimos than in Africans, indicating shorter distal limbs in the former two groups. However, the crural indices of Neandertals are even smaller—smaller than those of any modern humans. Mean brachial indices of European Neandertals are also smaller than those of any moderns, including Eskimos; but those of West Asian Neandertals fall above the mean for modern Europeans and Eskimos and only slightly below the modern North African mean. This difference

between the limb proportions of European and Asian Neandertals might reflect different levels of cold adaptation in the two groups. Perhaps the West Asian Neandertals were in the process of re-adapting to more tropical conditions. One might also interpret the longer forearms of the West Asian Neandertals as the result of genetic admixture with more southerly (African?) humans. Additionally, many see the differing limb proportions of Neandertals and early modern Europeans (Fig. 7.21) as evidence that the moderns were immigrants from more tropical regions, who moved into Europe and replaced the Neandertals (Holliday 1997a,b, 2000; Klein 1999, Pearson 2000a,b; Tattersall and Schwartz 2001).

Inference from climate to body proportions can also be worked in the other direction. T. Holliday (1997b) used regressions developed by Trinkaus (1981) and Ruff and Walker (1993) for predicting mean annual temperature from brachial and crural indices in recent humans to make inferences about the environments of European Neandertals. He concluded that they were probably living in climates with a mean annual temperature between 2.4 °C and −1.7 °C. These figures cannot be taken as precise estimates of Pleistocene temperatures, but they do show that European Neandertals were living in relatively cold conditions.

Other lines of evidence also suggest that the Neandertal body was molded by adaptation to cold environments. As Schaaffhausen (1858, 1888) and Boule (1911/13) observed long ago, Neandertals had less curved ribs than modern people. This pattern, which is most obvious in the lower ribs (Franciscus and Churchill 2002), suggests a deeper, more barrel-shaped chest (Smith 1976, Trinkaus 1983b, Churchill 1994). A deeper thorax would conform to Bergmann's Rule, which states that a bulkier build will conserve heat by decreasing the surface-to-volume ratio. An anteroposteriorly deep chest is also implied by the relatively long clavicles of Neandertals, in whom the ratio of clavicular to humeral length is much greater than in most modern samples (Trinkaus 1981). No doubt, this ratio is affected by the fact that humerus length (the denominator of the ratio) tends to be less in Neandertals than in modern Europeans (Holliday 1997b; Ruff et al. 2002). On the other hand, Eskimos also have short humeri ($\bar{x} = 308$ mm, compared to means of 307 mm for Neandertals, 335 mm for early modern Europeans, and 343 mm for Skhūl/ Qafzeh), in accordance with Allen's Rule; but Eskimos do not have elongated clavicles (Holliday 1997b). This supports the inference from long clavicles to deep chests in Neandertals.

Neandertal ribs are also more robust than those of modern humans, and they bear more extensive and pronounced muscle scars. Franciscus and Churchill (2002) infer that Neandertals had a powerful thorax, capable of moving large amounts of air in and out rapidly. This is

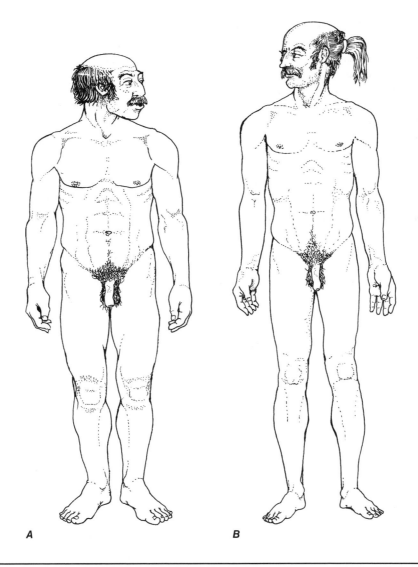

A   B

FIGURE 7.21

Differences in body form between a Neandertal (**left**) and an early modern European (**right**). Neandertals have relatively broader, more barrel-shaped trunks and stockier limbs with shorter distal segments. Early modern Europeans tend to be taller and to have a more linear body build, with narrower trunks, more slender limbs, and longer distal limb segments. (After Howells 1993.)

in keeping with the Neandertals' large nasal openings, muscular bodies, and impressive body masses. The powerfully muscled thorax (and big nasal passages) of Neandertals may have evolved to serve the respiratory demands of high activity levels in cold, arid environments. The large size of the Neandertal thorax implies that the enclosed heart and lungs were unusually big, probably for the same reasons (Churchill 2006).

Neandertal pelves were exceptionally broad. As Table 5.9 demonstrates, the bi-iliac breadth of the Kebara 2 Neandertal exceeds the modern human range; and the estimated pelvic breadth/stature ratio in this skeleton falls at the top of the modern European range, well out of the range of modern African values (see discussion in Chapter 5). This implies that the lower thorax of

Neandertals must also have been exceptionally broad (Ruff 1994, Ruff et al. 1993)—another piece of evidence for the large cross-sectional area of the Neandertal torso.

All these measurements and comparisons point to a strikingly cold-adapted body build for Neandertals (especially the European ones), with short limbs and a deep, wide trunk. This body form is broadly comparable to those of modern cold-adapted peoples (Leonard et al. 2002), but Neandertals carried this tendency further. Neandertal body form has been described as "hyperpolar" (Holliday 1997b) or "hyperarctic" (Ruff 1994). It is not clear how big an impact these traits would have had on a Neandertal's ability to withstand cold. From estimates of body mass and stature similar to those presented in Tables 6.1 and 7.15, Aiello and Wheeler (2003) estimate that the peculiarities of Neandertal body build would have compensated for only about a 1 °C drop in temperature (compared to early modern Europeans), all other things being equal. However, other things were probably not equal. Leonard and colleagues (Leonard 2002, Leonard et al. 2002) show that modern circumpolar populations have significantly higher basal metabolic rates (BMR) than more equatorial populations, due in part to the effects of temperature and day lengths on thyroid function. It seems safe to assume that European Neandertals would have had a similarly elevated BMR. Assuming a conservative 15% increase in Neandertal BMR, Aiello and Wheeler calculate that the distinctive body build of Neandertals (including the insulation value of their enhanced muscle mass) would have increased the level of their advantage over early moderns to 3.2 °C. This may seem a small advantage; but natural selection runs on small advantages. It would have been enough to spell the difference between life and death for some percentage of the Neandertal population over the course of a long, hard European winter during a glacial advance.

Nevertheless, other factors must have contributed to Neandertal survival under periglacial conditions. One contributing factor may have been a shift to a higher-energy diet. There is evidence from stable-isotope studies and archaeological evidence that Neandertals ate a lot of meat. This implies a diet rich in fat and high in calories (Richards et al. 2000, Tattersall and Schwartz 2000, Kuhn and Steiner 2006)—which is another similarity between Neandertals and modern Arctic populations (Leonard et al. 2002, Sorensen and Leonard 2001). Churchill (2006) calculates that an adult Neandertal would have required 3500–5000 kilocalories a day to survive in cold conditions. He suggests that the expanded thorax in Neandertals is less a reflection of Bergmann's Rule than an adaptation for providing the increased flow of oxygen needed to metabolize all those calories.

Although stocky bodies, short limbs, and wolfish diets must have helped to meet the harsh demands of the Neandertals' environments, they would not have been sufficient by themselves. Churchill (2006) calculates that if only the males in a hypothetical social group of ten Neandertals were hunting, they would have needed return rates from hunting that were about 35% higher than those recorded for wolves in North America in order to feed the caloric demands of the group. But unlike wolves, Neandertals were able to marshal the aid of extra-somatic adaptations—material culture—in confronting subfreezing temperatures. The archaeological record shows that Neandertals used both fire and artificial shelters (Trinkaus 1989). Aiello and Wheeler (2003) suggest further that Neandertals would have needed artificial body covering (clothing of some sort) with an insulation value at least equal to a modern business suit in order for them to survive away from fires and shelters

in subfreezing temperatures. They argue that even with all these cultural and biological adaptations to the cold, Neandertals would have had great difficulty surviving under the increasingly harsh conditions of the latter part of OIS 3, particularly during the winters. It seems clear that whatever assumptions we make, the Neandertal people of periglacial Europe must have led hard lives, in which they faced the continual threat of freezing or starvation during the coldest months of the year.

Some think that the differences in limb proportions between Neandertals and modern humans may have more to do with the functions of the limbs themselves than with adaptation to climate. Among carnivores, species with larger home ranges tend to have relatively longer distal limb segments (Kelly et al. 2006). It has been suggested that early modern Europeans were more mobile than Neandertals, and that the resulting need for greater locomotor efficiency might have been the selective factor that gave them longer lower limbs (Wolpoff 1989, 1992; Frayer et al. 1993). When Holliday and Falsetti (1995) tested this suggestion, using data from a sample of 19 modern hunter–gatherer groups from around the world, they found no relationship between mobility and relative lower-limb length even when climate was held constant. Another study (Weaver and Steudel-Numbers 2005) also concludes that the climatic hypothesis cannot be rejected. However, something besides just climate appears to have been influencing Neandertal limb proportions. In West Asian Neandertals, the lower limbs have short distal segments like those of European Neandertals, but the upper limbs do not (Table 7.16). This suggests that the proportions of the upper and lower limbs may have been responding to different selection pressures, presumably having something to do with their different functions.

Questions persist about how much can be inferred from body proportions. During early growth and development, the skeleton exhibits a certain plasticity in responding to environmental influences. For example, mice raised at low temperatures grow up with relatively, short limbs (Serrat et al. in press). Such metrics as the brachial and crural indices may respond to such influences. We might conjecture that frequent exposure to cold during development could retard growth in the length of the distal limb segments, resulting in lower brachial and crural indices in the adult. If that were so, the adaptive value of the adult morphology would not change, but our ability to make phylogenetic inferences from it would be severely compromised. However, other studies have shown that body proportions characteristic of specific modern human groups are established very early in ontogeny, and do not change even when overall growth patterns are altered (Holliday 1997a, 2000). These findings suggest that human body proportions are conservative and likely to be under rather tight genetic control.

Any attempt to draw phylogenetic inferences must reckon with the distinction between plesiomorphies and apomorphies. Some of the features that distinguish Neandertal body proportions from those of modern humans may lack phylogenetic implications because they are primitive for the genus *Homo*. On the basis of a composite reconstruction of a Neandertal skeleton (Sawyer and Maley 2005), the Neandertal thorax has been described as more bell-shaped than barrel-shaped—rather narrow at the top, but flaring out at the bottom. This shape, which reflects the pronounced lateral flare of the Neandertal ilia, is a derived form of that seen in the Nariokotome Ergaster (Chapter 5) and earlier hominins. As noted in Chapter 6, a wide pelvis is also seen in the Heidelbergs from Sima de los Huesos. Thus there is unquestionably a plesiomorphic component to the shape of the trunk in Neandertals. However, this does not negate the importance of trunk shape to Neandertal adaptation. As we have noted in other contexts, retained primitive features also contribute to the adaptations and capacities of any organism. Even if the features of limb proportions and body form that distinguish Neandertals from modern humans do not carry much of a phylogenetic signal, they still convey important information about the evolution of human adaptations in the Middle and Late Pleistocene.

Neck and Upper Limb

The cervical (neck) vertebrae of Neandertals have relatively long, thick, and horizontally oriented spinous processes. These bony levers serve as attachments for some of the epaxial neck muscles, which were probably larger in Neandertals (and other archaic *Homo*) than in modern humans. Archaic humans needed bigger neck muscles than ours, to help compensate for the more anterior position of the head's center of mass that results from having a more prognathic face. In Neandertals, these muscles would also have worked to steady the head against whatever pulls and jerks were producing the extraordinary wear on the labial surfaces of the front teeth. Boule (1911/13, 1921) incorrectly believed the cervical spinous processes to be ape-like in Neandertals. This misreading contributed to his reconstruction of their overall posture as ape-like. But Stewart (1962a) demonstrated that, while Neandertals' spinous processes lie at the extreme of recent human variation, they generally fall within the modern human range and are not like those of apes.

Neandertal scapulae tend to be large, providing a base for the big, powerful upper limb. They are characterized by robust muscle insertions and a relatively narrow, shallow glenoid fossa (the socket for the head of the humerus). The glenoid tends to face more laterally in Neandertals than in modern humans. Churchill and Trinkaus (1990) interpret these features as reflecting less complex back-to-front movements at the glenohumeral joint in Neandertals.

The most discussed peculiarity of the Neandertal scapula concerns the form of the **axillary border**, the lower lateral edge facing the axilla (armpit). Scapulae of modern humans usually have a crest that descends from just below the shoulder joint on the dorsal (back) side of the axillary border, with a distinct groove running ventral to this crest (Fig. 7.22). The groove serves as the

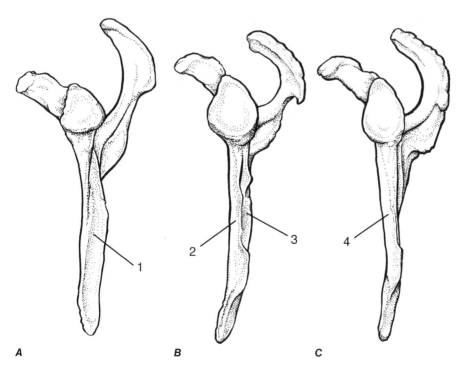

A *B* *C*

FIGURE 7.22

Schematic of scapular axillary-border variation. Left scapulae, viewed from the lateral (glenoid) aspect. **A**: Specimen with a dorsal axillary groove (1). This morph has its highest frequencies in Neandertals. **B**: Example of the bisulcate pattern, with both a small ventral (2) and a dorsal (3) groove. **C**: Specimen with a ventral axillary groove (4)—a very rare condition in Neandertals, but common in modern humans. (After Trinkaus 1983a.)

TABLE 7.17 ■ Axillary Border Morphology in Paleolithic, Mesolithic, and Recent Samples from Western Eurasia[a]

Sample	Dorsal Pattern	Bisulcate Pattern	Ventral Pattern
European Neandertals	64.7 (11)	23.5 (4)	11.8 (2)
Skhūl/Qafzeh	0.0	80.0 (4)	20.0 (1)
Early Upper Paleolithic	16.7 (3)	55.6 (10)	27.8 (5)
Late Upper Paleolithic	0.0	29.4 (5)	70.6 (12)
Mesolithic	10.6 (5)	21.3 (10)	68.1 (32)
Modern Europeans	0.4 (1)	23.8 (28)	75.8 (91)

[a]Values represent percentages; sample sizes are given in parentheses.
Source: After Frayer (1992b).

attachment area for the teres minor muscle, one of the rotator-cuff muscles that act to rotate the arm laterally. In recent human samples, this **ventral axillary groove** pattern tends to be predominant; but a second pattern, in which there are two smaller grooves on either side of a relatively central crest (the **bisulcate** pattern), is also common (Table 7.17). D. Gorjanović-Kramberger (1914) observed that many of the Krapina Neandertals had a third pattern, characterized by a ventral ridge and a **dorsal axillary groove** (see also Smith 1976). Stewart (1962b) noted the same pattern in Neandertal 1. Dorsal axillary grooves occur in more recent people, but at significantly lower frequencies (Table 7.17). E. Trinkaus has suggested that the difference in pattern might reflect an increased development of the teres minor muscle in Neandertals, in connection with the general muscular hypertrophy of the upper body (Trinkaus 1977, 1983a,b; Churchill 1994, 1996). The distinctive morphologies of the axillary groove and glenoid fossa in Neandertals may also be related to the greater depth of the thorax from front to back, which would be expected to affect muscle vectors, torques, and patterns of movement at the shoulder joint.

Like those of earlier *Homo*, Neandertal long bones generally have thicker cortical bone, stronger indicators of muscle attachments, and bigger joint surfaces than those of living humans. All of these features imply that Neandertals and other archaic humans were able to exert high levels of muscular force throughout the full range of movements at most joints. Thickening of the bony cortex deposits compact bone around the bone's surface, where it can most effectively resist bending stresses when the bone is loaded. In Neandertals, this thickening of the outer layer produces narrower medullary (marrow) cavities than those typically seen in recent humans (Fig. 7.26). Neandertal joint surfaces also tend to be larger than ours (relative to linear measurements of the bones), presumably because they had to

bear larger forces. However, when allometrically scaled to body mass, the joint surfaces of Neandertals are no larger than those of modern humans (Ruff et al. 1993).

The Neandertal humerus bears particularly strong markings of trunk and shoulder muscle attachments, especially for the deltoid and pectoralis major muscles. This is not unique to Neandertals, but the pattern of attachment of these and other muscles is somewhat distinctive in Neandertals (Trinkaus 1983a,b; Churchill 1994, 1996; Churchill and Smith 2000b). For example, the deltoid insertion on the humeri of Neandertals—and their predecessors from Sima de los Huesos—runs more nearly parallel to the shaft's long axis than it does in most modern humans, where it has a more oblique orientation (Carretero et al. 1997; see also Chapter 6). Neandertal humeral shafts tend to be shaped slightly differently from those of recent humans, with a more anteroposteriorly flattened (platymeric) midshaft (Table 6.4); but there is variation in this feature within both groups, and the two overlap with each other and with earlier *Homo* (Carretero et al. 1997, Churchill and Smith 2000b). A study of internal bone structure found that Neandertal males exhibited marked right-left asymmetry (Ben-Itzhak et al. 1988), suggesting that Neandertals (like modern humans) were typically right-handed. A recent multivariate study of the metrics of the distal humerus (Yokley and Churchill 2006) finds that Neandertals group separately from both modern humans and other archaic specimens such as Kabwe.

In addition to being relatively short, Neandertal ulnae are distinctive in having a trochlear notch (the ulnar surface of the elbow joint) that faces more anteriorly than in modern humans (Trinkaus 1983b, Churchill et al. 1996). This suggests greater loading of the arm in activities where the elbow was flexed—for instance, in thrusting with a spear. Neandertal radii are characterized by marked outward bowing of the midshaft and more medially oriented radial tuberosities (the attachment area for the arm's biceps muscle). Trinkaus and Churchill (1988) interpret these morphologies as yet more adaptations for producing powerful movements—specifically, flexion of the arm at the elbow (reflected in the radial-tuberosity orientation) and pronation and supination of the forearm (reflected in the curvature of the radial shaft). The outward bowing of the radial shaft would also have increased the surface area of the interosseous membrane between the radius and ulna, affording an increased surface of origin for more powerful flexors of the fingers and thumb.

The great strength of the Neandertal upper limb is also reflected in the anatomy of the hand (Musgrave 1970, Niewoehner 2006). Neandertal hands are similar to those of modern humans in their overall proportions and implied ranges of movement, but they differ in details that reflect muscular hypertrophy and distinctive patterns of loading at certain joints (Trinkaus 1983a,b). Many features of the hand skeleton of Neandertals suggest that they had a bone-crushing grip. Their carpal tunnels were bigger than those of modern people, implying that the tendons of the digital flexor muscles (which pass through the carpal tunnel) were thicker and more powerful. The same pattern is reflected in exaggerated muscle-attachment scars for various flexor tendons on the bones of Neandertal fingers. Muscle attachment areas on the carpal and first metacarpal bones show that the thumb could be rotated with unusual power, further enhancing gripping ability. A powerful grip is also suggested by the elongation of the distal phalanx of the thumb (Trinkaus and Villemeur 1991) and the expansion of the apical tufts (the bony fingertips) in Neandertals compared to most moderns. Studies of the carpometacarpal joints (Niewoehner et al. 1997, Niewoehner 2001) indicate that Neandertals had particularly forceful **transverse power grips**, which would be used in grasping something like a hammerstone or a baseball. By contrast, modern human hands are better adapted to withstand forces generated from actions involving an **oblique power grip**, used when grasping something with a handle (Churchill 2001). These anatomical differences may reflect significant differences in either technology or tool-use behavior, or both.

Pelvis and Lower Limb

Like their upper limbs, the lower limbs of Neandertals are characterized by big joint-surface areas, thick long-bone shafts, and other indications of powerful muscles exerting large forces. The Neandertal hipbone (*os coxae*) is generally robust. Like many other joint surfaces, the acetabulum is strikingly large in Neandertals, but the femoral head is no bigger than would be expected in a modern human of similar body mass (Ruff et al. 1993). The greater sciatic notch of the Neandertal hipbone seems to have had a pattern of sexual dimorphism like that of living humans, with the females having a wider, more open notch to augment the fore-and-aft diameter of the pelvic opening and birth canal (Smith 1980).

The Neandertal pubis—the part of the hipbone that extends around to the ventral midline—is distinctive and has prompted extensive debate. In Neandertals, the upper branch or **superior ramus** of the pubis, which forms the upper front edge of the obturator foramen (Fig. 7.23), is elongated and relatively thin compared to modern rami. This difference is established very early in Neandertal ontogeny (Tompkins and Trinkaus 1987). First noted by McCown and Keith (1939), the distinctive Neandertal pattern is evident in pelvic remains from Tabun, Shanidar, La Ferrassie, and Krapina (Stewart 1960, Smith 1976, Trinkaus 1976). Trinkaus (1984b) focused interest on this trait by suggesting that the

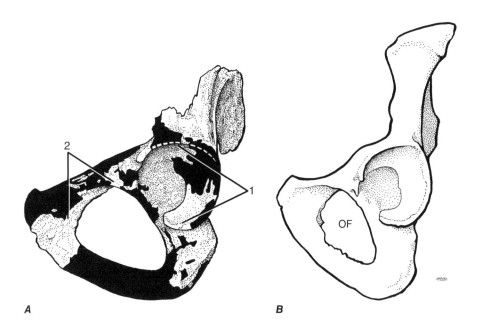

FIGURE 7.23

The left hipbone of the La Ferrassie 1 male Neandertal (**A**) and that of a recent human (**B**), viewed looking directly into the acetabula. The sex of the Neandertal specimen is determined from the shape of the greater sciatic notch (not shown). Note the large acetabulum (1) and the elongated superior pubic ramus (2) in the Neandertal compared to the modern specimen. OF, obturator foramen. Cf. Fig. 7.25.

elongated pubis reflected an expanded birth canal, which would allow for the birth of larger-headed infants. He saw this as an indication of a longer gestation period for Neandertals compared to modern humans, which would have had important implications for Neandertal life history.

No Neandertal pelvis was complete enough to actually measure the pelvic inlet or birth canal until the discovery of the Kebara 2 specimen in Israel in 1983. Analysis of this specimen indicated that the inlet was no larger than in modern humans (Rak and Arensburg 1987). This discovery cast doubt on Trinkaus's model but did not conclusively refute it, since Kebara 2 is a male and obviously did not give birth to any baby Neandertals. However, subsequent work by Walrath and Glantz (1996) indicates that the elongated superior pubic ramus in female Neandertals is exactly what one would expect to provide adequate pelvic inlet size for females from a skeletally large and heavy human population. Additionally, Rosenberg (1988) demonstrates that, among modern populations, females with the highest body weight have the longest pubes. She concludes that Neandertal pubic length and inlet size was probably proportional to the body mass of female Neandertals. These findings suggest that elongating the superior pubic ramus probably did enlarge pelvic inlet size for female Neandertals, but no more than would be predicted from their body build.

Rak (1990, 1993) contends that the pelvic inlet of Neandertals is shifted ventrally within the pelvic girdle (Fig. 7.24). This displacement supposedly moves the sacrum downward and forward to a point more nearly over an axis drawn through the two hip joints, thereby balancing the weight of the upper body more directly over the hip joints so that weight is more efficiently transferred from the trunk to the lower limb. In recent humans, the efficiency of weight transfer is optimized in a different way: the pelvis is tipped forward, rotating the top of the sacrum forward and the pubic symphysis downward and backward. Rak argues that the Neandertal morphology would produce a similar sacral displacement without the tilting. The increased length of the Neandertal pubic bones is explained partly by the ventral shift in the birth canal and partly by the fact that the distance between the two hip joints is greater in Neandertals, so that the pubis has to be elongated to reach to the midline. Rak (1993) suggests that the Neandertal morphology would produce certain (very slight) differences in posture and locomotion between Neandertals and modern humans and would also impair Neandertal abilities to absorb forces generated by locomotion. He sees these pelvic differences as a strong indicator that Neandertals represent a species distinct from the Skhul/Qafzeh folk and other modern humans.

Black (1999, 2004) offers a different biomechanical explanation of the Neandertal pubis. The outer surface of the pubic bone provides the surface of origin for the big **adductor** muscles on the inside of the thigh, which pull the thigh toward the midline when they contract and can act to stabilize the hip joint against various forces acting on the trunk. Black thinks that the Neandertal pubis may have been elongated to increase the attachment surface and thigh volume available for an

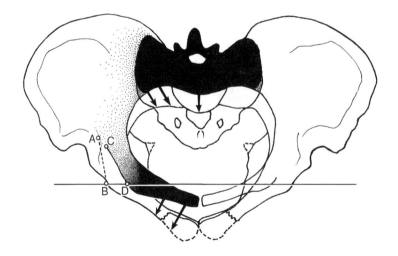

FIGURE 7.24

Schematic drawing of the Kebara 2 Neandertal pelvis from Israel compared to a recent human male specimen (*stippled*). The Neandertal pubis extends far in front of the solid line, which connects the anterior margins of the left and right acetabula. The dashed line on the left (line AB) represents the plane of the Neandertal acetabular opening, which is oriented more laterally than that of recent humans (line CD). (After Rak and Arensburg 1987.)

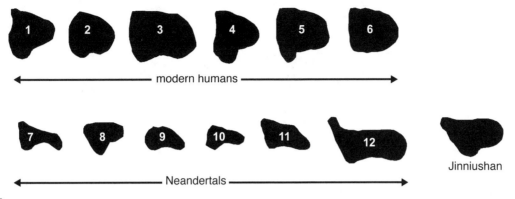

FIGURE 7.25

Cross sections of the superior pubic ramus in Pleistocene and Recent *Homo*. 1, Recent European male; 2, recent European female; 3, Skhūl 4; 4, Skhūl 9; 5, Oberkassel 2; 6, Qafzeh 9; 7, Shanidar 1; 8, Shanidar 4; 9, Tabūn 1 (left pubis);10, Tabūn 1 (right pubis); 11, Amud 1; 12, Kebara 2. The fossil from Jinniushan in China (Fig. 8.14) is contemporary with Neandertals and has been likened to them. (After Rosenberg 1998.)

adductor-muscle mass that was larger than in other humans. This enlargement of the adductors may have simply been part and parcel of the general beefing-up of the Neandertal musculature, or it may (as Black conjectures) have had a specific adaptive role in delivering and sustaining powerful thrusting forces during big-game hunting with spears.

The peculiarities of pubic morphology in Neandertals may also be related to their barrel-shaped trunk, with its exaggerated anteroposterior depth. No matter what the biomechanical consequences of pubic elongation may have been, it would also have had the effect of increasing the fore-and-aft diameter of the pelvis (Fig. 7.24). It seems logical to see such an increase as connected with a similar increase in thoracic depth, which is most pronounced at the pelvic end of the Neandertal ribcage (Smith 1992b). However, there are indications that the western Asian Neandertals had less barrel-shaped thoraxes than the European ones (Franciscus and Churchill 2002) but equally long and thin superior pubic rami. Evidently, the connection between upper trunk shape and the superior pubic ramus is not simple or straightforward.

Some aspects of Neandertal pubic morphology might simply be plesiomorphies. *Australopithecus* pelves have somewhat elongated pubic bones, although they do not manifest the thinning of the superior ramus seen in Neandertals. Unfortunately, there are no Ergaster or Erectine fossils that preserve the pubic region; but a long, thin superior ramus is seen in the Heidelberg pubes from Atapuerca (Sima de los Huesos), including that of the nearly complete Pelvis 1 (Chapter 6). The Jinniushan specimen from China, which is roughly contemporaneous with Neandertals (dating to OIS 6), has also been described as having a Neandertal-like superior pubic ramus (Rosenberg 1998, Rosenberg et al. 2006, Marchal 1999). Its cross-sectional profile is

much like those found among the Atapuerca people or some Asian Neandertals (Fig. 7.25). In summary, the morphology of the superior pubic ramus in Neandertals is probably related to more than one aspect of body build; but it does not appear to be linked specifically to gestation lengths or to be an exclusively Neandertal autapomorphy.

In addition to their large articular surfaces, Neandertal femora exhibit thick cortical bone in their shafts (Fig. 7.26), relatively low angles between the shaft and the neck, moderate anterior–posterior bowing of the shaft, and the absence of a pilaster (the ridge on the back of the middle half of the shaft that elevates the linea aspera from the remainder of the bone). The bowing of the shaft of the femur and the curvature of some other long bones in Neandertals led Virchow (1872) to claim that they suffered from rickets, a malady resulting from vitamin-D deficiency that causes irregular deformation of bone shafts. However, it is clear that Neandertals do not exhibit the histological signs or the extreme bone distortion characteristic of this condition (Ortner and Putschar 1981). The bowing of the Neandertal femur is probably an adaptive response to resist high anteroposterior bending stresses on the shaft produced by big, powerful thigh muscles. Trinkaus (1993c) argues similarly that the low neck-shaft angles seen in Neandertal femora are also related to loads on the lower limb resulting from elevated activity levels during growth. The markedly thick shafts of Neandertal femora and tibiae also reflect high levels of loading resulting from high lifelong levels of activity (Lovejoy and Trinkaus 1980, Ruff et al. 1993, Trinkaus and Ruff 1989, Trinkaus et al. 1999a). Again, these features are also primitive retentions, found in earlier members of the genus *Homo* (Chapters 5 and 6). Finally, 100% of all European Neandertals (but none of the Asian Neandertals) exhibit a **proximal lateral femoral flange** (Fig. 7.27), a lateral

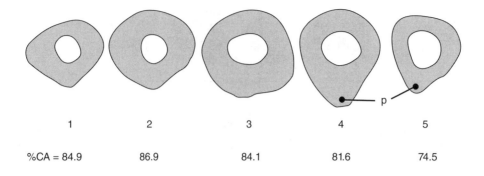

1	2	3	4	5
%CA = 84.9	86.9	84.1	81.6	74.5

FIGURE 7.26

Femoral cross-sectional contours at midshaft, illustrating variation in percent cortical area (% CA) and shape in Pleistocene and recent humans. 1, Habiline (KNM-ER 1481); 2, Erectine (Zhoukoudian 1); 3, Neandertal (La Chapelle-aux-Saints); 4, early modern human (Paviland 1); 5, recent modern human (Pecos Pueblo). Note the relatively broad midshaft of the Neandertal, the distinct pilasters (p) of the two modern femora, and the reduced cortical thickness of the recent femur. (After Ruff et al. 1993.)

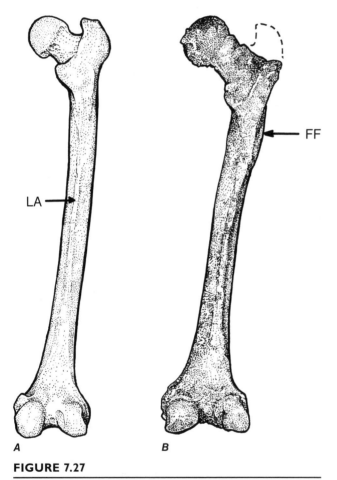

FIGURE 7.27

Posterior view of (**A**) the right femur of a recent human and (**B**) the Neandertal femur from Spy 1 (Belgium). Specimens are scaled to the same length. Note the larger articular surfaces of the Neandertal femur and its relatively broader shaft. This latter feature is particularly evident at the proximal shaft, where the proximal lateral femoral flange (FF) is developed on the Neandertal specimen. LA, linea aspera. (**A**, after Rohen et al. 1998.)

bulging of the upper end of the femoral shaft (Kidder et al. 1991). The functional significance of the flange is unknown, but its presence contributes to a distinctive mediolateral broadening of the proximal femur in European Neandertals.

Like the other elements of the limb skeleton, Neandertal foot bones are generally robust, implying high levels of activity-related stress. The proximal phalanges of the toes are relatively short, but have expanded joint surfaces (Trinkaus and Hilton 1996). Enlarged articular surfaces are also seen in other foot bones, particularly the talus; and the attachments for ligaments and muscles involved in plantarflexion and maintenance of the pedal arches are very rugose. According to Trinkaus (1983b), Neandertal feet are otherwise functionally similar to those of recent humans and do not indicate any significant differences in movement abilities or patterns. They seem well-suited to providing a stable base for Neandertal activities.

All these features of Neandertal limbs suggest that these ancient Europeans were powerfully built people, whose bodies were adapted to generating and resisting strong muscular forces. Although Klein (1999) raises the possibility that these features may lack functional significance and merely reflect genetic drift in an isolated European lineage, the bulk of the evidence makes this highly unlikely (Pearson et al. 2006). It is hard to see why drift would produce so many changes in different parts of the skeleton, all tending toward the same functional end. However, these changes do appear to have a genetic basis, inasmuch as many of the distinctively Neandertal traits show up in very young individuals (Vlček 1970, 1973; Smith 1976, Heim 1982, Trinkaus 1983b, Tompkins and Trinkaus 1987). This implies that these traits were not developed during life as mechanistic responses to severe functional demands, but rather were a part of the Neandertals' genetic program. Comparisons between the upper and lower limbs of Near Eastern Neandertals and early moderns indicate that the two groups had similar levels of lower-limb robustness (when scaled for differences in body mass), but that Neandertals had more robust upper limbs (Trinkaus et al. 1998). Evidently, both groups placed similar high demands on their lower limbs—demands that probably included extensive movement over rough terrain and activities relating to subsistence (and possibly to defense as well)—but for some reason, Neandertals needed even more powerful upper limbs than early modern humans. These differences may reflect distinctive Neandertal behaviors in using weapons or other implements. Pearson et al. (2006) interpret these differences in upper-limb robusticity differently. They find similarities in this regard between Neandertals and modern populations who practice intensive foraging over limited ranges, with both groups having more robust upper limbs than the early modern people from Skhūl/Qafzeh and Upper Paleolithic (Gravettian) sites in Europe. These differences suggest to them that the early moderns may have been foraging over more extensive geographic territories than Neandertals were.

❚ NEANDERTAL LIFE HISTORY

Dental development reflects patterns and timing of overall maturation in various primate species (Smith 1989a), and this may be true for Neandertals as well. From his studies of the Krapina Neandertal teeth, M. Wolpoff (1979) concluded that third molars erupted earlier in Neandertals than in modern humans. His assessment of relative wear in the molars led him to suggest that Neandertal third molars were in occlusion by age 14—four years earlier than in moderns. This is commensurate with the idea that Neandertals, like other archaic humans, may have developed more rapidly than people do today. Wolpoff's conclusion has been supported by other studies. Dean et al. (1986) inferred a more rapid development in Neandertals from their study of the Devil's Tower Neandertal child from Gibraltar. Tompkins (1996) concluded that third molars developed faster in Neandertals than in modern humans, but that this was also true of early modern Europeans. Studies on the rate of enamel formation indicate that Neandertal teeth formed more rapidly than in modern people (Dean 1985, Stringer et al. 1990) and that Neandertals reached adulthood by 15 years of age (Ramierez Rossi and Bermúdez de Castro 2004). However, Thompson and Nelson (2000) contend that the pattern of Neandertal growth is consistent with either slow overall body growth or rapid dental development.

F. Ramierez Rossi and J. Bermúdez de Castro (2004) used microscopy to study rates of enamel formation in teeth from a large series of Neandertals, Heidelbergs,

and modern humans from the Paleolithic and Meso-lithic. During development, the rate of enamel formation in humans and great apes fluctuates with a period of about nine days, resulting in tree-ring-like bands manifested as **striae of Retzius** (SR) in sections and ridges called **perikymata** on the surface of the tooth (FitzGerald 1998, Schwartz and Dean 2001, Dean et al. 2001). Ramierez Rossi and Bermúdez de Castro found that the spacing of perikymata was similar during early development in both the modern and fossil samples, but that later in development, the fossil samples exhibited fewer perikymata per millimeter of enamel thickness. Assuming that SR periodicity is constant, this implies that later enamel development was accelerated in the fossil teeth. In the case of Neandertals, teeth formed 15% faster than in modern humans. But the Neandertal teeth also formed faster than those of the earlier Heidelbergs. Ramierez Rossi and Bermúdez de Castro interpret these findings as documenting a specialized, autapomorphous pattern of accelerated growth and maturation in Neandertals, thereby supporting the recognition of Neandertals as a distinct and unique species. Ponce de León and Zollikofer (2001, 2006) and Thompson and Nelson (2000) also interpret the reported differences in Neandertal ontogenetic patterns as consistent with their placement in a separate species from *H. sapiens.*

A recent study on growth increments of the anterior teeth comparing Neandertals to modern Europeans, South Africans, and Inuits (Eskimos) challenges the findings of Ramierez Rossi and Bermúdez de Castro. On the basis of perikymata counts, Guatelli-Steinberg et al. (2005) conclude that anterior tooth formation in Neandertals was indeed faster than in Inuits, but slower than in South Africans and about the same as in Europeans. Thus, Neandertal anterior tooth formation times lie within the wide range found among modern humans. Guatelli-Steinberg and colleagues note that Ramierez Rossi and Bermúdez de Castro considered only one modern human sample, resulting in a misperception of Neandertal distinctiveness. However, a subsequent study indicated that the enamel growth curves in Neandertals were more linear than in any of the modern samples, indicating a different pattern of growth in Neandertals (Guatelli-Steinberg et al. 2006).

An earlier study of perikymata patterns by Mann et al. (1991), using less sophisticated technology and a smaller Neandertal sample (5 Krapina incisors), similarly found that Neandertal perikymata counts overlapped the modern human range almost completely. Mann's group went on to challenge the underlying supposition of a regular periodicity of enamel formation, arguing that bouts of disease and other factors resulting in growth irregularities make it impossible to assume that enamel always forms at so regular a rate.

Apart from enamel deposition rates, there are other signs that development was accelerated in Neandertals.

Several studies have shown that Neandertal children tend to exhibit larger vault and upper facial dimensions than similar-aged moderns (Vlček 1970, Smith 1976, Minugh-Purvis 1988, 1998; Minugh-Purvis et al. 2000)—assuming that we can assign accurate chronological ages to juvenile Neandertals, which is not necessarily the case (Tillier 1989). From some of these studies and many others (Vlček 1973, Smith 1976, Heim 1982, Tillier 1983, 1987, 1998, 1999; Tompkins and Trinkaus 1987, Maureille and Bar 1999, Ponce de Léon and Zollikofer 2001, Ruff et al. 2002, Trinkaus 2002), it has been established that even very young Neandertals show distinctive features of skeletal anatomy and shape, which distinguish them from similar modern children and foreshadow the morphology of Neandertal adults. In a pioneering study of Neandertal heterochrony, S. Leigh (1985) showed that the postnatal growth of the palate followed identical trajectories in Neandertals and modern humans, but that Neandertals started out larger at birth. He suggested that this difference might be due to a prenatal acceleration of growth in Neandertals.

Leigh's ideas were further developed by M. Green and F. Smith (Green 1990, Green and Smith 1991, Smith and Green 1991, Smith 1991). Their model of Neandertal craniofacial growth focused on the two cartilaginous plates or **synchondroses** in the midline of the cranial base, which separate the occipital bone from the sphenoid in front of it (the **basilar** synchondrosis) and divide the central stem of the sphenoid into two parts (the **mid-sphenoid** synchondrosis). Much of the longitudinal growth in the cranial base occurs at these plates, and accelerated growth at these points would result in a longer cranial base earlier in ontogeny. Green and Smith argued that rapid growth at the basilar (spheno-occipital) synchondrosis would also have caused the mid-sphenoid synchondrosis to fuse early. Since increasing basicranial flexion during human ontogeny involves growth at the mid-sphenoid synchondrosis, early fusion there would result in a less flexed cranial base. Green and Smith speculated that this longer and flatter cranial base would have impacted both vault and facial shape, and may have been responsible for many of the craniofacial differences between Neandertals and modern humans (cf. Lieberman et al. 2000).

Green and Smith went on to suggest that the main reason for prenatal growth acceleration in Neandertals was to produce bigger babies. Thermoregulatory stresses impose strong selection pressures on young human infants (Roberts 1978, Meban 1983). Assuming that Neandertal cultural adaptations to the rigors of a cold climate were not as effective as those of comparable modern people (see "Neandertal Diet and Behavior," below), Green and Smith inferred that selection on European Pleistocene Neandertals would have strongly favored larger infants, whose body build and overall size would maximize their resistance to the cold. Their

model implies that selection acting on the very young would have produced an ontogenetic pattern in Neandertals different from that of recent humans, and that this could account for many of the distinctive craniofacial traits of Neandertals. Green and Smith further speculated that such a change might be responsible for some of the postcranial peculiarities of Neandertals as well, if the underlying genetic mechanisms had global effects on endochondral bone growth.

More rapid maturation in Neandertals may also have been associated with a shorter lifespan. Few Neandertals seem to have survived beyond their peak reproductive years. Trinkaus (1995a; cf. Trinkaus and Tompkins 1990) concludes that disproportionate numbers of Neandertals died before reaching the age of 40. Among modern hunters and gatherers, about 38% of actual deaths occur among young adults, although a larger percentage (~60%) of the skeletons recovered by archaeologists fall into this age range because the bones of babies and children tend not to be preserved long after death. But around 80% of all known Neandertal individuals appear to have died as young adults. When the Neandertal numbers are adjusted to compensate for the expected poor preservation of baby bones, the survivorship curve for immature individuals falls within the range of modern human variation; but older adults continue to be underrepresented. Trinkaus suggests that this underrepresentation results from some combination of four factors: environmental stress that limited life expectancy, the effects of pooling Neandertal data from disparate times and places, a highly mobile lifestyle in which vulnerable old people were more likely to die on the trail (and thus not to get buried in cave shelters), and possibly differential disposal of the corpses of the aged. A developmental pattern involving faster maturation could have evolved in the context of a short life expectancy, with selection favoring earlier maturation to extend the reproductive period in the face of a probable early death.

Lifespan, age at maturity, and other life-history variables must have been reflected in the size, density, and demography of Neandertal populations. Unfortunately, estimating such population parameters in the remote past is very difficult and notoriously imprecise. From ecological modeling and interpretations of the pertinent archaeological record, F. Hassan (1981) calculated overall population size for Middle Paleolithic people (including early modern humans associated with Mousterian or Middle Stone Age artifacts) to have been slightly more than one million. This implies a population density of 0.03 persons per square kilometer. Both these numbers are about three times smaller than equivalent estimates for Upper Paleolithic people (Chapter 8). Evidently, Upper Paleolithic technologies and subsistence strategies could sustain a much larger human biomass than their precursors could. O'Connell (2006, p. 55) suggests that modern humans ultimately prevailed over Neandertals in Europe not because they were smarter or more capable, but rather because they developed a "... more expensive subsistence economy, one that succeeded only because they arrived in numbers large enough to prevent the resident Neanderthals from continuing to do as they had for tens of thousands of years before."

Certainly, the Neandertals' big bodies and elevated metabolic rates must have been energetically expensive, requiring a very large caloric intake. The expenditure side of any organism's energy budget can be reckoned up under three headings: normal maintenance or survival activities (including moving and various physiological functions), growth and development, and reproduction (Leonard 2002, Sorensen and Leonard 2001). What we know of Neandertal biology suggests that maintenance, survival, and growth would have required almost all of the energy the Neandertal diet could provide, leaving precious little surplus for reproduction. It seems correspondingly likely that Neandertals must have had lower long-term fertility rates than modern humans. This interpretation supports the archaeological estimates of low population density in Neandertals. It implies that Neandertals would have been a rare sight on the Pleistocene landscape.

▌ NEANDERTAL GENETICS

We saw in Chapter 1 that the cells of humans and other eukaryotes contain organelles called mitochondria, which appear to be the descendants of ancient bacteria. These symbiotic bacteria came on board at the beginning of eukaryote evolution to manage the processes of aerobic respiration in the oxygen-rich environment that resulted from the advent of photosynthesis. Although the mitochondria settled down in their new homes and became good citizens of the eukaryote cell, they remained true to their bacterial heritage by refusing to buy into the whole business of sex, with its haploid gametes and recombination. Mitochondria stubbornly retain their own DNA and reproduce by fission independently from the replication cycle of the rest of the cell. Both eggs and sperm cells contain mitochondria; but in most eukaryote species, the mitochondria carried into the egg by the sperm are almost always destroyed after fertilization and do not exchange DNA with the mitochondria in the egg (McVean 2001). The mitochondrial DNA (mtDNA) in the developing zygote is therefore inherited exclusively from the mother, and reflects only the maternal line in an organism's family tree.

Mitochondrial DNA has a distinct advantage for the paleontologist. Normal cells have only a single copy of their nuclear DNA, because there is only one nucleus. But because one cell can contain thousands of mitochondria, mtDNA has a much greater chance of surviving in ancient bone than nuclear DNA has. Fragments of mtDNA have been recovered from 12 Neandertal specimens to date. These mtDNA sequences are relatively short, with the longest being in the neighborhood of 380 nucleotide base pairs—about 2.3% of the 16,569 base pairs in the human mitochondrial genome. More recently, nuclear DNA has also been isolated from Neandertal fossils. Approximately a million base pairs of this (presumably Neandertal) nuclear DNA have been sequenced thus far (Green et al. 2006). This is a far smaller percentage (0.03%) of the nuclear genome, which contains some 3.2 *billion* base pairs. These studies thus give us a very narrow window on a tiny fraction of Neandertal genetic makeup and variability. Nevertheless, the opportunity to study Neandertal genetics has the potential to revolutionize our knowledge of these archaic people.

The first Neandertal mtDNA sample was isolated from the right humerus of the original Neandertal skeleton (Neandertal 1) under careful controls in S. Pääbo's laboratory and published by Krings et al. (1997). A 377-base-pair sequence was extracted from hypervariable region 1 of the D loop, the non-coding part of mtDNA that regulates its replication. This sequence was compared to a sample of 994 homologous sequences of recent humans as well as the standard recent human reference sequence. The major analysis involved pairwise comparisons, in which two individual sequences are compared randomly. When the pairwise comparisons involved recent human sequences only, the average difference between sequences was only eight base pairs (s.d. = 3.1, range = 1–24 differences). But comparisons between the Neandertal 1 and recent sequences yielded an average of 27 base-pair differences (s.d. = 2.2, range = 22–36 differences). As Fig. 7.28 shows, comparisons with chimpanzees yield many more differences on average, but this Neandertal sequence is nevertheless relatively distinctive when compared to those of living humans. Operating on the commonly held assumption that changes in this region of the mtDNA genome are not under natural selection, Krings and colleagues ran a molecular-clock analysis and calculated that the Neandertal lineage had been separate from the lineage(s) leading to modern people for about 600 Ky. The statement printed on the cover of the journal *Cell*, where the 1997 article was published, sums up the conclusion generally drawn from this study: "Neanderthals were not our ancestors."

Neandertal 1 yielded a second sequence, from hypervariable region 2 of the mtDNA genome (Krings et al. 1999). Other sequences were soon obtained from a Russian infant specimen from Mesmaiskaya (Ovchinnikov et al. 2000) whose Neandertal affinities have been questioned (Hawks and Wolpoff 2001b; cf. Hublin and Bailey 2006), an undiagnostic foot bone from the Croatian site of Vindija (Krings et al. 2000),

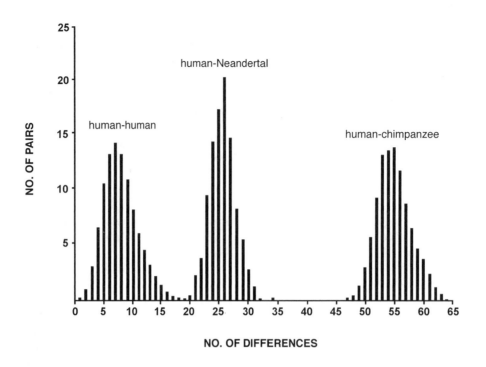

FIGURE 7.28

Distribution of pairwise comparisons of Neandertal–modern human, modern human–modern human, and Neandertal–chimpanzee mitochondrial DNA (redrawn from Krings et al. 1997). See text for discussion.

and a second right humerus from the original Neandertal site (Neandertal 2) that exhibits archaic features (Schmitz et al. 2002). Serre et al. (2004) published sequences from four additional Neandertal specimens: two from Vindija, and one each from the Engis 2 juvenile and the adult skeleton from La Chapelle-aux-Saints. These sequences are not as complete as the first four, but one of the new Vindija sequences is identical to the one published from the site in 2000. Three additional sequences were published in 2005–2006: one from El Sidrón 441 in Spain (Lalueza-Fox et al. 2005), one from Les Rochers-de-Villenueve in France (Beauval et al. 2005), and one from Scladina Cave in Belgium (Orlando et al. 2006). The Scladina Cave specimen is of note because it is significantly older (~100 Ky) than the other samples. These additional Neandertal sequences cluster with the original Neandertal 1 sequence when compared to recent humans, and all exhibit sequences not found in recent human samples. Serre and colleagues note that the observed diversity of Neandertal mtDNA is similar to that seen in recent humans, and lower than that of the great apes. All of these facts generally have strengthened the view that Neandertals represent a separate species from *Homo sapiens* and are not a part of *H. sapiens'* direct ancestry.

Isolating and authenticating ancient DNA is a tricky process. Contamination is always a major problem (Cooper et al. 2004). How can we make sure that DNA from a fossil is truly ancient, and not left on the fossil by recent contact with some other organism? One important sign is the condition of the DNA. Authentic ancient DNA is always significantly damaged and eroded (Serre et al. 2004), and long undegraded sequences can therefore be recognized as modern contaminants. DNA from a fossil can also be rejected as a contaminant if it is identical to that of some living species. For example, Serre et al. (2004) rejected DNA extracted from six fossils of the extinct cave bear (*Ursus spelaeus*) because it was identical to modern human sequences. This seems like a reasonable thing to do. But if this is a criterion of sample construction, it cannot also be an empirical discovery about the sample. If DNA from a human fossil does not count as ancient unless it lies outside the range of modern variation, then we should not be surprised to find that all ancient DNA lies outside the modern human range. In fact, some DNA sequences from Neandertal specimens have been rejected because they were the same as recent human sequences. Serre et al. (2004) note that thus far all Neandertal remains that have proved to contain usable mtDNA have yielded "Neandertal-type" sequences differing from any known from modern humans. Nevertheless, they acknowledge that "… in the absence of further technical improvements, it is impossible to produce undisputable human mtDNA sequences from ancient human remains." A "modern-type" fragment of the Neandertal mitochondrial genome could still be rejected as a contaminant if it were insufficiently different from recent human sequences.

Different authors have drawn conflicting conclusions from these mtDNA data. Knight (2003) compared the nucleotide sequences of the first four Neandertal hypervariable region 1's and two Neandertal hypervariable region 2's with homologous areas of nine recent humans representing diverse mitochondrial lineages. He found four highly consistent sites that grouped all Neandertals together to the exclusion of all modern humans. At one location, Knight identified a distinctive insertion in all four Neandertals that was lacking in the moderns. Noting the rarity of such a circumstance, Knight suggests that this insertion is a synapomorphy that defines the Neandertal mitochondrial lineage and demonstrates its separation from modern human lineages. On the other hand, G. Gutiérrez et al. (2002) constructed mtDNA trees using data from both hypervariable regions 1 and 2. Their analysis yields a bifurcated tree, in which one branch ends in 10 African haplotypes and a second branch comprises 367 other extant haplotypes (African and non-African) plus two Neandertals. This analysis contradicts the claim that Neandertal mitochondrial lineages are totally separate from those of all living humans.

It is often claimed that the Neandertal DNA data settle the debate over modern human origins in favor of the total-replacement version of the "Out of Africa" model. This conclusion seems premature. As we will see in the next chapter, certain facts about the genetics of modern human populations point to a different conclusion (Relethford 2001). There are also some unresolved issues concerning the Neandertal data. Some of these mtDNA sequences come from bones that may not be those of Neandertals. This caveat applies to the Russian baby and to all three specimens from Vindija. The Vindija specimens are fragments that may never be unequivocally identified as Neandertals on morphological grounds. (Two of them have not yet been morphometrically analyzed or published.) All four of these specimens are potentially from a time span when early modern humans may have been present in eastern Europe (Churchill and Smith 2000a, Trinkaus 2005, Smith et al. 2005), and so their "Neandertal-type" mtDNA sequences may actually come from early modern humans. We know that this is a real possibility because the single ancient mtDNA sample known from Australia, which was extracted from a specimen with modern human morphology, also yielded an mtDNA haplotype outside the modern human range (Chapter 8). Even if we assume that all these specimens are Neandertals, we know from the genetics of living populations that twelve Neandertals are not enough to estimate the range of Neandertal mtDNA diversity.

In some comparisons, Neandertals are best described as clustering at the periphery of modern human variation rather than falling entirely out of it. Some pairwise comparisons between Neandertals and recent humans are actually closer than some comparisons between recent humans (Fig. 7.28). Krings et al. (1997) indicate only about a 0.002% overlap between such comparisons, but it is still an overlap. Moreover, if the mtDNA diversity of recent humans plus Neandertals is added together, it is still less than that documented between the three subspecies of chimpanzees (*Pan troglodytes*) by Morin et al. (1994). This fact has led Wolpoff and colleagues (Wolpoff et al. 1997; Wolpoff and Caspari 1997) and Smith et al. (2005) to suggest that Neandertals might better be regarded as a subspecies of *H. sapiens* than a distinct species. Another issue is that Knight's study utilizes only nine modern human sequences. Again, this is not sufficient to claim that the inferred Neandertal synapomorphies are totally absent in modern people. Finally, we have no mtDNA from Neandertal contemporaries in Africa or Asia, and so we do not know whether the differences between Neandertals and living humans represent Neandertal autapomorphies or symplesiomorphies of archaic *Homo*.

A final complicating factor concerns the question of neutrality in the mtDNA genome. The widely accepted assumption that mtDNA does not undergo selection is critical not only to molecular-clock estimates of Neandertal divergence times, but also to the claim that the absence of Neandertal-type mtDNA in modern human populations demonstrates the extinction of the Neandertal nuclear genome (rather than selection against individuals with Neandertal-type mtDNA). However, several assessments, beginning with that of Spuhler (1989; see also Smith et al. 1989a), have challenged the assumption that mtDNA is selectively neutral. J. Hawks (2006) has marshaled the evidence that suggests that significant selection has impacted the human mtDNA genome during the last 30–50 Ky. Hawks notes that mitochondrial DNA variants in recent humans are associated with performance in athletics, certain brain disorders, chronic diseases relating to aging, and even overall longevity. Recent human mtDNA variation may also be correlated with climate and may affect dietary energetics and insulin metabolism. These facts indicate that mtDNA variants have phenotypic effects. Therefore, they cannot be exempt from natural selection. In addition, the distribution of variation in recent human mtDNA is not consistent with a neutral-evolution model. These issues are explored further in Chapter 8.

In 2006, the study of Neandertal genetics took a major new turn with the publication of two Neandertal nuclear DNA sequences, one of ~1 million base pairs (Green et al. 2006) and another of 65,250 base pairs (Noonan et al. 2006). Like the published mtDNA fragments, both of these nuclear sequences are significantly different from their counterparts in modern humans. Green et al. (2006) calculate that of the changes that have occurred in the modern human lineage since it diverged from chimpanzees, 8.2% took place after the divergence of a separate Neandertal lineage. Assuming a *Homo–Pan* split at 6.5 Mya, they place this human–Neandertal divergence at 516 Kya, give or take 10%. However, they also find that for single-nucleotide polymorphisms (SNPs) that distinguish humans from chimpanzees, the Neandertal sample has the derived (human) allele in about 30% of all cases—which is too high a frequency to allow for a decisive speciation event with reproductive isolation between modern humans and Neandertals at ~500 Kya. They conclude that gene flow may have continued for some time between modern humans and Neandertals (cf. Weiss and Smith 2007). Since the DNA fragments attributed to the Neandertal X chromosome are more divergent from modern humans than those mapped to other (autosomal) chromosomes, Green et al. (2006; p. 335) suggest that gene flow "… may have occurred predominately from modern human males into Neanderthals." Noonan et al. (2006) find that Neandertal and modern humans are 99.5% genetically identical. However, they also find that the differences imply a split between these lineages at ~370 Kya. Their data reveal no indication of a uniquely Neandertal genetic contribution to modern humans.

Both of these pioneering studies are based on DNA extracted from a single fossil specimen referred to as Vi 80, a small, nondescript fragment of what may be a tibia shaft from Vindija Cave in Croatia. The inventory number of this specimen is actually Vi 33.16. (The number "80" refers to the year it was found.) Direct AMS radiocarbon dating of the fossil yields an estimated age of 38.31 ± 2.1 Ky. Although mtDNA from this specimen places it among the Neandertals (Serre et al. 2004), the bone itself is not diagnostic morphologically, and we cannot be certain that it represents a typical Neandertal. Since the cranial remains of the Vindija hominins deviate from typical Neandertals in the direction of modern humans (see below), it may be that Vi 33.16—and the nuclear DNA derived from it—is sampling a hybrid Neandertal population with some admixture of early modern genes.

Two additional studies of Neandertal nuclear DNA have yielded significant results. In one study, Lalueza-Fox et al. (2007) identified a pigmentation gene called MC1R (melanocortin 1 receptor) in two Neandertal specimens, one each from El Sidrón (Spain) and Monti Lessini (Italy). The MC1R gene helps regulate the balance between **pheomelanin**, which produces red and yellow coloration, and the dark brown or black skin pigment **eumelanin**. Light pigmentation of the skin and red hair occur in recent humans bearing MC1R alleles that restrict the function of the receptor (Rees 2003). The two Neandertals yielded MC1R alleles that differ from known modern human variants by a single point mutation, but functional assessment indicates that the Neandertal variant would have limited MC1R activity to a level that results in pale skin and red hair in modern

people (Lalueza-Fox et al. 2007, Culotta 2007). This finding is not surprising. As we noted in Chapter 5, reduced skin pigmentation would have had crucial adaptive advantages for any ancient occupants of Europe or other high-latitude areas. It is reassuring to find this conclusion corroborated by actual genetic data.

The recent findings concerning the FOXP2 gene were discussed briefly earlier in this chapter. This gene codes for a protein that is expressed in the cortex of the developing brain and is thought to be a transcription factor governing the expression of other genes. In modern humans, mutations of the FOXP2 gene appear to cause disorders of neurological development that result in defective abilities to articulate, analyze, and understand speech (Lai et al. 2001). It is not yet clear how the uniquely human form of the FOXP2 gene actually differs in function from its homologs in other mammals. Still, the fact that Neandertals share the uniquely human FOXP2 sequence (Krause et al. 2007a) strongly suggests that they also shared a human capacity for language.

Krause et al. (2007a) conclude that the FOXP2 mutations shared by modern humans and Neandertals were probably fixed in their last common ancestor before the presumed split between the two populations, which they date to between 300 and 400 Kya. But there are other possibilities. For example, Neandertals might have acquired the modern FOXP2 variant through interbreeding with their modern-human contemporaries. Krause and his team acknowledge this possibility, but consider it unlikely because mtDNA and Y-chromosome data do not demonstrate evidence of gene flow between Neandertals and modern humans. On the other hand, studies of nuclear DNA indicate that such introgression may have occurred (Templeton 2002, 2005; Evans et al. 2006). Could the FOXP2 evidence be a further indication of Neandertal–modern human genetic exchange? It is even possible that the modern variant first emerged in Neandertals and entered the modern gene pool via interbreeding.

▌ NEANDERTAL TECHNOLOGY

As is the case for earlier hominins, our knowledge about the behavior of Neandertals is based largely on the stone tools they left behind. Compared to their predecessors, the Neandertals were clearly more expert stone-tool makers, who crafted a much wider range of lithic tools (Hayden 1993, Kuhn 1995, Roebroeks and Gamble 1999, Dibble and McPherron 2006). The cultural complex generally associated with Neandertals is known as the Mousterian, named for the French site of Le Moustier. A minor variant of Mousterian, the Micoquian, is found with some Neandertals—for example, at the Neander Valley type site. Some late Neandertals are apparently associated with Initial Upper Paleolithic industries, including the Châtelperronian at such French sites as Saint-Césaire and Arcy-sur-Cure (Grotte du Renne), the Szeletian in Central Europe, and the IUP variant at Vindija (Churchill and Smith 2000a, Svoboda 2006; but see Bar-Yosef 2006). Conversely, some Mousterian variants in western Asia and North Africa (e.g., the so-called Levalloiso-Mousterian) are associated with more modern human fossils. The following descriptions apply specifically to the Mousterian (Middle Paleolithic) industries of Eurasia that are found in association with Neandertal remains.

The Mousterian is a flake-based industry in which various techniques were used to pre-shape the flake before it was struck from the core. The textbook example is the so-called Levallois technique, which is found first in the Acheulean but is much more common in certain Mousterian variants. Pre-shaping the primary flake before it is removed yields flakes of more standardized form and probably better working edges, although it is not especially economical in conserving raw material. Mousterian assemblages are characterized by high frequencies of various types of side-scrapers, but also encompass denticulates, notches, flake-blades, and points. The percentages of these types are not the same in all Mousterian variants or contexts of use (Bordes 1961, 1968; Kuhn 1995, Mellars 1996; see also Roebroeks and Gamble 1999). Use-wear analysis of Mousterian tools shows that most of them were used for working various materials, including wood, bone, hides, and meat, and that the denticulates and notches were used more specifically for wood-working (Beyries 1988). In addition to their skill as stone knappers, Mousterian people were also adept at choosing high-quality raw materials for making their tools. Those raw materials generally came from nearby locations, indicating that Neandertals did not often range far in search of usable stone (Geneste 1988, Simek and Smith 1997, Klein 1999). But in some areas, the Mousterian artisans made use of raw materials from more than 100 km away from the site where the tools were found. Kuhn and Stiner (1992) conclude that Mousterian people developed very complex procurement strategies to obtain the raw material necessary for their high-quality lithic technology.

Bordes (1961), who developed the standard classification system for Mousterian tools, argued that the variability observed in Mousterian assemblages grouped them into four more or less distinct clusters. This clustering was not due, in his view, to differences in the activities carried on at different types of site; rather, it demonstrated the existence of different cultural traditions or ethnic groups—"tribes" perhaps—among the Neandertals (Bordes 1968). Bordes' analysis was limited essentially to southwestern France, where he believed that the four "tribes" defined on lithic variability existed as distinct units over a period of more than 60,000 years. A contrary interpretation of Mousterian variability was offered by Binford and Binford (1966, 1969), who argued that it simply represented seasonal or activity-related differences in the tool complexes produced by a single people. Dibble (1988, Dibble and Rolland 1992) notes that many of Bordes' tool categories may actually represent successive stages in the use and resharpening of a few basic types. This sequence of use and reworking is sometimes referred to as the **operational chain** (Fr. *châine operatoire*). Refinements of the chronology of Bordes' "tribal" groups also suggest some temporal patterning among them (Mellars 1986, 1996). In short, the significance of Mousterian lithic variability as originally defined by Bordes remains uncertain, but it is probably the result of a number of factors—perhaps including "ethnic variation."

Little is known about other aspects of Mousterian technology. A single, well-crafted wooden spear is known from the Mousterian site of Lehringen, Germany (Movius 1950), and two wooden "shovels" plus other indications of wood use are known from Abric Romani, Spain (Carbonell and Castro-Curel 1992). Wood must have been far more widely used in Mousterian technology than these few preserved artifacts would indicate. The well-made spears from the pre-Mousterian site of Schöningen in Germany (Thieme 1996, 1997) are strong evidence that this was true for earlier periods as well. The paucity of wooden artifacts in Middle Paleolithic (and earlier) contexts is probably due to factors of preservation. Although bone is plentiful in some Neandertal sites and used pieces of bone are not rare, skillfully shaped bone tools are generally lacking in Mousterian contexts (Mellars 1996, Münzel and Conard 2004). However, flat-based bone points known as Mladeč points are relatively common in late Mousterian assemblages in Central Europe (Monet-White 1996). A few bone tools, including a bone point and implements (retouchers) used to do fine retouch work on stone tools, are found in Middle Paleolithic levels in the Swabian Mountains of southwestern Germany (Münzel and Conard 2004). Sophisticated bone technology is directly associated with Neandertals in Initial Upper Paleolithic levels at the Grotte du Renne (Hublin et al. 1996) and Vindija (Karavanić and Smith 1998), and bone points are widespread in Initial Upper Paleolithic entities, like the Szeletian, in Central Europe (Svoboda 2006). All of these occurrences fall toward the end of the Neandertal time span.

Neandertals made extensive use of fire, and many Mousterian sites have evidence of well-defined hearths. These may be no more than thin deposits of charcoal, burned bones and ash, but many sites have evidence of hearths recurring at the same place over long periods of time within relatively thick Mousterian deposits (Bar-Yosef et al. 1992, Rigaud et al. 1995, Barton 2000, Schliegl et al. 2003). However, Mousterian hearths lack the sophisticated construction often found in their Upper Paleolithic counterparts (Perlés 1976, Trinkaus

1989). Although Mousterian tools were probably used to work hides, which suggests that Neandertals were capable of producing some form of clothing, there is no evidence of the bone needles and other artifacts that signal the presence of more efficient, tailored clothing in the Upper Paleolithic. Evidence of Mousterian structures is rare and does not suggest elaborate construction. According to Klein (2003), the only indisputable house ruin is from the Châtelperronian levels of the Grotte du Renne. All of this indicates that although Neandertals had developed certain cultural means of buffering themselves against their harsh environments, their expedients must have been less effective than those of such cold-adapted modern peoples as the Upper Paleolithic Europeans or the historical Lapps and Inuit.

▌DIET AND SUBSISTENCE BEHAVIOR

What evidence we have about Neandertal subsistence suggests that it was a particularly exaggerated form of what we referred to in Chapter 5 as the "Pleistocene human subsistence strategy," involving high activity levels, rapid metabolic turnover, and a lot of meat and other high-calorie foods in the diet. Not surprisingly, estimated daily energy expenditures for Neandertals are large and entail very high returns from foraging (Sorensen and Leonard 2001, Steegmann et al. 2002, Churchill 2006). Mousterian sites from western Europe to western Asia yield a wide variety of animal bones, mostly of herbivores of various body sizes. The faunal data show that Neandertals exploited a broader spectrum of species than their Upper Paleolithic successors did (Grayson and Delpech 2006). There is persuasive evidence of Neandertal exploitation of shellfish in Italy (Stiner 1994) and Gibraltar (Barton 2000). Among the most common prey animals found in Eurasian sites are aurochs (wild oxen), bison, red and fallow deer, reindeer, horses, wild goats and sheep, elephants (mammoth), and rhinoceros. Klein (1999) notes that these last two large species are generally rare and were presumably not regularly hunted. In Klein's view, the level of hunting skill needed to deal with mammoth and rhinoceros is not seen in the archaeological record until after the emergence of modern human behavior. Nevertheless, aurochs, bison, and red deer are large and formidable prey; and it is clear from the faunal remains at Mousterian sites that Neandertals relied heavily on the hunting of big game and the exploitation of other animal resources in meeting their energetic needs.

The inventories of faunal remains found in Mousterian sites vary with seasonality of occupation and other aspects of individual site environments (Klein 1973b, Straus 1992, Stiner 1994, Marean and Kim 1998, Marean 1998, Grayson and Delpech 2006). Some sites show considerable specialization on certain species—for example, the German site of Saltzgitter Lebenstedt, where Mousterian hunters preyed systematically on reindeer (Gaudzinski and Roebroeks 2000). Extensive exploitation of larger and more dangerous animals is evident at some Mousterian sites. Prey of this sort included bison at Mauran in France (Farizy et al. 1994), Merck's rhinoceros at Krapina (Gorjanović-Kramberger 1913, Miracle 1999), aurochs at La Borde in France (Jaubert et al. 1990), and perhaps horses at Vogelherd (Niven 2006). Large accumulations of mammoth bones are found at several Mousterian sites in the Ukraine (Klein 1973b; Hoffecker 2002). Klein suggests that Mousterian people might have been collecting these mammoth bones for use in building structures, rather than hunting the living animals; but at the other sites mentioned, the evidence of Neandertal predation on big, powerful ungulates is compelling. Although the remains of carnivores are rarely found in Mousterian sites, fossils of the extinct cave bear (*Ursus spelaeus*) remains are numerous at Krapina, Vindija, and many sites in other parts of Europe (Kurtén 1976, Malez 1978). This might be taken as evidence that Neandertals were hunting cave bear as well; and cut marks on cave bear bones from Mousterian and later sites in Germany do suggest that when humans and bears met, humans had the upper hand (Münzel et al. 2001, Münzel and Conard 2004). However, they may not have met often or on purpose. Cave bear bones are usually found in strata where evidence of human presence is minimal or lacking. Analysis of another Pleistocene cave-living bear species (*U. deningeri*) from a site in Turkey suggests that most of the bear remains found in Pleistocene caves are the result of natural deaths during hibernation (Stiner 1998). It is understandable that humans would have avoided these large, deadly animals when at all possible—and vice versa.

Questioning the prowess of Neandertal hunters, Binford (1985, 1989) argued that Neandertals were fundamentally scavengers. Although it is difficult to draw a sharp boundary between hunting and scavenging in the archaeological record, most subsequent analyses have repudiated Binford's conclusions (Conard 1992, Stiner 1994, Gaudzinski 1996, Marean 1998, Marean and Kim 1998, Speth and Tchernov 1998, Klein 1999, Conard and Prindville 2000, Gaudzinski and Roebroeks 2000, Jöris 2001, Münzel and Conard 2004, Bar-Yosef 2004, Grayson and Delpech 2006). Stable-isotope studies, comparing ^{15}N levels in Neandertal bones from Belgium, France, and Croatia with those of various animals with differing diets from the same regions, show Neandertals clustering with species that get their protein nearly exclusively from meat (Bocherens et al. 1991, 1999, 2001, 2005; Bocherens and Drucker 2006, Richards et al. 2000). Microwear patterns on Neandertal teeth strongly resemble those of the most carnivorous modern humans (e.g., Inuit, Fuegans) as opposed to people with more generalized diets (Lalueza-Fox et al. 1993, 1996).

Of course, none of this implies that Neandertals were exclusively carnivorous. Direct evidence of food-plant remains (seeds, fruits, tubers) has been recovered from the Neandertal fossil site of Kebara (Lev et al. 2005). Residues of starchy plants have also been detected on Mousterian tools (Hardy 2004). Plant foods must have provided essential nutrients for Neandertals, and starchy underground storage organs (USOs) and other carbohydrate-rich plant tissues could have made crucial contributions to their high caloric needs. However, the evidence suggests that Neandertals obtained most of their protein, and a lot of their calories as well, by eating meat. Kuhn and Stiner (2006) have argued that Neandertal subsistence activities were so focused on predation that females (and presumably children) also must have been active in the chase—not as hunters, but as beaters or in other capacities—and that this narrow focus on hunting would have limited the Neandertals' ability to exploit other resources. O'Connell (2006) thinks that this specialized subsistence pattern placed Neandertals at a disadvantage in competing with modern humans, who were able to make use of a wider range of food sources.

These facts and inferences argue against Binford's view of Neandertals as dedicated scavengers. That model is also improbable for theoretical reasons. As Richards et al. (2000) note, there are no specialized scavengers among mammals. The only land vertebrates narrowly adapted for scavenging are vultures and other birds that can rely on economical soaring flight and keen eyesight to survey large territories and locate carcasses quickly. Of course, all mammalian predators (including historical hunting peoples) steal or scavenge prey killed by other predators when the opportunity presents itself, and no doubt Neandertals did so as well. But all the evidence points to Neandertals as effective and efficient predators on a wide variety of large Pleistocene mammals. This interpretation is further supported by analyses of Neandertal skeletal injuries, which are similar to those seen in rodeo riders who work in close contact with large ungulates (Berger and Trinkaus 1995). Such injuries would not result from contact with carcasses; and attacks by primary predators trying to defend their kills from scavenging Neandertals would have left injuries of a different sort. However, Gardner (2001) found that Neandertal trauma patterns do not match those seen in farm workers who are injured by large farm animals. She suggests that the injuries seen in Neandertals probably resulted from other causes—falls while traversing difficult terrain, rock falls in caves, or interpersonal violence.

NEANDERTALS AND LANGUAGE

A great deal of human communication is carried on through body "language," inarticulate noises, and other modes of communication that we share with other animals. But practically all the behavioral capacities that we perceive as distinctively and essentially human—religion, art, music, poetry, philosophy, science—depend for their execution or their instruction on our ability to communicate abstract ideas and hypotheses by using our chewing and swallowing apparatus to modulate the buzzing noises produced by our larynx. We can encode the signals of language in other media, from books and computer chips to semaphore flags and hand gestures, when circumstances call for it; but for most of us, the vocal channel is primary and semi-instinctive. An ability to speak is the primary marker of human status.

This is not simply a matter of our parochial human-centered perceptions. Language has incomparable adaptive importance, more than any other distinctive feature of human anatomy or behavior. We can describe its major adaptive functions under four headings:

1. **Group Coordination.** All animals communicate with other members of their species, if only to make sure that sperm get released in near proximity to ova. Many social animals have evolved more complicated signal systems that can produce coordinated action by a whole group. At its most elaborate, nonlinguistic communication can produce coordinated group actions involving thousands of individuals, as in the complex economy of the beehive. But even bee communication is limited to evoking a restricted repertoire of instinctive, knee-jerk responses to a narrow range of stimuli.

Language gives human beings the power to bend group efforts to any task that is both imaginable and possible. Through verbal signals, we can coordinate the efforts of a whole group, tribe, or nation of large, smart, nimble-fingered apes and direct them in minute detail to change the world in any way we wish—reshaping landscapes, fighting a war, or building on a monumental scale. The causal chains of object modification that underlie human technology find a parallel in the operational chains of contract, threat, and command through which human societies disseminate verbal signals to coordinate mass action. Taken together, these unbounded, endlessly branching linkages of technique and command give *Homo sapiens* enormous power over the world.

2. **Contrafactual Representations.** The signals through which other animals communicate can transmit only certain limited kinds of information about the way things are at the moment (or were in the recent past, in the case of scent-marking). But the possession of language enables humans, and no other animals, to inform

each other in infinitely expansible detail about things that happened a month ago or that will happen tomorrow—or that may happen if certain steps are taken. We can therefore plan our actions in advance. This feature of language makes our group coordination possible. It allows us to think in detail about our objectives before we attempt to carry them out—to model hypothetical acts in our utterances, both as individuals and as councils, and judge and debate their possible outcomes.

3. Time-Binding. When other animals die, the experience and knowledge accumulated in the course of their lives dies with them. Through language, humans can cheat death by transmitting experience and knowledge from one brain to another. This ability is of immense adaptive importance to our species, because it means that individuals can access the experience and knowledge of their fellows, and that each generation can build on the knowledge and skills of its predecessors. The book you hold in your hands is one tiny part of this vast flow of intergenerational information.

4. The Framework of Consciousness. Language expands the scope of conceptual thought enormously by furnishing socially shared, negotiable labels for the categories of things and actions that we perceive in the world. It also allows us to represent ourselves to ourselves in the third person, modeling our own appearance and actions as they might appear to others. Language thereby gives human beings a degree of self-awareness denied to all other species (Bickerton 1990).

The incomparable adaptive value and great symbolic importance of language have long impelled scientists to search for evidence of its onset in the prehistoric record. Such evidence is hard to come by. Writing, which we can think of as a sort of fossilized speech, is a very recent phenomenon, dating back only a few thousand years to the rise of the first civilizations in India, Iraq, and Egypt. Language must predate the first direct evidence of its existence by many thousands or even by millions of years. The onset of language can therefore be detected only indirectly, from paleontological evidence of speech-related changes in anatomy or from archaeological evidence of speech-related behaviors. Unfortunately, the speech apparatus is made up almost wholly of tissues that do not fossilize. Its anatomy can be only vaguely inferred from the skeletal evidence furnished by fossils—via markings on bone where muscles or ligaments attached, or openings in bones for the passage of nerves and blood vessels, or the impressions made on the bones of the skull by the brain.

The language abilities of the Neandertals have been debated by scientists ever since the "Neandertal question" first came into focus in the 1860s. Most of this debate has centered on brain morphology. The brain is

certainly the most important organ involved in human language (Deacon 1997). People who are unable to hear or speak can develop fully human linguistic skills by making other noises, gestures, or marks on paper, as long as their brains are normal and intact. Whatever separates people from other animals in this regard is first and foremost a matter of the structure of the human brain. But all that we know about the brains of Neandertals is inferred from the inner surfaces of their braincases. Neandertal endocranial volumes, and therefore brain volumes, are comparable to those of modern humans (Tables 6.1 and 7.9), which implies that any differences between Neandertal and modern human brains in speech-related functions must have resided in their internal organization or "wiring." But as previous chapters have noted, extracting information about neural organization from hominin endocasts borders on the impossible.

Many earlier studies nevertheless sought to extract such information from the skulls of Neandertals. In what was perhaps the most influential of these early attempts, Boule and Anthony (1911) concluded that the endocast of the La Chapelle-aux-Saints cranium evinced reduced frontal lobes, an ape-like placement of the lunate sulcus, and other features implying a subhuman level of neural abilities. However, Holloway (1985) finds that the frontal volumes of Neandertal endocasts do not differ from those of modern humans when scaled correctly. He has also demonstrated that Neandertal endocasts exhibit the modern human-like pattern of petalias or hemispheric asymmetries (discussed earlier for Erectines) and that Broca's and Wernicke's areas, which play important roles in human speech and language comprehension, are similar on Neandertal endocasts and those of recent humans (Holloway 1983, 1985, 2000; Holloway and de La Coste-Lareymondie 1982). Holloway (1985) concludes that no significant differences between Neandertals and modern humans can be confidently identified from endocranial casts. But he also cautions that the vague sulcal patterns preserved on endocasts do not permit the accurate identification of any features of brain organization specifically related to language (cf. Gannon and Laitman 1993). In principle, we might be able to discern the linguistic capacities of the Neandertal brain if we had a complete Neandertal nuclear genome—and if we knew how the genome is related to brain structure, and if we knew exactly what brain structures are necessary and sufficient for language. At the moment, we know none of these things. However, the identification of the modern human variant of the FOXP2 gene in Neandertals suggests that the Neandertal brain may have had much the same sort of innate language "wiring" as ours.

Another potential source of anatomical information on language capabilities is the vocal tract. Most mammals cannot be trained to imitate human speech—

partly because they lack the necessary neural control over their chewing and swallowing apparatus, but also because that apparatus is not adapted to making the necessary sounds. (Parrots and other "talking" birds produce acoustically similar sound patterns using entirely different organs.) Certain peculiarities of human anatomy appear to be adaptations for producing the sounds of speech. Scientists have accordingly looked for evidence of these speech-related apomorphies of the human vocal tract in the bones of Neandertals and other ancient hominins.

The sounds of language originate with the buzzing noises made by the **larynx**. This complicated contraption of cartilages and muscles sits halfway down the neck below the back of the tongue (Fig. 7.29). Its main job is to serve as a sort of valve at the top of the **trachea** or windpipe, closing off the airway when we swallow to prevent food and drink from winding up in the lungs. The lips of the laryngeal opening, the **vocal folds**, can also be parted slightly and tensed so that they vibrate in the breeze expelled from the lungs, producing sound waves that we hear as a musical tone. If the air and the sound waves are allowed to escape through the mouth without obstruction, the result is a **vowel**. (Consonants are produced by transient complete or partial blockages of the airway by the lips, teeth, tongue, or vocal folds.) The differences between different vowel sounds result from changes in the shape and size of two air spaces, one above the tongue and one behind it. These spaces contribute resonant overtones or **formant frequencies** that reflect the size and shape of the airway. When the tongue changes shape or position, the formant frequen-

cies change, and we perceive a change in the vowel "color" of the tone emitted by the larynx. (Compare, for example, the different positions of your own tongue in pronouncing the vowels in English *heat* and *hot*.)

Nonhuman mammals encounter anatomical obstacles to pronouncing the complete spectrum of human vowels. Although most of them can breathe through their mouths if they have to (for example, in panting), it does not come easily to them. Their larynx sits further up and forward in their neck than ours does, and their **epiglottis**—a valve-like tab of cartilage that sticks up from the front of the larynx behind the tongue—protrudes up above the soft palate into the back end of the nasal passages. This arrangement efficiently channels exhaled air from the larynx out through the nose and over the maxilloturbinal surfaces, but it blocks the air from escaping into the mouth.

The two resonant spaces in the upper airway of nonhuman mammals are also not well differentiated from each other. Because other mammals lack the right-angled bend between the axis of the face and the axis of the neck that comes with our upright posture, they do not have one space above the tongue and one behind it; rather, they have one long, curved space above the tongue. They can use their tongue and lips to alter the shape of this space and produce a certain range of vowel sounds (as a cat does, for example, when it meows), but the range is smaller than ours and the vowels are less distinct. In human beings, the larynx is shifted to a position further down the neck, and the epiglottis does not ordinarily reach the soft palate. This shift produces a larger resonant space behind the tongue and allows

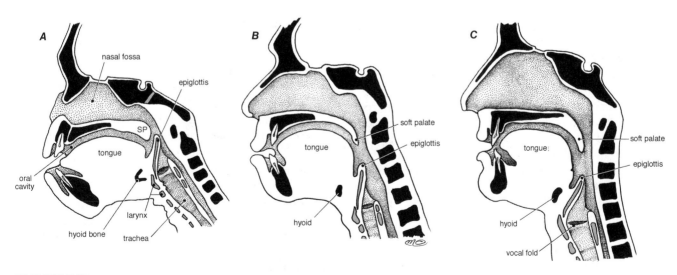

FIGURE 7.29

Diagrammatic mid-sagittal sections through the heads of (**A**) a chimpanzee, (**B**) the La Chapelle Neandertal as reconstructed by P. Lieberman, and (**C**) a recent human, showing the configurations of the vocal tract (*stipple*) and principal bones (*black*) and cartilages (*gray tone*). SP, soft palate. (**A**, after Crelin 1987; **B**, after Lieberman 1993; **C**, after Netter 1989.)

exhaled air to pass out more easily via the mouth, facilitating the production of both vowels and consonants. But it also makes it easier for swallowed food to "go down the wrong way" into the trachea. The descent of the larynx is therefore widely regarded as a specifically speech-related human apomorphy, which was selected for in spite of its deleterious effects on swallowing because of the overriding importance of vowels and consonants to the human way of life.

In an influential 1971 article, P. Lieberman and E. Crelin argued that Neandertals had lacked the human apomorphies of the supralaryngeal airway (Lieberman and Crelin 1971). Crelin had found that these apomorphies were not present in newborn modern humans, which resemble apes in having a relatively flat, unflexed cranial base and a high position of the larynx (Crelin 1969, 1973). Lieberman and Crelin compared the morphology of the human newborn with that of the adult Neandertal skull from La Chapelle-aux-Saints and concluded that there were strong similarities between the two, including a relatively long neurocranium (compared to the palate), overall mandibular shape (particularly the size of the ramus relative to the body of the mandible), the orientation of the styloid and pterygoid processes, and the lack of a chin. On the basis of these similarities, the Neandertal vocal tract was reconstructed using the recent-human newborn as a model.

A key variable in this reconstruction was the location of the **hyoid** bone. This small, U-shaped bone lies in the front of the neck between the oral end of the larynx and the lower edge of the mandible. Although it does not contact any other bone, it is connected to the mandible, tongue, larynx, cranial base, and sternum by a radiating fan of muscles and ligaments. Lieberman and Crelin inferred its location in La Chapelle by estimating the intersection of two structures that attach to the hyoid: the geniohyoid muscle (originating from the **mental spine** on the internal face of the mandibular symphysis) and the stylohyoid ligament (coursing from the **styloid process** of the temporal bone on the base of the skull). The resulting reconstruction placed the adult Neandertal hyoid—and therefore the attached larynx—high in the throat, much as in a human newborn or a chimpanzee. This move significantly reduced the resonant space behind the tongue. A computer simulation was then used to map the range of different vowel "colors" that could have been generated from the formant frequencies available in the reconstructed Neandertal vocal tract. Lieberman and Crelin (1971; Lieberman et al. 1972, Lieberman 1984, 1989) concluded that Neandertals would have had a smaller range of vowels than modern humans, and that this would have made it more difficult to hear and decipher the phonetic contrasts between their speech sounds. They also argued that Neandertal vocal-tract anatomy would have hampered or precluded the production of certain consonants (for example, the velar stops [k] and [g]).

Contrary to the impression that is sometimes given, Lieberman and Crelin never claimed that Neandertals lacked language or were incapable of speech. They granted that the Neandertal brain was sufficiently large and human-like to have allowed Neandertals to employ a language of some sort. But the absence of the human peculiarities of the supralaryngeal airway indicated to them that Neandertal speech was slow and clumsy (Lieberman et al. 1972, p. 303). Evidently, Neandertals had never been subjected to heavy natural selection for rapid, efficient decoding of sharply distinct speech sounds, and so language could not have had the overwhelming adaptive importance for them that it has for modern humans.

The conclusions of Lieberman and Crelin have met with a generally favorable reception from those who prefer to see *H. neanderthalensis* as a species separate from our own. However, other experts have offered a steady stream of criticisms, reviewed in detail by J. Spuhler (1977) and L. Schepartz (1993). Among the major issues raised are the following:

1. Several researchers have questioned the efficacy of using projections from the styloid process and the mental spine to locate the position of the hyoid (Morris 1974, Falk 1975, Wind 1976, Chan 1991). The orientation of the bands of soft tissue attached to these bony landmarks can be influenced by several factors that cannot be controlled for, and so the point where these projections intersect cannot be determined with the needed precision.

2. The supposed flatness of the cranial base in the La Chapelle-aux-Saints specimen exerted a strong influence on the Lieberman–Crelin argument and on subsequent work along the same lines (Laitman et al. 1979, 1992). However, this flatness is the product of an inaccurate reconstruction. A corrected reconstruction by Heim (1989) shows the base to have been more flexed. Both Heim and Frayer (1992b) conclude that basicranial flexion does not differentiate Neandertals from modern humans.

3. Even if the Neandertals had less basicranial flexion than modern humans, it would not have had the consequences claimed. Flexion of the cranial base is highly variable in modern humans, and does not appear to reflect differences in language capability (Arensburg et al. 1990, Burr 1976, LeMay 1975). A recent study tested the relationship between cranial-base flexion and hyoid/larynx position using a longitudinal series of modern human radiographs, and found no statistically significant relationship between them (Lieberman and McCarthy 1999; cf. Lieberman et al. 2000). The data from these studies indicate that

cranial-base flexion is related instead to the degree of facial projection.

4. P. Houghton (1993) notes that the figure of the La Chapelle skull used in the reconstruction is not oriented in the standard horizontal reference plane (the Frankfurt horizontal; see Appendix). Proper orientation necessitates changes in the relative position of the cervical vertebrae commensurate with upright posture. Making these changes shifts the styloid and pterygoid processes of La Chapelle from a newborn to an adult human position. Houghton concludes also that the other differences cited between Neandertals and moderns reflect the Neandertals' midfacial prognathism.

5. The Kebara 2 Neandertal hyoid bone (Arensburg et al. 1989) does not differ significantly in size or morphology from those of modern humans, and is quite different from the hyoids of apes (Arensburg et al. 1990, Arensburg and Tillier 1991, Schepartz 1993). The detailed similarities between Neandertal and modern human hyoids make it correspondingly less likely that the position and connections of the hyoid and larynx differed significantly between the two groups.

6. A new comparative simulation concludes that the Neandertal vowel repertoire was essentially identical to that of modern humans (Boë et al. 2002).

7. Recent comparative analyses of Neandertal and modern human mandibular morphology, especially the bony markings for the attachments and nerve of the mylohyoid muscle, suggest that the hyoid and larynx were in the same position in both Neandertals and moderns (Arensburg et al. 1990, Houghton 1993). It has been argued that the higher position of the hyoid posited by Lieberman and Crelin would inhibit, if not actually prohibit, swallowing in Neandertals (Falk 1975, Arensburg et al. 1990).

8. The hearing of modern humans is distinctive in being unusually sensitive to sounds in the 2- to 4-kHz range. This human peculiarity has been interpreted as a special adaptation for the perception of speech sounds. Computer simulations of the curve of auditory sensitivity for the Sima de los Huesos Heidelbergs, based on the anatomy of the middle and outer ears, generate an audiogram resembling that of modern humans and differing from that of chimpanzees (Martínez et al. 2004). This finding has been taken as evidence that specializations for human speech capabilities were present in the lineage leading to Neandertals.

Some investigators have sought evidence of language ability in other aspects of cranial anatomy. Keith (1915) argued that a capacity for speech could be inferred from two features of the mandible: (a) the presence of distinct mental spines on the inner aspect of the symphysis (where the tongue muscle genioglossus attaches) (b) and the development of an external mandibular buttress (the chin). According to Keith, the mental spines reflect the level of development of the tongue musculature, and the buttressing of the mandible on the external surface frees up space inside the mandible for speech-related movements of the tongue. Because Neandertals (and the Mauer mandible) lack both features, Keith inferred that they had subhuman capacities for speech.

Similar claims based on Keith's reasoning persisted in the scientific literature well into the 1960s. However, subsequent studies have shown that the expression of mental spines is variable in modern humans with normal speech (Hooton 1946, DuBrul and Reed 1960, Vallois 1961, 1962). There is also no reason to think that the chin has anything to do with talking. As noted elsewhere (Chapters 6 and 8), the protrusion of the modern human chin results from a rearward shift of the teeth relative to the cranial base—not, as Keith's analysis implies, from a forward shift of the bony buttress at the lower edge of the symphysis. As might be expected from the Neandertals' marked midfacial prognathism (Fig. 7.16), their mandibles in fact enclose a rather large space for the tongue.

The hypoglossal nerve, the motor nerve to the muscles of the tongue, passes through a small canal in the occipital bone near the foramen magnum. Kay et al. (1998) found that Neandertals and modern humans had on the average relatively larger hypoglossal canals than *Australopithecus africanus* or living apes. They conjectured that Neandertals and modern humans might have had an enlarged hypoglossal nerve, implying refined motor control of the tongue commensurate with spoken language. However, subsequent work demonstrated that some monkeys and some *Australopithecus* fossils also have big hypoglossal canals (Degusta et al. 1999). Further research (Jungers et al. 2003) showed that the size of the hypoglossal nerve cannot be inferred from hypoglossal canal size in hominoids. As far as is presently known, no aspect of skeletal anatomy can tell us whether Neandertals—or any other ancient humans—had human-like capacities for making and using language.

▌SYMBOLIC BEHAVIOR

If morphology tells us nothing about the language abilities of Neandertals, an alternative approach is to look for archaeological evidence of so-called **symbolic behaviors** that can be taken as reliable signs of the presence of language. The trouble is that we cannot be sure what those are. Practically every behavior that distinguishes us from other living animals has at one time or another been analyzed as a correlate of language. But because *Homo sapiens* is the only talking animal, all these supposed correlations are based on a sample size of one. It is hard to say which other uniquely human behaviors—making fire, making music, drawing pictures, and so on—are genuinely connected with linguistic ability and which ones are independent behavioral peculiarities of our species.

The two main candidates for the title of "symbolic behavior" are art and religion. It seems reasonable to think that religious beliefs—say, about the ceremonious treatment of the bodies of the dead—could not be either framed or communicated without recourse to language. And although there is no structural correspondence between art and language, both represent ways of thinking about and representing the world that involve abstraction and symbolism. The question of whether Neandertals had either art or religion has therefore been hotly contested.

There is precious little in the Mousterian archaeological record that can be regarded as evidence of artistic expression. There is better evidence from Initial Upper Paleolithic (Châtelperronian) sites associated with Neandertals (see below); but even this pales in comparison to the extensive, elaborate, and sophisticated art of the Upper Paleolithic proper in Europe (White 1986, 2003; Simek 1992, Hahn 1993, Klein 1999, 2003; Conard 2003). The supposedly artistic or symbolic objects that have been claimed from strictly Mousterian contexts (Hayden 1993) are few in number and unimpressive in appearance. Bones or teeth that appear to have been pierced or scored for hanging (perhaps around the neck, as primitive jewelry) are known from such sites as La Quina in France, Scalyn in Belgium, and Prolom II in Ukraine (Chase and Dibble 1987, Stepanchuk 1993, d'Errico and Villa 1997). Bones and pieces of stone from several Mousterian sites bear marks that may have been intentionally made by humans for no evident functional purpose, and a piece of mammoth molar shaped into an oval plate has been recovered from Tata in Hungary (Marshack 1976, 1989). Marquet and Lorblanchet (2003) claim the use of red ochre and other pigments, and a modified limestone slab from La Ferrassie (France), as further evidence of what they term the "birth" of Paleolithic art in the Mousterian. The centerpiece of Marquet and Lorblanchet's claims is a small chunk of worked flint with a bone splinter stuck through a natural hole in the stone, recovered from a secure Mousterian context at La Roche-Cotard (France) and dated to around 32 Kya. They interpret this object as an attempt to enhance a natural piece of stone to look like a human face. Not surprisingly, this interpretation is controversial.

Another controversial specimen is the broken shaft of a juvenile bear femur associated with Mousterian artifacts from Divje Babe I in Slovenia. Turk et al. (1995) identified four relatively evenly spaced holes on this shaft, leading them to interpret it as a bone flute. This item and most of the supposed Mousterian artworks have been explained away by other experts as non-artifacts, produced by animal activity and natural forces (Chase and Dibble 1987, 1990; d'Errico and Villa 1997, Klein 1999).

The most convincing art objects associated with Neandertals are probably the items of personal adornment (grooved and bored teeth and other ornamental pieces) from the Châtelperronian levels at Grotte du Renne at Arcy-sur-Cure (Hublin et al. 1996). J.-J. Hublin and colleagues suggest that these items reflect some form of influence on Neandertal culture by the makers of the Aurignacian culture, in which such items are relatively common (R. White 2003). This would indicate temporal overlap between the makers of the early Aurignacian (presumably moderns) and the Châtelperronian (Neandertals). However, Zilhão and colleagues (Zilhão et al. 2006; Zilhão and d'Errico 1999, 2003b; Zilhão et al. 2006) argue that the Châtelperronian always predates the Aurignacian in Europe. As they see it, these artifacts are not evidence of cultural interaction between these two groups, but an indigenous achievement of the Neandertals. Zilhão (2006) goes further and suggests that the presence of similar items in the Aurignacian actually represents cultural continuity between Neandertals and the early modern Europeans of the Aurignacian. Even earlier, Simek (1992) argued that much of the symbolic behavior of the Upper Paleolithic actually developed in the Middle Paleolithic and does not clearly differentiate the two. At the opposite pole of opinion, Bar-Yosef (2006) contends that the Châtelperronian is the first cultural complex brought into Europe by immigrating modern populations, and that it was not produced by Neandertals.

Evidence of ritual or religion among Neandertals is as sparse as evidence of Neandertal art. One early claim involved a supposed cave bear cult, inferred from the distribution of bear bones in such caves as Drachenloch in Switzerland, Petershöhle in Germany, and Le Régourdou in France (Bergounioux 1958). Cave bear skulls at these sites were supposedly placed in limestone slab boxes, or displayed on slabs with bear long bones. But subsequent analysis demonstrated that these arrays of bear remains are better interpreted as the result of denning and other activities of the bears themselves (Kurtén 1976; cf. Stiner 1998). There is no convincing indication of human modification of or interaction with these bones.

The final possible example of Neandertal symbolic behavior is the intentional interment or burial of the dead. It is often claimed that Neandertals were the first humans to bury their dead, but the issue is not a simple one. Not all Neandertal sites with human remains preserve any evidence of burial. Neandertal bones from several sites, including Krapina and Vindija in Croatia and Kůlna in the Czech Republic, bear cutmarks and other evidence of postmortem processing, and similar things appear to have gone on at some western European sites. For example, Defleur et al. (1999) and Rosas et al. (2006b) conclusively demonstrate processing of Neandertal bones at Moula-Guercy (France) and El Sidrón (Spain). But it is not clear whether this treatment of the dead among Neandertals should count as symbolic behavior or subsistence activity. T. White (1992, 2001) views these processed remains as the leftovers from cannibal feasts driven by hunger. It has even been suggested that the Neandertals were pushed into extinction by prion diseases (which include kuru and mad cow disease) that they contracted by feeding on the brains of their own dead (Chiarelli 2004). On the other hand, M. Russell (1987a,b) argued that the processing evident on the Krapina folk shows evidence of **secondary burial**—that is, the ceremonious reburial of human skeletons, which often involves bone-cleaning that can leave cutmarks. The most famous supposed case of ritual cannibalism in Neandertals, the Guattari (Monte Circeo) 1 skull (Blanc, 1961), has been plausibly reinterpreted as the result of hyena activity (Stiner 1991, White and Toth 1991, Mussi 2001).

The case for deliberate interment of Neandertal dead comes mainly from three other sorts of evidence: the recognition of burial pits at some sites, the flexed position in which several Neandertal bodies appear to have been laid to rest, and the state of preservation and completeness of many Neandertal skeletons. The most convincing evidence for deliberately excavated burial pits comes from the French sites of La Chapelle-aux-Saints, La Ferrassie, Le Régourdou, Le Roc du Marsal, and Saint-Césaire; from the Ukranian site of Kiik-Koba; and from Kebara in Israel (Vlček 1973, Vandermeersch 1976, Harrold 1980, Bar-Yosef et al. 1992, Defleur 1993). But even at sites where no evidence for a pit was noted, the completeness of the Neandertal specimens often argues for some form of intentional burial. A good example of this is the original Neander Valley specimen (Neandertal 1). The state of the 16 original bones found (Fig. 7.1) suggests that this was probably a virtually complete skeleton at the time of its discovery, although the circumstances of its recovery preclude any demonstration of a burial pit. The recent recognition of additional bones from the site, including complete hand and foot bones, many of which may come from this same individual (Schmitz et al. 2002, Smith et al. 2006), enhance the probability that this specimen was originally nearly complete and may represent a burial. However, Gargett (1989, 1999), who contends that there is no convincing evidence of deliberate burial at Neandertal sites, argues that the excellent state of preservation of many Neandertal specimens probably results from accidental "burial" by rock falls or other natural phenomena, not from intentional interment.

The comments affixed to Gargett's 1989 article (see also Straus 1989) demonstrate that there is good evidence for the intentional inhumation of Neandertals at some sites. But it is not clear whether this represents eschatology or housekeeping. Some authors regard Neandertal burials as evidence of Neandertal religion, the paragon of symbolic behavior (Bergounioux 1958, Defleur 1993). Others see them as just an expedient for covering up a stinking corpse (Klein 1999). The second interpretation seems less plausible to us. If a dead Neandertal was perceived as no more than an offensive hunk of dead meat, it could have been simply thrown out of the cave—or perhaps carried over to the next valley and dumped there, so as to avoid attracting hyenas and other unwelcome visitors. Taking the cadaver outside would have been less trouble than digging even a cursory grave in a cave floor, and more effective at reducing the stench.

The evidence for ceremonial burial would be more convincing if there were evidence that the Neandertals regularly buried artifacts or food with the dead. But as Harrold (1980) showed some years ago, Neandertal burials are simple affairs and rarely contain any evidence of grave goods. (This is true for the early modern graves from western Asia as well.) Supposed cases of Neandertal grave goods—for example, the flower pollen supposed to represent funeral bouquets placed atop the "flower burial" of Shanidar 4, or the goat skulls associated with the Teshik Tash juvenile—may be the products of natural processes rather than Neandertal funerary practices (Chase and Dibble 1987, Klein 1999).

In short, it is not clear what the diversity in Neandertals' treatment of their dead means. Interment may be a form of symbolic behavior; but it may also just be a sign of some emotional response—affection, respect, fear, or disgust—toward a lifeless body. Processing of Neandertal bones may have been part of a reburial ritual, but it might also reflect hunger-driven cannibalism, or something else altogether. Whatever it means, this diversity in treatment of human remains is at any rate another clear sign of the complexity and flexibility of Neandertal behavior. Not all Neandertals did the same thing every time and everywhere, not even with their dead!

The difficulty of figuring out what human corpses meant to Neandertals exemplifies the problems inherent in trying to infer complex behavior from fossils or Paleolithic archaeology. Unfortunately, it is often just these types of complex behavior that interest us the most. Documenting changes in tooth size or facial prognathism is relatively easy, but it has limited fascination outside the small circle of relevant specialists. Documenting language, religion, altruism, or social stratification stirs universal interest and gives us the feeling that we are getting at core elements in human nature and behavior. But saying anything definitive about the presence of these elements in the lives of pre-modern humans is frustratingly hard. This is not grounds for giving up the attempt. Progress continues to be made in these areas, and scientists should and will continue to seek ways to wring this type of information from the stones and bones found at ancient human sites. Even as things stand now, there are a great many things we can say with confidence about the biology and way of life of the Neandertals. This knowledge gives us an exceptionally detailed understanding of these very close and recent relatives of our own kind.

EARLY EUROPEAN NEANDERTALS

European hominins before 200 Kya had already begun to show Neandertal features (Chapter 6). Some regard the humans from such sites as Sima de los Huesos (SH) as *bona fide* Neandertals. However, the Heidelbergs from SH, Tautavel, and other European sites do not consistently exhibit all the distinctively Neandertal features. It is not clear where the line should be drawn between these people and the first true Neandertals. One of us has argued that the earliest Europeans to qualify unequivocally as Neandertals are the OIS 7 (~200 Ky) specimens from Fischer's and Kämpfe's quarries at Ehringsdorf, Germany (Smith 1984). Vlček (1993) rejects this idea, arguing that the Ehringsdorf fossils are more like modern humans than any Neandertals. His reconstruction of the Ehringsdorf H skull shows a posterior vault silhouette that has its maximum breadth of the vault rather high on the parietals, and thus lacks the distinctive oval *"en bombe"* shape seen in Neandertals. However, earlier reconstructions by both F. Weidenreich and O. Kleinschmidt are more Neandertal-like (Behm-Blancke 1958), with occipital buns and oval occipital contours (Figs. 7.30 and 7.31). Although Ehringsdorf H is difficult to reconstruct accurately, its suprainiac fossa and Neandertal-like supraorbital and mastoid regions demonstrate its Neandertal affinities, and the evidence that it is significantly different from Neandertals in overall cranial form is not compelling. The other Ehringsdorf specimens also fit comfortably within the Neandertals. For example, the femur (Ehringsdorf E) exhibits anterior-posterior bowing and lacks a pilaster. The adult (Ehringsdorf F) and subadult (Ehringsdorf G) mandibles conform to the distinctive Neandertal mandibular pattern, with large anterior teeth and receding symphyses, and are particularly similar to the mandibles from Krapina (Smith 1976).

As mentioned previously, several fragmentary or isolated specimens attributed to OIS stages 6 and 7 are now usually included with Neandertals. In many cases, the affinities of these specimens are difficult to determine, but none is clearly non-Neandertal and some of them show unquestionable Neandertal traits. For instance, one specimen from Biache-Saint-Vaast in northern France preserves a posterior cranial vault, BV1 (Fig. 7.32), plus a partial maxilla and some teeth. The vault

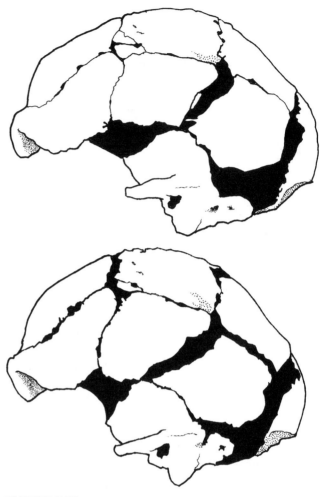

FIGURE 7.30

Two reconstructions of the Ehringsdorf H cranium (Germany) in lateral view. **Top**: Reconstruction by O. Kleinschmidt. **Bottom**: Reconstruction by F. Weidenreich. The differences between the two reveal the effects that a reconstructor's assumptions can have on the form of a reconstructed skull. However, the scope of those effects is limited. In this case, both reconstructions reveal occipital bunning and a projecting supraorbital torus, and both are consistent with Neandertal affinities for the specimen. (After Behm-Blancke 1958.)

shows distinctive Neandertal traits, including a distinct occipital bun, a suprainiac fossa, and a mastoid region that angles toward the midline and is less projecting than the occipitomastoid crest (Stringer et al. 1984). Remains of a supraorbital region and left partial cranial vault (BV 2) from this site also evince clear Neandertal affinities (Guipert et al. 2007).

The cranial remains from another early French site, Fontéchevade, are fragmentary but historically significant, because they were once promoted as evidence for the Pre-Sapiens model of modern human origins

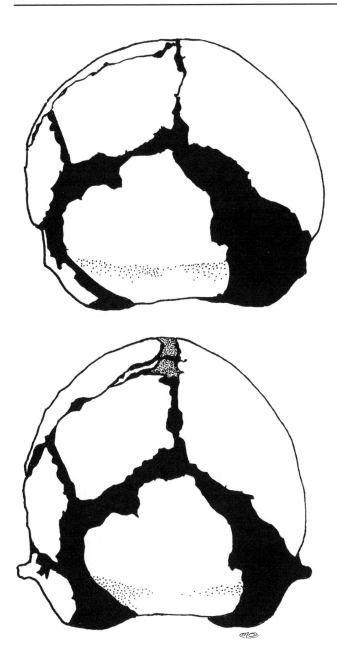

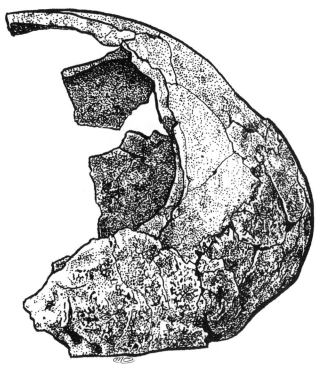

FIGURE 7.32

The Biache-Saint-Vaast cranium from France in lateral view, showing the occipital bun. (After Tattersall 1995b.)

FIGURE 7.31

Two reconstructions of the Ehringsdorf H cranium (Germany) in rear view. **Top**: Reconstruction by O. Kleinschmidt. **Bottom**: Reconstruction by F. Weidenreich. Both reconstructions reveal archaic cranial form, but Kleinschmidt's is more clearly Neandertal-like. (After Behm-Blancke 1958.)

(Chapter 6). Fontéchevade I (now 4) is a small piece of juvenile frontal bone from just above the nose. Fontéchevade II (now 5) is a partial adult vault consisting of a parietal and a frontal (lacking the supraorbital area). Vallois (1954, 1958) argued that both specimens lacked supraorbital tori and were therefore more "progressive" than Neandertals. However, Fontéchevade 4 is

a juvenile; and as noted above, juvenile Neandertals generally lack marked development of the torus. Vallois' claims that Fontéchevade 5 lacked a torus were based on the orientation of a small segment of frontal sinus, and have since been refuted (Trinkaus 1973, Tillier 1977, Stringer et al. 1984). The metrical pattern of Fontéchevade 5 does not differ from that of Neandertals (Corruccini 1975). Viewed from above, this vault has a strongly Neandertal-like shape, and the occipital view shows the oval contour characteristic of Neandertals, with the maximum breadth located low on the parietals. If these specimens had not been discovered at a time when the Pre-Sapiens model was still being advocated, they would probably have been considered Neandertals from the start.

Howell's (1951, 1957) version of the Pre-Neandertal model was rooted in his contention that the early, pre-Würm European Neandertals showed fewer Neandertal apomorphies than the later, "classic" Neandertals. The same ideas are inherent in the accretion model (Chapter 6). They are also evident in Condemi's (1992) analysis of the Saccopastore skulls from Italy. These two specimens are relatively late (OIS 5), and Saccopastore 2 is associated with a Mousterian industry (Mussi 2001). Nevertheless, Condemi regards them as representing an early, relatively unspecialized stage in the emergence of the Neandertal pattern. The main reason for thinking this is that Saccopastore 1, a complete cranium missing the supraorbital torus and most of the teeth, is small and lacks an occipital bun. But both it and the larger, less complete Saccopastore 2 have a distinctly Neandertal facial structure. Both skulls exhibit mid-facial prognathism and infraorbital morphology identical to later "classic" Neandertals in pattern, although they appear to be smaller in overall size. Their small size is evident in their estimated cranial capacities (1245–1300 cc), which fall below the Neandertal mean (1350). But other pre-Würm Neandertals also have small braincases (Table 7.9). Both Saccopastore specimens have a mastoid region of the Neandertal type, and Saccopastore 1 has a suprainiac fossa. Saccopastore 2 does not preserve the occipital, but it does preserve a Neandertal-like supraorbital torus. While Condemi considers the Saccopastore specimens as representing an early "stage" in the emergence of the most extreme Neandertal pattern, Stringer (1991) notes that the ranges of variation in Würm and pre-Würm Neandertals overlap so broadly that he is not inclined to recognize important differences between these samples. A similar view was outlined earlier by one of us (Smith 1976). The differences between the earlier and later samples seem to be due to allometry, not to the presence of fewer Neandertal traits in the pre-Würm specimens.

By far the largest sample of pre-Würm Neandertal remains comes from the rock-shelter site of Krapina in Croatia, which has yielded large collections of Mousterian tools (Simek and Smith 1997, Karavanić 2007) and about 1000 pieces of Neandertal skeletons (Gorjanović-Kramberger 1906, Smith 1976, Radovčić et al. 1988). The word "pieces" is advisable here, because all the larger skeletal elements are represented only by fragments (Radovčić et al. 1988), and many appear to have been intentionally smashed. T. White (2001) and others (Smith 1976) have posited that the bones were processed by humans, much as other animal remains at Neandertal sites were. Most of these fragmentary remains, which include many small and fragile bones, were evidently placed in a pit of some type soon after the death of their owners (Gorjanović-Kramberger 1905). Because the specimens were protected by being relatively rapidly covered up, the result is amazing preservation of lots of Neandertal fragments—in this case, without intentional inhumation of single skeletons. While the fragmentation is unfortunate, it is compensated for by relatively large sample sizes for important parts of the skeleton (supraorbital regions, mandibles, temporal bones, and teeth) and by the preservation of a cross section of ages and sexes in a single sample. Because of Gorjanović-Kramberger's careful excavations, the stratigraphic position of most major pieces is known (Radovčić 1988). He identified eight cultural levels at the site, all Mousterian, with a major concentration of human bones in Levels 3 and 4, which he called the "Homo Zone." There is a much smaller concentration of human bones in Level 8, while a few specimens are found in other levels as well (Smith 1976, Radovčić et al. 1988). The concentration of the tools and human remains in a few dense clumps suggests that there were relatively few periods of occupation at the site (Simek 1991, Simek and Smith 1997).

Gorjanović-Kramberger argued that the Krapina deposits were laid down fairly rapidly. He based this on his own observations regarding the rapid degeneration of the sandstone that formed the Krapina rock shelter and subsequently made up the sediment in which the human remains were deposited. His conclusions were questioned subsequently on the basis of the faunal evidence (see discussions in Smith 1976 and Malez 1978); but the ESR dates from the site indicate that all the deposits were accumulated around 130 Ky (Rink et al. 1995). These dates support Gorjanović-Kramberger's original interpretation, and place the site at the end of OIS 6 or the beginning of OIS 5e.

The Krapina hominins differ from typical Neandertals in some minor features, which recur throughout the sample. For example, all three of the "Homo Zone" specimens preserving nasal bones exhibit the same deviation of the internasal suture, which is unknown in other Neandertals (Smith and Smith 1986). Also unique among Neandertals is the lack of an anterior mastoid tubercle in the Krapina sample (Frayer 1992a,b), and there is a high incidence of anomalies in the mandibular fourth premolars (Wolpoff 1999). These shared peculiarities suggest that the Krapina hominins, especially those from and around the "Homo Zone," may be sampling a slightly distinctive local population—perhaps with a high degree of relatedness among the individuals represented.

While the Krapina "population" exhibits considerable variability, almost none of the variants fall outside the Neandertal range of variation (Gorjanović-Kramberger 1906, Hrdlička 1930, Smith 1976, Wolpoff 1979). H. Klaatsch and O. Hauser suggested that a second, more modern form of *Homo* was present at Krapina, but this idea was refuted by Gorjanović-Kramberger (Smith 1976, Radovčić 1988). Later analyses have confirmed his conclusion that all the Krapina hominins are Neandertals (Hrdlička 1930, Smith 1976, Brace 1979, Minugh-Purvis et al. 2000).

Krapina is another pre-Würm sample that is sometimes seen as representing an unspecialized or "progressive" grade of Neandertal evolution. However, a careful assessment of the total morphological pattern of the sample reveals the same thing for Krapina as for Saccopastore: Some Neandertal characteristics are less pronounced in these people because they were rather small. Most of the more complete crania are either small adults (Krapina 3 and 6) or juveniles (Krapina 1 and 2). The same is true for the mandibles and maxillae, although the teeth are surprisingly "megadont" compared to those of other Neandertal samples (Table 6.5; Smith 1976, Wolpoff 1979). But all the distinctive Neandertal features (except the anterior mastoid tubercle) are found in the Krapina hominins:

▶ All occipitals that preserve the appropriate area exhibit suprainiac fossae.

▶ The juvenile Krapina 2 skull, the only posterior vault complete enough to judge occipital bunning, shows distinct lambdoidal flattening and other features indicating the initial stages of bun formation.

▶ Maxillary incisors are large and strongly shovel-shaped.

▶ Mandibular molars exhibit extensive taurodontism.

▶ Supraorbital tori are thick and projecting, with expanded frontal sinuses limited to the torus itself.

▶ The infraorbital plate has a parasagittal orientation.

▶ Mandibles exhibit retromolar spaces, medial pterygoid tubercles, and receding symphyseal profiles, and lack evidence of distinct chin projection.

▶ The superior pubic rami are elongated and thinned.

▶ Ribs and clavicles indicate a deep (barrel-shaped) thorax.

▶ The femora have distinct proximal lateral femoral flanges.

In short, there is really nothing "progressive" about the Krapina specimens or any other pre-Würm Neandertals, because they do not differ significantly from later "classic" Neandertals. Additional support for this interpretation comes in the form of a stone endocast (with a few adhering bone fragments) and natural casts of a radius and ulna from the pre-Würm Slovakian site of Gánovce, assigned to OIS 5e. The endocast is long and low, with a cranial capacity of $1320\,cm^3$. Like the Krapina specimens, it exhibits distinct Neandertal features, including a marked occipital bun and a characteristic oval shape from the rear (Vlček 1969, Smith 1982). Analysis of the ~150 Ky La Chaise (Abri Bourgois-Delaunay) remains also reveals no differences in anatomical form from later Würm Neandertals (Condemi 2001).

The fossils of European Heidelbergs do seem to document a gradual consolidation in Europe of the distinctive Neandertal morphology, as implied by the "Accretion Model" (Chapter 6); but that process seems to have run its course by the time of these early Neandertals. These folk tend to be a bit small, but morphologically they are full-blown Neandertals. Apart from differences due to size, the ranges of variation in Würm and pre-Würm Neandertals overlap to such an extent that no meaningful distinctions can be drawn between them (Smith 1976, Stringer 1991).

WÜRM NEANDERTALS FROM WESTERN EUROPE

The inhabitants of Europe during the last (Würm) glacial maximum were the archetypal, textbook-case Neandertals that Howell (1951, 1952, 1957) dubbed "classic" Neandertals. Examples include the Monte Circeo (Guattari) specimens from Italy, the original Neandertal remains from Germany, the Spy sample from Belgium, the Gibraltar skulls, the southwestern French series (La Chapelle-aux-Saints, La Ferrassie, La Quina, Le Moustier, etc.), and a host of other, more recently discovered sites and specimens (Defleur et al. 1999, Quam et al. 2001, Arsuaga et al. 2006, Rosas et al. 2006b). Howell (1951, 1952, 1957) argued that these "classic" specimens exhibited Neandertal features at their most extreme because these people were the most thoroughly adapted of all Neandertals to a cold periglacial climate. Some of the most complete Würm specimens (La Chapelle, Guattari 1, La Ferrassie 1) are among the largest Neandertals known and have very large cranial capacities (Table 7.9). These specimens probably provide the basis for the often-repeated statement that Neandertals have bigger brains than modern humans. But when cranial capacities from smaller specimens (Gibraltar-Forbes Quarry, Spy 2, La Quina 5) are included, the mean cranial capacity for Würm European Neandertals is in fact smaller than that of the early modern Europeans (Table 7.9).

One of the most significant recently discovered sites in western Europe is El Sidrón in northern Spain (Rosas et al. 2006b). Dated to ~43 Kya by AMS radiocarbon, this site has yielded over 1300 fragments of human skeletal remains from at least eight individuals. Rosas and colleagues note that the morphological features of the El Sidrón sample conform to a Neandertal pattern. Such features include labially curved, shovel-shaped maxillary incisors; taurodont molars; receding mandibular symphyses; well-developed supraorbital tori; weakly projecting mastoid process; and suprainiac fossae. The postcranial bones are mostly hands and feet. Like those from Krapina, the human bones appear to have been processed by humans. Unlike Krapina, El Sidrón has yielded mainly human bones. Mousterian tools are also present but may not be directly associated with the human fossils.

Unlike most previously discovered sites, El Sidrón is being excavated with the aim of minimizing possible contamination of ancient DNA. This is a difficult undertaking, but it is paying dividends. El Sidrón has already yielded Neandertal mitochondrial DNA (Lalueza-Fox et al. 2005), as well as significant parts of the nuclear genome—including the modern human FOXP2 "language" gene (Krause et al. 2007a) and the MC1R variants that suggest at least some Neandertals had pale skin and red hair (Lalueza-Fox et al. 2007).

There are also a few Neandertals from Central Europe that are similar in age to these "classic" western European specimens (Jelínek 1969, Smith 1982, Smith and Trinkaus 1991). They include the Šaľa specimens from Slovakia; mandibles from Ochoz and Šipka and a partial maxilla and parietal from Kůlna (all from the Czech Republic); fragmentary child and adult remains from Subalyuk in Hungary; and a mandibular third molar from the site of Lakonis in Greece (Harvati et al. 2003). The Šipka specimen, the symphyseal region of a child's lower jaw, is historically important because it was one of the first Neandertal specimens to be found in association with Pleistocene fauna and Paleolithic tools (Table 7.1), and also because it figured prominently in Virchow's assessment of Neandertals as pathological. Šipka has been described as resembling modern humans in certain respects, but it is really too fragmentary to support these claims (Vlček 1969, Jelínek 1969, Smith 1982). Similar claims for "progressive" tendencies have been made for the Šaľa 1 frontal, but a recent reanalysis of this specimen concludes that its morphology is that of a standard Neandertal (Sládek et al. 2002). Kůlna is well dated to 50.5 ± 5 Ky by ESR dating (Rink et al. 1996). The half maxilla from the site exhibits a narrow nose and other features sometimes described as "transitional" to modern human morphology (Jelínek 1966, 1969; Smith 1982).

WESTERN AND CENTRAL ASIAN NEANDERTALS

Western Asian Neandertals are known from Israel, Syria, and Iraq, while the site of Teshik Tash in Uzbekistan yields the only unequivocal Neandertal from Central Asia proper (Hrdlicka 1939). However, Neandertals also have been recovered from the Crimean region of the Ukraine (Trinkaus 1984a), and perhaps from the Russian Caucasus region (Ovchinnikov et al. 2000, Hoffecker 2002). These finds indicate that Neandertals inhabited the whole western third of the Eurasian land mass at one time or another during their reign—and recent discoveries hint that their range, or at least their genetic influence, may have extended as far east as the fringes of Siberia.

Teshik Tash, a cave site located near Tashkent, has yielded a burial of an 8- or 9-year-old male surrounded by goat horns and skulls that appear to have been carefully placed around the pit (but see Gargett 1989). The site is essentially undated, but it contains Mousterian tools and a Pleistocene fauna (Ullrich 1955b). Despite its young age at death, the Teshik Tash skull already manifests a Neandertal-like cranial vault with a suprainiac fossa, characteristic Neandertal morphology of the mastoid and occipitomastoid eminence, an *en bombe*

occipital silhouette, and the beginnings of a continuous supraorbital torus. The facial skeleton also appears to be Neandertal-like in its midsagittal facial prognathism, infraorbital morphology, mandibular form, and nasal breadth. The upper incisors are shovel-shaped. Despite these facts, some have argued that Teshik Tash is not a typical Neandertal. Minugh-Purvis (1988) believes the face is not as Neandertal-like as the vault. Weidenreich (1945b) regarded Teshik Tash as similar to the Mt. Carmel hominins (both Tabun and Skhūl) and saw it as intermediate between Neandertals and modern people. Other Asian Neandertals seem to have less emphatically Neandertal morphology than the European Neandertals in some respects, and this may be true of Teshik Tash as well (but see Glantz and Ritzman 2004). Analysis of mitochondrial DNA from the Teshik Tash left femur reveals the presence of uniquely Neandertal sequences, although most of the sequences recovered from the specimen were contaminants from modern humans (Krause et al. 2007b).

Human remains associated with Mousterian tools have recently been recovered from Okladnikov Cave in the Altai region of southwestern Siberia. The most ancient of these fossils are several isolated teeth that have been attributed to Neandertals and dated to between 37,750 and 44,800 years ago using radiocarbon and uranium-series techniques (Krause et al. 2007b). The site has also yielded four postcranial specimens of more recent date: the distal halves of a subadult humerus and femur, probably from the same individual (~30–38 Ky), a piece of a distal adult humerus (~24 Ky), and an undated adult hand phalanx. These postcranial dates are uncalibrated radiocarbon estimates and therefore may be several thousand years too young, but the subadult individual probably falls within the temporal range of European Neandertals. The subadult postcranials and the adult hand bone are said to be non-modern in morphology; the adult humerus is of uncertain affiliation (Viola et al. 2006). Analysis of mtDNA from the three long-bone fragments showed that the subadult specimens carried diagnostic Neandertal sequences, while the adult specimen did not (Krause et al. 2007b). Again, the vast majority of sequences identified in all the Okladnikov Cave specimens fall within the modern human range and are presumably contaminants. Although we cannot yet feel confident that these fragmentary remains represent morphological Neandertals, they provide evidence that Neandertal genetic influence reached farther east than previously thought.

The only other significant Central Asian fossil found with a Mousterian industry, a right temporal bone from Darra-i-Kur in Afghanistan, has a mastoid-occipitomastoid region and temporal fossa of modern human form (Dupree 1972). Trinkaus (1984a) is probably right to call it a modern human. A Middle Paleolithic lower deciduous molar from Khudji, Tajikistan also is not diagnostically Neandertal (Trinkaus et al. 2000).

In the Crimea, Mousterian-associated Neandertal remains are found at the sites of Kiik-Koba and Zaskal'naya V and VI (Ullrich 1958, Vlček 1973, Trinkaus 1984a, Hoffecker 2002). Excavated between 1924 and 1926, Kiik-Koba revealed a disturbed burial pit containing the right leg, numerous right and left hand and foot bones, and a single tooth of an adult male Neandertal (Ullrich 1958). Less than half a meter away, another pit was found containing the flexed burial of a 5- to 7-month-old Neandertal child. This juvenile is very poorly preserved—but its morphology is unquestionably Neandertal (Vlček 1973). The Kiik-Koba remains are associated with the Mousterian—but as noted earlier, other Mousterian sites from western Asia have yielded fossils with fundamentally modern morphology. The child's skull from Sarosel'e, long thought to be such a fossil (Ullrich 1955a, Alexeyev 1976), is probably an intrusive burial from historical times (Marks et al. 1997).

The supposed Neandertals from the Caucasus are fragmentary and enigmatic. The dating and affinities of the infants from Mezmaiskaya infants are debated (see "Late Neandertals," below). A few other sites from the northern Caucasus have yielded bits and pieces attributed to Neandertals: the mandible and isolated teeth of another infant from Barakaevskaya Cave, some adult teeth and postcranial fragments from Monasheskaya Cave, and an isolated adult tooth from Matuzka Cave (Hoffecker 2002).

Further to the west, a larger and more informative Neandertal sample comes from Shanidar Cave in the Zagros Mountains of Iraq. Excavated by R. Solecki in the 1950s, this large cave yielded the remains of nine Neandertals in various states of completeness (Table 7.18), associated with a Mousterian industry (Solecki 1957, 1963). Several studies (Şenyürek 1957, Stewart 1977, Trinkaus 1983b, Franciscus and Churchill 2002) have demonstrated that the Shanidar people were unequivocal Neandertals, possessing almost all of the characteristic craniodental and postcranial features associated with the Neandertals of Europe. However, a few Neandertal peculiarities are missing or less developed in the Shanidar people. Their torsos appear to have been less barrel-shaped than those of typical Neandertals (Franciscus and Churchill 2002). The femora from Shanidar lack the proximal lateral flange seen on all the Neandertal femora from Europe. The only complete adult skull, Shanidar 1 (Sh 1), lacks an occipital bun and therefore exhibits parietal and occipital angles that are intermediate between those of European Neandertals and modern humans (Table 7.5). Other Asian Neandertals also lack these European apomorphies of the skull and femur (see below).

The Shanidar sample can be divided stratigraphically into an early group (Sh 2, 4, 6-9) and a later group

TABLE 7.18 ■ The Shanidar Neandertals

Specimen Number	Description	Group
1	Virtually complete skeleton of an adult male with extensive pathologies. Age: 30–45 years.	Late
2	Fragmentary cranium, complete mandible And left lower leg of an adult male. Age: 20–30 years.	Early
3	Largely complete male postcranial skeleton with four teeth. Age: 34–50 years.	Late
4	Fragmented cranium and mandible, with a largely complete postcranium (also fragmented and crushed) of a male. Age: 30–45 years.	Early
5	Partial cranium and postcranium of a male. Age: 35–50 years.	Late
6	Facial fragments and teeth with partial postcranial skeleton of a female. Age: ~25 years.	Early
7	Fragmentary cranium and postcranial remains of an infant: Age 8–9 months.	Early
8	Various bones (duplicate elements) associated with Sh 4 and 6. Probably from a single individual. Fragmentary cranium and postcrania of a female (?). Age: Young adult (?).	Early
9	Cervical and thoracic vertebrae of an infant. Age: 6–12 months.	Early

Source: After Trinkaus (1983b).

(Sh 1, 3, 5). The early specimens resemble the pre-Neandertal from Zuttiyeh (Chapter 6) in retaining more primitive features, particularly a more prognathic lateral face, while the later group shows the more typical Neandertal midsagittal facial prognathism (Trinkaus 1983b). Trinkaus (1982) suggests that the long, low frontals of Sh 1 and 5 may testify to some form of artificial cranial deformation.

The Shanidar 1 male skeleton shows a lot of pathologies and degenerative changes. Several scenarios have been put forward to explain them (Trinkaus 1984b). One possibility is that they all resulted from a massive crushing injury, perhaps suffered in a fall or a cave-in. Alternatively, Shanidar 1 may have suffered an injury to the left side of the skull (apparent in the asymmetry of the face and the evidence of a healed fracture on the left lateral orbit and anterior cranial vault), resulting in damage to the cerebral motor cortex on the left side. Because each cerebral hemisphere controls the opposite side of the body, left cortical damage could have produced right-side paralysis. Such paralysis might explain the severe injuries seen on Shanidar 1's right arm (multiple fractures and loss of the bone just above the elbow, resulting in disuse atrophy of the humerus) and compensatory degenerative changes in several postcranial joints. Shanidar 1 also had extreme anterior dental wear, resulting in what must have been painful exposure of the pulp cavities. To top it all off, he probably died in a rock fall. If we wanted to follow the African paleontologists' practice of giving pet names to important fossils, "Lucky" would be a fitting tag for Shanidar 1!

Fragmentary Neandertal remains are known from several other localities in western Asia, including a Mousterian-associated radius from Bistun Cave in Iran (Coon 1951) and other isolated pieces (Trinkaus 1984a). But the remaining Asian specimens of major importance all come from four Mousterian sites in the Levant: Amud, Kebara, and Tabūn (all in Israel) and Dederiryeh (in Syria). Apart from the Tabūn C2 mandible, the human fossils from these sites are clearly Neandertals; but they differ in some ways from the Neandertals of Europe (Keith and McCown 1937, McCown and Keith 1939, Howell 1958, 1960; Endo and Kimura 1970, Suzuki 1970a,b; Trinkaus 1984a, 1995b; Smith 1985, Simmons et al. 1991, Wolpoff 1999, Arensburg and Belfer-Cohen 1998, Kramer et al. 2001). The Amud 1 individual is one of the largest Neandertals known and has the largest Neandertal cranial capacity (1740 cm³) on record, whereas Tabun C1 is one of the smallest Neandertals, with a capacity of only 1271 cm³. Both crania exhibit most of the characteristic Neandertal features (McCown and Keith 1939, Howell 1960, Suzuki 1970a); but their vaults are more rounded, particularly posteriorly, and they lack occipital buns (Fig. 7.33). These differences from European Neandertals are reflected in smaller parietal and larger occipital angles (Table 7.5). Multivariate analysis indicates that the frontals of both skulls are quite close in form to those of early modern Near Easterners (Simmons et al. 1991, Smith et al. 1995).

The two Dederiyeh skulls, both from children who died at around two years of age, resemble those of other young Neandertals in showing incipient development of Neandertal shape patterns and characteristics (Dodo et al. 1998, 2002, Akazawa and Muhesen 2002). There is no trace of incipient bunning in these infant skulls, but bunning is also not evident in European Neandertals of similar developmental age (Vlček 1970). Rak et al. (1994) have demonstrated the Neandertal affinities of the occipital bone and mandible of the 10-month-old Amud 7 infant. They point specifically to the elongated foramen magnum, medial pterygoid tubercle, and various symphyseal traits as distinctive Neandertal features.

In the postcranial skeleton, the Tabūn C1, Amud 1, and Kebara 2 specimens exhibit various symplesiomorphies shared with European Neandertals and earlier

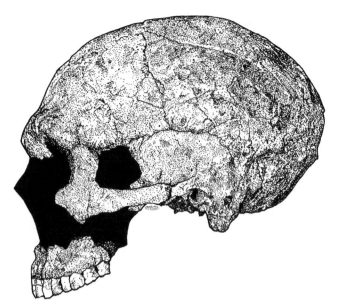

FIGURE 7.33

The Amud 1 cranium from Israel in lateral view. Although conforming to a Neandertal gestalt, the Amud cranium exhibits a higher cranial vault and a more prominent mastoid process than European Neandertals. The latter characteristic is exaggerated by the absence of the occipitomastoid region. Like other Near Eastern Neandertals, Amud 1 lacks occipital bunning. (After Delson 1985.)

humans. The Dederiyeh children's skeletons show that Neandertal traits, including the shortened distal limb segments, were established early in ontogeny (Dodo et al. 1998, Akazawa and Muhesen 2002). Kondo and Ishida (2002) observe that considerable variation is evident in these two young Neandertals. According to them, the Dederiyeh 2 skeleton is less robust than Dederiyeh 1 and overlaps in certain features with recent children of like age.

Trinkaus (1995b) compared the states of postcranial characters of Near Eastern Neandertals (including the Shanidar sample) with those of three other groups: the Skhūl/Qafzeh sample of early modern humans, European Neandertals, and earlier *Homo*. Of the 41 characters he examined, 32 comprised primitive features retained in both Neandertal groups and earlier humans, but not in Skhūl/Qafzeh. These symplesiomorphies of archaic *Homo* include a long and thin superior pubic ramus, low crural indices, broad bi-iliac breadth, a dorsal scapular axillary border, a low femoral neck-shaft angle, short stature, and the pattern of humeral shape and robusticity. Only one synapomorphy (absence of a gluteal tuberosity fossa) in Trinkaus's sample was shared by both archaic and modern Near Easterners to the exclusion of European Neandertals. Also, all Near Eastern Neandertals and Skhūl/Qafzeh lack the proximal lateral femoral flange that is ubiquitous among European Neandertals (Kidder et al. 1991). Three traits appear to be apomorphies peculiar to European Neandertals: a particularly deep, barrel-shaped chest, a very low brachial index, and exceptionally pronounced bowing of the radius. The first two of these may be adaptations to the cold climate of ice-age Europe. In sum, the Near Eastern Neandertals are more like European Neandertals than they are like the Skhūl/Qafzeh group; but they are not identical to their European cousins, and most of the non-modern postcranial traits that they share with them are also shared with some or all Erectines and Heidelbergs.

Only two character states in Trinkaus's analysis—enlarged thoracic and cervical vertebral canals—were uniquely shared by Skhūl/Qafzeh and the two Neandertal groups. These traits may be synapomorphies or parallelisms, reflecting the increase in relative brain size that Neandertals share with early modern humans (Tables 6.1 and 7.9). Enlargement of the brain would be expected to correlate with an enlargement of the upper end of the spinal cord and its bony conduit.

The Tabun C2 mandible (Fig. 7.34) presents more problems for our understanding of later human evolution than any other fossil from this part of the world. Like Neandertal mandibles, this robust, virtually complete mandible has big anterior teeth (Stefan and Trinkaus 1998a) and a retromolar space; but it also appears to exhibit a projecting chin (in side view). Its robustness and modern-looking symphysis distinguish it sharply from the smaller, more primitive Tabun C1 mandible. Its presence among the Tabun Neandertals helped

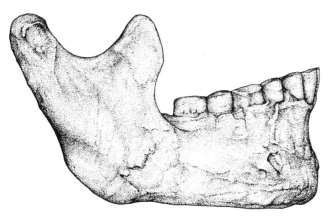

FIGURE 7.34

The Tabūn C2 mandible in lateral view. This specimen appears to have a mosaic of archaic traits, including specifically Neandertal ones, and more modern features. Anatomy of the mandibular symphysis suggests a modern-like chin structure, but breakage of the area precludes a conclusive demonstration of this. (Drawing courtesy of R. Quam.)

impel McCown and Keith (1939) to the conclusion that all the Mount Carmel hominins represent a single transitional population connecting Neandertals and modern humans. Conversely, Bar-Yosef and Pilbeam (1993) and Rak (1998) have claimed that C2 represents an early modern human. Rak bases this conclusion on three features of C2: the presence of a true chin, a mandibular notch of the modern human type, and a medial juncture of the mandibular notch crest with the condyle. He explains away the Neandertal-like retromolar gap of C2 by claiming that it is not a "true" retromolar space, but only a false semblance of one produced by a "pre-angular notch."

A different conclusion is reached by Quam and Smith (1998), who conclude that the C2 specimen has affinities to both Neandertal and early modern samples. Their multivariate analysis of 15 metric features does indeed group C2 with early moderns. However, this jaw also has a medial pterygoid tubercle, a retromolar space (*contra* Rak), a horizontal-oval mandibular foramen, and a posterior position of the mental foramina (under the P_4/M_1 septum). These traits, along with various features of tooth size and morphology (Stefan and Trinkaus 1998a), tend to affiliate C2 with Neandertals. Quam and Smith note that part of the basal area of the symphysis is missing in C2, which makes it hard to assess the meaning of its symphyseal morphology. Nevertheless, it shows evidence of a distinctly modern type of chin, with a mandibular incurvature, a basal projection at the symphysis, and the presence of an anterior marginal tubercle and mental fossa (particularly on the left side). These last two features are associated with the presence of a mental trigone ("true chin"), of a sort peculiar to modern humans (Chapter 8).

In short, the C2 mandible exhibits a mosaic of archaic and modern features. There are four possible explanations for this:

1. The human populations of the Levant may have been wholly confluent with those of eastern Africa during the period of the emergence of modern human morphology. Tabun C2 might then simply represent a transitional form, an evolutionary intermediate between archaic and modern mandibular morphology. The older age estimate for Tabun Level C (~170 Ky) would place Tabun C2 in the right span of time to be such an intermediate. Unfortunately, we do not have a comparable adult mandible from Africa during the "transitional" period to check out this possibility.

2. Tabun C2 may represent the effects of gene flow from Africa, reflecting interbreeding between archaic (Eurasian) and more modern (African) populations in a Near Eastern hybrid zone. If so, Tabun C2 furnishes the earliest known evidence for assimilation in Eurasia. The younger age estimate for Level C

(~120 Ky) fits with this interpretation, because essentially modern humans were already present in Africa at that time (Chapter 8).

3. C2 could be one of those "essentially modern humans," and its presence at Tabun could be evidence of the initial movements of modern people out of Africa. This interpretation is supported by Rak and others, but it does not explain the archaic and specifically Neandertal-like features of C2.

4. C2 may simply be an atypical representative of archaic *Homo* from the Middle East (Stefan and Trinkaus 1998a). By this reading, its modern-looking features would be interpreted as the products of parallel evolution, or as unusual variants within the local archaics' gene pool.

LATE NEANDERTALS

Neandertals survived in Europe long after more modern people were established in Africa and western Asia. Table 7.2 lists the seven sites in Europe where Neandertal fossils are dated to <36 Kya, along with several even younger sites containing Mousterian cultural complexes (Mellars 1996, d'Errico et al. 1998, Churchill and Smith 2000a). Some of these dates are controversial, and some of them may be underestimates (Conard and Bolus 2003, Zilhão and d'Errico 2003b). But most of these sites are assuredly less than 40 Ky old.

Some scholars think that the Iberian Peninsula represented the last refuge of the Neandertals (Hublin et al. 1995, d'Errico et al. 1998, Zilhão and d'Errico 1999, Zilhão 2000). J. Zilhão, F. d'Errico, and their co-workers suggest that Mousterian culture persisted south of the Ebro River in northern Spain until 28–30 Kya, overlapping with the Upper Paleolithic for 5–10 Ky with no evidence of interaction. The "Ebro frontier," in their view, represents the northern limit of this last bastion of the Neandertals. But as noted earlier, equally late dates for Neandertals come from Vindija in Croatia, and possibly from Mezmaiskaya in the Caucasus (Table 7.2). Evidently, Neandertals managed to survive quite late in several areas of Europe (Smith et al. 1999). Moreover, the archaeological data from the Iberian Peninsula itself do not demonstrate the existence of a clear-cut boundary at the Ebro "frontier" (Vaquero et al. 2006).

In western Europe, the most recent Neandertal fossils (other than the single tooth from the Portuguese site of Figueira Brava) date to 33–36 Kya (Table 7.2). The 33.8-Ky-old French site of Grotte du Renne (Arcy-sur-Cure) has yielded a series of fragmentary human remains associated with a Châtelperronian industry. A juvenile temporal bone from this site exhibits a Neandertal-like mastoid area and a bony labyrinth that is unlike that of modern humans (Hublin et al. 1996). The 29 isolated teeth recovered from the Châtelperronian layers at the

site represent at least six individuals and display a series of features that are common in Neandertals but rare in modern populations (Bailey and Hublin 2006a,b). The dimensions of the cheek teeth in this sample fall in the overlap range between Neandertals and early moderns, but the anterior teeth sort with the Neandertals. From Zafarraya in Spain, Hublin et al. (1995) report a pubic bone with a thin superior pubic ramus, a femur, a heavily damaged mandibular fragment, and a virtually complete mandible described as exhibiting characteristic Neandertal morphology in the symphysis and ramus. These fossils are all associated with the Mousterian in Level D at the site, which is dated to 33.2 Ky (Hublin et al. 1995, Hublin and Bailey 2006).

But some other late Neandertal remains deviate from typical Neandertal morphology in ways that hint at a transition to modern humans, or admixture or convergence with them. The best such specimen from western Europe is a fragmentary skull (Figs. 7.35 and 7.36) and partial postcranial skeleton from the French site of La

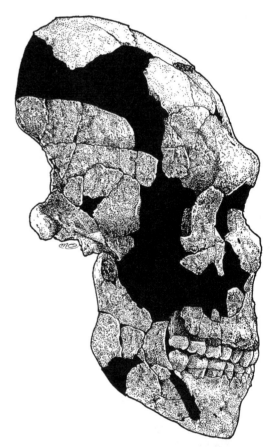

FIGURE 7.35

The late Neandertal cranium and mandible from Saint-Césaire, France. In this lateral view, the specimen evinces typical Neandertal morphology, except for a slight mandibular incurvature and a reduced retromolar space. (After Johanson and Edgar 1996.)

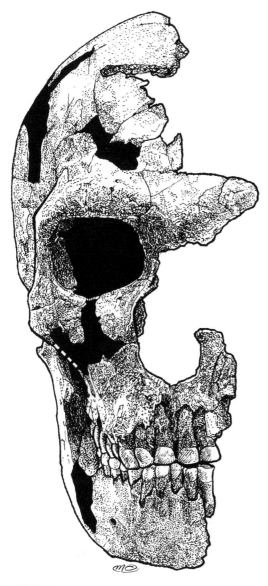

FIGURE 7.36

The late Neandertal cranium and mandible from Saint-Césaire, France in frontal view. The specimen exhibits typically Neandertal facial morphology in this view except for a relatively narrow nasal aperture. (After Johanson and Edgar 1996.)

Roche à Pierrot, Saint-Césaire, which is associated with the Châtelperronian and dates to around 36 Kya (Lévêque and Vandermeersch 1980, 1981). The Saint-Césaire 1 skull exhibits characteristically Neandertal infraorbital and mandibular morphology and midsagittal facial prognathism (Fig. 7.35). But the supraorbital torus is small (though Neandertal-like in shape and form), the anterior teeth are somewhat reduced, and the nose is very narrow for a Neandertal (Fig. 7.36). The postcranial remains generally align Saint-Césaire 1 with European Neandertals; but again, there are some "progressive,"

modern-looking details (Lévêque and Vandermeersch 1981, Stringer et al. 1984, Trinkaus et al. 1999a). A biomechanical analysis of the femoral diaphysis suggests a "hyperarctic" body build of the typical Neandertal sort. The femur is rounded in cross section and lacks a pilaster, while a moderate proximal lateral femoral flange is present. However, Trinkaus and colleagues also note that the mid-shaft cross section of the Saint-Césaire femur is more anterioposteriorly reinforced (elongated) than in other Neandertals, and falls close to early moderns in this feature. They interpret this reinforcement in Saint-Césaire and early modern Europeans as an adaptation for more extensive daily ranging than in typical Neandertals, although Saint-Césaire accomplished this differently (in that a pilaster is not developed). Trinkaus and colleagues also note that the shafts of the Saint-Césaire upper-limb bones are not as stoutly reinforced as those of other Neandertals, implying a behavioral pattern of reduced loading and less forceful movements of the arm—another resemblance to early moderns.

M. Wolpoff (1980a, 1999) sees the Neandertal remains from Hortus (France) as evincing transitional features not found in most Neandertals. Although not precisely dated, the site appears to be from the end of Würm 2 (OIS 4) and contains typical Mousterian artifacts (Piveteau and de Lumley 1963). Some 40 fragmentary human remains have been recovered. Wolpoff concludes that the Hortus teeth are reduced compared to those of other Neandertals—especially the anterior teeth, which are even smaller than those of other late Neandertals. The most complete specimen is Hortus 4, an adolescent mandible that is described as combining "… a poorly developed mental eminence with an anterior position for the ramus relative to the teeth" (Wolpoff 1980a, p. 300).

In central Europe, the Vindija sample takes center stage in any discussion of late Neandertals. Vindija is a large cave in the Hrvatsko Zagorje not far from the older site of Krapina. Excavations here in the 1970s and 1980s recovered a large series of fragmentary human remains from a long stratigraphic sequence. Although a few of these fossils come from the early Upper Paleolithic Level F, the most important finds come from the older levels of G3 and G1. These levels, dated respectively to ~38–42 Kya and 32–33 Kya (Table 7.2), are associated with a cultural assemblage described as having both Upper Paleolithic and Middle Paleolithic elements (Karavanić 1995, 2007; Karavanić and Smith 1998, 2000; Janković et al. 2006). The human fossils from these levels, all of which are Neandertals, have been extensively described and analyzed in a series of publications over the past quarter-century (Malez et al. 1980, Smith and Ranyard 1980, Wolpoff et al. 1981, Malez and Ullrich 1982, Smith et al. 1985, Smith and Ahern 1994, Ahern et al. 2004, Janković et al. 2006).

TABLE 7.19 ■ Brow Ridge Thickness Index (BTI) Data for Selected Hominin Samples[a]

Sample/Specimen	BTI	Sample N, SD
Krapina	85	11, 8
Vindija	79	6, 4
Early Upper Paleolithic	71	13, 5
Zuttiyeh	97	
Tabun C1	109	
Amud 1	92	
Skhūl/Qafzeh	84	3, —
Bodo	102 (cast)	
Kabwe 1	93	
Florisbad	93	
Ngaloba	90	
Jebel Irhoud 1	85	
Jebel Irhoud 2	81	

[a]Calculation of the BTI is explained in the text (from Smith et al. 1995).

Most of the Vindija Neandertals derive from Level G3, although the younger G1 level has also yielded a few specimens. No significant differences between the G1 and G3 fossils have been identified, so we will treat both Vindija groups together as a single sample. Like those from Krapina, the Vindija Neandertal remains are very fragmentary and may have been brought together by human processing. Although they are clearly the bones of Neandertals, they exhibit some systematic differences from other Neandertal samples. For example, the supraorbital tori of the adults are thinner and less projecting than those of other Neandertals (Table 7.6). They are also different in shape, with a pronounced thinning of the torus over the center of each orbit (Smith and Ranyard 1980, Smith et al. 1989b, Ahern et al. 2002). This thinning is reflected in values of the Brow Ridge Thickness Index (BTI), which is the mid-orbit thickness expressed as a percentage of the lateral thickness (Table 7.19). In this regard, the Vindija sample differs from earlier Neandertals and resembles early modern Europeans (Fig. 7.37). The only Vindija frontal for which the contour of the frontal squama can be assessed appears to have a higher frontal squama than most Neandertals (Malez et al. 1980, Wolpoff et al. 1981). A partial braincase from Vindija seems to show other "progressive" features—that is, a more rounded vault and less postorbital constriction—compared to Krapina (Ahern et al. 2004).

The other parts of the face in the Vindija Neandertals also show a pattern of reduction and change. Two adult maxillae both exhibit nasal breadths and alveolar heights that are more than two standard deviations below the Neandertal mean (Table 7.11; Smith 1992a). All four adult Vindija mandibles have more vertical symphyses than typical Neandertals, and two (Vi 206, Vi

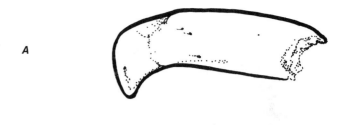

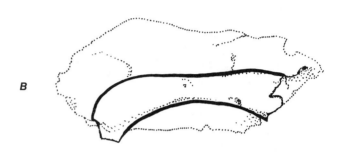

FIGURE 7.37

Schematic drawing of the supraorbital regions of three Croatian specimens. These specimens are viewed from slightly below the Frankfurt Horizontal in order to illustrate the shape differences. **A**, Krapina 28; **B**, Vindija 202; **C**, Velika Pećina. Once thought to be Pleistocene in age, Velika Pećina is now dated to the Holocene (Smith et al. 1999). However, its morphology is representative of early modern European supraorbital form. [After a drawing by M. O. Smith in Smith and Ranyard (1980).]

231) exhibit indications of an incipient mandibular incurvature and at least a *mentum osseum* (Fig. 7.38). The symphyseal angle of the Vindija sample (see Appendix) averages 87° (N = 3, SD = 1.6), while those of the Krapina and Western European Neandertals average 99.6° (N = 5, SD 1.8) and 95.5° (N = 8, SD = 6.5), respectively (Wolpoff et al. 1981).

It is important to emphasize that, apart from the aforementioned features, the crania, mandibles, and teeth of the Vindija sample clearly conform to the Nean-

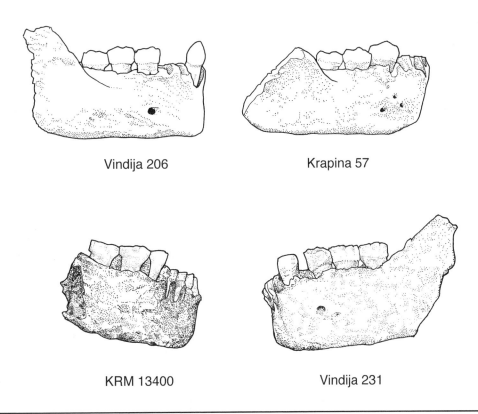

Vindija 206

Krapina 57

KRM 13400

Vindija 231

FIGURE 7.38

Symphyseal profiles in the Vindija 206 and 231 Neandertal mandibles from Croatia, compared to the earlier Krapina 57 mandible from Croatia and the Klasies River Mouth 13400 specimen from South Africa. A more vertical symphysis, an incipient *mentum osseum,* and a shallow mandibular incurvature are characteristic of the late Neandertals from Vindija. These differences from earlier Neandertals represent similarities to the KRM mandible, considered by many to be that of an anatomically modern human (Chapter 8).

dertal pattern. The same is true for the relatively few postcranial remains that can definitely be identified as human. A fragmentary scapula from Vindija has a Neandertal-like form with a dorsal axillary groove (Wolpoff et al. 1981), and a radial shaft exhibits characteristically Neandertal bowing and medial orientation of the radial tuberosity (Ahern et al. 2004). Nevertheless, a pattern of facial reduction is seen throughout the Vindija sample. It is not clear what it means. It does not reflect the inclusion of subadult individuals or the small body size of the Vindija people, as some have argued (Bräuer 1989, Stringer and Bräuer 1994). Care was taken to exclude subadults from the comparative analyses (Smith and Ranyard 1980, Wolpoff et al. 1981, Smith 1994, Ahern and Smith 2004), and the Vindija people were not small (Trinkaus and Smith 1995). Facial reduction in these late Neandertals therefore must reflect either gene flow from more modern populations elsewhere, or separate, parallel evolution of Neandertals in a modern direction. These possibilities apply to all Neandertals that show some "progressive" tendencies in their anatomy, including those from the Levant and Western Europe.

The most easterly of the possible late Neandertal sites is Mezmaiskaya Cave in the northern Caucasus. Level 3

at this site yielded a fragmentary skeleton of an infant (0–7 months of age), which has been directly carbon-dated (using AMS) to 29.2 Kya (Ovchinnikov et al. 2000; but see Higham et al. 2006). Golovanova et al. (1999) identify this infant as a Neandertal because it has short distal limb elements, a strongly developed occipitomastoid area, and an elongated foramen magnum. These authors also report cranial remains of a 1- to 2-year-old infant from the slightly younger Level 2 at the site. Both levels are associated with Mousterian artifacts, and the Level-3 infant was apparently buried in a pit. Mitochondrial DNA recovered from the Level-3 infant associates the specimen with Neandertals (Ovchinnikov et al. 2000).

However, the 29.2 Ky direct date for bone from the Level-3 infant is at odds with other reported radiocarbon dates for Levels 2 and 3 by Golovanova et al. (1999). These dates range from 32 to 41 Ky for the younger Level 2 and >45 Ky for Level 3. Thus it would seem that the infant is intrusive into these deposits, assuming the dating is correct. This is not necessarily a problem, as Neandertals could certainly have existed here at ~30 Ky as they did in other parts of Europe. However, the morphological affinities of so young an individual are difficult to determine (Creed-Miles et al. 1996, Hawks and

Wolpoff 2001b), and a detailed, comparative morphometric analysis of the Level-3 infant has yet to be published. Barriel and Tillier (2002) argue that the presumed Neandertal features actually vary widely in various populations and that the Mezmaiskaya infant could well be an intrusive modern human child. As Hawks and Wolpoff (2001b) point out, at least two of the features used to argue for Neandertal affinities for the Level-3 Mezmaiskaya infant (a pronounced occipitomastoid region and shorter distal limbs) are also found in the Lagar Velho 1 child from the Upper Paleolithic of Portugal (Chapter 8).

MAJOR ISSUES

1. **Neandertal Morphology: A Few Big Differences, or Many Small Ones?** Neandertals are morphologically distinctive, and a number of studies suggest that they show a pattern of ontogenetic development different from that of modern humans. As we have seen, various experts have described certain Neandertal peculiarities as adaptations—to cold climates, to the exigencies of life in a harsh environment, or to the use of the anterior teeth as tools. Other scientists think that the distinctiveness of Neandertals resulted more from genetic drift and isolation than from adaptation of any sort. Still others believe that at least some aspects of the Neandertal morphological gestalt result from mechanical responses of bone to environmentally-driven demands on the skeleton during life. More than one of these factors may have played a role in the making of Neandertals. But a more fundamental question needs to be asked: Can Neandertal morphology be explained by some sort of global developmental process affecting the entire organism, or do all the separate Neandertal features and complexes need to be explained separately?

Churchill (1998) refers to these two perspectives on Neandertal evolution respectively as "integrated" (or structuralist) and "particulate" (see also Rosas et al. 2006a). Integrated approaches date back to 1975, when Brothwell suggested that Neandertal morphology might be the product of an overall change in growth patterns, resulting from a shift in the hormones regulating adolescent growth and perhaps driven by selection favoring larger adults, who were better adapted to cold (Brothwell 1975). Similar heterochronic explanations have been advanced by later authors, including Lieberman (1998a, Lieberman et al. 2000) and Green and Smith (1991). The Green/Smith model, discussed earlier, posits an acceleration of prenatal growth to account for many elements of the distinctive Neandertal anatomical pattern. Green and Smith argued that most Neandertal characteristics could not be the result of an adolescent growth spurt or mechanistic responses to lifelong stresses on bone, because those characteristics are already developed in very young Neandertal children. According to their model, accelerated prenatal growth was favored because it yielded larger infants, who would have been better able to survive cold stress during the vulnerable first months after birth. Various characteristic Neandertal features of thorax shape, limb proportions, and limb bones may have resulted from this same process (Smith 1991, 1992b), because, as discussed previously, such features appear very early in Neandertal ontogeny. In this view, what is important for survival in Neandertals is the form of the infant, and most of the peculiarities of Neandertal adults are just side effects of prenatally accelerated growth.

An "integrated" explanation of this sort is appealing because it conforms to the principle of parsimony. Studies of Neandertal mitochondrial and nuclear DNA reveal differences between Neandertals and more modern humans; but those differences are trivial compared to those separating humans from any other living species. If each of the differences between Neandertals and moderns resulted from several changes in the structural part of the genome, the sum of all these changes would probably have to be rather extensive. On the other hand, a change in some aspect of the regulatory genome—say, one producing a change in the rate or timing of growth processes—could yield extensive, coordinated changes in many different parts of the body without equally extensive alterations in different parts of the genome. Such a change might still be an adaptive one driven by natural selection, but many or most of its effects would be incidental pleiotropies with no adaptive significance of their own. An explanation of Neandertal morphology in these terms would be preferred on grounds of parsimony. It would also avoid a number of famous theoretical problems implicit in the assumption that all features of an organism are adaptations produced by, and under the effective control of, natural selection (Gould and Lewontin 1979).

As we noted in examining similar claims about limb and pelvis proportions in *Australopithecus afarensis* (Chapter 4), those who offer explanations of this sort have an obligation to provide some empirical grounds for thinking that the supposedly "integrating" regulatory factors in fact exist. In an attempt to provide such empirical evidence, Churchill (1996, 1998) conducted a detailed comparative analysis of upper-body metrics in Neandertals, early modern humans (Skhūl/Qafzeh and EUP), and four recent human groups differing in body build and proportions. He found that patterns of correlation among the metric variables (covariance matrices) were homogeneous across the various groups, indicating the presence of consistent integrative factors, and that a model incorporating integration between body form and upper-limb traits fitted the data better than a model with no integration. Still, Churchill found that less than half of the total variation of features seen in the data could be explained by an integrated or structuralist model. He concludes that trait-specific selection pressures and/or developmental plasticity related to specific behavior patterns must have shaped many features of Neandertal morphology in general and of the upper body in particular. A similar study of covariance among cranial features by Lahr and Wright (1996) yielded similar results, revealing an underlying pattern of integration—but one that explained only about 40% of the total variation.

The results of these studies suggest that the complete Neandertal package cannot be explained with reference to a single ontogenetic shift. They also suggest that we will eventually be able to bundle certain elements of that package into pleiotropic complexes of correlated traits, produced by a relatively few underlying factors of genetic regulation and ontogenetic growth. It therefore seems likely that prenatal growth acceleration or some other integrational factor provided the basis for the Neandertal morphological pattern. However, it also seems clear that some Neandertal apomorphies will always have to be explained in particulate terms, as the products of individual selection pressures or activity-based effects. This is presumably true for some of the Neandertals' primitive features as well. Like many other aspects of human evolution, Neandertal morphological form appears to be too complex to be explained by a single factor.

2. *Homo neanderthalensis*, or *Homo sapiens neanderthalensis?* For many scientists, the anatomical, genetic, and even behavioral differences between Neandertals and modern humans demand that Neandertals be assigned to their own species, *Homo neanderthalensis.* The key issue here is how a species is to be defined in the fossil record (Chapter 2). Because Mayr's biological species concept (BSC) cannot be applied directly to extinct organisms, many paleontologists regard the morphological species concept (MSC) as the best theoretical substitute. Identifying Neandertals as a morphospecies in the context of cladistic theory requires that they exhibit a consistently distinctive morphology defined by uniquely derived or autapomorphic features. This requirement explains the ongoing quest for unique Neandertal apomorphies by cladistically oriented scholars, from Santa Luca (1978) down to the present. The preceding pages have assessed candidate Neandertal autapomorphies put forward by Schwartz, Tattersall, Rak, Hublin, Stringer, Lieberman, and many others. Some of these authors regard morphospecies labels merely as convenient labels for consistent clusters of traits, with no further biological implications. But most advocates of separate species status for *H. neanderthalensis* believe that the anatomical peculiarities of Neandertals are signs of their reproductive isolation, and that

careful application of the MSC will allow us to define a morphologically distinct Neandertal grouping that meets the criteria of the BSC as well.

Identifying Neandertals as a separate morphospecies may be useful in a heuristic sense (Smith 1994, 2002; Smith and Ahern 2006). But the leap from the MSC to the BSC seems unwarranted. Even many of those who regard Neandertals as a distinct species acknowledge that there may have been a small amount of gene flow between Neandertal and early modern populations. Many other researchers contend that there is significant evidence of morphological continuity—and therefore of genetic continuity—between Neandertals and early modern humans, in both Europe (Frayer 1992a,b, 1997; Frayer et al. 1993, Smith 1984, Wolpoff 1999, Wolpoff et al. 2001, 2004; Smith et al. 2005, Ahern 2006) and the Near East (Kramer et al. 2001, Simmons et al. 1991, Wolpoff 1999). Simmons (1994, 1999) and Smith et al. (1995, 2005) have argued specifically that evidence of assimilation is found in the Near East and at least in the central part of Europe. Voisin (2006) believes that assimilation probably did occur in these areas but not in western Europe, where Neandertals exhibited the most "extreme" morphological pattern. Voisin suggests a "speciation by distance" model, in which only the western European Neandertals were geographically remote enough from the areas of modern human origins to actually become reproductively isolated. Nevertheless, if intermixture occurred even sporadically all the way from central Europe to the Near East, there must have been a very extensive hybrid zone between Neandertal and early modern populations. Such a broad hybrid zone would cast doubt on whether the morphs involved really should be thought of as separate species (Mayr 1969, Wiley 1981).

3. Are Neandertals Ancestors of Modern Humans? - Strictly speaking, the status of *H. neanderthalensis* as a distinct species is logically separable from the issue of Neandertal ancestry for modern humans. Even if the genetic barrier between the two populations was entirely permeable, so that there was no foundation for recognizing Neandertals as a separate biological species, Neandertals might still have no living descendants—for example, if all the genes flowed from early moderns into a shrinking Neandertal population. Conversely, even if we recognize Neandertals as a separate species from *Homo sapiens*, that would not imply that Neandertals could not be ancestors of living humans. It depends on the geometry of the cladogram. For example, many of the proponents of species status for Neandertals believe that they emerged as the result of the splitting of *Homo heidelbergensis* into *H. neanderthalensis* in Europe and early *H. sapiens* in Africa. This phylogeny allows *H. heidelbergensis* to be specifically distinct from our own species and at the same time ancestral to it. However,

most cladistic systematists do not accord separate species status to two chronospecies arrayed in an ancestor-descendant sequence. From this perspective, *H. neanderthalensis* cannot be an ancestor of *H. sapiens* unless there was a cladogenetic (branching) event in which *H. neanderthalensis* gave rise to *H. sapiens* and another species. Since there is no evidence of any other species that might be derived from Neandertals, cladists generally reject any possibility of ancestral status for the Neandertals.

This way of thinking about species and speciation is intimately tied to a punctuationist view of the evolutionary process (Gould 1987, 1988; Eldredge and Tattersall 1982). In the punctuationist model, speciation occurs in small populations that become geographically (or perhaps ecologically or behaviorally) isolated from the main body of their species. New selection pressures, founder effect, and drift combine to fix evolutionary novelty more rapidly in these small peripheral populations than in the parent population. If the resulting genetic transformation of the peripheral population is sufficiently great, reproductive isolation will result. When contact between the two populations is reestablished, the two will not be able to interbreed, and will constitute separate biological species. This model is the one generally adopted by advocates of species status for Neandertals, who see Neandertals as evolving in nearly complete isolation in glacial Europe while modern humans developed their own apomorphies in Africa.

But if contact with Neandertals were reestablished before reproductive isolation had become complete, or if gene flow continued at a low level between the two populations throughout the process of regional differentiation, European and African humans would exhibit significant morphological differences but still be capable of exchanging genes. There are both theoretical and empirical reasons for thinking that this is what in fact happened. No matter how completely the Neandertal gene pool may have been geographically isolated from other human populations, it is doubtful that that isolation went on long enough for complete reproductive isolation to occur (Cartmill and Holliday 2006). A general survey of speciation times in eutherian mammals (Holliday 2006a) shows that geographical separations lasting about one million years on average are needed to produce complete reproductive isolation after sympatry is reestablished. And as we will see in the next chapter, the fossil records of Neandertals and early modern humans strongly suggest that geographical contact between the two populations was extensive and long-lasting. These facts are not in conflict with the demonstrable morphological distinctiveness of Neandertals from both western Asia and Europe. Significant

morphological change—even enough to demarcate a separate species using morphospecies criteria—can accumlate in parapatric populations without resulting in the reproductive isolation that defines biological species (Godfrey and Marks 1991).

However, the most persuasive empirical reason for doubting that complete reproductive isolation occurred in this case is the existence of specimens with mixed morphology. Some of these are fossils of early modern humans found in or at the edges of the Neandertal homeland, which exhibit features of the skeleton otherwise characteristic of Neandertals. These specimens will be taken up in the next chapter. But the evidence surveyed in this chapter demonstrates that a significant percentage of late Neandertals—from Western Europe (Saint-Césaire, Hortus), central Europe (Vindija), and perhaps the Caucasus (Mezmaiskaya)—evince cranial, dental, or postcranial morphology that deviates from typical or "classic" Neandertals in the direction of modern humans. Significantly, elements of material culture long thought to be unique to modern humans, including blade-tool technology and items of personal adornment, are associated with these late Neandertal survivors.

Many researchers would support the contention of Hublin and Bailey (2006) that the "progressive" features found among late Neandertals probably do not reflect any genetic connection to early modern humans, including post-contact hybridization or assimilation. But if this contention is sound, then these "progressive" features of behavior and morphology must have been evolving in parallel within the Neandertal population. Such parallel evolution would affirm the humanity of these ancient Europeans in a different sense. If Neandertals and modern humans were unable to interbreed and produce fertile offspring, then the "progressive" traits seen in some late Neandertal populations—larger and rounder braincases, more vertical foreheads, more vertical mandibular symphyses with incipient chins, narrower nasal apertures, shorter and smaller faces, reduced supraorbital tori with mid-orbital thinning, and evidence of Upper Paleolithic behavioral capacities—can only be interpreted as the products of evolutionary convergence with modern humans, driven by evolutionary forces much the same as those that produced similar changes somewhat earlier in Africa. No matter whether we interpret these traits as convergences or as tokens of gene flow out of Africa into Europe, we cannot avoid recognizing these late Neandertals as the same sort of creatures as ourselves: upright, talking apes with a capacity for technological innovation and symbolic behavior. Questions about interfertility take on a lesser importance in the light of that admission.

CHAPTER EIGHT

The Symbolic Ape: The Origin of Modern Humans

Signs, said the philosopher C. S. Peirce, are of three kinds: indexes, icons, and symbols. An **index** is related to the thing it stands for by a cause-and-effect connection, in the way that dark clouds are a sign of coming rain. An **icon** has some sort of actual resemblance to the thing it signifies, as a map resembles a territory. A **symbol** is a sign that has a purely arbitrary and conventional relation to its referent. In Peirce's sense, almost all words are symbols. The word "horse," for example, is not causally linked to horses, and it does not look or sound or smell like a horse. Its connection with horses is a matter of convention.

If words are symbols, then humans have been symbolic animals for as long as they have been talking. Symbol use in this sense probably dates back to a time before the advent of modern humans—almost certainly to the Neandertals, and probably to the Heidelbergs before them. But when most of us use the word "symbol," we mean something else. In ordinary usage, a "symbol" is an object or sign that stands for, represents, or *symbolizes* something that cannot be expressed in a few words and may not be readily expressible in language

at all. A cross is an example of this deeper sort of symbol. We say that a cross is a symbol of Christianity; but the cross does not stand for Christianity in the way that the word-symbol "Christianity" does. Its meaning is both deeper and vaguer. The symbolic cross encodes a set of attitudes and beliefs concerning the world and the meaning of human life, including the tenets of Christian faith, a system of rituals, and innumerable historical facts, stories, and legends. It not only connotes those attitudes, practices, and beliefs; it also affirms them. The display of the cross identifies a person, a household, or a nation as a participant in this belief system, and it says a vast amount about that person or group to those who understand its symbolic meanings.

Although some other animals can be taught to understand and even use human words, the symbolism of the cross is beyond their grasp. As the anthropologist Leslie White (1949) observed, no nonhuman animal can understand the difference between distilled water and holy water. Yet this sort of symbolism pervades almost every aspect of modern human behavior. Most of the things that we make and do are invested with complex, conventional, nonverbal meanings. The clothes we wear, the structures we make, the implements we use, and even the foods we concoct are not fashioned simply for utility; they also carry information about us, our group allegiances, and our convictions about the Good, the True, and the Beautiful.

It is hard to determine exactly when humans became symbolic animals in this sense. In the archaeological record, such unquestionably symbolic objects as paintings, sculpture, grave goods, and jewelry point to the presence of symbolic thought. But we cannot always be certain what constitutes a symbolic object. If the symbolic connotations of the Christian cross were unknown, what would we say about a cross uncovered in an archaeological dig? Could a functional, rather than symbolic, role for this object be excluded? Conversely, how can we be sure that an apparently utilitarian object did not have a symbolic significance for its makers?

For most of the past century, it has been widely assumed that symbolic behavior came into being at the onset of the Upper Paleolithic, about 40 Kya, as part of a single complex of apomorphies that also included language and modern human morphology. (For recent reviews, see Gamble 1986, 1999; Mellars 1996, 2004, 2006a; Straus 1992; Klein 1999; and Churchill and Smith 2000a.) Like many long-standing assumptions in paleo-anthropology, this notion reflects the Europe-centered origins of our science. The archaeological record in Europe seems to testify to a sharp break between European Upper Paleolithic (EUP) cultures—associated with modern human skeletons, and rich in symbolic artifacts—and earlier Middle Paleolithic cultures associated with more archaic *Homo* and lacking objects of art and personal adornment. And because words are referred to as "symbols," it has been an easy jump to the conclusion that the archaic Middle Paleolithic people lacked language because they did not leave these sorts of "symbolic" artifacts behind. But as more has been learned about early modern humans and their artifacts and life-ways in lands outside of Europe, it has become increasingly doubtful that all these novelties came into existence together as parts of a single package.

■ A "CREATIVE EXPLOSION"?

The advent of fully human forms of symbolic behavior in Europe is generally equated with the first appearance of the **Aurignacian** culture, which began perhaps as early as 39 Kya and was in full swing by 36 Kya. The earliest (Initial) Upper Paleolithic expressions in Europe were relatively simple, and some of them were probably produced by Neandertals (Chapter 7). But the Aurignacian is distinctly different. Compared to its predecessors, the Aurignacian is characterized by more finely crafted stone tools, a greater diversity of stone tool

types, the manufacture of standardized bone tools, and various forms of artistic expression including personal ornaments, carvings, and the famous cave paintings of France and Spain (Gamble 1986, 1999; Mellars 1996, 2004, 2006a; Klein 1999, Churchill and Smith 2000a; Conard et al. 2003a). The presence of art objects at Aurignacian sites is widely viewed as marking the advent of a new type of human intellect. Aurignacian and other EUP art has been interpreted as evidence of various sorts of symbolic thought, including magic and religion, the use of notational systems, and affirmations of personal and group identity (Marshack 1976, 1989; White 1986, 2003). The implications for many scholars of this so-called "creative explosion," as it was dubbed by J. Pfeiffer (1982), are summed up in the following quote from Le Gros Clark (1966, pp. 116–117):

> At the end of the Mousterian phase of Paleolithic culture, the Neandertal inhabitants of Europe were abruptly replaced by people of completely modern European type. There is reason to suppose this new population, the Aurignacians, having developed their distinctive culture elsewhere, probably in Asia, migrated into Europe, and with their superior social organization, quickly displaced Mousterian man and occupied his territory.

The "superior social organization" attributed to the makers of the Aurignacian is supposed to have been a byproduct of the onset of symbolic behavior. The new capacity for symbolism revolutionized human social behavior, because it made it possible to mark group identities and personal status distinctions, expand social networks, and share information using syntactic language of a fully modern sort. Or so the story goes. This story implies that cultural factors were what gave early modern people the competitive edge on their predecessors. But those who tell this sort of story generally assume that the cultural differences between the two were at bottom biological differences, and that the new behaviors were made possible by a new sort of brain organization in modern people.

The Le Gros Clark quote also expresses a biogeographic assumption: namely, that the Aurignacian "explosion" signals an immigration of alien modern humans into Europe from somewhere else. The British archaeologist P. Mellars is one of the leading proponents of this popular idea. Mellars (1989, 1993, 2002, 2004, 2006a,b,c) contends that the Aurignacian developed from the early Upper Paleolithic of the Near East after it was carried into Europe by modern-human immigrants. He sees the Aurignacian as a single, cohesive cultural entity, which radiates throughout much of Europe and tracks the spread of modern people across the continent.

Others see the development of the Aurignacian differently. M. Otte (2004) argues that the Aurignacian appeared first in Europe, with no real antcedents on the peripheries of Europe or in the Near East. This claim implies that the classic Aurignacian was not something imported by immigrants. The scope and unity of the "Aurignacian" as a cultural entity is also disputed (Straus 2003). N. Teyssandier (2006, p. 25) concludes that the Aurignacian "… includes distinct socio-cultural phenomena that may have different origins and histories." O. Bar-Yosef (2006, p. 475) bemoans the "… indiscriminate use of the term Aurignacian" and argues that this label should be restricted to a subset of "Aurignacian" assemblages sharing a narrowly defined suite of elements. His restricted definition excludes many assemblages in both Europe and West Asia that are often labeled as "Aurignacian."

Noting regional differences in the European "Aurignacian," Karavanić and Smith (1998) suggest that several European variants of the Aurignacian retain influences carried forward from the local Middle Paleolithic, exemplified by Vindija. Zilhão (2006) rejects the Vindija evidence, but he too discerns a pattern of indigenous European cultural contributions to the Aurignacian. As Zilhão sees it, the so-called Proto-Aurignacian commingles two different traditions of personal adornment—one (the use of modified marine shells) brought in from West Asia, and a second (the use of pierced and grooved teeth, bones, and fossils as beads) that has local antecedents in the Initial Upper Paleolithic of Europe. The fossil evidence suggests that this second tradition was invented by Neandertals, before modern humans came on the scene. Zilhão concludes that "Neanderthals and moderns mixed, and it matters" (Zilhão 2006, p. 183). Although his arguments are based mainly on the archaeological evidence, Zilhão notes that the occurrence of cultural continuity in Europe strengthens the case for some degree of biological continuity.

Like Zilhão, Bar-Yosef finds a broad continuity in Europe between the various IUP variants and classical Upper Paleolithic entities like the Aurignacian. But he draws a different biological conclusion. Bar-Yosef (2003) sees the Aurignacian as a native European development from IUP cultures brought in by intrusive modern humans. In his view, the IUP does not reflect any cultural or biological carryover from the Neandertals, but rather "… may simply indicate the rapid movement of modern humans accompanied by a high degree of cultural individualism" (Bar-Yosef 2006, p. 467). Bar-Yosef (2003) traces the origins of the IUP to the Ahmarian industries of the western Mediterranean. Unfortunately for his interpretation, all the human fossils associated with the European IUP are late Neandertals, not modern humans (Chapter 7).

In Africa, the use of shell ornaments and other aspects of "modern" behavior seem to have a longer time depth than in Europe. A striking example is the recent discovery at Blombos Cave in South Africa of 19 snail shells, each bearing a hole in the same spot opposite the mouth (d'Errico et al. 2005). These shell beads, which were probably strung as a necklace or bracelet, are dated to 75 Kya. They furnish the earliest clear archaeological evidence for personal adornment. McBrearty and Brooks (2000) argue that the advent of modern human behavior was not a sudden "revolution" but a gradual process, which began in southern Africa around 100 Kya. They point to East and South African evidence for sophisticated bone and stone technology, as well as for early symbolic behavior, well in excess of 50 Kya. As we will show in more detail later, there is increasingly strong evidence that distinctively modern human behavior began in Africa, where it emerged relatively gradually during the **Middle Stone Age**—the African equivalent of the European Middle Paleolithic (d'Errico 2003, Henshilwood and Marean 2003).

Klein (1998, 1999, 2000, 2003; Klein and Edgar 2002) also identifies Africa as the cradle of modern human behavior, but he believes that the origin of modernity was more punctuational and more recent. Rejecting the evidence for modern-type cultural behaviors in Africa as far back as 100 Kya (Yellen et al. 1995, McBrearty and Brooks 2000, d'Errico et al. 2005; cf. Wong 2005), Klein argues that such behaviors did not appear until around 50 Kya, when they were made possible by a fundamental neural reorganization in the human brain. He associates this supposed neural reorganization with the onset of language. Klein contends that the acquisition of language gave early modern Africans a decisive advantage over other hominins, allowing them to spread out of Africa and replace their archaic cousins in Eurasia—including some that had essentially modern skeletons but retained Middle Paleolithic cultural adaptations. Presumably, these behaviorally backward modern humans were handicapped by a more primitive brain structure that was unable to handle semantics and syntax.

Klein acknowledges that there is no evidence in the fossil record for this "great leap forward" in neurology. However, he suggests that a "… single mutation could underlie the fully modern capacity for speech" (Klein and Edgar 2002, pp. 271–272). Klein's candidate for this crucial mutation is the uniquely human variant of the FOXP2 gene, which seems to be part of the substrate for human language abilities (Lai et al. 2001, Enard et al. 2002). But as noted earlier (Chapter 7), Neandertals apparently shared the modern human FOXP2 variant (Krause et al. 2007a), and there is no good reason to think that their language abilities were subhuman. At present, neither paleontology nor genetics furnishes any real evidence for a neural revolution at 50 Kya.

Others have reached conclusions that are opposite from those of Klein. F. d'Errico (2003) argues that European Neandertals did not differ significantly in technological capabilities, complexity of subsistence strategies, or symbolic traditions from their more modern-looking contemporaries in Africa and the Near East. He notes that the florescence of "symbolic" behavior in the European Aurignacian is a unique case and that such behaviors do not characterize early modern humans elsewhere, including many who lived considerably later in time. D'Errico concludes that the archaeological record does not support a single point of origin for behavioral modernity.

Like d'Errico and Zilhão, Henshilwood and Marean (2003) think that the "creative explosion" in the European Aurignacian was a situation-specific case that does not furnish a model for what went on in other regions. This is particularly true of the unique art associated with this cultural complex. Henshilwood and Marean argue that the new behaviors seen in the Upper Paleolithic can be explained more parsimoniously without positing a macromutation that reorganized the human brain. With respect to resource utilization, they suggest that some resources may not have been exploited earlier because they are relatively difficult or dangerous to obtain. As long as there are "easier pickins," there is no reason to turn to these resources. However, changes in resource availability, or population pressures due to increasing population size or density, may lead to changes in resource procurement and associated changes in cultural adaptations. These changes may have more to do with environmental context than with ability or intellect. Finally, taphonomic factors may explain some apparent changes in behavior. In discussing the absence of sophisticated bone tools in the African Middle Stone Age, Henshilwood and Marean suggest that older archaeological complexes may simply not preserve as much as later ones. Middle Paleolithic people may have produced artistic or symbolic objects in media that did not withstand the ravages of time. Perhaps this is one reason why evidence of artistic expression occurs only at the end of the Neandertal reign in Europe (Chapter 7).

Many cultures in the ethnographic present produced complex symbolic representations that would be unlikely to last very long in an archaeological context. If the Lakota Sioux had lived 10 Kya, the complexities of their music, ritual, dance, religion, legends, costumes, ornaments, and graphic art would have been lost to history. The archaeology of late Pleistocene North America provides a striking contrast to the "creative explosion" seen in Aurignacian Europe. There were no hominins in the Americas until about 13 Kya, when fully modern humans entered North America via Alaska and quickly spread throughout both American continents. But although many North American sites have yielded remains of the stone tools and animal prey of these "Paleoindian" people, they left almost no enduring traces of their symbolic behavior—cave paintings, rock carvings, sculptures, personal ornaments, and so on (Griffin et al. 1988, Fiedel 1992, Zeitlin and Zeitlin 2002,

Martin 2005). This does not imply that they lacked such behavior or that they were less than completely human. These first Americans were just as capable of making symbolic objects as their civilized descendants who produced the pre-Columbian artworks of Peru and Mexico. Yet either they rarely made such things, or they made them of materials that decayed and vanished. The same thing is true of many other Paleolithic peoples as well.

No doubt, all hominins who made symbolic artifacts also had language. But we cannot infer from this that those who did not make them must have lacked language. This is like arguing that if all dogs have teeth, then all non-dogs must be toothless. The notion that language, human cognition, and symbolic artifacts are all byproducts of a single revolutionary transformation springs almost entirely from this simple logical error. Since animals with no capacity for art or religion have shown surprising abilities to master certain rudiments of language in the laboratory (Kako 1999), it seems reasonable to think that language ability preceded the onset of art, religion, and other sorts of symbolic behavior in the course of human evolution.

MODERN HUMAN ANATOMY—THE SKULL

Defining what counts as "modern" human anatomy is almost as hard as defining modern human behavior. Most analyses of anatomical modernity focus on the skull. However, modern human skulls vary considerably in both morphology and metrics (De Villiers 1968, Howells 1973b, Howell 1978, Lahr 1996, Habgood 2003), and most of the definitions of "modern human" that have been proposed are either too vague to be useful or too narrow to encompass the range of modern human variation.

Day and Stringer (1982) attempted to formulate a precise, quantitatively based definition of "modern human" or *Homo sapiens sensu stricto*. Many of the metric traits that they used are illustrated in the Appendix and discussed below and in Chapter 7 (Table 7.5). The cranial features that they identify as distinctively modern include:

- A short, high skull (ratios of basion–bregma height and of vertex radius to glabello-occipital length exceeding 0.70 and 0.64, respectively).

- A long, high parietal arch with narrow inferior and broad superior dimensions (parietal angle less than 138°, indicating a high, round cranial vault).

- A high frontal bone (frontal angle less than 133°).

- A supraorbital torus that is not continuous but divided into lateral and medial portions.

- An occipital bone that is uniformly curved rather than angulated (occipital angle no less than 114°).

- A mental eminence (chin) on the mandible.

Some Late Pleistocene crania that meet these criteria and are widely regarded as anatomically modern are in some sense not exactly "modern," because multivariate morphometric analyses place them outside the range of variation of recent humans. These aberrant early "moderns" include the Mladeč 1, 2, and 5 crania from the Czech Republic, the Qafzeh 6 and Skhūl 4, 5, and 9 crania from Israel, and the Border Cave 1 skull from South Africa (Kidder et al. 1992, Corruccini 1992). Several studies have shown that skeletal variation in Late Pleistocene "modern" humans exceeds the recent human range (Howells 1989, Stringer 1992, Lahr 1996, Crevecoeur 2007), particularly in samples older than about 35 Ky. But as Stringer (2002) and Lahr (1996) point out, there is no reason why definitions of "modern human" should be based on recent humans alone.

A more serious difficulty is that Stringer and Day's definition of *H. sapiens* would exclude some recent humans. It seems that what counts as "modern" human morphology depends to some degree on region of origin. For example, only a few early modern European crania fall out of the modern range morphometrically (Kidder et al. 1992); but many Australasian specimens do not meet all of Stringer and Day's criteria, particularly those involving the occipital and frontal bones (Wolpoff 1986b).

These regional differences in "modern" morphology may reflect differing amounts of local archaic contributions to regional modern-human gene pools (Smith 2002; cf. Wolpoff 1999, Habgood 2003). Perhaps archaic humans contributed more to the gene pools of early modern populations in Australasia than their counterparts in Europe did, with the result that many modern Australians do not conform very well to globally conceived definitions of modernity. Both Lahr and Stringer would reject this interpretation. But whatever the explanation, it seems clear that the boundary between archaic and modern humans must be defined differently for different regional populations. A recent Aboriginal Australian skull has to be regarded as modern in its regional context, even if it does not meet all of Stringer and Day's criteria.

Although an all-encompassing definition may be impossible, it is nevertheless possible to offer a polythetic description of modern human cranial anatomy and metrics. In lateral view (Fig. 8.1), the modern skull presents a high vault with relatively rounded contours and a high, vertical forehead. This is reflected in high vault-height indices for all modern human samples, including Pleistocene ones, compared to Neandertals (Table 7.4). The steeper forehead is reflected in lower frontal angles than those of Neandertals, and the more rounded lateral contours of the modern vault are demonstrated by higher occipital and lower parietal angles (Table 7.5). Nevertheless, Australian and Melanesian crania tend to have relatively lower vaults than crania

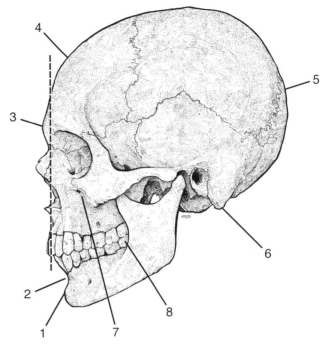

FIGURE 8.1

A recent human skull in lateral view. This modern specimen is **orthognathic**—that is, practically the whole face is located posterior to the anterior limit of the braincase (dashed line). Other distinctive modern features include (1) a projecting mental eminence or chin (cf. Fig. 8.2); (2) a well-excavated mandibular incurvature; (3) relatively weak development of the supraorbital superstructures; (4) a high, vertical forehead; (5) smoothly arched cranial contours; (6) a projecting mastoid process; (7) a canine fossa; and (8) absence of a retromolar space. (After Rohen et al. 1998.)

from other recent samples (Howells 1973b, 1989). Significantly, the archaic people from Australasia (Ngandong) are also low-vaulted compared to comparably ancient humans from Europe (Neandertals) and Africa, suggesting that a relatively low vault has long been characteristic of the Australasian region.

Other features associated with modern human skulls (Fig. 8.1) include a mastoid process that projects below the surrounding cranial base, an orthognathic face (a flat face located directly under the forehead, with some variation in the degree of alveolar prognathism), a forward-projecting "chin," and the general absence of a retromolar space (a gap between the back of the third molar and the front of the mandibular ramus). However, retromolar spaces are seen in some modern specimens (Franciscus and Trinkaus 1995), and are not characteristic of all archaic *Homo*.

The distinctive characteristics of the modern human chin (Fig. 8.2) were first defined by F. Weidenreich

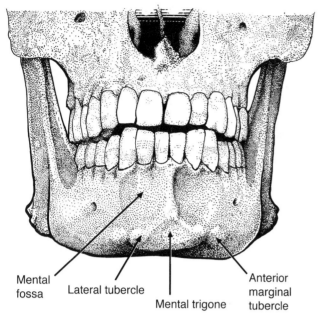

Mental
fossa Lateral tubercle Mental trigone Anterior
marginal
tubercle

FIGURE 8.2

Major features of the modern human chin. (Following
Weidenreich 1936b.)

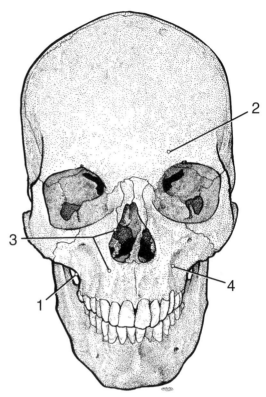

FIGURE 8.3

A recent human skull in frontal view. Noted features are:
(1) an angled zygomaticoalveolar margin (ZAM); (2)
absence of a supraorbital torus and generally relatively
weak development of the supraorbital superstructures;
(3) generally small nasal aperture and reduced facial
lengths (upper facial height, nasal height, alveolar
height); and (4) presence of a canine fossa. (After Rohen
et al. 1998.)

(1936b). They include an irregular triangle-shaped
central elevation, the **mental trigone**, bounded on
each side by distinct depressions called **mental fossae**.
The mental trigone is also bordered laterally by the
lateral tubercles, and even more laterally by the **anterior marginal tubercles**. Viewed from the side, the
basal projection of the trigone is separated from the top
of the alveolar process by a concavity, the **mandibular
incurvature** (*incurvatio mandibulae*). Schwartz and
Tattersall (2000) believe that the chin thus defined is an
autapomorphic feature that distinguishes *Homo sapiens*
proper from all earlier hominins. They argue that the
modern human chin is distinctively different from the
basal projections (the *mentum osseum*) seen in some
Neandertals and other archaic humans (Chapter 7),
which lack the mental trigone and associated structures.
Others have suggested that these structures are visible
in some archaic specimens (Mallegni and Trinkaus
1997) but are usually obscured by the forward thrust of
the alveolar part of the symphyseal region (Wolpoff
1975a, Smith 1976, 1983).

Modern human skulls generally lack supraorbital tori
(Fig. 8.3). However, tori are seen in some individuals and
are relatively common in certain populations—e.g.,
some Aboriginal Australian groups (Russell 1985). In
many skulls, what appears to be a torus is actually a
well-endowed modern brow ridge, a structure divided
into **superciliary arches** and **supraorbital trigones**
(Smith and Ranyard 1980). In such cases, the superciliary arch (located over the orbits just lateral to the

midline) is usually very thick and strongly arched; but
the bone flattens laterally into a variable plate-like structure, the supraorbital trigone. The Mladeč 5 cranium,
an early modern specimen from the Czech Republic,
illustrates this morphology (Fig. 8.4). At first glance, the
frontal bone seems to exhibit a pronounced torus, but
closer inspection shows that the supraorbital ridge is
divided into a prominent superciliary arch and a robust
but distinct supraorbital trigone.

Modern human faces tend to be significantly shorter
and narrower than their archaic predecessors (Tables
7.10 and 7.11), and the zygomaticoalveolar margin
(ZAM) forms a more pronounced angle between the
lower edge of the bony cheek and the side of the
alveolar region (Fig. 8.3). Medial to the suture between
the zygomatic and maxilla, the lower edge of the
bony cheek usually rises slightly before descending to
meet the alveolar process of the maxilla. The resulting
concavity, the **malar notch** (Fig. 3.34B), accentuates

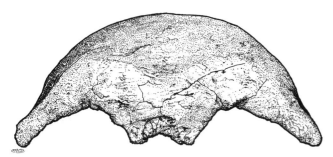

FIGURE 8.4

The brow ridges of the Mladeč 5 early modern human cranium from the Czech Republic. While this specimen has the most robust supraorbital region of early modern European crania, it does not have a supraorbital torus. Instead, the brow ridges are divided into thick, strongly arched superciliary arches medially and distinctly thinner supraorbital trigones laterally. This specimen is viewed from slightly below the Frankfurt Horizontal, in order to better reveal the shape of the supraorbital structures.

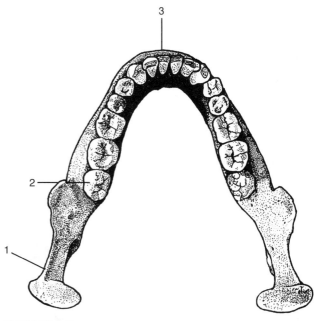

FIGURE 8.5

Occlusal view of a recent human mandible. The mandibular incisure crest (1) intersects the condylar neck laterally. Rak (1998; Rak et al. 2002) considers this a defining feature of modern humans. Because the anterior edge of the ramus is positioned lateral to the approximate center of the M3 (2), there is no retromolar space. The reduction and retraction of the anterior teeth result in a projecting chin (3).

the angulation of the ZAM. However, some modern skulls lack this notch (Fig. 8.3; Pope 1991). Modern human faces also characteristically exhibit a **canine fossa**, a vertical furrow or depression on the maxilla below and medial to the infraorbital foramen. Its function (if any) is not evident. In archaic *Homo*, the canine fossa tends to be found only in smaller and more gracile individuals. However, with the overall facial reduction that characterizes modern humans, the canine fossa emerges as a relatively constant feature, even in large and robust skulls.

There is considerable variability among modern human populations in the size and morphology of the teeth (Wolpoff 1971b, Cadien 1972). In general, modern front teeth are smaller than those of Neandertals and other late archaic hominins, but the cheek teeth are not (Fig. 8.5). We noted earlier (Chapter 7) that the size of the front teeth tends to be correlated with the breadth of the nasal aperture; and as we might expect, nasal breadth is also generally reduced in modern humans (especially in northern Eurasia). However, there is a lot of regional variation in nasal breadth, which appears to be correlated with climatic differences. Nasal breadths and indices (Table 7.11) are generally largest in recent populations living in warm, humid regions, while the nose opening tends to be narrower and higher in cold-adapted people (Coon 1962, Wolpoff 1968b, Howells 1973b). There is a significant correlation between nasal breadth and aspects of anterior palate breadth in recent humans (Glanville 1969), but it is not clear whether changes in anterior palate breadth are

side effects of changes in nasal breadth, or vice versa—or neither, or both. Variation in the size and shape of the nasal opening probably reflects a complex interplay of selection pressures, not just climatic factors (Wolpoff 1968b, Smith 1983, Yokley 2007).

Modern human braincases generally exhibit their greatest breadth high up on the parietals, giving them a house-shaped (French "*en maison*") silhouette in rear view (Fig. 8.6). In Neandertals and other archaic people, maximum vault breadths fall lower down on the parietals ("*en bombe*," Chapter 7) or even at the level of the cranial base. The maximum breadth of some modern human braincases actually lies at the cranial base as well; but in these cases, the bi-parietal breadth is only slightly smaller than the maximum breadth. Unlike Neandertals and other archaic *Homo*, modern humans (especially males) have an external occipital protuberance on the occipital bone, visible in both the rear and basal views (Fig. 8.6). Modern human occipitals also bear inferior and superior nuchal lines (with the superior line marking the boundary between the nuchal and occipital planes—Fig. 8.7), but they almost never have an occipital or nuchal torus.

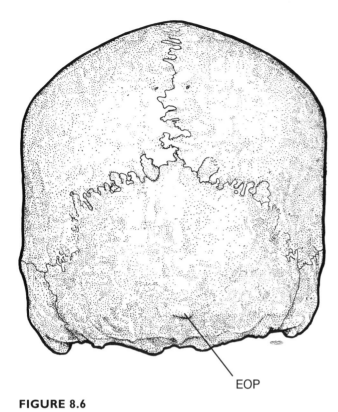

FIGURE 8.6

A recent human cranium in rear view. The sides of the vault are essentially parallel, and the maximum cranial breadth falls high up on the vault. EOP, external occipital protuberance. (After Rohen et al. 1998.)

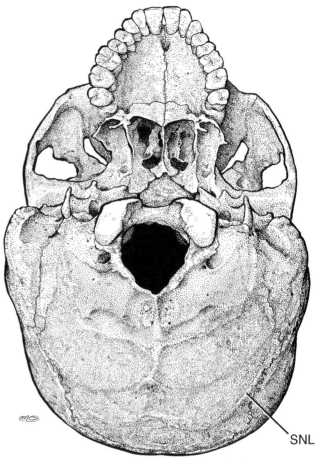

FIGURE 8.7

A recent human skull viewed from below. Compare Figs. 4.83, 5.12, and 7.12. SNL, superior nuchal line. (After Rohen et al. 1998.)

MODERN HUMAN ANATOMY—CRANIAL CAPACITY

Contrary to what might be expected, humans today have smaller braincases than many earlier human populations. Cranial capacity data for 122 recent human groups yield means of $1427 \pm 81.6 \, \text{cm}^3$ for males and $1272 \pm 82.9 \, \text{cm}^3$ for females, with a combined-sex range of 1070–1651 and mean of $1349 \pm 77.5 \, \text{cm}^3$ (Beals et al. 1984). Tobias (1971) reports a similar mean ($1345 \pm 200 \, \text{cm}^3$) for a mixed-sex sample of >1000 recent human skulls. These averages are some 60–200 cm^3 smaller than the means for late archaic humans from the western Old World—that is, Würm European Neandertals, West Asian Neandertals, and late archaic humans from Africa (the "African Transitional Group," below). However, *early* modern human means from these three regions are about the same as the sample means for late archaic *Homo* from the same region (Tables 7.9 and 8.4). In short, braincase size appears to reach a peak in early modern humans, and declines somewhat between the Pleistocene and the present.

Some of these differences may reflect differences in body size. They may also reflect patterns of climate change since the Pleistocene. As noted in earlier chapters, the thermodynamic effects of climate are major factors in explaining variation in recent human cranial capacities. For example, Beals et al. (1984) report a mixed-sex mean cranial capacity of $1415 \pm 51 \, \text{cm}^3$ for 19 Asian populations living in winter-frost conditions, whereas the mean for seven Asian populations inhabiting more tropical environs is $1284 \pm \text{cm}^3$. Virtually all available information indicates that this sort of regional brain-size variation does not correspond to differences in intelligence in recent humans (see Deacon 1997, pp. 146–164). It is probably related to the prevalence of larger body bulk in colder climates. Cranial-capacity reduction in recent humans may likewise reflect trends toward smaller and more gracile bodies in post-Pleistocene human populations (see below). Whatever its explanation, it does not imply that natural selection is working to make us all stupider.

MODERN HUMAN ANATOMY—THE POSTCRANIAL SKELETON

It is often claimed that the modern human postcranial skeleton is less robust than those of earlier humans. Our ideas about what is "modern" in the postcranium are based chiefly on comparisons with Neandertals, since little is known about other archaic *Homo* postcrania. Such comparisons may be misleading. Neandertals were exceptionally big, stocky, and muscular people, and many supposedly distinctive features of modern limbs appear to be allometric differences that reflect the larger body mass of Neandertals. For example, Neandertals have larger femoral joint surfaces than modern humans; but the differences disappear when properly scaled for body mass (Ruff et al. 1993). The limb bones of Neandertals (and other archaic *Homo*) tend to have pronounced and rugose muscle markings, but many modern skeletons have muscle attachment areas that are just as rugose. Relative to length, the long bones of modern limbs are more gracile than those of Neandertals, with thinner shafts, thinner cortical layers, and larger marrow cavities—on average. But again, there is a lot of overlap between the two groups in these metrics. For instance, the femoral robusticity index of modern human samples has a range that completely encompasses that of the Neandertals (Table 8.1). Some of the other postcranial features that are often cited as distinctively modern (Pearson 2000a) are not so distinctive if viewed in a broader context. For example, limb proportions clearly distinguish Neandertals from early modern Europeans or West Asians (Table 7.16); but they would not so clearly distinguish Neandertals from Lapps or Inuit. As noted earlier, these differences are probably related to climatic adaptation.

One genuine morphological peculiarity of the modern human femur is the pilaster. This ridge running down the back of the femur is capped by a linear rugosity, the linea aspera, to which powerful thigh muscles attach. The functional meaning of the pilaster is unclear. Archaic humans, which have a linea aspera without a pilaster (Figs. 7.26 and 7.27), presumably had thighs as muscular as those of any living people. The presence or absence of a pilaster may reflect genetic differences (e.g., different positioning of the muscles on the back of the thigh), environmental differences (differences in loading patterns on the shaft), or both.

As noted in Chapter 7, we do not yet know much about the relative contributions of environmental and genetic factors to the shaping of the postcranium. A lot of postcranial variation, both among recent samples and between recent and archaic ones, seems to have more to do with environmental influences during ontogeny than with natural selection or phylogenetic relationships (Trinkaus 1983a, 1989; Churchill 1994, 1996, 1998). Many of the postcranial differences between archaic and modern humans may simply reflect the response of the developing bones to differing levels of mechanical stress during growth (Ruff et al. 1993).

THE GEOCHRONOLOGY OF MODERN HUMAN ORIGINS

As recently as the middle 1980s, it was widely held that modern humans appeared roughly at the same time throughout the Old World, sometime between 40 and 35 Kya (Wolpoff 1980a, Smith 1985, Trinkaus and Smith 1985). When significantly earlier faunal and radiocarbon dates began to appear for modern human specimens from Israel and South Africa (Bar-Yosef and Vandermeersch 1981, Beaumont et al. 1978), plausible reasons were at first offered for rejecting them (Jelinek 1982, Trinkaus 1984a, Smith 1985, Smith et al. 1989a). But during the 1980s, the use of newer dating techniques, which could provide reliable radiometric dates for specimens as old as 200 Ky, began to reveal a different pattern in the emergence of modern human morphology across the Old World (see Mellars and Stringer 1989, Akazawa et al. 1992, Bräuer and Smith 1992, and Aitken et al. 1993). By the early 1990s, it had become clear that modern human morphology appeared much earlier in West Asia and Africa than in Europe. The currently accepted chronology indicates that modern human anatomy appeared first in East Africa, then spread to southern Africa and the Near East, and entered Europe last of all, probably after it had become established in North Africa, East Asia, and Australasia. At the moment, the earliest modern humans known from Europe date back to only ~35 Kya (Churchill and Smith 2000a, Svoboba 2000, Trinkaus et al. 2003, Conard et al. 2004, Wild et al. 2005, Rougier et al. 2007).

In Europe, the first evidence of anatomical modernity is coupled to the first clear signs of modern behavior. But this is not true anywhere else. The earliest anatomically modern specimens in Africa and the Near East occur in cultural contexts that differ little from those associated with earlier, more archaic humans in these

TABLE 8.1 ■ Indices of Robusticity (Femoral Length/ Midshaft Diameter) for Selected Late Pleistocene and Recent Human Samples

Sample	Mean	Range
Neandertals	7.1	6.8–7.4
Skhūl/Qafzeh	6.3	5.4–7.0
Early Upper Paleolithic Europe	6.3	5.6–7.0
Late Upper Paleolithic Europe	6.4	5.4–7.5
Mesolithic Europe	6.5	5.7–7.5
Medieval Hungarians	6.2	5.4–8.1

Source: Data courtesy of D. Frayer.

areas. Although there are some signs of modern behavioral innovations in Africa during the late Middle Stone Age (McBrearty and Brooks 2000, Henshilwood et al. 2002, d'Errico et al. 2005), they begin long after the first appearance of early modern human anatomy. In the Near East, the Mousterian cultural assemblages associated with early modern fossils show only minor differences from those found with local Neandertals (Bar-Yosef 2000), and they contain no unequivocal evidence for symbolic behavior. Obviously, there is no simple connection between morphological and behavioral modernity.

If Africa is indeed the cradle of modern human anatomy, there should be some evidence of an early shift toward the modern morphological pattern in the post-Erectine African fossil record, and that evidence should be earlier than other possible indications of "transitional" morphology. Does the African fossil record yield evidence of an early, distinctively transitional human population? The answer appears to be yes.

THE AFRICAN TRANSITION: BACKGROUND AND DATING

The reign of the Neandertals in Europe was underway by the beginning of OIS 6 around 200 Kya, and continued down to the onset of the Aurignacian some 35 Kya. African fossil humans overlapping the early part of this time span (~250–160 Kya) are considerably more like modern humans than Neandertals, though they retain some archaic features. These specimens, listed in Table 8.2, constitute the **African Transitional Group** (ATG). Even more modern-looking people appear in East Africa somewhat later on, though still during the period of

Neandertal ascendancy to the north. These later, more modern specimens include the >104-Ky-old Omo Kibish 1 specimen from Ethiopia (Day 1969, Bräuer 1984a) and the 160-Ky-old Herto specimens from the same country (White et al. 2003). The Klasies River Mouth (KRM) specimens from coastal South Africa have also been cited as early fully anatomically modern humans dating to ca. 100 Kya (Singer and Wymer 1982, Bräuer 1984a, 1989; Rightmire and Deacon 1991). Although their modernity has been overemphasized (Wolpoff and Caspari 1990, F. Smith 1993, 1994; Ahern and Smith 2004), the KRM fossils are certainly different from, and broadly contemporary with, European and West Asian Neandertals. The geographic distribution of all these finds is shown in Fig. 8.8. They constitute the main fossil evidence for the claim that modern human morphology first appeared in Africa (Bräuer 1984a, 1989, 1992, 2003; Klein 1999, Stringer and Andrews 1988).

The ATG (Table 8.2) is particularly crucial for this claim (Bräuer 2006, Rightmire 1978, 1984b; Smith 1985, 1993, 1997e, 2002). The anatomical features of this group are a mosaic of modern traits combined with more archaic ones resembling those of Heidelbergs or even Neandertals. The modern features mainly involve the facial skeleton, while the more primitive characteristics are found in the cranial vault. Unfortunately, faces and vaults are practically all that is known of the ATG humans. (Other ATG remains include one juvenile mandible, a few teeth, and two possible postcranial elements.) The meaning of the ATG fossils is also obscured by dating uncertainties.

As might be expected from all these question marks, the taxonomy of the ATG fossils is disputed. Members of the ATG have at various times been described as

TABLE 8.2 ■ Fossil Human Remains Assigned to the African Transitional Group (ATG)[a]

Site	Human Remains—Brief Description
Florisbad (S. Africa)	Fragmentary facial skeleton with third molar. Anterior cranial vault: anterior parietals and frontal with pronounced supraorbital torus.
Ngaloba (Tanzania) [Laetoli Hominid 18]	Complete cranial vault with supraorbital torus, lacking most of the cranial base, and a complete face (except for the teeth and zygomatic arches).
Eliye Springs (Kenya) [KNM-ES-11693]	Complete cranium with a damaged face (maxillae and anterior facial bones badly eroded) and lacking teeth.
Guomde (Kenya) [KNM-ER 3884]	Posterior cranial vault, supraorbital torus, and maxilla with a complete dentition. Also a possibly associated adult femur (KNM-ER 999).
Omo Kibish PHS (Ethiopia) [Omo Kibish 2]	Cranial vault with supraorbital torus, lacking the face and cranial base.
Jebel Irhoud (Morocco)	JI 1—Complete cranium with supraorbital torus, lacking most of the cranial base and dentition.
	JI 2—Cranial vault with supraorbital torus, lacking the face, the right half of the posterior vault, and most of the cranial base.
	JI 3—Mandible of an 8-year-old.
	JI 4—Right juvenile humerus shaft.

[a]All specimens are adults unless otherwise indicated.

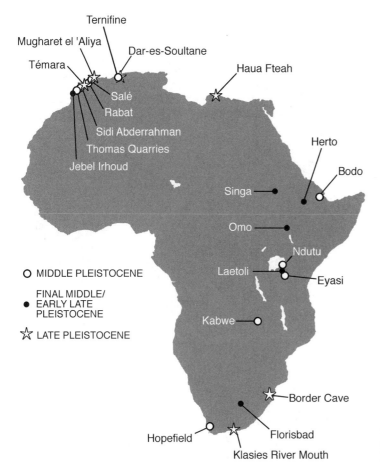

FIGURE 8.8

Map of important Middle and Late Pleistocene hominin fossil sites from Africa.

African Neandertals, archaic *Homo sapiens*, or modern humans. Some scholars have recently advocated resurrecting the taxon *Homo helmei* for this group. This taxonomic name, originally "*Homo (Africanthropus) helmei*," was proposed in 1935 by T. Dreyer for the Florisbad specimen. In 1996, C. Stringer suggested that *H. helmei*—that is, the ATG—falls in the modern human clade. However, M. Lahr and R. Foley (1998; Foley and Lahr 1997) contend that the range of *H. helmei* extended into Europe and regard it as the last common ancestor of Neandertals and modern humans.

The face and anterior vault from Florisbad, South Africa (Fig. 8.9) was the first of the ATG to be discovered. Found in 1932, the Florisbad skull was reconstructed and described by Dreyer (1935). A more recent reconstruction has been made by R. Clarke (1985a). The specimen was recovered from the eye of an ancient spring, in association with Middle Stone Age artifacts and fauna (Rightmire 1978, Clarke 1985a). Because the skull was found in the debris cone of the spring eye, its exact archaeological and geological context remain uncertain. This uncertainty has obfuscated the evolutionary position and significance of the specimen (Rightmire 1978). In 1974, R. Protsch assigned a date of 39 Kya to the Florisbad remains. This late date was based on

[14]C dates for a wooden implement from the site and an amino-acid racemization date on a hippopotamus tooth, both of which were assumed to be contemporaneous with the human skull. Protsch's date for Florisbad contributed to the impression that humans with this sort of transitional morphology persisted in Africa until relatively recent times. It provided part of the basis for the erroneous perspective on Late Pleistocene African hominin chronology and evolution proposed by Protsch (1975). However, direct ESR dating of the third molar from the Florisbad hominin has yielded an estimated age of 259 ± 35 Ky for the specimen (Grün et al. 1996). This age estimate for Florisbad fits into a recent pattern of increasingly older dates for Middle and Late Pleistocene human fossils from Africa (Grün and Stringer 1991, Smith et al. 1999, Grün 2006).

Dates and contexts are also uncertain for the other ATG members. The Eliye Springs cranium from Kenya (Bräuer and Leakey 1986) and the Omo Kibish 2 calotte from Ethiopia (Day 1969) are surface finds and lack primary geological context. If Omo Kibish 2 derives from Member 1 of the Kibish Formation, as has been argued (Butzer 1969), then it is between 196 and 104 Ky old (McDougall et al. 2005). The Guomde skull from Kenya was recovered from an area of unconformity in

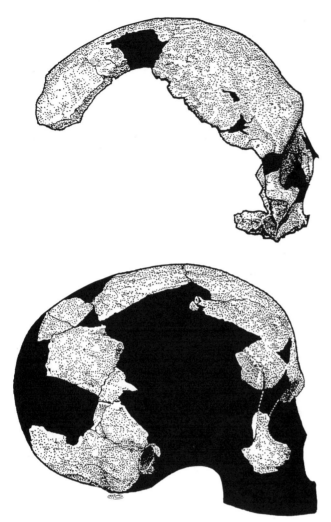

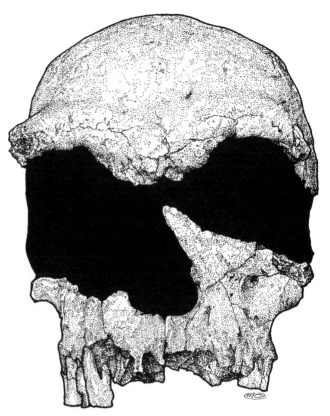

FIGURE 8.10

The Ngaloba (Laetoli Hominin 18) from Laetoli, Tanzania in frontal view. The marked supraorbital torus maintains an approximately constant thickness across the orbit. Note the relatively angled zygomaticoalveolar margins.

FIGURE 8.9

The Florisbad cranium (**top**) compared to Border Cave 1 (**bottom**), both from South Africa. A member of the African Transitional Group, Florisbad has a more developed supraorbital torus and a more receding frontal squama (forehead) than the early-modern specimen from Border Cave.

the Upper Chari Formation, where it contacts the Holocene Galana Boi Formation, and was originally thought to date to between 500 and 100 Kya (Bräuer et al. 1992). A later gamma-ray spectrometry date on the specimen resulted in an age estimate of 272 Ky +∞, −113 Ky (Bräuer et al. 1997). This estimate is compatible with the original date, but its enormous error range makes it of little value. Of the four Jebel Irhoud (JI) fossil remains from Morocco in North Africa, only one specimen (the juvenile humerus, JI 4) was recovered in a firm archaeological context, associated with the North African Mousterian (Hublin et al. 1987). The JI 1 and 2 adult skulls and the subadult JI 3 mandible may derive from the same deposits (Ennouchi 1963, 1968, 1969; Hublin

1993), but this is not certain. Assuming that all the JI specimens are of similar age, Grün and Stringer (1991) have assigned them to OIS 6—between 130 and 190 Kya—on the basis of ESR dates on mammalian teeth from the Mousterian deposits at the site. This date is supported by uranium-series and ESR dating of the JI 3 mandible itself, which yields an age of 160 ± 16 Ky (T. Smith et al. 2007). The Ngaloba or Laetoli Hominid 18 skull from Tanzania (Fig. 8.10) was recovered in general association with MSA artifacts and fauna, but there are questions about the security of the association (Masao 1992, Magori and Day 1983). The deposits from which the specimen is believed to derive have yielded age estimates of 129 ± 4 and 108 ± 30 Ky based on uranium-thorium dating of a giraffe vertebra (Hay 1987), and of 120 ± 30 Ky based on geological correlation to the radiometrically dated Ndutu Beds at Olduvai (Magori and Day 1983). More recently, amino-acid racemization dating of ostrich eggshells from the Ngaloba Beds has produced a date of ≥200 Kya for LH 18 (Manega 1993). But again, the association of the dated objects with the LH 18 specimen is uncertain. Putting together all these shaky estimates, we can say that the available evidence, such as it is, suggests that the members of the ATG are between 250 and 120 Ky in age.

THE AFRICAN TRANSITION: VAULT MORPHOLOGY

Cranial vaults in the ATG are long, relatively low, and impressively broad (Table 8.3). The mean vault-height index for this group is 56.5, almost identical to that of European Neandertals at 57.0 (Table 7.4). Viewed from the side, the frontal bones appear relatively low and flat in some ATG specimens (e.g., Ngaloba, Eliye Springs, Florisbad; see Fig. 8.9), but in others they seem more arched. For example, in Jebel Irhoud 1 and 2, the forehead is moderately steep (Fig. 8.11) and the frontal angles are 133° and 131°, respectively (Hublin 1993)—closer to those of modern humans than to those of Neandertals (Table 7.5). On the other hand, parietal angles in these two specimens are very large (150° and 145.5°, respectively: Hublin 1993), equaling or even exceeding those of Neandertals (Table 7.5). Omo Kibish 2 and LH 18 also have large parietal angles, of 147° and 143° respectively (Bräuer 1984a). Estimated cranial capacity in the ATG averages 1368 cm³ (Table 8.4)—well above the mean of the roughly contemporaneous Pre-Würm Neandertals, about equal to the overall Neandertal mean, and below the means for the Skhul/Qafzeh and European early-modern groups (Table 7.9).

The rear of the skull in the ATG is variable in some respects but more uniform in others. For example, none of the group has a suprainiac fossa, and in general they do not possess well-developed nuchal tori (Fig. 8.12). However, both Omo Kibish 2 and Eliye Springs do show remnants of the inferior torus margin. In all the ATG skulls, the maximum breadth of the neurocranium tends to fall on the lower part of the parietals, but the upper (biparietal) breadth is generally only slightly less. This

gives the skulls a contour in posterior view that is different from the oval shape of the Neandertals and looks somewhat more like (but certainly not identical to) the pentagonal silhouette typical of modern humans (Fig. 8.12). Both Jebel Irhoud specimens have occipital buns. Although the bun in JI 1 (Fig. 8.11) lacks the characteristically Neandertal lateral extension, it exhibits the lambdoidal flattening, relatively vertical upper occipital plane, and infratoral shelf that are seen in Neandertal buns. Like Neandertals, JI 1 also has a lowered anterior nuchal plane, which means that the occipitomastoid eminence is almost as inferiorly projecting as the mastoid

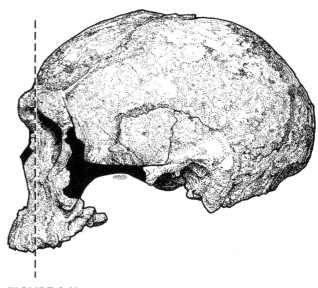

FIGURE 8.11

The Jebel Irhoud 1 cranium from Morocco in lateral view. The dashed line indicates the approximate position of the anterior braincase. Prognathism is reduced in this specimen, but it is still not fully orthognathic.

TABLE 8.3 ■ Selected Cranial Vault Measurements for ATG Specimens[a]

Specimen	Maximum Cranial Length	Maximum Cranial Breadth	Vertex Radius[b]	Minimum Frontal Breadth	Vault Height Index[c]
Ngaloba	207	132	115	103.5	56
Eliye Springs	(196)[d]	158	112	105.2	57
Omo 2	(215)[d]	145	118	108.9	55
Guomde	—	150	(128)[d]	(111)[d]	—
Jebel Irhoud 1	197	148	114	106.1	58
Jebel Irhoud 2	195	—	—	105.2	—

[a]All measurements by FHS. Omo 2 and Guomde measurements taken from casts.
[b]Vertex radius defined by Howells (1973b): see Appendix.
[c]100 × vertex radius/maximum cranial length.
[d]Both Omo 2 and Eliye Springs lack the glabellar region. Guomde does not preserve the exact point of vertex, and there are gaps between the reconstructed pieces of the anterior frontal. Thus, these measurements are estimates.

TABLE 8.4 ■ Cranial Capacity Estimates for Members of the African Transitional Group and the Earliest Modern Africans

Specimen	Group[a]	Cranial Capacity (cm³)
Ngaloba	ATG	1367
Omo 2	ATG	1435
Eliye Springs	ATG	>1300[b]
Guomde	ATG	1400[c]
Jebel Irhoud 1	ATG	1305[d]
Jebel Irhoud 2	ATG	1400[d]
Omo 1	EMA	1430
Singa	EMA (?)	1340[e]
Herto 1	EMA	1450[f]
ATG mean:		1368
EMA mean:		1407

[a]ATG, African Transitional Group; EMA, Early Modern African.
[b]From Bräuer and Leakey (1986).
[c]From Bräuer et al. 1992.
[d]From Holloway et al. 2004.
[e]From Stringer et al. (1985).
[f]From White et al. (2003).
Source: Data from Bräuer (1984a) unless otherwise specified.

process. However, the mastoid is not as strongly inclined inferomedially as in Neandertals, nor does it appear to be as small and gracile. JI 1 also lacks an anterior mastoid tubercle (Santa Luca 1978).

Some experts deny any similarity between Neandertal buns and those seen in the Jebel Irhoud skulls (Hublin 1993), while others see them as essentially similar (Bräuer 1984a). The bunning at Jebel Irhoud occurs on an overall vault form that is not like a Neandertal's, but is more like that of other ATG members to the south. T. Simmons and colleagues (Simmons and Smith 1991, Smith et al. 1995) claim that this difference in vault shape explains the differences between the JI and Neandertal buns. They suggest that the presence of a slightly modified Neandertal-style occipital bun in the JI specimens may reflect gene flow from Europe into Africa across the Strait of Gibraltar.

All members of the ATG exhibit marked supraorbital tori that are continuous across glabella, although there is considerable variation in size. A very thick torus is seen in Ngaloba (LH 18) (Fig. 8.10), Omo Kibish 2 (preserved only laterally), and Florisbad (Fig. 8.9). In the latter two specimens, the torus is not well separated from the squama and thus is not as distinctly obvious as that of Ngaloba. Jebel Irhoud 1 and Guomde have distinct tori (Fig. 8.13), but they are thinner than those

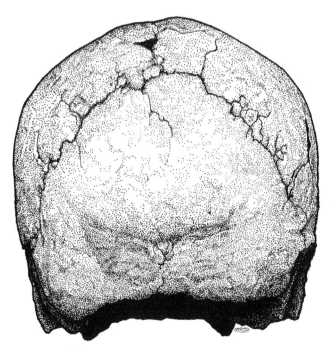

FIGURE 8.12

The Ngaloba cranium (Laetoli Hominin 18) from Laetoli, Tanzania in rear view. Note the essentially modern shape of the vault and the absence of a suprainiac fossa.

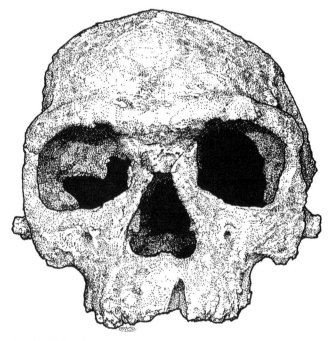

FIGURE 8.13

The Jebel Irhoud 1 cranium (Morocco) in frontal view. This skull exhibits a supraorbital torus with constant thickness across the orbits, an angled zygomaticoalveolar margin, a relatively small nasal aperture, and a short face.

of Ngaloba, Omo Kibish 2, and Florisbad. All the ATG tori are smaller than those of the African Heidelbergs, and the degree of reduction varies little across the entire torus (Smith 1992a, Smith et al. 1995). For example, the browridge thickness index for the ATG averages 87, similar to that seen in the early Neandertals, which means that most ATG members do not show the localized midorbital reduction characteristic of later European Neandertals or early modern Europeans (Table 7.19). An exception is Jebel Irhoud 2, which does overlap with the late European Neandertal form—another possible sign of gene flow across the Strait of Gibraltar. Eliye Springs would probably have had an impressive torus, judging from the thick bone at the front of the frontal squama, but the torus itself is not preserved.

THE AFRICAN TRANSITION: FACIAL MORPHOLOGY

Most of the ATG crania (Florisbad, Ngaloba, Guomde, Eliye Springs and Jebel Irhoud 1) preserve parts of the facial skeleton, but only JI 1 retains a complete face that is fully connected to the braincase (Figs. 8.11 and 8.13). Florisbad preserves much of the right side of the face, but no fully reliable measurements can be made on it and there is no clear, direct contact between the face and braincase. Ngaloba has a more complete face, but it too retains no clear connection with the braincase. Guomde's face is represented only by the alveolar parts of its maxillae and a bit of the face lateral to the nose. Although Eliye Springs has a partial face, mostly on the left, the external part is largely abraded away.

In both morphology and metrics, the known parts of ATG faces are much more like those of modern humans than those of Neandertals are. Again, only Jebel Irhoud 1 permits measurement of such dimensions as nasal and upper facial heights. These measure 53 mm and 76.8 mm respectively, both values lying about 2 standard deviations below the corresponding Neandertal means (Tables 7.10 and 7.11). A recent reconstruction of the Ngaloba cranium provides estimates of these values that are even smaller than those of JI 1 (Cohen 1996). Nevertheless, nasal-aperture breadth in the ATG averages 33 mm ($N = 3$), right at the Neandertal mean; and the ATG specimens have distinctively broad and generally flat interorbital regions, with interorbital breadth averaging 34.5 mm ($N = 4$)—much broader than in Neandertals or modern European samples (Table 7.10). These nasal traits of the ATG may be regional characteristics, since early and recent modern humans from Africa also tend to have broader and flatter noses than people from western Eurasia (Howells 1973b, White et al. 2003).

The faces of Ngaloba, Florisbad, and Jebel Irhoud 1 display canine fossae and inwardly inflected or angled zygomaticoalveolar margins (ZAM). Bräuer and Colleagues (1992) note the presence of a shallow canine fossa in Guomde as well. A canine fossa and an angled ZAM are characteristic of modern human skulls. They are not seen in Ergasters, Indonesian Erectines, Neandertals, and most Heidelbergs (including Kabwe 1). However, some other archaic *Homo*, including the Chinese Erectines, Steinheim, Ndutu, Kabwe 2, and the "*H. antecessor*" fragments from Atapuerca, show one or both of these traits. The fossa and angled ZAM therefore cannot be read simply as modern-human apomorphies. Rather, they should be taken as reflecting long-standing variability in the degree of "puffiness" or inflation of the human maxilla. But in the context of the late Pleistocene, these features are associated with modern human morphology—initially in the ATG, and subsequently with anatomical modernity in other parts of the world.

The deflation of the maxilla and the reduction of facial height are aspects of a general shrinkage of the face in the ATG as compared to more archaic humans. This shrinkage may also involve a reduction of facial prognathism, though this is less clear. For instance, although Fig. 8.9 seems to show that Florisbad is no more prognathic than the Border Cave 1 skull, the hafting of the face onto the anterior vault in Florisbad is open to alternative reconstructions, and Border Cave 1 really does not have a face at all. Again, only Jebel Irhoud 1 can provide reliable metrics.

All of the ATG specimens that preserve enough of the face to make a judgment appear to be relatively flat-faced, lacking the distinctive pattern of midsagittal facial (combining alveolar and upper facial) prognathism seen in Neandertals. Jebel Irhoud 1 has a subspinale angle of 125°—slightly below the mean of recent humans, but well above the mean of the midfacially prognathic Neandertals (Table 5.7). The question is whether the ATG morphology involves total facial prognathism, as in the Erectines and some Heidelbergs, or a pattern more like modern human orthognathism. The nasion radius (NAR) in JI 1 is 97 mm, and its NAR Index value (NAR as a percentage of cranial length) is 49. Both these values are considerably below the Neandertal mean and slightly below the early modern European means (Table 7.12). This shows that JI 1 does not have a strongly projecting face, even by modern standards, and certainly does not have the Neandertal pattern of prognathism. One reconstruction of Ngaloba suggests that this ATG specimen was more prognathic (Cohen 1996), but it is impossible to be sure since there is no bony connection between the face and cranial vault. If JI 1 is typical of the ATG, then prognathism was reduced in this transitional group—as we might expect, given the other signs of facial reduction in the ATG. However, this conclusion is based on a sample size of one.

THE AFRICAN TRANSITION: ADDITIONAL BONES, ARCHAEOLOGY, AND OTHER MATTERS

A relatively complete juvenile mandible (JI 3) and a juvenile humeral shaft (JI 4) have also been recovered from Jebel Irhoud. According to Hublin and Tillier (1981), the JI 3 specimen is from a roughly 8-year-old individual. It has large posterior teeth and a receding symphysis, but no specifically Neandertal features. Hublin and Tillier note that the specimen has a reduced retromolar space and lacks the characteristic Neandertal morphology of the condyle and surrounding area (Chapter 7). Hublin (1993) argues that JI 3 also shows incipient development of a mental trigone and other distinctive features of the modern human chin. Moreover, JI 3 appears to be the earliest hominin specimen that exhibits a pattern of dental eruption pattern and crown formation timing that specifically resembles those seen in modern humans, as contrasted to earlier members of the genus *Homo* (T. Smith et al. 2007). Thus the JI 3 mandible, like the crania from this site, presents a mosaic of archaic and modern features.

The JI 4 humerus is also rather archaic in form, with a thick shaft (and thus a reduced medullary cavity), anterior–posterior flattening of the shaft, and muscle insertions that are quite pronounced for so young an individual (Hublin et al. 1987). However, Hublin (1993; cf. Chapter 6) argues that these character states are simply primitive for *Homo*, and that neither these nor any other features of the Jebel Irhoud people represent signs of Neandertal influence.

Gamma-ray spectrometry dating has indicated a date for the KNM-ER 999 femur from Koobi Fora (Chapter 6) that is comparable to that attributed to the Guomde cranium (Bräuer et al. 1997). The Koobi Fora femur is robust but basically modern in form, with a pilaster, a modern shaft shape, and a high neck-shaft angle (Trinkaus 1993b). Its estimated length of 47–50 cm exceeds those of all archaic specimens (except Amud 1) and falls in the early-modern range. This suggests the possibility of a modern body form for the East African members of the ATG. However, this inference clashes with the more archaic morphology of the JI 4 humerus, and the large error factors in the gamma-ray spectrometry age estimate make it less than certain that ER 999 really is this early. It may also be that postcranial form, like cranial form, was variable in the ATG, exhibiting both archaic and more modern aspects.

Could the apparent shrinkage of the facial skeleton in the ATG be a byproduct of innovations in material culture that diminished the functional importance of the face and teeth? As noted above, there is no real evidence for this plausible-sounding idea. The archaeological context for the African Transitional Group is purely Middle Paleolithic—a Mousterian variant at Jebel Irhoud and the Middle Stone Age for the others. The North African Mousterian does not differ significantly from Eurasian Mousterian variants (Clark 1982). Middle Stone Age (MSA) assemblages from further south are broadly similar to the Eurasian and North African Mousterian in terms of lithic technology (tool types, use of prepared-core techniques, etc.), rarity of both formal bone tools and evidence of artistic expression, and time span (Clark 1982, 1988; Thackeray 1992, Masao 1992, Klein 1999).

Some possible innovations lacking in the Mousterian do turn up later in the MSA. These include the **Katanda** barbed points and other bone artifacts from the Congo (Yellen et al. 1995), engraved red ochre pieces and shell beads from Blombos Cave in South Africa (Henshilwood et al. 2002, d'Errico et al. 2005), stone bladelets and some modification of pigments at Pinnacle Point Cave in South Africa (Marean et al. 2007), more intensive use of aquatic resources (Klein 1999, Marean et al. 2007), and the brief existence of the advanced-looking Howieson's Poort industry (see below). But all these occur after the establishment of the "transitional" skeletal morphologies seen in the ATG. The Pinnacle Point evidence dates to only 164 Kya, and the other archaeological signs of innovation are at least 30 Ky younger (Marean et al. 2007). Whatever the technological requirements for the transition to a modern human morphology were, they must have already been present in typical Middle Paleolithic cultures. The archaeological record sheds no light on why the transition to modern human morphology began in Africa rather than Europe or Asia.

Nevertheless, the Katanda points, the Blombos and Pinnacle Point artifacts, and other early African innovations (see McBrearty and Brooks 2000, McBrearty and Stringer 2007) might be indicative of the onset of modern human behavior in Africa earlier than in other regions. The **Howieson's Poort** (HP) industry was once considered a sort of African IUP—a transitional industry bridging the gap between the MSA and the **Late Stone Age** (LSA), which is the equivalent of the Upper Paleolithic in Africa south of the Sahara (Clark 1970). Like LSA assemblages, HP lithics include backed forms that suggest hafting and are often made of raw material brought from some distance away. However, the HP lacks many other LSA characteristics, and its stratigraphic position and early dates (around 70 Kya) at several sites place it well within the span of the MSA, not at the end of it (Deacon 1989, 1995). Ambrose and Lorenz (1990) suggest that the transient innovations seen in the HP may reflect a temporary shift in subsistence ecology related to the onset of the last glacial advance. The Katanda points are very finely made, are associated with otherwise typical MSA lithic artifacts, and apparently date to between 90 and 150 Kya (Yellen et al. 1995), predating other occurrences of bone points in eastern Africa by over 40 Ky (Brooks et al. 1995). However, the animal bones from Katanda are more abraded than the bone points, suggesting transport by water (Klein 1999). Thus, the dates derived from the animal bones may not be applicable to the bone artifacts.

Taken together (and assuming the accuracy of the Katanda dating), the archaeological evidence from sub-Saharan Africa does seem to point to the emergence of some distinctively modern forms of human behavior during the later MSA. But these innovations occur within the context of the Middle Paleolithic, and they generally follow rather than precede or accompany the establishment of anatomical modernity in Africa. One possible interpretation of these facts is that Mousterian/MSA peoples everywhere were capable of such behaviors but did not realize that capability—at any rate, not in forms and media that were preserved in the archaeological record.

The so-called Amudian or Pre-Aurignacian lithic industry from western Asia provides a suggestive parallel to the Howieson's Poort phenomenon at an even earlier time horizon. The Amudian is characterized by the production of tool types (blades, burins, endscrapers, and backed knives) that are generally associated with Upper Paleolithic industries in this region (Jelinek 1982, 1992). But the Amudian is interstratified with the Acheuleo-Yabrudian, a Middle Paleolithic industry that ended at least 170 Kya (Bar-Yosef 1998, 2000), long before the emergence of the Upper Paleolithic. Like the Howieson's Poort industry, the Amudian appears to represent a transient phase of innovation in a sequence of more typical Middle Paleolithic lithic assemblages. If we regard these transient interludes as marking the onset of modern human behavior, then we are left with explaining where this behavior went after its relatively short, early appearance on the archaeological stage.

It may be, as Kuhn et al. (2004, p. 247) suggest, that the shift to Late Stone Age/Upper Paleolithic behavior was not a complete break with the Middle Paleolithic, "... but rather an extension and expansion of some subset of it" that emerged as a response to environmental changes of some sort during the late Pleistocene. If so, then the Amudian and Howieson's Poort industries may represent similar early responses to similar changes, over a shorter time span and a more circumscribed geographic range.

EAST ASIAN ARCHAIC HUMANS: BACKGROUND AND CONTEXT

Relatively few possible contemporaries of the Neandertals hail from the eastern half of Eurasia. The only candidates from Australasia and South Asia are the Ngandong, Sambungmacan, and Narmada remains (Chapter 6).

Although the Ngandong hominins have been called "tropical Neandertals" (von Koenigswald 1958), they lack many of the distinctive Neandertal features (Santa Luca 1980) and are no longer described in these terms. Mainland East Asia has yielded a few specimens that may be Neandertal contemporaries (Wolpoff et al. 1984, Pope 1992, Wu and Poirier 1995, Wu 1997), almost all of which come from China (Table 8.5). The most important of these are Changyang (found in 1956), Maba (in 1958), Xujiayao (in 1974), Dali (in 1978), and Jinniushan or Yinkou (in 1984). These Chinese fossils are generally similar to those from India and Indonesia, but they also possess some features that have been described as regionally distinctive for mainland human populations in East Asia (Wolpoff 1999).

Like the ATG fossils, many of the Chinese specimens are embroiled in contextual uncertainties. The most complete cranium is the Dali skull (Figs. 8.14 and 8.16), recovered from gravels in a terrace of the Lohe River along with Pleistocene fauna and a few nondescript artifacts (Wu and You 1979). Damage to the Dali maxilla suggests some water transport of the specimen, which always has to be reckoned with in dealing with fossils from river-terrace gravels. The association of the cranium and artifacts with each other and with the

TABLE 8.5 ■ Selected Archaic Humans from China

Site	Brief Description	Date (Kya)[a]
Changyang	Left maxilla portion with teeth	194 + 24/−20[b] 196 + 20/−17
Chaohu	Occipital and maxilla with teeth	160–200[c]
Dincun	Parietal (infant); three adult teeth	114 ± 2[d] 75 ± 0.6
Dali	Complete cranium with damaged lower face and no teeth	209 ± 23[c]
Jinniushan	Reconstructed skull and partial postcranial skeleton	200[e]
Maba	Upper face and anterior cranial vault	129–135[f]
Tongzi	Teeth	113–181[f]
Xujiayao	Cranial fragments, mandible and teeth	104–125[c]

[a]All dates in Kya. For additional details on dating and other specimens, see Wu and Poirier (1995).
[b]^{230}Th date (Yuan et al. 1986).
[c]Uranium-series dates: Chaohu (Chen et al. 1987), Dali (Chen et al. 1984).
[d]ESR dates (Wu and Poirier 1995)—also uranium-series dates (160–210: Chen et al. 1984).
[e]Based on ESR and uranium-series dates (Chen et al. 1994).
[f]Uranium-series date (Yuan et al. 1986).
Source: Data from Wu and Poirier (1995).

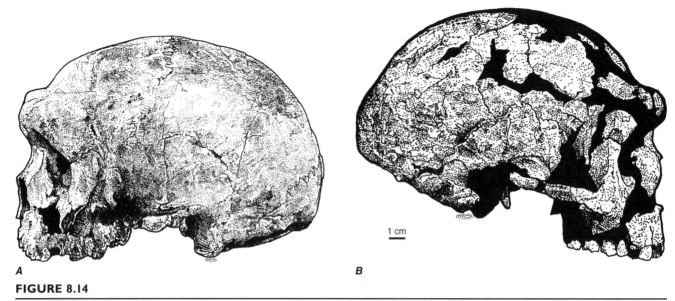

A   **B**

FIGURE 8.14

Lateral views of archaic crania from China. **A**, Dali; **B**, Jinniushan.

dated items from the site is therefore open to question. The context of the partial skeleton from Jinniushan seems more secure (Wu and Poirier 1995). It derives from a stratified limestone-cave sequence, where it was deposited in erosion pockets in older strata, along with Pleistocene faunal remains (but no artifacts). A third cranium, Maba, was also found in a limestone cave with Pleistocene fauna but no archaeological remains (Wu and Wu 1985). Tools and other clear indicators of human activity are lacking at the Chaohu and Changyang sites as well. Three other sites—Dincun, Xujiayao, and Tongzi—have yielded more fragmentary human remains from this time period in association with archaeological finds (Wu and Poirier 1995). However, even where these cultural remains are most extensive (at Xujiayao), there is no clear evidence of how the site was being used by humans.

▌EAST ASIAN ARCHAIC SITES AND SPECIMENS

Dali

Our knowledge of this cranium (Fig. 8.14) derives mainly from a series of studies by X. Wu and colleagues (Wu 1981, 1997; Wu and Wu 1985, Wu and Poirier 1995, Wolpoff et al. 1984). The braincase is long, broad, and strikingly low, with a basion–bregma height of 118 mm (Wu 1981)—below the means for the Ngandong and Neandertal samples. The frontal squama is strongly receding, albeit with a distinct frontal boss. The occipital is strongly angled but has a relatively even sagittal curvature. The bones of the vault are very thick. Cranial capacity is only 1120 cm³, slightly below the Ngandong mean (Chapter 6). Despite its rather primitive vault, Dali exhibits indications of braincase expansion of the sort

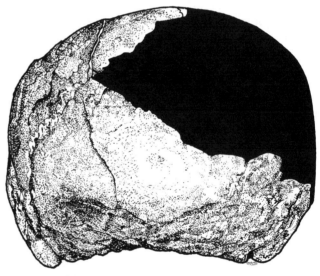

FIGURE 8.15

Dali (China), rear view. Maximum cranial breadth falls on the cranial base, at the level of the supramastoid ridges, although the biparietal breath is only slightly less.

seen in other post-Erectine *Homo*. The minimum frontal breath is expanded, the occipital plane is larger than the nuchal plane, and the maximum breadth of the braincase lies up on the parietals rather than down on the cranial base (Fig. 8.15). The posterior vault does not exhibit features characteristic of occipital bunning, and the mastoid process is relatively expanded and projecting. The nuchal torus is weakly developed and restricted to the medial aspect. Wu identifies Dali as the skull of a male around 30 years old.

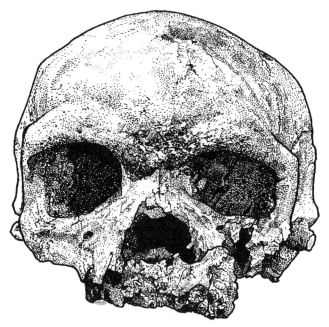

FIGURE 8.16

Dali (China), frontal view. The specimen exhibits a well-developed supraorbital torus, a broad interorbital area, and an apparently broad nose. Distortion to the left side and lower face is due to upward and posterior displacement of the palate and lower face. The right side of the specimen seems to show a horizontal orientation of the lateral part of the zygomaticoalveolar margin.

Viewed from the front (Fig. 8.16), Dali sports an impressive supraorbital torus, arched over each orbit and separated from the forehead by a sulcus. The torus is strongly projecting and very thick, thinning only slightly laterally. Below the torus, Dali has a broad, relatively flat face. The exact nature of the prognathism in the specimen is difficult to determine because of the damage to the lower part of the face, which has been mashed up into the middle of the face. The damage is most evident on the left, but the entire lower face is distorted. The Dali face nevertheless appears to have had at least a moderate degree of total facial prognathism. The nose opening and interorbital region are broad. Upper facial height and other facial lengths cannot be estimated, due to the facial damage. Pope (1991, 1992) points to the existence of a maxillary "notch" and a zygomaticoalveolar margin (ZAM) that seems to form a distinct angle between the alveolar and cheek regions. These features appear to be modern in a general sense, and are not unlike their counterparts in the African Transitional Group.

Jinniushan

This partial skeleton is probably that of a young adult female. It was first studied by R. Wu (1988) and subsequently by other researchers (Wolpoff et al. 1984, Wu and Wu 1985, Wu and Poirier 1995, Rosenberg 1998, Rosenberg et al. 2006). The skull is much more gracile than Dali's, with a higher, rounder cranial vault, thinner vault bones, smaller mastoids, and a much thinner supraorbital torus. Estimated cranial capacity is 1390 cm^3 (Wu and Poirier 1995, p. 122). The face is also gracile, but resembles Dali's in appearing relatively flat (total facial prognathism), with a broad nasal and interorbital area. According to Pope (1991), the infraorbital anatomy is advanced, again like Dali's. The postcranials comprise a left ulna and hipbone, several vertebrae and ribs, a patella, and 44 hand and foot bones (Wu and Poirier 1995). The ulna is relatively gracile, with smallish articular surfaces but with thick diaphyseal walls. Its dimensions yield a stature estimate of 165 cm. Rosenberg (1998) notes that the hipbone is large, with a large acetabulum. The superior pubic ramus is elongated and thinned, approaching the form seen in some Asian Neandertals (Fig. 7.25). Rosenberg et al. (2006) show that this specimen also resembles Neandertals in having a wide trunk, a large body mass (78 kg), and short limbs. From the site's location relatively far north in China, they infer that Jinniushan represents a cold-adapted population with an overall body form much like that of the similarly adapted European Neandertals.

Maba

Originally known as Mapa, this anterior vault and upper face was the first post-Erectine archaic skull reported from China (Woo and Peng 1959). It was initially thought to show strong similarities to Neandertals (see also Howells 1959 and Coon 1962), probably in part because the upper nose seemed to be more strongly projecting than did the remnant of the lateral face. More recent authors have stressed the forward orientation of the infraorbital surface and other features that relate this specimen to Dali (Pope 1992, 1997b; Wu and Poirier 1995, Wolpoff 1999). The vault has a thick, projecting supraorbital torus that arches over the orbits, a supratoral sulcus, and a receding forehead with a pronounced boss. Maba appears to be significantly more recent than the Jinniushan and Dali crania.

Other Cranial Pieces

Other, equally early sites in China have yielded more fragmentary cranial remains. These include occipital and maxillary fragments from Chaohu and a maxillary fragment from Changyang. The Chaohu occipital is broad and appears less angled than those of Chinese Erectines. Nevertheless, it has a distinct nuchal torus and a possible suprainiac fossa (Wu and Poirier 1995). The maxillae of these specimens are relatively small fragments, but Chaohu seems to have had a broad nose, and Changyang's alveolar process was apparently quite prognathic. Somewhat more recent cranial finds, more similar to Maba in age, include a small piece of relatively undiagnostic infant parietal from Dincun and some 15 cranial pieces from Xujiayao. The sample from the latter site is one of the largest from China, but unfortunately the specimens are very fragmentary. M. Wu (1980, Wu and Wu 1985) thinks that the Xujiayao parietal pieces suggest low vaults but lack the angular tori and other features characteristic for Erectines. The occipitals, though thick, are also less angled than those of Erectines. A complete temporal bone has a small mastoid and weak muscle markings. The only mandibular ramus recovered from Xujiayao is relatively broad.

Dentition

Teeth associated with this fossil sample derive mainly from Dincun, Tongzi, Jinniushan, Chaohu, Changyang, and Xujiayao (Zhang 1986, 1989; Wu and Wu 1985, Wu and Poirier 1995). All the teeth, both anterior and posterior, are generally similar in size to those of Erectine samples. The upper incisors are shovel-shaped and follow the East Asian pattern of shoveling as defined by Crummett (1994, 1995)—that is, they have straight incisive edges, marginal ridges, and a lingual tubercle at the base of the crown with finger-like projections extending above it (Fig. 5.14).

EAST ASIAN ARCHAICS—CONTINUITY OR SOMEONE NEW?

Advocates of Multiregional Evolution (MRE) contend that the East Asian remains exhibit several features supporting extensive regional continuity between archaic and modern human populations in this part of the world (Frayer et al. 1993, Pope 1988, 1991, 1992, 1997b, Thorne and Wolpoff 1992, Wolpoff 1989, 1999, Wolpoff et al. 1984, 2001, Wu and Poirier 1995, Wu 1997). For example, G. Pope (1991, 1992) identifies a strong malar notch as a distinctive East Asian feature. M. Wolpoff (1999, p. 577) lists the following character states as distinctly "Asian features" that link Jinniushan (and to some extent Yuxian) with a Zhoukoudian female:

▶ Arched supraorbitals with a sulcus separating them from the frontal squama.

▶ A tall, narrow keel on the frontal bone.

▶ A transversely flat, vertically oriented face.

▶ Marked facial breadth.

▶ A horizontal suture between the top of the nasal bones and the frontal bone.

▶ A high position for the malar notch.

▶ Incisor shoveling of the East Asian type (Fig. 5.14).

▶ M3 reduction.

However, others argue that these and other supposed "East Asian" features are either primitive retentions or are invalid as markers of regional continuity for other reasons (Bräuer 1989, Lahr 1994, 1996; Bräuer and Stringer 1997, Stringer and Bräuer 1994). In a detailed analysis of nine features that have been touted as evidence of regional continuity in East Asia (Table 8.8), Habgood (1992, 2003) finds that none of them qualify as distinctive regional traits, because they are either common in populations outside East Asia or are not ubiquitous enough in East Asian samples. We will return to Habgood's argument later.

The origin of the post-Erectine archaic humans from China is an important issue that has received little attention. They differ from local Erectines in many of the same ways that post-Erectine humans differ from their predecessors in other parts of the Old World. These shared differences must represent evidence of either (1) extensive gene flow connecting all the post-Erectine archaic populations from Africa over to China, or (2) a replacement event in which immigrating Heidelberg-grade archaic humans from elsewhere supplanted the Chinese Erectines, or (3) evolutionary parallelism in different regions of the Old World. Scholars who posit multiple speciation events during the evolution of Pleistocene *Homo* prefer the third explanation for the advanced-looking features of Neandertals (Chapter 7) and the Ngandong hominins (Chapter 6), but they have had little to say about the post-Erectine archaics from China. Johanson and Edgar (1996) suggest that Dali might represent a spread of Heidelbergs from western Eurasia into East Asia. However, they also make the surprising suggestion that Dali—perhaps the most archaic of the Neandertal contemporaries from East Asia—belongs to *Homo sapiens,* whereas other Heidelbergs and Neandertals are placed in separate species (Johanson and Edgar 1996). It is not clear why they assign Dali to *Homo sapiens,* or what this assignment is supposed to imply for the origin of *H. sapiens* from *H. heidelbergensis.*

In short, both the taxonomy and phylogenetic position of Dali and the other archaic post-Erectine humans from East Asia remain obscure. This obscurity reflects their rather scanty record and their problematic dates. It will probably not be cleared up until we have more and better-dated fossil finds representing this chapter of the human evolutionary story.

EARLY MODERN HUMANS: THE EAST AFRICAN RECORD

Two sites in Ethiopia have yielded the earliest hominin fossils that exhibit a distinctively modern, albeit robust, morphology. One of these sites is Locality KHS from the Omo Kibish Formation in southern Ethiopia. In 1967, this site yielded a single fragmentary cranium, mandible, and postcranial skeleton of an adult individual (Day 1969). The hominin specimen was found partially *in situ* (Butzer 1969) and was associated with lithic debris and fauna, neither of which is diagnostic (Howell 1978). A uranium-thorium date of 130 Kya was reported for Member 1 of the Kibish Formation (Butzer et al. 1969), but this date was on mollusk shell and its accuracy was later called into question (Smith 1985, 1993; Smith et al. 1989a). More recently, McDougall et al. (2005) have reported ^{40}Ar/^{39}Ar dates of 196 Kya from Member 1 below the level where the Omo Kibish 1 specimen was recovered, and of 104 Kya from the overlying Member 3. McDougall and colleagues argue that these deposits accumulated quickly (cf. Fleagle et al. 2002). They infer from this that the Omo Kibish 1 specimen must be close to 196 Ky old, because it was found just above the lower dated tuff. Estimates based on relative speed of accumulation of sediments are notoriously imprecise, but it is now clear that the Omo Kibish 1 specimen dates in excess—and perhaps well in excess—of 104 Kya.

The second site, Herto, is located in the Middle Awash of northern Ethiopia and was excavated in 1997. The Upper Herto Formation has yielded a series of hominin fossils dated by ^{40}Ar/^{39}Ar to a time range between 154 and 160 Kya (Clark et al. 2003, White et al. 2003). These finds comprise three skulls: a complete cranium of a robust adult, Herto 1 (BOU-VP-16/1); a less complete but even more robust adult, Herto 2 (BOU-VP-16/2); and a 6- to 7-year-old child's cranium, Herto 5 (BOU-VP-16/5). No postcranial remains or mandibles have been reported from these deposits. The hominin specimens are associated with a lithic assemblage containing both Acheulean and Middle Stone Age elements.

The Omo Kibish 1 skull retains only a few strategic pieces of the face and anterior braincase, but the rear of the vault is much more complete (Fig. 8.17). This rear portion shows that Omo 1 had projecting mastoid processes and a high cranial vault with contours that appear evenly rounded in lateral view. The maximum breadth of the vault is located high on the parietals, giving the specimen the typically modern house-shaped (*en maison*) vault form when viewed from the rear (Fig. 8.18). There is no indication of a nuchal torus or a suprainiac fossa, and the occipital plane is larger than the nuchal plane. Although the fragments of the frontal bone do not fit together precisely, the forehead appears to be relatively high. Cranial capacity is estimated at 1430 cm^3 (Table 8.4). Small pieces of face and mandible suggest an angled ZAM (zygomaticoalveolar margin) and the presence of a mental trigone and canine fossa. The brow ridges project noticeably, but they are not very thick and do not exhibit the distinct midorbital thinning characteristic of early modern specimens from Europe. Both descriptive and multivariate analyses of the Omo Kibish 1 skull (Day 1969, 1986a; Day and Stringer 1982, Bräuer 1984a, Habgood 2003, Schwartz and Tattersall

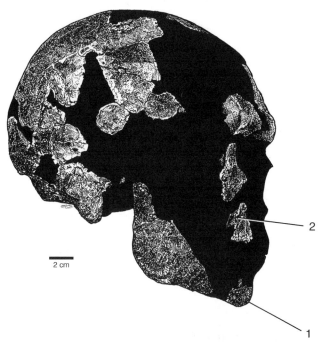

2 cm

FIGURE 8.17

The fragmentary Omo Kibish 1 skull (site KHS) from southern Ethiopia, lateral view. Although incomplete, Omo 1 exhibits modern-like cranial contours and a distinct chin (1). A small portion of maxilla suggests the presence of a canine fossa (2). (After Johanson and Edgar 1996.)

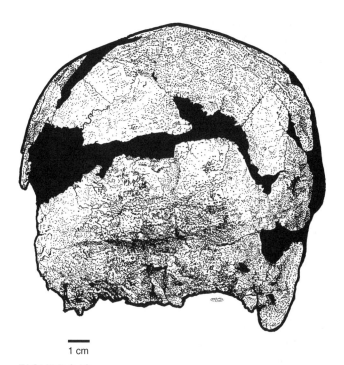

1 cm

FIGURE 8.18

The Omo Kibish 1 skull (site KHS), rear view. Note the distinctly modern shape of the vault from this perspective. (After Schwartz and Tattersall 2003.)

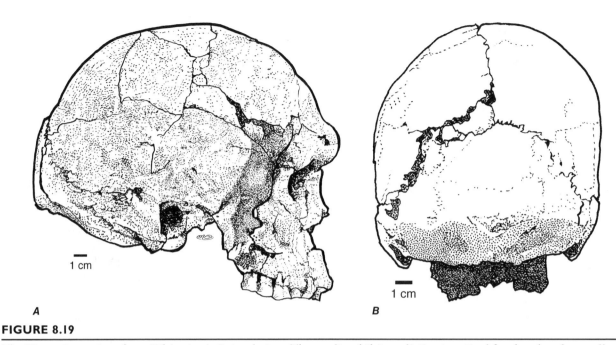

A

B

FIGURE 8.19

The Herto 1 cranium from Ethiopia. **A**: Lateral view. The vault exhibits a distinct vertical forehead and overall modern shape. The face is orthognathic and located totally under the braincase. The most archaic feature is the continued presence of a well-developed supraorbital torus. **B**: Semidiagrammatic posterior view, showing the modern human shape of the vault. (After White et al. 2003.)

2003) establish the modern form and relationships of this specimen, although the study by Day and Stringer does indicate close multivariate ties with Omo Kibish 2, which has a morphologically more primitive vault. The postcranial pieces of Omo Kibish 1 are very fragmentary, and only the femur has been analyzed. Kennedy (1984) shows that in all significant features, such as cortical bone thickness and possession of a pilaster, the Omo Kibish 1 femur falls squarely in the modern human range.

The Herto 1 specimen (Fig. 8.19) is a more complete cranium, though slightly distorted. It has a high cranial vault, with projecting mastoids and a moderately steep frontal (White et al. 2003). The occipital, however, is markedly angular. In fact, the occipital angle is 103°— more acute than most modern humans or even Neandertals (Table 7.5). Nevertheless, the occipital morphology is not equivalent to an occipital bun. There is a massive external occipital protuberance (a modern human feature), and there may be a form of suprainiac fossa that is common in early modern Europeans. The face is not prognathic, but there is a well-developed supraorbital torus that projects moderately and is thicker than that of Omo Kibish 1. Like that of Omo Kibish 1, the torus retains a relatively even thickness across the orbit, unlike the tori in Neandertals or the brow ridge form of early modern Europeans. White and colleagues also note that Herto 1 has a distinct canine fossa, an angled ZAM with a deep malar notch, and a para-coronally oriented infraorbital area. Viewed from the rear (Fig. 8.19B), the specimen is obviously distorted but still clearly exhibits a modern (house form) shape. Cranial capacity is slightly larger than Omo 1 at $1450\,cm^3$ (Table 8.4). The more fragmentary Herto 2 skull is more robust than Herto 1 (White et al. 2003).

The overall gestalt of the Herto and Omo Kibish 1 specimens clearly supports their designation as early modern humans. Like the earliest moderns in West Asia, these early modern Africans retain few primitive traits apart from the supraorbital torus. Although some of their other features, like the occipital morphology of Herto 1, are certainly robust, they are not specifically archaic. Otherwise, the salient features of these specimens fit the pattern seen in recent modern skulls. This is underlined by the third skull from Herto, the 6- to 7-year-old Herto 5 child's cranium (White et al. 2003). This specimen lacks the cranial superstructures seen in the adults, as would be expected for a child of this age. It has an undistorted globular vault with a high forehead, a rounded sagittal occipital contour, and a modern *"en maison"* shape from the rear. Herto 5 shows that a modern human cranial form was established early in ontogeny in these early Ethiopians.

As White and colleagues note, there are no features that clearly align these specimens specifically with recent Africans. Their closest similarities seem to be with other early modern specimens from the Near East and the members of the African Transitional Group (Stringer 2003). This pattern of affinities strengthens the argument for an early emergence of the modern human anatomical form in Africa, with recognizable roots extending back well over 200 Ky into the ATG.

Four other East African sites provide further evidence for modern people at an early date in this part of the world. A partial mandible from Diré Dawa (Port Epic) in Ethiopia is dated to >60 Kya on the basis of obsidian hydration and association with Middle Stone Age artifacts (Clark 1988). Small and badly eroded, the Diré Dawa mandible has been described by Bräuer (1984a) as a mosaic of modern features (vertical symphysis, no retromolar space) and archaic traits (absence of a chin). From the Mumba rock shelter in Tanzania come three isolated molars, which have been uranium-series dated to 130–109 Kya and described as modern in size and morphology (Bräuer and Mehlman 1988). Two more Ethiopian sites, Bouri and Aduma, have yielded cranial-vault bones that appear to be modern (Haile-Selassie et al. 2004a), although Trinkaus (2005) identifies a suprainiac fossa in the Abuma 3 occipital.

A more informative but uniquely problematic early-modern skull was recovered in 1924 from Singa in the Sudan (Vallois 1951). ESR analysis of two teeth from the site (Grün and Stringer 1991) has produced age estimates of 97 Ky (early uptake) and 160 Ky (linear uptake). There is also a uranium–thorium date of 133 ± 2 Kya on matrix adhering to the specimen (McDermott et al. 1996), which is generally regarded as a minimum for the skull. First described as an "ancestral Bushman" (Woodward 1938), the Singa specimen is a complete calvaria with some damage to the base (Stringer 1979, Stringer et al. 1985, Schwartz and Tattersall 2003) and an estimated cranial capacity of 1350 cm³ (Table 8.4). Like the Herto and Omo Kibish 1 skulls, Singa has a rather thick, protuberant supraorbital torus along with a more modern-shaped vault sporting large mastoid processes and prominent parietal bosses. It has long been conjectured that some aspects of its form may have a pathological basis (Brothwell 1974). One unusual pathology of the Singa skull is the absence of the bony labyrinth in the right temporal bone, probably as the result of an acoustic neuroma (Spoor et al. 1998). It is not clear how the pathology of the Singa specimen may have affected its morphology. But the discovery of similar, nonpathological skulls of comparable antiquity at Herto and Omo Kibish suggests that it deserves to be counted as a third very early modern cranium from Africa.

OUT OF [EAST] AFRICA: EARLY MODERN PEOPLE IN NORTH AND SOUTH AFRICA

Although the oldest fossils of modern humans now known are from East Africa, earlier arguments favoring

Africa as the locus of modern human origins pointed to South Africa as the source. Such arguments began in the late 1970s with P. Beaumont's claim that remains from Border Cave in South Africa established the presence of modern humans there more than 50 Kya (Beaumont 1980, Beaumont et al. 1978). During the 1980s, scientific attention shifted to the coastal South African site of Klasies River Mouth Caves, which has yielded even older specimens described as anatomically modern (Singer and Wymer 1982, Bräuer 1984a, Rightmire and Deacon 1991, Deacon and Shuurman 1992, Klein 1999). Isolated teeth and bony fragments from several other early sites in southern Africa (Table 8.6) have also been described as fully modern. The most notable are Equus Cave (Grine and Klein 1985), Sea Harvest (Grine and Klein 1993), Die Kelders (Grine et al. 1991), and Hoedjies Punt (Berger and Parkington 1995). While these fragmentary finds generally fall into the modern range of variation, they do not preserve much diagnostic anatomy. The key South African specimens are those from Border Cave and Klasies River Mouth.

The Border Cave cranial and mandibular specimens clearly conform to a modern human pattern (De Villiers 1973, 1976; Beaumont et al. 1978, Rightmire 1979b, 1984b; Bräuer 1984a, 1989; Smith 1985, Smith et al. 1989a, Stringer 1989, Stringer and Andrews 1988, Schwartz and Tattersall 2003). The Border Cave 1 skull is fragmentary but has a high forehead and rounded cranial contours (Fig. 8.9). Although little is left of the face, the preserved right zygomatic indicates a modern facial form. The only archaic-appearing feature of Border Cave is a moderately developed supraorbital torus, which maintains a relatively even thickness across the orbits (Smith 1992a). The adult Border Cave 2 and 5 mandibles are fully modern in size and shape and have chins defined by distinctive mental trigones. The Border Cave 3 infant skeleton also appears to be anatomically modern. However, its proximal ulnae retain an archaic morphology that resembles the Neandertal pattern described in Chapter 7 (Rightmire and Deacon 1991, Morris 1992, Pearson and Grine 1996). Its humeral shaft is robust, but within the range of modern variation (Pearson and Grine 1996).

Few dispute the modern morphology of the Border Cave remains. Although the dates associated with the MSA artifacts from the cave are extremely early, stretching back to ~195 Kya for the lowest deposits (Butzer et al. 1978), they too are not generally questioned. The problem here is the context of the human fossils themselves. None of the Border Cave specimens were recovered under controlled excavation conditions (Klein 1983, 1989). Border Cave 1 and 2 were found by guano prospectors in 1941–1942 (Cooke et al. 1945), and their stratigraphic context is uncertain. Border Cave 5 derives from a pit in the lower MSA deposits, and the Border Cave 3 infant was an apparently intrusive burial

TABLE 8.6 ■ Early Modern Human Sites and Specimens from South and North Africa

Site (Country)	Human Remains	Date (Kya)	Dating References
Border Cave (South Africa)	Adult cranium, infant skeleton, 2 adult mandibles, humerus, ulna, 2 metatarsals	>49 (^{14}C) 55–80 (ESR) 47–121[a] 74 ± 5	Beaumont et al. (1978) Grün and Stringer (1991) Miller et al. (1993) Grün et al. (2003)
Dar-es-Soltane II (Morocco)	Adult anterior skull (Dar-es-Soltane 5), child's skull, adolescent mandible	34–127[b]	Debénath et al. (1986), Hublin (1993)
Equus Cave (South Africa)	Mandibular fragment, 10 unassociated teeth	>27–71[c]	Klein (1999)
Haua Fteah (Libya)	Two mandible fragments	>47 (^{14}C) 30–150[d]	McBurney (1967)
Hoedjiespunt (South Africa)	Cranial fragments, teeth, postcranial fragments	71–300[e]	Berger and Parkington (1995); Stynder et al. (2001)
Die Kelders Cave I (South Africa)	Isolated teeth (>20), 2 hand phalanges	45–71[c]	Grine et al. (1991)
Klasies River Mouth Caves (South Africa)	Five partial mandibles; zygomatic; fragments of 2 maxillae, temporal, supraorbital region, and other cranial bones; isolated teeth; several postcranial pieces, including a radius and ulna.	60–115[c]	Grün and Stringer (1991), Deacon and Shuurman (1992), Klein (1999)
Mugharet el 'Aliya (Morocco)	Subadult maxillary fragment	34–127[b]	Debénath et al. (1986)
Nazlet Khater (Egypt)	Two partial skeletons	37.6 + 0.35/−0.31[f] 38 ± 6[g]	Vermeersch (2002) Crevecoeur (2007)
Sea Harvest (South Africa)	Premolar and phalanx	>40–127[h]	Grine and Klein (1993)
Taramsa Hill (Eqypt)	Juvenile partial skeleton	~50–80[i]	Vermeersch et al. (1998)
Témara (Smuggler's Cave) (Morocco)	Frontal, parietal, occipital and mandible fragments	34–127[b]	Debénath et al. (1986), Hublin (1993)
Zouhrah Cave (Morocco)	Partial mandible, canine	34–127[b]	Hublin (1993)

[a]These dates are based on isolucine epimerization of ostrich egg shell. The dates from two lower levels are 106 Ky and 145 Ky. Miller et al. (1993) believe that the actual age lies between these extremes, so the average is used here.
[b]These dates are based on association with the Aterian industry (Klein 1999). There are no chronometric dates from this site. t is not possible to determine where in this time range the human skeletal material falls, and the upper end of the range may be too old.
[c]Dated by ESR, faunal associations, and geology.
[d]Dated by association with the North African Levalloiso-Mousterian industry (Klein 1999).
[e]Dated by uranium series, faunal association, and geology.
[f]AMS [14]C.
[g]ESR date.
[h]Dated by ^{14}C and geology.
[i]OSL (optically stimulated luminescence) date.

in the same deposits. The adult postcranial specimens were collected from the dump left by the guano collectors (Morris 1992). Although Beaumont (1980) argues that matrix on some of the Border Cave specimens suggests that they come from the lower MSA levels, the human bones are more complete than the animal bones from the same levels and appear to have had a different taphonomic history (Klein 1983). The possibility that the human bones may be intrusive burials is supported further by an analysis of their crystallinity. Bone crystallinity tends to increase with age, and so it ought to be similar in bones of similar

age from the same deposit. The analysis by Sillen and Morris (1996) supports an MSA-level age for the humerus and ulna—which is consistent with these specimens' somewhat archaic morphology—but not for the other Border Cave fossils. Although ESR dating of dental enamel from Border Cave 5 yields an age of 74 ± 5 Ky (Grün et al. 2003, Grün 2006), this would still make the Border Cave remains considerably younger than Herto and Omo Kibish 1, and somewhat younger than even the samples from Klasies River Mouth and Skhūl/Qafzeh. Their relatively modern morphology is therefore not surprising.

The stratigraphic context and dating of the Klasies River Mouth (KRM) human remains are more reliable. These specimens (and the MSA cultural remains associated with them) were scientifically excavated, and the stratigraphic correlations among the site's various caves and rock shelters appear firm (Singer and Wymer 1982). Most of the human skeletal specimens come from the SAS member of the cave deposits, which overlies an older stratum (the LBS member) that has yielded uranium-disequilibrium dates of 98 and 110 Kya for its upper levels (Deacon 1989). Luminescence dating indicates an age of ≤130 Ky for the LBS member (Feathers 2002). The available data bracket the age of most of the KRM hominins between 60 and 110 Ky, apart from two maxillary fragments that come from older, underlying deposits (Deacon and Shuurman 1992).

The human fossils from Klasies River Mouth are fragmentary; in fact, there are no complete skeletal elements. White (1987) makes a strong case that the remains were processed for food, which would explain their fragmentation. Nevertheless, some of these bits and pieces retain potentially diagnostic morphology. The most diagnostic postcranial element from KRM is a well-preserved proximal ulna. This bone is archaic in the cross-sectional geometry of its shaft and its anterior-facing trochlear notch (as in Neandertals and the Border Cave ulna), although similar morphology is found in some early moderns from western Eurasia (Churchill et al. 1996).

The supposedly modern morphology is seen in the cranial pieces. These are particularly fragmentary, and most of them are not very informative. A small supraorbital fragment (from just above the nose and medial orbits) and a virtually complete right zygomatic bone convey some hints about KRM facial morphology. The zygomatic (KRM 16651) has been described as modern (Singer and Wymer 1982, Rightmire and Deacon 1991, Bräuer and Singer 1996), and it does appear to have had a horizontal lateral ZAM (zygomaticoalveolar margin), as in recent humans. But so do the members of the African Transitional Group. Other studies (F. Smith 1993, 1994; Janković and Smith 2005) show that the frontal process of this zygomatic, which forms the lower lateral margin of the orbit, has a thick, columnar cross section as in archaic *Homo* (Fig. 8.20). In this feature, the KRM zygomatic falls solidly within the

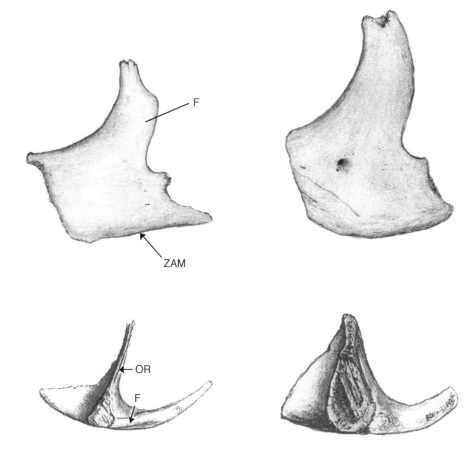

Recent   KRM 16651

FIGURE 8.20

The Klasies River Mouth left zygomatic bone KRM 16651, compared to a recent human specimen: facial surfaces (above) and vertical views (below). Note the distinct orbital (OR) and facial (F) plates in the recent specimen. By contrast, the post-orbital process of KRM 16651 exhibits a columnar-like form in which the area between these plates is filled in. The lateral aspect of the zygomaticoalveolar margin (ZAM) is approximately horizontal in both specimens. (Drawing by M. O. Smith.)

range of Neandertals (Janković and Smith 2005). Although KRM 16651 is no Neandertal, its morphology suggests that it is not fully modern either. This zygomatic is also quite large, comparable in size to those of such archaic African specimens as Kabwe 1 and Bodo (Smith 1993).

The supraorbital fragment, KRM 16425, evinces the broad interorbital region characteristic of African late Pleistocene specimens. Unlike earlier African archaics, it lacks a supraorbital torus (Smith 1993), which is the main reason for calling it "modern" (Rightmire and Deacon 1991). However, it is not certain that the specimen is fully adult. The development of the frontal sinuses in KRM 16425 suggests that it represents an older subadult rather than a fully adult individual (Ahern and Smith 2004). Adolescent Neandertals also lacked a strongly developed supraorbital torus, as the skull of the 14-year-old from Le Moustier demonstrates. If KRM 16425 represents a specimen at the same ontogenetic stage as Le Moustier 1, it might have developed a more pronounced torus later in life. A temporal fragment from the KRM site is generally modern in form (Grine et al. 1998), but the medial wall of its glenoid (mandibular) fossa is formed entirely by the temporal bone with no sphenoid contribution, as in Erectines and Neandertals (Smith 1976, Stringer 1984).

The key elements in the KRM sample are the five mandibles, all of which have been described as fully modern (Rightmire and Deacon 1991). However, some of them are not. The most striking thing about them is their relatively small size. The KRM 16424 mandible, the only one lacking a symphysis, is particularly small (Lam et al. 1996). The most commonly pictured KRM mandible is 41815, which has a distinct mental trigone and wholly modern symphyseal morphology. But the other three symphyses (KRM 21776, 13400, 14695) are less vertical, and only 21776 shows any clear indication of a mental trigone. These three symphyseal profiles do not differ much from those of some Neandertals (Fig. 7-38). Again, these fossils are certainly not Neandertals— but they are not modern humans, either (Lam et al. 1996). As one of us has previously noted (Smith 1994), it would be difficult to find a recent African sample in which 50% of the individuals lacked a chin structure.

Taken as a whole, the entire KRM sample actually appears less modern than the older fossils from Herto and Omo Kibish to the north. That fact requires some explanation. We believe that the KRM sample is the product of intermixture between local late-archaic people (resembling the ATG) and more fully modern populations moving southward out of East Africa (Smith et al. in press). Trinkaus (2005) has offered a similar interpretation. Increased variability is characteristic of "hybrid" samples (Ackerman 2006) because they sample more allelic variants than either parent stock, and hence

display a greater number of different genotypes and phenotypes in the F2 and subsequent generations. We suggest that the unusual degree of variability seen in the KRM sample reflects its "hybrid" status.

Similar things appear to have been going on in North Africa, where early modern specimens have been recovered from five sites (Table 8.6). It is difficult to establish the maximum age of any of these specimens, but those associated with Middle Paleolithic (Aterian or Levalloiso-Mousterian) tools could in principle date back as much as 130 Kya (Klein 1999). The sole certainty is that all of them are older than 30 or 40 Ky (Table 8.6). The only ones associated with Mousterian artifacts like those from Jebel Irhoud are two posterior pieces of mandible from Haua Fteah, Libya (McBurney 1967), which may or may not be from the same time range as Jebel Irhoud. The initial descriptions of the Haua Fteah fossils reported a relatively archaic morphology (Tobias 1967b), but neither specimen has any of the supposedly distinctive traits of Neandertal mandibles (coronoid notch, incisural crest, retromolar space, and so on) discussed in Chapter 7.

The other early modern fossils from North Africa are associated not with the Mousterian but with the Aterian, a Middle Paleolithic variant characterized by various tanged pieces and dating to roughly the same time span as the North African Mousterian. The most complete of these fossils is an anterior cranial vault and mandible from a cave in Morocco. This specimen, Dar-es-Soultane 5, is robust and broad but exhibits a basically modern form, reflected in its relatively high forehead, angled ZAM, large and robust mastoid process, and orthognathic face (Ferembach 1976a). According to Hublin (1993, 2000), the mandible is also robust but modern. Dar-es-Soultane 5 has a moderately developed supraorbital torus with a relatively even thickness across the orbits, as in other early modern Africans (Smith 1992a). Other Aterian-associated specimens come from Temara, also in Morocco. More fragmentary than Dar-es-Soultane 5, they include a modern-looking frontal and occipital (Ferembach 1976b). Unlike the other African specimens discussed previously, the Temara frontal has a lateral supraorbital region of the modern type, with a flattened supraorbital trigone (Hublin 1993). The associated mandible is equally modern, though it does not exhibit strong development of a chin.

The final specimen associated with the Aterian is the left side of an approximately 9-year-old child's maxilla from the Mughuret el 'Aliya near Tangier, Morocco. Its morphological pattern is quite different from that seen in the other Aterian-associated fossils. The specimen, also called the Tangier maxilla, preserves only the alveolar part of the maxilla, from the approximate midline at the front to the mesial part of the crypt for the developing permanent M2 at the rear. The permanent canine and premolars are also present in their crypts. Discovered by C. Coon in 1939, the Tangier maxilla was first described by M. Şenyürek (1940) as a North African representative of the Neandertals. Its Neandertal affinity was reaffirmed by a subsequent study (Myster and Smith 1990). Minugh-Purvis (1993) has argued that the Neandertal-like features of the specimen were due to its young age. However, it is much closer to comparable subadult Neandertals than to modern human children in some key features, including the very large size of its alveolar process, the orientation of its infraorbital region, and its lack of a canine fossa (Hutchinson 2000). Its teeth are also relatively large.

In short, the Tangier maxilla shows every indication of having come from a child that would have grown up to have the big, inflated face of an adult Neandertal. The specimen is too fragmentary to be confidently described as a Neandertal, but its strong similarity to these people from the North cannot be easily dismissed. Along with the variable Neandertal-like traits seen in the ATG fossils from Jebel Irhoud, the Tangier maxilla provides further evidence of Neandertal genetic influence on the early Late Pleistocene populations of North Africa. As at Klasies River Mouth, the marked variability among the fossils from these North African sites may reflect the genetic results of admixture between two distinctive populations—moderns expanding into North Africa from the south, and a much smaller Neandertal contingent entering Africa from the continent to the north.

A number of anatomically modern specimens, mostly less than 30 Ky in age, are associated with later, Upper Paleolithic cultural complexes in North Africa (Klein 1999). Probably the oldest is a partial skeleton from Taramsa Hill in Egypt, which seems to be 80–50 Ky in age (Vermeersch et al. 1998). This specimen is not yet fully studied, but appears to be morphologically similar to the Skhūl/Qafzeh hominins (described below). Two more partial skeletons from Nazlet Khater in Egypt have been dated to ~37–38 Kya (Crevecoeur 2007, Vermeersch 2002, Vermeersch et al. 1984a, 1984b). The skull and postcranial bones of the more complete specimen, Nazlet Khater 2, are robust but clearly modern (Crevecoeur and Trinkaus 2004), and the cranial capacity measures a full $1420\,cm^3$ (Holloway et al. 2004). Nevertheless, this skeleton retains a few primitive traits, including a rather low frontal, a short parietal arc, a wide mandibular ramus, and certain archaic-looking postcranial features (Crevecoeur 2007).

▌THE FIRST MODERN PEOPLE OUTSIDE AFRICA: THE NEAR EASTERN EVIDENCE

The earliest non-African human fossils that are generally considered modern come from two sites in Israel: the small cave of Mugharet es-Skhūl (Cave of the Kids) at Mt. Carmel, and the larger cave complex of Jebel Qafzeh located near Nazareth. The excavations at Skhūl, undertaken in 1931 and 1932 under the direction of D. Garrod, uncovered the remains of at least 10 individuals, including three juveniles and seven adults. All were found in Level B in association with Mousterian artifacts (Keith and McCown 1937, McCown and Keith 1939). Excavation at Qafzeh came in two phases. The first phase, begun in 1933 under the direction of R. Neuville, yielded two specimens (Qafzeh 1 and 2) associated with the Upper Paleolithic, and five specimens (Qafzeh 3-7) associated with the Mousterian (Neuville 1951, Vandermeersch 1981). Neuville excavated in the cave's interior, where the Mousterian specimens were recovered from Level L. The second excavation period at Qafzeh ran intermittently from 1966 to 1975 under the direction of B. Vandermeersch (1981), who excavated mainly in the cave entrance and the terrace immediately in front of it. His diggings produced Qafzeh hominins 8 through 18, all associated with the Mousterian and (except for some isolated teeth) all derived from Levels XVII–XXIV in the cave entrance.

Ever since the discovery of these caves, arguments about the dating of the Skhūl-Qafzeh (SQ) sample have played an important role in changing views on modern human origins (Trinkaus 1984a). Some archaeological studies suggested that the Skhūl and Qafzeh remains are about equally old—40 to 45 Ky—and not as ancient as those from Tabūn (Jelinek 1982, 1992). These late dates were cited as support for the view that the SQ people could have evolved from the Neandertals of the Near East and were no older than the earliest modern Europeans (Wolpoff 1980a, Trinkaus 1984a, Smith 1985). However, other archaeological analyses (Neuville 1951) and studies of microfauna and sedimentology suggested an older age for Qafzeh, perhaps about 70 Ky (Bar-Yosef and Vandermeersch 1981). The turning point in this debate came in 1988, when TL analyses on burned flints from Qafzeh Levels XVII–XXIII gave a mean date of 92 ± 5 Kya (Valladas et al. 1988). This date was supported by ESR dating of mammal teeth from Levels XV–XXI, which yielded age estimates of ~100 Ky and 120 Ky using early and linear uptake models respectively (Grün and Stringer 1991). Uranium-series dates at Qafzeh yield a wide range of age estimates—from 34 to 120 Ky, depending on what we assume about the impact of leaching on $^{234}U/^{238}U$ ratios (Grün 2006). The current state of our knowledge is summarized in Fig. 7.4.

Dating of Skhūl has proven even more confusing, but the current data are still suggestive of an age similar to that of Qafzeh. Early- and linear-uptake models of ESR dating have provided estimates of 81 Ky and 101 Ky (Stringer et al. 1989, Grün and Stringer 1991) for Skhūl Level B, while uranium-series dates are 46–88 Kya on early uptake and 66–102 Kya on linear uptake (McDermott et al. 1993). TL dates range from 99 to

167 Kya, but cluster around 119 Kya (Mercier et al. 1993). McDermott et al. (1993) raise the possibility that there are two widely separated time periods represented at Skhūl. Some recent support for this notion comes from direct uranium-series dating on two of the hominin specimens: Skhūl 9 has been dated to 131 ± 2 Kya, while Skhūl 2 clocks in at only 32 ± 0.8 Kya (Grün 2006).

Despite all the variation in these dates, a reasonable summary age estimate for the SQ specimens is 80–100 Ky, although some of them may be younger. This makes them significantly older than the earliest modern humans from sites in Europe or further east in Asia.

The chronological pattern that has emerged from these studies seems to document the geographical spread of modern human morphology from an African center of origin, first into the Levant and later into Europe and the rest of Asia. The new SQ chronology is one of the main factors that has furthered the elaboration and acceptance of a monocentric "out of Africa" model of modern human origins (Smith et al. 1999). The Recent African Origins interpretation of the SQ evidence portrays the SQ hominins as fundamentally modern in form. That claim goes back to the initial analyses of the Skhūl specimens (Keith and McCown 1937) and to F.C. Howell's influential papers in the 1950s (Howell 1957, 1958, 1960), which categorized the Skhūl sample as a different human "type" from the Levantine Neandertals. The first detailed studies of the Qafzeh sample (Vallois and Vandermeersch 1972, Vandermeersch 1981) drew a similar categorical distinction between the Levantine Neandertals and the entire SQ sample. This distinction has become widely accepted (Trinkaus 1984a, Rak 1993, 1998; Stringer 1994, Lahr 1996, Vandermeersch 1997, Klein 1999, Hublin 2000, Holliday 2000, Tattersall and Schwartz 2001, Habgood 2003).

But from early on, an alternative view of the SQ sample has emphasized its variability rather than its discreteness—especially in the case of Skhūl. As mentioned in Chapter 7, Keith and McCown (1937) initially recognized two "types" of humans at Mt. Carmel: Neandertals at Tabūn and moderns at Skhūl. However, more detailed analyses led them to argue that the Mt. Carmel people constituted a single, highly variable population. In their 1939 monograph, McCown and Keith noted that while the Tabūn C1 specimen tended to show a Neandertal morphological pattern and the Skhūl specimens a more modern one, not all the characteristics of individual specimens placed them decisively in one category or the other (McCown and Keith 1939). Although the extremes of the range could be considered different "types," McCown and Keith argued that variation in the expression of individual characteristics bridged the gap between the samples, making it impossible to draw an unambiguous morphological line between the Neandertal and modern groups.

Some later researchers have reached similar conclusions. A recent analysis of the Levantine hominins and

their African contemporaries by Kramer et al. (2001) identifies 32 characteristics (24 postcranial and 8 cranial) that reflect whether a specimen is more Neandertal-like or modern. As expected, the Tabūn C1 specimen falls at the Neandertal end of the spectrum; yet four of its characters actually exhibit the most "modern" state found in this sample. Conversely, the Skhūl hominins generally fall toward the modern human end of the sample, but five of them show the most Neandertal-like states of specific characters. Skhūl 5, which is the most modern-like in 16 traits, is the most Neandertal-like in five others. Other morphometric studies likewise show that the Mt. Carmel sample—and indeed the Levantine hominin sample as a whole—is not unequivocally separable into distinct Neandertal and modern groups (Simmons et al. 1991, Arensburg and Belfer-Cohen 1998, Wolpoff and Lee 2001).

There are several relatively complete adult skulls from Skhūl and Qafzeh. Perhaps the best-known is the robust Skhūl 5 cranium (Fig. 8.21), but other adult

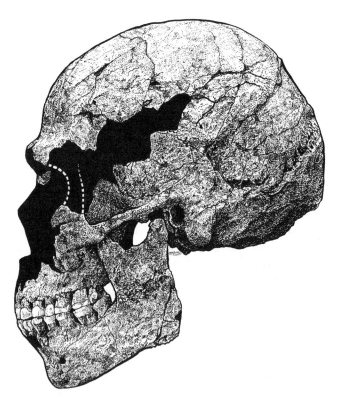

FIGURE 8.21

The Skhūl 5 skull (Mt. Carmel, Israel), lateral view. The adult Skhūl hominins have modern vault shapes, but maintain some facial prognathism and distinct supraorbital tori. Skhūl 5 shows considerable alveolar prognathism, a retromolar space, and a distinct chin structure. Like other Near Eastern early moderns, Skhūl 5 lacks evidence for an occipital bun. The midfacial region is heavily reconstructed, and the anatomy of this area is unknown in this specimen. (After Johanson and Edgar 1996.)

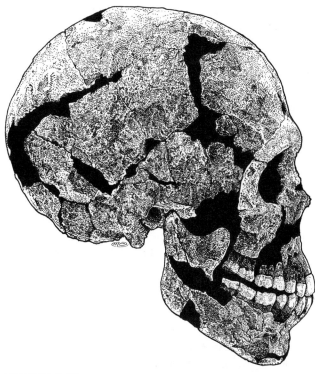

FIGURE 8.22

Lateral view of the Qafzeh 9 late-adolescent skull from Israel. The vault of this specimen is clearly modern, with a high forehead and lack of occipital bunning. Qafzeh 9 exhibits a distinct chin but otherwise orthognathic face. (After Johanson and Edgar 1996.)

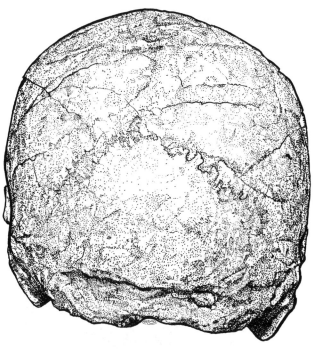

FIGURE 8.23

The Skhūl 5 cranium (Mt. Carmel, Israel), rear view. Note the lack of a suprainiac fossa and the modern shape of the vault.

skulls—particularly Skhūl 4, Skhūl 9, Qafzeh 6, and Qafzeh 9 (Fig. 8.22)—are equally important. Overall, these skulls exhibit distinctly modern cranial vaults with high, vertical foreheads, rounded cranial contours, prominent mastoid processes, and no occipital buns. From the rear (Fig. 8.23), the vault of Skhūl 5 presents the parallel-sided, house-form shape that characterizes modern humans, with no evidence of a suprainiac fossa. The general impression of modern cranial vault form conveyed by these specimens holds for the entire SQ adult sample, although there is significant variation (for example, Skhūl 9 has a suprainiac fossa). Studies of the SQ juveniles indicate that their affinities are also with modern humans rather than Neandertals (Minugh-Purvis 1998, Tillier 1999, Minugh-Purvis et al. 2000). Yet one supposed Neandertal apomorphy, the anterior mastoid tubercle, is more common in the SQ skulls than in European Neandertals (Table 7.7).

Metric data generally confirm the relatively modern form of the SQ vaults. In their vault-height index values (Table 7.4) and frontal angles (Table 7.5), SQ skulls differ from Neandertals and group with European Upper Paleolithic and recent European samples, reflecting their vertical foreheads and high cranial vaults. The

modern form of the SQ braincases is partly a correlate of their extremely large size. The mean endocranial volume of the five SQ adult skulls is an impressive 1554 cm³—a larger average than we see for any other human population sampled in this book (Tables 7.9, 8.4). Even the early modern Europeans, despite their later dates and more northerly provenience, have smaller braincases on the average. Intriguingly, we see the same difference among Neandertals: the Near Eastern ones have braincase volumes about 5% larger than late (Würm) Neandertals from Europe (Table 7.9)—the opposite of what we might expect on the basis of latitudinal effects.

The SQ faces also generally conform to a modern pattern. Though broad, they are shorter and less prognathic than Neandertal faces (Figs. 8.21 and 8.22). One reflection of their reduced prognathism is their relatively low nasion projection (Table 7.7). The SQ crania also lack the characteristic infraorbital "puffiness" of Neandertals. Although Qafzeh 9 and Skhūl 5 have somewhat oblique orientations of the ZAM (Fig. 8.24), Skhūl 4, Skhūl 9, and Qafzeh 6 have more indented, modern-looking ZAMs; and Skhūl 4, Skhūl 9, Qafzeh 6, and Qafzeh 9 all appear to have had canine fossae.

Although not specifically Neandertal-like, the supraorbital region is one of the more primitive parts of the SQ crania. Adult specimens from both sites tend to have a distinct supraorbital torus (Fig. 8.21), which thins lat-

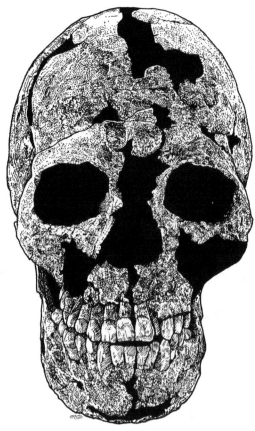

FIGURE 8.24

The Qafzeh 9 late-adolescent skull from Israel, frontal view. The face of Qafzeh 9 is heavily reconstructed and distorted. Unlike other Skhūl/Qafzeh hominins, Qafzeh 9 lacks a supraorbital torus and has a modern-like brow ridge. The anterior teeth are quite large, but the other features of the face conform to a modern form. (After Johanson and Edgar 1996.)

erally but projects as strongly as the tori of European and Near Eastern Neandertals (Simmons et al. 1991). The torus tends to be continuous across the top of the nasal opening, and is often separated from the frontal squama by a supratoral sulcus. None of the SQ specimens (except for Qafzeh 9) exhibit the midorbital thinning characteristic of European early modern humans. In fact, the browridge thickness index for SQ is about the same as for early Central European Neandertals (Table 7.19). The outlier here is Qafzeh 9, which has a brow ridge much like those of early modern skulls from Upper Paleolithic Europe. In an analysis of frontal-bone morphometrics, Simmons et al. (1991) find that Qafzeh 9 forms a distinct cluster with the more recent (Upper Paleolithic) Qafzeh 1 and 2 frontals, whereas the other SQ frontals group with Near Eastern Neandertals and Zuttiyeh.

As Table 7.14 demonstrates, the SQ mandibles are quite large. According to Schwartz and Tattersall (2000, 2003), three Qafzeh specimens—the Qafzeh 8 and 9 adults and the Qafzeh 11 subadult—have bony chins with the distinctive morphology of modern humans (Fig. 8.2). However, the other SQ mandibles do not. Instead, they have a "variably teardrop-shaped, mound-like structure that thins before reaching the inferior margin" of the symphysis (Schwartz and Tattersall 2003, p. 359). Schwartz and Tattersall argue that this morphology, along with that of the supraorbital torus and the presence of retromolar spaces in the Skhūl mandibles (Fig. 8.21), establishes the Skhūl fossils and part of the Qafzeh sample as a different morph, potentially a different species, from the other Qafzeh specimens (mainly Qafzeh 9—Figure 8.22) and modern humans. Rak (1998, Rak et al. 2002) asserts, however, that all of the mandibular features of the entire SQ sample align them with modern humans.

Multivariate cranial analyses by C. Stringer and others (Stringer 1974, 1978, 1994; Bräuer 1984a, 1989; Lahr 1996, Habgood 2003) demonstrate the distinctiveness of the SQ group from Neandertals. Most experts today accept Stringer's conclusion that the SQ population is part of a modern-human clade that excludes Neandertals (Fig. 8.25). However, some multivariate analyses using both metric and non-metric data conclude that SQ cannot be separated as a distinct group from the Neandertals of the Levant (Simmons et al. 1991, Kramer et al. 2001, Wolpoff and Lee 2001). Other studies question the unequivocal identification of certain Skhūl specimens as anatomically modern humans (Corruccini 1992, 1994; Kidder et al. 1992). While the SQ sample does seem to have closer affinities with more recent humans than with Neandertals, there are distinctly archaic features in it as well. Not surprisingly, adherents of conflicting schools of thought draw divergent conclusions from these facts. Proponents of the "Out of Africa" model see the SQ sample as documenting a species-replacement event, in which invasive modern humans from Africa displaced their Neandertal cousins from the Middle East. Regional-continuity theorists, on the other hand, stress the variability and archaic features of the SQ crania, which they see as reflecting evolutionary continuity or genetic admixture between Neandertals and modern humans.

Postcranial data may provide an independent test of the degree of separation between the SQ and Near Eastern Neandertal groups (Holliday 2000). In a comprehensive study of postcranial features, Trinkaus (1995b) noted several characteristic Neandertal traits that are absent in the SQ sample. Unlike Neandertals, the SQ people are quite tall, averaging some 16–17 cm taller than either European or Near Eastern Neandertals and ~7 cm taller than even the early modern Europeans (Table 7.15). Much like the early modern European sample, the SQ sample exhibits only bisulcate and

CRANIAL CHANGES PRIMARILY IN –

–vault shape, face size   –vault size, face shape

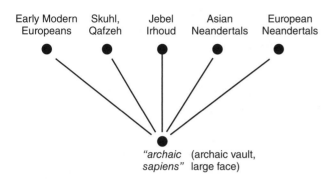

Early Modern Europeans   Skuhl, Qafzeh   Jebel Irhoud   Asian Neandertals   European Neandertals

"archaic sapiens"   (archaic vault, large face)

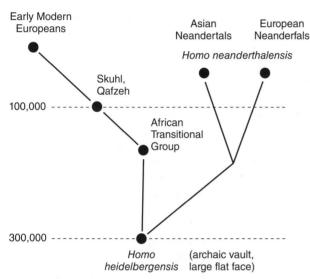

Early Modern Europeans

Asian Neandertals   European Neanderfals

Homo neanderthalensis

Skuhl, Qafzeh

100,000

African Transitional Group

300,000

Homo heidelbergensis   (archaic vault, large flat face)

FIGURE 8.25

Schematic representation of the relationships between Neandertals, Skhūl/Qafzeh hominins, early modern Europeans, the African Transitional Group (including Jebel Irhoud), and their ancestors. (After Stringer 1994.)

TABLE 8.7 ■ Placement of Skhūl/Qafzeh and Near Eastern Neandertals in Recent Human Populations, Based on Holliday's (2000, p. 62) Discriminant-Function Analysis of Postcranial Body-Shape Features

Specimen	EUR %	NAF %	SAF %
Qafzeh 3	5.1	37.1	57.8*
Qafzeh 8	8.3	6.4	85.3*
Qafzeh 9	7.5	33.0	59.6*
Skhūl 4	0.4	29.0	70.6*
Skhūl 5	99.7*	0.3	0
Skhūl 6	24.4	30.4	45.2*
Amud 1	96.7*	3.2	0
Kebara 2	83.5*	16.5	0
Shanidar 1	12.4	57.3*	28.9
Shanidar 2	87.0*	8.5	4.5
Shanidar 3	30.6	68.4*	0.1
Shanidar 4	91.7*	7.5	0.8
Shanidar 5	98.3*	1.7	0.1
Shanidar 6	97.1*	2.1	0.8
Tabūn C1	52.0*	39.4	8.5

Numerical values represent the degree of confidence with which a specimen can be assigned to EUR (European), NAF (North African) or SAF (sub-Saharan African) modern human groups. The recent human group to which each fossil specimen is most plausibly assigned is indicated by boldface type and an asterisk.

ventral sulcus patterns of the scapula, with no sign of the dorsal sulcus pattern that predominates among European Neandertals (Table 7.17). Higher femoral neck-shaft angles in SQ compared to the Neandertals suggest less mechanical loading of the lower limb (Trinkaus 1993c). Although the Skhūl femora have robust shafts (McCown and Keith 1939), they bear pilasters and are otherwise similar in cross section to modern human femora. The SQ femora lack the proximal-lateral femoral flange (Fig. 7.27) characteristic of European Neandertals, although in this regard they also resemble Near Eastern Neandertals and differ from early modern Europeans (Kidder et al. 1991). Some features of the SQ

hand bones—reduction in the mechanical advantage of the thumb flexors and wrist muscles, less robust muscle attachments, narrower fingertips—distinguish them from Neandertals and align them with modern humans (Niewoehner 2001). Several authors see these features as indicating that the SQ folk are a different morph from the Neandertals, a morph capable of more oblique grips and finer manipulation (Trinkaus 1983a,b; Trinkaus and Villemeur 1991, Niewoehner 2001, Churchill 2001).

Holliday (2000) argues that body form should provide the most decisive test of the species-replacement interpretation of the SQ fossils. He reasons that if the SQ postcrania exhibit a tropical body form—as opposed to the nontropical pattern of recent Near Easterners (Eveleth and Tanner 1976) or the stocky, cold-adapted body form of Neandertals—then they can be plausibly interpreted as tropically adapted (and presumably African) immigrants into the Near East. Holliday's multivariate analysis, which incorporates measurements of relevant joint diameters, body breadths, and bone lengths (femur, tibia, clavicle, radius and humerus), compares SQ, Neandertals, and later fossils to modern humans from Europe, North Africa, and sub-Saharan Africa. Holliday finds that the overall SQ body form is phenetically closest to that of sub-Saharan Africans and very different from that of typical Neandertals (Table 8.7). This finding supports the "Out of Africa" model.

However, Holliday's study also reveals a significant exception to the generally tropical SQ body-form pattern—namely, Skhūl 5, which has short distal limb segments like those of cold-adapted modern people (Table 7.16). The estimated brachial (70.5) and crural (80.0) indices for this specimen are much lower than those of other SQ individuals, falling into the Neandertal range (Trinkaus 1981) and placing it closest to the European group in Holliday's analysis. As Holliday notes, the Skhūl 5 postcranials are heavily reconstructed, making accurate measurement of bone lengths problematic. But this uncertainty should not have affected most of the measures used in his analysis, which places Skhūl 5 in the European group with the highest confidence of any fossil assignment. Wolpoff (1999) cites the body form of Skhūl 5 as evidence against sub-Saharan affinities for the entire SQ sample. An alternative interpretation, discussed below, is that the Skhūl 5 body form and other atypical variants in the SQ sample are signs of a limited contribution by local archaic humans to an essentially modern gene pool derived from migrants out of sub-Saharan Africa.

Prehistorians have also looked for cultural and behavioral differences between the SQ people and Near Eastern Neandertals. Bar-Yosef (1992, 1994, 1995, 2000) has detailed the lithic differences between the Tabūn B type Mousterian, which is associated with Near Eastern Neandertals at Kebara and Amud, and Tabūn C type, associated with the hominins at Skhūl and Qafzeh. Both use Levallois technology and are dominated by side scrapers, but Tabūn B type Mousterian has more points. In general, the differences are no greater than those between different Mousterian types in Europe (Mellars 1996, Shea 2003). They probably reflect ecological differences between different phases of human occupation in the Levant, not fundamental differences in technological ability between competing species of *Homo*. Shea (1988, 1997) suggests that the higher incidence of points in the Tabūn B assemblages may be associated with more frequent ambush hunting by the Neandertals. From comparisons of the faunal remains at Tabūn B and C sites, Shea and Lieberman (Lieberman and Shea 1994, Lieberman 1998b) infer that the Levantine Neandertals inhabited their sites more continuously during multiple seasons of the year. Shea (2003) concludes that the ability of the Asian Neandertals to remain in a single site year-round may reflect a superior physiological adaptation to the markedly cooler climates that prevailed in the region between 75 and 50 Kya, allowing them to outcompete the more tropically adapted SQ people during this Levantine cold snap.

Shea sees the Tabūn C Mousterian as showing signs of modern human behavior not found in the Neandertal sites (Tabūn B). He notes, however, that these signs (use of grave goods, red ochre, and long-distance transport of marine shells) disappear from the Levant during the succeeding cold-snap period of Neandertal resurgence, and do not return until the permanent establishment of "fully modern" behavior at ~50 Kya. Supposed differences in burial practice between the SQ humans and the Levantine Neandertals (Harrold 1980) have been denied by Bar-Yosef (2000), who concludes that there are no demonstrable differences in the burials of these two groups.

AFRICAN AND CIRCUM-MEDITERRANEAN GENE FLOW AND MODERN HUMAN ORIGINS

Anatomically modern humans (AMH) first appear in East Africa (Herto, Omo Kibish KHS) around 160 Kya. By at least 80 Kya and perhaps as early as 120 Kya, AMH had spread to the Levant (Qafzeh and Skhūl). If the ESR date for Border Cave is accurate, AMH were in southern Africa by 70 Kya. The North African dates are less certain, but estimates for Dar-es-Soultane II and Témara suggest that early modern people were established in Morocco sometime between 43 and 127 Kya (Table 8.6).

The Recent African Origins (RAO) model depicts all these occurrences as documenting the origin and spread of a new species of *Homo*. In particular, the Skhūl/Qafzeh group is portrayed by RAO proponents as an intrusion of this new species from Africa into western Asia, where it contended against and ultimately replaced the Neandertals. This replacement may have followed a long period of fluctuating population boundaries in the Near East, so that during some periods Neandertals inhabited the Levant while at other times AMH pushed up from the south and displaced them (Bar-Yosef 1992, 1994; Bar-Yosef and Vandermeersch 1993, Holliday 2000, Shea 2003). In the end, however, AMH prevailed in western Asia, and Neandertals faded into extinction without contributing to the biological makeup of modern populations. Or so the story goes. Although there has been little discussion about the potential spread of morphological modernity within Africa, the punctuationist models of evolution favored by many RAO proponents imply that AMH should have arisen locally within Africa, as a peripheral isolate from a larger population of archaic humans. Presumably, this new species should then have replaced archaic peoples throughout Africa, in the same way that SQ-type humans are supposed to have replaced Neandertals in western Asia.

In evaluating this species-replacement model, the key question is whether there is any evidence of interbreeding or hybridization between modern humans and various archaic groups, either in Africa or in western Asia. Many proponents of one or another theory of modern human origins think this is an easy question, and offer confident (though conflicting) answers. We think it is an extremely difficult question, and our answers are correspondingly tentative.

Interbreeding between two genetically different populations may not produce viable offspring. In such cases, the two populations are reproductively isolated, and belong to unequivocally different species. Even if interpopulation matings can produce viable offspring, the populations may still be effectively isolated, either because pre-zygotic isolating mechanisms prevent such matings from occurring, or because post-zygotic isolating mechanisms render the hybrids infertile, as in the case of mules. In such cases, we would not expect to find evidence of hybridization in the fossil record, since hybrids would be nonexistent or vanishingly rare animals with zero fitness. (As far as we know, there are no fossil mules.) But as we know from many current examples (Chapter 2), matings between genetically, morphologically, or karyologically distinct populations may produce hybrid offspring with only slightly or moderately reduced fertility, or even with enhanced fertility (Levin 2002). In such cases, the two populations may be referred to as species, semispecies, or subspecies, depending on the fertility of the hybrids, the geographical extent of the zones where they occur, and the tastes and philosophy of the taxonomist.

Although hybridization between semispecies is a common phenomenon, it is extremely difficult to find fossil evidence of it. Karyological evidence is never available, because chromosomes do not fossilize. Direct genetic evidence from fossil DNA is scrappy and obtainable only from relatively recent specimens. The incidence of interbreeding between ancient African and Near Eastern humans can be assessed only from the evidence of skeletal morphology. Such evidence is not likely to take the form of fossil specimens manifesting a globally intermediate, half-and-half morphology. Two hybridizing species or semispecies need not contribute equally to either the gene pool or the morphological traits of the hybrid population. If one of the parent populations is considerably more abundant on the landscape than the other, the rarer one's contribution to the hybrid group may be proportionally very small. Over the long run, such contributions could be quickly swamped by those of the more abundant parent population, vanishing from the gene pool altogether or remaining identifiable only by the persistence of specific morphological details. Even when such telltale details persist in the hybrid population to bear witness to the contributions of the rarer parent type, they will be expressed in the context of a genome derived principally from the other parent population, and therefore may differ from their original morphological expression. In such cases, it may be impossible to develop a scientific consensus concerning their homology.

The archaic features of the Skhūl/Qafzeh (SQ) people are similar overall to those seen in the early moderns

from Ethiopia. Both groups retain large and robust faces, robust vaults, prominent supraorbital tori, and slight but persistent facial prognathism. Thus, the SQ sample can be plausibly interpreted as manifesting a primitive early-AMH morphological pattern, derived from the earlier East African specimens. But a few traits of the SQ sample might represent genetic contributions from Neandertals. These are:

- The overall shape of the cranial vault (Kidder et al. 1992, Corruccini 1992), particularly of the frontal bone (Simmons et al. 1991, Smith et al. 1995).
- A high incidence of anterior mastoid tubercles (Table 7.7).
- Absence of the proximal lateral femoral flange in the SQ sample as well as in Near Eastern Neandertals (Kidder et al. 1991).
- The presence of retromolar spaces in the SQ mandibles, as exemplified by Skhūl 5 (Fig. 8.21) and Qafzeh 9 (Fig. 8.22).
- The limb proportions of Skhūl 5.

On a closer look, some of these possible Neandertal inheritances seem less plausible. Some SQ individuals fall outside the recent human range in overall vault-shape metrics because they retain widespread primitive traits that have been lost in later AMH (Kidder et al. 1992). But these symplesiomorphies do not link the SQ skulls specifically to Neandertals. Similar remarks apply to the frontal-bone data. For example, the SQ specimens (except Qafzeh 9) all exhibit supraorbital tori that are thickest medially and thinnest laterally, as in Neandertals. But so do the African Transitional Group (ATG) specimens and early modern Africans. Moreover, the SQ and African tori all tend to thin evenly in their mid-orbital and lateral aspects (Smith 1992a), without the localized mid-orbital thinning that characterizes late European Neandertals and early modern Europeans (Tables 7.6 and 7.19). Again, the SQ morphology appears to be a plesiomorphy that can be easily derived from that seen in the ATG or in early African AMH. But by the same token, the pattern of mid-orbital thinning peculiar to Neandertals and early moderns from Europe can be read as a regional synapomorphy, reflecting genetic continuity between archaic and modern peoples in the Neandertal heartland (see below).

The femoral flange can be interpreted in two ways, depending on what we think about character-state polarity. Because we do not know what the state of this character was in the ATG or early modern Africans, it is not known whether presence or absence of the flange is the derived condition. In Europe, both Neandertals and early moderns have the flange; in the Near East, both Neandertals and early moderns (SQ) lack it (Kidder et al. 1991). If the presence of the flange is primitive, then its absence in the SQ people could be a synapomorphy acquired through hybridization with the local Neandertals. Conversely, if the flange is primitively absent, its presence in early modern Europeans may be a sign of local hybridization between Neandertals and invasive AMH from further south. But no matter which is the primitive state, the distribution of the femoral flange provides evidence for regional continuity somewhere—if not in the Levant, then in Europe.

Another apomorphous character potentially linking the SQ group to Neandertals is the presence of a retromolar space. Apart from Neandertals, the only hominins showing this gap between M3 and the mandibular ramus are populations suspected of having some genetic connections with Neandertals—that is, the Neandertals' potential ancestors among European Heidelbergs, some early modern Europeans, the SQ group, and the earlier Tabūn C2 mandible. The presence of a retromolar space may therefore be a tie to the Neandertal lineage, reflecting a shared stage in a particular sequence of reduction in prognathism (Fig. 7.16). Rak (1998, Rak et al. 2002) argues that the retromolar space in Tabūn C2 is not homologous with that of Neandertals. This seems unlikely, but Rak's claim illustrates the sort of dispute that recurs whenever attempts are made to identify Neandertal contributions to other populations. As usual, simply listing verbal descriptions of shared properties does not settle the matter, because different morphologies can often be described in the same terms, and no two morphologies are ever identical.

The apparently cold-adapted limb proportions of Skhūl 5 might also be claimed as a possible sign of Neandertal contributions to the gene pool of the otherwise long-limbed, tropically-adapted SQ folk. Some researchers have made such claims about the limb skeleton of an Upper Paleolithic child from Lagar Velho in Portugal (see below). Others dismiss the Lagar Velho skeleton as just an aberrant individual variant, "simply a chunky Gravettian child" (Tattersall and Schwartz 1999). Whichever position one adopts, it is important to notice that these arguments cut both ways. If the tropical body build of typical early modern people in Europe and the Levant is taken as evidence for a recent arrival from Africa, then by the same token the cold-adapted, Eskimo-like limb proportions of Skhūl 5 and Lagar Velho ought to count as evidence for the presence of Neandertal genes in these same gene pools.

Most of the debate over admixture in the Near East has centered on these arguably Neandertal-like traits in the SQ sample. However, similar questions deserve to be asked about possible signs of hybridization in the Levantine Neandertals. Such hybridization has been offered as an explanation for the puzzling mixture of archaic (including specifically Neandertal) and modern features that some see in the Tabūn C2 mandible (McCown and Keith 1939, Quam and Smith 1998). If this mandible is a product of Neandertal-modern interbreed-

ing, then modern humans must have spread northward very soon after their first appearance in Africa. (The oldest secure date for AMH in Africa is 160 Kya at Herto; Tabūn Level C is dated to 170–120 Kya.) This may well be what happened. The parallel case of the early Erectines from Dmanisi shows that new variants of *Homo* can spread out of Africa very rapidly, and early modern humans may well have been even more mobile and invasive than primitive Erectines.

Other Near Eastern Neandertals share the basic morphological pattern of European Neandertals, but they are more like modern humans in some ways. They lack occipital buns and proximal lateral femoral flanges. They have higher brachial indices. Their cranial vaults are higher and more rounded than those of European Neandertals, a difference reflected in their higher vault-height indices, parietal angles, and occipital angles (Tables 7.4 and 7.5). The more "progressive" or modern-looking morphology of the West Asian Neandertals has been noted by many investigators (McCown and Keith 1939, Suzuki 1970b, Arensburg and Belfer-Cohen 1998). It may explain why some analysts find it impossible to sort Near Eastern Neandertals and early AMH into distinctly separate groups (Simmons et al. 1991, Kramer et al. 2001, Wolpoff and Lee 2001). Perhaps the Levantine Neandertals, not the SQ people, provide the best evidence for hybridization in the Near Eastern Pleistocene.

As noted earlier, distinctively Neandertal-like features also crop up among early modern humans from North Africa, including the large size and "puffiness" of the Tangier maxilla and the occipital buns of the Jebel Irhoud crania. Simmons and colleagues (Simmons et al. 1991, Simmons and Smith 1991, Smith et al. 1995) have suggested that such traits may reflect gene flow from European Neandertal populations—in this case, across the Strait of Gibraltar. Since Erectines appear to have reached Flores before 800 Kya, which would have required crossing an even greater expanse of open water (Chapter 5), crossings of the Strait of Gibraltar by Neandertals are at least a theoretical possibility. However, much as in the Near East, evidence of archaic genetic influence wanes and disappears in North Africa before the end of the Middle Paleolithic.

All these facts suggest that interactions between human populations around the Mediterranean during the late Pleistocene were probably complex, with genes flowing in both directions during this initial period of contact between Neandertals and early modern humans. At this time, both the southwestern and eastern Mediterranean may have been hybrid zones between parapatric hominin semispecies. Other hybrid zones involving other archaic populations may have occurred further south. The morphological mosaic seen in the Klasies River Mouth (KRM) sample suggests that similar admixture went on between early moderns spreading into

South Africa and local representatives of the ATG. However, by ~75 Kya (Grün et al. 2003), the KRM mosaic had been replaced by more visibly modern people—the same course of events seen in the north of Africa.

MODERN HUMAN ORIGINS IN EAST ASIA

The fossil evidence of early anatomically modern humans (AMH) on the Asian mainland east of the Levant is sketchy (Brown 1992). For most of this area, the earliest remains of AMH are too recent to be relevant to the issue of modern human origins. The best evidence comes from China, which has yielded a few possibly relevant specimens older than 35 Ky.

Perhaps the oldest of these is a partial skeleton from Liujiang in Guangxi Province in southern China. This specimen supposedly derives from a breccia just inside the mouth of a limestone cave. U-series dates yield an age of >68 Ky and perhaps as much as 139 Ky for the breccia (Shen et al. 2002). Unfortunately, the relationship of the dated material to the human skeletal remains is not clear (Brown 1999, Habgood 2003). Salawusu (originally Sjara-osso-gol) in Mongolia has yielded a series of human remains (Licent et al. 1926) dated to between 37–50 Kya on the basis of U-series analysis of possibly associated fauna (Yuan et al. 1983). However, all of these fossils were recovered from erosion surfaces in the banks of the Salawusu River, and the relationship between the dated fauna and the human bones is not certain (Wu and Poirier 1995). The Ziyang (previously Tzeyang) cranium was once described as China's earliest known AMH (Woo 1958), and it might be just that—but again, the dates are uncertain. [14]C dates of 37.4 ± 3 Ky and 39.3 ± 2.5 Ky have been obtained from mineralized trees associated with fauna and tools found 100 m from the Ziyang skull (Li and Zhang 1984). The fauna from this neighboring site is supposedly similar to that found with Ziyang; but another site nearby has yielded ^{14}C dates of only 7485 ± 130 years (Wu and Poirier 1995), and it is not clear that the older site truly corresponds in age to the human skull.

The most reliably dated early modern human from China is a partial skeleton from Tianyuan Cave near Zhoukoudian, AMS-dated to 34.4 ± 0.5 Kya. The Tianyuan 1 skeleton preserves a partial mandible and some 30 postcranial elements (Tong et al. 2004, Shang et al. 2007). The mandible exhibits essentially modern morphology, including a distinctly modern chin. The postcrania are also modern, with pilasters on the femora and other features typical of early modern humans. The crural index (partly based on reconstructed bones) is high, suggesting recent tropical ancestry. Shang et al. (2007) conclude that the origins of AMH in this region must have involved substantial gene flow from early modern populations to the south and west. But they also interpret certain features of the Tianyuan 1 teeth, hand bones, and tibia as inheritances from admixture with local archaic humans.

Both the Liujang and Ziyang cranial vaults are fundamentally modern in shape, with steep foreheads, rounded vault contours, and maximum breadths located high on the parietals (Wu and Zhang 1985, Wu and Poirier 1995). Although both specimens have moderately pronounced superciliary arches, their supraorbital regions conform to a modern pattern. (So do those of three frontal specimens from Salawusu.) The Liujiang occiput bears a slight protrusion, reminiscent of the hemibuns seen in early European AMH (see below). Both Liujiang and Ziyang have distinct frontal keeling, and Ziyang has an angular torus. Besides the Tianyuan skeleton, postcranial bones from these sites include a partial pelvis, lower vertebral column, and two femoral shafts from Liujiang, and two femora and a tibia from Salawusu. They all look fundamentally modern (Wu and Poirier 1995).

The faces of these Chinese skulls are also modern-looking, with canine fossae and angled ZAMs. They exhibit some regionally distinctive features seen throughout East Asia today (Table 8.8). Liujiang's face is described as being transversely flat, with an anteriorly placed frontal process of the zygomatic, and it has the flat, non-protrusive nasal bones characteristic of later East Asians (Habgood 1992). Liujiang, Ziyang, and Salawusu all exhibit shovel-shaped maxillary incisors that conform to the East Asian pattern defined by Crummett (1994, 1995).

Modern human remains were recovered from the Upper (Shandingdong) Cave at Zhoukoudian in 1933 and 1934 (Pei 1934). As is the usual case for Chinese fossils, their age is uncertain. The Upper Cave fauna points to an Upper Pleistocene age (Wu and Zhang 1985). However, the association of the human remains with the faunal sample is unclear (Brown 1999). Two radiocarbon dates on animal bones from the site yield ages of 10,470 ± 360 and 18,865 ± 420 years. Some authors (Wu and Zhang 1985, Kamminga and Wright 1989) argue on stratigraphic grounds that the human remains are closer to the younger date. On the other hand, there is an older radiocarbon date of 33.2 ± 2 Kya from lower levels of the Upper Cave deposits. Brown (1999) believes that the more recent dates may well derive from contaminated samples, and that the cultural levels are probably closer to this older estimate. Brown suggests an age range of 24–29 Kya for the human remains.

The Upper Cave fossils sample at least eight individuals, represented by three skulls (UC 101, 102, and 103), a maxillary fragment, four mandibles, a radius, six femora, and a patella (Wu and Zhang 1985, Wu and Poirier 1995). The crania come from Level 4 in the lower chamber of the Upper Cave complex, but other human and cultural

TABLE 8.8 ■ **Analysis of Presence/Absence of East Asian Regional Continuity Features in East Asian and Comparative Samples**[a]

Crania	Feature								
	1	2	3	4	5	6	7	8	9
Sample: Erectines and Ergasters									
KNM-ER 3733	—	—	—	—	?	?	Y	Y?	—?
KNM-ER 3883	—	—	Y?	—?	?	?	?	?	—
Olduvai OH 9	—	?	?	?	?	?	?	?	?
Sangiran 17	—	—	—?	Y	?	?	—	—	Y/—
Ngandong 11	—	—	?	?	?	?	?	?	?
Zhoukoudian (reconstruction)	Y	Y	Y	Y	?	Y	Y	Y	Y
Sample: East Asia									
Jinniushan	Y	—?	Y?	Y	Y	Y	?	Y?	Y
Dali	Y	Y?	Y?	Y	?	Y	Y?	Y	Y
Maba	Y	Y	Y	Y	?	?	?	?	—
Liujiang	—	Y	Y	Y	Y	Y?	Y	Y	Y
Upper Cave 101	Y	—	—	Y	—?	—?	Y	Y	Y
Upper Cave 102	Y	Y	—	Y	?	Y?	Y	Y	Y
Upper Cave 103	Y	Y	Y	Y	?	Y?	Y	Y	Y
Minatogawa 1	Y	—	—	Y	?	Y?	Y	Y	?
Sample: Australasia									
Wajak 1	Y	Y	Y	Y	?	Y?	Y	—	Y
Wajak 2	?	Y	Y	Y	?	?	?	?	?
Keilor	Y	Y	Y	Y	?	Y?	Y	Y?	Y
Kow Swamp 1[b]	Y	—	—?	—	?	—?	Y	—	Y
Kow Swamp 5[b]	Y	—	—	—	?	—	Y	—	Y
Kow Swamp 15[b]	Y	—	?	—	?	—	Y	—	Y
Sample: Africa									
Jebel Irhoud 1	Y	—	—	Y	?	—?	?	Y	Y
Bodo 1	Y	—	—	Y	?	—?	?	Y	Y
Ndutu	—?	—	—	?	?	—?	?	?	?
Kabwe 1	Y	—	—	—	?	Y/—	—	—	Y
Omo Kibish 1	—?	—?	?	?	?	?	?	?	Y
Border Cave 1	Y	Y	?	Y?	?	?	?	?	Y
Florisbad	—	Y?	?	Y?	?	?	?	—?	Y
Sample: Western Asia									
Zuttiyeh	—	Y	?	Y	?	—?	—	?	—
Skhūl 4	—	?	?	Y	?	?	?	—	Y
Skhūl 5	—	—	?	Y	?	?	—	—	Y?
Qafzeh 9	—	?	?	Y	?	?	—	—	Y
Amud 1	—	—	—	—	?	?	?	?	—
Sample: Europe									
Petralona	Y	—	—	—	?	Y/—?	—?	—	Y
Arago 21	Y	—	—	Y/—?	?	Y/—	—	—	Y
Steinheim	—	—	—	—	?	—?	—	—	Y
La Ferrassie 1	—	—	?	—	?	?	—	—	Y
Cro-Magnon 1	—	—	—	Y	?	?	—	Y	Y
Oberkassel 1	Y	—	—	Y	?	?	—	Y	Y
Oberkassel 2	Y	—	—	Y	?	?	—	Y?	—
Dolní Věstonice 3	—	—	—	Y	?	?	—	Y?	—
Předmostí 3	Y	—	—	Y	?	?	—	Y?	—
Předmostí 4	—	—	—	Y	?	?	—	Y?	—
Mladeč 1	—	—	—	Y	?	?	—	Y	—

[a]Numbers for features correspond to those listed in the text. Y = present, — = absent, ? = morphology unknown, identification questionable, or region not known. Y/— = conflicting assessments by different observers.
[b]Thorne (1976) judged that the Kow Swamp crania lack sagittal keeling.
Source: From Habgood (2003).

remains are found in Levels 1–3. The human remains are associated with a few undiagnostic lithic pieces, a necklace and other items of personal adornment, and at least one bone needle (Pei 1939). The three Upper Cave skulls have figured prominently in discussions of the origin of modern East Asians (Habgood 2003). Weidenreich (1939b) argued that they represented three different modern human "racial types": a young adult female (UC 102) with Melanesian features, an adult female (UC 103) exhibiting "Eskimoid" characteristics, and a male (UC 101) with features relating it to Upper Paleolithic Europeans and "primitive Mongoloids" (i.e., Native Americans and other vaguely Asian-looking peoples). Discounting Weidenreich's picture of a multiethnic confluence in the Pleistocene, later authors have regarded these three skulls as merely sampling normal variation within a single local population (Wu and Zhang 1985, Kamminga and Wright 1989, Kamminga 1992, Habgood 2003). Unfortunately, the Upper Cave material vanished in 1941 along with the ZKD Erectine fossils (Chapter 5).

The modern morphology of the Upper Cave skulls is confirmed by several multivariate analyses of their metrics, although their affinities to various modern human groups are debated (see below). Both UC 102 and 103 show evidence of postmortem distortion and may have been artificially deformed. Despite the distortion, the vaults are distinctly high and rounded, with relatively high foreheads and maximum breadths lying high on the parietals (in rear view). The faces have canine fossae and angled ZAMs with varying degrees of malar notch development (Pope 1991). The male cranium has a prominent, but modern-shaped, brow ridge. According to Wolpoff (1999), the Upper Cave postcranial bones and mandibles are also modern in appearance.

A similar description applies to the remains from the Minatogawa limestone quarry on Okinawa. These fossils, comprising the skeleton of an adult male (Minatogawa 1) and three more fragmentary female skeletons, are the only early modern specimens known from the insular parts of East Asia. The Minatogawa fossils are probably bracketed by radiocarbon dates of 18.25 ± 0.65 and 16.6 ± 0.3 Kya (Suzuki and Hanihara 1982). Though robust, the nearly complete male skeleton is anatomically modern (Suzuki 1982, Brown 1999, Wolpoff 1999, Habgood 2003).

In a multivariate analysis of Pleistocene and recent skulls from the eastern part of the Old World, Brown (1999) found that the Minatogawa 1, Upper Cave 101, and Liujiang crania grouped together outside the range of modern East Asian populations, but were somewhat closer to Australian, Eskimo, and early Japanese (Jomon and Ainu) samples. Other multivariate analyses have also failed to find strong links between modern "Mongoloids" and these early modern fossils from China and Okinawa (Kamminga and Wright 1989, Howells 1989, 1995; Wright 1995, Neves and Pucciarelli 1998, Cun-

ningham and Westcott 2002). Habgood (2003, p. 229) concludes that the Upper Cave, Minatogawa 1, and Liujiang crania represent a "... morphologically variable early East Asian population that displays some so-called 'Mongoloid features' but has not developed the modern 'Mongoloid' configuration, which does not appear until the mid-Holocene."

As noted earlier, several scientists (Weidenreich 1939b, 1947a; Wu 1997, Wolpoff 1999, Wolpoff et al. 2001) have pointed to fossils from mainland Asia as providing strong evidence for regional continuity. Since 1984, this fossil evidence has been cited in support of the Multiregional Evolution model (Wolpoff et al. 1984). However, the early modern fossils from East Asia are essentially modern in their morphology, with little or no indication of lingering archaic features of the sort seen in the African Transitional Group. The evidence for continuity (rather than replacement) in East Asia is seen in specific anatomical details that are said to be distinctive of both archaic and modern humans from this area. The regional distinctiveness of these details has been called into question by Habgood (1992, 2003), who compared the occurrence of nine such features in archaic peoples (Neandertals, Heidelbergs, and the ATG) and early modern humans from East Asia, Africa, Australasia, Europe, and West Asia. To determine the primitive state of each character, Habgood also included Erectines and Ergasters in his sample. The features he examined were:

1. Sagittal keeling (frontal and parietal) with parasagittal depressions.

2. A nondepressed nasal root.

3. Nonprojecting, perpendicularly oriented nasal bones.

4. Flat upper faces with forward-jutting zygomatic bones.

5. Shovel-shaped incisors, especially upper laterals.

6. A relatively horizontal course of the frontonasal and frontomaxillary sutures.

7. The presence of a marginal process below the zygomaticofrontal suture.

8. An angular rather than rounded junction of the zygomatic bone and the zygomatic process of the maxilla, forming a deep notch.

9. Rounding of the infraorbital margins of the orbit.

The results of this analysis are presented in Table 8.8. Habgood concludes that all of these character states are present too often in other samples (or not consistently enough in East Asia) to be considered distiinctively East Asian regional features. Nevertheless, Habgood finds some evidence for possible continuity in East Asia—not in the distribution of individual features, but rather in the fact that some of them tend to co-occur far more frequently in East Asia than elsewhere. He notes that features 2, 3, 6, and 8 are found together as a cluster in many ancient and modern East Asian skulls, but "… are not all commonly found on single individuals outside east Asia" (Habgood 2003, p. 239). He suggests that this combination (together with feature 4) may represent a single functional complex associated with facial flatness, particularly in the upper face. Overall facial flatness is primitive within *Homo* in connection with a pattern of total facial prognathism (Chapter 5). Habgood argues that strong retention or exaggeration of facial flatness and its correlated traits may represent a marker of regional continuity throughout Middle and Late Pleistocene human evolution in East Asia. Advocates of Multiregional Evolution hold that this complex, along with the East Asian maxillary incisor shoveling pattern and other morphological features, clearly reflects long-term genetic continuity in East Asia (Wolpoff 1989, 1999; Wolpoff et al. 1984, Thorne and Wolpoff 1992, Frayer et al. 1993). Although Habgood dismisses "incisor shoveling" as a regional feature because it occurs commonly elsewhere, his assessment does not take into account the distinctive regional differences in patterns of shovel-shaping (Crummett 1994, 1995). It is the *pattern* of shoveling that Wolpoff (1999) identifies as an East Asian regional feature, not just the occurrence of shoveling of any sort whatever.

As usual, a lot of the debate over these issues would be resolved by better dating and more fossils. If the generally shaky chronological estimates just discussed are close to correct, there is a gap of 60 or 70 Ky between the latest archaic East Asians and these early AMH. Perhaps transitional forms, something like the African Transitional Group, inhabited East Asia during this period and have yet to be discovered. Considering the geographic immensity of East Asia, the record for both archaic hominins and early AMH is very meager, and we can feel sure that the existing evidence does not tell the whole story. For what it is worth, that evidence suggests that modern morphology appeared relatively rapidly in East Asia and that continuity from archaics to moderns can be traced only in a few anatomical details. But as in western Asia, there is no evidence in the archaeological record for any abrupt change in human behavior in East Asia during this period. Rather, the various regional lithic assemblages of the Middle and Upper Pleistocene in China reflect a long and relatively slow *in situ* evolution, with no sign of significant outside influences (Qiu 1985, Jia and Huang 1985, Pope 1992).

▌ MODERN HUMAN ORIGINS IN AUSTRALASIA

Australasia, as defined here, comprises two regions. The first of these are the islands off Southeast Asia that lie on the relatively shallow Sunda Shelf. As discussed in Chapter 5, many of these islands would have been connected to the Asian mainland during Pleistocene drops in sea level. At the glacial maxima around 3 Mya, 1.25 Mya, and 65–45 Kya, sea level would have fallen by ≥100 m (Bellwood 1992, 1997). During these periods, Sundaland would have covered some four million square kilometers and included such islands as Java, Sumatra, Borneo, and parts of the Philippines (Fig. 8.26). The second region is Sahulland, comprising Australia, New Guinea, Tasmania and other smaller islands. These were also joined together when sea level fell. Smaller clusters of islands were connected during lesser drops in sea level, which were not uncommon during the Pleistocene (Chapter 5). However, even when sea levels were at their lowest, Sahul and Sunda were never joined. Crossing Wallace's Line between Bali and Lombok in Indonesia would always have required negotiating at least 50 km of open water. Even after making it across this barrier, people would have had to cross several other stretches of open sea as they island-hopped to Sahulland (Fig. 8.26).

Several sites on the Asian side of Wallace's Line document the early appearance of modern people in this region. We have already discussed the Ngandong and Sambungmacan samples, which represent the only archaic humans (pre-modern but post-, or very late, Erectines) in Australasia. Potential early modern human remains have been recovered from three other localities in Sundaland: Wajak (Java), Tabon Cave (Palawan Island) and Niah Cave (Borneo). These remains are all anatomically modern, but it is not clear how early they are.

The Wajak (formerly Wadjak) specimens were recovered in 1888, 1889, and 1890 (Dubois 1922), and were initially hailed as the first fossil hominin discoveries outside of Europe. They include two robust crania with pronounced brow ridges and large teeth, as well as some postcranial pieces. For many years Wajak was considered to be late Pleistocene in age, and the skulls figured prominently in discussions of the peopling of Australasia because of similarities drawn between them and recent native Australian crania (Weidenreich 1945c). Today they are pertinent mainly for historical reasons, since analysis of the fauna consigns them to a much later (Mesolithic or Neolithic) period in time (van den Brink 1982). This late date is compatible with the morphology of the skulls (Storm and Nelson 1992). Wolpoff et al. (1984) present a succinct discussion of the Wajak remains and the controversies surrounding their interpretation.

In 1959, Harrison reported (Harrison 1959) recovery of a partial cranium at a depth of 2.74 m from the Great Cave of Niah, a complex, multicomponent site on Borneo (Harrison 1957; see Figure 8.26). Postcranial remains were also recovered. An age of around 40 Ky has been claimed for the cranium, which became known as the

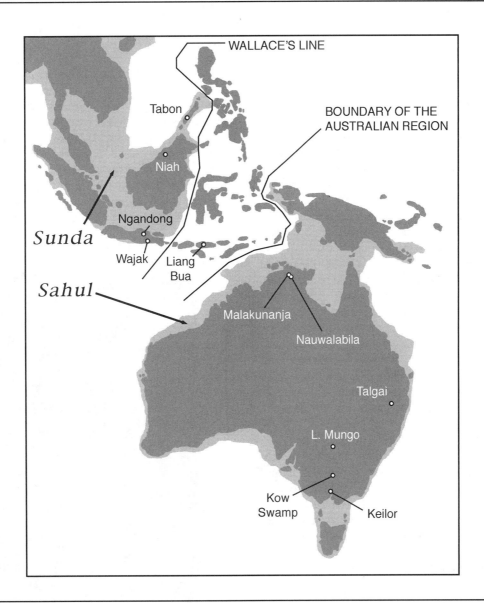

FIGURE 8.26

Late Pleistocene sites (plus Ngandong) pertinent to the origin of modern people in Australasia. The lighter gray tone indicates relatively shallow (continental shelf) areas, exposed when sea levels fell during glacial maxima.

"deep skull of Niah." This estimate is based on standard ^{14}C dates of $39{,}820 \pm 1012$ and $32{,}630 \pm 700$ years BP (before the present) on charcoal and burned bone located stratigraphically some 20 cm above the skull (Harrison 1957, 1959; Kennedy 1979). P. Bellwood (1997) and others have suggested that the specimen is an intrusive, possibly Holocene, burial and that Harrison's stratigraphic analysis was flawed. However, subsequent research at the site indicates that the "deep skull" is not intrusive, and uranium-series dating of the specimen itself accords it an age of at least 35.2 ± 2.6 Kya (Barker et al. 2007). A date somewhere between 35 and 40 Kya seems plausible.

The "deep skull" is not well preserved and required significant reconstruction. It has been variously identi-

fied as a 16-year-old adolescent (Brothwell 1960) or a young adult aged 20–30 years (Birdsell 1979). It has relatively thin vault bones, a high forehead, extremely slight brow ridge development, rounded cranial contours, a deep nasal root, a broad and short face, a broad nasal aperture, a moderately developed canine fossa, and a broad, deep palate (Kennedy 1979, Habgood 2003). The overall gestalt of the specimen is clearly modern. As with the mainland Asian material, there has been some debate concerning the biological affinities of the Niah skull. Brothwell (1960) noted its affinities to Tasmanians. Howells (1973c) found that it clustered with Tasmanians, Melanesians, and Australians. However, X. Wu's (1987) comprehensive analysis found Niah to be more closely related to East Asiatic early moderns like Liujiang and Minatogawa than to native Australians.

Tabon Cave on Palawan Island in the Philippines (Fig. 8.26) has yielded an array of fragmentary human specimens, including cranial, mandibular, and postcranial pieces. These fossils have been stratigraphically correlated to dates of 22–24 Kya on other material from the cave deposits; but the relevance of these dates is debated, because older and younger fossils may have been mixed together by animals digging in the area where the human remains were recovered (Howells 1973c, Dizon et al. 2002). A uranium–thorium date on a human specimen from the site yields an age estimate of 31 ± 8 Kya (Dizon et al. 2002, Pawlik and Ronquillo 2003). The best-known specimen from Tabon is a gracile frontal bone with a distinct forehead and moderately developed brow ridges (Howells 1973c, Wolpoff et al. 1984, Dizon et al. 2002). The interorbital area is broad, with a slightly depressed nasion and narrow nasal bones that have a "pinched" appearance (Dizon et al. 2002, Habgood 2003). Two mandibular fragments from Tabon are robust, and one has a three-rooted molar (Macintosh 1978). All the remains are modern, but (again) their affinities with living human populations are debated. Howells (1973c) suggested that Tabon represents a non-Mongoloid population with close affinities to the Ainu and Tasmanians. Others have noted similarities to Australians (Macintosh 1978, Wolpoff et al. 1984, Habgood 2003), and most researchers find resemblances to later Asians as well.

The first colonists of Australia appear to have been anatomically modern humans, able to build and navigate seagoing vessels that could cross the water gap between the Sunda and Sahul landmasses. The date of their arrival is debated. Early estimates based on archaeological evidence indicated that people had probably first reached Australia sometime between 35 and 60 Kya (Jones 1979, 1989, 1992). In the early 1990s, thermoluminescence (TL) and optically stimulated luminescence (OSL) dating techniques (Roberts et al. 1990, 1994) produced age estimates of 50–60 Ky for early artifacts from two sites, Malakunanja and Nauwalabila, along the northwestern edge of the Australian continent where immigrants from Sundaland would have been expected to make landfall (Fig. 8.26). These early dates have been questioned by O'Connell and Allen (1998), who note the possibility of considerable disturbance at both sites, argue that the dated materials may not be associated with the artifacts, and point out that an extensive ^{14}C chronology indicates an age of only some 35 Ky for the arrival of the first Australians.

Additional support for the early dates comes from a complex array of chronometric techniques applied to fossils and sediments from a site near Lake Mungo in southeastern Australia. U-series and ESR dates taken directly from an early modern human skeleton from this site, Willandra Lakes Hominid 3 (WLH 3, or Mungo 3: Fig. 8.27), yield an age estimate of 62 ± 6 Ky (Thorne et al. 1999). This estimate is in general agreement with an OSL date of 61 ± 2 Kya on the sediment in which the Mungo (WLH) 3 burial was placed, and a U-series date of 81 ± 21 Lya on the calcite matrix adhering to the bones.

Critics of these results (Bowler and Magee 2000, Gillespie and Roberts 2000) argue that the early dates are inconsistent with the stratigraphy of the locality and with a TL date of 43 ± 3 Kya for the deposits. Although

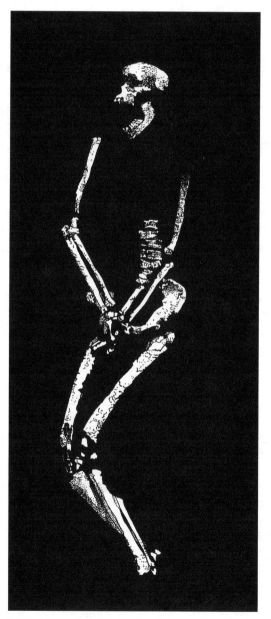

FIGURE 8.27

The Lake Mungo (Willandra Lakes) Hominin 3 burial. Note the relatively elongated distal limb segments.

these criticisms were rebutted (Grün et al. 2000), Bowler et al. (2003) subsequently reported OSL dates that indicate an age of 38–42 Ky for WLH 3. Stone flakes from the area are bracketed between 50,100 ± 2400 and 45,700 ± 2300 years BP, which may represent a date for the first occupation of the Willandra Lakes area. But even if we accept this as a date for the arrival of humans in southeastern Australia, the first immigrants into the continent must have landed in the northwest a few thousand years earlier. Therefore, no matter which dates we accept, the early moderns known from Sundaland (Wajak, Tabon Cave, Niah Cave) are probably too recent to represent the source population for the first Australians.

What was that source population (or populations)? The most widely held theories posit multiple waves of colonization from different sources (Brown 1997, Laitman and Thorne 2000, Habgood 2003). From his studies of variation in living Australians, Birdsell (1969, 1979) concluded that three successive waves of immigrants—Asian Negritos, non-"Mongoloid" Asians, and a third group possibly originating in India—had entered Australia and interbred in varying proportions to form various Australian populations. Thorne (1976, 1980; Laitman and Thorne 2000) proposes a dual-source hypothesis, based mainly on the pertinent human fossil record from Australia and East Asia. In Thorne's model, native Australians are the descendants of two groups of immigrants: a robust, archaic-looking stock from Java, and a more gracile morph from Pleistocene China.

The first early Australian human fossil to be discovered was a distorted juvenile cranium found in 1886 at Talgai in Queensland, eastern Australia. Because it lacks an archaeological context, Talgai's antiquity has always been open to question (Brown 1997). Smith (1918) described the Talgai cranial vault as similar to those of native Australians, but the distortion of the skull led him to the mistaken conclusion that the lower face and canines were persistently ape-like. They are not (Hellman 1934). Nevertheless, Talgai is distinctly robust for a juvenile, with a characteristically Australian vault form and the facial form, prognathism pattern, pronounced brow ridge, and large teeth that Thorne attributes to the robust (Java-like) morph of early Australians.

The next fossil Australian to be unearthed showed a different morphological pattern. Recovered near Melborne in 1940, the Keilor skull is considered to date to at least 12 Kya. Other material from the site has yielded a maximum radiocarbon age of 38.8 Ky (O'Connell and Allen 1998). Keilor is more gracile than Talgai, with little prognathism, weak brow ridge development, and a less robust face. Weidenreich (1945c) nevertheless viewed it as a "Wadjak type," connecting living Australians to the Ngandong group of Javan archaics. Thorne (1977, 1980), however, sees Keilor as more like Chinese Paleolithic specimens, especially Liujiang, and thus as

evidence for an East Asian contribution to the genesis of native Australians. The morphological dichotomy represented by these two skulls, discovered in the early days of Australian paleoanthropology, underlies the ongoing debate about the peopling of Australia.

In 1925, a robust cranium was unearthed in southeastern Australia at the edge of Kow Swamp, near Cohuna in New South Wales (Keith 1931). Subsequent excavation of Kow Swamp, beginning in 1967, has uncovered the bones of 30 individuals in various degrees of completeness (Thorne and Macumber 1972, Thorne 1976, 1977, 1980). Radiocarbon-based age estimates for the Kow Swamp sample range from 9.5 to 13 Ky (Brown 1997). The Kow Swamp crania generally exhibit the characteristics of Thorne's robust (Java) group (Figs. 8.28, 8.29)—so much so that Thorne has described them as demonstrating "... the survival of *Homo erectus* features in Australia until as recently as 10,000 years ago" (Thorne and Macumber 1972, p. 316). Other specimens that are usually placed in Thorne's "robust" group come from sites at Lake Nitchie, Cossack, and Willandra Lakes (Thorne 1977, Freedman and Lofgren 1979, Webb 1989).

The Willandra Lakes sample, also from New South Wales, first came to light in 1969 with the discovery of the cremated remains of a female (Mungo 1 or WLH 1) and fragments of a possible male (Mungo 2 or WLH 2). The sample now comprises the (mostly very fragmen-

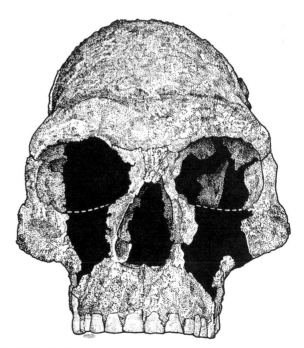

FIGURE 8.28

The Kow Swamp 1 early modern Australian cranium: frontal view. Note the robust face and supraorbital torus.

tary) remains of 135 individuals, many of which have been cremated (Webb 1989). All of these specimens were deposited before 15 Kya, when the lake deposits ceased forming, but the precise dating for most of them is unknown (Webb 1989). The dating controversy for WLH (Mungo) 3 has already been discussed. The other specimens are younger. AMS dates for WLH 1 range from $16,940 \pm 635$ to $25,120 \pm 1380$ BP. Direct dating of five other WLH specimens yields age estimates from around 11,600 to 18,600 years BP (Webb 1989). Several of these specimens, including WLH 1 and WLH 3, fall into Thorne's gracile (China) group (Bowler et al. 1972, Bowler and Thorne 1976). WLH 3, for example, has a higher, more rounded cranial vault and weaker brow-ridge development than typical "robust" specimens like Kow Swamp 5 (Brown 2000). But as at Kow Swamp, there is variation in the sample, and the WLH 50 calvaria (Fig. 8.29) is often pointed to as a typical representative of the robust group. Some think that WLH 50 is at least 30 Kya and probably far older. Preliminary ESR and gamma-ray spectrometry dating have yielded age estimates of 29 and 14 Ky, respectively (Habgood 2003).

Wolpoff (1999) has argued that although WLH 3 is less robust than many other early Australians (including WLH 50), it is in many ways more like robust Australians and even Ngandong than like early modern specimens from China. For example, Wolpoff describes the WLH 3 face as showing marked alveolar prognathism (as in the Kow Swamp skulls) and an everted temporal process of the zygomatic, "... giving the face a pentagon-like profile as seen from the front (like Sangiran 17 and many more recent Australians)" (Wolpoff 1999, p. 739). He also notes that WLH 3 exhibits a lateral frontal trigone (see Chapter 6)—a feature limited to the Ngandong skulls among archaic hominins. WLH 3 is also significant because a mitochondrial DNA sample from the specimen has been interpreted as an ancient, non-African-derived modern human mtDNA lineage (see below). Finally, WLH 3 appears to have the linear body build typical of tropical humans (Fig. 8.27). This has less significance as a marker of replacement versus continuity in this part of the world than in western Eurasia, since archaic people in Australasia would presumably have had more tropical body proportions to begin with. However, such a body build is not evident in the Ngandong tibia (Chapter 6).

The most critical single specimen in recent discussions of native Australian origins is Willandra Lakes Hominid 50, which comprises a robust calvaria and some additional fragments. Despite their importance, the WLH 50 remains have yet to be completely described, and only the calvaria is well known (Fig. 8.29). The maximum breadth of the long, broad cranial vault is located very low down, at the level of the cranial base. This is a primitive feature, characteristic of more archaic humans. However, the biparietal breadth is only slightly smaller, and the braincase is expansive, with a capacity

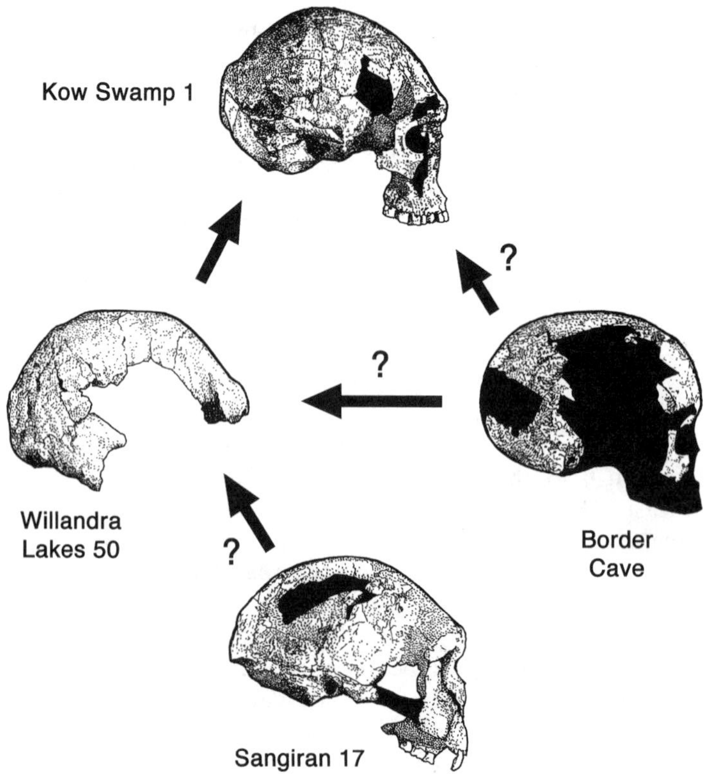

FIGURE 8.29

Diagram showing the general similarities in vault form between Sangiran 17, Willandra Lakes Hominin 50, and Kow Swamp 1. These similarities contrast with the form seen in Border Cave 1 and suggest a local component in the emergence of early modern Australasians. (After Frayer et al. 1993.)

estimated by P. Brown at 1540 cm³ (Habgood 2003, p. 185). The frontal bone of WLH 50 is long and low, with a prominent brow ridge that is incipiently divided into superciliary arches and supraorbital trigones.

There is no reason to regard WLH 50 as anything but a modern human (Hawks et al. 2000b), and the skull exhibits a number of features that are common on more recent Australian crania (Brown 1989, Webb 1989, Wolpoff 1999, Habgood 2003). However, Hawks et al. (2000b) contend that other characteristics of WLH 50 specifically link it to Ngandong. For example, although the lateral parts of both brow ridges are missing on WLH 50, Hawks and his coauthors argue that the existing bony contours show that lateral frontal trigones would have been present. They also suggest that the nuchal torus area would have been Ngandong-like, though its morphology is unfortunately obscured by surface erosion. In a multivariate (discriminant-function) metric analysis, they find that WLH 50 falls closer to Ngandong than to other potential ancestors (Skhūl/Qafzeh and Late Pleistocene Africans). This finding is supported by their analyses of pairwise differences of nonmetric data for individuals within these samples. Hawks and his co-workers conclude that these analyses support a dual derivation of early modern humans in Australia, from Africa and S/Q on the one hand and from Ngandong on the other (Hawks et al. 2000b, Wolpoff et al. 2001). These studies imply that the archaic Indonesian contribution was comparable in impact to the influences from more modern populations derived from the West.

Others reject these conclusions. A multivariate analysis by Stringer (1998) shows a distinct separation between Ngandong and WLH 50 and places WLH 50 closer to S/Q than to modern native Australians (at least on the first principal component). Brown (1992) argues that overall cranial shape in WLH 50 has little to do with Ngandong. Brown notes, for example, that WLH 50 has a considerably higher vault with a much less-developed supraorbital region and different occipital morphology. Habgood (2003) asserts that the comparison summarized in Figure 8.29 would yield different impressions of sample affinities if other specimens were used—for example, if WLH 3 or Keilor were substituted for WLH 50, or if Qafzeh 6 was used rather than Border Cave. Lastly, Webb (1989, 1990) claims that the unusual thickening of the WLH 50 braincase is pathological, reflecting the impact of anemia (but cf. Hawks et al. 2000b). Except for Habgood, none of these authors sees any evidence of significant contributions from archaic Indonesian humans to the origin of early modern Australasians.

The characteristics that are claimed to reflect regional continuity in Australia vary from study to study, and the conclusions drawn vary with them. Habgood (2003) analyzed the following traits taken from sources arguing for continuity (see Table 8.9 for references):

1. Flatness of the frontal bone viewed in the sagittal plane.
2. Posterior placement of the minimum frontal breadth (well behind the orbits).
3. Relatively horizontal orientation of the inferior supraorbital border.
4. Distinct pre-bregmatic eminence.
5. Low position of maximum parietal breadth (at or near the parietomastoid angle).
6. Marked facial prognathism.
7. Prominent zygomaxillary (malar) tuberosity.
8. Eversion of the lower border of the zygomatic.
9. Rounding of the inferior-lateral border of the orbit.
10. Lower border of the nasal aperture with no distinct line dividing the nasal floor from the subnasal face of the maxilla.
11. Marked curvature of the posterior alveolar plane of the maxilla (corresponding to the mandibular "curve of Spee").
12. The pattern of facial and dental reduction (especially of the posterior teeth).

Tallying the distribution of these features in other *Homo* (Table 8.9), Habgood concluded that all but characters 7 and 11 were primitive retentions from Erectines, Ergasters, and/or Heidelbergs. Several were also frequently present in non-Australian samples (particularly 1, 3, 5–7, and 11). As in his analysis of the East Asian evidence, Habgood concludes that none of these features by itself evinces regional continuity, but that a cluster of co-occurring features (Nos. 1, 2, 6, 7, and 8) provides some evidence for continuity between Indonesian archaics and modern Australian humans.

In a different analysis of continuity, Wolpoff et al. (2001) compared 16 nonmetric cranial characters for WLH 50 and various fossil specimens from Ngandong, Africa, and the Levant (Table 8.10). Only one of these characters (the pre-bregmatic eminence) is included in the Habgood analysis. They found that WLH 50 differed from the Ngandong specimens in only 3.7 features on average, versus averages of 7.3 and 9.3 differences in comparisons with S/Q and African crania respectively. In a similar study of 17 features of the mandible, Kramer (1991) likewise found that eight features link modern Australians to the Indonesian Erectine sample, while only three uniquely link the Australians to modern African specimens.

Many of the candidate features of regional continuity in Australasia are primitive traits, supposedly transmitted from Indonesian archaic peoples to modern human immigrants passing through Java en route to Australia. An orthodox cladistic systematist might protest that such features are symplesiomorphies, which cannot be

TABLE 8.9 ■ P. Habgood's (2003, p. 193) Assessment of "Regional Continuity Features" for Australasia[a]

Crania	Features											
	1	2	3	4	5	6	7	8	9	10	11	12
Sample: Erectines and Ergasters												
KNM—ER 3733	—	—	—	—?	Y	Y	—?	—	—	Y	?	?
KNM—ER 3883	Y	—	—	—	Y	?	—	—?	Y/—	?	?	?
Olduvai 9	Y	—	Y	?	Y	?	?	?	?	?	?	?
Sangiran 17	Y	Y	Y	Y/—	Y	Y	Y/—	Y	Y/—	Y	?	?
Ngandong 11	Y	Y	Y	Y	Y	?	?	?	?	?	?	?
Zhoukoudian[b]	Y/—	—	—	—	Y	Y	—	—	Y	Y	—	Y
Sample: Australia												
WLH 50	Y	Y	Y	—	Y	?	Y	Y	Y	?	?	?
Kow Swamp 1	Y	Y	Y	—	Y	Y	Y	Y	Y	Y	Y	Y
Kow Swamp 5	Y	Y	Y	?	?	Y	Y	Y	Y	Y	?	Y
Kow Swamp 15	—	Y?	Y	?	?	Y	Y	Y	Y	Y	Y?	Y
Cohuna	Y	Y	Y	Y	Y	Y	Y	Y	Y	Y	Y	Y
Talgai	Y	Y?	Y?	?	Y	Y	?	Y?	Y?	Y	Y?	Y
Cossack	Y	Y/—	Y	—	Y	Y	Y	Y?	Y	?	Y	Y
Mossgiel	Y	Y	Y	—?	Y	Y?	Y	Y	Y	?	Y?	Y
Lake Nitchie	—	—	Y	—	Y	Y	Y	Y	Y	Y	Y	Y
Keilor	—	—	Y	—	Y	Y	Y?	Y	Y	Y	?	Y
WLH 1	—	—	Y	—	—	?	—	?	Y	?	?	?
WLH 3	—	—	Y	—	Y	?	Y	?	?	?	?	?
Lake Tandou	—	—	Y	—	Y?	Y	Y?	Y	Y	?	Y	Y
Sample: Sundaland												
Niah	—	?	?	—	?	Y	—	—	Y	Y	?	?
Wajak 1	—	—	Y?	—	Y	Y	—	—	Y	Y	?	?
Wajak 2	?	?	Y?	?	?	?	?	?	?	Y	—	Y
Sample: East Asia												
Maba	—	—	—	?	?	?	?	?	?	?	?	?
Liujiang	—	—	Y?	—	—	Y	—?	—	Y	Y	—	Y
Upper Cave 101	—	—	Y	—	Y	Y	—?	—	Y	Y	—	Y
Upper Cave 102	Y	—	Y?	Y	Y	Y	—?	—	Y	Y	?	?
Upper Cave 103	—	—	Y	—	Y	Y	—?	Y	Y	Y	?	?
Sample: Africa South of the Sahara												
Bodo	—?	—	Y	?	?	Y	—	Y	Y	Y?	?	?
Ndutu	?	—	—	?	Y	Y	?	—	?	?	?	?
Kabwe 1	Y	—	Y	—	Y	Y	Y	—	Y	Y	Y	Y
Laetoli 18	Y	—	Y	—	Y	Y	?	?	?	Y	?	?
Omo Kibish 1	—	—	?	—	?	?	—?	?	Y	?	?	?
Border Cave 1	—	—	—	—	?	?	Y?	?	Y	?	?	?
Florisbad	—	—	—	—	?	?	—?	?	Y	—	?	?
Sample: North Africa—Western Asia												
Jebel Irhoud 1	—	—	—?	—	Y	Y	—?	—	Y	Y	?	?
Wadi Halfa[c]	Y/—	—	—	—	Y/—	Y	Y	Y	Y?	?	—	Y
Zuttiyeh	—	—	—	—	?	?	—?	—	—	?	?	?
Skhūl 4	—	—	?	?	Y	Y	—	—	Y	Y	—	Y
Skhūl 5	—	—	Y	—	—	Y	—	—	Y?	?	—	Y
Jebel Qafzeh 9	—	—	Y	?	—	Y	—?	—	Y	Y	—	Y
Amud 1	Y	—	—	—	—	Y	—?	—	—	—?	—	Y

TABLE 8.9 ■ Continued

Crania	Features											
	1	2	3	4	5	6	7	8	9	10	11	12
Sample: Europe												
Petralona	Y	—	Y	—	Y	Y	—?	—	Y	—	—	Y
Arago 21	Y	Y	—	—	Y	Y	—	—	Y	—	—	Y
Steinheim	Y	—	—	—	Y	Y	—	—	Y	?	—?	Y
La Ferrassie 1	Y	—	—	—	—	Y	—	—	Y	—	—	Y
Cro-Magnon 1	—	—	Y	—	—	—	?	—	Y	?	?	?
Oberkassel 1	—	—	Y	—	—	—	Y	Y	Y	?	—	?
Oberkassel 2	—	—	Y	—	—	—	—	—	—	?	—	Y
Dolní Věstonice 3	—	—	Y	—	—	—	—	—	—	?	Y?	Y
Předmostí 3	—	—	—	—	—	—	—	—	—	—	—	Y
Předmostí 4	—	—	—	—	—	—	—	—	—	—	—	Y
Mladeč 1	—	—	—	—	—	—	—	—	—	—	?	?

[a]"Regional Continuity Features" as proposed by Thorne and Wolpoff (1981), Wolpoff et al. (1984), and Frayer et al. (1993). Numbers for features correspond to those listed in the text. Y = present;— = absent; ? = morphology unknown, identification questionable, or region not known. Y/— = conflicting assessments by different observers.
[b]Weidenreich's Zhoukoudian reconstruction.
[c]Observations on a sample of Wadi Halfa specimens.

TABLE 8.10 ■ Pairwise Comparison of WLH 50 to Indonesian, African, and Levantine Fossil Specimens[a]

Ngandong 5	2	Omo Kibish 1	6
Ngandong 9	3	Qafzeh 6	7
Ngandong 10	3	Qafzeh 9	7
Ngandong 11	3	Skhūl 5	7
Ngandong 4	4	Omo Kibish 2	8
Ngandong 1	5	Jebel Irhoud 1	10
Skhūl 9	5	Ngaloba	10
Ngandong 6	6	Singa	12
Jebel Irhoud 2	6		

[a]The number given for each specimen represents the number of differences between it and WLH 50 (from Wolpoff et al. 2001, p. 295).

used as evidence of biological relationships. This objection would be relevant if we were dealing with differences between species, but it is not relevant here. If a primitive trait is not retained in the earliest modern humans from Africa, its presence in later modern humans from Australia is strong evidence against the claim that those early Africans are the exclusive progenitors of all modern humans. Symplesiomorphies lingering on the Eurasian periphery provide a good reason for thinking that local archaic populations must have contributed to the genetic makeup of early modern populations in those peripheral areas.

Three more points need to be made about the origin of modern people in Australasia. First, the only post-Erectine, pre-modern fossils known from this large area of the Old World are the Ngandong specimens—which lack faces. Our ability to understand the pattern of modern human emergence in Australasia is correspondingly limited. It will remain limited until we know more about post-Erectine populations in this region. Second, it is difficult to determine whether the peopling of Australia resulted from multiple waves of immigration from several centers or a much more limited incursion from a single source. Ordinarily, we would expect such migrations to involve multiple waves of immigrants; but for Australia, the difficulties that early humans would have faced in crossing from Sunda to Sahul must have markedly restricted the opportunities for immigration. At this point, it is impossible to be certain how many waves or sources of origin were involved.

Finally, Australia and Tasmania offer the opportunity to assess the effects of geographic isolation on human populations during the late Pleistocene. Throughout most of the early period of Sahul's occupation, connections between its inhabitants and the rest of the world must have been severely limited. Yet the native Australians remained conspecific with the rest of *Homo sapiens*. Even the Tasmanians, who were isolated from other Sahul populations by the Bass Strait for at least 8 Ky and possibly for as much as 13 Ky (Wolpoff 1999), were fully able to interbreed with modern humans from the other side of the planet. In fact, genetic studies (Pardoe 1991) show that the Tasmanians were no more different genetically from other Australians than they would have been expected to be if the Bass Strait had been perpetually dry land. The Australian–Tasmanian example suggests that it would take a much longer period of complete peripheral isolation to produce speciation in the genus *Homo*. We will return to this issue at the end of this book.

EUROPE: THE LAST FRONTIER

As we noted earlier, the onset of the Upper Paleolithic industry called the Aurignacian is generally taken as an archaeological marker of the appearance of anatomically modern humans in Europe. The Aurignacian is characterized by many new forms of material culture, including distinctive types of blades, beaked burins, a variety of end scrapers including keeled or carinated scrapers, split-based bone points and other standardized bone tools, and especially by its impressive cave paintings and carvings. These striking differences between the Aurignacian and Mousterian have been documented and emphasized by many archaeologists (Bordes 1961, 1968, 1984; Gamble 1986, 1999; Klein 1999, Mellars 1989, 1993, 2002, 2004; White 1986, 2003; Zilhão and d'Errico 2003a,b). The apparently sharp break between the two cultural horizons has long been one of the main supports for the idea that immigrating moderns abruptly replaced the European Neandertals. But the break between the two cultures is not as sharp as was once thought; and recent discoveries have cast doubt on almost all of the supposed associations between early Aurignacian artifacts and the bones of anatomically modern humans.

Radiometric dates for the Aurignacian range from 42.8 to 30.2 Kya (Churchill and Smith 2000a). The earliest dated Aurignacian assemblages come from El Castillo, Spain (five AMS dates from 39.8 to 42.2 Kya); Geißenklösterle, Germany (AMS and TL dates, 33.1 to 40.2 Ky); Trou Magrite, Belgium (AMS, 41 Kya); Willendorf, Austria (^{14}C dates of 39.5 and 41.7 Ky); and Samuilica Cave, Bulgaria (^{14}C, 42.8 Kya). The early part of this time range overlaps with various Initial Upper Paleolithic (IUP) industries (Table 8.11). Where these IUP industries are associated with human remains, those remains are either Neandertals (at the French sites of Arcy-sur-Cure, Grotte du Renne, and Saint-Césaire) or nondiagnostic but arguably Neandertal-like fragments at various sites in Central Europe (Glen and Kaczanowski 1982, Smith 1984, Allsworth-Jones 1990, Churchill and Smith 2000a). Of these early Aurignacian dates, only the El Castillo dates are associated with human remains—and these too are nondiagnostic (see below).

If we accept the conventional view that the Aurignacian was made exclusively by modern humans, these dates imply an overlap of Neandertals and early moderns for several thousand years in Europe. As noted in Chapter 7, Zilhão and d'Errico (1999, 2003b; d'Errico et al. 1998) question this overlap, claiming that the early Aurignacian dates (>36.5 Kya) and many of the IUP dates are inaccurate for various reasons. In their view, the IUP cultures

TABLE 8.11 ■ Age Ranges for Initial Upper Paleolithic Industries Based on Available Radiometric Date Estimates[a]

Industry (Location)	ND[b]	Range (Ky)
Châtelperronian (France and Northeastern Spain)	11	31–45.1
Uluzzan (Italian Penninsula)	2	32–33, >31
Leaf Point Early Upper Paleolithic (Northern Europe)	4	27.2–~45
Szeletian (Central Europe)	5	40.2–43
Bachokirian (Central Europe)	14	33.8–46
Bohunician (Central Europe)	2	34.5–43

[a]Bohunician dates from Svoboda (2003); other data from Churchill and Smith (2000a, pp. 66–68).
[b]ND, number of dates.

always predate the Aurignacian. If this chronology is correct, then the personal-adornment objects, bone implements, and other "advanced" tools found with Neandertals at sites like Arcy-sur-Cure (Hublin et al. 1996, R. White 2001) and possibly Vindija (Karavanić and Smith 1998) must represent independent inventions by the Neandertals themselves—not, as some have suggested, "borrowings" copied by Neandertals from modern human immigrants living in the neighborhood.

In denying extensive temporal overlap between Neandertals and moderns, Zilhão and d'Errico reject the claim that the Châtelperronian is stratigraphically younger than the Aurignacian at some sites in the Perigord of France (Laville et al. 1980). However, such interstratification has been demonstrated recently at the Châtelperronian type-site (Gravina et al. 2005). This finding supports the original interpretation of the cultural sequences at the Perigord sites. It provides further evidence for a considerable period of overlap of Neandertals and early modern populations in this part of Europe—assuming, that is, that the Aurignacian is always reliably associated with early modern humans.

It is now generally conceded that the Châtelperronian and other pre-Aurignacian IUP complexes in Europe derive from European Mousterian antecedents. But it has also been suggested that some aspects of the Aurignacian derive from IUP antecedents in Central Europe. Svoboda (2003) argues that although the Bohunician IUP cannot be a direct precursor of the Aurignacian, some role for it in the origins of the Aurignacian cannot be excluded. Otte and Kozlowski (2003) point to the Bachokirian IUP as a potential Aurignacian ancestor. Since there are no known precursors for the Aurignacian outside of Europe, it may well be that the emergence of the Aurignacian is an Eastern/Central European phenomenon (Teyssandier 2005, 2006; Teyssandier et al. 2006). The early dates for the Aurignacian from Geißenklosterle, and other data from the Swabian Jura of Germany, have led some to suggest that Aurignacian peoples first migrated up through the Danube Valley and spread from there throughout central and Western Europe (Conard 2003, Conard et al. 2003b, Conard and Bolus 2003).

There is little evidence for the biological identity of the makers of the Aurignacian, particularly in its early manifestations. It is widely assumed that the Aurignacian is connected with modern humans. This assumption originated with a supposed association between Aurignacian artifacts and modern human remains in the famous Cro-Magnon rock shelter in southwestern France. But recent discoveries have called this and other such associations into question. Table 8.12 lists a series of European AMH remains once thought to be associ-

ated with the Aurignacian, or at least to date from the Aurignacian time span. Re-dating of these remains, mostly by direct AMS dating of the human remains themselves, has yielded surprising results. The modern human specimen from Velika Pećina in Croatia, once thought to date back 34 Ky, has been shown to be Holocene in age (Smith et al. 1999). The Cro-Magnon fossils and the lesser-known La Rochette remains, also from France, are less than 28 Kya and are now thought to fall into the Gravettian cultural period following the Aurignacian (Henry-Gambier 2002, Ortscheidt 2002). The Zlatý Kůň remains from the Czech Republic have been redated from >30 Kya to the late Upper Paleolithic, around 13 Kya. Other influential specimens formerly dated to >30 Kya (Vogelherd, Hahnöfersand) have been shown to be less than 10 Ky old. The only significant AMH fossils from Europe still thought to be possibly as old as 30 Ky are the Mladeč remains from the Czech Republic. The Mladeč sample therefore becomes uniquely important for our understanding of the affinities of the earliest modern Europeans.

The site of Mladeč, known as Lautsch in German, is a complex cave system located in the Moravian Karst region. The site has been known since 1829, but the first systematic excavations were carried out in 1881 and 1882 by J. Szombathy, who uncovered human fossil remains in the main cave (Szombathy 1925). Subsequent excavations by other scientists between 1903 and 1922 yielded a large series of human remains from other parts of the cave system. Svoboda (2000) has shown that the Mladeč site was not a habitation site; the human remains were probably dropped into the cave from above through a vertical chimney. However, there is general consensus that the human remains are associated with the artifacts recovered from the site, which include a number of massive-based (or Mladeč) bone points. These points are common in the early Upper Paleolithic, particularly in central Europe, but they are also found in late Mousterian and Initial Upper Paleolithic contexts as well (Svoboda 2005).

At least 137 human skeletal elements were recovered from Mladeč, but unfortunately most of them were destroyed at the end of World War II in a fire at Mikulov Castle on the Czech-Austrian border (Smith 1997f). Szombathy (1925) provided some descriptions and illustrations of selected pieces, but most of the specimens were never thoroughly studied. What we know of the Mladeč hominins is based mainly on some 30-odd pieces recovered by Szombathy and housed in Vienna (the Mladeč 1 and 2 crania, the Mladeč 3 child's skull, the Mladeč 8 face, and several postcranial fragments), plus some additional hand and cranial bones and one cranium (Mladeč 5) that miraculously survived the Mikulov fire. These specimens have been discussed and analyzed by many paleoanthropologists (Jelínek 1969, 1976, 1983; Wolpoff 1980a, 1999; Smith 1982, 1984, 2002; Frayer 1978, 1986; Frayer et al. 1993, Bräuer and Broeg 1998,

TABLE 8.12 ■ Selected European Human Fossils Suggested to Be Associated with the Aurignacian or to Derive from the Aurignacian Time Period, Compared with Recent Direct Dates on the Fossils

Site	Previous Date (Basis of Dating)	Current Direct Date (Reference)
Mladeč (Czech Republic)	34–35 Kya[a]	29.7 Kya[f] (Wild et al. 2005)
Vogelherd (Germany)	30.2–31.9 Kya[b]	4.7 Kya[g] (Conard et al. 2004)
Velika Pećina (Croatia)	>33.9 Kya[c]	5 Kya[h] (Smith et al. 1999)
Hahnöfersand (Germany)	36.3 Kya, 35 Kya[d]	7 Kya[i] (Terberger et al. 2001)
Cro-Magnon (France)	30–35 Kya[e]	27.7 Kya[j] (Henry-Gambier 2002)
Zlatý Kůň (Czech Republic)	30–35 Kya[e]	12.9 Kya[h] (Svoboda et al. 2002)

[a]Carbon-14 dating of calcium carbonate crust that supposedly sealed the human remains at the "Dome of the Dead" locality at Mladeč (Svoboda et al. 2002).

[b]Carbon-14 dating from Aurignacian level where human remains were recovered in 1931 (see Churchill and Smith 2000a).

[c]Carbon-14 dating from Upper Paleolithic (Aurignacian ?) level where human remains were recovered in 1961 (see Smith et al. 1999).

[d]Carbon-14 and amino-acid racemization dates, respectively, determined by R. Protsch (see Terberger et al. 2001).

[e]Date inferred from presumed association with early Aurignacian.

[f]Mean of six dates ranging from 26.3 Kya to 31.5 Kya from six different Mladeč human specimens (Wild et al. 2005, p. 334).

[g]Mean of 12 dates ranging from 3.6 Kya to 5.2 Kya from five different Vogelherd human specimens (Conard et al. 2004, p. 200).

[h]Direct single date on the fossil specimen.

[i]Average of two dates reported in Terberger et al. 2001 (pp. 521, 523).

[j]Not a direct date on a Cro-Magnon human specimen but a date on *Littorina* shell collected in 1868 in association with the human remains from the Cro-Magnon rock shelter.

Wolpoff et al. 2001, Schwartz and Tattersall 2002, Bräuer et al. 2004, Smith et al. 2005, Frayer et al. 2006, Minugh-Purvis et al. 2006, Trinkaus et al. 2006a, Wolpoff et al. 2006).

The Mladeč 1 skull is robust but anatomically modern in form, with rounded vault contours, a well-developed forehead, a relatively orthognathic face, modern supra-orbital form, distinct canine fossae, and angled zygomaticoalveolar margins (ZAM). The vault is somewhat low for an Early Upper Paleolithic specimen (Table 7.4); the brow ridges are moderately developed; and there is a small occipital hemibun (see below). Overall, the skull appears quite similar to the "Old Man" of Cro-Magnon (Figs. 8.32–8.34)—a similarity noted by Szombathy (1925), who regarded the Mladeč remains as representatives of the "Cro-Magnon race." The Mladeč 2 cranium is even more gracile and more fully modern in vault form. The Mladeč 8 lower face is large and has a very broad nasal aperture (Table 7.11), but otherwise is not distinctively like Neandertals. Judging from the photographs published by Szombathy (1925), Mladeč mandibles had modern chin structures and generally resembled other Upper Paleolithic specimens. Although some Mladeč limb bones seem to have had expanded joint surfaces, the surviving postcranials are basically modern, with femoral pilasters and relatively thin long-bone walls (Smith 1984, Trinkaus et al. 2006a).

However, both Mladeč femora have proximal lateral femoral flanges, like those of European Neandertals (Chapter 7). Several researchers have noted other similarities between Mladeč 5 and Neandertals (Jelínek 1969, 1983; Wolpoff 1980a, 1999; Smith 1982, 1984; Frayer 1986, Frayer et al. 1993, Frayer et al. 2006, Wolpoff et al. 2001). Mladeč 5 (Fig. 8.30) has a low cranial vault, with a vault-height index falling at the European Neandertal mean of 57 (Table 7.4). Although its mastoid processes are not as small or as medially inclined as those of Neandertals, they do not protrude as much below the cranial base as those of typical male Upper Paleolithic or recent skulls (Table 7.8). As in Neandertals, this is probably due to the inferior projection of the surrounding occipitomastoid region. As noted in Chapter 7, the inferior projection of this part of the cranial base is one of the features associated with the development of an occipital bun. Mladeč 5 has at least limited development of all of those features, and it accordingly sports one of the best-developed occipital buns found among Early Upper Paleolithic (EUP) specimens. This bun clearly differs from those of Neandertals. The occipital buns of Mladeč 5 and other modern humans are restricted to the area near the midline of the skull, rather than extending laterally as in Neandertals, and they tend to be positioned lower on the vault. They have been dubbed **hemibuns** (Smith 1982) to distinguish them from the true occipital buns of Neandertals. Although Mladeč 5's hemibun is the most laterally extended of any EUP spec-

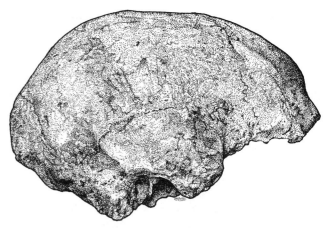

FIGURE 8.30

The Mladeč 5 cranium (Czech Republic), lateral view. This specimen exhibits a high forehead but a relatively low vault and an occipital bun.

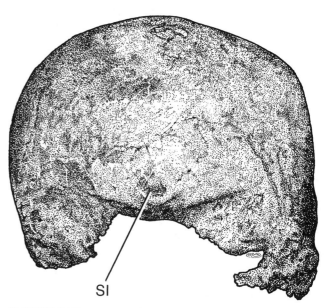

SI

FIGURE 8.31

The Mladeč 5 cranium (Czech Republic), posterior view. From the rear, Mladeč 5 has a modern-shaped vault and a small, centrally restricted suprainiac fossa (SI).

imen, it is not that of a Neandertal. Mladeč 5 also has a small occipital depression that can be interpreted as a suprainiac fossa (Fig. 8.31), but this feature too is not identical to the suprainiac fossae of Neandertals. It is smaller and more circular, and it is restricted to the midline area. Viewed from the rear (Fig. 8.31), Mladeč 5 exhibits a basically modern "*en maison*" vault shape. And although its supraorbital region is impressively

TABLE 8.13 ■ Pairwise Differences Between Mladeč Male Crania and Those of Skhūl, Qafzeh, and European Neandertals

Specimen	Number of Differences from Mladeč 5 (30 Features)	Number of Differences from Mladeč 6 (22 Features)
Skhūl 4	8	11
Qafzeh 6	10	6
Skhūl 9	11	11
Spy 2	12	8
La Chapelle	13	7
Qafzeh 9	15	14
La Ferrassie 1	15	9
Guattari 1	15	7
Skhūl 5	23	15

Source: Wolpoff et al. (2001, p. 295).

large, exceeding that of most Neandertals in several dimensions, it is thoroughly modern in having distinct superciliary arches and supraorbital trigones (Fig. 8.4).

Wolpoff et al. (2001) focused on Mladeč 5 and another male skull, Mladeč 6, in arguing the case for regional continuity in Europe. As in their study of WLH 50 from Australia, they scored a series of nonmetric features in the Mladeč skulls and compared them to a series of Skhūl/Qafzeh (S/Q) and Neandertal males. By their reckoning (Table 8.13), Mladeč 5 shows about the same average number of differences from Neandertals (14.8) as it does from the S/Q specimens (14.0); and Mladeč 6 exhibits markedly fewer average differences from Neandertals (7.8) than from S/Q (11.6). They conclude, just as they did in the Australian case, that a dual origin for early modern Europeans—partly from invading moderns, and partly from local archaics—cannot be ruled out.

The Neandertal-like aspects of the Mladeč male crania have been called into question, particularly by G. Bräuer and his colleagues H. (Bräuer and Broeg 1998, Bräuer et al. 2004, 2006). They argue that the hemibuns of the Mladeč and other EUP specimens are not related to the Neandertal pattern, but resemble the bunlike occipital protuberances found in earlier African (Jebel Irhoud, Ngaloba) and recent North African and other early modern specimens. A similar argument is made for the suprainiac fossa. Bräuer and Broeg emphasize the essentially modern form of the Mladeč 5 brow ridge, and argue that the Mladeč and other EUP crania are fundamentally modern and not like Neandertals in vault shape and the anatomy of the face, nuchal torus, and occipitomastoid region.

There is no doubt that there are systematic differences between the states of these characters in EUP people and those in Neandertals. But morphological

differences do not rule out homology or genetic continuity between the Neandertal and EUP conditions. We will return to this issue below.

Apart from the Mladeč remains, most of the other human fossils that can still be linked to the early phases of the Aurignacian are fragmentary and/or undiagnostic (Churchill and Smith 2000a). For example, at El Castillo in Spain, one tooth and three small cranial fragments from an adult and a partial mandible of a child are associated with Aurignacian tools in a deposit dated to between 39.8 and 42.2 Kya (see above). Unfortunately, the human remains were never fully described and are now lost. The surviving records indicate that none of the teeth from this site are diagnostic; and the child's mandible, while lacking a distinct chin, cannot be unequivocally attributed to either modern humans or Neandertals (Garralda 1997, Churchill and Smith 2000a). Other sites have even more fragmentary or isolated specimens or suffer from even greater questions of context. The only human fossils securely associated with Aurignacian assemblages are the fragmentary remains from the French sites of Brassempouy, La Quina, and Les Rois. While they are all modern in morphology, they also are all carbon-dated to less than 34 Kya (Dujardin 2003, Henry-Gambier et al. 2004).

Some other fossil specimens have been directly or indirectly dated to the Aurignacian time span, but are not associated with Aurignacian artifacts. These include a partial maxilla from Kent's Cavern, England, dated to ~31 Kya (Stringer 1990b), a fragmentary (and undescribed) tibia and fibula from Kostenki in Russia (Richards et al. 2001), and the Muierii 1 and Cioclinova 1 crania from Romania, dated respectively to 30 ± 0.8 and 29 ± 0.7 Kya (or 28.5 ± 0.17 Kya) (Păunescu 2001, Soficaru et al. 2007). The Kent's Cavern maxilla is fragmentary, has not been thoroughly analyzed, and may be undiagnostic. The Cioclinova calvaria resembles the Mladeč skulls in combining an overall modern form with a hemibun, moderate brow ridges, and a suprainiac fossa (Smith 1984, Soficaru et al. 2007).

More significant, however, are the remains from the Pestera cu Oase (Oase Cave), Romania, particularly a robust adult mandible directly dated by AMS to $34,950 + 990/-890$ Kya (Trinkaus et al. 2003a). This specimen, Oase 1, is the oldest modern-human specimen known from Europe. It has all the features associated with a modern chin (albeit with a rather vertical symphysis), and the overall gestalt of its morphology clearly places Oase 1 in the Upper Paleolithic (early modern) morphological group. However, the specimen does have a small medial pterygoid tubercle and horizontal-oval mandibular foramina, both of which could indicate some connection to Neandertals. A fragmentary cranium, Oase 2, is also modern but exhibits such possibly Neandertal-derived features as a slight

TABLE 8.14 ■ Important Human Skeletal Remains Associated with the Gravettian in Europe

Site (Location)	Skeletal Remains	Date (Reference)
Brno 2 (Czech Republic)	Adult calvaria, facial and mandibular pieces	$23,680 \pm 200$[a] (Pettit and Trinkaus 2000)
Cro-Magnon (France)	Adult skull and partial skeleton (Cro-Magnon 1: the "Old Man"; adult cranium and postcranial remains (Cro-Magnon 2); partial cranium and postcranial remains (Cro-Magnon 3); additional cranial and postcranial remains.	$27,680 \pm 270$[b] (Henry-Gambier 2002)
Dolní Věstonice/Pavlov (Czech Republic)	73 specimens (mostly isolated bones or teeth) but including: Pavlov 1, DV 3, DV 13, DV 14, DV 15, and DV 16 associated skeletons; and DV 1, DV 2, and DV 11 calottes.	25,500–27,000[c] (Svoboda et al. 1996; Trinkaus and Svoboda 2006a)
Grotte des Enfants (Italy)	Fragmentary skeletons of two juvenile individuals (Grotte des Enfants 1,2), two adult skulls and postcrania (Grotte des Enfants 4,5), skull and postcrania of a Subadult (Grotte des Enfants 5).	≤28 Kya[d] (Shea and Brooks 2000)
Lagar Velho (Portugal)	Partial skull and postcranial skeleton of a child.	24.5 Kya[e] (Pettit et al. 2002)
Předmostí (Czech Republic)	Mass burial of 18 individuals, plus remains of 11 others, of various ages and degrees of completeness.	26,000–27,000[f] (Svoboda et al. 1996)
Sungír (Russia)	Skeletons of adult male (Sungír 2) and two children (Sungír 3,4); partial cranium of an adult female (Sungír 1)	23.6 Kya[g] (Pettit and Bader 2000)

[a] Direct ^{14}C date on human specimen.
[b] Carbon-14 date on materials associated with human remains.
[c] Range based on a series of ^{14}C dates associated with the human remains.
[d] Estimated date based on archaeological associations.
[e] Mean of four ^{14}C dates on material associated with the human specimen.
[f] Based on two ^{14}C dates on material associated with the human remains.
[g] Mean of three ^{14}C dates directly on the human specimens.

hemibun and upper midfacial projection (Trinkaus et al. 2003b, Trinkaus et al. 2006b, Rougier et al. 2007).

There is a much larger sample of human remains associated with the Gravettian (Pavlovian in Central Europe), the industrial/cultural complex that follows the Aurignacian in most of Europe. The Gravettian dates from ~28 to ~21 Kya. Although its stone and bone tools differ from those of the Aurignacian (Gamble 1986, Klein 1999), the Gravettian is equally characterized by elaborate aesthetic expression, including the famous "Venus" figurines found in many sites throughout Europe. Table 8.14 lists some of the important specimens and samples of human skeletal remains associated with this complex. All of these specimens derive from burials containing grave goods or other indications of symbolic behavior (Zilhão and Trinkaus 2002, Trinkaus and Svoboda 2006b). For example, the Lagar Velho 1 burial, a 4.5- to 5-year-old child, was partially covered with red ochre and was interred with perforated shells and deer teeth (Vanhaeren and d'Errico 2002). The Pavlovian specimens from Předmostí were buried in a mass grave in a 4-m × 2.5-m oval pit covered by limestone rubble. The Brno 2 burial contained, among other grave goods, a large ivory "puppet" made of several separate pieces (Svoboda et al. 1996). In addition to the mass grave at Předmostí, there are multiple burials at Dolní

Věstonice and Surgír. The former site yields a famous triple burial, DV 13, 14, & 15 (Svoboda et al. 1996, Sládek et al. 2000, Trinkaus and Svoboda 2006a). Sungír has two children buried head-to-head. The Sungír 2 male burial and the Sungír 3–4 double burial are decorated with over 10,000 beads that appear to have been attached to "tailored" clothing (Klein 1999).

Since its discovery in 1868, the Cro-Magnon 1 skeleton has been used as the representative specimen for the earliest modern Europeans (Boule 1921, Smith and Spencer 1997). As noted above, new dating demonstrates that Cro-Magnon is not as old as previously thought (Henry-Gambier 2002; Table 8.14). Often referred to as "Le Vieillard" or the old man, Cro-Magnon 1 preserves a complete, edentulous cranium, a mandible and partial postcranial skeleton. The skull is large but modern in form, with a relatively high vault (Table 7.4) and forehead and a rounded posterior vault contour in lateral view (Fig. 8.32). There is at most a slight hint of a hemibun. The face is orthognathic, and its modern form is reflected in the presence of canine fossae, an angled ZAM, well-developed brow ridges, and a relatively high and narrow nose (Figs. 8.32 and 8.33). The mandible (not illustrated) has a distinct chin. The vault exhibits the modern "*en maison*" form in rear view (Fig. 8.34) and lacks any development of a suprainiac fossa.

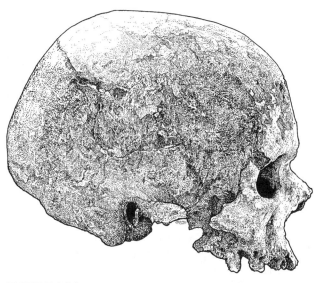

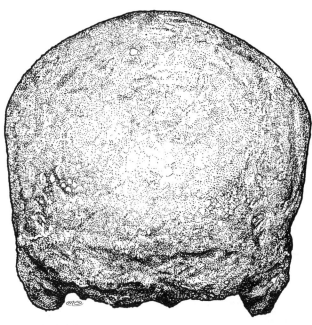

FIGURE 8.32

The Cro-Magnon 1 cranium (France), lateral view. This specimen, long considered typical of the earliest modern humans, exhibits a fully modern morphological pattern: high vault with a vertical forehead and rounded vault contours, orthognathic face, canine fossa, and weak development of the supraorbital superstructures.

FIGURE 8.34

The Cro-Magnon 1 cranium (France), posterior view. Note the modern vault shape and the absence of a supra-iniac fossa.

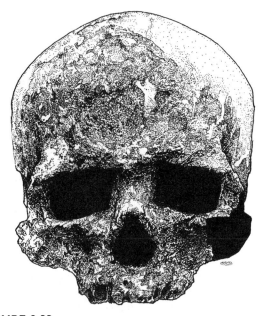

FIGURE 8.33

The Cro-Magnon 1 cranium (France), frontal view. Note the angled zygomaticoalveolar margin, the relatively small nasal aperture, and absence of a supraorbital torus. The teeth were lost postmortem.

The Cro-Magnon 2 skull, probably that of a female, is more gracile. Although the Cro-Magnon 3 male vault has a distinct hemibun and more marked brow ridges, the morphological pattern of the Cro-Magnon sample is undeniably modern (Camps and Olivier 1970, Stringer et al. 1984, Gambier 1989, 1997). The habitual use of Cro-Magnon 1 as the "type specimen" for the earliest modern Europeans exaggerates the differences between them and the Neandertals; but the differences between the Gravettian (and the late Aurignacian) people and the Neandertals are nonetheless real and consistent.

The Pavlovian specimens from the Czech Republic are also fundamentally modern in overall form (Tables 7.4 and 7.10), differing little in their basic description from Cro-Magnon 1 or the Mladeč specimens (Matiegka 1934, 1938; Jelínek 1969, Smith 1982, 1984; Van Vark et al. 1992, Svoboda et al. 1996, Sládek et al. 2000, Franciscus and Vlček 2006, Franciscus et al. 2006, Trinkaus and Svoboda 2006b). However, many scholars have seen certain more or less Neandertal-like characteristics in these central European specimens (Jelínek 1969, Wolpoff 1980a, 1999; Smith 1982, 1984; Frayer 1992a,b, 1997). Virtually all of the skulls have relatively pronounced hemibuns; and while these tend to be most pronounced in the large robust males (e.g., Brno 2, Předmostí 3, Pavlov 1), they are also evident in the females (e.g., Předmostí 4, Dolní Věstonice 3). Most of these specimens also show development of a feature which may be a form of suprainiac fossa. The males have particularly large and prominent brow ridges (albeit of a modern form). The upper midfaces of this sample also tend to be more forwardly projecting than those of their non-European contemporaries or later European specimens. Some of the Pavlovian mandibles (e.g., Předmostí 3) have retromolar spaces.

Table 7.7 presents comparative data on four morphological details that have been proposed as markers of genetic continuity between Neandertals and early modern Europeans. The distribution of suprainiac fossae, horizontal-oval mandibular foramina, and pronounced upper facial (nasion) projection does suggest some degree of morphological continuity in Europe, because the earlier and supposedly ancestral modern humans outside of Europe (Skhūl/Qafzeh) show markedly lower frequencies of these Neandertal-like features. The same inference can be drawn from a comparison of occipital buns, which are present in 81.9% of Neandertals and 68.4% of early modern Europeans but absent in the Skhūl/Qafzeh group (Smith et al. 2005). It seems significant that these Neandertal-like features tend to disappear, or become markedly reduced, over time in Europe. This pattern of gradual reduction appears to reflect evolutionary processes that are also discernible in the genetic data (see below). But this conclusion assumes that these anatomical traits are homologous and genetically based; and as we will see, those assumptions are debated.

The Gravettian postcranials also conform to a modern anatomical pattern (Matiegka 1938, Camps and Olivier 1970, Smith 1984, Sládek et al. 2000, Zilhão and Trinkaus 2002, Holliday 2006b, Trinkaus and Svoboda 2006b). The femora have pilasters; pubic rami are not elongated; and upper limbs do not show Neandertal features. However, many femora do have proximal lateral flanges (Kidder et al. 1991), and the axillary borders of early Upper Paleolithic (EUP) scapulae exhibit frequencies of sulcus positions (Table 7.17) that are intermediate between those of Neandertals (predominantly dorsal) and modern Europeans (mainly ventral). EUP peoples were taller and lankier than Neandertals (Table 7.15), with relatively narrow and linear trunks and high brachial and crural indices (Table 7.16). These body and limb proportions are more similar to those of tropical peoples today than to Neandertals or even to recent Europeans. They suggest a recent tropical derivation of the EUP populations (Holliday 1997a, 2000, 2006b; Holliday and Falsetti 1995, Pearson 2000a,b).

At the end of 1998 and beginning of 1999, a burial containing the partial skeleton of a 4.5- to 5-year-old child was recovered by J. Zilhão from the Lagar Velho site in the Lapido Valley of Portugal (Zilhão and Trinkaus 2002). This skeleton, Lagar Velho (LV) 1, includes several skull and mandibular fragments and much of the postcranial skeleton (Duarte et al. 1999, 2002). The postcranial remains exhibit a morphological pattern

that suggest a combination of Neandertal and modern human features, and it has been claimed that LV 1 provides evidence for hybridization between Neandertals and early moderns (Duarte et al. 1999, Trinkaus and Zilhão 2002). Zilhão and E. Trinkaus (2002), the main proponents of this position, are careful to point out that most of the anatomical features of LV 1 are clearly those of a modern human child. For example, the overall vault contours conform to those of appropriate-aged modern children and differ from those of Neandertals. The mastoid processes are large and inferiorly projecting. There is no incipient supraorbital torus like those seen in Neandertal juveniles. The nasal opening is modern in size and morphology. The interorbital breadth is narrow. The lower front teeth are small and modern in their morphology. Like modern humans of comparable age, but unlike Neandertals and most mammals, LV 1 shows no evidence of a suture between the maxilla and the primitively separate bone (the premaxilla: Fig. 3.8) containing the sockets of the upper incisors. The short clavicle and superior pubic ramus indicate a slender body build, and the femora were not as robust as those of Neandertals.

However, a few characteristics associate LV 1 uniquely with Neandertals as opposed to modern humans. Trinkaus (2002) argues that the LV 1 cranium is distinctively Neandertal-like in its large, projecting occipito-mastoid crest and the presence of markedly developed semispinalis capitis muscle-attachment fossae that are separated on either side of the midline. He also points to the presence of a suprainiac fossa and a markedly receding symphysis as characters probably reflecting a Neandertal ancestry. However, limb proportions provide the main evidence for a Neandertal connection (Duarte et al. 1999, Ruff et al. 2002). The LV 1 child has notably short distal limb elements relative to proximal elements—a pattern unknown in other EUP children but characteristic of Neandertals. Since limb proportions are established relatively early in ontogeny (Chapter 7), it is unlikely that LV 1 would have simply "grown out" of this condition.

Trinkaus, Zilhão, and colleagues interpret this mosaic pattern of characteristics in the LV 1 child as indication of admixture between Neandertals and early modern people. The Neandertal contribution would appear to have been much less than the modern contribution, although exact estimates of the extent of each contribution are not possible. This interpretation implies that some Neandertal genes were assimilated into early modern human populations in Europe. However, by 24.5 Kya, there were no Neandertals; so the actual interbreeding would have had to have occurred not during the Gravettian, but earlier in the Upper Paleolithic time span. Judging from other Gravettian remains, Neandertal features must have been relatively rare in the population by 24.5 Kya. Thus it seems extremely fortunate that the fossil record contains an individual like Lagar Velho 1, which reflects so much Neandertal biological influence.

"Extremely fortunate," of course, is another way of saying "extremely unlikely," and the hybrid or assimilation interpretation for LV 1 has been vigorously challenged. In a commentary on the original publication on LV 1, Tattersall and Schwartz (1999) dismiss LV 1 as "simply a chunky Gravettian child." They claim that the Neandertal-like skull characteristics of LV 1 are not really shared derived features, and that its Neandertal-like limb proportions arose independently as an adaptation to the cold last-glacial environments of coastal Portugal. Stringer (2001) also argues that the limb proportions of LV 1 could have occurred as the result of short-term genetic adaptation to these conditions. Defenders of the Neandertal-hybridization hypothesis (Trinkaus and Zilhão 2002, Ruff et al. 2002) counter that this is unlikely as there is no evidence of such an adaptation in other Gravettian children or adults, including those that lived in even colder climates. Furthermore, they note that LV 1 resembles tropical peoples and other EUP specimens in having a narrow and linear thorax, so that its body proportions exhibit a mosaic of disparately adapted traits. Ruff et al. (2002) also demonstrate that the LV 1 limb proportions are probably not the result of nutritional influences.

It bears repeating that the overall morphology of all these early Aurignacian and Gravettian Europeans is modern and distinctively different from that of the Neandertals. Given the chronological patterns associated with the appearance of early modern humans, their presence in Europe is most plausibly explained as the result of immigration of early modern populations from areas further south. But exactly what happened when these immigrant populations contacted the indigenous Neandertals remains a matter for debate. The evidence from Europe, as well as that from other regions of Eurasia, suggests some degree of archaic contribution to these early modern populations. However, the European evidence provides perhaps the clearest indication that this contribution was relatively small.

Since modern humans were present in West Asia around 100 Kya at Qafzeh and Skhūl, why did it take them an additional 65 Ky to penetrate into south central Europe? One key reason is that the European Neandertals appear to have been very well adapted to Pleistocene Europe, both biologically and behaviorally. They were unquestionably highly intelligent, resourceful people, who must have represented a formidable barrier to the systematic immigration of early moderns. The initial immigrants may have found that barrier especially hard to overcome if they were, as their body form suggests, physiologically adapted to tropical environments. Early moderns may have been unable to compete successfully with Neandertals in Pleistocene Europe as long as they possessed only a Middle Paleolithic level of technology, which is what we find at the earliest modern-human sites in the Middle East. However, in Aurignacian and later Upper Paleolithic sites, we find evidence of more efficient cultural buffers, including better pyro-technology and tailored clothing (Trinkaus 1989), that would have made early modern peoples more competitive. Perhaps the entry of early moderns into Europe had to await the emergence of these and other such cultural innovations.

A strictly biological factor that may also have retarded modern human immigration into Europe is skin color. As noted in Chapter 5, pale, depigmented skin would have been a necessity for successful human adaptation to locales above $40°$ N latitude (Loomis 1967, Jablonski and Chaplin 2000). Early moderns emerging from Africa may simply have been too darkly pigmented to thrive in Europe. This problem would have been exacerbated by their need to cover their skins with insulating clothing to survive in cold northern environments. As long as they retained dark skins adapted to tropical environments, early modern humans in Europe would probably have suffered from an increased incidence of rickets. This would have reduced their viability and thus their ability to compete successfully with Neandertals—at least some of whom have been shown by studies of ancient DNA to have been pale-skinned (Lalueza-Fox et al. 2007).

The mutated MC1R gene from which Neandertal depigmentation is inferred is also mutated in light-skinned modern Europeans. Though the two variants differ only in a single point mutation, they are different alleles, and some hold that neither is likely to have given rise to the other (Culotta 2007). If the modern European allele was independently acquired, then it may well have taken some time for mutation and natural selection to equip modern humans living on the southern fringe of Europe with the pale skins that they needed to compete with Neandertals on their home turf.

RECENT HUMAN GENETICS AND MODERN HUMAN ORIGINS

Some of the most important data concerning modern human origins come from analyses of genetic patterns in living human populations. Three such sources of genetic information have been brought to bear on this issue:

1. **Nuclear genes** (DNA), which code for most of the well-known polymorphisms in humans (and also encompass microsatellite DNA and *Alu* insertions).

2. **Mitochondrial genes** (DNA), which are passed on only along female lineages and thus are clonally inherited (Chapter 7).

3. **Y-chromosome features**, which are nuclear DNA passed on only through male lines (on the Y sex chromosome) and thus have different inheritance patterns from other nuclear DNA.

Almost all genetic data from these sources suggest that recent humans exhibit surprisingly little genetic diversity for such a widespread species. Many scientists (see discussion in Relethford 2001) hold that the reason for this limited variability in the human genome is that all modern humans are descended from a relatively recent ancestral population that passed through a genetic bottleneck (Chapter 2), in which population size was drastically reduced—perhaps to no more than 10,000 individuals (Harpending et al. 1998). This constriction in the size of the population would have strongly reduced genetic diversity as well, and the low diversity in modern human populations can thus be interpreted as a lingering effect of the hypothesized ancestral bottleneck.

It is not clear what the cause of this recent bottleneck was. Some think that the eruption of Mt. Toba in Indonesia may have caused a "volcanic winter" that lasted for many years and resulted in extreme climatic change (Ambrose 1998). However, this event is somewhat late (~70 Kya) to be associated with modern human origins, and it is not clear that it would have negatively impacted cold-adapted archaic people like the Neandertals. The date and severity of the supposed bottleneck are also debated. Hawks and colleagues (2000a, Hawks and Wolpoff 2001b) believe that it occurred around 2 Mya with the origin of *Homo erectus*, and contend that the geneticists' estimates of population size on the order of 10,000 are probably orders of magnitude too low (see also Templeton 1997, 1998). But many others argue that the bottleneck occurred in Africa between 100 and 200 Kya, in connection with a speciation event that produced modern *Homo sapiens sensu stricto*. Such a constriction in population size would be expected as part of the segregation of a peripheral isolate implied in the theory of allopatric speciation. In this view, all recent humans trace their genetic roots back to a small, genetically isolated ancestral population in Africa. Since this population and the associated genetic bottleneck were relatively recent, not enough time has elapsed for mutation to reestablish expected levels of genetic diversity in *Homo sapiens*. And because human populations in other parts of the world were established by migration of relatively small groups out of Africa, their genetic diversity would have been further limited by founder effect. Therefore, the greatest degree of diversity would be expected in Africa, where human populations accumulating new genetic diversity longer than anywhere else. Such a finding would provide strong support for the Recent African Origin (RAO) model of modern human origins (Relethford 2001, Pearson 2004).

On the whole, this expectation is supported by the genetic data. By the 1970s, accumulating data on nuclear genes had suggested a relatively recent common origin for all modern people (Howells 1976), and large-scale studies of genetic diversity in protein and blood-group loci were interpreted as showing an African origin of this diversity (Nei and Roychoudhury 1974, Jones 1982). In 1982, a study of variation in over 100 nuclear genes led M. Nei and A. Roychoudhury to conclude that European and Asian populations split from Africans at 110 ± 34 Kya and that the split between European and Asian populations did not occur until 41 ± 15 Kya. They further noted that the various populations analyzed within Europe were genetically very closely related (except for the Lapps), and that the same was true for most populations within Asia. These facts supported the idea that populations in Europe and Asia radiated very recently. Although another large-scale study indicated an initial split between Asians on the one hand and Europeans plus Africans on the other (Cavalli-Sforza and Bodmer 1971), later work by Cavalli-Sforza and co-workers (1993, Cavalli-Sforza 1995) used a large series of nuclear DNA data to support an African origin.

A study of mitochondrial DNA by Cann et al. (1987) was the first to establish a firm genetic basis for the RAO model. Cann and colleagues collected mtDNA from 147 individuals worldwide and identified 133 different variants or haplotypes (Fig. 8.35). The parsimony-based molecular phylogenies derived from this analysis revealed the following major points (Fig. 8.35):

► The mtDNA haplotypes clustered into two major groups. One group entirely comprises African haplotypes, and the other one includes other African and all non-African haplotypes.

► The greatest degree of mtDNA diversity is found in Africa.

► The root (last common ancestor) of all haplotypes was an African haplotype.

► An mtDNA mutation rate of 2–4% per million years was established by calculations based on "relatively certain" divergence times estimated for specific living populations. From this, Cann and co-workers estimated that the last common ancestor of all living humans lived between 140 and 290 Kya, and that Eurasian populations diverged from Africans between 90 and 180 Kya.

► No mtDNA haplotype exists in Eurasia that cannot be traced back to this African origin. Thus there is no extant human mtDNA haplotype that appears to derive from archaic Eurasian populations (e.g., Neandertals).

On the basis of these findings, Cann and colleagues (see also Cann 1988, Stoneking and Cann 1989) advo-

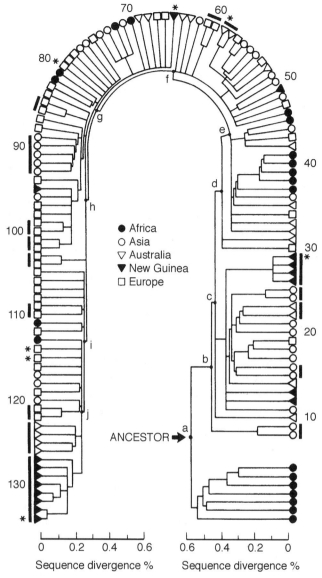

FIGURE 8.35

Dendrogram reflecting the relationships between the mitochondrial DNA haplotypes identified by Cann and colleagues (1987) in recent humans. See text for discussion. (After Cann et al. 1987, by permission.)

cated a total-replacement model of human origins in which archaic Eurasians were wholly supplanted by African immigrants. Of course, their data did not imply that these migrating Africans must have been anatomically modern people. But when the mtDNA and other genetic data were combined with patterns evident in the relevant parts of the human fossil record, the obvious conclusion for many was that modern humans had originated in Africa as a new species, and then spread into Eurasia without interbreeding with the local archaics (Stringer and Andrews 1988).

Critics noted several problems with the original Cann et al. (1987) paper, including some questionable assumptions about mutation rates and their inclusion of African-Americans in their "African" sample. In response, Vigilant et al. (1991) undertook an improved mtDNA study, sampling native Africans rather than African-Americans and using chimpanzees as an outgroup to allow a more precise estimate of the age of the last common ancestor of recent human mtDNA haplotypes. The Vigilant group's results were strikingly similar to those of the 1987 study and were widely regarded as a decisive confirmation of the Cann group's conclusions. However, doubts were soon raised concerning the African rooting of the mtDNA tree in African populations, because it was possible to generate trees rooted in non-African populations that fit the mtDNA data just as well (Hedges et al. 1992, Templeton 1992). Using an improved method, Penny et al. (1995) argued that the Vigilant group's data are indeed best interpreted as demonstrating an African root. The same conclusion was reached by Ingram et al. (2000) from a study of the entire human mitochondrial genome.

Nevertheless, these genetic data remain open to other interpretations. The fact that Africa exhibits more variation in mtDNA haplotypes and nuclear polymorphisms is generally taken as an indication that modern human populations have been evolving longer in Africa than in other regions of the world—and that modern humans therefore must have originated in Africa. But studies by several authors (Rogers and Harpending 1992, Rogers and Jorde 1995, Relethford and Harpending 1995, Harpending and Relethford 1997, Relethford 1998a,b, 2001) offer an alternative explanation in terms of differential population size. Some evidence suggests that Africa had a larger effective population size than other areas of the Old World during most of the Pleistocene (Relethford 2001). This is to be expected if the ancestral human adaptation was to African environments, with other parts of the Old World providing relatively more marginal habitat. Because larger populations generate and sustain greater genetic diversity, such a population-size differential could well explain the origin and persistence of high levels of genetic diversity in sub-Saharan Africa. Relethford (2001) argues that this differential could lead to the appearance of total replacement when African populations spread into Eurasia, even if there was some degree of local archaic contribution to modern gene pools in areas outside of Africa. None of these studies rule out an African origin of all modern human genetic variation, but they demonstrate that the data are compatible with other accounts.

Other questions about the mtDNA studies concern the assumption of neutrality. The prevailing interpretations of the mtDNA data—and of many nuclear DNA polymorphisms as well—assume that changes at these loci are evolutionarily neutral, so that all variation is due entirely to mutation rate and drift. Once mutation rates are determined, this assumption allows the use of a molecular clock to estimate the date of the last common ancestor. But if natural selection is involved, then the pattern of change cannot be tied to mutation rates alone. If many mutations at any locus were removed by selection, then divergence times estimated on the basis of an assumed neutrality would be too recent. Differential patterns of selection could also produce patterns of genetic relationships like that seen in Figure 8.35. An early indication that selection might be impacting mtDNA patterns in humans was provided by Excoffier and Langaney (1989), who showed that distribution of mtDNA haplotypes does not conform to a neutral model. Subsequent claims for selective impact on all or part of the mtDNA genome have been made by other researchers (Templeton 1996, Wise et al. 1997, Hey 1997, Hawks et al. 2000a, Elson et al. 2004, Hawks 2006). It is important to remember that recombination does not occur in mtDNA, so all alleles are effectively linked. Thus selection anywhere in the mtDNA genome could impact frequency patterns for any other part of it.

Studies of nuclear DNA polymorphisms, including those on the Y chromosome, have also been interpreted as suggesting a recent origin of human genetic variation in Africa (Cavalli-Sforza et al. 1993, Cavalli-Sforza 1995, Underhill et al. 1997, Hammer et al. 1998, Thomson et al. 2000; see discussions in Relethford 2001 and Pearson 2004). However, studies on some nuclear DNA traits give results that are not easily reconciled with a recent, monocentric origin of all modern human genetic variation. Potentially much older divergence times have been estimated for allelic variants of the ZFX gene on the X chromosome (between 162 and 932 Kya; Huang et al. 1998) and the PDHA1 gene (between 1.35 and 2.16 Mya; Harris and Hey 1999). Harding and colleagues (1997, Harding 1997) analyzed 16 haplotypes of beta-globin DNA and concluded that the last common ancestor would have lived ~800 Kya. These analyses also found ancient haplotype lines going back in Asia well in excess of 200 Ky. This implies the presence of gene flow between Africa and Asia extending much farther back in time than the Recent African Origin model permits. Hammer et al. (1998) found that while divergence dates derived from mtDNA and Y-chromosome studies were similar, the Y-chromosome data indicated an early spread from Africa followed by gene flow from Asia back into Africa within the past 200 Ky—another indication of a gene-flow pattern not allowed for in the RAO model.

Finally, studies of the microcephalin (MCPH1) gene indicate introgression from an archaic, non-African source into the modern human gene pool. Evans et al. (2006) identify a specific group of closely related variants at this locus, the haplotype D group, which arose from a common ancestor about 37 Kya. They conclude that the ancestral D-group mutation arose from a lineage that diverged from the modern human one significantly earlier, entered the modern human gene pools through admixture with the representatives of this lineage, and rose quickly to a frequency of about 70% by positive selection. The positive selection stems from the fact that, although the exact function of MCPH1 is unclear, it is critical to normal brain development and function. Evans and colleagues note that finding a gene that reflects relatively low levels of admixture is generally difficult, because weakly selected or neutral alleles from such a source would tend to be lost through drift or other random factors. MCPH1 is an exception, because of the strong positive selection acting on it. They conclude that their work on microcephalin supports the possibility of low levels of admixture between modern and archaic humans at around 37 Kya.

The evidence presented above indicates that genetic replacement was not complete and that there are some non-African contributions to the modern human gene pools. Of course, this evidence is derived from far fewer genetic systems than the number indicating an African origin. Several geneticists and population biologists have accordingly espoused what is called the "Mostly Out of Africa" model (Rogers and Harpending 1992, Rogers and Jorde 1995, Harpending and Relethford 1997, Jorde et al. 1998, Relethford 2001). This model suggests that the majority of human genetic diversity is traceable back to an African root, but that the exceptions noted above provide evidence of non-African contributions to the recent human gene pool. As Jorde et al. (1998, p. 134) put it: "(a)n African origin with some mixing of populations appears to be the most likely possibility." In defending this "mostly out of Africa" perspective, J. Relethford notes that "(e)ven if we could demonstrate only a very small (genetic) contribution outside of Africa, the fundamental evolutionary process is still different from speciation and replacement" (Relethford 2001, p. 209).

A. Templeton's analyses of human mtDNA patterns have provided additional strong evidence contradicting the Recent African Origin (RAO) model. Templeton (1993) applied a method that allowed him to compare the patterns of mtDNA distribution expected under two models: (a) a model of recurrent gene flow between regions, and (b) a model involving complete replacement. His analysis revealed several population expansions from different points of origin, but no evidence of a single worldwide replacement expansion originating from Africa. Templeton argued that the evidence best fits a pattern in which human populations were connected by continuous gene flow, albeit with restrictions based on distances between groups, throughout the past several hundred thousand years. However, his analysis indicates that this pattern of recurrent gene flow occurred out of Africa into Asia, with a later expansion into Europe that may have been a replacement.

Templeton originally considered these results as supporting multiregional evolution. But in a later study, he slightly modified this view. In 2002, he presented an analysis of 10 human genetic systems (mtDNA, six nuclear DNA sequences, a Y-chromosome sequence, and two X-linked sequences). He argued that the data were not commensurate with a single, recent migration out of Africa but rather three major "out of Africa" migrations (Fig. 8.36). According to Templeton, two earlier migrations established lines of alleles in various areas of Eurasia that were not derived from the recent migration defined by the RAO model (Templeton 2005). These ancient alleles are also found in recent Eurasian populations, implying that not all alleles originated from the most recent migrations out of Africa—an event that Templeton (2002, 2005) dates to ~130 Kya. Templeton acknowledges this most recent series of migrations from Africa as the principal influence on the genetic structure of recent human populations, but he emphasizes that the data do not fit a total replacement of earlier

humans during this final expansion. His statistical analysis strongly rejects a complete replacement of Eurasian archaics ($P < 10^{-17}$) (Templeton 2005). Rather, he concludes that the genetic data are compatible with the assimilation model (Smith et al. 1989), the multiregional model with expansions followed by admixture (Wolpoff et al. 1994b), and the "Mostly out of Africa" model (Relethford 2001).

ANCIENT DNA IN EARLY MODERN HUMANS

Fossils of early modern humans from Europe and Australia have yielded DNA fragments that are relevant to the issue of modern human origins. But unfortunately, no such evidence is yet available for early modern Africans, members of the African Transitional Group, or archaic hominins outside of Europe. The first study which claimed to provide genetic analysis of an early modern human specimen was that of Scholz et al. (2000). They used the southern blot hybridization technique, which is supposed to quantify similarities across

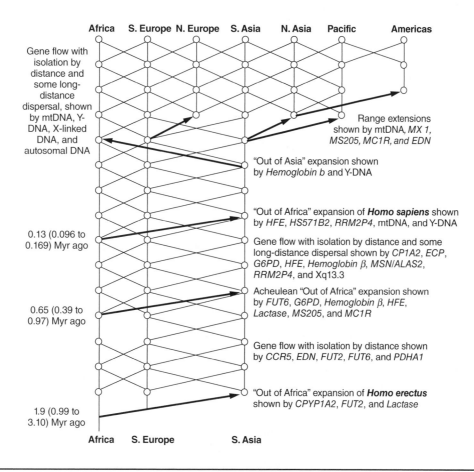

FIGURE 8.36

Schematic showing the three major "Out of Africa" genetic expansions and subsequent "Out of Asia" events identified by A. Templeton. These data support a "mostly out of Africa" model for human genetic evolution. (After Templeton 2005).

the entire genome rather than in specific segments thereof. The results of this study showed that the two Neandertals studied (a Krapina clavicle and a parietal from the German site of Warendorf) differed from the single modern human examined (the Vogelherd 3 humerus) by at least a factor of two. From this Scholz and colleagues claimed that even early modern humans were distinctly different from Neandertals. However, even if the southern blot technique is an appropriate analytical tool (and no other laboratories have employed it to analyze fossil hominin DNA), the conclusions do not follow, because it has since been discovered that the Vogelherd material does not represent early modern Europeans. Direct dating of the fossils (Conard et al. 2004) demonstrates that all the Vogelherd material (including Vogelherd 3) is all less than 5 Ky old.

Fortunately, mtDNA sequences are known from a few genuinely Upper Paleolithic skeletons. In 2003, Spanish and Italian scientists extracted mtDNA sequences from two Gravettian-aged skeletons from the site of Paglicci Cave in southern Italy (Caramelli et al. 2003), which were directly dated to 23 ± 0.35 and 24.7 ± 0.43 Kya. Genetic distance analysis by these authors showed that the mtDNA sequences of the two Gravettian individuals clustered with those of more recent modern humans and were distinctly different from the three Neandertal sequences known at that time.

In a pivotal study published in the following year, Serre et al. (2004) provided mtDNA sequences for five additional Upper Paleolithic human specimens from Europe: Mladeč 2 and 25c from the Czech Republic, and Cro-Magnon, La Madeleine, and Abri Pataud from France. The sequences of all five of these specimens fell within the modern human range of mtDNA sequences, and lacked the distinctive differences from modern sequences documented in Neandertals (Chapter 7). Serre and co-workers concluded, as did the two earlier studies, that there was no direct evidence of any Neandertal contribution to the early modern European gene pool.

However, the Serre group took the argument one step further—and it is a very important step. They showed that using the available early-modern sequences, a Neandertal contribution of ≥25% to the early-modern gene pool in Europe can be statistically excluded, but not a contribution of less than 25%. To exclude a Neandertal contribution of 10%, 50 additional early modern European mtDNA sequences would be needed; and to exclude a 5% Neandertal contribution, a sample of sequences greater than the number of Upper Paleolithic skeletons currently known would be necessary! Serre and colleagues point out that if the early modern population sizes were much larger than those of the Neandertals, which seems likely, these percentages would be lower. But although the genetic evidence currently available does not demonstrate a Neandertal contribution to early modern European gene pools, it does not exclude

such a contribution. A contribution of, say, 6% to 15% would not be insignificant; and such a level of contribution cannot be excluded based on currently available genetic data.

The only ancient DNA recovered outside of Europe that is pertinent to the issue of modern human origins comes from Australia. Adcock et al. (2001a) extracted mtDNA from several early Australian specimens, including WLH (Mungo) 3, three other Willandra Lakes individuals (WLH 4, 15, 55) and six Kow Swamp individuals. All of these individuals yielded mtDNA sequences that clustered with recent humans—except WLH 3. Although WLH 3 is anatomically modern, its mtDNA sequence does not fall in the range of living Australians or other modern samples. Dated to around 40 or 60 Kya (see above), WLH 3 is the oldest modern human specimen that has yielded an mtDNA sequence. Adcock and colleagues suggest that the WLH 3 sequence diverged before the purported African last common ancestor of recent human mtDNA. Relethford (2001) cites the WLH 3 sequence as an excellent example of the phenomenon of mtDNA lineage extinction, because this haplotype is not found in recent humans. He notes that if the WLH 3 mtDNA haplotype could go extinct in early modern human populations, so could Neandertal lineages that might have been assimilated into early modern European gene pools.

The Adcock et al. study has been questioned on methodological grounds (including the fact that the sequence has not been duplicated in another laboratory), and doubts have been expressed about contamination of the mtDNA recovered from the fossils (Colgan 2001, Cooper et al. 2001, Groves 2001, Trueman 2001). These questions were subsequently addressed by Adcock et al. (2001b,c), but the veracity of the WLH 3 mtDNA sequence continues to be a point of contention.

MODERN HUMAN ORIGINS: THE MODELS VS. THE FACTS

Aiello (1993) reviewed four models that have been put forward to explain modern human origins: the African-replacement model, Bräuer's African-hybridization-and-replacement model, the assimilation model, and the multiregional model. Aiello regards the first two as basically similar, although the second allows for greater degrees of hybridization between modern and archaic humans. Aiello notes a similar close relationship between the assimilation model (AM) and the multiregional-evolution (MRE) model. Advocates of both AM and MRE hold that local archaic peoples contributed significantly to early modern human populations in various regions of Eurasia. Both resist the assignment of any of these archaic groups, including the Neandertals, to species other than *Homo sapiens*. But unlike AM, the MRE model posits a much greater degree of regional continuity, and denies that the spread of modern alleles (and ultimately populations) from an African source was the catalyst for the emergence of anatomically modern people throughout both Europe and Asia.

The Recent African Origin Model

The RAO model gives rise to four testable predictions (Frayer et al. 1993):

1. Human skeletal anatomy and genetic data should be relatively homogeneous, reflecting a recent common ancestry of all modern humans.

2. There should be evidence of an early transition to modern human anatomy in Africa, and the earliest appearance of the modern human anatomical complex should be found in Africa. The appearance of early modern people in Eurasia should be geographically and temporally patterned in a manner compatible with a spread outward from an African center.

3. There should be no evidence of an evolutionary transition between archaic and modern people anywhere in Eurasia at any time.

4. There should be no evidence of any continuity between archaic and modern populations anywhere in Eurasia. In other words, there should be no morphological or genetic traits in early modern Eurasians that are more readily derived from local archaics than from the presumed African ancestors of all modern humans.

How do these predictions stack up against the facts?

Prediction 1. As predicted by RAO, modern human skeletal anatomy exhibits a relative homogeneity, reflected in multivariate cranial analyses and morphological studies of various skeletal elements and processes. Although some "modern" fossil skulls do not fall within the recent human range, supporters of RAO note that such specimens are relatively few, and experts of all persuasions generally recognize them as modern humans from their overall morphological gestalt. Regional variations in recent human skeletal anatomy (e.g., in body builds and limb proportions) can be explained as the result of relatively recent ecogeographic adaptation or perhaps genetic drift, all of which would have occurred over the last few tens of thousands of years. Genetic variation among living humans is also much less than that seen in many other species, including chimpanzees, and there is considerable evidence that our genetic homogeneity reflects our descent from a population that had passed through a rather extreme genetic bottleneck. Most (if not all) of the genetic data point to Africa as the place where the ancestors of

modern humans emerged from that bottleneck. Supporters of RAO correctly insist that these facts conform to Prediction 1 of their model.

Prediction 2. As discussed previously in this chapter, the African Transitional Group (ATG) evinces a mosaic cranial morphology (an archaic vault combined with a more modern face) that looks like an intermediate waypoint in the evolution of modern human anatomy. This intermediate condition in the ATG appears to predate evidence of any such transitional morphology outside of Africa. Moreover, the fossil hominin finds from Herto and Omo Kibish KHS represent the earliest dated specimens of anatomically modern humans known at the current time. All these facts support Prediction 2. So does the apparent geographical and temporal pattern in the appearance of modern humans. After their initial appearance in East Africa, anatomically modern people appear first in the Near East (the sites of Skhūl and Qafzeh) at ~100 Kya, and are not seen in Europe until ~35 Kya. The earliest appearance of modern humans in eastern Asia and Australasia is not well dated, but there is no evidence that they were there before 60 Kya, and that date is probably too early. The available data thus support Prediction 2 of RAO.

Prediction 3. Supporters of RAO point to the transition sampled by the ATG as the only compelling example of a morphological transition from archaic to modern hominin morphology. Other potentially intermediate fossils are dismissed as the products of convergent or parallel evolution. For example, supporters of RAO believe that the Ngandong hominins and specimens like Willandra Lakes hominin 50 do not represent evidence of a transition in Australasia. They view Ngandong as a sample of late-surviving Asian Erectines that have evolved certain "advanced" features such as bigger braincases in parallel with the true human lineage. Conversely, WLH 50 is considered a robust modern specimen, and its similarities to Ngandong and earlier Erectines (Fig. 8.29) are discounted as homoplasies. In analyzing fossils from West Asia and Europe, supporters of RAO stress the distinctions between Neandertals and early moderns and argue that no samples or specimens can be recognized as transitional between the two. For example, the seemingly modern features seen in the Vindija sample of late Neandertals are dismissed as parallelisms, or rejected on the grounds that the sample is biased toward female and subadult individuals. Proponents of RAO thus can make a strong case that the data support prediction 3, but their arguments for this conclusion are more debatable.

Prediction 4. Are there any isolated features in undeniably modern Eurasian humans that provide evidence of continuity with local archaics? This question is even more hotly debated. Supporters of RAO have offered strong critiques of every such feature that has been proposed, insisting that the features being compared are not homologous (Bräuer et al. 2004), or that their use as indicators of regional continuity is problematic because their genetic basis is unknown (Lieberman 1995). Pearson (2004) argues that if D. Frayer's (1992b) data on the prevalence of regional-continuity features in Europe are used to calculate the level of genetic admixture, they yield estimated percentages of Neandertal contribution to modern European populations ranging from 21.5% to 79.9%. Since the genetic data (see above) demonstrate no evidence of Neandertal contribution to modern Europeans, he concludes that these estimates are unrealistically high—implying that these features cannot be invoked as indicators of regional continuity in Europe. As the proponents of RAO see it, the available data are therefore compatible with Prediction 4.

Alternative Views—Multiregional Evolution

There are certainly other ways to view the relevant genetic and anatomical data. The best known of these alternatives is the Multiregional Evolution model (MRE), first articulated by M. Wolpoff, X. Wu, and A. Thorne in 1984 (Wolpoff et al. 1984). This model depicts Middle and Late Pleistocene human evolution as a latticework of regional lineages connected by gene flow that has united all members of the genus *Homo* as a single polytypic species during this entire span of time (Fig. 6.3). How would MRE explain the data pertinent to the four predictions of RAO?

Prediction 1. According to MRE proponents, the morphological homogeneity of humans today is more apparent than real. However homogeneous recent human skeletal anatomy may appear, the persistence of regionally distinctive ancient traits in modern populations rules out a single, monocentric origin—or even a "mostly" single origin—for modern human morphology. For example, although Wolpoff (1989) acknowledges that there may have been less local continuity in Europe than in East Asia or Australasia, he and his colleagues (Wolpoff et al. 2004) believe that the anatomical evidence cannot exclude a 50% Neandertal contribution to the early modern European gene pool.

Proponents of MRE also offer different interpretations of the genetic data that have been claimed as evidence for RAO. They stress three previously noted facts: (1) not all genetic systems point to a recent African origin; (2) patterns of differential effective population size, rather than a recent population bottleneck, may explain the concentration of genetic diversity in Africa; and (3) evolution in mitochondrial DNA may not be neutral. The last point is particularly important for MRE supporters, since selection on mitochondrial DNA vari-

ants would make it impossible to use a molecular clock to attach a date to the last common ancestor of modern humans (or more precisely, to the last common ancestor of our mitochondria). Such selection might also help to explain why recent human mtDNA diversity is lower than in the Late Pleistocene, as evidenced by the extinct mtDNA variants reported from ancient Australians.

Prediction 2. MRE certainly recognizes the African Transitional Group (ATG) as evidence for the evolution of African Heidelbergs into early modern Africans, but MRE supporters would not agree that this sample documents the initial emergence of "modernity" as a unitary morphological complex. As regards the earliest modern humans, MRE "... predicts that there will be no single origin for these populations that can be located on genetic or morphological grounds" (Wolpoff 1989, p. 93). Additionally, the designation of certain non-African samples as "anatomically modern" is challenged by MRE. For example, the recognition of the Skhūl/Qafzeh (S/Q) sample as anatomically modern is a crucial part of the RAO model; but supporters of MRE argue that this sample cannot be clearly differentiated from Near Eastern Neandertals. To MRE proponents, the S/Q hominins represent another local population in transition between archaic and modern morphology, not a fully modern population of African origin. Thus MRE does not recognize the distinct chronogeographic pattern of early modern human spread touted by RAO, and the model denies that modern human biology emerges as a complex in Africa. Rather, MRE asserts that the individual traits characteristic of modern people emerge in different regions and coalesce into the recent human pattern in specific regions at different times, depending upon the direction and magnitude of gene flow, genetic drift, and the regional selective environment.

Prediction 3. In addition to the ATG and the S/Q sample, MRE recognizes transitional specimens and samples in other regions of Eurasia as well. In Europe, the best evidence for such a transition is the Level G sample from Vindija. Although much later than the ATG and S/Q samples, this sample has been claimed as evidence of a Neandertal population changing morphologically in the direction of early modern humans. These morphological changes in the Vindija hominins are real and not simply the result of sample bias, as RAO proponents contend (Chapter 7). Supporters of MRE accordingly view Vindija and other late Neandertals (e.g., Hortus, Saint-Césaire) as evidence of a European transition to modern human morphology. And although the relevant fossil records in East Asia and Australasia are rather meager, MRE interprets them as documenting unbroken sequences connecting Erectines and recent populations in these regions as well. Wolpoff (1989, p. 84) summarizes this argument succinctly:

The pattern of hominid evolution at the eastern periphery still appears much as Weidenreich described it. It involves a morphological gradient spread over the eastern periphery between north and south ends that are unique and distinguishable as far back into the past as the earliest evidence for habitation can be found, but also involves geographically and morphologically intermediate specimens that show there was never populational isolation. Because of the marked differences in morphological features that show continuity at the north and south ends of the eastern periphery, *this pattern clearly discounts complete replacement as a valid explanation of Pleistocene hominid evolution.*

Prediction 4. MRE supporters have identified many regionally distinctive traits that supposedly demonstrate archaic contributions to local early-modern human populations. For East Asia and Southeast Asia, as many as nine and 12 major features respectively have been cited as traits documenting regional continuity. As Thorne (2000) notes, the degree of continuity would be expected to differ from region to region, and within the same region over time. Although proponents of MRE concede that not all, or even most, of these features are unique to any one region (Tables 8.8 and 8.9), they contend that high rates of co-occurrence of these features in specific regions designate them as evidence of local continuity between archaic and early modern humans. Fewer such features have been identified for Europe (Table 7.7). Nevertheless, it has been argued that those features may imply a Neandertal contribution of as much as 50% to the early-modern European gene pool (Wolpoff et al. 2004; cf. Pearson 2004). The level of archaic contributions in East Asia and Australasia would therefore presumably have been even higher.

Alternative Views—The Assimilation Model

Wolpoff et al. (2000, p. 132) have argued that the only two models that apply to the study of modern human origins are RAO and MRE, and that "... there is no process that can lie between them or be a compromise." By their definition, a model that posits anything other than total genetic replacement throughout Eurasia is, by definition, a variant of the MRE model. However, intermediate "variants" have been proposed that involve premises that Wolpoff and other MRE proponents reject. At least three such intermediate models have been put forward: G. Bräuer's "Replacement with Hybridization" model, the "Mostly Out of Africa" model, and the Assimilation model. The first and third of these models are primarily based on the fossil record, while the second is grounded in the genetic data.

Bräuer's model (Bräuer 2003, 2006; Bräuer and Broeg 1998, Bräuer et al. 2004) acknowledges the possibility of "hybridization" between modern and late archaic humans, and rejects the erection of species boundaries between them. However, Bräuer also contends that such hybridization probably did not occur, and that if it did, its effects were negligible. In its practical application to the fossil data, Bräuer's model thus has a strong flavor of replacement—if not of species, at least of local populations. The "Mostly Out of Africa" model is supported by several geneticists (Rogers and Harpending 1992, Rogers and Jorde 1995, Jorde et al. 1998, Harpending and Relethford 1997, Relethford 2001, Templeton 2002, 2005). Relethford (2001) likens this model to Thorne's "center and edge" concept, applied here to modern human origins rather than to Erectine evolution. Relethford accepts a recent African origin for most modern human traits, both genetic and morphological, because of the probability of a larger effective population size in Africa. As he notes, this model fits the paleontological evidence as well. However, he argues that the spread of moderns out of Africa is not likely to have resulted in a total replacement of all archaic Eurasians, because of the "... findings of some non-African influence in some genes and DNA sequences as well as the fossil record for regional continuity" (Relethford 2001, p. 205).

The Assimilation Model (AM) was first formally proposed by Smith and colleagues in 1989, although aspects of it had been put forward in earlier work (Trinkaus and Smith 1985, Smith 1985). Like MRE, AM acknowledges an important role for gene flow between archaic and modern humans in Eurasia and rejects a separate specific status for Neandertals and other Middle and Late Pleistocene archaic hominin groups. But when the AM was first articulated, it clearly differed from MRE on three important points: (1) a relatively recent African center of origin for anatomically modern humans, (2) the pattern of spread of early modern human morphology; and (3) the extent of regional continuity across Eurasia.

In contrast to MRE, AM accepts that modern human morphology evolved as a complex in Africa. The dating and morphological pattern of the African Transitional Group shows that this transition was underway by 250 Kya. Although the ATG is spread throughout Africa, it seems likely that fully modern humans initially appeared in East Africa, if we accept the dating of Omo Kibish (KHS) and Herto to >104 and <196 Kya. From this origin, a complicated pattern of population migration and demic diffusion spread modern human morphology into the rest of Africa as well as into the Levant (at ~100 Kya), and then later into Asia, Europe and the rest of the world. While this is all fundamentally compatible with RAO (see Prediction 2), AM posits that indigenous archaic populations were not totally replaced throughout Eurasia—or for that matter in the African peripheries—by the radiating moderns. Rather, interactions between these early moderns and aboriginal archaic peoples included some degree of biological admixture—at least some of the time and in some places.

Thus both MRE and AM agree that there is evidence of regional continuity across the archaic/modern human boundary throughout Eurasia. The distinction here is a matter of degree. While MRE is not always specific as to the extent of continuity in Europe, Australasia, and East Asia, it implies that regional archaic contributions to early moderns in those regions may have been on the order of 50% (Wolpoff et al. 2004) or more, and that these large contributions are reflected in relatively large numbers of regionally distinctive features. By contrast, AM proponents have consistently argued that evidence of continuity is found only in limited numbers of anatomical details. The data presented earlier in this chapter suggest that there are significantly fewer compelling regional-continuity features in Europe, Australasia, and East Asia than MRE proponents have argued. How, then, would AM assess the four predictions discussed for RAO and MRE?

Prediction 1. AM accepts that the relative morphological homogeneity of modern humans reflects the emergence of the modern-human morphological complex in one region and its subsequent spread to other regions *as a complex*. Nevertheless, in southern and northern Africa, East Asia, Australasia, and Europe, early modern populations retain a few morphological characteristics that evidently derive from local archaic peoples. Such features include occipital buns, suprainiac fossae, and nasion projection in Europe; Asian-type incisor shovelling, malar notches, and features of the upper nose in East Asia; and the complex of five features identified by Habgood (1989, 2003) as a signal of regional continuity in Australasia. Of course, the reliability of such features as indicators of genetic continuity is hotly contested by proponents of the RAO model (see below).

AM also fits the arguments made by the geneticists who advocate a "Mostly Out of Africa" model of modern human origins. Although most recent human genetic variation has a relatively recent African origin, such nuclear DNA polymorphisms such as the beta-globin cluster, ZFX, PDHA1, and MCPH1 show distributions that are not compatible with the wholly African origin envisioned by advocates of RAO. While this evidence derives from relatively few genes, it does suggest that a total replacement of all archaic peoples is unlikely.

Genetic data provide no evidence of the level of Eurasian regional genetic continuity that one might expect with MRE. However, this is not necessarily a decisive objection to the MRE model. The predominance of African haplotypes in living humans may result from the fact that Africa had a larger effective population size throughout most of the Pleistocene than Eurasia (Rele-

thford 1998a, 2001; Relethford and Harpending 1995). Following the initial expansion of modern humans out of Africa, continued gene flow from the larger African populations would be expected to have enforced greater proportions of African genes everywhere else. In time, continuing gene flow out of Africa could have produced a genetic picture indistinguishable from one of near-total population replacement. Even if a great deal of local genetic continuity was at first present in early modern human populations in Europe or Asia, the local gene pools may eventually have been swamped by gene flow out of the larger populations in Africa. This sort of process may have been going on for a long time, if Templeton (2002, 2005) has correctly identified allelic signals of previous migrations out of Africa, assimilated into the gene pools established by the last big African emigration around 130 Kya.

The seven samples of ancient mtDNA recovered from early modern Europeans reveal no indication of a Neandertal contribution. But the mtDNA data cannot exclude a Neandertal contribution of <25% to early modern European gene pools (Serre et al. 2004). This is significantly less than the ~50% Neandertal contribution that some MRE proponents think the morphological evidence allows for (Wolpoff et al. 2004), but it is perfectly compatible with the Assimilation Model. If the Neandertal genetic contribution to early modern European populations were on the order of 10%, then subsequent gene flow, genetic swamping, and drift could easily have removed almost all evidence of that contribution within a few thousand years. Similar remarks apply to other regions. This may explain why so few ancient non-African contributions can be discerned in today's populations.

The single early modern mtDNA haplotype from Australia comes from Willandra Lakes 3. The WLH 3 individual is certainly morphologically modern, but the haplotype recovered falls out of the modern range. This means that this mtDNA lineage subsequently must have gone extinct, since it is not found in living Australians or other modern peoples. If this lineage could go extinct in 40 Ky, so too could mtDNA lineages contributed to early modern Europeans by Neandertals—especially if the number of those lineages was small, as the Assimilation Model assumes.

Prediction 2. In regard to the data bearing on this prediction, AM basically agrees with the RAO: modern human morphology was established first in East Africa and then spread into Eurasia. But unlike the advocates of RAO, AM supporters view the Klasies River Mouth hominins from South Africa not as samples of this early modern population in Africa, but as an admixed sample reflecting hybridization between late archaic natives of South Africa and early-modern immigrants from East Africa. AM proponents argue that similar things went on as early modern populations expanded into North Africa and Eurasia as well.

Prediction 3. Again like the RAO, the Assimilation Model posits an original transition to modern human morphology in Africa, reflected in the fossils belonging to the African Transitional Group. However, like MRE, AM recognizes transitional samples and specimens in Eurasia as well. From an AM perspective, the Tabūn C2 mandible represents early evidence of such a transition, incorporating a mosaic of modern and archaic (and perhaps even some specifically Neandertal) anatomical features (Quam and Smith 1998). If the TL chronology for the near East is correct, Tabūn C2 is contemporary with some of the ATG, and might be viewed as simply indicating an extension of the ATG into the Levant. RAO proponents would accept this interpretation. But another interpretation, which would be more attractive to AM supporters, is that the otherwise basically archaic-looking Level-C hominins from Tabūn represent a Neandertal population that has assimilated some gene flow from African populations to the south.

Assimilation theorists see similar evidence of admixed Neandertal samples in Europe, particularly the Level-G sample from Vindija. They argue that the modern-like changes seen in this population, occurring in the context of a basically Neandertal morphological gestalt, reflect the beginnings of modern human gene flow into Europe at around 40–45 Kya. At this point in time, the extent of gene flow was not extensive, and there is no fossil evidence that any modern human immigrants had actually entered Europe. However, a relatively minor amount of gene flow into Europe from modern populations in adjacent areas would account for the changes in morphological pattern seen in Vindija and some other late Neandertal samples.

Unfortunately, similar late-archaic samples are not known from East Asia and Australasia, with the possible exception of the Ngandong fossils. The Ngandong hominins are more advanced than their Erectine predecessors in certain features (mainly related to increase in brain size) and can in this sense be broadly described as "transitional" between Erectines and early moderns. However, there is little evidence that they sample a truly transitional population linking their predecessors and successors in the region. Similar remarks apply to Chinese humans from this time period. In these parts of the world, the best evidence for continuity comes from the presence of regional-continuity features in the early modern specimens, not from transitional populations with intermediate morphology (like the ATG).

Prediction 4. AM and MRE advocates agree that there is evidence for such regional-continuity features in the earliest modern people from the peripheries of the Old World—Europe, East Asia, and Australasia. However,

AM recognizes fewer of those features, implying significantly less archaic-modern continuity in these areas than is generally accepted by proponents of MRE. The small percentage of non-African alleles detected in recent Eurasian gene pools implies a correspondingly small contribution to those populations from local archaic sources. It is certainly possible, however, that genetic continuity was greater in the past, and that many ancient alleles have been eliminated by subsequent demographic phenomena or evolutionary processes.

Morphological evidence of continuity is also not extensive. The most plausible indicators of a Neandertal contribution to early modern European populations are high frequencies of occipital bunning and suprainiac fossae, the amount of upper midfacial projection, the morphological pattern of the supraorbital torus, and possibly proximal-lateral femoral flanges. The body proportions of the Lagar Velho 1 specimen also suggest the persistence of Neandertal genes in these populations. In East Asia and Australasia, other regionally distinctive clusters of co-occurring features imply some degree of continuity between local archaics and early moderns. The case for assimilation in southern Africa is based on the unusual mix of archaic and modern features (and the unusually high degree of size variability) seen in the Klasies River Mouth sample.

Many researchers reject such characteristics as indicators of continuity. They argue that these supposed "regional continuity" features are not homologous, inaccurately defined, not regionally distinctive, or not known to be genetically determined. These are valid issues, and it is hard to settle them decisively in the present state of our knowledge. Nevertheless, some of these objections can be addressed to some degree. For example, occipital bunning in Neandertals may not be homologous with that seen in early modern Europeans. However, in this case homology cannot be rejected on morphological grounds, since the differences between Neandertals and early modern buns can be explained as side effects of other shape differences in the cranial vault. Thus, applying the word "bunning" to both morphologies is not merely a case of imprecise definition, as has been claimed (Bräuer and Broeg 1998). It is also true that occipital hemibuns (the modern variant) are commonly found in some recent populations from other parts of the world. But the high frequency of this feature in early modern Europeans, and its rarity in other early modern samples, provide a compelling reason for thinking that the genes underlying this morphology entered the modern gene pool through Europe.

The genetic control of such features is a major issue in these debates. As Lieberman (1995) correctly insists, we do not know for certain that any of these features is genetically based. If they are not, they cannot provide evidence for regional continuity. But this argument cuts both ways. The RAO model is grounded in the converse assumption, that the facial changes that distinguish modern humans from their archaic predecessors reflect genetic change of some sort. In fact, we have no more direct evidence for this assumption than we have for the genetic basis of many "regional continuity" features. If we reject occipital bunning, suprainiac fossae, or other shared plesiomorphies as evidence for regional continuity because their genetic basis is unknown, then we should be equally hesitant about accepting shared apomorphies as evidence for the African affinities of early modern humans until the genetic basis of those apomorphies is better understood.

ASSIMILATION AND INTERACTIONS BETWEEN MODERN AND ARCHAIC HUMANS

According to the Assimilation Model, modern human morphology evolved as a complex in East Africa around 160 Kya, or slightly earlier. Population movements and demic diffusion carried this distinctive morphological complex into southern Africa and the Levant just over 100 Kya. This complex subsequently came to predominate throughout Eurasia and Australasia, and it was ultimately carried by modern human colonists into areas devoid of archaic forerunners, like the Americas. The controversy surrounding modern human origins centers on exactly what happened when early modern migrants encountered indigenous archaic folk outside of East Africa.

In our view, the mosaic of features exhibited by the Klasies River Mouth human sample testifies to the assimilation of local southern African archaic elements (essentially members of the African Transitional Group or *Homo helmei*) into an expanding early modern human population (or perhaps vice versa). In the Levant, similar evidence of assimilation is seen in the morphology of both Tabūn C2 and the local Neandertals. As shown in Chapter 7, Levantine Neandertals are more like early modern humans than European Neandertals are in various features of the cranial vault. The Assimilation Model interprets the similarities between Levantine Neandertals and the Skhūl/Qafzeh group as indications of interbreeding. After ~40 Kya, signs of early-modern genetic influence begin to appear in Europe as well, manifested in the "progressive" tendencies of late Neandertal samples and specimens that fall in the 40- to 30-Ky time range.

As gene flow and population movements continued, early modern morphology became increasingly dominant throughout Africa and the peripheries of Eurasia. By around 75 Kya (assuming the correctness of the most recent date for Border Cave), fully modern humans were established in southern Africa. In North Africa, there are no samples comparable to the Klasies River

Mouth hominins, and dating uncertainties make it unclear when early moderns first appeared in this area. For now, all we can say is that it happened sometime between 127 and 34 Kya. In the European periphery, the first fully modern humans date back only some 35 Ky. In the East Asian and Australasian peripheries, the date for the first early moderns probably lies between 60 and 40 Kya.

According to the Assimilation Model, what ultimately happened in all these cases was equivalent to Levin's (2002) process of "extinction by hybridization." Although Levin defines this phenomenon with reference to interbreeding species, it applies equally well to regionally defined populations or semispecies. If the archaic peoples were much rarer on the landscape than their modern counterparts in regions of sympatry or parapatry, or if population movements were almost invariably out of Africa into the Eurasian periphery, then the more abundant and perpetually reinforced African populations would eventually have overwhelmed the local archaic peoples. Once interbreeding was established in even a few parts of the smaller populations' ranges, a behavioral conduit for movement of genes from the larger, more modern, African-derived populations into the smaller ones would have been laid down, leading ultimately to the genetic or allelic swamping of the smaller local populations. Within a relatively few generations—perhaps within a span of a millennium or two—the regional archaic populations would have become morphologically "extinct." They would not have become extinct in the sense applicable to interspecific competition; rather, they would simply have been assimilated into early modern populations. Genetic and morphological traces of that assimilation lingered for a while in the resulting hybrid populations, but were eventually obliterated—not necessarily by the pressures of natural selection, but by patterns of gene flow, population movements, and relative gene pool sizes.

Evidence for the relative population sizes assumed in this scenario is derived from both the archaeological record and ecological modeling. The most complete assessment of human population sizes in the Pleistocene has been provided by F. Hassan (1981). Hassan estimated both the population density and overall population size (based on archaeological data) for Middle Paleolithic and Upper Paleolithic groups. He concluded that Middle Paleolithic people (meaning any people associated with Mousterian or Middle Stone Age cultural assemblages, thus including some early moderns) had a total population size of slightly more than one million and a density of 0.03 persons per square kilometer. The population size estimated for Upper Paleolithic peoples was around six times greater, but their population density was only about three times greater (0.10 persons per km²), because Upper Paleolithic people inhabited almost twice as much land area as Middle Paleolithic people did—at least by Hassan's calculations. Of course, these estimates are only approximations, but they show a demographic patterning that supports the scenario of assimilation. There appear to have been far more moderns than archaics per square mile, and far more square miles inhabited exclusively by anatomically modern Upper Paleolithic humans. Similar figures calculated for Neandertal population size on the basis of estimated physiological, subsistence, and life-history factors suggest that the disproportion may have been even greater in Europe (Leonard 2002, O'Connell 2006, Sorensen and Leonard 2001).

It is not known why modern morphology appeared first in East Africa. As we noted earlier, there is no evidence for any association of the new morphology with behavioral changes, either as causes or effects of the morphological novelties. While some cultural innovations may appear in the African MSA that are not known for the Eurasian Mousterian, those innovations are too late to have been a prime mover in the emergence of modern morphology. Perhaps such late innovations eventually helped to drive the subsequent one-way flow of modern populations out of Africa (McBrearty and Brooks 2000). This factor was apparently not involved in the earliest stages of outflow, since the early moderns from the Levant show no trace of such innovations. But once such behavioral innovations had appeared in Africa, the ecological and demographic factors enumerated by Hassan would have led to an expansion of the African population and helped to fuel a series of outmigrations. The result was inevitable; given consistent one-way migration from Africa into other parts of the Old World, African genes and morphology ultimately and necessarily prevailed. But in the Assimilation Model, this process was not the abrupt, genocidal replacement envisioned by RAO theorists. Rather, it was a dissolution by dilution of the local archaic populations. Swamped by an influx of modern humans and African genes over a period of several thousand years, the local gene pools were eventually diluted to the extent that their regionally distinctive morphologies and genetic contributions are virtually imperceptible in recent humans. Like the Delaware or the Tasmanians in more recent times, they disappeared in the same way that a cup of wine disappears when 40 gallons of water are stirred into it.

The available evidence suggests that Europe was probably the last major region of the Old World to hold out against these pressures of assimilation. Why was this the case? One possibility is that the Neandertals were so well-adapted to their European setting that modern humans could not make inroads into Europe until Upper Paleolithic technological innovations (e.g., tailored clothing, better pyrotechnology) gave them a more effective cultural buffer against that environment. Once moderns were able to adapt to Pleistocene Europe,

their demographic advantages would have begun to take hold. Zubrow (1989) calculates that a 2% difference in mortality in favor of early moderns compared to Neandertals would have led to the latter's extinction in 30 generations.

Early modern humans equipped with Upper Paleolithic technology may have had ecological as well as demographic and technological advantages over Neandertals. There is some evidence that they were exploiting a broader range of dietary resources. Some isotopic data suggest that early moderns were including fish, mollusks, and birds in their diet, whereas Neandertals were almost exclusively hunting terrestrial herbivores (Richards et al. 2000, 2001). However, other isotopic studies (Drucker and Bocherens 2004) and archaeological data (Bar-Yosef 2004) have failed to reveal significant differences between Neandertal and early modern European subsistence patterns. If the two groups did rely on identical resources for subsistence, that would have made the competition between them all the more intense. Finlayson (2004) argues that Neandertals disappeared because they could not adapt to the radically changing climates of 30–40 Kya. Dating uncertainties make it difficult to pinpoint whether climate played a major role in the disappearance of Neandertals (Tzedakis et al. 2007), and Neandertals and their precursors had of course survived earlier fluctuations in the Pleistocene climates of Europe. But they had never done so in the face of competition from Upper Paleolithic invaders. While climatic instability may well have had a negative impact on the Neandertals, there is no reason to doubt that the causes of their disappearance were inextricably linked to the influx of modern people.

But there are also no compelling grounds for thinking that the disappearance of the Neandertals represented a species-replacement event. Klein (1999, p. 518) argues that "... the Cro-Magnon invasion of Europe would have differed fundamentally from the historic European invasion of the Americas or Australia," because the "Cro-Magnons" and Neandertals belonged to different species and had fundamentally different capacities for culture. As we see it, the archaeological evidence from Africa, the Levant, and Europe provides no clear indications of such differences in capacity; and the biological and morphological evidence suggests that there was at least a limited amount of gene flow between early modern populations and their archaic antecedents throughout the Old World. As noted earlier, a "hybrid zone" between Neandertals and early moderns apparently extended at least from Central Europe into the Near East, and possibly into North Africa as well. The breadth of this zone strongly suggests that Neandertals and early moderns were not different species. From an AM perspective, the European invasions of Australia and the Americas actually furnish an excellent model for the complex population dynamics involved in the spread of early modern humans and their encounters with indigenous archaic folk. What happened and continues to happen to indigenous American and Australian populations, both biologically and culturally, parallels what AM represents as having happened to the Neandertals and other archaic populations in the late Pleistocene.

One of us (FS) has long been a frequent visitor to the Eastern Band of the Cherokee, who inhabit the Qualla Boundary in western North Carolina. Many of these people are not readily identifiable to the casual observer as Cherokees. In general, they look like and have the same behaviors as their non-Cherokee neighbors. Although significant differences in gene frequencies and Native American phenotypic traits persist, biological swamping of the Cherokee by the far more numerous European-derived Americans is undeniable from both the genetic and morphological evidence. Cultural assimilation has been even more thorough. While some vestiges of Cherokee uniqueness remain, most of the items of material culture seen on the Qualla Boundary are the same as those found in surrounding areas not occupied by Cherokee. Future excavations of early 21st-century sites on the Qualla Boundary would disclose few traces of pre-contact Cherokee culture. Skeletal studies might reveal more of the Cherokee uniqueness to future archaeologists; but the effects of biological swamping would certainly be apparent, and distinctive morphological indicators would probably be relatively rare. How much of the Cherokee biological distinctiveness will remain if this process of assimilation continues for another thousand years? We suspect that precious little would remain and that what remained would be virtually impossible to detect in a biological or archaeological study.

We believe that this mirrors what happened to the Neandertals and other archaic Eurasians. As opponents of such an idea would point out, Neandertals and Asian archaics were far more different morphologically from the early modern Europeans than the Cherokees were from the Europeans they encountered in the 17th and 18th centuries. And if Cherokees and Europeans are taken as exemplars of human subspecies or "races," then it is natural to think that Neandertals and early moderns should be separated at a higher level of the Linnaean hierarchy—presumably that of the species—at least in terms of the morphological species concept.

But as hinted earlier, the equation of modern human "races" with subspecies may be a misconception. Our concepts of human subspecies have traditionally centered on the geographically correlated differences that exist between recent human populations or "races." But as RAO proponents are fond of pointing out, those differences are surprisingly small compared to the differences between comparable regional populations in other mammalian species. Perhaps we need to adjust our concepts of "race" accordingly. It might be biologi-

cally sounder to think of all humans today as constituting a single "race" or subspecies. If so, then the differences between the regionally distinctive archaic populations of the Pleistocene—between Neandertals and the ATG, or between Dali and Ngandong—presumably exemplify the degree of difference to be expected between human subspecies.

We can accordingly describe the entire human population today as representing the African subspecies *Homo sapiens sapiens*, with slight amounts of local admixture from other subspecies that once existed in Europe (*Homo sapiens neanderthalensis*), Asia, and Australasia. The other subspecies have disappeared due to genetic swamping and assimilation. Most of the *H. s. sapiens* gene pool is African, because Africa was the original home of this subspecies and the initial surge in the size of its population began there. *Homo sapiens sapiens* is ubiquitous throughout the world today because beginning about 100 Kya, successive waves of outmigrants from Africa demographically and genetically swamped the indigenous archaic people of Eurasia (and parts of Africa as well). But while Neandertals and other archaic populations were not the major contributors to the gene pool of *H. s. sapiens*, they appear to have made minor contributions to that gene pool in the Pleistocene. Therefore, they should not be separated from the African race at a higher taxonomic level than the subspecies.

As noted earlier, there are theoretical reasons for doubting that the regional human populations of the Pleistocene could have been sufficiently isolated from each other long enough for the process of allopatric speciation to run its course. Some empirical evidence in support of this view has been marshaled by Holliday (2006a). Holliday assembled 328 cases of known hybridization between eutherian mammal species for which viable hybrid offspring are documented. In 156 of these cases, it is not known whether the offspring are fertile; but in the other 172 cases the offspring are demonstrably or probably fertile (although at least 26 follow Haldane's Rule that hybrids of one sex, usually the males, are sterile). Using estimates from the fossil record, Holliday determined that the most rapid attainment of partial reproductive isolation was between the genera *Bos* and *Bison* at 1 My, and the most rapid attainment of complete isolation was between horses and zebras/donkeys (genus *Equus*) at ~2 My (Cartmill and Holliday 2006). Estimates based on molecular clocks indicate that the shortest time needed to achieve partial isolation (between lions and tigers, genus *Panthera*) was 1.55 My, and complete isolation (between red-fronted and Thompson's gazelles, genus *Gazella*) required a minimum of 1.4 My. A generation length of 4–5 years was used to estimate how many generations it took for partial and full reproductive isolation to develop in this eutherian sample. The *minimum* numbers are 200,000–

250,000 generations for partial isolation (*Bos/Bison*) and 280,000-350,000 generations for full isolation (*Gazella* spp.). Various estimates date the divergence of Neandertals from modern humans to between 250 and 700 Kya. No matter what assumptions we make about geographical ranges or behavioral isolating mechanisms, generation lengths in later human evolution were surely at least 10 years. Thus, Neandertals and modern humans could not have been separated from each other for more than 70,000 human generations—less than half of the number of generations documented as necessary for speciation in eutherian mammals. It is therefore parsimonious to assume that they were not reproductively isolated.

In our view, the available facts bearing on modern human origins are best explained by the Assimilation Model. However, the competing models cannot be decisively rejected. If the features used to indicate regional continuity prove not to be genetically determined, and if other explanations are found for the few genetic indicators of continuity in Eurasia, then AM (and MRE) would be fatally undermined. On the other hand, if future research shows that archaic Eurasian contributions to early modern populations were substantially larger than the current evidence implies, or if new fossil evidence shows that the spread of modern morphology in the late Pleistocene follows no clear chronological or geographical pattern, then MRE will emerge as a better

alternative than AM. It is our hope and belief that all but one—or perhaps all—of these incompatible models will eventually go down under the accumulating weight of empirical evidence. The evolution of scientific ideas, like the evolution of organic life, progresses through the survival of the fittest variants and the disappearance of the less adequate ones.

In the history of life on this planet, the story of the human lineage is uniquely complex, because its final chapters involve elements and factors that are not found in the stories of any other organisms. Most of our story is shared with those of millions of other species. But over the last few moments of geological time, our line has emerged from creatures not unlike many other eutherian mammals to become something with no real parallels in the story of evolution—a bipedal, brainy, migrating, talking, and symbolic ape. We have tried to capture the essence of that transformation and the excitement of struggling to understand it. We have assuredly failed to some extent on both accounts, because of our limited knowledge of the facts and because of the biases that we bring to their interpretation. Our final hope is that the readers of this book can transcend our limitations to arrive at new ideas and questions of their own and thereby help to move paleoanthropology toward a deeper understanding of what our biological history tells us about being human.

Appendix: Cranial Measurements

In the chapters of this book that deal with fossil hominin remains, cranial metrics are used extensively to aid in both descriptions and comparisons. We present here the definitions of these measurements and the anatomical points at which the measurements are taken (Tables A.1 to A.4). The named points are illustrated in Figs. A.1–A.3. Angles are illustrated in Fig. A.4. The method for measuring the mandibular symphyseal angle is taken from G. Olivier (1969). Other definitions are adapted from the fourth edition of R. Martin's *Lehrbuch der Anthropologie* (Bräuer 1988) and from W. W. Howells's *Cranial Variation in Man* (Howells 1973b).

The standard orientation of skulls for measurement and comparison is in the ear-eye plane, formalized by the Frankfurt Agreement of 1884 as a plane drawn tangent to the upper edge of the external acoustic meatuses and the lower margins of the orbits (cf. Bräuer and Knußmann 1988). The terms *anterior, posterior, superior,* and *inferior* are defined with reference to this so-called **Frankfurt Horizontal** (Fig. A.1). The other two common reference planes are the **midsagittal plane**, drawn through the midline of the skull, and the **coronal plane** (any plane perpendicular to both the Frankfurt and midsagittal planes). All the terms defined below refer to the external surface of the skull.

The only dental metric used in this book is tooth breadth—the maximum labiolingual (lip-to-tongue) or buccolingual (cheek-to-tongue) diameter of the tooth crown (Martin's measurement 81-1). Most of the postcranial measurements that we discuss are maximum long-bone lengths. Other postcranial measurements used are defined in the text or in cited sources. Further details can be found in Bräuer (1988).

TABLE A.1 ■ **Definitions and Standard Abbreviations for Craniometric Points**

Point (Abbreviation)	
Asterion (ast)	The caudal end of the suture between the parietal and temporal bones (Figs. A.1 and A.3).
Bregma (b)	The anterior end of the **sagittal suture** (between the two parietal bones: Figs. A.1 and A.2).
Basion (ba)	The **midsagittal** (midline) point on the front edge of the foramen magnum (Fig. A.4).
Coronale (co)	Most lateral point on the coronal suture (between the parietal and frontal bones: Fig. A.2).
Ectoconchion	The intersection of (a) the most anterior surface of the lateral orbital border and (b) a horizontal plane bisecting the maximum vertical heights of the left and right orbital openings.
Euryon (eu)	Most lateral point on the cranial vault (Figs. A.2 and A.3).
Frontomalare temporale (fmt)	Most lateral point on the zygomaticofrontal suture (Figs. A.1 and A.2).
Frontotemporale (ft)	Most anterior point on the superior temporal line on the frontal bone (Fig. A.1).
Glabella (g)	Most anterior midsagittal point on the frontal bone (Fig. A.1).
Gnathion (gn)	The most inferior midsagittal point on the mandibular symphysis (Fig. A.1).
Infradentale (id)	The most anterior midsagittal point on the alveolar process of the mandible between the lower central incisors (Fig. A.1).
Inion (i)	The point on the midsagittal plane where both superior nuchal lines meet (Figs. A.1 and A.3).
Lambda (l)	The caudal end of the sagittal suture (Figs. A.1 and A.3).
Metopion (m)	The point on the frontal bone furthest from the nasion-bregma chord (Fig. A.1).
Mastoideale (ms)	The most inferior point on the mastoid process (Figs. A.1 and A.3).
Maxillofrontale (mf)	The point where the **maxillofrontal suture** (between the frontal and maxilla) crosses the rim of the orbit (Fig. A.2).
Nasion (n)	The midsagittal point on the suture between the nasal and frontal bones (Fig. A.2).
Nasospinale (ns)	The midsagittal point on a horizontal line drawn perpendicular to the midsagittal plane through the lowest point on the rim of the nasal (piriform) aperture (Fig. A.1 and A.2).
Opisthion (o)	The midsagittal (midline) point on the posterior edge of the foramen magnum (Fig. A.4).
Opisthocranion (op)	The most posterior point on the cranial vault (Fig. A.1).
Orbitale (or)	The most inferior point on the orbital rim (where it intersects the Frankfurt plane: Fig. A.1).
Pogonion (pg)	The most anterior midsagittal point on the protuberance of the chin, below the concavity or mandibular incurvature beneath the anterior incisors (Fig. A.1). If there is no mandibular incurvature, a point on the anterior face of the mandible just above the inferior margin of the symphysis is used instead in measurements involving pogonion. This point is selected to correspond as closely as possible to the point described above.
Porion (po)	The uppermost point on the rim of the external auditory meatus (where it intersects the Frankfurt plane: Fig. A.1).
Prosthion (pr)	The most anterior midsagittal point on the alveolar process of the maxilla lying between the upper central incisors (Figs. A.1 and A.2).
Stephanion (st)	The point where the superior temporal line crosses the coronal suture (Figs. A.1 and A.2).
Subspinale (ss)	The most posterior point seen on the profile of the maxilla below the anterior nasal spine (Fig. A.1).
Vertex (v)	The highest point on the cranial vault (Fig. A.1).
Zygomaxillare anterius (zma)	The intersection of the zygomaxillary suture and the limit of the attachment of the masseter muscle on the facial surface (Fig. A.2).

TABLE A.2 ■ Linear Measurements of the Cranium and Mandible[a]

Measurement	Definition
Maximum cranial length (M1)	Glabella to opisthocranion
Maximum cranial breadth (M8)	Euryon to euryon (see notes 1 and 2)
Parietal breadth	(See note 2)
Minimum frontal breadth (M9)	Frontotemporale to frontotemporale
Maximum frontal breadth (M10)	Coronale to coronale (see note 3)
Biasterionic breadth (M12)	Asterion to asterion (maximum occipital breadth)
Mastoid height (M19a)	Porion to the level of mastoideale, measured vertically
Auricular height (M20)	Bregma to porion (projected to the midsagittal plane)
Occipital plane length	Lambda to inion
Nuchal plane length	Inion to opisthion
Bifrontal breadth (M43)	Frontomalare temporale to frontomalare temporale
Upper facial height (M48)	Nasion to prosthion
Alveolar height (M48.1)	Nasospinale to prosthion
Interorbital breadth (M50)	Maxillofrontale to maxillofrontale
Orbital breadth (M51)	Maxillofrontale to ectoconchion
Nasal breadth (M54)	Greatest breadth of the piriform aperture
Nasal height (M55)	Nasion to nasospinale
Mandibular symphysis height (M69)	Gnathion to infradentale
Mandibular minimum ramus breadth (M71a)	Minimum anterior–posterior dimension of the ramus
Mandibular corpus height	Vertical distance between the lowest point on the corpus and the highest point on the alveolar process, measured in a coronal plane through the mental foramen
Mandibular corpus breadth	Maximum medial–lateral dimension measured in a coronal plane through the mental foramen
Mandibular symphyseal thickness: base	Anterior–posterior thickness of the symphysis at the level of pogonion
Mandibular symphyseal thickness: alveolar process	Minimum anterior–posterior thickness of the symphysis above the level of pogonion

[a]All measurements are chords (straight-line distances between two points). The "M" numbers refer to the numbers assigned to the measurements by R. Martin (Bräuer 1988).

Notes:
1. Measurements between the same points indicate a breadth measurement from the point on one side of the body to the same point on the opposite side.
2. In post-Erectine hominins, parietal breadth is the same thing as maximum cranial breadth (M8). In apes and early hominins, the maximum breadth of the braincase falls further down, on the temporal bone (Figs. 4.84, 5.11, and 5.22B). In these cases, parietal breadth is measured separately at the approximate midpoint of the coronal suture.
3. Coronale often coincides with stephanion.

TABLE A.3 ■ Radius Measurements[a]

Radius (Martin Number)	Description
Vertex radius (M94)	Transmeatal axis to vertex
Nasion radius (M95)	Transmeatal axis to nasion
Prosthion radius (M97)	Transmeatal axis to prosthion
Zygomaxillary radius (M102)	Transmeatal axis to zygomaxillare anterius

[a]W. W. Howells (1973b) popularized the use of radii in craniometric analyses. Each radius is defined as the minimum distance from the **transmeatal axis**—a transverse axis drawn through the centers of both external auditory meatuses—to the point in question (prosthion, nasion, etc.). Martin numbers are from Bräuer (1988).

TABLE A.4 ■ Angles Used in This Volume[a]

Angle (Martin Number)	Description
Subspinale angle (M76a)	Angle between the lines drawn from subspinale to each zygomaxillare anterius. For details, see Howells (1973b). Howells abbreviation: SSA.
Frontal angle (M32-5)	Angle between the chords drawn from metopion (Fig. A.1) to nasion and bregma. For details, see Howells (1973b). Howells abbreviation: FRA.
Parietal angle (M33e)	Angle formed by the chords from lambda and bregma to the point on the sagittal suture furthest from the bregma-lambda chord (parietal chord). For details, see Howells (1973b). Howells abbreviation: PAA.
Occipital angle (M33d)	Angle formed by the chords from lambda and opisthion to the midsagittal point on the occipital bone furthest from the opisthion-lambda chord. For details, see Howells (1973b). Howells abbreviation: OCA.
Mandibular symphyseal angle	Angle between the pogonion-infradentale chord and the alveolar plane. For details, see Olivier (1969).

[a]Martin numbers are from Bräuer (1988). Except for the subspinale angle, angles are illustrated in Fig. A.4.

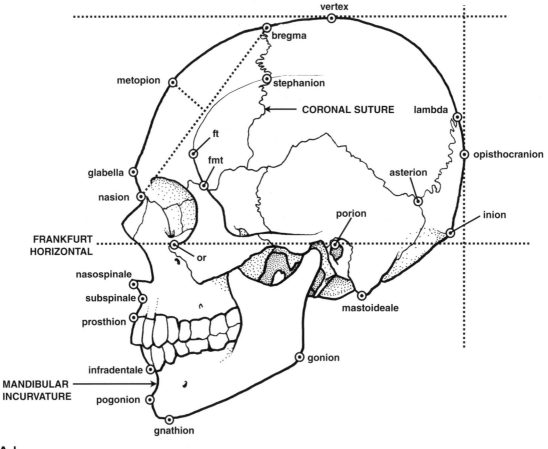

FIGURE A.1

Craniometric points: lateral aspect. Abbreviations: fmt, frontomalare temporale; ft, frontotemporale; or, orbitale.

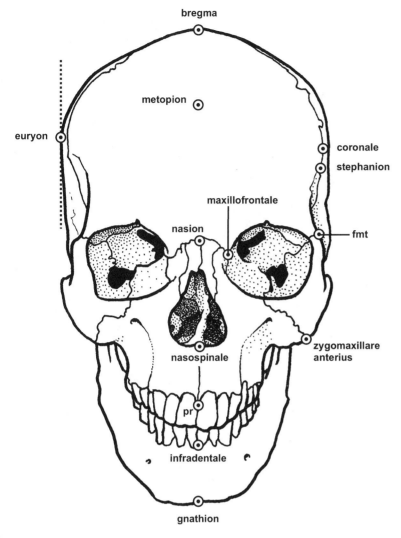

FIGURE A.2

Craniometric points: frontal aspect. pr, prosthion. Other abbreviations as in Fig. A.1.

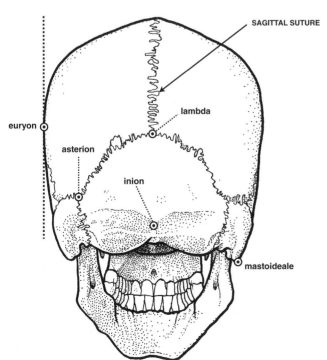

FIGURE A.3

Craniometric points: posterior aspect.

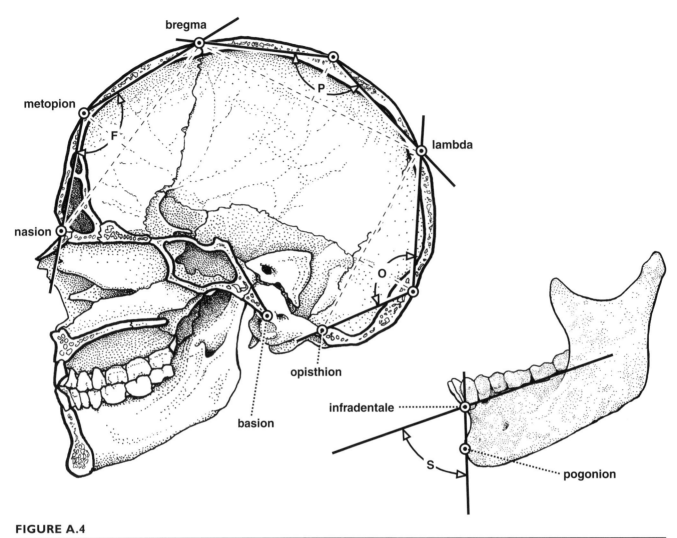

FIGURE A.4

Craniometric angles, illustrated on a hemisected skull (medial aspect) and mandible (lateral aspect). F, frontal angle; O, occipital angle; P, parietal angle; S, mandibular symphyseal angle.

Bibliography

EDITED VOLUMES

Aitken, M., C. Stringer, and P. Mellars, Editors (1993) *The Origins of Modern Humans and the Impact of Chronometric Dating*. Princeton: Princeton University Press.

Akazawa, T., K. Aoki, and T. Kimura, Editors (1992) *The Evolution and Dispersal of Modern Humans in Asia*. Tokyo: Hokusen-Sha.

Akazawa, T., K. Aoki, and O. Bar-Yosef, Editors (1998) *Neandertals and Modern Humans in Western Asia*. New York: Plenum.

Akazawa, T., and S. Muhesen, Editors (2002) *Neanderthal Burials. Excavations of the Dediryeh Cave, Afrin, Syria*. Kyoto: International Research Center for Japanese Studies.

Arsuaga, J., J. Bermúdez De Castro, and E. Carbonell, Editors (1997a) *The Sima de los Huesos Hominid Site. Journal of Human Evolution* **33**:105–421.

Bar-Yosef, O., and D. Pilbeam, Editors (2000) *The Geography of Neandertals and Modern Humans in Europe and the Greater Mediterranean*. Cambridge: Harvard Peabody Museum Bulletin Number 8.

Begun, D., C. Ward, and M. Rose, Editors (1997) *Function, Phylogeny, and Fossils: Miocene Hominoid Evolution and Adaptations*. New York: Plenum.

Beinhauer, K., R. Kraatz, and G. Wagner, Editors (1996) Homo erectus heidelbergensis *von Mauer*. Kolloquium I. Sigmaringen: Thorbecke.

Bermúdez De Castro, J., E. Carbonell, and J. Arsuaga, Editors (1999a) *Gran Dolina Site: TD6 Aurora Stratum (Burgos, Spain). Journal of Human Evolution* **37**:309–700.

Brain, C., Editor (1993) *Swartkrans: A Cave's Chronicle of Early Man*. Pretoria: Transvaal Museum Monograph Number 8.

Brantingham, P., S. Kuhn, and K. Kerry, Editors (2004) *The Early Upper Paleolithic Beyond Western Europe*. Berkeley: University of California Press.

Bräuer, G., and F. Smith, Editors (1992) *Continuity or Replacement? Controversies in* Homo sapiens *Evolution*. Rotterdam: Balkema.

Butzer, K., and G. Isaac, Editors (1975) *After the Australopithecines*. The Hague: Mouton.

Ciochon, R., and R. Corruccini, Editors (1983) *New Interpretations of Ape and Human Ancestry*. New York: Plenum.

Clark, G., and K. Willermet, Editors (1997) *Conceptual Issues in Modern Human Origins Research*. New York: Aldine De Gruyter.

Conard, N., Editor (2006) *When Neanderthals and Modern Humans Met*. Tübingen: Kerns Verlag.

Coppens, Y., F. C. Howell, G. Isaac, and R. Leakey, Editors (1976) *Earliest Man and Environments in the Lake Rudolf Basin.* Chicago: University of Chicago Press.

Coppens, Y., and B. Sénut, Editors (1991) *Origine(s) de la Bipèdie chez les Hominidés.* Paris: Centre Nationale de la Recherche Scientifique.

Corruccini, R., and R. Ciochon, Editors (1994) *Integrative Paths to the Past. Paleoanthropological Advances in Honor of F. Clark Howell.* Englewood Cliffs, NJ: Prentice-Hall.

Crow, T., Editor (2002) *The Speciation of Modern* Homo sapiens. Oxford: Oxford University Press.

Delson, E., Editor (1985) *Ancestors: The Hard Evidence.* New York: Liss.

Delson, E., I. Tattersall, J. Van Couvering, and A. Brooks, Editors (2000) *Encyclopedia of Human Evolution and Prehistory*, 2nd edition. New York: Garland.

Dibble, H., and A. Monet-White, Editors (1988) *Upper Pleistocene Prehistory of Western Eurasia.* Philadelphia: University of Pennsylvania Press.

Ereshefsky, M., Editor (1992) *The Units of Evolution: Essays on the Nature of Species.* Cambridge: MIT Press.

Fleagle, J., and R. Kay, Editors (1994) *Anthropoid Origins.* New York: Plenum.

Franzen, J., Editor (1994) *100 Years of Pithecanthropus. The* Homo erectus *Problem. Courier Forschungs-Institut Senckenberg* **171**:1–361.

Gebo, D., Editor (1993) *Postcranial Adaptation in Nonhuman Primates.* DeKalb: Northern Illinois University Press.

Grine, F., Editor (1988) *Evolutionary History of the "Robust" Australopithecines.* New York: Aldine de Gruyter.

Hartwig, W., Editor (2002) *The Primate Fossil Record.* Cambridge: University of Cambridge Press.

Harvati, K., and T. Harrison, Editors (2006) *Neanderthals Revisited: New Approaches and Perspectives.* Dordrecht: Springer.

Howells, W. W., Editor (1967) *Ideas on Human Evolution. Selected Essays 1949–1961,* college edition. New York: Atheneum.

Kimbel, W., and L. Martin, Editors (1993) *Species, Species Concepts and Primate Evolution.* New York: Plenum.

Leakey, M. G., and R. Leakey, Editors (1978) *Koobi Fora Research Project, Volume 1. The Fossil Hominids and an Introduction to Their Context, 1968–1974.* Oxford: Clarendon Press.

Lillegraven, J., Z. Kielan-Jaworowska, and W. Clemens, Jr., Editors (1979) *Mesozoic Mammals—The First Two-Thirds of Mammalian History.* Berkeley: University of California Press.

Luckett, W., and F. Szalay, Editors (1975) *Phylogeny of the Primates: A Multidisciplinary Approach.* New York: Plenum.

Maglio, V., and H. Cooke, Editors (1978) *Evolution of African Mammals.* Cambridge: Harvard University Press.

McGrew, W., L. Marchant, and T. Nishida, Editors (1996) *Great Ape Societies.* Cambridge: Cambridge University Press.

Mellars, P., Editor (1990) *The Emergence of Modern Humans: An Archaeological Perspective.* Ithaca: Cornell University Press.

Mellars, P., and C. Stringer, Editors (1989) *The Human Revolution: Behavioral and Biological Perspectives on the Origins of Modern Humans.* Edinburgh: University of Edinburgh Press.

Nitecki, M., and D. Nitecki, Editors (1994) *Origins of Anatomically Modern Humans.* New York: Plenum.

Ovey, C., Editor (1964) *The Swanscombe Skull.* London: Royal Anthropological Institute (Occasional Paper Number 20).

Ravosa, M., and M. Dagosto, Editors (2007) *Primate Origins: Adaptations and Evolution.* New York: Springer.

Roebroeks, W., and C. Gamble, Editors (1999) *The Middle Palaeolithic Occupation of Europe.* Leiden: University of Leiden Press.

Roebroeks, W., and T. van Kolfschoten, Editors (1995) *The Earliest Occupation of Europe.* Leiden: University of Leiden Press.

Ross, C., and R. Kay, Editors (2004) *Anthropoid Origins: New Visions.* New York: KluwerAcademic.

Schmitz, R., Editor (2006) *Neanderthal 1856–2006.* Mainz: Verlag Philipp von Zabern.

Sigmon, B., and J. Cybulski, Editors (1981) Homo erectus: *Papers in Honor of Davidson Black.* Toronto: University of Toronto Press.

Smith, F., and F. Spencer, Editors (1984) *The Origins of Modern Humans. A World Survey of the Fossil Evidence.* New York: Liss.

Spencer, F., Editor (1997) *History of Physical Anthropology. An Encyclopedia.* Two volumes. New York: Garland.

Strasser, E., J. Fleagle, A. Rosenberger, and H. McHenry, Editors (1998) *Primate Locomotion: Recent Advances.* New York: Plenum.

Stringer, C., Editor (1981) *Aspects of Human Evolution.* London: Taylor & Francis.

Stringer, C., R. N. Barton, and C. Finlayson, Editors (2000) *Neanderthals on the Edge.* Oxford: Oxbow Books.

Suzuki, H., and F. Takai, Editors (1970) *The Amud Man and His Cave Site.* Tokyo: University of Tokyo Press.

Teschler-Nicola, M., Editor (2006) *Early Modern Humans at the Moravian Gate. The Mladeč Caves and Their Remains.* Vienna: Springer.

Tobias, P., Editor (1985) *Hominid Evolution: Past, Present and Future.* New York: Alan R. Liss.

Trinkaus, E., Editor (1989) *The Emergence of Modern Humans. Biocultural Adaptations in the Later Pleistocene.* Cambridge: University of Cambridge Press.

Trinkaus, E., and J. Svoboda, Editors (2006a) *Early Modern Human Evolution in Central Europe. The People of Dolní Věstonice and Pavlov.* Oxford: Oxford University Press.

Tuttle, R., Editor (1975) *Paleoanthropology: Morphology and Paleoecology.* The Hague: Mouton.

van Andel, T., and W. Davies, Editors (2003) *Neanderthals and Modern Humans in the European Landscape During the Last Glaciation.* Cambridge: McDonald Institute for Archaeological Research.

von Koeningswald, G. H. R., Editor (1958) *Hundert Jahre Neanderthaler.* Utrecht: Kemink en Zoon.

Vrba, E., G. Denton, T. Partridge, and L. Burckle, Editors (1995) *Paleoclimate and Evolution, with Emphasis on Human Origins.* New Haven: Yale University Press.

Walker, A., and R. Leakey, Editors (1993a) *The Nariokotome* Homo erectus *Skeleton.* Cambridge: Harvard University Press.

Wood, B., L. Martin, and P. Andrews, Editors (1986) *Major Topics in Primate and Human Evolution.* Cambridge: Cambridge University Press.

Wu, R., and J. Olsen, Editors (1985) *Paleoanthropology and Paleolithic Archaeology in the People's Republic of China.* New York: Academic Press.

Zilhao, J., and F. d'Errico, Editors (2003) *The Chronology of the Aurignacian and of the Transitional Technocomplexes.* Lisbon: *Trabalhos de Arqueologia* Number 33.

Zilhão, J., and E. Trinkaus, Editors (2002) *Portrait of the Artist as a Young Child. The Gravettian Human Skeleton from the Abrigo do Lagar Velho and Its Archaeological Context.* Lisbon: *Trabalhos de Arqueologia* Number 22.

ARTICLES AND CHAPTERS

Most journals are cited in the bibliography using the abbreviations given below. Abbreviations are not used for journals with single-word titles, such as *Nature* or *Science,* nor for less frequently cited journals.

Journal Abbreviations: *AA, American Anthropologist; AAS, Acta Anthropologica Sinica; AIPH, Archives de l'Institut de Paléontologie Humaine; AJHB, American Journal of Human Biology; AJHG, American Journal of Human Genetics; AJP, American Journal of Primatology; AJPA, American Journal of Physical Anthropology; AK, Archäologisches Korrespondenzblatt; AO, Archaeology in Oceania; ANYAS, Annals of the New York Academy of Sciences; APAMNH, Anthropological Papers of the American Museum of Natural History; AR, Anatomical Record; ARA, Annual Review of Anthropology; AS, American Scientist; AZ, Anthropologischer Anzeiger; BARIS, British Archaeological Reports International Series; BGSC, Bulletin of the Geological Society of China; BMSAP, Bulletin et Mémoire de la Société de l'Anthropologie de Paris; CA, Current Anthropology; CFS, Courier Forschungsinstitut Senckenberg; CRAS, Comptes Rendus de l'Académie des Sciences de Paris; CRE, Comptes Rendus du Paleo; CSHS, Cold Springs Harbor Symposia on Quantitative Biology; EAZ, Ethnologische-Archäologische Zeitschrift; ERAUL, Études et Recherches Archéologiques de l'Université de Liège; EG, Eiszeitalter und Gegenwart; FP, Folia Primatologica; HB, Human Biology; HEC, Human Ecology; HE, Human Evolution; IJO, International Journal of Osteoarcheology; IJP, International Journal of Primatology; JA, Journal of Anatomy; JAA, Journal of Anthropological Archaeology; JAR, Journal of Anthropological Research; JAS, Journal of Archaeological Science; JHE, Journal of Human Evolution; JM, Journal of Morphology; JME, Journal of Molecular Evolution; JRAI, Journal of the Royal Anthropological Institute; JWP, Journal of World Prehistory; MAGV, Mitteilungen der anthropologischen Gesellschaft in Wien; MBE, Molecular Biology and Evolution; MGU, Mitteilungen der Gesellschaft für Urgeschichte (Blaubeuren); MPE, Molecular Phylogenetics and Evolution; MQRSA, Modern Quaternary Research in Southeast Asia; MTM, Memoirs of the Transvaal Museum; NG, National Geographic; NH, Natural History; PAPS, Proceedings of the American Philosophical Society; PNAS, Proceedings of the National Academy of Sciences (USA); PPS, Proceedings of the Prehistoric Society; PTRSL, Philosophical Transactions of the Royal Society of London, Series B; QI, Quaternary International; QR, Quaternary Research, SA, Scientific American; SAJS, South African Journal of Science; SJA, Southwest Journal of Anthropology; TAPS, Transactions of the American Philosophical Society; VP, Vertebrata PalAsiatica; YPA, Yearbook of Physical Anthropology; ZE, Zeitschrift für Ethnologie; ZMA, Zeitschrift für Morphologie und Anthropologie.*

For chapters cited below that were published in multiply-referenced edited volumes, the source volume is referenced by giving the name(s) of the editor(s), date of the publication, and page numbers of the cited chapter. For example: P. Brown's 1997 publication on "Australian paleoanthropology" is cited as being "In: Spencer 1997:138–145." This refers to F. Spencer's edited volume *History of Physical Anthropology. An Encyclopedia,* published in 1997. The full citation for Spencer is given in the "Edited Volumes" section of the bibliography (see above).

A

Abbate, E., et al. (1998) A one-million-year-old *Homo* cranium from the Danakil (Afar) Depression of Eritrea. *Nature* **393**:458–460.

Abitbol, M. (1996) *Birth and Human Evolution: Anatomical and Obstetrical Mechanics in Primates.* Westport, CT: Bergin & Garvey.

Ackermann, R., J. Rogers, and J. Cheverud (2006) Identifying the morphological signature of hybridization in primate and human evolution. *JHE* **51**:632–645.

Adam, K. (1954) Die zeitliche Stellung des urmenschen Fundschicht von Steinheim an der Murr innerhalb des Pleistozäns. *EG* **4/5**:18–21.

Adams, D., A. Gold, and A. Burt (1978) Rise in female-initiated sexual activity at ovulation and its suppression by oral contraceptives. *New England Journal of Medicine* **299**:1145–1150.

Adcock, G., E. Dennis, S. Easteal, G. Huttley, L. Jerimn, W. Peacock, and A. Thorne (2001a) Mitochondrial DNA sequences in ancient Australians: Implications for modern human origins. *PNAS* **98**:537–542.

Adcock, G., E. Dennis, S. Easteal, G. Huttley, L. Jerimn, W. Peacock, and A. Thorne (2001b) Response to "Human origins and ancient DNA." *Science* **292**:1656.

Adcock, G., E. Dennis, S. Easteal, G. Huttley, L. Jerimn, W. Peacock, and A. Thorne (2001c) Lake Mungo 3: A response to recent critiques. *AO* **36**:170–174.

Ahern, J. (1998) Underestimating intraspecific variation: The problem with excluding Sts 19 from *Australopithecus africanus*. *AJPA* **105**:461–480.

Ahern, J. (2005) Foramen magnum position variation in *Pan troglodytes, Homo sapiens*, and Plio–Pleistocene hominids: Implications for recognizing the earliest hominids. *AJPA* **127**:267–276.

Ahern, J. (2006) Non-metric variation in recent humans as a model for understanding Neanderthal–early modern human differences: just how "unique" are Neanderthal unique traits? In: Harvati and Harrison 2006: 255–268.

Ahern, J., I. Karavanić, M. Paunović, I. Janković, and F. Smith (2004) New discoveries and interpretations of hominid fossils and artifacts from Vindija Cave, Croatia. *JHE* **46**:27–67.

Ahern, J., S. Lee, and J. Hawks (2002) The late Neandertal supraorbital fossils from Vindija Cave, Croatia: A biased sample? *JHE* **43**:419–432.

Ahern, J., and F. Smith (2004) Adolescent archaics or adult moderns? Le Moustier 1 as a model for estimating age at death of fragmentary supraorbital fossils in the modern human origins debate. *Homo* **55**:1–19.

Aiello, L. (1993) The fossil evidence for modern human origins in Africa: A revised view. *AA* **95**:73–96.

Aiello, L., and C. Dean (1990) *Human Evolutionary Anatomy*. New York: Academic Press.

Aiello, L., and P. Wheeler (1995) The expensive-tissue hypothesis: The brain and digestive system in human and primate evolution. *CA* **36**:199–221.

Aiello, L., and P. Wheeler (2003) Neanderthal thermoregulation and the glacial climate. In: van Andel and Davies 2003: 147–166.

Aiello, L., B. Wood, C. Key, and M. Lewis (1999) Morphological and taxonomic affinities of the Olduvai ulna (OH 36). *AJPA* **109**:89–110.

Aigner, J. (1986) The age of Zhoukoudian Locality 1: The newly proposed O^{18} correspondences. *Anthropus (Brno)* **23**:157–173.

Aitken, M., J. Huxtable, and N. Debenham (1986) Thermoluminescence dating in the Paleolithic: burned flint, stalagmitic calcite, and sediment. In: A. Tuffreau and J. Sommé, Editors. *Chronostratigraphie et faciés culturels du Paléolithique inférieur et moyen dans l'Europe du Nord-Ouest*. Paris: Supplément au Bulletin de l'Association Française pour l'Étudie du Quaternaire. 7–14.

Alemseged, Z., Y. Coppens, and D. Geraads (2002) Hominid cranium from Omo: Description and taxonomy of Omo-323-1976-896. *AJPA* **117**:103–112.

Alemseged, Z., F. Spoor, W. Kimbel, R. Bobé, D. Geraads, D. Reed, and J. Wynn (2006) A juvenile early hominin skeleton from Dikika, Ethiopia. *Nature* **443**:296–301.

Alemseged, Z., J. Wynn, W. Kimbel, D. Reed, D. Geraads, and R. Bobé (2005) A new hominin from the Basal Member of the Hadar Formation, Dikika, Ethiopia, and its geological context. *JHE* **49**:499–514.

Alexander, R. (2004) Bipedal animals, and their differences from humans. *JA* **204**:321–330.

Alexandersen, V. (1967) The pathology of the jaws and the temporomandibular joint. In: D. Brothwell and A. Sandison, Editors. *Diseases in Antiquity*. Springfield: Charles C Thomas.

Alexeev, V. (1986) *The Origin of the Human Race*. Moscow: Progress.

Alexeyev, V. (1976) Position of the Staroselye find in the hominid system. *JHE* **5**:413–421.

Allen, J. (1877) The influence of physical conditions in the genesis of species. *Radical Review* **1**:108–140.

Allin, E. (1975) Evolution of the mammalian middle ear. *JM* **147**:403–438.

Allin, E., and J. Hopson (1992) Evolution of the auditory system in Synapsida ("mammal-like reptiles" and primitive mammals) as seen in the fossil record. In: D. Webster, R. Fay, and A. Popper, Editors. *The Evolutionary Biology of Hearing*. New York: Springer-Verlag, 587–614.

Allman, J. (2000) *Evolving Brains*. New York: Scientific American Library.

Allsworth-Jones, P. (1990) The Szeletian and the stratigraphic succession in Central Europe and adjacent areas: Main trends, recent results, and problems for resolution. In: Mellars 1980: 160–242.

Alperson, N., O. Barzilai, D. Dag, G. Hartman, and Z. Matskevich (2000) The age and context of the Tabun I skeleton: A reply to Schwartz et al. *JHE* **38**:849–853.

Altermann, W., and J. Kazmierczak (2003) Archaean microfossils: A reappraisal of early life on Earth. *Research in Microbiology* **154**:611–617.

Ambrose, S. (1998) Late Pleistocene human population bottlenecks, volcanic winter, and differentiation of modern humans. *JHE* **34**:623–651.

Ambrose, S., and K. Lorenz (1990) Social and ecological models for the Middle Stone Age in southern Africa. In: Mellars 1980: 3–33.

An, Z., and C. Ho (1989) New magnetostratigraphic dates of Lantian *Homo erectus*. *QR* **32**:213–221.

Andrews, P. (1981) Species diversity and diet in monkeys and apes during the Miocene. In: Stringer 1981: 25–61.

Andrews, P. (1984) An alternative explanation of the characters used to define *Homo erectus*. *CFS* **69**:167–178.

Andrews, P. (1992) Evolution and environment in the Hominoidea. *Nature* **360**:641–646.

Andrews, P., and J. Cronin (1982) The relationships of *Sivapithecus* and *Ramapithecus* and the evolution of the orang-utan. *Nature* **297**:541–546.

Andrews, P., and E. Delson (2000) Pliopithecidae. In: Delson et al. 2000: 577–578.

Ankel, F. (1965) Der Canalis sacralis als Indikator für die Länge der Caudalregion der Primaten. *FP* **3**:263–276.

Anonymous (1971) Confusion over fossil man. *Nature* **232**:294–295

Antón, S. (1994) Mechanical and other perspectives on Neandertal craniofacial morphology. In: Corruccini and Ciochon 1994: 677–695.

Antón, S. (1997) Developmental age and taxonomic affinity of the Mojokerto child, Java, Indonesia. *AJPA* **102**:497–514.

Antón, S. (2002) Evolutionary significance of cranial variation in Asian *Homo erectus*. *AJPA* **118**:301–323.

Antón, S. (2003) Natural history of *Homo erectus*. *YPA* **46**:126–170.

Antón, S. (2004) The face of Olduvai Hominid 12. *JHE* **46**:337–347.

Antón, S., W. Leonard, and M. Robertson (2002a) An ecomorphological model of the initial hominid dispersal from Africa. *JHE* **43**:773–785.

Antón, S., S. Márquez, and K. Mowbray (2002b) Sambungmacan 3 and cranial variation in Asian *Homo erectus*. *JHE* **43**: 555–562.

Antunes, M., and A. Cunha (1992) Neanderthalian remains from Figueira Brava cave, Portugal. *Géobios* **25**:681–692.

Arambourg, C. (1954) L'hominidien fossile de Ternifine (Algérie). *CRAS* **239**:893–895.

Arambourg, C. (1955) A recent discovery in human paleontology: *Atlanthropus* of Ternifine (Algeria). *AJPA* **13**:191–201.

Arambourg, C. (1963) L'*Atlanthropus mauritanicus*. *AIPH* **32**:37–190.

Arambourg, C., and P. Biberson (1956) The fossil human remains from the Paleolithic site of Sidi Abderrahmen (Morocco). *AJPA* **14**:467–490.

Arambourg, C., and Y. Coppens (1967) Sur la découverte, dans le pléistocène inferieur de la vallée de l'Omo (Ethiopie), d'une mandibule d'Australopithécien. *CRAS* **265**:589–590.

Arambourg, C., and Y. Coppens (1968) Découverte d'un australopithécien nouveau dans les gisements de l'Omo (Ethiopie). *SAJS* **64**:58–59.

Arcadi, A. (2006) Species resilience in Pleistocene hominids that traveled far and ate widely: An analogy to the wolf-like canids. *JHE* **51**:383–394.

Archibald, D. (1996) Fossil evidence for a late Cretaceous origin of "hoofed" mammals. *Science* **272**:1150–1153.

Arensburg, B., and A. Belfer-Cohen (1998) Sapiens and Neandertals. Rethinking the Levantine Middle Paleolithic hominids. In: Akazawa et al. 1998: 311–322.

Arensburg, B., L. Schepartz, A. Tillier, B. Vandermeersch, and Y. Rak (1990) A reappraisal of the anatomical basis for speech in Middle Paleolithic hominids. *AJPA* **83**:137–146.

Arensburg, B., and A. Tillier (1991) Speech and the Neanderthals. *Endeavor* **15**:26–28.

Arensburg, B., A. Tillier, B. Vandermeersch, H. Duday, L. Schepartz, and Y. Rak (1989) A Middle Paleolithic human hyoid bone. *Nature* **338**:758–760.

Argue, D., D. Dolon, C. Groves, and R. Wright (2006) *Homo floresiensis*: Microcephalic, pygmoid, *Australopithecus* or *Homo*? *JHE* **51**:360–374.

Armstrong, E. (1983) Relative brain size and metabolism in mammals. *Science* **220**:1302–1304.

Arnason, U., G. Gullberg, S. Gretarsdottir, B. Ursing, and A. Janke (2000) The mitochondrial genome of the sperm whale and a new molecular reference for estimating eutherian divergence dates. *JME* **50**:569–578.

Arrese, C., M. Archer, and L. Beazley (2002) Visual capabilities in a crepuscular marsupial, the honey possum (*Tarsipes rostratus*): A visual approach to ecology. *Journal of Zoology (London)* **256**:151–158.

Arrese, C., S. Dunlop, A. Harman, C. Braekevelt, W. Ross, J. Shand, and L. Beazley (1999) Retinal structure and visual acuity in a polyprotodont marsupial, the fat-tailed dunnart (*Sminthopsis crassicaudata*). *Brain, Behavior and Evolution* **53**:111–126.

Arsuaga, J., C. Lorenzo, J. Carretero, A. Gracia, I. Martínez, N. García, J. Bermúdez de Castro, and E. Carbonell (1999a). A complete human pelvis from the Middle Pleistocene of Spain. *Nature* **399**:255–258.

Arsuaga, J., I. Martínez, A. Gracia, J. Cerretero, and E. Carbonell (1993) Three new human skulls from the Sima de los Huesos Middle Pleistocene site in Sierra de Atapuerca, Spain. *Nature* **362**:534–537.

Arsuaga, J., I. Martínez, A. Gracia, J. Carretero, C. Lorenzo, N. García, and A. Ortega (1997b) Sima de los Huesos (Sierra de Atapuerca, Spain). The site. In: Arsuaga et al. 1997a: 109–127.

Arsuaga, J., I. Martínez, A. Gracia, and C. Lorenzo (1997c) The Sima de los Huesos crania (Sierra de Atapuerca, Spain). A comparative study. In: Arsuaga et al. 1997a: 219–281.

Arsuaga, J., I. Martínez, C. Lorenzo, A. Gracia, A. Muñoz, O. Alonso, and J. Gallego (1999b) The human cranial remains from Gran Dolina Lower Pleistocene site (Sierra de Atapuerca, Spain). In: Bermúdez de Castro et al. 1999a: 431–457.

Arsuaga, J., V. Villaverde, R. Quam, I. Martínez, J. Carretero, C. Lorenzo, and A. Gracia (2006) New Neandertal remains from Cova Negra (Valencia, Spain). *JHE* **54**:31–58.

Ascenzi, A. I. Biddittu, P. Cassoli, A. Serge, and E. Serge-Naldi (1996) A calvarium of late *Homo erectus* from Ceprano, Italy. *JHE* **31**:409–423.

Ascenzi, A., F. Mallegni, G. Manzi, A. Serge, and E. Serge Naldini (2000) A re-appraisal of Ceprano calvaria affinities with *Homo erectus* after the new reconstruction. *JHE* **39**:443–450.

Asfaw, B. (1983) A new hominid parietal from Bodo, Middle Awash Valley, Ethiopia. *AJPA* **61**:367–371.

Asfaw, B. (1987) The Belohdelie frontal: New evidence of early hominid cranial morphology from the Afar of Ethiopia. *JHE* **16**:611–624.

Asfaw, B., Y. Beyene, G. Suwa, R. Walter, T. White, G. Wolde-Gabriel, and T. Yemane (1992) The earliest Acheulean from Konso-Gardula. *Nature* **360**:732–735.

Asfaw, B., W. Gilbert, Y. Beyene, W. Hart, P. Renne, G. Wolde-Gabriel, E. Vrba, and T. White (2002) Remains of *Homo erectus* from Bouri, Middle Awash, Ethiopia. *Nature* **416**:317–320.

Asfaw, B., T. White, C. O. Lovejoy, B. Latimer, S. Simpson, and G. Suwa (1999) *Australopithecus garhi*: A new species of early hominid from Ethiopia. *Science* **284**:629–635.

Ashton, E. H., and S. Zuckerman (1950) Some quantitative dental characters of fossil anthropoids. *Philosophical Transactions of the Royal Society B*, **234**:485–520.

Atsalis, S., J. Schmid, and P. Kappeler (1996) Metrical comparisons of three species of mouse lemur. *JHE* **31**:61–68.

Ayala, F. (1982) Gradualism versus punctualism in speciation: reproductive isolation, morphology, genetics. In: C. Barigozzi, Editor. *Mechanisms of Speciation*. New York: Liss. 51–66.

Ayala, F. (1985) Reduction in biology: A recent challenge. In: D. Depew and B. Weber, Editors. *Evolution at a Crossroads: The New Biology and the New Philosophy of Science*. Cambridge: MIT Press. 65–79.

B

Baba, H., and F. Aziz (1992) Human tibial fragment from Sambungmachan, Java. In: Akazawa et al. 1992: 349–361.

Baba, H., F. Aziz, Y. Kaifu, G. Suwa, R. Kono, and T. Jacob (2003) *Homo erectus* calvarium from the Pleistocene of Java. *Science* **299**:1384–1388.

Baba, H., F. Aziz, S. Narasaki, Sudijono, Y. Kaifu, A. Suprijo, M. Hyodo, E. Susanto, and T. Jacob (2000) A new hominid incisor from Sangiran, Central Java. *JHE* **38**:855–862.

Bada, J. (1985) Amino acid racemization dating of fossil bones. *Annual Review of Earth and Planetary Science* **13**:241–268.

Bacon, A.-M. (1991) La fosse intertrochantérienne: étude morphofonctionelle et évolution au sein des primates simiiformes. In: Coppens and Sénut 1991: 89–98.

Bahuchet, S., D. McKey, and I. de Ganne (2004) Wild yams revisited: Is independence from agriculture possible for rain forest hunter–gatherers? *HEC* **19**:213–243.

Bailey, R., G. Head, M. Jenike, B. Owen, R. Rechtman, and E. Zechenter (1989) Hunting and gathering in tropical rain forest: Is it possible? *AA* **91**:59–82.

Bailey, R., and T. Headland (2004) The tropical rain forest: Is it a productive environment for human foragers? *HEC* **19**:261–285.

Bailey, S. (2002) A closer look at Neanderthal postcanine dental morphology: The mandibular dentition. *AR (New Anatomist)* **269**:148–156.

Bailey, S. (2004) A morphometric analysis of maxillary molar crowns of Middle–Late Pleistocene hominins. *JHE* **47**:183–198.

Bailey, S., and J.-J. Hublin (2006a) Dental remains from the Grotte du Renne at Arcy-sur-Cure (Yonne). *JHE* **50**:485–508.

Bailey, S., and J.-J. Hublin (2006b) Did Neanderthals make the Châtelperronian assemblage from La Grotte du Renne (Arcy-sur-Cure, France)? In: Havarti and Harrison 2006: 191–210.

Balout, L., P. Biberson, and J. Tixier (1967) L'Achuléen de Ternifine (Algérie): Gisement de l'Atlanthrope. *Anthropologie* **71**:217–238.

Balter, M. (2005) Small but smart? Flores hominid shows signs of advanced brain. *Science* **307**:1386–1389.

Balter, M., and A. Gibbons (2000) A glimpse of humans' first journey out of Africa. *Science* **288**:948–950.

Balzeau, A., and D. Grimaud-Hervé (2006) Cranial base morphology and cranial base pneumatization in Asian *Homo erectus*. *JHE* **51**:350–359.

Barker, G., H. Barton, M. Bird, et al. (2007) The "human revolution" in lowland tropical Southeast Asia: The antiquity and behavior of anatomically modern humans at Niah Cave (Sarawak, Borneo). *JHE* **52**:243–261.

Barrickman, N., M. Bastian, K. Isler, and C. van Schaik (2008) Life history costs and benefits of encephalization: A comparative test using data from long-term studies of primates in the wild. *JHE* **54**:568–590.

Barriel, V., and A.-M. Tillier (2002) L'enfant de Maezmaiskaya (Caucase) examiné par une double perspective paléogénétique et paléoanthropologique. *BMSAP* 163–191.

Barton, N. (2000) Mousterian hearths and shellfish: Late Neanderthal activities on Gibraltar. In: Stringer et al. 2000: 211–220.

Bartstra, G. (1983) The fauna from Trinil, type locality of *Homo erectus*: A reinterpretation. *Geologie en Mijnbouw* **62**:329–336.

Bartstra, G. (1984) Dating the Pacitanian: Some thoughts. *CFS* **69**:253–258.

Bartstra, G. (1985) Sangiran, the stone implements of Ngebung and the Paleolithic of Java. *MQRSA* **9**:99–113.

Barstra, G., S. Soegondho, and A. Van der Wijk (1988) Ngandong man: Age and artifacts. *JHE* **17**:325–337.

Barut-Kusimba, S., and F. Smith (2001) Acheulean. In: P. Peregrine and M. Ember, Editors. *Encyclopedia of Prehistory*, Volume 1. Africa. New York: Kluwer Academic/Plenum Publishers, pp. 1–22.

Bar-Yosef, O. (1988) The date of the Southwest Asian Neanderthals. In: M. Otte, Editor. *L'Homme de Neanderthal*, Volume 3. Liège: Université de Liège, pp. 31–38.

Bar-Yosef, O. (1992) Middle Paleolithic human adaptations in the Mediterranean Levant. In: Akazaka et al. 1992: 189–216.

Bar-Yosef, O. (1994) The contributions of southwest Asia to the study of the origin of modern humans. In: Nitecki and Nitecki 1994: 23–66.

Bar-Yosef, O. (1995) The Lower and Middle Paleolithic in the Mediterranean Levant: Chronology and cultural entities. In: H. Ulrich, Editor. *Man and Environment in the Paleolithic*. Liège: Université de Liège.

Bar-Yosef, O. (1998) The chronology of the Middle Paleolithic of the Levant. In: Akazawa et al. 1998: 39–56.

Bar-Yosef, O. (2000) The Middle and Early Upper Paleolithic in southwest Asia. In: Bar-Yosef and Pilbeam 2000: 107–156.

Bar-Yosef, O. (2002) The Upper Paleolithic revolution. *ARA* **31**:363–393.

Bar-Yosef, O. (2003) Away from home: prehistoric colonizations, exchanges and diffusions in the Mediterranean basin. In: B. Vandermeersch, Editor. *Echanges et diffusions dans la préhistoire méditerranéenne*. Paris: CTHS, pp. 71–81.

Bar-Yosef, O. (2004) Eat what is there: Hunting and gathering in the world of Neanderthals and their neighbours. *IJO* **14**:333–342.

Bar-Yosef, O. (2006) Neanderthals and modern humans: A different interpretation. In: Conard 2006: 467–482.

Bar-Yosef, O., and N. Goren-Inbar (1993) The lithic assemblages of 'Ubeidiya: a Lower Paleolithic site in the Jordon Valley. *Oedem* **45**:1–266.

Bar-Yosef, O., and J. Callander (1999) The woman from Tabun: Garrod's doubts in historical perspective. *JHE* **37**:879–885.

Bar-Yosef, O., and D. Pilbeam (1993) Dating hominid remains. *Nature* **366**:415.

Bar-Yosef, O., and B. Vandermeersch (1981) Notes concerning the possible age of the Mousterian layers in Qafzeh Cave. In: P. Sanlaville and J. Cauvin, Editors. *Préhistoire du Levant*. Paris: Editions du CNRS. 281–285.

Bar-Yosef, O., and B. Vandermeersch (1993) Modern humans in the Levant. *SA* **267**:94–100.

Bar-Yosef, O., B. Vandermeersch, B. Arensberg, A. Belfer-Cohen, P. Goldberg, L. Meignen, Y. Rak, J. Speth, E. Tchernov, A.-M. Tillier, and S. Weiner (1992) The excavations in Kebara Cave, Mt. Carmel. *CA* **33**:497–550.

Basmajian, J. (1982) *Primary Anatomy*, 8th edition. Baltimore: Williams and Wilkins.

Bassinot, F., L. Labeyrie, E. Vincent, X. Quidelleur, N. Shackleton, and Y. Lancelot (1994) The astronomical theory of climate and the age of the Brunhes–Matuyama magnetic reversal. *Earth and Planetary Science Letters* **126**: 91–108.

Bateman, A. (1948) Intra-sexual selection in *Drosophila*. *Heredity* **2**:349–368.

Baum, D., and M. Donoghue (1995) Choosing among alternative "phylogenetic" species concepts. *Systematic Botany* **20**:560–573.

Beall, C., and A. Steegmann (2000) Human adaptation to climate: temperature, ultraviolet radiation and altitude. In: S. Stinson et al., Editors. *Human Biology. An Evolutionary and Biocultural Perspective*. New York: Wiley-Liss. 163–224.

Beals, K., C. Smith, and S. Dodd (1984) Brain size, cranial morphology, climate and time machines. *CA* **25**:301–330.

Beard, K. (1988) The phylogenetic significance of strepsirrhinism in Paleogene primates. *IJP* **9**:83–96.

Beard, K. (1991) Vertical postures and climbing in the morphotype of Primatomorpha: implications for locomotor evolution in primate history. In: Coppens and Sénut 1991: 79–87.

Beard, K. (1993) Origin and evolution of gliding in early Cenozoic Dermoptera (Mammalia, Primatomorpha). In: R. MacPhee, Editor. *Primates and Their Relatives in Phylogenetic Perspective*. New York: Plenum, pp. 63–90.

Beard, K. (1998) A new genus of Tarsiidae (Mammalia: Primates) from the middle Eocene of Shanxi Province, China, with notes on the historical biogeography of tarsiers. *Bulletin of the Carnegie Museum of Natural History* **34**:260–277.

Beard, K. (2002) Basal anthropoids. In: Hartwig 2002: 133–149.

Beard, K. (2004) *The Hunt for the Dawn Monkey: Unearthing the Origins of Monkeys, Apes, and Humans*. Berkeley: University of California Press.

Beard, K., J.-J. Jaeger, Y. Chaimanee, J. Rossie, A. Soe, S. Tun, L. Marivaux, and B. Marandat (2005) Taxonomic status of purported primate frontal bones from the Eocene Pondaung Formation of Myanmar. *JHE* **49**:468–481.

Beard, K., L. Krishtalka, and R. Stucky (1991) First skulls of the Early Eocene primate *Shoshonius cooperi* and the anthropoid–tarsier dichotomy. *Nature* **349**:64–67.

Beard, K., and R. MacPhee (1994) Cranial anatomy of *Shoshonius* and the antiquity of the Anthropoidea. In: Fleagle and Kay 1994: 55–97.

Beard, K., T. Qi, M. Dawson, B. Wang, and C. Li (1994) A diverse new primate fauna from middle Eocene fissure-fillings in southeastern China. *Nature* **368**:604–609.

Beard, K., and J. Wang (2004) The eosimiid primates (Anthropoidea) of the Heti Formation, Yuanqu Basin, Shanxi and Henan Provinces, People's Republic of China. *JHE* **46**:401–432.

Beaumont, P. (1980) On the age of the Border Cave hominids 1–5. *Paleontologia Africana* **23**:21–33.

Beaumont, P., H. de Villiers, and J. Vogel (1978) Modern man in sub-Saharan Africa prior to 49,000 years BP: A review and evaluation with particular reference to Border Cave. *SAJS* **74**:409–414.

Beauval, C., B. Maureille, F. Lecrampe-Cuyaubére, D. Serre, D. Peressinotto, J.-G. Bordes, D. Cochard, I. Couchoud, D.

Dubrasquet, V. Laroulandie, A. Lenoble, J.-B. Mallye, S. Pasty, J. Primault, N. Rohland, S. Pääbo, and E. Trinkaus (2005) A late Neandertal femur from Les Rochers-de-Villeneuve, France. *PNAS* **102**:7085–7090.

Beckner, M. (1959) *The Biological Way of Thought.* New York: Columbia University Press.

Begun, D. (1993) Knuckle-walking ancestors. *Science* **259**:294.

Begun, D. (2002a) The Pliopithecoidea. In: Hartwig 2002: 221–240.

Begun, D. (2002b) European hominoids. In: Hartwig 2002: 339–368.

Begun, D., and A. Walker (1993) The endocast. In: Walker and Leakey 1993a: 326–358.

Begun, D., Ward, C., and M. Rose (1997) Events in hominoid evolution. In: Begun et al. 1997: 389–415.

Behm-Blancke, G. (1958) Umwelt, Kultur und Morphologie des Eem-interglazialen Menschen von Ehringsdorf bei Weimar. In: von Koenigswald 1958: 141–150.

Belitsky, S., N. Goren-Inbar, and E. Werker (1991) A Middle Pleistocene wooden plank with man-made polish. *JHE* **20**:349–353.

Belmaker, M., E. Tchernov, S. Condemi, and O. Bar-Yosef (2002) New evidence for hominid presence in the Lower Pleistocene of the Southern Levant. *JHE* **43**:43–56.

Bellomo, R. (1994) Methods of determining early hominid behavioral activities associated with the controlled use of fire at FxJj 20 Main, Koobi Fora, Kenya. *JHE* **27**:173–195.

Bellwood, P. (1992) Southeast Asia before history. In: N. Tarling, Editor. *The Cambridge History of Southeast Asia.* Cambridge: Cambridge University Press, pp. 55–136.

Bellwood, P. (1997) *Prehistory of the Indo-Malaysian Archipelago.* Honolulu: University of Hawaii Press.

Benefit, B., and M. McCrossin (2002) The Victoriapithecidae, Cercopithecoidea. In: Hartwig 2002: 241–253.

Ben-Itzhak, S., P. Smith, and R. Bloom (1988) Radiographic study of the humerus in Neandertals and *Homo sapiens. AJPA* **77**:231–242.

Berckhemmer, F. (1933) Ein Menschen-Schädel aus den diluvialen Schottern von Steinhein a. d. Murr. *AZ* **10**:318–321.

Berger, A. (1992) Astronomical theory of paleoclimates and the last glacial–interglacial cycle. *Quaternary Science Review* **11**:571–581.

Berger, A., et al. (1984) *Milankovitch and Climate: Understanding the Response to Astronomical Forcing.* Part 1. Boston: Reidel.

Berger, L., D. de Ruiter, C. Steininger, and J. Hancox (2003) Preliminary results of excavations at the newly investigated Coopers D deposit, Gauteng, South Africa. *SAJS* **99**:276–278.

Berger, L., R. Lacruz, and D. de Ruiter (2002) Revised age estimates of *Australopithecus*-bearing deposits at Sterkfontein, South Africa. *AJPA* **119**:192–197.

Berger, L., and J. Parkington (1995) A new Pleistocene hominid-bearing locality at Hoedjiespunt, South Africa. *AJPA* **98**:601–609.

Berger, L., and P. Tobias (1996) A chimpanzee-like tibia from Sterkfontein, South Africa and its implication for the interpretation of bipedalism in *Australopithecus africanus. JHE* **30**:343–348.

Berger, T., and E. Trinkaus (1995) Patterns of trauma among the Neandertals. *JAS* **22**:841–852.

Bergmann, C. (1847) Über die Verhältnisse der Warmeökonomie der Thiere zu ihre Grösse. *Göttingen Studien* **1**: 595–708.

Bergounioux, F. (1958) "Spiritualité" de l'homme de Néandertal. In: von Koenigswald 1958: 151–166.

Bermúdez De Castro, J., A. Durand, and S. Ipiña (1993) Sexual dimorphism in the human dental sample from the SH site (Sierra de Atapuerca, Spain): A statistical approach. *JHE* **24**:43–56.

Bermúdez De Castro, J., and M. Nicholás (1997) Paleodemography of the Atapuerca-SH Middle Pleistocene hominid sample. In: Arsuaga et al. 1997a: 333–355.

Bermúdez De Castro, J., J. Arsuaga, E. Carbonell, A. Rosas, I. Martínez, and M. Mosquera (1997) A hominid from the Lower Pleistocene of Atapuerca, Spain: Possible ancestor to Neandertals and modern humans. *Science* **276**:1392–1395.

Bermúdez De Castro, J., A. Rosas, and M. Nicolás (1999b) Dental remains from Atapuerca TD6 (Gran Dolina site, Burgos, Spain). In: Bermúdez de Castro et al. 1999a: 523–566.

Bernor, R. (1983) Geochronology and zoogeography of Miocene hominoids. In: Ciochon and Corruccini 1983: 21–64.

Beynon, A., and B. Wood (1986) Variation in enamel thickness and structure in East African hominids. *AJPA* **70**:177–193.

Beyries, S. (1988) Functional variability of lithic sets in the Middle Paleolithic. In: Dibble and Monet-White 1988: 213–224.

Bickerton, D. (1990) *Language and Species.* Chicago: University of Chicago Press.

Biegert, J. (1963) The evaluation of characteristics of the skull, hands, and feet for primate taxonomy. In: S. Washburn, Editor. *Classification and Human Evolution.* Chicago: Aldine, pp. 116–145.

Bilsborough, A. (1976) Pattern of evolution in Middle Pleistocene hominids. *JHE* **5**:423–439.

Binford, L. (1981) *Bones: Ancient Men and Modern Myths.* New York: Academic Press.

Binford, L. (1985) Human ancestors: Changing views of their behavior. *JAA* **4**:292–347.

Binford, L. (1987) American Association of Physical Anthropologists Annual Luncheon Address, April 1986: The hunting hypothesis, archaeological methods, and the past. *YPA* **30**:1–9.

Binford, L. (1989) Isolating the transition to cultural adaptations: An organizational approach. In: Mellars and Stringer 1989: 18–41.

Binford, L., and S. Binford (1966) A preliminary analysis of functional variability in the Mousterian of Levallois faces. *AA* **68**:238–295.

Binford, L., and S. Binford (1969) Stone tools and human behavior. *SA* **220**:70–84.

Binford, L., and Ho, C. (1985) Taphonomy at a distance: Zhoukoudian, "The cave home of Beijing man." *CA* **26**:413–442.

Binford, L., M. Mills, and N. Stone (1988) Hyena scavenging behavior and its implications for the interpretation of faunal assemblages from FLK 22 (the Zinj floor) at Olduvai Gorge. *JAA* **7**:99–135.

Birdsell, J. (1967) Preliminary data on the trihybrid origin of the Australian Aborigines. *Archaeology and Physical Anthropology in Oceania* **2**:100–155.

Birdsell, J. (1979) Physical anthropology in Australia today. *ARA* **8**:417–430.

Bischoff, J., J. Fitzpatrick, L. León, J. Arsuaga, C. Falgueres, J. Bahain, and T. Bullen (1997) Geology and preliminary dating of the hominid-bearing sedimentary fill of the Sima de los Huesos Chamber, Cueva Mayor of the Sierra de Atapuerca, Burgos, Spain. In: Arsuaga et al. 1997a: 129–154.

Bischoff, J., D. Shamp, A. Armburu, J. Arsuaga, E. Carbonell, and J. Bermúdez De Castro (2003) The Sima de los Huesos hominids date to beyond U/Th equilibrium (>350 kyr) and perhaps to 400–500 kyr: New radiometric dates. *JAS* **30**:275–280.

Bischoff, J., R. Williams, R. Rosenbauer, A. Aramburu, J. Arsuaga, N. Garcia, and G. Cuenca-Bescós (2007) High resolution U-series dates from the Sima de los Huesos hominids yields $600^{+\infty}/_{-66}$ kyrs: Implications for the evolution of the early Neanderthal lineage. *JAS* **34**:763–770.

Black, D. (1933) The Croonian Lecture: On the discovery, morphology and environment of *Sinanthropus pekinensis*. *PTRSL* **223**:57–120.

Black, M. (1999) The "trunk torsion hypothesis" and Neandertal superior pubic ramal morphology. *JHE* **36**:A2.

Black, M. (2004) *Mechanical determinants of form in the pubis of middle-to-late Pleistocene Homo*. Ph.D. thesis, Duke University.

Blackwell, B., and H. Schwarcz (1986) U-series analyses of the lower travertine at Ehringsdorf, DDR. *QR* **25**:215–222.

Blanc, A. (1961) Some evidence for the ideologies of early man. In: S. Washburn, Editor. *The Social Life of Early Man*. Chicago: Aldine. 119–136.

Blanco, R., and W. Jones (2005) Terror birds on the run: A mechanical model to estimate its maximum running speed. *PRSL* **272**:1769–1773.

Bloch, J., and D. Boyer (2002) Grasping primate origins. *Science* **298**:1606–1610.

Bloch, J., and D. Boyer (2007) New skeletons of Paleocene–Eocene Plesiadapiformes: A diversity of arboreal positional behaviors in early primates. In: Ravosa and Dagosto 2007: 535–581.

Bloch, J., and M. Silcox (2001) New basicrania of Paleocene–Eocene *Ignacius*: Reevaluation of the plesiadapiform–dermopteran link. *AJPA* **116**:184–198.

Bloch, J., and M. Silcox (2006) Cranial anatomy of the Paleocene plesiadapiform *Carpolestes simpsoni* (Mammalia, Primates) using ultra high-resolution X-ray computed tomography, and the relationship of plesiadapiforms to Euprimates. *JHE* **50**:1–35.

Bloch, J., Silcox, M., Boyer, D., and Sargis, E. (2007) New Paleocene skeletons and the relationship of plesiadapiforms to crown-clade primates. *PNAS* **104**:1159–1164.

Blumenscheine, R., C. Peters, F. Masao, R. Clarke, A. Deino, R. Hay, C. Swisher, I. Stanistreet, G. Ashley, L. McHenry, N. Sikes, N. van der Merwe, J. Tactikos, A. Cushing, D. Deocampo, J. Njau, and J. Ebert (2003) Late Pliocene *Homo* and hominid land use from western Olduvai Gorge, Tanzania. *Science* **299**:1217–1221.

Boaz, N. (1988) Status of *Australopithecus afarensis*. *YPA* **31**:85–113.

Boaz, N. (1994) Significance of the Western Rift for hominid evolution. In: Corruccini and Ciochon 1994: 321–343.

Boaz, N., and F. C. Howell (1977) A gracile hominid cranium from upper Member G of the Shungura Formation, Ethiopia. *AJPA* **46**:93–108.

Bocherens, H., and D. Drucker (2006) Dietary competition between Neanderthals and modern humans: Insights from stable isotopes. In: Conard 2006: 129–143.

Bocherens, H., M. Fizet, A. Mariotti, B. Lange-Badre, B. Vandermeersch, J. Borel, and G. Bellon (1991) Isotopic biochemistry (^{13}C, ^{15}N) of fossil vertebrate collagen: Application of the study of a past food web including Neanderthal man. *JHE* **20**:481–492.

Bocherens, H., D. Billiou, A. Mariotti, M. Patou-Mathis, M. Otte, D. Bonjean, and M. Toussaint (1999) Palaeoenvironmental and palaeodietary implications of isotopic biochemistry on last interglacial Neanderthal and mammal bones in Otte Scladina Cave (Belgium). *JAS* **26**:599–607.

Bocherens, H., D. Billiou, A. Mariotti, M. Toussaint, M. Patou-Mathis, D. Bonjean, and M. Otte (2001) New isotopic evidence for dietary habits of Neandertals from Belgium. *JHE* **40**:497–505.

Bocherens, H., D. Drucker, D. Billion, M. Patou-Mathis, and B. Vandermeersch (2005) Isotope evidence for diet and subsistence pattern of the Saint-Césaire 1 Neanderthal: Review and use of a multi-source mixing model. *JHE* **49**: 71–87.

Boë, L.-J., J.-L. Heim, K. Honda, and S. Maeda (2002) The potential Neandertal vowel space was as large as that of modern humans. *Journal of Phonetics* **30**:465–484.

Boesch, C. (1996) Social grouping in Taï chimpanzees. In: McGrew et al. 1996: 101–113.

Bogduk, N., and L. Twomey (1987) *Clinical Anatomy of the Lumbar Spine*. Melbourne: Churchill Livingstone.

Bogin, B. (1990) The evolution of human childhood. *Bioscience* **40**:16–25.

Boins, L., D. Geraads, G. Guérin, A. Haga, J. Jaeger, and S. Sen (1984) Découverte d'un hominidé fossile dans le Pléistocène de la République de Djibouti. *CRAS D* **299**:1097–1100.

Bonifay, E. (1991) Les premiers industries du Sud-Est de la France et du Massif-Central. In: E. Bonifay, and B. Vandermeersch, Editors. *Les premiers Europèens*. Paris: Éditions du Comité Travaux Historiques et Sceintifiques. 63–80.

Bonis, L. De, J.-J. Jaeger, B. Coiffait, and P.-E. Coiffait (1988) Découverte du plus ancien primate Catarrhinien connu dans l'Éocène supérieur d'Afrique du Nord. *CRAS Sér.* **2, 306**: 929–934.

Bordes, F. (1961) Mousterian cultures in France. *Science* **134**:803–810.

Bordes, F. (1968) *The Old Stone Age*. New York: McGraw-Hill.

Bordes, F. (1984) *Leçons sur le paléolithique. Tome II. Le paléolithique en Europe*. Paris: Editions du C.N.R.S.

Boule, M. (1911/13) L'homme fossile de La Chapelle-aux-Saints. *Annales de Paléontologie* **6**:111–172; **7**:21–192; **8**:1–70; 209–278.

Boule, M. (1921) *Les hommes fossiles: Éléments de paléontologie humaine*. Paris: Masson.

Boule, M., and R. Anthony (1911) L'encephale de l'homme fossile de la Chapelle-aux-Saints. *L'Anthropologie* **22**:129–196.

Boule, M., and H.-V. Vallois (1952) *Les hommes fossiles. Eléments de paléontologie humaine*, 4th edition. Paris: Masson et Cie. 1957 English translation by M. Bullock. New York: Dryden Press.

Bowler, J., H. Johnson, J. Olley, J. Prescott, R. Roberts, W. Shawcross, and N. Spooner (2003) New ages for human occupation and climatic change at Lake Mungo, Australia. *Nature* **421**:837–840.

Bowler, J., and J. Magee (2000) Redating Australia's oldest human remains: A skeptic's view. *JHE* **38**:719–726.

Bowler, J., and A. Thorne (1976) Human remains from Lake Mungo: Discovery and excavation of Lake Mungo III. In: R. Kirk, and A. Thorne, Editors. *The Origin of the Australians*. Canberra: Australian Institute of Aboriginal Studies. 127–138.

Bowler, J., A. Thorne, and H. Polach (1972) Pleistocene man in Australia: Age and significance of the Mungo skeleton. *Nature* **240**:48–50.

Bowler, P. (1983) *The Eclipse of Darwinism: Anti-Darwinian Evolution Theories in the Decades around 1900*. Baltimore: Johns Hopkins University Press.

Bowler, P. (1986) *Theories of Human Evolution: A Century of Debate 1844–1944*. Baltimore: Johns Hopkins University Press.

Bown, T., and M. Kraus (1979) Origin of the tribosphenic molar and metatherian and eutherian dental formulae. In: Lillegraven et al. 1979: 172–181.

Brace, C. L. (1962) Cultural factors in the evolution of the human dentition. In: M. Montagu, Editor. *Culture and the Evolution of Man*. New York: Oxford University Press, pp. 343–354.

Brace, C. L. (1964) The fate of the "Classic" Neanderthals: A consideration of hominid catastrophism. *CA* **5**:3–43.

Brace, C. L. (1967) *The Stages of Human Evolution*. Englewood Cliffs, NJ: Prentice-Hall.

Brace, C. L. (1976) Monkey business and bird brains. In: E. Giles, and J. Friedlaender, Editors. *The Measures of Man*. Cambridge: Peabody Museum Press, pp. 54–71.

Brace, C. L. (1979) Krapina "classic" Neanderthals, and the evolution of the European face. *JHE* **8**:527–550.

Brace, C. L. (1988) Punctuationalism, cladistics and the legacy of Medieval Neoplatonism. *HE* **3**:121–138.

Brace, C. L. (1989) Medieval thinking and the paradigms of paleoanthropology. *AA* **91**:442–446.

Brace, C. L. (1995) *The Stages of Human Evolution*, 5th edition. Englewood Cliffs, NJ: Prentice-Hall.

Brace, C. L., P. Mahler, and R. Rosen (1973) Tooth measurement and the rejection of the taxon "*Homo habilis*." *YPA* **16**:50–68.

Bradley, B., D. Doran-Sheehy, D. Lukas, C. Boesch, and L. Vigilant (2004) Dispersed male networks in western gorillas. *Current Biology* **14**:510–513.

Bradley, B., M. Robbins, E. Williamson, H. Steklis, N. Steklis, N. Eckhardt, C. Boesch, and L. Vigilant (2005) Mountain gorilla tug-of-war: Silverbacks have limited control over reproduction in multimale groups. *PNAS* **102**:9418–9423.

Brain, C. (1981) *The Hunters or the Hunted? An Introduction to African Cave Taphonomy*. Chicago: University of Chicago Press.

Brain, C. (1985) Cultural and taphonomic comparisons of hominids from Swartkrans and Sterkfontein. In: Delson 1985: 72–75.

Brain, C. (1993) The occurrence of burnt bones at Swartkrans and their implications for the control of fire by early hominids. In: Brain 1993: 229–242.

Brain, C., S. Chrucher, J. Clark, F. Grine, P. Shipman, R. Susman, A. Turner, and V. Watson (1988) New evidence of early hominids, their culture and environment from the Swartkrans Cave, South Africa. *SAJS* **84**:828–835.

Brain, C., and P. Shipman (1993) The Swartkrans bone tools. In: Brain 1993: 195–215.

Brain, C., and A. Sillen (1988) Evidence from the Swartkrans Cave for the earliest use of fire. *Nature* **336**:464–466.

Bramble, D., and D. Lieberman (2004) Endurance running and the evolution of *Homo*. *Nature* **432**:345–352.

Brasier, M., O. Green, A. Jephcoat, A. Kleppe, M. Van Kranendonk, J. Lindsay, A. Steele, and N. Grassineau (2002) Questioning the evidence for Earth's oldest fossils. *Nature* **416**:76–81.

Brasier, M., O. Green, J. Lindsay, N. McLoughlin, A. Steele, and C. Stoakes (2005) Critical testing of Earth's oldest putative fossil assemblage from the ~3.5 Ga Apex chert, Chinaman Creek, Western Australia. *Precambrian Research* **140**:55–102.

Bräuer, G. (1984a) A craniological approach to the origin of anatomically modern *Homo sapiens* in Africa and implications for the appearance of modern Europeans. In: Smith and Spencer 1984: 327–410.

Bräuer, G. (1984b) The "Afro-European Sapiens hypothesis" and hominid evolution in East Asia during the Middle and Upper Pleistocene. *CFS* **69**:145–165.

Bräuer, G. (1988) Osteometrie. In: R. Knußmann, Editor. *Anthropologie: Handbuch der vergleichenden Biologie des Menschen*, Volume I. Stuttgart: Gustav Fischer Verlag, 160–232.

Bräuer, G. (1989) The evolution of modern humans: A comparison of the African and non-African evidence. In: Mellars and Stringer 1989: 123–154.

Bräuer, G. (1992) Africa's place in the evolution of *Homo sapiens*. In: Bräuer and Smith 1992: 83–98.

Bräuer, G. (1994) How different are Asian and African *Homo erectus?* In: Franzen 1994: 301–318.

Bräuer, G. (2003) Der Ursprung lag in Afrika. *Spektrum der Wissenschaft* March:38–46.

Bräuer, G. (2006) The African origin of modern humans and the replacement of the Neanderthals. In: Schmitz 2006: 337–372.

Bräuer, G., and H. Broeg (1998) On the degree of Neandertal-modern continuity in the earliest Upper Paleolithic crania from the Czech Republic: evidence from nonmetrical features. In: K. Omoto, and P. Tobias, Editors. *The Origins of Past Modern Humans; Toward Reconciliation*. Singapore: World Scientific, pp. 106–125.

Bräuer, G., H. Broeg, and C. Stringer (2006) Earliest Upper Paleolithic crania from Mladeč, Czech Republic and the question of Neanderthal-modern continuity: Metrical evidence from the fronto-facial region. In: Harvati and Harrison 2006: 269–279.

Bräuer, G., M. Collard, and C. Stringer (2004) On the reliability of recent tests of the Out of Africa hypothesis for modern human origins. *AR* **279**A: 701–707.

Bräuer, G., and R. Knußmann (1988) Grundlagen der Anthropometrie. In: R. Knußmann, Editor. *Anthropologie: Handbuch der vergleichenden Biologie des Memschen*, Volume I. Stuttgart: Gustav Fischer Verlag, pp. 129–160.

Bräuer, G., and R. Leakey (1986) The ES-11693 cranium from Eliye Springs, West Turkana, Kenya. *JHE* **15**:289–312.

Bräuer, G., R. Leakey, and E. Mbua (1992) A first report of the ER-3884 cranial remains from Ileret/East Turkana, Kenya. In: Bräuer and Smith 1992: 111–119.

Bräuer, G., and E. Mbua (1992) *Homo erectus* features used in cladistics and their variability in Asian and African hominids. *JHE* **22**:79–108.

Bräuer, G., and M. Mehlman (1988) Hominid molars from a Middle Stone Age level at Mumba Rock shelter, Tanzania. *AJPA* **75**:69–76.

Bräuer, G., and M. Schultz (1996) The morphological affinities of the Plio–Pleistocene mandible from Dmanisi, Georgia. *JHE* **30**:445–481.

Bräuer, G., and R. Singer (1996) The Klasies zygomatic bone: Archaic or modern? *JHE* **30**:161–165.

Bräuer, G., and C. Stringer (1997) Models, polarization and perspectives on modern human origins. In: Clark and Willermet 1997: 191–201.

Bräuer, G., Y. Yokoyama, C. Falguères, and E. Mbua (1997) Modern human origins backdated. *Nature* **386**:337–338.

Bridgeland, D. (1994) *Quaternary of the Thames. Geological Conservation Review Series 7*. London: Chapman, and Hall.

Bridgeland, D., P. Gibbard, P. Harding, R. Kemp, and G. Southgate (1985) New information and results from recent excavations at Barnfield, Swanscombe. *Quaternary Newsletter* **46**:25–38.

Bromage, T., F. Schrenk, et al. (1995) Paleoanthropology of the Malawi Rift: An early hominind mandible from the Chiwondo Beds, northern Malawi. *JHE* **28**:78–108.

Bromham, L., M. Phillips, and D. Penny (1999) Growing up with dinosaurs: Molecular dates and the mammalian radiation. *Trends in Ecology and Evolution* **14**:113–118.

Brooks, A., D. Helgren, J. Kramer, A. Franklin, W. Hornyak, J. Keating, R. Klein, W. Rink, H. Schwarcz, J. Leith Smith, K. Stewart, N. Todd, J. Verniers, and J. Yellen (1995) Dating and context of three Middle Stone Age sites with bone points in the Upper Semliki Valley, Zaire. *Science* **268**:548–553.

Broom, R. (1936a) A new fossil anthropoid skull from South Africa. *Nature* **138**:486–488.

Broom, R. (1938) The Pleistocene anthropoid apes of South Africa. *Nature* **142**:377–379.

Broom, R. (1949) Another new type of fossil ape-man. *Nature* **163**:57.

Broom, R., and J. Robinson (1949) A new type of fossil man. *Nature* **164**:322–323.

Broom, R., and J. Robinson (1950) Further evidence of the structure of the Sterkfontein ape-man *Plesianthropus*. *MTM* **4**:7–84.

Brose, D., and M. Wolpoff (1971) Early Upper Paleolithic man and late Middle Paleolithic tools. *AA* **73**:1156–1194.

Brothwell, D. (1960) Upper Pleistocene human skull from Niah Caves. *Sarawak Museum Journal* **9**:323–349.

Brothwell, D. (1974) The Upper Pleistocene Singa skull: A problem in paleontological interpretation. In: W. Bernhard, and A. Kandler, Editors. *Bevölkerungsbiologie*. Stuttgart: Fischer, pp. 534–545.

Brothwell, D. (1975) Adaptive growth rate changes as a possible explanation for the distinctiveness of the Neanderthalers. *JAS* **2**:161–163.

Brown, B. (1994) Comparative dental anatomy of African *Homo erectus*. In: Franzen 1994: 175–184.

Brown, B., and A. Walker (1993) The dentition. In: Walker and Leakey 1993a: 161–192.

Brown, F. (1994) Development of Pliocene and Pleistocene chronology of the Turkana Basin, East Africa, and its relation to other sites. In: Corruccini and Ciochon 1994: 285–312.

Brown, F., J. Harris, R. Leakey, and A. Walker (1985) Early *Homo erectus* skeleton from west Lake Turkana, Kenya. *Nature* **316**:788–792.

Brown, F., and I. McDougall (1993) Geological setting and age. In: Walker and Leakey 1993a: 9–20.

Brown, K., and D. Mack (1978) Food sharing among captive *Leontopithecus rosalia*. *FP* **29**:268–290.

Brown, P. (1989) *Coobol Creek. A Morphological and Metrical Analysis of the Crania, Mandibles and Dentitions of a Prehistoric Australian Human Population. Terra Australis 13*. Canberra: Australian National University.

Brown, P. (1992) Recent human evolution in East Asia and Australasia. *PTRSL* **337**:235–242.

Brown, P. (1994) Cranial vault thickness in Asian *Homo erectus* and *Homo sapiens*. In: Franzen 1994: 33–46.

Brown, P. (1997) Australian paleoanthropology. In: Spencer 1997: 138–145.

Brown, P. (1999) The first modern East Asians? Another look at Upper Cave 101, Liujiang, and Minatogawa 1. In: K. Omoto, Editor. *Interdisciplinary Perspectives of the Origins of the Japanese*. Kyoto: Center for Japanese Studies. 105–130.

Brown, P. (2000) Australian Pleistocene variation and the sex of Lake Mungo 3. *JHE* **38**:743–749.

Brown, P., T. Sutikna, M. Morwood, R. Soejono, Jatmiko, E. Saptomo, and Rokus Awe Due (2004) A new small-bodied hominin from the Late Pleistocene of Flores, Indonesia. *Nature* **431**:1056–1061.

Bruere, A., and P. Ellis (1979) Cytogenetics and reproduction of sheep with multiple centric fusions (Robertsonian translocations). *Journal of Reproduction and Fertility* **57**:363–375.

Bruner, E., and G. Manzi (2006) Saccopastore 1: The earliest Neanderthal? A new look at an old cranium. In: Harvati and Harrison 2006: 23–36.

Brunet, M. (2002) Reply to Wolpoff et al. *Nature* **419**:582.

Brunet, M., A. Beauvilain, Y. Coppens, E. Heintz, A. Moutaye, and D. Pilbeam (1995) The first australopithecine 2,500 kilometres west of the Rift Valley (Chad). *Nature* **378**:273–240.

Brunet, M., A. Beauvilain, Y. Coppens, E. Heintz, A. Moutaye, and D. Pilbeam (1996) *Australopithecus bahrelghazali*, une novelle espèce d'Hominidé ancien de la région de Koro Toro (Tchad). *CRAS série IIa*, **332**:907–913.

Brunet, M., F. Guy, D. Pilbeam, D. Lieberman, A. Likius, H. Mackaye, M. Ponce de Léon, C. Zollikofer, and P. Vignaud (2002) A new hominid from the Upper Miocene of Chad, Africa. *Nature* **418**:145–152.

Brunet, M., F. Guy, D. Pilbeam, D. Lieberman, A. Likius, H. Mackaye, M. Ponce de León, C. Zollikofer, and P. Vignaud (2005) New material of the earliest hominid from the Upper Miocene of Chad. *Nature* **434**:752–755.

Budd, G. (2001) Why are arthropods segmented? *Evolution and Development* **3**:332–342.

Bunn, H. (1981) Archaeological evidence for meat-eating by Plio-Pleistocene hominids from Koobi Fora and Olduvai Gorge. *Nature* **291**:574–577.

Bunn, H. (1991) A taphonomic perspective in the archaeology of human origins. *ARA* **20**:433–467.

Bunn, H. (1994) Early Pleistocene hominid foraging strategy along the ancestral Omo River at Koobi Fora, Kenya. *JHE* **27**:247–266.

Bunn, H., and E. Kroll (1986) Systematic butchery by Plio/Pleistocene hominids at Olduvai Gorge, Tanzania. *CA* **27**:431–452.

Burling, R. (1992) The crucial mutation for language. *Journal of Linguistic Anthropology* **2**:81–91.

Burr, D. (1976) Further evidence concerning speech in Neandertal man. *Man* **11**:104–110.

Bush, E., E. Simons, and J. Allman (2004) High-resolution computed tomography study of the cranium of a fossil anthropoid primate, *Parapithecus grangeri*: New insights into the evolutionary history of primate sensory systems. *AR* **281**A: 1083–1087.

Bush, G. (1975) Modes of animal speciation. *Annual Review of Ecology and Systematics* **6**:334–364.

Busk, G. (1865) On a very ancient cranium from Gibraltar. *Report of the British Association for the Advancement of Science (1864 Meeting)*: 91–92.

Butler, P., and J. Mills (1959) A contribution to the odontology of *Oreopithecus*. *Bulletin of the British Museum of Natural History (Geology)* **4**:3–26.

Butler, R. F. 1992. *Paleomagnetism: Magnetic Domains to Geologic Terranes*. Oxford: Blackwell Scientific Publications.

Butlin, R. (1989) Reinforcement of premating isolation. In: D. Otte, and J. Endler, Editors. *Speciation and Its Consequences*. Sunderland, MA.: Sinauer Associates. 158–179.

Butzer, K. (1969) Geological interpretation of two Pleistocene hominid sites in the Lower Omo Basin. *Nature* **222**: 1133–1135.

Butzer, K., P. Beaumont, and J. Vogel (1978) Lithostratigraphy of Border Cave, KwaZulu, South Africa: A Middle Stone Age sequence beginning c. 195,000 B.P. *JAS* **5**:317–341.

Butzer, K., F. Brown, and D. Thurber (1969) Horizontal sediments of the lower Omo Valley: Kibish Formation. *Quaternaria* **110**:15–29.

C

Cachel, S., and J. Harris (1995) Ranging patterns, land-use and subsistence in *Homo erectus*, from the perspective of evolutionary ecology. In: J. Bower, and S. Sartono, Editors. *Evolution and Ecology of* Homo erectus. Leiden: Pithecanthropus Centennial Foundation, pp. 51–66.

Cachel, S., and J. Harris (1996) The paleobiology of *Homo erectus*: Implications for understanding the adaptive zone of this species. In: G. Pwiti, and R. Soper, Editors. *Aspects of African Archaeology*. Harare: University of Zimbabwe Publications, pp. 3–9.

Cachel, S., and J. Harris (1998) The lifeways of *Homo erectus* inferred from archaeology and evolutionary ecology: A perspective from East Africa. In: M. Petraglia, and R. Korisettar, Editors. *Early Human Behaviour in Global Context*. London: Routledge, pp. 108–132.

Cadien, J. (1972) Dental variation in man. In: S. Washburn, and P. Dolhinow, Editors. *Perspectives on Human Evolution*. Volume **2**. New York: Holt, Rinehart, and Winston, pp. 199–222.

Campbell, B. (1965) The nomenclature of the Hominidae. *Occasional Papers of the Royal Anthropological Institute* **22**:1–33.

Campbell, N. (1993) *Biology*, 3rd edition. Redwood City: Benjamin/Cummings.

Camps, G., and G. Olivier (1970) *L'homme de Cro-Magnon*. Paris: Arts et Métier Graphiques.

Cann, R. (1988) DNA and human origins. *ARA* **17**:127–143.

Cann, R., M. Stoneking, and A. Wilson (1987) Mitochondrial DNA and human evolution. *Nature* **325**:31–36.

Cantu, R., and C. Richardson (1997) *Mule Deer Management in Texas*. Austin: Texas Parks and Wildlife.

Caramelli, D., C. Lalueza-Fox, C. Vernesi, M. Lari, A. Casoli, F. Mallegni, B. Chiarelli, I. Dupanloup, J. Bertranpetit, G. Barbujani, and G. Bertorelli (2003) Evidence for a genetic discontinuity between Neandertals and 24,000-year-old anatomically modern Europeans. *PNAS* **100**:6593–6597.

Carbonell, E., J. Bermúdez de Castro, J. Arsuaga, J. Díez, A. Rosas, G. Cuenca-Bescós, R. Sala, M. Mosquera, and X. Rodríguez (1995) Lower Pleistocene hominids and artifacts from Atapuerca-TD6 (Spain). *Science* **269**:826–832.

Carbonell, E., J. Bermúdez de Castro, J. Parés, A. Pérez-González, G. Cuenca-Bescós, A. Ollé, M. Mosquera, R. Huguet, J. van der Made, A. Rosas, R. Sala, J. Vallverdú, N. García, D. Granger, M. Martinón-Torres, X. Rodríguez, G. Stock, J. M. Vergès, E. Allué, F. Burjach, I. Cáceres, A. Canals, A. Benito, C. Díez, M. Lozano, A. Mateos, M. Navazo, J. Rodríguez, J. Rosell, and J. Arsuaga (2008) The first hominin of Europe. *Nature* **452**:465–469.

Carbonell, E., and Z. Castro-Curel (1992) Paleolithic wooden artifacts from the Abric Romani (Capellades, Barcelona, Spain). *JAS* **19**:707–719.

Carbonell, E., M. Gárcia-Antón, C. Mallol, M. Mosquera, A. Ollé, X. Rodríguez, M. Sahnouni, R. Sala, and J. Vergès (1999) The TD6 level lithic industry from Gran Dolina, Atapuerca (Burgos, Spain): production and use. In: Bermúdez de Castro et al. 1999a: 653–693.

Carbonell, E., and Rodríguez, X. (1994) Early Middle Pleistocene deposits and artifacts in the Gran Dolina site (TD 4) of the "Sierra de Atapuerca" (Burgos, Spain). *JHE* **26**:291–311.

Carretero, J., J. Arsuaga, and C. Lorenzo (1997) Clavicles, scapulae, and humeri from the Sima de los Huesos site (Sierra de Atapuerca, Spain). In: Arsuaga et al. 1997a: 357–408.

Carretero, J., C. Lorenzo, and J. Arsuaga (1999) Axial and appendicular skeleton of *Homo antecessor*. In: Bermúdez de Castro et al. 1999a: 459–499.

Carrier, D. (1984) The energetic paradox of human running and hominid evolution. *CA* **25**:484–495.

Carroll, R. (1988) *Vertebrate Paleontology and Evolution*. New York: Freeman.

Cartelle, C., and W. Hartwig (1996) A new extinct primate among the Pleistocene megafauna of Bahia, Brazil. *PNAS* **93**:6405–6409.

Cartmill, M. (1970) *The orbits of arboreal mammals: A reassessment of the arboreal theory of primate evolution*. Ph.D. Dissertation, University of Chicago.

Cartmill, M. (1972) Arboreal adaptations and the origin of the order Primates. In: R. Tuttle, Editor. *The Functional and Evolutionary Biology of Primates*. Chicago: Aldine-Atherton, pp. 97–122.

Cartmill, M. (1974a) *Daubentonia, Dactylopsila*, woodpeckers, and klinorhynchy. In: R. Martin, G. Doyle, and A. Walker, Editors. *Prosimian Biology*. London: Duckworth, London, pp. 655–670.

Cartmill, M. (1974b) Rethinking primate origins. *Science* **184**:436–443.

Cartmill, M. (1974c) Pads and claws in arboreal locomotion. In: F. Jenkins, Jr., Editors. *Primate Locomotion*. New York: Academic Press. 45–83.

Cartmill, M. (1975a) Strepsirrhine basicranial structures and the affinities of the Cheirogaleidae. In: Luckett and Szalay 1975: 313–354.

Cartmill, M. (1975b) *Primate Origins*. Minneapolis: Burgess.

Cartmill, M. (1980) Morphology, function and evolution of the anthropoid postorbital septum. In: R. Ciochon, and A. Chiarelli, Editors. *Evolutionary Biology of the New World Monkeys and Continental Drift*. New York: Plenum, pp. 243–274.

Cartmill, M. (1981) Hypothesis testing and phylogenetic reconstruction. *Zeitschrift für zoologische Systematik und Evolutionsforschung* **19**:73–96.

Cartmill, M. (1982) Assessing tarsier affinities: Is anatomical description phylogenetically neutral? *Geobios, mémoire special* **6**:279–287.

Cartmill, M. (1990) Human uniqueness and theoretical content in paleoanthropology. *International Journal of Primatology* 11:173–192.

Cartmill, M. (1992) New views on primate origins. *EA* 1:105–111.

Cartmill, M. (1993) *A View to a Death in the Morning: Hunting and Nature through History.* Cambridge: Harvard University Press.

Cartmill, M. (1994) Anatomy, antinomies, and the problem of anthropoid origins. In: Fleagle and Kay 1994: 549–566.

Cartmill, M. (1999) Revolution, evolution, and Kuhn: A response to Chamberlain and Hartwig. *EA* 8:45–47.

Cartmill, M. (2002) Paleoanthropology: Science or mythological charter? *JAR* 58:183–201.

Cartmill, M. (in press) Primate classification and diversity. In: M. Platt and A. Ghazanfar, Editors. *Primate Neuroethology.* Oxford University Press.

Cartmill, M., and T. Holliday (2006) Species taxa, characters, and symplesiomorphies. *AJPA Supplement* 42: 74.

Cartmill, M., W. Hylander, and J. Shafland (1987) *Human Structure.* Cambridge: Harvard University Press.

Cartmill, M., and R. Kay (1978) Craniodental morphology, tarsier affinities, and primate suborders. In: D. Chivers and K. Joysey, Editors. *Recent Advances in Primatology*, Volume 3. London: Academic Press, pp. 205–214.

Cartmill, M., P. Lemelin, and D. Schmitt (2002) Support polygons and symmetrical gaits in mammals. *Zoological Journal of the Linnean Society* 136:401–420.

Cartmill, M., P. Lemelin, and D. Schmitt (2007) Primate gaits and primate origins. In: Ravosa and Dagosto 2007: 403–436.

Cartmill, M., R. MacPhee, and E. Simons (1981) Anatomy of the temporal bone in early anthropoids, with remarks on the problem of anthropoid origins. *AJPA* 56:3–21.

Cartmill, M., and K. Milton (1977) The lorisiform wrist joint and the evolution of "brachiating" adaptations in the Hominoidea. *AJPA* 47:249–272.

Cartmill, M., D. Pilbeam, and G. Isaac (1986) One hundred years of paleoanthropology. *AS* 74:410–420.

Caspari, R. (1991) *The evolution of the posterior cranial vault in the Central European Upper Pleistocene.* Ph.D. Dissertation, University of Michigan. Ann Arbor: University Microfilms.

Cathey, J., J. Bickham, and J. Patton. (1998) Introgressive hybridization and nonconcordant evolutionary history of maternal and paternal lineages in North American deer. *Evolution* 52:1224–1229.

Cavagna, G., N. Heglund, and C. Taylor (1977) Mechanical work in terrestrial locomotion: two basic mechanisms for minimizing energy expenditure. *American Journal of Physiology* 233: R243–R261.

Cavalla-Sforza, L. (1995) *The Great Human Diasporas: The History of Diversity and Evolution.* Reading, MA: Addison-Wesley.

Cavalli-Sforza, L., and W. Bodmer (1971) *The Genetics of Human Populations.* San Francisco: Freeman.

Cavalla-Sforza, L., P. Menozzi, and A. Piazza (1993) Demic expansions and human evolution. *Science* 259:639–646.

Cela-Conde, C., and C. Altaba (2002) Multiplying genera versus moving species: a new taxonomic proposal for the family Hominidae. *SAJS* 98:229–232.

Cela-Conde, C., and F. Ayala (2003) Genera of the human lineage. *PNAS* 100:7684–7689.

Cerling, T. (1992) Development of grasslands and savannas in east Africa during the Neogene. *Paleogeography, Paleoclimatology, Paleoecology* 97:241–247.

Chaimanee, Y., V. Suteethorn, P. Jintasakul, C. Vidthayanon, B. Marandat, and J.-J. Jaeger (2004) A new orang-utan relative from the Late Miocene of Thailand. *Nature* 427: 439–441.

Chambers, R. (1844) *Vestiges of the Natural History of Creation.* London: Churchill.

Chan, L. (1991) The deduction of the level of the human larynx from bony landmarks: its relevance to the evolution of speech. *HE* 6:249–261.

Chandley, A., R. Short, and W. Allen (1975) Cytogenetic studies of three equine hybrids. *Journal of Reproduction and Fertility* 23(Suppl):356–370.

Chappell, J., and N. Shackleton (1986) Oxygen isotopes and sea level. *Nature* 324:137–140.

Charles-Dominique, P. (1975) Nocturnality and diurnality: An ecological interpretation of these two modes of life by an analysis of the higher vertebrate fauna in tropical forest ecosystems. In: W. Luckett and F. Szalay, Editors. *Phylogeny of the Primates: A Multidisciplinary Approach.* New York: Plenum. 69–88.

Charles-Dominique, P., and R. Martin (1970) Evolution of lorises and lemurs. *Nature* 227:257–260.

Chase, P., and H. Dibble (1987) Middle Paleolithic symbolism: A review of current evidence and interpretations. *JAA* 6:263–296.

Chase, P., and H. Dibble (1990) On the emergence of modern humans. *CA* 38:58–59.

Chavaillon, J., C. Brahimi, and Y. Coppens (1974) Première découverte d'hominidé dans l'un des sites acheuléens de Melka-Kunturé (Éthiopie). *CRAS* 278:3299–3302.

Chen, J-Y., D-Y. Huang, and C-W. Li (1999) An Early Cambrian craniate-like chordate. *Nature* 402:518–522.

Chen, T., S. Yuan, and S. Gao (1984) The study of uranium series dating of fossil bones and an absolute age sequence for the main Paleolithic sites of north China. *AAS* 3:259–268.

Chen, T., S. Yuan, S. Gao, and Y. Hu (1987) Uranium series dating of the Hexian and Chaoxian human fossil sites. *AAS* 6:249–254.

Chen, T., Q. Yang, and E. Wu (1994) Antiquity of *Homo sapiens* in China. *Nature* 368:55–56.

Chen, T., Q. Yang, Y. Hu, W. Bao, and T. Li (1997) ESR dating of tooth enamel from Yuxian *Homo erectus* site, China. *Quaternary Science Reviews* **16**:455–458.

Chiarelli, B. (2004) Spongiform encephalopathy, cannibalism and Neanderthals' extinction. *HE* **19**:81–91.

Chomsky, N. (1964) *Language and Mind*. New York: Harcourt, Brace, and World.

Chomsky, N. (1975) *Reflections on Language*. New York: Random House.

Chow, M., and T. Rich (1982) *Shuotherium dongi*, n. gen. and sp., a therian with pseudo-tribosphenic molars from the Jurassic of Sichuan, China. *Australian Mammalogy* **5**:127–142.

Chu, K., J. Qi, Z. Yu, and Y. Anh (2004) Origin and phylogeny of chloroplasts revealed by a simple correlation analysis of complete genomes. *MBE* **21**:200–206.

Churchill, S. (1994) *Human upper body evolution in the Eurasian Late Pleistocene*. Ph.D. Dissertation, University of New Mexico, Albuquerque.

Churchill, S. (1996) Particulate versus integrated evolution of the upper body in late Pleistocene humans: A test of two models. *AJPA* **100**:559–583.

Churchill, S. (1998) Cold adaptation, heterochrony and Neandertals. *EA* **7**:46–61.

Churchill, S. (2001) Hand morphology, manipulation, and tool use in Neandertals and early modern humans of the Near East. *PNAS* **98**:2953–2955.

Churchill, S. (2006) Bioenergetic perspectives on Neanderthal thermoregulatory and activity budgets. In: Harvati and Harrison 2006: 113–133.

Churchill, S., O. Pearson, F. Grine, E. Trinkaus, and T. Holliday (1996) Morphological affinities of the proximal ulna from Klasies River Main Site: Archaic or modern? *JHE* **31**:213–237.

Churchill, S., and F. Smith (2000a) Makers of the earliest Aurignacian of Europe. *YPA* **43**:61–115.

Churchill, S., and F. Smith (2000b) A modern human humerus from the early Aurignacian of Vogelherdhöhle (Stetten, Germany), *AJPA* **112**:251–273.

Churchill, S., and E. Trinkaus (1990) Neandertal scapular glenoid morphology. *AJPA* **83**:147–160.

Cifelli, R. (2000) Cretaceous mammals of Asia and North America. *Paleontological Society of Korea Special Publication* **4**:49–84.

Cifelli, R. (2001) Early mammalian radiations. *Journal of Paleontology* **75**:1214–1226.

Cifelli, R., and C. de Muizon (1997) Dentition and jaw of *Kokopellia juddi*, a primitive marsupial or near marsupial from the medial Cretaceous of Utah. *Journal of Mammalian Evolution* **4**:241–258.

Ciochon, R. (1991) The ape that was: Asian fossils reveal humanity's giant cousin. *Natural History* **November**:54–62.

Ciochon, R., and A. Chiarelli (1980) Paleobiogeographic perspectives on the origin of the Platyrrhini. In: R. Ciochon and A. Chiarelli, Editors. *Evolutionary Biology of the New World Monkeys and Continental Drift*. New York: Plenum, pp. 459–493.

Ciochon, R., and R. Corruccini (1976) Shoulder joint of Sterkfontein *Australopithecus*. *SAJS* **72**:80–82.

Ciochon, R., P. Gingerich, G. Gunnell, and E. Simons (2001) Primate postcrania from the late middle Eocene of Myanmar. *PNAS* **99**:7672–7677.

Ciochon, R., and G. Gunnell (2004) Eocene large-bodied primates from Myanmar and Thailand: Morphological considerations and phylogenetic affinities. In: Ross and Kay 2004: 249–282.

Clark, J. D. (1969) *Kalambo Falls Prehistoric Site*, Volume I. Cambridge: Cambridge University Press.

Clark, J. D. (1970) *The Prehistory of Africa*. New York: Praeger.

Clark, J. D. (1982) The cultures of the Middle Paleolithic/Middle Stone Age. In: J. Clark, Editor. *The Cambridge History of Africa*. Volume I. Cambridge: Cambridge University Press, pp. 248–341.

Clark, J. D. (1988) The Middle Stone Age of East Africa and the beginnings of regional identity. *JWP* **2**:235–305.

Clark, J. D. (1994) The Acheulean industrial complex in Africa and elsewhere. In: Corruccini and Ciochon 1994: 451–469.

Clark, J. D., Y. Beyenne, G. WoldGabriel, W. Hart, P. Renne, H. Gilbert, A. Aefluer, G. Suwa, S. Katoh, K. Ludwig, J. Boisserie, B. Asfaw, and T. White (2003) Stratigraphic, chronological and behavioural context of Pleistocene *Homo sapiens* from Middle Awash, Ethiopia. *Nature* **423**:747–752.

Clark, J. D., J. de Heinzelin, K. Schick, W. Hart, T. White, G. WoldeGabriel, R. Walter, G. Suwa, B. Asfaw, E. Vrba, and Y. H-Selassie (1994) African *Homo erectus*: Old radiometric ages and young Oldowan assemblages in the Middle Awash Valley, Ethiopia. *Nature* **264**:1907–1910.

Clark, J. D., K. Oakley, L. Wells, and J. McCleland (1950) New studies on Rhodesian Man. *JRAI* **77**:1–32.

Clark, J. G. D. (1977) *World Prehistory: A New Perspective*. Cambridge: Cambridge University Press.

Clarke, R. (1976) New cranium of *Homo erectus* from Lake Ndutu, *Tanzania*. *Nature* **262**:485–487.

Clarke, R. (1977) A juvenile cranium and some adult teeth of early *Homo* from Swartkrans, Transvaal. *SAJS* **73**:46–49.

Clarke, R. (1979) Early hominid footprints from Tanzania. *SAJS* **75**:148–149.

Clarke, R. (1985a) A new reconstruction of the Florisbad cranium, with notes on the site. In: Delson 1985: 301–305.

Clarke, R. (1985b) Early Acheulean with *Homo habilis* at Sterkfontein. In: Tobias 1985: 287–298.

Clarke, R. (1988) A new *Australopithecus* cranium from Sterkfontein and its bearing on the ancestry of *Paranthropus*. In: Grine 1988: 285–292.

Clarke, R. (1994a) Advances in understanding the craniofacial anatomy of South African early hominids. In: Corruccini and Ciochon 1994: 205–222.

Clarke, R. (1994b) The significance of the Swartkrans *Homo* to the *Homo erectus* problem. In: Franzen 1994: 185–193.

Clarke, R. (1998) First ever discovery of a well-preserved skull and associated skeleton of *Australopithecus*. *SAJS* **94**: 460–463.

Clarke, R. (2000) A corrected reconstruction and interpretation of the *Homo erectus* calvaria from Ceprano, Italy. *JHE* **39**:433–442.

Clarke, R. (2002a) On the unrealistic "revised age estimates" for Sterkfontein. *SAJS* **98**:415–419.

Clarke, R. (2002b) Newly revealed information on the Sterkfontein Member 2 *Australopithecus* skeleton. *SAJS* **98**: 523–526.

Clarke, R., and F. C. Howell (1972) Affinities of the Swartkrans 847 cranium. *AJPA* **37**:319–336.

Clarke, R., F. C. Howell, and C. Brain (1970) More evidence of an advanced hominid at Swartkrans. *Nature* **225**: 1219–1222.

Clarke, R., and P. Tobias (1995) Sterkfontein Member 2 foot bones of the oldest South African hominid. *Science* **269**:521–524.

Clemens, W., and Z. Kielan-Jaworowska (1979) Multituberculata. In: Lillegraven et al. 1979: 99–149.

Clemente, C. (1997) *Anatomy: A Regional Atlas of the Human Body*, 4th edition. Baltimore: Williams and Wilkins.

Clutton-Brock, T. (1980) Primates, brains, and ecology. *Journal of Zoology* **190**:309–323.

Coffing, K., C. Feibel, M. Leakey, and A. Walker (1994) Four-million-year-old hominids from East Lake Turkana, Kenya. *AJPA* **93**:55–65.

Cohen, P. (1996) Fitting a face to Ngaloba. *JHE* **30**:373–379.

Cole, M., and T. Cole (1994) Metric variation in the supraorbital region in Northern Plains Indians. In: D. Owsley, and R. Jantz, Editors. *Skeletal Biology in the Great Plains*. Washington, DC: Smithsonian Institution, pp. 209–217.

Cole, T., and F. Smith (1987) An odontometric assessment of variability in *Australopithecus afarensis*. *HE* **2**:221–234.

Colgon, D. (2001) Commentary on G. J. Adcock et al., 2001, Mitochondrial DNA sequences in ancient Australians: Implications for modern human origins. *AO* **36**:168–169.

Collard, M., and L. Aiello (2000) From forelimbs to two legs. *Nature* **404**:339–340.

Conard, N. (1992) Tönchesberg and its position in the prehistory of northern Europe. *Monographien Römisch-Germanisches Zentralmuseum* **20**.

Conard, N. (2003) Paleolithic ivory sculptures from south-western Germany and the origin of art. *Nature* **426**:830–832.

Conard, N., and M. Bolus (2003) Radiocarbon dating and the appearance of modern humans and timing of cultural innovations in Europe: New results and new challenges. *JHE* **44**:331–371.

Conard, N., G. Dippon, and P. Goldberg (2003b) Chronostratigraphy and archaeological context of the Aurignacian deposits at Geißenklösterle. In: Zilhão and d'Errico 2003: 165–176.

Conard, N., P. Grootes, and F. Smith (2004) Unexpectedly recent dates for the human remains from Vogelherd. *Nature* **430**:198–201.

Conard, N., L. Niven, K. Mueller, and A. Stewart (2003a) The chronostratigraphy of the Upper Paleolithic deposits at Vogelherd. *MGU* **12**:73–86.

Conard, N., and T. Prindville (2000) Middle Paleolithic hunting economies in the Rhineland. *IJO* **10**:286–309.

Condemi, S. (1988) Caractères plesiomorphes et apomorphes de l'os temporal des Néandertaliens Européens Würmiens. In: *L'Homme de Néandertal*. Volume 3: *L'Anatomie*. Liège: Études et Recherches Archéologiques de l'Université de Liège pp. 49–52.

Condemi, S. (1992) *Les hommes fossiles de Saccopastore et leurs relations phylogénétiques*. Paris: CNRS.

Condemi, S. (1996) Does the human fossil specimen from Reilingen (Germany) belong to the *Homo erectus* or to the Neanderthal lineage? *Anthropologie* **34**:69–77.

Condemi, S. (2001) *Les Néandertaliens de La Chaise*. Paris: Comité des Travaux Historiques et Scientifiques.

Conroy, G. (2002) Speciosity in the early *Homo* lineage: Too many, too few, or just about right? *JHE* **43**:759–766.

Conroy, G., C. Jolly, D. Cramer, and J. Kalb (1978) Newly discovered fossil hominid skull from the Afar Depression, Ethiopia. *Nature* **276**:67–70.

Conroy, G., M. Vannier, and P. Tobias (1990) Endocranial features of *Australopithecus africanus* revealed by 2- and 3-D computed tomography. *Science* **247**:838–841.

Conroy, G., G. Weber, H. Seidle, P. Tobias, A. Kane, and B. Brunsden (1998) Endocranial capacity in an early hominid cranium from Sterkfontein, South Africa. *Science* **280**: 1730–1731.

Constant, D., and F. Grine (2001) A review of taurodontism with new data on indigenous southern African populations. *Archives of Oral Biology* **46**:1021–1026.

Conway, B., J. McNabb, and N. Ashton, Editors (1996) *Excavations at Barnfield Pit, Swanscombe, 1968–1972*. London: British Museum.

Cook, J., C. Stringer, A. Currant, H. Schwarcz, and A. Wintle (1982) A review of the chronology of the European Middle Pleistocene hominid record. *YPA* **25**:19–65.

Cooke, H. B., B. Malan, and L. Wells (1945) Fossil man in the Lebombo Mountains, South Africa: The "Border Cave," Ingwavuma District, Zululand. *Man* **45**:6–13.

Coolidge, H. (1933) *Pan paniscus*. Pygmy chimpanzee from south of the Congo River. *AJPA* **18**:1–57.

Coon, C. (1951) *Cave Explorations in Iran 1949*. Philadelphia: University Museums Monographs.

Coon, C. (1962) *The Origin of Races*. New York: A.A. Knopf.

Cooper, A., A. Drummond, and E. Willerslev (2004) Ancient DNA: Would the real Neandertal please stand up? *Current Biology* **14**:431–433.

Cooper, A., A. Rambaut, V. Macaulay, E. Willerslev, A. J. Hansen, and C. B. Stringer (2001) Human origins and ancient DNA. *Science* **292**:1655–1656.

Coppens, Y. (1994) East Side Story: The origin of humankind. *SA* **270**:88–95.

Coppens, Y. (1999) Introduction. In: T. Bromage and F. Schrenk, Editors. *African Biogeography, Climate Change, and Human Evolution*. New York: Oxford University Press, pp. 13–18.

Coqueugniot, H., J. Hublin, F. Veillon, F. Houët, and T. Jakob (2004) Early brain growth in *Homo erectus* and implications for cognitive ability. *Nature* **431**:299–302.

Corruccini, R. (1974) Calvarial shape relationships between fossil hominids. *YPA* **18**:89–109.

Corruccini, R. (1975) Metrical analysis of Fontéchevade II. *AJPA* **42**:95–98.

Corruccini, R. (1978) Comparative osteometrics of the hominoid wrist joint with special reference to knuckle-walking. *JHE* **7**:307–321.

Corruccini, R. (1992) Metrical reconsideration of the Skhul IV and IX and Border Cave 1 crania in the context of modern human origins. *AJPA* **87**:433–445.

Corruccini, R. (1994) Reaganomics and the fate of the progressive Neandertals. In: Corruccini and Ciochon 1994: 697–708.

Corruccini, R., and H. McHenry (2001) Knuckle-walking hominid ancestors. *JHE* **40**:507–511.

Coutselinis, A., C. Dritsas, and T. Pitsios (1991) Expertise médico-légale du crâne Pléistocène LA01/S2 (Apidima II) Apidima, Laconie, Grèce. *L'Anthropologie (Paris)* **95**:401–408.

Covert, H. (1986) Biology of early Cenozoic primates. In: D. Swindler and J. Erwin, Editors. *Comparative Primate Biology*, Volume 1. New York: Alan R. Liss, pp. 335–359.

Covert, H. (2002) The earliest fossil primates and the evolution of prosimians: Introduction. In: Hartwig 2002: 13–20.

Covert, H. (2005) Does overlap among the adaptive radiations of omomyoids, adapoids, and early anthropoids cloud our understanding of anthropoid origins? In: Ross and Kay 2005: 139–155.

Covert, H., and B. Williams (1991) The anterior lower dentition of *Washakius insignis* and adapid–anthropoidean affinities. *JHE* **21**:463–467.

Covert, H., and B. Williams (1994) Recently recovered specimens of North American Eocene omomyids and adapids and

their bearing on debates about anthropoid origins. In: Fleagle and Kay 1994: 29–54.

Coyne, J. (1992) Genetics and speciation. *Nature* **355**:511–515.

Coyne, J., and H. Orr (1989) Patterns of speciation in *Drosophila*. *Evolution* **43**:362–381.

Creed-Miles, M., A. Rosas, and R. Kruszynski (1996) Issues in the identification of Neandertal derivative traits at early post-natal stages. *JHE* **30**:147–153.

Crelin, E. (1969) *Anatomy of the Newborn: An Atlas*. Philadelphia: Lea & Febiger.

Crelin, E. (1973) *Functional Anatomy of the Newborn*. New Haven: Yale University Press.

Crelin, E. (1987) *The Human Vocal Tract: Anatomy, Function, Development, and Evolution*. New York: Vantage.

Crespi, B., and C. Semeniuk (2004) Parent–offspring conflict in the evolution of vertebrate reproductive mode. *American Naturalist* **163**:635–653.

Crevecoeur, I. (2007) The Early Upper Paleolithic human remains of Nazlet Khater 2, Egypt. *AJPA Supplement* **44**:93.

Crevecoeur, I., and E. Trinkaus (2004) From the Nile to the Danube: A comparison of the Nazlet Khater 2 and Oase 1 early modern human mandibles. *Anthropologie* **42**:229–239.

Crompton, A. (1985) Origin of the mammalian temporomandibular joint. In: D. Carlson, J. McNamara, and K. Ribbens, Editors. *Developmental Aspects of Craniomandibular Joint Disorders*. Craniofacial Growth Series, Monograph #16, Center for Human Growth and Development, University of Michigan, Ann Arbor, pp. 1–18.

Crompton, A., and W. Hylander (1986) Changes in mandibular function following the acquisition of a dentary-squamosal jaw articulation. In: N. Hotton, P. MacLean, J. Roth, and E. Roth, Editors. *The Ecology and Biology of Mammal-like Reptiles*. Washington, DC: Smithsonian Institution Press, pp. 263–282.

Crompton, A., and F. Jenkins, Jr. (1978) Mesozoic mammals. In: Maglio and Cooke 1978: 46–55.

Crompton, A., and F. Jenkins, Jr. (1979) Origin of mammals. In: Lillegraven et al. 1979: 59–73.

Crompton, A., and P. Parker (1978) Evolution of the mammalian masticatory apparatus. *AS* **66**:192–201.

Crompton, R., Y. Li, W. Wang, M. Günther, and R. Savage (1998) The mechanical effectiveness of erect and "bent-hip, bent-knee" bipedal walking in *Australopithecus afarensis*. *JHE* **35**:55–74.

Crummett, T. (1994) *The evolution of shovel shaping: Regional and temporal variation in human incisor morphology*. Ph.D. Dissertation, University of Michigan. Ann Arbor: University Microfilms.

Crummett, T. (1995) The 3 dimensions of shovel shaping. In: J. Moggi-Cecchi, Editor. *Proceedings of the 9th International Symposium on Dental Anthropology*. Florence: International

Institute for the Study of Man, Angelo Pontecorboli Editione, pp. 305–313.

Cuenca-Bescós, G., C. Laplana Conesa, J. Canudo, and J. Arsuaga (1997) Small mammal remains from Sima de los Huesos. In: Arsuaga et al. 1997a: 175–190.

Cuenca-Bescós, G., G. Laplana, and J. Canudo (1999) Biochronological implications of the Arvicolidae (Rodentia, Mammalia) from the Lower Pleistocene hominid-bearing level of Trinchera Dolina 6 (TD6, Atapuerca, Spain). In: Bermúdez de Castro et al. 1999a: 353–373.

Culotta, E. (2007) Ancient DNA reveals Neandertals with red hair, fair complexions. *Science* 318:546–547.

Cunningham, D., and D. Wescott (2002) Within-group human variation in the Asian Pleistocene: The three Upper Cave crania. *JHE* **42**:627–638.

Curnoe, D. (2001) Early *Homo* from southern Africa: A cladistic perspective. *SAJS* **97**:186–190.

Curnoe, D. (2002) Fossil human taxonomic hypotheses examined with topology-dependent permutation tail probability (T-PTP) tests. *SAJS* **98**:310–312.

Curnoe, D. (2006) Odontometric systematic assessment of the Swartkrans SK 15 mandible. *Homo* **57**:263–294.

Curnoe, D., A. Thorne, and J. A. Coate (2006) Timing and tempo of primate speciation. *Journal of Evolutionary Biology* **19**:59–65.

Curnoe, D., and P. Tobias (2006) Description, new reconstruction, comparative anatomy, and classification of the Sterkfontein Stw 53 cranium, with discussions about the taxonomy of other southern African early *Homo* remains. *JHE* **50**:36–77.

Cuvier, G. (1831) *A Discourse on the Revolutions of the Surface of the Globe: and the Changes Thereby Produced in the Animal Kingdom*. Philadelphia: Carey & Lea.

Czarnetzki, A. (1989) Ein archaischer Hominidencalvariarest aus einer Kiesgrube in Reilingen, Rhein Neckar-Kreis. *Quartär* **39/40**:191–201.

Czarnetzki, A., T. Jakob, and C. Pusch (2003) Paleopathological and variant conditions of the *Homo heidelbergensis* type specimen (Mauer, Germany). *JHE* **44**:479–495.

D

Daegling, D. (1993) Functional morphology of the human chin. *EA* **1**:170–177.

Daegling, D., and F. Grine (1999) Terrestrial foraging and dental microwear in *Papio ursinus*. *Primates* **40**:559–572.

Dagosto, M. (1988) Implications of postcranial evidence for the origin of euprimates. *JHE* **17**:35–56.

Dagosto, M. (1993) Postcranial anatomy and locomotor behavior in Eocene primates. In: Gebo 1993: 150–174.

Dagosto, M., and D. Gebo (1994) Postcranial anatomy and the origin of the Anthropoidea. In: Fleagle and Kay 1994: 567–593.

Dagosto, M., D. Gebo, and K. Beard (1999) Revision of the Wind River faunas, Early Eocene of central Wyoming. Part 14. Postcranium of *Shoshonius cooperi* (Mammalia: Primates). *Annals of the Carnegie Museum* **68**:175–211.

Dahlberg, F., Editor (1981) *Woman the Gatherer*. New Haven: Yale University Press.

D'Août, K., P. Aerts, D. De Clercq, K. De Meester, and L. Van Elsacker (2002) Segment and joint angles of the hind limb during bipedal and quadrupedal walking of the bonobo (*Pan paniscus*). *AJPA* **119**:37–51.

D'Août, K., E. Vereecke, K. Schoonaert, D. De Clercq, L. van Elsacker, and P. Aert (2004) Locomotion in bonobos (*Pan paniscus*): Differences and similarities between bipedal and quadrupedal terrestrial walking, and a comparison with other locomotor modes. *JA* **204**:353–361.

Dart, R. (1925) *Australopithecus africanus*: The man-ape of South Africa. *Nature* **115**:195–199.

Dart, R. (1926) Taungs and its significance. *NH* **26**:315–327.

Dart, R. (1948) The Makapansgat protohuman *Australopithecus prometheus*. *AJPA* **6**:259–283.

Dart, R. (1949) The predatory implemental technique of *Australopithecus*. *AJPA* **7**:1–38.

Dart, R. (1953) The predatory transition from ape to man. *International Anthropological and Linguistic Review* **1**:201–217.

Dart, R. (1955) Cultural status of the South African man-apes. *Annual Report of the Smithsonian Institution* 1955:317–338.

Dart, R. (1957) The osteodontokeratic culture of *Australopithecus prometheus*. *Transvaal Museum Memoirs* **10**:1–105.

Dart, R., and D. Craig (1959) *Adventures with the Missing Link*. New York: Harper and Bros., New York. 1961 reprint, New York: Viking Press.

Darwin, C. (1859) *The Origin of Species by Means of Natural Selection: or, The Preservation of Favoured Races in the Struggle for Life*. New York: Modern Library reprint (4th edition), 1936.

Darwin, C. (1871) *The Descent of Man and Selection in Relation to Sex*. London: John Murray. Modern Library reprint, New York, 1936.

Darwin, C. (1883) *The Variation of Animals and Plants under Domestication*, 2nd edition. New York: Appleton.

Darwin, E. (1794) *Zoonomia; or, The Laws of Organic Life*, 3rd (1801) edition. London: J. Johnson.

Day, M. (1965) *Guide to Fossil Man: A Handbook of Human Paleontology*. London: Cassell.

Day, M. (1969) Omo human skeletal remains. *Nature* **222**:1135–1138.

Day, M. (1971) Postcranial remains of *Homo erectus* from Bed IV, Olduvai Gorge, Tanzania. *Nature* **232**:383–387.

Day, M. (1984) The postcranial remains of *Homo erectus* from Africa, Asia, and possibly Europe. *CFS* **69**:113–121.

Day, M. (1985) Hominid locomotion—From Taung to the Laetoli footprints. In: Tobias 1985: 115–127.

Day, M. (1986b) Bipedalism: Pressures, origins and modes. In: Wood et al. 1986: 188–202.

Day, M. (1991) Bipedalism and prehistoric footprints. In: Coppens and Sénut 1991: 199–213.

Day, M. (1995) Continuity and discontinuity in the postcranial remains of *Homo erectus*. *ERAUL* **62**:181–190.

Day, M., R. Leakey, A. Walker, and B. Wood (1975) New hominids from East Rudolf, Kenya, I. *AJPA* **42**:461–476.

Day, M., and T. Molleson (1973) The Trinil femora. In: M. Day, Editor. *Human Evolution. Symposia of the Society for the Study of Human Biology*, Volume 11. London: Taylor and Francis, pp. 127–154.

Day, M., and C. B. Stringer (1982) A reconsideration of the Omo-Kibish remains and the *erectus*–*sapiens* transition. In: M. de Lumley, Editor. *L'Homo erectus et la place de l'homme de Tautavel parmi les hominidés*. Nice: CNRS. 814–846.

Day, M., and E. Wickens (1980) Laetoli Pliocene hominid footprints and bipedalism. *Nature* **286**:385–387.

Deacon, H. (1989) Late Pleistocene paleoecology and archaeology in the southern Cape, South Africa. In: Mellars and Stringer 1989: 547–564.

Deacon, H. (1995) An unsolved mystery at the Howieson's Poort name site. *South African Archaeological Bulletin* **50**: 110–120.

Deacon, H., and R. Shuurman (1992) The origins of modern people: The evidence from Klasies River. In: Bräuer and Smith 1992: 121–129.

Deacon, T. (1997) *The Symbolic Species: The Co-Evolution of Language and the Brain*. New York: Norton.

Dean, D., and E. Delson (1995) *Homo* at the gates of Europe. *Nature* **373**:472–473.

Dean, D., J. Hublin, R. Holloway, and R. Ziegler (1998) On the phylogenetic position of the pre-Neandertal specimen from Reilingen, Germany. *JHE* **34**:485–508.

Dean, M. (1985) Root cone angle of the permanent mandibular teeth of modern man and certain fossil hominids. *AJPA* **68**:233–238.

Dean, M. (1986) *Homo* and *Paranthropus*: Similarities in the cranial base and developing dentition. In: Wood et al. 1986: 249–265.

Dean, M. (1988) Growth processes in the cranial base of hominoids and their bearing on morphological similarities that exist in the cranial base of *Homo* and *Paranthropus*. In: Grine 1988: 107–112.

Dean, M., C. Stringer, and T. Bromage (1986) Age at death of the Neanderthal child from Devil's Tower (Gibraltar) and the implications for studies of general growth and development in Neanderthals. *AJPA* **70**:301–309.

Dean, M., M. Leakey, D. Reid, F. Schrenk, G. Schwartz, C. Stringer, and A. Walker (2001) Growth processes in teeth distinguish modern humans from *Homo erectus* and earlier hominins. *Nature* **414**:628–631.

Debénath, A., J.-P. Raynal, J. Roche, J.-P. Texier, and D. Ferenbach (1986) Stratigraphie, habitat, typologie, et devenir de l'Atérien marocain: Données récentes. *Anthropologie* **90**:233–246.

Defleur, A. (1993) *Les sepultures moustériennes*. Paris: Editions du CNRS.

Defleur, A., O. Dutour, H. Vallades, and B. Vandermeersch (1993) Cannibalism among the Neanderthals. *Nature* **362**:214.

Defleur, A., T. White, P. Valensi, L. Slimak, and É. Crégut-Bonnoure (1999) Neanderthal cannibalism at Moula-Guercy, Ardéche, France. *Science* **286**:128–131.

De Gregorio, B., and T. Sharp (2006) The structure and distribution of carbon in 3.5 Ga Apex chert: Implications for the biogenicity of Earth's oldest putative microfossils. *American Mineralogist* **91**:784–789.

DeGusta, D. (2002) Comparative skeletal pathology and the case for conspecific care in Middle Pleistocene hominids. *JAS* **29**:1435–1438.

DeGusta, D. (2003) Aubesier 11 is not evidence of Neanderthal conspecific care. *JHE* **45**:91–94.

DeGusta, D., W. Gilbert, and S. Turner (1999) Hypoglossal canal size and hominid speech. *PNAS* **96**:1800–1804.

de Heinzelin, J., J. Clark, T. White, W. Hart, P. Renne, G. WoldeGabriel, Y. Beyene, and E. Vrba (1999) Environment and behavior of 2.5-million-year-old Bouri hominids. *Science* **284**:625–635.

Deinard, A., G. Sirugo, and K. Kidd (1998) Hominoid phylogeny: Inferences from a subterminal minisatellite analyzed by repeat expansion detection (RED). *JHE* **35**:313–317.

Deino, A., L. Tauxe, M. Monaghan, and A. Hill (2002) ^{40}Ar/^{39}Ar geochronology and paleomagnetic stratigraphy of the Lukeino and Lower Chemeron Formations at Tabarin and Kapcheberek, Tugen Hills, Kenya. *JHE* **42**:117–140.

de la Torre, I. (2004) Omo revisited: Evaluating the technological skills of Pliocene hominids. *CA* **45**:439–465.

Delfino, V., and E. Vacca (1993) An archaic human skeleton discovered at Altamura [Bari, Italy]. *Rivista di Antropologia* **71**:249–257.

Deloison, Y. (1991) Les Australopithèques marchaient-ils comme nous? In: Coppens and Sénut 1991: 177–186.

Delson, E. (1988) Chronology of South African australopith site units. In: Grine 1988: 317–324.

Delson, E. (2000a) Cercopithecoidea. In: Delson et al. 2000: 171–172.

Delson, E. (2000b) *Dendropithecus*-group. In: Delson et al. 2000: 206–207.

Delson, E., K. Harvati, D. Reddy, L. Marcus, K. Mowbray, G. Sawyer, T. Jacob, and S. Márquez (2001) The Sambungmacan 3 *Homo erectus* calvaria: A comparative morphometric and morphological study. *AR* **262**:380–397.

de Lumley, H. (1975) Cultural evolution in France in its paleoecological setting during the Middle Pleistocene. In: Butzer and Isaac 1975: 745–808.

de Lumley, H. (1976) Les premières industries humains en Provence. In: H. de Lumley, Editor. *La préhistoire française.* Paris: Centre National de la Recherche Scientifique. 765–794.

de Lumley, H., and A. Sonakia (1985) Contexte stratigraphique et archéologique de l'homme de Narmada, Hathnora, Madhya Pradesh, India. *L'Anthropologie* **89**:3–12.

de Lumley, H., A. Fournier, J. Krzepkowska, and A. Eschassoux (1988) L'industrie du Pléistocène inférieur de la grotte du Vallonet, Roquebrune-Cap-Martin, Alpes-Maritimes. *L'Anthropologie* **92**:502–614.

de Lumley, M.-A. (1981) Les Anténéandertaliens en Europe. In: Sigmon and Cybulski 1981: 115–132.

Demes, B. (1987) Another look at an old face: Biomechanics of the Neandertal facial skeleton reconsidered. *JHE* **16**:297–304.

de Menocal, P. (1995) Plio–Pleistocene African climate. *Science* **270**:53–59.

Dennell, R. (1997) The world's oldest spears. *Nature* **385**: 787–788.

Dennell, R., H. Rendell, and E. Hailwood (1988) Late Pliocene artifacts from northern Pakistan. *CA* **29**:495–498.

Dennell, R., H. Rendell, L. Hurcombe, and F. Hailwood (1994) Archaeological evidence for hominids in northern Pakistan before one million years ago. In: Franzen 1994: 151–155.

Dennell, R., and W. Roebroeks (1996) The earliest colonization of Europe: The short chronology revisited. *Antiquity* **70**: 535–542.

Dennell, R., and W. Roebroeks (2005) An Asian perspective on early human dispersal from Africa. *Nature* **438**:1099–1014.

Denton, G. (1999) Cenozoic climate change. In: T. Bromage and F. Schrenk, Editors. *African Biogeography, Climate Change, and Human Evolution.* New York: Oxford University Press. 94–114.

Derr, J. N., D. Hale, D. Ellsworth, and J. Bickham (1991) Fertility in an F1 male hybrid of white-tailed deer (*Odocoileus virginianus*) × mule deer (*O. hemionus*). *Journal of Reproduction and Fertility* **93**:111–117.

d'Errico, F. (2003) The invisible frontier. A multiple species approach for the origin of behavioral modernity. *EA* **12**: 188–202.

d'Errico, F., C. Henshilwood, M. Vanhaeren, and K. van Niekerk (2005) *Nassarius kraussianus* shell beads from Blombos Cave: Evidence for symbolic behavior in the Middle Stone Age. *JHE* **48**:3–24.

d'Errico, F., and P. Villa (1997) Holes and grooves: The contribution of microscopy to the problem of art origins. *JHE* **33**:1–31.

d'Errico, F., J. Zilhão, M. Julien, D. Baffier, and J. Pelgrin (1998) Neanderthal acculturation in Western Europe? A critical review of the evidence and its interpretation. *CA* **39**:1–44.

De Villiers, H. (1968) *The Skull of the South African Negro.* Johannesburg: Witwatersrand University Press.

De Villiers, H. (1973) Human skeletal remains from Border Cave, Ingwavuma District, KwaZulu, South Africa. *Annals of the Transvaal Museum* **28**:229–256.

De Villiers, H. (1976) A second adult human mandible from Border Cave, Inwavuma District, Kwazulu, South Africa. *SAJS* **72**:212–215.

De Vos, J., and P. Sondaar (1994) Dating hominid sites in Indonesia. *Science* **266**:1726–1727.

De Vos, J., P. Sondaar, G. van der Bergh, and F. Aziz (1994) The *Homo* bearing deposits of Java and its ecological context. In: Franzen 1994: 129–140.

de Waal, F. (2001) Apes from Venus: Bonobos and human social evolution. In: F. de Waal, Editor. *Tree of Origin: What Primate Behavior Can Tell Us about Human Social Evolution.* Cambridge: Harvard University Press, pp. 39–68.

Diamond, J., and A. Bond (2003) A comparative analysis of social play in birds. *Behaviour* **140**:1091–1115.

Dibble, H. (1988) Tyopological aspects of reduction and intensity of utilization of lithic resources in the French Mousterian. In: Dibble and Monet-White 1988: 181–197.

Dibble, H., and N. Rolland (1992) On assemblage variation in the Middle Paleolithic of western Europe: History, perspectives and a new synthesis. In: H. Dibble and P. Mellars, Editors. *The Middle Paleolithic: Adaptation, Behavior, and Variability.* Philadelphia: University of Pennsylvania Museum Monographs 72, pp. 1–28.

Dibble, H., and S. McPherron (2006) The missing Mousterian. *CA* **47**:777–803.

Dixson, A. (1983) Observations on the evolution and behavioral significance of "sexual skin" in female primates. *Advances in the Study of Behavior* **13**:63–106.

Dizon, E., F. Détroit, F. Sémah, C. Falguères, S. Hameau, W. Ronquillo, and E. Cabanis (2002) Notes on the morphology and age of the Tabon Cave fossil *Homo sapiens. CA* **43**: 660–666.

Dobson, S. (2005) Are the differences between Stw 431 (*Australopithecus africanus*) and A. L. 288-1 (*A. afarensis*) significant? *JHE* **49**:143–154.

Dobson, S., and E. Trinkaus (2002) Cross-sectional geometry and morphology of the mandibular symphysis in Middle and Late Pleistocene *Homo. JHE* **43**:67–87.

Dobzhansky, T. (1937) *Genetics and the Origin of Species.* New York: Columbia University Press.

Dobzhansky, T. (1951) *Genetics and the Origin of Species*, 3rd edition. New York: Columbia University Press.

Dodo, Y., O. Kondo, S. Muhesen, and T. Akazawa (1998) Anatomy of the Neandertal infant skeleton from Dederiyeh Cave, Syria. In: Akazawa et al. 1998: 323–338.

Dodo, Y., O. Kondo, and T. Nara (2002) The skull of the Neanderthal child of burial no. 1. In Akazawa and Muhesen 2002: 93–137.

Doran, D., and A. McNeilage (2001) Subspecific variation in gorilla behavior: the influence of ecological and social factors. In: M. Robbins, P. Sicotte, and K. Stewart, Editors. *Mountain Gorillas: Three Decades of Research at Karisoke*, Cambridge University Press, Cambridge, pp. 123–149.

Doty, R. (2001) Olfaction. *Annual Review of Psychology* **52**: 423–452.

Drake, R., and G. Curtis (1987) K–Ar geochronology of the Laetoli fossil localities. In: M. Leakey and J. Harris, editors. *Laetoli: A Pliocene Site in Northern Tanzania*. Oxford: Clarendon Press, Oxford, pp. 48–52.

Drayna, D. (2005) Founder mutations. *SA* **293**:78–85.

Dreyer, M. (1935) The Florisbad skull. *SAJS* **32**:601–602.

Drucker, D., and H. Bocherens (2004) Carbon and nitrogen stable isotopes as tracers of change in diet breadth during Middle and Upper Paleolithic in Europe. *IJO* **14**:162–177.

Duarte, C., J. Maurício, P. Pettitt, P. Souto, E. Trinkaus, E. van der Plicht, and J. Zilhão (1999) The early Upper Paleolithic human skeleton from the Abrigo do Lagar Velho (Portugal) and modern human emergence in Iberia. *PNAS* **96**:7604–7609.

Duarte, C., S. Hillson, T. Holliday, and E. Trinkaus (2002) The Lagar Velho 1 human skeletal inventory. In: Zilhão and Trinkaus 2002: 221–255.

Dubois, E. (1894) *Pithecanthropus erectus, eine menschenähnliche Übergangsform aus Java*. Batavia: Landesdruckerei.

Dubois, E. (1899a) Upon the brain-cast of *Pithecanthropus erectus*. *Proceedings of the International Congress of Zoology* (Cambridge, 1898): 78–95.

Dubois, E. (1899b) *Pithecanthropus erectus*: A form from the ancestral stock of mankind. *Annual Report of the Smithsonian Institution* (1898): 445–459.

Dubois, E. (1922) The proto-Australasian fossil man of Wadjak. *Verhandelingen der Koninklijke Akademie van Wetenschappen te Amsterdam* **23**:1013–1051.

Dubois, E. (1935) On the gibbon-like appearance of *Pithecanthropus erectus*. *Verhandelingen der Koninklijke Akademie van Wetenschappen te Amsterdam* **38**:578–585.

DuBrul, E., and C. Reed (1960) Skeletal evidence of speech? *AJPA* **18**:153–156.

Ducrocq, S. (1999) *Siamopithecus eocaenus*, a late Eocene anthropoid primate from Thailand: Its contribution to the evolution of anthropoids in Southeast Asia. *JHE* **36**:613–635.

Dujardin, V. (2003) Sondages à La Quina aval (Gardes-le-Pontaroux, Charente). *Antiquités Nationales* **33**:21–26.

Duncan, A., J. Kappelman, and L. Shapiro (1994) Metatarsophalangeal joint function and positional behavior in *Australopithecus afarensis*. *AJPA* **93**:67–81.

Dunsworth, H., and A. Walker (2002) Early genus *Homo*. In: Hartwig 2002: 419–435.

Dupont, É. (1866) Étude sur les fouilles scientifiques exécutées pendant l'hiver de 1865–1866 dans les cavernes des bords de la Lesse. *Bulletin de l'Académie Royal de Belgique* **22**:44–45.

Dupree, L., Editor (1972) *Prehistoric Research in Afghanistan (1959–1966)*. TAPS 62.

Durband, A. (2002) Squamotympanic fissure in the Ngandong and Sambungmacan hominids: A reply to Delson et al. *AR* **266**:138–141.

Durband, A. (2003) A re-examination of purported "Meganthropus" cranial remains. *AJPA Supplement* **36**:91.

E

Eckhardt, R., K. Galik, and A. Kuperavage (2005) Questions about *Orrorin* femur: Response. *Science* **307**:845.

Eckhardt, R., A. Kuperavage, D. Frayer, and M. Henneberg (2007) More than meets the eye: LB1, the transforming hominin. *AJPA Supplement* **44**:104.

Egeland, C. P., M. Domínguez-Rodrigo, and R. Barba (2007) Geological and paleoecological overview of Olduvai Gorge. In: M. Domínguez-Rodrigo, R. Barba, and C. P. Egeland, Editors. *Deconstructing Olduvai: A Taphonomic Study of the Bed I Sites*, Springer, New York, pp. 33–38.

Eizirik, E., W. Murphy, M. Springer, and S. O'Brien (2004) Molecular phylogeny and dating of early primate divergences. In: Ross and Kay 2004: 45–64.

Eldredge, N. (1989) *Macroevolutionary Dynamics: Species, Niches and Adaptative Peaks*. New York: McGraw-Hill.

Eldredge, N. (1993) What, if anything, is a species? In: Kimbel and Martin 1993: 3–20.

Eldredge, N., and S. Gould (1972) Punctuated equilibrium: An alternative to phyletic gradualism. In: T. Schopf, Editor. *Models in Paleobiology*. San Fransisco: Freeman. 82–115.

Eldredge, N., and I. Tattersall (1982) *The Myths of Human Evolution*. New York: Columbia University Press.

Elftman, H., and J. Manter (1935) The evolution of the human foot, with especial reference to the joints. *Journal of Anatomy* **70**:56–67.

Ellenberger, W., H. Baum, and H. Dittrich (1956) *An Atlas of Animal Anatomy for Artists*. New York: Dover.

Elliot, D. (1913) *A Review of the Primates*, Volume 3. New York: American Museum of Natural History.

Elliot, M., and B. Crespi (2008) Placental invasiveness and brain-body allometry in eutherian mammals. *Journal of Evolutionary Biology* **21**:1763–1778.

Elson, J., D. Turnbull, and N. Howell (2004) Comparative genomics and the evolution of human mitochondrial DNA: Assessing the effects of selection. *AJHG* **74**:229–238.

Enard, W., M. Przeworski, S. Fisher, C. Lai, V. Wiebe, T. Kitano, A. Monaco, and S. Pääbo (2002) Molecular evolution of FOXP2, a gene involved in speech and language. *Nature* **418**:869–872.

Endo, B., and T. Kimura (1970) Postcranial skeleton of the Amud man. In: Suzuki and Takai 1970: 231–409.

Engel, M. (2000) A new interpretation of the oldest fossil bee (Hymenoptera: Apidae). *Novitates* **3296**:1–11.

Enlow, D. (1975) *Handbook of Facial Growth.* Philadelphia: W.B. Saunders.

Ennouchi, E. (1963) Les néandertaliens du Jebel Irhoud (Maroc). *CRAS* **256**:2459–2460.

Ennouchi, E. (1968) Le deuxième crâne de l'homme d'Irhoud. *Annales de Paléontologie* **54**:117–128.

Ennouchi, E. (1969) Présence d'un enfant néandertalien au Jebel Irhoud (Maroc). *Annales de Paléontologie* **55**:251–265.

Erikson, G. (1963) Brachiation in the New World monkeys. *Symposia of the Zoological Society of London* **10**:135–164.

Etkin, W. (1954) Social behavior and the evolution of man's mental faculties. *American Naturalist* **88**:129–142.

Etler, D. (1996) The fossil evidence for human evolution in Asia. *ARA* **25**:275–301.

Etler, D., and T. Li (1994) New archaic human fossil discoveries in China and their bearing on hominid species definition during the Middle Pleistocene. In: Corruccini and Ciochon 1994: 639–675.

Evans, P., N. Mekel-Bobrov, E. Vallender, R. Hudson, and B. Lahn (2006) Evidence that the adaptive allele of the brain size gene microcephalin introgressed into *Homo sapiens* from an archaic *Homo* lineage. *PNAS* **103**:18178–18183.

Eveleth, P., and J. Tanner (1976) *Worldwide Variation in Human Growth.* Cambridge: Cambridge University Press.

Evin, J. (1990) Validity of the radiocarbon dates beyond 35,000 years B.P. *Paleogeography, Paleoclimatology, Paleoecology* **80**:71–78.

Excoffier, L., and A. Langaney (1989) Origin and differentiation of human mitochondrial DNA. *AJHG* **44**:73–85.

F

Falguères, C., J. Bahain, Y. Yokoyama, J. Arsuaga, J. Bermúdez de Castro, E. Carbonell, J. Bischoff, and J. Dolo (1999) Earliest humans in Europe: The age of TD6 Gran Dolina, Atapuerca, Spain. In: Bermúdez de Castro et al. 1999a:343–352.

Falk, D. (1975) Comparative anatomy of the larynx in man and the chimpanzee: Implications for language in Neanderthal. *AJPA* **43**:123–132.

Falk, D. (1983) Cerebral cortices of East African early hominids. *Science* **221**:1072–1074.

Falk, D. (1986) Evolution of cranial blood drainage in hominids: Enlarged occipital/marginal sinuses and emissary foramina. *AJPA* **70**:311–324.

Falk, D. (1988) Enlarged occipital/marginal sinuses and emissary foramina: Their significance in hominid evolution. In: Grine 1988: 85–96.

Falk, D., and G. Conroy (1983) The cranial venous sinus system in *Australopithecus afarensis. Nature* **306**:779–781.

Falk, D., C. Hildebolt, K. Smith, M. Morwood, T. Sutikna, P. Brown, Jatmiko, E. Saptomo, B. Brunsden, and F. Prior (2005) The brain of *Homo floresiensis. Science* **308**:242–245.

Falk, D., C. Hildebolt, K. Smith, M. Morwood, T. Sutikna, P. Brown, Jatmiko, E. Saptomo, B. Brunsden, and F. Prior (2006) Response to comment on "The brain of *Homo floresiensis.*" *Science* **312**:999.

Falk, D., C. Hildebolt, K. Smith, M. Morwood, T. Sutikna, J. Jatmiko, E. Saptomo, H. Imhof, H. Seidler, and F. Prior (2007) Brain shape in human microcephalics and *Homo floresiensis. PNAS* **104**:2513–2518.

Falk, D., C. Hildeboldt, K. Smith, M. Morewood, T. Sutikna, Jatmiko, E. Saptomo, H. Imhoff, H. Seidler, B. Brunsden, and F. Prior (2007) LB1 is not a microcephalic. *AJPA Supplement* **44**:106–107.

Farizy, C., F. David, and J. Jaubert (1994) *Hommes et Bisons du Paléolithique Moyen à Mauran (Haute-Garonne).* Paris: Editions du CNRS.

Feathers, J. (2002) Luminescence dating in less than ideal conditions: Case studies from Klasies River main site and Duinefontein, South African. *JAS* **29**:177–194.

Feibel, C. (1995) Geological context and the ecology of *Homo erectus* in Africa. In: J. Bower and S. Sartono, Editors. *Evolution and Ecology of Homo erectus.* Leiden: Pithecanthropus Centennial Foundation, pp. 67–74.

Feibel, C., F. Brown, and I. MacDougall (1989) Stratigraphic context of the fossil hominids from the Omo Group deposits: Northern Turkana Basin, Kenya and Ethiopia. *AJPA* **78**:595–622.

Feldesman, M., J. Klecker, and J. Lundy (1990) Femur/stature ratio and estimates of stature in Mid- and Late Pleistocene hominids. *AJPA* **83**:359–372.

Feldesman, M., and J. Lundy (1988) Stature estimates for some African Plio-Pleistocene fossil hominids. *JHE* **17**:583–596.

Felsenstein, J. (1978) The number of evolutionary trees. *Systematic Zoology* **27**:27–33.

Fenton, C., and M. Fenton (1952) *Giants of Geology.* Garden City, NY: Doubleday.

Ferembach, D. (1976a) Les restes humains de la grotte de Dar-es-Soultane 2 (Maroc). *BMSAP* **3**:183–193.

Ferembach, D. (1976b) Les restes humains de Temara (Campagne 1975). *BMSAP* **3**:175–180.

Ferembach, D. (1997) Geoffroy Saint-Hilaire. In: F. Spencer, editor. *History of Physical Anthropology: An Encyclopedia*, Volume 1. New York: Garland Publishing, pp. 421–423.

Fiedel, S. (1992) *Prehistory of the Americas*, 2nd edition. Cambridge: Cambridge Univiversity Press.

Fiedler, L. (1996) Die Hornsteinartefakte von Mauer. In: Beinhauer et al. 1996: 155–159.

Finlayson, C. (2004) *Neanderthals and Modern Humans. An Ecological and Evolutionary Perspective*. Cambridge: University of Cambridge Press.

Fiscella, G., and F. Smith (2006) Ontogenetic study of the supraorbital region in modern humans: A longitudinal test of the spatial model. *AZ* **64**:147–160.

FitzGerald, C. (1998) Do enamel microstructures have regular time dependency? Conclusions from the literature and a large-scale study. *JHE* **35**:371–386.

Fleagle, J. (1999) *Primate Adaptation and Evolution*, 2nd edition. New York: Academic Press.

Fleagle, J., Z. Assefa, F. Brown, I. MacDougall, and S. Yirga (2002) Paleoanthropology of the Kibish Formation, Ethiopia. *AJPA Supplement* **34**:70–71.

Fleagle, J., D. Rasmussen, S. Yirga, T. Bown, and F. Grine (1991) New hominid fossils from Fejej, southern Ethiopia. *JHE* **21**:145–152.

Fleagle, J., J. T. Stern, Jr., W. Jungers, R. Susman, A. Vangor, and J. Wells (1981) Climbing: A biomechanical link with brachiation and with bipedalism. *Symposia of the Zoological Society of London* **48**:359–375.

Fleagle, J., and M. Tejedor (2002) Early platyrrhines of southern South America. In: Hartwig 2002: 161–173.

Flint, R. (1971) *Glacial and Quaternary Geology*. New York: Wiley.

Fogarty, M., and F. Smith (1987) Late Pleistocene climatic reconstruction in North Africa and the emergence of modern Europeans. *HE* **2**:311–319.

Foley, R. (1991) How many species of hominid should there be? *JHE* **20**:413–427.

Foley, R. (2001) The evolutionary consequences of increased carnivory in hominids. In: C. Stanford and H. Bunn, Editors. *Meat-Eating and Human Evolution*. New York: Oxford University Press. 305–331.

Foley, R., and M. Lahr (1997) Mode 3 technologies and the evolution of modern humans. *Cambridge Archaeological Journal* **7**:3–36.

Folk, E., and H. Semken (1991) The evolution of sweat glands. *International Journal of Biometeorology* **35**:180–186.

Fox, D., D. Fisher, and L. Leighton (1999) Reconstructing phylogeny with and without temporal data. *Science* **284**: 1816–1819.

Fox, R., G. Youzwyshyn, and D. Krause (1992) Post-Jurassic mammal-like reptile from the Paleocene. *Nature* **358**: 233–235.

Fraipont, C. (1936) Les hommes fossiles d'Engis. *AIPH* **16**: 1–52.

Fraipont, J., and M. Lohest (1886) La race humaine de Néanderthal ou de Cannstadt, en Belgique: Recherches ethnographiques sur les ossements humaines découverts dans les dépôts quaternaires d'une grotte à Spy et détermination de leur âge géologique. Note préliminaire. *Bulletin de l'Académie royal de Belgique* **12**(3ᵉ ser.):741–784.

Fraipont, J., and M. Lohest (1887) La race humaine de Néanderthal ou de Cannstatdt en Belgique: Recherches ethnographiques sur les ossements humaines découverts dans les dépôts quaternaires d'une grotte à Spy et détermination de leur âge géologique. *Archives de Biologie* **7**:587–757.

Franciscus, R. (1999) Neandertal nasal structures and upper respiratory tract "specialization." *PNAS* **96**:1805–1809.

Franciscus, R. (2003) Internal nasal floor configuration in *Homo* with special reference to the evolution of Neandertal facial form. *JHE* **44**:701–729.

Franciscus, R. (2005) The midfacial morphology. In: Zilhão and Trinkaus 2002: 297–311.

Franciscus, R., and S. Churchill (2002) The costal skeleton of Shanidar 3 and a reappraisal of Neandertal thoracic morphology. *JHE* **42**:303–356.

Franciscus, R., and E. Trinkaus (1988) Nasal morphology and the emergence of *Homo erectus*. *AJPA* **75**:517–527.

Franciscus, R., and E. Trinkaus (1995) Determinants of retromolar space presence in Pleistocene *Homo* mandibles. *JHE* **28**:577–595.

Franciscus, R., and E. Vlček (2006) The cranial remains. In: Trinkaus and Svoboda 2006a: 63–152.

Franciscus, R., E. Vlček, and E. Trinkaus (2006) The mandibular remains. In: Trinkaus and Svoboda 2006a: 156–178.

Frayer, D. (1978) *Evolution of the Dentition in Upper Paleolithic and Mesolithic Europe. University of Kansas Publications in Anthropology*. Number 10.

Frayer, D. (1986) Cranial variation at Mladeč and the relationship between Mousterian and Upper Paleolithic hominids. V. Novotný, Editor. *Fossil Man: New Facts—New Ideas. Anthropos* **23**:243–256.

Frayer, D. (1992a) The presence of Neandertal features in post-Neandertal Europeans. In: Bräuer and Smith 1992: 179–188.

Frayer, D. (1992b) Evolution at the European edge: Neanderthal and Upper Paleolithic relationships. *Préhistoire Européen* **2**: 9–69.

Frayer, D. (1997) Perspectives on Neanderthals as ancestors. In: Clark and Willermet 1997: 220–234.

Frayer, D., J. Jelínek, M. Oliva, and M. Wolpoff (2006) Aurignacian male crania, jaws and teeth from the Mladeč caves,

Moravia, Czech Republic. In: Teschler-Nicola 2006: 185–272.

Frayer, D., and M. Wolpoff (1985) Sexual dimorphism. *AAR* **14**:429–473.

Frayer, D., M. Wolpoff, A. Thorne, F. Smith, and G. Pope (1993) Theories of modern human origins: The paleoanthropological test. *AA* **95**:14–50.

Freedman, L., W. Blumer, and M. Lofgren (1991) Endocranial capacity of Western Australian Aboriginal crania: Comparisons and association with stature and latitude. *AJPA* **84**:399–405.

Freedman, L., and M. Lofgren (1979) The Cossack skull and a dihybrid origin of the Australian Aborigines. *Nature* **282**:298–300.

Fridrich, J. (1989) *Prezletice: A Lower Paleolithic Site in Central Bohemia (Excavations 1969–1985)*. Prague: National Museum.

Fryer, G., and T. Iles (1972) *The Cichlid Fishes of the Great Lakes of Africa: Their Biology and Evolution*. Edinburgh: Oliver and Boyd.

Futuyma, D. (1998) *Evolutionary Biology*, 3rd edition. Sunderland, MA: Sinauer Associates.

G

Gabounia, L., M. de Lumley, A. Vekua, D. Lordkipanidze, and H. de Lumley (2002) Découverte d'un nouvel hominidé á Dmanissi (Transcaucasie, Géorgie). *CRE* **1**:243–253.

Gabunia, L. (1992) Die menschliche Unterkiefer von Dmanisi (Georgien, Kaukasus). *Jahrbuch des Romisch-Germanischen Zentral Museums (Mainz)* **39**:185–208.

Gabunia, L., and A. Vekua (1995) A Plio-Pleistocene hominid mandible from Dmanisi, East Georgia, Caucasus. *Nature* **373**:509–512.

Gabunia, L., O. Jöris, A. Justus, D. Lordkipanidze, A. Muschelišvili, M. Nioradze, C. Swisher, and A. Vekus (1999) Neue Hominidenfunde des Altpaläolithischen Fundplatzes Dmanisi (Georgien, Kaukasus) im Kontext aktueller Grabungsergebnisse. *AK* **29**:451–488.

Gabunia, L., A. Vekua, and D. Lordkipanidze (2000a) The environmental contexts of early human occupation of Georgia (Transcaucasia). *JHE* **38**:785–802.

Gabunia, L., A. Vekua, D. Lordkipanidze, C. Swisher, R. Ferring, A. Justus, M. Nioradze, M. Tvalchrelidze, S. Antón, G. Bosinski, O. Jöris, M. de Lumley, G. Majsuradze, and A. Mouskhelishvili (2000b) Earliest Pleistocene hominid cranial remains from Dmanisi, Republic of Georgia: Taxonomy, geological setting, and age. *Science* **288**:1019–1025.

Gabunia, L., S. Antón, D. Lordkipanizde, A. Vekua, A. Justus, and C. Swisher (2001) Dmanisi and dispersal. *EA* **10**:158–170.

Galik, K., B. Sénut, M. Pickford, D. Gommery, J. Treil, A. Kuperavage, and R. Eckhardt (2004) External and internal morphology of the BAR 1002′00 *Orrorin tugenensis* femur. *Science* **305**:1450–1453.

Gallardo, M., J. Bickham, R. Honeycutt, R. Ojeda, and N. Kohler (1999) Discovery of tetraploidy in a mammal. *Nature* **401**:341.

Gambaryan, P. (1974) *How Mammals Run: Anatomical Adaptations*. New York: Wiley.

Gambier, D. (1989) Fossil hominids from the early Upper Palaeolithic (Aurignacian) of France. In: Mellars and Stringer 1989: 194–211.

Gambier, D. (1997) Modern human origins at the beginnings of the Upper Paleolithic in France: Anthropological data and perspectives. In: Clark and Willermet 1997: 117–131.

Gamble, C. (1986) *The Paleolithic Settlement of Europe*. Cambridge: University of Cambridge Press.

Gamble, C. (1999) *The Paleolithic Societies of Europe*. Cambridge: University of Cambridge Press.

Gannon, P., and J. Laitman (1993) Can we see language areas on human brain endocasts? *AJPA Supplement* **16**:91.

Gao, J. (1975) Australopithecine teeth associated with *Gigantopithecus*. *VP* **13**:81–88.

García, N., J. Arsuaga, and T. Torres (1997) The carnivore remains from the Sima de los Huesos Middle Pleistocene site (Sierra de Atapuerca, Spain). In: Arsuaga et al. 1997a: 155–174.

Gardner, J. (2001) An analysis of the pathology of the Krapina Neandertals. *AJPA Supplement* **32**:68.

Garey, M., and D. Johnson (1979) *Computers and Intractability: A Guide to the Theory of NP-Completeness*. San Francisco: Freeman.

Gargett, R. (1989) Grave shortcomings: The evidence for Neanderthal burial. *CA* **30**:157–190.

Gargett, R. (1999) Middle Paleolithic burial is not a dead issue: The view from Qafzeh, Saint-Césaire, Kebara, Amud, and Dederiyeh. *JHE* **37**:27–90.

Garralda, M. (1997) The human paleontology of the Middle to Upper Paleolithic Transition on the Iberian peninsula. In: Clark and Willermet 1997: 148–160.

Gatesy, S., and A. Biewener (1991) Bipedal locomotion: Effects of speed, size and limb posture in birds and humans. *Journal of Zoology (London)* **224**:127–147.

Gaudzinski, S. (1996) On bovid assemblages and their consequences for the knowledge of subsistence patterns in the middle paleolithic. *PPS* **62**:19–39.

Gaudzinski, S., and W. Roebroeks (2000) Adults only. Reindeer hunting at the Middle Paleolithic site Salzgitter Lebenstedt, Northern Germany. *JHE* **38**:497–521.

Gavan, J. (1953) Growth and development of the chimpanzee: A longitudinal and comparative study. *HB* **25**:93–143.

Gebo, D. (1986) Anthropoid origins: The foot evidence. *JHE* **15**:421–430.

Gebo, D. (1992) Plantigrady and foot adaptation in African apes: Implications for hominid origins. *AJPA* **89**:29–58.

Gebo, D. (1996) Climbing, brachiation, and terrestrial quadrupedalism: Historical precursors of hominid bipedalism. *AJPA* **101**:55–92.

Gebo, D. (2002) Adapiformes: Phylogeny and adaptation. In: Hartwig 2002: 21–43.

Gebo, D. (2004) A shrew-sized origin for primates. *YPA* **47**:40–62.

Gebo, D., and M. Dagosto (2004) Anthropoid origins: Postcranial evidence from the Eocene of Asia. In: Ross and Kay 2004: 369–380.

Gebo, D., M. Dagosto, K. Beard, T. Qi, and J. Wang (2000a) The oldest known anthropoid postcranial fossils and the early evolution of higher primates. *Nature* **404**:276–278.

Gebo, D., M. Dagosto, K. Beard, and T. Qi (2000b) The smallest primates. *JHE* **38**:585–594.

Gebo, D., M. Dagosto, K. Beard, and T. Qi (2001) Middle Eocene primate tarsals from China: Implications for haplorhine evolution. *AJPA* **116**:83–107.

Gebo, D., M. Dagosto, K. Beard, and J. Wang (1999) A first metatarsal of *Hoanghonius stehlini* from the Late Middle Eocene of Shanxi Province, China. *JHE* **37**:801–806.

Gebo, D., L. MacLatchy, R. Kityo, A. Deino, J. Kingston, and D. Pilbeam (1997) A new hominoid genus from the Early Miocene of Uganda. *Science* **276**:401–404.

Gebo, D., and G. Schwartz (2006) Foot bones from Omo: Implications for hominind evolution. *AJPA* **129**:499–511.

Gebo, D., E. Simons, D. Rasmussen, and M. Dagosto (1994) Eocene anthropoid postcrania from the Fayum, Egypt. In: Fleagle and Kay 1994: 203–233.

Gee, H. (1996) Box of bones "clinches" identity of Piltdown paleontology hoaxer. *Nature* **381**:261–262.

Geneste, J. (1988) Économie des resources lithiques dans le Moustérien du sud-ouest de la France. *ERAUL* **33**:75–97.

Geoffroy Saint-Hilaire, E. 1830. *Principes de philosophie zoologique discutés en Mars 1830 au sein de l''Académie Royale des Sciences.* Paris: Pichot et Didier.

Geraads, D., and E. Tchernov (1983) Fémurs humains du Pléistocène moyen de Gesher Benot Ya'acov (Israël). *L'Anthropologie* **87**:138–141.

Geraads, D., P. Berrio, and H. Roche (1980) La faune et l'industrie des sites à *Homo erectus* des carrières Thomas (Maroc): Précisions sur l'âge de ces hominidés. *CRAS* D **291**:195–198.

Geraads, D., J. Hublin, J. Jaeger, Haiyang Tong, Sevket Sen, and P. Toubeau (1986) The Pleistocene hominid site of Ternifine, Algeria: New results on the environment, age, and human industries. *QR* **25**:380–386.

Gibbons, A. (1997) A new face for human ancestry. *Science* **276**:1331–1333.

Gibbons, A. (2001) The riddle of coexistence. *Science* **291**: 1725–1729.

Gibbons, A. (2002) Glasnost for hominids: Seeking access to fossils. *Science* **297**:1464–1468.

Gibbons, A. (2004) Oldest human femur wades into controversy. *Science* **305**:1885.

Gibbons, A. (2005) Facelift supports skull's status as oldest member of the human family. *Science* **308**:179–191.

Gilbert, J., F. Ribot, C. Fernandez, B. Martínez, R. Caporicci, and D. Campillo (1989) Anatomical study: Comparison of the cranial fragment from Venta Micena (Orce, Spain) with fossil and extant mammals. *JHE* **4**:283–305.

Gilbert, J., D. Campillo, B. Martínez, F. Sánchez, R. Caporrici, C. Jimenez, C. Fernandez, and F. Ribot (1991) Nouveaux restes d'hominidés dans les gisements d'Orce et de Cueva Victoria, Espagne. In: E. Bonifay and B. Vandermeersch, Editors. *Les premiers Européens.* Paris: Éditions du Comité des Traveaux Historiques et Scientifiques, pp. 273–282.

Gilbert, J., F. Sánchez, A. Malgosa, and B. Martínez Navarro (1994) Découvertes de restes humains dans les gisements d'Orce (Granada, Espagne). *CRAS* **319**:963–968.

Gilbert, J., D. Campillo, J. Argués, E. Garcia-Olivares, C. Borja, and J. Lowenstein (1998) Hominid status of the Orce cranial remains reasserted. *JHE* **34**:203–217.

Gilbert, W., T. White, and B. Asfaw (2003) *Homo erectus, Homo ergaster, Homo "cepranensis"* and the Daka cranium. *JHE* **45**:255–259.

Gillespie, C. (1959) *Genesis and Geology.* New York: Harper, and Row.

Gillespie, R., and R. Roberts (2000) On the reliability of age estimates for human remains at Lake Mungo. *JHE* **38**:727–732.

Gingerich, P. (1977) Correlation of tooth size and body size in living hominoid primates, with a note on relative brain size in *Aegyptopithecus* and *Proconsul*. *AJPA* **47**:395–398.

Gingerich, P. (1980) Eocene Adapidae, paleobiogeography, and the origin of South American Platyrrhini. In: R. Ciochon and A. Chiarelli, Editors. *Evolutionary Biology of the New World Monkeys and Continental Drift.* New York: Plenum, 123–138.

Gingerich, P. (1984) Primate evolution: Evidence from the fossil record, comparative morphology, and molecular biology. *YPA* **27**:57–72.

Gingerich, P. (1985) Species in the fossil record: Concepts, trends, and transitions. *Paleobiology* **11**:27–41.

Gingerich, P. (1990) African dawn for primates. *Nature* **346**:411.

Gingerich, P., D. Dashzeveg, and D. Russell (1991) Dentition and systematic relationships of *Altanius orlovi* (Mammalia, Primates) from the early Eocene of Mongolia. *Geobios* **24**:637–646.

Gingerich, P., M. ul Haq, I. Zalmout, I. Khan, and M. Malkani (2001) Origin of whales from early artiodactyls: Hands and feet of Eocene Protocetidae from Pakistan. *Science* **293**: 2239–2242.

Gingerich, P., and M. Uhen (1994) Time of origin of primates. *JHE* **27**:443–445.

Girondot, M., and J. Garcia (1998) Senescence and longevity in turtles: What telomeres tell us. In: D. Miaud and R. Guyetant, Editors. *Current Studies in Herpetology.* Le Bouget du Lac, France: Societa Europaea Herpetologica, pp. 133–137.

Gisis, I., and O. Bar-Yosef (1974) New excavations in Zuttiyeh Cave. *Paléorient* **2**:175–180.

Glantz, M., and T. Ritzman (2004) A reanalysis of the Neandertal status of the Teshik–Tash child. *AJPA Supplement* **38**:100–101.

Glanville, E. (1969) Nasal shape, prognathism and adaptation in man. *AJPA* **30**:29–37.

Glen, E., and K. Kaczanowski (1982) Human remains. In: J. Kozlowski, Editor. *Excavations in the Bacho Kiro Cave (Bulgaria): Final Report.* Warsaw: Panstwowe Wydawnictwo Naukowe, pp. 75–79.

Godfrey, L., and W. Jungers (2002) Quaternary fossil lemurs. In: Hartwig 2002: 97–121.

Godfrey, L., and W. Jungers (2003) The extinct sloth lemurs of Madagascar. *EA* **12**:252–263.

Godfrey, L., and J. Marks (1991) The nature and origin of primate species. *YPA* **34**:39–68.

Godinot, M. (1992) Apport à la systématique de quatre genres d'Adapiformes (primates, Éocène). *CRAS* **314**:237–242.

Godinot, M. (1994) Early North African primates and their significance for the origin of Simiiformes (= Anthropoidea). In: Fleagle and Kay 1994: 235–295.

Gohau, G. (1990) *A History of Geology.* New Brunswick: Rutgers University Press.

Goldberg, P., S. Weiner, O. Bar Yosef, Q. Xu, and J. Liu (2001) Site formation processes at Zhoukoudian, China. *JHE* **41**:483–530.

Goldizen, A. (1986) Tamarins and marmosets: Communal care of offspring. In: B. Smuts, D. Cheney, R. Seyfarth, R. Wrangham and T. Struhsaker, Editors. *Primate Societies.* Chicago: University of Chicago Press, pp. 34–43.

Goldschmidt, R. (1940) *The Material Basis of Evolution.* New Haven: Yale University Press.

Golovanova, L., J. Hoffecker, V. Kharitonov, and G. Romanova (1999) Mezmaiskaya Cave: A Neanderthal occupation in the northern Caucasus. *CA* **40**:77–86.

Gommery, D., B. Sénut, and A. Keyser (2002) Description d'un bassin fragmentaire de *Paranthropus robustus* du site Plio-Pléistocène de Drimolin (Afrique du Sud). *Geobios* **35**:265–281.

Goodall, J. (1963) Feeding behaviour of wild chimpanzees: A preliminary report. *Symposia of the Zoological Society of London* **10**:39–47.

Goodall, J. (1986) *The Chimpanzees of Gombe: Patterns of Behavior.* Cambridge: Harvard University Press.

Goodall, J., A. Bandora, E. Bergmann, C. Busse, H. Matama, E. Mpongo, A. Pierce and D. Riss (1979) Intercommunity interactions in the chimpanzee population of the Gombe National Park. In: D. Hamburg and E. McCown, Editors. *The Great Apes.* Menlo Park: Benjamin/Cummings.

Gore, R. (2002) The first pioneer? *NG* **202**:xxxii–1.

Gorjanović-Kramberger, D. (1902) Der paläolithische Mensch und seine Zeitgenossen aus dem Diluvium von Krapina in Kroatien, III. *MAGW* **32**:189–216.

Gorjanović-Kramberger, D. (1905) Der paläolithische Mench und seine Zeitgenossen aus dem Diluvium von Krapina in Kroatien, IV. *MAGW* **35**:197–229.

Gorjanović-Kramberger, D. (1906) *Der diluviale Mensch von Krapina in Kroatien. Ein Beitrag zur Paläoanthropologie.* Wiesbaden: Kreidel.

Gorjanović-Kramberger, D. (1907) Die Kronen und Wurzeln der Mahlzähne des *Homo primigenius* und ihre genetische Bedeutung. *AZ* **31**:97–138.

Gorjanović-Kramberger, D. (1908) Der prizmatische Molarwurzeln rezenter und diluvialer Menschen. *AZ* **32**:401–413.

Gorjanović-Kramberger, D. (1913) *Život i kultura dilijalnoga čovkeka iz Krapine u Hrvatskoj.* Zagreb: Djela Jugoslavenske akademije znanosti i umjetnosti.

Gorjanović-Kramberger, D. (1914) Der Axillarand des Schulterblattes des Menschen von Krapina. *Glasnik Hrvatskog prirodoslovnog Društva* **26**:231–257.

Gould, S. J. (1982a) The meaning of punctuated equilibrium and its role in validating a hierarchical approach to macroevolution. In: R. Milkman, Editor. *Perspectives on Evolution.* Sunderland, MA: Sinauer Associates, pp. 83–104.

Gould, S. J. (1982b) Darwinism and the expansion of evolutionary theory. *Science* **216**:380–387.

Gould, S. J. (1986) Evolution and the triumph of homology, or why history matters. *AS* **74**:60–69.

Gould, S. J. (1987) Bushes all the way down. *NH* **96**:12–19.

Gould, S. J. (1988) A novel notion of Neanderthal. *NH* **97**:16–21.

Gould, S. J. (1989) *Wonderful Life.* New York: Norton.

Gould, S. J. (1990) Men of the thirty-third division. *NH* **99**:12–24.

Gould, S. J., and N. Eldredge (1977) Punctuated equilibria: The tempo and mode of evolution reconsidered. *Paleobiology* **3**:115–151.

Gould, S. J., and R. Lewontin (1979) The spandrels of San Marco and the Panglossian paradigm: A critique of the adaptationist programme. *Proceedings of the Royal Society of London, Series B* **205**:581–598.

Gower, C. (1923) A contribution to the morphology of the *apertura piriformis. AJPA* **6**:27–36.

Gowlett, J., J. Harris, D. Walton, and B. Wood (1981) Early archaeological sites, further hominid remains and traces of fire from Chesowanja, Kenya. *Nature* **294**:125–129.

Gradstein, F., J. Ogg, and A. Smith, Editors. (2005) *Geologic Time Scale 2004*. Cambridge: Cambridge University Press.

Graham, J., N. Aguilar, R. Dudley, and C. Gans (1995) Implications of the late Paleozoic oxygen pulse for physiology and evolution. *Nature* **375**:117–120.

Graham, J., N. Aguilar, R. Dudley, and C. Gans (1997) The late Paleozoic atmosphere and the ecological and evolutionary physiology of tetrapods. In: S. Sumida and K. Martin, Editors. *Amniote Origins: Completing the Transition to Land*. San Diego: Academic Press, pp. 141–167.

Grant, J. C. B. (1972) *An Atlas of Anatomy*, 6th edition. Baltimore: Williams and Wilkins.

Grant, P. (1986) *Ecology and Evolution of Darwin's Finches*. Princeton: Princeton University Press.

Grant, P., and B. Grant (1992) Hybridization of bird species. *Science* **256**:193–197.

Grausz, H., R. Leakey, A. Walker, and C. Ward (1988) Associated cranial and postcranial bones of *Australopithecus boisei*. In: Grine 1988: 127–132.

Gravina, B., P. Mellars, and C. Bronk Ramsey (2005). Radiocarbon dating of interstratified Neanderthal and early modern human occupations at the Chatelperronian type-site. *Nature* **438**:51–56.

Gray, J. (1959) *How Animals Move*. Harmondsworth: Pelican Books.

Grayson, D., and F. Delpech (2006) Was there increasing dietary specialization across the Middle-to-Upper Paleolithic transition in France? In: Conard 2006: 377–417.

Green, D., A. Gordon, and B. Richmond (2007) Limb-size proportions in *Australopithecus afarensis* and *Australopithecus africanus*. *JHE* **52**:187–200.

Green, M. (1990) *Neandertal craniofacial growth: An ontogenetic model*. M.A. thesis. Knoxville: University of Tennessee.

Green, M., and F. Smith (1991) Heterochrony and Neandertal craniofacial morphogenesis. *AJPA Supplement* **12**:82.

Green, R., J. Krause, S. Ptak, A. Briggs, M. Ronan, Lei Du, M. Egholm, J. Rothberg, M. Paunovic, and S. Pääbo (2006) Analysis of one million base pairs of Neanderthal DNA. *Nature* **444**:330–336.

Greenfield, L. (1979) On the adaptive pattern of "Ramapithecus." *AJPA* **50**:527–548.

Greenfield, L. (1990) Canine reduction in early man: A critique of three mechanical models. *HE* **5**:213–226.

Grehan, J. (2006) Mona Lisa smile: The morphological enigma of human and great ape evolution. *AR* **298B**:139–157.

Griffin, J. B., D. Meltzer, B. Smith, and W. Sturtevant (1988) A mammoth fraud in science. *American Antiquity* **53**:578–582.

Griffin, T., R. Main, and C. Farley (2004) Biomechanics of quadrupedal walking: How do four-legged animals achieve inverted pendulum-like movements? *Journal of Experimental Biology* **207**:3545–3558.

Grimaldi, D. (1999) The co-radiations of pollinating insects and angiosperms in the Cretaceous. *Annals of the Missouri Botanical Garden* **86**:373–406.

Grine, F. (1984) Comparison of the deciduous dentitions of African and Asian hominids. *CFS* **69**:69–82.

Grine, F. (1985) Australopithecine evolution: The deciduous dental evidence. In: Delson 1985: 153–167.

Grine, F. (1986) Dental evidence for dietary differences in *Australopithecus*: A quantitative analysis of permanent molar microwear. *JHE* **15**:783–822.

Grine, F. (1987) Quantitative analysis of occlusal microwear in *Australopithecus* and *Paranthropus*. *Scanning Microscopy* **1**: 647–656.

Grine, F. (1989) New hominid fossils from the Swartkrans Formation (1979–1986 excavations): Craniodental specimens. *AJPA* **79**:409–449.

Grine, F. (2000) Swartkrans. In: Delson et al. 2000: 681–682.

Grine, F., B. Demes, W. Jungers, and T. Cole (1993) Taxonomic affinity of the early *Homo* cranium from Swartkrans. *AJPA* **92**:411–426.

Grine, F., W. Jungers, P. Tobias, and O. Pearson (1995) Fossil *Homo* femur from Berg Aukas, northern Namibia. *AJPA* **97**:151–185.

Grine, F., and R. Klein (1985) Pleistocene and Holocene human remains from Equus Cave, South Africa. *Anthropology* **8**: 55–98.

Grine, F., and R. Klein (1993) Late Pleistocene human remains from the Sea Harvest site, Saldanha Bay, South Africa. *SAJS* **89**:145–152.

Grine, F., R. Klein, and T. Volman (1991) Dating, archaeology, and human fossil remains from the Middle Stone Age levels of Die Kelders, South Africa. *JHE* **21**:363–395.

Grine, F. E., and Martin, L. B. 1988. Enamel thickness and development in *Australopithecus* and *Paranthropus*. In: Grine 1988: 3–42.

Grine, F., O. Pearson, R. Klein, and P. Rightmire (1998) Additional human remains from Klasies River Mouth, South Africa. *JHE* **35**:95–107.

Grine, F., P. Ungar, M. Teaford, and S. El-Zaatari (2006) Molar microwear in *Praeanthropus afarensis*: Evidence for dietary stasis through time and under diverse paleoecological conditions. *JHE* **51**:297–319.

Groves, C. (1989) *A Theory of Primate and Human Evolution*. Oxford: Clarendon Press.

Groves, C. (2001) Lake Mungo 3 and his DNA. *AO* **36**: 166–167.

Groves, C., and V. Mazák (1975) An approach to the taxonomy for the Hominidae: Gracile Villafranchian hominids of Africa. *Casopis pro Mineralogii A Geologii* **20**:225–246.

Grün, R. (1996) A re-analysis of electron spin resonance dating results associated with the Petralona hominid. *JHE* **30**: 227–241.

Grün, R. (2006) Direct dating of human fossils. *YPA* **49**:2–48.

Grün, R., P. Beaumont, P. Tobias, and S. Eggins (2003) On the age of Border Cave 5 human mandible. *JHE* **45**:155–167.

Grün, R., J. Brink, N. Spooner, L. Taylor, C. Stringer, R. Franciscus, and A. Murray (1996) Direct dating of Florisbad hominid. *Nature* **382**:500–501.

Grün, R., P. Huang, W. Huang, F. McDermott, A. Thorne, C. Stringer, and G. Yan (1998) ESR and U-series analysis of teeth from the paleoanthropological site of Hexian, Anhui Province, China. *JHE* **34**:555–564.

Grün, R., P. Huang, X. Wu, C. Stringer, A. Rhorne, and M. McCulloch (1997) ESR analysis of teeth from the paleoanthropological site of Zhoukoudian, China. *JHE* **32**:83–91.

Grün, R., J. Maroto, S. Eggins, C. Stringer, S. Robertson, L. Taylor, G. Mortimer, and M. McCulloch (2006) ESR and U-series analyses of enamel and dentine fragments of the Banyoles mandible. *JHE* **50**:347–358.

Grün, R., H. Schwarcz, and C. Stringer (1991) ESR dating of teeth from Garrod's Tabun Cave collections. *JHE* **20**:231–248.

Grün, R., N. Spooner, A. Thorne, G. Mortimer, J. Simpson, M. McCulloch, A. Taylor, and D. Curnoe (2000) Age of the Lake Mungo 2 skeleton: Reply to Bowler and Magee and to Gillespie and Roberts. *JHE* **38**:733–741.

Grün, R., and C. Stringer (1991) Electron spin resonance dating and the evolution of modern humans. *Archaeometry* **33**:153–199.

Grün, R., and C. Stringer (2000) Tabun revisited: Revised ESR chronology and new ESR and U-series analyses of dental material from Tabun C1. *JHE* **39**:601–612.

Grün, R., and A. Thorne (1997) Dating the Ngandong humans. *Science* **276**:1575–1576.

Guatelli-Steinberg, D., and J. Irish (2005) Brief communication: Early hominin variation in first molar dental trait frequencies. *AJPA* **128**:477–484.

Guatelli-Steinberg, D., D. Reid, T. Bishop, and C. Larsen (2005) Anterior growth periods in Neandertals were comparable to those of modern humans. *PNAS* **102**:14197–14202.

Guatelli-Steinberg, D., D. Reid, and T. Bishop (2006) Did the lateral enamel of Neandertal anterior teeth grow differently from modern humans? *JHE* **52**:72–84.

Guipert, G., B. Mafart, A. Tuffreau, and M.-A. de Lumley (2007) 3D reconstruction and study of a new late Middle Pleistocene hominid: Biache-Saint-Vaast 2, Nord, France. *AJPA Supplement* **44**:121–122.

Gunnell, G., and W. Miller (2001) Origin of Anthropoidea: dental evidence and recognition of early anthropoids in the fossil record, with comments on the Asian anthropoid radiation. *AJPA* **114**:177–191

Gunnell, G., and K. Rose (2002) Tarsiiformes: Evolutionary history and adaptation. In: Hartwig 2002: 45–82.

Gunz, P., and K. Harvati (2006) The Neanderthal "chignon": Variation, integration and homology. *JHE* **52**:262–274.

Gutierréz, G., D. Sanchez, and A. Marin (2002) A reanalysis of the ancient mitochondrial DNA sequences recovered from Neandertal bones. *MBE* **19**:1359–1366.

H

Habgood, P. (1989) The origin of anatomically modern humans in Australia. In: Mellars and Stringer 1989: 245–273.

Habgood, P. (1992) The origin of anatomically modern humans in East Asia. In: Bräuer and Smith 1992: 273–288.

Habgood, P. (2003) A morphometric investigation of the origin(s) of anatomically modern humans. *BARIS* **1176**:1–313.

Haeckel, E. (1866) *Generelle Morphologie der Organismen*, two volumes. Berlin: Reimer.

Haeckel, E. (1868) *Natürliche Schöpfungsgeschichte*. Berlin: Reimer.

Hahn, J. (1993) Aurignacian art in Central Europe. In: H. Knecht, A. Pike-Tay, and R. White, Editors. *Before Lascaux: The Complex Record of the Early Upper Paleolithic*. Boca Raton: CRC Press, pp. 229–241.

Haig, D. (1993) Genetic conflicts in human pregnancy. *Quarterly Review of Biology* **68**:495–531.

Haile-Selassie, Y. (2001) Late Miocene hominids from the Middle Awash, Ethiopia. *Nature* **412**:178–181.

Haile-Selassie, Y., B. Asfaw, and T. White (2004a) Hominind cranial remains from Upper Pleistocene deposits at Aduma, Middle Awash, Ethiopia. *AJPA* **123**:1–10.

Haile-Selassie, Y., G. Suwa, and T. White (2004b) Late Miocene teeth from Middle Awash, Ethiopia, and early hominid dental evolution. *Science* **303**:1503–1505.

Haileab, B., and F. Brown (1992) Turkana Basin–Middle Awash correlations of the Sangabtole and Hadar formations. *JHE* **22**:453–468.

Hallam, T. (2004) *Catastrophes And Lesser Calamities: The Causes of Mass Extinctions*. Oxford: Oxford University Press.

Hammer, M., T. Karafet, A. Rasanayangam, E. Wood, T. Altheide, T. Jenkins, R. Griffiths, A. Templeton, and S. Zegura (1998) Out of Africa and back again: Nested cladistic analysis of human Y chromosome variation. *MBE* **15**:427–441.

Hammond, M. (1982) The expulsion of the Neandertals from human ancestry. Marcelin Boule and the social context of scientific research. *Social Studies of Science* **12**:1–36.

Hamrick, M., and S. Inouye (1995) Thumbs, tools and early humans. Technical Comment. *Science* **268**:586–587.

Han, T., and B. Runnegar (1992) Megascopic eukaryotic algae from the 2.1-billion-year-old Negaunee iron-formation, Michigan. *Science* **257**:232–235.

Hanson, E. (1933) Social regression in the Orinoco and Amazon Basins: Notes on a journey in 1931 and 1932. *Geographical Review* **23**:578–598.

Harada, M., H. Schneider, M. Schneider, I. Sampaio, J. Czelusniak, J., and M. Goodman (1995) DNA evidence on the phylogenetic systematics of New World monkeys: Support for the sister-grouping of *Cebus* and *Saimiri* from two unlinked nuclear genes. *MPE* **4**:331–349.

Harcourt-Smith, W., and L. Aiello (2004) Fossils, feet and the evolution of human bipedal locomotion. *JA* **204**:403–416.

Harding, R. (1997) Lines of descent from mitochondrial Eve: An evolutionary look at coalescence. In: P. Donnelly and S. Tavaré, Editors. *Progress in Population Genetics and Human Evolution*. New York: Springer, pp. 15–31.

Harding, R., S. Fullerton, R. Griffiths, J. Bond, M. Cox, J. Schneider, D. Moulin, and J. Clegg (1997) Archaic African and Asian lineages in the genetic ancestry of modern humans. *AJHG* **60**:772–789.

Hardy, B. (2004) Neanderthal behavior and stone tool function at the Middle Paleolithic site of La Quina, France. *Antiquity* **78**:547–565.

Harley, D. (1982) Models of human evolution. *Science* **217**:296.

Harmon, E. (2006) Size and shape variation in *Australopithecus afarensis* proximal femora. *JHE* **51**:217–227.

Harpending, H., M. Batzer, M. Gurven, L. Jorde, A. Rogers, and S. Sherry (1998) Genetic traces of ancient demography. *PNAS* **95**:1961–1967.

Harpending, H., and J. Relethford (1997) Population perspectives on human origins research. In: Clark and Willermet 1997: 361–368.

Harris, A., and J. Hey (1999) X chromosome evidence for ancient human histories. *PNAS* **96**:3320–3324.

Harris, J., and S. Capaldo (1993) The earliest stone tools: Their implications for an understanding of the activities and behaviour of late Pliocene hominids. In: A. Berthelet and J. Chavaillon, Editors. *The Use of Tools by Human and Non-human Primates*. Oxford: Clarendon Press. 196–220.

Harrison, R. 1993. *Hybrid Zones and the Evolutionary Process*. New York: Oxford University Press.

Harrison, R., and W. Montagna (1969) *Man*. New York: Appleton-Century-Crofts.

Harrison, T. (1986) New fossil anthropoids from the middle Miocene of East Africa and their bearing on the origin of the Oreopithecidae. *AJPA* **71**:265–284.

Harrison, T. (1987) The phylogenetic relationships of the early catarrhine primates: A review of the current evidence. *JHE* **16**:41–80.

Harrison, T. (1989) New estimates of cranial capacity, body size, and encephalization in *Oreopithecus bambolii*. *AJPA* **78**: 237.

Harrison, T. (2002) Late Oligocene to middle Miocene catarrhines from Afro-Arabia. In: Hartwig 2002: 311–338.

Harrison, T. (2005) The zoogeographic and phylogenetic relationships of early catarrhine primates in Asia. *Anthropological Science* **113**:43–51.

Harrison, T., and L. Rook (1997) Enigmatic anthropoid or misunderstood ape? The phylogenetic status of *Oreopithecus bambolii* reconsidered. In: Begun et al. 1997: 327–362.

Harrison, T. (1957) The Great Cave of Niah. *Man* **57**:161–166.

Harrison, T. (1959) New archaeological and ethnological results from Niah Cave, Sarawak. *Man* **59**:1–8.

Harrold, F. (1980) A comparative analysis of Eurasian Paleolithic burials. *WA* **12**:195–211.

Hartwig, W., and D. Meldrum (2002) Miocene platyrrhines of the northern Neotropics. In: Hartwig 2002: 175–188.

Harvati, K. (2003) Quantitative analysis of Neanderthal temporal bone morphology using three-dimensional geometric morphometrics. *AJPA* **120**:323–338.

Harvati, K., S. Frost, and K. McNulty (2004) Neanderthal taxonomy reconsidered: Implications of 3D primate models of intra- and interspecific differences. *PNAS* **101**:1147–1152.

Harvati, K., E. Panagopoulou, and K. Krakanas (2003) First Neanderthal remains from Greece: The evidence from Lakonis. *JHE* **45**:465–473.

Harvati, K., and T. Weaver (2006a) Human cranial anatomy and the differential preservation of population history and climate signatures. *AR* **288A**:1225–1233.

Harvati, K., and T. Weaver (2006b) Reliability of cranial morphology in reconstructing Neanderthal phylogeny. In: Harvati and Harrison 2006: 239–254.

Hashimoto, C., Y. Tashiro, D. Kimura, T. Enomoto, E. Ingmanson, G. Idani, and T. Furuichi (1998) Habitat use and ranging of wild bonobos (*Pan paniscus*) at Wamba. *IJP* **19**:1045–1060.

Hassan, F. (1981) *Demographic Archaeology*. New York: Academic.

Hatley, T., and J. Kappelman (1980) Pigs, bears, and Plio-Pleistocene hominids: A case for the exploitation of below-ground food resources. *Human Ecology* **8**:371–387.

Hauser, M., N. Chomsky, and W. Fitch (2002) The faculty of language: What is it, who has it, and how did it evolve? *Science* **298**:1569–1580.

Häusler, M. (2002) New insights into the locomotion of *Australopithecus africanus* based on the pelvis. *EA* **11**(Suppl. 1):53–57.

Häusler, M., S. Martelli, and T. Boeni (2002) Vertebrae numbers of the early hominid lumbar spine. *JHE* **43**:621–643.

Häusler, M., and H. McHenry (2004) Body proportions of *Homo habilis* reviewed. *JHE* **46**:433–465.

Häusler, M., and P. Schmid (1995) Comparison of the pelves of Sts 14 and AL 288-1: Implications for birth and sexual dimorphism in australopithecines. *JHE* **29**:363–383.

Hawkes, K., J. O'Connell, and N. Blurton Jones (1997) Hadza women's time allocation, offspring provisioning, and the evolution of post-menopausal lifespans. *CA* **38**:551–578.

Hawkes, K., J. O'Connell, N. Blurton Jones, H. Alvarez, and E. Charnov (1998) Grandmothering, menopause and the evolution of human life histories. *PNAS* **95**:1336–1339.

Hawks, J. (2004). How much can cladistics tell us about early hominid relationships? *AJPA* **125**:207–219.

Hawks, J. (2006) Selection on mitochondrial DNA and the Neanderthal problem. In: Harvati and Harrison 2006: 221–238.

Hawks, J., K. Huntley, S.-H. Lee, and M. Wolpoff (2000a) Population bottlenecks and Pleistocene human evolution. *MBE* **17**:2–22.

Hawks, J., S. Oh, K. Hunley, S. Dobson, G. Cabana, P. Dayalu, and M. Wolpoff (2000b) An Australasian test of the recent African origin theory using the WLH-50 calvarium. *JHE* **39**:1–22.

Hawks, J., and M. Wolpoff (2001a) The accretion model of Neandertal evolution. *Evolution* **55**:1474–1485.

Hawks, J., and M. Wolpoff (2001b) Brief communication: Paleoanthropology and the population genetics of ancient genes. *AJPA* **114**:269–272.

Hay, R. (1987) Geology of the Laetoli area. In: M. Leakey, and J. Harris, Editors. *Results of the Laetoli Expeditions 1975–1981*. Oxford: Oxford University Press, pp. 23–47.

Hay, R., and M. Leakey (1982) The fossil footprints of Laetoli. *SA* **246**(2):50–57.

Hayden, B. (1993) The cultural capacities of Neandertals: A review and reevaluation. *JHE* **24**:113–146.

Hedges, R., R. Housley, C. Bronk-Ramsey, and G. Van Klinken (1994) Radiocarbon dates from the Oxford AMS system: Archaeometry datelist 9. *Archaeometry* **36**:337–374.

Hedges, S., S. Kumar, K. Tamura, and M. Stoneking (1992) Human origins and analysis of mitochondrial DNA sequences. *Science* **255**:737–739.

Heesy, C. (2004) On the relationship between orbit orientation and binocular visual field overlap in mammals. *AR* **281A**:1104–1110.

Heesy, C. (2005) Function of the mammalian postorbital bar. *Journal of Morphology* **264**:363–380.

Heesy, C. (2008) Ecomorphology of orbit orientation and the adaptive significance of binocular vision in primates and other mammals. *Brain, Behavior and Evolution* **71**:54–67.

Heesy, C., and C. Ross (2004). Mosaic evolution of activity pattern, diet, and color vision in haplorhine primates. In: Ross and Kay 2004: 665–698.

Heesy, C., C. Ross, and B. Demes (2007) Oculomotor stability and the functions of the postorbital bar and septum. In: Ravosa and Dagosto 2006: 257–284.

Heim, J. (1976) Les hommes fossiles de La Ferrassie I: Le gisement, les squelettes adultes (crâne et squelette du tronc). *AIPH* **35**:1–331.

Heim, J. (1978) Contribution du massif facial a la morphogenèse du crâne Néandertalien. In: *Les origines humaines et les époches de l'intelligence*. Paris: Masson, pp. 183–215.

Heim, J. (1982) *Les enfants néandertaliens de la Ferrassie*. Paris: Masson.

Heim, J. (1989) La nouvelle reconstruction du crâne de La Chapelle-aux-Saints. Methode et results. *BMSAP* **1**:95–118.

Heinrich, R., M. Rose, R. Leakey, and A. Walker (1993) Hominid radius from the Middle Pliocene of Lake Turkana, Kenya. *AJPA* **92**:139–148

Hellman, M. (1934) The form of the Talgai palate. *AJPA* **19**:1–15.

Hennig, W. (1966) *Phylogenetic Systematics*. Urbana: University of Illinois Press.

Henri-Martin, G. (1964) La dernière occupation moustérienne de La Quina (Charente). *CRAS* **258**:3533–3535.

Henry-Gambier, D. (2002) Les fossils de Cro-Magnon (Les Eyzier-de-Tayac, Dordogne): Nouvelles données sur leur position chronologique et leur attribution culturelle. *BMSAP* **14**:89–112.

Henry-Gambier D., B. Maureille, and R. White (2004). Vestiges humains des niveaux de l'Aurignacien ancien du site de Brassempouy (Landes). *BMSAP* **16**:49–87

Henshilwood, C., F. d'Errico, R. Yates, Z. Jacobs, C. Tribolo, G. Duller, N. Mercier, J. Sealy, H. Valladas, I. Watts, and A. Wintle (2002) Emergence of modern human behavior: Middle Stone Age engravings from South Africa. *Science* **295**:1278–1279.

Henshilwood, C., and C. Marean (2003) The origin of modern human behavior. *CA* **44**:627–651.

Hesse, H., and H. Ullrich (1966) Schädel der "Homo mousteriensis Hauseri" wiederfunden. *Biologische Rundschau* **4**:158–160.

Hey, J. (1997) Mitochondrial and nuclear genes present conflicting portraits of human origins. *MBE* **14**:166–172.

Higham, T., C. Bronk-Ramsey, I. Karavanic, F. Smith, and E. Trinkaus (2006) Revised direct radiocarbon dating of the Vindija G_1 Upper Paleolithic Neandertals. *PNAS* **103**:553–557.

Hill, A. (1985) Early hominid from Baringo, Kenya. *Nature* **315**:222–224.

Hill, A., S. Ward, A. Deino, G. Curtis, and R. Drake (1992) Earliest *Homo*. *Nature* **355**:719–722.

Hill, K. (1982) Hunting and human evolution. *JHE* **11**:521–544.

Hillenius, W. (1992) The evolution of nasal turbinates and mammalian endothermy. *Paleobiology* **18**:17–29.

Hinton, R. (1981) Changes in articular eminence morphology with dental function. *AJPA* **54**:439–455.

Hoffecker, J. (2002) *Desolate Landscapes: Ice Age Settlement in Eastern Europe*. New Brunswick: Rutgers University Press.

Hohmann, G., and B. Fruth (2003) Culture in bonobos? Between-species and within-species variation in behavior. *CA* **44**:563–571.

Holliday, T. (1997a) Body proportions in Late Pleistocene Europe and modern human origins. *JHE* **32**:423–447.

Holliday, T. (1997b) Postcranial evidence of cold adaptation in European Neandertals. *AJPA* **104**:245–258.

Holliday, T. (2000) Evolution at the crossroads: modern human emergence in Western Asia. *AA* **102**:54–68.

Holliday, T. (2003) Species concepts, reticulation, and human evolution. *CA* **44**:653–673.

Holliday, T. (2006a) Neanderthals and modern humans: An example of a mammalian syngameon? In: Harvati and Harrison 2006: 281–297.

Holliday, T. (2006b) Body proportions. In: Trinkaus and Svoboda 2006: 224–232.

Holliday, T., and A. Falsetti (1995) Lower limb length of European early modern humans in relation to mobility and climate. *JHE* **29**:141–153.

Holloway, R. (1962) A note on sagittal cresting. *AJPA* **20**:527–530.

Holloway, R. (1972) New australopithecine endocast, SK 1585, from Swartkrans, South Africa. *AJPA* **37**:173–185.

Holloway, R. (1975) Early hominid endocasts: Volumes, morphology, and significance for hominid evolution. In: Tuttle 1975: 393–415.

Holloway, R. (1980) Indonesian "Solo" (Ngandong) endocranial reconstructions: Some preliminary observations and comparisons with Neandertal and *Homo erectus* groups. *AJPA* **53**:285–295.

Holloway, R. (1981a) The Indonesian *Homo erectus* brain endocasts revisited. *AJPA* **55**:503–521.

Holloway, R. (1981b) Volumetric and asymmetry determinations on recent hominid endocasts: Spy I and II, Djebel Irhoud I, and the Salé *Homo erectus* specimens, with some notes on Neanderthal brain size. *AJPA* **55**:385–393.

Holloway, R. (1982) *Homo erectus* brain endocasts: Volumetric and morphological observations with some comments on cerebral asymmetries. In: H. de Lumley, Editor. Homo erectus *et la place de l'homme de Tautavel parmi les hominidés fossiles*. Nice: Premier Congrès de Paléontologie Humaine, pp. 355–369.

Holloway, R. (1983) Human paleontological evidence relevant to language behavior. *Human Neurobiology* **2**:105–114.

Holloway, R. (1985) The poor brain of *Homo sapiens neanderthalensis*, see what you please. … In: Delson 1985: 319–324.

Holloway, R. (1988) "Robust" australopithecine brain endocasts: Some preliminary observations. In: Grine 1988: 97–105.

Holloway, R. (2000) Brain. In: Delson et al. 2000: 141–149.

Holloway, R., D. Broadfield, and M. Yuan (2004) *The Human Fossil Record*. Volume 3. *Brain Endocasts—The Paleoneurological Evidence*. New York: Wiley-Liss.

Holloway, R., P. Brown, P. Schoenemann, and J. Monge (2006) The brain endocast of *Homo floresiensis*: Microcephaly and other issues. *AJPA Supplement* **42**:105.

Holloway, R., and M. de La Coste-Lareymondie (1982) Brain endocast asymmetry in pongids and hominids: Some preliminary finds on the paleontology of cerebral dominance. *AJPA* **58**:101–110.

Hooker, J., D. Russell, and A. Phélizon (1999) A new family of Plesiadapiformes (Mammalia) from the Old World Lower Paleogene. *Palaeontology* **42**:377–407.

Hooton, E. (1946) *Up From the Ape*. New York: Macmillan.

Hopwood, A. (1932) The age of "Oldoway Man." *Man* **32**:192–195.

Hopwood, A. (1933) Miocene primates from British East Africa. *Annals and Magazine of Natural History* (Series 10) **11**:96–98.

Hotton, N., E. Olson, and D. Beerbower (1997) Amniote origins and the discovery of herbivory. In: S. Sumida and K. Martin, Editors. *Amniote Origins: Completing the Transition to Land*. San Diego: Academic Press, pp. 207–264.

Hou, Y., R. Potts, B. Yuan, Z. Guo, A. Deino, W. Wei, J. Clark, G. Xie, and W. Huang (2000) Mid-Pleistocene Acheulean-like stone technology of the Bose Basin, South China. *Sciences* **287**:1622–1626

Houghton, P. (1993) Neandertal supralaryngeal vocal tract. *AJPA* **90**:139–146.

Howell, F. C. (1951) The place of Neanderthal man in human evolution. *AJPA* **9**:376–416.

Howell, F. C. (1952) Pleistocene glacial ecology and the evolution of "Classic Neandertal" man. *SJA* **8**:377–410.

Howell, F. C. (1957) The evolutionary significance of variation and varieties of "Neanderthal" man. *Quarterly Review of Biology* **32**:330–347.

Howell, F. C. (1958) Upper Pleistocene men of the Southwest Asian Mousterian. In: Von Koenigswald 1958: 185–198.

Howell, F. C. (1960) European and northwest African Middle Pleistocene hominids. *CA* **1**:195–232.

Howell, F. C. (1978) Hominidae. In: Maglio and Cooke 1978: 154–248.

Howell, F. C. (1994a) Thoughts on Eugène Dubois and the "*Pithecanthropus*" stage. In: Franzen 1994: 11–20.

Howell, F. C. (1994b) A chronostratigraphic and taxonomic framework of the origins of modern humans. In: Nitecki and Nitecki 1994: 253–319.

Howell, F. C. (1996) Thoughts on the study and interpreataion of the human fossil record. In: W. Meikle, F. C. Howell, and N. Jablonski, Editors. *Contemporary Issues in Human Evolution*. San Franscisco: California Academy of Sciences, pp. 1–45.

Howell, F. C. (1999) Paleo-demes, species clades, and extinctions in the Pleistocene hominin record. *JAR* **55**:191–243.

Howell, F. C., and Y. Coppens (1976) An overview of Hominidae from the Omo succession, Ethiopia. In: Coppens et al. 1976: 522–532.

Howell, F. C., P. Haesaerts, and J. de Heinzelin (1987) Depositional environments, archaeological occurrences, and hominids from Members E and F of the Shungura Formation (Omo Basin, Ethiopia). *JHE* **16**:665–700.

Howells, W. W. (1946) *Mankind So Far*. New York: Doubleday.

Howells, W. W. (1959) *Mankind in the Making*. New York: Doubleday.

Howells, W. W. (1966) *Homo erectus*. *SA* **215**:46–53.

Howells, W. W. (1973a) *The Evolution of the Genus* Homo. Reading, MA: Addison-Wesley.

Howells, W. W. (1973b) *Cranial Variation in Man. Papers of the Peabody Museum of Archaeology and Ethnology* 67. Cambridge: Harvard University Press.

Howells, W. W. (1973c) *The Pacific Islanders*. New York: Charles Schribner's Sons.

Howells, W. W. (1974) Neanderthals: Names, hypotheses, and scientific method. *AA* **76**:24–38.

Howells, W. W. (1976) Explaining modern man: Evolutionists versus migrationists. *JHE* **5**:577–596.

Howells, W. W. (1980) *Homo erectus*—who, when, and where: a survey. *YPA* **23**:1–23.

Howells, W. W. (1985) Taung: A mirror for American anthropology. In: Tobias 1985: 19–24.

Howells, W. W. (1989) *Skull Shapes and the Map. Craniometric Analyses in the Dispersion of Modern* Homo. *Papers of the Peabody Museum of Archaeology and Ethnology* 79. Cambridge: Harvard University Press.

Howells, W. W. (1993) *Getting There: The Story of Human Evolution*. Washington, DC: Compass Press.

Howells, W. W. (1995) *Who's Who in Skulls. Ethnic Identification of Crania from Measurements. Papers of the Peabody Museum of Archaeology and Ethnology* 83. Cambridge: Harvard University Press.

Hrdlička, A. (1920) Shovel-shaped teeth. *AJPA* **3**:429–465.

Hrdlička, A. (1927) The Neanderthal phase of man. *JRAI* **57**:249–274.

Hrdlička, A. (1930) *The Skeletal Remains of Early Man. Smithsonian Miscellaneous Collections* **83**:1–379.

Hrdlička, A. (1939) Important Paleolithic find in Central Asia. *Science* **90**:996–998.

Hu, Y., J. Meng, Y. Wang, and C. Li (2005) Large Mesozoic mammals fed on young dinosaurs. *Nature* **433**:149–152.

Huang, P., S. Jin, R. Liang, Z. Lu, L. Zheng, Z. Yuan, Z. Fang, and B. Cai (1991) Study of ESR dating for burying age of the first skull of Peking Man and chronological scale of the cave deposits in Zhoukoudian site. *AAS* **10**:107–115.

Huang, W., R. Ciochon, G. Yumiin, R. Larick, F. Quiren, H. Schwarcz, C. Younge, J. de Vos, and W. Rink (1995) Early *Homo* and associated artifacts in Asia. *Nature* **378**:275–278.

Huang, W., Y. Fu, B. Chang, X. Gu, L. Jorde, and W. Li (1998) Sequence variaition in ZFX introns in human populations. *MBE* **15**:138–142.

Hublin, J.-J. (1978) Quelques caractères apomorphes du crâne néandertalien et leur interprétation phylogénique. *CRAS* **287**: 923–926.

Hublin, J.-J. (1980) La Chaise Suard, Engis 2, et La Quina H 18: Développement de la morphologie externe chez l'enfant prénéandertalien et néandertalien. *CRAS* **291**:669–672.

Hublin, J.-J. (1982) Les Anténéandertaliens: Presapiens ou Prénéandertaliens. *Geobios, mémoire spécial* **6**:345–357.

Hublin, J.-J. (1985) Human fossils from the North African Middle Pleistocene and the origin of *Homo sapiens*. In: Delson 1985: 283–288.

Hublin, J.-J. (1986) Some comments on the diagnostic features of *Homo erectus*. *Anthropos (Brno)* **23**:175–187.

Hublin, J.-J. (1988a) Les presapiens Européens. In: *L'Homme de Néandertal*, Volume 3. *L'Anatomie*. Liège: 75–80.

Hublin, J.-J. (1988b) Les plus anciens representants de la lignée Prenéandertalienne. In: *L'Homme de Néandertal*, Volume 3. *L'Anatomie*. Liège: 81–94.

Hublin, J.-J. (1988c) Caractères derivés de la region occipito-mastoïdenne chez les Néandertaliens. In: *L'Homme de Néandertal*, Volume 3. *L'Anatomie*. Liège: 67–73.

Hublin, J.-J. (1993) Recent human evolution in Northwest Africa. In: Aitken et al. 1993: 118–131.

Hublin, J.-J. (1998) Climatic changes, paleogeography, and the evolution of the Neandertals. In: Akazawa et al. 1998: 295–310.

Hublin, J.-J. (2000) Modern-nonmodern hominid interactions: A Mediterranean perspective. In: Bar-Yosef and Pilbeam 2000: 157–182.

Hublin, J.-J., and S. Bailey (2006) Revisiting the last Neanderthals. In: Conard 2006: 105–128.

Hublin, J.-J., C. Ruiz, P. Lara, M. Fontugne, and J. Reyss (1995) The Mousterian site of Zafarraya (Andalucia, Spain): Dating and implications on the paleolithic peopling process of Western Europe. *CRAS* **312**(série II a):931–937.

Hublin, J.-J., F. Spoor, M. Braun, F. Zonneveld, and S. Condemi (1996) A late Neanderthal associated with Upper Paleolithic artifacts. *Nature* **381**:224–226.

Hublin, J.-J., and A. Tillier (1981) The Mousterian juvenile mandible from Irhoud (Morocco): A phylogenetic interpretation. In: Stringer 1981: 176–185.

Hublin, J.-J., A. Tillier, and J. Tixier (1987) L'humèrus d'enfant Moustèrien (Homo 4) du Jebel Irhoud (Maroc) dan son contexte archèologique. *BMSAP* **4**(series 14):115–142.

Hubrecht, A. (1897) *The Descent of the Primates*. New York: Charles Scribner's Sons.

Huchon, D., O. Madsen, M. Sibbald, M. K. Ament, K., M. Stanhope, F. Catzeflis, W. de Jong, and E. Douzery (2002) Rodent phylogeny and a timescale for the evolution of Glires: Evidence from an extensive taxon sampling using three nuclear genes. *MBE* **19**:1053–1065.

Huffman, O. (2001) Geological context and age of the Perning/Modjokerto *Homo erectus*, East Java. *JHE* **40**:353–362.

Huffman, O., Y. Zaim, J. Kappelman, D. Ruez, J. De Vos, Y. Rizal, F. Aziz, and C. Hertler (2006) Relocation of the 1936 Modjokerto skull discovery site near Perning, East Java. *JHE* **50**:431–451.

Hughes, A., and P. Tobias (1977) A fossil skull probably of the genus *Homo* from Sterkfontein, Transvaal. *Nature* **265**: 310–312.

Hunt, G. (2000) Human-like, population-level specialization in the manufacture of pandanus tools by New Caledonian crows *Corvus moneduloides*. *Proceedings of the Royal Society of London B* **267**:403–413.

Hunt, K. (1991) Mechanical implications of chimpanzee positional behavior. *AJPA* **86**:521–536.

Hunt, K., and V. Vitzthum (1986) Dental metric assessment of the Omo fossils: Implications for the phyletic position of *Australopithecus africanus*. *AJPA* **71**:141–156.

Hürzeler, J. (1948) Zur Stammesgeschichte der Necrolemuriden. *Schweizerischen paläontologischen Abhandlungen* **55**:1–46.

Hürzeler, J. (1968) Questions et réflexions sur l'histoire des anthropomorphes. *Annales de Paléontologie (Vertebrés)* **54**(**2**): 13–41, 195–233.

Hutchinson, V. (2000) *Circum-Mediterranean population dynamics in the Late Pleistocene: Evidence from the Tangier (Morocco) maxilla.* M.A. Thesis. Northern Illinois University.

Hutterer, K. (1985) The Pleistocene archaeology of Southeast Asia in regional context. *MQRSA* **9**:1–23.

Hutton, J. (1795) *Theory of the Earth: With Proofs and Illustrations*, reprint edition. Weinheim: H.R. Engelmann (J. Cramer), 1959.

Huxley, T. (1863) *Evidence as to Man's Place in Nature.* London: Williams, and Norgate.

Hylander, W. (1975a) The human mandible: Lever or link? *AJPA* **43**:227–242.

Hylander, W. (1975b) Incisor size and diet in anthropoids with special reference to Cercopithecidae. *Science* **189**:1095–1098.

Hylander, W. (1977) The adaptive significance of Eskimo craniofacial morphology. In: A. Dahlberg and T. Graber, Editors. *Orofacial Growth and Development.* The Hague: Mouton. 129–169.

Hylander, W. (1984) Stress and strain in the mandibular symphysis of primates: A test of competing hypotheses. *AJPA* **64**: 1–46.

Hylander, W., P. Picq, and K. Johnson (1991a) Masticatory-stress hypotheses and the supraorbital region of primates. *AJPA* **86**: 1–36.

Hylander, W., P. Picq, and K. Johnson (1991b) Function of the supraorbital region of primates. *Archives of Oral Biology* **36**: 273–281.

Hylander, W., and C. Vinyard (2006) The evolutionary significance of canine reduction in hominids: Functional links between jaw mechanics and canine length. *PaleoAnthropology* 2006: A2.

Hyodo, M., N. Nakaya, A. Urabe, H. Saegusa, X. Shunrong, Y. Jiyun, and J. Xuepin (2002) Paleomagnetic dates of hominid remains from Yuanmou, China and other Asian sites. *JHE* **27**:27–41.

I

Imbrie, J., and K. Imbrie (1979) *Ice Ages: Solving the Mystery.* New York: Macmillan.

Ingram, J. (1978) Parent-infant interactions in the common marmoset (*Callithrix jacchus*). In: D. Kleiman, Editor. *The Biology and Conservation of the Callitrichidae.* Washington, DC: Smithsonian Institution Press. 281–291.

Ingram, M., H. Kaessmann, S. Pääbo, and U. Gyllensten (2000) Mitochondrial genome variation and the origin of modern humans. *Nature* **408**:708–713.

Isaac, G. (1982) Models of human evolution. *Science* **217**:295.

Isaac, G. (1984) The archaeology of human origins: Studies of the Lower Pleistocene in East Africa 1971–1981. In: F. Wendorf and A. Close, Editors. *Advances in World Archaeology.* New York: Academic Press, pp. 1–87.

Ishida, H., Y. Kunimatsu, T. Takano, Y. Nakano, and M. Nakatsukasa (2004) *Nacholapithecus* skeleton from the Middle Miocene of Kenya. *JHE* **46**:69–103.

Ishida, H., and M. Pickford (1997) A new late Miocene hominoid from Kenya: *Samburupithecus kiptalami* gen. et sp. nov. *CRAS, Sciences de la terre et des planètes* **325**:823–829.

Isler, K., and C. van Schaik (2006) Costs of encephalization: The energy trade-off hypothesis tested on birds. *JHE* **51**: 228–243.

Itihara, M., N. Watanabe, D. Kadar, and H. Kumai (1994) Quaternary stratigraphy of the hominid bearing formations in the Sangiran area, Central Java. In: Franzen 1994: 123–128.

J

Jabbour, R., G. Richards, and J. Anderson (2002) Mandibular condyle traits in Neanderthals and other *Homo*: A comparative, correlative and ontogenetic study. *AJPA* **119**:144–155.

Jablonski, N. (2002) Fossil Old World monkeys: The late Neogene radiation. In: Hartwig 2002: 255–299.

Jablonski, N., and G. Chaplin (2000) The evolution of human skin coloration. *JHE* **39**: 57–106.

Jacob, T. (1973) Paleoanthropological discoveries in Indonesia with special reference to the finds of the last two decades. *JHE* **2**:473–486.

Jacob, T. (1975) Morphology and paleoecology of early man in Java. In: Tuttle 1975: 311–325.

Jacob, T. (1978) The puzzle of Solo man. *MQRSA* **4**:31–40.

Jacob, T. (1980) The Pithecanthropus of Indonesia: Phenotype, genetics and ecology. In: L. Könnigson, Editor. *Current Argument on Early Man*. Oxford: Pergamon, pp. 170–179.

Jacob, T., E. Indriati, R. Soejono, K. Hsu, D. Frayer, R. Eckhardt, A. Kuperavange, A. Thorne, and M. Henneberg (2006) Pygmoid Australasian *Homo sapiens* skeletal remains from Liang Bua, Flores: Population affinities and pathological abnormalities. *PNAS* **103**:13421–13426.

Jacob, T., R. Soejono, L. Freeman, and F. Brown (1978) Stone tools from Mid-Pleistocene sediments in Java. *Science* **202**: 885–887.

Jacobs, B., J. Kingston, and L. Jacobs (1999) The origin of grass-dominated ecosystems. *Annals of the Missouri Botanical Garden* **86**:590–643.

Jaeger, J.-J., U. Soe, U. Aung, M. Benammi, Y. Chaimanee, R.-M. Ducrocq, T. Tun, U. Thein, and S. Ducrocq (1998) New Myanmar middle Eocene anthropoids: An Asian origin for catarrhines? *CRAS* **321**:953–959.

James, W. (1960) *The Jaws and Teeth of Primates*. London: Pitman Medical Publishing.

Janis, C. (1993) Tertiary mammal evolution in the context of changing climates, vegetation, and tectonic events. *Annual Reviews of Ecology and Systematics* **24**:467–500.

Janković, I. (2004) Neandertals—150 years later. *CAN* **28**(supplement 2):379–401.

Janković, I., I. Karavanic, J. Ahern, D. Brajković, J. Mauch Lenardić, and F. Smith (2006) Vindija Cave and the modern human peopling of Europe. *CAN* **30**:457–466.

Janković, I., and F. H. Smith (2005) Comparative morphometrics of Neandertal zygomatic bones. *AJPA Supplement* **40**: 121–122.

Jaubert, J., M. Lorblanchet, H. Laville, R. Slott-Moller, A. Turq, and J. Brugal (1990) *Les chasseurs d'aurochs de la Borde: Un site du Paléolithique moyen (Livenon, Lot)*. Paris: Editions de la Maison des Sciences de l'Homme.

Jeffery, N., and F. Spoor (2002) Brain size and the human cranial base: A prenatal perspective. *AJPA* **118**:324–340.

Jelinek, A. (1982) The Tabun cave and Paleolithic man in the Levant. *Science* **216**:1369–1375.

Jelinek, A. (1992) Problems in the chronology of the Middle Paleolithic and the first appearance of early modern *Homo sapiens* in southwest Asia. In: Akazawa et al. 1992: 253–275.

Jelínek, J. (1966) Jaw of an intermediate type of Neanderthal man from Czechoslovakia. *Nature* **212**:701–702.

Jelínek, J. (1969) Neanderthal man and *Homo sapiens* in Central and Eastern Europe. *CA* **10**:457–503.

Jelínek, J. (1976) The *Homo sapiens neanderthalensis* and *Homo sapiens sapiens* relationship in central Europe. *Anthropologie* **14**:79–81

Jelínek, J. (1983) The Mladeč finds and their evolutionary importance. *Anthropologie* **21**: 57–64.

Jenkin, F. (1867) "The origin of species." *The North British Review* **46**:277–318.

Jenkins, F., Jr. (1971) The postcranial skeleton of African cynodonts. *Bulletin of the Peabody Museum (Yale)* **36**:1–216.

Jenkins, F., Jr. (1974) Tree shrew locomotion and the origins of primate arborealism. In: F. Jenkins, Jr., Editors. *Primate Locomotion*. New York: Academic Press, pp. 85–115.

Jenkins, F., Jr., and F. Parrington (1976) The postcranial skeletons of the Triassic mammals *Eozostrodon*, *Megazostrodon* and *Erythrotherium*. *PTRSL* **273**:387–431.

Jerison, H. (1973) *Evolution of the Brain and Intelligence*. New York: Academic Press.

Ji, Q., Z.-X. Luo, C.-X. Yuan, J. Wible, J.-P. Zhang, and J. Georgi (2002) The earliest known eutherian mammal. *Nature* **416**:816–820.

Jia, L., and W. Huang (1985) The Late Paleolithic of China. In: Wu and Olsen 1985: 211–223.

Jia, L., and W. Huang (1990) *The Story of Peking Man*. Oxford: Oxford University Press.

Jiggins, C., W. McMillan, W. Neukirchen, and J. Mallet (1996) What can hybrid zones tell us about speciation? The case of *Heliconius erato* and *H. himera* (Lepidoptera: Nymphalidae). *Biological Journal of the Linnaean Society* **59**: 221–242.

Johanson, D. (1980) Early African hominid phylogenesis: A re-evaluation. In: L.-K. Konigsson, Editor. *Current Argument on Early Man*. Oxford: Pergamon Press, pp. 31–69.

Johanson, D., and B. Edgar (1996) *From Lucy to Language*. New York: Simon and Schuster.

Johanson, D., C. O. W. Kimbel, T. White, S. Ward, M. Bush, B. Latimer, and Y. Coppens (1982a) Morphology of the Pliocene partial skeleton (A.L. 288-1) from the Hadar Formation, Ethiopia. *AJPA* **57**:403–451.

Johanson, D., F. Masao, G. Eck, T. White, R. Walter, W. Kimbel, B. Asfaw, P. Manega, P. Ndessokia, and G. Suwa (1987) New partial skeleton of *Homo habilis* from Olduvai Gorge, Tanzania. *Nature* **327**:205–209.

Johanson, D., and M. Taieb (1976) Plio-Pleistocene hominid discoveries in Hadar, Ethiopia. *Nature* **260**:293–297.

Johanson, D., and T. White (1979) A systematic assessment of early African hominids. *Science* **203**:321–330.

Johanson, D., T. White, T., and Y. Coppens (1978) A new species of the genus *Australopithecus* from the Pliocene of eastern Africa. *Kirtlandia* **28**:1–14.

Johanson, D., T. White, and Y. Coppens (1982b) Dental remains from the Hadar Formation, Ethiopia: 1974–1977 collections. *AJPA* **57**:545–603.

Johnson, B., and G. Miller (1997) Archaeological applications of amino acid racemization. *Archaeometry* **39**:265–287.

Johnson, K., M. McCrossin, and B. Benefit (2000) Circular shapes do not an ape make: Comments on interpretation of the inferred *Morotopithecus* scapula. *AJPA Supplement* **30**: 189–190.

Johnson, S., and L. Shapiro (1998) Positional behavior and vertebral morphology in a telines and cebines. *AJPA* **105**: 333–354.

Johnson, T., C. Scholz, M. Talbot, K. Kelts, R. Ricketts, G. Ngobi, K. Beuning, I. Ssemmanda, and J. McGill (1996) Late Pleistocene desiccation of Lake Victoria and rapid evolution of cichlid fishes. *Science* **273**:1091–1093.

Jolly, A. (1966) *Lemur Behavior*. Chicago: University of Chicago Press.

Jolly, A. (2003) *Lemur catta*, ring-tailed lemur, *maky*. In: S. Goodman and J. Benstead, Editors. *The Natural History of Madagascar*. Chicago: University of Chicago Press, pp. 1329–1331.

Jolly, C. (1970a) The seed eaters: A new model of hominid differentiation based on a baboon analogy. *Man* **5**:5–26.

Jolly, C. (1970b) *Hadropithecus*: A lemuroid small-object feeder. *Man* (new series) **5**:619–626.

Jolly, C. (2001) A proper study of mankind: Analogies from the papionin monkeys and their implications for human evolution. *YPA* **44**:177–204.

Jolly, C., and J. Phillips-Conroy (2003) Testicular size, mating system, and maturation schedules in wild anubis and hamadryas baboons. *IJP* **24**:125–142.

Jolly, C., T. Woolley-Barker, S. Beyene, T. Disotell, and J. Phillips-Conroy (1997) Intergeneric hybrid baboons. *IJP* **18**:597–627.

Jones, J. (1982) How different are human races? *Nature* **293**: 188–190.

Jones, R. (1979) The fifth continent: Problems concerning the human colonization of Australia. *ARA* **8**:445–466.

Jones, R. (1989) East of Wallace's Line: Issues and problems in the colonization of the Australian continent. In: Mellars and Stringer 1989: 743–782.

Jones, R. (1992) The human colonisation of the Australian continent. In: Bräuer and Smith 1992: 289–301.

Jorde, L., M. Bamshad, and A. Rogers (1998) Using mitochondrial and nuclear DNA markers to reconstruct human evolution. *BioEssays* **20**:126–136.

Jöris, O. (2001) Der Spätmittelpaläolithische Fundplatz Buhlen (Grabungen 1966–69). Stratigraphie, Steinartifakte, und Fauna des oberen Fundplatzes. Bonn: *Universitätsforschungen zur Prähistorischen Archäologie* 73.

Jouffroy, F. (1991) La "main sans talon" du primate bipède. In: Coppens and Sénut, 1991: 21–35.

Jungers, W. (1977) Hindlimb and pelvic adaptations to vertical climbing in *Megaladapis*, a giant subfossil prosimian from Madagascar. *YPA* **20**:508–524.

Jungers, W. (1982) Lucy's limbs: Skeletal allometry and locomotion in *Australopithecus afarensis*. *Nature* **297**:676–678.

Jungers, W. (1988a) New estimates of body size in australopithecines. In: Grine 1988: 115–125.

Jungers, W. (1988b) Relative joint size and hominoid locomotor adaptations with implications for the evolution of hominoid bipedalism. *JHE* **17**:247–265.

Jungers, W. (1991) A pygmy perspective on body size and shape in *Australopithecus afarensis* (AL 288-1, "Lucy"). In: Coppens and Sénut 1991: 215–224.

Jungers, W., and F. Grine (1986) Dental trends in the australopithecines: The allometry of mandibular molar dimensions. In: Wood et al. 1986: 203–219.

Jungers, W., A. Pokempner, R. Kay, and M. Cartmill (2003) Hypoglossal canal size in living hominoids and the evolution of human speech. *HB* **75**:473–484.

Jungers, W., and J. Stern (1983) Body proportions, skeletal allometry and locomotion in the Hadar hominids: A reply to Wolpoff. *JHE* **12**:673–684.

K

Kako, E. (1999) Elements of syntax in the systems of three language-trained animals. *Animal Learning and Behavior* **27**:1–14.

Kalb, J., E. Oswald, S. Tebedge, A. Mebrate, E. Tola, and D. Peak (1982) Geology and stratigraphy of Neogene deposits, Middle Awash Valley, Ethiopia. *Nature* **298**:17–25.

Kamminga, J. (1992) New interpretations of the Upper Cave, Zhoukoudian. In: Akazawa et al. 1992: 379–400.

Kamminga, J., and R. Wright (1989) The Upper Cave at Zhoukoudian and the origins of the Mongoloids. *JHE* **17**:739–767.

Kano, T. (1996) Male rank order and copulation rate in a unit-group of bonobos at Wamba, Zaïre. In: McGrew et al. 1996: 135–145.

Kaplan, J., J. Phillips-Conroy, M. Fontenot, C. Jolly, L. Fairbanks, and J. Mann (1999) Cerebrospinal fluid monoaminergic metabolites differ in wild anubis and hybrid (*Anubis hamadryas*) baboons: Possible relationships to life history and behavior. *Neuropsychopharmacology* **20**:517–524.

Kappelman, J., M. Alçiçek, N. Kazanci, M. Schultz, M. Özkul, and Ş. Şen (2008) Brief communication: First *Homo erectus* from Turkey and implications for migrations into temperate Eurasia. *AJPA* **135**:110–116.

Kappelman, J., E. Simons, and C. Swisher (1992) New age-determinations for the Eocene–Oligocene boundary sediments in the Fayum Depression, Northern Egypt. *Journal of Geology* **100**:647–667.

Kappelman, J., C. Swisher, J. Fleagle, S. Yirga, T. Bown, and M. Feseha (1996) Age of *Australopithecus afarensis* from Fejej, Ethiopia. *JHE* **30**:139–146.

Karavanić, I. (1995) Upper Paleolithic occupation levels and late occurring Neandertals at Vindija Cave (Croatia) in the context of Central Europe and the Balkans. *JAR* **51**:9–35.

Karavanić, I. (2007) Le Moustérien en Croatie. *L'Anthropologie* 111:321–345.

Karavanic, I., and F. Smith (1998) The Middle/Upper Paleolithic interface and the relationship of Neanderthals and early modern humans in the Hrvatsko Zagorje, Croatia. *JHE* 34:223–248.

Karavanic, I., and F. Smith (2000) More on the Neanderthal problem: the Vindija case. *CA* 41:838–840.

Kay, R. (1975) The functional adaptations of primate molar teeth. *AJPA* 43:195–216.

Kay, R. (1977) The evolution of molar occlusion in the Cercopithecidae and early catarrhines. *AAPA* 46:327–352.

Kay, R. (1981) The nut-crackers: a new theory of the adaptations of the Ramapithecinae. *AJPA* 55:141–151.

Kay, R. (1982) *Sivapithecus simonsi*, a new species of Miocene hominoid, with comments on the phylogenetic status of the Ramapithecinae. *IJP* 3:113–173.

Kay, R. (1984) On the use of anatomical features to infer foraging behavior in extinct primates. In: P. Rodman and J. Cant, Editors. *Adaptations for Foraging in Nonhuman Primates: Contributions to an Organismal Biology of Prosimians, Monkeys, and Apes.* New York: Columbia University Press. pp. 21–53.

Kay, R. (1985) Dental evidence for the diet of *Australopithecus*. *ARA* 14:315–341.

Kay, R. (2005) A synopsis of the phylogeny and paleobiology of Amphipithecidae, South Asian Middle and Late Eocene primates. *Anthropological Science* 113:33–42.

Kay, R., and M. Cartmill (1977) Cranial morphology and adaptations of *Palaechthon nacimienti* and other Paromomyidae (Plesiadapoidea, ?Primates), with a description of a new genus and species. *JHE* 6:19–53.

Kay, R., M. Cartmill, and M. Balow (1998) The hypoglossal canal and the origin of human vocal behavior. *PNAS* 95: 5417–5419.

Kay, R., and F. Grine (1988) Tooth morphology, wear and diet in *Australopithecus* and *Paranthropus* from southern Africa. In: Grine 1988: 427–447.

Kay, R., and K. Hiiemae (1974) Jaw movement and tooth use in recent and fossil primates. *AJPA* 40:227–256.

Kay, R., C. Ross, and B. Williams (1997) Anthropoid origins. *Science* 275:797–804.

Kay, R., D. Schmitt, C. Vinyard, J. Perry, M. Takai, N. Shigehara, and N. Egi (2004a) The paleobiology of Amphipithecidae, South Asian late Eocene primates. *JHE* 46:3–25.

Kay, R., and E. Simons (1983) A reassessment of the relationship between later Miocene and subsequent Hominoidea. In: Ciochon and Corruccini 1983: 577–624.

Kay, R., J. Thewissen, and A. Yoder (1992) Cranial anatomy of *Ignacius graybullianus* and the affinities of the Plesiadapiformes. *AJPA* 89:477–498.

Kay, R., B. Willams, C. Ross, M. Takai, and N. Shigehara (2004b) Anthropoid origins: A phylogenetic analysis. In: Ross and Kay 2004: 91–135.

Kay, R., and P. Ungar (1997) Dental evidence for diet in some Miocene catarrhines with comments on the effects of phylogeny on the interpretation of adaptation. In: Begun et al. 1997: 131–151.

Keeley, L., and N. Toth (1981) Microwear polishes on early stone tools from Koobi Fora, Kenya. *Nature* 293:494–465.

Keith, A. (1915) *The Antiquity of Man*, two volumes. London: Williams & Norgate.

Keith, A. (1923) Man's posture: Its evolution and disorders. *British Medical Journal* 1:451–454, 499–502, 545–548, 587–590, 624–626, 669–672.

Keith, A. (1925) The fossil anthropoid ape from Taungs. *Nature* 115:234–235.

Keith, A. (1927) A report on the Galilee skull. In: F. Turville-Petre, Editor. *Researches in Prehistoric Galilee, 1925–1926.* London: British School of Archaeology in London.

Keith, A. (1931) *New Discoveries Relating to the Antiquity of Man.* London: Williams & Norgate.

Keith, A. (1947) *A New Theory of Human Evolution.* New York: Philosophical Library.

Keith, A., and T. McCown (1937) Mount Carmel Man: His bearing on the ancestry of modern races. In: G. MacCurdy, Editor. *Early Man.* Philadelphia: Lippincott, pp. 44–52.

Kelley, J. (2002) The hominoid radiation in Asia. In: Hartwig 2002: 369–384.

Kennedy, G. (1983a) Some aspects of femoral morphology in *Homo erectus*. *JHE* 12:587–616.

Kennedy, G. (1983b) A morphometric and taxonomic assessment of a hominine femur from the Lower Member, Koobi Fora, Lake Turkana. *AJPA* 61:429–436.

Kennedy, G. (1984) The emergence of *Homo sapiens*: The postcranial evidence. *Man* 19:94–110.

Kennedy, K. (1979) The deep skull of Niah: An assessment of twenty years of speculation concerning its evolutionary significance. *Asian Perspectives* 20:32–50.

Kennedy, K. (2000) *God-Apes and Fossil Men: Paleoanthropology in South Asia.* Ann Arbor: University of Michigan Press.

Kennedy, K., A. Sonakia, J. Chiment, and K. Verma (1991) Is the Narmada hominid an Indian *Homo erectus*? *AJPA* 86: 475–496.

Kermack, K., F. Mussett, and H. Rigney (1981) The skull of *Morganucodon*. *Zoological Journal of the Linnaean Society* 71:1–158.

Kern, H. M., Jr., and Straus, W. L., Jr. 1949. The femur of *Plesianthropus transvaalensis*. *AJPA* 7:53–77.

Keyser, A. (2000) The Drimolen skull: The most complete australopithecine cranium and mandible to date. *SAJS* 96:189–193.

Keyser, A., C. Menter, J. Moggi-Cecchi, T. Pickering, and L. Berger (2000) Drimolen: A new hominid-bearing site in Gauteng, South Africa. *SAJS* **96**:193–197.

Kidder, J., and A. Durband (2004) A re-evaluation of the metric diversity within *Homo erectus*. *JHE* **46**:299–315.

Kidder, J., R. Jantz, and F. Smith (1992) Defining modern humans: A multivariate approach. In: Bräuer and Smith 1992: 157–177.

Kidder, J., F. Smith, and R. Jantz (1991) The origin of modern *Homo sapiens* in Europe and Southwest Asia: proximal femoral evidence. *AJPA Supplement* **12**:104.

Kielan-Jaworowska, Z., R. Cifelli, and Z.-X. Luo (2002) Dentition and relationships of the Jurassic mammal *Shuotherium*. *Acta Palaeontologica Polonica* **47**:479–486.

Kimbel, W. (1984) Variation in the pattern of cranial venous sinuses and hominid phylogeny. *AJPA* **63**:243–263.

Kimbel, W. (1988) Identification of a partial cranium of *Australopithecus afarensis* from the Koobi Fora Formation, Kenya. *JHE* **17**:647–656.

Kimbel, W., D. Johanson, and Y. Coppens (1982) Pliocene hominid cranial remains from the Hadar Formation, Ethiopia. *AJPA* **57**:403–452.

Kimbel, W., D. Johanson, and Y. Rak (1994) The first skull and other discoveries of *Australopithecus afarensis* at Hadar, Ethiopia. *Nature* **368**:449–451.

Kimbel, W., D. Johanson, and Y. Rak (1997) Systematic assessment of a maxilla of *Homo* from Hadar, Ethiopia. *AJPA* **103**:235–262.

Kimbel, W., C. Lockwood, C. Ward, M. Leakey, Y. Rak, and D. Johanson (2006) Was *Australopithecus anamensis* ancestral to *A. afarensis*? A case of anagenesis in the hominin fossil record. *JHE* **51**:134–152.

Kimbel, W., and Y. Rak (1993) The importance of species taxa in paleoanthropology and an argument for the phylogenetic concept of the species category. In: Kimbel and Martin 1993: 461–484.

Kimbel, W., Y. Rak, and D. Johanson (2004) *The Skull of Australopithecus afarensis*. New York: Oxford University Press.

Kimbel, W., R. Walter, D. Johanson, et al. (1996) Late Pleistocene *Homo* and Oldowan tools from the Hadar Formation (Kada Hadar Member), Ethiopia. *JHE* **31**:549–561.

Kimbel, W., and T. White (1988) Variation, sexual dimorphism and the taxonomy of *Australopithecus*. In: Grine 1988: 175–192.

Kimbel, W., T. White, and D. Johanson (1984) Cranial morphology of *Australopithecus afarensis*: a comparative study based on a composite reconstruction of the adult skull. *AJPA* **64**:337–388.

Kimbel, W., T. White, and D. Johanson (1985) Craniodental morphology of the hominids from Hadar and Laetoli: Evidence of "*Paranthropus*" and *Homo* in the mid-Pliocene of eastern Africa? In: Delson 1985: 120–137.

Kimbel, W., T. White, and D. Johanson (1988) Implications of KNM-ER 17000 for the evolution of "robust" australopithecines. In: Grine 1988: 259–268.

King, W. (1864) The reputed fossil man of Neanderthal. *Quarterly Journal of Science* **1**:88–97.

Kirk, E. (2003) *Evolution of the primate visual system*. Ph.D. Dissertation, Duke University.

Kirk, E., M. Cartmill, R. Kay, and P. Lemelin (2003) Comment on "Grasping primate origins." *Science* **300**:741.

Kirk, E., P. Lemelin, M. Hamrick, D. Boyer, and J. Bloch (2008) Intrinsic hand proportions of euarchontans and other mammals: Implications for the locomotor behavior of plesiadapiforms. *JHE* **55**:278–299.

Kirk, E., and E. Simons (2001) Diets of fossil primates from the Fayum Depression of Egypt: A quantitative analysis of molar shearing. *JHE* **40**:203–229.

Klein, C., and N. Buekes (1992) Time distribution, stratigraphy, sedimentologic setting, and geochemistry of Precambrian iron-formations. In: J. W. Schopf and C. Klein, Editors. *The Proterozoic Biosphere: A Multidisciplinary Study*. New York: Cambridge University Press. 139–146.

Klein, R. (1973a) Geological antiquity of Rhodesian man. *Nature* **244**:311–312.

Klein, R. (1973b) *Ice Age Hunters of the Ukraine*. Chicago: University of Chicago Press.

Klein, R. (1983) The stone age prehistory of southern Africa. *ARA* **12**:25–48.

Klein, R. (1988) The archaeological significance of animal bones from Acheulean sites in southern Africa. *African Archaeological Review* **6**:3–25.

Klein, R. (1989) Biological and behavioral perspectives on modern human origins in southern Africa. In: Mellars and Stringer 1989: 529–546.

Klein, R. (1998) Why anatomically modern people did not disperse from Africa 100,000 years ago. In: Akazawa et al. 1998: 509–521.

Klein, R. (1999) *The Human Career. Human Biological and Cultural Origins*, 2nd edition. Chicago: University of Chicago Press.

Klein, R. (2000) Archaeology and the evolution of human behavior. *EA* **9**:17–36.

Klein, R. (2003) Whither the Neanderthals? *Science* **299**: 1525–1527.

Klein, R., and B. Edgar (2002) *The Dawn of Human Culture*. New York: John Wiley & Sons.

Kleklamp, J., A. Riedel, C. Harper, and H.-J. Kretschmann (1987) A quantitative study of Australian Aboriginal and Caucasian brains. *JA* **150**:191–210.

Knight, A. (2003) The phylogenetic relationship of Neandertal and modern human mitochondrial DNAs based on informative nucleotide sites. *JHE* **44**:1–6.

Knoll, A. (2003) *Life on a Young Planet: The First Three Billion Years of Evolution on Earth*. Princeton: Princeton University Press.

Koch, T. (1973) *Anatomy of the Chicken and Domestic Birds*. Ames: Iowa State University Press.

Kochetkova, V. (1978) *Paleoneurology*. Washington: Winston & Sons.

Köhler. M., and S. Moyà-Solà (1997) Ape-like or hominid-like? The positional behavior of *Oreopithecus bambolii* reconsidered. *PNAS* **94**:11747–11750.

Kondo, O., and H. Ishida (2002) Postcranial bones of the Neanderthal child of burial no. 2. In: Akazawa and Muhesen 2002: 299–321.

Kool, K. (1993) The diet and feeding behavior of the silver leaf monkey (*Trachypithecus auratus sondaicus*) in Indonesia. *IJP* **14**:667–700.

Korey, K. (1990) Deconstructing reconstruction: The OH 62 humerofemoral index. *AJPA* **83**:25–33.

Kornet, D. (1993) Permanent splits as speciation events: A formal reconstruction of the internodal species concept. *Journal of Theoretical Biology* **164**:407–435.

Kortlandt, A. (1972) *New Perspectives on Ape and Human Evolution. Amsterdam*: Stichting voor Psychobiologie, University of Amsterdam.

Kramer, A. (1991) Modern human origins in Australasia: Replacement or evolution? *AJPA* **86**:455–473.

Kramer, A. (1993) Human taxonomic diversity in the Pleistocene: Does *Homo erectus* represent multiple hominid species? *AJPA* **91**:161–171.

Kramer, A. (1994) A critical appraisal for the existence of Southeast Asian australopithecines. *JHE* **36**:3–21.

Kramer, A., T. Crummett, and M. Wolpoff (2001) Out of Africa and into the Levant: replacement or admixture in western Asia. *QI* **75**:51–63.

Kramer, A., S. Donnelly, J. Kidder, S. Ousley, and S. Olah (1995) Craniometric variation in large-bodied hominoids: Testing the single-species hypothesis for *Homo habilis*. *JHE* **29**:443–462.

Kramer, A., T. Djubiantono, F. Aziz, J. Bogard, R. Weeks, D. Weinand, W. Hames, J. Elam, A. Durband, and Agus (2005) The first hominid fossil recovered from West Java, Indonesia. *JHE* **48**:661–667.

Kramer, A., and L. Konigsberg (1994) The phyletic position of Sangiran 6 as determined by multivariate analysis. In: Franzen 1994: 105–114.

Krantz, G. (1975) An explanation for the diastema of Javan Erectus skull IV. In: Tuttle 1975: 361–372.

Krause, D. (2001) Fossil molar from a Madagascan marsupial. *Nature* **412**:497–498.

Krause, J., C. Lalueza-Fox, L. OrLando, W. Enard, R. Green, H. Burbano, J.-J. Hublin, C. Hänni, J. Fortea, M. de La Rasilla, J. Bertranpetit, A. Rosas, and S. Pääbo (2007a) The derived FOXP2 variant of modern humans was shared with Neandertals. *Current Biology* **17**:1908–1912.

Krause, J., L. Orlando, D. Serre, B. Viola, K. Prüfer, M. Richards, J.-J. Hublin, C. Hänni, A. Derevienko, and S. Pääbo (2007b) Neandertals in Central Asia and Siberia. *Nature* **442**:902–904.

Krause, W., and F. Jenkins, Jr. (1983) The postcranial skeleton of North American multituberculates. *Bulletin of the Museum of Comparative Zoology* **150**:199–246.

Krings, M., A. Stone, R. Schmitz, H. Kraininzki, M. Stoneking, and S. Pääbo (1997) Neanderthal DNA sequences and the origin of modern humans. *Cell* **90**:19–30.

Krings, M., H. Geisert, R. Schmitz, H. Kraininzki, and S. Pääbo (1999) DNA sequence of the mitochondrial hypervariable region II from the Neandertal type specimen. *PNAS* **96**:5581–5585.

Krings, M., C. Capelli, F. Tsenchentscher, H. Geisert, S. Meyer, A. von Haeseler, K. Grossschmidt, G. Possnert, M. Paunović, and S. Pääbo (2000) A view of Neandertal genetic diversity. *Nature Genetics* **26**:144–146.

Krogman, W. (1950) Classification of fossil men: Concluding remarks of the chairman. *CSHS* **15**:119–121.

Kroll, E. (1994) Behavioral implications of Plio-Pleistocene archaeological site structure. *JHE* **27**:107–138.

Kroll, E., and G. Isaac (1984) Configurations of artifacts and bones at early Pleistocene sites in East Africa. In: H. Hietala, Editor. *Intrasite Spacial Analysis in Archaeology*. Cambridge: Cambridge University Press, pp. 4–31.

Kuhn, S. (1995) *Mousterian Lithic Technology: An Ecological Perspective*. Princeton: Princeton University Press.

Kuhn, S., P. Brantingham, and K. Kerry (2004) The early Upper Paleolithic and the origins of modern human behavior. In: Brantingham et al. 2004: 242–248.

Kuhn, S., and M. Stiner (1992) New research on Riparo Mochi, Balzi Rossi (Liguria): preliminary results. *Quaternaria Nova* **2**:77–90.

Kuhn, S., and M. Steiner (2006) What's a mother to do? The division of labor among Neandertals and modern humans in Eurasia. *CA* **47**:953–980.

Kuman, K., and R. Clarke (2000) Stratigraphy, artifact industries and hominid associations for Sterkfontein Member 5. *JHE* **38**:827–847.

Kumazawa, Y., and M. Nishida (1999) Complete mitochondrial DNA sequences of the green turtle and blue-tailed mole skink: Statistical evidence for archosaurian affinity of turtles. *MBE* **16**:784–792.

Kummer, B. (1991). Biomechanical foundations of the development of human bipedalism. In: Coppens and Sénut 1991: 1–8.

Kuroda, S., T. Nishihara, S. Suzuki, and R. Oko (1996) Sympatric chimpanzees and gorillas in the Ndoki Forest, Congo. In: McGrew et al. 1996:71–81.

Kurtén, B. (1976) *The Cave Bear Story*. New York: Columbia University Press.

L

Laden, G., and R. Wrangham (2005) The rise of the hominids as an adaptive shift in fallback foods: Plant underground storage organs (USOs) and australopith origins. *JHE* **49**:482–498.

Lague, M., and W. Jungers (1996) Morphometric variation in Plio-Pleistocene hominid distal humeri. *AJPA* **101**:401–427.

Lahr, M. (1994) The multiregional hypothesis of modern human origins: A reassessment of its morphological basis. *JHE* **26**:23–56.

Lahr, M. (1996) *Evolution of Modern Human Diversity*. Cambridge: Cambridge University Press.

Lahr, M. (1997) The evolution of modern human cranial diversity. In: Clark and Willermet 1997: 304–318.

Lahr, M., and R. Foley (1994) Multiple dispersals and modern human origins. *EA* **3**:48–60.

Lahr, M., and R. Foley (1998) Towards a theory of modern human origins: geography, demography and diversity in recent human evolution. *YPA* **41**:137–176.

Lahr, M., and R. Foley (2004) Human evolution writ small. *Nature* **431**:1043–1044.

Lahr, M., and R. Wright (1996) The question of robusticity and the relationship between cranial size and shape in *Homo sapiens*. 31:157–191.

Lai, C., S. Fisher, J. Hurst, F. Vargha-Khadem, and A. Monaco (2001) A fork-head domain gene is mutated in a severe speech and language disorder. *Nature* **413**:519–523.

Laitman, J, R. Heimbuch, and E. Crelin (1979) The basicranium of fossil hominids as an indicator of their upper respiratory systems. *AJPA* **51**:15–34.

Laitman, J., J. Reidenberg, and P. Gannon (1992) Fossil skulls and hominid vocal tracts: New approaches to charting the evolution of human speech. In: J. Wind et al. Editors. *Language Origins: A Multidisciplinary Approach*. Dordrecht: Kluwer, pp. 395–407.

Laitman, J., J. Reidenberg, J. Marquez, and P. Gannon (1996) What the nose knows: New understandings of Neandertal upper respiratory tract specializations. *PNAS* **93**:10543–10545.

Laitman, J., and A. Thorne (2000) Australia. In: Delson et al. 2000: 107–112.

Lalueza-Fox, C., and A. Pérez-Pérez (1993) The diet of the Neanderthal child Gibraltar 2 (Devil's Tower) through the study of the vestibular striation pattern. *JHE* 24:29–41.

Lalueza-Fox, C., A. Pérez-Pérez, and D. Turbon (1996) Dietary inferences through buccal microwear analysis of Middle and Late Upper Pleistocene human fossils. *AJPA* **100**:367–387.

Lalueza-Fox, C., H. Römpler, D. Caramelli, C. Stäubert, G. Catalono, D. Hughes, N. Rohland, E. Pilli, L. Longo, S. Condemi, M. de la Rasilla, J. Fortea, A. Rosas, M. Stoneking, M. Schöneberg, J. Bertranpetit, and M. Hofreiter (2007) A melanocortin 1 receptor allele suggests varying pigmentation among Neandertals. *Science* **318**:1453–1455.

Lalueza-Fox, C., M. Sampietro, D. Caramelli, Y. Puder, M. Lari, F. CaLafell, C. Martinez-Maza, M. Bastir, J. Fortea, M. de laRasilla, J. Bertranpetit, and A. Rosas (2005) Neandertal evolutionary genetics: Mitochondrial DNA data from the Iberian peninsula. *MBE* **4**:1077–1081.

Lam, Y., O. Pearson, and C. Smith (1996) Chin morphology and sexual dimorphism in the fossil hominid mandible sample from Klasies River Mouth. *AJPA* **100**:545–557.

Lamarck, J.-B. 1809. *Philosophie Zoologique*. Paris: Dentu.

Lancaster, J. (1975) *Primate Behavior and the Emergence of Human Culture*. New York: Holt, Rinehart, and Winston.

Lancaster, J. (1978) Carrying and sharing in human evolution. *Human Nature* 1:83–89.

Langbroek, M., and W. Roebroeks (2000) Extraterrestrial evidence on the age of the hominids from Java. *JHE* **38**:595–600.

Langdon, J., J. Bruckner, and H. Baker (1991) Pedal mechanics and bipedalism in early hominids. In: Coppens and Sénut 1991: 159–167.

Larick, R., and R. Ciochon (1996) The first Asians. *Archaeology* **49**:51–53.

Larson, S. (1998) Parallel evolution in the hominoid trunk and forelimb. *EA* **6**:87–89.

Larson, S., W. Jungers, M. Morwood, T. Sutikna, Jatmiko, E. Saptomo, R. Due, and T. Djubiantono (2007) Misconceptions about the postcranial skeleton of *Homo floresiensis*. *AJPA Supplement* **44**:151.

Latham, A., and H. Schwarcz (1992) The Petralona hominid site: uranium-series reanalysis of "layer 10" calcite and associated paleomagnetic analyses. *Archaeometry* **34**:135–140.

Latimer, B. (1991) Locomotor adaptations in *Australopithecus afarensis*: The issue of arboreality. In: Coppens and Senut 1991: 169–176.

Latimer, B., and C. O. Lovejoy (1989) The calcaneus of *Australopithecus afarensis* and its implications for the evolution of bipedality. *AJPA* **78**:369–386.

Latimer, B., and C. O. Lovejoy (1990a) Hallucal tarsometatarsal joint in *Australopithecus afarensis*. *AJPA* **82**:125–133.

Latimer, B., and C. O. Lovejoy (1990b) Metatarsophalangeal joints of *Australopithecus afarensis*. *AJPA* **83**:13–23.

Latimer, B., and J. Ohman (2001) Axial dysplasia in *Homo erectus*. *JHE* **40**:12.

Latimer, B, J. Ohman, and C. O. Lovejoy (1987) Talocrural joint in African hominids: implications for *Australopithecus afarensis*. *AJPA* **74**:155–175.

Latimer, B., and C. Ward (1993) The thoracic and lumbar vertebrae. In: Walker and Leakey 1993a: 266–293.

Laville, H., J.-P. Rigaud, and J. Sackett (1980) *Rockshelters of the Perigord*. New York: Academic Press.

Leakey, L. (1934) *Adam's Ancestors*. London: Metheun.

Leakey, L. (1935) *The Stone Age Races of Kenya*. London: Oxford University Press.

Leakey, L. (1936) Fossil human remains from Kanam and Kanjera, Kenya Colony. *Nature* **138**:643.

Leakey, L. (1953) *Adam's Ancestors: The Evolution of Man and His Culture*, 4th edition. London: Methuen.

Leakey, L. (1959) A new fossil skull from Olduvai. *Nature* **184**:491–493.

Leakey, L. (1961) New finds at Olduvai Gorge. *Nature* **189**:649–650.

Leakey, L., P. Tobias, and J. Napier (1964) A new species of the genus *Homo* from Olduvai Gorge. *Nature* **202**:7–9.

Leakey, M. D. (1971a) Discovery of postcranial remains of *Homo erectus* and associated artifacts in Bed IV at Olduvai Gorge, Tanzania. *Nature* **232**:380–383.

Leakey, M. D. (1971b) *Olduvai Gorge: Excavations in Beds I and II, 1960–1963*. Cambridge: Cambridge University Press.

Leakey, M. D. (1975) Cultural practices in the Olduvai sequence. In: Butzer and Isaac 1975: 477–494.

Leakey, M. D., R. Clarke, and L. Leakey (1971) New hominid skull from Bed I, Olduvai Gorge, Tanzania. *Nature* **232**:308–312.

Leakey, M. D., and R. Hay (1979) Pliocene footprints in the Laetolil Beds at Laetoli, northern Tanzania. *Nature* **278**:317–323.

Leakey, M. D., and R. Hay (1982) The chronological position of the fossil hominids of Tanzania. In: M. de Lumley, Editor. *L'Homo erectus et la place de l'homme de Tautavel parmi les hominidés fossiles*. Nice: Premier Congrés International de Paléontologie Humaine, pp. 753–765.

Leakey, M. G., C. Feibel, I. McDougall, and A. Walker (1995a) New four-million-year-old hominid species from Kanapoi and Allia Bay, Kenya. *Nature* **376**:565–571.

Leakey, M. G., C. Feibel, I. McDougall, C. Ward, and A. Walker (1998) New specimens and confirmation of an early age for *Australopithecus anamensis*. *Nature* **393**:62–66.

Leakey, M.G., F. Spoor, F. Brown, P. Gathogo, C. Kiarie, L. Leakey, and I. McDougall (2001) New hominin genus from eastern Africa shows diverse middle Pliocene lineages. *Nature* **410**:433–440.

Leakey, M. G., P. Ungar, and A. Walker (1995b) A new genus of large primate from the late Oligicene of Lothidok, Turkana District, Kenya. *JHE* **28**:519–531.

Leakey, M. G., and A. Walker (1997) *Afropithecus*: Function and phylogeny. In: Begun et al. 1997: 225–239.

Leakey, R. (1971) Further evidence of lower Pleistocene hominids from East Rudolf, North Kenya. *Nature* **231**:241–245.

Leakey, R. (1972) Further evidence of lower Pleistocene hominids from East Rudolf, North Kenya, 1971. *Nature* **237**:264–269.

Leakey, R. (1973a) Evidence for an advanced Plio-Pleistocene hominid from East Rudolf, Kenya. *Nature* **242**:447–450.

Leakey, R. (1973b) Further evidence of lower Pleistocene hominids from East Rudolf, North Kenya, 1972. *Nature* **242**:170–173.

Leakey, R. (1974) Further evidence of lower Pleistocene hominids from East Rudolf, North Kenya, 1973. *Nature* **248**:653–656.

Leakey, R. (1976) New hominid finds from the Koobi Fora Formation in Northern Kenya. *Nature* **261**:574–576.

Leakey, R., M. G. Leakey, and A. Behrensmeyer (1978) The hominid catalogue. In: Leakey and Leakey 1978: 86–90.

Leakey, R., M. Leakey, and A. Walker (1988) Morphology of *Afropithecus turkanensis* from Kenya. *AJPA* **76**:289–307.

Leakey, R., and A. Walker (1976) *Australopithecus, Homo erectus*, and the single species hypothesis. *Nature* **261**:572–574.

Leakey, R., and A. Walker (1985) Further hominids from the Plio-Pleistocene of Koobi Fora, Kenya. *AJPA* **67**:135–163.

Leakey, R., and A. Walker (1988) New *Australopithecus boisei* specimens from East and West Lake Turkana, Kenya. *AJPA* **76**:1–24.

Leakey, R., and B. Wood (1973) New evidence of the genus *Homo* from East Rudolf, Kenya II. *AJPA* **39**:355–368.

Leakey, R., and B. Wood (1974) New evidence of the genus *Homo* from East Rudolf, Kenya IV. *AJPA* **41**:237–244.

Lebel, S., and E. Trinkaus (2002) Middle Pleistocene human remains from the Bau d'Aubesier. *JHE* **43**:659–685.

Lebel, S., E. Trinkaus, M. Faure, P. Fernandez, C. Guérin, D. Richter, N. Mercier, H. Valladas, and G. Wagner (2001) Comparative morphology and paleobiology of Middle Pleistocene human remains from the Bau d'Aubesier, Vaucluse, France. *PNAS* **98**:11097–11102.

Lee, R. (1968) What hunters do for a living, or how to make out on scarce resources. In: R. Lee and I. DeVore, Editors. *Man the Hunter*. Chicago: Aldine. 30–48.

Lee, R. (1969) !Kung Bushman subsistence: An input–output analysis. In: P. Vayda, Editor. *Environment and Cultural Behavior*. Garden City, NY: Natural History Press, pp. 47–79.

Lee, S.-H. (2005) Brief communication: is variation in the cranial capacity of the Dmanisi sample too high to be from a single species? *AJPA* **127**:263–266.

Lee, S-H., and M. Wolpoff (2003) The pattern of evolution in Pleistocene human brain size. *Paleobiology* **29**:186–196.

Lee-Thorp, J., J. Thackeray, and N. van der Merwe (2000) The hunters and the hunted revisited. *JHE* **39**:565–576.

Lee-Thorp, J., and M. Sponheimer (2006) Contributions of biogeochemistry to understanding hominin dietary ecology. *YPA* **49**:131–148.

Lee-Thorp, J., M. Sponheimer, and N. van der Merwe (2003) What do stable isotopes tell us about hominid dietary and ecological niches in the Pliocene? *IJO* **13**:104–113.

Le Gros Clark, W. (1938) The endocranial cast of the Swanscombe skull bones. *JRAI* **68**:61–67.

Le Gros Clark, W. (1950) Hominid characteristics of the australopithecine dentition. *JRAI* **80**:37–54.

Le Gros Clark, W. (1954) Reason and fallacy in the study of fossil man. *The Advancement of Science* **43**:280–292.

Le Gros Clark, W. (1958) Bones of contention. *Man* **88**:131–145.

Le Gros Clark, W. (1959) *The Antecedents of Man: An Introduction to the Evolution of the Primates*. Edinburgh: Edinburgh University Press.

Le Gros Clark, W. (1964) *The Fossil Evidence for Human Evolution*, 2nd edition. Chicago: University of Chicago Press.

Le Gros Clark, W. (1966) *History of the Primates*, 5th edition. Chicago: Phoenix.

Le Gros Clark, W. (1967) *Man–Apes or Ape–Men? The Story of Discoveries in Africa*. New York: Holt, Rinehart, and Winston.

Le Gros Clark, W., and L. Leakey (1951) The Miocene Hominoidea of East Africa. *Fossil Mammals of Africa* **1**:1–117.

Leigh, S. (1985) Ontogenetic and static allometry of the Neandertal and modern hominid palate. *AJPA* **66**:195.

Leigh, S. (1992) Cranial capacity evolution in *Homo erectus* and early *Homo sapiens*. *AJPA* **87**:1–13.

Leigh, S. (2001) Evolution of human growth. *EA* **10**:223–236.

Leigh, S. (2006) Brain ontogeny and life history in *Homo erectus*. *JHE* **50**:104–108.

LeMay, M. (1975) The language capability of Neanderthal man. *AJPA* **42**:9–14.

Lemelin, P., and D. Schmitt (2007) Origins of grasping and locomotor adaptations in primates: Comparative and experimental approaches using a primate model. In: Ravosa and Dagosto 2007: 329–380.

Lenneberg, E. (1964) The capacity for language acquisition. In: J. Fodor and J. Katz, Editors. *The Structure of Language*. Englewood Cliffs, NJ: Prentice-Hall, pp. 579–603.

Lenneberg, E. (1967) *Biological Foundations of Language*. New York: Wiley.

Leonard, W. (2000) Human nutritional evolution. In: S. Stinson, B. Bogin, R. Huss-Ashmore, and D. O'Rourke, Editors. *Human Biology: An Evolutionary and Biocultural Perspective*. New York: Wiley-Liss. 295–343.

Leonard, W. (2002) Food for thought. *SA* **287**:106–115.

Leonard, W., and M. Robertson (1994) Evolutionary perspectives on human nutrition: the influence of brain and body size on diet and metabolism. *AJHB* **6**:77–88.

Leonard, W., and M. Robertson (1995) Energetic efficiency of human bipedality. *AJPA* **97**:335–338.

Leonard, W., and M. Robertson (1997a) Rethinking the energetics of bipedality. *CA* **38**:304–309.

Leonard, W., and M. Robertson (1997b) Comparative primate energetics and hominid evolution. *AJPA* **102**:265–281.

Leonard, W., M. Sørensen, V. Galloway, G. Spencer, M. Mosher, L. Osipova, and V. Spitsyn (2002) Climatic influences on basal metabolic rates among circumpolar populations. *AJHB* **14**:609–620.

Lepland, A., M. van Zuilen, G. Arrhenius, M. Whitehouse, and C. Fedo (2005) Questioning the evidence for Earth's earliest life—Akilia revisited. *Geology* **33**:77–79.

Lester, L. (1994) The history of life. *Creation Research Society Quarterly* **31**:(2): insert.

Leutenegger, W., and B. Shell (1987) Variability and sexual dimorphism in canine size of *Australopithecus* and extant hominoids. *JHE* **16**:359–367.

Lev, E., M. Kislev, and O. Bar-Yosef (2005) Mousterian vegetal food in Kebara Cave, Mt. Carmel. *JAS* **32**:475–484.

Lévêque, F., and B. Vandermeersch (1980) Découverte de restes humaines dans un niveau castelperronien à Saint-Césaire (Charente-Maritime) *CRAS* **291**:187–189.

Lévêque, F., and B. Vandermeersch (1981) Le néandertalien de Saint-Césaire. *La Recherche* **12**:242–244.

Levin, D. (2002) Hybridization and extinction. *AS* **90**:254–261.

Lewis, G. (1934) Preliminary notice of new man-like ape from India. *American Journal of Science* **27**:161–179.

Lewis, O. (1972) Evolution of the hominoid wrist joint. In: R. Tuttle, Editor. *The Functional and Evolutionary Biology of Primates*. Chicago: Aldine-Atherton, pp. 207–222.

Lewis, O. (1981) Functional morphology of the joints of the evolving foot. *Symposia of the Zoological Society of London* **46**:169–188.

Li, T., and D. Etler (1992) New Middle Pleistocene hominid crania from Yunxian in China. *Nature* **357**:404–407.

Li, X., and S. Zhang (1984) Paleoliths discovered in Ziyang locality B. *AAS* **3**:215–224.

Licent, E., P. Teilhard de Chardin, and D. Black (1926) On a presumably Pleistocene human tooth from the Sjara-osso-gol (Southeastern Ordos) deposits. *Bulletin of the Geological Society of China* **5**:285–290.

Lieberman, D. (1995) Testing hypotheses about recent human evolution from skulls. *CA* **36**:159–197.

Lieberman, D. (1996) How and why humans grow thin skulls: Experimental evidence for systemic cortical robusticity. *AJPA* **101**:217–236.

Lieberman, D. (1998a) Sphenoid shortening and the evolution of modern human cranial shape. *Nature* **393**:158–162.

Lieberman, D. (1998b) Neandertal and early modern human mobility patterns. In: Akazawa et al. 1998: 263–275.

Lieberman, D. (2001) Another face in our family tree. *Nature* **410**:419–420.

Lieberman, D. (2005) Further fossil finds from Flores. *Nature* **437**:957–958.

Lieberman, D. (2007) Homing in on early *Homo*. *Nature* **449**:291–292.

Lieberman, D., and R. McCarthy (1999) The ontogeny of cranial base angulation in humans and chimpanzees and its implications for reconstructing pharyngeal dimensions. *JHE* **36**:487–517.

Lieberman, D., D. Pilbeam, and B. Wood (1988) A probabilistic approach to the problem of sexual dimorphism in *Homo habilis*: A comparison of KNM-ER 1470 and KNM-ER 1813. *JHE* **17**:503–511.

Lieberman, D., D. Raichlen, H. Pontzer, D. Bramble, and E. Cutright-Smith (2006) The human gluteus maximus and its role in running. *Journal of Experimental Biology* **209**: 2143–2155.

Lieberman, D., C. Ross, and M. Ravosa (2000) The primate cranial base: Ontogeny, function, and integration. *YPA* **43**:117–169.

Lieberman, D., and J. Shea (1994) Behavioral differences between archaic and modern humans in the Levant. *AA* **96**:300–332.

Lieberman, D., B. Wood, and D. Pilbeam (1996) Homoplasy and early *Homo*: An analysis of the evolutionary relationships of *Homo habilis* sensu stricto and *Homo rudolfensis*. *JHE* **30**:97–120.

Lieberman, P. (1984) *The Biology and Evolution of Language*. Cambridge: Harvard University Press.

Lieberman, P. (1989) The origins of some aspects of human speech and cognition. In: Mellars and Stringer 1989: 391–414.

Lieberman, P. (1993) *Uniquely Human: The Evolution of Speech, Thought, and Selfless Behavior*. Cambridge: Harvard University Press.

Lieberman, P., and E. Crelin (1971) On the speech of Neanderthal man. *Linguistic Inquiry* **2**:203–222.

Lieberman, P., E. Crelin, and D. Klatt (1972) Phonetic ability and related ability of the newborn, adult human, Neanderthal man, and the chimpanzee. *AA* **74**:287–307.

Lillegraven, J. (1969) Latest Cretaceous mammals of upper part of Edmonton Formation of Alberta, Canada, and review of marsupial–placental dichotomy in mammalian evolution. *University of Kansas Paleontological Contributions* **50**: 1–122.

Linne, K. (1735) *Systema Naturae sive Regna Tria Naturae systematice proposita per Classes, Ordines, Genera, et Species*. Leiden: T. Haak. 1964 facsimile reprint, Nieuwkoop: de Graaf.

Lipson, S., and D. Pilbeam (1982) *Ramapithecus* and hominoid evolution. *JHE* **11**:545–548.

Liu, C.-W. J-Y. Chen, and T-E. Hua (1998) Precambrian sponges with cellular structures. *Science* **279**:879–882.

Liu, T., S. Stenger, H. Li, L. Wenzel, B. Tan, S. Krutzik, M. Ochoa, J. Schauber, K. Wu, C. Meinkin, D. Kamen, M. Wagner, R. Bals, A. Steinmeyer, U. Zugel, R. Gallo. D. Eisenberg, M. Hewison, B. Hollis, J. Adams, B. Bloom, and R. Modlin (2006) Toll-loke receptor triggering of a vitamin D-mediated human antimicrobial response. *Science* **311**:1770–1773.

Livingstone, F. (1962) On the non-existence of human races. *CA* **3**:279–281.

Livingstone, F. (1992) Gene flow in the Pleistocene. *HB* **64**:67–80.

Lockwood, C., W. Kimbel, and D. Johanson (2000) Temporal trends and metric variation in the mandibles and dentition of *Australopithecus afarensis*. *JHE* **39**:23–55.

Lockwood, C., B. Richmond, W. Jungers, and W. Kimbel (1996) Randomization procedures and sexual dimorphism in *Australopithecus afarensis*. *JHE* **31**:537–548.

Lockwood, C., and P. Tobias (2002) Morphology and affinities of new hominin cranial remains from Member 4 of the Sterkfontein Formation, Gauteng Province, South Africa. *JHE* **42**:389–450.

Loomis, W. (1967) Skin-pigment regulation of vitamin-D biosynthesis in man. *Science* **157**:501–506.

Lordkipanidze D., T. Jashashvili, A. Vekua, M. Ponce de León, C. Zollikofer, G. Rightmire, H. Pontzer, R. Ferring, O. Oms, M. Tappen, M. Bukhsianidze, J. Agusti, R. Kahlke, G. Kiladze, B. Martienz-Navarro, A. Mouskhelishvili, M. Niordaze, and L. Rook (2007) Postcranial evidence from early *Homo* from Dmanisi, Georgia. *Nature* **449**:305–310.

Lordkipanidze, D., A. Vekua, R. Ferring, G. Rightmire, J. Agustill, G. Kiladze, A. Mouskhelishvili, M. Nioradze, M. Ponce de León, M. Tappen, and C. Zollikofer (2005) The earliest toothless hominin skull. *Nature* **434**:717–718.

Lorenzo, C., J. Arsuaga, and J. Carretero (1999) Hand and foot remains from the Gran Dolina Early Pleistocene site (Sierra de Atapuerca, Spain). In: Bermúdez de Castro et al. 1999a: 501–522.

Lovejoy, C. O. (1970) The taxonomic status of the "Meganthropus" mandibular fragments from the Djetis Beds of Java. *Man* **5**:228–236.

Lovejoy, C. O. (1974) The gait of australopithecines. *YPA* **17**: 147–161.

Lovejoy, C. O. (1975) Biomechanical perspectives on the lower limb of early hominids. In: R. Tuttle, Editor. *Primate Functional Morphology and Evolution*. The Hague: Mouton, pp. 291–326.

Lovejoy, C. O. (1981) The origin of man. *Science* **211**:341–350.

Lovejoy, C. O. (1982a) Diaphyseal biomechanics of the locomotor skeleton of Tautavel Man with comments on the evolution of skeletal changes in Late Pleistocene man. In: H. deLumley, Editor. *L'Homo erectus et la place de l'homme de Tautavel parmi les hominidés fossiles*. Nice: Premier Congrès International de Paléontologie Humaine, pp. 447–470.

Lovejoy, C. O. (1982b) Models of human evolution. *Science* **217**:304–305.

Lovejoy, C. O. (1988) Evolution of human walking. *SA* **259**:118–126.

Lovejoy, C. O. (1993) Modeling human origins: are we sexy because we're smart, or smart because we're sexy? In: D. Rasmussen, Editor. *The Origin and Evolution of Humans and Humanness.* Boston: Jones and Bartlett pp. 1–28.

Lovejoy, C. O. (2005a) Natural history of human gait and posture. Part 1. Spine and pelvis. *Gait and Posture* **21**:95–112.

Lovejoy, C. O. (2005b) Natural history of human gait and posture. Part 2. Hip and thigh. *Gait and Posture* **21**:113–124.

Lovejoy, C. O. (2007) Natural history of human gait and posture. Part 3. Knee. *Gait and Posture* **25**:325–341.

Lovejoy, C. O., M. Cohn, and T. White (1999) Morphological analysis of the mammalian postcranium: a developmental perspective. *PNAS* **96**:13247–13252.

Lovejoy, C. O., K. Heiple, and A. Burstein (1973) The gait of *Australopithecus. AJPA* **38**:757–780.

Lovejoy, C. O., and E. Trinkaus (1980) Strength and robusticity of the Neandertal tibia. *AJPA* **53**:465–470.

Lowe, A. J., and Abbott, R. J. (2004) Reproductive isolation of a new hybrid species, *Senecio eboracensis* Abbott & Lowe (Asteraceae). *Heredity* **92**:386–395.

Luckett, W. (1975) Ontogeny of the fetal membranes and placenta: Their bearing on primate phylogeny. In: Luckett and Szalay 1975: 157–182.

Luckett, W. (1976) Cladistic relationships among primate higher categories: Evidence of the fetal membranes and placenta. *FP* **25**:245–276.

Luckett, W. (1977) Ontogeny of amniote fetal membranes and their application to phylogeny. In: M. Hecht, P. Goody, and B. Hecht, Editors. *Major Patterns in Vertebrate Evolution.* New York: Plenum. 439–516.

Luo, Z.-X., Q. Ji, J. Wible, and C.-X. Yuan (2003) An Early Cretaceous tribosphenic mammal and metatherian evolution. *Science* **302**:1934–1940.

Luo, Z.-X., , Z. Kielan-Jaworowska, and R. Cifelli (2001a) Dual origin of tribosphenic mammals. *Nature* **409**:53–57.

Luo, Z.-X., Crompton, A. W., and Sun, A-L. (2001b). A new mammaliaform from the Early Jurassic of China and evolution of mammalian characteristics. *Science* **292**:1535–1540.

Lyell, C. (1830) *Principles of Geology: Being an Attempt to Explain the Former Changes of the Earth's Surface by Reference to Causes Now in Operation.* London: John Murray.

M

MacArthur, R., and E. Wilson (1967) *The Theory of Island Biogeography.* Princeton: Princeton University Press.

Macintosh, N. (1978) The Tabon Cave mandible. *Archaeology and Physical Anthropology in Oceania* **13**:143–159.

MacLarnon, A. (1993) The vertebral canal. In: Walker and Leakey 1993a: 359–390.

MacLatchy, L. (1996) Another look at the australopithecine hip joint. *JHE* **31**:455–476.

MacLatchy, L., D. Gebo, R. Kityo, and D. Pilbeam (2000) Postcranial functional morphology of *Morotopithecus bishopi*, with implications for the evolution of modern ape locomotion. *JHE* **39**:159–183.

MacPhee, R., and M. Cartmill (1986) Basicranial structures and primate systematics. In: D. Swindler, Editor. *Comparative Primate Biology, Volume 1: Systematics, Evolution, and Anatomy.* New York: Alan R. Liss, pp. 219–275.

MacPhee, R., and I. Horovitz (2002) Extinct Quaternary platyrrhines of the Greater Antilles and Brazil. In: Hartwig 2002:189–200.

Madar, S., M. Rose, J. Kelley, L. MacLatchy, and D. Pilbeam (2001) New *Sivapithecus* postcranial specimens from the Siwaliks of Pakistan. *JHE* **42**:705–752.

Maggioncalda, A. (1995) *The socioendocrinology of orangutan growth, development, and reproduction—An analysis of endocrine profiles of juvenile, developing adolescent, developmentally arrested adolescent, adult, and aged captive male orangutans.* Ph.D. thesis, Duke University.

Magori, F., and M. Day (1983) Laetoli Hominid 18: An early *Homo sapiens* skull. *JHE* **12**:747–753.

Maier, W. (1980) Nasal structures in Old and New World primates. In: R. Ciochon and A. Chiarelli, Editors. *Evolutionary Biology of the New World Monkeys and Continental Drift.* New York: Plenum, pp. 219–241.

Malez, M. (1976) Excavation of the Villafranchian site Šandalja I near Pula (Jugoslavija). In: K. Valoch, Editor. *Les premiers industries de l'Europe.* Nice: IX Congress UISPP, Colloque VIII, pp. 104–123.

Malez, M. (1978) Stratigrafski, paleofaunski, paleolitski odnosi krapinog nalažišta. In: M. Malez, Editor. *Krapinski Pračovjek i Evolucija Hominida.* Zagreb: Jugoslavenska Akademija Znanosti I Umjetnosti. 61–102.

Malez, M., F. Smith, D. Rukavina, and J. Radovčic (1980) Upper Pleistocene fossil hominids from Vindija, Croatia, Yugoslavia. *CA* **21**:365–367.

Malez, M., and H. Ullrich (1982) Neuere paläoanthropologische Untersuchungen am Material aus der Höhle Vindija (Kroatien, Jugoslawien). *Palaeontologia Jugoslavica* **29**:1–44.

Mallegni, F. and E. Trinkaus (1997) A reconsideration of the Archi I Neandertal mandible. *JHE* **33**:651–668.

Mallet, J. (2001) The speciation revolution. *Journal of Evolutionary Biology* **14**:887–888.

Manega, P. (1993) *Geochronology, geochemistry and isotopic study of the Plio-Pleistocene hominid sites and the Ngorongoro volcanic highlands in Northern Tanzania.* Ph.D. thesis. University of Colorado.

Mania, D., U. Mania, and E. Vlček (1993) Zu den Funden der Hominiden-Reste aus dem mittelpleistozänen Travertin von Bilzingsleben von 1987–1993. *EAZ* **34**:511–524.

Mania, D., U. Mania, and E. Vlček (1995) Dernières découvertes de restes humains dans le travertin du pléistocène moyen de

Bilzingsleben: état actuel des recherches. *L'Anthropologie (Paris)* **99**:42–54.

Mania, D., and E. Vlček (1987) *Homo erectus* from Bilzingsleben (GDR)—his culture and his environment. *Anthropologie (Brno)* **26**:1–45.

Mann, A. (1972) Hominid and cultural origins. *Man* **7**:379–386.

Mann, A. (1975) Some paleodemographic aspects of the South African Australopithecines. Philadelphia: University of Pennsylvania Publications in Anthropology, No. 1.

Mann, A., J. Monge, and M. Lampl (1991) Investigation into the relationship between perikymata counts and crown formation times. *AJPA* **86**:175–188.

Mann, A., and M. Weiss (1996) Hominoid phylogeny and taxonomy: A consideration of the molecular and fossil evidence in an historical perspective. *MPE* **5**:169–181.

Manzi, G., E. Bruner, and P. Passarello (2003) The one-million-year-old cranium from Bouri (Ethiopia): A consideration of its *H. erectus* affinities. *JHE* **44**:731–736.

Manzi, G., F. Mallegni, and A. Ascenzi (2001) A cranium for the earliest Europeans: phylogenetic position of the hominid from Ceprano, Italy. *PNAS* **98**:10011–10016.

Marchal, F. (1999) L'os coxal du genre *Homo* au Pléistocène inférieur et moyen. *L'Anthropologie (Paris)* **103**:223–235.

Marean, C. (1998) A critique of the evidence of scavenging by Neandertals and early modern humans: New data from Kobeh Cave (Zagros Mountains, Iran) and Die Kelders 1 Layer 10 (South Africa). *JHE* **35**:111–136.

Marean, C., M. Bar-Matthews, J. Bernatchez, E. Fisher, P. Goldberg, A. Herries, Z. Jacobs, A. Jerardino, P. Karkanas, T. Minichillo, P. Nilssen, E. Thompson, I. Watts, and H. Williams (2007) Early human use of marine resources and pigment in South Africa during the Middle Pleistocene. *Nature* **449**:905–908.

Marean, C., and S. Kim (1998) Mousterian large mammal remains from Kobeh Cave: Behavioral implications for Neandertals and early modern humans. *CA* **39**:579–613.

Marivaux, L., Y. Chaimanee, S. Ducrocq, B. Marandat, J. Sudre, A. Soe, S. Tun, W. Htoon, and J.-J. Jaeger (2003) The anthropoid status of a primate from the late middle Eocene Pondaung formation (Central Myanmar): Tarsal evidence. *PNAS* **100**: 13173–13178.

Marko, P. (2002) Fossil calibration of molecular clocks and the divergence times of geminate species separated by the Isthmus of Panama. *MBE* **19**:2005–2021.

Marks, A., Y. Demidenko, K. Monigal, V. Usik, C. Ferring, A. Burke, J. Rink, and C. McKinney (1997) Staroselje and the Staroselje child: New excavations, new results. *CA* **38**:112–123.

Marks, J. (1994) Blood will tell (won't it?): A century of molecular discourse in anthropological systematics. *AJPA* **94**:59–80.

Marks, J. (2005) Phylogenetic trees and evolutionary forests. *EA* **14**:49–53.

Marquet, J., and M. Lorblanchet (2003) A Neanderthal face? The proto-figure from La Rochet-Cotard, Langeais (Indreet Loire, France). *Antiquity* **77**:661–670.

Márquez, S., K. Mowbray, G. Sawyer, T. Jacob, and A. Silvers (2001) New fossil hominid calvaria from Indonesia—Sambungmacan 3. *AR* **262**:344–368.

Marshack, A. (1976) Implications of the Paleolithic symbolic evidence for the origin of language. *AS* **64**:136–145.

Marshack, A. (1989) Evolution of the human capacity: The symbolic evidence. *YPA* **32**:1–34.

Marshall, L. (1988) Land mammals and the Great American Interchange. *AS* **76**:380–388.

Marston, A. (1936) Preliminary note on a new fossil human skull from Swanscombe, Kent. *Nature* **138**:200–201.

Marston, A. (1937) The Swanscombe skull. *JRAI* **67**:339–406.

Martin, H. (1923) *L'homme fossile de la Quina.* Paris: O. Doin.

Martin, P. (2005) *Twilight of the Mammoths: Ice Age Extinctions and the Rewilding of America.* Berkeley: Univiversity of California Press.

Martin, R. (1973) Comparative anatomy and primate systematics. *Symposia of the Zoological Society of London* **33**:301–337.

Martin, R. (1981) Relative brain size and basal metabolic rate in terrestrial vertebrates. *Nature* **293**:57–60.

Martin, R. (1983) Human brain evolution in an ecological context. 52nd James Arthur Lecture on the Evolution of the Human Brain. New York: American Museum of Natural History.

Martin, R. (1989) Evolution of the brain in early hominids. *Ossa* **14**:49–62.

Martin, R. (1990) *Primate Origins and Evolution: A Phylogenetic Reconstruction.* London: Chapman & Hall.

Martin, R. (1991) New fossils and primate origins. *Nature* **349**:19–20.

Martin, R. (1993) Primate origins: plugging the gaps. *Nature* **363**:223–234.

Martin, R. (1996) Scaling of the mammalian brain: The maternal energy hypothesis. *News in Physiological Sciences* **11**:149–156.

Martin, R. (1998) Comparative aspects of human brain evolution: Scaling, energy costs, and confounding variables. *Memoirs of the California Academy of Sciences* **24**:35–68.

Martin, R. (2004) Chinese lantern for early primates. *Nature* **427**:22–23.

Martin, R., A. Maclarnon, J. Phillips, L. Dussubieux, P. Williams, and W. Dobyns (2006a) Comment on "The brain of LB1, *Homo floresiensis*." *Science* **312**:999.

Martin, R., A. Maclarnon, J. Phillips, and W. Dobyns (2006b) Flores hominid: New species or microcephalic dwarf? *AR* **288A**:1123–1145.

Martínez, I., M. Rosa, J-I. Arsuaga, P. Jarabo, R. Quam, C. Lorenzo, A. Gracia, J.-M. Carretero, J.-M. Bermúdez de Castro, and E. Carbonell (2004) Auditory capacities in Middle Pleistocene humans from the Sierra de Atapuerca in Spain. *PNAS* **101**:9976–9981.

Martínez-Navarro, B., and P. Palmqvist (1995) Presence of the African machairodont *Megantereon whitei* (Broom 1937) (Felidae, Carnivora, Mammalia) in the Lower Pleistocene site of Venta Micena (Orce, Granada, Spain) with some considerations on the origin, evolution and dispersal of the genus. *JAS* **22**:569–582.

Masao, F. (1992) The Middle Stone Age with reference to Tanzania. In: Bräuer and Smith 1992: 99–109.

Matiegka, J. (1934) *Homo Předmostensis: Fossilni Člověk z Předmostí na Moravě. 1. Lebky*. Prague: Ceská Academie Věd I Uměni.

Matiegka, J. (1938) *Homo Předmostensis: Fossilni Člověk z Předmostí na Moravě. 2. Ostané Částí Kostrové*. Prague: Česká Academie Věd I Uměni.

Maureille, B. (2002) La redécouverte du noveau-né Néandertalien Le Moustier 2. *Paleo* **14**:221–238.

Maureille, B., and D. Bar (1999) The premaxilla in Neandertal and early modern children: Ontogeny and morphology. *JHE* **37**:137–152.

Maureille, B., and F. Houët (1997) Nouvelles données sur des caractéristiques dérivées du massif facial supérieur des Néandertaliens. *Archaeologie et Préhistoire* **108**:89–98.

Maynard Smith, J. (1966) Sympatric speciation. *American Naturalist* **100**:637–650.

Maynard Smith, J., and R. Savage (1959) The mechanics of mammalian jaws. *School Science Review* **141**:289–301.

Mayr, E. (1942) *Systematics and the Origin of Species*. New York: Columbia University Press.

Mayr, E. (1950) Taxonomic categories in fossil hominids. *CSHS* **15**:109–118.

Mayr, E. (1969) *Principles of Systematic Zoology*. New York: McGraw-Hill.

Mayr, E. (1976) *Evolution and the Diversity of Life*. Cambridge: Harvard University.

Mayr, E. (1982) Processes of speciation in animals. In: C. Barigozzi, Editor. *Mechanisms of Speciation*. New York: Liss, pp. 1–19.

McBrearty, S., and A. Brooks (2000) The revolution that wasn't: A new interpretation of the origin of modern human behavior. *JHE* **39**:453–563.

McBrearty, S., and C. Stringer (2007) The coast in colour. *Nature* **449**:793–794.

McBurney, C., Editor (1967) *The Haua Fteah (Cyrenaica) and the Stone Age of the South-East Mediterranean*. Cambridge: Cambridge University Press.

McCarthy, R., M. Holmes, L. Lucas, and K. O' Donnell (2007) Taxonomy of Middle Pleistocene humans: What is *Homo heidelbergensis* anyway? *AJPA Supplement* **44**:167.

McCown, T., and A. Keith (1939) *The Stone Age of Mount Carmel, Volume 2: The Fossil Human Remains from the Levailloiso-Mousterian*. Oxford. Clarendon Press.

McCrossin, M. (2005) Review of *Geology and Paleontology of the Miocene Sinap Formation, Turkey*, edited by M. Fortelius, J. Kappelman, S. Sen, and R. Bernor. *AJPA* **128**:488–490.

McCrossin, M., and B. Benefit (1997) On the relationships and adaptations of *Kenyapithecus*, a large-bodied hominoid from the Middle Miocene of eastern Africa. In: Begun et al. 1997: 241–267.

McCrossin, M., B. Benefit, S. Gitau, A. Palmer, and K. Blue (1998) Fossil evidence for the origins of terrestriality among Old World higher primates. In: Strasser et al. 1998: 353–396.

McCullom, M., F. Grine, S. Ward, and W. Kimbel (1993) Subnasal morphological variation in extant hominoids and fossil hominids. *JHE* **24**:87–111.

McDermott, F., R. Grün, C. Stringer, and C. Hawkesworth (1993) Mass-spectrometric U-series dates for Israeli Neanderthal/early modern hominid sites. *Nature* **363**:252–255.

McDermott, F., C. Stringer, R. Grün, C. Williams, V. Din, and C. Hawkesworth (1996) New Late Pleistocene uranium–thorium and ESR dates for the Singa hominid (Sudan). *JHE* **31**:507–516.

McDougall, I., F. Brown, and J. Fleagle (2005) Stratigraphic placement and age of modern humans from Kibish, Ethiopia. *Nature* **433**:733–736.

McFadden, P. (1980) An overview of padeomagnetic chronology with special reference to the south African hominid sites. *Palaeontologia Africana* **23**:35–40.

McGrew, W. (2001a) The nature of culture: Prospects and pitfalls of cultural primatology. In: F. de Waal, Editor. *Tree of Origin: What Primate Behavior Can Tell Us about Human Social Evolution*. Cambridge: Harvard University Press, pp. 229–254.

McGrew, W. (2001b) The other faunivory: Primate insectivory and early human diet. In: Stanford and Bunn 2001: 160–178.

McGrew, W., P. Baldwin, and C. Tutin (1988) Diet of wild chimpanzees (*Pan troglodytes verus*) at Mt. Assirik, Senegal: I. Composition. *AJP* **16**:213–226.

McHenry, H. (1975) A new pelvic fragment from Swartkrans and the relationship between the robust and gracile australopithecines. *AJPA* **43**:245–262.

McHenry, H. (1984) Relative cheek-tooth size in *Australopithecus*. *AJPA* **64**:297–306.

McHenry, H. (1985) Implications of postcanine megadontia for the origin of *Homo*. In: Delson 1985: 178–183.

McHenry, H. (1988) New estimates of body weight in early hominids and their significance to encephalization and megadontia in "robust" australopithecines. In: Grine 1988: 133–148.

McHenry, H. (1991a) First steps? Analyses of the postcranium of early hominids. In: Coppens and Sénut 1991: 133–141.

McHenry, H. (1991b) Petite bodies of the "robust" australopithecines. *AJPA* **86**:445–454.

McHenry, H. (1991c) Sexual dimorphism in *Australopithecus afarensis*. *JHE* **20**:21–32.

McHenry, H. (1992) Body size and proportions in early hominids. *AJPA* **87**:407–431.

McHenry, H. (1994a) Behavioral ecological implications of early hominid body size. *JHE* **27**:77–87.

McHenry, H. (1994b) Tempo and mode in human evolution. *PNAS* **91**:6780–6786.

McHenry, H. (1994c) Early hominid postcrania: Phylogeny and function. In: Corruccini and Ciochon 1994: 251–268.

McHenry, H. (1996a) Homoplasy, clades, and hominid phylogeny. In: E. Meikle, F. C. Howell, and N. Jablonski, Editors. *Contemporary Issues in Human Evolution*. San Francisco: California Academy of Sciences, pp. 77–92.

McHenry, H. (1996b) Sexual dimorphism in fossil hominids and its socioecological implications. In: J. Steele and S. Shennan, Editors. *The Archaeology of Human Ancestry: Power, Sex and Tradition*. London: Routledge, pp. 91–109.

McHenry, H., and L. Berger (1998a) Body proportions in *Australopithecus afarensis* and the origins of the genus *Homo*. *JHE* **35**:1–22.

McHenry, H., and L. Berger (1998b) Limb lengths in *Australopithecus* and the origins of the genus *Homo*. *SAJS* **94**:447–450.

McHenry, H., and R. Corruccini (1978) The femur in early human evolution. *AJPA* **49**:473–488.

McHenry, H., and R. Corruccini (1983) The wrist of *Proconsul africanus* and the origin of hominoid postcranial adaptations. In: Ciochon and Corruccini 1983: 353–367.

McHenry, H., and A. Jones (2006) Hallucial divergence in early hominids. *JHE* **50**:534–539.

McKee, J. (1989) Australopithecine anterior pillars: reassessment of the functional morphology and phylogenetic relevance. *AJPA* **80**:1–9.

McKee, J. (1996) Faunal evidence and Sterkfontein Member 2 foot bones of early hominid. *Science* **271**:1301.

McKee, J. (1999) The autocatalytic nature of hominid evolution in African Plio-Pleistocene environments. In: T. Bromage and F. Schrenk, Editors. *African Biogeography, Climate Change, and Human Evolution*. New York: Oxford University Press, pp. 57–67.

McKee, J., J. Thackeray, and L. Berger (1995) Faunal assemblage seriation of southern African Pliocene and Pleistocene fossil deposits. *AJPA* **96**:235–250.

McMahon, T. (1985) The role of compliance in mammalian running gaits. *Journal of Experimental Biology* **115**:263–282.

McMenamin, M., and D. McMenamin (1989) *The Emergence of Animals: The Cambrian Breakthrough*. New York: Columbia University Press.

McVean, G. (2001) What do patterns of genetic variability reveal about mitochondrial recombination? *Heredity* **87**:613–620.

Meban, C. (1983) The surface area and volume of the human fetus. *JA* **137**:271–278.

Mehlman, M. (1984) Archaic *Homo sapiens* from Lake Eyasi, Tanzania: Recent misinterpretations. *JHE* **13**:487–501.

Mehlman, M. (1987) Provenience, age and association of archaic *Homo sapiens* crania from Lake Eyasi, Tanzania. *JAS* **14**: 133–162.

Meldrum, D., and R. Wunderlich (1998) Midfoot flexibility in ape foot dynamics, early hominid footprints and bipedalism. *AJPA Supplement* **26**:161.

Mellars, P. (1986) A new chronology for the Mousterian period. *Nature* **322**:410–411.

Mellars, P. (1989) Major issues in the emergence of modern humans. *CA* **30**:349–385.

Mellars, P. (1993) Archaeology and the population dispersal hypothesis of modern human origins in Europe. In: Aitken et al. 1993: 196–216.

Mellars, P. (1996) *The Neanderthal Legacy: An Archaeological Perspective from Western Europe*. Princeton: Princeton University Press.

Mellars, P. (2002) Archaeology and the origins of modern humans: European and African perspectives. In: Crow 2002: 31–47.

Mellars, P. (2004) Neanderthals and the modern human colonization of Europe. *Nature* **432**:461–465.

Mellars, P. (2006a) A new radiocarbon revolution and the dispersal of modern humans in Eurasia. *Nature* **438**:931–935.

Mellars, P. (2006b) Going east: New genetic and archaeological perspectives on the modern human colonization of Eurasia. *Science* **313**:796–800.

Mellars, P. (2006c) Archaeology and the dispersal of modern humans in Europe: Deconstructing the "Aurignacian." *EA* **15**:167–182.

Meng, J., Y. Hu, Y. Wang, and C. Li (2003) The ossified Meckel's cartilage and internal groove in Mesozoic mammaliaforms: Implications to the origin of the definitive mammalian middle ear. *Zoological Journal of the Linnean Society* **138**:431–448.

Menter, C., A. Keyer, et al. (1999) First record of hominind teeth from the Plio-Pleistocene site of Gondolin, South Africa. *JHE* **37**:299–307.

Mercier, N. H. Valladas, J. Joron, J. Reyss, F. Lévêque, and B. Vandermeersch (1991) Thermoluminescence dating of the late Neanderthal remains from Saint-Césaire. *Nature* **351**:737–739.

Mercier, N., H. Vallada, O. Bar-Yosef, C. Stringer, and J. Joron (1993) Thermoluminescence date for the Mousterian burial site of Es-Skhul, Mt. Carmel. *JAS* **20**:169–174.

Mercier, N., H. Valladas, G. Valladas, J. Reyss, A. Jelinek, L. Meignen, and J. Joron (1995) TL dates of burnt flints from Jelinek's excavations at Tabun and their implications. *JAS* **22**:495–510.

Meyer, M. (2005) *Functional biology of the* Homo erectus *axial skeleton from Dmanisi, Georgia*. Ph.D. thesis, University of Pennsylvania.

Meyer, M., D. Lordkipanidze, and A. Vekua (2006) Evidence for the anatomical capacity for spoken language in *Homo erectus*. *AJPA Supplement* **42**:130.

Millard, A., and A. Pike (1999) Uranium series dating of the Tabun Neanderthal: A cautionary note. *JHE* **36**:581–586.

Miller, E., and E. Simons (1997) Dentition of *Proteopithecus sylviae*, an archaic anthropoid from the Fayum, Egypt. *PNAS* **94**:13760–13764.

Miller, G., P. Beaumont, A. Jull, and B. Johnson (1993) Pleistocene geochronology and palaeothermometry from protein diagenesis in ostrich eggshells: Implications for the evolution of modern humans. In: Aitken et al.: 49–68.

Miller, J. (1991) Does brain size variability provide evidence of multiple species in *Homo habilis*? *AJPA* **84**:385–398.

Miller, J. (2000) Craniofacial variation in *Homo habilis:* An analysis of the evidence for multiple species. *AJPA* **112**:103–128.

Miller, M., G. Christensen, and H. Evans (1964) *Anatomy of the Dog*. Philadelphia: W. B. Saunders.

Minugh-Purvis, N. (1988) *Pattern of craniofacial growth and development in Upper Pleistocene hominids*. Ph.D. thesis, University of Pennsylvania.

Minugh-Purvis, N. (1993) Reexamination of the immature hominid maxilla from Tangier, Morocco. *AJPA* **92**:449–461.

Minugh-Purvis, N. (1998) The search for the earliest modern Europeans: A comparison of the Es-Skhul I and Krapina I juveniles. In: Akazawa et al. 1998: 339–352.

Minugh-Purvis, N., J. Radovčić, and F. Smith (2000) Krapina 1: A juvenile Neandertal from the early Late Pleistocene of Croatia. *AJPA* **111**:393–424.

Minugh-Purvis, N., T. Viola, and M. Teschler-Nicola (2006) The Mladeč 3 infant. In: Teschler-Nicola 2006: 357–384.

Miracle, P. (1999) Rhinos and beavers and bears, oh my! Zooarchaeological perspectives on the Krapina fauna 100 years after Gorjanović. *The Krapina Neandertals and Human Evolution in Central Europe*. Program and Abstracts, p. 34.

Mishra, S., T. Venkaatesan, S. Rajaguru, and K. Somalyajulu (1995) Earliest Acheulean industry from peninsular India. *CA* **36**: 847–851.

Moggi-Cecchi, J., P. Tobias, and A. Beynon (1998) The mixed dentition and associated skull fragments of a juvenile fossil hominid from Sterkfontein, South Africa. *AJPA* **106**:425–465.

Molleson, T., and K. Oakley (1966) Relative antiquity of the Ubeidiya hominid. *Nature* **209**:1268.

Monahan, C. (1996) New zoological data from Bed II, Olduvai Gorge, Tanzania: Implications for hominid behavior in the Early Pleistocene. *JHE* **31**:93–128.

Monet-White, A. (1996) *Le Paléolithique en ancienne Yougoslavie*. Grenoble: Millon.

Moore, J. (1996) Savanna chimpanzees, referential models and the last common ancestor. In: McGrew et al. 1996: 275–292.

Moore, R. (1953) *Man, Time and Fossils*. New York: A. A. Knopf.

Morant, G. (1938) The form of the Swanscombe skull. *JRAI* **68**:67–97.

Morbeck, M. (1975) *Dryopithecus africanus* forelimb. *JHE* **4**:39–46.

Morbeck, M. (1983) Miocene hominoid discoveries from Rudabánya: Implications from the postcranial skeleton. In: Ciochon and Corruccini 1983: 369–404.

Morgan, E. (1982) *The Aquatic Ape*. London: Souvenir.

Morin, P., J. Moore, R. Chakraborty, Li Jin, J. Goodall, and D. Woodruff (1994) Kin selection, social structure, gene flow, and the evolution of chimpanzees. *Science* **265**:1193–1201.

Morris, A. (1992) Biological relationships between Upper Pleistocene and Holocene populations in southern Africa. In: Bräuer and Smith 1992: 131–143.

Morris, D. (1974) Neandertal speech. *Linguistic Inquiry* **5**:144–150.

Morton, D. (1922) Evolution of the human foot. I. *AJPA* **5**:305–336.

Morwood, M., P. O'Sullivan, F. Aziz, and A. Raza (1998) Fission track ages of stone tools and fossils on the east Indonesian island of Flores. *Nature* **392**:173–179.

Morwood, M., F. Aziz, Nasruddin, D. Hobbs, P. O'Sullivan, and A. Raza (1999) Archaeological and paleontological research in central Flores, east Indonesia: Results of fieldwork 1997–1998. *Antiquity* **73**:273–286.

Morwood, M., R. Soejono, R. Roberts, T. Sutikna, C. Turney, K. Westaway, W. Rink, J.-X. Zhao, G. van den Bergh, Rokus Awe Due, D. Hobbs, M. Moore, M. Bird, and L. Fifield (2004) Archaeology and age of a new hominin from Flores in eastern Indonesia. *Nature* **431**:1087–1091.

Morwood, M., P. Brown, Jatmiko, T. Sutikna, E. Wahyu Saptomo, E. Westaway, Rokus Awe Due, R. Roberts, T. Maeda, S. Wasisto, and T. Djubiantono (2005) Further evidence for small-bodied hominins from the Late Pleistocene of Flores. *Nature* **437**:1012–1017.

Moss, M., and R. Young (1960) A functional approach to craniology. *AJPA* **18**:281–292.

Moulton, D. G. (1977) Minimum odorant concentrations detectable by the dog and their implications for olfactory receptor sensitivity. In: D. Müller-Schwarze and M. Mozell, Editors. *Chemical Signals in Vertebrates*. New York: Plenum, pp. 455–464.

Movius, H. (1944) Early man and Pleistocene stratigraphy in southern and eastern Asia. *Papers of the Peabody Museum* **19**: 1–25.

Movius, H. (1948) The lower Paleolithic cultures of southern and eastern Asia. *TAPS* **38**:329–420.

Movius, H. (1950) A wooden spear of third interglacial age from Lower Saxony. *SJA* **6**:139–142.

Movius, H. (1969) Lower Paleolithic cultures in southern and eastern Asia. Reprinted in: W. W. Howells, Editor. *Early Man in the Far East. Studies in Physical Anthropology*, No. 1. New York: Humanities Press, pp. 17–82.

Moy-Thomas, J., and R. Miles (1971) *Paleozoic Fishes*, 2nd edition. Philadelphia: W. B. Saunders.

Moyá-Solá, S., and M. Köhler (1996) A *Dryopithecus* skeleton and the origins of great ape locomotion. *Nature* **372**:156–159.

Moyá-Solá, S., and M. Köhler (1997a) The phylogenetic relationships of *Oreopithecus bambolii* Gervais, 1872. *Comptes rendus de l'Académie des sciences de Paris. Sciences de la terre et des planètes* **324**:141–148.

Moyá-Solá, S., and M. Köhler (1997b) The Orce skull: Anatomy of a mistake. *JHE* **33**:91–97.

Moyá-Solá, S., M. Köhler, D. Alba, I. Casanovas-Vilar, and J. Galindo (2004) *Pierolapithecus catalaunicus*, a new Middle Miocene great ape from Spain. *Science* **306**:1339–1344.

Müller-Beck, H. (1996) Zur Datierung der Steinartefakte von Mauer, Grahenrain. In: Beinhauer et al. 1996: 147–151.

Müntzing, A. (1930) Über Chromosomenvermehrung in *Galeopsis*-Kreuzungen und ihre phylogenetische Bedeutung. *Hereditas* **14**:153–172.

Münzel, S., and N. Conard (2004) Change and continuity in subsistence during the Middle and Upper Palaeolithic in the Ach Valley of Swabia (South-west Germany). *IJO* **14**:225–243.

Münzel, S., K. Langguth, N. Conard, and H. P. Uerpmann (2001) Höhlenbärenjagd auf der schwäbischen Alb vor 30.000 Jahren. *AK* **31**:317–328.

Murrill, R. (1975) A comparison of the Rhodesian and Petralona upper jaws in relation to other Pleistocene hominids. *ZMA* **66**:176–187.

Murrill, R. (1981) *Petralona Man*. Springfield: Charles C Thomas.

Musgrave, J. (1970) How dexterous was Neandertal man? *Nature* **223**:538–541.

Musil, R., and K. Valoch (1968) Stránská Skála: Its meaning for Pleistocene studies. *CA* **9**:534–539.

Mussi, M. (1995) The earliest occupation of Europe: Italy. In: Roebroeks and van Kolfschoten 1995: 27–49.

Mussi, M. (2001) *Earliest Italy*. New York: Kluwer Academic/ Plenum.

Muybridge, E. (1887) *Animals in Motion*. 1957 reprint. New York: Dover.

Muybridge, E. (1913) *The Human Figure in Motion*. 1955 reprint. New York: Dover.

Myster, S., and F. H. Smith (1990) The taxonomic dilemma of the Tangier maxilla: A metric and non-metric assessment. *AJPA* **81**:273–274.

N

Nagel, A. (1989) Development of temperature regulation in the common white-toothed shrew, *Crocidura russula. Comparative Biochemistry and Physiology*, A, **92**:409–413.

Nakagawa, S., N. Gemmell, and T. Burke (2004) Measuring vertebrate telomeres: Applications and limitations. *Molecular Ecology* **13**:2523–2533.

Nakatsukasa, M., A. Yamanaka, D. Shimizu, T. Takano, Y. Kunimatsu, Y. Nakano, and H. Ishida (2003) Definitive evidence for tail loss in *Nacholapithecus*, an East African Miocene hominoid. *JHE* **45**:179–186.

Nakatsukasa, M., H. Tsujikawa, Y. Kunimatsu, D. Shimizu, and H. Ishida (1998) A newly discovered *Kenyapithecus* skeleton and its implications for the evolution of positional behavior in Miocene East African hominoids. *JHE* **34**:657–664.

Napier, J. (1961) Prehensility and opposability in the hands of primates. *Symposia of the Zoological Society of London* **5**:115–132.

Napier, J. (1962) Fossil hand bones from Olduvai Gorge. *Nature* **196**:409–411.

Napier, J., and P. Napier (1967) *Handbook of Living Primates*. New York: Academic Press.

Narita, Y., and S. Kuratani (2005) Evolution of the vertebral formulae in mammals: A perspective on developmental constraints. *Journal of Experimental Zoology (Molecular and Developmental Evolution)* **304B**:91–106.

Nei, M., and A. Roychoudhury (1974) Gene differences between Caucasian, Negro, and Japanese populations. *Science* **177**:434–435.

Nei, M., and A. Roychoudhury (1982) Genetic relationships and evolution of human races. Evolutionary *Biology* **15**:1–59.

Nekaris, K. (2005) Foraging behaviour of the slender loris (*Loris lydekkerianus lydekkerianus*): Implications for theories of primate origins. *JHE* **49**:289–300.

Nelson, S. (2003) *The Extinction of Sivapithecus: Faunal and Environmental Changes Surrounding the Disappearance of a Miocene Hominoid in the Siwaliks of Pakistan*. Boston: Brill Academic Publishers.

Netter, F. (1989) *Atlas of Human Anatomy*. Summit, NJ: CIBA-Geigy Corporation.

Neuville, R. (1951) Paléolithique et Mésolithique du désert de Judée. *AIPH* **24**:1–270.

Nevell, L., A. Gordon, and B. Wood (2007) *Homo floresiensis* and *Homo sapiens* size-adjusted cranial shape variation. *AJPA Supplement* **44**:175–176.

Neves, W., and H. Pucciarelli (1998) The Zhoukoudian Upper Cave 101 skull as seen from the Americas. *JHE* **34**:219–222.

Newman, R. (1970) Why man is such a sweaty and thirsty naked animal: A speculative review. *HB* **42**:12–27.

Ni, X., Y. Wang, Y. Hu, and C. Li (2004) A euprimate skull from the early Eocene of China. *Nature* **427**:65–68.

Niewoehner, W. (2001) Behavioral implications from the Skhul/Qafzeh early modern human hand remains. *PNAS* **98**:2979–2984.

Niewoehner, W. (2006) Neanderthal hands in their proper perspective. In: Harvati and Harrison 2006: 157–190.

Niewoehner, W., A. Weaver, and E. Trinkaus (1997) Neandertal capitate-metacarpal articular morphology. *AJPA* **103**:219–233.

Niklas, K. (1994) *Plant Allometry: The Scaling of Form and Process.* Chicago: University of Chicago Press.

Niklas, K., and B. Tiffney (1994) The quantification of plant biodiversity through time. *PTRSL* **345**:35–44.

Nishida, T., and M. Hiraiwa-Hasegawa (1986) Chimpanzees and bonobos: Cooperative relationships among males. In: B. Smuts, D. Cheney, R. Seyfarth, R. Wrangham and T. Struhsaker, Editors. *Primate Societies.* Chicago: University of Chicago Press, Chicago. 165–177.

Niven, L. (2006) *The Palaeolithic Occupation of Vogelherd Cave.* Tübingen: Kerns Verlag.

Noonan, J., G. Coop, S. Kudaravalli, D. Smith, J. Krause, J. Alessi, Feng Chen, D. Platt, S. Pääbo, J. Pritchard, and E. Rubin (2006) Sequencing and analysis of Neanderthal genomic DNA. *Science* **314**:1113–1118.

Norscia, I., V. Carrai, and S. Borgognini-Tarli (2006) Influence of dry season and food quality and quantity on behavior and feeding strategy of *Propithecus verreauxi* in Kirindy, Madagascar. *IJP* **27**:1001–1022.

O

Oakley, K. (1975) A reconsideration of the date of the Kanam jaw. *JAS* **2**:151–152.

Oakley, K., P. Andrews, L. Keeley, and J. Clark (1977) A reappraisal of the Clacton spearpoint. *Proceedings of the Prehistoric Society* **43**:13–30.

Oakley, K., and A. Montagu (1949) A reconsideration of the Galley Hill skeleton. *Bulletin of the British Museum (Natural History) Geology* **1**:25–48.

O'Brien, S. (2003) *Tears of the Cheetah, and Other Tales from the Genetic Frontier.* New York: St. Martin's Press.

O'Connell, J. (2006) How did modern humans displace Neanderthals? Insights from hunter–gatherer ethnography and archaeology. In: Conard 2006: 43–64.

O'Connell, J., and J. Allen (1998) When did humans first arrive in Greater Australia and why is it important to know? *EA* **6**:132–146.

O'Connell, J., K. Hawkes, and N. Blurton Jones (1999) Grandmothering and the evolution of *Homo erectus*. *JHE* **36**:461–485.

O'Connell, J., K. Hawkes, K. Lupo, and N. Blurton Jones (2002) Male strategies and Plio-Pleistocene archaeology. *JHE* **43**:831–872.

Ohman, J., T. Krochta, C. O. Lovejoy, R. Mensforth, and B. Latimer (1997) Cortical bone distribution in the femoral neck of hominoids: Implications for the locomotion of *Australopithecus afarensis*. *AJPA* **104**:117–131.

Ohman, J., C. O. Lovejoy, and T. White (2005) Questions about *Orrorin* femur. *Science* **307**:45.

Olivier, G. (1969) *Practical Anthropology.* Springfield: Charles C Thomas.

Olson, J. (2006) Photosynthesis in the Archean Era. *Photosynthesis Research* **88**:109–117.

Olson, T. (1981) Basicranial morphology of the extant hominoids and Pliocene hominids: The new material from the Hadar Formation, Ethiopia and its significance in early human evolution and taxonomy. In: Stringer 1981:99–128.

Olson, T. (1985) Cranial morphology and systematics of the Hadar Formation hominids and "*Australopithecus*" *africanus*. In: Delson 1985: 102–119.

Orban-Segebarth, R., and F. Procureur (1983) Tooth size of *Meganthropus paleojavanicus:* An analysis of distance between some fossil hominids and a modern human population. *JHE* **12**:711–720.

Orgel, L. (1998) The origin of life—A review of facts and speculations. *Trends in Biochemical Sciences* **23**:491–495.

Orlando, L., P. Darlu, M. Toussaint, D. Bonjean, M. Otte, and C. Hänni (2006) Revisiting Neandertal diversity with a 100,000 year old mtDNA sequence. *Current Biology* **16**:R400–R402.

Orms, O., J. Parés, B. Martínez-Navarro, J. Agustí, I. Torro, G. Martínez-Fernández, and A. Turq (2000) Early human occupation of western Europe: Paleomagnetic data for two Paleolithic sites in Spain. *PNAS* **97**:10666–10670.

Ortner, D., and W. Putschar (1981) *Identification of Pathological Conditions in Human Skeletal Remains.* Washington, DC: Smithsonian Institution Press.

Ortscheidt, J. (2002) Datation d'un vestige humain provenant de La Rochette (Saint Léon-sur-Vézère, Dordogne) par la méthode du carbone 14 en spectrométrie de masse. *Paléo* **14**:239–240.

Osborn, H. (1926) Why Central Asia? *NH* **26**:263–269.

Osborn, H. (1927) Recent discoveries relating to the origin and antiquity of Man. *Science* **65**:481–488.

Osgood, W., and C. Herrick (1921) *A Monographic Study of the American Marsupial,* Caenolestes, *with a Description of the Brain of* Caenolestes. Chicago: Field Museum of Natural History.

Otte, M. (2004) The Aurignacian in Asia. In: Brantingham et al. 2004: 144–150.

Otte, M., and J. Kozlowski (2003) Constitution of the Aurignacian throughout Eurasia. In: Zilhão and d'Errico 2003: 19–27.

Ovchinnikov, I., A. Götherström, G. Romanova, V. Kharitonov, K. Lindén, and W. Goodwin (2000) Molecular analysis of Neanderthal DNA from the Northern Caucasus. *Nature* **404**:490–493.

Oxnard, C. (1963) Locomotor adaptations in the primate forelimb. *Symposia of the Zoological Society of London* **10**:165–182.

Oxnard, C. (1969) Aspects of the mechanical efficiency of the hands of higher primates as demonstrated by two-dimensional photoelasticity. *AR* **163**:239.

Oxnard, C. (1973) *Form and Pattern in Human Evolution: Some Mathematical, Physical, and Engineering Approaches.* Chicago: University of Chicago Press.

Oxnard, C. (1975a) *Uniqueness and Diversity in Human Evolution: Morphometric Studies of Australopithecines.* Chicago: University of Chicago Press.

Oxnard, C. (1975b) The place of the australopithecines in human evolution: Grounds for doubt? *Nature* **258**:389–395.

Oyen, O., R. Rice, and M. Cannon (1979) Browridge structure and function in extant primates. *AJPA* **51**:83–96.

P

Paciulli, L. (1995) Ontogeny of phalangeal curvature and positional behavior in chimpanzees. *AJPA Supplement* **20**:165.

Paley, W. (1802) *Natural Theology: or, Evidences of the Existence and Attributes of the Deity.* London: J. Faulder, London. 1854 U.S. edition, New York: American Tract Society.

Palmquist, P. (1997) A critical re-evaluation of the evidence for the presence of hominids in Lower Pleistocene times at Venta Micea, southern Spain. *JHE* **33**:83–89.

Pap, I., A. Tillier, B. Arensberg, and M. Chech. (1996) The Subalyuk Neanderthal remains (Hungary): A reexamination. *Annals of the Hungarian Natural History Museum* **88**:233–270.

Pardoe, C. (1991) Isolation and evolution in Tasmania. *CA* **32**:1–21.

Parés, J., and A. Péres-González (1999) Magnetochronology and stratigraphy at Gran Dolina section, Atapuerca (Burgos, Spain). In: Bermúdez de Castro et al. 1999a: 325–342.

Parfitt, S., R. Barendregt, M. Breda, I. Candy, M. Collins, G. Coope, P. Durbidge, M. Field, J. Lee, A. Lister, R. Mutch, K. Penkman, R. Preece, J. Rose, C. Stringer, R. Symmons, J. Whittaker, J. Wymer, and A. Stuart (2005) The earliest record of human activity in Europe. *Nature* **438**:1008–1012.

Parrington, F. (1946) On the cranial anatomy of cynodonts. *Proceedings of the Zoological Society of London* **116**:181–197.

Partridge, T. (1978) Re-appraisal of lithostratigraphy of Sterkfontein hominid site. *Nature* **275**:282–287.

Partridge, T. (2002) On the unrealistic "revised age estimates" for Sterkfontein. *SAJS* **98**:419–120.

Partridge, T., D. Granger, M. Caffee, and R. Clarke (2003) Lower Pliocene hominid remains from Sterkfontein. *Science* **300**:607–612.

Partridge, T., J. Shaw, D. Heslop, and R. Clarke (1999) The new hominid skeleton from Sterkfontein, South Africa: age and preliminary assessment. *Journal of Quaternary Science* **14**:293–298.

Patterson, B. (1966) A new locality for early Pleistocene fossils in northwestern Kenya. *Nature* **212**:577–581.

Patterson, B., and W. W. Howells (1967) Hominid humeral fragment from early Pleistocene of northwestern Kenya. *Science* **156**:64–66.

Patterson, H. (1978) More evidence against speciation by reinforcement. *SAJS* **74**:369–371.

Patterson, H. (1992) The recognition concept of species. In: Ereshefsky 1992: 139–158.

Păunescu, A. (2001) *Paleoliticul şi mezoliticul din spatiul transilvan.* Bucuresti: Editura AGIR.

Pavlov, P., J. Svendsen, and S. Indrelid (2001) Human presence in the European Arctic nearly 40,000 years ago. *Nature* **413**:64–67.

Pawlik, A., and W. Ronquillo (2003) The Paleolithic in the Phillippines. *Lithic Technology* **28**:79–93.

Pearson, O. (2000a) Postcranial remains and the origin of modern humans. *EA* **9**:229–247.

Pearson, O. (2000b) Activity, climate, and postcranial robusticity: Implications for modern human origins and scenarios of adaptive change. *CA* **41**:569–607.

Pearson, O. (2004) Has the combination of genetic and fossil evidence solved the riddle of modern human origins? *EA* **13**:145–159.

Pearson, O., R. Cordero, and A. Busby (2006) How different were Neanderthals' habitual activities? A comparative analysis with diverse groups of recent humans. In: Havarti and Harrison 2006: 135–156.

Pearson, O., and F. Grine (1996) Morphology of the Border Cave ulna and humerus. *SAJS* **92**:231–236.

Pei, W. (1934) A preliminary report on the Late Paleolithic Cave of Choukoutien. *BGSC* **13**:327–357.

Pei, W. (1939) The Upper Cave industry of Choukoutien. *Paleontologia Sinica*, new series C, **10**:1–57.

Penck, A., and E. Brückner (1909) *Die Alpen in Eiszeitalter.* Leipzig.

Pennington, J. (1989) *Bowes and Church's Food Values of Portions Commonly Used*, 15th edition. New York: Harper Collins.

Penny, D., M. Steel, P. Waddell, and M. Hendy (1995) Improved analyses of mtDNA sequences support a recent African origin for *Homo sapiens*. *MBE* **12**:863–882.

Peretto, C., Editor (1991) *Isernia la Pineta, nuovi contributi scientifici.* Isernia: Instituto regionale per gli studi storici del Molise, V. Cuoco.

Pérez, P., A. Gracia, I. Martínez, and J. Arsuaga (1997) Paleopathological evidence of the cranial remains from the Sima de los Huesos Middle Pleistocene site (Sierra de Atapuerca, Spain). Description and preliminary inferences. In: Arsuaga et al. 1997a: 409–421.

Perlés, C. (1976) Le feu. In: H. deLumley, Editor. *La Préhistoire Français*, Volume 1. Paris: CNRS, pp. 679–683.

Pernkopf, E. (1963) *Atlas of Topographical and Applied Human Anatomy*, Volume. I. Philadelphia: W. B. Saunders.

Pettigrew, J. (1986) Flying primates? Megabats have the advanced pathway from eye to midbrain. *Science* **231**:1304–1306.

Pettitt, P., and N. Bader (2000) Direct AMS radiocarbon dates for the Sungir mid Upper Palaeolithic burials. *Antiquity* **74**:269–270.

Pettitt, P., and E. Trinkaus (2000) Direct radiocarbon dating of the Brno 2 Gravettian human remains. *Anthropologie (Brno)* **38**:149–150.

Pettitt, P., H. van der Plicht, C. Ramsey, A. Soares, and J. Zilhão (2002) The radiocarbon chronology. In: Zilhão and Trinkaus 2002: 132–138.

Pfeiffer, J. (1982) *The Creative Explosion*. New York: Harper & Row.

Pickering, T. (2001) Taphonomy of the Swartkrans hominid post-crania and its bearing on issues of meat-eating and fire management. In: C. Stanford and H. Bunn, Editors. *Meat-Eating and Human Evolution*. Oxford: Oxford University Press. 33–51.

Pickering, T., R. Clarke, and J. Heaton (2004) The context of Stw 573, an early hominid skull and skeleton from Sterkfontein Member 2: Taphonomy and paleoenvironment. *JHE* **46**:279–297.

Pickford, M., D. Johanson, C. O. Lovejoy, T. White, and J. Aronson (1983) A hominoid humeral fragment from the Pliocene of Kenya. *AJPA* **60**:337–346.

Pickford, M., S. Moyá-Solá, and M. Köhler (1997) Phylogenetic implications of the first African Middle Miocene hominoid frontal bone from Otavi, Namibia. *CRAS, Sciences de la terre et des planètes* **325**:459–466.

Pickford, M., and B. Sénut (2001) The geological and faunal context of Late Miocene hominid remains from Lukeino, Kenya. *CRAS* **332**:145–152.

Pickford, M., B. Sénut, D. Gommery, and J. Treil (2002) Bipedalism in *Orrorin tugenensis* revealed by its femora. *CRAS Palévol* **1**:191–203.

Pilbeam, D. (1972) *The Ascent of Man: An Introduction to Human Evolution*. New York: Macmillan.

Pilbeam, D. (1982) New hominoid skull material from the Miocene of Pakistan. *Nature* **295**:232–234.

Pilbeam, D. (1996) Genetic and morphological records of the Hominoidea and hominid origins: A synthesis. *Molecular Phylogenetics and evolution* **5**:155–168.

Pilbeam, D. (1997) Research on Miocene hominoids and hominid origins: The last three decades. In: Begun et al. 1997: 13–28.

Pilbeam, D. (2002) Perspectives on the Miocene Hominoidea. In: Hartwig 2002: 303–310.

Pilbeam, D., M. Rose, J. Barry, and S. Shah (1990) New *Sivapithecus* humeri from Pakistan and the relationship of *Sivapithecus* and *Pongo*. *Nature* **348**:237–239.

Pilgrim, G. (1910) Notices of new mammalian genera and species from the Tertiaries. *Records of the Geological Survey of India* **40**:63–71.

Pineda, D., J. Gonzalez, P. Callaerts, K. Ikeo, W. Gehring, and E. Salo (2000) Searching for the prototypic eye genetic network: *Sine oculis* is essential for eye regeneration in planarians. *PNAS* **97**:4525–4529.

Piveteau, J. (1957) *Traité de paléontologie, Volume 7: Primates*. Paris: Masson.

Piveteau, J., and M. de Lumley (1963) Découverte de restes neandértaliens dans la grotte de l'Hortus (Valflaunès, Hérault). *CRAS* **256**:40–44.

Plavcan, J. (2001) Sexual dimorphism in primate evolution. *YPA* **44**:25–53.

Plavcan, J., C. Lockwood, W. Kimbel, M. Lague, and E. Harmon (2005) Sexual dimorphism in *Australopithecus afarensis* revisited: How strong is the case for a human-like pattern of dimorphism? *JHE* **48**:313–320.

Plavcan, J., and C. van Schaik (1992) Intrasexual competition and canine dimorphism in anthropoid primates. *AJPA* **87**:461–477.

Plavcan, J., and C. van Schaik (1997a) Interpreting hominid behavior on the basis of sexual dimorphism. *JHE* **32**:345–374.

Plavcan, J., and C. van Schaik (1997b) Intrasexual competition and body weight dimorphism in anthropoid primates. *AJPA* **103**:37–68.

Pollock, J., and R. Mullin (1987) Vitamin C biosynthesis in prosimians: Evidence for the anthropoid affinity of *Tarsius*. *AJPA* **73**:65–70.

Ponce de Léon, M., and C. Zollikofer (1999) New evidence from Le Moustier 1: Computer assisted reconstruction and morphometry of the skull. *AR* **254**:474–489.

Ponce de Léon, M., and C. Zollikofer (2001) Neanderthal cranial ontogeny and its implications for late hominid diversity. *Nature* **412**:534–538.

Ponce de Léon, M., and C. Zollikofer (2006) Neanderthals and modern humans—chimps and bonobos: Similarities and differences in development and evolution. In: Harvati and Harrison 2006: 71–88.

Pope, G. (1988) Recent advances in Far Eastern paleoanthropology. *ARA* **17**:43–77.

Pope, G. (1989) Bamboo and human evolution. *NH* **98**:48–57.

Pope, G. (1991) Evolution of the zygomaxillary region in the genus *Homo* and its relevance to the origin of modern humans. *JHE* **21**:189–213.

Pope, G. (1992) Craniofacial evidence for the origin of modern humans in China, *YPA* **35**:243–298.

Pope, G. (1997a) Java. In: Spencer 1997: 544–553.

Pope, G. (1997b) Paleoanthropological research traditions in the Far East. In: Clark and Willermet 1997: 269–282.

Pope, G., and J. Cronin (1984) The Asian Hominidae. *JHE* **13**:377–396.

Pope, G., S. Barr, A. Macdonald, and S. Nakabanlang (1986) Earliest radiometrically dated artifacts from Southeast Asia. *CA* **27**:275–279.

Popesko, P. (1977) *Atlas of Topographic Anatomy of the Domestic Animals*, 2nd edition. Philadelphia: W. B. Saunders.

Porat, N., M. Chazan, H. Schwarcz, and L. Horowitz (2002) Timing of the Lower to Middle Paleolithic boundary: new dates from the Levant. *JHE* **43**:107–122.

Potts, R. (1988) *Early Hominid Activities at Olduvai*. New York: Aldine de Gruyter.

Potts, R., A. Behrensmeyer, A. Deino, P. Ditchfield, and J. Clark (2004) Small Mid-Pleistocene hominin associated with East African Acheulean technology. *Science* **305**:76–78.

Potts, R., and P. Shipman (1981) Cutmarks made by stone tools on bones from Olduvai Gorge, Tanzania. *Nature* **291**:577–580.

Poulianos, A. (1971) Petralona: A Middle Pleistocene cave in Greece. *Archaeology* **24**:6–11.

Preuschoft, H., and H. Witte (1991) Biomechanical reasons for the evolution of hominid body shape. In: Coppens and Sénut 1991: 59–77.

Prins, H., and J. Reitsma (1989) Mammalian biomass in an African equatorial rain forest. *Journal of Animal Ecology* **58**:851–861.

Protsch, R. (1974) Florisbad: Its paleoanthropology, chronology and archaeology. *Homo* **25**:68–78.

Protsch, R. (1975) The absolute dates of Upper Pleistocene sub-Saharan fossil hominids and their place in human evolution. *JHE* **4**:297–322.

Pycraft, W., G. Elliot Smith, M. Yearsley, J. Carter, R. Smith, A. Hopwood, D. Bate, and W. Swinton (1928) *Rhodesian Man and Associated Remains*. London: British Museum (Natural History).

Q

Qiu, Z.. (1985) The Middle Paleolithic of China. In: Wu and Olsen 1985: 187–210.

Qiu, Z., Y. Gu, Y. Zhang, and S. Zhang (1973) Newly discovered *Sinanthropus* remains and stone artifacts at Choukoutien. *VP* **11**:109–131.

Quam, R., and F. Smith (1998) A reassessment of the Tabun C2 mandible. In: Akazawa et al. 1998: 405–421.

Quam, R., J-L. Arsuaga, J.-M. Bermúdez de Castro, J. Díez, G. Lorenzo, N. Garcia, and A. Ortega (2001) Human remains from Valdegoba Cave (Huérmeces, Burgos, Spain). *JHE* **41**:385–435.

Quatrefages de Breau, J., and J. Hamy (1882) *Crania ethnica: Les crânes des races humains*. Paris: Balliére.

R

Radovčić, J. (1988) *Gorjanović-Kramberger and Krapina Early Man*. Zagreb: Hrvatski Prirodoslovni Muzej and Školska Knjiga.

Radovčić, J., F. Smith, E. Trinkaus, and M. Wolpoff (1988) *The Krapina Hominids: An Illustrated Catalog of Skeletal Collection*. Zagreb: Mladost.

Rafferty, K. (1998) Structural design of the femoral neck in primates. *JHE* **34**:361–383.

Rak, Y. (1983) *The Australopithecine Face*. New York: Academic Press.

Rak, Y. (1985) Sexual dimorphism, ontogeny and the beginning of differentiation of the robust australopithecine clade. In: Tobias 1985: 233–237.

Rak, Y. (1986) The Neanderthal: A new look at an old face. *JHE* **15**:151–164.

Rak, Y. (1988) On variation in the masticatory system of *Australopithecus boisei*. In: Grine 1988: 193–198.

Rak, Y. (1990) On the difference between two pelvises of Mousterian context from the Qafzeh and Kebara caves, Israel. *AJPA* **81**:323–332.

Rak, Y. (1991) Lucy's pelvic anatomy: its role in bipedal gait. *JHE* **20**:283–290.

Rak, Y. (1993) Morphological variation in *Homo neanderthalensis* and *Homo sapiens* in the Levant: A biogeographic model. In: Kimbel and Martin 1993: 523–536.

Rak, Y. (1998) Does any Mousterian cave present evidence of two hominid species? In: Akazawa et al. 1998: 353–366.

Rak, Y., and B. Arensburg (1987) Kebara 2 Neanderthal pelvis: First look at a complete inlet. *AJPA* **73**:227–231.

Rak, Y., A. Ginzburg, and E. Geffen (2002) Does *Homo neanderthalensis* play a role in modern human ancestry? The mandibular evidence. *AJPA* **119**:199–204.

Rak, Y., W. Kimbel, and E. Hovers (1994) A Neandertal infant from Amud Cave, Israel. *JHE* **26**:313–324.

Rak, Y., W. Kimbel, and D. Johanson (1996) The crescent of foramina in *Australopithecus afarensis* and other early hominids. *AJPA* **101**:93–99.

Ramierez Rossi, F., and J. Bermúdez de Castro (2004) Surprising rapid growth in Neanderthals. *Nature* **428**:936–939.

Ranieri, S., and S. Washburn (1981) Who brought home the bacon? *New York Review of Books* **28**(14):59.

Rasmussen, D. (1986) Anthropoid origins: A possible solution to the Adapidae–Omomyidae paradox. *JHE* **15**:1–12.

Rasmussen, D. (1990) Primate origins: Lessons from a neotropical marsupial. *AJP* **22**:263–277.

Rasmussen, D. (2002) Early catarrhines of the African Eocene and Oligocene. In: Hartwig 2002: 203–220.

Rasmussen, D., G. Conroy, and E. Simons (1998) Tarsier-like locomotor specializations in the Oligocene primate *Afrotarsius*. *PNAS* **95**:14848–14850.

Rasmussen, D., and E. Simons (1988) New specimens of *Oligopithecus savagei*, early Oligocene primate from the Fayum, Egypt. *FP* **51**:182–208.

Rasmussen, D., and R. Sussman (2007) Parallelisms among primates and possums. In: Ravosa and Dagosto 2007: 775–803.

Raup, D. (1991) *Extinction: Bad Genes or Bad Luck?* New York: W. W. Norton.

Rauum, R., K. Sterner, C. Noviello, C. Stewart, and T. Disotell (2005) Catarrhine divergence times estimated from complete mitochondrial genomes: concordance with fossil and nuclear DNA evidence. *JHE* **48**:237–2257.

Raven, P. (1976) Systematics and plant population biology. *Systematic Botany* 1:284–316.

Ravosa, M. (1988) Browridge development in Cercopithecidae: A test of two models. *AJPA* **76**:535–555.

Ravosa, M. (1991a) Ontogenetic perspectives on mechanical and nonmechanical models of primate circumorbital morphology. *AJPA* **85**:95–112.

Ravosa, M. (1991b) Interspecific perspective on mechanical and nonmechanical models of primate circumorbital morphology. *AJPA* **86**:369–396.

Ravosa, M., V. Noble, W. Hylander, K. Johnson, and E. Kowalski (2000) Masticatory stress, orbital orientation and the function of the primate postorbital bar. *JHE* **38**:667–693.

Ravosa, M., and D. Savakova (2004) Euprimate origins: The eyes have it. *JHE* **46**:357–364.

Rees, J. (2003) Genetics of hair and skin color. *Annual Review of Genetics* **37**:67–90.

Relethford, J. (1998a) Genetics of modern human origins and diversity. *ARA* **27**:1–23.

Relethford, J. (1998b) Mitochondrial DNA and ancient population growth. *AJPA* **105**:1–7.

Relethford, J. (1999) Models, predictions and the fossil record of modern human origins. *EA* **8**:7–10.

Relethford, J. (2001) *Genetics and the Search for Modern Human Origins.* New York: Wiley-Liss.

Relethford, J., and H. Harpending (1995) Ancient differences in population size can mimic a recent African origin of modern humans. *CA* **36**:667–674.

Rendell, H., and R. Dennell (1985) Dated Lower Paleolithic artifacts from northern Pakistan. *CA* **26**:393.

Reno, P., R. Meindl, M. McCollum, and C. O. Lovejoy (2003) Sexual dimorphism in *Australopithecus afarensis* was similar to that of modern humans. *PNAS* **100**:9404–9409.

Reno, P., D. DeGusta, M. Serrat, R. Meindl, T. White, R. Eckhardt, A. Kuperavage, K. Galik, and C. O. (2005a) Plio-Pleistocene hominid limb proportions: Evolutionary reversals or estimation errors? *CA* **46**:575–588.

Reno, P., R. Meindl, M. McCollum, and C. O. Lovejoy (2005b) The case is unchanged and remains robust: *Australopithecus afarensis* exhibits only moderate skeletal dimorphism. *JHE* **40**:279–288.

Reynolds, T. (1985) Stresses on the limbs of quadrupedal primates. *AJPA* **67**:351–362.

Richards, G., R. Jabbour, and J. Anderson (2003) *Medial Mandibular Ramus.* BARIS **1138**: 1–113.

Richards, G., and A. Plourde (1995) Reconsideration of the "Neandertal" infant, Amud 7. *AJPA* Supplement 20: 180–181.

Richards, M., P. Pettitt, M. Stiner, and E. Trinkaus (2001) Stable isotope evidence for increasing dietary breadth in the European mid-Upper Paleolithic. *PNAS* **98**:6528–6532.

Richards, M., P. Pettitt, E. Trinkaus, F. Smith, M. Paunovic, and I. Karavanić (2000) Neanderthal diet and Vindija and Neanderthal predation: The evidence from stable isotopes. *PNAS* **97**:7663–7666.

Richmond, B. (1997) Ontogeny of phalangeal curvature and locomotor behavior in lar gibbons. *AJPA* Supplement **24**:197.

Richmond, B., D. Begun, and D. Strait (2001) Origin of human bipedalism: the knuckle-walking hypothesis revisited. *YPA* **44**:70–105.

Richmond, B., and D. Strait (2000) Evidence that humans evolved from a knuckle-walking ancestor. *Nature* **404**:382–385.

Ricklan, D. (1987) Functional anatomy of the hand of *Australopithecus africanus*. *JHE* **16**:643–664.

Ricklan, D. (1990) The precision grip in *Australopithecus africanus*: anatomical and behavioral correlates. In: G. Sperber, Editor. *From Apes to Angels: Essays in Anthropology in Honor of Philip V. Tobias.* New York: Wiley-Liss, pp. 177–183.

Rigaud, J., J. Simek, and T. Ge (1995) Mousterian fires from the Grotte XVI (Dordogne, France). *Antiquity* **69**:902–912.

Rightmire, G. P. (1976) Relationships of Middle and Upper Pleistocene hominids from sub-Saharan Africa. *Nature* **260**:238–240.

Rightmire, G. P. (1978) Florisbad and human population succession in southern Africa. *AJPA* **48**:475–486.

Rightmire, G. P. (1979a) Cranial remains of *Homo erectus* from Beds II and IV, Olduvai Gorge, Tanzania. *AJPA* **51**:99–116.

Rightmire, G. P. (1979b) Implications of the Border Cave skeletal remains for later Pleistocene human evolution. *CA* **48**:23–35.

Rightmire, G. P. (1980) Middle Pleistocene hominids from Olduvai Gorge, Northern Tanzania. *AJPA* **53**:225–241.

Rightmire, G. P. (1981) Patterns in the evolution of *Homo erectus*. *Paleobiology* **7**:241–246.

Rightmire, G. P. (1983) The Lake Ndutu cranium and early *Homo sapiens* in Africa. *AJPA* **61**:245–254.

Rightmire, G. P. (1984a) Comparison of *Homo erectus* from Africa and Southeast Asia. *CFS* **69**:83–98.

Rightmire, G. P. (1984b) *Homo sapiens* in Sub-Saharan Africa. In: Smith and Spencer 1984: 295–325.

Rightmire, G. P. (1986) Stasis in *Homo erectus* defended. *Paleobiology* **12**:324–325.

Rightmire, G. P. (1988) *Homo erectus* and later Middle Pleistocene humans. *ARA* **17**:239–259.

Rightmire, G. P. (1990) *The Evolution of* Homo erectus. Cambridge: Cambridge University Press.

Rightmire, G. P. (1992) *Homo erectus:* Ancestor or evolutionary side branch? *EA* **1**:43–49.

Rightmire, G. P. (1994) The relationship of *Homo erectus* to later Middle Pleistocene hominids. In: Franzen 1994: 319–326.

Rightmire, G. P. (1996) The human cranium from Bodo, Ethiopia: Evidence for speciation in the Middle Pleistocene? *JHE* **31**:21–39.

Rightmire, G. P. (1998) Human evolution in the Middle Pleistocene: The role of *Homo heidelbergensis*. *EA* **6**:218–227.

Rightmire, G. P., and H. Deacon (1991) Comparative studies of late Pleistocene human remains from Klasies River Mouth, South Africa. *JHE* **20**:131–156.

Rightmire, G. P., D. Lordkipanidze, and A. Vekua (2006) Anatomical descriptions, comparative studies and evolutionary significance of the hominin skulls from Dmanisi, Republic of Georgia. *JHE* **50**:115–141.

Rightmire, G. P., A. Van Arsdale, and D. Lordkipanidze (2008) Variation in the mandibles from Dmanisi, Georgia. *JHE* **54**:904–908.

Rink, W., H. Schwarcz, F. Smith, and J. Radovčić (1995) ESR ages for Krapina hominids. *Nature* **378**:24.

Rink, W., H. Schwarcz, K. Valoch, L. Seitl, and C. Stringer (1996) ESR dating of Micoquian industry and Neanderthal remains at Kůlna Cave, Czech Republic. *JAS* **23**:889–901.

Roberts, D. (1978) *Climate and Human Variability*, 2nd edition. Menlo Park: Cummings.

Roberts, M., C. Stringer, and S. Parfitt (1994) A hominid tibia from Middle Pleistocene sediments at Boxgrove, UK. *Nature* **369**:311–313.

Roberts, R., R. Jones, and M. Smith (1990) Thermoluminescence dating of a 50,000-year-old human occupation site in northern Australia. *Nature* **345**:153–156.

Roberts, R., R. Jones, N. Spooner, M. Head, A. Murray, and M. Smith (1994) The human colonisation of Australia: Optical dates of 53,000 and 60,000 years bracket human arrival at Deaf Adder Gorge, Northern Territory. *Quaternary Science Reviews* **13**:575–586.

Robertson, D., M. McKenna, O. Toon, S. Hope, and J. Lillegraven (2004) Survival in the first hours of the Cenozoic. *Geological Society of America Bulletin* **116**:760–768.

Robinson, J. (1953a) *Meganthropus*, australopithecines, and hominids. *AJPA* **11**:1–38.

Robinson, J. (1953b) The nature of *Telanthropus capensis*. *Nature* **171**:33.

Robinson, J. (1954a) The genera and species of the Australopithecinae. *AJPA* **12**:181- 200.

Robinson, J. (1954b) Prehominid dentition and hominid evolution. *Evolution* **8**:324–334.

Robinson, J. (1956) *The Dentition of the Australopithecinae*. Pretoria: *Transvaal Museum Memoir* 9.

Robinson, J. (1960) Affinities of the new Olduvai australopithecine. *Nature* **186**:456–457.

Robinson, J. (1965) *Homo "habilis"* and the australopithecines. *Nature* **205**:121–124.

Robinson, J. (1966) The distinctiveness of Homo habilis. *Nature* **209**:957–960.

Robinson, J. (1967) Variation and the taxonomy of the early hominids. In: T. Dobzhansky, M. Hecht and W. Steere, Editors. *Evolutionary Biology*, Volume I. New York: Appleton–Century–Crofts, pp. 69–100.

Robinson, J. (1968) The origin and adaptive radiation of the australopithecines. In: G. Kurth, Editor. *Evolution and Hominisation*. Stuttgart: Fischer. 150–175.

Robinson, J. (1972) *Early Hominid Posture and Locomotion*. Chicago: University of Chicago Press.

Roebroeks, W. (2002) Hominid behavior and the earliest occupation of Europe: An exploration. *JHE* **41**:437–461.

Roebroeks, W. (2005) Life on the Costa del Cromer. *Nature* **438**:921–922.

Roebroeks, W., N. Conard, and T. van Kolfshoten (1992) Dense forests, cold steppes and the Palaeolithic settlement of northern Europe. *CA* **33**:551–586.

Roebroeks, W., and C. Gamble, Editors (1999) *The Middle Palaeolithic Occupation of Europe*. Leiden: University of Leiden Press.

Roebroeks, W., and T. van Kolfschoten (1994) The earliest occupation of Europe: A short chronology. *Antiquity* **68**: 489–503.

Roebroeks, W., and T. van Kolfschoten (1995) The earliest occupation of Europe: A reappraisal of artifactual and chronological evidence. In: Roebroeks and van Kolfschoten 1995: 297–315.

Rogers, A., and H. Harpending (1992) Population growth makes waves in the distribution of pairwise genetic differences. *MBE* **9**:552–569.

Rogers, A., and L. Jorde (1995) Genetic evidence on modern human origins. *HB* **67**:1–36.

Rogers, M., J. Harris, and C. Feibel (1994) Changing pattern of land use by Plio/Pleistocene hominids in the Lake Turkana Basin. *JHE* **27**:139–158.

Rohen, J., C. Yokochi, and E. Lütjen-Drecoll (1998) *Color Atlas of Anatomy*, 4th edition. Baltimore: Williams and Wilkins.

Rolland, N. (1992) The Paleolithic colonization of Europe: An archaeological and biogeographic perspective. *Trabajos de Prehistoria* **49**:69–111.

Rollins, D. (1990) Managing desert mule deer. Texas Agricultural Extension Service Bulletin 1636. College Station: Texas A&M University.

Romer, A. (1956) *Osteology of the Reptiles*. Chicago: University of Chicago Press.

Romer, A. (1970a) The Chanares (Argentina) Triassic reptile fauna. VI. A chiniquodontid cynodont with an incipient squamosal-dentary jaw articulation. *Breviora* **344**:1–18.

Romer, A. (1970b) *The Vertebrate Body*, 4th edition. Philadelphia: W. B. Saunders.

Rook, L., L. Bondioli, F. Casali, M. Rossi, M. Köhler, S. Moyá-Solá, and R. Macchiarelli (2004) The bony labyrinth of *Oreopithecus bambolii*. *JHE* **46**:347–354.

Rosas, A. (1987) Two new mandibular fragments from Atapuerca/Ibeas (SH site). A reassessment of the affinities of the Ibeas mandibular sample. *JHE* **16**:417–427.

Rosas, A. (1995) Seventeen new mandibular specimens from the Atapuerca/Ibeas Middle Pleistocene hominids sample. *JHE* **28**:533–559.

Rosas, A., M. Bastir, C. Martínez-Maza, A. García-Tabernero, and C. Lalueza-Fox (2006a) Inquiries into Neanderthal craniofacial development and evolution: "accretion" vs. "organismic" models. In: Havarti and Harrison 2006: 37–70.

Rosas, A., and J. Bermúdez De Castro (1998) The Mauer mandible and the evolutionary significance of *Homo heidelbergensis*. *Geobios* **31**:687–697.

Rosas, A., and J. Bermúdez De Castro (1999) The ATD6-5 mandibular specimen from Gran Dolina (Atapuerca, Spain). Morphological study and phylogenetic implications. In: Bermúdez de Castro et al. 1999a: 567–590.

Rosas, A., C. Martínez-Maza, M. Bastir, A. García-Tabernero, C. Lalueza-Fox, R. Huguet, J. Ortiz, R. Julia, V. Soler, T. De Torres, E. Martínez, J. Cañaveras, S. Sánchez-Moral, S. Cuezva, J. Lario, D. Santamaria, M. De la Rasilla, and J. Fortea (2006b) Paleobiology and comparative morphology of a late Neandertal sample from El Sidrón, Asturias, Spain. *PNAS* **103**:19266–19271.

Rose, K. (1995) The earliest primates. *EA* **3**:159–173.

Rose, K., and T. Bown (1984) Gradual phyletic evolution at the generic level in early Eocene omomyid primates. *Nature* **309**:250–252.

Rose, K., and T. Bown (1991) Additional fossil evidence on the differentiation of the earliest euprimates. *PNAS* **88**:98–101.

Rose, K., M. Godinot, and T. Bown (1994) The early radiation of euprimates and the initial diversification of Omomyidae. In: Fleagle and Kay 1994: 1–28.

Rose, M. (1975) Functional proportions of primate lumbar vertebral bodies. *JHE* **4**:21–38.

Rose, M. (1991) The process of bipedalization in hominids. In: Coppens and Sénut 1991: 37–48.

Rose, M. (1997) Functional and phylogenetic features of the forelimb in Miocene hominoids. In: Begun et al. 1997: 79–100.

Rosenberg, K. (1988) The functional significance of Neandertal pubic length. *CA* **29**:595–617.

Rosenberg, K. (1992) The evolution of modern human childbirth. *YPA* **35**:89–124.

Rosenberg, K. (1998) Morphological variation in West Asian postcrania: Implications for obstetric and locomotor behavior. In: Akazawa et al. 1998: 367–379.

Rosenberg, K., Lü Ziné, and C. Ruff (2006) Body size, body form and encephalization in a Middle Pleistocene archaic human from northern China. *PNAS* **103**:3352–3556.

Rosenberg, K., and W. Trevathan (1996) Bipedalism and human birth: the obstetrical dilemma revisited. *EA* **4**:161–168.

Rosenberg, K., and W. Trevathan (2001) The evolution of human birth. *SA* **285**(November):72–77.

Ross, C. (2004) The tarsier fovea: Functionless vestige or nocturnal adaptation? In: Ross and Kay 2004: 477–537.

Ross, C., and H. Covert (2000) The petrosal of *Omomys carteri* and the evolution of the primate basicranium. *JHE* **39**:225–251.

Ross, C., M. Henneberg, M. Ravosa, and S. Richard (2004) Curvilinear, geometric, and phylogenetic modeling of basicranial flexion: Is it adaptive, is it constrained? *JHE* **46**:185–213.

Ross, C., and R. Kay (2004) Anthropoid origins: Retrospective and prospective. In: Ross and Kay 2004: 701–737.

Ross, C., and M. Ravosa (1993) Basicranial flexion, relative brain size, and facial kyphosis in *Homo sapiens* and some fossil hominids. *AJPA* **91**:305–324.

Ross, C., B. Williams, and R. Kay (1998) Phylogenetic analysis of anthropoid relationships. *JHE* **35**:221–306.

Rossie, J. (2005) Anatomy of the nasal cavity and paranasal sinuses in *Aegyptopithecus* and early Miocene African catarrhines. *AJPA* **126**:250–267.

Rossie, J., X. Ni, and K. Beard (2006) Cranial remains of an Eocene tarsier. *PNAS* **103**:4381–4285.

Rothschild, B., I. Hershkovitz, and C. Rothschild (1995) Origin of yaws in the Pleistocene. *Nature* **378**:343–344.

Rougier, H., S. Milota, R. Rodrigo, M. Gherase, L. Sarcina, O. Moldovan, J. Zilhão, S. Constantin, R. Franciscus, C. Zollikofer, M. Ponce de León, and E. Trinkaus (2007) Pestera cu Oase 2 and the cranial morphology of early modern Europeans. *PNAS* **104**:1165–1170.

Ruff, C. (1988) Hindlimb articular surface allometry in Hominoidea and *Macaca*, with comparisons to diaphyseal scaling. *JHE* **17**:687–714.

Ruff, C. (1991) Climate, body size and body shape in hominid evolution. *JHE* **21**:81–105.

Ruff, C. (1993) Climatic adaptation and hominid evolution: The thermoregulatory imperative. *EA* **2**:53–60.

Ruff, C. (1994) Morphological adaptation to climate in modern and fossil hominids. *YPA* **37**:65–107.

Ruff, C. (1995) Biomechanics of the hip and birth in early *Homo*. *AJPA* **98**:527–574.

Ruff, C. (1998) Evolution of the hominid hip. In: Strasser et al. 1998: 449–469.

Ruff, C., E. Trinkaus, A. Walker, and C. Larsen (1993) Postcranial robusticity in *Homo* I: Temporal trends and mechanical interpretations. *AJPA* **91**:21–53.

Ruff, C., E. Trinkaus, and T. Holliday (1997) Body mass and encephalization in Pleistocene *Homo*. *Nature* **387**: 173–176.

Ruff, C., E. Trinkaus, and T. Holliday (2002) Body proportions and size. In: Zilhão and Trinkaus 2002: 365–391.

Ruff, C., and A. Walker (1993) Body size and body shape. In: Walker and Leakey 1993a: 234–265.

Runnegar, B. (1992) Evolution of the earliest animals. In: J. Schopf, Editor. *Major Events in the History of Life*. Boston: Jones and Bartlett. 65–93.

Russell, D., P. Louis, and D. Savage (1967) Primates of the French early Eocene. *University of California Publications in the Geological Sciences* **73**:1–46.

Russell, M. (1985) The supraorbital torus: "A most remarkable peculiarity." *CA* **26**:337–360.

Russell, M. (1987a) Bone breakage in the Krapina hominid collection. *AJPA* **72**:373–380.

Russell, M. (1987b) Mortuary practices at the Krapina Neandertal site. *AJPA* **72**:381–398.

Ruvolo, M. (1997) Molecular phylogeny of the hominoids: Inferences from multiple independent DNA sequence data sets. *MBE* **14**:248–265.

Ryan, A. S., and Johanson, D. C. 1989. Anterior dental microwear in *Australopithecus afarensis*: Comparisons with human and nonhuman primates. *JHE* **18**:235–268.

S

Saban, R. (1977) The place of Rabat man (Kébibat, Morocco) in human evolution. *CA* **18**:518–524.

Sacher, G. (1975) Maturation and longevity in relation to cranial capacity in hominid evolution. In: Tuttle 1975:417–441.

Saller, K. (1962) Die Ofnet-Funde in neuer Zusammensetzung. *ZMA* **52**:1–51.

Sanders, W. (1990) Weight transmission through the lumbar vertebrae and sacrum in australopithecines. *AJPA* **81**:289.

Sanders, W. (1995) *Function, allometry and evolution of the australopithecine lower precaudal spine*. Ph.D. dissertation, New York University.

Sanders, W. (1998) Comparative morphometric study of the australopithecine vertebral series Stw-H8/H41. *JHE* **34**:249–302.

Sanders, W., and M. Bodenbender (1994) Morphometric analysis of lumbar vertebra UMP 67–28: Implications for spinal function and and phylogeny of the Miocene Moroto hominoid. *JHE* **26**:203–237.

Sandrock, O., Y. Dauphin, O. Kullmer, R. Abel, F. Schrenk, and C. Denys (1999) Malema: Preliminary taphonomic analysis of an African locality. *CRAS* **328**:133–139

Santa Luca, A. (1978) A re-examination of presumed Neanderthal fossils. *JHE* **7**:619–636.

Santa Luca, A. (1980) *The Ngandong Fossil Hominids. Yale University Publications in Anthropology* **78**:1–175.

Sarich, V. (1971) A molecular approach to the question of human origins. In: P. Dolhinow and V. Sarich, Editors. *Background for Man: Readings in Physical Anthropology*. Boston: Little, Brown, pp. 60–81.

Sarich, V., and A. Wilson (1967) Immunological time scale for human evolution. *Science* **158**:1200–1203.

Sarmiento, E. (1987) *Oreopithecus* and its significance in the origin of the Hominoidea. *American Museum Novitates* **2881**: 1–44.

Sarmiento, E. (1996) Quadrupedalism in the hominid lineage: 11 years after. *AJPA Supplement* **22**:208.

Sarmiento, E. (1998) Generalized quadrupeds, committed bipeds and the shift to open habitats: an evolutionary model of hominid divergence. *American Museum Novitates* **3520**: 1–78.

Sarmiento, E., and L. Marcus (2000) The os navicular of humans, great apes, OH 8, Hadar, and *Oreopithecus*: function, phyogeny, and multivariate analyses. *American Museum Novitates* **3288**:1–38.

Sartono, S. (1971) Observations on a new skull of *Pithecanthropus erectus* (Pithecanthropus VIII) from Sangiran, Central Java. *Verhandelingen der Koninkijke Nederlandse Akademie van Wetenschappen* **74**:185–194.

Sartono, S. (1975) Implications arising from Pithecanthropus VIII. In: Tuttle 1975: 327–360.

Sartono, S. (1982) Sagittal cresting in *Meganthropus paleojavanicus* von Koenigswald. *MQRSA* **7**:201–210.

Sawada, Y., M. Pickford, B. Sénut, T. Itaya, M. Hyodo, T. Miura, C. Kashine, T. Chujo, and H. Fujii (2002) The age of *Orrorin tugenensis*, an early hominid from the Tugen Hills, Kenya. *CRAS Palévol* **1**:293–303.

Sawyer, G., and B. Maley (2005) Neanderthal reconstructed. *AR* **283**B:23–31.

Schaaffhausen, H. (1858) Zur Kenntniß der ältensten Rassenschädel. *Müllers Archiv* **5**:453–478.

Schaaffhausen, H. (1880) Funde in der Schipkahöhle in Mähren. *Verhandlungen des naturhistorisches Vereins der preussischen Rheinlande und Westfalens* **73**:260–264.

Schaaffhausen, H. (1888) *Der Neanderthaler Fund*. Bonn: Marcus.

Schaefer, M. (1999) Brief communication: Foramen magnum–carotid foramina relationship: Is it useful for species designation? *AJPA* **110**:467–471.

Schepartz, L. (1993) Language and modern human origins. *YPA* **36**:91–126.

Schick, K. (1994) The Movius line reconsidered. In: Corruccini and Ciochon 1994: 569–596.

Schick, K., N. Toth, W. Qi, J. Clark, and D. Ettler (1991) Archaeological perspectives in the Nihewan Basin, China. *JHE* **21**:13–26.

Schliegl, S., P. Goldberg, H. Pfretschner, and N. Conard (2003) Paleolithic burnt bone horizons from the Swabian Jura: Distinguishing between *in situ* fireplaces and dumping areas. *Geoarchaeology* **18**:541–565.

Schmid, P. (1983) Eine Rekonstruktion des Skelettes von A.L. 288-1 (Hadar) und deren Konsequenzen. *FP* **40**:283–306.

Schmid, P. (1991) The trunk of the australopithecines. In: Coppens and Sénut 1991: 225–234.

Schmidt-Nielsen, K., W. Bretz, and C. Taylor (1970) Panting in dogs: Unidirectional air flow over evaporative surfaces. *Science* **169**:1102–1104.

Schmitt, D. (1993) Forelimb mechanics as a function of substrate type during quadrupedalism in two anthropoid primates. *JHE* **26**:441–457.

Schmitt, D. (1998) Forelimb mechanics during arboreal and terrestrial quadrupedalism in Old World monkeys. In: Strasser et al. 1998:175–200.

Schmitt, D. (1999) Compliant walking in primates. *Journal of Zoology (London)* **248**:149–160.

Schmitt, D. (2003) Insights into the evolution of human bipedalism from experimental studies of humans and other primates. *Journal of Experimental Biology* **206**:1437–1448.

Schmitt, D., and S. Larson (1995) Heel contact as a function of substrate type and speed in primates. *AJPA* **96**:39–50.

Schmitt, D., J. Stern, and S. Larson (1996) Compliant gait in humans: Implications for substrate reaction forces during australopithecine bipedalism. *AJPA Supplement* **22**:209.

Schmitz, J., and H. Zischler (2004) Molecular cladistic markers and the infraordinal phylogenetic relationships of primates. In: Ross and Kay 2004: 65–77.

Schmitz, R. (1997) Neandertal (Feldhofer Grotte). In: Spencer 1997: 710–711.

Schmitz, R. (2003) Interdiziplinäre Untersuchungen an den Neuenfunden aus dem Neandertal. *MGU* **12**:25–45.

Schmitz, R. (2005) Die Entdeckung des fossilen Menschen im 18. und 19. Jahrhundert. *MGU* **14**:25–35.

Schmitz, R. (2006) The rediscovery of the sediments from the Neanderthal type specimen's site. In: Schmitz 2006: 55–60.

Schmitz, R., D. Serre, G. Bonani, S. Feine, F. Hillgruber, H. Krainitzki, S. Pääbo, and F. Smith (2002) The Neandertal type site revisited: Interdisciplinary investigations of skeletal remains from the Neander Valley, Germany. *PNAS* **99**: 13342–13347.

Schmitz, R., and J. Thissen (2000) *Neandertal. Die Geschichte geht weiter.* Heidelberg: Spektrum.

Schoenenmann, P., and J. Allen (2006) Scaling of body and brain weight within modern and fossil hominids: Implications for the Flores specimen. *AJPA Supplement* **42**:159–160.

Schoetensack, O. (1908) *Der Unterkiefer des* Homo heidelbergensis *aus den Sanden von Mauer bei Heidelberg.* Leipzig: Engelmann.

Scholz, M., L. Bachmann, G. Nicholson, J. Bachmann, I. Giddings, B. Rüschoff-Thale, A. Czarnetzki, and C. Pusch (2000) Genomic differentiation of Neanderthals and anatomically modern man allows a fossil DNA-based classification of morphologically indistinguishable hominid bones. *AJHG* **66**:1927–1932.

Schopf, J. W. (1999) *Cradle of Life: The Discovery of Earth's Earliest Fossils.* Princeton: Princeton University Press.

Schrenk, F., T. Bromage, et al. (1993) Oldest *Homo* and Pliocene biogeography of the Malawi Rift. *Nature* **365**:833–835.

Schülke, O., P. Kappeler, and H. Zischler (2004) Small testes size despite high extrapair paternity in the pair-living nocturnal primate *Phaner furcifer. Behavioral Ecology and Sociobiology* **55**:293–301.

Schultz, A. (1961) Vertebral column and thorax. In: H. Hofer, A. Schultz, and D. Starck, Editors. *Primatologia: Handbuch der Primatenkunde*, Volume IV, Part 5. Basel: S. Karger, pp. 1–66.

Schultz, A. (1969) *The Life of Primates.* London: Weidenfeld and Nicholson.

Schuman, E., and C. Brace (1954) Metric and morphologic variations in the Liberian chimpanzee; comparisons with anthropoid and human dentitions. *HB* **26**:239–268.

Schwalbe, G. (1901) Der Neandertalschädel. *Bonner Jahrbuch* **106**:1–72.

Schwalbe, G. (1904) *Die Vorgeschichte des Menschen.* Braunschweig: Vieweg.

Schwalbe, G. (1906) *Studien zur Vorgeschichte des Menschen.* Stuttgart: Schweizerbart'sche.

Schwarcz, H. (1992) Uranium series dating in paleoanthropology. *EA* **1**:56–62.

Schwarcz, H., W. Buhay, R. Grün, H. Valladas, E. Tchernov, O. Bar-Yosef, and B. Vandermeersch (1989) ESR dating of the Neanderthal site, Kebara cave, Israel. *JAS* **16**:653–661.

Schwarcz, H., and R. Grün (1992) Electron spin resonance (ESR) dating of the origin of modern man. *PTRSL* **337**:145–148.

Schwarcz, H., R. Grün, and P. Tobias (1994) ESR dating studies of the australopithecine site of Sterkfontein, South Africa. *JHE* **26**:175–181.

Schwarcz, H., R. Grün, B. Vandermeersch, O. Bar-Yosef, H. Valladas, and E. Tchernov (1988) ESR dates for the hominid burial site of Qafzeh in Israel. *JHE* **17**:733–737.

Schwarcz, H., and A. Latham (1984) Uranium-series age determinations of travertines from the site of Vértesszőllős, Hungary. *JAS* **11**:326–336.

Schwarcz, H., and W. Rink (1998) Progress in ESR and U-series chronology of the Levantine Paleolithic. In: Akazawa et al. 1998: 57–67.

Schwarcz, H., J. Simpson, and C. Stringer (1998) The Neanderthal skeleton from Tabun: U-series data by gamma ray spectrometry. *JHE* **35**:635–645.

Schwartz, G. (2000) Taxonomic and functional aspects of the patterning of enamel thickness distribution in extant large-bodied hominoids. *AJPA* **111**:221–244.

Schwartz, G., and C. Dean (2001) The ontogeny of canine dimorphism in extant hominoids. *AJPA* **115**:269–283.

Schwartz, J. (1984) The evolutionary relationships of man and orang-utans. *Nature* **308**:501–505.

Schwartz, J. (1990) *Lufengpithecus* and its potential relationship to an orang-utan clade. *JHE* **19**:591–605.

Schwartz, J. (1997) *Lufengpithecus* and hominoid phylogeny: Problems in delineating and evaluating phylogenetically relevant characters. In: Begun et al. 1997: 363–388.

Schwartz, J. (2000) Taxonomy of the Dmanisi crania. *Science* **289**:55.

Schwartz, J. (2003) Another perspective on hominid diversity. *Science* **301**:763–764.

Schwartz, J. (2004) Getting to know *Homo erectus*. *Science* **305**:53–54.

Schwartz, J., and I. Tattersall (1996) Significance of some previously unrecognized apomorphies in the nasal region of *Homo neanderthalensis*. *PNAS* **93**:10852–10854.

Schwartz, J., and I. Tattersall (2000) The human chin revisited: What is it and who has it? *JHE* **38**:367–409.

Schwartz, J., and I. Tattersall (2002) *The Human Fossil Record*, Volume 1: *Terminology and Craniodental Morphology of Genus Homo (Europe)*. New York: Wiley-Liss.

Schwartz, J., and I. Tattersall (2003) *The Human Fossil Record*, Volume 2: *Craniodental Morphology of the Genus Homo (Africa and Asia)*. New York: Wiley-Liss.

Schwartz, J. H., and I. Tattersall (2005) *The Human Fossil Record*, Volume 4: *Craniodental Morphology of Early Hominids (Genera Australopithecus, Paranthropus, Orrorin) and Overview*. New York: Wiley-Liss.

Schwartz, J., I. Tattersall, and N. Eldredge (1978) Phylogeny and classification of the Primates revisited. *YPA* **21**:95–133.

Schwartz, J., I. Tattersall, W. Huang, Y. Gu, R. Ciochon, R. Larrick, F. Quiren, J. de Vos, H. Schwarcz, W. Rink, and C. Younge (1996) Whose teeth? *Nature* **381**:201–202.

Seiffert, E., E. Simons, and Y. Attia (2003) Fossil evidence for an ancient divergence of lorises and galagos. *Nature* **422**: 421–424.

Seiffert, E., E. Simons, and C. Simons (2004) Phylogenetic, biogeographic, and adaptive implications of new fossil evidence bearing on crown anthropoid origins and early stem catarrhine evolution. In: Ross and Kay 2004: 157–181.

Seiffert, E., E. Simons, W. Clyde, J. Rossie, Y. Attia, T. Bown, P. Chatrath, and M. Mathison (2005) Basal anthropoids from Egypt and the antiquity of Africa's higher primate radiation. *Science* **310**:300–304.

Seilacher, A. (1994) Early life on Earth: Late Proterozoic fossils and the Cambrian explosion. In: S. Bengtson, J. Bergstrom, G. Vidal, and A. Knoll, Editors. *Early Life on Earth*. New York: Columbia University Press. 389–400.

Sémah, F. (1984) The Sangiran Dome in the Javan Plio-Pleistocene chronology. *CFS* **69**:245–252.

Sémah, F., A. Sémah, T. Djubiantono, and H. Simanjuntak (1992) Did they also make stone tools? *JHE* **23**:439–446.

Semaw, S. 2000 The world's oldest stone artefacts from Gona, Ethiopia: Their implications for understanding stone technology and patterns of human evolution between 2.6–1.5 million years ago. *JAS* **27**:1197–1214.

Semaw, S., S. Simpson, J. Quade, P. Renne, R. Butler, W. McIntosh, N. Levin, M. Dominguez-Rodrigo, and M. Rogers (2005) Early Pliocene hominids from Gona, Ethiopia. *Nature* **433**:301–305.

Sénut, B. (1980) New data on the humerus and its joints in Plio-Pleistocene hominids. *Collegium Antropologicum* **4**:87–94.

Sénut, B. (1981a) *L'humérus et ses articulations chez les hominidés Plio-Pléistocènes*. Paris: Centre Nationale de la Recherche Scientifique.

Sénut, B. (1981b) Humeral outlines in some hominoid primates and in Plio-Pleistocene hominids. *AJPA* **56**:275–283.

Sénut, B. 1996. Pliocene hominid systematics and phylogeny. *SAJS* **92**:165–166.

Sénut, B., and M. Pickford (2004) La dichotomie grands singes—homme revisitée. *CRAS Palévol* **3**:265–276.

Sénut, B., M. Pickford, D. Gommery, P. Mein, K. Cheboi, and Y. Coppens (2001) First hominid from the Miocene (Lukeino Formation, Kenya). *CRAS* **332**:137–144.

Sénut, B., and C. Tardieu (1985) Functional aspects of Plio-Pleistocene hominid limb bones: implications for taxonomy and phylogeny. In: Delson 1985: 193–201.

Şenyürek, M. (1940) Fossil man in Tangier. *Papers of the Peabody Museum of Archaeology and Ethnology*, Volume XVI, Part 3.

Şenyürek, M. (1957) The skeleton of the fossil infant found in Shanidar Cave, northern Iraq. *Anatolia* **2**:49–55.

Sergi, S. (1953) I profaneranthropi di Swanscombe e di Fontéchevade. *Rendiconti dell'Academia Nazionale dei Lincei* **15**:601–608. Reprinted as: Morphological position of the "Prophaneranthropi" (Swanscombe and Fontéchevade). In: Howells 1967: 507–520.

Serrat, M., D. King, and C.O. Lovejoy (in press) Temperature regulates limb length in homeotherms by directly modulating cartilage growth. *PNAS*.

Serre, D., A. Langaney, M. Chech, M. Teschler-Nikola, M. Paunović, P. Mennecier, M. Hofreiter, G. Possnert, and S. Pääbo (2004) No evidence of Neandertal mtDNA contribution to early modern humans. *PLOS (Public Library of Science) Biology* **2**:313–317.

Shackleton, N. (1967) Oxygen isotope analyses and Pleistocene temperatures reassessed. *Nature* **215**:259–265.

Shackleton, N. (1975) The stratigraphic record of deep-sea cores and its implication for the assessment of glacials, interglacials, stadials, and interstadials in the Mid-Pleistocene. In: Butzer and Isaac 1975: 1–24.

Shackleton, N. (1987) Oxygen isotopes, ice volume and sea level. *Quarternary Science Reviews* **6**:183–190.

Shackleton, N. (1995) New data on the evolution of Pliocene climatic variability. In: Vrba et al. 1995: 242–248.

Shang, H., H. Tong, S. Zhang, F. Chen, and E. Trinkaus (2007) An early modern human from Tianyuan Cave, Zhoukoudian, China. *PNAS* **104**:6573–6578.

Shapiro, H. (1974) *Peking Man*. New York: Simon and Schuster.

Shapiro, L., C. Seiffert, L. Godfrey, W. Jungers, E. Simons, and G. Randria (2005) Morphometric analysis of lumbar vertebrae in extinct Malagasy strepsirrhines. *AJPA* **128**:823–839.

Sharman, G., and P. Pilton (1964) Life history and reproduction of the red kangaroo (*Megaleia rufa*). *Proceedings of the Zoological Society of London* **142**:29–48.

Shea, J. (1988) Spear points from the Middle Paleolithic of the Levant. *Journal of Field Archaeology* **15**:441–450.

Shea, J. (1997) Middle Paleolithic spear point technology. In: H. Kneckt, Editor. *Projectile Technology*. New York: Plenum, pp. 79–106.

Shea, J. (2003) Neandertals, competition, and the origin of modern human behavior in the Levant. *EA* **12**:173–187.

Shea, J., and A. Brooks (2000) Grimaldi. In: Delson et al. 2000: 297–298.

Shen, G., T. Ku, H. Cheng, R. Edwards, Z. Yuan, and Q. Wang (2001) High-precision U-series dating of Locality 1 at Zhoukoudian, China. *JHE* **41**:679–688.

Shen, G., W. Wang, Q. Wang, J. Zhou, K. Collerson, C. Zhou, and P. Tobias (2002) U-series dating of Liujiang hominid site, Guangxi, Southern China. *JHE* **43**:817–829.

Shigehara, N., M. Takai, R. Kay, A. Soe, S. Tun, T. Tsubamato, and T. Thein (2002) The upper dentition and face of *Pondaungia cotteri* from central Myanmar. *JHE* **43**:143–146.

Shipman, P. (1988) Diet and subsistence strategies at Olduvai Gorge. In: B. Kennedy and G. le Moine, Editors. *Diet and Subsistence: Current Archaeological Perspectives*. Calgary: Archaeological Association of the University of Calgary, pp. 3–11.

Shipman, P. (1989) Altered bones from Olduvai Gorge, Tanzania: Techniques, problems, and implications of their recognition. In: R. Bonnichsen and M. Sorg, Editors. *Bone Modification*. Orono: Center for the Study of the First Americans, University of Maine, pp. 317–334.

Shipman, P. (1994) *The Evolution of Racism*. New York: Simon and Schuster.

Shipman, P. (2001) *The Man Who Found the Missing Link*. New York: Simon and Schuster.

Shipman, P., and P. Storm (2002) Missing links: Eugène Dubois and the origins of paleoanthropology. *EA* **11**:108–116.

Shreeve, J. (1996) Sunset on the savanna. *Discover* **17**(7): 116–125.

Shu, D-G., H-L. Luo., S. Conway Morris, X-L. Zhang, S-X. Hu, L. Chen, J. Han, M. Zhu, Y. Li, and L-Z. Chen (1999) Lower Cambrian vertebrates from South China. *Nature* **402**: 42–46.

Shubin, N., E. Daeschler, and F. Jenkins, Jr. (2006) The pectoral fin of *Tiktaalik roseae* and the origin of the tetrapod limb. *Nature* **440**:764–771.

Shubin, N., C. Tabin, and S. Carroll (1997) Fossils, genes and the evolution of animal limbs. *Nature* **388**:639–649.

Sigmon, B. (1975) Functions and evolution of hominoid hip and thigh musculature. In: R. Tuttle, Editor. *Primate Functional Morphology and Evolution*. The Hague: Mouton, pp. 237–252.

Silcox, M., J. Bloch, E. Sargis, and D. Boyer (2005) Euarchonta (Dermoptera, Scandentia, Primates). In: K. Rose and D. Archibald, Editors. *The Rise of Placental Mammals*. Baltimore: Johns Hopkins University Press, pp. 127–144.

Sillen, A. 1992. Strontium–calcium ratios (Sr/Ca) of *Australopithecus robustus* and *Homo* sp. in Swartkrans. *JHE* **23**:495–516.

Sillen, A. (1993) Was *Australopithecus robustus* an omnivore? *SAJS* **89**:71–72.

Sillen, A., G. Hall, and R. Armstrong (1995) Strontium/calcium ratios (Sr/Ca) and strontium isotopic ratios ($^{87}Sr/^{86}Sr$) of *Australopithecus robustus* and *Homo* from Swartkrans. *JHE* **28**:277–285.

Sillen, A., and A. Morris (1996) Diagenesis of bone from Border Cave: Implications for the age of the Border Cave hominids. *JHE* **31**:499–506.

Simek, J. (1991) Stone tool assemblages from Krapina (Croatia, Yugoslavia). In: A. Monet-White and S. Holen, Editors. *Raw Material Economics among Prehistoric Hunter–Gatherers*. University of Kansas Publications in Anthropology, Volume 19, pp. 59–71.

Simek, J. (1992) Neandertal cognition and the Middle to Upper Paleolithic transition. In: Bräuer and Smith 1992: 231–245.

Simek, J., and F. Smith (1997) Chronological changes in the stone tool assemblages from Krapina (Croatia). *JHE* **32**:561–575.

Simmons, T. (1994) Archaic and modern *Homo sapiens* in the contact zones: evolutionary schematics and model predictions. In: Nitecki and Nitecki 1994: 201–225.

Simmons, T. (1999) Migration and contact zones in modern human origins: Baboon models for hybridization snd species recognition. *Anthropologie* **36**:101–109.

Simmons, T., A. Falsetti, and F. Smith (1991) Frontal bone morphometrics of southwest Asian Pleistocene hominids. *JHE* **20**:249–269.

Simmons, T., and F. Smith (1991) Human population relationships in the Late Pleistocene. *CA* **32**:623–627.

Simons, E. (1961a) Notes on Eocene tarsioids and a revision of some Necrolemurinae. *Bulletin of the British Museum (Natural History), Geological Series* **5**:43–69.

Simons, E. (1961b) The phyletic position of *Ramapithecus. Postilla Yale Peabody Museum* **57**:1–9.

Simons, E. (1964a) The early relatives of man. *SA* **211**: 50–64.

Simons, E. (1964b) On the mandible of *Ramapithecus. PNAS* **51**:528–535.

Simons, E. (1972) *Primate Evolution.* New York: Macmillan.

Simons, E. (1987) New faces of *Aegyptopithecus* from the Oligocene of Egypt. *JHE* **16**:273–290.

Simons, E. (1989) Description of two genera and species of Late Eocene Anthropoidea from Egypt. *PNAS* **86**:9956–9960.

Simons, E. (1990) Discovery of the oldest known anthropoidean skull from the Paleogene of Egypt. *Science* **247**:1567–1569.

Simons, E. (1992) Diversity in the early Tertiary anthropoidean radiation in Africa. *PNAS* **94**:180–184.

Simons, E. (1997) Preliminary description of the cranium of *Proteopithecus sylviae,* an Egyptian late Eocene anthropoidean primate. *PNAS* **94**:14970–14975.

Simons, E. (2004) The cranium and adaptations of *Parapithecus grangeri,* a stem anthropoid from the Fayum Oligocene of Egypt. In: Ross and Kay 2004: 183–204.

Simons, E., and T. Bown (1985) *Afrotarsius chatrathi,* new genus, new species: First tarsiiform primate (Tarsiidae?) from Africa. *Nature* **313**:475–477.

Simons, E., and D. Pilbeam (1965) A preliminary revision of the Dryopithecinae (Pongidae, Anthropoidea). *FP* **3**:81–152.

Simons, E., J. Plavcan, and J. Fleagle (1999) Canine sexual dimorphism in Egyptian Eocene African primates: *Catopithecus* and *Proteopithecus. PNAS* **96**:2559–2562.

Simons, E., and D. Rasmussen (1994) A whole new world of ancestors: Eocene anthropoideans from Africa. *EA* **3**:128–139.

Simons, E., T. Ryan, and Y. Attia (2007) A remarkable female cranium of the early Oligocene anthropoid *Aegyptopithecus zeuxis* (Catarrhini, Propliopithecidae). *PNAS* **104**:8731–8736.

Simpson, G. G. (1944) *Tempo and Mode in Evolution.* New York: Columbia University Press.

Simpson, G. G. (1953) *The Major Features of Evolution.* New York: Columbia University Press.

Simpson, G. G. (1961) *Principles of Animal Taxonomy.* New York: Columbia University Press.

Simpson, G. G. (1963) The meaning of taxonomic statements. In: S. Washburn, Editor. *Classification and Human Evolution.* New York: Aldine de Gruyter.

Sinclair, A., M. Leakey, and N. Norton-Griffiths (1986) Migration and hominid bipedalism. *Nature* **324**:307–308.

Singer, R., and J. Wymer (1982) *The Middle Stone Age at Klasies River Mouth in South Africa.* Chicago: University of Chicago Press.

Skelton, R., H. McHenry, and G. Drawhorn (1986) Phylogenetic analysis of early hominids. *CA* **27**:21–43.

Skelton, R., and H. McHenry (1992) Evolutionary relationships among early hominids. *JHE* **23**:309–349.

Skinner, M. (1991) Bee brood consumption: An alternative explanation for hypervitaminosis A in KNM-ER 1808 (*Homo erectus*) from Koobi Fora, Kenya. *JHE* **20**:493–503.

Skinner, M., A. Gordon, and N. Collard (2006) Mandibular size and shape variation in the hominins at Dmanisi, Republic of Georgia. *JHE* **51**:36–49.

Skinner, M., and G. Sperber (1982) *Atlas of the Radiographs of Early Man.* New York: Liss.

Sládek, V., E. Trinkaus, S. Hillson, and T. Holliday (2000) *The People of the Pavlovian.* Brno: The Dolní Věstonice Studies, Volume 5.

Sládek, V., E. Trinkaus, A. Šefčáková, and R. Halouza (2002) Morphological affinities of the Šaľa frontal bone. *JHE* **43**:787–815.

Smith, B. H. (1989a) Dental development as a measure of life history in primates. *Evolution* **43**:683–688.

Smith, B. H. (1989b) Growth and develoopment and its significance for early hominid behavior. *Ossa* **14**:63–96.

Smith, B. H. (1991) Dental development and the evolution of life history in Hominidae. *AJPA* **86**:157–174.

Smith, B. H. (1993) The physiological age of KNM-WT 15000. In: Walker and Leakey 1993a: 195–220.

Smith, F. (1976) *The Neandertal Remains from Krapina: A Descriptive and Comparative Study.* Knoxville: University of Tennessee Department of Anthropology Reports of Investigation 15.

Smith, F. (1978) Evolutionary significance of the mandibular foramen area in Neandertals. *AJPA* **48**:523–531.

Smith, F. (1980) Sexual differences in European Neandertal crania with special reference to the Krapina remains: Problems and analysis. *JHE* **9**:359–375.

Smith, F. (1982) Upper Pleistocene hominid evolution in South-Central Europe: a review of the evidence and analysis of trends. *CA* **23**:667–703.

Smith, F. (1983) Behavioral interpretation of changes in craniofacial morphology across the archaic/modern *Homo sapiens* transition. *BARIS* **164**:143–163.

Smith, F. (1984) Fossil hominids from the Upper Pleistocene of central Europe and the origin of modern Europeans. In: Smith and Spencer 1984: 137–209.

Smith, F. (1985) Continuity and change in the origin of modern *Homo sapiens. ZMA* **75**:197–222.

Smith, F. (1991) The Neandertals: Evolutionary dead ends or ancestors of modern people? *JAR* **47**:219–238.

Smith, F. (1992a) The role of continuity in modern human origins. In: Bräuer and Smith 1992: 145–156.

Smith, F. (1992b) On excavations at Kebara Cave. *CA* **33**:540–541.

Smith, F. (1993) Models and realities in modern human origins: The African fossil evidence. In: Aitken et al. 1993: 234–248.

Smith, F. (1994) Samples, species and speculations in the study of modern human origins. In: Nitecki and Nitecki 1994: 227–249.

Smith, F. (1997a) Modern human origins. In: Spencer 1997: 661–672.

Smith, F. (1997b) Neandertals. In: Spencer 1997: 711–722.

Smith, F. (1997c) Gibraltar. In: Spencer 1997: 435–438.

Smith, F. (1997d) Virchow, Rudolf (1821–1902). In: Spencer 1997: 1094–1095.

Smith, F. (1997e) Modern human origins. In: J. O. Vogel, Editor. *Encyclopedia of Precolonial Africa*. Walnut Creek: Altamira Press. 257–266.

Smith, F. (1997f) Mladeč. In: Spencer 1997: 659–660.

Smith, F. (2002) Migrations, radiations, and continuity: Patterns in the evolution of Middle and Late Pleistocene humans. In: Hartwig 2002: 437–456.

Smith, F., and J. Ahern (1994) Additional cranial remains from Vindija Cave, Croatia. *AJPA* **93**:275–280.

Smith, F., and J. Ahern (2006) Problems with species identification in the human fossil record with special emphasis on the Neandertal question. *AJPA Supplement* **42**:166.

Smith, F., D. Boyd, and M. Malez (1985) Additional Upper Pleistocene human remains from Vindija Cave, Croatia, Yugoslavia. *AJPA* **68**:375–383.

Smith, F., A. Falsetti, and S. Donnelly (1989a) Modern human origins. *YPA* **32**:35–68.

Smith, F., A. Falsetti, and T. Simmons (1995) Circum–Mediterranean biological connections and pattern of Late Pleistocene human evolution. In: H. Ullrich, Editor. *Man and Environment in the Paleolithic*. Liège: ERAUL, pp. 197–207.

Smith, F., and M. Green (1991) Heterochrony, life history and Neandertal morphology. *AJPA Supplement* **12**:164.

Smith, F., V. Hutchinson, and I. Janković (in press) Assimilation and modern human origins in the African peripheries. In: S. Reynolds, and A. Gallagher, Editors. *African Genesis: Perspectives on Hominid Evolution*. Johannesburg: University of the Witwatersrand Press.

Smith, F., I. Janković, and I. Karavanić (2005) The assimilation model, modern human origins in Europe, and the extinction of the Neandertals. *QI* **137**:7–19.

Smith, F., and S. Paquette (1989) The adaptive basis of Neandertal facial form, with some thoughts on the nature of modern human origins. In: Trinkaus 1989: 181–210.

Smith, F., and G. Ranyard (1980) Evolution of the supraorbital region in Upper Pleistocene fossil hominids from South-Central Europe. *AJPA* **53**:589–609.

Smith, F., J. Simek, and M. Harrill (1989b) Geographic variation in supraorbital torus reduction during the later Pleistocene (c. 80,000–15,000 BP). In: Mellars and Stringer 1989: 172–193.

Smith, F., and M. Smith (1986) On the significance of anomalous nasal bones in the Neandertals from Krapina. In: V. Novotný, Editor. *Fossil Man: New Facts—New Ideas. Anthropos* **23**:217–226.

Smith, F., M. Smith, and R. Schmitz (2006) Human skeletal remains from the 1997 and 2000 excavations of cave deposits derived from Kleine Feldhofer Grotte in the Neander Valley, Germany. In: Schmitz 2006: 187–246.

Smith, F., and F. Spencer (1997) Cro-Magnon. In: Spencer 1997: 298–301.

Smith, F., and E. Trinkaus (1991) Les origines de l'homme moderne en Europe centrale; un cas de continuité. In: J.-J. Hublin and A.-M. Tillier, Editors. *Aux Origines de l'Homo sapiens*. Paris: Presse Universitaires de France, pp. 251–290.

Smith, F., E. Trinkaus, P. Pettitt, I. Karavanic, and M. Paunović (1999) Direct radiocarbon dates for Vindija G1 and Velika Pećina late Plesistocene hominid remains. *PNAS* **96**: 12281–12286.

Smith, G. Elliot (1925) The fossil anthropoid ape from Taungs. *Nature* **115**:235.

Smith, G. Elliot (1927) *The Evolution of Man*. London: Oxford University Press.

Smith, K. (1992) The evolution of the mammalian pharynx. *Zoological Journal of the Linnean Society (London)* **104**:313–349.

Smith, S. (1918) The fossil human skull found at Talgai, Queensland. *PTRSL* **208**:351–387.

Smith, T., P. Tafforeau, D. Reid, R. Grün, S. Eggins, M. Boutakiout, and J.-J. Hublin (2004) Earliest evidence of modern human life history in North African early *Homo sapiens*. *PNAS* **104**: 6138–6133.

Soficaru, A., A. Doboş, and E. Trinkaus (2006) Early modern humans from Peştera Muierii, Baia de Fier, Romania. *PNAS* **103**:17196–17201.

Soficaru, A. C. Petrea, A. Doboş, and E. Trinkaus (2007) The human cranium from Peştera Cioclinova Uscată, Romania: Context, age, taphonomy, morphology and paleopathology. *CA* **48**:611–619.

Sohn, S., and M. Wolpoff (1993) The Zuttiyeh face: A view from the East. *AJPA* **91**:325–348.

Sokal, R., and T. Crovello (1992) The biological species concept: A critical evaluation. In: Ereshefsky 1992: 27–55.

Sokal, R., and P. Sneath (1963) *Principles of Numerical Taxonomy*. San Francisco: Freeman.

Solecki, R. (1957) Shanidar Cave. *SA* **197**:58–64.

Solecki, R. (1963) Prehistory in the Shanidar Valley, northern Iraq. *Science* **139**:179–193.

Sollas, W. (1907) On the cranial and facial characters of the Neanderthal race. *PTRSL* **199**:321–337.

Sonakia, A. (1985) Early *Homo* from Narmada Valley, India. In: Delson 1985: 334–338.

Sondaar, P., G. van den Bergh, B. Mumbroto, F. Aziz, J. de Vos, and U. Batu (1994) Middle Pleistocene faunal turnover and colonization of Flores (Indonesia) by *Homo erectus*. *CRAS* **319**:1255–1262.

Sonntag, C. (1924) *The Morphology and Evolution of the Apes and Man.* London: John Bale, Sons and Danielsson.

Sorensen, M., and W. Leonard (2001) Neandertal energetics and foraging efficiency. *JHE* **40**:483–495.

Spencer, F. (1984) The Neandertals and their evolutionary significance: A brief historical survey. In: Smith and Spencer 1984: 1–49.

Spencer, F. (1986) *Ecce Homo. An Annotated Bibliographic History of Physical Anthropology.* Westport, CT: Greenwood Press.

Spencer, F. (1990) *Piltdown: A Scientific Forgery.* Oxford: Oxford University Press.

Spencer, F. (1997a) Engis. In: Spencer 1997: 360–362.

Spencer, F. (1997b) Piltdown. In: Spencer 1997: 821–825.

Spencer, F., and F. Smith (1981) The significance of Aleš Hrdlička's "Neanderthal phase of man." *AJPA* **56**:435–456.

Spencer, M., and B. Demes (1993) Biomechanical analysis of masticatory configuration in Neandertals and Inuits. *AJPA* **91**:1–20.

Speth, J., and E. Tchernov (1998) The role of hunting and scavenging in Neandertal procurement strategies. In: Akazawa et al. 1998: 223–239.

Sponheimer, M., and J. Lee-Thorp (1999) Isotopic evidence for the diet of an early hominid, *Australopithecus africanus. Science* **283**:368–370.

Sponheimer, M., J. Loudon, D. Codron, M. Howells, J. Pruetz, J. Codron, J., D. de Ruiter, and J. Lee-Thorp (2006) Do "savanna" chimpanzees consume C_4 resources? *JHE* **51**:128–133.

Spoor, F., T. Garland, G. Krovitz, T. Ryan, M. Silcox, and A. Walker (2007a) The primate semicircular canal system and locomotion. *PNAS* **104**:10808–10812.

Spoor, F., J.-J. Hublin, and O. Kondo (2002) The bony labyrinth of the Dederiyeh child. In: Akazawa and Muhesen 2002: 215–220.

Spoor, F., J.-J. Hublin, M. Braun, and F. Zonneveld (2003) The bony labyrinth of Neanderthals. *JHE* **44**:141–165.

Spoor, F., M. Leakey, P. Gathogo, F. Brown, S. Antón, I. McDougall, I. Kiarie, F. Manthi, and L. Leakey (2007b) Implications of new early *Homo* fossils from Ileret, east of Lake Turkana, Kenya. *Nature* **448**:688–691.

Spoor, F., C. Stringer, and F. Zonneveld (1998) Rare temporal bone pathology of the Singa calvaria from Sudan. *AJPA* **107**:41–50.

Spoor, F., B. Wood, and F. Zonneveld (1994) Implications of early hominid labyrinthine morphology for the evolution of human bipedal locomotion. *Nature* **369**:645–648.

Spoor, F., and F. Zonneveld (1994) The bony labyrinth in *Homo erectus*: A preliminary report. In: Franzen 1994: 251–256.

Spoor, F., and F. Zonneveld (1998) Comparative review of the human bony labyrinth. *YPA* **41**:211–251.

Spuhler, J. (1977) Biology, speech and language. *ARA* **6**:509–561.

Spuhler, J. (1989) Evolution of mitochondrial DNA in human and other organisms. *AJHB* **1**:509–528.

Stanford, C. (1996) The hunting ecology of wild chimpanzees: Implications for the evolutionary ecology of Pliocene hominids. *CA* **98**:96–113.

Stanford, C. (1999) *The Hunting Apes: Meat Eating and the Origins of Human Behavior.* Princeton: Princeton University Press.

Stanford, C. (2001a) The ape's gift: Meat-eating, meat-sharing, and human evolution. In: F. de Waal, Editor. *Tree of Origin: What Primate Behavior Can Tell Us about Human Social Evolution.* Cambridge: Harvard University Press, pp. 95–117.

Stanford, C. (2001b) A comparison of social meat-foraging by chimpanzees and human foragers. In: C. Stanford and H. Bunn, Editors. *Meat-Eating and Human Evolution.* New York: Oxford University Press, pp. 122–140.

Stanley, S. (1989) *Earth and Life Through Time*, 2nd edition. New York: Freeman.

Stapledon, O. (1935) *Odd John: A Story Between Jest and Earnest.* London: Methuen.

Stebbins, G., and A. Day (1967) Cytogenetic evidence for long continued stability in the genus *Plantago. Evolution* **21**:409–428.

Steegmann, A. T. (1970) Cold adaptation and the human face. *AJPA* **32**:234–250.

Steegmann, A. T., F. Cerny, and T. Holliday (2002) Neandertal cold adaptation: physiological and enegetic factors. *AJHB* **14**:566–583.

Stefan, V., and E. Trinkaus (1998a) Discrete trait and morphometric affinities of the Tabun 2 mandible. *JHE* **34**:443–468.

Stefan, V., and E. Trinkaus (1998b) La Quina 9 and Neandertal mandibular variability. *BMSAP* **10**:293–324.

Steiper, M., and M. Ruvolo (2003) New World monkey phylogeny based on X-linked G6PD DNA sequences. *MPE* **27**:121–130.

Stepanchuk, V. (1993) Prolom II, a Middle Paleolithic cave in the eastern Crimea with non-utilitarian bone artifacts. *PPS* **9**:17–37.

Stephan, H. (1972) Evolution of primate brains: a comparative anatomical investigation. In: R. Tuttle, Editor. *The Functional and Evolutionary Biology of Primates.* Chicago: Aldine-Atherton, pp. 155–174.

Stephan, H., and O. andy (1969) Quantitative comparative neuroanatomy of primates: An attempt at a phylogenetic interpretation. *Annals of the New York Academy of Sciences* **167**:370–387.

Stern, J. (1975) Before bipedality. *YPA* **19**:59–68.

Stern, J. (2000) Climbing to the top: A personal memoir of *Australopithecus afarensis. EA* **9**:113–133.

Stern, J., and R. Susman (1981) Electromyography of the gluteal muscles in *Hylobates, Pongo,* and *Pan*: Implications for the evolution of hominid bipedality. *AJPA* **55**:153–166.

Stern, J. T., and R. Susman (1983) The locomotor anatomy of *Australopithecus afarensis*. *AJPA* **60**:279–317.

Stern, J., and R. Susman (1991) "Total morphological pattern" versus the "magic trait": conflicting approaches to the study of early hominid bipedalism. In: Coppens and Sénut 1991: 99–111.

Stewart, J. (1997) Morphology and evolution of the egg of oviparous amniotes. In: S. Sumida and K. Martin, Editors. *Amniote Origins: Completing the Transition to Land*. San Diego: Academic Press, pp. 291–326.

Stewart, K., and A. Harcourt (1986) Gorillas: Variation in female relationships. In: B. Smuts, D. Cheney, R. Seyfarth, R. Wrangham, and T. Struhsaker, Editors. *Primate Societies*. Chicago: University of Chicago Press, pp. 155–164.

Stewart, T. D. (1959) Restoration and study of the Shanidar 1 Neanderthal skeleton in Baghdad, Iraq. *Yearbook of the American Philosophical Society, 1958*, pp. 274–278.

Stewart, T. D. (1960) Form of the pubic bone in Neanderthal man. *Science* **131**:1437–1438.

Stewart, T. D. (1962a) Neanderthal cervical vertebrae with special attention to the Shanidar Neanderthals from Iraq. *Biblioteca Primatologica* **1**:130–154.

Stewart, T. D. (1962b) Neanderthal scapulae with special attention to the Shanidar Neanderthals from Iraq. *Anthropos* **57**:781–800.

Stewart, T. D. (1964) A neglected primitive feature of the Swanscombe skull. In: Ovey 1964: 151–160.

Stewart, T. D. (1977) The Neanderthal skeletal remains from Shanidar Cave, Iraq: A summary of the findings to date. *PAPS* **121**:122–165.

Stiner, M. (1991) A taphonomic perspective on the origins of the faunal remains from Grotta Guatteri (Latium, Italy). *CA* **32**:103–117.

Stiner, M. (1992) The place of hominids among predators: interspecific comparisons of food procurement and transport. In: J. Hudson, Editor. *From Bones to Behavior*. Carbondale: Southern Illinois University Press, pp. 38–61.

Stiner, M. (1994) *Honor Among Thieves*. Princeton: Princeton University Press.

Stiner, M. (1998) Mortality analysis of Pleistocene bears and its paleoanthropological relevance. *JHE* **34**:303–326.

Stoneking, M., and R. Cann (1989) African origin of human mitochondrial DNA. In: Mellars and Stringer 1989: 17–30.

Storm, P., and A. Nelson (1992) The many faces of Wadjak man. *Archaeology in Oceania* **27**:37–46.

Stovall, J., L. Price, and A. Romer (1966) The postcranial skeleton of the giant Permian pelycosaur *Cotylorhynchus romeri*. *Bulletin of the Museum of Comparative Zoology (Harvard)* **135**:1–30.

Strait, S. (2001) Dietary reconstruction of small-bodied omomyoid primates. *Journal of Vertebrate Paleontology* **21**:322–334.

Strait, D., and F. Grine (1999) Cladistics and early hominid phylogeny. *Science* **285**:1210.

Strait, D., F. Grine, and M. Moniz (1997) A reappraisal of early hominid phylogeny. *JHE* **32**:17–82.

Straus, L. (1989) Grave reservations: More on Paleolithic burial evidence. *CA* **30**:633–634.

Straus, L. (1992) *Iberia before the Iberians: The Stone Age Prehistory of Cantabrian Spain*. Albuquerque: University of New Mexico Press.

Straus, L. (2003) "The Aurignacian"? Some thoughts. In: Zilhão and d'Errico 2003: 11–17.

Straus, W., and A. Cave (1957) Pathology and posture of Neanderthal man. *Quarterly Review of Biology* **32**:348–363.

Strauss, H., D. Des Marais, J. Hayes, and R. Summons (1992) The carbon-isotopic record. In: J. W. Schopf and C. Klein, Editors. *The Proterozoic Biosphere: A Multidisciplinary Study*. New York: Cambridge University Press, pp. 117–127.

Stringer, C. (1974a) A multivariate study of the Petralona skull. *JHE* **3**:397–404.

Stringer, C. (1974b) Population relationships of later Pleistocene hominids: A multivariate study of available crania. *JAS* **1**:317–342.

Stringer, C. (1978) Some problems in Middle and Upper Pleistocene hominid relationships. In: D. Chivers and K. Joysey, Editors. *Recent Advances in Primatology*, Volume 3: *Evolution*. London: Academic Press, pp. 395–418.

Stringer, C. (1979) A re-evaluation of the fossil human calvaria from Singa, Sudan. *Bulletin of the British Museum of Natural History (Geology)* **32**:77–83.

Stringer, C. (1983) Some further notes on the morphology and dating of the Petralona skull. *JHE* **12**:731–742.

Stringer, C. (1984) The definition of *Homo erectus* and the existence of the species in Africa and Europe. *CFS* **69**:131–143.

Stringer, C. (1986a) The credibility of *Homo habilis*. In: Wood 1986: 266–294.

Stringer, C. (1986b) An archaic character in the Broken Hill innominate E. 719. *AJPA* **71**:115–120.

Stringer, C. (1989) Documenting the origin of modern humans. In: Trinkaus 1989: 67–96.

Stringer, C. (1990a) The emergence of modern humans. *SA* **263**:98–104.

Stringer, C. (1990b) British Isles. In: R. Orban, Editor. *Hominid Remains: An Update. British Isles and Eastern Germany*. Brussels: Université Libre de Bruxelles, pp. 1–40.

Stringer, C. (1991) A metrical study of the Guattari and Saccopastore crania. *Quaternaria Nova* **1**:621–638.

Stringer, C. (1992) Reconstructing recent human evolution. *PTRSL B* **337**:217–224.

Stringer, C. (1993) Secrets of the pit of the bones. *Nature* **362**:502–503.

Stringer, C. (1994) Out of Africa: A personal history. In: Nitecki and Nitecki 1994: 149–174.

Stringer, C. (1996) The Boxgrove tibia: Britain's oldest hominid and its place in the Middle Pleistocene record. In: C. Gamble and A. Lawson, Editors. *The English Paleolithic Reviewed.* Trust for Wessex Archaeology, pp. 52–56.

Stringer, C. (1998) A metrical study of the WLH-50 calvaria. *JHE* **34**:327–332.

Stringer, C. (2001) What happened to the Neandertals? *General Anthropology* **5**:4–7.

Stringer, C. (2002) Modern human origins: progress and prospects. *PTRSL* **357**:563–579.

Stringer, C. (2003) Out of Ethiopia. *Nature* **423**:692–695.

Stringer, C., and P. Andrews (1988) Genetic and fossil evidence for the origin of modern humans. *Science* **239**:1263–1268.

Stringer, C., and G. Bräuer (1994) Models, misreading and bias. *AA* **96**:416–424.

Stringer, C., L. Cornish, and P. Stuart-Macadam (1985) Preparation and further study of the Singa skull from the Sudan. *Bulletin of the British Museum of Natural History (Geology)* **38**:347–358.

Stringer, C., M. Dean, and R. Martin (1990) A comparative study of cranial and dental development within a recent British sample and among Neanderthals. In: J. DeRousseau, Editor. *Primate Life History and Evolution.* New York: Wiley, pp. 115–152.

Stringer, C., and C. Gamble (1993) *In Search of the Neanderthals: Solving the Puzzle of Human Origins.* London: Thames & Hudson.

Stringer, C., R. Grün, H. Schwarcz, and P. Goldberg (1989) ESR dates for the hominid burial site of Es-Skhul in Israel. *Nature* **338**:756–758.

Stringer, C., F. C. Howell, and J. Melentis (1979) The significance of the fossil hominid skull from Petralona, Greece. *JAS* **6**:235–253.

Stringer, C., and J. Hublin (1999) New age estimates for the Swanscombe hominid, and their significance for human evolution. *JHE* **37**:873–877.

Stringer, C., J. Hublin, and B. Vandermeersch (1984) The origin of anatomically modern humans in western Europe. In: Smith and Spencer 1984: 51–135.

Stringer, C., L. Humphrey, and T. Crompton (1997) Cladistic analysis of dental traits in recent humans using a fossil outgroup. *JHE* **32**:389–402.

Stringer, C., E. Trinkaus, M. Roberts, S. Parfitt, and R. Macphail (1998) The Middle Pleistocene human tibia from Boxgrove. *JHE* **34**:509–547.

Stubblefield, S., R. Warren, and B. Murphy (1986) Hybridization of free-ranging white-tailed and mule deer in Texas. *Journal of Wildlife Management* **50**:688–690.

Stynder, D., J. Moggi-Cecchi, L. Berger, and J. Parkington (2001) Human mandibular incisors from the Middle Pleistocene locality of Hoedjiespunt, South Africa. *JHE* **41**:369–383.

Sun, G., D. Dilcher, S. Zheng, S., and Z. Zhou (1998) In search of the first flower: A Jurassic angiosperm, *Archaefructus*, from Northeast China. *Science* **282**:1692–1695.

Susman, R. (1988a) New postcranial remains from Swartkrans and their bearing on the functional morphology and behavior of *Paranthropus robustus*. In: Grine 1988: 149–172.

Susman, R. (1988b) Hand of *Paranthropus robustus* from Member 1, Swartkrans: Fossil evidence for tool behavior. *Science* **240**:781–784.

Susman, R. (1989) New hominid fossils from the Swartkrans Formation (1979–1986 excavations): Postcranial specimens. *AJPA* **79**:451–474.

Susman, R. (1991) Species attribution of the Swartkrans thumb metacarpals: Reply to Drs. Trinkaus and Long. *AJPA* **86**:549–552.

Susman, R. (1993) Hominid postcranial remains from Swartkrans. In: Brain 1993: 117–136.

Susman, R. (1994) Fossil evidence for early hominid tool use. *Science* **265**:1570–1573.

Susman, R. (1998) Hand function and tool behavior in early hominids. *JHE* **35**:23–46.

Susman, R., and T. Brain (1988) New first metatarsal (SKX 5017) from Swartkrans and the gait of *Paranthropus robustus*. *AJPA* **77**:7–16.

Susman, R., J. Stern, and W. Jungers (1985) Locomotor adaptations in the Hadar hominids. In: Delson 1985: 184–192.

Sussman, R. (1991) Primate origins and the evolution of angiosperms. *AJP* **23**:209–223.

Suwa, G. (1988) Evolution of the "robust" australopithecines in the Omo succession: Evidence from mandibular premolar morphology. In: Grine 1988: 199–222.

Suwa, G., B. Asfaw, Y. Beyene, T. White, S. Katoh, S. Nagaoka, H. Nakaya, K. Uzawa, P. Renne, and G. WoldeGabriel (1997) The first skull of *Australopithecus boisei*. *Nature* **389**: 489–492.

Suwa, G., R. Kono, S. Katoh, B. Asfaw, and Y. Beyene (2007) A new species of great ape from the late Miocene epoch in Ethiopia. *Nature* **448**:921–924.

Suwa, G., T. White, and F. C. Howell (1996) Mandibular postcanine dentition from the Shungura Formation, Ethiopia: Crown morphology, taxonomic allocations and Plio-Pleistocene hominid evolution. *AJPA* **101**:247–282.

Suzuki, H. (1970a) The skull of the Amud man. In: Suzuki and Takai 1970: 123–206.

Suzuki, H. (1970b) General conclusions. In: Suzuki and Takai 1970: 425–426.

Suzuki, H. (1982) Skulls of the Minatogawa man. In: H. Suzuki and K. Hanihara, Editors. *The Minatogawa Man.* Toyko: University of Tokyo Press. 7–49.

Suzuki, H., and K. Hanihara, Editors (1982) *The Minatogawa Man.* Tokyo: University of Tokyo Press.

Svoboda, J. (1987) Lithic industries of the Arago, Bilzingsleben, and Vértesszőllős hominids: Comparison and evolutionary interpretation. *CA* **28**:219–227.

Svoboda, J. (2000) The depositional context of the Early Upper Paleolithic human fossils from Koněprusy (Zlatý kůň) and Mladeč Caves, Czech Republic. *JHE* **38**:523–536.

Svoboda, J. (2003) The Bohunician and the Aurignacian. In: Zilhão and d'Errico 2003: 123–131.

Svoboda, J. (2005) The Neandertal extinction in eastern Central Europe. *QI* **137**:69–75.

Svoboda, J. (2006) The Danube Gate to Europe: Patterns of chronology, settlement archaeology, and demography of late Neandertals and early modern humans of the Middle Danube. In: Conard 2006: 233–267.

Svoboda, J., V. Ložek, and E. Vlček (1996) *Hunters between East and West. The Paleolithic of Moravia.* New York: Plenum.

Svoboda, J., J. van der Plicht, and V. Kuželka (2002) Upper Paleolithic and Mesolithic human fossils from Moravia and Bohemia (Czech Republic): Some new ^{14}C dates. *Antiquity* **76**:957–962.

Sweet, W., and P. Donoghue (2001) Conodonts: past, present, future. *Journal of Paleontology* **75**:1174–1184.

Swindler, D., and C. Wood (1973) *An Atlas of Primate Gross Anatomy: Baboon, Chimpanzee, and Man.* Seattle: University of Washington Press.

Swisher, C., G. Curtis, T. Jacob, A. Getty, A. Suprijo, and Widiasmoro (1994) Age of the earliest known hominids in Java, Indonesia. *Science* **263**:1118–1121.

Swisher, C., W. Rink, S. Antón, H. Schwarcz, G. Curtis, A. Suprijo, and Widiasmoro (1996) Latest *Homo erectus* of Java: Potential contemporaneity with *Homo sapiens* in Southeast Asia. *Science* **274**:1870–1874.

Szabo, B., and D. Collins (1975) Age of fossil bones from British interglacial sites. *Nature* **254**:680–682.

Szalay, F. (1968) The beginnings of primates. *Evolution* **22**:19–36.

Szalay, F. (1972) Paleobiology of the earliest primates. In: R. Tuttle, Editor. *The Functional and Evolutionary Biology of Primates.* Chicago: Aldine-Atherton, pp. 3–35.

Szalay, F. (1976) Systematics of the Omomyidae (Tarsiiformes, Primates): taxonomy, phylogeny, and adaptations. *Bulletin of the American Museum of Natural History* **156**:157–450.

Szalay, F. (2007) Ancestral locomotor modes, placental mammals, and the origin of euprimates: Lessons from history. In: Ravosa and Dagosto 2007: 457–487.

Szalay, F., and A. Berzi (1973) Cranial anatomy of *Oreopithecus*. *Science* **180**:183–185.

Szalay, F., and E. Delson (1979) *Evolutionary History of the Primates.* New York: Academic Press.

Szilvássy, J., H. Kritscher, and E. Vlček (1986) Die Bedeutung röntgenologischer Methoden für die anthropologische Unter-suchung ur- und frühgeschichtliche Gräberfelder. *Annalen des Naturhistorischen Museums zu Wien* **89**:313–352.

Szombathy, J. (1925) Die diluvialen Menschenreste aus der Fürst–Johannes–Höhle bei Lautsch in Mähren. *Die Eiszeit* **2**:1–34, 73–95.

T

Tague, R., and C. O. Lovejoy (1986) The obstetric pelvis of AL 288-1 (Lucy). *JHE* **15**:237–255.

Takahata, Y., H. Ihobe, and G. Idani (1996) Comparing copulations of chimpanzees and bonobos: Do females exhibit proceptivity or receptivity? In: McGrew et al. 1996: 146–155.

Takai, M., N. Shigehara, N. Egi, and T. Tsubamoto (2003) Endo-cranial cast and morphology of the olfactory bulb of *Amphipi-thecus mogaungensis* (latest middle Eocene of Myanmar). *Primates* **44**:137–144.

Takai, M., C. Sein, T. Tsubamoto, N. Egi, M. Maung, and N. Shigehara (2005) A new eosimiid from the latest middle Eocene in Pondaung, central Myanmar. *Anthropological Science* **113**:17–25.

Tamrat, E., N. Thouveny, M. Taïeb, and N. Opdyke (1995) Revised magnetostratigraphy of the Plio-Pleistocene sedimentary sequence of the Olduvai Formation (Tanzania). *Paleogeography, Paleoclimatology, Paleoecology* **114**:273–283.

Tan, Y., A. Yoder, N. Yamashita, and W.-H. Li (2005) Evidence from opsin genes rejects nocturnality in ancestral primates. *PNAS* **102**:14712–14716.

Tanner, N. (1981) *On Becoming Human: A Model of the Transition from Ape to Human and the Reconstruction of Early Human Social Life.* Cambridge: Cambridge University Press.

Tanner, N., and A. Zihlman (1976) Women in evolution. Part I: Innovation and selection in human origins. *Signs: Journal of Women in Culture and Society* **1**:585–608.

Tappen, N. (1973) Structure of bone in the skulls of Neanderthal fossils. *AJPA* **38**:93–98.

Tappen, N. (1980) The vermiculate surface pattern in the brow ridges of australopithecines and other very ancient hominids. *AJPA* **52**:512–528.

Tardieu, C. (1979) Aspects biomécaniques de l'articulation du genou chez les Primates. *Bulletin de la Societé d'Anatomie de Paris* **4**:66–86.

Tardieu, C. (1983) *L'articulation du genou: Analyse morpho-fonctionelle chez les Primates et les Hominidés fossiles.* Paris: Centre Nationale de la Recherche Scientifique.

Tardieu, C. (1986) Evolution of the knee intra-articular menisci in primates and some fossil hominids. In: J. Else and P. Lee, Editors. *Primate Evolution*, Volume 1, Cambridge: Cambridge University Press, pp. 183–190.

Tardieu, C. (1991) Étude comparative des déplacements du centre de gravité du corps pendant la marche par une nou-velle méthode d'analyse tridimensionelle. Mise à l'épreuve d'une hypothèse évolutive. In: Coppens and Senut 1991: 49–58.

Tardieu, C., and H. Preuschoft (1996) Ontogeny of the knee joint in humans, great apes and fossil hominids: Pelvi-Femoral relationships during postnatal growth in humans. *FP* **66**:68–81.

Tattersall, I. (1973) Cranial anatomy of Archaeolemurinae (Lemuroidea, Primates). *APAMNH* **52**(1):1–110.

Tattersall, I. (1986) Species recognition in human paleontology. *JHE* **15**:165–175.

Tattersall, I. (1992) Species concepts and species identification in human evolution. *JHE* **22**:341–349.

Tattersall, I. (1995a) *The Fossil Trail*. Oxford: Oxford University Press.

Tattersall, I. (1995b) *The Last Neanderthal: The Rise, Success and Mysterious Extinction of Our Closest Human Relatives*. New York: Macmillan.

Tattersall, I. (1996) Paleoanthropology and perception. In: W. Meickle, F. C. Howell, and N. Jablonski, Editors. *Contemporary Issues in Human Evolution*. San Francisco: California Academy of Sciences.

Tattersall, I. (1998) *Becoming Human: Evolution and Human Uniqueness*. New York: Harcourt Brace.

Tattersall, I. (2000) Paleoanthropology and evolutionary biology. *Anthropologie* **38**:165–168.

Tattersall, I., and E. Delson (1999) Primates. In: R. Singer, Editor. *Encyclopedia of Paleontology*. Chicago: Fitzroy Dearborn, pp. 949–959.

Tattersall, I., and G. Sawyer (1996) The skull of "Sinanthropus" from Zhoukoudian: A new reconstruction. *JHE* **31**:311–314.

Tattersall, I., and J. Schwartz (1999) Hominids and hybrids: the place of Neanderthals in human evolution. *PNAS* **96**:7117–7119.

Tattersall, I., and J. Schwartz (2000) Diet and the Neanderthals. *Acta Universitatis Carolinae Medica* **41**:29–36.

Tattersall, I., and J. Schwartz (2001) *Extinct Humans*. Boulder: Westview Press.

Tavaré, S., C. Marshall, O. Will, C. Soligo, and R. Martin (2002) Using the fossil record to estimate the age of the last common ancestor of extant primates. *Nature* **416**:726–729.

Teaford, M., and P. Ungar (2000) Diet and the evolution of the earliest human ancestors. *PNAS* **97**:13506–13511.

Teleki, G. (1973) *The Predatory Behavior of Wild Chimpanzees*. Lewisburg: Bucknell University Press.

Teleki, G. (1981) The omnivorous diet and eclectic feeding habits of chimpanzees in Gombe National Park, Tanzania. In: R. Harding and G. Teleki, Editors. *Omnivorous Primates: Gathering and Hunting in Human Evolution*. New York: Columbia University Press, pp. 303–343.

Templeton, A. (1992) Human origins and analysis of mitochondrial DNA sequences. *Science* **255**:737.

Templeton, A. (1993) The "Eve" hypothesis: A genetic critique and reanalysis. *AA* **95**:51–72.

Templeton, A. (1996) Contingency tests of neutrality using intra/interspecific gene trees: The rejection of neutrality for the evolution of the mitochondrial cytochrome oxidase II gene in the hominoid primates. *Genetics* **144**:1263–1270.

Templeton, A. (1997) Testing the Out of Africa replacement hypothesis with mitochondrial DNA data. In: Clark and Willermet 1997: 329–360.

Templeton, A. (1998) Human races: A genetic and evolutionary perspective. *AA* **100**:632–650.

Templeton, A. (2002) Out of Africa again and again. *Nature* **416**:45–51.

Templeton, A. (2005) Haplotype trees and modern human origins. *YPA* **48**:33–59.

Terberger, T., M. Street, and G. Bräuer (2001) Der menschliche Schädelrest aus der Elbe bei Hahnöfersand und seine Bedeutung für die Steinzeit Norddeutschlands. *AK* **31**:521–526.

Tetreault, N., A. Hakeem, and J. Allman, (2004) The distribution and size of retinal ganglion cells in *Microcebus murinus, Cheirogaleus medius,* and *Tarsius syrichta*: Implications for the evolution of sensory systems in primates. In: Ross and Kay 2004: 463–475.

Teyssandier, N. (2005) Neue Perspektiven zu den Anfängen des Aurignacien. *MGU* **14**:11–24.

Teyssandier, N. (2006) Questioning the first Aurignacian: Mono or multi cultural phenomenon during the formation of the Upper Paleolithic in Central Europe and the Balkans. *Anthropologie* **44**:9–29.

Teyssandier, N., M. Bolus, and N. Conard (2006) The early Aurignacian in Central Europe and its place in a European perspective. In: O. Bar-Yosef and J. Zilhão, Editors. *Toward a Definition of the Aurignacian*. Lisbon: Instituto português de arqueologia (Trabalhos de arqueologia), pp. 241–256.

Thackeray, A. (1992) The Middle Stone Age south of the Limpopo River. *JWP* **6**:385–440.

Thackeray, J., J. Braga, J. Treil, N. Niksch, and J. Labuschagne (2002) "Mrs Ples" (Sts 5) from Sterkfontein: An adolescent male? *SAJS* **98**:21–22.

Theunissen, B. (1988) *Eugène Dubois and the Ape-Man from Java: The History of the First "Missing Link" and Its Discoverer*. Dordrecht: Kluwer Academic.

Theunissen, B. (1997) Dubois, (Marie) Eugène (François Thomas) (1858–1940). In: Spencer 1997: 353–354.

Thewissen, J., M. Williams, L. Roe, and S. Hussain (2001) Skeletons of terrestrial cetaceans and the relationship of whales to artiodactyls. *Nature* **413**:277–281.

Thieme, H. (1996) Altpaläolithische Wurfspeere aus Schöningen, Niedersachsen: Ein Vorbericht. *AK* **26**:377–393.

Thieme, H. (1997) Lower Paleolithic hunting spears from Germany. *Nature* **385**:807–810.

Thieme, H. (1999) Altpaläolithische Holzgeräte aus Schöningen Lkr. Helmstedt. Bedeutsame Funde zur Kulturentwicklung des frühen Menschen. *Germania* **77**:451–487.

Thoma, A. (1966) L'occipital de l'homme Mindélien de Vértesszőllős. *L'Anthropologie* **70**:495–534.

Thoma, A. (1967) Human teeth from the Lower Paleolithic of Hungary. *ZMA* **58**:152–180.

Thoma, A. (1969) Biometrische Studie über das Occipitale von Vértesszőllős. *ZMA* **60**:229–241.

Thompson, J., and A. Nelson (2000) The place of Neandertals in the evolution of hominid growth and development. *JHE* **38**:475–495.

Thomson, R., J. Pritchard, P. Shen, P. Oefner, and M. Feldman (2000) Recent common ancestry of human Y chromosome: Evidence from DNA sequence data. *PNAS* **97**:7360–7365.

Thomson, W. (1862) On the age of the sun's heat. *Macmillan's Magazine* **5**:288–293.

Thomson, W. (1864) On the secular cooling of the Earth. *Transactions of the Royal Society of Edinburgh* **23**:167–169.

Thorne, A. (1976) Morphological contrasts in Pleistocene Australians. In: R. Kirk and A. Thorne, Editors. *The Origin of the Australians*. Canberra: Australian Institute of Aboriginal Studies, pp. 95–112.

Thorne, A. (1977) Separation or reconciliation? Biological clues to the development of Australian society. In: J. Allen et al., Editors. *Sunda and Sahul: Prehistoric Studies in Southeast Asia, Melanesia, and Australia*. New York: Academic Press, pp. 187–204.

Thorne, A. (1980) The longest link: Human evolution in Southeast Asia and the settlement of Australia. In: J. Fox et al., Editors. *Indonesia: Australian Perspectives*. Canberra: Australian National University, pp. 35–43.

Thorne, A. (1981) The center and the edge: The significance of Australian hominids to African paleoanthropology. In: R. Leakey and B. Ogot, Editors. *Proceedings of the 8th Panafrican Congress of Prehistory and Quaternary Studies*. Nairobi: TILLMIAP, pp. 180–181.

Thorne, A. (2000) Modern human origins: Multiregional evolution. In: Delson et al. 2000: 425–429.

Thorne, A., R. Grün, G. Mortimer, N. Spooner, J. Simpson, M. McCulloch, A. Taylor, and D. Curnoe (1999) Australia's oldest human remains: Age of the Lake Mungo 3 skeleton. *JHE* **36**:591–612.

Thorne, A., and P. Macumber (1972) Discoveries of late Pleistocene man at Kow Swamp, Australia. *Nature* **238**:316–319.

Thorne, A., and M. Wolpoff (1981) Regional continuity in Australasian Pleistocene hominid evolution. *AJPA* **55**:337–349,

Thorne, A., and M. Wolpoff (1992) The multiregional evolution of humans. *SA* **266**:76–83.

Thouveny, N., and E. Bonifay (1984) New chronological data on European Plio-Pleistocene faunas and hominid occupation sites. *Nature* **308**:355–358.

Tillier, A.-M. (1977) La pneumatisation du massif cranio-facial chez les hommes actuels et fossiles. *BMSAP* **4**:177–198, 287–316.

Tillier, A.-M. (1980) Les dents d'enfant de Ternifine (Pléistocène moyen d'Algérie). *L'Anthropologie (Paris)* **84**:413–421.

Tillier, A.-M. (1983) Le crâne d'enfant d'Engis: Un example de distribution des caractères juvéniles, primitifs, et néanderthaliens. *Bulletin de la Société Royale Belgique d'Anthropologie et Préhistoire* **94**:51–75.

Tillier, A.-M. (1987) L'enfant néandertalien de la Quina H 18 et l'ontogénie des Néandertaliens. In: B. Vandermeersch (Editor). *Prèhistoire de Pointou-Charentes: Problèmes Actuels*. Paris: Comité des Travaux Historiques et Scientifiques, pp. 201–206.

Tillier, A.-M. (1989) The evolution of modern humans: evidence from young Mousterian individuals. In: Mellars and Stringer 1989: 286–297.

Tillier, A.-M. (1998) Ontogenetic variation in Late Pleistocene *Homo sapiens* from the Near East. In: Akazawa et al. 1998: 381–389.

Tillier, A.-M. (1999) *Les enfants Mousteriéns de Qafzeh*. Paris: Cahiers de Paléoanthropologie.

Tobias, P. (1960) The Kanam jaw. *Nature* **185**:946–947.

Tobias, P. (1962) Early members of the genus *Homo* in Africa. In: G. Kurth, Editor. *Evolution und Hominization*. Stuttgart: Gustav Fischer, pp. 194–195.

Tobias, P. (1964) The Olduvai Bed I hominine with special reference to its cranial capacity. *Nature* **202**:3–4.

Tobias, P. (1966) Fossil hominid remains from 'Ubediya, Israel. *Nature* **211**:130–133.

Tobias, P. (1967a) *Olduvai Gorge,* Volume 2: *The Cranium and Maxillary Dentition of* Australopithecus (Zinjanthropus) boisei. Cambridge: Cambridge University Press.

Tobias, P. (1967b) The hominid skeletal remains of Haua Fteah. In: C. McBurney, Editor. *The Haua Fteah (Cyrenaica) and the Stone Age of the South-East Mediterranean*. Cambridge: Cambridge University Press, pp. 337–352.

Tobias, P. (1971) *The Brain in Human Evolution*. New York: Columbia University Press.

Tobias, P. (1978) The earliest Transvaal members of the genus *Homo* with another look at some problems of hominid taxonomy and systematics. *ZMA* **69**:225–265.

Tobias, P. (1988) Numerous apparently synapomorphic features in *Australopithecus robustus, Australopithecus boisei*, and *Homo habilis*: Support for the Skelton–McHenry–Drawhorn hypothesis. In: Grine 1988: 293–308.

Tobias, P. (1991) *Olduvai Gorge,* Volume 4: *The Skulls, Endocasts and Teeth of* Homo habilis. Cambridge: Cambridge University Press.

Tobias, P., and R. Clarke (1996) Faunal evidence and Sterkfontein Member 2 foot bones of early hominid: Response. *Science* **271**:1301–1302.

Tobias, P., and G. von Koenigswald (1964) A comparison between the Olduvai hominines and those of Java and some implications for hominid phylogeny. *Nature* **204**:515–518.

Tocheri, M., W. Jungers, S. Larson, C. Orr, T. Sutikna, Jatmiko, E. Saptomo, R. Due, T. Djubiantono, and M. Morwood (2007) Morphological affinities of the wrist of *Homo floresiensis*. *Paleoanthropology* **207**:A33.

Tokuriki, M. (1973) Electromyographic and joint-mechanical studies in quadrupedal locomotion. I. Walk. *Japanese Journal of Veterinary Science* **35**:433–446.

Tompkins, R. (1996) Relative dental development of Upper Pleistocene hominids compared to human population variation. *AJPA* **99**:103–118.

Tompkins, R., and E. Trinkaus (1987) La Ferrassie 6 and the development of Neandertal pubic morphology. *AJPA* **73**:233–239.

Tong, H., H. Shang, S. Zhang, and F. Chen (2004) A preliminary report on the newly found Tianyuan Cave, a Late Pleistocene human fossil site near Zhoukoudian. *Chinese Science Bulletin* **49**:853–857.

Toth, N., and K. Schick (1986) The first million years: The archaeology of protohuman culture. *Advances in Archaeological Method and Theory* **9**:1–96.

Toth, N., K. Schick, E. Savage-Rumbaugh, R. Sevcik, and D. Rumbaugh (1993) *Pan* the tool maker: Investigations into the stone tool-making and tool-using capabilities of a bonobo (*Pan paniscus*). *JAS* **20**:81–91.

Toussaint, M., G. Macho, P. Tobias, T. Partridge, and A. Hughes (2003) The third partial skeleton of a late Pliocene hominin (Stw 431) from Sterkfontein, South Africa. *SAJS* **99**:215–223.

Trinkaus, E. (1973) A reconsideration of the Fontéchevade fossils. *AJPA* **39**:25–35.

Trinkaus, E. (1976) The morphology of European and Southwest Asian Neandertal pubic bones. *AJPA* **44**:95–104.

Trinkaus, E. (1977) A functional interpretation of the axillary border of the Neandertal scapula. *JHE* **6**:231–234.

Trinkaus, E. (1978) Dental remains from the Shanidar adult Neanderthals. *JHE* **7**:369–382.

Trinkaus, E. (1981) Neanderthal limb proportions and cold adaptation. In: Stringer 1981: 187–224.

Trinkaus, E. (1982) Artificial cranial deformation of the Shanidar 1 and 5 crania. *CA* **23**:198–199.

Trinkaus, E. (1983a) Neandertal postcrania and the adaptive shift to modern humans. *BARIS* **164**:165–200.

Trinkaus, E. (1983b) *The Shanidar Neandertals*. New York: Academic Press.

Trinkaus, E. (1984a) Western Asia. In: Smith and Spencer 1984: 251–293.

Trinkaus, E. (1984b) Neandertal pubic morphology and gestation length. *CA* **25**:508–514.

Trinkaus, E. (1985) Pathology and posture of the La Chapelle-aux-Saints Neandertal. *AJPA* **67**:19–41.

Trinkaus, E. (1987a) Bodies, brawn, brains, and noses: human ancestors and human predation. In: M. Nitecki and D. Nitecki, Editors. *The Evolution of Human Hunting*. New York: Plenum, pp. 107–145.

Trinkaus, E. (1987b) The Neandertal face: Evolutionary and functional perspectives on a recent hominid face. *JHE* **16**:429–443.

Trinkaus, E. (1989) The Upper Pleistocene transition. In: Trinkaus 1989: 42–66.

Trinkaus, E. (1992) Cladistics and later Pleistocene human evolution. In: Bräuer and Smith 1992: 1–7.

Trinkaus, E. (1993a) Variation in the position of the mandibular mental foramen and the identification of Neandertal apomorphies. *Rivista di Antropologia* **71**:259–274.

Trinkaus, E. (1993b) A note on the KNM-ER 999 hominid femur. *JHE* **24**:493–504.

Trinkaus, E. (1993c) Femoral neck-shaft angles of the Qafzeh-Skhul early modern humans and activity levels among immature Near Eastern Middle Paleolithic hominids. *JHE* **25**:393–416.

Trinkaus, E. (1993d) Comment. *CA* **34**:620–622.

Trinkaus, E. (1995a) Neanderthal mortality patterns. *JAS* **22**:121–142.

Trinkaus, E. (1995b) Near Eastern late archaic humans. *Paléorient* **21**:9–23.

Trinkaus, E. (2002) The cranial morphology. In: Zilhão and Trinkaus 2002: 256–286.

Trinkaus, E. (2003) Neandertal faces were not long; modern human faces are short. *PNAS* **100**:8142–8145.

Trinkaus, E. (2004) Eyasi 1 and the suprainiac fossa. *AJPA* **124**:28–32.

Trinkaus, E. (2005) Early modern humans. *ARA* **34**:207–230.

Trinkaus, E., and S. Churchill (1988) Neandertal radial tuberosity orientation. *AJPA* **75**:15–21.

Trinkaus, E., S. Churchill, C. Ruff, and B. Vandermeersch (1999a) Long bone shaft robusticity and body proportions of the Saint-Césaire 1 Châtelperronian Neandertal. *JAS* **26**:753–773.

Trinkaus, E., and C. Hilton (1996) Neandertal pedal proximal phalanges: diaphyseal loading patterns. *JHE* **30**:399–425.

Trinkaus, E., and M. LeMay (1982) Occipital bunning among later Pleistocene hominids. *AJPA* **57**:27–35.

Trinkaus, E., and J. Long (1990) Species attribution of the Swartkrans Member 1 first metacarpals: SK 84 and SKX 5020. *AJPA* **83**:419–424.

Trinkaus, E., S. Milota, R. Rodrigo, G. Mircea, and O. Modovan (2003b) Early modern human cranial remains from the Peştera cu Oase, Romania. *JHE* **45**:245–253.

Trinkaus, E., O. Moldovan, S. Milota, A. Bîlgăr, L. Sarcina, S. Athreya, S. Bailey, R. Rodrigo, G. Mircea, T. Higham, C. Ramsey, and J. Van der Plicht (2003a) An early modern human from the Peştera cu Oase, Romania. *PNAS* **100**:11231–11236.

Trinkaus, E., V. Ramov, and S. Lauklin (2000) Middle Paleolithic human deciduous incisor from Khudji, Tajikistan. *JHE* **38**:575–583.

Trinkaus, E., and C. Ruff (1989) Diaphyseal cross-sectional morphology and biomechanics of the Fond-de-Forêt femur and the Spy 2 femur and tibia. *Bulletin de la Societé Royal Belgique d'Anthropologie et Préhistoire* **100**:33–42.

Trinkaus, E., C. Ruff, and S. Churchill (1998) Upper limb versus lower limb loading patterns among Near Eastern Middle Paleolithic hominids. In: Akazawa et al. 1998: 391–404.

Trinkaus, E., and P. Shipman (1993) *The Neandertals.* New York: Knopf.

Trinkaus, E., and F. Smith (1985) The fate of the Neandertals. In: Delson 1985: 325–333.

Trinkaus, E., and F. Smith (1995) Body size of the Vindija Neandertals. *JHE* **28**:201–208.

Trinkaus, E., F. Smith, T. Stockton, and L. Shackelford (2006a) The human postcranial remains from Mladeč. In: Teschler-Nikola 2006: 385–445.

Trinkaus, E., C. Stringer, C. Ruff, R. Hennessy, M. Roberts, and S. Parfitt (1999b) Diaphyseal cross-sectional geometry of the Boxgrove 1 Middle Pleistocene human tibia. *JHE* **37**:1–25.

Trinkaus, E., and J. Svoboda (2006b) The paleobiology of t he Pavlovian people. In: Trinkaus and Svoboda 2006a: 459–465.

Trinkaus, E., and R. Tomkins (1990) The Neandertal life cycle: Probability and perceptibility of contrasts with recent humans. In: J. DeRousseau, Editor. *Primate Life History and Evolution.* New York: Wiley-Liss, pp. 153–180.

Trinkaus, E., and I. Villemeur (1991) Mechanical advantages of the Neandertal thumb in flexion: A test of a hypothesis. *AJPA* **84**:249–260.

Trinkaus, E., and J. Zilhão (2002) Phylogenetic implications. In: J. Zilhão and E. Trinkaus 2000: 497–518.

Trinkaus, E., J. Zilhão, H. Rougier, R. Rodrigo, S. Milota, M. Gherase, L. Sarcină, O. Moldovan, I. Băltean, V. Codrea, S. Bailey, R. Franciscus, M. Ponce de León, and C. Zollikofer (2006b) The Peştera cu Oase and early modern humans in southeastern Europe. In: Conard 2006: 145–164.

Trivers, R. L. (1972). Parental investment and sexual selection. In: B. Campbell, Editor. *Sexual Selection and the Descent of Man 1871–1971.* Chicago: Aldine Press, pp. 136–179.

Trotter, M., and G. Gleser (1958) A re-evaluation of estimation of stature based on measurements during life and of long bones after death. *AJPA* **16**:79–123.

Trueman, J. (2001) Does the Lake Mungo 3 mt DNA evidence stand up to analysis? *AO* **36**:163–165.

Tuffreau, A., and P. Antoine (1995) The earliest occupation of Europe: Continental northwest Europe. In: Roebroeks and van Kolfschoten 1995: 297–315.

Turk, I., J. Dirjec, and B. Kavur (1995) The oldest musical instrument in Europe discovered in Slovenia? *Raprave IV. Razreda SAZU* **36**:287–293.

Turq, A., B. Martínez-Navarro, P. Palmquist, A. Arribas, J. Agustí, and J. Rodriguez-Vidal (1996) Le Plio-Pléistocène de la région d'Orce, province de Grenade, Espagne: Bilan et perspectives de recherché. *Paléo* **8**:161–204.

Turville-Petre, F., Editor (1927) *Researches in Prehistoric Galilee, 1925–1926.* London: British School of Archaeology in Jerusalem.

Tutin, C., R. Ham, L. White, and M. Harrison (1997) The primate community of the Lopé Reserve, Gabon: Diets, responses to fruit scarcity, and effects on biomass. *AJP* **42**:1–24.

Tuttle, R. (1967) Knuckle-walking and the evolution of hominoid hands. *AJPA* **26**:171–206.

Tuttle, R. (1969) Knuckle-walking and the problem of human origins. *Science* **166**:953–961.

Tuttle, R. (1974) Darwin's apes, dental apes, and the descent of man: normal science in evolutionary anthropology. *CA* **15**:389–398.

Tuttle, R. (1975) Parallelism, brachiation, and hominoid phylogeny. In: Luckett and Szalay 1975: 447–480.

Tuttle, R. (1981) Evolution of hominid bipedalism and prehensile capabilities. *PTRSL* **292**:89–94.

Tuttle, R. (1985) Ape footprints and Laetoli footprints: a response to the SUNY claims. In: Tobias 1985: 129–133.

Tuttle, R. (1987) Kinesiological inferences and evolutionary implications from Laetoli bipedal trails G-1, G-2/3, and A. In: M. D. Leakey and J. Harris, Editors. *Laetoli: A Pliocene Site in Northern Tanzania.* Oxford: Clarendon Press. 503–523.

Tuttle, R. (1990) The pitted pattern of Laetoli feet. *NH* **90**:60–65.

Tuttle, R. (1994) Up from electromyography: Primate energetics and the evolution of human bipedalism. In: Corruccini and Ciochon 1994: 269–284.

Tuttle, R., J. Basmajian, and H. Ishida (1975) Electromyography of the gluteus maximus muscle in gorillas and the evolution of hominid bipedalism. In: R. Tuttle, Editor. *Primate Functional Morphology and Evolution.* The Hague: Mouton, pp. 253–269.

Tuttle, R. D. Webb, and M. Baksh (1990) Further progress on the Laetoli trails. *JAS* **17**:347–362.

Tuttle, R., D. Webb, and N. Tuttle (1991) Laetoli footprint trails and the evolution of hominid bipedalism. In: Coppens and Sénut 1991: 187–198.

Tyler, D. (1994) The taxonomic status of "*Meganthropus.*" In: Franzen 1994: 115–121.

Tyler, D. (1995) The current picture of hominid evolution in Java. *AAS* **14**:315–323.

Tyler, D. (2001) "Meganthropus" cranial remains from Java. *HE* **16**:81–101.

Tzedakis, P., K. Hughes, I. Cacho, and K. Harvati (2007) Placing late Neanderthals in a climatic context. *Nature* **449**:206–208.

U

Ullrich, H. (1955a) Paläolithische Menschenreste aus der Sowjetunion. I. Das Mousterien-Kind von Staroselje (Krim). *ZMA* **47**:91–98.

Ullrich, H. (1955b) Paläolithische Menschenreste aus der Sokjetunion. II. Das Kinderskelett aus der grotte Teschik-Tasch. *ZMA* **47**:99–112.

Ullrich, H. (1958) Neandertalfunde aus der Sowjetunion. In: von Koeningswald 1958: 72–106.

Underhill, P., L. Jin, A. Lin, S. Mehdi, T. Jenkins, D. Vollrath, R. Davis, L. Cavalli-Sforza, and P. Oefner (1997) Direction of numerous Y chromosome biallelic polymorphisms by denaturing high-performance liquid chromatography. *Genome Research* **7**:996–1005.

Ungar, P., K. Fennell, K. Gordon, and E. Trinkaus (1997) Neanderthal incisor beveling. *JHE* **32**:407–421.

Ungar, P., and F. Grine (1991) Incisor size and wear in *Australopithecus africanus* and Paranthropus robustus. *JHE* **20**:313–340.

Ungar, P., F. Grine, M. Teaford, and S. El Zaatari (2006) Dental microwear and diets of early African *Homo*. *JHE* **50**:78–95.

V

Valentine, J. (2004) *On the Origin of Phyla*. Chicago: University of Chicago Press.

Valladas, H., J. Geneste, J. Joron, and J. Chadelle (1986) Thermoluminescence dating of Le Moustier (Dordogne, France). *Nature* **322**:452–454.

Valladas, H., J. Reyss, J. Joron, G. Valladas, O. Bar-Yosef, and B. Vandermeersch (1988) Thermoluminescence dating of Mousterian "proto-Cro-Magnon" remains from Israel and the origin of modern man. *Nature* **331**:614–615.

Valladas, H., J. Joron, G. Valladas, B. Arensburg, O. Bar-Yosef, A. Belfer-Cohen, P. Goldberg, H. Laville, L. Meignen, Y. Rak, E. Tchernov, A. Tillier, and B. Vandermeersch (1987) Thermoluminescence dates for the Neanderthal burial site at Kebara Cave in Israel. *Nature* **330**:159–160.

Vallois, H. (1951) La mandibule humaine fossile de la grotte du Port Épic près Diré Daoua (Abyssinie). *L'Anthropologie* **55**:231–238.

Vallois, H. (1954) Neandertals and presapiens. *JRAI* **84**:111–130.

Vallois, H. (1958) L'origine de l'*Homo sapiens*. In: La grotte de Fontéchevade. II. Anthropologie. *AIPH* **29**:7–164. Reprinted as: The origin of *Homo sapiens*. In: Howells 1967: 473–499.

Vallois, H. (1961) The social life of early man: The evidence of skeletons. In: S. Washburn, Editor. *The Social Life of Early Man*. Chicago: Aldine, pp. 214–235.

Vallois, H. (1962) Language articulé et squelette. *Homo* **13**:114–121.

Vallois, H. (1969) Le temporal néandertalien H27 de La Quina: étude anthropologique. *L'Anthropologie* **73**:365–400.

Vallois, H., and B. Vandermeersch (1972) Le crâne moustérien de Qafzeh (Homo VI). Étude anthropologique. *L'Anthropologie* **76**:71–96.

Valoch, K. (1984) Early Paleolithic in Moravia, Czechoslovakia. *PPS* **50**:63–69.

Valoch, K. (1987) The early Paleolithic site Stranská Skála I near Brno (Czechoslovakia). *Anthropologie* **25**:125–142.

Valoch, K. (1995) The earliest occupation of Europe: Eastern Central and Southeastern Europe. In: Roebroeks and van Kolfschoten 1995: 67–84.

Van Couvering, J. (2000a) Glaciation. In: Delson et al. 2000: 289–292.

Van Couvering, J. (2000b) Pleistocene. In: Delson et al. 2000: 567–570.

Van Couvering, J. (2000c) Fejej. In: Delson et al. 2000: 267–268.

van den Brink, L. (1982) On the mammal fauna of the Wadjak Cave, Java (Indonesia). *MQRSA* **7**:177–193.

Vandermeersch, B. (1976) Les sépultures néandertaliennes. In: H. deLumley, Editor. *La Préhistoire Français, Volume 1*. Paris: CNRS. 725–727.

Vandermeersch, B. (1981) *Les hommes fossiles de Qafzeh (Israël)*. Paris: CNRS.

Vandermeersch, B. (1982) The first *Homo sapiens sapiens* in the Near East. *BARIS* **151**:297–299.

Vandermeersch, B. (1985) The origin of the Neanderthals. In: Delson 1985: 306–309.

Vandermeersch, B. (1989) The evolution of modern humans: Recent evidence from southwest Asia. In: Mellars and Stringer 1989: 155–164.

Vandermeersch, B. (1997) The Near East and Europe: continuity or discontinuity? In: Clark and Willermet 1997: 107–116.

Vandermeersch, B., and E. Trinkaus (1995) The postcranial remains of the Régourdou 1 Neandertal: The shoulder and arm remains. *JHE* **28**:439–476.

van der Merwe, N., J., Thackeray, J. Lee-Thorp, and J. Luyt (2003) The carbon isotope ecology and diet of *Australopithecus africanus* at Sterkfontein, South Africa. *JHE* **44**:581–597.

Vanhaeren, M., and F. d'Errico (2002) The body ornaments associated with the burial. In: Zilhão and Trinkaus 2002: 154–186.

van Schaik, C., M. Ancrenaz, G. Borgen, B. Galdikas, C. Knott, I. Singleton, A. Suzuki, S. Utami, and M. Merrill (2003) Orangutan cultures and the evolution of material culture. *Science* **299**:102–105.

van Schaik, C., R. Deaner, and M. Merrill (1999) The conditions for tool use in primates: Implications for the evolution of material culture. *JHE* **36**:719–741.

van Schaik, C., and J. Van Hooff (1996) Toward an understanding of the orangutan's social system. In: McGrew et al. 1996: 3–15.

Van Valen, L. (1965) Morphological variation and width of the ecological niche. *American Naturalist* **94**:377–390.

Van Vark, G., A. Bilsborough, and W. Henke (1992) Affinities of European Upper Paleolithic *Homo sapiens* and later human evolution. *JHE* **23**:401–417.

Vaquero, M. J. Maroto, A. Arrizabalaga, J. Baena, E. Baquedano, E. Carríon, J. Jordà, M. Martinón, M. Menéndez, R. Montes, and J. Rosell (2006) The Neandertal-modern human meeting in Iberia: A critical view of the cultural, geographical and chronological data. In: Connard 2006: 419–439.

Vekua, A., D. Lordkipanidze, G. Rightmire, J. Agusti, R. Ferring, G. Maisuradze, A. Mouskhelisvili, M. Nioradze, M. Ponce de Leon, M. Tappen, M. Tvalchrelidze, and C. Zollikofer (2002) A new skull of early *Homo* from Dmanisi, Georgia. *Science* **297**:85–89.

Vermeersch, P. (2002) Two Upper Paleolithic burials at Nazlet Khater. In: P. Vermeersch, Editor. *Paleolithic Quarrying Sites in Upper and Middle Egypt*. Leuven: Leuven University Press, pp. 273–282.

Vermeersch, P., M. Paulissen, G. Gijselings, M. Otte, A. Thoma, P. van Peer, and R. Lauwers (1984a) 33,000-year-old chert mining site and related *Homo* in the Egyptian Nile Valley. *Nature* **309**:342–344.

Vermeersch, P., G. Gijselings, and M. Paulissen (1984b) Discovery of the Nazlet Khater man, Upper Egypt. *JHE* **13**:281–286.

Vermeersch, P., E. Paulissen, S. Stokes, C. Charlier, P. van Peer, C. Stringer, and W. Lindsay (1998) A Middle Paleolithic burial of a modern human at Taramsa Hill, Egypt. *Antiquity* **72**:475–484.

Vigilant, L., M. Stoneking, H. Harpending, K. Hawkes, and A. Wilson (1991) African populations and the evolution of human mitochondrial DNA. *Science* **253**:1503–1507.

Vignaud, P., P. Duringer, H. Mackaye, A. Likius, C. Blondel, J-R. Boisserie, L. De Bonis, V. Eisenmann, M-E. Etienne, D. Geraads, F. Guy, T. Lehmann, F. Lihoreau, N. Lopez-Martinez, C. Mourer-Chauviré, O. Otero, J-C. Rage, M. Schuster, L. Viriot, A. Zazzo, and M. Brunet (2002) Geology and palaeontology of the Upper Miocene Toros-Menalla hominid locality, Chad. *Nature* **418**:152–155.

Villa, P. (1983) *Terra Amata and the Middle Pleistocene Archaeological Record of Southern France. University of California Publications in Anthropology* **13**:1–303.

Villa, P. (1996) Book Review. The first Italians. Le industrie litiche del giacimento paleolitico di Isernia La Pineta. *Lithic Technology* **21**:71–79.

Vinyard, C., and F. Smith (1997) Morphometric relationships between the suprorbital region and frontal sinus in Melanesian crania. *Homo* **48**:1–21.

Vinyard, C., and F. Smith (2001) Morphometric testing of structural hypotheses of the supraorbital region in modern humans. *ZMA* **83**:23–41.

Viola, T., M. Teschler-Nikola, O. Kullmer, K. Schäfer, A. Derevianko, and H. Seidler (2006) The Okladnikov Cave hominids—The easternmost Neanderthals? In: *150 Years of Neandertal Discoveries. Terra Nostra* **2006/2**:139.

Virchow, R. (1872) Untersuchung des Neanderthal-Schädels. *ZE* **4**:157–165.

Virchow, R. (1882) Der Kiefer aus der Schipka-Höhle und der Kiefer von La Naulette. *ZE* **14**:277–310.

Vlček, E. (1967) Die Sinus frontales bei europäischen Neandertalern. *AZ* **30**:166–189.

Vlček, E. (1969) *Neandertaler der Tschechoslowakei*. Prague: Academia.

Vlček, E. (1970) Étude comparative onto-phylogénétique de l'enfant du Pech de l'Azé par rapport á d'autres enfants Néandertalien. *AIPH* **33**:149–178.

Vlček, E. (1973) Postcranial skeleton of a Neandertal child from Kiik-Koba, U.S.S.R. *JHE* **2**:537–544.

Vlček, E. (1978) A new discovery of *Homo erectus* in Central Europe. *JHE* **7**:239–251.

Vlček, E. (1993) *Fossile Menschenfunde von Weimar-Ehringsdorf*. Stuttgart: Theiss.

Vogel, G. (1999) Chimps in the wild show stirrings of culture. *Science* **284**:2070–2073.

Vogel, P. (2004) The current molecular phylogeny of eutherian mammals challenges previous interpretations of placental evolution. *Placenta* **26**:591–596.

Voisin, J. (2006) Speciation by distance and temporal overlap: A new approach to understanding Neanderthal evolution. In: Harvati and Harrison 2006: 299–314.

Vollbrecht, J. (1992) *Das Altpaläolithikum aus den unteren Schichten in Kärlich*. Masters Thesis. Köln: Universität Köln.

Volman, T. (1984) Early prehistory of southern Africa. In: R. Klein, Editor. *Southern African Prehistory and Paleoenvironments*. Rotterdam: Balkema, pp. 169–220.

von Berg, A., S. Condemi, and M. Frechen (2000) Die Schädelkalotte eines Neandertalers von Ochtendung/Osteifel—Archäologie, Paläoanthropologie und Geologie. *Eiszeitalter und Gegenwart* **50**:56–58.

von Dornum, M., and M. Ruvolo (1999) Phylogenetic relationships of the New World monkeys (Primates, Platyrrhini) based on nuclear G6PD DNA sequences. *MPE* **11**:459–476.

von Koenigswald, G. (1936) Erste Mitteilungen über einen fossilen Hominiden aus dem Altpleistozän Ostjavas. *Koninklijke Nederlandse Akademie van Wetenschappen* **39**:1000–1009.

von Koenigswald, G. (1940) Neue *Pithecanthropus*-Funde 1936–1938: Ein Beitrag zur Kenntnis der Praehominiden. *Wetenschappelijke Mededelingen—Dienst van dem Mijnbouw in Nederlandsch-Oost Indië* **28**:1–232.

von Koenigswald, G. (1950) Fossil homininds of the Lower Pleistocene of Java. *18th International Geological Congress (London, 1948)* **9**:59–61.

von Koenigswald, G. (1956) *Meeting Prehistoric Man.* London: Thames and Hudson.

von Koenigswald, G. (1958) Der Solo-Mensch von Java: Ein tropischer Neanderthaler. In: von Koenigswald 1958: 21–26.

von Koenigswald, G. (1975) Early man in Java—Catalogue and problems. In: Tuttle 1975: 303–309.

von Koenigswald, G., and F. Weidenreich (1939) The relationship between *Pithecanthropus* and *Sinanthropus*. *Nature* **144**:926–929.

von Koenigswald, W. (1992) Zur Ökologie und Biostratigraphie der beiden plesitozänen Faunen von Mauer bei Heidelberg. In: K. Beinhauer, and G. Wagner, Editors. *Schichen—85 Jahre Homo erectus heidelbergensis von Mauer.* Mannheim/Heidelberg: Edition Braus/Reiß-Museum der Stadt Mannheim, pp. 101–110.

von Koenigswald, W., and T. van Kolfschoten (1996) The *Mimomys-Arvicola* boundary and the enamel thickness quotient (SDQ) of *Arvicola* as stratigraphic markers in the Middle Pleistocene. In: C. Turner, Editor. *The Early Middle Pleistocene in Europe.* Rotterdam: Balkema, pp. 211–226.

Vrba, E. (1980) Evolution, species, and fossils: how does life evolve? *SAJS* **76**:61–84.

Vrba, E. (1984) Evolutionary pattern and process in the sister-group Alcelaphini-Aepycerotini (Mammalia: Bovidae). In: N. Eldredge and S. Stanley, Editors. *Living Fossils.* New York: Springer, pp. 62–79.

Vrba, E. (1985) Early hominids in southern Africa: Updated observations of chronological and ecological background. In: Delson 1985: 195–200.

Vrba, E. (1995) The fossil record of African antelopes (Mammalia, Bovidae) in relation to human evolution and paleoclimate. In: Vrba et al. 1995: 385–424.

W

Wagner, G. (1996) Überlegungen zur Zeitstellung der Fundschicht des *Homo erectus heidelbergensis* in den Sanden von Mauer. In: Beinhauer et al. 1996: 53–56.

Walker, A. (1976) Remains attributable to *Australopithecus* in the East Rudolf succession. In: Coppens et al. 1976: 484–489.

Walker, A. (1981) The Koobi Fora hominids and their bearing on the genus *Homo.* In: Sigmon and Cybulski 1981: 193–215.

Walker, A. (1993) Perspectives on the Nariokotome discovery. In: Walker and Leakey 1993a: 411–430.

Walker, A. (1997) *Proconsul*: Function and phylogeny. In: Begun et al. 1997: 209–224.

Walker, A., and R. Leakey (1978) The hominids of East Turkana. *SA* **239**:54–66.

Walker, A., and R. Leakey (1988) The evolution of *Australopithecus boisei.* In: Grine 1988: 247–258.

Walker, A., and R. Leakey (1993b) The skull. In: Walker and Leakey 1993a: 63–94.

Walker, A., and R. Leakey (1993c) The postcranial bones. In: Walker and Leakey 1993a: 95–160.

Walker, A., R. Leakey, J. Harris, and F. Brown (1986) 2.5-Myr *Australopithecus boisei* from west of Lake Turkana, Kenya. *Nature* **322**:517–522.

Walker A., and C. Ruff (1993) The reconstruction of the pelvis. In: Walker and Leakey 1993a: 221–233.

Walker, A., M. Zimmerman, and R. Leakey (1982) A possible case of hypervitaminosis A in *Homo erectus. Nature* **296**:248–250.

Walker, J., R. Cliff, and A. Latham (2006) U-Pb isotopic evidence for the age of Littlefoot, Sterkfontein, South Africa. *PaleoAnthropology* **2006**:A10.

Wallace, A. (1890) *The Malay Archipelago.* 1962 Edition. New York: Dover.

Wallace, J. (1975) Did La Ferrassie I use his teeth as a tool? *CA* **16**:393–401.

Walrath, D., and M. Glantz (1996) Sexual dimorphism in the pelvic midplane and its relationship to Neandertal reproductive patterns. *AJPA* **100**:89–100.

Walter, R., P. Manega, R. Hay, R. Drake, and G. Curtis (1991) Laser fusion ^{40}Ar/^{39}Ar dating of Bed I, Olduvai Gorge, Tanzania. *Nature* **354**:145–149.

Wang, W., R. Crompton, Y. Li, and M. Gunther (2003) Optimum ratio of upper to lower limb lengths in hand-carrying of a load under the assumption of frequency coordination. *Journal of Biomechanics* **36**:249–252.

Ward, C. (1993) Torso morphology and locomotion in *Proconsul nyanzae. AJPA* **92**:291–328.

Ward, C. (1997) Functional anatomy and phyletic implications of the hominoid trunk and hindlimb. In: Begun et al. 1997: 101–130.

Ward, C. (1998) *Afropithecus, Proconsul,* and the primitive hominoid skeleton. In: E. Strasser, J. Fleagle, A. Rosenberger, and H. McHenry, Editors. *Primate Locomotion: Recent Advances.* New York: Plenum, pp. 337–352.

Ward, C. (2002) Interpreting the posture and locomotion of *Australopithecus afarensis*: Where do we stand? *YPA* **45**:185–215.

Ward, C., M. G. Leakey, and A. Walker (1999) The new hominid species *Australopithecus anamensis. EA* **7**:197–205.

Ward, C., M. G. Leakey, and A. Walker (2001) Morphology of *Australopithecus anamensis* from Kanapoi and Allia Bay, Kenya. *JHE* **41**:255–368.

Ward, S. (1997) The taxonomy and phylogenetic relationships of *Sivapithecus* revisited. In: Begun et al. 1997: 269–290.

Ward, S., and B. Brown (1986) The facial skeleton of *Sivapithecus indicus*. In: D. Swindler, Editor. *Comparative Primate Biology*, Volume I: *Systematics, Evolution, and Anatomy*. New York: Alan R. Liss, pp. 413–452.

Ward, S., and D. Duren (2002) Middle and late Miocene African hominoids. In: Hartwig 2002: 385–397.

Ward, S., and A. Hill (1987) Pliocene hominid partial mandible from Tabarin, Baringo, Kenya. *AJPA* **72**:21–37.

Ward, S., and W. Kimbel (1983) Subnasal morphology and the systematic position of *Sivapithecus*. *AJPA* **61**:157–171.

Ward, S., and D. Pilbeam (1983) Maxillofacial morphology of Miocene hominoids from Africa and Indo-Pakistan. In: Ciochon and Corruccini 1983: 211–238.

Washburn, S. (1947) The relation of the temporal muscle to the form of the skull. *AR* **99**:239–248.

Washburn, S. (1951) The analysis of primate evolution with particular reference to the origin of man. *Symposia on Quantitative Biology* **15**:67–77.

Watts, D. (1996) Comparative socio-ecology of gorillas. In: McGrew et al. 1996: 16–28.

Weaver, T., and K. Steudel-Numbers (2005) Does climate or mobility explain the differences in body proportions between Neandertals and their Upper Paleolithic successors? *EA* **14**:218–223.

Webb, S. (1989) *The Willandra Lakes Hominids*. Canberra: Australian National University.

Webb, S. (1990) Cranial thickening in an Australian hominid as a possible paleoepidemiological indicator. *AJPA* **82**:403–411.

Weber, J., A. Czarnetzki, and C. Pusch (2005) Comment on: The brain of LB 1, *Homo floresiensis*. *Science* **310**:236.

Weidenreich, F. (1936a) Observations on the form and the proportions of the endocranial casts of *Sinanthropus pekinensis* and the great apes: A comparative study of brain size. *Paleontologia Sinica Series D* **7**(Fascicle 4): 1–50.

Weidenreich, F. (1936b) The mandibles of *Sinanthropus pekinensis*: A comparative study. *Paleontologia Sinica Series D* **7**(Fascicle 3):1–162.

Weidenreich, F. (1937) The dentition of *Sinanthropus pekinensis*. *Paleontologia Sinica New Series D* **1**:1–180.

Weidenreich, F. (1939a) Six lectures of *Sinanthropus pekinensis* and related problems. *BGSC* **19**:1–110.

Weidenreich, F. (1939b) On the earliest representatives of modern mankind recovered on the soil of East Asia. *Bulletin of the Natural History Society, Peking* **13**(3):161–174.

Weidenreich, F. (1940) The torus occipitalis and related structures and their transformation in the course of human evolution. *BGSC* **19**:379–558.

Weidenreich, F. (1941) The extremity bones of *Sinanthropus pekinensis*. *Paleontologia Sinica New Series D* **5**:1–150.

Weidenreich, F. (1943) The skull of *Sinanthropus pekinensis*: A comparative study on a primitive hominid skull. *Paleontologia Sinica New Series D* **10**:iii–484.

Weidenreich, F. (1945a) Giant early man from Java and South China. *APAMNH* **40**:1–134.

Weidenreich, F. (1945b) The Paleolithic child from the Teshik Tash cave in southern Uzbekistan (Central Asia). *AJPA* **3**:237–249.

Weidenreich, F. (1945c) The Keilor skull: A Wadjak type from southeast Australia. *AJPA* **3**:225–236.

Weidenreich, F. (1946) *Apes, Giants and Man*. Chicago: University of Chicago Press.

Weidenreich, F. (1947a) Facts and speculations concerning the origin of *Homo sapiens*. *AA* **49**:187–203.

Weidenreich, F. (1947b) The trend of human evolution. *Evolution* **1**:221–236.

Weidenreich, F. (1951) Morphology of Solo Man. *APAMNH* **43**:205–290.

Weiner, J. (1955) *The Piltdown Forgery*. Oxford: Oxford University Press.

Weiner, J., K. Oakley, and W. Le Gros Clark (1953) The solution to the Piltdown problem. *Bulletin of the British Museum (Natural History)* **2**:141–146.

Weiner, M., J. Ricci, J. Phelan, T. Plummer, S. Gaulda, R. Potts, and T. Bromage (2006) Pathology of an archaic *Homo* mandible from Kanam, Kenya. Meeting abstracts, American Association for Dental Research.

Weinert, H. (1936) Der Urmenschschädel von Steinheim. *ZMA* **35**:413–518.

Weinert, H., W. Bauermeister, and A. Remane (1940) *Africanthropus njarasensis*. Beschreibung und phylogenetische Einordnung des ersten Affenmenschen aus Ostafrika. *ZMA* **38**:252–308.

Weir, A., J. Chappell, and A. Kacelnik (2002) Shaping of hooks in New Caledonian crows. *Science* **297**:981.

Weiss, K., and F. Smith (2007) Out of the veil of death rode the one million! Neandertals and their genes. *BioEssays* **29**:105–110.

Wheeler, P. (1984) The evolution of bipedality and the loss of functional body hair in hominids. *JHE* **13**:91–98.

Wheeler, P. (1985) The influence of bipedalism on the energy and water budgets of early hominids. *JHE* **21**:117–136.

Wheeler, P. (1991a) The loss of functional body hair in man: The influence of thermal environment, body form, and bipedalism. *JHE* **14**:23–28.

Wheeler, P. (1991b) The thermoregulatory advantages of hominid bipedalism in open equatorial environments. *JHE* **21**:107–115.

Wheeler, P. (1992a) The influence of the loss of functional body hair on the water budgets of early hominids. *JHE* **23**:379–388.

Wheeler, P. (1992b) The thermoregulatory advantages of large body size for hominids foraging in savannah environments. *JHE* **23**:351–362.

Wheeler, P. (1993) The influence of stature and body form on hominid energy and water budgets; a comparison of *Australopithecus* and early *Homo* physiques. *JHE* **24**:13–28.

White, F. (1996) Comparative socio-ecology of *Pan paniscus*. In: McGrew et al. 1996: 29–41.

White, F. (1998) Seasonality and socioecology: The importance of variation in fruit abundance to bonobo sociality. *IJP* **19**:1013–1027.

White, L. (1994) Biomass of rain forest mammals in the Lopé Reserve, Gabon. *Journal of Animal Ecology* **63**:499–512.

White, L. A. (1949) *The Science of Culture: A Study of Man and Civilization*. New York: Farrar, Straus and Cudahy.

White, M. (1968) *Modes of Speciation*. San Francisco: Freeman.

White, R. (1986) *Dark Caves, Bright Visions*. New York: American Museum of Natural History.

White, R. (2001) *Personal ornaments from the Grotte du Renne at Arcy-sur-Cure*. Athena Review **2**:41–46.

White, R. (2003) *Prehistoric Art: The Symbolic Journey of Humankind*. New York: Abrams.

White, T. (1977) *The anterior mandibular corpus of early African Hominidae: functional significance of shape and size*. Ph.D. Dissertation, University of Michigan. Ann Arbor: University Microfilms.

White, T. (1985) The hominids of Hadar and Laetoli: An element-by-element comparison of the dental samples. In: Delson 1985: 138–152.

White, T. (1987) Cannibals at Klasies? *Saggitarius* **2**:6–9.

White, T. (1988) The comparative biology of "robust" *Australopithecus*: Clues from context. In: Grine 1988: 449–483.

White, T. (1992) *Prehistoric Cannibalism at Mancos Canyon 5MTURMR-2346*. Princeton: Princeton University Press.

White, T. (2000) Olduvai Gorge. In: Delson et al. 2000: 486–489.

White, T. (2001) Once we were cannibals. *SA* **285**:58–65.

White, T. (2002) Earliest hominids. In: Hartwig 2002: 407–417.

White, T. (2003) Early hominids—Diversity or distortion? *Science* **299**:1994–1997.

White, T., B. Asfaw, D. DeGusta, H. Gilbert, G. Richards, G. Suwa, and F. C. Howell (2003) Pleistocene *Homo sapiens* from Middle Awash, Ehiopia. *Nature* **423**:742–747.

White, T., and D. Johanson (1982) Pliocene hominid mandibles from the Hadar Formation, Ethiopia: 1974–1977 collections. *AJPA* **57**:501–544.

White, T., and D. Johanson (1989) The hominid composition of Afar Locality 333: some preliminary observations. In: G. Giacobini, Editor. *HOMINIDAE: Proceedings of the 2nd International Congress of Human Paleontology*. Milan: Editoriale Jaca Book, pp. 97–101.

White, T., D. Johanson, and W. Kimbel (1981) *Australopithecus africanus*: Its phylogenetic position reconsidered. *SAJS* **77**:445–470.

White, T., D. Johanson, and W. Kimbel (1983) *Australopithecus africanus*: its phylogenetic position reconsidered. In: Ciochon and Corruccini 1983: 721–780.

White, T., and G. Suwa, G. 1987. Hominid footprints at Laetoli: Facts and interpretations. *AJPA* **72**:485–514.

White, T., G. Suwa, and B. Asfaw (1994) *Australopithecus ramidus*, a new species of early hominid from Aramis, Ethiopia. *Nature* **371**:306–312.

White, T., G. Suwa, and B. Asfaw (1995) Corrigendum: *Australopithecus ramidus*, a new species of early hominid from Aramis, Ethiopia. *Nature* **375**:88.

White, T., G. Suwa, W. Hart, G. WoldeGabriel, J. de Heinzelin, J. Clark, B. Asfaw, and E. Vrba (1993) New discoveries of *Australopithecus* at Maka in Ethiopia. *Nature* **366**:261–265.

White, T., G. Suwa, S. Simpson, and B. Asfaw (2000) Jaws and teeth of *Australopithecus afarensis* from Maka, Middle Awash, Ethiopia. *AJPA* **111**:45–68.

White, T., and N. Toth (1991) The question of ritual cannibalism at Grotta Guattari. *CA* **32**:118–138.

White, T., G. WoldeGabriel, B. Asfaw, S. Ambrose, Y. Beyene, R. Bernor, J-R. Boisserie, B. Currie, H. Gilbert, Y. Haile-Selassie, W. Hart, L. Hlusko, F. C. Howell, R. Kono, T. Lehmann, A. Louchart, C. O. Lovejoy, P. Renne, H. Saegusa, E. Vrba, H. Wesselmann, and G. Suwa (2006) Asa Issie, Aramis and the origin of *Australopithecus*. *Nature* **440**:883–889.

Whiten, A., J. Goodall, W. McGrew, T. Nishida, V. Reynolds, Y. Sugiyama, C. Tutin, R. Wrangham, and C. Boesch (1999) Cultures in chimpanzees. *Nature* **399**:682–685.

Wible, J., and H. Covert (1987) Primates: cladistic diagnosis and relationships. *JHE* **16**:1–20.

Widianto, H., and V. Zeitoun (2003) Morphological description, biometry, and phylogenetic position of the skull of Nagwi I (East Java, Indonesia). *IJO* **13**:339–351.

Widianto, H., A. Sémah, T. Djubiantono, and F. Sémah (1994) A tentative reconstruction of the cranial human remains of Hanoman I from Bukuran, Sangiran (Central Java). In: Franzen 1994: 47–59.

Wieland, C. 1994. Birds of a feather don't breed together! *Creation Magazine* **16**:10–12.

Wild, E., M. Teschler-Nikola, W. Kutschera, P. Steier, E. Trinkaus, and W. Wanek (2005) Direct dating of early Upper Palaeolithic human remains from Mladeč. *Nature* **435**:332–335.

Wildman, D., C. Chen, O. Erez, L. Grossman, M. Goodman, and R. Romero (2006) Evolution of the mammalian placenta revealed by phylogenetic analysis. *PNAS* **103**:3203–3208.

Wildman, D., M. Uddin, G. Liu, L. Grossman, and M. Goodman (2003) Implications of natural selection in shaping 99.4% nonsynonymous DNA identity between humans and chimpanzees: Enlarging genus *Homo*. *PNAS* **100**:7181–7188.

Wiley, E. O. (1981) *Phylogenetics: The Theory and Practice of Phylogenetic Systematics*. New York: John Wiley & Sons.

Wiley, E. O. (1992) The evolutionary species concept reconsidered. In: Ereshefsky 1992: 79–92.

Wiley, E. O., and R. Mayden (2000) The evolutionary species concept. In: Q. Wheeler, and R. Meier, Editors. *Species Concepts and Phylogenetic Theory: A Debate*. New York: Columbia University Press, pp. 70–89.

Wimmer, R., S. Kirsch, G. Rappold, and W. Schempp (2002) Direct evidence for the *Homo–Pan* clade. *Chromosome Research* **10**:55–61.

Wind, J. (1976) Phylogeny of the human vocal tract. *ANYAS* **280**:612–630.

Wise, C., M. Sraml, D. Rubinzstein, and S. Easteal (1997) Comparative nuclear and mitochondrial genome diversity in humans and chimpanzees. *MBE* **14**:707–716.

Wobst, M. (1976) Local relationships in Paleolithic society. *JHE* **5**:49–58.

WoldeGabriel, G., Y. Haile-Selassie, P. Renne, W. Hart, S. Ambrose, B. Asfaw, G. Heisken, and T. White (2001) Geology and paleontology of the late Miocene Middle Awash Valley, Afar Rift, Ethiopia. *Nature* **412**:175–178.

WoldeGabriel, G., T. White, G. Suwa, P. Renne, J. De Heinzelin, W. Hart, and G. Helken (1994) Ecological and temporal placement of early Pliocene hominids at Aramis, Ethiopia. *Nature* **371**:330–333.

Wolfe, L., J. Gray, J. Robinson, L. Lieberman, and E. Peters (1982) Models of human evolution. *Science* **217**:302.

Wolpoff, M. (1968a) "Telanthropus" and the single species hypothesis. *AA* **70**:447–493.

Wolpoff, M. (1968b) Climatic influence on skeletal nasal aperture. *AJPA* **29**:405–423.

Wolpoff, M. (1969) Cranial capacity and the taxonomy of Olduvai Hominid 7. *Nature* **223**:182–183.

Wolpoff, M. (1970a) Taxonomy and cranial capacity of Olduvai Hominid 7. *Nature* **227**:747.

Wolpoff, M. (1970b) The evidence for multiple hominid taxa at Swartkrans. *AA* **72**:576–607.

Wolpoff, M. (1971a) Vertesszöllos and the presapiens theory. *AJPA* **35**:209–216.

Wolpoff, M. (1971b) *Metric Trends in Hominid Dental Evolution*. Cleveland: Case Western Reserve University Studies in Anthropology, Number 2.

Wolpoff, M. (1971c) Interstitial wear. *AJPA* **34**:205–228.

Wolpoff, M. (1971d) Competitive exclusion among Lower Pleistocene hominids: The single species hypothesis. *Man* **6**:601–614.

Wolpoff, M. (1974) Sagittal cresting in the South African australopithecines. *AJPA* **40**:397–408.

Wolpoff, M. (1975a) Some aspects of human mandibular evolution. In: J. McNamara, Editor. *Determinants of Mandibular Form and Growth*. Ann Arbor: Center for Human Growth and Development, pp. 1–64.

Wolpoff, M. (1975b) Sexual dimorphism in the australopithecines. In: Tuttle 1975: 245–284.

Wolpoff, M. (1978) Some aspects of canine size in the australopithecines. *JHE* **7**:115–126.

Wolpoff, M. (1979) The Krapina dental remains. *AJPA* **50**:67–114.

Wolpoff, M. (1980a) *Paleoanthropology*. New York: Knopf.

Wolpoff, M. (1980b) Cranial remains of Middle Pleistocene European hominids. *JHE* **9**:339–358.

Wolpoff, M. (1983) Lucy's little legs. *JHE* **12**:443–453.

Wolpoff, M. (1984) Evolution in *Homo erectus*: The question of stasis. *Paleobiology* **10**:389–406.

Wolpoff, M. (1986a) Stasis in the interpretation of evolution in *Homo erectus*: A reply to Rightmire. *Paleobiology* **12**:325–328.

Wolpoff, M. (1986b) Describing anatomically modern *Homo sapiens*: A distinction without a definable difference. In: V. Novotný, Editor. *Fossil Man: New Facts—New Ideas. Anthropos* **23**:41–53.

Wolpoff, M. (1989) Multiregional evolution: The fossil alternative to Eden. In: Mellars and Stringer 1989: 62–108.

Wolpoff, M. (1992) Theories of modern human origins. In: Bräuer and Smith 1992: 25–63.

Wolpoff, M. (1999) *Paleoanthropology*. Second edition. New York. McGraw-Hill.

Wolpoff, M., and R. Caspari (1990) On Middle Paleolithic/Middle Stone Age hominid taxonomy. *CA* **31**:394–395.

Wolpoff, M., and R. Caspari (1997) *Race and Human Evolution*. New York: Simon & Schuster.

Wolpoff, M., and D. Frayer (2005) Unique ramus anatomy for Neandertals? *AJPA* **128**:245–251.

Wolpoff, M., D. Frayer, and J. Jelínek (2006) Aurignacian female crania and teeth from the Mladeč caves, Moravia, Czech Republic. In: Teschler-Nokola 2006: 273–340.

Wolpoff, M., J. Hawks, and R. Caspari (2000) Multiregional, not multiple origins. *AJPA* **112**:129–136.

Wolpoff, M., J. Hawks, D. Frayer, and K. Huntley (2001) Modern human ancestry at the peripheries: A test of the replacement theory. *Science* **291**:293–297.

Wolpoff, M., J. Hawks, B. Sénut, M. Pickford, and J. Ahern (2006) An ape or *the* ape: Is the Toumaï cranium TM 266 a hominid? *PaleoAnthropology* **2006**:36–50.

Wolpoff, M., and S-H. Lee (2001) The late Pleistocene human species of Israel. *BMSAP* **13**:291–310.

Wolpoff, M., B. Mannheim, A. Mann, J. Hawks, R. Caspari, K. Rosenberg, D. Frayer, G. Gill, and G. Clark (2004) Why not the Neandertals? *WA* **36**:527–546.

Wolpoff, M., B. Sénut, M. Pickford, and J. Hawks (2002) *Sahelanthropus* or "Sahelpithecus"? *Nature* **419**:581–582.

Wolpoff, M., F. Smith, and D. Frayer (1997) Neandertals are a race of *Homo sapiens*. *JHE* **32**:A25.

Wolpoff, M., F. Smith, M. Malez, J. Radovčic, and D. Rukavina (1981) Upper Pleistocene human remains from Vindija cave, Croatia, Yugoslavia. *AJPA* **54**:499–545.

Wolpoff, M., A. Thorne, J. Jelínek, and Y. Zhang (1994a) The case for sinking *Homo erectus*. 100 years of Pithecanthropus is enough! In: Franzen 1994: 341–361.

Wolpoff, M., A. Thorne, F. Smith, D. Frayer, and G. Pope (1994b) Multiregional evolution: A world-wide source for modern human populations. In: Nitecki and Nitecki 1994: 175–199.

Wolpoff, M., X. Wu, and A. Thorne (1984) Modern *Homo sapiens* origins: A general theory of hominid evolution involving the fossil evidence from East Asia. In: Smith and Spencer 1984: 411–483.

Wong, K. (2005) The morning of the modern mind. *SA* **292**:86–95.

Woo (Wu), J. (R.) (1958) Tzeyang Paleolithic man—Earliest representative of modern man in China. *AJPA* **16**:459–465.

Woo (Wu), J. (R.) (1964) Mandible of *Sinanthropus lantianensis*. *CA* **5**:98–101.

Woo (Wu), J. (R.) (1966) The skull of Lantian man. *CA* **7**:83–86.

Woo (Wu), J. (R.), and T. Chao (1959) New discovery of *Sinanthropus* mandible from Choukoutien. *Vertebrata PalAsiatica* **3**:169–172.

Woo (Wu), J. (R.), and Peng, R. (1959) Fossil human skull of early Palaeoanthropic stage found at Mapa, Shaokuan, Kwangtung province. *Paleovertebrata et Paleoanthropologia* **1**:159–164.

Wood, B. (1984) The origin of *Homo erectus*. *CFS* **69**:99–111.

Wood, B. (1985) A review of the definition, distribution, and relationships of *Australopithecus africanus*. In: Tobias 1985: 227–232.

Wood, B. (1991) *Koobi Fora Research Project*, Volume 4: *Hominid Cranial Remains*. Oxford: Clarendon Press.

Wood, B. (1992) Origin and evolution of the genus *Homo*. *Nature* **355**:783–790.

Wood, B. (1994) Taxonomy and evolutionary relationships of *Homo erectus*. In: Franzen 1994: 159–165.

Wood, B. (1999a) '*Homo rudolfensis*' Alexeev, 1986–fact or phantom? *JHE* **36**:115–118.

Wood, B. (1999b) Plio-Pleistocene hominins from the Baringo region, Kenya. In: P. Andrews and P. Banham, Editors. *Late Cenozoic Environments and Hominid Evolution: A Tribute to Bill Bishop*. London: Geological Society, pp. 113–122.

Wood, B. (2002) Hominid revelations from Chad. *Nature* **418**:133–135.

Wood, B., and M. Collard (1999) The human genus. *Science* **284**:65–71.

Wood, B., and M. Ellis (1986) Evidence for dietary specialization in the "robust" australopithecines. In: V. Novotný and A. Mizerová, Editor. *Fossil Man: New Facts—New Ideas. Papers in Honour of Jan Jelínek's Life Anniversary. Anthropos* **23**:101–124.

Wood, B., and P. Quinney (1996) Assessing the pelvis of AL 288-1. *JHE* **31**:563–568.

Wood, B., and B. Richmond (2000) Human evolution: taxonomy and paleobiology. *Journal of Anatomy* **196**:19–60.

Wood, B., and D. Strait (2004) Patterns of resource use in early *Homo* and *Paranthropus*. *JHE* **46**:119–162.

Wood, B., C. Wood, and L. Konigsberg (1994) *Paranthropus boisei*: an example of evolutionary stasis? *AJPA* **95**:117–136.

Wood, J. (1982) Models of human evolution. *Science* **217**: 297–298.

Woodward, A. Smith (1921) A new cave-man from Rhodesia, South Africa. *Nature* **108**:371–372.

Woodward, A. Smith (1938) A fossil skull of an ancestral Bushman from the Anglo-Egyptian Sudan. *Antiquity* **12**:193–195.

Wootten, M., M. Kadaba, and G. Cochran (1990) Dynamic electromyography. II. Normal patterns during gait. *Journal of Orthopaedic Research* **8**:259–265.

Wrangham, R. (2001) Out of the *Pan*, into the fire: How our ancestors' evolution depended on what they ate. In: F. de Waal, Editor. *Tree of Origin: What Primate Behavior Can Tell Us about Human Social Evolution*. Cambridge: Harvard University Press, pp. 119–143.

Wrangham, R., C. Chapman, A. Clark-Arcadi, and G. Isabirye-Basuta (1996) Social ecology of Kanyawara chimpanzees: Implications for understanding the costs of great ape groups. In: McGrew et al. 1996: 45–57.

Wrangham, R., N. Conklin-Brittain, and K. Hunt (1998) Dietary response of chimpanzees and cercopithecines to seasonal variation in fruit abundance. I. Antifeedants. *IJP* **19**:949–970.

Wrangham, R., and D. Peterson (1996) *Demonic Males: Apes and the Origins of Human Violence*. New York: Houghton Mifflin.

Wright, R. (1995) The Zhoukoudian Upper Cave skull 101 and multiregionalism. *JHE* **29**:181–183.

Wu, L., Z. Zhang, and X. Wu (2005) Middle Pleistocene human cranium from Tangshan (Nanjing), southeast China: A new reconstruction and comparisons with *Homo erectus* from Eurasia and Africa. *AJPA* **127**:253–262.

Wu, M. (1980) Human fossils discovered at Xujiayao site in 1977. *Vertebrata PalAsiatica* **18**:229–238.

Wu, R. (1985) New Chinese *Homo erectus* and recent work at Zhoukoudian. In: Delson 1985: 245–248.

Wu, R. (1988) The reconstruction of the fossil human skull from Jinniushan, Yinkou, Lianing Province and its main features. *AAS* **7**:97–101.

Wu, R., and X. Dong (1985) *Homo erectus* in China. In: Wu and Olson 1985: 79–89.

Wu, R., and S. Li (1983) Peking man. *SA* **248**:78–86.

Wu, X. (1981) A well-preserved cranium of an archaic type of early *Homo sapiens* from Dali, China. *Scienta Sinica* **24**:530–538.

Wu, X. (1987) Relations between Upper Paleolithic men in China and their southern neighbors in Niah and Tabon. *AAS* **6**:180–183.

Wu, X. (1997) On the decent of modern humans in East Asia. In: Clark and Willermet 1997: 283–293.

Wu, X., and F. Poirier (1995) *Human Evolution in China*. Oxford: Oxford University Press.

Wu, X., and M. Wu (1985) Early *Homo sapiens* in China. In: Wu and Olsen 1985: 91–106.

Wu, X., and Y. You (1979) A preliminary observation of the Dali man site. *VP* **17**:294–303.

Wu, X., and Z. Zhang (1985) *Homo sapiens* remains from Late Paleolithic and Neolithic China. In: Wu and Olsen 1985: 91–106.

Wunderlich, R., A. Walker, and W. Jungers (1999) Rethinking the positional behavior of *Oreopithecus*. *AJPA Supplement* **28**: 282.

Würges, K. (1986) Artefakte aus den ältesten Quartärsedimenten (Schichten A–C) der Tongrube Kärlich, Kreis Mayen-Koblenz/Neuwieder Becken. *AK* **16**:1–6.

Würges, K. (1991) Neue altpaläolithische Funde aus der Tongrube Kärlich, Kreis Mayen-Koblenz/Neuwieder Becken. *AK* **21**:449–455.

Wymer, J. (1955) A further fragment of the Swanscombe skull. *Nature* **176**:426–427.

Wynn, J. (2000) Paleosols, stable carbon isotopes, and paleoenvironmental interpretation of Kanapoi, northern Kenya. *JHE* **39**:411–432.

X

Xiao, S., Y. Zhang, and A. Knoll (1998) Three-dimensional preservation of algae and animal embryos in a Neoproterozoic phosphorite. *Nature* **391**:553–558.

Y

Yamagiwa, J., T. Maruhashi, T. Yumoto, and N. Mwanza (1996) Dietary and ranging overlap in sympatric gorillas and chimpanzees in Kahuzi-Biega National Park, Zaire. In: McGrew et al. 1996: 82–98.

Yamashita, N. (1998) Functional dental correlates of food properties in five Malagasy lemur species. *AJPA* **106**:169–188.

Yates, F., and M. Healey (1951) Statistical methods in anthropology. *Nature* **168**:1116–1117.

Yellen, J., A. Brooks, A. Cornelissan, M. Mehlman, and K. Stewart (1995) A Middle Stone Age worked bone industry from Katanda, Upper Semliki Valley, Zaire. *Science* **268**:553–556.

Yi, S., and G. Clark (1983) Observations on the lower Paleolithic of Northeast Asia. *CA* **24**:181–292.

Yoder, A. (1992) *The phylogenetic affinities of the Cheirogaleidae: A molecular and morphological analysis*. Ph.D. thesis, Duke University.

Yoder, A., R. Rasoloarison, S. Goodman, J. Irwin, S. Atsalis, M. Ravosa, and J. Ganzhorn (2001) Remarkable species diversity in Malagasy mouse lemurs (Primates, Microcebus). *PNAS* **97**:11325–11330.

Yoder, A., and Z. Yang (2000) Estimation of primate speciation dates using local molecular clocks. *MPE* **17**:1081–1090.

Yokley, T. (1999) Variation in Neandertal internal nasal morphology: Evidence from Krapina and Vindija. *AJPA Supplement* **28**:283.

Yokley, T. (2007) *The functional and adaptive significance of anatomical variation in recent and fossil human nasal passages*. Ph.D. thesis, Duke University.

Yokley, T., and S. Churchill (2006) Archaic and modern human distal humeral morphology. *JHE* **51**:603–616.

Young, N., and L. MacLatchy (2004) The phylogenetic position of *Morotopithecus*. *JHE* **46**:163–184.

Yuan, S., T. Chen, and S. Gao (1983) Uranium series dating of "Ordos Man" and "Sjara-osso-gol Culture." *AAS* **2**:90–94.

Yuan, S., T. Chen, and S. Gao (1986) Uranium series chronological sequence of some Paleolithic sites in south China. *AAS* **5**:179–190.

Z

Zapfe, H. (1958) The skeleton of *Pliopithecus (Epipliopithecus) vindobonensis* Zapfe and Hürzeler. *AJPA* **16**:441–457.

Zapfe, H. (1960) Die Primatenfunde aus der miozäne Spaltenfüllung von Neudorf an der March (Děvinskà Nová Ves), Tschechoslovakei. *Abhandlungen der schweizerischen paläontologischen Gesellschaft* **78**:1–293.

Zeitlin, R., and J. Zeitlin (2000) The Paleoindian and Archaic cultures of Mesoamerica. In: R. Adams and M. MacLeod, Editors. *The Cambridge History of the Native Peoples of the Americas*, Volume 2: *Mesoamerica*. Cambridge: Cambridge University Press, pp. 45–121.

Zhang, S. (1985a) The Early Paleolithic of China. In: Wu and Olsen 1985: 147–186.

Zhang, Y. (1984) The "Australopithecus" of West Hubei and some early Pleistocene hominids of Indonesia. *AAS* **3**:85–92.

Zhang, Y. (1985b) *Gigantopithecus* and *"Australopithecus"* in China. In: Wu and Olssen 1985: 69–78.

Zhang, Y. (1986) The dental remains of early *Homo sapiens* found in China. *AAS* **5**:103–113.

Zhang, Y. (1989) Tooth wear in early *Homo sapiens* from Chaohu and the hypothesis of use of anterior teeth as tools. *AAS* **8**:314–319.

Zhang, Z. (1997) A new Paleoproterozoic clastic-facies microbiota from the Changzhougou Formation, Changcheng Group, Jixian, north China. *Geological Magazine* **134**:145–150.

Zhao, J., K. Hu, K. Collerson et al. (2001) Thermal ionization mass spectrometry U-series dating of a hominind site from near Nanjing, *China. Geology* **29**:27–30.

Zhou, C., Z. Lin, and Y. Wang (2000) Climatic cycles investigated by sediment analysis in Peking Man's cave, Zhoukoudian, China. *JAS* **27**:101–109.

Zhu, R., K. Hoffman, R. Potts, C. Deng, Y. Pan, B. Guo, C. Shi, Z. Guo, B. Yuan, Y. Hou, and W. Huang (2001) Earliest presence of humans in northeast Asia. *Nature* **413**:413–417.

Ziegler, R., and D. Dean (1998) Mammalian fauna and biostratigraphy of the pre-Neandertal site of Reilingen, Germany. *JHE* **34**:469–484.

Zihlman, A. (1976) Sexual dimorphism and its behavioral implications in early hominids. In: P. Tobias and Y. Coppens, Editors. *Les Plus Anciens Hominidés*. Paris: Éditions du Centre National de la Recherche Scientifique, pp. 268–293.

Zihlman, A. (1978) Women in Evolution. Part II. Subsistence and social organization among early hominids. *Signs: Journal of Women in Culture and Society* **4**:4–20.

Zihlman, A. (1979) Pygmy chimpanzee morphology and the interpretation of early hominids. *South African Journal of Science* **75**:165–168.

Zihlman, A. (1981) Women as shapers of the human adaptation. In: F. Dahlberg, Editor. *Woman the Gatherer*. New Haven: Yale University Press, pp. 75–120.

Zihlman, A. (1985) *Australopithecus afarensis:* Two sexes or two species? In: Tobias 1985, pp. 213–220.

Zihlman, A. (1987) American Association of Physical Anthropologists Annual Luncheon Address, April 1985: Sex, sexes, and sexism in human origins. *YPA* **30**:11–19.

Zihlman, A. (1990) Knuckling under: Controversy over hominid origins. In: G. Sperber, Editor. *From Apes to Angels: Essays in Anthropology in Honor of Philip V. Tobias*. New York: Wiley-Liss, pp. 185–196.

Zihlman, A., and L. Brunker (1979) Hominid bipedalism: Then and now. *YPA* **22**:132–162.

Zihlman, A., J. Cronin, D. Cramer, and V. Sarich (1978) Pygmy chimpanzee as a possible prototype for the common ancestor of humans, chimpanzees and gorillas. *Nature* **275**:744–746.

Zilhão, J. (2000) The Ebro frontier: A model for the late extinction of Iberian Neanderthals. In: Stringer et al. 2000: 111–121.

Zilhão, J. (2006) Neandertals and moderns mixed, and it matters. *EA* **15**:183–195.

Zilhão, J., and F. d'Errico (1999) The chronology and taphonomy of the earliest Aurignacian and its implications for the understanding of Neandertal extinction. *JWP* **13**:1–68.

Zilhão, J., and F. d'Errico (2003a) An Aurignacian "Garden of Eden" in southern Germany? An alternative interpretation of the Geissenklösterle and a critique of the *Kulturpumpe* model. *Paleo* **15**:69–86.

Zilhão, J., and F. d'Errico (2003b) The chronology of the Aurignacian and transitional technocomplexes. Where do we stand? In: Zilhão and d'Errico 2003: 313–349.

Zilhão, J., F. d'Errico, J-G. Bordes, A. Lenoble, J-P. Texier, and J-P. Rigaud (2006) Analysis of Aurignacian interstratification at the Châtelperronian-type site and implications for the behavioral modernity of Neadertals. *PNAS* **103**:12643–12648.

Zilhão, J., and E. Trinkaus (2002) Social implications. In: Zilhão and Trinkaus 2002: 519–541.

Zollikofer, C., M. Ponce de Léon, D. Lieberman, F. Guy, D. Pilbeam, A. Likius, H. Mackaye, P. Vignaud, and M. Brunet (2005) Virtual cranial reconstruction of *Sahelanthropus tchadensis. Nature* **434**:755–759.

Zubrow, E. (1989) The demographic modeling of Neanderthal extinction. In: Mellars and Stringer 1989: 212–231.

Zuckerman, S. (1954) Correlation of change in the evolution of higher primates. In: J. Huxley, A. Hardy, and F. Ford, Editors. *Evolution as a Process*. 1963 reprint. New York: Collier, pp. 347–401.

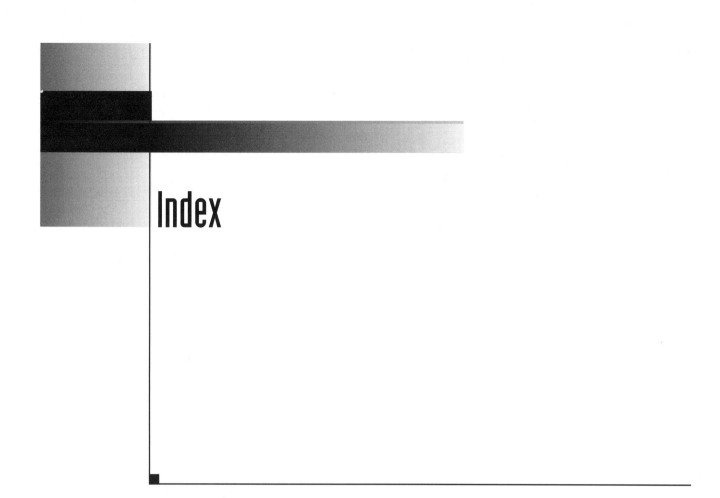

Index